WORLD **CHECKLIST** AND **BIBLIOGRAPHY OF**

Campanulaceae

Campanulaceae

Thomas G. Lammers

PLANTS PEOPLE
POSSIBILITIES

First published 2007
by Royal Botanic Gardens, Kew
Richmond, Surrey, TW9 3AB, UK
www.kew.org

ISBN 978 1 84246 186 0

British Library Cataloguing in Publication Data
A catalogue record for this book is available from the British Library

World Checklists and Bibliographies, 7
[The first six in this series were published respectively as Magnoliaceae (October 1996), Fagales (May 1998), Coniferae (August 1998), Euphorbiaceae (with Pandaceae) (February 2000), Sapotaceae (December 2001), and Araceae (with Acoraceae) (July 2002), Araliaceae (2003)

Address of the author:
Department of Biology and Microbiology, University of Wisconsin Oshkosh, Oshkosh, Wisconsin 54901, U.S.A.

Design by Jeff Eden, Publishing & Media Resources, Royal Botanic Gardens, Kew
Typesetting and page layout: Margaret Newman

For information or to purchase all Kew titles please visit
www.kewbooks.com or email publishing@kew.org

All proceeds to go to support Kew's work in saving the world's plants for life

Cover photograph: *Platycodon grandiflorus*, Thomas G. Lammers

Printed in the United Kingdom and USA by Lightning Source

Contents

Preface

My involvement with the family Campanulaceae dates back to the summer of 1981, when I first visited Dr. Tod F. Stuessy in his lab at The Ohio State University to discuss undertaking graduate work with him. In describing potential dissertation projects, he mentioned that several members of his lab had studied insular endemics, and that such groups provided a great deal of grist for the research mill. "For example", he said, handing me an open copy of Sherwin Carlquist's book, *Hawaii: A Natural History*, "look at how the Lobelioideae have radiated in the Hawaiian Islands." I stared slack-jawed at Carlquist's gorgeous color photographs of *Clermontia*, *Cyanea*, and the rest, with their strange tree-like habits, huge flowers, and fleshy fruits. I had no idea such things existed; the lobelioids I knew were wildflowers such as *Lobelia cardinalis*, or the petite Bedding Lobelia of garden shops. I resolved then and there to learn all I could about this fascinating group of plants. From this initial point of contact, I fanned out and expanded my research into the whole of the Campanulaceae, spending the next quarter century researching diverse aspects of their classification, nomenclature, evolution, and biogeography.

In January 1994, I was invited to prepare the treatment of Campanulaceae for the Asterales volume of Klaus Kubitzki's *The Families and Genera of Vascular Plants* (FGVP). This series was designed as an encyclopedic reference work, summarizing a great deal of fundamental biological data about each family, providing a key to its genera, and a description of each genus. One bit of information to be provided was the number of species in each genus.

One would suppose that in the opening decade of the Twenty-first Century, it would not be difficult for biologists to answer so basic a question as "How many kinds of ____ are there?" After all, our scientists have split the atom, landed instruments on Mars, and decoded our DNA; surely they can tell us how many different kinds of cockatoo or titmouse are to be found on this planet. But the truth is that such questions seldom can be answered with any degree of confidence, often not even within an order of magnitude. This sad fact is true not only for tiny, cryptic, and difficult-to-study organisms such as microbes and deep-sea invertebrates, but even for large conspicuous organisms such as the flowering plants.

As I prepared treatments of the various genera of Campanulaceae for the FGVP manuscript, I was shocked at the disparities I found in the numbers of species reported by various current reference works. For example, the family as a whole was said to include between 1800 and 2200 species, apportioned among 60 to 82 genera. Resolving these incongruities was complicated by the fact that these references seldom explained how their figures were derived. They seemed, like Athena, to have sprung full-blown from the forehead of Zeus. One suspects that many were simply inherited from earlier reference works, with some mark-up to cover inflation.

I concluded that I would have to just *count* the numbers of species in each genus myself if I did not wish to perpetuate bad information. Consequently, in April 1994, I began to compile a simple checklist of the species recognized in current taxonomic literature. As word of this project circulated, I was subsequently invited to expand the checklist and publish it as part of Kew's *World Checklist and Bibliography* series, which commenced publication in 1996.

Work on the two projects progressed in an alternating and reciprocal fashion for several years. A rough idea of the genera to be recognized was required for the first draft of the species checklist, while a greater understanding of the species involved suggested alterations in the generic classification. As the work progressed, numerous nomenclatural difficulties were discovered that required attention (see references under **Special**). More importantly, new taxonomic data from various sources, particularly DNA sequences, accumulated at a rapid pace, which dictated changes in the treatment.

Although the rate of data acquisition continues unabated, publishing deadlines for the FGVP contribution dictated that a line be drawn in the sand and the manuscript completed. This occurred in 2003 and the treatment has now been published (Lammers 2007, under **General**). After submission of the FGVP manuscript, work on the species checklist accelerated until its completion in late 2005.

It is my intention that these two publications be used hand-in-hand. The FGVP treatment provides fundamental information about the family and its constituent taxa down to the rank of genus, while the present checklist then provides a thorough accounting of the species within each genus, as well as a comprehensive guide to the literature and (as an appendix) an outline of a revised classification of the family that may be used *pro tempore*. By taking the two works as a coordinated whole, we have the closest approach to a comprehensive monograph of the Campanulaceae since the treatment by Candolle (1839, under **General**) for the *Prodromus*. All that are wanting are keys to identify the species and descriptions of the species, and these may often be found in the works cited in the bibliography.

It is realized that this is *far* from the final word on the classification of the Campanulaceae. Ongoing molecular phylogenetic research in several labs has already pointed to serious problems with some of the generic circumscriptions embraced here. Unfortunately, such research has not as yet included enough exemplars to permit a comprehensive revision, and it may be many years before a clear picture emerges. It is my hope that the availability of this checklist will hasten the production of a sound generic classification, by providing a comprehensive accounting of the species to be apportioned among the genera.

When I was an undergraduate at Iowa State University, just beginning to contemplate the life of a taxonomist, Dr. Duane Isely offered me the following bit of advice: "No one ever wrote a monograph sitting on a barstool." The past twelve years of effort on this project are a testimonial to the veracity of Isely's maxim. As a corollary, I would add that "no one ever wrote a monograph alone." This volume would not have been possible without the generous assistance of a great many people, whose kindness I gratefully acknowledge.

The most difficult part of preparing this checklist was the acquisition of the protologues for the more than 13,000 names included herein. The majority of the requisite books and serials were to be found in the libraries of the Field Museum of Natural History, the Lloyd Library in Cincinnati, Miami University, the Missouri Botanical Garden, the New York Botanical Garden, and The Ohio State University, and I gratefully acknowledge the abundant help extended to me by the staffs of these institutions. Many other works, however, are extremely rare in the world's libraries and could not be located in these collections. Fortunately, a host of taxonomists and librarians around the globe (including the superb interlibrary loan staff at the University of Wisconsin Oshkosh) responded positively to my pleas for assistance. I cannot begin to name all of the individuals who provided me with copies of scarce publications, but each should know the extreme debt of gratitude I feel; thanks to them, it was possible for me to gather and examine the protologues for nearly 99% of the names included in the checklist.

The discussions of the various genera and their relationships benefited greatly from my ability to briefly mention preliminary results of ongoing molecular phylogenetic researches in the family; I extend my heartfelt appreciation to Eric Knox (Indiana University) and Alexandre Antonelli (Göteborg University), who generously permitted me to do so. The entries for the numerous Chinese species were greatly enhanced by the detailed distributional data provided to me by the *Flora of China* staff at Missouri Botanical Garden. Financial support for critical aspects of the finalization of this work during the summer of 2004 were supported by the University of Wisconsin Oshkosh Faculty Development Board via Project FDR-171.

Finally, I must thank the many people at Kew, who invited me to publish this checklist under their aegis and stuck with me on the circuitous and often rocky path to completion, especially David G. Frodin, Rafaël Govaerts, Simon Owens, Gina Fullerlove, Margaret Newman, Michelle Payne, and Lydia White.

THOMAS G. LAMMERS
UNIVERSITY OF WISCONSIN OSHKOSH
5 JANUARY 2007

Introduction

Aims and scope

This volume is a synonymized checklist of the genera, species, subspecies, and named hybrids assigned to the family Campanulaceae.[1] Eighty-four genera are recognized herein; these encompass a total of 2319 species, 391 subspecies (including autonyms), and 27 named hybrids. Another 8135 names are treated as synonyms. For each accepted taxon, information is provided on its geographic distribution, life-form, and sporic chromosome number (if known). Pertinent bibliographic references are integrated throughout this checklist.

Names

This checklist attempts to include every validly published name at or below specific rank, as well as all validly published generic names, together with the author(s) and the place and date of publication for each. Citation of authors follows *Authors of plant names* (Brummitt & Powell 1992); of books, *Taxonomic literature* (ed. 2; Stafleu & Cowan 1976-88) and its supplements (Stafleu & Mennega 1992-2000); and of periodicals, *Botanico-periodicum-Huntianum/supplementum* (Bridson 1991). Accepted names are set in Roman type, synonyms in italics.

Acceptance of taxa was based on assessments of current taxonomic literature, as well as the author's own experience with the family over the past 25 years. The goal has been to make this checklist reflect prevailing practice in the world taxonomic community, so far as possible, rather than to promulgate massive changes. That is not to say that big changes are not on the horizon. In just the past five years, molecular data cladistically analyzed have made it painfully clear that some generic circumscriptions embraced here are patently unnatural by anyone's definition. Were it merely a matter of a bit of paraphyly, a traditional taxonomist might be able to look the other way. But the data available thus far intimate that large genera such as *Campanula, Centropogon, Lobelia,* and *Wahlenbergia* are in all likelihood *poly*phyletic, a situation no taxonomist can countenance. Unfortunately, while an overall picture is emerging, too many important details are not yet in focus. The data are as yet insufficient for the sort of major overhaul required; too many critical taxa have not yet been included in the analyses. It is hoped that the availability of this checklist will help to hasten the arrival of a sound classification of the family.

The backbone of this checklist is an alphabetical listing of generic names; for clarity and precision, the nomenclatural type of each is indicated. For each accepted genus, information is given regarding the subfamily to which it belongs (Lammers 2007), the number of accepted species listed here, its overall geographical distribution, and chromosome number(s) (if known). Additional notes cover relationships to other genera and infrageneric classification.

The names of accepted species are listed alphabetically within the genus to which they belong. For named hybrids, the putative or confirmed parental formula follows the place of publication. Synonyms are listed under the accepted name to which they belong; heterotypic synonyms are listed chronologically, one to a paragraph, with derivative homotypic synonyms following in the same paragraph. The basionym or replaced synonym of an accepted name is denoted by an asterisk (*). For ease of reference, all synonyms under a given generic name are

---

Notes

[1] Campanulaceae Juss., nom. cons., has been circumscribed variously by different authors; the circumscription adopted here (following Cronquist 1981 and Lammers 2007) includes the genera sometimes segregated as Cyananthaceae J. Agardh; Cyphiaceae A. DC.; Cyphocarpaceae Reveal & Hoogl.; Jasionaceae Dumort.; Lobeliaceae R. Br., nom. cons.; and Nemacladaceae Nutt., but excludes those comprising Pentaphragmataceae J. Agardh, nom. cons., and Sphenocleaceae Mart. ex A. DC., nom. cons.

listed alphabetically at the end of the list of accepted species, with an indication of the accepted species to which they belong following an equal sign (===). Names of uncertain application are also included in these synonymy lists; they are denoted with a question mark following the equal sign (=== ?). The full bibliographic entries for these unattributed names appear alphabetically in a summary section following the last genus. Names excluded from Campanulaceae to other families are treated similarly in a second summary section. A third section accounts for the sole name based on fossil material, while a fourth lists hybrids that have been named, but which do not have a legitimate name currently.

Some species have been divided into infraspecific taxa. In my opinion, the only infraspecific taxa that merit recognition are groups of conspecific populations that are not only morphologically distinguishable from each other, but are also geographically coherent in some fashion, typically allopatric or parapatric. Such taxa are not as clearly demarcated as congeneric species, often showing intergradation in a zone of contact. Groups of populations that meet these criteria have been accorded both subspecific and varietal rank by various authors. However, this is very nearly the only definition ever applied to the former rank, while the latter has been defined in diverse and often conflicting ways (Raven et al. 1974). For this reason, I have consistently used the rank of subspecies in my research, and this is the sole infraspecific rank recognized in this checklist.

Bibliography

The bibliography is selective rather than exhaustive. The original place of publication of each generic name is cited, as well as those books and papers judged to be of some import and utility to taxonomists: e.g., monographs, complete or partial revisions and synopses, major floristic treatments, phylogenetic analyses, discussions of major characters or evolutionary trends, and high quality plant portraits. Works believed to be of especial significance and usefulness are bulleted (•). Each work cited has been annotated with regard to its general contents and useful features; the language(s) in which it was written is noted in abbreviated form. Literature dealing with the family as a whole or with all its representatives in a given geographic region are listed in a block preceding the first genus; literature pertaining to a single genus is listed following the introductory paragraph for that genus.

Geographical distribution

In reporting distributions, a distinction is made between what is believed to be a taxon's indigenous range and naturalized occurrences. Assessing the former from the literature generally presents no difficulties, but the latter can. Although few Campanulaceae are weedy in the way that, e.g., many Chenopodiaceae or Poaceae are weedy, numerous members of the family are cultivated as ornamentals in regions outside their natural ranges (see references under **Cultivation**). Their ecological status in these areas may range from completely unable to survive without human intervention to behaving like a component of the native vegetation. Unfortunately, authors vary considerably in their criteria for where to draw the line on this continuum between "cultivated" and "naturalized". I have tried to err on the side of caution in reporting naturalized distributions, looking for evidence that a species is indeed capable of spreading from gardens and thriving unassisted for an indefinite period of time. Similarly, I have adopted a conservative approach in deciding which species to regard as truly in cultivation. Many species are cultivated in major botanic gardens and by fanciers and aficionados of the unusual. However, I have denoted as 'cult.' only those species that are commercially available as seeds or nursery stock, and that are grown by the average gardener. No attempt is made to indicate where these species are cultivated; they might reasonably be expected anywhere that growing conditions permit.

The native and naturalized distribution of a taxon is furnished in two ways. The first is a general statement in narrative form, such as one might find in any flora or monograph; statements for genera are much simpler and more generalized than those for species and subspecies. When the presence of a taxon in a given region or location is not certainly known, a question mark is used; when an exact location within a country or region is not known, the question mark is placed in parentheses. Examples include:

S. Italy (Appennines)
E. Mexico (San Luis Potosí, Veracruz)
Brazil? [presence in Brazil not certainly known]
Brazil (?) [exact location in Brazil not known]
Venezuela & Colombia
Cuba, Hispaniola, Puerto Rico
Assam to Laos, Sri Lanka, Andaman Is., Nicobar Is. & Jawa
Continental Europe to W. Siberia & Turkey; cult.; naturalized in Great Britain &
    North America

This narrative statement is then followed by the appropriate TDWG geographical codes (Brummitt 2001) expressed to that system's third level. In the TDWG codes, a two-digit number indicates the first two levels: continent (e.g., 2 = Africa), and the region within that continent (e.g., 27 = Southern Africa). The three-letter code following the two-digit number indicate the third-level, the unit, which is typically a country, state, island, or other comparable area (e.g., 27 NAM = Namibia). 'ALL' is used if a species or subspecies is known to occur in every unit within a given region, while uncertainty regarding distribution within a region is expressed by '+'. Naturalization is denoted by setting the three-letter unit codes in lower case; if all occurrences within a given region are the result of naturalization, the two-digit number for the region is placed in parentheses. Examples include:

83 ECU	[S. America: Ecuador]
77 TEX	[SC. U.S.A.: Texas]
27 ALL	[S. Africa (throughout)]
84+	[Brazil (more exact distribution not known)]
76 CAL 85 AGS CLS	[SW. U.S.A.: California; S. South America: Argentina and Chile]
36 CHC CHI CHM CHN CHQ CHS 37 MON 38 KOR	[Asia: China, Mongolia, Korea]
(10) grb (12) fra 13 ALB ITA YUG	[SE. Europe: Albania, Italy, Yugoslavia; naturalized in Great Britain and France]
(36) (38) (50) 71 72 73 74 75 76 77 78 79 80 (81) 83 84 85	[of a genus: N. & S. America; naturalized in Caribbean, Asia, and Australia]

Life-forms

The terminology for life-forms, definitions of which follow, is based on the system of Raunkiaer (1934), as elaborated in other volumes of this series. Campanulaceae are remarkable for their diversity of life forms; of these, hemicryptophyte is perhaps the commonest.

Main categories

Phanerophyte (phan.)
    Stems: woody and indefinitely persistent
    Buds: normally 3 m or more above ground
    e.g.: very large shrubs and trees such as *Cyanea leptostegia*, *Lobelia petiolata*

Nanophanerophyte (nanophan.)
    Stems: woody and indefinitely persistent
    Buds: above soil level but normally less than 3 m above ground
    e.g.: shrubs such as *Azorina vidalii*, *Clermontia micrantha*

Herbaceous phanerophyte (herb. phan.)
   Stems: herbaceous and persisting for several years
   Buds: above soil level
   e.g.: giant rosette plants such as *Lobelia exaltata, L. telekii*.

Chamaephyte (cham.)
   Stems: herbaceous and/or woody and persistent
   Buds: on or just above soil level but never more than 0.5 m above ground
   e.g.: perennial herbs such as *Lobelia chamaepitys, Lysipomia sphagnophila*

Hemicryptophyte (hemicr.)
   Stems: herbaceous, often dying back after the growing season but with buds or growth at
   soil level surviving
   Buds: just on or below soil level
   e.g.: perennial herbs such as *Campanula persicifolia, Jasione crispa*

Geophyte [not abbreviated]
   Hemicryptophytes that survive unfavorable seasons in the form of a rhizome, bulb, tuber,
   or root bud
   Buds: below soil level
   e.g.: perennial herbs such as *Cyphia elata, Ostrowskia magnifica*

Therophyte (ther.)
   Plants surviving unfavorable seasons in the form of seeds, completing their life-cycle
   during the favorable season
   e.g., annuals such as *Campanula erinus, Nemacladus capillaris*

Biennial [not abbreviated]
   Therophytes completing their life-cycle in two favorable seasons, passing the intervening
   unfavorable season as a hemicryptophyte
   e.g., *Campanula medium, Michauxia campanuloides*

Aquatic plants

Helophyte (hel.)
   Hemicryptophytes growing in soil saturated with water, or in water and with leaf- and
   flower-bearing shoots held above water
   Buds: just on or below soil level
   e.g.: *Lobelia boykinii, Siphocampylus verticillatus*

Hydrophyte [inclusive of the two following subcategories]
   Plants with vegetative shoots entirely in water, the leaves usually submersed and/or
   floating; flower-bearing parts may emerge above the water
   Buds: permanently or temporarily on the bottom of the water

Hydrohemicryptophyte (hydrohemicr.): aquatic hemicryptophytes
   e.g.: *Lobelia dortmanna*

Hydrotherophyte (hydrother.): aquatic therophytes
   e.g.: *Howellia aquatilis, Lobelia aquatica*

Climbing plants

   Climbing plants are denoted by prefixing 'cl.' to one of the main categories. If the taxon
   only sometimes climbs, this abbreviation is placed in parentheses. Examples include:
   Climbing phanerophytes (cl. phan.); e.g.: *Centropogon lianeus*
   Climbing nanophanerophytes (cl. nanophan.); e.g.: *Centropogon ferrugineus*
   Climbing tuberous geophytes (cl. tuber geophyte); e.g.: *Canarina canariensis*
   Sometimes climbing nanophanerophytes ((cl.) nanophan.); e.g.: *Burmeistera pallida*

References

References cited above but not found below are listed in the following section, under **General**.

Bridson, G. (1991). Botanico-periodicum-Huntianum/supplementum. 1068 pp. Pittsburgh: Hunt Institute for Botanical Documentation.
Brummitt, R. K. (2001). World geographical scheme for recording plant distributions (ed. 2). xvi , 137 pp. Pittsburgh: Hunt Institute for Botanical Documentation.
Brummitt, R. K. & C. E. Powell (1992). Authors of plant names. 732 pp. Kew: Royal Botanic Gardens.
Raunkiaer, C. (1934). The life forms of plants and statistical plant geography. xvi, 632 pp., illus. London: Oxford University Press.
Raven, P. H., S. G. Shetler & R. L. Taylor (1974). Proposals for the simplification of infraspecific terminology. Taxon 23: 828-831.
Stafleu, F. & R. S. Cowan (1976-1988). Taxonomic literature (ed. 2), 7 vols. Utrecht: Bohn, Scheltema & Holkema.
Stafleu, F. & E. A. Mennega (1992-2000). Taxonomic literature, supplement. 6 vols. Königstein: Koeltz.

Abbreviations

al.	Latin *alii:* others
Bos.	Bosnian
Bul.	Bulgarian
C.	central
Cat.	Catalan
Ch.	Chinese
cham.	chamaephyte [life-form]
cl.	climber, climbing
Co.	county
cult.	Latin *cultus:* cultivated
Cz.	Czech
Distr.	district
Du.	Dutch
E.	east(ern)
En.	English
etc.	Latin *et cetera:* and the rest
e.g.	Latin *exempli gratia:* for example
Fr.	French
Ge.	German
hel.	helophyte [life form]
hemicr.	hemicryptophyte [life form]
herb.	herbaceous
hort.	Latin *hortorum:* of gardens; or *hortulanorum:* of horticulturalists
Hu.	Hungarian
hydrohemicr.	hydrohemicryptophyte [life-form]
hydrother.	hydrotherophyte [life-form]
I(s).	island(s)
ICBN	*International Code of Botanical Nomenclature*
i.e.	Latin *id est:* that is

It.	Italian
Ja.	Japanese
Ko.	Korean
La.	Latin
Medit.	Mediterranean
Mt(s).	mountain(s)
N.	north(ern)
nanophan.	nanophanerophyte [life form]
nom. cons.	Latin *nomen conservandum:* conserved name [ICBN]
nom. illeg.	Latin *nomen illegitimum:* illegitimate name [ICBN]
nom. inval.	Latin *nomen invalidum:* name not validly published [ICBN]
nom. nud.	Latin *nomen nudum:* name published without a description [ICBN]
nom. rejic.	Latin *nomen rejiciendum:* rejected name [ICBN]
nom. superfl.	Latin *nomen superfluum:* name superfluous when published [ICBN]
nov.	Latin *novus:* new
Pen.	peninsula(r)
pers. comm.	personal communication
phan.	phanerophyte [life-form]
Pol.	Polish
Por.	Portugese
p.p.	Latin *pro parte:* partly
Prov.	province
q.v.	Latin *quod vide:* which see
Rep.	Republic
Ru.	Russian
S.	south(ern)
Se.	Serbian
sic	Latin *sic:* thus
s.l.	Latin *sensu lato:* in the broad sense
Sp.	Spanish
sp.	species
sphalm.	Latin *sphalmate:* by mistake
s.s.	Latin *sensu stricto:* in the narrow sense
subtrop.	subtropical
Sw.	Swedish
syn.	Latin *synonymon:* synonym
temp.	temperate
ther.	therophyte [life-form]
trop.	tropical
W.	west(ern)
?	not known, doubtful
*	basionym or replaced synonym (before a synonym)
×	nothotaxon (before a generic name or specific epithet)
+	range more than as indicated but not certainly known

Campanulaceae

84 genera, 2319 species, 391 subspecies, and 27 named hybrids, divided among five subfamilies: Campanuloideae Burnett (1045 spp.), Cyphioideae Walp. (64 spp.), Cyphocarpoideae Miers (3 spp.), Lobelioideae Burnett (1192 spp.), and Nemacladoideae Lammers (15 spp.). Cosmopolitan, with representatives on six continents and many oceanic archipelagoes, from the tropics to the frigid zones. However, the subfamilies are not evenly distributed. Campanuloideae and Lobelioideae are very widespread but largely non-overlapping: the former are found primarily in the temperate zones of the Old World, while the latter are distributed predominantly in the world's tropical and subtropical zones. The other three subfamilies (sometimes treated as Cyphioideae s. l., though the name Cyphocarpoideae has priority) are restricted to rather small areas: Cyphioideae to southern and tropical Africa, Cyphocarpoideae to northern Chile, and Nemacladoideae to southwestern North America. The only region to have significant numbers of both Campanuloideae and Lobelioideae (as well almost all the Cyphioideae) is southern Africa. Other major centers of diversity are the Andes of South America for Lobelioideae and Eurasia from the Mediterranean to the Caucasus for Campanuloideae. Also noteworthy is the presence of numerous endemic Lobelioideae on the small but highly isolated Hawaiian Islands.

General

Adanson, M. (1763). Les campanules. Campanulae. In Familles des plantes 2: 132-134. Paris: Vincent. Fr. — Original description of family, but prior to 1789 starting data for family nomenclature; circumscribed in manner consistent with that employed here.

de Jussieu, A. L. (1789). Campanulaceae, les Campanulacées. In Genera plantarum secundum ordines naturales disposita: 163-166. Paris: Vidua Herissant. La. — Formal establishment of family, with generic descriptions; circumscribed more broadly than Adanson (1763), as it also includes *Ceratostema* Juss. (Ericaceae), *Forgesia* Comm. ex Juss. (Escalloniaceae), *Gesneria* L. (Gesneriaceae), and *Scaevola* L. (Goodeniaceae).

de Jussieu, A. L. (1811). Mémoire sur les Lobeliacées et les Stylidiées, nouvelles familles des plantes. Ann. Mus. Natl. Hist. Nat. 18: 1-19 + pl. 1-3, illus. Fr. — Advocates recognition of *Lobelia* as distinct family, but only gives name in French form.

Bonpland, A. (1813). Tab. 7. In Description des plantes rares cultivées à Malmaison et à Navarre: 19. Paris: P. Didot. Fr. — Validates Jussieu's description of family name Lobeliaceae.

• de Candolle, A. (1830). Monographie des Campanulées. viii + 384 pp. + pl. 1-20, illus. Paris: Veuve Desray. Fr., La. — Fundamental monograph for Campanuloideae, with key to genera, descriptions, supra- and infrageneric classifications, full nomenclature, and specimen citations.

• Presl, C. (1836). Prodromus monographiae Lobeliacearum. 52 pp. Prague: Theophilus Haase. La. — Synonymized checklist of Lobelioideae with descriptions of genera and supra- and infrageneric classifications.

• de Candolle, A. (1839). Lobeliaceae, Campanulaceae, Cyphiaceae. In A. P. de Candolle, Prodromus systematis naturalis regni vegetabilis 7: 339-501, 784-793. Paris: Treuttel & Würtz. La. — Last comprehensive monograph of family, with descriptions, supra- and infrageneric classifications, full nomenclature, and specimen citations; segregation of *Cyphia* as distinct family.

Nuttall, T. (1842). Description and notices of new or rare plants in the natural orders Lobeliaceae, Campanulaceae, Vaccinieae, Ericaceae, collected in a journey over the continent of North America, and during a visit to the Sandwich Islands, and Upper California. Trans. Amer. Philos. Soc. (n.s.) 8: 251-272. En. — Establishment of new genus *Nemacladus* as distinct family.

Horaninow, P. (1847). Lobeliaceae, Campanulaceae. In Characteres essentiales familiarum ac tribuum regni vegetabilis et amphorganici: 99-101. St. Petersburg: K. Wienhöberian. La. — Descriptions of families and tribes with lists of included genera.

Miers, J. (1848). Contributions to the botany of South America. *Cyphocarpus*. London J. Bot. 7: 59-64. En. — Establishment of new genus *Cyphocarpus* as subfamily Cyphocarpoideae.

Bentham, G. (1875). Notes on the gamopetalous orders belonging to the campanulaceous and oleaceous groups. Bot. J. Linn. Soc. 15: 1-16. En. — Diverse taxonomic notes preparatory to Bentham (1876).

• Bentham, G. (1876). Campanulaceae. In G. Bentham & J. D. Hooker, Genera plantarum 2: 541-564. London: L. Reeve. La. — Generic monograph with descriptions and supra- and infrageneric classifications; family divided into three tribes: Campanuleae Dumort., Cyphieae Sond., Lobelieae Rchb.

• Schönland, S. (1889). Campanulaceae. In A. Engler & K. Prantl, Die natürlichen Pflanzenfamilien IV. 5: 40-70, illus. Leipzig: Wilhelm Engelmann. Ge. — Generic monograph with keys, descriptions, and supra- and infrageneric classifications; ranked Bentham's (1876) tribes as subfamilies.

Engler, A. (1897). Campanulaceae. In A. Engler & K. Prantl, Die natürlichen Pflanzenfamilien, Nachtr.: 319-320. Leipzig: Wilhelm Engelmann. Ge. — Supplement to Schönland (1889).

Greene, E. L. (1904). Affinities of the Cichoriaceae. Leafl. Bot. Observ. Crit. 1: 59-63. En. — Hypothesizes that Campanulaceae ancestral to cichorioid Asteraceae.

Briquet, J. (1930). Le nombre des carpelles dans la fleur des campanules. Compt. Rend. Séances Soc. Phys. Genève 47: 20-24. Fr. — Discussion of variation in carpel number among Campanuloideae.

Rosén, W. (1932). Zur Embryologie der Campanulaceen und Lobeliaceen. Acta Horti Gothob. 7: 31-42, illus. Ge. — Descriptive and comparative embryology.

Sugiura, T. (1942). Studies on the chromosome numbers in Campanulaceae I. Campanuloideae-Campanuleae. Cytologia 12: 418-434, illus. En. — Summary of current cytological knowledge.

• Wimmer, F. E. (1943). Campanulaceae-Lobelioideae, I. Teil. In A. Engler & L. Diels, Das Pflanzenreich IV. 276b: i-viii +1-260 + 4 pl., illus. Leipzig: Wilhelm Engelmann. Ge., La. — Monograph of tribe Delisseeae C. Presl, with keys, descriptions, supra- and infrageneric classifications, full nomenclature, and specimen citations.

Rosén, W. (1949). Endosperm development in Campanulaceae and closely related families. Bot. Not. 1949: 137-147, illus. En. — Follow-up to Rosén (1932), emphasizing endosperm developmental patterns.

Subramanyam, K. (1951). Interrelationships of Campanulatae. Half-Yearly J. Mysore Univ. (n.s.), B 12: 331-339, illus. En. — Relationships of Campanuloideae and Lobelioideae to each other and to other families, based largely on embryological data.

Erdtman, G. (1952). Campanulaceae. In Pollen morphology and taxonomy: 90-94, illus. Stockholm: Almqvist & Wiksell. En. — Descriptive and comparative palynology.

• Wimmer, F. E. (1953). Campanulaceae-Lobelioideae, II. Teil. In A. Engler & L. Diels, Das Pflanzenreich IV. 276b: i-viii, 261-813 + 11 pl., illus. Berlin: Akademie-Verlag. Ge., La. — Monograph of tribe Lobelieae Rchb. with keys, descriptions, supra- and infrageneric classifications, full nomenclature, and specimen citations; includes supplement to Delisseeae monograph (Wimmer 1943).

• Gadella, T. W. J. (1966). Some notes on the delimitation of genera in the Campanulaceae. Proc. Kon. Ned. Akad. Wetensch. C 69: 502-521, illus. En. — Review of problems of generic circumscription in Campanuloideae, based largely on cytological data.

Chapman, J. L. (1967). Comparative palynology in Campanulaceae. Trans. Kansas Acad. Sci. 69: 197-200, illus. En. — Descriptive palynological survey; family divided into three groups: 3-12-porate (most Campanuloideae); 3-colporate and reticulate (Lobelioideae, *Jasione*) or spinuliferous (Nemacladoideae); and 6-9-col(por)ate (E. Asian Campanuloideae).

• Wimmer, F. E. (1968). Campanulaceae-Lobelioideae supplementum et Campanulaceae-Cyphioideae. In A. Engler & L. Diels, Das Pflanzenreich IV. 276c: i-x, 815-1024 + Taf. 1-69, illus. Berlin: Akademie-Verlag. Ge., La. — Monograph of Cyphioideae s. l. with keys, descriptions, supra- and infrageneric classifications, full nomenclature, and specimen citations; includes supplement to Lobelioideae monograph (Wimmer 1943, 1953).

Carlquist, S. (1969). Wood anatomy of Lobelioideae (Campanulaceae). Biotropica 1: 47-72, illus. En. — Stem anatomy in woody taxa; woodiness hypothesized to be derived within family and due to paedomorphosis.

Knobloch, I. W. (1972). Intergeneric hybridization in flowering plants. Taxon 21: 97-103. En. — Lists "×Fockeanthus" H. R. Wehrh., nom. inval., as only intergeneric hybrid in family.

Dunbar, A. (1973). Pollen ontogeny in some species of Campanulaceae. A study by electron microscopy. Bot. Not. 126: 277-315, illus. En. — Descriptive study of gametogenesis in *Campanula* and *Jasione*.

Mabberley, D. J. (1974). Pachycauly, vessel-elements, islands and the evolution of arborescence in "herbaceous" families. New Phytol. 73: 967-975, illus. En. — Criticizes Carlquist's (1969) hypothesis that woodiness in Campanulaceae due to paedomorphosis; sees woodiness as primitive in family.

• Dunbar, A. (1975). On pollen of Campanulaceae and related families with special reference to the surface ultrastructure I. Campanulaceae subfam. Campanuloidae [sic]. Bot. Not. 128: 73-101, illus. En.; II. Campanulaceae subfam. Cyphioidae [sic] and subfam. Lobelioidae [sic]; Goodeniaceae; Sphenocleaceae. Bot. Not. 128: 102-118, illus. En. — Descriptive and comparative palynology; Campanuloideae and Lobelioideae quite distinct, with Cyphioideae s. l. as a "connecting link".

Dunbar, A. & H. G. Wallentinus (1976). On pollen of Campanulaceae III. A numerical taxonomic investigation. Bot. Not. 129: 69-72. En. — Cluster analysis of palynological data divides family into six groups and suggests Cyphioideae s. l. is not a natural group.

Shulkina, T. V. (1978). [Life forms in Campanulaceae Juss., their geographical distribution and connection with taxonomy.] Bot. Žurn. 63: 153-169, illus., maps Ru. — Biogeographic correlations with life-forms; three centers of distribution for Campanuloideae: Mediterranean, S. Africa, E. Asia.

• Kolakovskii, A. A. (1980). [New evidence on the systematics of Campanulaceae.] Soobšč. Akad. Nauk Gruzinsk. SSR 98: 653-656, illus. Ru. — Variation in details of fruit anatomy among Old World Campanuloideae suggests need for revised generic classification.

• Shulkina, T. V. (1980). The significance of life-form characters for systematics, with special reference to the family Campanulaceae. Pl. Syst. Evol. 136: 233-246, illus. En. — Detailed exposition on utility of life-form as taxonomic character among Campanuloideae.

Shulkina, T. V. & S. E. Zikov (1980). [The anatomical structure of the stem in the family Campanulaceae s. str. in relation to the evolution of life forms.] Bot. Žurn. 65: 627-638, illus. Ru. — Anatomy correlates with growth form in genera surveyed.

• Cronquist, A. (1981). Family Campanulaceae A. L. de Jussieu 1789 nom. conserv., the Bellflower Family. In An integrated system of classification of flowering plants: 983-986, illus. New York: Columbia University Press. En. — Extremely detailed description of family.

Belyaev, A. A. (1984a). [Seed anatomy in some representatives of the family Campanulaceae.] Bot. Žurn. 69: 585-594, illus. Ru. — Comprehensive survey of testal structure in Campanuloideae and Lobelioideae; 11 distinct testal types recognized.

Belyaev, A. A. (1984b). [Surface ultrastructure and some morphological characteristics of seeds in the representatives of the family Campanulaceae.] Bot. Žurn. 69: 890-898 + 2 pl., illus. Ru. — Comprehensive survey of testal surface in Campanuloideae and Lobelioideae; 10 distinct testal patterns recognized.

• Dunbar, A. (1984). Pollen morphology in Campanulaceae IV. Nord. J. Bot. 4: 1-19, illus. En. — Descriptive and comparative palynology; two main types recognized: (1) porate, often with spinules, and (2) 3-col(por)ate, never with spinules.

Shulkina, T. V. (1984). [Main evolutionary patterns of life-forms in the Campanulaceae s. str. family] Bot. Žurn. 69: 3-13, illus. Ru. — Patterns of divergence in life-forms among various genera.

Belyaev, A. A. (1986). [The features of anatomy and surface ultrastructure of the seed coat in some representatives of critical genera of the Campanulaceae family.] Bot. Žurn. 71: 1371-1376 + 2 pl., illus. Ru. — Testal structure and appearance in select taxa.

Bigazzi, M. (1986). Ultrastructural and cytochemical observations on fibrillar intranuclear inclusions in the family Campanulaceae. Caryologia 39: 199-210, illus. En. — Protein inclusions with fibrillar substructure found in nuclei of mesophyll cells in leaves of some Campanuloideae, but not in others nor in any Lobelioideae surveyed.

- Kolakovskii, A. A. (1986). [Carpology of the Campanulaceae and problems in their taxonomy.] Bot. Žurn. 71: 1155-1166 + 4 pl., illus. Ru. — Distinguishes 55 fruit types among Campanuloideae, divisible into two major categories: with or without an axicorn.

- Kolakovskii, A. A. (1987). [System of the Campanulaceae family from the Old World.] Bot. Žurn. 72: 1572-1581. Ru. — Revised classification, based largely on fruit structure described previously (Kolakovskii 1986), dividing genera treated here as Campanuloideae into four subfamilies: Prismatocarpoideae Kolak., Canarinoideae Kolak., Wahlenbergioideae Kolak., and Campanuloideae.

- Shulkina, T. V. (1988). [Architectural models in the Campanulaceae s. str. family, their geography and possible ways of transformation.] Bot. Žurn. 73: 3-16, illus., maps. Ru. — Campanuloideae quite diverse in architectural models, especially the herbaceous taxa.

Erbar, C. & P. Leins (1989). On the early floral development and the mechanisms of secondary pollen presentation in *Campanula*, *Jasione* and *Lobelia*. Bot. Jahrb. Syst. 111: 29-55, illus. En. — Relation of floral development to portioned discharge of pollen.

Leins, P. & C. Erbar (1990). On the mechanism of secondary pollen presentation in the Campanulales-Asterales complex. Bot. Acta 103: 87-92, illus. En. — Structure of floral organs involved in portioned pollen release.

Gudkova, I. Y. & G. P. Borshchenko (1991). [The serological study of the Campanulaceae. The phylogenetic relations in the tribe Phyteumateae.] Bot. Žurn. 76: 809-817. Ru. — Protein serology supports close relationships among *Asyneuma*, *Cryptocodon*, *Petromarula*, *Phyteuma*, and *Sergia*.

- Lammers, T. G. (1992). Circumscription and phylogeny of the Campanulales. Ann. Missouri Bot. Gard. 79: 388-413. En. — Synthesis of diverse data to explore relationships of Campanulaceae to other families.

- Lammers, T. G. (1993). Chromosome numbers of Campanulaceae. III. Review and integration of data for subfamily Lobelioideae. Amer. J. Bot. 80: 660-675. En. — Synthesis of all published chromosome number reports; base number $x = 7$, polyploidy extensive, aneuploidy in some genera.

- Cosner, M. E., R. K. Jansen & T. G. Lammers (1994). Phylogenetic relationships in the Campanulales based on *rbc*L sequences. Pl. Syst. Evol. 190: 79-95, illus. En. — Molecular data support monophyly of family and its division into five subfamilies rather than three.

- Kolakovskii, A. A. (1994). [The conspectus of the system of the Old World Campanulaceae.] Bot. Žurn. 79(1): 109-124, maps. Ru. — Update of Kolakovskii's (1987) classification.

Erbar, C. & P. Leins (1995). Portioned pollen release and the syndromes of secondary pollen presentation in the Campanulales-Asterales complex. Flora 190: 323-338, illus. En. — Mechanisms for discharging pollen in small portions, with phylogenetic hypotheses to explain their diversity.

Gustafsson, M. H. G. (1995). Petal venation in Asterales and related orders. Bot. J. Linn. Soc. 118: 1-18, illus. En. — Character is diverse among Campanulaceae, but most taxa form dense reticulum with definite midvein.

- Gustafsson, M. H. G. & K. Bremer (1995). Morphology and phylogenetic interrelationships of the Asteraceae, Calyceraceae, Campanulaceae, Goodeniaceae, and related families (Asterales). Amer. J. Bot. 82: 250-265. En. — Phylogenetic analysis of morphological data supports monophyly of Campanulaceae and its division into five subfamilies.

- Hong, D. Y. (1995). The geography of the Campanulaceae: on the distribution centres. Acta Phytotax. Sin. 33: 521-536, maps. Ch. — Hypothesizes that Campanuloidae arose in SW. China by end of Cretaceous and that Cape of Good Hope and Mediterranean are secondary diversification centers; classifies subfamily into six informal groups: *Cyananthus* Group, *Platycodon* Group, *Ostrowskia* Group, *Wahlenbergia* Group, *Jasione* Group, and *Campanula* Group.

Erbar, C. & P. Leins (1996). Distribution of the character states "early sympetaly" and "late sympetaly" within the "Sympetalae Tetracyclicae" and presumably allied groups. Bot. Acta 109: 427-440, illus. En. — Sympetaly develops very early in floral ontogeny in family.

• Stace, H. M. & S. H. James (1996). Another perspective on cytoevolution in Lobelioideae (Campanulaceae). Amer. J. Bot. 83: 1356-1364. En. — Disagree with Lammers' (1993) hypothesis that subfamily has $x = 7$; argue that $x = 14$ or even 21, and that lower numbers are result of descending aneuploidy.

Takhtajan, A. (1996). Campanulales Reichenbach 1828. In Diversity and classification of flowering plants: 405-412. New York: Columbia University Press. En. — Five subfamilies of Campanulaceae elevated to familial rank; genera treated here as Campanuloideae divided into four subfamilies: "Cyananthoideae" nom. nud., Ostrowskioideae Takht., Canarinoideae Kolak., and Campanuloideae.

• Tobe, H. & N. R. Morin (1996). Embryology and circumscription of Campanulaceae and Campanulales: a review of literature. J. Pl. Res. 109: 425-435. En. — Synthesis of available embryological data supports broad circumscription of family.

Bremer, K. & M. H. G. Gustafsson (1997). East Gondwana ancestry of the sunflower alliance of families. Proc. Natl. Acad. Sci. USA 94: 9188-9190. En. — Phylogenetic analysis of molecular data supports monophyly of Campanulaceae, Campanuloideae, and Lobelioideae; and separation of Cyphocarpoideae and Nemacladoideae from Cyphioideae.

Lammers, T. G. (1998). Nemacladoideae, a new subfamily of Campanulaceae. Novon 8: 36-37. En. — Establishment of formal five-subfamily classification, based on data by Cosner et al. (1994).

Kårehed, J., J. Lundberg, B. Bremer & K. Bremer (1999). Evolution of the Australasian families Alseuosmiaceae, Argophyllaceae, and Phellinaceae. Syst. Bot. 24: 660-682, illus. En. — Phylogenetic analysis of molecular data supports monophyly of Campanulaceae.

Soltis, D. E., P. E. Soltis, M. W. Chase, M. E. Mort, D. C. Albach, M. Zanis, V. Savolainen, W. H. Hahn, S. B. Hoot, M. F. Fay, M. Axtell, S. M. Swensen, L. M. Prince, W. J. Kress, K. C. Nixon & J. S. Farris (2000). Angiosperm phylogeny inferred from 18S rDNA, *rbcL*, and *atpB* sequences. Bot. J. Linn. Soc. 133: 381-461. En. — Phylogenetic analysis of molecular data supports monophyly of Campanulaceae.

Albach, D. C., P. S. Soltis, D. E. Soltis & R. G. Olmstead (2001). Phylogenetic analysis of asterids based on sequences of four genes. Ann. Missouri Bot. Gard. 88: 163-212. En. — Phylogenetic analysis of molecular data supports monophyly of Campanulaceae.

Buss, C. C., T. G. Lammers & R. R. Wise (2001). Seed coat morphology and its systematic implications in *Cyanea* and other genera of Lobelioideae (Campanulaceae). Amer. J. Bot. 88: 1301-1308, illus. En. — Seed surface structure used to resolve various taxonomic and evolutionary questions.

• Eddie, W. M. M., T. Shulkina, J. Gaskin, R. C. Haberle & R. K. Jansen (2003). Phylogeny of Campanulaceae s. str. inferred from ITS sequences of nuclear ribosomal DNA. Ann. Missouri Bot. Gard. 90: 554-575. En. — Phylogenetic analysis of molecular data supports recognition of two major alliances: platycodonoid clade, largely in Asia, with colpate or colporate pollen; and campanuloid (N. Hemisphere) + wahlenbergioid (S. Hemisphere) clade with porate pollen.

• Lundberg, J. & K. Bremer (2003). A phylogenetic study of the order Asterales using one morphological and three molecular data sets. Int. J. Pl. Sci. 164: 553-578. En. — Combined phylogenetic analyses support monophyly of Campanulaceae, Campanuloideae, and Lobelioideae; and separation of Nemacladoideae from Cyphioideae.

• Shulkina, T. V., J. F. Gaskin & W. M. M. Eddie (2003). Morphological studies toward an improved classification of Campanulaceae s. str. Ann. Missouri Bot. Gard. 90: 576-591. En. — Synthesis of diverse morphological data supports recognition within Campanuloideae of two lineages: (1) plants primarily in Asia with elongate epicotyl and internodes in seedling, fundamentally opposite phyllotaxy, rhythmic seasonal growth with long dormancy, sympodial branching, and colpate or colporate pollen; and (2) plants from throughout world with shortened epicotyl and internodes in seedling, alternate phyllotaxy, continuous growth in non-flowering period, both sympodial and monopodial growth, and porate pollen.

- Lammers, T. G. (2007). Campanulaceae. In K. Kubitzki (ed.), The families and genera of vascular plants, vol. 8 Asteridae (J. Kadereit & C. Jeffrey, eds.): 26-56, illus. Leipzig: LE-TeX Jelonek, Schmidt & Voeckler. En. — Generic monograph, with keys, descriptions, and full nomenclature.

Special

The following papers were published explicitly as precursors to the present work, to validate requisite nomenclatural changes.

Lammers, T. G. (1995). Transfer of the southern African species of *Lightfootia,* nom. illeg., to *Wahlenbergia* (Campanulaceae, Campanuloideae). Taxon 44: 333-339. En. — Provision of names for 39 species of *Wahlenbergia.*

Lammers, T. G. (1996). Proposal to conserve the name *Edraianthus* against *Pilorea* (Campanulaceae). Taxon 45: 563-564. En. — Ensuring continued use of familiar name.

Lammers, T. G. (1997). Proposal to conserve the name *Laurentia* (Campanulaceae) with a conserved type. Taxon 46: 795-799. En. — Failed attempt to permit continued use of familiar name.

Lammers, T. G. (1998). New names and new combinations in Campanulaceae. Novon 8: 31-35. En. — Fifteen nomenclatural innovations for species of *Campanula, Cyanea, Delissea, Lobelia,* and *Wahlenbergia.*

Lammers, T. G. (1999a). Nomenclatural consequences of the synonymization of *Hypsela reniformis* (Campanulaceae: Lobelioideae). Novon 9: 73-76. En. — Provision of name in *Lobelia* for type of *Hypsela,* and transfer of its three other species to *Isotoma.*

Lammers, T. G. (1999b). A new *Lobelia* from Mexico, with additional new combinations in world Campanulaceae. Novon 9: 381-389, illus. En. — Nine nomenclatural innovations for species and subspecies of *Campanula, Centropogon, Codonopsis, Cyanea, Cyclocodon, Delissea,* and *Lobelia.*

Serra, L., M. B. Crespo & T. G. Lammers (1999). *Wimmerella,* a new South African genus of Lobelioideae (Campanulaceae). Novon 9: 414-418. En. — Provision of names for 10 species formerly assigned to *Laurentia.*

Lammers, T. G. (2001). Further nomenclatural innovations in world Campanulaceae. Novon 11: 67-68. En. — Four nomenclatural innovations for species of *Codonopsis, Lobelia,* and *Wahlenbergia.*

Lammers, T. G. (2006). Validation of names at subspecific rank in North American Campanulaceae. Novon 16: 69-73. En. — Thirteen nomenclatural innovations for subspecies in *Diastatea, Downingia, Githopsis, Heterotoma, Lobelia, Palmerella,* and *Triodanis.*

Europe

Grenier, J. C. & D. A. Godron (1853). Lobéliacées, Campanulacées. In Flore de France 2: 395-422. Paris: J. B. Baillière. Fr. — Flora with descriptions and full nomenclature.

Willkomm, M. (1868). Lobeliaceae Juss., Campanulaceae Juss. In M. Willkomm & J. Lange, Prodromus florae Hispanicae 2: 278-298. Stuttgart: E. Schweizerbart. La. — Flora of Spain with keys, descriptions, full nomenclature, and specimen citations.

Nyman, C. F. (1879). Campanulaceae Juss., Lobeliaceae Juss. In Conspectus florae Europaeae: 475-487. Örebro: Officina Bohliniana. La. — Checklist for Europe, with synonymy.

Arcangeli, G. (1882). Campanulaceae. In Compendio della flora Italiana: 447-457. Torino. It. — Flora of Italy, with key to genera and brief descriptions.

Nyman, C. F. (1889-90). Campanulaceae Juss., Lobeliaceae Juss. In Conspectus florae Europaeae, suppl. 2: 206-213, 368. Örebro: Officina Bohliniana. La. — Supplement to Nyman (1879).

Willkomm, M. (1893). Lobeliaceae, Campanulaceae. In Supplementum prodromi florae Hispanicae: 125-131. Stuttgart: E. Schweizerbart. La. — Supplement to Willkomm (1868).

Dalla Torre, K. W. & L. G. Sarnthein (1912). Campanulaceae. In Flora der gefürsteten Graftschaft Tirol, des Landes Vorarlberg und des fürstentnumes Liechtenstein 6: 431-478. Innsbruck: Wagner. Ge. — Checklist for W. Austria, with full nomenclature and specimen citations.

Hayek, A. (1912). Campanulaceae Juss. In Flora von Steiermark 2(1): 438-470. Berlin: Gebrüder Borntraeger. Ge. — Flora for C. Austria, with keys, descriptions, synonymy, and specimen citations.

Jávorka, S. (1925). Campanulaceae. In Magyar flóra: 1070-1084. Budapest: Studium. Hu. — Flora for Hungary, with keys and synonymy.

Hayek, A. (1930). Prodromus florae peninsulae Balcanicae, Campanulaceae, Lobeliaceae. Repert. Spec. Nov. Regni Veg. 30(2): 521-567. En. — Floristic treatment for Balkans, with keys, descriptions, and full nomenclature.

Fournier, P. (1946). Campanulacées, Lobéliacées. In Les quatre flores de la France (ed. 2): 906-922, illus. Paris: Paul Lechevalier. Fr. — Flora with keys and synonymy.

• Fedorov, A. A. (1957). Campanulaceae Juss., Lobeliaceae Juss. In V. L. Komarov, Flora URSS 24: 126-453, 459-475, illus., map. Leningrad: Academia Scientiae URSS. Ru. — Flora for former Soviet Union, with keys, descriptions, supra- and infrageneric classifications, and full nomenclature.

• Tutin, T. G. (1976). Campanulaceae. In T. G. Tutin, V. H. Heywood, N., A. Burges, D. M. Moore, D. H. Valentine, S. M. Walters & D. A. Webb (eds.), Flora Europaea 4: 74-102. Cambridge: Cambridge University Press. En. — Flora of Europe with keys, descriptions, and full nomenclature.

Franco, J. (1984). Campanulaceae. In Nova flora de Portugal (continente e Açores) 2: 323-333, 569. Lisbon: Sociedade Astória. Por. — Flora with keys, descriptions, and full nomenclature.

• Greuter, W., H. M. Burdet & G. Long (1984). Campanulaceae. In Med-checklist 1: 120-154, map. Genève: Conservatoire et Jardin Botanique. En. — Checklist of species in countries bordering Mediterranean, with full nomenclature.

Meikle, R. D. (1985). Campanulaceae. In Flora of Cyprus 2: 1047-1059, illus. Kew: Royal Botanic Gardens. En. — Flora with keys, descriptions, full nomenclature, and specimen citations.

Clapham, A. R., T. G. Tutin & D. M. Moore (1987). Campanulaceae (incl. Lobeliaceae). In Flora of the British Isles (ed. 3): 421-425, illus. Cambridge: Cambridge University Press. En. — Flora with keys and descriptions.

Stace, C. (1991). Campanulaceae - Bellflower Family. In New flora of the British Isles: 756-765. Cambridge: Cambridge University Press. En. — Flora with keys and brief descriptions.

Kerguélen, M. (1993). *Campanula, Jasione, Phyteuma*. In Index synonimique de la flore de France: 34-35, 99, 136-137, xv. Paris: Muséum National d'Histoire Naturelle. Fr. — Checklist with synonymy.

Bolòs, O. & J. Vigo (1996). Campanulàcies (incl. Lobeliàcies). Flora dels països Catalans 3: 646-677, illus., maps. Barcelona: Barcino. Cat. — Flora for SE. Spain, with keys.

• Sales, F., I. C. Hedge, S. Castroviejo & J. J. Aldasoro, eds. (2001). Campanulaceae. In S. Castroviejo (ed.), Flora Iberica 14: 104-175, illus. Madrid: Real Jardín Botánico. Sp. — Flora of Spain with keys, descriptions, and full nomenclature.

Africa

Webb, P. B. & S. Berthelot (1844). Campanulaceae. Bartl. In Histoire naturelle des Iles Canaries 2(3): 1-9. Paris: Béthune. La. — Flora of Canary Islands, with descriptions, full nomenclature, and partial suprageneric classification.

Bolle, C. (1861). Addenda ad floram Atlantidis, praecipue insularum Canariensium Gorgadumque V. Bonplandia 9: 50-55. La. — Update of Webb & Berthelot (1844), with descriptions and full nomenclature.

• Sonder, O. W. (1865). Campanulaceae, Juss. In W. H. Harvey & O. W. Sonder, Flora Capensis 3: 530-605. Dublin: Hodges Smith. En. — Flora with keys, descriptions, full nomenclature, and specimen citations.

Melliss, J. C. (1875). Lobeliaceae, Campanulaceae. In St. Helena: a physical, historical, and topographical description of the island: 292-293 + pl. 46-49, illus. London: L. Reeve. En. — Includes botanical notes; four native species plus two cultivated.

Battandier, J. A. (1889). Lobéliacées Jussieu, Campanulacées Jussieu. In Flore de l'Algérie, Dicotylédones: 569-576. Paris: F. Savy. Fr. — Flora with keys and descriptions.

Marloth, R. (1932). Campanulaceae. In The flora of South Africa 3: 208-215 + pl. 54, illus. Capetown: Darter. En. — Generalized descriptive account, with keys to tribes and genera and diverse comments on natural history and taxonomy.

Turrill, W. B. (1949). On the flora of St. Helena. Kew Bull. 1948: 358-362. En. — Discussion of status of endemic flora.

Adamson, R. S. (1950). Campanulaceae R. Br. In R. S. Adamson & T. M. Salter, Flora of the Cape Peninsula: 740-759. Cape Town: Juta. En. — Flora with keys and descriptions.

Markgraf, F. (1950). Die Campanulaceen von Südwestafrika. Bot. Jahrb. Syst. 75: 206-220. Ge. — Floristic treatment for Namibia, with keys, descriptions, and full nomenclature.

• Quézel, P. (1953). Les Campanulacées d'Afrique du Nord. Feddes Repert. Spec. Nov. Regni Veg. 56: 1-65, illus., maps. Fr. — Floristic treatment with keys and descriptions

• Wimmer, F. E. (1953). Flore de Madagascar et des Comores (plantes vasculaires). 186. Lobéliacées (Lobeliaceae). 41 pp., illus. Paris: Firmin-Didot. Fr. — Flora for Madagascar and Comoros, with keys, descriptions, full nomenclature, and specimen citations.

• Hedberg, O. (1957). Afroalpine vascular plants. A taxonomic revision. Symb. Bot. Upsal. 15: 1-411 + pl. I-XII, illus. (Campanulaceae, pp. 184-196, 333-335). En. — Floristic/ecological treatment with keys, full nomenclature, and specimen citations.

Hepper, F. N. (1963). Campanulaceae. In J. Hutchinson & J. M. Dalziel, Flora of west tropical Africa (ed. 2) 2: 309-311, illus. London: Crown Agents for Overseas Governments. En. — Flora with keys, descriptions, full nomenclature, and specimen citations.

Quézel, P. & S. Santa (1963). Campanulacées. In Nouvelle flore de l'Algérie et des régions désertiques méridionales: 895-902, illus. Paris: Centre National de la Recherche Scientifique. Fr. — Flora with keys.

• Wimmer, F. E. (1963). Lobeliaceae. In J. Hutchinson & J. M. Dalziel, Flora of west tropical Africa (ed. 2) 2: 311-315, illus. London: Crown Agents for Overseas Governments. En. — Flora with keys, descriptions, full nomenclature, and specimen citations.

Roessler, H. (1966). Campanulaceae, Lobeliaceae. In H. Merxmüller, Prodromus einer Flora von Südwestafrika 136: 1-9, 137: 1-4. Lehre: J. Cramer. Ge. — Flora of Namibia, with keys, brief descriptions, full nomenclature, and specimen citations.

Burger, W. C. (1967). Campanulaceae, Lobeliaceae. In Families of flowering plants in Ethiopia: 122, 205, illus. Stillwater: Oklahoma State University Press. En. — Family description and list of genera.

Badré, F. & T. Cadet (1972). Lobeliaceae des Mascareignes. Adansonia (n.s.) 11: 667-683, illus. Fr. — Floristic treatment for Mauritius, Réunion, and Rodrigues, with keys, descriptions, full nomenclature, and specimen citations.

Täckholm, V. (1974). Campanulaceae A. Juss. In Students' flora of Egypt (ed. 2): 520-522, illus. Cairo: Cairo University. En. — Flora with keys and brief descriptions.

• Dyer, R. A. (1975). Campanulaceae, Lobeliaceae. In The genera of southern African flowering plants 1: 638-646, 734. Pretoria: Department of Agricultural Technical Services. En. — Generic flora with keys, descriptions, and full nomenclature.

• Badré, F. (1976). Flore des Mascareignes. 111. Campanulacées. 18 pp., illus. Paris: ORSTOM. Fr. — Flora for Mauritius, Réunion, and Rodrigues, with keys, descriptions, and full nomenclature.

• Thulin, M. (1976). Flora of tropical east Africa, Campanulaceae. 39 pp., illus., map. Rotterdam: Balkema. En. — Flora with keys, descriptions, full nomenclature, and specimen citations.

Ghafoor, A. (1977). Flora of Libya. 49. Campanulaceae. 10 pp., illus., map. Tripoli: Al Faateh University. En. — Flora with keys, descriptions, full nomenclature, and specimen citations.

• Thulin, M. (1977). Flora d'Afrique centrale (Zaire-Rwanda-Burundi), Spermatophytes, Campanulaceae. 49 pp., illus. Meise: Jardin Botanique National de Belgique. Fr. — Flora with keys, descriptions, full nomenclature, and specimen citations.

Perez de Paz, J. (1978). Estudio palinologico de las Campanulaceas endemicas de la region Macaronesica. Bot. Macarones. 6: 27-42, illus. Por. — Descriptive account of pollen morphology.

• Thulin, M. (1978). Flore de Madagascar et des Comores. 187. Campanulacées. 24 pp., illus., map. Paris: Muséum National d'Histoire Naturelle. Fr. — Flora for Madagascar and Comoros, with keys, descriptions, full nomenclature, and specimen citations.

Thulin, M. (1983a). Some tropical African Lobeliaceae. Chromosome numbers, new taxa and comments on taxonomy and nomenclature. Nord. J. Bot. 3: 371-382, illus., map. En. — Preparatory material for subsequent floristic treatments in the region.

• Thulin, M. (1983b). Campanulaceae, Lobeliaceae. In E. Launert (ed.), Flora Zambesiaca 7(1): 87-157, illus. Kew: Royal Botanic Garden. En. — Flora of Mozambique, Malawi, Zambia, Zimbabwe and Botswana, with keys, descriptions, full nomenclature and specimen citations.

• Greuter, W., H. M. Burdet & G. Long (1984). Campanulaceae. In Med-checklist 1: 120-154, map. Genève: Conservatoire et Jardin Botanique. En. — Checklist of species in countries bordering Mediterranean, with full nomenclature.

• Thulin, M. (1984). Flora of tropical east Africa, Lobeliaceae. 59 pp., illus., map. Rotterdam: Balkema. En. — Flora with keys, descriptions, full nomenclature, and specimen citations.

• Thulin, M. (1985). Flora d'Afrique Centrale (Zaire-Rwanda-Burundi), Spermatophytes. Lobeliaceae. 65 pp., illus. Meise: Jardin Botanique National de Belgique. Fr. — Flora with keys, descriptions, full nomenclature, and specimen citations.

Cronk, Q. C. B. (1987). The history of endemic flora of St Helena: a relictual series. New Phytol. 105: 509-520. En. — Discussion of evolutionary relationships and biogeography.

Kolberg, H., W. Giess, M. A. N. Müller & B. Strohbach (1992). Campanulaceae, Lobeliaceae. In List of Namibian plant species. Dinteria 22: 73-74. En. — Checklist with synonymy.

• Welman, W. G. (1993). Campanulaceae, Lobeliaceae. In T. H. Arnold & B. C. de Wet (eds.), Plants of southern Africa: names and distribution. Mem. Bot. Surv. S. Africa 62: 686-698. En. — Checklist of species, with nomenclator and distributional data.

• Tebbs, B. R. (1994). Campanulaceae. In J. R. Press and M. J. Short (eds.), Flora of Madeira: 329-333 + pl. 42, illus. London: HMSO. — Flora with keys and descriptions.

• Figueiredo, E. (1995). Flora de Cabo Verde. Plantas vasculares. 86. Campanulaceae. 11 pp., illus. Lisbon: Praia. Por. — Flora with key and descriptions.

Fennane, M. & M. Ibn Tattou (1998). Catalogue des plantes vasculaires rares, menacées ou endémiques du Maroc. Bocconea 8: 5-243 (Campanulaceae, 58-60). Fr. — Notes with full nomenclature on spp. endemic to or imperiled in Morocco.

• Goldblatt, P. & J. Manning (2000). Campanulaceae. In Cape plants. A conspectus of the Cape flora of South Africa. Strelitzia 9: 388-402, 706-708. En. — Checklist with key to genera and brief species descriptions.

• Welman, W. G. (2000). Campanulaceae, Lobeliaceae. In O. A. Leistner (ed.), Seed plants of southern Africa: families and genera. Strelitzia 10: 200-202, 339-340. En. — Generic flora with keys and descriptions.

Dobignard, A. (2002). Contributions à la connaissance de la flore du Maroc et de l'Afrique du Nord. Nouvelle série. I. J. Bot. Soc. Bot. France 20: 5-43, illus., maps. Fr. — Floristic notes for Morocco and N. Africa, with full nomenclature and specimen citations.

Asia

* Hooker, J. D. & T. Thomson (1858). Praecursores ad floram Indicam: being sketches of the natural families of Indian plants, with remarks on their distribution, structure, and affinities. Series I. Stylidieae, Goodenovieae, et Campanulaceae (including Lobeliaceae). J. Proc. Linn. Soc., Bot. 2: 1-29. En. — Floristic treatment with descriptions, full nomenclature, and specimen citations.

von Trautvetter, E. R. (1879). Catalogus Campanulacearum Rossicarum. Trudy Imp. S.-Petersburgsk. Bot. Sada 6: 43-104. La. — Checklist for Russia with full nomenclature.

Merrill, E. D. (1923). Campanulaceae. In An enumeration of Philippine flowering plants 3: 586-589. Manila: Bureau of Science. En. — Checklist with full nomenclature and specimen citations.

Tsoong, P. C. (1935). Preliminary study on Chinese Campanulaceae. Contr. Inst. Bot. Natl. Acad. Peiping 3: 61-118 + pl. IX-XIV, illus. En. — Checklist with key to genera, full nomenclature, and specimen citations.

Hara, H. (1952). Campanulaceae [nom. conserv.] Jussieu. In Enumeratio spermatophytarum Japonicarum 2: 87-103. Tokyo: Iwanami Shoten. Ja., La. — Checklist for Japan with full nomenclature.

* Fedorov, A. A. (1957). Campanulaceae Juss., Lobeliaceae Juss. In V. L. Komarov, Flora URSS 24: 126-453, 459-475, illus., map. Leningrad: Academia Scientiae URSS. Ru. — Flora for former Soviet Union, with keys, descriptions, supra- and infrageneric classifications, and full nomenclature.

Kitamura, S. (1960). Campanulaceae. In Flora of Afghanistan: 377-383, illus. En. Tokyo: Kyoto University. — Checklist with full nomenclature.

Moeliono, B. & P. Tuyn (1960). Campanulaceae. In C. G. G. J. van Steenis (ed.), Flora Malesiana (ser. I) 6(1): 107-141, illus., maps. Djakarta: Noordhoff-Kolff. En. — Flora of Malesia and Papuasia with keys, descriptions, and full nomenclature.

Backer, C. A. & R. C. Bakhuizen van den Brink, Jr. (1965). Campanulaceae, Lobeliaceae. In Flora of Java (spermatophytes only): 446-452. Groningen: P. Noordhoff. En. — Flora with keys, descriptions, and synonymy.

* Rechinger, K. H. & H. Schiman-Czeika (1965). Campanulaceae. In K. H. Rechinger (ed.), Flora Iranica 13: 1- 51 + tab. 1-12, illus. Graz: Akademische Druck- und Verlagsanstalt. Ge., La. — Flora of Iran, Afghanistan, and adjacent portions of Pakistan, Iraq, Azerbaijan, and Turkmenistan, with keys, descriptions, full nomenclature, and specimen citations.

Van Thuan, N. (1969). Campanulaceae. In A. Aubréville (ed.), Flore de Cambodge, du Laos et du Vietnam 9: 1-55, illus. Paris: Muséum National d'Histoire Naturelle. Fr. — Flora of Indochina, with keys, descriptions, and full nomenclature.

Gusseinov, S. A. (1970). [On the Campanulaceae of Daghestan.] Bot. Žurn. (Moscow & Leningrad) 55: 1498-1501, illus. Ru. — Checklist for Dagestan in N. Caucasus.

Hara, H. (1971). Campanulaceae. In H. Hara (ed.), The flora of eastern Himalaya, second report: 129-131, illus. Tiokyo: University of Tokyo Press. En. — Checklist with full nomenclature and specimen citations.

Huang, T. C. (1972). Campanulaceae. In Pollen flora of Taiwan: 72-73 + pl. 29-31, illus. Taipei: National Taiwan University. En. — Descriptive palynology, including a key to genera.

van Steenis, C. G. G. J. (1972). Campanulaceae. In C. G. G. J. van Steenis (ed.), Flora Malesiana (ser. I) 6(6): 928. Djakarta: Noordhoff-Kolff. En. — Supplement to Moeliono & Tuyn (1960).

Kao, M. T. & C. E. DeVol (1974). The Campanulaceae of Taiwan. Taiwania 19: 123-147, illus. En. — Floristic treatment with keys, descriptions, full nomenclature, and specimen citations.

Ying, S. S. (1975). A list of the alping [sic] plants of Taiwan (II). Campanulaceae. Quart. J. Chin. Forest. 8: 127. En. — Checklist of high-elevation species.

* Damboldt, J. (1976). Materials for a flora of Turkey XXXII: Campanulaceae. Notes Roy. Bot. Gard. Edinburgh 35: 39-52, illus. En. — Preparatory material for Damboldt (1978).

Walker, E. H. (1976). Campanulaceae. Kikyô Ka; Bellflower Family. In Flora of Okinawa and the southern Ryukyu Islands: 1005-1009. Washington: Smithsonian Institution Press. En. — Flora with keys, descriptions, full nomenclature, and specimen citations.

Maekawa, F., H. Hara & T. Tuyama (1977). [Campanulaceae.] In Makino's new illustrated flora of Japan: 612-617, illus. Tokyo: Hokuryukan. Ja. — Flora treatment with descriptions and synonymy.

• Damboldt, J. (1978). Lobeliaceae, Campanulaceae. In P. H. Davis (ed.), Flora of Turkey and the East Aegean Islands 6: 1-89, illus., maps. Edinburgh: University Press. En. — Flora with keys, descriptions, full nomenclature, and specimen citations.

• Feinbrun-Dothan, N. (1978). Campanulaceae. In Flora Palaestina 3: 277-285 + pl. 473-489, illus. Jerusalem: Israel Academy of Sciences & Humanities. En. — Flora for Israel and Jordan, with keys, descriptions and full nomenclature.

Kitagawa, M. (1979). Campanulaceae. In Neo-lineamenta florae Manshuricae: 599-608. Vaduz: J. Cramer. En. — Checklist for Manchuria, with full nomenclature.

Wu, C. Y. (1979). Lobeliaceae. In C.Y. Wu & J. Chen (eds.), Flora Yunnanica 2: 521-536, illus. Beijing: Science Press. Ch. — Flora for Yunnan Prov., China, with keys, descriptions and full nomenclature.

Kolakovskii, A. A. (1980). Campanulaceae. In Flora Abchazii 1: 125-142. Tblisi: Metsniereba. Ru. — Flora of Abkhaziya in Transcaucasus, with keys, descriptions, and synonymy.

Hara, H. (1982). Campanulaceae. In H. Hara, A. O. Chater & L. H. J. Williams, An enumeration of the flowering plants of Nepal 3: 49-54, 220. London: British Museum (Natutral History). En. — Checklist with full nomenclature and specimen citations.

Haridasan, V. K. & P. K. Mukherjee (1982). Phytogeographical notes on the Indian Campanulaceae. J. Econ. Taxon. Bot. 3: 739-749. En. — Compilation of distributional data and geographic relationships of all species in India.

• Cramer, L. H. (1983). Campanulaceae, Lobeliaceae. In A revised handbook to the flora of Ceylon 4: 160-177. New Delhi: Amerind. En. — Flora of Sri Lanka with keys, descriptions, full nomenclature, and specimen citations.

• Hong, D. Y., ed. (1983). Campanulales. In Flora Reipublicae Popularis Sinicae 73(2): i-viii + 1-208, illus. Beijing: Science Press. Ch. — Flora of China, with keys, descriptions, supra- and infrageneric classifications, and full nomenclature.

• Greuter, W., H. M. Burdet & G. Long (1984). Campanulaceae. In Med-checklist 1: 120-154, map. Genève: Conservatoire et Jardin Botanique. En. — Checklist of species in countries bordering Mediterranean, with full nomenclature.

• Nasir, E. (1984). Campanulaceae. In E. Nasir & S. I. Ali (eds.), Flora of Pakistan 155: 1-22, illus., map. Islamabad: Pakistan Agricultural Research Council. En. — Flora with keys, descriptions, full nomenclature, and specimen citations.

Ohwi, J. (1984). Campanulaceae. In Flora of Japan (in English): 848-854. Washington: Smithsonian Institution Press. En. — Flora with keys, descriptions, and synonymy.

Hong, D. Y. (1985). Campanulaceae. In C. Y. Wu (ed.) , Flora Xizangica 4: 560-601, illus. Beijing: Science Press. Ch. — Flora for Xinjiang and Qinghai Prov., China, with keys, descriptions and full nomenclature.

Keng, H. (1985). Annotated list of seed plants of Singapore (IX). Campanulaceae. Gard. Bull. Singapore 38: 166-167. — Floristic treatment with key, descriptions, and synonymy.

• Davis, P. H., R. R. Mill & K. Tan (1988). Campanulaceae. In Flora of Turkey and the East Aegean Islands, supplement: 177-181. Edinburgh: University Press. En. — Supplement to Damboldt (1978).

Haridasan, V. K. & P. K. Mukherjee (1988a). Seed surface features of some members of the Indian Campanulaceae. Phytomorphology 37: 277-285, illus. En. — Survey of seed coat morphology in select taxa.

• Haridasan, V. K. & P. K. Mukherjee (1988b). Lobeliaceae. In P. K. Hajra & M. Sanjappa (eds.), Fascicles of flora of India 19: 39-63, illus. Calcutta: Botanical Survey of India. En. — Flora with keys, descriptions, and full nomenclature.

Lee, S. T., Y. J. Chung & J. K. Lee (1988). A palynotaxonomic study of the Korean Campanulaceae. Korean J. Pl. Taxon. 18: 115-131, illus. Ko. — Survey of pollen surface morphology in select taxa.

Lee, T. B. [Campanulaceae]. (1989). In Illustrated flora of Korea: 719-725. Hyangmunsa: Soul T'ukpylosi. Ko. — Flora produced in South Korea, with descriptions and synonymy.

Huang, S. H. (1991). Campanulaceae nom. conserv. In C.Y. Wu & J. Chen (eds.), Flora Yunnanica 5: 452-509, illus. Beijing: Science Press. Ch. — Flora of Yunnan Prov., China, with keys, descriptions and full nomenclature.

Hong, D. Y. & L. M. Ma (1992). Campanulaceae. In H. W. Li & Z. Y. Zhu (eds.), Flora Sichuanica 10: 515-586, illus. Beijing: Science Press. Ch. — Flora of Sichuan Prov., China, with keys, descriptions and full nomenclature.

Lammers, T. G. (1992). Systematics and biogeography of the Campanulaceae of Taiwan. In C. I. Peng (ed.), Phytogeography and botanical inventory of Taiwan: 43-61. Taipei: Academia Sinica. En. — Discussion of evolution and biography of indigenous species.

Ohba, H. & S. Akiyama (1992). Campanulaceae. In The alpine flora of the Jaljale Himal, East Nepal: 62-63. Tokyo: University of Tokyo. En. — Checklist with full nomenclature and specimen citations.

Heller, D. & C. C. Heyn (1993). Campanulaceae. In Conspectus florae orientalis 8: 1-14, maps. Jerusalem: Israel Academy of Sciences & Humanities. En. — Checklist for Middle East with full nomenclature.

• Shimizu, T. (1993). Campanulaceae. In K. Iwatsuki, T. Yamazaki, D. E. Boufford & H. Ohba (eds.), Flora of Japan 3a: 405-418. Tokyo: Kodansha. En. — Flora with keys, descriptions, and full nomenclature.

Ying, T. S., Y. L. Zhang & D. E. Boufford (1993). Campanulaceae. In The endemic genera of seed plants of China: 297-303, illus., maps. Beijing: Science Press. En. — Detailed descriptions of *Echinocodon* and *Homocodon*.

Zhao, Y. Z. (1993). Campanulaceae. In Y. C. Ma (ed.), Flora Intramongolica (ed. 2) 4: 434-476, illus. Huhhot: Intramongolica Popularis. Ch. — Floristic treatment for Nei Mongol Prov., China, with keys, descriptions and full nomenclature.

• Czerepanov, S. K. (1995). Campanulaceae Juss., Lobeliaceae R. Br. In Vascular plants of Russia and adjacent states (the former USSR): 153-157, 312, map. Cambridge: Cambridge University Press. En. — Checklist with synonymy.

• Oganesian, M. (1995). Synopsis of Caucasian Campanulaceae. Candollea 50: 275-308, map. En. — Checklist of species in Caucasus, with full nomenclature.

Turner, I. M. (1995). A catalogue of the vascular plants of Malaya. Campanulaceae. Gard. Bull. Singapore 47: 157-158. En. — Checklist with synonymy.

Yoo, K. O. & W. T. Lee (1995). Seed morphology of Korean Campanulaceae. Korean J. Pl. Taxon. 25: 61-70, illus. Ko. — Survey of seed coat morphology in select taxa.

• Haridasan, V. K. & P. K. Mukherjee (1996). Campanulaceae. In P. K. Hajra & M. Sanjappa (eds.), Fascicles of flora of India 22: 25-118, illus. Calcutta: Botanical Survey of India. En. — Flora with keys, descriptions, and full nomenclature.

• Lammers, T. G. (1998). Campanulaceae. In T. C. Huang (ed.), Flora of Taiwan (ed. 2) 4: 775-802, 1169-1171, illus. Taipei: National Taiwan University. En. — Flora with keys, descriptions, full nomenclature, and specimen citations.

Bhellum, B. L. & R. Mangotra (1999). A contribution to the flora of the district Doda of Jammu and Kashmir state (family Campanulaceae Juss.). Indian J. Forest. 22: 187-188. En. — Checklist for area in N. India, with keys, full nomenclature, and specimen citations.

Im, R. J. (1999). Campanulaceae Juss., Lobeliaceae Jucc [sic.]. In Flora Coreana 7: 12-45, illus. Pyongyang: Science & Technology Publishing House. Ko. — Flora produced in North Korea, with keys, descriptions, and full nomenclature.

• Güner, A. & B. Yildiz (2000). Campanulaceae. In A. Güner, N. Özhatay, T. Ekim & K. H. C. Başer (eds.), Flora of Turkey and the East Aegean Islands, Supplement 2: 177-181. Edinburgh: University Press. En. — Second supplement to Damboldt (1978); see Davis et al. (1988) for first supplement.

Grubov, V. I. (2001). Campanulaceae Juss. In Key to the vascular plants of Mongolia: 608-611. Enfield: Science Publishers. En. — Checklist with keys.

Kress, W. J., R. A. DeFilipps, E. Farr & D. Y. Y. Kyi (2003). A checklist of the trees, shrubs, herbs, and climbers of Myanmar. Campanulaceae. Contr. U.S. Natl. Herb. 45: 185. En. — Checklist.

Australasia

Brown, R. (1810). Campanulaceae. In Prodromus florae Novae Hollandiae et Insulae Van-Diemen: 559-565. London: J. Johnson. La. — Flora with descriptions and full nomenclature.

Endlicher, S. (1836). Lobeliaceae, Campanulaceae. In Bemerkungen über die Flora der Südseeinseln. Ann. Wiener Mus. Naturgesch 1: 170. La. — Checklist of species of Australasia, Hawaiian and Society Is., with full nomenclature and specimen citations.

Hooker, J. D. (1844). Lobeliaceae, Juss. In The botany of the antarctic voyage of H.M. discovery ships *Erebus* and *Terror* in the years 1839-1843. I. Flora antarctica: 41-44 + tab. XXIX, illus. London: Reeve Bros. En. — Flora of Antipodean Is. with descriptions.

Hooker, J. D. (1852). Lobeliaceae, Juss.; Campanulaceae, Juss. In The botany of the antarctic voyage of H.M. discovery ships *Erebus* and *Terror* in the years 1839-1843. II. Flora Novae-Zelandiae 1: 156-160. London: Reeve Bros. En. — Flora of New Zealand with descriptions and full nomenclature.

Hooker, J. D. (1856). Lobeliaceae, Campanulaceae. In The botany of the antarctic voyage of H.M. discovery ships *Erebus* and *Terror* in the years 1839-1843. III. Flora Tasmaniae 1: 236-240 + pl. LXIX-LXXI, illus. London: Reeve Bros. En. — Flora of Tasmania with descriptions and full nomenclature.

• Bentham, G. & F. Mueller (1868). Campanulaceae. In Flora Australiensis: a description of the plants of the Australian Territory 4: 121-138. London: L. Reeve. En. — Flora with keys, descriptions, full nomenclature, and specimen citations.

Mueller, F. (1882). Campanulaceae. In Systematic census of Australian plants 1: 85. Melbourne: M'Carron Bird. En. — Checklist with full nomenclature.

• Allan, H. H. (1961). Campanulaceae, Lobeliaceae. In Flora of New Zealand 1: 787-794, 796-802, 976, 1031-1032. Wellington: R. E. Owen. En. — Flora with keys, descriptions and full nomenclature.

Willis, J. H. (1972). Campanulaceae, Lobeliaceae. In A handbook to the plants of Victoria 2: 624-633. Carlton: Melbourne University Press. En. — Flora with keys and full nomenclature.

• Carolin, R. C. (1981). Campanulaceae. In J. Jessop (ed.), Flora of central Australia: 349-351, illus. Sydney: Reed. En. — Flora with keys and descriptions.

Carolin, R. C. (1982). Campanulaceae, Lobeliaceae. In N. C. W. Beadle, O. D. Evans & R. C. Carolin, Flora of the Sydney Region (ed. 3): 432-435, illus. Sydney: Reed. En. — Flora for region in New South Wales, with keys and generic descriptions.

Green, J. W. (1985). Census of the vascular plants of Western Australia (ed. 2): 156. Perth: Western Australian Herbarium. En. — Checklist with full nomenclature.

• Stanley, T. D. & E. M. Ross (1986). Campanulaceae. In Flora of south-eastern Queensland 2: 474-480, illus. Brisbane: Queensland Department of Primary Industries. En. — Flora with keys and descriptions.

• Toelken, H. R. (1986). Campanulaceae. In Flora of South Australia (ed. 4) 3: 1367-1383, illus. Adelaide: Government Printing Division. En. — Flora with keys, descriptions, and full nomenclature.

Wheeler, J. R. (1987). Campanulaceae, Lobeliaceae. In N. G. Marchant, J. R. Wheeler, B. L. Rye, E. M. Bennett, N. S. Lander & T. D. Macfarlane, Flora of the Perth region 2: 598-606, illus. Perth: Western Australian Herbarium. En. — Flora for region in Western Australia, with keys and descriptions.

Forbes, S. J. & J. H. Ross (1988). Campanulaceae. In A census of the vascular plants of Victoria (ed. 2): 46. Melbourne: National Herbarium of Victoria. En. — Checklist with synonymy and full nomenclature.

Warcup, J. H. (1988). Mycorrhizal associations and seedling development in Australian Lobelioideae (Campanulaceae). Austral. J. Bot. 36: 461-472, illus. En. — Perennials solely vesicular-arbuscular (VA), annuals both VA and ectomycorrhizal.

• Webb, C. J., W. R. Sykes & P. J. Garnock-Jones (1988). Campanulaceae. In Flora of New Zealand 4: 457-461, illus. Wellington: R. E. Owen. En. — Coverage of naturalized species, with keys, descriptions, and full nomenclature.

- Smith, P. J. (1992). Campanulaceae. In G. J. Harden (ed.), Flora of New South Wales 3: 117-123, illus. Sydney: New South Wales University Press. En. — Flora with keys and descriptions.

Wheeler, J. R. (1992). Lobeliaceae. In J. R. Wheeler (ed.), B. L. Rye, B. L. Koch & A. J. G. Wilson, Flora of the Kimberley region: 871-874, illus. Perth: Western Australian Herbarium. En. — Flora for region in Western Australia, with keys and descriptions.

Wheeler, J. R. & P. J. Smith (1992). Campanulaceae. In J. R. Wheeler (ed.), B. L. Rye, B. L. Koch & A. J. G. Wilson, Flora of the Kimberley region: 869-871, illus. Perth: Western Australian Herbarium. En. — Flora for region in Western Australia, with key and descriptions.

- Wiecek, B. (1992). Lobeliaceae. In G. J. Harden (ed.), Flora of New South Wales 3: 123-131, illus. Sydney: New South Wales University Press. En. — Flora with keys and descriptions.

Pacific

Endlicher, S. (1836). Lobeliaceae, Campanulaceae. In Bemerkungen über die Flora der Südseeinseln. Ann. Wiener Mus. Naturgesch 1: 170. La. — Checklist of species of Australasia, Hawaiian and Society Is., with full nomenclature and specimen citations.

Gray, A. (1861). Notes on Lobeliaceae, Goodeniaceae, &c. of the collection of the U.S. South Pacific Exploring Expedition. Proc. Amer. Acad. Arts 5: 146-152. En. — Taxonomic commentary, including realignment of circumscriptions for endemic Hawaiian genera.

Mann, H. (1867-68). Enumeration of Hawaian [sic] plants, Lobeliaceae. Proc. Amer. Acad. Arts 7: 177-187. En. — Flora with descriptions, full nomenclature, and specimen citations.

Drake del Castillo, E. (1886). Campanulaceae. In Illustrationes florae insularum Maris Pacifici: 217-223. Paris: G. Masson. Fr. — Checklist for Hawaiian and Society Is. with full nomenclature and specimen citations

- Hillebrand, W. (1888). Lobeliaceae. In Flora of the Hawaiian Islands: 234-264. London: Williams & Norgate. En. — Flora with keys, descriptions, full nomenclature, and specimen citations.

Drake del Castillo, E. (1893). Campanulacées. In Flore de la Polynésie Française: 112-114. Paris: G. Masson. Fr. — Flora of Society Is., with keys, descriptions, full nomenclature, and specimen citations.

Rock, J. F. (1913). Campanulaceae. Tribe Lobelioideae. In The indigenous trees of the Hawaiian Islands: 469-494, 508-512, illus. Honolulu: published privately. En. — Floristic treatment of arborescent species, with keys, descriptions, full nomenclature, and specimen citations.

- Rock, J. F. (1919). A monographic study of the Hawaiian species of the tribe Lobelioideae family Campanulaceae. Mem. Bernice Pauahi Bishop Mus. 7(2): i-xvi + 1-395, illus. En. — Floristic treatment with keys, descriptions, infrageneric classifications, full nomenclature, and specimen citations.

Selling, O. H. (1947). Studies in Hawaiian pollen statistics II. The pollens of the Hawaiian phanerogams. Lobeliaceae. Spec. Publ. Bernice Pauahi Bishop Mus. 38: 318-320 + pl. 49, illus. En. — Descriptive palynology.

Degener, O. & I. Degener (1957). Lobeliaceae. *Lobelia* family. Flora Hawaiiensis, fam. 339. 2 pp. Honolulu: published privately. En. — Description of family and key to genera in Hawaiian Is. This flora comprises thousands of individually published pages, issued between 1932 and 1966. In addition to this introductory sheet, many of the species were illustrated and described. These are not listed here, but may be found in the index by Mill et al., Taxon 34: 249-250 (1985).

Stone, B. C. (1967). A review of the endemic genera of Hawaiian plants. Bot. Rev. (Lancaster) 33: 216-259. En. — Brief descriptive treatment with full nomenclature.

Degener, O. & I. Degener (1983). The *Lobelia* family (339). In Plants of Hawaii national parks illustrative of plants and customs of the South Seas (rev. ed.): 454-465, illus. Honolulu: published privately. En. — Popular account of family in Hawaiian Is.

• Lammers, T. G. (1990). Campanulaceae Bellflower Family. In W. L. Wagner, D. R. Herbst & S. H. Sohmer, Manual of the flowering plants of Hawai'i: 420-489, illus. Honolulu: University of Hawaii Press. En. — Flora of Hawaiian Is. with keys, descriptions, and synonymy for all species.

St. John, H. (1990). Lobeliaceae (*Lobelia* family). In J. R. Wichman & H. St. John, A chronicle and flora of Niihau: 135-137 + 1 pl., illus. Lāwa'i: National Tropical Botanical Garden. En. — Flora for a Hawaiian island, with key and full nomenclature.

Smith, A. C. (1991). Campanulaceae. In Flora vitiensis nova 5: 243-247. Lāwa'i: National Tropical Botanical Garden. En. — Flora with keys, descriptions, full nomenclature, and specimen citations.

Kartesz, J. T. (1994). Campanulaceae. In A synonymized checklist of the vascular flora of the United States, Canada, and Greenland (ed. 2): 186-191. Portland: Timber Press. En. — Includes species in Hawaiian Islands.

North America

Grisebach, A. H. R. (1861). Lobeliaceae. In Flora of the British West Indian islands: 385-388. London: L. Reeve. En. — Flora with descriptions.

Gray, A. (1878). Lobeliaceae, Campanulaceae. In Synoptical flora of North America 2(1): 1-14, 394. New York: Ivison, Blakeman, Taylor. En. — Flora with descriptions and synonymy.

Gray, A. (1886). Lobeliaceae, Campanulaceae. In Synoptical flora of North America (ed. 2) 2(1): 393-396. New York: Ivison, Blakeman, Taylor. En. — Supplement to Gray (1878).

Standley, P. C. (1938). Flora of Costa Rica, Lobeliaceae. Publ. Field Mus. Nat. Hist., Bot. ser. 18: 1405-1418. En. — Floristic treatment with descriptions, full nomenclature, and specimen citations.

• McVaugh, R. (1943). Campanulales, Campanulaceae, Lobelioideae. In North American Flora 32A: 1-134. New York: New York Botanical Garden. En. — Flora with keys, descriptions, and full nomenclature.

Moscoso, R. M. (1943). Campanulaceae, Lobeliaceae. In Catalogus florae Domingensis: 645-648. New York: published privately. La. — Checklist for Dominican Rep., with full nomenclature.

Marie-Victorin, J. L. C. (1944). Mise au point sur les Lobéliacées de l'ile de Cuba. Contr. Inst. Bot. Univ. Montreal 49: 3-18, illus. Fr. — Synopsis with keys and full nomenclature

McVaugh, R. (1945). Notes on North American Campanulaceae. Bartonia 23: 36-40. En. — Diverse taxonomic and floristic comments.

Fernald, M. L. (1950). Campanulaceae (Bluebell family). In Gray's manual of botany (ed. 8): 1350-1357, illus. New York: American Book. En. — Flora for NE. USA and adjacent Canada, with keys, descriptions, and synonymy.

• McVaugh, R. (1951). Campanulaceae Juss. In C. L. Lundell, Flora of Texas 3: 331-366. Renner: Texas Research Foundation. En. — Flora with keys, descriptions, full nomenclature, and specimen citations.

Gleason, H. A. (1952). Campanulaceae, the Harebell Family; Lobeliaceae, the Lobelia Family. In The new Britton and Brown illustrated flora of the northeastern United States and adjacent Canada 3: 314-322, illus. New York: New York Botanical Garden. En. — Flora with keys, descriptions, and synonymy.

Hitchcock, C. L., A. Cronquist & M. Ownbey (1959). Campanulaceae. In Vascular plants of the Pacific Northwest 4: 482-495, illus. Seattle: University of Washington Press. En. — Flora with keys, descriptions, and full nomenclature.

• Liogier, E. E. (1962). Campanulaceae. In Flora de Cuba 5: 165-173. Rio Piedras: Editorial Universitaria. Sp. — Flora with keys and descriptions.

Radford, A. E., H. E. Ahles & C. R. Bell (1964). Campanulaceae. In Manual of the vascular flora of the Carolinas: 1001-1008, illus., maps. Chapel Hill: University of North Carolina Press. En. — Flora with keys and descriptions.

Böcher, T. W., K. Holmen & K. Jakobsen (1968). Campanulaceae. In The flora of Greenland: 172-174, illus. Copenhagen: P. Haase & Sons. En. — Flora with key and descriptions.

Hultén, E. (1968). Campanulaceae (Bluebell Family). Flora of Alaska and neighboring territories: 847-851, illus., maps. Stanford: Stanford University Press. En. — Flora with key and descriptions.

Munz, P. A. & D. D. Keck (1970). Campanulaceae. Bellflower Family. In A California flora: 1061-1073, illus. Berkeley: University of California Press. En. — Flora with keys and descriptions.

• Adams, C. D. (1972). Campanulaceae. In Flowering plants of Jamaica: 734-737. Mona: University of the West Indies. En. — Flora with keys, descriptions, full nomenclature, and specimen citations.

Correll, D. S. & H. B. Correll (1972). Campanulaceae Juss. Bluebell Family. In Aquatic and wetland plants of southwestern United States: 1571-1585, illus. Washington: Environmental Protection Agency. En. — Flora with keys, descriptions, and synonymy.

Hitchcock, C. L. & A. Cronquist (1973). Campanulaceae. In Flora of the Pacific Northwest: 457-461, illus. Seattle: University of Washington Press. En. — Flora with keys, generic descriptions, and synonymy.

Welsh, S. l. (1974). Campanulaceae, Bellflower Family. In Anderson's flora of Alaska and adjacent parts of Canada: 66-69, illus. Provo: Brigham Young University Press. En. — Flora with keys and descriptions.

• Nash, D. L. (1976). Flora of Guatemala, Campanulaceae. Fieldiana Bot. 24: 396-431, illus. En. — Floristic treatment with keys, descriptions, and full nomenclature.

• Wilbur, R. L. (1977). Flora of Panama, Campanulaceae. Ann. Missouri Botanical Gard. 63: 593-655, illus. En. — Floristic treatment with keys, descriptions, full nomenclature, and specimen citations.

Fournet, J. (1978). Lobeliaceae. In Flore illustée des phanérogames de Guadeloupe et de Martinique: 594-601, illus. Paris: Institut National de la Recherche Agronomique. Fr. — Flora with keys, descriptions, and synonymy.

Scoggin, H. J. (1979). Campanulaceae (Bluebell or Harebell Family), Lobeliaceae (Lobelia Family). In The flora of Canada 1434-1441. Ottawa: National Museums of Canada. En. — Flora with keys.

Wiggins, I. L. (1980). Campanulaceae. Bluebell family. In Flora of Baja California: 229-235, illus. Stanford: Stanford University Press. En. — Flora with keys and generic descriptions.

Godfrey, R. K. & J. W. Wooten (1981). Campanulaceae (Bellflower Family). In Aquatic and wetland plants of southeastern United States, Dicotyledons: 734-751, illus. En. — Athens: University of Georgia Press. Flora with keys, descriptions, and synonymy.

Correll, D. S. & H. B. Correll (1982). Campanulaceae Juss. Bluebell Family. In Flora of the Bahama Archipelago: 1433-1437, illus. Vaduz: J. Cramer. En. — Flora with description and full nomenclature.

Cronquist, A., A. H. Holmgren, N. H. Holmgren, J. L. Reveal & P. K. Holmgren (1984). Campanulaceae, the Bellflower Family. In Intermountain flora 4: 515-527, illus. New York: New York Botanical Garden. En. — Flora with keys, descriptions, and full nomenclature.

Brooks, R. E. (1986). Campanulaceae Juss., the Bellflower Family. In Great Plains Flora Association, Flora of the Great Plains: 808-816. Lawrence: University Press of Kansas. En. — Flora for C. North America, with keys, generic descriptions, and synonymy.

• Rosatti, T. J. (1986). The genera of Sphenocleaceae and Campanulaceae in the southeastern United States. J. Arnold Arbor. 67: 1-64, illus. En. — Floristic treatment with key, descriptions, suprageneric classification, and full nomenclature.

Shetler, S. G. & N. R. Morin (1986). Seed morphology in North American Campanulaceae. Ann. Missouri Bot. Gard. 73: 653-688, illus. En. — Detailed study of seed coat structure and appearance and its taxonomic implications.

Kelly, D. L. (1988). The threatened flowering plants of Jamaica. Biol. Conserv. 46: 201-216, maps. En. — Discussion of species considered rare or imperiled.

• Howard, R. A. (1989). Lobeliaceae. Flora of the Lesser Antilles 6: 497-507, illus. Jamaica Plain: Arnold Arboretum. En. — Flora with keys, descriptions, and full nomenclature.

- Morin, N. (1993). Campanulaceae. In J. C. Hickman (ed.), The Jepson manual, higher plants of California: 459-468, illus. Berkeley: University of California Press. En. — Flora with keys, descriptions, and synonymy.

  Kartesz, J. T. (1994). Campanulaceae. In A synonymized checklist of the vascular flora of the United States, Canada, and Greenland (ed. 2): 186-191. Portland: Timber Press. En. — Checklist of species in North America north of Mexico, with synonymy.

- Liogier, H. (1995). Campanulaceae. In La flora de la Española 7: 440-459, illus. San Pedro de Macorís: Ediciones de la Universidad Central del Este. Sp. — Flora of Hispaniola with keys, descriptions, and full nomenclature.

- Liogier, H. (1997). Campanulaceae. In Descriptive flora of Puerto Rico and adjacent islands 5: 209-217, illus. San Juan: Editorial de la Universidad de Puerto Rico. En. — Flora with keys, descriptions, and full nomenclature.

- Rzedowski, J. & G. Calderón de Rzedowski (1997). Flora del Bajio y de Regiones Adyacentes. 58. Campanulaceae. 67 pp., illus., maps. Pátzcuaro: Instituto de Ecología. Sp. — Flora for a portion of C. Mexico, with keys, descriptions, full nomenclature, and specimen citations.

  Douglas, G. W. (1998). Campanulaceae. In G. W. Douglas, G. B. Straley, D. Meidinger & J. Pojar, Illustrated flora of British Columbia 2: 198-211, illus. Victoria: British Columbia Ministry of Forests. En. — Flora with keys.

  Wunderlin, R. P. (1998). Campanulaceae (Bellflower Family). In Guide to the vascular plants of Florida: 588-591. Gainesville: University Press of Florida. En. — Flora with keys.

  Magee, D. W. & H. E. Ahles (1999). Campanulaceae (Bluebell Family). In Flora of the Northeast: 971-976, illus., maps. Amherst: University of Massachusetts Press. En. — Flora for New England and adjacent New York, with keys, descriptions, and synonymy.

  Crow, G. E. & C. B. Hellquist (2000). Campanulaceae / Bellflower Family. In Aquatic and wetland plants of northeastern North America 1: 365-369, illus. Madison: University of Wisconsin Press. En. — Flora with keys and synonymy.

- Wilbur, R. L. (2001). Campanulaceae Juss. In W. D. Stevens, C. U. Ulloa, A. Pool & O. M. Montiel (eds.), Flora de Nicaragua 1: 557-564. St. Louis: Missouri Botanical Garden Press. Sp. — Flora with keys, descriptions, and full nomenclature.

- Gleason, H. A. & A. Cronquist (2004). Campanulaceae, the Bellflower family. In Manual of the vascular plants of northeastern United States and adjacent Canada (ed. 2, 7th printing): 496-500. New York: New York Botanical Garden. En. — Flora with keys, descriptions, and synonymy.

  Lammers, T. G. (2004). Campanulaceae. In N. Smith, S. A. Mori, A. Henderson, D. W. Stevenson & S. V. Heald (eds.), Flowering plants of the Neotropics: 78-80, illus. Princeton: Princeton University Press. En. — Descriptive account of family.

South America

  Hooker, J. D. (1846). Lobeliaceae, Juss. In The botany of the antarctic voyage of H.M. discovery ships *Erebus* and *Terror* in the years 1839-1843. I. Flora antarctica: 325. London: Reeve Bros. En. — Flora of Tierra del Fuego & Falkland Is., with descriptions.

  Gay, C. (1849). Lobeliaceas, Campanulaceas. In Historia fisica y politica de Chile, Botanica 4: 319-344. Santiago: Museo Historia Natural. Sp. — Flora with descriptions.

- Kanitz, A. (1878). Lobeliaceae. In C. F. P. von Martius, Flora Brasiliensis 6(4): 130-158 + tab. 39-45, illus. Munich: published privately. La. — Flora with keys, descriptions, full nomenclature, and specimen citations.

  Philippi, F. (1881). Lobeliaceae, Campanulaceae. In Catalogus plantarum vascularium Chilensium: 232-236. Santiago: Imprenta Nacional. La. — Checklist of species in Chile, with synonymy and full nomenclature.

  Hemsley, W. B. (1884). Report on the botany of the Juan Fernandez, the south-eastern Moluccas, and the Admirality Islands, Campanulaceae. In Report on the scientific results of the voyage of H.M.S. *Challenger* during the years 1873-1876, botany 1(3): 45-46 + pl. LVI, illus. London: HMSO. En. — Floristic account for Juan Fernández Is., with full nomenclature and specimen citations.

• Kanitz, A. (1885). Campanulaceae. In C. F. P. von Martius, Flora Brasiliensis 6(4): 178-188 + tab. 48-49, illus. Munich: published privately. La. — Flora with keys, descriptions, full nomenclature, and specimen citations.

Zahlbruckner, A. (1897). Revisio Lobeliacearum Boliviensium hucusque cognitarum. Bull. Torrey Bot. Club 24: 371-388. La. — Floristic treatment with keys, descriptions, full nomenclature, and specimen citations.

Reiche, K. (1902). Zur Kenntnis der Bestäubung chilenischer Campanulaceen und Goodeniaceen. Verh. Deutsch. Wiss. Vereins Santiago de Chile 4: 509-522. Ge. — Pollination biology among Chilean spp.

Reiche, K. (1905). Estudios criticos sobre la flora de Chile, Campanuláceas. Anales Univ. Chile 117: 206-208, 451-464, 481-482. Sp. — Flora with keys, descriptions, and full nomenclature.

Gleason, H. A. (1921). A rearrangement of the Bolivian species of *Centropogon* and *Siphocampylus*. Bull. Torrey Bot. Club 48: 189-201. En. — Regional synopsis with key, brief descriptions, and full nomenclature.

Skottsberg, C. (1921). The phanerogams of the Juan Fernandez Islands, Campanulaceae. In C. Skottsberg (ed.), The natural history of Juan Fernandez and Easter Island 2: 175-180, illus. Uppsala: Almqvist & Wiksells. En. — Flora with key, descriptions, full nomenclature, and specimen citations.

Lanjouw, J. (1934). Campanulaceae. In A. Pulle (ed.), Flora of Suriname 4(1): 302-305. Amsterdam: Koninklijke Vereeniging Indisch Instituut. En. — Flora with key, descriptions, and full nomenclature.

• Wimmer, F. E. (1935). Conspectus Lobelioidearum in Brasilia, Paraguay, Uruguay, Argentinia crescentium. Revista Sudamer. Bot. 2: 89-107. Ge. — Key to the species in non-Andean South America.

Wimmer, F. E. (1937). Flora of Peru, Campanulaceae. Bellflower family. Field Mus. Nat. Hist., Bot. Ser. 13: 383-489. En. — Floristic treatment with keys, descriptions, full nomenclature, and specimen citations.

Pittier, H., T. Lasser, L. Schnee, Z. Luces de Febres & V. Badillo (1947). Campanulaceae. In Catalogo de la flora Venezolana 2: 472-475. Caracas: Vargas. Sp. — Checklist of species in Venezuela, with key to genera.

Skottsberg, C. (1956). Derivation of the flora and fauna of Juan Fernandez and Easter Island, Campanulaceae. In C. Skottsberg (ed.), The natural history of Juan Fernandez and Easter Island 1: 210. Uppsala: Almqvist & Wiksells. En. — Biogeographic and evolutionary relationships of native species.

Foster, R. C. (1958). Campanulaceae. In A catalogue of the ferns and flowering plants of Bolivia. Contr. Gray Herb. 184: 200-202. En. — Checklist of species, with synonymy and full nomenclature.

Fabris, H. A. (1965). Campanulaceae. In A. L. Cabrera (ed.), Flora de la provincia de Buenos Aires: 407-412, illus. Buenos Aires: Instituto Nacional de Tecnologia Agropecuaria. Sp. — Flora for area in Argentina, with keys, descriptions, full nomenclature, and specimen citations.

Moore, D. M. (1968). Lobeliaceae. In The vascular flora of the Falkland Islands: 113-114. London: British Antarctic Survey. En. — Flora with descriptions and full nomenclature.

Wiggins, I. L. (1971). Campanulaceae. Bellflower Family. In I. L. Wiggins & D. M. Porter, Flora of the Galápagos Islands: 296-299, illus. Stanford: Stanford University Press. En. — Flora with descriptions and full nomenclature.

Bacigalupo, N. M. (1974). Campanulaceae, Campanuláceas. In A. Burkart (ed.), Flora ilustrada de Entre Rios (Argentina) 6B: 95-101, illus. Buenos Aires: Instituto Nacional de Tecnologia Agropecuaria. Sp. — Flora with keys, descriptions, full nomenclature, and specimen citations.

Cabrera, A. L. & E. M. Zardini (1978). Campanulaceae. In Manual de la flora de los alrededores de Buenos Aires (ed. 2): 603-605, illus. Buenos Aires: Editorial Acme. Sp. — Flora for area in Argentina, with keys and descriptions.

Navas Bustamante, L. E. (1979). Lobeliaceae. In Flora de la Cuenca de Santiago de Chile 3: 162-164, 414-415, illus. Santiago: Editiones de la Universidad de Chile. Sp. — Flora with key, descriptions, and full nomenclature.

• Jeppesen, S. (1981). Campanulaceae, Lobeliaceae. In G. Harling & B. Sparre (eds.), Flora of Ecuador 14: 1-170, illus., map. Stockholm: Swedish Natural Science Research Council. En. — Flora with keys, descriptions, full nomenclature, and specimen citations.

Moore, D. M. (1983). Campanulaceae, Lobeliaceae. In Flora of Tierra del Fuego: 208-210, 214, illus., maps. Oswestry: Anthony Nelson. En. — Flora for S. Argentina and Chile, with keys and brief descriptions.

• Fromm Trinta, E. & E. Santos (1989). Flora ilustrada Catarinense, Campanuláceas. 75 pp., illus., maps. Itajaí: Herbario "Barbosa Rodrigues". Por. — Flora for Santa Catarina in S. Brazil, with descriptions, full nomenclature, and specimen citations.

de Godoy, S. A. P. (1992). Flora da Serra do Cipó, Minas Gerais: Campanulaceae. Bol. Bot. Univ. São Paulo 13: 241-257, illus. Por. — Floristic treatment for region in SE. Brazil, with descriptions, full nomenclature, and specimen citations.

• Lammers, T. G., D. Kama & N. Morin (1993). Campanulaceae. In L. Brako & J. L. Zarruchi, Catalogue of the flowering plants and gymnosperms of Peru: 310-326. St. Louis: Missouri Botanical Garden. En. — Checklist.

Pontiroli, A. (1993). Campanulaceae. In A. L. Cabrera (ed.), Flora de la provincia de Jujuy República Argentina 9: 515-537, illus. Buenos Aires: Instituto Nacional de Tecnologia Agropecuaria. Sp. — Flora for NW. Argentina, with keys, descriptions, full nomenclature, and specimen citations.

Jung-Mendaçolli, S. L. & L. P. Cabral (1997). Campanulaceae. In M. M. R. F. Melo (ed.), Flora fanerogâmica da Ilha do Cardoso 5: 69-72, illus. São Paulo: Instituto de Botânica. Por. — Flora for region in SE. Brazil, with descriptions, full nomenclature, and specimen citations.

• Stein, B. A. (1998). Campanulaceae. In P. E. Berry, B. K. Holst & K. Yatskievych (eds.), Flora of the Venezuelan Guayana 4: 122-129, illus. St. Louis: Missouri Botanical Garden Press. En. — Flora with keys, descriptions, and full nomenclature.

• Chiapella, J. (1999). Campanulaceae. In M. N. Correa (ed.), Flora Patagonica 6: 479-488, illus. Buenos Aires: Instituto Nacional de Tecnologia Agropecuaria. Sp. — Flora for S. Argentina, with keys, descriptions, full nomenclature, and specimen citations.

• Chiapella, J. (1999). Campanulaceae. In F. O. Zuloaga & O. Morrone (eds.), Catálogo de las plantas vasculares de la República Argentina 2: 495-500. St. Louis: Missouri Botanical Garden Press. Sp. — Checklist with synonymy and distributional data.

• Jørgensen, P. M. (1999). Campanulaceae. In P. M. Jørgensen & S. León-Yánez (eds.), Catalogue of the vascular plants of Ecuador: 374-382. St. Louis: Missouri Botanical Garden Press. En. — Checklist.

Gracie, C. A. (2002). Campanulaceae (Bellflower Family). In S. A. Mori, G. Cremers, C. A. Gracie, J. J. de Granville, S. V. Heald, M. Hoff & J. D. Mitchell (2002). Guide to the plants of central French Guiana 2: 183-184 + pl. 38a-b, illus. New York: New York Botanical Garden. En. — Flora with descriptions.

Lammers, T. G. (2004). Campanulaceae. In N. Smith, S. A. Mori, A. Henderson, D. W. Stevenson & S. V. Heald (eds.), Flowering plants of the Neotropics: 78-80, illus. Princeton: Princeton University Press. En. — Descriptive account of family.

Cultivation

Voss, A. (1894). Campanulaceae, Glockenblumengewächse. In A. Siebert & A. Voss, Vilmorin's Blumengärterei Beschreibung, Kultur und Verwendung (ed. 3) 1: 560-578, illus. Berlin: Paul Parey. Ge. — Species in cultivation, with keys, descriptions, and synonymy.

Bailey, L. H. (1949). Campanulaceae. Bellflower family; Lobeliaceae. Lobelia family. In Manual of cultivated plants most commonly grown in the continental United States and Canada (rev. ed.): 958-972, illus. New York: MacMillan. En. — Descriptive account of species in cultivation, with keys and synonymy.

- Bailey, L. H. (1953). The garden of bellflowers in North America. xiii + 155 pp., illus. New York: MacMillan. En. — Semi-popular account of select species in cultivation, with keys, descriptions, and synonymy.
  Lawrence, G. H. M. (1955). Bellflowers for your garden. Horticulture 33: 234-235, 250-251, 253, illus. En. — Popular account of select species in cultivation.
- Eddie, W. M. M. & T. J. Ayers (2000). Campanulaceae. In J. Cullen, J. C. M. Alexander, C. D. Brickell, J. R. Edmondson, P. S. Green, V. H. Heywood, P. M. Jørgensen, S. L. Jury, S. G. Knees, H. S. Maxwell, D. M. Willer, N. K. B. Robson, S. M. Walters & P. F. Yeo (eds.), The European garden flora 6: 465-502, illus. Cambridge: Cambridge University Press. En. — Account of species cultivated in Europe, both out-of-doors and under glass, with keys, descriptions, and synonymy.
  Lammers, T. G. (2005). Campanulaceae. In G. W. Staples and D. R. Herbst, A tropical garden flora: 216-218, illus. Honolulu: Bishop Museum Press. En. — Account of species cultivated in Hawaiian Is., with key, descriptions, and synonymy.

Adenophora

Campanuloideae, 67 species, eastern Asia, from western Siberia to the Russian Far East, south to Himalaya and Vietnam, with one species extending well into Europe and another endemic to Krym. There is no comprehensive infrageneric classification; Fedorov (1957), Baranov (1963), Hong (1983), and Tu et al. (1998) all have proffered systems, but only for the species in their respective regions. Preliminary analysis of ITS data (Ge et al. 1998) suggests that recognition of two sections (sect. *Adenophora* and sect. *Microdiscus* Fed.) may prove best. Despite these plants' great overall resemblence to various species of *Campanula*, a preliminary molecular phylogeny of the family (Eddie et al. 2003, under **General**) indicates that *Adenophora* (with the addition of *Hanabusaya*, q.v.) forms a discrete clade within the "Rapunculus clade". The genus has not been monographed in its entirety since Candolle (1839, under **General**); the most comprehensive account currently available is Hong's (1983) floristic treatment for China. $2n$ = 34, 36, 68, 72, 102. Type [designated by the author]: *Campanula verticillata* Pall.

Fischer, F. E. L. (1823). Adumbratio generis *Adenophorae*. Mém. Soc. Imp. Naturalistes Moscou 6: 165-169. La. — Establishment of genus as distinct from *Campanula*.

Korshinsky, S. (1894). Untersuchungen über die russischen *Adenophora*-Arten. Mém. Acad. Imp. Sci. Saint Pétersbourg (ser. 7) 42(2): 1-41. Ge. — Floristic treatment for Russia with key and descriptions; postulates rampant hybridization of a few true species as cause of variability and polymorphism.

Borbás, V. de (1904). Az *Adenophora* kritikája (Recensio *Adenophorum*). Magyar Bot. Lapok 3: 189-196. Hu. — Synopsis with infrageneric classification and descriptions; suggests best treated as subgenus of *Campanula*.

Kitagawa, M. (1935). Species generis *Adenophorae* Manshuricae australes & Jeholenses. In S. Tokunaga (ed.), Report of the first scientific expedition to Manchoukuo 4(2): 106-115, illus. Tokyo: Waseda University. En. — Floristic treatment for NE. China with full nomenclature and specimen citations.

Reverdatto, V. V. (1935). Conspectus specierum generis *Adenophorae* Fisch. florae Jenisseensis. Sist. Zametki Mater. Gerb. Krylova Tomsk. Gosud. Univ. Kujbyševa 3-4: 1-6, illus. Ru. — Floristic treatment for C. Asia with key, descriptions, and specimen citations.

Kitamura, S. (1936). Notes on some adenophoras of Japan. Acta Phytotax. Geobot. 5: 204-210. Ja., La. — Miscellaneous taxonomic notes.

Milne-Redhead, E. (1936). *Adenophora morrisonensis*. Bot. Mag. 159: tab. 9444 + 3 pp., illus. En. — Portrait of subsp. *morrisonensis*, with description and notes on cultivation.

Kitamura, S. (1941). On *Adenophora triphylla* DC. Acta Phytotax. Geobot. 10: 305-311. Ja., La. — Infraspecific classification of this widespread and polymorphic species.

Sealy, J. R. (1941). *Adenophora coelestis*. Bot. Mag. 163: tab. 9626 + 4 pp., illus. En. — Portrait with description and notes on cultivation.

Bailey, L. H. (1953). *Adenophora*. In The garden of bellflowers in North America: 116-126, illus. New York: MacMillan. En. — Account of species cultivated in North America, with key and descriptions.

• Fedorov, A. A. (1957). *Adenophora* Fisch. In V. L. Komarov, Flora URSS 24: 343-372, illus. Leningrad: Academia Scientiae URSS. Ru. — Flora for former Soviet Union, with key, descriptions, infrageneric classification, and full nomenclature.

• Baranov, A. I. (1963). Materials to the monograph of the species of *Adenophora* of N. E. China. Quart. J. Taiwan Mus. 16: 143-179, illus. En. — Flora with keys, descriptions, full nomenclature, and infrageneric classification; prompted by disagreement with Fedorov (1957).

Kitagawa, M. (1965). Brief notes on *Adenophora verticillata* Fischer. J. Jap. Bot. 40: 319-320. En. — Brief taxonomic notes on this widespread polymorphic species.

Shimizu, T. (1965). A taxonomic and cytological note on *Adenophora* of Taiwan. J. Jap. Bot. 40: 372-377, illus. En. — Floristic treatment with key, full nomenclature, specimen citations, and chromosome counts.

Shimizu, T. (1966). Investigations on Japanese *Adenophora* I. Observations on a population of *A. petrophila*. J. Fac. Sci. Shinsu Univ. 1: 27-40, illus. Ja. — Variation and cytology in *A. nikoensis*.

Thuan, N. V. (1969). *Adenophora* Fisch. In Flore du Camdodge du Laos et du Vietnam 9: 22-23. Paris: Muséum National d'Histoire Naturelle. Fr. — Flora treatment with descriptions.

Ponomarchuk, G. I. (1971). [Sistematika i nomenklatura *Adenophora remotiflora* (Siebold et Zucc.) Miq.] Bjull. Glavn. Bot. Sada 79: 57-61, illus. Ru. — Synonymization of *A. trachelioides* under *A. remotiflora*.

Tutin, T. G. (1973). *Adenophora lilifolia* (L.) Besser. Bot. J. Linn. Soc. 67: 280. En. — Short note pointing out correct spelling and author citation.

Ponomarchuk, G. I. (1974). On the taxonomic independence of *Adenophora verticillata* Fisch. Bot. Žurn. 59: 234-237. Ru. — Argues for recognition as distinct from *A. triphylla*; detailed synonymies.

Kanai, H. (1978). An approach to a logical identification of Japanese plants: 1. Identification chart of Japanese species of the genus *Adenophora* (Campanulaceae). Bull. Natl. Sci. Mus., Tokyo, B 4: 155-159. En. — Identification guide in the form of a taxon × character matrix.

Zhao, Y. Z. (1980). [A preliminary report on the taxonomy study of the genus *Adenophora* Fisch. of Inner Mongolia.] Acta Sci. Nat. Univ. Intramongolicae 11: 54-61, illus. Ch. — Floristic treatment for Nei Monggol Prov., China, with key.

Shimizu, T. & J. Okazaki (1982). Investigations on Japanese *Adenophora* II. Chromosome numbers and pollen grains of some species. Acta Phytotax. Geobot. 33: 328-335, illus. Ja. — Cytological and palynological data in diverse species.

• Hong, D. Y. (1983). *Adenophora* Fisch. In Flora Reipublicae Popularis Sinicae 73(2): 92-139, illus. Beijing: Science Press. Ch. — Flora of China, with key, descriptions, infrageneric classification, and full nomenclature.

Fu, C. X. & M. Y. Liu (1986). Study on the taxonomy of *Adenophora* Fischer in Heilongjiang Province. Nat. Sci. J. Harbin Normal Univ. 2: 41-52. Ch. — Floristic treatment.

Fu, C. X. & M. Y. Liu (1987). Chromosome studies of 10 Chinese species of *Adenophora* Fisch. Acta Phytotax. Sin. 25: 180-188, illus. Ch. — Cytological data; polyploidy, aneuploidy, and changes in karyotype all present.

Okazaki, J. (1988). A taxonomic study of the *Adenophora nikoensis* complex (Campanulaceae) I. *A. uryuensis* and its identity. Acta Phytotax. Geobot. 39: 94-104, illus., maps. Ja. — Statistical analysis of morphology and cytology support recognition of *A. uryuensis*.

Lee, J. K. & S. T. Lee (1990). *Adenophora racemosa* (Campanulaceae), a new species from Korea. Korean J. Pl. Taxon. 20: 121-126, illus. En. — Includes key to Korean species of sect. *Platyphyllae* (Borbás) Fed.

Qiu, J. Z. & D. Y. Hong (1990). Notes on two species of the genus *Adenophora* from Liaoning Prov. and their chromosome numbers. Acta Phytotax. Sin. 28: 397-402, illus., map. Ch. — Elevation of *A. polyantha* var. *contracta* to specific rank; data supports recognition of *A. pinifolia*.

Huxley, A. J., ed. (1992). *Adenophora* Fisch. In The new Royal Horticultural Society dictionary of gardening 1: 59-60. London: MacMillan. En. — Account of the species in cultivation.

Lancaster, R. (1993). Plants that should be better known: *Adenophora stricta* var. [sic] *sessilifolia*. The Garden 118: 418-419, illus. En. — Introduction of *Adenophora stricta* subsp. *sessilifolia* to horticulture in Britain.

• Okazaki, J. (1993). *Adenophora* Fisch. In K. Iwatsuki, T. Yamazaki, D. E. Boufford & H. Ohba, Flora of Japan 3a: 406-410. Tokyo: Kodansha. En. — Flora with infrageneric classification, key, and descriptions.

Qiu, J. Z. & D. Y. Hong (1993). A biosystematic study on *Adenophora gmelinii* complex (Campanulaceae). Acta Phytotax. Sin. 31: 17-41, illus. Ch. — Experimental hybridizations used to test reproductive isolation.

Ge, S. & D. Y. Hong (1994). Biosystematic studies on *Adenophora potaninii* Korsh. complex (Campanulaceae). I. Phenotypic plasticity. Acta Phytotax. Sin. 32: 489-503, illus. Ch. — Analysis of variation in plants grown in common garden showed much environmentally induced variation in characters thought diagnostic.

Ge, S. & D. Y. Hong (1994). Biosystematic studies on *Adenophora potaninii* Korsh. complex(Campanulaceae). II. Crossing experiment. Cathaya 6: 15-26. En. — Experimental hybridizations support reducing *A. biformifolia* to synonymy under *A. wawreana,* and *A. bockiana* and *A. polydentata* to synonymy under *A. potaninii;* and treating *A. wawreana* and *A. potaninii* as conspecific subspecies (though requisite combination not effected).

Lee, J. K. & S. T. Lee (1994). A taxonomic study of the genus *Adenophora* in Korea. J. Nat. Sci. Sung Kyun Kwan Univ. 45(2): 15-32. Ko. — Floristic treatment with keys, descriptions, infrageneric classification, full nomenclature, and specimen citations.

Ge, S. & D. Y. Hong (1995). Biosystematic studies on *Adenophora potaninii* Korsh. complex (Campanulaceae) III. Genetic variation and taxonomic value of morphological characters. Acta Phytotax. Sin. 33: 433-443, illus. Ch. — Variation in many characters (once thought diagnostic) shown to be continuous.

Ge, S. & D. Y. Hong (1996). Genetic analysis of isozyme loci in *Adenophora potaninii* Korsh. Acta Bot. Sin. 38: 431-438 + pl. I-II, illus. Ch. — Preliminary electrophoretic survey of eight enzyme sytstems.

• Ge, S., B. A. Schaal & D. Y. Hong (1997). A reevaluation of the status of *A. lobophylla* based on ITS sequence, with reference to the utility of ITS sequence in *Adenophora.* Acta Phytotax. Sin. 35: 385-395, cladograms. Ch. — First application of DNA sequencing in genus suggests that *A. lobophylla* allied to *A. himalayana* rather than *A. potaninii* complex.

Lee, S. T., J. K. Lee & S. T. Kim (1997). A new species of *Adenophora* (Campanulaceae) from Korea. J. Pl. Res. 110: 77-80, illus. En. — Description of *A. erecta,* with key to species of sect. *Remotiflorae* (Fed. ex A. I. Baranov) P. F. Tu & G. J. Xu in China and Korea.

Zu, Y. G., W. H. Zhang, S. X. Wu, J. J. Zhou & F. J. Yang (1997). Comparative studies of sexual reproduction and asexual propagation between populations of *Adenophora lobophylla* and *A. potaninii.* Acta Bot. Sin. 39: 1065-1072, illus. Ch. — Poor seed germination and seedling survivorship may account for rarity of *A. lobophylla.*

Wang, K. Q. & S. Ge (1998). A karyotype study on five species of *Adenophora.* Acta Bot. Yunnan. 20: 58-62. Ch. — Karyotype analysis of five Chinese species.

• Ge, S., B. Schaal & D. Y. Hong (1998). A preliminary phylogenetic analysis of the genus *Adenophora* (Campanulaceae) based on ITS sequence data [abstract]. Amer. J. Bot. 85 (6, Suppl.): 172. En. — Molecular data support division of genus into two sections, sect. *Adenophora* and sect. *Microdiscus* Fed.

Ge, S. & D. Y. Hong (1998). Biosystematic studies on *Adenophora potaninii* Korsh. complex (Campanulaceae) IV. Allozyme variation and differentiation. Acta Phytotax. Sin. 36: 481-489. Ch. — Enzyme electrophoresis supports recognition of *A. lobophylla* and merger of *A. wawreana* and *A. potaninii.*

• Tu, P. F., H. B. Chen, G. J. Xu & L. S. Xu (1998). Classification and evolution of the genus *Adenophora* Fisch. in China. Acta Bot. Boreal.-Occid. Sin. 18: 613-621, illus. Ch. — Promulgation of revised infrageneric classification; includes non-cladistic phylogeny.

Hall, T. & Y. Harvey (1999). *Adenophora triphylla.* Bot. Mag. 16: 90-97 + pl. 365, illus. En. — Portrait with description, extensive (though not complete) synonymy, and key to two varieties recognized, with comments on morphological variation and cultivation.

• Krestovskaja, T. (2002). Genus *Adenophora* Fisch. (Campanulaceae) in Asia Centrali. Novosti Sist. Vyssh. Rast. 34: 235-243. Ru. — Floristic treatment for C. Asia, with key, full nomenclature, and specimen citations.

Adenophora Fisch., Mém. Soc. Imp. Naturalistes Moscou 6: 165 (1823). *Campanula* sect. *Adenophora* (Fisch.) D. Dietr., Syn. Pl. 1: 755 (1839). *Campanula* subg. *Adenophora* (Fisch.) Borbás, Magyar Bot. Lapok 3: 190 (1904).
Europe, temp. & trop. Asia. 11 13 14 30 31 32 36 37 38 40 41.
    Floerkea Spreng., Anleit. Kenntn. Gew. (ed. 2) 2: 523 (1818), as 'Flörkea;' non Willd., Ges. Naturf. Freunde Berlin Neue Schriften 3: 448 (1801).
    Campanula sect. *Floerkeae* Colla, Herb. Pedem. 4: 19 (1835).

Adenophora amurica C. X. Fu & M. Y. Liu, Bull. Bot. Res. North-East. Forest. Inst. 6: 159 (1986).
Manchuria (Heilongjiang). 36 CHM. Hemicr. or geophyte. *2n* = 68.

Adenophora aurita Franch., J. Bot. (Morot) 9: 366 (1895).
SC. China (Sichuan). 36 CHC. Hemicr. or geophyte.
Adenophora watsonii W. W. Sm., Notes Roy. Bot. Gard. Edinburgh 8: 175 (1914), as 'watsoni'.

Adenophora borealis D. Y. Hong & Y. Z. Zhao, Fl. Reipubl. Popul. Sin. 73(2): 187 (1983).
N. China (Hebei, Nei Mongol). 36 CHI CHN. Hemicr. or geophyte. *2n* = 34.
Adenophora borealis var. *oreophila* Y. Z. Zhao, Acta Sci. Nat. Univ. Intramongolicae 11: 57 (1980).

Adenophora brevidiscifera D. Y. Hong, Fl. Reipubl. Popul. Sin. 73(2): 184 (1983).
SC. China (Sichuan). 36 CHC. Hemicr. or geophyte.

Adenophora capillaris Hemsl., J. Linn. Soc., Bot. 26: 10 (1889).
C. China. 36 CHC CHN. Hemicr. or geophyte.

subsp. **capillaris**
C. China (Guizhou, Hubei, Shandong, Sichuan). 36 CHC CHN. Hemicr. or geophyte.
Adenophora capillaris var. *tenuifolia* Diels, Bot. Jahrb. Syst. 29: 605 (1901).
Adenophora longisepala P. C. Tsoong, Contr. Inst. Bot. Natl. Acad. Peiping 3: 74 (1935).

subsp. **leptosepala** (Diels) D. Y. Hong, Fl. Reipubl. Popul. Sin. 73(2): 136 (1983).
SC. China (Sichuan, Yunnan). 36 CHC. Hemicr. or geophyte.
* *Adenophora leptosepala* Diels, Notes Roy. Bot. Gard. Edinburgh 5: 175 (1912).
Adenophora leptosepala var. *linearifolia* C. Y. Wu, Rep. Yunnan Trop. Subtrop. Fl. Res. Inst. 1: 69 (1965).
Adenophora urceolata C. Y. Wu, Rep. Yunnan Trop. Subtrop. Fl. Res. Inst. 1: 74 (1965).

Adenophora changaica Gubanov & Kamelin, Bjull. Moskovsk. Obšč. Isp. Prir., Otd. Biol. 93(5): 111 (1988).
Mongolia. 37 MON. Hemicr. or geophyte.

Adenophora coelestis Diels, Notes Roy. Bot. Gard. Edinburgh 5: 173 (1912).
SC. China (Sichuan, Yunnan). 36 CHC. Hemicr. or geophyte. *2n* = 102.
Adenophora ornata Diels, Notes Roy. Bot. Gard. Edinburgh 5: 174 (1912).
Adenophora megalantha Diels, Notes Roy. Bot. Gard. Edinburgh 5: 175 (1912).
Adenophora pachyrhiza Diels, Notes Roy. Bot. Gard. Edinburgh 5: 176 (1912).
Adenophora bulleyana var. *alba* C. Y. Wu, Rep. Yunnan Trop. Subtrop. Fl. Res. Inst. 1: 65 (1965).
Adenophora coelestis var. *stenophylla* Diels ex C. Y. Wu, Rep. Yunnan Trop. Subtrop. Fl. Res. Inst. 1: 66 (1965).
Adenophora huangae C. Y. Wu, Rep. Yunnan Trop. Subtrop. Fl. Res. Inst. 1: 68 (1965).
Adenophora ornata var. *alba* C. Y. Wu, Rep. Yunnan Trop. Subtrop. Fl. Res. Inst. 1: 71 (1965).
Adenophora raphanorrhiza C. Y. Wu, Rep. Yunnan Trop. Subtrop. Fl. Res. Inst. 1: 72 (1965).

Adenophora contracta (Kitag.) J. Z. Qiu & D. Y. Hong, Acta Phytotax. Sin. 28: 399 (1990).
Manchuria (Liaoning). 36 CHM. Hemicr. or geophyte.
* *Adenophora polyantha* var. *contracta* Kitag. in S. Tokun., Rep. First Sci. Exped. Manchoukuo 4(2): 112 (1935).

Adenophora cordifolia D. Y. Hong, Fl. Reipubl. Popul. Sin. 73(2): 185 (1983).
C. China (Henan). 36 CHS. Hemicr. or geophyte.

Adenophora divaricata Franch. & Sav., Enum. Pl. Jap. 2: 423 (1879). *Adenophora polymorpha* var. *divaricata* (Franch. & Sav.) Makino, Bot. Mag. (Tokyo) 12 (Jap.): 57 (1898).
Russian Far East to Japan (Honshu, Shikoku) & N. China (Hebei, Shandong, Shanxi). 31 AMU KHA PRM 36 CHM CHN 38 JAP KOR. Hemicr. or geophyte. $2n = 34$.
Adenophora divaricata f. *alternifolia* H. Hara, J. Jap. Bot. 10: 371 (1934).
Adenophora tyosenensis Nakai ex Chung, Nom. Pl. Kor. Herb.: 125 (1949).
Adenophora divaricata f. *angustifolia* A. I. Baranov, Quart. J. Taiwan Mus. 16: 163 (1963).

Adenophora elata Nannf., Acta Horti Gothob. 5: 16 (1930).
N. China (Hebei, Nei Mongol, Shanxi). 36 CHI CHN. Hemicr. or geophyte. $2n = 68$.
Adenophora wutaiensis Hurus., Bot. Mag. (Tokyo) 62: 46 (1949).

Adenophora erecta S. T. Lee, J. K. Lee & S. T. Kim, J. Pl. Res. 110: 77 (1997).
South Korea (Ullung Do). 38 KOR. Hemicr. or geophyte.

Adenophora fusifolia Y. N. Lee, Bull. Korea Pl. Res. 4: 10 (2004).
South Korea. 38 KOR. Hemicr. or geophyte.
Adenophora fusifolia f. *albiflora* Y. N. Lee, Bull. Korea Pl. Res. 4: 13 (2004).

Adenophora gmelinii (Biehler) Fisch., Mém. Soc. Imp. Naturalistes Moscou 6: 167 (1823), as 'gmelini'.
Mongolia to SE. Siberia, Korea & NE. China (Hebei, Nei Mongol, Shanxi). 30 BRY CTA 31 AMU 36 CHI CHM CHN 37 MON 38 KOR. Hemicr. or geophyte. $2n = 34, 68$.
* *Campanula gmelinii* Biehler, Pl. Nov. Herb. Spreng.: 14 (1807), as 'gmelini'. *Adenophora polymorpha* var. *gmelinii* (Biehler.) Trautv. ex Herder, Trudy Imp. S.-Peterburgsk. Bot. Sada 1: 307 (1873), as 'gmelini'. *Adenophora communis* var. *gmelinii* (Biehler) Trautv., Trudy Imp. S.-Peterburgsk. Bot. Sada 6: 99 (1879), as 'gmelini'.

subsp. **gmelinii**
Mongolia to SE. Siberia, Korea & NE. China (Hebei, Nei Mongol, Shanxi). 30 BRY CTA 31 AMU 36 CHI CHM CHN 37 MON 38 KOR. Hemicr. or geophyte. $2n = 34, 68$.
Campanula erysimoides Vest ex Schult. in Roem. & Schult., Syst. Veg. 5: 102 (1819). *Adenophora coronopifolia* var. *erysimoides* (Vest ex Schult.) Steud., Nomencl. Bot. (ed. 2) 1: 25 (1840). *Adenophora erysimoides* (Vest ex Schult.) Kitag., Rep. Inst. Sci. Res. Manchoukuo 2: 298 (1938).
Campanula coronopifolia Fisch. ex Schult. in Roem. & Schult., Syst. Veg. 5: 157 (1819). *Adenophora coronopifolia* (Fisch. ex Schult.) Fisch., Mém. Soc. Imp. Naturalistes Moscou 6: 167 (1823). *Campanula fischeriana* Spreng., Syst. Veg. 4(2): 77 (1827). *Adenophora polymorpha* var. *coronopifolia* (Fisch. ex Schult.) Trautv. ex Herder, Trudy Imp. S.-Peterburgsk. Bot. Sada 1: 309 (1873). *Adenophora communis* var. *coronopifolia* (Fisch. ex Schult.) Trautv., Trudy Imp. S.-Peterburgsk. Bot. Sada 6: 98 (1879). *Adenophora gmelinii* var. *coronopifolia* (Fisch. ex Schult.) Y. Z. Zhao, Acta Sci. Nat. Univ. Intramongolicae 11: 59 (1980).
Campanula rabelaisiana Schult. in Roem. & Schult., Syst. Veg. 5: 158 (1819). *Adenophora rabelaisiana* (Schult.) G. Don in Loudon, Hort. Brit.: 75 (1830).
Adenophora lamarckii var. *angustifolia* A. DC., Monogr. Campan.: 357 (1830). *Campanula liliifolia* var. *pycnodonta* Borbás, Magyar Bot. Lapok 3: 196 (1904); non *Campanula liliifolia* var. *angustifolia* M. Bieb., Fl. Taur.-Cauc. 3: 140 (1819).
Adenophora gmelinii var. *stylosa* A. DC., Monogr. Campan.: 363 (1830).
Campanula coronopifolia var. *odontosepala* Borbás, Magyar Bot. Lapok 3: 192 (1904).
Campanula gmelinii var. *pomponiifolia* Fisch. ex Borbás, Magyar Bot. Lapok 3: 192 (1904).
Adenophora coronopifolia var. *longisepala* Reverd., Sist. Zametki Mater. Gerb. Krylova Tomsk. Gosud. Univ. Kujbyševa 3-4: 4 (1935).
Adenophora coronopifolia var. *pubicalycina* Reverd., Sist. Zametki Mater. Gerb. Krylova Tomsk. Gosud. Univ. Kujbyševa 3-4: 5 (1935).
Adenophora coronopifolia var. *puberula* Reverd., Sist. Zametki Mater. Gerb. Krylova Tomsk. Gosud. Univ. Kujbyševa 3-4: 5 (1935).

Adenophora pachyphylla Kitag., Rep. Inst. Sci. Res. Manchoukuo 2: 297 (1938).
Adenophora gmelinii var. *pachyphylla* (Kitag.) Y. Z. Zhao, Acta Sci. Nat. Univ.
Intramongolicae 11: 59 (1980).
Adenophora biloba Y. Z. Zhao, Ann. Bot. Fenn. 41: 381 (2004).
Adenophora coronopifolia var. *rotundiflora* Y. N. Lee, Bull. Korea Pl. Res. 4: 14 (2004).

subsp. **hailinensis** J. Z. Qiu & D. Y. Hong, Acta Phytotax. Sin. 31: 38 (1993).
Manchuria (Heilongjiang). 36 CHM. Hemicr. or geophyte.

subsp. **nystroemii** J. Z. Qiu & D. Y. Hong, Acta Phytotax. Sin. 31: 38 (1993).
N. China (Hebei, Nei Mongol, Shanxi). 36 CHI CHN. Hemicr. or geophyte.

Adenophora golubinzevaeana Reverd., Sist. Zametki Mater. Gerb. Krylova Tomsk. Gosud.
Univ. Kujbyševa 3-4: 4 (1935).
C. Siberia (Krasnoyarsk). 30 KRA. Hemicr. or geophyte.
Adenophora golubinzevaeana var. *pilosa* Reverd., Sist. Zametki Mater. Gerb. Krylova
Tomsk. Gosud. Univ. Kujbyševa 3-4: 4 (1935).

Adenophora grandiflora Nakai, Bot. Mag. (Tokyo) 23: 188 (1909).
Korea. 38 KOR. Hemicr. or geophyte.

Adenophora hatsushimae Kitam., Acta Phytotax. Geobot. 9: 244 (1940).
Japan (Kyushu). 38 JAP. Hemicr. or geophyte.

Adenophora himalayana Feer, Bot. Jahrb. Syst. 12: 618 (1890).
E. Kazakhstan to C. China & Tibet. 32 KAZ KGZ TZK 36 CHC CHN CHQ CHT CHX 40
NEP WHM. Hemicr. or geophyte.

subsp. **alpina** (Nannf.) D. Y. Hong, Fl. Reipubl. Popul. Sin. 73(2): 132 (1983).
W. Himalaya to C. China (Gansu, Shaanxi, Sichuan). 36 CHC CHN 40 WHM. Hemicr.
or geophyte.
Adenophora tsinlingensis Pax & K. Hoffm., Repert. Spec. Nov. Regni Veg. Beih. 12: 498
(1922).
* *Adenophora alpina* Nannf., Acta Horti Gothob. 5: 14 (1930).

subsp. **himalayana**
E. Kazakhstan to W. China (Gansu, Qinghai, Sichuan, Xinjiang) & Tibet. 32 KAZ KGZ
TZK 36 CHC CHN CHQ CHT CHX 40 NEP WHM. Hemicr. or geophyte.
Adenophora smithii Nannf., Acta Horti Gothob. 5: 21 (1930).
Adenophora smithii f. *crispa* Nannf., Acta Horti Gothob. 5: 22 (1930).

Adenophora hubeiensis D. Y. Hong, Fl. Reipubl. Popul. Sin. 73(2): 186 (1983).
C. China (Hubei). 36 CHC. Hemicr. or geophyte.

Adenophora hunanensis Nannf. in Hand.-Mazz., Symb. Sin. 7: 1070 (1936).
S. & E. China. 36 CHC CHN CHS. Hemicr. or geophyte.

subsp. **huadungensis** D. Y. Hong, Fl. Reipubl. Popul. Sin. 73(2): 186 (1983).
E. China (Anhui, Fujian, Jiangsu, Jiangxi, Zhejiang). 36 CHS. Hemicr. or geophyte.

subsp. **hunanensis**
S. & E. China (Guangdong, Guangxi, Guizhou, Hebei, Henan, Hubei, Hunan, Jiangxi,
Shaanxi, Shanxi, Sichuan). 36 CHC CHN CHS. Hemicr. or geophyte.

Adenophora × **izuensis** H. Ohba & S. Watan., J. Jap. Bot. 73: 81 (1998), pro sp. A.
remotiflora × A. triphylla.
Japan (Honshu). 38 JAP. Hemicr. or geophyte.

Adenophora jacutica Fed. in Kom., Fl. URSS 24: 470 (1957).
NE. Siberia (Yakutiya). 30 YAK. Hemicr. or geophyte.

Adenophora jasionifolia Franch., J. Bot. (Morot) 9: 365 (1895).
SC. China (Sichuan, Yunnan), Tibet. 36 CHC CHT. Hemicr. or geophyte. $2n = 34$.
Adenophora forrestii Diels, Notes Roy. Bot. Gard. Edinburgh 5: 174 (1912).
Adenophora forrestii var. *intercedens* Pax & K. Hoffm., Repert. Spec. Nov. Regni Veg.,
Beih. 122: 500 (1922).
Adenophora pumila P. C. Tsoong, Contr. Inst. Bot. Natl. Acad. Peiping 3: 76 (1935).
Adenophora forrestii var. *handeliana* Nannf. in Hand.-Mazz., Symb. Sin. 7: 1070 (1936).
Adenophora atuntzensis C. Y. Wu, Rep. Yunnan Trop. Subtrop. Fl. Res. Inst. 1: 64 (1965).
Adenophora microcodon C. Y. Wu, Rep. Yunnan Trop. Subtrop. Fl. Res. Inst. 1: 70 (1965).

Adenophora kayasanensis Kitam., Acta Phytotax. Geobot. 5: 247 (1936).
Korea. 38 KOR. Hemicr. or geophyte.

Adenophora khasiana (Hook. f. & Thomson) Oliv. ex Collett & Hemsl., J. Linn. Soc., Bot.
28: 80 (1890).
Tibet to SC. China (Sichuan, Yunnan) & Myanmar. 36 CHC CHT 40 ASS EHM 41 MYA.
Hemicr. or geophyte. $2n = 34$.
* *Campanula khasiana* Hook. f. & Thomson, J. Proc. Lin. Soc., Bot. 2: 25 (1858).
Adenophora bulleyana Diels, Notes Roy. Bot. Gard. Edinburgh 5: 175 (1912).
Adenophora diplodonta Diels, Notes Roy. Bot. Gard. Edinburgh 5: 176 (1912).
Adenophora albescens C. Y. Wu, Rep. Yunnan Trop. Subtrop. Fl. Res. Inst. 1: 63 (1965).
Adenophora chionantha C. Y. Wu, Rep. Yunnan Trop. Subtrop. Fl. Res. Inst. 1: 65 (1965).
Adenophora dimorphophylla C. Y. Wu, Rep. Yunnan Trop. Subtrop. Fl. Res. Inst. 1: 67 (1965).
Adenophora bulleyana var. *angustifolia* C. Y. Wu, Rep. Yunnan Trop. Subtrop. Fl. Res.
Inst. 1: 65 (1965).
Adenophora likiangensis C. Y. Wu, Rep. Yunnan Trop. Subtrop. Fl. Res. Inst. 1: 69 (1965).
Adenophora roseiflora C. Y. Wu, Rep. Yunnan Trop. Subtrop. Fl. Res. Inst. 1: 73 (1965).

Adenophora koreana Kitam., Acta Phytotax. Geobot. 5: 205 (1936).
Korea. 38 KOR. Hemicr. or geophyte.

Adenophora kulunensis Y. Z. Zhao, Acta Phytotax. Sin. 44: 615 (2006).
Inner Mongolia (Nei Monggol). 36 CHI. Hemicr. or geophyte.
* *Adenophora urceolata* Y. Z. Zhao, Acta Bot. Fenn. 39: 335 (2002); non C. Y. Wu, Rep.
Yunnan Trop. Subtrop. Fl. Res. Inst. 1: 74 (1965).

Adenophora lamarckii Fisch., Mém. Soc. Imp. Naturalistes Moscou 6: 168 (1823).
Campanula lamarckii (Fisch.) D. Dietr., Syn. Pl. 1: 755 (1839). *Adenophora polymorpha* var.
lamarckii (Fisch.) Herder, Trudy Imp. S.-Peterburgsk. Bot. Sada 1: 309 (1873). *Adenophora
communis* var. *lamarckii* (Fisch.) Trautv., Trudy Imp. S.-Peterburgsk. Bot. Sada 6: 98 (1879).
Adenophora lilifolia var. *lamarckii* (Fisch.) Krylov, Fl. Alt. 3: 702 (1904).
Kazakhstan to S. Siberia, Mongolia & W. China (Xinjiang); Korea. 30 ALT IRK 32 KAZ 36
CHX 37 MON 38 KOR. Hemicr. or geophyte.
Adenophora lamarckii var. *longifolia* Nakai, Rep. Veg. Daisetsu Mts.: 13 (1930).

Adenophora liliifolia (L.) A. DC., Monogr. Campan. 358 (1830).
E. Europe to W. Siberia & W. China (Xinjiang); cult. 11 AUT CZE GER HUN POL SWI 13
ITA ROM YUG 14 BLR BLT RUC RUE RUS UKR 30 ALT WSB 32 KAZ 36 CHX. Hemicr.
or geophyte. $2n = 34, 102$.
* *Campanula liliifolia* L., Sp. Pl.: 165 (1753), as 'lilifolia'.
Campanula alpini L., Sp. Pl. (ed. 2): 1669 (1763). *Campanula rhomboidalis* var. *alpini* (L.)
L., Syst. Veg. (ed. 13): 173 (1774).
Campanula stylosa Lam., Encycl. 1: 580 (1785). *Campanula liliifolia* var. *stylosa* (Lam.)
Besser, Cat. Hort. Cremeneci: 27 (1816). *Adenophora stylosa* (Lam.) Fisch., Mém. Soc.
Imp. Naturalistes Moscou 6: 168 (1823). *Adenophora liliifolia* var. *stylosa* (Lam.)
Korsh., Mém. Acad. Imp. Sci. Saint Pétersbourg (ser. 7) 42(2): 40 (1894).

Campanula periplocifolia Lam., Encycl. 1: 580 (1785). *Campanula stylosa* var. *periplocifolia* (Lam.) Pers., Syn. Pl. 1: 189 (1805). *Adenophora periplocifolia* (Lam.) A. DC., Monogr. Campan.: 361 (1830), as 'periplocaefolia'.

Campanula umbrosa F. Dietr., Vollst. Lex. Gärtn. 1: 675 (1802).

Campanula suaveolens Schrad. ex Hornem., Suppl. Hort. Bot. Hafn.: 23 (1819). *Campanula liliifolia* var. *suaveolens* (Schrad. ex Hornem.) Schult. in Roem. & Schult., Syst. Veg. 5: 110 (1819). *Adenophora suaveolens* (Schrad. ex Hornem.) Rchb., Iconogr. Bot. Exot. 1: 22 (1824). *Adenophora liliifolia* var. *suaveolens* (Schrad. ex Hornem.) Steud., Nomencl. Bot. (ed. 2) 1: 25 (1840).

Campanula liliifolia var. *hybrida* Schult. in Roem. & Schult., Syst. Veg. 5: 110 (1819).

Campanula intermedia Schult. in Roem. & Schult., Syst. Veg. 5: 110 (1819). *Adenophora intermedia* (Schult.) Sweet, Hort. Brit.: 249 (1826); non Ledeb., Index Seminum Horti Dorpat.: 1 (1824).

Campanula fischeri Schult. in Roem. & Schult., Syst. Veg. 5: 116 (1819). *Adenophora fischeri* (Schult.) G. Don in Loudon, Hort. Brit.: 75 (1830).

Campanula spreta Schult. in Roem. & Schult., Syst. Veg. 5: 123 (1819). *Adenophora liliifolia* var. *spreta* (Schult.) A. DC., Monogr. Campan.: 359 (1830).

Campanula liliifolia var. *angustifolia* M. Bieb., Fl. Taur.-Cauc. 3: 140 (1819).

Adenophora communis Fisch., Mém. Soc. Imp. Naturalistes Moscou 6: 168 (1823).

Adenophora polymorpha Ledeb., Fl. Altaic. 1: 246 (1829).

Adenophora liliifolia var. *infundibuliformis* A. DC., Monogr. Campan.: 359 (1830). *Adenophora liliifolia* f. *infundibuliformis* (A. DC.) Voss in Siebert & Voss, Vilm. Blumengärtn. (ed. 3) 1: 560 (1894).

Campanula cordata Tausch, Flora 25: 288 (1842); non Vis., Stirp. Dalmat. Spec.: 5 (1826); nec Peterm., Fl. Lips. Excurs.: 187 (1838) *Adenophora cordata* B. D. Jacks., Index Kew. 1: 39 (1893).

Adenophora suaveolens var. *latifolia* Schur, Enum. Pl. Transsilv.: 432 (1866).

Adenophora polymorpha var. *integerrima* Trautv., Bull. Soc. Imp. Naturalistes Moscou 39: 406 (1866). *Adenophora communis* var. *integerrima* (Trautv.) Trautv., Trudy Imp. S.-Peterburgsk. Bot. Sada 6: 99 (1879). *Adenophora liliifolia* var. *integerrima* (Trautv.) Korsh., Mém. Acad. Imp. Sci Saint Pétersbourg (ser. 7) 42(2): 40 (1894).

Adenophora setulosa Borbás, Magyar Orv. Termész. Vándorgyül. Tört. Vázl. Munk. 23: 204 (1887).

Campanula mikoi Borbás, Magyar Bot. Lapok 3: 192 (1904).

Campanula perpallens Borbás, Magyar Bot. Lapok 3: 193 (1904).

Campanula alpini var. *villosula* Borbás, Magyar Bot. Lapok 3: 194 (1904). *Adenophora liliifolia* f. *villosula* (Borbás) Hayek, Repert. Spec. Nov. Regni Veg. Beih. 30(2): 550 (1930).

Campanula alpini var. *asperula* Borbás, Magyar Bot. Lapok 3: 195 (1904).

Campanula alpini var. *botryantha* Borbás, Magyar Bot. Lapok 3: 195 (1904).

Campanula alpini var. *hungarica* Borbás, Magyar Bot. Lapok 3: 195 (1904).

Campanula liliifolia var. *hirtula* Borbás, Magyar Bot. Lapok 3: 196 (1904).

Campanula liliifolia var. *polyadenia* Borbás, Magyar Bot. Lapok 3: 196 (1904).

Campanula liliifolia var. *setulosa* Borbás, Magyar Bot. Lapok 3: 196 (1904).

Campanula liliifolia subvar. *subverticillata* Borbás, Magyar Bot. Lapok 3: 196 (1904). *Adenophora liliifolia* subf. *subverticillata* (Borbás) Hayek, Repert. Spec. Nov. Regni Veg. Beih. 30(2): 550 (1930).

Adenophora liliifolia var. *alba* Nakai in T. Mori, Enum. Pl. Cor.: 337 (1922). *Adenophora liliifolia* f. *alba* (Nakai) Nakai, Bull. Natl. Sci. Mus. Tokyo 31: 110 (1952).

Adenophora liliifolia var. *angusta* Nakai in T. Mori, Enum. Pl. Cor.: 337 (1922).

Adenophora liliifolia var. *pocsii* Soó, Acta Bot. Acad. Sci. Hung. 18: 174 (1973).

Adenophora liliifolioides Pax & K. Hoffm., Repert. Spec. Nov. Regni Veg. Beih. 12: 499 (1922). C. China (Gansu, Shaanxi, Sichuan), Tibet. 36 CHC CHN CHT. Hemicr. or geophyte.

Adenophora gracilis Nannf., Acta Horti Gothob. 5: 17 (1930).

Adenophora lobophylla D. Y. Hong, Fl. Reipubl. Popul. Sin. 73(2): 186 (1983).
SC. China (Sichuan). 36 CHC. Hemicr. or geophyte. *2n* = 34.

Adenophora longipedicellata D. Y. Hong, Fl. Reipubl. Popul. Sin. 73(2): 185 (1983).
SC. China (Guizhou, Hubei, Sichuan). 36 CHC. Hemicr. or geophyte.

Adenophora maximowicziana Makino, Bot. Mag. (Tokyo) 20: 38 (1906).
Japan (Shikoku). 38 JAP. Hemicr. or geophyte.

Adenophora micrantha D. Y. Hong, Fl. Reipubl. Popul. Sin. 73(2): 185 (1983).
Inner Mongolia (Nei Mongol). 36 CHI. Hemicr. or geophyte.
Adenophora suolunensis P. F. Tu & X. F. Zhao, Acta Bot. Boreal.-Occid. Sin. 18: 616 (1998).

Adenophora microsperma Y. Y. Qian, Guihaia 18: 9 (1998).
SC. China (Yunnan). 36 CHC. Hemicr. or geophyte.

Adenophora morrisonensis Hayata, Mat. Fl. Formosa: 165 (1911).
Taiwan. 38 TAI. Hemicr. or geophyte. *2n* = 34.

subsp. **morrisonensis**
Taiwan. 38 TAI. Hemicr. or geophyte. *2n* = 34.

subsp. **uehatae** (Yamam.) Lammers, Bot. Bull. Acad. Sin. 33: 285 (1992).
Taiwan. 38 TAI. Hemicr. or geophyte. *2n* = 34.
* *Adenophora uehatae* Yamam., J. Trop. Agric. Soc. Formos. 4: 484 (1932). *Adenophora coelestis* var. *uehatae* (Yamam.) Masam., Trans. Nat. Hist. Soc. Taiwan 29: 271 (1939).

Adenophora nikoensis Franch. & Sav., Enum. Pl. Jap. 2: 423 (1879).
Japan (Honshu). 38 JAP. Hemicr. or geophyte. *2n* = 34, 68, 102.
Adenophora lamarckii f. *multiloba* Takeda, Bot. Mag. (Tokyo) 24: 107 (1910). *Adenophora nipponica* f. *multiloba* (Takeda) Kitam., Acta Phytotax. Geobot. 5: 207 (1936). *Adenophora polymorpha* f. *multiloba* (Takeda) Makino & Nemoto, Fl. Japan:109 (1925), as 'multifida'. *Adenophora nikoensis* f. *multifida* H. Hara, J. Jap. Bot. 13: 467 (1937); non *Adenophora nikoensis* f. *multiloba* Honda, Bot. Mag. (Tokyo) 50: 670 (1936).
Adenophora nikoensis f. *linearifolia* Takeda, Bot. Mag. (Tokyo) 24: 107 (1910).
Adenophora nikoensis f. *macrocalyx* Takeda, Bot. Mag. (Tokyo) 24: 108 (1910).
Adenophora petrophila H. Hara, J. Jap. Bot. 11: 826 (1935). *Adenophora nikoensis* var. *petrophila* (H. Hara) H. Hara, J. Jap. Bot. 13: 468 (1937). *Adenophora pereskiifolia* f. *petrophila* (H. Hara) T. Shimizu, Acta Phytotax. Geobot. 17: 86 (1958). *Adenophora pereskiifolia* var. *petrophila* (H. Hara) T. Shimizu, Acta Phytotax. Geobot. 33: 335 (1982).
Adenophora nipponica Kitam., Acta Phytotax. Geobot. 5: 206 (1936). *Adenophora nikoensis* f. *nipponica* (Kitam.) H. Hara, J. Jap. Bot. 13: 466 (1937).
Adenophora nipponica var. *stenophylla* Kitam., Acta Phytotax. Geobot. 5: 207 (1936). *Adenophora nikoensis* f. *stenophylla* (Kitam.) H. Hara, J. Jap. Bot. 13: 467 (1937).
Adenophora nikoensis f. *globiflora* Hiyama, J. Jap. Bot. 26: 157 (1951).
Adenophora teramotoi Hurus., J. Jap. Bot. 27: 199 (1952). *Adenophora nikoensis* var. *teramotoi* (Hurus.) Okazaki & T. Shimizu in K. Iwats., T. Yamaz., Boufford & H. Ohba, Fl. Japan 3a: 409 (1993).
Adenophora teramotoi var. *hispidula* T. Shimizu, J. Fac. Textile Sci. Technol. Shinsu Univ., A 36(12): 66 (1963). *Adenophora nikoensis* f. *hispidula* (T. Shimizu) Yonek., J. Jap. Bot. 80: 329 (2005); non T. Shimizu, New Alp. Fl. Japan 1: 311 (1982).
Adenophora nikoensis f. *hispidula* T. Shimizu, New Alp. Fl. Japan 1: 311 (1982).
Adenophora nikoensis var. *persicaria* Ohwi, Acta Phytotax. Geobot. 33: 334 (1982).

Adenophora ningxianica D. Y. Hong ex S. Ge & D. Y. Hong, Novon 9: 46 (1999).
N. China (Gansu, Nei Mongol). 36 CHI CHN. Hemicr. or geophyte.

Adenophora omeiensis Y. Z. Zhu, Acta Bot. Yunnan. 13: 141 (1991).
SC. China (Sichuan). 36 CHC. Hemicr. or geophyte.

Adenophora palustris Kom., Trudy Imp. S.-Peterburgsk. Bot. Sada 18: 426 (1901).
Manchuria to Japan (Honshu). 36 CHM 38 JAP KOR. Hemicr. or geophyte. $2n = 102$.
Adenophora palustris f. *leucantha* Nakai ex Matsum., Nippon Shokubutsumeii (ed. 9) 2: 83 (1916). *Adenophora palustris* var. *leucantha* (Nakai ex Matsum.) Nakai in T. Mori, Enum. Pl. Cor.: 337 (1922).
Adenophora palustris var. *linearis* Nakai ex Matsum., Nippon Shokubutsumeii (ed. 9) 2: 83 (1916).

Adenophora paniculata Nannf., Acta Horti Gothob. 5: 19 (1930).
N. China (Hebei, Henan, Nei Mongol, Shaanxi, Shandong, Shanxi). 36 CHI CHN. Hemicr. or geophyte.
Adenophora paniculata var. *psilosa* Kitag. in S. Tokun., Rep. First Sci. Exped. Manchoukuo 4(2): 109 (1935).
Adenophora paniculata var. *pilosa* Kitag. in S. Tokun., Rep. First Sci. Exped. Manchoukuo 4(2): 110 (1935).
Adenophora paniculata var. *dentata* Y. Z. Zhao, Acta Sci. Nat. Univ. Intramongolicae 11:58 (1980).
Adenophora paniculata var. *petiolata* Y. Z. Zhao, Acta Sci. Nat. Univ. Intramongolicae 11: 58 (1980).

Adenophora pereskiifolia (Fisch. ex Schult.) Fisch. ex G. Don in Loudon, Hort. Brit.: 75 (1830), as 'pereskiaefolia'.
SE. Siberia to Kuril Is., Japan (Hokkaido, Honshu) & Manchuria. 30 BRY CTA 31 AMU KHA KUR 36 CHM 37 MON 38 JAP KOR. Hemicr. or geophyte. $2n = 34, 68, 72$.
* *Campanula pereskiifolia* Fisch. ex Schult. in Roem. & Schult., Syst. Veg. 5: 116 (1819), as 'pereskiaefolia'. *Adenophora polymorpha* var. *pereskiifolia* (Fisch. ex Schult.) Makino, Bot. Mag. (Tokyo) 12 (Jap.): 56 (1898), as 'pereskiaefolia'.
Adenophora latifolia Fisch., Mém. Soc. Imp. Naturalistes Moscou 6: 168 (1823). *Adenophora polymorpha* var. *latifolia* (Fisch.) Trautv., Bull. Soc. Imp. Naturalistes Moscou 39: 406 (1866). *Adenophora communis* var. *latifolia* (Fisch.) Trautv., Trudy Imp. S.-Peterburgsk. Bot. Sada 6: 99 (1879).
Adenophora polymorpha var. *verticillata* Franch. & Sav., Enum. Pl. Jap. 2: 422 (1879).
Campanula rhomboidea Borbás, Magyar Bot. Lapok 3: 192 (1904); non Falk, Topogr. Beitr. 2: 128 (1786).
Adenophora curvidens Nakai, Bot. Mag. (Tokyo) 29: 6 (1915). *Adenophora perskiifolia* var. *curvidens* (Nakai) Kitag., Lin. Fl. Manshur.: 417 (1939).
Adenophora moiwana Nakai, Bot. Mag. (Tokyo) 36: 126 (1922). *Adenophora pereskiifolia* var. *moiwana* (Nakai) H. Hara, J. Jap. Bot. 26: 281 (1951).
Adenophora manshurica Nakai, Veg. Apoi: 13, 70 (1930). *Adenophora divaricata* var. *manshurica* (Nakai) Kitag. in S. Tokun., Rep. First Sci. Exped. Manchoukuo 4(2): 106 (1935). *Adenophora divaricata* f. *manshurica* (Nakai) Kitag., Neo-Lineam. Fl. Manshur.: 600 (1979).
Adenophora latifolia f. *albiflora* Nakai, Bull. Natl. Sci. Mus. Tokyo 31: 110 (1952). *Adenophora manshurica* f.*albiflora* (Nakai) U. C. La, Fl. Coreana 7: 23 (1999).
Adenophora onoi Tatew. & Kitam., Acta Phytotax. Geobot. 5: 209 (1936).
Adenophora moiwana var. *heterotricha* Nakai ex H. Hara, Bot. Mag. (Tokyo) 51: 895 (1937). *Adenophora pereskiifolia* var. *heterotricha* (Nakai ex H. Hara) H. Hara, J. Jap. Bot. 26: 281 (1951).
Adenophora ishiyamae Miyabe & Tatew., Trans. Sapporo Nat. Hist. Soc. 15: 207 (1938).
Adenophora yokoyamae Miyabe & Tatew., Trans. Sapporo Nat. Hist. Soc. 15: 209 (1938).
Adenophora pereskiifolia f. *puberula* Kitag., Rep. Inst. Sci. Res. Manchoukuo 5: 158 (1941).
Adenophora pereskiifolia f. *linearifolia* T. Shimizu, Acta Phytotax. Geobot. 17: 87 (1958).
Adenophora pereskiifolia f. *stenophylla* Toyok. & Nosaka, Acta Phytotax. Geobot. 18: 196 (1960).
Adenophora pereskiifolia var. *yamadae* Toyok. & Nosaka, Acta Phytotax. Geobot. 18: 196 (1960).

Adenophora pereskiifolia subsp. *subalpina* A. I. Baranov, Quart. J. Taiwan Mus. 16: 161
(1963). *Adenophora polyantha* subsp. *subalpina* (A. I. Baranov) Kitag., Neo-Lineam. Fl.
Manshur.: 603 (1979).
Adenophora pereskiifolia f. *linearifolia* A. I. Baranov, Quart. J. Taiwan Mus. 16: 162
(1963); non T. Shimizu, Acta Phytotax. Geobot. 17: 87 (1958).
Adenophora pereskiifolia var. *alternifolia* Fuh ex Y. Z. Zhao, Acta Sci. Nat. Univ.
Intramongolicae 11: 57 (1980). *Adenophora pereskiifolia* subsp. *alternifolia* (Fuh ex Y.
Z. Zhao) C. X. Fu & M. Y. Liu, Acta Phytotax. Sin. 25: 186 (1987).
Adenophora pereskiifolia var. *angustifolia* Y. Z. Zhao in Ma, Fl. Intramongolica (ed. 2) 4:
846 (1993).

Adenophora petiolata Pax & K. Hoffm., Repert. Spec. Nov. Regni Veg. Beih. 12: 499 (1922).
NC. China (Gansu, Henan, Shaanxi, Shanxi). 36 CHN. Hemicr. or geophyte. 2*n* = 34.

Adenophora pinifolia Kitag. in S. Tokun., Rep. First Sci. Exped. Manchoukuo 4(2): 110 (1935).
Manchuria (Liaoning). 36 CHM. Hemicr. or geophyte.

Adenophora polyantha Nakai, Bot. Mag. (Tokyo) 23: 188 (1909).
N. China (Gansu, Hebei, Henan, Jiangsu, Liaoning, Ningxia, Nei Mongol, Shaanxi,
Shandong, Shanxi), Korea. 36 CHI CHM CHN CHS 38 KOR. Hemicr. or geophyte. 2*n* =
34, 68.
Campanula chanetii H. Lév., Repert. Spec. Nov. Regni Veg. 9: 450 (1911). *Adenophora
chanetii* (H. Lév.) D. F. Chamb., Notes Roy. Bot. Gard. Edinburgh 35: 248 (1977).
Adenophora scabridula Nannf., Acta Horti Gothob. 5: 20 (1930).
Adenophora scabridula var. *viscida* P. C. Tsoong, Contr. Inst. Bot. Natl. Acad Peiping 3:
80 (1935).
Adenophora obovata Kitam., Acta. Phytotax. Geobot. 5: 247 (1936).
Adenophora polyantha var. *media* Nakai & Kitag. in S. Tokun., Rep. First Sci. Exped.
Manchoukuo 4(1): 57 (1934).
Adenophora polyantha var. *glabricalyx* Kitag. in S. Tokun., Rep. First Sci. Exped.
Manchoukuo 4(2): 111 (1935).
Adenophora polyantha f. *eriocaulis* Kitag. in S. Tokun., Rep. First Sci. Exped. Manchoukuo
4(2): 111 (1935).
Adenophora polyantha f. *densipila* Kitag. in S. Tokun., Rep. First Sci. Exped. Manchoukuo
4(2): 112 (1935).
Adenophora polyantha var. *scabricalyx* Kitag. in S. Tokun., Rep. First Sci. Exped.
Manchoukuo 4(2): 112 (1935). *Adenophora polyantha* subsp. *scabricalyx* (Kitag.) J. Z.
Qiu & D. Y. Hong, Acta Phytotax. Sin. 31: 39 (1993).
Adenophora polyantha var. *rhombica* Y. Z. Zhao in Ma, Fl. Intramongolica (ed. 2) 4: 847
(1993).

Adenophora potaninii Korsh., Mém. Acad. Imp. Sci. Saint Pétersbourg (ser. 7) 42(3): 39
(1894), as 'potanini'.
C. China (Gansu, Ningxia, Qinghai, Shaanxi, Shanxi, Sichuan). 36 CHC CHI CHN CHQ.
Hemicr. or geophyte. 2*n* = 34, 102.
Adenophora bockiana Diels, Bot. Jahrb. Syst. 29: 605 (1901).
Adenophora polydentata P. F. Tu & G. J. Xu, J. China Pharm. Univ. 21: 16 (1990).

Adenophora probatovae A. E. Kozhevn., Bot. Žurn. (Moscow & Leningrad) 89: 1211 (2004).
Russian Far East (Primorye). 31 PRM. Hemicr. or geophyte.
**Adenophora verticillata* var. *primorskensis* A. E. Kozhevn., Sosud. Rast. Sovet. Dal'nego
Vostoka 8: 280 (1996).

Adenophora racemosa J. K. Lee & S. T. Lee, Korean J. Pl. Taxon. 20: 122 (1990).
Korea. 38 KOR. Hemicr. or geophyte.

Adenophora remotidens Hemsl., J. Linn. Soc., Bot. 26: 12 (1889).
Korea. 38 KOR. Hemicr. or geophyte.

Adenophora remotiflora (Siebold & Zucc.) Miq., Ann. Mus. Bot. Lugduno-Batavum 2:
193 (1866).
Russian Far East to Japan (Honshu, Kyushu, Shikoku) & E. China (Anhui, Hebei, Jiangsu,
Shandong, Shanxi, Zhejiang). 31 PRM 36 CHM CHN CHS 38 JAP KOR. Hemicr. or
geophyte. $2n = 34, 36$.
* *Campanula remotiflora* Siebold & Zucc., Abh. Math.-Phys. Cl. Königl. Bayer. Akad. Wiss.
4: 180 (1846).
Adenophora trachelioides Maxim., Prim. Fl. Amur.: 186 (1859).
Adenophora isabellae Hemsley, J. Bot. 14: 207 (1876).
Adenophora trachelioides var. *cordatifolia* Deb., Acta Soc. Linn. Bordeaux 33: 66 (1879).
Adenophora remotiflora var. *cordatifolia* (Deb.) Zahlbr., Ann. K. K. Naturhist. Hofmus.
10 (Notizen): 56 (1895).
Adenophora remotiflora f. *longifolia* Kom., Fl. Manshur. 3: 557 (1907).
Adenophora remotiflora f. *cordata* Kom., Fl. Manshur. 3: 557 (1907).
Adenophora remotiflora f. *angustifolia* Makino in Iinuma, Sômoku-dzusetsu (ed. 3) 1: 184
(1907). *Adenophora remotiflora* var. *angustifolia* (Makino) Honda, Bot. Mag. (Tokyo)
51: 95 (1937).
Adenophora remotiflora var. *hirsuta* Honda, Bot. Mag. (Tokyo) 45: 138 (1931).
Adenophora remotiflora f. *leucantha* Honda, Bot. Mag. (Tokyo) 57: 107 (1943).
Adenophora remotiflora f. *oreophila* Hiyama, J. Jap. Bot. 22: 56 (1948).
Adenophora trachelioides f. *puberula* A. I. Baranov, Quart. J. Taiwan Mus. 16: 155 (1963).
Adenophora remotiflora f. *albiflora* Y. N. Lee, Korean J. Bot. 24: 29 (1981).
Adenophora trachelioides subsp. *giangsuensis* D. Y. Hong, Fl. Reipubl. Popul. Sin. 73(2):
186 (1983).
Adenophora remotiflora var. *hirticalyx* S. T. Lee, Y. J. Chung & J. K. Lee, Korean J. Pl.
Taxon. 20: 192 (1990).

Adenophora rupestris Reverd., Sist. Zametki Mater. Gerb. Krylova Tomsk. Gosud. Univ.
Kujbyševa 3-4: 2 (1935).
S. Siberia (Irkutsk). 30 IRK. Hemicr. or geophyte.

Adenophora rupincola Hemsl., J. Linn. Soc., Bot. 26: 12 (1889).
C. China (Hubei, Jiangxi, Sichuan). 36 CHC CHS. Hemicr. or geophyte.
Adenophora pubescens Hemsl., J. Linn. Soc., Bot. 26: 12 (1889).

Adenophora sinensis A. DC., Monogr. Campan.: 354 (1830). *Campanula sinensis* (A. DC.) D.
Dietr., Syn. Pl. 1: 755 (1839). *Adenophora polymorpha* var. *sinensis* (A. DC.) Pampan.,
Nuovo Giorn. Bot. Ital. (n.s.) 17: 732 (1910).
SE. China (Anhui, Fujian, Guangdong, Hunan, Jiangxi). 36 CHS. Hemicr. or geophyte.
Adenophora sinensis var. *glabra* A. DC., Monogr. Campan.: 354 (1830).

Adenophora stenanthina (Ledeb.) Kitag., Lin. Fl. Manshur.: 418 (1939).
S. Siberia to N. China. 30 ALT BRY CTA IRK TUV 31 AMU 36 CHI CHN CHQ 37 MON.
Hemicr. or geophyte. $2n = 34$.
* *Campanula stenanthina* Ledeb., Mém. Acad. Imp. Sci. St. Pétersbourg Hist. Acad. 5: 525
(1814).

subsp. **stenanthina**
S. Siberia to N. China (Gansu, Hebei, Nei Mongol, Ningxia, Shaanxi, Shanxi). 30 ALT
BRY CTA IRK TUV 31 AMU 36 CHI CHN 37 MON. Hemicr. or geophyte. $2n = 34$.
Campanula coronata Ker Gawl., Bot. Reg. 2: tab. 149 (1816). *Adenophora coronata* (Ker
Gawl.) A. DC., Monogr. Campan.: 363 (1830).
Campanula sajanensis Pall. ex Schult. in Roem. & Schult., Syst. Veg. 5: 102 (1819); non
Fisch., Cat. Jard. Gorenki: 52 (1808).

Campanula marsupiiflora Schult. in Roem. & Schult., Syst. Veg. 5: 116 (1819).
Adenophora marsupiiflora (Schult.) Fisch., Mém. Soc. Imp. Naturalistes Moscou 6: 167 (1823). *Adenophora verticillata* var. *marsupiiflora* (Schult.) Trautv., Trudy Imp. S.-Peterburgsk. Bot. Sada 6: 96 (1879). *Adenophora polymorpha* var. *marsupiiflora* (Schult.) Franch., Pl. David. 1: 192 (1883).
Adenophora intermedia Ledeb., Index Seminum Horti Dorpat.: 1 (1824).
Adenophora montana Turcz., Bull. Soc. Imp. Naturalistes Moscou 21(2): 484 (1848).
Adenophora marsupiiflora var. *pilosa* Korsh., Mém. Acad. Imp. Sci. Saint Pétersbourg (ser. 7) 42(2): 30 (1894).
Adenophora marsupiiflora var. *korshinskiana* Diels, Bot. Jahrb. Syst. 29: 605 (1901).
Adenophora marsupiiflora var. *jaluensis* Kom., Fl. Manshur. 3: 560 (1907).
Adenophora insolens Reverd., Sist. Zametki Mater. Gerb. Krylova Tomsk. Gosud. Univ. Kujbyševa 3-4: 3 (1935).
Adenophora marsupiiflora var. *crispata* Turcz. ex Kitag. in S. Tokun., Rep. First Sci. Exped. Manchoukuo 4(2): 108 (1935). *Adenophora crispata* (Turcz. ex Kitag.) Kitag., Lin. Fl. Manshur.: 415 (1939). *Adenophora stenanthina* f. *crispata* (Turcz. ex Kitag.) Kitag., Rep. Inst. Sci. Res. Manchoukuo 4: 97 (1940). *Adenophora stenanthina* var. *crispata* (Turcz. ex Kitag.) Y. Z. Zhao, Acta Sci. Nat. Univ. Intramongolicae 11: 59 (1980).
Adenophora collina Kitag., Rep. Inst. Sci. Res. Manchoukuo 4: 98 (1940). *Adenophora stenanthina* var. *collina* (Kitag.) Y. Z. Zhao, Acta Sci. Nat. Univ. Intramongolicae 11: 59 (1980).
Adenophora collina f. *latifolia* Kitag., Rep. Inst. Sci. Res. Manchoukuo 4: 98 (1940).
Adenophora stenanthina var. *angustilanceifolia* Y. Z. Zhao, Acta Sci. Nat. Univ. Intramongolicae 11: 58 (1980), as 'angusti-lanceifolia'.
Adenophora pratensis Y. Z. Zhao in Ma, Fl. Intramongolica (ed. 2) 4: 847 (1993).

subsp. **sylvatica** D. Y. Hong, Fl. Reipubl. Popul. Sin. 73(2): 187 (1983).
NW. China (Gansu, Qinghai). 36 CHN CHQ. Hemicr. or geophyte.

Adenophora stenophylla Hemsl., J. Linn. Soc., Bot. 26: 10 (1889).
Mongolia & Manchuria. 36 CHM 37 MON. Hemicr. or geophyte. $2n = 34$.
Adenophora stenophylla var. *denudata* Kitag., Bot. Mag. (Tokyo) 48: 618 (1934).
Adenophora mongolica A. I. Baranov, Acta Soc. Harbin Invest. Nat. Ethnogr., Bot. 12: 35 (1954).

Adenophora stricta Miq., Ann. Mus. Bot. Lugduno-Batavum 2: 192 (1866). *Adenophora polymorpha* var. *stricta* (Miq.) Makino, Bot. Mag. (Tokyo) 12 (Jap.): 57 (1898). $2n = 34$.
China (Anhui, Gansu, Guangxi, Guizhou, Henan, Hubei, Hunan, Jiangsu, Jiangxi, Shaanxi, Sichuan, Yunnan, Zhejiang), Korea; cult.; naturalized in Japan. 36 CHC CHN CHS 38 jap KOR. Hemicr. or geophyte.

subsp. **confusa** (Nannf.) D. Y. Hong, Fl. Reipubl. Popul. Sin. 73(2): 105 (1983).
SC. China (Yunnan); cult. 36 CHC. Hemicr. or geophyte.
* *Adenophora confusa* Nannf. in Hand.-Mazz., Symb. Sin. 7: 1068 (1936).

subsp. **henanica** P. F. Tu & G. J. Xu, J. China Pharm. Univ. 21: 17 (1990).
E. China (Henan). 36 CHS. Hemicr. or geophyte.

subsp. **sessilifolia** D. Y. Hong, Fl. Reipubl. Popul. Sin. 73(2): 185 (1983).
C. & S. China (Gansu, Guangxi, Guizhou, Henan, Hubei, Hunan, Shaanxi, Sichuan, Yunnan). 36 CHC CHN CHS. Hemicr. or geophyte.

subsp. **stricta**
E. China (Anhui, Hunan, Jiangsu, Jiangxi, Zhejiang) & Korea; naturalized in Japan. 36 CHS 38 jap KOR. Hemicr. or geophyte. $2n = 34$.
Adenophora sinensis var. *pilosa* A. DC., Monogr. Campan.: 354 (1830). *Campanula axilliflora* Borbás, Magyar Bot. Lapok. 3: 191 (1904); non *Campanula pilosa* Pall. ex Schult. in Roem. & Schult., Syst. Veg. 5: 148 (1819). *Adenophora axilliflora* (Borbás) Borbás ex Prain, Index Kew., Suppl. 3: 4 (1908).

Adenophora stricta var. *glabrior* Miq., Ann. Mus. Bot. Lugduno-Batavum 2: 193
(1866).
Adenophora argyi H. Lév., Bull. Acad. Int. Géogr. Bot. 23: 292 (1914).
Adenophora rotundifolia H. Lév., Bull. Acad. Int. Géogr. Bot. 23: 292 (1914).
Adenophora stricta var. *lancifolia* Honda, Bot. Mag. (Tokyo) 50: 608 (1936).
Adenophora stricta var. *nanjingensis* P. F. Tu & G. J. Xu, J. China Pharm. Univ. 21: 17
(1990).
Adenophora stricta var. *qinglongshanica* P. F. Tu & G. J. Xu, J. China Pharm. Univ. 21:
17 (1990).

Adenophora sublata Kom., Bot. Mater. Gerb. Glavn. Bot. Sada SSSR 6: 13 (1926).
Russian Far East. 31 KHA PRM. Hemicr. or geophyte. $2n = 68$.

Adenophora takedae Makino, Bot. Mag. (Tokyo) 20: 37 (1906), as 'takedai'.
Japan (Honshu). 38 JAP. Hemicr. or geophyte. $2n = 34$.
Adenophora howozana Takeda, Bot. Mag. (Tokyo) 24: 108 (1910). *Adenophora takedae* var.
howozana (Takeda) Sugimoto ex Honda, Nom. Pl. Jap.: 335 (1939).

Adenophora taquetii H. Lév., Repert. Spec. Nov. Regni Veg. 12: 22 (1913).
Korea. 38 KOR. Hemicr. or geophyte.

Adenophora tashiroi (Makino & Nakai) Makino & Nakai, J. Coll. Sci. Imp. Univ. Tokyo 31:
66 (1911).
Korea (Cheju Do) & Japan (Kyushu). 38 JAP KOR. Hemicr. or geophyte.
Adenophora polymorpha var. *tashiroi* Makino & Nakai, Bot. Mag. (Tokyo) 23: 21 (1909).

Adenophora taurica (Sukacz.) Juz., Bot. Mater. Gerb. Bot. Inst. Komarova Akad. Nauk SSSR
13: 301 (1950).
Krym. 14 KRY. Hemicr. or geophyte.
Adenophora liliifolia subsp. *taurica* Sukacz. in Popl., Spisok Rast. Krymsk. Zapov.:
93 (1931).

Adenophora tricuspidata (Fisch. ex Schult.) A. DC., Monogr. Campan.: 355 (1830).
S. Siberia to Manchuria (Heilongjiang) & Inner Mongolia (Nei Mongol). 30 BRY CTA IRK
TVA 31 AMU KHA PRM 36 CHI CHM. Hemicr. or geophyte. $2n = 34$.
* *Campanula tricuspidata* Fisch. ex Schult. in Roem. & Schult., Syst. Veg. 5: 158 (1819).
Adenophora denticulata Fisch., Mém. Soc. Imp. Naturalistes Moscou 6: 167 (1823).
Campanula denticulata (Fisch.) Spreng., Syst. Veg. 1: 735 (1824); non Burch., Trav. S.
Africa 1: 538 (1822). *Adenophora polymorpha* var. *denticulata* (Fisch.) Trautv. ex
Herder, Trudy Imp. S.-Peterburgsk. Bot. Sada 1: 308 (1873). *Adenophora communis* var.
denticulata (Fisch.) Trautv., Trudy Imp. S.-Peterburgsk. Bot. Sada 6: 97 (1879).
Adenophora turczaninowii Feer, Bot. Jahrb. Syst. 12: 619 (1890), as 'turczaninowi'.
Adenophora denticulata f. *angustifolia* Korsh., Trudy Imp. S.-Peterburgsk. Bot. Sada 12:
365 (1893).
Adenophora richteri Borbás, Magyar Bot. Lapok 1: 253 (1902). *Campanula richteri*
(Borbás) Borbás, Magyar Bot. Lapok 3: 193 (1904).
Adenophora denticulata var. *ciliata* Reverd., Sist. Zametki Mater. Gerb. Krylova Tomsk.
Gosud. Univ. Kujbyševa 3-4: 4 (1935).
Adenophora denticulata var. *linifolia* Reverd., Sist. Zametki Mater. Gerb. Krylova Tomsk.
Gosud. Univ. Kujbyševa 3-4: 4 (1935).

Adenophora triphylla (Thunb.) A. DC., Monogr. Campan.: 365 (1830).
E. Asia, from Russian Far East & Japan to Indochina. 30 BRY CTA 31 AMU KHA KUR
PRM SAK 36 CHI CHM 38 JAP KOR NNS TAI 41 LAO VIE. Hemicr. or geophyte. $2n =$
34, 36.

Campanula verticillata Pall., Reise Russ. Reich. 3: 719 (1776); non Hill, Veg. Syst. 8: 10 (1765). *Adenophora verticillata* Pall. ex Fisch., Mém. Soc. Imp. Naturalistes Moscou 6: 167 (1823). *Adenophora radiatifolia* Nakai, Bull. Natl. Sci. Mus. Tokyo 31: 110 (1952).

* *Campanula triphylla* Thunb. in Murray, Syst. Veg. (ed. 14): 211 (1784). *Adenophora verticillata* var. *triphylla* (Thunb.) Regel, Tent. Fl. Ussur.: 108 (1861). *Adenophora verticillata* f. *triphylla* (Thunb.) Makino, Bot. Mag. (Tokyo) 22: 168 (1908).

Campanula tetraphylla Thunb. in Murray, Syst. Veg. (ed. 14): 211 (1784). *Adenophora tetraphylla* (Thunb.) Fisch. ex Kitag., Lin. Fl. Manshur.: 418 (1939). *Adenophora triphylla* var. *tetraphylla* (Thunb.) Makino, New Ill. Fl. Japan: 83 (1940).

Adenophora pereskiifolia var. *japonica* Regel, Index Seminum Hortus Petrop. 1864 (suppl.): 17 (1865). *Adenophora triphylla* var. *japonica* (Regel) H. Hara, J. Jap. Bot. 26: 281 (1951).

Adenophora verticillata var. *angustifolia* Regel, Tent. Fl. Ussur.: 108 (1861). *Adenophora verticillata* f. *angustifolia* (Regel) Makino, Bot. Mag. (Tokyo) 12 (Jap.): 59 (1898). *Adenophora triphylla* var. *angustifolia* (Regel) Kitam., Acta Phytotax. Geobot. 10: 308 (1941). *Adenophora radiatifolia* var. *angustifolia* (Regel) Nakai, Bull. Natl. Sci. Mus. Tokyo 31: 111 (1952).

Adenophora verticillata var. *subintegrifolia* Regel, Tent. Fl. Ussur. 108 (1861). *Adenophora verticillata* f. *subintegrifolia* (Regel) Makino, Bot. Mag. (Tokyo) 12 (Jap.): 60 (1898).

Adenophora verticillata var. *latifolia* Miq., Ann. Mus. Bot. Lugduno-Batavum 2: 192 (1866).

Adenophora verticillata var. *triphylla* Miq., Ann. Mus. Bot. Lugduno-Batavum 2: 192 (1866); non (Thunb.) Regel, Tent. Fl. Ussur.: 108 (1861).

Adenophora verticillata var. *hirsuta* F. Schmidt, Reis. Amur-Land., Bot.: 155 (1868). *Adenophora verticillata* f. *hirsuta* (F. Schmidt) Makino, Bot. Mag. (Tokyo) 20: 39 (1906). *Adenophora verticillata* subf. *hirsuta* (F. Schmidt) Makino, Bot. Mag. (Tokyo) 22: 167 (1908). *Adenophora thunbergiana* f. *hirsuta* (F. Schmidt) Kudô, Rep. Veg. N. Saghalien: 225 (1924). *Adenophora triphylla* f. *pilosissima* Kitam., Acta Phytotax. Geobot. 10: 311 (1941); non *Adenophora triphylla* f. *hirsuta* Kitam., Acta Phytotax. Geobot. 10: 308 (1941). *Adenophora radiatifolia* var. *hirsuta* (F. Schmidt) Nakai, Bull. Natl. Sci. Mus. Tokyo 31: 111 (1952). *Adenophora tetraphylla* var. *hirsuta* (F. Schmidt) D. F. Chamb., Notes Roy. Bot. Gard. Edinburgh 35: 249 (1977).

Adenophora verticillata var. *canescens* Franch. & Sav., Enum. Pl. Jap. 2: 422 (1879). *Adenophora triphylla* f. *canescens* (Franch. & Savat.) Kitam., Acta Phytotax. Geobot. 10: 310 (1941).

Adenophora verticillata var. *alternifolia* Franch. & Sav., Enum. Pl. Jap. 2: 422 (1879). *Adenophora verticillata* lusus *alternifolia* (Franch. & Sav.) Makino, Bot. Mag. (Tokyo) 20: 39 (1906). *Adenophora verticillata* subf. *alternifolia* (Franch. & Sav.) Makino, Bot. Mag. (Tokyo) 22: 167 (1908).

Adenophora verticillata var. *brevidens* Franch. & Sav., Enum. Pl. Jap. 2: 422 (1879).

Adenophora verticillata f. *crenata* Franch. & Sav., Enum. Pl. Jap. 2: 422 (1879).

Adenophora verticillata f. *dentata* Franch. & Sav., Enum. Pl. Jap. 2: 422 (1879).

Adenophora verticillata f. *incisa* Franch. & Sav., Enum. Pl. Jap. 2: 422 (1879).

Adenophora verticillata var. *oppositifolia* Franch. & Sav., Enum. Pl. Jap. 2: 422 (1879).

Adenophora verticillata f. *serrulata* Maxim. ex Franch. & Sav., Enum. Pl. Jap. 2: 422 (1879).

Adenophora verticillata var. *pilosissima* Engl., Bot. Jahrb. Syst. 6: 68 (1885).

Adenophora verticillata var. *media* Korsh., Trudy Imp. S.-Peterburgsk. Bot. Sada 12: 365 (1893). *Adenophora tetraphylla* var. *media* (Korsh.) A. I. Baranov, Quart. J. Taiwan Mus. 16: 159 (1963).

Adenophora verticillata var. *angustifolia* Korsh., Trudy Imp. S.-Peterburgsk. Bot. Sada 12: 365 (1893); non Regel, Tent. Fl. Ussur.: 108 (1861). *Adenophora tetraphylla* var. *angustifolia* A. I. Baranov, Quart. J. Taiwan Mus. 16: 159 (1963).

Adenophora verticillata var. *denticulata* Korsh., Trudy Imp. S.-Peterburgsk. Bot. Sada 12: 365 (1893).

Adenophora verticillata var. *maritima* Korsh., Trudy Imp. S.-Peterburgsk. Bot. Sada 12: 365 (1893).

Adenophora verticillata prole *princeps* Korsh., Trudy Imp. S.-Peterburgsk. Bot. Sada 12: 365 (1893). *Adenophora verticillata* f. *princeps* (Korsh.) Kom., Fl. Manshur. 3: 567 (1907). *Adenophora triphylla* f. *princeps* (Korsh.) Kitam., Acta Phytotax. Geobot. 10: 309 (1941). *Adenophora tetraphylla* var. *princeps* (Korsh.) A. I. Baranov, Quart. J. Taiwan Mus. 16: 159 (1963).

Adenophora verticillata subf. *glabra* Makino, Bot. Mag. (Tokyo) 22: 168 (1908). *Adenophora verticillata* var. *glabra* (Makino) Makino in Makino & Nemoto, Fl. Japan: 111 (1925). *Adenophora thunbergiana* f. *glabra* (Makino) H. Hara, Bot. Mag. (Tokyo) 51: 897 (1937). *Adenophora triphylla* f. *glabra* (Makino) Kitam., Acta Phytotax. Geobot. 10: 310 (1941).

Adenophora verticillata var. *linearis* Hayata, Fl. Mont. Formos.: 148 (1908). *Adenophora verticillata* f. *linearis* (Hayata) Matsum., Index Pl. Jap. 2(2): 614 (1912). *Adenophora triphylla* f. *linearis* (Hayata) Kitam., Acta Phytotax. Geobot. 10: 308 (1941).

Adenophora polymorpha var. *rhombifolia* H. Lév., Repert. Spec. Nov. Regni Veg. 12: 22 (1913). *Adenophora radiatifolia* var. *rhombifolia* (H. Lév.) Nakai, Bull. Natl. Sci. Mus. Tokyo 31: 110 (1952).

Adenophora verticillata var. *abbreviata* H. Lév., Repert. Spec. Nov. Regni Veg. 12: 22 (1913). *Adenophora radiatifolia* var. *abbreviata* (H. Lév.) Nakai, Bull. Natl. Sci. Mus. Tokyo 31: 110 (1952). *Adenophora tetraphylla* var. *abbreviata* (H. Lév.) D. F. Chamb., Notes Roy. Bot. Gard. Edinburgh 35: 249 (1977).

Adenophora thunbergiana Kudô, Rep. Veg. N. Saghalien: 224 (1924).

Adenophora hakusanensis Nakai, Bot. Mag. (Tokyo) 42: 14 (1928). *Adenophora thunbergiana* var. *hakusanensis* (Nakai) H. Hara, Bot. Mag. (Tokyo) 51: 897 (1937). *Adenophora triphylla* var. *hakusanensis* (Nakai) Kitam., Acta Phytotax. Geobot. 10: 311 (1941).

Adenophora obtusifolia Merr., Sunyatsenia 1: 82 (1930).

Adenophora kurilensis Nakai, Bot. Mag. (Tokyo) 44: 10 (1930). *Adenophora triphylla* var. *kurilensis* (Nakai) Kitam., Acta Phytotax. Geobot. 10: 311 (1941).

Adenophora thunbergiana var. *lancifolia* H. Hara, J. Jap. Bot. 10: 371 (1934). *Adenophora thunbergiana* f. *lancifolia* (H. Hara) H. Hara, Bot. Mag. (Tokyo) 51: 896 (1937). *Adenophora triphylla* f. *lancifolia* (H. Hara) Kitam., Acta Phytotax. Geobot. 10: 310 (1941).

Adenophora pulchra Kitam., Acta Phytotax. Geobot. 5: 204 (1936).

Adenophora insularis Kitam., Acta Phytotax. Geobot. 5: 209 (1936). *Adenophora triphylla* var. *insularis* (Kitam.) Kitam., Acta Phytotax. Geobot. 10: 310 (1941).

Adenophora puellaris Honda, Bot. Mag. (Tokyo) 50: 669 (1936). *Adenophora triphylla* var. *puellaris* (Honda) H. Hara, Enum. Sperm. Jap. 2: 93 (1952).

Adenophora thunbergiana f. *violacea* H. Hara, Bot. Mag. (Tokyo) 51: 897 (1937). *Adenophora triphylla* f. *violacea* (H. Hara) T. Shimizu, New Alp. Fl. Japan 1: 311 (1982).

Adenophora thunbergiana f. *brevidens* H. Hara, Bot. Mag. (Tokyo) 51: 897 (1937).

Adenophora kurilensis f. *albiflora* Tatew., Trans. Sapporo Nat. Hist. Soc. 16: 6 (1939). *Adenophora triphylla* f. *albiflora* (Tatew.) H. Hara, Enum. Sperm. Jap. 2: 93 (1952).

Adenophora thunbergiana f. *totoki* Hiyama ex Honda, Nom. Pl. Jap.: 336 (1939).

Adenophora triphylla f. *latifolia* Kitam., Acta Phytotax. Geobot. 10: 308 (1941).

Adenophora triphylla f. *hirsuta* Kitam., Acta Phytotax. Geobot. 10: 308 (1941).

Adenophora triphylla subsp. *aperticampanulata* Kitam., Acta Phytotax. Geobot. 10: 309 (1941).

Adenophora triphylla f. *pilosa* Kitam., Acta Phytotax. Geobot. 10: 309 (1941).

Adenophora verticillata f. *hirsuta* Hatus., ?; non (F. Schmidt) Makino, Bot. Mag. (Tokyo) 20: 39 (1906). *Adenophora radiatifolia* f. *hirsuta* (Hatus.) Nakai, Bull. Natl. Sci. Mus. Tokyo 31: 111 (1952).

Adenophora triphylla f. *leucantha* H. Hara, Enum. Sperm. Jap. 2: 93 (1952).

Adenophora triphylla f. *albiflora* Tawada, Biol. Mag. 7(5): 30 (1968); non (Tatew.) H. Hara, Enum. Sperm. Jap. 2: 93 (1952).

Adenophora triphylla f. *procumbens* T. Shimizu, Acta Phytotax. Geobot. 33: 334 (1982).

Adenophora tetraphylla var. *integrifolia* Y. Z. Zhao in Ma, Fl. Intramongolica (ed. 2) 4: 847 (1993).

Adenophora uryuensis Miyabe & Tatew., Trans. Sapporo Nat. Hist. Soc. 14: 80 (1935).
  Adenophora pereskiifolia var. *uryuensis* (Miyabe & Tatew.) Toyok. & Nosaka, Acta Phytotax. Geobot. 18: 195 (1960).
  Japan (Hokkaido). 38 JAP. Hemicr. or geophyte. $2n = 34$.
    Adenophora uryuensis f. *angustifolia* S. Watan., Hokuriku J. Bot. 5: 121 (1956).
      Adenophora pereskiifolia f. *angustifolia* (S. Watan.) Toyok. & Nosaka, Acta Phytotax. Geobot. 18: 195 (1960).

Adenophora wawreana Zahlbr., Ann. K. K. Naturhist. Hofmus. 10 (Notizen): 56 (1895).
  Manchuria (Liaoning) & N. China (Hebei, Henan, Nei Mongol, Shanxi). 36 CHI CHM CHN CHS. Hemicr. or geophyte. $2n = 34$.
    Adenophora wawreana f. *foliosa* Zahlbr., Ann. K. K. Naturhist. Hofmus. 10 (Notizen): 56 (1895).
    Adenophora wawreana f. *oligotricha* Kitag. in S. Tokun., Rep. First Sci. Exped. Manchoukuo 4(2): 114 (1935).
    Adenophora wawreana f. *polytricha* Kitag. in S. Tokun., Rep. First Sci. Exped. Manchoukuo 4(2): 115 (1935).
    Adenophora wawreana var. *lanceifolia* Y. Z. Zhao, Acta Sci. Nat. Univ. Intramongolicae 11: 57 (1980).
    Adenophora biformifolia Y. Z. Zhao, Acta Sci. Nat. Univ. Intramongolicae 11: 57 (1980).

Adenophora wilsonii Nannf. in Hand.-Mazz., Symb. Sin. 7: 1075 (1936).
  C. China (Guizhou, Hubei, Shaanxi, Sichuan). 36 CHC CHN. Hemicr. or geophyte.

Adenophora wulingshanica D. Y. Hong, Fl. Reipubl. Popul. Sin. 73(2): 187 (1983).
  NE. China (Hebei, Nei Monggol). 36 CHI CHN. Hemicr. or geophyte. $2n = 34$.
    Adenophora elata f. *verticillata* Kitag. in S. Tokun., Rep. First Sci. Exped. Manchoukuo 4(2): 107 (1935).
    Adenophora wulingshanica var. *alterna* Y. Z. Zhao in Ma, Fl. Intramongolica (ed. 2) 4: 847 (1993).

Adenophora xifengensis (P. F. Tu & Y. S. Zhou) P. F. Tu & Y. S. Zhou, Acta Bot. Boreal.-Occid. Sin. 18: 617 (1998). Hemicr. or geophyte.
  N. China (Gansu). 36 CHN.
    * *Adenophora stenanthina* subsp. *xifengensis* P. F. Tu & Y. S. Zhou, J. China Pharm. Univ. 21: 16 (1990).

Synonyms:
Adenophora albescens C. Y. Wu === **Adenophora khasiana** (Hook. f. & Thomson) Feer
Adenophora alpina Nannf. === **Adenophora himalayana** subsp. **alpina** (Nannf.) D. Y. Hong
Adenophora argyi H. Lév. === **Adenophora stricta** Miq. subsp. **stricta**
Adenophora atuntzensis C. Y. Wu === **Adenophora jasionifolia** Franch.
Adenophora axilliflora (Borbás) Borbas ex Prain === **Adenophora stricta** Miq. subsp. **stricta**
Adenophora biformifolia Y. Z. Zhao === **Adenophora wawreana** Zahlbr.
Adenophora biloba Y. Z. Zhao === **Adenophora gmelinii** (Biehler) Fisch. subsp. **gmelinii**
Adenophora bockiana Diels === **Adenophora potaninii** Korsh.
Adenophora borealis var. *oreophila* Y. Z. Zhao === **Adenophora borealis** D. Y. Hong & Y. Z. Zhao
Adenophora bulleyana Diels === **Adenophora khasiana** (Hook. f. & Thomson) Feer
Adenophora bulleyana var. *alba* C. Y. Wu === **Adenophora coelestis** Diels
Adenophora bulleyana var. *angustifolia* C. Y. Wu === **Adenophora khasiana** (Hook. f. & Thomson) Feer
Adenophora capillaris var. *tenuifolia* Diels === **Adenophora capillaris** Hemsl. subsp. **capillaris**
Adenophora chanetii (H. Lév.) D. F. Chamb. === **Adenophora polyantha** Nakai
Adenophora chionantha C. Y. Wu === **Adenophora khasiana** (Hook. f. & Thomson) Feer
Adenophora coelestis var. *stenophylla* Diels ex C. Y. Wu === **Adenophora coelestis** Diels
Adenophora coelestis var. *uehatae* (Yamam.) Masam. === **Adenophora morrisonensis** subsp. **uehatae** (Yamam.) Lammers

Adenophora collina Kitag. === **Adenophora stenanthina** (Ledeb.) Kitag. subsp. **stenanthina**

Adenophora collina f. *latifolia* Kitag. === **Adenophora stenanthina** (Ledeb.) Kitag. subsp. **stenanthina**

Adenophora communis Fisch. === **Adenophora liliifolia** (L.) A. DC.

Adenophora communis var. *coronopifolia* (Fisch. ex Schult.) Trautv. === **Adenophora gmelinii** (Biehler) Fisch. subsp. **gmelinii**

Adenophora communis var. *denticulata* (Fisch.) Trautv. === **Adenophora tricuspidata** (Fisch. ex Schult.) A. DC.

Adenophora communis var. *gmelinii* (Biehler) Trautv. === **Adenophora gmelinii** (Biehler) Fisch.

Adenophora communis var. *integerrima* (Trautv.) Trautv. === **Adenophora liliifolia** (L.) A. DC.

Adenophora communis var. *lamarckii* (Fisch.) Trautv. === **Adenophora lamarckii** Fisch.

Adenophora communis var. *latifolia* (Fisch.) Trautv. === **Adenophora pereskiifolia** (Fisch. ex Schult.) Fisch. ex G. Don

Adenophora confusa Nannf. === **Adenophora stricta** subsp. **confusa** (Nannf.) D. Y. Hong

Adenophora cordata B. D. Jacks. === **Adenophora liliifolia** (L.) A. DC.

Adenophora coronata (Ker Gawl.) A. DC. === **Adenophora stenanthina** (Ledeb.) Kitag. subsp. **stenanthina**

Adenophora coronopifolia (Fisch. ex Schult.) Fisch. === **Adenophora gmelinii** (Biehler) Fisch. subsp. **gmelinii**

Adenophora coronopifolia var. *erysimoides* (Vest ex Schult.) Steud. === **Adenophora gmelinii** (Biehler) Fisch. subsp. **gmelinii**

Adenophora coronopifolia var. *longisepala* Reverd. === **Adenophora gmelinii** (Biehler) Fisch. subsp. **gmelinii**

Adenophora coronopifolia var. *puberula* Reverd. === **Adenophora gmelinii** (Biehler) Fisch. subsp. **gmelinii**

Adenophora coronopifolia var. *pubicalycina* Reverd. === **Adenophora gmelinii** (Biehler) Fisch. subsp. **gmelinii**

Adenophora coronopifolia var. *rotundiflora* Y. N. Lee === **Adenophora gmelinii** (Biehler) Fisch. subsp. **gmelinii**

Adenophora crispata (Turcz. ex Kitag.) Kitag. === **Adenophora stenanthina** (Ledeb.) Kitag. subsp. **stenanthina**

Adenophora curvidens Nakai === **Adenophora pereskiifolia** (Fisch. ex Schult.) Fisch. ex G. Don

Adenophora denticulata Fisch. === **Adenophora tricuspidata** (Fisch. ex Schult.) A. DC.

Adenophora denticulata f. *angustifolia* Korsh. === **Adenophora tricuspidata** (Fisch. ex Schult.) A. DC.

Adenophora denticulata var. *ciliata* Reverd. === **Adenophora tricuspidata** (Fisch. ex Schult.) A. DC.

Adenophora denticulata var. *linifolia* Reverd. === **Adenophora tricuspidata** (Fisch. ex Schult.) A. DC.

Adenophora dimorphophylla C. Y. Wu === **Adenophora khasiana** (Hook. f. & Thomson) Feer

Adenophora diplodonta Diels === **Adenophora khasiana** (Hook. f. & Thomson) Feer

Adenophora divaricata f. *alternifolia* H. Hara === **Adenophora divaricata** Franch. & Sav.

Adenophora divaricata f. *angustifolia* A. I. Baranov === **Adenophora divaricata** Franch. & Sav.

Adenophora divaricata f. *manshurica* (Nakai) Kitag. === **Adenophora pereskiifolia** (Fisch. ex Schult.) Fisch. ex G. Don

Adenophora divaricata var. *manshurica* Nakai === **Adenophora pereskiifolia** (Fisch. ex Schult.) Fisch. ex G. Don

Adenophora elata f. *verticillata* Kitag. === **Adenophora wulingshanica** D. Y. Hong

Adenophora erysimoides (Vest ex Schult.) Kitag. === **Adenophora gmelinii** (Biehler) Fisch. subsp. **gmelinii**

Adenophora fischeri (Schult.) G. Don === **Adenophora liliifolia** (L.) A. DC.

Adenophora forrestii Diels === **Adenophora jasionifolia** Franch.

Adenophora forrestii var. *handeliana* Nannf. === **Adenophora jasionifolia** Franch.

Adenophora forrestii var. *intercedens* Pax & K. Hoffm. === **Adenophora jasionifolia** Franch.

Adenophora fusifolia f. *albiflora* Y. N. Lee === **Adenophora fusifolia** Y. N. Lee

Adenophora gmelinii var. *coronopifolia* (Fisch. ex Schult.) Y. Z. Zhao === **Adenophora gmelinii** (Biehler) Fisch. subsp. **gmelinii**

Adenophora gmelinii var. *pachyphylla* (Kitag.) Y. Z. Zhao === **Adenophora gmelinii** (Biehler) Fisch. subsp.**gmelinii**

Adenophora gmelinii var. *stylosa* A. DC. === **Adenophora gmelinii** (Biehler) Fisch. subsp. **gmelinii**

Adenophora golubinzevaeana var. *pilosa* Reverd. === **Adenophora golubinzevaeana** Reverd.

Adenophora gracilis Nannf. === **Adenophora liliifolioides** Pax & K. Hoffm.

Adenophora hakusanensis Nakai === **Adenophora triphylla** (Thunb.) A. DC.

Adenophora howozana Takeda === **Adenophora takedae** Makino

Adenophora huangae C. Y. Wu === **Adenophora coelestis** Diels

Adenophora insolens Reverd. === **Adenophora stenanthina** (Ledeb.) Kitag. subsp. **stenanthina**

Adenophora insularis Kitam. === **Adenophora triphylla** (Thunb.) A. DC.

Adenophora intermedia Ledeb. === **Adenophora stenanthina** (Ledeb.) Kitag. subsp. **stenanthina**

Adenophora intermedia (Schult.) Sweet === **Adenophora liliifolia** (L.) A. DC.

Adenophora isabellae Hemsl. === **Adenophora remotiflora** (Siebold & Zucc.) Miq.

Adenophora ishiyamae Miyabe & Tatew. === **Adenophora pereskiifolia** (Fisch. ex Schult.) G. Don

Adenophora kurilensis Nakai === **Adenophora triphylla** (Thunb.) A. DC.

Adenophora kurilensis f. *albiflora* Tatew. === **Adenophora triphylla** (Thunb.) A. DC.

Adenophora lamarckii var. *angustifolia* A. DC. === **Adenophora gmelinii** (Biehler) Fisch. subsp. **gmelinii**

Adenophora lamarckii f. *multiloba* Takeda === **Adenophora nikoensis** Franch. & Sav.

Adenophora lamarckii var. *longifolia* Nakai === **Adenophora lamarckii** Fisch.

Adenophora latifolia Fisch. === **Adenophora pereskiifolia** (Fisch. ex Schult.) Fisch. ex G. Don

Adenophora latifolia f. *albiflora* Nakai === **Adenophora pereskiifolia** (Fisch. ex Schult.) Fisch. ex G. Don

Adenophora leptosepala Diels === **Adenophora capillaris** subsp. **leptosepala** (Diels) D. Y. Hong

Adenophora leptosepala var. *linearifolia* C. Y. Wu === **Adenophora capillaris** subsp. **leptosepala** (Diels) D. Y. Hong

Adenophora likiangensis C. Y. Wu === **Adenophora khasiana** (Hook. f. & Thomson) Oliv. ex Collett & Hemsl.

Adenophora liliifolia f. *alba* (Nakai) Nakai === **Adenophora liliifolia** (L.) A. DC.

Adenophora liliifolia f. *infundibuliformis* (A. DC.) Voss === **Adenophora liliifolia** (L.) A. DC.

Adenophora liliifolia f. *villosula* (Borbás) Hayek === **Adenophora liliifolia** (L.) A. DC.

Adenophora liliifolia subf. *subverticillata* (Borbás) Hayek === **Adenophora liliifolia** (L.) A. DC.

Adenophora liliifolia subsp. *taurica* Sukacz. === **Adenophora taurica** (Sukacz.) Juz.

Adenophora liliifolia var. *alba* Nakai === **Adenophora liliifolia** (L.) A. DC.

Adenophora liliifolia var. *angusta* Nakai === **Adenophora liliifolia** (L.) A. DC.

Adenophora liliifolia var. *infundibuliformis* A. DC. === **Adenophora liliifolia** (L.) A. DC.

Adenophora liliifolia var. *integerrima* (Trautv.) Korsh. === **Adenophora liliifolia** (L.) A. DC.

Adenophora liliifolia var. *lamarckii* (Fisch.) Krylov === **Adenophora lamarckii** Fisch.

Adenophora liliifolia var. *pocsii* Soó === **Adenophora liliifolia** (L.) A. DC.

Adenophora liliifolia var. *spreta* (Schult.) A. DC. === **Adenophora liliifolia** (L.) A. DC.

Adenophora liliifolia var. *stylosa* (Lam.) Korsh. === **Adenophora liliifolia** (L.) A. DC.

Adenophora liliifolia var. *suaveolens* (Schrad. ex Hornem.) Steud. === **Adenophora liliifolia** (L.) A. DC.

Adenophora longisepala P. C. Tsoong === **Adenophora capillaris** Hemsl. subsp. **capillaris**

Adenophora manshurica Nakai === **Adenophora pereskiifolia** (Fisch. ex Schult.) Fisch. ex G. Don

Adenophora manshurica f.*albiflora* (Nakai) U. C. La === **Adenophora pereskiifolia** (Fisch. ex Schult.) Fisch. ex G. Don

Adenophora marsupiiflora (Schult.) Fisch. === **Adenophora stenanthina** (Ledeb.) Kitag. subsp. **stenanthina**

Adenophora marsupiiflora var. *crispata* Turcz. ex Kitag. === **Adenophora stenanthina** (Ledeb.) Kitag. subsp. **stenanthina**

Adenophora marsupiiflora var. *jaluensis* Kom. === **Adenophora stenanthina** (Ledeb.) Kitag. subsp. **stenanthina**

Adenophora marsupiiflora var. *korshinskiana* Diels === **Adenophora stenanthina** (Ledeb.) Kitag. subsp. **stenanthina**

Adenophora marsupiiflora var. *pilosa* Korsh. === **Adenophora stenanthina** (Ledeb.) Kitag. subsp. **stenanthina**

Adenophora megalantha Diels === **Adenophora coelestis** Diels

Adenophora microcodon C. Y. Wu === **Adenophora jasionifolia** Franch.

Adenophora moiwana Nakai === **Adenophora pereskiifolia** (Fisch. ex Schult.) Fisch. ex G. Don

Adenophora moiwana var. *heterotricha* Nakai ex H. Hara === **Adenophora pereskiifolia** (Fisch. ex Schult.) Fisch. ex G. Don

Adenophora mongolica A. I. Baranov === **Adenophora stenophylla** Hemsl.

Adenophora montana Turcz. === **Adenophora stenanthina** (Ledeb.) Kitag. subsp. **stenanthina**

Adenophora nikoensis f. *globiflora* Hiyama === **Adenophora nikoensis** Franch. & Sav.

Adenophora nikoensis f. *hispidula* T. Shimizu === **Adenophora nikoensis** Franch. & Sav.

Adenophora nikoensis f. *hispidula* (T. Shimizu) Yonek. === **Adenophora nikoensis** Franch. & Sav.

Adenophora nikoensis f. *linearifolia* Takeda === **Adenophora nikoensis** Franch. & Sav.

Adenophora nikoensis f. *macrocalyx* Takeda === **Adenophora nikoensis** Franch. & Sav.

Adenophora nikoensis f. *multifida* H. Hara === **Adenophora nikoensis** Franch. & Sav.

Adenophora nikoensis f. *nipponica* (Kitam.) H. Hara === **Adenophora nikoensis** Franch. & Sav.

Adenophora nikoensis f. *stenophylla* (Kitam.) H. Hara === **Adenophora nikoensis** Franch. & Sav.

Adenophora nikoensis var. *persicaria* Ohwi === **Adenophora nikoensis** Franch. & Sav.

Adenophora nikoensis var. *petrophila* (Hara) H. Hara === **Adenophora nikoensis** Franch. & Sav.

Adenophora nikoensis var. *teramotoi* (Hurus.) Okazaki & T. Shimizu === **Adenophora nikoensis** Franch. & Sav.

Adenophora nipponica Kitam. === **Adenophora nikoensis** Franch. & Sav.

Adenophora nipponica f. *multiloba* (Takeda) Kitam. === **Adenophora nikoensis** Franch. & Sav.

Adenophora nipponica var. *stenophylla* Kitam. === **Adenophora nikoensis** Franch. & Sav.

Adenophora obovata Kitam. === **Adenophora polyantha** Nakai

Adenophora obtusifolia Merr. === **Adenophora triphylla** (Thunb.) A. DC.

Adenophora onoi Tatew. & Kitam. === **Adenophora pereskiifolia** (Fisch. ex Schult.) Fisch. ex G. Don

Adenophora ornata Diels === **Adenophora coelestis** Diels

Adenophora ornata var. *alba* C. Y. W === **Adenophora coelestis** Diels

Adenophora pachyphylla Kitag. === **Adenophora gmelinii** (Biehler) Fisch. subsp. **gmelinii**

Adenophora pachyrhiza Diels === **Adenophora coelestis** Diels

Adenophora palustris f. *leucantha* Nakai ex Matsum. === **Adenophora palustris** Kom.

Adenophora palustris var. *leucantha* (Nakai ex Matsum.) Nakai === **Adenophora palustris** Kom.

Adenophora palustris var. *linearis* Nakai ex Matsum. === **Adenophora palustris** Kom.

Adenophora paniculata var. *dentata* Y. Z. Zhao === **Adenophora paniculata** Nannf.

Adenophora paniculata var. *petiolata* Y. Z. Zhao === **Adenophora paniculata** Nannf.

Adenophora paniculata var. *pilosa* Kitag. === **Adenophora paniculata** Nannf.

Adenophora paniculata var. *psilosa* Kitag. === **Adenophora paniculata** Nannf.

Adenophora pereskiifolia f. *angustifolia* (S. Watan.) Toyok. & Nosaka === **Adenophora uryuensis** Miyabe & Tatew.

Adenophora pereskiifolia f. *linearifolia* A. I. Baranov === **Adenophora pereskiifolia** (Fisch. ex Schult.) G. Don

Adenophora pereskiifolia f. *linearifolia* T. Shimizu === **Adenophora pereskiifolia** (Fisch. ex Schult.) Fisch. ex G. Don

Adenophora pereskiifolia f. *petrophila* (H. Hara) T. Shimizu === **Adenophora nikoensis** Franch. & Sav.

Adenophora pereskiifolia f. *puberula* Kitag. === **Adenophora pereskiifolia** (Fisch. ex Schult.) G. Don

Adenophora pereskiifolia f. *stenophylla* Toyok. & Nosaka === **Adenophora pereskiifolia** (Fisch. ex Schult.) Fisch. ex G. Don

Adenophora pereskiifolia subsp. *alternifolia* (Fuh ex Y. Z. Zhao) C. X. Fu & M. Y. Liu === **Adenophora pereskiifolia** (Fisch. ex Schult.) Fisch. ex G. Don

Adenophora pereskiifolia subsp. *subalpina* A. I. Baranov === **Adenophora pereskiifolia** (Fisch. ex Schult.) Fisch. ex G. Don

Adenophora pereskiifolia var. *alternifolia* Fuh ex Y. Z. Zhao === **Adenophora pereskiifolia** (Fisch. ex Schult.) Fisch. ex G. Don

Adenophora pereskiifolia var. *angustifolia* Y. Z. Zhao === **Adenophora pereskiifolia** (Fisch. ex Schult.) Fisch. ex G. Don

Adenophora pereskiifolia var. *curvidens* (Nakai) Kitag. === **Adenophora pereskiifolia** (Fisch. ex Schult.) Fisch. ex G. Don

Adenophora pereskiifolia var. *heterotricha* (Nakai ex H. Hara) H. Hara === **Adenophora pereskiifolia** (Fisch. ex Schult.) Fisch. ex G. Don

Adenophora pereskiifolia var. *japonica* Regel === **Adenophora triphylla** (Thunb.) A. DC.

Adenophora pereskiifolia var. *moiwana* (Nakai) H. Hara === **Adenophora pereskiifolia** (Fisch. ex Schult.) Fisch. ex G. Don

Adenophora pereskiifolia var. *petrophila* (H. Hara) T. Shimizu === **Adenophora nikoensis** Franch. & Sav.

Adenophora pereskiifolia var. *uryuensis* (Miyabe & Tatew.) Toyok. & Nosaka === **Adenophora uryuensis** Miyabe & Tatew.

Adenophora pereskiifolia var. *yamadae* Toyok. & Nosaka === **Adenophora pereskiifolia** (Fisch. ex Schult.) Fisch. ex G. Don

Adenophora periplocifolia (Lam.) A. DC. === **Adenophora liliifolia** (L.) A. DC.

Adenophora petrophila H. Hara === **Adenophora nikoensis** Franch. & Sav.

Adenophora polyantha f. *densipila* Kitag. === **Adenophora polyantha** Nakai

Adenophora polyantha f. *eriocaulis* Kitag. === **Adenophora polyantha** Nakai

Adenophora polyantha subsp. *scabricalyx* (Kitag.) J. Z. Qiu & D. Y. Hong === **Adenophora polyantha** Nakai

Adenophora polyantha subsp. *subalpina* (A. I. Baranov) Kitag. === **Adenophora pereskiifolia** (Fisch. ex Schult.) Fisch. ex G. Don

Adenophora polyantha var. *contracta* Kitag. === **Adenophora contracta** (Kitag.) J. Z. Qiu & D. Y. Hong

Adenophora polyantha var. *glabricalyx* Kitag. === **Adenophora polyantha** Nakai

Adenophora polyantha var. *media* Nakai & Kitag. === **Adenophora polyantha** Nakai

Adenophora polyantha var. *rhombica* Y. Z. Zhao === **Adenophora polyantha** Nakai

Adenophora polyantha var. *scabricalyx* Kitag. === **Adenophora polyantha** Nakai

Adenophora polydentata P. F. Tu & G. J. Xu === **Adenophora potaninii** Korsh.

Adenophora polymorpha Ledeb. === **Adenophora liliifolia** (L.) A. DC.

Adenophora polymorpha f. *integrisepala* Herder === ?

Adenophora polymorpha f. *multiloba* (Takeda) Makino & Nemoto === **Adenophora nikoensis** Franch. & Sav.

Adenophora polymorpha f. *raddeana* Herder === ?

Adenophora polymorpha var. *alternifolia* Franch. & Sav. === ?

Adenophora polymorpha var. *calicina* Franch. & Sav. === ?

Adenophora polymorpha var. *chanetii* H. Lév. === ?

Adenophora polymorpha var. *coronopifolia* (Fisch. ex Schult.) Trautv. ex Herder === **Adenophora gmelinii** (Biehler) Fisch. subsp. **gmelinii**

Adenophora polymorpha var. *denticulata* (Fisch.) Trautv. ex Herder === **Adenophora tricuspidata** (Fisch. ex Schult.) A. DC.

Adenophora polymorpha var. *divaricata* (Franch. & Sav.) Makino === **Adenophora divaricata** Franch. & Sav.

Adenophora polymorpha var. *gmelinii* (Biehler) Trautv. ex Herder === **Adenophora gmelinii** (Biehler) Fisch.

Adenophora polymorpha var. *integerrima* Trautv. === **Adenophora liliifolia** (L.) A. DC.

Adenophora polymorpha var. *lamarckii* (Fisch.) Trautv. ex Herder === **Adenophora lamarckii** Fisch.

Adenophora polymorpha var. *latifolia* (Fisch.) Trautv. === **Adenophora pereskiifolia** (Fisch. ex Schult.) Fisch. ex G. Don

Adenophora polymorpha var. *marsupiiflora* (Schult.) Franch. === **Adenophora stenanthina** (Ledeb.) Kitag. subsp. **stenanthina**

Adenophora polymorpha var. *pereskiifolia* (Fisch. ex Schult.) Makino === **Adenophora pereskiifolia** (Fisch. ex Schult.) Fisch. ex G. Don

Adenophora polymorpha var. *rhombifolia* H. Lév. === **Adenophora triphylla** (Thunb.) A. DC.

Adenophora polymorpha var. *sinensis* (A. DC.) Pampan. === **Adenophora sinensis** A. DC.

Adenophora polymorpha var. *stricta* (Miq.) Makino === **Adenophora stricta** Miq.

Adenophora polymorpha var. *tashiroi* Makino & Nakai === **Adenophora tashiroi** (Makino & Nakai) Makino & Nakai

Adenophora polymorpha var. *urticifolia* Franch. & Sav. === ?

Adenophora polymorpha var. *verticillata* Franch. & Sav. === **Adenophora pereskiifolia** (Fisch. ex Schult.) Fisch. ex G. Don

Adenophora pratensis Y. Z. Zhao === **Adenophora stenanthina** (Ledeb.) Kitag. subsp. **stenathina**

Adenophora prenanthoides Prain ex Diels === ?

Adenophora pubescens Hemsl. === **Adenophora rupincola** Hemsl.

Adenophora puellaris Honda === **Adenophora triphylla** (Thunb.) A. DC.

Adenophora pulchra Kitam. === **Adenophora triphylla** (Thunb.) A. DC.

Adenophora pumila P. C. Tsoong === **Adenophora jasionifolia** Franch.

Adenophora rabelaisiana (Schult.) G. Don === **Adenophora gmelinii** (Biehler) Fisch. subsp. **gmelinii**

Adenophora radiatifolia Nakai === **Adenophora triphylla** (Thunb.) A. DC

Adenophora radiatifolia f. *hirsuta* (Hatus.) Nakai === **Adenophora triphylla** (Thunb.) A. DC.

Adenophora radiatifolia var. *abbreviata* (H. Lév.) Nakai === **Adenophora triphylla** (Thunb.) A. DC.

Adenophora radiatifolia var. *angustifolia* (Regel) Nakai === **Adenophora triphylla** (Thunb.) A. DC.

Adenophora radiatifolia var. *hirsuta* (F. Schmidt) Nakai === **Adenophora triphylla** (Thunb.) A. DC.

Adenophora radiatifolia var. *rhombifolia* (H. Lév.) Nakai === **Adenophora triphylla** (Thunb.) A. DC.

Adenophora raphanorrhiza C. Y. Wu === **Adenophora coelestis** Diels

Adenophora remotiflora f. *albiflora* Y. N. Lee, Korean J. Bot. 24: 29 (1981). === **Adenophora remotiflora** (Siebold & Zucc.) Miq.

Adenophora remotiflora f. *angustifolia* Makino === **Adenophora remotiflora** (Siebold & Zucc.) Miq.

Adenophora remotiflora f. *cordata* Kom. === **Adenophora remotiflora** (Siebold & Zucc.) Miq.

Adenophora remotiflora f. *leucantha* Honda === **Adenophora remotiflora** (Siebold & Zucc.) Miq.

Adenophora remotiflora f. *longifolia* Kom. === **Adenophora remotiflora** (Siebold & Zucc.) Miq.

Adenophora remotiflora f. *oreophila* Hiyama === **Adenophora remotiflora** (Siebold & Zucc.) Miq.

Adenophora remotiflora var. *cordatifolia* (Deb.) Zahlbr. === **Adenophora remotiflora** (Siebold & Zucc.) Miq.

Adenophora remotiflora var. *hirsuta* Honda === **Adenophora remotiflora** (Siebold & Zucc.) Miq.

Adenophora remotiflora var. *hirticalyx* S. T. Lee, Y. J. Chung & J. K. Lee === **Adenophora remotiflora** (Siebold & Zucc.) Miq.

Adenophora reticulata G. Don ex W. H. Baxter & Wooster === ?

Adenophora richteri Borbás === **Adenophora tricuspidata** (Fisch. ex Schult.) A. DC.

Adenophora roseiflora C. Y. Wu === **Adenophora khasiana** (Hook. f. & Thomson) Feer

Adenophora rotundifolia H. Lév. === **Adenophora stricta** Miq. subsp. **stricta**

Adenophora scabridula Nannf. === **Adenophora polyantha** Nakai

Adenophora scabridula var. *viscida* P. C. Tsoong === **Adenophora polyantha** Nakai

Adenophora setulosa Borbás === **Adenophora liliifolia** (L.) A. DC.

Adenophora sinensis var. *glabra* A. DC. === **Adenophora sinensis** A. DC.

Adenophora sinensis var. *pilosa* A. DC. === **Adenophora stricta** Miq. subsp. **stricta**

Adenophora smithii Nannf. === **Adenophora himalayana** Feer subsp. **himalayana**

Adenophora smithii f. *crispa* Nannf. === **Adenophora himalayana** Feer subsp. **himalayana**

Adenophora stenanthina f. *crispata* (Turcz. ex Kitag.) Kitag. === **Adenophora stenanthina** (Ledeb.) Kitag. subsp. **stenanthina**

Adenophora stenanthina subsp. *xifengensis* P. F. Tu & Y. S. Zhou === **Adenophora xifengensis** (P. F. Tu & Y. S. Zhou) P. F. Tu & Y. S. Zhou

Adenophora stenanthina var. *angustilanceifolia* Y. Z. Zhao === **Adenophora stenanthina** (Ledeb.) Kitag. subsp. **stenanthina**

Adenophora stenanthina var. *collina* (Kitag.) Y. Z. Zhao === **Adenophora stenanthina** (Ledeb.) Kitag. subsp. **stenanthina**

Adenophora stenanthina var. *crispata* (Turcz. ex Kitag.) Y. Z. Zhao === **Adenophora stenanthina** (Ledeb.) Kitag. subsp. **stenanthina**

Adenophora stenophylla var. *denudata* Kitag. === **Adenophora stenophylla** Hemsl.

Adenophora stricta var. *glabrior* Miq. === **Adenophora stricta** Miq. subsp. **stricta**

Adenophora stricta var. *lancifolia* Honda === **Adenophora stricta** Miq. subsp. **stricta**

Adenophora stricta var. *nanjingensis* P. F. Tu & G. J. Xu === **Adenophora stricta** Miq. subsp. **stricta**

Adenophora stricta var. *qinglongshanica* P. F. Tu & G. J. Xu === **Adenophora stricta** Miq. subsp. **stricta**

Adenophora stylosa (Lam.) Fisch. === **Adenophora liliifolia** (L.) A. DC.

Adenophora suaveolens (Schrad. ex Hornem.) Rchb. === **Adenophora liliifolia** (L.) A. DC.

Adenophora suaveolens var. *latifolia* Schur === **Adenophora liliifolia** (L.) A. DC.

Adenophora suolunensis P. F. Tu & X. F. Zhao === **Adenophora micrantha** D. Y. Hong

Adenophora takedae var. *howozana* (Takeda) Sugimoto ex Honda === **Adenophora takedae** Makino

Adenophora tatewakiana Hurus. === ?

Adenophora teramotoi Hurusawa === **Adenophora nikoensis** Franch. & Sav.

Adenophora teramotoi var. *hispidula* T. Shimizu === **Adenophora nikoensis** Franch. & Sav.

Adenophora tetraphylla (Thunb.) Fisch. ex Kitag. === **Adenophora triphylla** (Thunb.) A. DC.

Adenophora tetraphylla var. *abbreviata* (H. Lév.) D. F. Chamb. === **Adenophora triphylla** (Thunb.) A. DC.

Adenophora tetraphylla var. *angustifolia* A. I. Baranov === **Adenophora triphylla** (Thunb.) A. DC.

Adenophora tetraphylla var. *hirsuta* (F. Schmidt) D. F. Chamb. === **Adenophora triphylla** (Thunb.) A. DC.

Adenophora tetraphylla var. *integrifolia* Y. Z. Zhao === **Adenophora triphylla** (Thunb.) A. DC.

Adenophora tetraphylla var. *media* (Korsh.) A. I. Baranov === **Adenophora triphylla** (Thunb.) A. DC.

Adenophora tetraphylla var. *princeps* (Korsh.) A. I. Baranov === **Adenophora triphylla** (Thunb.) A. DC.

Adenophora thunbergiana Kudô === **Adenophora triphylla** (Thunb.) A. DC.

Adenophora thunbergiana f. *brevidens* H. Hara === **Adenophora triphylla** (Thunb.) A. DC.

Adenophora thunbergiana f. *glabra* (Makino) H. Hara === **Adenophora triphylla** (Thunb.) A. DC.

Adenophora thunbergiana f. *hirsuta* (F. Schmidt) Kudô === **Adenophora triphylla** (Thunb.) A. DC.

Adenophora thunbergiana f. *lancifolia* (H. Hara) H. Hara === **Adenophora triphylla** (Thunb.) A. DC.

Adenophora thunbergiana f. *violacea* H. Hara === **Adenophora triphylla** (Thunb.) A. DC.

Adenophora thunbergiana f. *totoki* Hiyama ex Honda === **Adenophora triphylla** (Thunb.) A. DC.

Adenophora thunbergiana var. *hakusanensis* (Nakai) H. Hara === **Adenophora triphylla** (Thunb.) A. DC.

Adenophora thunbergiana var. *lancifolia* H. Hara === **Adenophora triphylla** (Thunb.) A. DC.

Adenophora trachelioides Maxim. === **Adenophora remotiflora** (Siebold & Zucc.) Miq.

Adenophora trachelioides f. *puberula* A. I. Baranov === **Adenophora remotiflora** (Siebold & Zucc.) Miq.

Adenophora trachelioides subsp. *giangsuensis* D. Y. Hong === **Adenophora remotiflora** (Siebold & Zucc.) Miq.

Adenophora trachelioides var. *cordatifolia* Deb. === **Adenophora remotiflora** (Siebold & Zucc.) Miq.

Adenophora triphylla f. *albiflora* (Tatew.) H. Hara === **Adenophora triphylla** (Thunb.) A. DC.

Adenophora triphylla f. *albiflora* Tawada === **Adenophora triphylla** (Thunb.) A. DC.

Adenophora triphylla f. *canescens* (Franch. & Sav.) Kitam. === **Adenophora triphylla** (Thunb.) A. DC.

Adenophora triphylla f. *glabra* (Makino) Kitam. === **Adenophora triphylla** (Thunb.) A. DC.

Adenophora triphylla f. *hirsuta* Kitam. === **Adenophora triphylla** (Thunb.) A. DC.

Adenophora triphylla f. *lancifolia* (H. Hara) Kitam. === **Adenophora triphylla** (Thunb.) A. DC.

Adenophora triphylla f. *latifolia* Kitam. === **Adenophora triphylla** (Thunb.) A. DC.

Adenophora triphylla f. *linearis* (Hayata) Kitam. === **Adenophora triphylla** (Thunb.) A. DC.

Adenophora triphylla f. *pilosa* Kitam. === **Adenophora triphylla** (Thunb.) A. DC.

Adenophora triphylla f. *pilosissima* Kitam. === **Adenophora triphylla** (Thunb.) A. DC.

Adenophora triphylla f. *princeps* (Korsh.) Kitam. === **Adenophora triphylla** (Thunb.) A. DC.

Adenophora triphylla f. *procumbens* T. Shimizu === **Adenophora triphylla** (Thunb.) A. DC.

Adenophora triphylla f. *violacea* (H. Hara) T. Shimizu === **Adenophora triphylla** (Thunb.) A. DC.

Adenophora triphylla subsp. *aperticampanulata* Kitam. === **Adenophora triphylla** (Thunb.) A. DC.

Adenophora triphylla var. *angustifolia* (Regel) Kitam. === **Adenophora triphylla** (Thunb.) A. DC.

Adenophora triphylla var. *hakusanensis* (Nakai) Kitam. === **Adenophora triphylla** (Thunb.) A. DC.

Adenophora triphylla var. *insularis* (Kitam.) Kitam. === **Adenophora triphylla** (Thunb.) A. DC.

Adenophora triphylla var. *japonica* (Regel) H. Hara === **Adenophora triphylla** (Thunb.) A. DC.

Adenophora triphylla var. *kurilensis* (Nakai) Kitam. === **Adenophora triphylla** (Thunb.) A. DC.

Adenophora triphylla var. *puellaris* (Honda) H. Hara === **Adenophora triphylla** (Thunb.) A. DC.

Adenophora triphylla var. *tetraphylla* (Thunb.) Makino === **Adenophora triphylla** (Thunb.) A. DC.

Adenophora tsinlingensis Pax & K. Hoffm. === **Adenophora himalayana** subsp. **alpina** (Nannf.) D. Y. Hong

Adenophora turczaninowii Feer === **Adenophora tricuspidata** (Fisch. ex Schult.) A. DC.

Adenophora tyosenensis Nakai ex Chung === **Adenophora divaricata** Franch. & Sav.

Adenophora uehatae Yamam. === **Adenophora morrisonensis** subsp. **uehatae** (Yamam.) Lammers

Adenophora urceolata C. Y. Wu === **Adenophora capillaris** subsp. **leptosepala** (Diels) D. Y. Hong

Adenophora urceolata Y. Z. Zhao === **Adenophora kulunensis** Y. Z. Zhao

Adenophora uryuensis f. *angustifolia* S. Watan. === **Adenophora uryuensis** Miyabe & Tatew.

Adenophora uyemurae Kudo === **Campanula uyemurae** (Kudô) Miyabe & Tatew.

Adenophora verbenifolia Bunge ex Heynh. === **?**

Adenophora verticillata Pall. ex Fisch. === **Adenophora triphylla** (Thunb.) A. DC

Adenophora verticillata f. *angustifolia* (Regel) Makino === **Adenophora triphylla** (Thunb.) A. DC.

Adenophora verticillata f. *crenata* Franch. & Sav. === **Adenophora triphylla** (Thunb.) A. DC.

Adenophora verticillata f. *dentata* Franch. & Sav. === **Adenophora triphylla** (Thunb.) A. DC.

Adenophora verticillata f. *hirsuta* Hatus. === **Adenophora triphylla** (Thunb.) A. DC.

Adenophora verticillata f. *hirsuta* (F. Schmidt) Makino === **Adenophora triphylla** (Thunb.) A. DC.

Adenophora verticillata f. *incisa* Franch. & Sav. === **Adenophora triphylla** (Thunb.) A. DC.

Adenophora verticillata f. *linearis* (Hayata) Matsum. === **Adenophora triphylla** (Thunb.) A. DC.

Adenophora verticillata f. *princeps* (Korsh.) Kom. === **Adenophora triphylla** (Thunb.) A. DC.

Adenophora verticillata f. *serrulata* Maxim. ex Franch. & Sav. === **Adenophora triphylla** (Thunb.) A. DC.

Adenophora verticillata f. *subintegrifolia* (Regel) Makino === **Adenophora triphylla** (Thunb.) A. DC.

Adenophora verticillata f. *triphylla* (Thunb.) Makino === **Adenophora triphylla** (Thunb.) A. DC.

Adenophora verticillata lusus *alternifolia* (Franch. & Sav.) Makino === **Adenophora triphylla** (Thunb.) A. DC.

Adenophora verticillata prole *princeps* Korsh. === **Adenophora triphylla** (Thunb.) A. DC.

Adenophora verticillata subf. *alternifolia* (Franch. & Sav.) Makino === **Adenophora triphylla** (Thunb.) A. DC.

Adenophora verticillata subf. *glabra* Makino === **Adenophora triphylla** (Thunb.) A. DC.

Adenophora verticillata var. *abbreviata* H. Lév. === **Adenophora triphylla** (Thunb.) A. DC.

Adenophora verticillata var. *alternifolia* Franch. & Sav. === **Adenophora triphylla** (Thunb.) A. DC.

Adenophora verticillata var. *angustifolia* Korsh. === **Adenophora triphylla** (Thunb.) A. DC.

Adenophora verticillata var. *angustifolia* Regel === **Adenophora triphylla** (Thunb.) A. DC.

Adenophora verticillata var. *brevidens* Franch. & Sav. === **Adenophora triphylla** (Thunb.) A. DC.

Adenophora verticillata var. *canescens* Franch. & Sav. === **Adenophora triphylla** (Thunb.) A. DC.

Adenophora verticillata var. *denticulata* Korsh. === **Adenophora triphylla** (Thunb.) A. DC.

Adenophora verticillata var. *glabra* (Makino) Makino === **Adenophora triphylla** (Thunb.) A. DC.

Adenophora verticillata var. *hirsuta* F. Schmidt === **Adenophora triphylla** (Thunb.) A. DC.

Adenophora verticillata var. *latifolia* Miq. === **Adenophora triphylla** (Thunb.) A. DC.

Adenophora verticillata var. *linearis* Hayata === **Adenophora triphylla** (Thunb.) A. DC.

Adenophora verticillata var. *maritima* Korsh. === **Adenophora triphylla** (Thunb.) A. DC.

Adenophora verticillata var. *marsupiiflora* (Schult.) Trautv. === **Adenophora stenanthina** (Ledeb.) Kitag. subsp. **stenanthina**

Adenophora verticillata var. *media* Korsh. === **Adenophora triphylla** (Thunb.) A. DC.

Adenophora verticillata var. *oppositifolia* Franch. & Sav. === **Adenophora triphylla** (Thunb.) A. DC.

Adenophora verticillata var. *pilosissima* Engl. === **Adenophora triphylla** (Thunb.) A. DC.

Adenophora verticillata var. *subintegrifolia* Regel === **Adenophora triphylla** (Thunb.) A. DC.

Adenophora verticillata var. *triphylla* Miq. === **Adenophora triphylla** (Thunb.) A. DC.

Adenophora verticillata var. *triphylla* (Thunb.) Regel === **Adenophora triphylla** (Thunb.) A. DC.

Adenophora watsonii W. W. Sm. === **Adenophora aurita** Franch.

Adenophora wawreana f. *foliosa* Zahlbr. === **Adenophora wawreana** A. Zahlbr.

Adenophora wawreana f. *oligotricha* Kitag. === **Adenophora wawreana** A. Zahlbr.

Adenophora wawreana f. *polytricha* Kitag. === **Adenophora wawreana** A. Zahlbr.

Adenophora wawreana var. *lanceifolia* Y. Z. Zhao === **Adenophora wawreana** A. Zahlbr.

Adenophora wulingshanica var. *alterna* Y. Z. Zhao === **Adenophora wulingshanica** D. Y. Hong

Adenophora wutaiensis Hurusawa === **Adenophora elata** Nannf.

Adenophora yokoyamae Miyabe & Tatew. === **Adenophora pereskiifolia** (Fisch. ex Schult.) G. Don

Aikinia

This name was proposed as a replacement for illegitimate *Schultesia* (q.v.), but is itself a later homonym of *Aikinia* Wall. (Poaceae) and *Aikinia* R. Br. (Gesneriaceae). If it becomes desireable to segregate *Wahlenbergia hederacea* (q.v.) from the remainder of that genus, as the phylogeny of Eddie et al. (2003, under **General**) intimates, a new name will be required. Based on: *Schultesia* Roth non Spreng.

Fourreau, P. (1869). Catalogue des plantes du cours du Rhone (suite). Ann. Soc. Linn. Lyon (n.s.) 17: 89-200. Fr. — Validation of name.

Synonyms:
Aikinia Salisb. ex Fourr. === **Wahlenbergia** Schrad. ex Roth
Aikinia hederacea (L.) Fourr. === **Wahlenbergia hederacea** (L.) Rchb.

Annaea

Campanula hieracioides was segregated on the basis of the funnel-like summit of its capsule and the apparent shift in dehiscence from pores near the summit to clefts near the base. Type [designated by the author]: *Annaea hieracioides* (Kolak.) Kolak.

Kolakovskii, A. A. (1979). [*Annaea*: a new genus of Campanulaceae from Abkhazian limestones.] Soobšč. Akad. Nauk Gruzinsk. SSR 94: 161-164, illus. Ru. — Establishment of genus.
Kolakovskii, A. A. & R. V. Lakoba (1993). [Capsule morphology and anatomy of genus *Annaea* Kolak. (Campanulaceae) endemical for Caucasus.] Bjull. Moskovsk. Obšč. Isp. Prir., Otd. Biol. 98: 135-137, illus. Ru. — Detailed studies of fruit structure.

Synonyms:
Annaea Kolak. === **Campanula** L.
Annaea hieracioides (Kolak.) Kolak. === **Campanula hieracioides** Kolak.

Apetahia

Lobelioideae, 4 species, French Polynesia; two are endemic to the Marquesas (though found on different islands), a third to the Tubuai (Austral) Islands, and the last to the Society Islands. On the basis of morphology, the genus seems most closely related to *Sclerotheca*, likewise endemic to Polynesia; two of the species here placed in *Apetahia* were formerly assigned to that genus. The treatment here follows that of Florence (1997). 2*n* = 28. Type [by monotypy]: *Apetahia raiateensis* Baill.

Baillon, H. (1882). Sur l'*apetahi* de Raiatea. Bull. Mens. Soc. Linn. Paris 1: 310-311. Fr. — Establishment of genus.
Brown, F. B. H. (1935). Flora of southeastern Polynesia III. Dicotyledons. *Sclerotheca* De Candolle. Bernice P. Bishop Mus. Bull. 130: 325-328, illus. En. — Floristic treatment of species on Tubuai and Society Is., with key, descriptions, and specimen citations.
Wimmer, F. E. (1953). Subtribus Apetahiinae E. Wimm. In A. Engler & L. Diels, Das Pflanzenreich IV. 276b: 728-730, illus. Berlin: Akademie-Verlag. Ge., La. — Monograph of monogeneric subtribe, with key, descriptions, full nomenclature, and specimen citations.
Wimmer, F. E. (1968). Subtrib. Apetahiinae E. Wimm. In A. Engler & L. Diels, Das Pflanzenreich IV. 276c: 892. Berlin: Akademie-Verlag. Ge., La. — Supplement to Wimmer (1953).
Thorne, R. F. (1972). Major disjunctions in the geographic ranges of seed plants. Quart. Rev. Biol. 47: 365-411, maps. En. — *Apetahia* one of just 14 genera confined to Pacific Basin, which show major distributional disjunctions within region.
• Dodd, E. (1976). Polynesia's sacred isle. 224 pp., illus. New York: Dodd Mead. En. — Popular account of environment and culture of Raiatea, in which *A. raiateensis* figures prominently.
Whistler, W. A. (1982). A naturalist in the South Pacific in search of the *apetahi*. Bull. Pacific Trop. Bot. Gard. 12: 1-4 + cover photo, illus. En. — Popular account of *A. raiateensis* and its habitat.

- Florence, J. (1997). New species of *Plakothira* (Loasaceae), *Melicope* (Rutaceae), and *Apetahia* (Campanulaceae) from the Marquesas Islands. Allertonia 7: 238-253, illus. En. — Synopsis of genus, including emendation of generic boundaries based on a clarification of fruit anatomy and a key to all species.

Apetahia Baill., Bull. Mens. Soc. Linn. Paris 1: 310 (1882).
  SC. Pacific. 61.

Apetahia longistigmata (F. Br.) E. Wimm., Ann. Naturhist. Mus. Wien 56: 373 (1948).
  Marquesas (Nuku Hiva, Hiva Oa, Ua Huka, Ua Pou, Tahuata). 61 MRQ. Nanophan.
  * *Sclerotheca longistigmata* F. Br., Fl. S.E. Polynesia: 327 (1935).

Apetahia margaretae (F. Br.) E. Wimm., Pflanzenr. IV.276b: 18 (1943).
  Tubuai Is. (Rapa-iti). 61 TUB. Nanophan.
  * *Sclerotheca margaretae* F. Br., Fl. S.E. Polynesia: 326 (1935).

Apetahia raiateensis Baill., Bull. Mens. Soc. Linn. Paris 1: 310 (1882).
  Society Is. (Raiatea). 61 SCI. Nanophan. 2*n* = 28.

Apetahia seigelii J. Florence, Allertonia 7: 248 (1997).
  Marquesas (Fatu Hivu). 61 MRQ. Nanophan.

Astrocodon

Astrocodon was erected to accomodate a pair of eastern Asian species (now merged), which were said to resemble *Adenophora* in overall habit but to lack the diagnostic nectary of that genus. Two species of northwestern North America were added subsequently. Type [designated by the author]: *Astrocodon kruhseanus* (Fisch. ex Regel & Tiling) Fed.

> Fedorov, A. A. (1957). *Astrocodon* Fed., gen. nov. In V. L. Komarov, Flora URSS 24: 375-379, 471, illus. Leningrad: Academia Scientiae URSS. Ru. — Establishment of genus as distinct from *Campanula*.

Synonyms:
Astrocodon Fed. === **Campanula** L.
Astrocodon auritus (Greene) A. P. Khokhr. === **Campanula aurita** Greene
Astrocodon expansus (Rudolph) Fed. === **Campanula expansa** Rudolph
Astrocodon kruhseanus (Fisch. ex Regel & Tiling) Fed. === **Campanula expansa** Rudolph
Astrocodon piperi (Howell) A. P. Khokhr. === **Campanula piperi** Howell

Asyneuma

Campanuloideae, 33 species, throughout much of Asia and extending into Europe and northern Africa. Although Fedorov (1957) divided the genus into two sections (each comprising two series) and Rechinger and Schiman-Czeika (1965) added a third section, Damboldt (1970) declined to employ a formal classification, instead dividing the species among nine informal groups. A preliminary molecular phylogeny (Eddie et al. 2003, under **General**) suggests that *Asyneuma*, *Petromarula*, *Physoplexis*, and *Phyteuma* form a clade within the "Rapunculus clade". The most recent monograph is that of Damboldt (1970). 2*n* = 20, 22, 24, 28, 30, 32, 34, 48, 56, 68. Type [by monotypy]: *Asyneuma canescens* (Waldst. & Kit.) Griseb. & Schenk.

> Sims, J. (1807). *Phyteuma campanuloides*. Three-flowered phyteuma. Bot. Mag. 26: tab. 1015 + 1 pg., illus. En. — Portrait of *A. campanuloides* with description.

Loddiges, C. (1822). *Phyteuma virgata*. Bot. Cab. 7: pl. 667 + 1 pg., illus. En. — Portrait of *A. virgatum* subsp. *virgatum* with notes on cultivation.

Grisebach, A. & A. Schenk (1852). Iter hungaricum a. 1852 susceptum. Arch. Naturgesch. 18: 291-362. Ge. — Establishment of genus as distinct from *Campanula* and *Phyteuma*.

Bornmüller, J. (1921). Ein Beitrag zur Kenntnis der Gattung *Asyneuma* Griseb. Beih. Bot. Centralbl. 38(2): 333-351. Ge. — Synopsis of genus and discussion of relationships to other genera, with full nomenclature and specimen citations.

Stapf, O. (1922). *Podanthum floribundum*. Bot. Mag. 148: tab. 8936 + 3 pp., illus. En. — Portrait of *A. limoniifolium* subsp. *pestalozzae,* with description, notes on relationships to other genera, and cultivation.

McVaugh, R. (1945). Notes on North American Campanulaceae. 1. The genus *Asyneuma* in the North American flora. Bartonia 23: 36. En. — Transfers *Campanula prenanthoides* (q.v.) of western North America to *Asyneuma* on basis of floral morphology, a move refuted by Damboldt (1970).

• Fedorov, A. A. (1957). *Asyneuma* Griseb. et Schenk. In V. L. Komarov, Flora URSS 24: 396-420, illus. Leningrad: Academia Scientiae URSS. Ru. — Flora of former Soviet Union, with key, descriptions, infrageneric classification, and full nomenclature.

• Rechinger, K. H. & H. Schiman-Czeika (1965). *Asyneuma*. In K. H. Rechinger (ed.), Flora Iranica 13: 39-47, illus. Graz: Akademische Druck- und Verlagsanstalt. Ge., La. — Flora with key, descriptions, infrageneric classification, full nomenclature, and specimen citations.

Damboldt, J. (1968). Vorarbeiten zu einer Revision der Gattung *Asyneuma* (Campanulaceae). I. Willdenowia 5: 35-54, illus. Ge. — Preparatory material for Damboldt (1970).

Damboldt, J. (1969). Vorarbeiten zu einer Revision der Gattung *Asyneuma* (Campanulaceae). II. Zur systematischen Stellung von *Asyneuma scoparium* (Boiss. et Hausskn.) Bornm. Willdenowia 5: 241-243, illus. Ge. — Removal of *A. scoparium* to *Campanula,* based on corolla morphology.

Contandriopoulos, J. (1970). Contribution a l'etude cytotaxonomique des Campanulacees du Proche-Orient, II. – Le *Asyneuma* de Turqui. Bull. Soc. Bot. France 117: 209-220. Fr. — Chromosome counts.

• Damboldt, J. (1970). Revision der Gattung *Asyneuma*. Boissiera 17: 1-128 + taf. A-B, illus., maps. Ge. — Monograph with keys, descriptions, full nomenclature, and specimen citations.

• Damboldt, J. (1976). *Asyneuma* Griseb. & Schenk. In T. G. Tutin, V. H. Heywood, N., A. Burges, D. M. Moore, D. H. Valentine, S. M. Walters & D. A. Webb (eds.), Flora Europaea 4: 95. Cambridge: Cambridge University Press. En. — Flora with key and descriptions.

• Damboldt, J. (1978). *Asyneuma* Griseb. & Schenk. In P. H. Davis (ed.), Flora of Turkey and the East Aegean Islands 6: 65-81, maps. Edinburgh: University Press. En. — Flora with key and descriptions.

Hong, D. Y. (1983). *Asyneuma* Griseb. et Schenk. In Flora Reipublicae Popularis Sinicae 73(2): 140-142. Beijing: Science Press. Ch. — Flora of China, with key, descriptions, and full nomenclature.

Nasir, E. (1984). *Asyneuma*. In E. Nasir & S. I. Ali (eds.), Flora of Pakistan 155: 5-6, illus. Islamabad: Pakistan Agricultural Research Council. En. — Flora with key and descriptions.

Oganesian, M. E. (1984). [The genus *Asyneuma* Griseb. et Schenk in Armenia.] Biol. Žurn. Arnenii 37: 166-168. Ru. — Floristic treatment with chromosome numbers.

• Ayers, T. J. (1991). The cultivated species of *Phyteuma* and *Asyneuma* (Campanulaceae). Baileya 23: 126-138, illus. En. — Account of species in cultivation, with keys and descriptions.

• Gudkova, I. Y. & G. P. Borshchenko (1991). [The serological study of the Campanulaceae. The phylogenetic relations in the tribe Phyteumateae.] Bot. Žurn. 76(6): 809-817. Ru. — Protein serology supports close relationship to *Cryptocodon, Petromarula, Phyteuma,* and *Sergia,* and a somewhat more distant relationship to *Cylindrocarpa*.

Huxley, A. J., ed. (1992). *Asyneuma*. In The new Royal Horticultural Society dictionary of gardening 1: 278. London: MacMillan. En. — Account of species in cultivation.

Parolly, G. (2000). Notes on two neglected Turkish *Asyneuma* taxa (Campanulaceae). Willdenowia 30: 67-75, illus., maps. En. — Transfer of poorly understood *Campanula juncea* to *Asyneuma* and clarification of its nature.

Yildiz, B. (2000). *Asyneuma* Griseb. & Schenk. In A. Güner, N. Özhatay, T. Ekım & K. H. C. Başer (eds.), Flora of Turkey and the East Aegean Islands, supplement 2: 176-181. Edinburgh: University Press. En. — Supplement to Damboldt (1978); includes revised key to all species in Turkey.

Lakušić, D. & F. Conti (2004). *Asyneuma pichleri* (Campanulaceae), a neglected species of the Balkan Peninsula. Pl. Syst. Evol. 247: 23-36, illus., maps. En. — Advocates a broader circumscription than Damboldt (1970), by including *Campanula trichocalycina* and its segregate, *C. pichleri*.

Asyneuma Griseb. & Schenk, Arch. Naturgesch. 18: 335 (1852).
Europe, N. Africa & temp. Asia. 11 13 14 20 31 32 33 34 36 38 40 41.
Phyteuma sect. *Podanthum* G. Don, Gen. Hist. 3: 748 (1834). *Trachanthelium* Kit. ex Schur., Enum. Pl. Transsilv.: 431 (1866). *Podanthum* (G. Don) Boiss., Fl. Orient. 3: 945 (1875).
Trochocodon Candargy, Bull. Soc. Bot. France 44: 148 (1897).
Asyneumopsis Contandr., Quézel & Pamukç., Ann. Univ. Provence Sci. 46: 57 (1972).

Asyneuma amplexicaule (Willd.) Hand.-Mazz., Ann. K. K. Naturhist. Hofmus. 27: 431 (1913).
Turkey to Caucasus (Abkhaziya, Gruziya, Armenia, Nakhichevan, Azerbaijan) & Iran. 33 TCS 34 IRN IRQ TUR. Cham. or hemicr. $2n = 24, 34, 48$.
* *Phyteuma amplexicaule* Willd., Sp. Pl. 1: 925 (1798), as 'amplexicaulis'. *Campanula amplexicaulis* (Willd.) Boiss., Diagn. Pl. Orient. (ser. 1) 11: 77 (1849); non Michx., Fl. Bor.-Amer. 1: 108 (1803). *Podanthum amplexicaule* (Willd.) Boiss., Fl. Orient. 3: 948 (1875).

subsp. **amplexicaule**
Turkey to Caucasus (Abkhaziya, Gruziya, Armenia, Nakhichevan, Azerbaijan) & Iran. 33 TCS 34 IRN IRQ TUR. Cham. or hemicr. $2n = 24, 34$.
Phyteuma kotschyi Boiss., Diagn. Pl. Orient. (ser. 1) 11: 71 (1849). *Campanula cilicia* Boiss., Diagn. Pl. Orient. (ser. 1) 11: 77 (1849). *Podanthum amplexicaule* var. *kotschyi* (Boiss.) Boiss., Fl. Orient. 3: 948 (1875). *Asyneuma amplexicaule* var. *kotschyi* (Boiss.) Bornm., Beih. Bot. Centralbl. 38(2): 339 (1921).
Phyteuma kotschyi var. *obtusicrenum* Boiss., Diagn. Pl. Orient. (ser. 1) 11: 72 (1849).
Phyteuma amplexicaule var. *majus* K. Koch, Linnaea 23: 630 (1850).
Phyteuma amplexicaule var. *minus* K. Koch, Linnaea 23: 630 (1850).
Podanthum amplexicaule var. *angustifolium* Boiss., Fl. Orient. 3: 948 (1875). *Asyneuma amplexicaule* var. *angustifolium* (Boiss.) Bornm., Beih. Bot. Centralbl. 38(2): 339 (1921).
Asyneuma talyschense Fed. in Kom., Fl. URSS 24: 473 (1957).

subsp. **aucheri** (A. DC.) Bornm., Beih. Bot. Centralbl. 38(2): 339 (1921).
Turkey to Iran. 34 IRN IRQ TUR. Cham. or hemicr. $2n = 24, 48$.
* *Phyteuma aucheri* A. DC. in DC., Prodr. 7: 456 (1839). *Campanula firma* Boiss., Diagn. Pl. Orient. (ser. 1) 11: 77 (1849); non *Campanula aucheri* A. DC. in DC., Prodr. 7: 460 (1839). *Podanthum amplexicaule* var. *aucheri* (A. DC.) Hausskn. ex Bornm., Beih. Bot. Centralbl. 20(2): 177 (1906).

Asyneuma anthericoides (Janka) Bornm., Beih. Bot. Centralbl. 38(2): 339 (1921).
Balkans. 13 BUL ROM YUG. Hemicr. $2n = 22$.
* *Podanthum anthericoides* Janka, Természetrajzi Füz. 2: 30 (1878). *Phyteuma anthericoides* (Janka) Nyman, Consp. Fl. Eur.: 484 (1879).
Podanthum grandiflorum Velen., Fl. Bulg.: 373 (1891). *Asyneuma grandiflorum* (Velen.) Bornm., Beih. Bot. Centralbl. 38(2): 341 (1921). *Podanthum anthericoides* var. *grandiflorum* (Velen.) Stoj. & Stef., Fl. Bulg. (ed. 2): 985 (1933).
Podanthum serbicum Formánek, Verh. Naturf. Vereines Brünn 32: 288 (1894).

Asyneuma anthericoides var. *dobrogense* Borza, Bul. Inform. Grăd. Bot. Univ. Cluj 3: 32 (1923). *Asyneuma anthericoides* f. *dobrogense* (Borza) Hayek, Repert. Spec. Nov. Regni Veg. Beih. 30(2): 555 (1930).
Podanthum anthericoides f. *glabrata* Stoj. & Acht., Izv. Carsk. Prir. Inst. Sofija 9: 143 (1936).
Podanthum anthericoides f. *villosa* Stoj. & Acht., Izv. Carsk. Prir. Inst. Sofija 9: 143 (1936).

Asyneuma argutum (Regel) Bornm., Beih. Bot. Centralbl. 38(2): 339 (1921).
Kazakhstan to Pakistan & W. Himalaya. 32 KAZ KGZ TZK UZB 34 AFG 40 PAK WHM. Hemicr.
* *Phyteuma argutum* Regel, Bull. Soc. Imp. Naturalistes Moscou 40(3): 183 (1867). *Podanthum argutum* (Regel) O. Fedtsch. & B. Fedtsch., Bot. Žurn. (St. Petersburg) 1: 53 (1906).

subsp. **argutum**
Kazakhstan to Pakistan & W. Himalaya. 32 KAZ TZK UZB 34 AFG 40 PAK WHM. Hemicr.
Phyteuma attenuatum Franch., Mission Capus Pl. Turkestan: 114 (1883). *Podanthum attenuatum* (Franch.) O. Fedtsch. & B. Fedtsch., Bot. Žurn. (St. Petersburg) 1: 53 (1906). *Asyneuma attenuatum* (Franch.) Bornm., Beih. Bot. Centralbl. 38(2): 340 (1921).
Podanthum trautvetteri B. Fedtsch., Bot. Žurn. (St. Petersburg) 1: 53 (1906). *Phyteuma trautvetteri* (B. Fedtsch.) B. Fedtsch., Rast. Turkestan.: 721 (1915). *Asyneuma trautvetteri* (B. Fedtsch.) Bornm., Beih. Bot. Centralbl. 38(2): 350 (1921).
Podanthum argutum var. *elegans* O. Fedtsch., Bot. Žurn. (St. Petersburg) 1: 53 (1906).
Podanthum argutum var. *foliosum* O. Fedtsch., Bot. Žurn. (St. Petersburg) 1: 53 (1906).
Asyneuma strictum Wendelbo, Nytt Mag. Bot. 1: 60 (1952). *Asyneuma thomsonii* var. *strictum* (Wendelbo) Kitam., Fl. Afghan.: 377 (1960).
Asyneuma ramosum Pavlov, Vestn. Akad. Nauk Kazahsk. SSR 8: 135 (1954).
Asyneuma thomsonii var. *nuristanica* Kitam., Fl. Afghan.: 377 (1960).

subsp. **baldshuanicum** (O. Fedtsch.) Damboldt, Boissiera 17: 43 (1970).
Tadzhikistan, Kirgizstan, Afganistan. 32 KGZ TZK 34 AFG. Hemicr.
* *Podanthum argutum* var. *baldshuanicum* O. Fedtsch., Bot. Žurn. (St. Petersburg) 1: 53 (1906). *Asyneuma baldshuanicum* (O. Fedtsch.) Fed. in Kom., Fl. URSS 24: 400 (1957).
Asyneuma debile Fed. in Kom., Fl. URSS 24: 472 (1957).

subsp. **pavlovii** Damboldt, Boissiera 17: 44 (1970).
Kazakhstan. 32 KAZ. Hemicr.

Asyneuma babadaghense Yildiz & Kit Tan, Willdenowia 18: 72 (1988), as 'babadaghensis'.
Turkey. 34 TUR. Cham. or hemicr.

Asyneuma campanuloides (M. Bieb. ex Sims) Bornm., Beih. Bot. Centralbl. 38(2): 340 (1921).
Caucasus (Severo-Osetiya, Dagestan, Abkhaziya, Gruziya, Armenia, Azerbaijan). 33 NCS TCS. Hemicr. or geophyte. $2n = 34$.
* *Phyteuma campanuloides* M. Bieb. ex Sims, Bot. Mag. 26: tab. 1015 (1807). *Phyteuma sibthorpianum* var. *campanuloides* (M. Bieb. ex Sims) Steud., Nomencl. Bot. (ed. 2) 2: 331 (1841). *Campanula marschalliana* Boiss., Diagn. Pl. Orient. (ser. 1) 11: 77 (1849). *Podanthum campanuloides* (M. Bieb. ex Sims) Boiss., Fl. Orient. 3: 949. 1875.
Phyteuma campanuloides var. *amplexifolium* K. Koch, Linnaea 23: 629 (1850).
Asyneuma campanuloides var. *gracilis* Manden. & Kuth., Zametki Sist. Geogr. Rast. 18: 64 (1955).

Asyneuma canescens (Waldst. & Kit.) Griseb. & Schenk, Arch. Naturgesch. 18: 335 (1852).
EC. Europe to Ukraine. 11 AUT CZE HUN 13 ALB BUL GRC ROM YUG 14 UKR. Hemicr. $2n = 30, 32, 34, 56$.
* *Phyteuma canescens* Waldst. & Kit., Descr. Icon. Pl. Hung. 1: 12 (1799). *Phyteuma lanceolatum* var. *canescens* (Waldst. & Kit.) Pers., Syn. Pl. 1: 194 (1805). *Campanula*

canescens (Waldst. & Kit.) Roth, Enum. Pl. Phaen. Germ. 1: 716 (1827).
Trachanthelium canescens (Waldst. & Kit.) Schur., Enum. Pl. Transsilv.: 431 (1866).
Podanthum canescens (Waldst. & Kit.) Boiss., Fl. Orient. 3: 950 (1875).

subsp. **canescens**
EC. Europe to Ukraine. 11 AUT CZE HUN 13 ALB BUL GRC ROM YUG 14 UKR. Hemicr. $2n = 30, 32, 34, 56$.
Phyteuma salignum Waldst. & Kit. ex Besser, Prim. Fl. Galiciae Austriac. 1: 368 (1809). *Phyteuma salicifolium* Kit. ex Schult., Oestr. Fl. (ed. 2) 1: 400 (1814). *Phyteuma canescens* var. *salicifolium* Steud., Nomencl. Bot. (ed. 2) 2: 331 (1841). *Campanula salicifolia* Boiss., Diagn. Pl. Orient. (ser. 1) 11: 76 (1849). *Podanthum salicifolium* Rupr., Bull. Acad. Imp. Sci. Saint-Pétersbourg 11: 203 (1867). *Phyteuma canescens* var. *salignum* (Waldst. & Kit. ex Besser) Nyman, Consp. Fl. Eur.: 483 (1879). *Podanthum canescens* var. *salicifolium* (Steud.) Fomin in N. M. Kusn., N. Bush & Fomin, Mater. Fl. Kavkaza 4(6): 136 (1906). *Asyneuma salicifolium* Flerow, Spisok Rast. Sev. Kavkaz. Dag.: 543 (1938), as 'salicifolia'. *Asyneuma salignum* (Waldst. & Kit. ex Besser) Fed. in Kom., Fl. URSS 24: 408 (1957). *Asyneuma canescens* subsp. *salicifolium* (Steud.) Soó, Acta Bot. Acad. Sci. Hung. 12: 366 (1966).
Phyteuma foliosum Kit., Linnaea 32: 426 (1863); non Bellardi ex Colla, Herb. Pedem. 4: 33 (1835). *Phyteuma canescens* subsp. *foliosum* Nyman, Consp. Fl. Eur.: 483 (1879). *Asyneuma canescens* var. *foliosum* (Nyman) Borza, Bul. Inform. Grăd. Bot. Univ. Cluj 3: 32 (1923).
Trachanthelium canescens var. *transsilvanicum* Schur, Enum. Pl. Transsilv.: 431 (1866).
Podanthum canescens subsp. *rhodopeum* Formánek, Verh. Naturf. Vereins Brünn 36: 46 (1898).
Podanthum canescens var. *nudiflorum* Davidov, Compt. Rend. Gymnas. Samokov 1908-1909: 38 (1909). *Asyneuma canescens* f. *nudiflorum* (Davidov) Hayek, Repert. Spec. Nov. Regni Veg. Beih. 30(2): 555 (1930).
Podanthum canescens f. *laevis* Rohlena, Sitzungsber. Königl. Böhm. Ges. Wiss. Prag, Math.-Naturwiss. Cl. 1912: 39 (1912). *Phyteuma canescens* f. *laevis* (Rohlena) Rohlena, Sitzungsber. Königl. Böhm. Ges. Wiss. Prag, Math.-Naturwiss. Cl. 1912: 39 (1912).

subsp. **cordifolium** (Bornm.) Damboldt, Boissiera 17: 57 (1970).
Macedonia. 13 YUG. Hemicr.
* *Asyneuma cordifolium* Bornm., Beih. Bot. Centralbl. 38(2): 333 (1921).

Asyneuma chinense D. Y. Hong, Fl. Reipubl. Popul. Sin. 73(2): 188 (1983).
S. China (Guangxi, Guizhou, Hubei, Sichuan, Yunnan). 36 CHC CHS. Hemicr.

Asyneuma comosiforme Hayek & Janch., Oesterr. Bot. Z. 70: 20 (1921).
Albania. 13 ALB. Hemicr.

Asyneuma compactum Damboldt, Boissiera 17: 58 (1970).
Turkey. 34 TUR. Hemicr. $2n = 34$.
* *Campanula compacta* Boiss. & Heldr. in Boiss., Diagn. Pl. Orient. (ser. 1) 11: 71 (1849); non Hegetschw., Fl. Schweiz.: 232 (1839).
Asyneuma compactum subsp. *glabrescens* Contandr., Quézel & Pamukç., Ann. Univ. Provence Sci. 46: 57 (1972).
Asyneuma compactum var. *eriocarpum* Parolly, Willdenowia 30: 72 (2000).

Asyneuma davisianum Yildiz & Kit Tan, Willdenowia 18: 68 (1988).
Turkey. 34 TUR. Cham. or hemicr.

Asyneuma ekimianum Kit Tan & Yildiz, Willdenowia 18: 74 (1988).
Turkey. 34 TUR. Hemicr.

subsp. **beritense** Kit Tan & Yildiz, Willdenowia 18: 76 (1988), as 'beritensis'.
Turkey. 34 TUR. Hemicr.

subsp. **ekimianum**
Turkey. 34 TUR. Hemicr.

subsp. **sivasicum** Kit Tan & Yildiz, Willdenowia 18: 76 (1988).
Turkey. 34 TUR. Hemicr.

Asyneuma filipes (Nábělek) Damboldt, Boissiera 17: 59 (1970).
Turkey & Iraq. 34 IRQ TUR. Hemicr. or geophyte.
* *Asyneuma lobelioides* var. *filipes* Nábělek, Spisy Přír. Fak. Masarykovy Univ. 70: 8 (1926).
Asyneuma lobelioides f. *glaberrima* Nábělek, Spisy Přír. Fak. Masarykovy Univ. 70: 8 (1926).
Asyneuma lobelioides f. *scabrida* Nábělek, Spisy Přír. Fak. Masarykovy Univ. 70: 8 (1926).
Asyneuma lobelioides var. *capitatum* Nábělek, Spisy Přír. Fak. Masarykovy Univ. 70: 8 (1926).

Asyneuma fulgens (Wall.) Briq., Candollea 4: 334 (1931).
Tibet to Myanmar, Sri Lanka & S. India (Karnataka, Tamil Nadu). 36 CHT 40 ASS EHM
IND NEP SRL 41 MYA. Ther.
* *Campanula fulgens* Wall. in Roxb., Fl. Ind. 2: 99 (1824).

Asyneuma giganteum (Boiss.) Bornm., Beih. Bot. Centralbl. 38(2): 341 (1921).
E. Aegean Is. (Rodhos). 34 EAI. Biennial.
* *Podanthum giganteum* Boiss., Fl. Orient. 3: 946 (1875). *Phyteuma giganteum* (Boiss.)
Nyman, Consp. Fl. Eur.: 483 (1879).

Asyneuma ilgazense Yildiz & Kit Tan, Willdenowia 18: 67 (1988), as 'ilgazensis'.
Turkey. 34 TUR. Hemicr.

Asyneuma isauricum Contandr., Quézel & Pamukç., Ann. Univ. Provence Sci. 46: 56 (1972).
Turkey. 34 TUR. Cham. or hemicr.

Asyneuma japonicum (Miq.) Briq., Candollea 4: 335 (1931).
Russian Far East to Japan (Honshu, Kyushu), Korea & Manchuria. 31 AMU KHA PRM 36
CHM 38 JAP KOR. Hemicr. or geophyte.
* *Phyteuma japonicum* Miq., Ann. Mus. Bot. Lugduno-Batavum 2: 192 (1866), as
'iaponicum'. *Campanula japonica* (Miq.) Vatke, Linnaea 38: 705 (1874).

Asyneuma junceum Parolly, Willdenowia 30: 71 (2000).
Turkey. 34 TUR. Cham. or hemicr. *2n* = 34.
* *Campanula juncea* Wettst., Denkschr. Kaiserl. Akad. Wiss., Math.-Naturwiss. Kl. 50: 117
(1885), as 'iuncea;' non Hill, Syst. Veg. 8: 14 (1765); nec (Buek) D. Dietr., Syn. Pl. 1:
756 (1839).

Asyneuma limoniifolium (L.) Janch., Mitt. Naturwiss. Vereines Univ. Wien (n.s.) 4: 35 (1906).
Italy to Turkey. 13 ALB BUL GRC ITA TUE YUG 34 EAI TUR. Cham. or hemicr. *2n* = 24.
* *Campanula limoniifolium* L., Sp. Pl. (ed. 2): 239 (1762), as 'limonifolium'. *Phyteuma
limoniifolium* (L.) Sm. in Sibth. & Sm., Fl. Graec. Prodr. 1: 144 (1806), as 'limonifolium'.
Podanthum limoniifolium (L.) Boiss., Fl. Orient. 3: 951 (1875), as 'limonifolium'.

subsp. **limoniifolium**
Italy to Turkey. 13 ALB BUL GRC ITA TUE YUG 34 EAI TUR. Cham. or hemicr. *2n* = 24.
Phyteuma repandum Sm. in Sibth. & Sm., Fl. Graec. Prodr. 1: 143 (1806). *Campanula
repanda* (Sm.) Boiss., Diagn. Pl. Orient. (ser. 1) 11: 76 (1849). *Podanthum limoniifolium*
var. *alpinum* Boiss., Fl. Orient. 3: 951 (1875). *Phyteuma limoniifolium* var. *repandum*
(Sm.) Nyman, Consp. Fl. Eur.: 484 (1879). *Podanthum limoniifolium* var. *repandum*
(Sm.) Halácsy, Consp. Fl. Graec. 2: 277 (1902). *Asyneuma limoniifolium* var. *alpinum*
(Boiss.) Bornm., Beih. Bot. Centralbl. 38(2): 344 (1921). *Asyneuma repandum* (Sm.)
Rothm., Bot. Jahrb. Syst. 73: 443 (1944).
Campanula phyteumoides Sibth. ex Zuccagni, Cent. Observ. Bot. 1: 16 (1806).

Phyteuma stylosum Schrank., Pl. Rar. Hort. Monac.: pl. 49 (1819).

Phyteuma strictum Sims, Bot. Mag. 47: tab. 2145 (1820), as 'stricta'. *Phyteuma limoniifolium* var. *strictum* (Sims) K. Koch, Linnaea 23: 629 (1850).

Phyteuma collinum Guss., Pl. Rar.: 97 (1826).

Phyteuma tenuifolium A. DC., Monogr. Campan.: 201 (1830). *Podanthum tenuifolium* (A. DC.) Boiss., Fl. Orient 3: 952 (1875). *Asyneuma tenuifolium* (A. DC.) Bornm., Beih. Bot. Centralbl. 38(2): 350 (1921).

Phyteuma stylidioides Boiss., Diagn. Pl. Orient. (ser. 1) 11: 73 (1849). *Campanula stylidioides* (Boiss.) Boiss., Diagn. Pl. Orient. (ser. 1) 11: 77 (1849).

Podanthum limoniifolium var. *canescens* Boiss., Fl. Orient. 3: 951 (1875). *Asyneuma limoniifolium* var. *canescens* (Boiss.) Bornm., Beih. Bot. Centralbl. 38(2): 344 (1921). *Asyneuma limoniifolium* f. *canescens* (Boiss.) Hayek, Repert. Spec. Nov. Regni Veg. Beih. 30(2): 556 (1930).

Podanthum otites Boiss., Fl. Orient. 3: 952 (1875). *Phyteuma otites* (Boiss.) Trautv., Trudy Imp. S.-Peterburgsk. Bot. Sada 6: 49 (1879). *Asyneuma otites* (Boiss.) Bornm., Beih. Bot. Centralbl. 38(2): 347 (1921).

Podanthum limoniifolium var. *ramosum* Hausskn., Mitt. Thüring. Bot. Vereins (n.s.) 7: 63 (1895). *Asyneuma limoniifolium* subf. *ramosum* (Hausskn.) Hayek, Repert. Spec. Nov. Regni Veg. Beih. 30(2): 556 (1930).

Trochocodon spicatus Candargy, Bull. Soc. Bot. France 44: 148 (1897).

Podanthum psaridis Heldr. ex Halácsy, Consp. Fl. Graec. 2: 277 (1902). *Asyneuma psaridis* (Heldr. ex Halácsy) Bornm., Beih. Bot. Centralbl. 38(2): 348 (1921). *Asyneuma limoniifolium* subsp. *psaridis* (Heldr. ex Halácsy) Hayek, Repert. Spec. Nov. Regni Veg. Beih. 30(2): 556 (1930).

Podanthum limoniifolium f. *heterophyllum* Rohlena, Sitzungsber. Königl. Böhm. Ges. Wiss. Prag, Math.-Naturwiss. Cl. 1912: 88 (1912). *Asyneuma limoniifolium* f. *heterophyllum* (Rohlena) Hayek, Repert. Spec. Nov. Regni Veg. Beih. 30(2): 556 (1930).

Podanthum rhodopaeum Davidov, Trav. Soc. Bulg. Sci. Nat. 8: 93 (1915).

Asyneuma parviflorum Turrill, Bull. Misc. Inform. Kew 1924: 349 (1924), as 'parviflora'.

Podanthum kellererianum Stef. in Stoj. & Stef., Fl. Bulg. (ed. 2): 984 (1933). *Asyneuma kellererianum* (Stef.) Stef., Magyar Bot. Lapok 33: 2 (1934).

subsp. **pestalozzae** (Boiss.) Damboldt, Boissiera 17: 80 (1970).
  Turkey. 34 TUR. Cham. or hemicr. 2*n* = 24.
  * *Phyteuma pestalozzae* Boiss., Diagn. Pl. Orient. (ser. 1) 11: 74 (1849). *Campanula pestalozzae* (Boiss.) Boiss., Diagn. Pl. Orient. (ser. 1) 11: 77 (1849); non *Campanula pestalozzae* Boiss., Diagn. Pl. Orient. (ser. 1) 11: 62 (1849). *Campanula willdenowiana* Boiss., Diagn. Pl. Orient. (ser. 2) 3: 117 (1856); non Schult. in Roem. & Schult., Syst. Veg. 5: 107 (1819).

  Podanthum lobelioides f. *stenophyllum* Bornm., Beih. Bot. Centralbl. 24(2): 477 (1909). *Asyneuma lobelioides* f. *stenophylla* (Bornm.) Bornm., Beih. Bot. Centralbl. 38(2): 345 (1921).

  Asyneuma lobelioides f. *aspera* Bornm., Beih. Bot. Centralbl. 38(2): 346 (1921).

  Podanthum floribundum Stapf, Bot. Mag. 138: tab. 8936 (1922). *Asyneuma floribundum* (Stapf) P. H. Davis, Kew Bull. 1949: 109 (1949).

  Asyneuma amanum Rech. f., Ann. Naturhist. Mus. Wien 57: 88 (1950).

Asyneuma linifolium (Boiss. & Heldr.) Bornm., Beih. Bot. Centralbl. 38(2): 345 (1921).
  Turkey. 34 TUR. Hemicr. or geophyte. 2*n* = 34, 68.
  * *Phyteuma linifolium* Boiss. & Heldr. in Boiss., Diagn. Pl. Orient. (ser. 1) 11: 75 (1849). *Campanula stenophylla* Boiss. & Heldr. in Boiss., Diagn. Pl. Orient. (ser. 1) 11: 77 (1849); non *Campanula linifolia* L., Amoen. Acad. 4: 479 (1759). *Podanthum linifolium* (Boiss. & Heldr.) Boiss., Fl. Orient. 3: 954 (1875).

subsp. **eximium** (Rech. f.) Damboldt, Boissiera 17: 83 (1970).
  Turkey. 34 TUR. Hemicr. or geophyte.
  * *Asyneuma eximium* Rech. f., Ann. Naturhist. Mus. Wien. 57: 86 (1950).

subsp. **glabrum** Kit Tan & Yildiz, Willdenowia 18: 72 (1988).
Turkey. 34 TUR. Hemicr. or geophyte.

subsp. **linifolium**
Turkey. 34 TUR. Hemicr. or geophyte. *2n* = 34, 68.

subsp. **nallihanicum** Kit Tan & Yildiz, Willdenowia 18: 70 (1988).
Turkey. 34 TUR. Hemicr. or geophyte.

subsp. **podanthoides** (Boiss. & Hausskn.) Damboldt, Notes Roy. Bot. Gard. Edinburgh
35: 49 (1976).
Turkey. 34 TUR. Hemicr. or geophyte.
* *Campanula podanthoides* Boiss. & Hausskn., Fl. Orient 3: 938 (1875).

Asyneuma lobelioides (Willd.) Hand.-Mazz., Ann. K. K. Naturhist. Hofmus. 27: 431 (1913).
Turkey & Caucasus (Armenia, Azerbaijan). 33 TCS 34 TUR. Cham. or hemicr. *2n* = 24.
* *Phyteuma lobelioides* Willd., Phytographia 1: 6 (1794). *Campanula hesperidifolia* Boiss.,
Diagn. Pl. Orient. (ser. 1) 11: 76 (1849); non *Campanula lobelioides* L. f., Suppl. Pl.:
140 (1782). *Campanula lobelioides* (Willd.) Vatke, Linnaea 38: 704 (1874); non
Campanula lobelioides L. f., Suppl. Pl.: 140 (1782). *Podanthum lobelioides* (Willd.)
Boiss., Fl. Orient. 3: 953 (1875).
Phyteuma lanceolatum Desf., Ann. Mus. Natl. Hist. Nat. 11: 55 (1808); non Vill., Hist. Pl.
Dauphiné 2: 517 (1787), as 'lanceolata;' nec Willd., Sp. Pl. 1: 924 (1798), as
'lanceolata'.
Phyteuma gracile Boiss. & Heldr. in Boiss., Diagn. Pl. Orient. (ser. 1) 11: 74 (1849).
Campanula gracilis (Boiss. & Heldr.) Boiss. & Heldr. in Boiss., Diagn. Pl. Orient. (ser.
1) 11: 77 (1849); non G. Forst., Fl. Ins. Austr.: 84 (1786); nec Avé-Lall., Pl. Ital. Bor.:
10 (1829).
Podanthum lobelioides var. *urceolatum* Fomin in N. M. Kusn., N. Bush & Fomin, Mater.
Fl. Kavkaza 4(6): 138 (1906). *Podanthum urceolatum* (Fomin) Schischk. in Grossh. &
Schischk., Pl. Orient. Exsic. 1-8: 38 (1924). *Asyneuma urceolatum* (Fomin) Fed. in
Kom., Fl. URSS 24: 416 (1957).
Asyneuma lobelioides f. *nana* Bornm., Beih. Bot. Centralbl. 38(2): 346 (1921).
Podanthum woronowii Fomin, Věstn. Tiflissk. Bot. Sada 10: 35 (1908). *Asyneuma
woronowii* (Fomin) Bornm., Beih. Bot. Centralbl. 38(2): 351 (1921).

Asyneuma lycium (Boiss.) Bornm., Beih. Bot. Centralbl. 38(2): 346 (1921).
Turkey. 34 TUR. Hemicr. or geophyte. *2n* = 34
* *Phyteuma lycium* Boiss., Diagn. Pl. Orient. (ser. 1) 11: 76 (1849). *Campanula lycia* (Boiss.)
Boiss., Diagn. Pl. Orient. (ser. 1) 11: 77 (1849). *Podanthum lycium* (Boiss.) Boiss., Fl.
Orient. 3: 954 (1875).

Asyneuma macrodon (Boiss. & Hausskn.) Bornm., Beih. Bot. Centralbl. 38(2): 346 (1921).
Iran. 34 IRN. Hemicr.
* *Podanthum macrodon* Boiss. & Hausskn. in Boiss., Fl. Orient. 3: 956 (1875).

Asyneuma michauxioides (Boiss.) Damboldt, Notes Roy. Bot. Gard. Edinburgh 35: 49 (1976).
Turkey. 34 TUR. Biennial. *2n* = 24, 30, 34.
* *Campanula michauxioides* Boiss., Diagn. Pl. Orient. (ser. 1) 4: 35 (1844).
Campanula michauxioides var. *dilacerata* Bornm., Mitt. Thüring. Bot. Vereins (n.s.) 20:
33 (1905).
Campanula michauxioides var. *hirtifolia* Huber-Mor., Bauhinia 2: 201 (1963).
Asyneuma davisii Quézel, Contandr. & Pamukç., Candollea 25: 351 (1970).

Asyneuma persicum (A. DC.) Bornm., Beih. Bot. Centralbl. 38(2): 347 (1921).
Turkey to Iran. 34 TUR IRQ IRN. Hemicr. or geophyte. *2n* = 24, 34.
* *Campanula persica* A. DC. in DC., Prodr. 7: 483 (1839). *Podanthum persicum* (A. DC.)
Boiss., Fl. Orient. 3: 956 (1875).

Campanula multicaulis Boiss., Diagn. Pl. Orient. (ser. 1) 7: 19 (1846). *Podanthum persicum* var. *pumilum* Boiss., Fl. Orient. 3: 956 (1875). *Podanthum persicum* var. *multicaule* (Boiss.) Bornm., Beih. Bot. Centralbl. 20(2): 178 (1906). *Asyneuma persicum* var. *pumilum* (Boiss.) Bornm., Beih. Bot. Centralbl. 38(2): 347 (1921). *Asyneuma multicaule* (Boiss.) Rech. f. & Schiman-Czeika in Rech. f., Fl. Iran. 13: 42 (1965).
Phyteuma asperum Boiss., Diagn. Pl. Orient. (ser. 1) 11: 72 (1849). *Campanula aspera* (Boiss.) Boiss., Diagn. Pl. Orient. (ser. 1) 11: 77 (1849); non Moench, Suppl. Meth.: 188 (1802). *Campanula boissieri* Vatke, Linnaea 38: 703 (1874). *Podanthum asperum* (Boiss.) Boiss., Fl. Orient. 3: 955 (1875). *Podanthum persicum* var. *asperum* (Boiss.) Bornm., Beih. Bot. Centralbl. 20(2): 178 (1906). *Asyneuma asperum* (Boiss.) Rech. f. & Schiman-Czeika in Rech. f., Fl. Iran. 13: 41 (1965).
Podanthum persicum f. *subsimplex* Bornm., Beih. Bot. Centralbl. 20(2): 178 (1906).

Asyneuma pichleri (Vis.) D. Lakušić & F. Conti, Pl. Syst. Evol. 247: 27 (2004).
Balkans. 13 ALB BUL GRC YUG. Hemicr.
* *Campanula pichleri* Vis., Mem. Reale Ist. Veneto Sci. 16: 106 (1872).
Asyneuma pichleri var. *subalpina* D. Lakušić & F. Conti, Pl. Syst. Evol. 247: 28 (2004).

Asyneuma pulchellum (Fisch. & C. A. Mey.) Bornm., Beih. Bot. Centralbl. 38(2): 348 (1921).
Turkey to Caucasus (Nakhichevan, Azerbaijan) & Iran. 33 TCS 34 IRN IRQ TUR. Biennial. $2n = 24$.
* *Phyteuma pulchellum* Fisch. & C. A. Mey., Index Sem. Hort. Petrop. 1: 35 (1835). *Campanula pulchella* (Fisch. & C. A. Mey.) Boiss., Diagn. Pl. Orient. (ser. 1) 11: 76 (1849); non Salisb., Prodr. Stirp. Chap. Allerton: 127 (1796). *Podanthum pulchellum* (Fisch. & C. A. Mey.) Boiss., Fl. Orient. 3: 947 (1875).

Asyneuma pulvinatum P. H. Davis, Kew Bull. 1949: 108 (1949).
Turkey. 34 TUR. Cham.

Asyneuma rigidum (Willd.) Grossh., Opred. Rast. Kavkaz.: 424 (1949).
Algeria; Turkey to Caucasus, Iran & Sinai. 20 ALG 33 NCS TCS 34 IRN IRQ LBS SIN TUR. Hemicr. $2n = 24, 28$
* *Phyteuma rigidum* Willd., Sp. Pl. 1: 925 (1798), as 'rigida'. *Campanula rigida* (Willd.) Boiss., Diagn. Pl. Orient. (ser. 1) 11: 76 (1849); non Stokes, Bot. Mat. Med. 1: 333 (1812). *Podanthum lanceolatum* var. *rigidum* (Willd.) Boiss., Fl. Orient. 3: 952 (1875). *Asyneuma lanceolatum* var. *rigidum* (Willd.) Hand.-Mazz., Ann. K. K. Naturhist. Hofmus. 27: 431 (1913).

subsp. **aurasiacum** (Batt. & Trab.) Damboldt, Boissiera 17: 97 (1970).
Algeria. 20 ALG. Hemicr.
* *Podanthum aurasiacum* Batt. & Trab., Bull. Soc. Bot. France 39: lx (1892). *Campanula aurasiaca* (Batt. & Trab.) Batt. & Trab., Fl. Algérie Tunisie: 220 (1905). *Asyneuma aurasiacum* (Batt. & Trab.) Bornm., Beih. Bot. Centralbl. 38(2): 340 (1921).

subsp. **graminifolium** Huber-Mor., Bauhinia 6: 286 (1978).
Turkey. 34 TUR. Hemicr.

subsp. **rigidum**
Turkey to Caucasus (Gruziya, Nakhichevan, Armenia, Azerbaijan, Dagestan) & Iran. 33 NCS TCS 34 IRN IRQ TUR. Hemicr. $2n = 28$.
Phyteuma lanceolatum Willd., Sp. Pl. 1: 924 (1798), as 'lanceolata;' non Vill., Hist. Pl. Dauphiné 2: 517 (1787), as 'lanceolata'. *Campanula fontanesiana* Boiss., Diagn. Pl. Orient. (ser. 1) 11: 76 (1849). *Podanthum lanceolatum* Boiss., Fl. Orient. 3: 951 (1875). *Asyneuma lanceolatum* Hand.-Mazz., Ann. K. K. Naturhist. Hofmus. 27: 431 (1913).
Phyteuma canescens var. *humile* K. Koch, Linnaea 23: 630 (1850).

Campanula controversa Boiss., Diagn. Pl. Orient. (ser. 2) 3: 115 (1856). *Podanthum controversum* (Boiss.) Boiss., Fl. Orient. 3: 949 (1875). *Podanthum lanceolatum* var. *controversum* (Boiss.) Bornm., Mitt. Thüring. Bot. Vereins (n.s.) 20: 38 (1905). *Asyneuma controversum* (Boiss.) Bornm., Beih. Bot. Centralbl. 38(2): 341 (1921). *Asyneuma lanceolatum* subsp. *controversum* (Boiss.) Bornm., Beih. Bot. Centralbl. 38(2): 341 (1921).
Campanula tauricola Boiss. & Balansa in Boiss., Diagn. Pl. Orient. (ser. 2) 3: 116 (1856).
Podanthum supinum Wettst., Sitzungsber. Kaiserl. Akad. Wiss., Math.-Naturwiss. Cl., Abt. 1 98: 377 (1889). *Asyneuma supinum* (Wettst.) Bornm., Beih. Bot. Centralbl. 38(2): 350 (1921).
Podanthum lanceolatum var. *flagellatum* Hausskn. & Bornm., Mitt. Thüring. Bot. Vereins (n.s.) 20: 33 (1905). *Asyneuma lanceolatum* f. *flagellatum* (Hausskn. & Bornm.) Bornm., Beih. Bot. Centralbl. 38(2): 342 (1921).
Asyneuma lanceolatum f. *stenophyllum* Bornm., Beih. Bot. Centralbl. 38(2): 343 (1921).
Asyneuma eldivenum Czeczott, Acta Soc. Bot. Poloniae 9: 42 (1932). *Phyteuma eldivenum* (Czeczott) Czeczott, Acta Soc. Bot. Poloniae 9: 42 (1932).
Phyteuma grossheimii Karjagin ex Grossh., Opred. Rast. Kavkaz.: 425 (1949).
Asyneuma densifolium Rech. f., Ann. Naturhist. Mus. Wien 57: 87 (1950).
Asyneuma haradjanii Rech. f., Ann. Naturhist. Mus. Wien 57: 89. (1950).

subsp. **sibthorpianum** (Schult.) Damboldt, Boissiera 17: 103 (1970).
Turkey. 34 TUR. Hemicr. $2n$ = 24.
* *Phyteuma ellipticum* Sm. in Sibth. & Sm., Fl. Graec. Prodr. 1: 143 (1806); non Vill., Hist. Pl. Dauphiné 2: 517 (1787). *Phyteuma sibthorpianum* Schult. in Roem. & Schult., Syst. Veg. 5: 84 (1819). *Phyteuma campanuloides* var. *sibthorpianum* (Schult.) A. DC., Monogr. Campan.: 206 (1830). *Campanula sibthorpiana* (Schult.) Boiss., Diagn. Pl. Orient. (ser. 1) 11: 77 (1849). *Podanthum sibthorpianum* (Schult.) Boiss., Fl. Orient. 3: 950 (1875). *Asyneuma sibthorpianum* (Schult.) Bornm., Beih. Bot. Centralbl. 38(2): 349 (1921).
Podanthum sibthorpianum var. *tmoleum* Boiss., Fl. Orient. 3: 950 (1875).

subsp. **sinai** (A. DC.) Damboldt, Boissiera 17: 105 (1970).
Sinai, Lebanon, Syria. 34 LBS SIN. Hemicr. $2n$ = 28.
* *Phyteuma sinai* A. DC. in DC., Prodr. 7: 455 (1839). *Campanula sinai* (A. DC.) Boiss., Diagn. Pl. Orient. (ser. 1) 11: 76 (1849). *Podanthum lanceolatum* var. *alpinum* Boiss., Fl. Orient. 3: 952 (1875). *Asyneuma lanceolatum* var. *alpinum* (Boiss.) Bornm., Beih. Bot. Centralbl. 38(2): 343 (1921).

Asyneuma thomsonii (Hook. f.) Bornm., Beih. Bot. Centralbl. 38(2): 350 (1921).
Tadzhikistan to Pakistan & W. Himalaya. 32 TZK 34 AFG 40 PAK WHM. Hemicr. $2n$ = 32.
* *Campanula thomsonii* Hook. f., J. Proc. Linn. Soc., Bot. 2: 25 (1858). *Phyteuma thomsonii* (Hook. f.) C. B. Clarke in Hook. f., Fl. Brit. India 3: 438 (1881), as 'thomsoni'.

Asyneuma trichocalycinum (Ten.) K. Malý, Glasn. Zemaljsk. Muz. Bosni Hercegovini 20: 555 (1908).
Italy to N. Africa. 13 ITA KRI SIC 20 ALG. Hemicr. $2n$ = 32.
* *Campanula trichocalycina* Ten., Fl. Napol. 1: xv (1811). *Podanthum trichocalycinum* (Ten.) Boiss., Fl. Orient. 3: 955 (1875). *Phyteuma trichocalycinum* (Ten.) Tanfani in Parl., Fl. Ital. 8: 63 (1888).
Campanula alburnica V. Brig., Stirp. Rar. 1: 3 (1816).
Campanula minae Strobl, Flora 66: 549 (1883).
Podanthum trichocalycinum var. *densiflora* Hausskn., Mitt. Thüring. Bot. Vereins (n.s.) 7: 63 (1895).

Asyneuma trichostegium (Boiss.) Bornm., Beih. Bot. Centralbl. 38(2): 350 (1921).
Turkey. 34 TUR. Hemicr.
* *Podanthum trichostegium* Boiss., Fl. Orient. 3: 954 (1875).

Asyneuma virgatum (Labill.) Bornm., Beih. Bot. Centralbl. 38(2): 351 (1921).
  E. Aegean Is. (Khios) to Iran. 33 TCS 34 EAI IRN LBS TUR. Biennial. $2n = 20, 24$.
   * *Campanula virgata* Labill., Icon. Pl. Syr. 2: 11 (1791). *Phyteuma virgatum* (Labill.) Willd.,
      Sp. Pl. 1: 924 (1798), as 'virgata'. *Podanthum virgatum* (Labill.) Boiss., Fl. Orient. 3:
      946 (1875).

  subsp. **cichoriiforme** (Boiss.) Damboldt, Boissiera 17: 113 (1970).
    E. Aegean Is. (Khios) & Turkey. 34 EAI TUR. Biennial. $2n = 20$.
    * *Phyteuma cichoriiforme* Boiss., Diagn. Pl. Orient. (ser. 1) 4: 37 (1844), as 'cichoriforme'.
       Campanula cichoriiformis (Boiss.) Boiss., Diagn. Pl. Orient. (ser. 1) 11: 77 (1849), as
       'cichoriformis'. *Podanthum cichoriiforme* (Boiss.) Boiss., Fl. Orient. 3: 947 (1875), as
       'cichoriforme'. *Asyneuma cichoriiforme* (Boiss.) Bornm., Beih. Bot. Centralbl. 38(2):
       340 (1921), as 'cichoriforme'.

  subsp. **mazanderanicum** (Rech. f.) Damboldt, Boissiera 17: 115 (1970).
    Iran. 34 IRN. Biennial. $2n = 20$.
    * *Asyneuma mazanderanicum* Rech. f., Österr. Akad. Wiss., Math.-Naturwiss. Kl., Anz. 87:
       197 (1950).

  subsp. **virgatum**
    Turkey to Armenia, Iran & Lebanon. 33 TCS 34 IRN LBS TUR. Biennial. $2n = 20, 24$.
    Phyteuma cappadocicum Boiss., Diagn. Pl. Orient. (ser. 1) 11: 73 (1849). *Campanula
       cappadocica* (Boiss.) Boiss., Diagn. Pl. Orient. (ser. 1) 11: 77 (1849). *Podanthum
       cappadocicum* (Boiss.) Boiss., Fl. Orient. 3: 946 (1875). *Asyneuma cappadocicum*
       (Boiss.) Bornm., Beih. Bot. Centralbl. 38(2): 340 (1921).
    Phyteuma leianthum Trautv., Trudy Imp. S.-Peterburgsk. Bot. Sada 3: 274 (1874).
       Podanthum leianthum (Trautv.) Boiss., Fl. Orient., Suppl.: 334 (1888). *Asyneuma
       leianthum* (Trautv.) Bornm., Beih. Bot. Centralbl. 38(2): 343 (1921).
    Podanthum brachylobum Boiss., Fl. Orient. 3: 947 (1875). *Asyneuma brachylobum* (Boiss.)
       Bornm., Beih. Bot. Centralbl. 38(2): 340 (1921).
    Asyneuma virgatum f. *peltatum* Witasek, Ann. K. K. Naturhist. Hofmus. 20: 419 (1905).
    Podanthum sintenisii Hausskn. ex Bornm., Mitt. Thüring. Bot. Vereins (n.s.) 20: 35
       (1905). *Podanthum brachylobum* var. *sintenisii* (Hausskn. ex Bornm.) Bornm., Mitt.
       Thüring. Bot. Vereins (n.s.) 20: 37 (1905). *Phyteuma sintenisii* (Hausskn. ex Bornm.)
       Bornm., Mitt. Thüring. Bot. Vereins (n.s.) 20: 38 (1905). *Asyneuma sintenisii*
       (Hausskn. ex Bornm.) Bornm., Beih. Bot. Centralbl. 38(2): 349 (1921).
    Podanthum obtusifolium Hausskn. ex Bornm., Mitt. Thüring. Bot. Vereins (n.s.) 20: 37
       (1905). *Phyteuma obtusifolium* (Hausskn. ex Bornm.) Bornm., Mitt. Thüring. Bot.
       Vereins (n.s.) 20: 38 (1905); non Freyn, Oesterr. Bot. Z. 41: 56 (1891). *Asyneuma
       obtusifolium* (Hausskn. ex Bornm.) Bornm., Beih. Bot. Centralbl. 38(2): 346 (1921).
    Podanthum aizoon Hausskn. ex Bornm., Mitt. Thüring. Bot. Vereins (n.s.) 20: 37 (1905).
       Phyteuma aizoon (Hausskn. ex Bornm.) Bornm., Mitt. Thüring. Bot. Vereins (n.s.) 20:
       38 (1905). *Asyneuma obtusifolium* var. *aizoon* (Hausskn. ex Bornm.) Bornm., Beih.
       Bot. Centralbl. 38(2): 346 (1921).
    Asyneuma cilicicum Contandr., Quézel & Pamukç., Ann. Univ. Provence Sci. 46: 56 (1972).
    Asyneumopsis stipitata Contandr., Quézel & Pamukç., Ann. Univ. Provence Sci. 46: 58.
       1972, as 'stipitatum'.

Synonyms:
Asyneuma amanum Rech. f. === **Asyneuma limoniifolium** subsp. **pestalozzae** (Boiss.)
  Damboldt
Asyneuma amplexicaule var. *kotschyi* (Boiss.) Bornm. === **Asyneuma amplexicaule** (Willd.)
  Hand.-Mazz. subsp. **amplexicaule**
Asyneuma amplexicaule var. *angustifolium* (Boiss.) Bornm. === **Asyneuma amplexicaule**
  (Willd.) Hand.-Mazz. subsp. **amplexicaule**
Asyneuma anhuiense B. A. Shen === **Triodanis perfoliata** subsp. **biflora** (Ruiz & Pav.) Lammers
Asyneuma anthericoides f. *dobrogense* (Borza) Hayek === **Asyneuma anthericoides** (Janka)
  Bornm.

Asyneuma anthericoides var. *dobrogense* Borza === **Asyneuma anthericoides** (Janka) Bornm.

Asyneuma asperum (Boiss.) Rech. f. & Schiman-Czeika === **Asyneuma persicum** (A. DC.) Bornm.

Asyneuma attenuatum (Franch.) Bornm. === **Asyneuma argutum** (Regel) Bornm. subsp. **argutum**

Asyneuma aurasiacum (Batt. & Trab.) Bornm. === **Asyneuma rigidum** subsp. **aurasiacum** (Batt. & Trab.) Damboldt

Asyneuma baldshuanicum (O. Fedtsch.) Fed. === **Asyneuma argutum** subsp. **baldshuanicum** (O. Fedtsch.) Damboldt

Asyneuma brachylobum (Boiss.) Bornm. === **Asyneuma virgatum** (Labill.) Bornm. subsp. **virgatum**

Asyneuma campanuloides var. *gracilis* Manden. & Kuth. === **Asyneuma campanuloides** (M. Bieb. ex Sims) Bornm.

Asyneuma canescens f. *nudiflorum* (Davidov) Hayek === **Asyneuma canescens** (Waldst. & Kit.) Griseb. & Schenk subsp. **canescens**

Asyneuma canescens subsp. *salicifolium* (Steud.) Soó === **Asyneuma canescens** (Waldst. & Kit.) Griseb. & Schenk subsp. **canescens**

Asyneuma canescens var. *foliosum* (Nyman) Borza === **Asyneuma canescens** (Waldst. & Kit.) Griseb. & Schenk subsp. **canescens**

Asyneuma cappadocicum (Boiss.) Bornm. === **Asyneuma virgatum** (Labill.) Bornm. subsp. **virgatum**

Asyneuma cichoriiforme (Boiss.) Bornm. === **Asyneuma virgatum** subsp. **cichoriiforme** (Boiss.) Damboldt

Asyneuma cilicicum Contandr., Quézel & Pamukç. === **Asyneuma virgatum** (Labill.) Bornm. subsp. **virgatum**

Asyneuma compactum subsp. *glabrescens* Contandr., Quézel & Pamukç. === **Asyneuma compactum** Damboldt

Asyneuma compactum var. *eriocarpum* Parolly === **Asyneuma compactum** Damboldt

Asyneuma canescens var. *foliosum* (Kit.) Borza === **Asyneuma canescens** (Waldst. & Kit.) Griseb. & Schenk subsp. **canescens**

Asyneuma controversum (Boiss.) Bornm. === **Asyneuma rigidum** (Willd.) Grossh. subsp. **rigidum**

Asyneuma cordifolium Bornm. === **Asyneuma canescens** subsp. **cordifolium** (Bornm.) Damboldt

Asyneuma davisii Quézel, Contandr. & Pamukç. === **Asyneuma michauxioides** (Boiss.) Damboldt

Asyneuma debile Fed. === **Asyneuma argutum** subsp. **baldshuanicum** (O. Fedtsch.) Damboldt

Asyneuma densifolium Rech. f. === **Asyneuma rigidum** (Willd.) Grossh. subsp. **rigidum**

Asyneuma ekimianum subsp. *beritensis* Kit Tan & Yildiz === **Asyneuma ekimianum** Kit Tan & Yildiz

Asyneuma ekimianum subsp. *sivasicum* Kit Tan & Yildiz === **Asyneuma ekimianum** Kit Tan & Yildiz

Asyneuma eldivenum Czeczott === **Asyneuma rigidum** (Willd.) Grossh. subsp. **rigidum**

Asyneuma eximium Rech. f. === **Asyneuma linifolium** subsp. **eximium** (Rech. f.) Damboldt

Asyneuma floribundum (Stapf) P. H. Davis === **Asyneuma limoniifolium** subsp. **pestalozzae** (Boiss.) Damboldt

Asyneuma grandiflorum (Velen.) Bornm. === **Asyneuma anthericoides** (Janka) Bornm.

Asyneuma haradjanii Rech. f. === **Asyneuma rigidum** (Willd.) Grossh. subsp. **rigidum**

Asyneuma kellerianum (Stef.) Stef. === **Asyneuma limoniifolium** (L.) Janch. subsp. **limoniifolium**

Asyneuma lanceolatum Hand.-Mazz. === **Asyneuma rigidum** (Willd.) Grossh. subsp. **rigidum**

Asyneuma lanceolatum f. *flagellatum* (Hausskn. & Bornm. ex Bornm.) Bornm. === **Asyneuma rigidum** (Willd.) Grossh. subsp. **rigidum**

Asyneuma lanceolatum f. *stenophyllum* Bornm. === **Asyneuma rigidum** (Willd.) Grossh. subsp. **rigidum**

Asyneuma lanceolatum subsp. *controversum* (Boiss.) Bornm. === **Asyneuma rigidum** (Willd.) Grossh. subsp. **rigidum**

Asyneuma lanceolatum var. *alpinum* (Boiss.) Bornm. === **Asyneuma rigidum** subsp. **sinai** (A. DC.) Damboldt

Asyneuma lanceolatum var. *rigidum* (Willd.) Hand.-Mazz. === **Asyneuma rigidum** (Willd.) Grossh.

Asyneuma leianthum (Trautv.) Bornm. === **Asyneuma virgatum** (Labill.) Bornm. subsp. **virgatum**

Asyneuma limonifolium f. *canescens* (Boiss.) Hayek === **Asyneuma limoniifolium** (L.) Janch. subsp. **limoniifolium**

Asyneuma limoniifolium f. *heterophyllum* (Rohlena) Hayek === **Asyneuma limoniifolium** (L.) Janch. subsp. **limoniifolium**

Asyneuma limoniifolium subf. *ramosum* (Hausskn.) Hayek === **Asyneuma limoniifolium** (L.) Janch. subsp. **limoniifolium**

Asyneuma limoniifolium subsp. *psaridis* (Heldr. ex Halácsy) Hayek === **Asyneuma limoniifolium** (L.) Janch. subsp. **limoniifolium**

Asyneuma *limoniifolium* var. *alpinum* (Boiss.) Bornm. === **Asyneuma limoniifolium** (L.) Janch. subsp. **limoniifolium**

Asyneuma *limoniifolium* var. *canescens* (Boiss.) Bornm. === **Asyneuma limoniifolium** (L.) Janch. subsp. **limoniifolium**

Asyneuma lobelioides f. *aspera* Bornm. === **Asyneuma limoniifolium** subsp. **pestalozzae** (Boiss.) Damboldt

Asyneuma *lobelioides* f. *glaberrima* Nábělek === **Asyneuma filipes** (Nábělek) Damboldt

Asyneuma *lobelioides* f. *nana* Bornm. === **Asyneuma lobelioides** (Willd.) Hand.-Mazz.

Asyneuma lobelioides f. *scabrida* Nábělek === **Asyneuma filipes** (Nábělek) Damboldt

Asyneuma lobelioides f. *stenophylla* (Bornm.) Bornm. === **Asyneuma limoniifolium** subsp. **pestalozzae** (Boiss.) Damboldt

Asyneuma lobelioides var. *capitatum* Nábělek === **Asyneuma filipes** (Nábělek) Damboldt

Asyneuma lobelioides var. *filipes* Nábělek === **Asyneuma filipes** (Nábělek) Damboldt

Asyneuma mazanderanicum Rech. f. === **Asyneuma virgatum** subsp. **mazanderanicum** (Rech. f.) Damboldt

Asyneuma multicaule (Boiss.) Rech. f. & Schiman-Czeika === **Asyneuma persicum** (A. DC.) Bornm.

Asyneuma obtusifolium (Hausskn. ex Bornm.) Bornm. === **Asyneuma virgatum** (Labill.) Bornm. subsp. **virgatum**

Asyneuma obtusifolium var. *aizoon* (Hausskn. ex Bornm.) Bornm. === **Asyneuma virgatum** (Labill.) Bornm. subsp. **virgatum**

Asyneuma otites (Boiss.) Bornm. === **Asyneuma limoniifolium** (L.) Janch. subsp. **limoniifolium**

Asyneuma parviflorum Turrill === **Asyneuma limoniifolium** (L.) Janch. subsp. **limoniifolium**

Asyneuma persicum var. *pumilum* (Boiss.) Bornm. === **Asyneuma persicum** (A. DC.) Bornm.

Asyneuma pichleri var. *subalpina* D. Lakušić & F. Conti === **Asyneuma pichleri** (Vis.) D. Lakušić & F. Conti

Asyneuma prenanthoides (Durand) McVaugh === **Campanula prenanthoides** Durand

Asyneuma psaridis (Heldr. ex Halácsy) Bornm. === **Asyneuma limoniifolium** (L.) Janch. subsp. **limoniifolium**

Asyneuma psilostachya (Boiss. & Kotschy) Bornm. === **Campanula psilostachya** Boiss. & Kotschy

Asyneuma ramosum Pavlov === **Asyneuma argutum** (Regel) Bornm. subsp. **argutum**

Asyneuma regelii (Trautv.) Bornm. === **Sergia regelii** (Trautv.) Fed.

Asyneuma repandum (Sm.) Rothm. === **Asyneuma limoniifolium** (L.) Janch. subsp. **limoniifolium**

Asyneuma salicifolium Flerow === **Asyneuma canescens** (Waldst. & Kit.) Griseb. & Schenk subsp. **canescens**

Asyneuma salignum (Waldst. & Kit. ex Besser) Fed. === **Asyneuma canescens** (Waldst. & Kit.) Griseb. & Schenk subsp. **canescens**

Asyneuma scoparium (Boiss. & Hausskn.) Bornm. === **Campanula scoparia** (Boiss. & Hausskn.) Damboldt

Asyneuma sibthorpianum (Schult.) Bornm. === **Asyneuma rigidum** subsp. **sibthorpianum** (Schult.) Damboldt

Asyneuma sintensii (Hausskn. ex Bornm.) Bornm. === **Asyneuma virgatum** (Labill.) Bornm. var. **virgatum**

Asyneuma strictum Wendelbo === **Asyneuma argutum** (Regel) Bornm. subsp. **argutum**

Asyneuma strigillosum (Boiss.) Bornm. === **Campanula strigillosa** Boiss.

Asyneuma supinum (Wettst.) Bornm. === **Asyneuma rigidum** (Willd.) Grossh. subsp. **rididum**

Asyneuma talyschense Fed. === **Asyneuma amplexicaule** (Willd.) Hand.-Mazz. subsp. **amplexicaule**

Asyneuma tenuifolium (A. DC.) Bornm. === **Asyneuma limoniifolium** (L.) Janch. subsp. **limoniifolium**

Asyneuma thomsonii var. *nuristanica* Kitam. === **Asyneuma argutum** (Regel) Bornm. subsp. **argutum**

Asyneuma thomsonii var. *strictum* (Wendelbo) Kitam. === **Asyneuma argutum** (Regel) Bornm. subsp. **argutum**

Asyneuma trautvetteri (B. Fedtsch.) Bornm. === **Asyneuma argutum** (Regel) Bornm. subsp. **argutum**

Asyneuma urceolatum (Fomin) Fed. === **Asyneuma lobelioides** (Willd.) Hand.-Mazz.

Asyneuma virgatum f. *peltatum* Witasek === **Asyneuma virgatum** (Labill.) Bornm. subsp. **virgatum**

Asyneuma woronowii (Fomin) Bornm. === **Asyneuma lobelioides** (Willd.) Hand.-Mazz.

Asyneumopsis

This genus and its sole species were based upon a specimen that falls readily within the range of variation of *Asyneuma virgatum* subsp. *virgatum* (Damboldt 1978, under **Asyneuma**). Type [by monotypy]: *Asyneumopsis stipitata* Contandr., Quézel & Pamukç.

Cantandriopoulos, J., P. Quézel & A. Pamukçuoğlu (1972). Campanulacées nouvelles du pourtour méditerranéen oriental. Ann. Univ. Provence sci. 46: 53-61 + pl. I-II + cliché 1-2, illus. Fr. — Establishment of genus as distinct from *Asyneuma*.

Synonyms:
Asyneumopsis Contandr., Quézel & Pamukç. === **Asyneuma** Griseb. & Schenk.
Asyneumopsis stipitata Contandr., Quézel & Pamukç. === **Asyneuma virgatum** (Labill.) Bornm. subsp. **virgatum**

Azorina

Campanuloideae, 1 species, the Azores. A preliminary phlyogeny based on nucleic acid sequences (Eddie et al. 2003, under **General**) offers some support for assignment of this species to *Campanula*. Sealey (1968) and Franco (1984) offer the best available treatments of the genus. 2*n* = 56. Type [by monotypy]: *Azorina vidalii* (H. C. Watson) Feer.

Watson, H. C. (1844). *Campanula vidalii,* H. C. Watson. Hooker's Icon. Pl. 7: tab. 684 + 1 pg., illus. En. — Portrait from dried specimen, with description.

Hooker, W. J. (1853). *Campanula vidalii*. Vidal's Bell-flower. Bot. Mag. 79: tab. 4748 + 2 pp., illus. En. — Portrait from cultivated plant, with description.

Feer, H. (1890). Beiträge zur Systematik und Morphologie der Campanulaceen. II.A. Drei neue monotype Gattungen. Bot. Jahrb. Syst. 12: 610-617 + Taf. VII-VIII, illus. Ge. — Establishment of genus as distinct from *Campanula*.

• Sealey, J. R. (1968). *Campanula vidalii*. Bot. Mag. 176: tab. 527 + 5 pp., illus. En. — Portrait with description.

Vasilevskaya, V. K. & T. V. Shulkina (1976). [The morphological and anatomical structure of the arborescent plant *Azorina vidalii* Feer (*Campanula vidalii* Wats.).] Trudy Moskovsk. Obšč. Isp. Prir. 42: 131-140, illus. Ru. — Detailed study of life-form and stem anatomy.

Perez de Paz, J. (1978). Estudio palinologico de las Campanulaceas endemicas de la region Macaronesica. Bot. Macarones. 6: 27-42, illus. Por. — Pollen morphology.

• Franco, J. (1984). *Azorina* Feer. Nova flora de Portugal (continente e Açores) 2: 327. Lisbon: Sociedade Astória. Por. — Flora with description.

Azorina Feer, Bot. Jahrb. Syst. 12: 611 (1890). *Campanula* sect. *Azorina* (Feer) Kuntze in T. Post & Kuntze, Lex. Gen. Phan.: 95 (1904).
Macaronesia. 21.

Azorina vidalii (H. C. Watson) Feer, Bot. Jahrb. Syst. 12: 612 (1890).
Azores (Corvo, Flores, Pico, São Jorge, Terceira, São Miguel, Santa Maria); cult. 21 AZO.
Nanophan. $2n = 56$.
  * *Campanula vidalii* H. C. Watson, Hooker's Icon. Pl. 7: pl. 684 (1844).
  Campanula vidalii f. *alba* Sealy, Bot. Mag. 176: pl. 527 (1968).

Baclea

This name though earlier cannot be used for *Pseudonemacladus* (q.v.) because it is a later homonym of *Baclea* Fourn. ex Baill. (Asclepiadaceae). Type [by monotypy]: *Baclea oppositifolia* (B. L. Rob.) Greene.

Greene, E. L. (1893). Two rare lobeliaceous plants. Erythea 1: 237-238. En. — Establishment of genus as distinct from *Nemacladus*.

Synonyms:
Baclea Greene === **Pseudonemacladus** McVaugh
Baclea oppositifolia (B. L. Rob.) Greene === **Pseudonemacladus oppositifolius** (B. L. Rob.) McVaugh

Benaurea

Rafinesque (1837) apparently was unaware of the earlier name *Musschia* (q.v.) when he created this genus to house *Campanula aurea* L. f. of Madeira. Type [by monotypy]: *Benaurea sempervirens* Raf.

Rafinesque, C. S. (1837). Flora Telluriana, pars secunda. Philadelphia: published privately. En. — Establishment of genus as distinct from *Campanula*.

Synonyms:
Benaurea Raf. === **Musschia** Dumort.
Benaurea sempervirens Raf. === **Musschia aurea** (L. f.) Dumort.

Berenice

Campanuloideae, 1 species, Réunion in the western Indian Ocean. On the basis of gross morphology, *Berenice* was originally assigned to Saxifragaceae s. l. (Argophyllaceae). However, data from subsequent studies of pollen morphology and anatomy (Erdtman & Metcalfe 1963, Badré et al. 1975) were more consistent with assignment to Campanulaceae. The recent discovery of a close phylogenetic relationship between Argophyllaceae and allied families on the one hand and the Campanulales on the other (Gustafsson & Bremer 1995, Kårehed et al. 1999, Lundberg & Bremer 2003; all under **General**) casts new interest on the relationships of this genus, which has yet to be included in any phylogenetic analysis. Type [by monotypy]: *Berenice arguta* Tul.

Tulasne, L. R. (1857) Florae madagascariensis fragmenta. Fragmentum alterum (1). Ann. Sci. Nat., Bot. (sér. 4) 8: 44-163. La. — Establishment of genus in Saxifragaceae, tribe Escallonieae.

Engler, A. (1930). *Berenice* Tul. In Die natürlichen Pflanzenfamilen (ed. 2) 18a: 217. Ge. — Description, as part of monograph of Saxifragaceae.

Bernardi, L. (1961). La mort de *Berenice*. Mus. Genève (n.s.) 18: 12-14. Fr. — Suggests genus extinct.

• Erdtman, G. & C. R. Metcalfe (1963). Affinities of certain genera *incertae sedis* suggested by pollen morphology and vegetative anatomy. III. The campanulaceous affinity of *Berenice arguta* Tulasne. Kew Bull. 17: 253-256 + pl. 2, illus. En. — Assignment of genus to Campanulaceae, in subfamily Campanuloideae.

• Badré, F., T. Cadet, G. Cusset & M. Hideux (1975). Position systématique, étude morphologique et palynologique du genre *Berenice*. Adansonia (n.s.) 15: 139-146, illus. Fr. — Further evidence for classification among Campanuloideae; not extinct as previously reported.

• Badré, F. (1976). *Berenice* Tul. In Flore des Mascareignes. 111. Campanulacées: 16-17, illus. Paris: ORSTOM. Fr. — Flora with description and full nomenclature.

Belyaev, A. A. (1986). [The features of anatomy and surface ultrastructure of the seed coat in some representatives of critical genera of the Campanulaceae family.] Bot. Žurn. 71: 1371-1376 + 2 pl., illus. Ru. — Description of testa.

Berenice Tul., Ann. Sci. Nat., Bot. (sér. 4) 8: 156 (1857).
Mascarenes. 29.

Berenice arguta Tul., Ann. Sci. Nat., Bot. (sér. 4) 8: 157 (1857).
Réunion. 29 REU. Nanophan.

Bolelia

This name was created as an avowed substitute for illegitimate *Clintonia* (q.v.), but was formally rejected in 1905 in favor of the later name *Downingia*. Based on: *Clintonia* Dougl. ex Lindl., non Raf.

Rafinesque, C. S. (1832). On 3 n. sp. of *Clintonia*. Atlantic J.: 120. En. — Validation of name.

Greene, E. L. (1890). Some genera of Rafinesque. *Bolelia*. Pittonia 2: 124-127. En. — Resurrection of name on grounds of priority, with synopsis of all species.

Greene, E. L. (1906). Doctor Torrey and *Downingia*. Torreya 6: 145-147. En. — Nomenclatural discussion regarding priority of various names for *Downingia*, justifying use of *Bolelia*.

Greene, E. L. (1910). The genus *Downingia*. Leafl. Bot. Observ. Crit. 2: 43-45. En. — Further nomenclatural discussion, repudiating his earlier use of *Bolelia* and taking up *Downingia*.

Synonyms:
Bolelia Raf. === **Downingia** Torr.

Bolelia bicornuta (A. Gray) Greene === **Downingia bicornuta** A. Gray
Bolelia brachyantha Rydb. === **Downingia laeta** (Greene) Greene
Bolelia concolor (Greene) Greene === **Downingia concolor** Greene
Bolelia concolor var. *tricolor* (Greene) Jeps. === **Downingia concolor** Greene subsp. **concolor**
Bolelia cuspidata Greene === **Downingia cuspidata** (Greene) Rattan
Bolelia elegans (Dougl. ex Lindl.) Greene === **Downingia elegans** (Dougl. ex Lindl.) Torr.
Bolelia humilis Greene === **Downingia pusilla** (A. DC.) Torr.
Bolelia insignis (Greene) Greene === **Downingia insignis** Greene
Bolelia laeta Greene === **Downingia laeta** (Greene) Greene
Bolelia montana (Greene) Greene === **Downingia montana** Greene
Bolelia ornatissima (Greene) Greene === **Downingia ornatissima** Greene
Bolelia pulchella (Lindl.) Greene === **Downingia pulchella** (Lindl.) Torr.
Bolelia pulchella f. *alba* Voss === **Downingia pulchella** (Lindl.) Torr.
Bolelia pulchella f. *atrocinerea* Voss === **Downingia pulchella** (Lindl.) Torr.
Bolelia pulchella f. *atropurpurea* Voss === **Downingia pulchella** (Lindl.) Torr.
Bolelia pusilla (A. DC.) Greene === **Downingia pusilla** (A. DC.) Torr.
Bolelia tricolor (Greene) Greene === **Downingia concolor** Greene subsp. **concolor**

Brachycodon

This genus was erected for species considered to be morphologically intermediate between *Campanula* and *Legousia*. Unfortunately, the name is a later homonym of *Brachycodon* (Benth.) Progel (Gentianaceae), and was replaced by *Brachycodonia*. Type [designated by the author]: *Brachycodon fastigiatus* (Dufour ex Schult.) Fed.

> Fedorov, A. A. (1957). *Brachycodon* Fed. In V. L. Komarov, Flora URSS 24: 341-343, illus. Leningrad: Academia Scientiae URSS. Ru. — Establishment of genus as distinct from *Campanula*.

Synonyms:
Brachycodon Fed. === **Campanula** L.
Brachycodon fastigiatus (Dufour ex Schult.) Fed. === **Campanula fastigiata** Dufour ex Schult.

Brachycodonia

The sole species of this genus is treated by most authors as *Campanula* subg. *Brachycodonia* (Fed.) Damboldt. Based on: *Brachycodon* Fed., non (Benth.) Progel.

> Karyagin, I. I., ed. (1961). Flora Azerbajdžana, vol. 8. Baku: Akademii Nauk Azerbaidzhanskoi SSR. Ru. — Validation of name.

Synonyms:
Brachycodonia Fed. === **Campanula** L.
Brachycodonia fastigiata (Dufour ex Schult.) Fed. === **Campanula fastigiata** Dufour ex Schult.

Brighamia

Lobelioideae, 2 species, Hawaiian Islands. Although earlier authors perceived a relationship to *Isotoma* or *Hippobroma* (Rock 1919, Stone 1967, St. John 1969, Lammers 1989), or regarded the genus as highly isolated within the subfamily (Wimmer 1953), molecular data (Givnish et al. 1995; E. Knox, pers. comm.; A. Antonelli, pers. comm.) and seed morphology (Buss et al. 2001) support a sister-relationship with *Delissea* in the Hawaiian clade. The treatment here is that of Lammers (1989, 1990). $2n = 28$. Type [by monotypy]: *Brighamia insignis* A. Gray.

Gray, A. (1861). Notes on Lobeliaceae, Goodeniaceae, &c. of the collections of the U.S. South Pacific Exploring Expedition. Proc. Amer. Acad. Arts 5: 146-152. En. — First mention of genus, as "a striking new *Isotoma*? of Kauai or Nihau [sic] in Remy's collection".

Mann, H. (1869). Notes on *Alsinidendron, Platydesma,* and *Brighamia,* new genera of Hawaiian plants; with an analysis of the Hawaiian flora. Mem. Boston Soc. Nat. Hist. 1: 529-541 + pl. 21-23, illus. En. — Establishment of genus by Asa Gray.

Hillebrand, W. (1888). *Brighamia,* Gray. In Flora of the Hawaiian Islands: 235. London: Williams & Norgate. En. — Flora with descriptions, full nomenclature, and specimen citations.

• Rock, J. F. (1919). A monographic study of the Hawaiian species of the tribe Lobelioideae family Campanulaceae, *Brighamia* A. Gray. Mem. Bernice Pauahi Bishop Mus. 7(2):149-152, illus. En. — Floristic treatment with descriptions, full nomenclature, and specimen citations.

Selling, O. H. (1947). Studies in Hawaiian pollen statistics II. The pollens of the Hawaiian phanerogams. Lobeliaceae. Spec. Publ. Bernice Pauahi Bishop Mus. 38: 318-320 + pl. 49, illus. En. — Descriptive palynology.

• Wimmer, F. E. (1953). Subtribus Brighamiinae E. Wimm. In A. Engler & L. Diels, Das Pflanzenreich IV. 276b: 730-731, illus. Berlin: Akademie-Verlag. Ge., La. — Monograph of monogeneric subtribe with descriptions, full nomenclature, and specimen citations.

Stone, B. C. (1957). Rediscovery of a rare lobelioid, *Brighamia insignis* f. *citrina,* in Kauai, Hawaiian Islands. Bull. Torrey Bot. Club 84: 175-177. En. — Notes on conservation status.

Stone, B. C. (1967). A review of the endemic genera of Hawaiian plants. *Brighamia* A. Gray. Bot. Rev. (Lancaster) 33: 243. En. — Brief descriptive treatment with generic description and full nomenclature.

• Wimmer, F. E. (1968). Subtrib. Brighamiinae E. Wimm. In A. Engler & L. Diels, Das Pflanzenreich IV. 276c: 892. Berlin: Akademie-Verlag. Ge., La. — Supplement to Wimmer (1953).

St. John, H. (1969). Monograph of the genus *Brighamia* (Lobeliaceae). Hawaiian plant studies 29. Bot. J. Linn. Soc. 62: 187-204, illus., maps — Monograph with key, descriptions, and specimen citations.

Perlman, S. (1979). *Brighamia* in Hawaii. Bull. Pacific Trop. Bot. Gard. 9: 1-2 + cover, illus. En. — Popular account emphasizing conservation concerns including lack of pollinators.

Christensen, C. (1979). Propagating Kauai's *Brighamia.* Bull. Pacific Trop. Bot. Gard. 9: 2-4 + cover, illus. En. — Practical account of earliest attempts at cultivation and *ex situ* conservation of *B. insignis.*

Lucas, S. A. (1981). Recent introductions of ornamental value. Bull. Pacific Trop. Bot. Gard. 11(1): 8-13, illus. En. — Introduction of *B. rockii* to cultivation.

Rauh, W. (1981). *Brighamia insignis.* A curious succulent of the lobelia family from the Hawaiian Islands. Cact. Succ. J. (Los Angeles) 53: 219-220, illus. En. — Introduction of genus to horticulture.

Rowley, G. D. (1983). *Brighamia:* succulent endemic of Hawaii. Brit. Cact. Succ. J. 1: 9-11, illus. En. — Descriptive account aimed at popularizing genus with horticulturalists.

Gregory, S. E. (1983). *Brighamia* at Kew. Brit. Cact. Succ. J. 1: 74, illus. En. — Account of *B. insignis* in cultivation.

Johnson, M. A. (1984). The chromosome number of *Brighamia citrina* var. *napaliensis;* a rare succulent endemic from Hawaii. Brit. Cact. Succ. J. 1: 106-107. En. — First determination of chromosome number for genus.

Johnson, M. A. (1986). *Brighamia citrina* var. *napaliensis.* Kew Mag. 3: 68-72 + pl. 60, illus, map. En. — Portrait with description, information on micropropagation, and conservation status.

• Lammers, T. G. (1989). Revision of *Brighamia* (Campanulaceae: Lobelioideae), a caudiciform succulent endemic to the Hawaiian Islands. Syst. Bot. 14: 133-138. En. — Monograph with key, descriptions, full nomenclature, and specimen citations.

Lammers, T. G. (1990). *Brighamia* A. Gray. In W. L. Wagner, D. R. Herbst & S. H. Sohmer, Manual of the flowering plants of Hawai'i: 422-424, illus. Honolulu: University of Hawaii Press. En. — Flora with key, descriptions, and synonymy.

Stolzenburg, W. (1992). The lonesome flower. Nature Conservancy 42(2): 28-29, illus. En. — Description of *in situ* conservation efforts involving manual pollination.

Givnish, T. J., K. J. Sytsma, J. F. Smith, & W. J. Hahn (1995). Molecular evolution, adaptive radiation, and geographic speciation in *Cyanea* (Campanulaceae, Lobelioideae). In W. L. Wagner & V. A. Funk (eds.), Hawaiian biogeography: evolution on a hot spot archipelago: 288-337, illus., maps. Washington: Smithsonian Institution Press. En. — Molecular phylogenetics indicate *Brighamia* is sister to *Delissea*.

• Gemmill, C. E. C., T. A. Ranker, D. Ragone, S. P. Perlman & K. R. Wood (1998). Conservation genetics of the endangered endemic Hawaiian genus *Brighamia* (Campanulaceae). Amer. J. Bot. 85: 528-539 + cover photo, illus, map. En. — Enzyme electrophoretic survey showed two species highly distinct genetically ($\bar{I}$ = 0.56); includes summary of conservation status for all populations.

Buss, C. C., T. G. Lammers & R. R. Wise (2001). Seed coat morphology and its systematic implications in *Cyanea* and other genera of Lobelioideae (Campanulaceae). Amer. J. Bot. 88: 1301-1308, illus. En. — Testal pattern supports sister relationship of *Brighamia* and *Delissea*.

• Hannon, D. P. & S. Perlman (2002). The genus *Brighamia*. Cact. Succ. J. (Los Angeles) 74: 67-76 + cover photo, illus. En. — Comprehensive information on cultivation and conservation status.

Lammers, T. G. (2005). *Brighamia*. In G. W. Staples & D. R. Herbst, A Tropical Garden Flora: 217, illus. Honolulu: Bishop Museum Press. En. — Account of genus in Hawaiian horticulture, with descriptions.

Brighamia A. Gray, Proc. Amer. Acad. Arts 7: 185 (1867).
Hawaiian Is. 63.

Brighamia insignis A. Gray, Proc. Amer. Acad. Arts 7: 185 (1867), as 'insiginis'.
Hawaiian Is. (Ni'ihau, Kaua'i); cult. 63 HAW. Phan. or nanophan. $2n = 28$.
Brighamia insignis f. *citrina* C. N. Forbes & Lydgate, Occas. Pap. Bernice Pauahi Bishop Mus. 6: 203 (1917). *Brighamia citrina* (C. N. Forbes & Lydgate) H. St. John, Pacific Sci. 12: 182 (1958).
Brighamia citrina var. *napaliensis* H. St. John, Bot. J. Linn. Soc. 62: 194 (1969).

Brighamia rockii H. St. John, Bot. J. Linn. Soc. 62: 196 (1969).
Hawaiian Is. (Molokai, Lanai?, Maui). 63 HAW. Phan. or nanophan.
Brighamia rockii f. *longiloba* H. St. John, Bot. J. Linn. Soc. 62: 201 (1969).
Brighamia remyi H. St. John, Bot. J. Linn. Soc. 62: 201 (1969).

Synonyms:
Brighamia citrina (C. N. Forbes & Lydgate) H. St. John === **Brighamia insignis** A. Gray
Brighamia citrina var. *napaliensis* H. St. John === **Brighamia insignis** A. Gray
Brighamia insignis f. *citrina* C. N. Forbes & Lydgate === **Brighamia insignis** A. Gray
Brighamia remyi H. St. John === **Brighamia rockii** H. St. John
Brighamia rockii f. *longiloba* H. St. John === **Brighamia rockii** H. St. John

Burmeistera

Lobelioideae, 102 species, Neotropics from Guatemala to Peru. Wimmer (1943) divided the species among two sections: *Burmeistera* and *Barbatae* E. Wimm. The latter was further divided into two subsections, but Stein (1987) transferred both species of subsect. *Aequilatae* E. Wimm. to *Siphocampylus*. Analyses of molecular data (E. Knox, pers. comm.; A. Antonelli, pers. comm.) support the monophyly of the genus and indicate that it forms a clade with *Centropogon* and *Siphocampylus*. The most recent monograph was that by Wimmer (1943, 1953, 1968). Type [by monotypy]: *Burmeistera ibaguensis* Triana.

Triana, J. (1855). Nuevos jeneros i especies de plantas para la flora Neo-granadina. 28 pp. Bogota: published privately. Sp. — Establishment of genus.

Karsten, H. (1856). Plantae Columbianae. *Burmeistera* Karst. et Triana. Linnaea 28: 444-446. La. — Reiteration of generic description, with addition of six more species to genus.

Gleason, H. A. (1925). Studies on the flora of northern South America – IV. The genus *Burmeistera*. Bull. Torrey Bot. Club 52: 93-104. En. — Synopsis of genus with key and some descriptions.

• Wimmer, F. E. (1932). *Burmeistera*, eine umstrittene Pflanzengattung und ihre Arten. Repert. Spec. Nov. Regni Veg. 30: 1-52 + tab. CXXIII-CXXVI, illus. Ge., La. — Monograph with key, descriptions, full nomenclature, and specimen citations.

Standley, P. C. (1938). Flora of Costa Rica, *Burmeistera* Karst. & Triana. Field Mus. Nat. Hist., Bot. Ser. 18: 1406-1409. En. — Floristic treatment with descriptions.

• McVaugh, R. (1943). *Burmeistera* Triana. In North American Flora 32A: 127-134. New York: New York Botanical Garden. En. — Flora with key, descriptions, and full nomenclature.

• Wimmer, F. E. (1943). *Burmeistera* Triana. In A. Engler & L. Diels, Das Pflanzenreich IV. 276b: 122-161, illus. Leipzig: Wilhelm Engelmann. Ge., La. — Monograph with keys, descriptions, infrageneric classification, full nomenclature, and specimen citations.

• Wimmer, F. E. (1953). *Burmeistera*. In A. Engler & L. Diels, Das Pflanzenreich IV. 276b: 767-768. Berlin: Akademie-Verlag. Ge., La. — Supplement to Wimmer (1943).

Wimmer, F. E. (1968). *Burmeistera* Triana. In A. Engler & L. Diels, Das Pflanzenreich IV. 276c: 834-838 + Taf. 13, illus. Berlin: Akademie-Verlag. Ge., La. — Second supplement to Wimmer (1943).

• Nash, D. L. (1976). Flora of Guatemala, *Burmeistera* Triana. Fieldiana, Bot. 24: 402-404, illus. En. — Floristic treatment with descriptions.

• Wilbur, R. L. (1976). A synopsis of the Costa Rican species of *Burmeistera* (Campanulaceae: Lobelioideae). Bull. Torrey Bot. Club 102: 225-231. En. — Floristic treatment with key and full nomenclature.

• Wilbur, R. L. (1977). Flora of Panama, *Burmeistera*. Ann. Missouri Bot. Gard. 63: 595-612, illus. En. — Floristic treatment with key and descriptions.

• Jeppesen, S. (1981). *Burmeistera* Triana. In G. Harling & B. Sparre (eds.), Flora of Ecuador 14: 12-48, illus., map. Stockholm: Swedish Natural Science Research Council. En. — Flora with keys, descriptions, full nomenclature, and specimen citations.

• Wilbur, R. L. (1981). Additional Panamanian species of *Burmeistera* (Campanulaceae: Lobelioideae). Ann. Missouri Bot. Gard. 68: 167-171. En. — Supplement to Wilbur (1977), with revised key.

Lozano C., G. & G. Galeano G. (1986). Una nueva especie de *Burmeistera* (Campanulaceae) de Colombia. Caldasia 15: 53-56, illus. Sp. — Description of *B. multipinnatisecta*, and synonymization of *B. pinnatisecta* (see Luteyn 1986) under *B. pteridoides*.

Luteyn, J. L. (1986). A new *Burmeistera* (Campanulaceae: Lobelioideae) from western Colombia. Syst. Bot. 11: 474-476, illus. En. — Description of *B. pinnatisecta*, with key to species with pinnate leaves.

Stein, B. A. (1987). *Siphocampylus oscitans* (Campanulaceae: Lobelioideae), a new name for *Burmeistera weberbaueri* from Peru. Ann. Missouri Bot. Gard. 74: 491-493. En. — Discussion of species of *Siphocampylus* with open anther orifice like that of *Burmeistera*.

Stein, B. A. (1987). Synopsis of the genus *Burmeistera* (Campanulaceae: Lobelioideae) in Peru. Ann. Missouri Bot. Gard. 74: 494-496. En. — Floristic treatment with key and full synonymy.

• Lammers, T. G. (1998). Review of the neotropical endemics *Burmeistera*, *Centropogon*, and *Siphocampylus* (Campanulaceae: Lobeliodeae), with description of 18 new species and a new section. Brittonia 50: 233-262. En. — Summary of current knowledge of genus and its subordinate taxa.

Lammers, T. G. & P. J. M. Maas (1998). First report of the genus *Burmeistera* (Campanulaceae) from Honduras. Sida 18: 363-364. En. — Major range extension for genus.

Muchhala, N. (2003). Exploring the boundary between pollination syndromes: bats and hummingbirds as pollinators of *Burmeistera cyclostigmata* and *B. tenuiflora* (Campanulaceae). Oecologia 134: 373-380. En. — Field studies of pollination in Costa Rica show these species are both chiropterophilous and ornithophilous.

Batterman, M. R. W. & T. G. Lammers (2004). Branched foliar trichomes of Lobelioideae (Campanulaceae) and the infrageneric classification of *Centropogon*. Syst. Bot. 29: 448-458, illus. En. — Scanning electron micrographs of *B. arbusculifera,* only known species with branched trichomes.

Muchhala, N. (2006). The pollination biology of *Burmeistera* (Campanulaceae): specialization and syndromes. Amer. J. Bot. 93: 1081-1089 + cover, illus. En. — Field studies of pollination in Ecuador demonstrate both chiropterophily and ornithophily in the genus.

Burmeistera Triana, Nuev. Gen. Esp.: 13 (1855). *Centropogon* sect. *Burmeistera* (Triana) Schönl., Nat. Planzenfam. IV.5: 65 (1889).
Neotropics. 80 82 83.

Burmeistera acuminata H. Karst., Linnaea 28: 445 (1856).
Colombia. 83 CLM. Nanophan.

Burmeistera aeribacca E. Wimm., Brittonia 8: 108 (1955).
Colombia. 83 CLM. Nanophan.

Burmeistera almedae Wilbur, Bull. Torrey Bot. Club 102: 229 (1976).
Costa Rica. 80 COS. Nanophan. or cham.

Burmeistera anderssonii Jeppesen in Harling & Sparre, Fl. Ecuador 14: 26 (1981).
Ecuador. 83 ECU. Nanophan.

Burmeistera arbusculifera Lammers, Novon 12: 207 (2002).
Ecuador. 83 ECU. Cl. phan. or cl. nanophan.

Burmeistera asclepiadea Gleason, Bull. Torrey Bot. Club 52: 103 (1925).
Colombia. 83 CLM. Nanophan. or cham.

Burmeistera aspera E. Wimm., Pflanzenr. IV.276b: 160 (1943).
Colombia & Ecuador. 83 CLM ECU. Cl. nanophan.

Burmeistera asplundii Jeppesen in Harling & Sparre, Fl. Ecuador 14: 28 (1981).
Ecuador. 83 ECU. Cl. nanophan. or cl. cham.

Burmeistera auriculata Muchhala & Lammers, Novon 15: 177 (2005).
Ecuador. 83 ECU. Nanophan.

Burmeistera borjensis Jeppesen in Harling & Sparre, Fl. Ecuador 14: 29 (1981).
Columbia & Ecuador. 83 CLM ECU. Nanophan. or (cl.) cham.

Burmeistera brachyandra E. Wimm., Pflanzenr. IV.276b: 129 (1943).
Ecuador. 83 ECU. Nanophan. or cham.

Burmeistera breviflora (Gleason) E. Wimm., Repert. Spec. Nov. Regni Veg. 29: 52 (1931).
Colombia. 83 CLM. Nanophan.
    * *Centropogon breviflorus* Gleason, Bull. Torrey Bot. Club 51: 448 (1924).

Burmeistera brighamioides Lammers, Novon 12: 208 (2002).
Ecuador. 83 ECU. Nanophan. or cham.

Burmeistera caldasensis (Gleason) E. Wimm., Repert. Spec. Nov. Regni Veg. 29: 52 (1932).
  Colombia. 83 CLM. Nanophan. or cham.
  * *Centropogon caldasensis* Gleason, Bull. Torrey Bot. Club 51: 446 (1924).

Burmeistera carnosa Gleason, Bull. Torrey Bot. Club 52: 102 (1925).
  Colombia. 83 CLM. Nanophan. or cham.
  Burmeistera carnosa var. *metastasia* E. Wimm., Brittonia 8: 109 (1955).

Burmeistera ceratocarpa Zahlbr., Repert. Spec. Nov. Regni Veg. 13: 534 (1915).
  Colombia & Ecuador. 83 CLM ECU. Nanophan. or cham.
  Burmeistera ceratocarpa var. *dentata* E. Wimm., Pflanzenr. IV.276b: 154 (1943).

Burmeistera chiriquiensis Wilbur, Ann. Missouri Bot. Gard. 63: 597 (1977).
  Costa Rica & Panama. 80 COS PAN. Nanophan. or cham.

Burmeistera chirripoensis Wilbur, Bull. Torrey Bot. Club 102: 228 (1976).
  Costa Rica. 80 COS. Cham.

Burmeistera coleoides (Vatke) E. Wimm., Repert. Spec. Nov. Regni Veg. 30: 35 (1932).
  South America (?). 82+ or 83+. Nanophan. or cham.
  * *Siphocampylus coleoides* Vatke, Linnaea 38: 733 (1874). *Centropogon coleoides* (Vatke)
    Zahlbr., Ann. K. K. Naturhist. Hofmus. 6: 439 (1891).

Burmeistera connivens Gleason, Bull. Torrey Bot. Club 52: 100 (1925).
  Colombia. 83 CLM. Cham.
  Burmeistera luteola E. Wimm., Repert. Spec. Nov. Regni Veg. 29: 57 (1931).

Burmeistera corei E. Wimm., Brittonia 8: 109 (1955).
  Colombia. 83 CLM. Nanophan.

Burmeistera crassifolia (E. Wimm.) E. Wimm., Repert. Spec. Nov. Regni Veg. 30: 33 (1932).
  Ecuador. 83 ECU. Nanophan. or cham.
  * *Centropogon crassifolius* E. Wimm., Repert. Spec. Nov. Regni Veg. 19: 389 (1924).
    Centropogon crassifolius var. *ovatifolius* E. Wimm., Repert. Spec. Nov. Regni Veg. 19: 389
    (1924). *Burmeistera crassifolia* var. *ovatifolia* (E. Wimm.) E. Wimm., Repert. Spec.
    Nov. Regni Veg. 30: 33 (1932).

Burmeistera crebra McVaugh, Brittonia 9: 31 (1957).
  Costa Rica. 80 COS. Cham.

Burmeistera crispiloba Zahlbr., Repert. Spec. Nov. Regni Veg. 13: 528 (1915).
  Ecuador. 83 ECU. Nanophan.
  Burmeistera montana E. Wimm., Repert. Spec. Nov. Regni Veg. 30: 23 (1932).
  Burmeistera succulenta var. *breviloba* E. Wimm., Pflanzenr. IV.276c: 836 (1968).

Burmeistera curviandra E. Wimm., Brittonia 8: 107 (1955).
  Colombia. 83 CLM. Nanophan.

Burmeistera cyclostigmata Donn. Sm., Bot. Gaz. (Crawfordsville) 20: 291 (1895).
  Costa Rica to Ecuador. 80 COS PAN 83 CLM ECU. (Cl.) nanophan. or cham.
  Burmeistera cyclostigmata var. *suerrensis* Donn. Sm., Bot. Gaz. 24: 394 (1897). *Burmeistera
    suerrensis* (Donn. Sm.) E. Wimm., Repert. Spec. Nov. Regni Veg. 30: 14 (1932).
  Burmeistera cerasifera Zahlbr., Repert. Spec. Nov. Regni Veg. 13: 532 (1915).
  Burmeistera suerrensis var. *almirantensis* E. Wimm., Repert. Spec. Nov. Regni Veg. 30: 14
    (1932).
  Burmeistera millei E. Wimm., Repert. Spec. Nov. Regni Veg. 30: 15 (1932).

Burmeistera mindoana E. Wimm., Repert. Spec. Nov. Regni Veg. 30: 27 (1932).
Burmeistera cylindrocarpa var. *candida* E. Wimm., Brittonia 8: 109 (1955).

Burmeistera cylindrocarpa Zahlbr., Repert. Spec. Nov. Regni Veg. 13: 533 (1915).
Ecuador. 83 ECU. Nanophan. or cham.

Burmeistera darienensis Wilbur, Ann. Missouri Bot. Gard. 63: 600 (1977).
Panama. 80 PAN. Nanophan.

Burmeistera dendrophila E. Wimm., Repert. Spec. Nov. Regni Veg. 30: 23 (1932).
Panama. 80 PAN. Nanophan. or cham.

Burmeistera dichlora E. Wimm., Repert. Spec. Nov. Regni Veg. 29: 56 (1931).
Colombia. 83 COL. Nanophan.

Burmeistera domingensis Jeppesen in Harling & Sparre, Fl. Ecuador 14: 20 (1981).
Ecuador. 83 ECU. Nanophan. or cham.

Burmeistera dukei Wilbur, Ann. Missouri Bot. Gard. 63: 601 (1977).
Panama. 80 PAN. Cham.

Burmeistera estrellana E. Wimm., Repert. Spec. Nov. Regni Veg. 30: 24 (1932).
Costa Rica. 80 COS. Cham.

Burmeistera fimbriata Lammers, Novon 12: 209 (2002).
Colombia. 83 CLM. Cham.

Burmeistera formosa (E. Wimm.) Jeppesen in Harling & Sparre, Fl. Ecuador 14: 22 (1981).
Ecuador. 83 ECU. Nanophan.
  * *Centropogon formosus* E. Wimm., Repert. Spec. Nov. Regni Veg. 22: 200 (1926).

Burmeistera fruticosa E. Wimm., Pflanzenr. IV.276c: 838 (1968).
Colombia. 83 CLM. Nanophan.

Burmeistera fuscoapicata E. Wimm., Repert. Spec. Nov. Regni Veg. 29: 55 (1931), as
'fusco-apicata'.
Colombia. 83 CLM. Cham.

Burmeistera glabrata (Kunth) Benth. & Hook. f. ex B. D. Jacks., Index Kew. 1: 361 (1893).
Colombia & Ecuador. 83 CLM ECU. (Cl.) nanophan. or cham.
  * *Lobelia glabrata* Kunth. in Humb., Bonpl. & Kunth, Nov. Gen. Sp. 3: 307 (quarto), 240
    (folio) (1819). *Siphocampylus glabratus* (Kunth) G. Don, Gen. Hist. 3: 702 (1834).
    Centropogon glabratus (Kunth) Planch. & Oerst., Vidensk. Meddel. Dansk. Naturhist.
    Foren. Kjøbenhavn 1857: 157 (1857).
  Burmeistera subcrenata E. Wimm., Pflanzenr. IV.276b: 160 (1943), as 'suberenata'.

Burmeistera glauca (E. Wimm.) Gleason, Bull. Torrey Bot. Club 52: 98. 1925.
Panama. 80 PAN. Nanophan. or cham.
  * *Centropogon glaucus* E. Wimm., Repert. Spec. Nov. Regni Veg. 19: 251 (1924).

Burmeistera globosa E. Wimm., Repert Spec. Nov. Regni Veg. 29: 54 (1931).
Colombia. 83 CLM. Nanophan.

Burmeistera hammelii Wilbur, Ann. Missouri Bot. Gard. 68: 168 (1981).
Panama. 80 PAN. Cham.

Burmeistera hippobromoides Lammers, Novon 12: 210 (2002).
Colombia. 83 CLM. Nanophan.

Burmeistera holm-nielsenii Jeppesen in Harling & Sparre, Fl. Ecuador 14: 35 (1981).
Ecuador. 83 ECU. Nanophan. or cl. cham.

Burmeistera huacamayensis Jeppesen in Harling & Sparre, Fl. Ecuador 14: 22 (1981).
Ecuador. 83 ECU. Nanophan.

Burmeistera ibaguensis Triana, Nuev. Gen. Esp.: 13 (1855).
Colombia. 83 CLM. Nanophan.

Burmeistera ignimontis E. Wimm., Repert. Spec. Nov. Regni Veg. 30: 17 (1932).
Ecuador. 83 ECU. Nanophan. or cham.
  Burmeistera ovalis E. Wimm., Repert. Spec. Nov. Regni Veg. 30: 18 (1932). *Burmeistera
  ignimontis* var. *ovalis* (E. Wimm.) E. Wimm., Pflanzenr. IV.276b: 131 (1943).
  Burmeistera cuyujensis Jeppesen in Harling & Sparre, Fl. Ecuador 14: 32 (1981).

Burmeistera intii Gómez-Laur. & L. D. Gómez, Brenesia 25-26: 311 (1988).
Costa Rica. 80 COS. Nanophan. or cham.

Burmeistera kalbreyeri E. Wimm., Repert. Spec. Nov. Regni Veg. 30: 12 (1932).
Colombia. 83 CLM. Cham.

Burmeistera killipii Gleason, Bull. Torrey Bot. Club 52: 99 (1925).
Colombia. 83 CLM. Nanophan.

Burmeistera kirkbridei Wilbur, Ann. Missouri Bot. Gard. 63: 603 (1977).
Panama. 80 PAN. Nanophan.

Burmeistera knaphusii Lammers, Novon 12: 212 (2002).
Ecuador. 83 ECU. Cl. phan.

Burmeistera lacerata H. Karst., Linnaea 28: 445 (1856).
Colombia. 83 CLM. Nanophan.

Burmeistera loejtnantii Jeppesen in Harling & Sparre, Fl. Ecuador 14: 23 (1981).
Ecuador. 83 ECU. Nanophan.

Burmeistera longifolia Gleason, Bull. Torrey Bot. Club 52: 97 (1925).
Colombia. 83 CLM. Cham.

Burmeistera lutosa E. Wimm., Repert. Spec. Nov. Regni Veg. 29: 56 (1931).
Ecuador. 83 ECU. Nanophan. or cham.
  Burmeistera villosula E. Wimm., Repert. Spec. Nov. Regni Veg. 29: 57 (1931).

Burmeistera maculata E. Wimm., Pflanzenr. IV.276b: 160 (1943).
Colombia. 83 CLM. Nanophan.

Burmeistera marginata H. Karst., Linnaea 28: 445 (1856).
Colombia. 83 CLM. Nanophan.

Burmeistera mcvaughii Wilbur, Ann. Missouri Bot. Gard. 68: 169 (1981).
Panama. 80 PAN. Nanophan.

Burmeistera microphylla Donn. Sm., Bot. Gaz. (Crawfordsville) 25: 146 (1898).
Costa Rica & Panama; Ecuador & Peru. 80 COS PAN 83 ECU PER. Nanophan. or cham.
Centropogon aurobarbatus E. Wimm., Repert. Spec. Nov. Regni Veg. 19: 251 (1924).
Burmeistera aurobarbata (E. Wimm.) E. Wimm., Repert. Spec. Nov. Regni Veg. 30: 35 (1932).
Burmeistera aurobarbata var. *valdecuspidata* Suesseng., Bot. Jahrb. Syst. 72: 287 (1942).
Burmeistera aurobarbata var. *cuspidata* E. Wimm., Pflanzenr. IV.276b: 151 (1943).

Burmeistera montipomum E. Wimm., Pflanzenr. IV.276b: 137 (1943).
Colombia. 83 CLM. Cl. phan. or cl. nanophan.

Burmeistera morii Wilbur, Ann. Missouri Bot. Gard. 63: 605 (1977).
Panama. 80 PAN. Nanophan. or cham.

Burmeistera multiflora Zahlbr., Repert. Spec. Nov. Regni Veg. 13: 530 (1915).
Colombia & Ecuador. 83 CLM ECU. Nanophan. or cham.
Burmeistera armadana E. Wimm., Repert. Spec. Nov. Regni Veg. 29: 54 (1931).

Burmeistera multipinnatisecta Lozano & Galeano, Caldasia 15: 53 (1986).
Colombia. 83 CLM. Cham.

Burmeistera mutisiana (Kunth) E. Wimm., Repert. Spec. Nov. Regni Veg. 29: 52 (1931).
Colombia. 83 CLM. Nanophan.
* *Lobelia mutisiana* Kunth in Hunb., Bonpl. & Kunth, Nov. Gen. Sp. 3: 308 (quarto), 240 (folio) (1819). *Centropogon mutisianus* (Kunth) Gleason, Bull. Torrey Bot. Club 51: 446 (1924). *Siphocampylus mutisianus* (Kunth) G. Don, Gen. Hist. 3: 703 (1834).

Burmeistera nuda (E. Wimm.) E. Wimm., Pflanzenr. IV.276b: 142 (1943).
Colombia. 83 CLM. Nanophan.
* *Burmeistera xerampelina* var. *nuda* E. Wimm., Repert. Spec. Nov. Regni Veg. 29: 59 (1931).

Burmeistera oblongifolia E. Wimm., Brittonia 8: 108 (1955).
Ecuador. 83 ECU. Nanophan.

Burmeistera obtusifolia E. Wimm., Repert. Spec. Nov. Regni Veg. 29: 58 (1931).
Costa Rica & Panama. 80 COS PAN. Nanophan.

Burmeistera ondocarpa E. Wimm., Pflanzenr. IV.276b: 160 (1943).
Colombia. 83 CLM. Nanophan.

Burmeistera oyacachensis Jeppesen in Harling & Sparre, Fl. Ecuador 14: 40 (1981).
Ecuador. 83 ECU. Nanophan.

Burmeistera pallida (Drake) E. Wimm., Repert. Spec. Nov. Regni Veg. 30: 50 (1932).
Venezuela, Ecuador, Peru. 82 VEN 83 ECU PER. (Cl.) nanophan.
* *Centropogon pallidus* Drake, J. Bot. (Morot) 3: 239 (1889).
Burmeistera ramosa E. Wimm., Repert. Spec. Nov. Regni Veg. 30: 16 (1932). *Burmeistera ramosa* var. *campii* E. Wimm., Brittonia 8: 108 (1955).

Burmeistera panamensis Wilbur, Ann. Missouri Bot. Gard. 63: 605 (1977).
Panama. 80 PAN. Nanophan.

Burmeistera parviflora E. Wimm. ex Standl., Fl. of Costa Rica: 1408 (1938).
Costa Rica & Panama. 80 COS PAN. Nanophan. or cham.

Burmeistera pastoralis E. Wimm., Repert. Spec. Nov. Regni Veg. 29: 58 (1931).
Colombia. 83 CLM. Nanophan.
>   *Burmeistera pastoralis* var. *glabrifolia* E. Wimm., Repert. Spec. Nov. Regni Veg. 29: 59
>   (1931).

Burmeistera pennellii Gleason, Bull. Torrey Bot. Club 52: 100 (1925).
Colombia. 83 CLM. Cham.

Burmeistera pinnatisecta Luteyn, Syst. Bot. 11: 474 (1986).
Colombia. 83 CLM. Nanophan. or cham.

Burmeistera pirrensis Wilbur, Ann. Missouri Bot. Gard. 63: 607 (1977).
Panama. 80 PAN. Nanophan. or cham.

Burmeistera pomifera H. Karst., Linnaea 28: 445 (1856).
Colombia. 83 CLM. Nanophan. or cham.

Burmeistera prunifolia McVaugh, J. Washington Acad. Sci. 39: 162 (1949).
Colombia. 83 CLM. Nanophan.

Burmeistera pteridoides McVaugh, Ann. Missouri Bot. Gard. 52: 400 (1965).
Colombia. 83 CLM. Nanophan. or cham.

Burmeistera puberula E. Wimm., Repert. Spec. Nov. Regni Veg. 38: 3 (1935).
Colombia. 83 CLM. Cham.

Burmeistera racemiflora Lammers, Brittonia 50: 258 (1998).
Ecuador. 83 ECU. Phan. or nanophan.

Burmeistera refracta E. Wimm., Repert. Spec. Nov. Regni Veg. 30: 11 (1932).
Ecuador & Peru. 83 ECU PER. Nanophan. or cham.

Burmeistera resupinata Zahlbr., Repert. Spec. Nov. Regni Veg. 13: 530 (1915).
Ecuador. 83 ECU. Nanophan. or cham.
>   *Burmeistera resupinata* var. *heilbornii* E. Wimm., Pflanzenr. IV.276b: 140 (1943).

Burmeistera rivina E. Wimm., Repert. Spec. Nov. Regni Veg. 38: 4 (1935).
Colombia. 83 CLM. Cham.

Burmeistera rostrata Jeppesen in Harling & Sparre, Fl. Ecuador 14: 25 (1981).
Columbia. 83 CLM. Nanophan.

Burmeistera rubrosepala (E. Wimm.) E. Wimm., Repert. Spec. Nov. Regni Veg. 30: 10 (1932).
Ecuador. 83 ECU. Nanophan.
>   * *Centropogon rubrosepalus* E. Wimm., Repert. Spec. Nov. Regni Veg. 19: 252 (1924).

Burmeistera smaragdi Lammers, Novon 12: 213 (2002).
Ecuador. 83 ECU. Nanophan.

Burmeistera sodiroana Zahlbr., Repert. Spec. Nov. Regni Veg. 13: 534. 1915.
Ecuador. 83 ECU. Nanophan. or cham.
>   *Burmeistera leucocarpa* Zahlbr., Repert. Spec. Nov. Regni Veg. 13: 529 (1915).
>   *Burmeistera leucocarpa* var. *dentata* E. Wimm., Repert. Spec. Nov. Regni Veg. 30: 25 (1932).

Burmeistera succulenta H. Karst., Linnaea 28: 445 (1856).
Venezuela to Peru. 82 VEN 83 CLM ECU PER. (Cl.) nanophan. or (cl.) cham.

Burmeistera succulenta var. *latisepala* E. Wimm., Repert. Spec. Nov. Regni Veg. 29: 55 (1931).
Burmeistera succulenta var. *meiophylla* Zahlbr. ex E. Wimm., Pflanzenr. IV.276b: 143 (1943).

Burmeistera sylvicola Zahlbr., Repert. Spec. Nov. Regni Veg. 13: 532 (1915).
Colombia. 83 CLM. Cham.

Burmeistera tambensis E. Wimm., Pflanzenr. IV.276b: 155 (1943).
Colombia. 83 CLM. Nanophan. or cham.

Burmeistera tenuiflora Donn. Sm., Bot. Gaz. (Crawfordsville) 25: 147 (1898).
Costa Rica & Panama. 80 COS PAN. Nanophan. or cham.

Burmeistera tomentosula H. Karst., Linnaea 28: 445 (1856).
Colombia. 83 CLM. Nanophan.

Burmeistera toroensis Wilbur, Ann. Missouri Bot. Gard. 63: 608 (1977).
Panama. 80 PAN. Nanophan. or cham.

Burmeistera truncata Zahlbr., Repert. Spec. Nov. Regni Veg. 13: 531 (1915).
Ecuador. 83 ECU. Nanophan.

Burmeistera utleyi Wilbur, Ann. Missouri Bot. Gard. 63: 609 (1977).
Panama. 80 PAN. Nanophan. or cham.

Burmeistera variabilis (Gleason) E. Wimm., Repert. Spec. Nov. Regni Veg. 29: 52 (1931).
Colombia. 83 CLM. Cham.
  * *Centropogon variabilis* Gleason, Bull. Torrey Bot. Club 51: 446 (1924).
  Centropogon gravidus Gleason, Bull. Torrey Bot. Club 51: 447 (1924). *Burmeistera gravida* (Gleason) E. Wimm., Repert. Spec. Nov. Regni Veg. 29: 52 (1931). *Burmeistera variabilis* var. *gravida* (Gleason) E. Wimm., Pflanzenr. IV.276b: 158 (1943).

Burmeistera venezuelensis Lammers, Brittonia 50: 260 (1998).
Venezuela. 82 VEN. Cham.

Burmeistera virescens (Benth.) Benth. & Hook. ex Hemsl., Biol. Centr.-Amer., Bot. 2: 263 (1881).
Guatemala & Honduras. 80 GUA HON. Nanophan. or cham.
  * *Lobelia virescens* Benth., Pl. Hartweg.: 77 (1841). *Centropogon virescens* (Benth.) Planch. & Oerst., Vidensk. Meddel. Dansk Naturhist. Foren. Kjøbenhavn 1857: 157 (1857).
  Centropogon lignescens E. Wimm., Repert. Spec. Nov. Regni Veg. 19: 252 (1924). *Burmeistera lignescens* (E. Wimm.) E. Wimm., Repert. Spec. Nov. Regni Veg. 30: 34 (1932).
  Burmeistera lignescens var. *martialis* E. Wimm., Repert. Spec. Nov. Regni Veg. 30: 34 (1932).
  Burmeistera virescens var. *subdentata* E. Wimm., Pflanzenr. IV.276b: 135 (1943).

Burmeistera vulgaris E. Wimm., Repert. Spec. Nov. Regni Veg. 30: 27 (1932).
Costa Rica to Ecuador. 80 COS PAN 83 CLM ECU. Nanophan. or cham.
  Burmeistera macrocalyx E. Wimm., Repert. Spec. Nov. Regni Veg. 30: 19 (1932).
  Burmeistera pittieri E. Wimm., Repert. Spec. Nov. Regni Veg. 30: 19 (1932).
  Burmeistera pittieri var. *decorans* E. Wimm., Repert. Spec. Nov. Regni Veg. 30: 20 (1932).

Burmeistera xerampelina E. Wimm., Repert. Spec. Nov. Regni Veg. 29: 59 (1931).
Colombia. 83 CLM. Cham.
  Burmeistera xerampelina f. *lilacina* E. Wimm., Pflanzenr. IV.276b: 147. 1943.

Burmeistera zurquiensis Wilbur, Bull. Torrey Bot. Club 102: 230 (1976).
   Costa Rica. 80 COS. Nanophan. or cham.

Synonyms:
Burmeistera armadana E. Wimm. === **Burmeistera multiflora** Zahlbr.
Burmeistera asteriscus E. Wimm. === **Centropogon peruvianus** (E. Wimm.) McVaugh
Burmeistera aurobarbata (E. Wimm.) E. Wimm. === **Burmeistera microphylla** Donn. Sm.
Burmeistera aurobarbata var. *cuspidata* E. Wimm. === **Burmeistera microphylla** Donn. Sm.
Burmeistera aurobarbata var. *valdecuspidata* Suesseng. === **Burmeistera microphylla** Donn. Sm.
Burmeistera carnosa var. *metastasia* E. Wimm. === **Burmeistera carnosa** Gleason
Burmeistera ceratocarpa var. *dentata* E. Wimm. === **Burmeistera ceratocarpa** Zahlbr.
Burmeistera cerasifera Zahlbr. === **Burmeistera cyclostigmata** Donn. Sm.
Burmeistera crassifolia var. *ovatifolia* (E. Wimm.) E. Wimm. === **Burmeistera crassifolia** (E.
   Wimm.) E. Wimm.
Burmeistera cuyujensis Jeppesen === **Burmeistera ignimontis** E. Wimm.
Burmeistera cyclostigmata var. *suerrensis* Donn. Sm. === **Burmeistera cyclostigmata** Donn. Sm.
Burmeistera cylindrocarpa var. *candida* E. Wimm. === **Burmeistera cyclostigmata** Donn. Sm
Burmeistera gravida (Gleason) E. Wimm. === **Burmeistera variabilis** (Gleason) E. Wimm.
Burmeistera ignimontis var. *ovalis* (E. Wimm.) E. Wimm. === **Burmeistera ignimontis** E. Wimm.
Burmeistera leucocarpa Zahlbr. === **Burmeistera sodiroana** Zahlbr.
Burmeistera leucocarpa var. *dentata* E. Wimm. === **Burmeistera sodiroana** Zahlbr.
Burmeistera lignescens (E. Wimm.) E. Wimm. === **Burmeistera virescens** (Benth.) Benth. &
   Hook. ex Hemsl.
Burmeistera lignescens var. *martialis* E. Wimm. === **Burmeistera virescens** (Benth.) Benth. &
   Hook. ex Hemsl.
Burmeistera luteola E. Wimm. === **Burmeistera connivens** Gleason
Burmeistera macrocalyx E. Wimm. === **Burmeistera vulgaris** E. Wimm.
Burmeistera macrocarpa (Zahlbr.) E. Wimm. === **Centropogon macrocarpus** Zahlbr.
Burmeistera matthiaei (A. DC.) Benth. & Hook. f. ex B. D. Jacks. === **Siphocampylus
   matthiaei** A. DC.
Burmeistera millei E. Wimm. === **Burmeistera cyclostigmata** Donn. Sm.
Burmeistera mindoana E. Wimm. === **Burmeistera cyclostigmata** Donn. Sm.
Burmeistera montana E. Wimm. === **Burmeistera crispiloba** Zahlbr.
Burmeistera ovalis E. Wimm. === **Burmeistera ignimontis** E. Wimm.
Burmeistera pastoralis var. *glabrifolia* E. Wimm. === **Burmeistera pastoralis** E. Wimm.
Burmeistera peruviana E. Wimm. === **Centropogon peruvianus** (E. Wimm.) McVaugh
Burmeistera pittieri E. Wimm. === **Burmeistera vulgaris** E. Wimm.
Burmeistera pittieri var. *decorans* E. Wimm. === **Burmeistera vulgaris** E. Wimm.
Burmeistera ramosa E. Wimm. === **Burmeistera pallida** (Drake) E. Wimm.
Burmeistera ramosa var. *campii* E. Wimm. === **Burmeistera pallida** (Drake) E. Wimm.
Burmeistera resupinata var. *heilbornii* E. Wimm. === **Burmeistera resupinata** Zahlbr.
Burmeistera splendens E. Wimm. === **Siphocampylus splendens** (E. Wimm.) Jeppesen ex B.
   A. Stein
Burmeistera subcrenata E. Wimm. === **Burmeistera glabrata** (Kunth) Benth. & Hook. f. ex B.
   D. Jacks.
Burmeistera succulenta var. *breviloba* E. Wimm. === **Burmeistera crispiloba** Zahlbr.
Burmeistera succulenta var. *latisepala* E. Wimm. === **Burmeistera succulenta** H. Karst.
Burmeistera succulenta var. *meiophylla* Zahlbr. ex E. Wimm. === **Burmeistera succulenta**
   H. Karst.
Burmeistera suerrensis (Donn. Sm.) E. Wimm. === **Burmeistera cyclostigmata** Donn. Sm.
Burmeistera suerrensis var. *almirantensis* E. Wimm. === **Burmeistera cyclostigmata** Donn. Sm.
Burmeistera tricolorata E. Wimm. === **Siphocampylus rusbyanus** Britt.
Burmeistera variabilis var. *gravida* (Gleason) E. Wimm. === **Burmeistera variabilis** (Gleason)
   E. Wimm.
Burmeistera villosula E. Wimm. === **Burmeistera lutosa** E. Wimm.

Burmeistera virescens var. *subdentata* E. Wimm. === **Burmeistera virescens** (Benth.) Benth. &
 Hook. ex Hemsl.
Burmeistera weberbaueri Zahlbr. === **Siphocampylus oscitans** B. A. Stein
Burmeistera xerampelina f. *lilacina* E. Wimm. === **Burmeistera xerampelina** E. Wimm.
Burmeistera xerampelina var. *nuda* E. Wimm. === **Burmeistera nuda** (E. Wimm.) E. Wimm.

Byrsanthes

The name *Byrsanthes,* like *Canonanthus* (q.v.), applies to the small group of species treated by
most authors as *Siphocampylus* subsect. *Nivei* McVaugh. Although it has been formally
rejected to protect *Byrsanthus* Guill. (Flacourtiaceae), this seems unnecessary under the
current ICBN (Art. 53 Ex. 10). Type [designated by Pfeiff., Nomencl. Bot. 1: 509 (1873)]:
Lobelia nivea Willd.

> Presl, C. *Byrsanthes.* In Prodromus monographiae Lobeliacearum: 41-42. Prague:
> Theophilus Haase. La. — Establishment of genus as distinct from *Lobelia.*
> McVaugh, R. (1949). Studies in South American Lobelioideae (Campanulaceae) with
> special reference to Colombian species. II. A new subsection of *Siphocampylus* sect.
> *Eusiphocampylus.* Brittonia 6: 453-458. En. — Treatment of group at infrageneric rank,
> with key, descriptions, full nomenclature, and specimen citations.

Synonyms:
Byrsanthes C. Presl === **Siphocampylus** Pohl
Byrsanthes halliana C. Presl === **Siphocampylus affinis** (Mirb.) McVaugh
Byrsanthes humboldtiana C. Presl === **Siphocampylus niveus** (Willd. ex Schult.) Vatke
Byrsanthes nivea (Willd. ex Schult.) A. DC. === **Siphocampylus niveus** (Willd. ex Schult.)
 Vatke

Calcaratolobelia

These species for many years resided in *Heterotoma* (q.v.), a genus defined by the presence of
a nectar spur. Ayers (1990) provided morphological evidence that the genus was biphyletic
with this circumscription, and removed all species but the type to *Lobelia.* Wilbur (1997)
disagreed with the inclusion of spurred species in *Lobelia,* and erected this genus to house
them. A more recent but very preliminary analysis (Koopman & Ayers 2005) suggests that
all spurred Lobelioideae form a clade embedded among species of *Lobelia.* Type [designated
by the author]: *Calcaratolobelia cordifolia* (Hook. & Arn.) Wilbur.

> Ayers, T. J. (1990). Systematics of *Heterotoma* (Campanulaceae) and the evolution of nectar
> spurs in the New World Lobelioideae. Syst. Bot. 15: 296-327, illus., maps. En. — Removal
> of these species from *Heterotoma* to *Lobelia,* with key and descriptions for all species.
> Wilbur, R. L. (1997). *Calcaratolobelia* (Campanulaceae): a new genus of spurred lobelioids
> from Mexico and Central America. Sida 17: 555-564. En. — Establishment of genus as
> distinct from *Heterotoma* and *Lobelia.*
> Koopman, M. M. & T. J. Ayers (2005). Nectar spur evolution in the Mexican lobelias
> (Campanulaceae: Lobelioideae). Amer. J. Bot. 92: 558-562. En. — Preliminary
> molecular phylogeny of spurred Lobelioideae.

Synonyms:
Calcaratolobelia Wilbur === **Lobelia** L.
Calcaratolobelia aurita (Brandegee) Wilbur === **Lobelia aurita** (Brandegee) T. J. Ayers
Calcaratolobelia cordifolia (Hook. & Arn.) Wilbur === **Lobelia cordifolia** Hook. & Arn.
Calcaratolobelia flexuosa (C. Presl) Wilbur === **Lobelia flexuosa** (C. Presl) A. DC.
Calcaratolobelia flexuosa var. *intermedia* (Hemsl.) Wilbur === **Lobelia flexuosa** subsp.
 intermedia (Hemsl.) Lammers

Calcaratolobelia gibbosa (S. Watson) Wilbur === **Lobelia endlichii** (E. Wimm.) T. J. Ayers
Calcaratolobelia goldmanii (Fern.) Wilbur === **Lobelia goldmanii** (Fern.) T. J. Ayers
Calcaratolobelia knoblochii (T. J. Ayers) Wilbur === **Lobelia knoblochii** T. J. Ayers
Calcaratolobelia macrocentron (Benth.) Wilbur === **Lobelia macrocentron** (Benth.) T. J. Ayers
Calcaratolobelia margarita (E. Wimm.) Wilbur === **Lobelia margarita** E. Wimm.
Calcaratolobelia mcvaughii (T. J. Ayers) Wilbur === **Lobelia mcvaughii** T. J. Ayers
Calcaratolobelia pringlei (B. L. Rob.) Wilbur === **Lobelia gypsophila** T. J. Ayers
Calcaratolobelia tenella (Turcz.) Wilbur === **Lobelia volcanica** T. J. Ayers
Calcaratolobelia villaregalis (T. J. Ayers) Wilbur === **Lobelia villaregalis** T. J. Ayers

Campanopsis

In Kuntze's (1891, 1898) absolutist view, the epithet *Campanopsis* had priority over *Wahlenbergia*, despite its difference in rank; that view is not sanctioned under the ICBN (Art. 11.2). Based on: *Campanula* sect. *Campanopsis* R. Br. Type [designated by Lammers, Taxon 44: 334 (1995)]: *Campanula gracilis* G. Forst.

> Kuntze, O. (1891). *Campanopsis.* In Revisio generum plantarum vascularium omnium atque cellularium multarum secundum leges nomenclaturae internationales 2: 378-379. Leipzig: Arthur Felix. Ge. — Establishment of genus.
> Kuntze, O. (1898). *Campanopsis.* In Revisio generum plantarum vascularium omnium atque cellularium multarum secundum leges nomenclaturae internationales 3(2): 185-186. Leipzig: Arthur Felix. Ge. — Supplement to Kuntze (1891).

Synonyms:
Campanopsis (R. Br.) Kuntze === **Wahlenbergia** Schrad. ex Roth
Campanopsis sect. *Cervicina* (Delile) Kuntze === **Wahlenbergia** Schrad. ex Roth
Campanopsis sect. *Edraianthus* (A. DC.) Kuntze === **Edraianthus** A. DC.
Campanopsis acaulis (E. Mey. ex A. DC.) Kuntze === **Wahlenbergia acaulis** E. Mey. ex A. DC.
Campanopsis adpressa (L. f.) Kuntze === **Wahlenbergia adpressa** (L. f.) Sond.
Campanopsis androsacea (A. DC.) Kuntze === **Wahlenbergia androsacea** A. DC.
Campanopsis androsacea var. *arenaria* (A. DC.) Kuntze === **Wahlenbergia androsacea** A. DC.
Campanopsis angustifolia (Roxb.) Kuntze === **Wahlenbergia angustifolia** (Roxb.) A. DC.
Campanopsis annularis (A. DC.) Kuntze === **Wahlenbergia annularis** A. DC.
Campanopsis arenaria (A. DC.) Kuntze === **Wahlenbergia androsacea** A. DC.
Campanopsis arguta (Hook. f.) Kuntze === **Wahlenbergia krebsii** subsp. **arguta** (Hook. f.) Thulin
Campanopsis banksiana (A. DC.) Kuntze === **Wahlenbergia banksiana** A. DC.
Campanopsis berteroi (Hook. & Arn.) Kuntze === **Wahlenbergia berteroi** Hook. & Arn.
Campanopsis bojeri (A. DC.) Kuntze === **Wahlenbergia undulata** (L. f.) A. DC.
Campanopsis bowkerae (Sond.) Kuntze === **Wahlenbergia bowkerae** Sond.
Campanopsis brasiliensis (Cham.) Kuntze === **Wahlenbergia brasiliensis** Cham.
Campanopsis burchelii (A. DC.) Kuntze === **Wahlenbergia burchelii** A. DC.
Campanopsis caffra (A. DC.) Kuntze === **?**
Campanopsis caledonica (Sond.) Kuntze === **Wahlenbergia undulata** (L. f.) A. DC.
Campanopsis campanuloides (Delile) Kuntze === **Wahlenbergia campanuloides** (Delile) Vatke
Campanopsis capensis (L.) Kuntze === **Wahlenbergia capensis** (L.) A. DC.
Campanopsis capillacea (L. f.) Kuntze === **Wahlenbergia capillacea** (L. f.) A. DC.
Campanopsis cartilaginea (Hook. f.) Kuntze === **Wahlenbergia cartilaginea** Hook. f.
Campanopsis cernua (Thunb.) Kuntze === **Wahlenbergia cernua** (Thunb.) A. DC.
Campanopsis costata (A. DC.) Kuntze === **Wahlenbergia costata** A. DC.
Campanopsis dalmatica (A. DC.) Kuntze === **Edraianthus dalmaticus** (A. DC.) A. DC.
Campanopsis debilis (H. Buek) Kuntze === **Wahlenbergia debilis** H. Buek
Campanopsis decipiens (A. DC.) Kuntze === **Wahlenbergia decipiens** A. DC.

Campanopsis denudata (A. DC.) Kuntze === **Wahlenbergia denudata** A. DC.

Campanopsis dicentrifolia (C. B. Clarke) Kuntze === **Codonopsis dicentrifolia** (C. B. Clarke) W. W. Sm.

Campanopsis dichotoma (A. DC.) Kuntze === **Wahlenbergia dichotoma** A. DC.

Campanopsis divergens (A. DC.) Kuntze === **Wahlenbergia divergens** A. DC.

Campanopsis ecklonii (H. Buek) Kuntze === **Wahlenbergia ecklonii** H. Buek

Campanopsis emirnensis (A. DC.) Kuntze === **Gunillaea emirnensis** (A. DC.) Thulin

Campanopsis epacridea (Sond.) Kuntze === **Wahlenbergia epacridea** Sond.

Campanopsis etbaica (Schweinf.) Kuntze === **Wahlenbergia lobelioides** subsp. **nutabunda** (Guss.) Murb.

Campanopsis exilis (A. DC.) Kuntze === **Wahlenbergia exilis** A. DC.

Campanopsis fernandeziana (A. DC.) Kuntze === **Wahlenbergia fernandeziana** A. DC.

Campanopsis flaccida (A. DC.) Kuntze === **?**

Campanopsis graminifolia (L.) Kuntze === **Edraianthus graminifolius** (L.) A. DC. subsp. **graminifolia**

Campanopsis hederacea (L.) Kuntze === **Wahlenbergia hederacea** (L.) Rchb.

Campanopsis hilsenbergii (A. DC.) Kuntze === **Wahlenbergia madagascariensis** A. DC.

Campanopsis homallanthina (Ledeb.) Kuntze === **Campanula expansa** Rudolph

Campanopsis huillana (A. DC.) Kuntze === **Gunillaea emirnensis** (A. DC.) Thulin

Campanopsis humilis (A. DC.) Kuntze === **Wahlenbergia lobelioides** subsp. **riparia** (A. DC.) Thulin

Campanopsis inconspicua (A. DC.) Kuntze === **?**

Campanopsis ingrata (A. DC.) Kuntze === **Wahlenbergia ingrata** A. DC.

Campanopsis inhambanensis (Klotzsch) Kuntze === **Wahlenbergia androsacea** A. DC.

Campanopsis kitaibelii (A. DC.) Kuntze === **Edraianthus graminifolius** (L.) A. DC.

Campanopsis krebsii (Cham.) Kuntze === **Wahlenbergia krebsii** Cham.

Campanopsis linarioides (Lam.) Kuntze === **Wahlenbergia linarioides** (Lam.) A. DC.

Campanopsis linarioides var. *longifolia* Kuntze === **Wahlenbergia linarioides** (Lam.) A. DC.

Campanopsis linifolia (Roxb.) Kuntze === **Wahlenbergia linifolia** (Roxb.) A. DC.

Campanopsis lobelioides (L. f.) Kuntze === **Wahlenbergia lobelioides** (L. f.) Schrad. ex Link

Campanopsis madagascariensis (A. DC.) Kuntze === **Wahlenbergia madagascariensis** A. DC.

Campanopsis mannii (Vatke) Kuntze === **Wahlenbergia silenoides** Hochst. ex A. Rich.

Campanopsis marginata (Thunb. ex Murray) Kuntze === **Wahlenbergia marginata** (Thunb. ex Murray) A. DC.

Campanopsis marginata var. *polymorpha* Kuntze === **Wahlenbergia marginata** (Thunb. ex Murray) A. DC.

Campanopsis marginata var. *rigida* Kuntze === **Wahlenbergia marginata** (Thunb. ex Murray) A. DC.

Campanopsis meyeri (A. DC.) Kuntze === **Wahlenbergia meyeri** A. DC.

Campanopsis montana (A. DC.) Kuntze === **Craterocapsa montana** (A. DC.) Hilliard & B. L. Burtt

Campanopsis namaquana (Sond.) Kuntze === **Wahlenbergia namaquana** Sond.

Campanopsis nudicaulis (A. DC.) Kuntze === **Wahlenbergia androsacea** A. DC.

Campanopsis oocarpa (Sond.) Kuntze === **Wahlenbergia oocarpa** Sond.

Campanopsis oppositifolia (A. DC.) Kuntze === **Wahlenbergia madagascariensis** A. DC.

Campanopsis oxyphylla (A. DC.) Kuntze === **Wahlenbergia oxyphylla** A. DC.

Campanopsis paniculata (L. f.) Kuntze === **Wahlenbergia paniculata** (L. f.) A. DC.

Campanopsis patula (A. DC.) Kuntze === **Wahlenbergia patula** A. DC.

Campanopsis pauciflora (A. DC.) Kuntze === **Wahlenbergia pauciflora** A. DC.

Campanopsis peduncularis (Wall. ex A. DC.) Kuntze === **Wahlenbergia peduncularis** (Wall. ex A. DC.) Hook. f. & Thomson

Campanopsis pilosa (H. Buek) Kuntze === **Wahlenbergia pilosa** H. Buek

Campanopsis procumbens (L. f.) Kuntze === **Wahlenbergia procumbens** (L. f.) A. DC.

Campanopsis procumbens var. *bolusiana* Kuntze === **?**

Campanopsis procumbens var. *oppositifolia* (A. DC.) Kuntze === **Wahlenbergia madagascariensis** A. DC.

Campanopsis prostrata (E. Mey. ex A. DC.) Kuntze === **Wahlenbergia prostrata** E. Mey. ex A. DC.

Campanopsis pumilio (Port. ex Schult.) Kuntze === **Edraianthus pumilio** (Port. ex Schult.) A. DC.

Campanopsis pusilla (Hochst. ex A. Rich.) Kuntze === **Wahlenbergia pusilla** Hochst. ex A. Rich.

Campanopsis ramulosa (E. Mey. ex A. DC.) Kuntze === **Wahlenbergia debilis** H. Buek

Campanopsis riparia (A. DC.) Kuntze === **Wahlenbergia lobelioides** subsp. **riparia** (A. DC.) Thulin

Campanopsis robusta (A. DC.) Kuntze === **Wahlenbergia robusta** (A. DC.) Sond.

Campanopsis roxburghii (A. DC.) Kuntze === **Wahlenbergia roxburghii** A. DC.

Campanopsis saxicola (R. Br.) Kuntze === **Wahlenbergia saxicola** (R. Br.) A. DC.

Campanopsis serpyllifolia (Vis.) Kuntze === **Edraianthus serpyllifolius** (Vis.) A. DC.

Campanopsis silenoides (Hochst. ex A. Rich.) Kuntze === **Wahlenbergia silenoides** Hochst. ex A. Rich.

Campanopsis spinulosa (A. DC.) Kuntze === ?

Campanopsis stellarioides (Cham.) Kuntze === **Wahlenbergia stellarioides** Cham.

Campanopsis stellarioides var. *angusta* (Sond.) Kuntze === **Wahlenbergia stellarioides** Cham.

Campanopsis stellarioides var. *integrifolia* (A. DC.) Kuntze === **Wahlenbergia stellarioides** Cham.

Campanopsis tenuifolia (Waldst. & Kit.) Kuntze === **Edraianthus tenuifolius** (Waldst. & Kit.) A. DC.

Campanopsis tenuis (A. DC.) Kuntze === **Wahlenbergia tenuis** A. DC.

Campanopsis tuberosa (Hook. f.) Kuntze === **Wahlenbergia tuberosa** Hook. f.

Campanopsis undulata (L. f.) Kuntze === **Wahlenbergia undulata** (L. f.) A. DC.

Campanopsis undulata f. *albiflora* Kuntze === **Wahlenbergia undulata** (L. f.) A. DC.

Campanopsis undulata f. *azurea* Kuntze === **Wahlenbergia undulata** (L. f.) A. DC.

Campanopsis undulata f. *flavida* Kuntze === **Wahlenbergia undulata** (L. f.) A. DC.

Campanopsis undulata f. *lilacina* Kuntze === **Wahlenbergia undulata** (L. f.) A. DC.

Campanopsis wyleyana (Sond.) Kuntze === **Wahlenbergia wyleyana** Sond.

Campanopsis zeyheri (H. Buek) Kuntze === **Wahlenbergia krebsii** Cham. subsp. **krebsii**

Campanula

Campanuloideae, 420 species, throughout the Arctic and North Temperate Zones and extending south into eastern Africa, southern Asia, and northern Mexico. There is no concensus regarding infrageneric classification. Numerous names are available at these ranks, but no system has been promulgated that takes into account *all* members of the genus; probably the best developed and most comprehensive is that of Damboldt (1976). In a preliminary phylogenetic analysis of nucleic acid sequence data (Eddie et al. 2003, under **General**), most of the species here assigned to *Campanula* fall into two discrete clades; each contains species of other genera. The "Campanulaceae s. str. clade" comprises the type species of *Azorina*, *Campanula*, *Edraianthus*, *Feeria*, *Roucela*, and *Trachelium*, as well as all sampled species referable to *Diosphaera*, *Hemisphaera*, *Michauxia*, and *Symphyandra*. The "Rapunculus clade" includes the type species of *Campanulastrum*, *Hanabusaya*, *Heterocodon*, *Legousia*, *Petromarula*, *Physoplexis*, and *Phyteuma*, plus all sampled species resferable to *Adenophora*, *Asyneuma*, *Githopsis*, and *Neocodon*. These two major clades form a pentachotomy with three smaller clades: (1) the type species of *Echinocodonia*, *Gadellia*, and *Musschia*, plus *Campanula peregrina;* (2) all sampled species of *Jasione* including its type, and (3) *Wahlenbergia hederacea*. Although the results of this analysis are extremely interesting and foreshadow profound changes in the classification of *Campanula*, only about 10% of the species in the genus were included. It simply would not be prudent to effect formal taxonomic changes until many more species have been sequenced. As a result, the very broad and traditional circumscription of *Campanula* employed previously (Lammers 2007, under **General**) is *per force* maintained here. There has been no comprehensive monograph since that of Candolle (1839, under **General**); however, the floristic accounts by Fedorov (1957), Shetler (1963), Fedorov & Kovanda (1976), Damboldt (1978), and Hong (1983) taken together do provide keys and descriptions for most species in the world. *2n* = 14, 16, 18, 20, 22, 24, 26, 28, 30, 32, 34, 36, 40, 46, 48, 50, 52, 54, 56, 58, 60, 68, 70, 72, 80, 84, 90, 102. Type [designated by Hitchc. & M. L. Green in Anon., Nomencl. Prop. Brit. Bot. 184 (1929)]: *Campanula latifolia* L. This same species was designated earlier [Britton & A. Br., Ill. Fl. N. US (ed. 2) 3: 294 (1913)], but that choice is superceded under ICBN Art. 10.5(b).

Linnaeus, C. (1753). *Campanula*. In Species plantarum: 163-169. Stockholm: Laurentius Salvius. La. — Establishment of genus in Pentandria Monogynia.

Linnaeus, C. (1754). *Campanula*. In Genera plantarum, ed. 5: 77. Stockholm: Laurentius Salvius. La. — Description of genus to accompany preceding.

Sims, J. (1801). *Campanula pumila*. Dwarf campanula. Bot. Mag. 15: tab. 512 + 1 pg., illus. En. — Portrait of *C. cespitosa* with description.

Sims, J. (1802). *Campanula azurea*. Azure campanula. Bot. Mag. 15: tab. 551 + 1 pg., illus. En. — Portrait of *C. rhomboidalis* with description.

Andrews, H. (1804). *Campanula laciniata*. Bot. Repos. 6: pl. 385 + 1 pg., illus. En. — Portrait with description.

Sims, J. (1806). *Campanula collina*. Sage-leaved bell-flower. Bot. Mag. 24: tab. 927 + 1 pg., illus. En. — Portrait of *C. collina* subsp. *collina* with description.

Sims, J. (1810). *Campanula thyrsoidea* [sic]. Long-spiked bell-flower. Bot. Mag. 32: tab. 1290 + 2 pp., illus. En. — Portrait of *C. thyrsoides* subsp. *thyrsoides* with description.

Sims, J. (1815). *Campanula punctata*. Spotted bell-flower. Bot. Mag. 41: tab. 1723 + 1 pg., illus. En. — Portrait with description.

Ker Gawler, J. B. (1817). *Campanula sarmatica*. Marschall's bell-flower. Bot. Reg. 3: tab. 237 + 2 pp., illus. En. — Portrait of *C. sarmatica* subsp. *sarmatica* with description.

Loddiges, C. (1821a). *Campanula scheuchzeri*. Bot. Cab. 5: pl. 485 + 1 pg., illus. En. — Portrait.

Loddiges, C. (1821b). *Campanula aggregata*. Bot. Cab. 6: pl. 505 + 1 pg., illus. En. — Portrait of *C. glomerata* subsp. *glomerata*.

Ker Gawler, J. B. (1822). *Campanula glomerata; β dahurica*. The clustered bell-flower of Dauria. Bot. Reg. 8: tab. 620 + 2 pp., illus. En. — Portrait of *C. glomerata* subsp. *speciosa* with description.

Sims, J. (1826a). *Campanula infundibuliformis*. Funnel-shaped bell-flower. Bot. Mag. 53: tab. 2632 + 2 pp., illus. En. — Portrait of *C. rapunculoides* subsp. *rapunculoides* with description.

Sims, J. (1826b). *Campanula speciosa*. Shewy bell-flower. Bot. Mag. 53: tab. 2649 + 2 pp., illus. En. — Portrait of *C. glomerata* subsp. *speciosa* with description.

Sims, J. (1826c). *Campanula ruthenica*. Taurian bellflower. Bot. Mag. 53: tab. 2653 + 2 pp., illus. En. — Portrait of *C. bononiensis* with description.

Hooker, W. J. (1834). *Campanula macrantha, β. polyantha*. Large-flowered giant bell-flower; many-blossomed var. Bot. Mag. 61: tab. 3347 + 2 pp., illus. En. — Portrait of *C. latifolia* subsp. *latifolia* with description.

von Schlechtendal, D. F. L. (1842). Ueber eine neue Gattung des Campanulaceae. Linnaea 16: 374-376. Ge. — Discovery of anonymous manuscript describing new genus "Depierrea;" Schlechtendal does not accept it, considering plant to be monstrous form of *C. rotundifolia* or an allied species.

Webb, P. B. (1848). *Campanula* (*Medium*) *jacobaea*, Chr. Sm. Hooker's Icon. Pl. 8: tab. 762 + 1 pg., illus. En. — Portrait with description.

Hooker, W. J. & J. Smith (1851). *Campanula colorata*. Deep-coloured bell-flower. Bot. Mag. 77: tab. 4555 + 2 pp., illus. En. — Portrait of *C. pallida* with description.

Planchon, J. E. (1851-52). *Campanula persicifolia* var. *coronata*. Fl. Serres Jard. Eur. 7: 151-152 + pl. 699, illus. Fr. — Portrait and discussion of a "double-flowered" mutant.

Vukotinović, L. F. (1854). Monographia generis *Campanula*. Specierum ad hocce pertinentium secundum principia historiae naturalis concinnata. Linnaea 26: 323-344. La. — Attempt to thoroughly revise classification and nomenclature of genus according to a system that seemingly disregards priority.

Hooker, J. D. (1858). *Campanula strigosa*. Strigose bell-flower. Bot. Mag. 84: tab. 5068 + 2 pp., illus. En. — Portrait with description.

Hooker, J. D. (1868). *Campanula isophylla*. Ligurian Campanula. Bot. Mag. 94: tab. 5745 + 2 pp., illus. En. — Portrait with description.

Van Houtte, L. (1869-70). *Campanula (?) soldanellaeflora plena*. Fl. Serres Jard. Eur. 18: 93 + pl. 1880, illus. Fr. — Portrait and discussion of unusual mutant of *C. rotundifolia*.

Timbal-Lagrave, E. (1873). Étude sur quelques campanules des Pyrénées. Mém. Acad. Sci. Toulouse (sér. 7) 5: 259-276 + pl. I-II, illus. Fr. — Floristic treatment with descriptions and full nomenclature.

Hooker, J. D. (1878). *Campanula macrostyla*. Bot. Mag. 104: tab. 6394 + 2 pp., illus. En. — Portrait with description.

Hooker, J. D. (1880). *Campanula fragilis*. Bot. Mag. 106: tab. 6504 + 2 pp., illus. En. — Portrait of *C. fragilis* subsp. *fragilis* with description.

Hooker, J. D. (1881a). *Campanula allionii*. Bot. Mag. 107: tab. 6588 + 2 pp., illus. En. — Portrait of *C. alpestris* with description.

Hooker, J. D. (1881b). *Campanula tommasiniana*. Bot. Mag. 107: tab. 6590 + 2 pp., illus. En. — Portrait with description.

Hooker, J. D. (1883). *Campanula jacobaea*. Bot. Mag. 109: tab. 6703 + 2 pp., illus. En. — Portrait with description.

Barnes, C. R. (1885). The process of fertilization in *Campanula americana* L. Bot. Gaz. (Crawfordsville) 10: 349-354 + pl. 10, illus. En. — Early embryological study.

Barnes, C. R. (1886). The fertilization of *Campanula americana*. Bot. Gaz. (Crawfordsville) 11: 99. En. — Supplement to Barnes (1885).

Barton, B. W. (1886). Notes on *Campanula medium*. Bot. Gaz. (Crawfordsville) 11: 208-211. En. — Experiments with insect pollination.

Feer, M. H. (1890). Recherches litteraires et synonymiques sur quelques campanules. I. - Sur trois campanules d' Espagne (*C. lusitanica* L., *affinis* R. et S., *arvatica* Lag.). J. Bot. (Morot) 4: 1-21. Fr. — Comments on relationships of three Spanish species.

Hooker, J. D. (1893). *Symphyandra hofmanni*. Bot. Mag. 119: tab. 7298 + 2 pp., illus. En. — Portrait of *C. hofmannii* with description.

Hooker, J. D. (1894). *Campanula excisa*. Bot. Mag. 120: tab. 7358 + 2 pp., illus. En. — Portrait with description.

Hooker, J. D. (1898). *Symphyandra wanneri*. Bot. Mag. 124: tab. 7594 + 2 pp., illus. En. — Portrait of *C. wanneri* with description.

Hooker, J. D. (1900). *Campanula mirabilis*. Bot. Mag. 126: tab. 7714 + 2 pp., illus. En. — Portrait with description.

• Pritchard, M. (1902). The genus *Campanula*. J. Roy. Hort. Soc. 27: 98-123, illus. En. — Synopsis of species in cultivation, divided among nine informal groups.

Witasek, J. (1906). Studien über einige Arten aus der Verwandtschaft der *Campanula rotundifolia* L. Magyar Bot. Lapok 5: 236-249. Ge. — Discussion of classification of *C. rotundifolia* and its allies.

• Beddome, R. H. (1907). An annotated list of the species of *Campanula*. J. Roy. Hort. Soc. 32: 196-221. En. — Nomenclator for genus, with emphasis on cultivated species.

Léveillé, H. (1908). A propos des *Campanula glomerata* et *cervicaria*, par M. Karl Ortlepp. Monde Pl. 10: 5-6. Fr. — Brief note on variation and relationships.

Wright, C. H. (1910). *Campanula beauverdiana*. Bot. Mag. 136: tab. 8299 + 2 pp., illus. En. — Portrait of *C. stevenii* subsp. *beauverdiana* with description.

Turrill, W. B. (1912). *Campanula arvatica*. Bot. Mag. 138: tab. 8431 + 2 pp., illus. En. — Portrait of *C. arvatica* subsp. *arvatica* with description.

Beyer, R. (1917). Ueber einige neue Bastarde und Abarten in der Gattung *Campanula* aus den Kottischen Alpen. Verh. Bot. Vereins Prov. Brandenburg 58: 108-119. Ge. — Description of several interspecific hybrids.

Brown, N. E. (1917). *Campanula ephesia*. Bot. Mag. 143: tab. 8715 + 2 pp., illus. En. — Portrait of *C. tomentosa* with description.

Turrill, W. B. (1919). *Campanula sulphurea*. Bot. Mag. 145: tab. 8827 + 2 pp., illus. En. — Portrait with description.

Lathouwers, M. V. (1922). Recherches expérimentales sur l'hérédité chez "*Campanula medium* L". Mém. Acad. Roy. Belgique, Cl. Sci. 8° (sér. 10) 4: 1-33 + pl. I-III, illus. Fr. — Determination of genetic basis of various morphological characters in a cultivated species.

Anonymous (1929). Dehiscence in *Campanula*. Amer. Bot. (Binghampton) 35: 23-24. En. — Brief note on capsule morphology and its effect on seed dispersal.

• Hruby, J. (1930). Campanulastudien innerhalb der *Vulgares* und ihrer Verwandten. Magyar Bot. Lapok 29: 152-276 + Tab. IV-VII. Ge. — Monograph of a group of polymorphic European species, with keys, descriptions, synonymy, and specimen citations.

• Hruby, J. (1933). Campanulastudien. II. Mitteilung. Magyar Bot. Lapok 29: 152-276 + Tab. IV-VII. Ge. — Supplement to Hruby (1930).

Milne-Redhead, E. (1934). *Campanula propinqua grandiflora*. Bot. Mag. 157: tab. 9349 + 3 pp., illus. En. — Portrait of *C. reuteriana*, with description.

Winton, A. L. (1935). Roots of the bellflower family (Campanulaceae). In The structure and composition of foods 2: 111. New York: John Wiley & Sons. En. — Root anatomy of *C. rapunculus*.

Milne-Redhead, E. (1936). *Campanula formanekiana*. Bot. Mag. 159: tab. 9436 + 4 pp., illus. En. — Portrait with description.

Senior, R. M. (1936). The campanulas of North America. Natl. Hort. Mag. 15: 97-102, illus. En. — Popular account of species indigenous to North America.

Wood, C. W. (1936). The *Campanula rotundifolia* tangle. Amer. Bot. (Binghampton) 42: 8-11. En. , Commentary on classification of a widespread and polymorphic species.

Alexander, E. J. (1939). *Campanula divaricata*. Addisonia 21: 13-14 + pl. 679, illus. En. — Portrait with description.

Milne-Redhead, E. (1939a). *Campanula incurva*. Bot. Mag. 161: tab. 9556 + 2 pp., illus. En. — Portrait with description and notes on cultivation.

Milne-Redhead, E. (1939b). *Campanula affinis*. Bot. Mag. 162: tab. 9568 + 4 pp., illus. En. — Portrait of *C. speciosa* subsp. *affinis* with description.

Guinochet, M. (1942). Recherches de taxonomie experimentale sur la flore des Alpes et de la region mediterraneene occidentale. II. Sur quelques formes du *Campanula rotundifolia* L. sens. lat. Bull. Soc. Bot. France 89: 70-75, 153-156. Fr. — Morphological variation in a widespread and polymorphic species.

McVaugh, R. (1942). A new *Campanula* from Idaho. Bull. Torrey Bot. Club 69: 241-243. En. — Includes key to species of *Campanula* in Rocky Mountains south of Canada.

Nakai, T. (1944). Notulae ad plantas Asiae Orientalis (XXXI). J. Jap. Bot. 20: 185-187. Ja. — Discussion of E. Asian representative of *C. glomerata*.

Lucas, E. V., L. Crane & E. M. Edwardes, transl. (1945). Rapunzel. In Grimm's Fairy Tales by the Brothers Grimm: 129-133. New York: Grosset & Dunlap. En. — *C. rapunculus* in folklore.

Smirnov, P. A. (1947). [About *Campanula steveni* of the authors of the Middle Russian flora]. Bjull. Moskovsk. Obšč. Isp. Prir., Otd. Biol. 52(3): 55-63. Ru. — Discussion of taxa in C. Asia allied to *C. stevenii*.

• Kharadze, A. L. (1949). Tentamen systematicae specierum caucasicarum g. *Campanulae* L. sectione *Medium* A.D.C. Zametki Sist. Geogr. Rast. 15: 13-33. Ru. — Floristic treatment with descriptions and full nomenclature.

Davis, P. H. (1950). *Tracheliopsis antilibanotica* P. H. Davis. Hooker's Icon. Pl. 35: tab. 3498 + 5 pp., illus. En. — Portrait of *C. antilibanotica* with description, and discussion of merits of recognizing *Diosphaera* and *Tracheliopsis*.

• Hruby, J. (1950). Campanulastudien innerhalb der *Vulgares* und ihrer Verwandten in Mitteleuropa. Mitt. Florist.-Soziol. Arbeitsgem. 2: 78-93. Ge. — Update of Hruby (1930, 1934) with keys and brief descriptions.

• Crook, H. C. (1951). Campanulas, their cultivation and classification. 256 pp., illus. London: Country Life. En. — Account of species in cultivation.

Turrill, W. B. (1952). *Campanula reiseri*. Bot. Mag. 169: tab. 178 + 3 pp., illus. En. — Portrait with description and notes on cultivation.

• Bailey, L. H. (1953). *Campanula* – the true bellflowers; *Symphyandra*. In The garden of bellflowers in North America: 8-111, 114-116, illus. New York: MacMillan. En. — Account of species cultivated in North America, with key and descriptions.

Kharadze, A. (1953). De *Campanulis* Caucasicis notulae criticae. Zametki Sist. Geogr. Rast. 17: 99-109. Ru. — Taxonomic notes on several Caucasian species.

Lawrence, G. H. M. (1955). Bellflowers for your garden. Horticulture 33: 234-235, 250-251, 253, illus. En. — Popular account of certain species in cultivation.

Turrill, W. B. (1955). *Campanula portenschlagiana*. Bot. Mag. 170: tab. 256 + 3 pp., illus. En. — Portrait with description.

Kharadze, A. (1956). Ad cognitionem specierum Caucasicarum generis *Campanula* L. Zametki Sist. Geogr. Rast. 19: 33-38. Ru. — Taxonomic notes on several Caucasian species.

Reiter, L. (1956). Einschlusse im Zellkern von *Campanula*. Protoplasma 45: 507-509, illus. Ge. — Unique protein inclusions in cell nuclei.

Turrill, W. B. (1956). *Campanula davisii*. Bot. Mag. 171: tab. 283 + 3 pp., illus. En. — Portrait with description.

• Fedorov, A. A. (1957). *Campanula* L., *Symphyandra* A. DC., *Brachycodon* Fed., *Popoviocodonia* Fed., *Astrocodon* Fed. In V. L. Komarov, Flora URSS 24: 131-343, 372-379, illus., map. Leningrad: Academia Scientiae URSS. Ru. — Flora for former Soviet Union, with keys, descriptions, infrageneric classifications, and full nomenclature.

Senior, R. M. (1959). Attractive perennial campanulas. Natl. Hort. Mag. 38: 33-41, illus. En. — Popular account of species in cultivation.

Turrill, W. B. (1959). *Campanula poscharskyana*. Bot. Mag. 172: tab. 334 + 4 pp., illus. En. — Portrait with description and notes on cultivation.

Alexander, E. J. (1960a). *Campanula cochlearifolia*. Addisonia 24: 3 + pl. 770, illus. En. — Portrait with description.

Alexander, E. J. (1960b). *Campanula latifolia macrantha*. Addisonia 24: 9 + pl. 773, illus. En. — Portrait of *C. latifolia* subsp. *latifolia* with description.

Dress, W. J. (1961). *Campanula punctata*. Baileya 9: 89-91, illus. En. — History of its use in horticulture.

• Gadella, T. W. J. (1962). Some cytological observations in the genus *Campanula*. Proc. Kon. Ned. Akad. Wetensch. C 65: 269-278. En. — Application of data on chromosome number variation to classification and evolution of genus.

Podlech, D. (1962). Beitrag zur Kenntnis der Subsektion *Heterophylla* (Witas.) Fed. der Gattung *Campanula* L. Ber. Deutsch. Bot. Ges. 75: 237-244. Ge. — Discussion of European species allied to *C. rotundifolia* L.

Shetler, S. G. (1962). Notes on the life history of *Campanula americana*, the tall bellflower.Michigan Bot. 1: 9-14, illus., map. En. — Details on pollination as it relates to floral structure.

Phitos, D. (1963). Beiträge zur Kenntnis der *Campanula rupestris*-Gruppe. Phyton (Horn) 10: 124-127. Ge. — Brief synopsis of group of Mediterranean species.

• Shetler, S. G. (1963). A checklist and key to the species of *Campanula* native or commonly naturalized in North America. Rhodora 65: 319-337. En. — Synopsis with key.

• Gadella, T. W. J. (1964a). Some cytological observations in the genus *Campanula* II. Proc. Kon. Ned. Akad. Wetensch. C 66: 270-283. En. — Application of data on chromosome number variation to classification and evolution of genus.

• Gadella, T. W. J. (1964b). Cytotaxonomic studies in the genus *Campanula*. Wentia 11: 1-104. En. — Synthesis of large body of cytological data and its application to questions of phylogeny and relationships in genus.

• Phitos, D. (1964). Trilokuläre *Campanula*-Arten der Ägäis. Oesterr. Bot. Z. 111: 208-230, illus., maps. Ge., La. — Monograph of group of E. Mediterranean species, with key, descriptions, and full nomenclature.

• Damboldt, J. (1965). Zytotaxonomische revision der isophyllen *Campanulae* in Europa. Bot. Jahrb. Syst. 84: 302-358. Ge. — Revision (based on chromosome numbers) of small group of species.

- Phitos, D. (1965). Die quinquelokulären *Campanula*-Arten. Oesterr. Bot. Z. 112: 449-498, maps. Ge., La. — Monograph of a group of Mediterranean species, with key, descriptions, and full nomenclature.
- Podlech, D. (1965). Revision der europäischen und nordafrikanischen Vertreiter der subsect. *Heterophylla* (Wit.) Fed. der Gattung *Campanula* L. Feddes Repert. 71: 50-187, illus. Ge. — Monograph with key, descriptions, full nomenclature, and specimen citations.
- Rechinger, K. H. & H. Schiman-Czeika (1965). *Symphyandra, Campanula, Diosphaera*. In K. H. Rechinger (ed.), Flora Iranica 13: 2, 7-39, illus. Graz: Akademische Druck- und Verlagsanstalt. Ge., La. — Flora with keys, descriptions, infrageneric classification, full nomenclature, and specimen citations.

Meikle, R. D. (1967). *Campanula hawkinsiana*. Bot. Mag. 176: tab. 505 + 4 pp., illus. En. — Portrait with description, discussion of its relationships, and notes on cultivation.

Hubac, J.-M. (1969). Premier essai d'étude de croisements experimentaux a l'aide des methodes numeriques de la taxonomie chez *Campanula rotundifolia* L., s.l. Compt.-Rend. Hebd. Séances Mém. Soc. Biol. 163: 336-344. Fr. — Integration of morphometric analysis and biosystematics.

Kovanda, M. (1970a). Polyploidy and variation in the *Campanula rotundifolia* complex. Part I. (General). Rozpr. Ceskoslov. Akad. Vûd. 80(2): 1-95. En. — Regional revision of complex and polymorphic species group.

Kovanda, M. (1970b). Polyploidy and variation in the *Campanula rotundifolia* complex. Part II. (Taxonomic). 1. Revision of the groups *Saxicolae, Lanceolatae*, and *Alpicolae* in Czechoslovakia and adjacent regions. Folia Geobot. Phytotax. 5: 171-208, illus., maps. En. — Regional revision of complex and polymorphic species group.

Podlech, D. (1970). Ergänzungen zur Revision der europäischen und nordafrikanischen Vertreiter der subsect. *Heterophylla* (Wit.) Fed. der Gattung *Campanula* L. Mitt. Bot. Staatssamml. München 8: 211-217. Ge. — Supplement to Podlech (1965).

Bielawska, H. (1971). Cytogenetic relationships among tetraploid representatives of *Campanula rotundifolia* L. s. str. Genet. Polon. 12: 223-225. En. — Cytological study of a widespread and polymorphic species.

Geslot A. & J. Medus (1971). Morphologie pollinique et nombre chromosomique dans la sous-section *Heterophylla* du genre *Campanula*. Canad. J. Genet. Cytol. 13: 888-894. Fr. — Palynological and cytological study of *C. rotundifolia* and its allies.

Huxley, A. J. (1971). *Campanula rupestris*. Bull. Alpine Gard. Soc. Gr. Brit. 39: 263-265, illus. En. — Portrait.

Tacik, T. (1971). Observationes ad *Campanulae* subsectionem *Heterophyllae* (Nyman ex Witasek) Fedorov e Polonia meridionali pertinentes. Pars 1. *Campanula rotundifolia* L. s.l. Fragm. Florist. Geobot. 17: 221-236. La. — Regional synopsis for a widespread and polymorphic species, with descriptions and full nomenclature for infraspecific taxa.

Bielawska, H. (1972). Cytogenetic relationships among some palearctic and nearctic tetraploid taxa of the *Campanula rotundifolia* L. group. Acta Soc. Bot. Poloniae 41: 289-300. En. — Cytological study of a widespread and polymorphic species.

Davis, P. H. (1972). *Campanula myrtifolia*. Bot. Mag. 179: tab. 637 + 3 pp., illus. En. — Portrait with description, notes on cultivation, and comments on the merits of recognizing *Diosphaera* and *Tracheliopsis*.

De Mezey, A. A. (1972). *Campanula elatines* L. Bull. Alpine Gard. Soc. Gr. Brit. 40: 224, 227, 229, illus. En. — Portrait.

Michel, M. (1972). Quelques campanules pour le jardin de rocaille. Bull. Mens. Soc. Linn. Lyon 41: 49-54, illus. Fr. — Popular account of species suitable for rock gardens.

Musch, A. & T. W. J. Gadella (1972). Some notes on the hybrid between *Campanula isophylla* Mor. and *C. pyramidalis* L. Acta Bot. Neerl. 21: 605-608, illus. En. — Description of artificial hybrid of intermediate chromosome number between two allopatric species.

Thaler, I. & E. Gailhofer-Denng (1972). Die feinstruktur der Zellkerneinschlusse von *Campanula moesiaca, C. persicifolia* und *C. trachelium*. Phyton (Horn) 14: 217-221, illus. Ge. — Ultrastructure of unique protein intranuclear inclusions.

Bielawska, H. (1973). Self-fertilization in *Campanula rotundifolia* L. s.l. group. Acta Soc. Bot. Poloniae 42: 253-264. En. — Breeding system variation.

Contandriopoulos, J., P. Quézel & J. Zaffran (1973). A propos des Campanules du groupe *aizoon* en Grèce méridionale et en Crète. Bull. Soc. Bot. France 120: 331-340, illus., maps. Fr. — Taxonomic, cytologic, and ecologic studies of small group of species.

Geslot, A. (1973). Contribution a l'etude cytotaxinomique de *Campanula rotundifolia* dans les Pyrenees francaises et espagnoles. Phyton (Horn) 15: 127-143. Fr. — Cytological study of a widespread and polymorphic species.

Hubac, J.M. (1973). Relations entre les dimensions du pollen et le niveau de polyploidie dans le complexe du *Campanula rotundifolia* L. s.l. Bull. Soc. Bot. France 119: 279-285. Fr. — Palynological and cytological study of a widespread and polymorphic species.

• Kharadze, A. L. & L. B. Serdyukova (1973). Generis *Campanula* L. nonnulla taxa superspecifica. Zametki Sist. Geogr. Rast. 30: 36-39, illus. Ru. — Revised classification of small group of species in Caucasus.

Contandriopoulos, J., P. Quézel & J. Zaffran (1974). A propos des campanules du groupe *aizoon* en Grèce méridionale et en Crète. Bull. Soc. Bot. France 120: 331-334, illus., map. Fr. — Synopsis of small group of E. Mediterranean species.

Geslot, A. & J. Medus (1974). Quelques remarques sur les relations entre morphologie pollinique et polyploidie dans le genre *Campanula* sous-section *Heterophylla*. Rev. Palaeobot. Palynol. 17: 233-243. Fr. — Use of pollen features to infer ploidy level in *C. rotundifolia* and is allies.

Shulkina, T. V. (1974). [Morphology of seedlings in some representatives of the genus *Campanula* L. in the USSR flora]. Bot. Žurn. 59: 439-447, illus. Ru. — Use of seedling morphology in taxonomy.

Hauser, M. L. (1975). Zytotaxonomische Untersuchungen an *Campanula patula* L. s.l. und *C. rapunculus* L. in der Schweiz und in Österreich. Veröff. Geobot. Inst. ETH Stiftung Rübel Zürich 53: 1-73, illus., maps. Ge. — Regional survey of cytological and morphological variation in two closely allied species.

Kovanda, M. (1975). *Campanula tatrae,* the correct name for *Campanula polymorpha.* Preslia 47: 26-30, En. — Clarification of an overlooked name.

Shulkina, T. V. (1975). [Peculiarities of seasonal development in some species of the genus *Campanula* L.] Bot. Žurn. 60: 109-117, illus. Ru. — Data on growth and reproduction.

• Damboldt, J. (1976). Materials for a flora of Turkey XXXII: Campanulaceae. Notes Roy. Bot. Gard. Edinburgh 35: 39-52, illus. En. — Preparatory material for Damboldt (1978); includes synopsis of revised infrageneric classification.

• Fedorov, A. A. & M. Kovanda (1976). *Campanula* L. In T. G. Tutin, V. H. Heywood, N., A. Burges, D. M. Moore, D. H. Valentine, S. M. Walters & D. A. Webb (eds.), Flora Europaea 4: 74-93. Cambridge: Cambridge University Press. En. — Flora with keys, descriptions, infrageneric classification, and full nomenclature.

Hunt, D. R. (1976). *Campanula ephesia*. Bot. Mag. 181: 33-35 + tab. 712, illus. En. — Portrait of *C. tomentosa* with description..

• Kharadze, A. L. (1976). Genus *Campanula* L. s.l. in Caucaso (Conspectus). Zametki Sist. Geogr. Rast. 32: 46-56. Ru. — Synopsis of species in Caucasus with full classification.

• Serdyukova, L. B. (1976a). Species generis *Campanula* L. e Transcaucasia australi sect. *Symphyandriformes* (Fom.) Char. Zametki Sist. Geogr. Rast. 32: 41-45, illus. Ru. — Synopsis of small group of species in Caucasus.

Serdyukova, L. B. (1976b). [Ad cognitionem inflorescentiarum structurae in generibus *Campanula* L. et *Symphyandra* A. DC.] Zametki Sist. Geogr. Rast. 33: 13–19, illus. Ru. — Morphology of inflorescences and their variation.

Shulkina, T. V. (1976). Campanulas native to the USSR. J. Scott. Rock Gard. Club 15: 126-133. En. — Popular account of species suitable for cultivation.

• Tutin, T. G. (1976). *Symphyandra* A. DC. In T. G. Tutin, V. H. Heywood, N., A. Burges, D. M. Moore, D. H. Valentine, S. M. Walters & D. A. Webb (eds.), Flora Europaea 4: 93. Cambridge: Cambridge University Press. En. — Flora with key, descriptions, and full nomenclature.

Thulin, M. (1976). *Campanula keniensis* Thulin sp. nov., and notes on allied species. Bot. Not. 128: 350-356, illus. En. — Synopsis of *C. edulis* complex of trop. Africa, with full nomenclature.

Crook, H. C. (1977). The genus *Symphyandra*. Bull Alpine Gard. Soc. Gr. Brit. 45: 246-254, illus. En. — Popular account with brief descriptions, published posthumously.

• Kovanda, M. (1977). Polyploidy and variation in the *Campanula rotundifolia* complex. Part II. (Taxonomic). 2. Revision of the groups *Vulgares* and *Scheuchzerianae* in Czechoslovakia and adjacent regions. Folia Geobot. Phytotax. 12: 23-89, illus., maps. En. — Regional revision of polymorphic species group, with descriptions and specimen citations.

Łańcucka-Środoniowa, M. (1977). New herbs described from the Tertiary of Poland. Acta Palaeobot. 18: 37-44, illus. En. — Miocene seeds described as *C. palaeopyramidalis*.

• Shulkina, T. V. (1977). [The types of life forms and their importance for the taxonomy of *Campanula* L.] Bot. Žurn. 62: 1102 — 1114, illus. Ru. — Application of life-form data to taxonomy.

• Damboldt, J. (1978). *Campanula* L., *Symphyandra* A. DC. In P. H. Davis (ed.), Flora of Turkey and the East Aegean Islands 6: 2-65, illus., maps. Edinburgh: University Press. En. — Flora with keys, descriptions, infrageneric classification, full nomenclature, and specimen citations.

• Feinbrun-Dothan, N. (1978). *Campanula* L. In Flora Palaestina 3: 277-283 + pl. 474-485, illus. Jerusalem: Israel Academy of Sciences & Humanities. En. — Flora with key, descriptions, and infrageneric classification.

Serdyukova, L. B. (1978). [Synopsis botanicae-geographicae specierum Caucasicarum nonnullarum generum *Campanula* L. et *Symphyandra* A. DC.] Zametki Sist. Geogr. Rast. 35: 55-62, maps. Ru. — Floristic treatment for Caucasus.

Łańcucka-Środoniowa, M. (1979). Macroscopic plant remains from the freshwater Miocene of the Nowy Sącz Basin (West Carpathians, Poland). Acta Palaeobot. 20: 3-117, illus. En. — Additional Miocene seed fossils referable to *Campanula* sp.

Shetler, S. G. (1979). Pollen-collecting hairs of *Campanula* (Campanulaceae): 1. Historical review. Taxon 28: 205-215. En. — Structure and function of invaginating hairs on style.

Geslot, A. (1980). Biometrie des graines et nombres chromosomiques dans la sous section *Heterophylla* du genre *Campanula* (Campanulaceae). Pl. Syst. Evol. 134: 193-206. Fr. — Use of pollen features to infer ploidy level in *C. rotundifolia* and its allies.

Kazanfarova, V. K. (1980). Taksonomiya i geografiya vidov podsektsii trilokulares roda *Campanula* L. severo vostochnogo Azerbaidzhana. Izv. Akad. Nauk Azerbajdžana SSR, Ser. Biol. 1: 23-28. Ru. — Floristic treatment for Azerbaijan.

Kolakovskii, A. A. & L. B. Serdyukova. (1980). [Ad systematicam specierum Caucasicarum generis *Campanula* L. notitiae novae.] Zametki Sist. Geogr. Rast. 36: 44-57, illus. Ru. — Synopsis of group of species in Caucasus.

• Morin, N. (1980). Systematics of the annual California campanulas (Campanulaceae). Madroño 27: 149-163, illus., maps. En. — Mongraph including key, descriptions, full nomenclature, and specimen citations.

Shulkina, T. V. (1980). [Geographical distribution of bluebell life forms of the section *Campanula* of the USSR flora.] Bjull. Moskovsk. Obšč. Isp. Prir., Otd. Biol. 85: 73, illus., map. Ru. — Application of life-form data to taxonomy.

Contandriopoulos, J. (1981). Contribution à l'étude cytotaxonomique du genre *Campanula* L. en Afrique du nord et centrale. Bol. Soc. Brot. 53: 887-906, illus., map. Fr. — Chromosome counts.

Oganesian, M. E. (1981). [The genera *Campanula* and *Symphyandra* (Campanulaceae) in southern Transcaucasia.] Bot. Žurn. 66: 399-408. Ru. — Synopsis for S. Transcaucasus, with key.

Serdyukova, L. B. (1982). Notula de nonnulis speciebus generis *Campanula* L., e sect. *Sibiricae* (Fomin) Char. Zametki Sist. Geogr. Rast. 38: 18-21, map. Ru. — Discussion of taxa in the Caucasus allied to *C. sibirica*.

• Shetler, S. G. (1982). Variation and evolution of the Nearctic Harebells (*Campanula* subsect. *Heterophylla*). 516 pp., illus., maps. Vaduz: J. Cramer. En. — Detailed biosystematic study of New World populations of *C. rotundifolia*, dividing them among four eco-geographical races (here treated as species).

Geslot, A. (1983). Contribution à la connaissance biosystématique des campanules de la sous-section *Heterophylla* (Wit.) Fed.: I. – Étude de la biologie de la reproduction des campanules pyrénéennes à feuilles hétéromorphes. Rev. Gén. Bot. 90: 185-220. Fr. — Detailed regional analysis of breeding systems

• Hong, D. Y. (1983). *Campanula* L. In Flora Reipublicae Popularis Sinicae 73(2): 78-92, illus. Beijing: Science Press. Ch. — Flora of China, with key, descriptions, infrageneric classification, and full nomenclature.

Janzon, L.-A. (1983). Pollination studies of *Campanula persicifolia* (Campanulaceae) in Sweden. Grana 22: 153-165, illus., map. En. — Pollinators diverse, but a megachilid bee and a fly were most reliable here.

Laane, M. M., B. E. Croff & R. Wahlstrøm (1983). Cytotype distribution in the *Campanula rotundifolia* complex in Norway and cytomorphological characteristics of diploid and tetraploid groups. Hereditas 99: 21-48, illus., map. En. — Chromosome number variation correlated with morphology.

Shulkina, T. V. (1983). Campanulas in the USSR. J. Scott. Rock Gard. Club 18: 270-271. En. — Popular account of species suitable for cultivation.

Shulkina, T. V. & E. A. Zemskova (1983). [Chromosome numbers and life-forms of certain *Campanula* (Campanulaceae) species criticae. Bot. Žurn. 68: 866-875, illus., map. Ru. — Life-forms and cytological data for several problematic species.

Baskin, J. M. & C. C. Baskin (1984). The ecological life cycle of *Campanula americana* in northcentral Kentucky. Bull. Torrey Bot. Club 111: 329-337. En. — Time of seed germination determines whether plants behave as winter annuals or biennials.

• Contandriopoulos, J. (1984a). Polyphylétisme des campanules annuelles. Bull. Soc. Bot. France 131: 315-324. Fr. — Evidence for multiple origins of therophytes from perennial ancestors.

• Contandriopoulos, J. (1984b). Differentiation and evolution of the genus *Campanula* in the Mediterranean region. In W. F. Grant (ed.), Plant Biosystematics: 141-158. Toronto: Academic Press. En. — Explicit hypotheses of evolution in genus, based on cytological data.

Geslot, A. (1984a). Contribution à l'étude biosystématique des campanules de la sous-section *Heterophylla* (Wit.) Fed. II. – Recherche des barrières de stérilité entre taxons pyrénéens. Rev. Gén. Bot. 91: 123-152. Fr. — Experimental hybridizations reveal considerable reproductive isolation within group.

Geslot, A. (1984b). Caryologie des *Campanula* subsect. *Heterophylla* (Wit.) Fed.: nouvelles numerations chromosomiques dans les Pyrenees. Phyton (Horn) 24: 173-191, maps. Fr. — Detailed cytological study of *C. rotundifolia* and its allies.

Nasir, E. (1984). *Campanula*. In E. Nasir & S. I. Ali (eds.), Flora of Pakistan 155: 6-21, illus. Islamabad: Pakistan Agricultural Research Council. En. — Flora with key and descriptions.

Serdyukova, L. B. (1984). [Concerning the species of the section *Latilimbus* (Fed.) Char. of the genus *Campanula* L.]. Soobšč. Akad. Nauk Gruzinsk. SSR 113: 137-140. Ru. — Monograph of small group of species.

• Carlström, A. (1986). A revision of the *Campanula drabifolia* complex (Campanulaceae). Willdenowia 15: 375-387, maps. En. — Treatment of group of Mediterranean species, with key, descriptions, full nomenclature, and specimen citations.

Inoue, K. & M. Amano (1986). Evolution of *Campanula punctata* Lam. in the Izu Islands: changes of pollinators and evolution of breeding systems. Pl. Spec. Biol. 1: 89-97, illus., map. En. — Role of pollinators in divergence of *C. microdonta*.

Serdyukova, L. B. (1986). [Notula de speciebus nonnullis generis *Campanula* e sectione *Latilimbus* (Fed.) Charadze (Campanulaceae) florae Georgiae.] Zametki Sist. Geogr. Rast. 41: 58-61. Ru. — Taxonomic treatment of small group of species.

Nurmi, J. (1987). Chromosome numbers and variation of the *Campanula rotundifolia* complex in northwestern Europe. Symb. Bot. Upsal. 27: 235-239, maps. En. — Biosystematic study of a widespread and polymorphic species.

Thulin, M. (1987). Notes on *Campanula dimorphantha* (Campanulaceae). Nord. J. Bot. 7: 419-420, map. En. — Asian *C. benthamii* conspecific with African *C. dimorphantha*.

Inoue, K. (1988). Pattern of breeding-system change in the Izu Islands in *Campanula punctata:* bumblebee-absence hypotheses. Pl. Spec. Biol. 3: 125-128, map. En. — *C. punctata* self-incompatible and bombid-pollinated, *C. microdonta* self-compatible and halictid-pollinated.

Inoue, K. & I. Washitani (1989). Geographical variation in thermal germination responses in *Campanula punctata* Lam. Pl. Spec. Biol. 4: 69-74. En. — Effects of temperature on seed germination in this and related *C. microdonta*.

Kovanda, M. & M. Ančev (1989). The *Campanula rotundifolia* complex in Bulgaria. Preslia 61: 193-207 + tab. I-III, illus., map. En. — Regional synopsis with key, full nomenclature, and specimen citations.

• Lewis, P. & M. Lynch (1989). Campanulas. 149 pp., illus. London: Christopher Helm. En. — Account of species in cultivation.

Richardson, T. E. & A. G. Stephenson (1989). Pollen removal and pollen deposition affect the duration of the staminate and pistillate phases in *Campanula rapunculoides*. Amer. J. Bot. 76: 532-538. En. — Experiments on dichogamy.

Inoue, K. & T. Kawahara (1990). Allozyme differentiation and genetic structure in island and mainland Japanese populations of *Campanula punctata* (Campanulaceae). Amer. J. Bot. 77: 1440-1448, map. En. — Variation in allozyme profiles supports recognition of *C. microdonta*.

Soljan, D. (1990). Morfološka diferencjacija populacija vrste *Campanula portenschlagiana* Schultes in Roemer et Schultes. Glasn. Zemaljsk. Muz. Bosne Hercegovine Sarajevu, Prir. Nauke (n.s.) 29: 39-50, illus. Serb. — Analysis of morphological variation, with key to infraspecific taxa.

Nyman, Y. (1991). Crossing experiments within the *Campanula dichotoma* group (Campanulaceae). Pl. Syst. Evol. 177: 185-192, map. En. — Experimental hybridizations support recognition of *C. afra*, *C. dichoma*, *C. kremeri*, and *C. occidentalis* as distinct species.

• Huxley, A. J., ed. (1992). *Campanula, Symphyandra.* In The new Royal Horticultural Society dictionary of gardening 1: 485-495, illus; 4: 414. London: MacMillan. En. — Account of species in cultivation.

Kelly, J. (1992). Campanulas. The Garden 117: 325-330, illus. En. — Popular account of species in cultivation.

• Nowicke, J. W., S. G. Shetler & N. Morin (1992). Exine structure of pantoporate *Campanula* (Campanulaceae) species. Ann. Missouri Bot. Gard. 79: 65-80, illus. En. — Only members of entire family with pantoporate pollen are five North American species.

Nyman, Y. (1992a). Pollination mechanisms in six *Campanula* species (Campanulaceae). Pl. Syst. Evol. 181: 97-108. En. — Comparative studies of reproductive biology in four Mediterranean annuals and two European perennials.

Nyman, Y. (1992b). Reproduction in *Campanula afra* (Campanulaceae): mating system and the role of the pollen collecting hairs. Pl. Syst. Evol. 183: 33-41. En. — Experimental studies of mating behhavior.

Phillips, S. (1992). Vintage vegetables. The Garden 117: 507-511. En. — Brief account of culinary usage of *C. rapunculus*.

Kolakovskii, A. A. & R. V. Lakoba (1993). [Capsule morphology and anatomy of genus *Annaea* Kolak. (Campanulaceae) endemical for Caucasus.] Bjull. Moskovsk. Obšč. Isp. Prir., Otd. Biol. 98: 135-137, illus. Ru. — Detailed studies of fruit structure in *C. hieracioides* Kolak.

Lobin, W., O. Kriesten & T. Leyens (1993). Die Kapverdische Glockenblume (*Campanula jacobaea* Webb): eine für Botanische Garten attraktive endemische Art von den Kapverdischen Inseln. Palmengarten 57: 120-121, illus. Ge. — Portrait and popular account.

Morin, N. (1993). *Campanula*. In J. C. Hickman (ed.), The Jepson manual. Higher plants of California: 459-460, illus. Berkeley: University of California Press. En. — Flora with key and descriptions.

Nyman, Y. (1993a). The pollen-collecting hairs of *Campanula* (Campanulaceae). I. Morphological variation and the retractive mechanism. Amer. J. Bot. 80: 1427-1436, illus. En. — Structure and function of invaginating hairs on style.

Nyman, Y. (1993b). The pollen-collecting hairs of *Campanula* (Campanulaceae). II. Function and adaptive significance in relation to pollination. Amer. J. Bot. 80: 1437-1443, illus. En. — Role in pollination of invaginating hairs on style.

- Oganesian, M. E. (1993). [Synopsis of the species of the subgenus *Scapiflorae* of the genus *Campanula* (Campanulaceae).] Bot. Žurn. 78(3): 145-157. Ru. — Synopsis of group of species in W. Asia, with full nomenclature.

- Shimizu, T. (1993). *Campanula* L. In K. Iwatsuki, T. Yamazaki, D. E. Boufford & H. Ohba, Flora of Japan 3a: 410-412. Tokyo: Kodansha. En. — Flora with key, descriptions, and infrageneric classification.

Cresswell, J. E. & A. W. Robertson (1994). Discrimination by pollen-collecting bumblebees among differentially rewarding flowers of an alpine wildflower, *Campanula rotundifolia* (Campanulaceae). Oikos 69: 304-308. En. — Field studies of bumblebee pollination in Switzerland.

Emig, W. & P. Leins (1994). Ausbreitungsbiologische Untersuchungen in der Gattung *Campanula* L. I. Vergleichende Windkanalexperimente zur Samenportionierung bei *C. trachelium* L., *C. sibirica* L. und *C. glomerata* L. Bot. Jahrb. Syst. 116: 243-257, illus. Ge. — Wind-tunnel experiments on seed dispersal.

- Oganisian, M. E. (1994). [Synopsis of the species of the subgenus *Scapiflorae* of the genus *Campanula* (Campanulaceae): key for determination of the species.] Bot. Žurn. 79: 95-104. Ru. — Identification guide for small group of species.

Contandriopoulos, J. (1995). Relations phylogenetiques et cytobiologiques entre le complexe de *Campanula edulis* Forssk. et *C. bordesiana* Maire (Campanulaceae). Bull. Mus. Natl. Hist. Nat., B, Adansonia 17: 59-67, illus. Fr. — Morphological polymorphism in North African *C. edulis* complex not correlated with chromosome number variation.

Krahulcová, A., L. Kirschnerová & M. Kovanda (1995). Polyploid *Campanula patula* in the Czech Republic. Preslia 67: 107-115, map. En. — Correlation of morphology and chromosome number in a widespread and polymorphic species.

Leyens, T. & W. Lobin (1995). *Campanula* (Campanulaceae) on the Cape Verde Islands – two species or only one? Willdenowia 25: 215-228, illus., map. En. — Floristic treatment with key, descriptions, full nomenclature, and specimen citations.

- Oganesian, M. (1995). Synopsis of Caucasian Campanulaceae. Candollea 50: 275-308, map. En. — Checklist of species in the Caucasus, with revised infrageneric classification and full nomenclature.

Emig, W. & P. Leins (1996). Ausbreitungsbiologische Untersuchungen in der Gattung *Campanula* L. II. Die Bedeutung der Kapselmorphologie und der Samenausgestaltung für das Ausbreitungsverhalten. Bot. Jahrb. Syst. 118: 505-528, illus. Ge. — Wind-tunnel experiments of seed dispersal as related to seed morphology and capsule structure.

Inoue, K., M. Maki & M. Masuda (1996). Evolution of *Campanula* flowers in relation to insect pollination on islands. In D. G. Lloyd & S. C. H. Barrett, eds., Floral biology: 377-400. New York: Chapman & Hall. En. — Shift in pollinators correlated with changes in floral morphology.

Kovanda, M. (1996). Chromosome number studies in *Campanula patula*. Biologia (Bratislava) 51: 383-386. En. — Cytological examination of numerous populations of widespread species; most diploid, though tetraploids found as well.

Lewis, P. (1998). Belles of the border. The Garden 123: 436-439, illus. En. — Detailed account of named cultivars of *C. persicifolia*.

- Lewis, P. & M. Lynch (1998). Campanulas. A gardener's guide. 176 pp., illus. Portland: Timber Press. En. — Account of species currently in cultivation.

Viktorov, V. P. & A. G. Elenevskii (1998). [On the taxonomy of *Campanula dichotoma* and *C. propinqua* (Campanulaceae).] Bot. Žurn. 83(10): 57-58. — Synonymization of latter name under former.

• Eddie, W. M. M. & M. J. Ingrouille (1999). Polymorphism in the Aegean "five-loculed" species of the genus *Campanula*, section *Quinqueloculares* (Campanulaceae). Nordic J. Bot. 19: 153-169, illus., maps. En. — Morphometric analysis emphasizes close relationships and mophological overlap among species in group.

Tyler, T. (1999). Fyra sorters toppklocka *Campanula glomerata* i Norden. Svensk. Bot. Tidskr. 93: 139-144, illus., maps. Sw. — Discussion of morphological variation in Scandinavian representatives of a widespread and polymorphic Eurasian species, including those naturalized or persisting after cultivation.

Güner, A. & R. Wilford (2000). *Campanula betulifolia.* Curtis's Bot. Mag. 17: 85-91, illus. En. — Portrait.

Park, K. R. & H. J. Jung (2000). [Isozyme and morphological variation in *Campanula punctata* and *C. takesimana* (Campanulaceae).] Korean J. Pl. Taxon. 30: 1-16. Ko. — Multivariate analysis of morphology supports recognition of two species; electrophoretic data suggest *C. takesimana* is derived from *C. punctata.*

Viktorov, V. P. (2000). [Morphology and the principal evolutionary trends of the inflorescences in the genus *Campanula* (Campanulaceae).] Bot. Žurn. 85: 80-90, illus. Ru. — Evolution of inflorescence structure.

Galloway, L. F. (2001). The effect of maternal and paternal environments on seed characters in the herbaceous plant *Campanula americana* (Campanulaceae). Amer. J. Bot. 88: 832-840. En. — Environmental variation affects seed characteristics.

Good-Avila, S. V., F. Frey & A. G. Stephenson (2001). The effect of partial self-incompatibility on the breeding system of *Campanula rapunculoides* L. (Campanulaceae) under conditions of natural pollination. Int. J. Pl. Sci. 162: 1081-1087. En. — Experimental studies of breeding system in a naturalized species.

Krestovskaya, T. V. (2001). [Genus *Campanula* L. (Campanulaceae) in Asia Centrali.] Novosti Sist. Vyssh. Rast. 33: 218-222. Ru. — Floristic treatment for C. Asia.

• Oganesian, M. E. (2001). [Review of the sections *Symphyandriformes* and *Cordifoliae* of the subgenus *Campanula* (*Campanula,* Campanulaceae).] Bot. Žurn. 86(11): 134-150. Ru. — Synopsis of group of species in W. Asia, with key.

Blionis, G. J. & D. Vokou (2002). Structural and functional divergence of *Campanula spatulata* subspecies on Mt Olympos (Greece). Pl. Syst. Evol. 232: 89-105. En. — Field studies of subsp. *spatulata* and subsp. *spruneriana* along an elevational gradient suggest reproductive isolation at least locally.

Galloway, L. F., T. Cirigliano & K. Gremski (2002). The contribution of display size and dichogamy to potential geitonogamy in *Campanula americana.* Int. J. Pl. Sci. 163: 133-139. En. — Field studies of relationship between floral morphology and pollination biology.

• Viktorov, V. (2002). [Conspectus taxonomicus generis *Campanula* L. (Campanulaceae) Rossiae et civitatum collimitanearum. Novosti Sist. Vyssh. Rast. 34: 197-234. Ru. — Synopsis of genus in former Soviet Union.

Good-Avila, S. V., T. Nagel, D. W. Vogler & A. G. Stephenson (2003). Effects of inbreeding on male function and self-fertility in the partially self-incompatible herb *Campanula rapunculoides* (Campanulaceae). Amer. J. Bot. 90: 1736-1745. En. — Field studies of breeding system impacts on fitness.

• Sáez, L. & J. J. Aldasoro (2003). A taxonomic revision of *Campanula* L. subgenus *Sicyocodon* (Feer) Damboldt and subgenus *Megalocalyx* Damboldt (Campanulaceae). Bot. J. Linn. Soc. 141: 215-241, illus., maps. En. — Monograph of pair of closely related groups of Mediterranean therophytes (sometimes accorded generic rank), with key, descriptions, full nomenclature, and specimen citations.

Kovačić, S. (2004). The genus *Campanula* L. (Campanulaceae) in Croatia, circum-Adriatic and west Balkan region. Acta Bot. Croat. 63: 171-202, map. En. — Floristic account, with detailed historical discussion, including infrageneric classification.

Campanula L., Sp. Pl.: 163 (1753).

Arctic & N. Temp. Zones; naturalized in Antarctic Zone. 10 11 12 13 14 20 21 23 24 25 30 31 32 33 34 35 36 37 38 40 41 70 71 72 73 74 75 76 77 78 79 (85) (90).

Roucela Dumort., Comment. Bot.: 14 (1822). *Campanula* subg. *Roucela* (Dumort.) Damboldt, Notes Roy. Bot. Gard. Edinburgh 35: 43 (1976).

Marianthemum Schrank, Denkschr. Königl.-Baier Bot. Ges. Regensburg 2: 34 (1822). *Campanula* sect. *Marianthemum* (Schrank) Schur, Enum. Pl. Transsilv.: 432 (1866).

Rapuntia Chevall., Fl. Gén. Env. Paris (ed. 2) 2: 526 (1836).

Symphyandra A. DC., Monogr. Campan.: 365 (1830).

Erinia Noulet, Fl. Bass. Sous-Pyrén.: 404 (1837).

Decaprisma Raf., Fl. Tellur. 2: 78 (1837).

Pentropis Raf., Fl. Tellur. 2: 80 (1837).

Loreia Raf., Fl. Tellur. 3: 82 (1837).

Medium Spach, Hist. Nat. Vég. 9: 553 (1838).

Cenekia Opiz in Bercht., Oekon.-techn. Fl. Böhm. 2(2): 9 (1839).

Medium Opiz in Bercht., Oekon.-techn. Fl. Böhm. 2(2): 9 (1839); non Spach, Hist. Nat. Vég. 9: 553 (1838).

Nenningia Opiz in Bercht., Oekon.-techn. Fl. Böhm. 2(2): 9 (1839).

Trachelioides Opiz in Bercht., Oekon.-techn. Fl. Böhm. 2(2): 9 (1839).

Weitenwebera Opiz in Bercht., Oekon.-techn. Fl. Böhm. 2(2): 9 (1839).

Quinquelocularia K. Koch, Linnaea 23: 630 (1850).

Sicyocodon Feer, Bot. Jahrb. Syst. 12: 613 (1890). *Campanula* sect. *Sicyocodon* (Feer) Kuntze in T. Post & Kuntze, Lex. Gen. Phan.: 95 (1904). *Campanula* subg. *Sicyocodon* (Feer) Damboldt, Notes Roy. Bot. Gard. Edinburgh 35: 43 (1976), as 'Sicyodon'.

Diosphaera Buser, Bull. Herb. Boissier 2: 519 (1894). *Campanula* sect. *Diosphaera* (Buser) Kuntze in T. Post & Kuntze, Lex. Gen. Phan.: 95 (1904).

Tracheliopsis Buser, Bull. Herb. Boissier 2: 524 (1894). *Campanula* sect. *Tracheliopsis* (Buser) Kuntze in T. Post & Kuntze, Lex. Gen. Phan.: 95 (1904).

Campanulastrum Small, Fl. S.E. US: 1141 (1903).

Rotantha Small, Man. S.E. Fl.: 1289 (1933); non Baker, J. Linn. Soc., Bot. 25: 317 (1890).

Petkovia Stef., God. Sofiisk. Univ. Agron.-Lesoved. Fak. 14: 100 (1936).

Brachycodon Fed. in Kom., Fl. URSS 24: 468 (1957); non Progel in Mart., Fl. Bras. 6(1):229 (1865). *Brachycodonia* Fed. in Karyagin, Fl. Azerbajdž. 8: 163 (1961). *Campanula* subg. *Brachycodonia* (Fed.) Damboldt, Notes Roy. Bot. Gard. Edinburgh 35: 44 (1976).

Astrocodon Fed. in Kom., Fl. URSS 24: 471 (1957).

Popoviocodonia Fed. in Kom., Fl. URSS 24: 470 (1957).

Annaea Kolak., Soobšč. Akad. Nauk Gruzinsk. SSR 94: 163 (1979). *Campanula* subg. *Annaea* (Kolak.) Ogan., Candollea 50: 299 (1995). *Campanula* sect. *Annaea* (Kolak.) Viktorov, Novosti Sist. Vyssh. Rast. 34: 231 (2002).

Gadellia Schulkina, Novosti Sist. Vyssh. Rast. 16: 177 (1979). *Campanula* sect. *Gadellia* (Schulkina) Viktorov, Novosti Sist. Vyssh. Rast. 34: 232 (2002).

Pseudocampanula Kolak., Soobšč. Akad. Nauk Gruzinsk. SSR 97: 414 (1980). *Campanula* subg. *Pseudocampanula* (Kolak.) Ogan., Candollea 50: 299 (1995). *Campanula* sect. *Pseudocampanula* (Kolak.) Viktorov, Novosti Sist. Vyssh. Rast. 34: 221 (2002).

Mzymteila Kolak., Soobšč. Akad. Nauk Gruzinsk. SSR 103: 150 (1981).

Hyssaria Kolak., Soobšč. Akad. Nauk Gruzinsk. SSR 103: 151 (1981).

Neocodon Kolak. & Serdyuk., Zametki Sist. Geogr. Rast. 40: 27 (1984).

Hemisphaera Kolak., Okhrana Prir. Gruz. 12: 161 (1984).

Echinocodon Kolak., Soobšč. Akad. Nauk Gruzinsk. SSR 121: 387 (1986); non D. Y. Hong, Acta Phytotax. Sin. 22: 183 (1984). *Echinocodonia* Kolak., Bot. Žurn. (Moscow & Leningrad) 79: 114 (1994).

Campanula acutiloba Vatke, Linnaea 38: 709 (1874).

Turkey to Iran. 34 IRN IRQ TUR. Hemicr.

Campanula cissophylla Boiss. & Hausskn. in Boiss., Pl. Orient. Nov. 1: 2 (1875).

Campanula cissophylla var. *brachycalyx* Bornm., Mitt. Thüring. Bot. Vereins (n.s.) 20: 32 (1905).

Campanula afganica Pomel, Nouv. Mat. Fl. Atl.: 257 (1875). *Campanula atlantica* Coss. & Durieu ex Batt., Fl. Algérie, Dicotyl.: 573 (1889).
N. Africa. 20 ALG TUN. Hemicr.
 Campanula atlantica var. *glabra* Bonnet, J. Bot. (Morot) 7: 197 (1893).
 Campanula atlantica var. *guergourensis* Batt., Fl. Algérie, Suppl. Phan.: 62 (1910).

Campanula afra Cav., Anales Ci. Nat. 3: 21 (1800). *Campanula dichotoma* subsp. *afra* (Cav.) Maire, Cavanillesia 2: 174 (1930).
W. Mediterranean. 12 SPA 20 MOR ALG TUN. Ther. 2*n* = 24.
 Campanula leptosiphon Pau & Sennen in Sennen, Diagn. Nouv.: 195 (1936).
 Campanula hirsutissima Sennen & Mauricio in Sennen, Diagn. Nouv.: 297 (1936).

Campanula aghrica Kit Tan & Sorger, Notes Roy. Bot. Gard. Edinburgh 41: 528 (1984).
Turkey. 34 TUR. Hemicr. or geophyte.

Campanula aizoides Zaffran ex Greuter, Fl. Rep. Cretan Area: 15 (1972). *Campanula aizoon* subsp. *aizoides* (Zaffran ex Greuter) Fed., Bot. J. Linn. Soc. 67: 281 (1973).
Greece & Kriti. 13 GRC KRI. Biennial. 2*n* = 16.

Campanula aizoon Boiss. & Spruner in Boiss., Diagn. Pl. Orient. (ser. 1) 4: 34 (1844).
Greece. 13 GRC. Biennial. 2*n* = 16.

Campanula akguelii Altan, Bull. Pure Appl. Sci., Modinagar, B 17: 71 (1998), as 'akgülii'.
Turkey. 34 TUR. Hemicr. or geophyte.

Campanula akhdarensis A. G. Mill. & Whitc., Notes Roy. Bot. Gard. Edinburgh 41: 109 (1983).
Oman. 35 OMA. Hemicr.

Campanula alaskana (A. Gray) Wight ex J. P. Anderson, Proc. Iowa Acad. Sci. 25: 447 (1920).
Alaska (E. Aleutians) to NW. U.S.A. (Washington). 70 ALU ASK YUK 71 BRC 73 WAS. Hemicr. or geophyte. 2*n* = 34, 68, 102.
 * *Campanula rotundifolia* var. *alaskana* A. Gray, Syn. Fl. N. Amer. (ed. 2) 2 (1): 395 (1886).
 Campanula rotundifolia var. *hirsuta* Macoun, Cat. Canad. Pl. 5: 338 (1890).
 Campanula latisepala Hultén, Fl. Alaska Yukon: 1460 (1948).
 Campanula latisepala var. *dubia* Hultén, Fl. Alaska Yukon: 1460 (1948).

Campanula alata Desf., Fl. Atlant. 1: 178 (1798).
W. Mediterranean. 12 POR SPA 20 ALG TUN. Hemicr. 2*n* = 36.
 Campanula primulifolia Brot., Fl. Lusit. 1: 288 (1804), as 'primulaefolia'. *Echinocodon primulifolius* (Brot.) Kolak., Soobšč. Akad. Nauk Gruzinsk. SSR 121: 387 (1986), as 'primulifolium'.

Campanula albanica Witasek, Magyar Bot. Lapok. 5: 246 (1906). *Campanula linifolia* subsp. *albanica* (Witasek) Hayek, Repert. Spec. Nov. Regni Veg. Beih. 30(2): 537 (1930). *Campanula linifolia* var. *albanica* (Witasek) Hruby, Magyar Bot. Lapok 29: 226 (1930).
Balkans. 13 ALB GRC YUG. Hemicr. or geophyte. 2*n* = 34, 102.

subsp. **albanica**
 Balkans. 13 ALB GRC YUG. Hemicr. or geophyte. 2*n* = 34, 102.
 Campanula rotundifolia subsp. *hellenica* Hayek, Repert. Spec. Nov. Regni Veg. Beih. 30(2): 540 (1930). *Campanula hellenica* (Hayek) Podlech, Ber. Deutsch. Bot. Ges. 75: 239 (1962).

Campanula linifolia f. *simplex* Hruby, Magyar Bot. Lapok 29: 226 (1930).
Campanula linifolia f. *ramulosa* Hruby, Magyar Bot. Lapok 29: 226 (1930).

subsp. **sancta** (Hayek) Podlech, Feddes Repert. 71: 85 (1965).
N. Greece (Mt. Athos). 13 GRC. Hemicr. or geophyte. *2n* = 34.
* *Campanula rotundifolia* subsp. *sancta* Hayek, Repert. Spec. Nov. Regni Veg. Beih. 30(2): 540 (1930).
Campanula balcanica subf. *simplex* Hruby, Magyar Bot. Lapok 29: 243 (1930).

Campanula aldanensis Fed. & Karav. in Kom., Fl. URSS 24: 466 (1957).
E. Siberia (Yakutiya). 30 YAK. Hemicr. or geophyte.

Campanula alliariifolia Willd., Sp. Pl. 1: 910 (1798), as 'alliariaefolia'. *Medium alliariifolium* (Willd.) Spach, Hist. Nat. Vég. 9: 557 (1838), as 'alliariaefolium'.
Turkey & Caucasus; cult.; naturalized in Great Britain. (10) grb 33 ALL 34 TUR. Hemicr. *2n* = 34, 68.

subsp. **alliariifolia**
Turkey & Caucasus (Abkhaziya, Armenia, Azhariya, Chechnya, Dagestan, Gruziya, Kabardino-Balkariya, Karacheyevo-Cherkessiya, Krasnodar, Severo-Ossetiya); cult.; naturalized in Great Britain. (10) grb 33 ALL 34 TUR. Hemicr. or geophyte. *2n* = 34, 68, 96.
Campanula lamiifolia Adam in F. Weber & D. Mohr, Beitr. Naturk. 1: 48 (1805).
Campanula macrophylla Sims, Bot. Mag. 23: tab. 912 (1806). *Campanula alliariifolia* var. *macrophylla* (Sims) A. DC., Monogr. Campan.: 241 (1830). *Campanula lamiifolia* var. *macrophylla* (Sims) A. DC. in DC., Prodr. 7: 464 (1839).
Campanula gundelia K. Koch, Linnaea 23: 634 (1850).
Campanula alliariifolia var. *cordata* Trautv., Trudy Imp. S.-Peterburgsk. Bot. Sada 6: 67 (1879).
Campanula makaschvilii E. A. Busch, Bot. Mater. Gerb. Bot. Inst. Komarova Akad. Nauk SSSR 7: 128 (1938).
Campanula alliariifolia subsp. *ochroleuca* Kem.-Nat., Trudy Tbilissk. Bot. Inst. 2:136 (1937). *Campanula ochroleuca* (Kem.-Nat.) Kem.-Nat., Zametki Sist. Geogr. Rast. 15: 10 (1949).
Campanula alliariifolia var. *alpestris* Kem.-Nat., Trudy Tbilissk. Bot. Inst. 2: 136 (1937). *Campanula ochroleuca* var. *alpestris* (Kem.-Nat.) Kem.-Nat., Zametki Sist. Geogr. Rast. 15: 10 (1949).
Campanula alliariifolia var. *rupestris* Kem.-Nat., Trudy Tbilissk. Bot. Inst. 2: 136(1937). *Campanula ochroleuca* var. *rupestris* (Kem.-Nat.) Kem.-Nat., Zametki Sist. Geogr. Rast. 15: 10 (1949).
Campanula alliariifolia var. *silvatica* Kem.-Nat., Trudy Tbilissk. Bot. Inst. 2: 136(1937). *Campanula ochroleuca* var. *silvatica* (Kem.-Nat.) Kem.-Nat., Zametki Sist. Geogr. Rast. 15: 10 (1949).
Campanula leskovii Fed., Bot. Mater. Gerb. Bot. Inst. Komarova Akad. Nauk SSSR 15: 378 (1953).
Campanula kirpicznikovii Fed., Bot. Mater. Gerb. Bot. Inst. Komarova Akad. Nauk SSSR 15: 379 (1953).

subsp. **letschchumensis** (Kem.-Nat.) Ogan., Candollea 50: 286 (1995).
Caucasus (Gruziya). 33 TCS. Hemicr.
* *Campanula letschchumensis* Kem.-Nat., Zametki Sist. Geogr. Rast. 15: 11 (1949).

Campanula alpestris All., Auct. Syn. Stirp. Horti Taur.: 11 (1773).
SW. Alps. 12 FRA 13 ITA. Hemicr. *2n* = 34.
Campanula allionii Vill., Prosp. Hisp. Pl. Dauphiné 22 (1779).
Campanula trilocularis Turra, Fl. Ital. Prodr.: 64 (1780).
Campanula nana Lam., Encycl. 1: 585 (1785).

Campanula alphonsii Wall. ex A. DC., Monogr. Campan.: 294 (1839).
S. India (Kerala, Tamil Nadu). 40 IND. Hemicr.

Campanula alpina Jacq., Enum. Stirp. Vindob.: 36 (1762). *Marianthemum alpinum* (Jacq.)
Schur, Sert. Fl. Transsilv.: 48 (1853). *Campanula macrorrhiza* var. *polycaulis* Vuk., Linnaea
26: 330 (1854). S. Europe. 11 AUT CZE GER POL 13 ALB BUL ITA ROM YUG 14 UKR.
Hemicr. or geophyte. 2*n* = 34.

   subsp. **alpina**
      E. Alps & Carpathians. 11 AUT CZE GER POL 13 ITA ROM YUG 14 UKR. Hemicr. or
         geophyte. 2*n* = 34.
      Campanula haynaldii Szontágh, Verh. K. K. Zool.-Bot. Ges. Wien 13: 1070 (1863), as
         'haynaldi'. *Campanula alpina* var. *haynaldii* (Szontágh) Nyman, Consp. Fl. Eur.: 477
         (1879), as 'haynaldi'.
      Campanula alpina var. *albiflora* Schur, Enum. Pl. Transsilv.: 433 (1866).
      Campanula alpina var. *frigida* Schur, Enum. Pl. Transsilv.: 433 (1866).
      Campanula alpina var. *calyculata* Schur, Enum. Pl. Transsilv.: 433 (1866).
      Campanula ciblesii Prodan, Bul. Sti. Acad. Republ. Populare Române, Sect. Biol. Sti.
         Agric., Ser. Bot. 9: 323 (1957).

   subsp. **orbelica** (Pančić) Urum., Spisy Bulg. Akad. Nauk 28: 147 (1923).
      Albania, Serbia, Bulgaria. 13 ALB BUL YUG. Hemicr. or geophyte.
      * *Campanula orbelica* Pančić, Elem. Fl. Bulg.: 48 (1883).

Campanula alsinoides Hook. f. & Thomson, J. Proc. Linn. Soc., Bot. 2: 24 (1858).
Pakistan & W. Himalaya. 40 PAK WHM. Hemicr.

Campanula americana L., Sp. Pl.: 164 (1753). *Specularia americana* (L.) Morgan ex J. James,
J. Cincinnati Soc. Nat. Hist. 7: 74 (1884). *Campanulastrum americanum* (L.) Small, Fl. S.E.
US: 1141 (1903).
E. North America (South Dakota to Ontario, New York, Florida & Oklahoma). 72 ONT 74
   ILL IOW KAN MIN MSO NEB OKL SDA WIS 75 INI MIC NWJ NWY OHI PEN WVA 78
   ARK FLA GEO KTY LOU MRY MSI NCA SCA TEN VRG. Biennial or ther. 2*n* = 34, 58, 102.
Phyteuma americanum Hill, Syst. Veg. 8: 15 (1765), as 'americana'.
Campanula planiflora Lam., Encycl. 1: 580 (1785).
Campanula nitida Aiton, Hort. Kew. 1: 221 (1789).
Campanula asteroides Lam., Tab. Encycl. 2: 55 (1796).
Campanula obliqua Jacq., Pl. Horti Schoenbr. 3: 47 (1798).
Campanula declinata Moench, Suppl. Meth.: 187 (1802).
Campanula acuminata Michx., Fl. Bor.-Amer. 1: 108 (1803).
Campanula americana var. *subulata* A. DC., Monogr. Campan.: 314 (1830).
Campanula illinoensis Fresen., Sem. Hort. Bot. Francofurt.: 4 (1836). *Campanula americana*
   var. *illinoensis* (Fresen.) Farw., Annual Rep. Michigan Acad. Sci. 20: 191 (1919).
Campanula americana f. *tubuliflora* Fernald, Rhodora 44: 458 (1942).
Campanula americana f. *albiflora* M. L. Grant, Proc. Iowa Acad. Sci. 60: 148 (1953).

Campanula amorgina Rech. f., Beih. Bot. Centralbl. 54 (abteil. B): 646 (1936).
S. Aegean Is. (Amorgós). 13 GRC. Hemicr. or geophyte.

Campanula anchusiflora Salisb. ex Sm. in Sibth. & Sm., Fl. Graec. Prodr. 1: 141 (1806).
Campanula rupestris subsp. *anchusiflora* (Salisb. ex Sm.) Hayek, Repert. Spec. Nov. Regni
Veg. Beih. 30(2): 525 (1930), as 'anchusaeflora'.
Greece. 13 GRC. Biennial. 2*n* = 34.

Campanula andina Rupr., Bull. Acad. Imp. Sci. Saint-Pétersbourg 11: 215 (1867). *Campanula
bayerniana* var. *andina* (Rupr.) Trautv., Trudy Imp. S.-Peterburgsk. Bot. Sada 2: 563 (1873).

Caucasus (Azerbaijan, Dagestan). 33 ALL. Hemicr. or geophyte.
Campanula gumbetica Boiss., Fl. Orient. 3: 914 (1875).
Campanula andina var. *alexeenkoi* Fomin, Trudy Tiflissk. Bot. Sada 6(2): 5 (1902).

Campanula andrewsii A. DC., Monogr. Campan.: 220 (1830). *Campanula rupestris* subsp. *andrewsii* (A. DC.) Hayek, Repert. Spec. Nov. Regni Veg. Beih. 30(2): 525 (1930).
Greece. 13 GRC. Biennial. $2n = 34$.

subsp. **andrewsii**
Greece. 13 GRC. Biennial. $2n = 34$.

subsp. **hirsutula** Phitos, Oesterr. Bot. Z. 112: 455 (1965).
Greece. 13 GRC. Biennial. $2n = 34$.

Campanula angustiflora Eastw., Proc. Calif. Acad. Sci. (ser. 3) 1: 132 (1898).
W. U.S.A. (N. & C. California). 76 CAL. Ther. $2n = 30$.

Campanula antalyensis Ayasligil & Kit Tan, Notes Roy. Bot. Gard. Edinburgh 42: 71 (1984).
Turkey. 34 TUR. Biennial.

Campanula antilibanotica (P. H. Davis) Greuter & Burdet, Willdenowia 11: 39 (1981).
Lebanon & Syria. 34 LBS. Hemicr.
* *Tracheliopsis antilibanotica* P. H. Davis, Hooker's Icon. Pl. 35: pl. 3498 (1950).

Campanula aparinoides Pursh, Fl. Amer. Sept. 1: 159 (1814).
E. North America (Saskatchewan to Nova Scotia & Georgia); naturalized in NW. U.S.A. & Finland. (10) fin 71 MAN SAS 72 NBR NSC ONT QUE 73 COL was WYO 74 ILL IOW MIN MSO NEB NDA SDA WIS 75 ALL 78 ALA DEL GEO KTY MRY NCA SCA TEN VRG. Hemicr. $2n = 34$.
Campanula aparinoides var. *multiflora* A. DC., Monogr. Campan.: 290 (1830).
Campanula aparinoides var. *erinoides* A. W. Wood, Class-book Bot.: 479 (1861).
Campanula aparinoides var. *rosea* Coleman, Kent Sci. Inst. Misc. Publ. 2: 24 (1874).
Campanula aparinoides var. *grandiflora* Holz., Minnesota Bot. Stud. 1: 566 (1896).
Campanula uliginosa Rydb. in Britt., Man. Fl. N. States: 885 (1901). *Campanula aparinoides* var. *uliginosa* (Rydb.) Gleason, Phytologia 4: 25 (1952).

Campanula ardonensis Rupr., Bull. Acad. Imp. Sci. Saint-Pétersbourg 11: 212 (1867).
Campanula tridentata var. *ardonensis* (Rupr.) Trautv., Trudy Imp. S.-Peterburgsk. Bot. Sada 6: 58 (1879). *Hemisphaera ardonensis* (Rupr.) Kolak. & Serdyuk., Okhrana Prir. Gruz. 12: 162 (1984).
Caucasus (Severo-Osetiya). 33 NCS. Hemicr. or geophyte.
Campanula ardonensis var. *latifolia* Fomin in N. M. Kusn., N. Bush & Fomin, Mater. Fl. Kavkaza 4(6): 67 (1905).

Campanula argaea Boiss. & Balansa in Boiss., Diagn. Pl. Orient. (ser. 2) 6: 119 (1859).
Turkey. 34 TUR. Biennial. $2n = 34$.

Campanula argentea Lam., Encycl. 1: 584 (1785).
Turkey. 34 TUR. Hemicr.
Campanula myosotidifolia Boiss. in Tchich., Asie Min., Bot. 2: 392 (1860).

Campanula argyrotricha Wall. ex A. DC. in DC., Prodr. 7: 473 (1839), as 'argyrothrica'.
Pakistan & Himalaya. 40 EHM IND NEP PAK WHM. Hemicr.

Campanula ariana Podlech, Mitt. Bot. Staatssamml. München 7: 109 (1968).
Afghanistan. 34 AFG. Hemicr. $2n = 28$.

Campanula aristata Wall. in Roxb., Fl. Ind. 2: 98 (1824).
Afghanistan to Himalaya & C. China (Gansu, Qinghai, Shaanxi, Sichuan, Yunnan). 34
AFG 36 CHC CHN CHT 40 EHM IND NEP PAK WHM. Hemicr.
Wahlenbergia cylindrica Pax & K. Hoffm., Repert. Spec. Nov. Regni Veg. Beih. 12: 501
(1922). *Campanula cylindrica* (Pax & K. Hoffm.) Nannf., Acta Horti Gothob. 5: 24
(1930).
Campanula aristata var. *longisepala* C. Marquand, J. Linn. Soc., Bot. 48: 196 (1929).

Campanula armena Steven, Mém. Soc. Imp. Naturalistes Moscou 3: 256 (1812).
Symphyandra armena (Steven) A. DC., Monogr. Campan.: 367 (1830).
Turkey to Caucasus (Armenia, Azerbaijan), Turkmenistan & Iran. 32 TKM 33 TCS 34 IRN
TUR. Hemicr. or geophyte. $2n = 34$.
Symphyandra armena var. *serratosepala* Fomin in N. M. Kusn., N. Bush & Fomin, Mater.
Fl. Kavkaza 4(6): 152 (1906).
Symphyandra daralaghezica Grossh., Trudy Azerbajdžansk. Otd. Zakavkazsk. Fil. Akad.
Nauk SSSR 1: 57 (1933). *Symphyandra armena* subsp. *daralaghezica* (Grossh.) Fed. in
Takht. & Fed., Fl. Erevana: 267 (1972). *Campanula daralaghezica* (Grossh.) Kolak. &
Serdyuk., Zametki Sist. Geogr. Rast. 36: 46 (1980).

Campanula arvatica Lag., Varied. Ci. 2(4): 40 (1805). *Wahlenbergia arvatica* (Lag.) Sweet,
Hort. Brit. (ed. 2) 593 (1830). *Wahlenbergia hederacea* var. *arvatica* (Lag.) Steud., Nomencl.
Bot. (ed. 2) 2: 782 (1841).
Spain. 12 SPA. Hemicr. or geophyte. $2n = 28$.

subsp. **adsurgens** (Levier & Leresche) Damboldt, Ber. Deutsch. Bot. Ges. 79: 305 (1966).
NW. Spain. 12 SPA. Hemicr. or geophyte. $2n = 28$.
* *Campanula adsurgens* Levier & Leresche, J. Bot. 17: 199 (1879).

subsp. **arvatica**
N. Spain. 12 SPA. Hemicr. or geophyte. $2n = 28$.
Campanula acutangula Leresche & Levier, J. Bot. 17: 198 (1879).

Campanula asperuloides (Boiss. & Orph.) Engl. in Engl. & Prantl, Nat. Pflanzenfam.,
Nachtr. 1: 319 (1897).
Greece. 13 GRC. Hemicr. $2n = 34$.
* *Trachelium asperuloides* Boiss. & Orph. in Boiss., Diagn. Pl. Orient. (ser. 2) 3: 117 (1856).
Diosphaera asperuloides (Boiss. & Orph.) Buser, Bull. Herb. Boissier 2: 523 (1894).

subsp. **asperuloides**
Greece. 13 GRC. Hemicr. $2n = 34$.
Diosphaera asperuloides var. *denudata* Buser, Bull. Herb. Boissier 2: 524 (1894).
Diosphaera asperuloides var. *vestita* Buser, Bull. Herb. Boissier 2: 524 (1894).

subsp. **taygetea** (Quézel & Contandr.) Greuter & Burdet, Willdenowia 11: 39 (1981).
Greece. 13 GRC. Hemicr. $2n = 34$.
* *Trachelium taygeteum* Quézel & Contandr., Taxon 16: 240 (1967).

Campanula atlantis Gattef., Maire & Weiller, Bull. Soc. Hist. Nat. Afrique N. 28: 367 (1937).
Morocco. 20 MOR. Hemicr. $2n = 24$.

Campanula aurita Greene, Pittonia 1: 221 (1888). *Astrocodon auritus* (Greene) A. P. Khokhr.,
Analiz Fl. Kolymsk. Nagor'ya: 122 (1989). $2n = 34$.
Alaska & W. Canada. 70 ASK NWT YUK 71 BRC. Hemicr.

Campanula austroxinjiangensis Y. K. Yang, J. K. Wu & J. Z. Li, Acta Phytotax. Sin. 30: 92
(1992), as 'austro-xinjiangensis'.
NW. China (Xinjiang). 36 CHX. Hemicr.

Campanula autraniana Albov, Bull. Herb. Boissier 2: 115 (1894).
Caucasus (Gruziya). 33 TCS. Hemicr. or geophyte. $2n = 34$.

Campanula axillaris Boiss. & Balansa in Boiss., Diagn. Pl. Orient. (ser. 2) 3: 109 (1856).
Turkey. 34 TUR. Hemicr.

Campanula baborensis Quézel, Feddes Repert. Spec. Nov. Regni Veg. 56: 19 (1953).
Algeria & Saudi Arabia. 20 ALG 35 SAU. Hemicr.

Campanula balansae Boiss. & Hausskn. in Boiss., Fl. Orient. 3: 931 (1875).
Turkey. 34 TUR. Ther. $2n = 16$.

Campanula balfourii R. Wagner & Vierh., Oesterr. Bot. Z. 56: 301 (1906).
Socotra. 24 SOC. Ther.

Campanula barbata L., Syst. Nat. (ed. 10) 2: 926 (1759). *Medium barbatum* (L.) Spach, Hist.
Nat. Vég. 9: 555 (1838). *Marianthemum barbatum* (L.) Schur, Sert. Fl. Transsilv.: 48 (1853).
Campanula macrorrhiza var. *pogonopetala* Vuk., Linnaea 26: 330 (1854).
Europe. 10 NOR 11 AUT CZE GER POL SWI 12 FRA SPA 13 ITA YUG. Hemicr. or
geophyte. $2n = 34$.
Campanula firmiana Vand., Fasc. Pl.: 9 (1771). *Campanula barbata* var. *firmiana* (Vand.)
Steud., Nomencl. Bot. (ed. 2) 1: 266 (1840).
Campanula barbata var. *pusilla* Gaudin, Fl. Helv. 2: 163 (1828). *Campanula barbata* f.
pusilla (Gaudin) Hayek & Hegi in Hegi, Ill. Fl. Mitt.-Eur. 6: 336 (1915).
Campanula barbata var. *uniflora* A. DC., Monogr. Campan.: 247 (1830).
Campanula barbata var. *strictopedunculata* Thomas ex Rchb., Icon. Fl. Germ. Helv. 19:
110 (1859).

Campanula baumgartenii Becker, Fl. Frankfurt 1: 264 (1827), as 'baumgarteni'. *Campanula
rotundifolia* var. *baumgartenii* (Becker) Nyman, Consp. Fl. Eur.: 479 (1879). *Campanula
rotundifolia* subsp. *baumgartenii* (Becker) Rouy, Fl. France 10: 80 (1908), as 'baumgarteni'.
C. Europe. 11 AUT GER 12 FRA 13 YUG. Hemicr. or geophyte. $2n = 68$.

subsp. **baumgartenii**
France & Germany. 11 GER 12 FRA. Hemicr. or geophyte. $2n = 68$.
Campanula rotundifolia var. *reniformis* Pers., Syn. Pl. 1: 188 (1805). *Campanula pusilla*
var. *reniformis* (Pers.) Steud., Nomencl. Bot.: 144 (1821). *Campanula rotundifolia* var.
lancifolia Mert. & W. D. J. Koch in Röhl., Deutschl. Fl. (ed. 3) 2: 156 (1826).
Campanula lancifolia Witasek, Abh. K. K. Zool.-Bot. Ges. Wien 1(3): 84 (1902); non
Roxb., Fl. Ind. 2: 96 (1824).

subsp. **beckiana** (Hayek) Podlech, Feddes Repert. 71: 122 (1965).
Austria & Slovenia. 11 AUT 13 YUG. Hemicr. or geophyte. $2n = 68$.
Campanula rotundifolia var. *major* Neilr., Fl. Wien: 298 (1846); non A. DC., Monogr.
Campan.: 282 (1830).
Campanula rotundifolia var. *multiflora* Neilr., Fl. Nied.-Oesterr.: 448 (1858).
* *Campanula beckiana* Hayek, Fl. Steiermark 2(1): 455 (1912).

Campanula bayerniana Rupr., Bull. Acad. Imp. Sci. Saint-Pétersbourg 11: 214 (1867).
Caucasus (Armenia, Nackhichevan, Nagorno Karabakh) & Iran. 33 TCS 34 IRN. Hemicr.
or geophyte.
Campanula bayerniana var. *trautvetteri* Fomin, Trudy Tiflissk. Bot. Sada 6(2): 7 (1902).
Campanula choziatowskyi Fomin, Trudy Tiflissk. Bot. Sada 4(3): 38 (1904). *Campanula
bayerniana* subsp. *choziatowskyi* (Fomin) Ogan., Bot. Žurn. (Moscow & Leningrad) 66:
402 (1981).
Campanula elegantissima Grossh., Trudy Azerbajdžana Otd. Zakavkazsk. Fil. Akad. Nauk
SSSR 1: 56 (1933).
Campanula betulifolia var. *glaberrima* Parsa, Kew Bull. 1948: 209 (1948).

Campanula takhtadzhianii Fed., Bot. Mater. Gerb. Bot. Inst. Komarova Akad. Nauk SSSR 15: 374 (1953).

Campanula savalanica Fed., Bot. Mater. Gerb. Bot. Inst. Komarova Akad. Nauk SSSR 15: 376 (1953).

Campanula bellidifolia Adam in F. Weber & D. Mohr, Beitr. Naturk. 1: 47 (1805). *Campanula tridentata* var. *bellidifolia* (Adam) Trautv., Trudy Imp. S.-Peterburgsk. Bot. Sada 4: 388 (1876). *Hemisphaera bellidifolia* (Adam) Kolak. & Serdyuk., Okhrana Prir. Gruz. 12: 163 (1984). Turkey, Caucasus (Krasnodar, Karacheyevo-Cherkessiya, Kabardino-Balkariya, Severo-Osetiya, Chechnya, Dagestan, Adzhariya, Gruziya, Armenia, Nakhichevan, Nagorno Karabakh, Azerbaijan), Iran. 33 ALL 34 IRN TUR. Hemicr. or geophyte. $2n = 34$.

subsp. **argunensis** (Rupr.) Viktorov, Bot. Žurn. (Moscow & Leningrad) 86(9): 118 (2001). Caucasus (Severo-Osetiya, Chechnya, Dagestan, Gruziya, Azerbaijan). 33 ALL. Hemicr. or geophyte.

Campanula pubiflora Rupr., Bull. Acad. Imp. Sci. Saint-Pétersbourg 11: 207 (1867). *Campanula tridentata* var. *pubiflora* (Rupr.) Trautv., Trudy Imp. S.-Peterburgsk. Bot. Sada 2: 562 (1873).

* *Campanula argunensis* Rupr., Bull. Acad. Imp. Sci. Saint-Pétersbourg 11: 209 (1867). *Campanula tridentata* var. *argunensis* (Rupr.) Trautv., Trudy Imp. S.-Peterburgsk. Bot. Sada 4: 388 (1876). *Hemisphaera argunensis* (Rupr.) Kolak. & Serdyuk., Okhrana Prir. Gruz. 12: 162 (1984). *Campanula saxifraga* subsp. *argunensis* (Rupr.) Ogan., Bot. Žurn. (Moscow & Leningrad) 78(3): 148 (1993).

Campanula tridentata var. *petrophila* Trautv., Trudy Imp. S.-Peterburgsk. Bot. Sada 10: 120 (1890); non (Rupr.) Trautv., Trudy Imp. S.-Peterburgsk. Bot. Sada 6: 58 (1879).

Campanula doluchanovii Kharadze, Zametki Sist. Geogr. Rast. 13: 54 (1947), as 'doluchanovi'. *Hemisphaera doluchanovii* (Kharadze) Kolak., Okhrana Prir. Gruz. 12: 158 (1984).

subsp. **aucheri** (A. DC.) Viktorov, Bot. Žurn. (Moscow & Leningrad) 86(9): 118 (2001). Turkey, Caucasus (Adzhariya, Gruziya, Armenia, Nakhichevan, Nagorno Karabakh), Iran. 33 ALL 34 IRN TUR. Hemicr. or geophyte. $2n = 34$.

* *Campanula aucheri* A. DC. in DC., Prodr. 7: 460 (1839). *Hemisphaera aucheri* (A. DC.) Kolak., Okhrana Prir. Gruz. 12: 163 (1984). *Campanula saxifraga* subsp. *aucheri* (A. DC.) Ogan., Bot. Žurn. (Moscow & Leningrad) 78(3): 149 (1993).

Campanula pilosa var. *pontica* K. Koch, Linnaea 23: 638 (1850).

Campanula alpigena K. Koch, Linnaea 23: 638 (1850). *Hemisphaera alpigena* (K. Koch) Kolak., Okhrana Prir. Gruz. 12: 162 (1984).

Campanula pallidiflora Rupr., Bull. Acad. Imp. Sci. Saint-Pétersbourg 11: 179 (1867).

Campanula gilanica Rupr., Bull. Acad. Imp. Sci. Saint-Pétersbourg 11: 182 (1867). *Campanula ruprechtii* Boiss., Fl. Orient. 3: 905 (1875), as 'ruprechti'. *Campanula tridentata* var. *gilanica* (Rupr.) Trautv., Trudy Imp. S.-Peterburgsk. Bot. Sada 4: 387 (1876). *Hemisphaera ruprechtii* Kolak. & Serdyuk., Okhrana Prir. Gruz. 12: 167 (1984).

Campanula saxifraga var. *transcaucasica* Rupr., Bull. Acad. Imp. Sci. Saint-Pétersbourg 11: 184 (1867).

Campanula fallax Rupr., Bull. Acad. Imp. Sci. Saint-Pétersbourg 11: 204 (1867).

Campanula affinis Fisch. ex Rupr., Bull. Acad. Imp. Sci. Saint-Pétersbourg 11: 205 (1867); *non* Schult. in Roem. & Schult., Syst. Veg. 5: 140 (1819).

Campanula hygrophila Rupr., Bull. Acad. Imp. Sci. Saint-Pétersbourg 11: 209 (1867).

Campanula hygrophila var. *obovata* Rupr., Bull. Acad. Imp. Sci. Saint-Pétersbourg 11: 209 (1867).

Campanula aucheri var. *compacta* Fomin in N. M. Kusn., N. Bush & Fomin, Mater. Fl. Kavkaza 4(6): 61 (1905). *Campanula armazica* Kharadze, Zametki Sist. Geogr. Rast. 13: 51 (1947); non *Campanula compacta* Hegetschw., Fl. Schweiz: 232 (1838-39); nec Boiss. & Heldr. in Boiss., Diagn. Pl. Orient. (ser. 1) 11: 71 (1849). *Hemisphaera armazica* (Kharadze) Kolak., Okhrana Prir. Gruz. 12: 163 (1984).

Campanula froedinii Rech. f., Österr. Akad. Wiss., Math.-Naturwiss. Kl., Anz. 87: 195 (1950).

subsp. **bellidifolia**

Caucasus (Severo-Osetiya). 33 NCS. Hemicr. or geophyte. $2n = 34$.

Campanula adamii M. Bieb., Fl. Taur.-Cauc. 1: 155 (1808), as 'adami'.

Campanula bellidifolia var. *longisepala* Fomin in N. M. Kusn., N. Bush & Fomin, Mater. Fl. Kavkaza 4(6): 65 (1905).

Campanula sosnowskyi Kharadze, Zametki Sist. Geogr. Rast. 13: 109 (1947). *Hemisphaera sosnowskyi* (Kharadze) Kolak., Okhrana Prir. Gruz. 12: 167 (1984).

subsp. **besenginica** (Fomin) Viktorov, Novosti Sist. Vyssh. Rast. 34: 223 (2002).

Caucasus (Karacheyevo-Cherkessiya). 33 NCS. Hemicr.

* *Campanula besenginica* Fomin, Trudy Tiflissk. Bot. Sada 6(2): 8 (1902). *Hemisphaera besengica* (Fomin) Kolak. & Serdyuk., Okhrana Prir. Gruz. 12: 163 (1984).

subsp. **meyerana** (Rupr.) Viktorov, Bot. Žurn. (Moscow & Leningrad) 86(9): 119 (2001).

Caucasus (Azerbaijan). 33 TCS. Hemicr. or geophyte.

* *Campanula meyerana* Rupr., Bull. Acad. Imp. Sci. Saint-Pétersbourg 11: 207 (1867). *Hemisphaera meyerana* (Rupr.) Kolak. & Serdyuk., Okhrana Prir. Gruz. 12: 166 (1984), as 'meyeriana'. *Campanula saxifraga* subsp. *meyerana* (Rupr.) Ogan., Bot. Žurn. (Moscow & Leningrad) 78(3): 148 (1993).

Campanula fominii Grossh., Trudy Azerbajdžana Otd. Zakavkazsk. Fil. Akad. Nauk SSSR 1: 56 (1933), as 'fomini'. *Hemisphaera fominii* (Grossh.) Kolak. & Serdyuk., Okhrana Prir. Gruz. 12: 166 (1984).

subsp. **saxifraga** (M. Bieb.) Viktorov, Bot. Žurn. (Moscow & Leningrad) 86(9): 118 (2001).

Caucasus (Krasnodar, Karacheyevo-Cherkessiya, Kabardino-Balkariya, Severo-Osetiya, Gruziya). 33 ALL. Hemicr. or geophyte.

* *Campanula saxifraga* M. Bieb., Fl. Taur.-Cauc. 1: 155 (1808). *Campanula tridentata* var. *saxifraga* Trautv., Trudy Imp. S.-Peterburgsk. Bot. Sada 2: 562 (1873). *Hemisphaera saxifraga* (M. Bieb.) Kolak., Okhrana Prir. Gruz. 12: 167 (1984).

Campanula saxifraga var. *leptorhiza* Sommier & Levier, Enum. Pl. Cauc.: 318 (1900).

Campanula saxifraga f. *stenophylla* Fomin in N. M. Kusn., N. Bush & Fomin, Mater. Fl. Kavkaza 4(6): 55 (1905).

Campanula bertolae Colla, Herb. Pedem. 4: 24 (1835). *Campanula rotundifolia* var. *bertolae* (Colla) Nyman, Consp. Fl. Eur., Suppl. 2: 208 (1889). *Campanula rotundifolia* subsp. *bertolae* (Colla) Vacc., Cat. Pl. Vall. Aoste: 607 (1911). *Campanula macrorhiza* var. *bertolae* (Colla) Hruby, Magyar Bot. Lapok 33: 150 (1934).

SW. Alps. 13 ITA. Hemicr. or geophyte. $2n = 102$.

Campanula re Colla, Herb. Pedem. 4: 25 (1835). *Campanula rotundifolia* var. *re* (Colla) Nyman, Consp. Fl. Eur., Suppl. 2: 208 (1889). *Campanula bertolae* var. *re* (Colla) Gola, Mem. Reale Accad. Sci. Torino (ser. 2) 60: 237 (1909). *Campanula rotundifolia* subvar. *re* (Colla) Hruby, Magyar Bot. Lapok 29: 163 (1930). *Campanula macrorhiza* subvar. *re* (Colla) Hruby, Magyar Bot. Lapok 33: 150 (1934).

Campanula betonicifolia Sm. in Sibth. & Sm., Fl. Graec. Prodr. 1: 141 (1806).

Turkey. 34 TUR. Biennial. $2n = 34$.

Campanula cariensis A. DC. in DC., Prodr. 7: 459 (1839).

Campanula betonicifolia var. *multicaulis* K. Koch, Linnaea 19: 28 (1845).

Campanula betonicifolia var. *micrantha* K. Koch, Linnaea 19: 29 (1845).

Campanula betonicifolia var. *byzantina* K. Koch, Linnaea 23: 632 (1850).

Campanula betulifolia K. Koch, Linnaea 23: 635 (1850), as 'betulaefolia'.

Turkey. 34 TUR. Hemicr. or geophyte.

Campanula denticulata Boiss. & A. Huet, Diagn. Pl. Orient. (ser. 2) 3: 107 (1856); non Burch., Trav. S. Africa 1: 538 (1822); nec (Fisch.) Spreng., Syst. Veg. 1: 735 (1824).

Campanula betulifolia var. *exappendiculata* Albov, Trudy Imp. S.-Peterburgsk. Bot. Sada 13: 116 (1893).

Campanula finitima Fomin, Věstn. Tiflissk. Bot. Sada 1: 15 (1905). *Symphyandra finitima* (Fomin) Fomin in N. M. Kusn., N. Bush & Fomin, Mater. Fl. Kavkaza 4(6): 151 (1906).

Campanula bipinnatifida P. H. Davis, Notes Roy. Bot. Gard. Edinburgh 22: 78 (1956). Turkey. 34 TUR. Hemicr.

Campanula bluemelii Halda, Preslia 61: 321 (1989). Turkey. 34 TUR. Hemicr. or geophyte.

Campanula bohemica Hruby in Domin & Podp., Klíč Úplné Květ. Republ. Českoslov.: 534 (1928). *Campanula baumgartenii* subsp. *bohemica* (Hruby) Tacik in Pawłowskiego & Jasiewicz, Fl. Polska 12: 78 (1971). Czech Rep. (Sudeten Mts.). 11 CZE. Hemicr. or geophyte. 2*n* = 68.

subsp. **bohemica**
   Czech Rep. (W. Sudeten Mts.) 11 CZE. Hemicr. or geophyte. 2*n* = 68.
   Campanula rotundifolia var. *grandiflora* Wimm., Fl. Schles.: 241 (1840).
   Campanula corcontica Šourek, Preslia 25: 7 (1953).

subsp. **gelida** (Kovanda) Kovanda, Folia Geobot. Phytotax. 12: 81 (1977).
   Czech Rep. (E. Sudeten Mts.) 11 CZE. Hemicr. or geophyte. 2*n* = 68.
  * *Campanula gelida* Kovanda, Folia Geobot. Phytotax. 3: 408 (1968).

Campanula bononiensis L., Sp. Pl. 165 (1753). *Campanula racemosa* var. *arctiflora* Vuk., Linnaea 26: 32 (1854).
   France to W. Siberia, Kazakhstan & Iran; naturalized in NE. U.S.A. 11 AUT CZE GER HUN POL SWI 12 FRA 13 ALB BUL GRC ITA ROM YUG 14 BLR BLT KRY RUC RUE RUS UKR 30 ALT WSB 32 KAZ 34 IRN TUR (75) mai. Hemicr. 2*n* = 34, 68.
   Campanula pyramidata Gilib., Hist. Pl. Europe 2: 209 (1798).
   Campanula simplex DC. in Lam. & DC., Fl. Franç. (ed. 3) 3: 730 (1805). *Cenekia simplex* (DC.) Opiz, Seznam: 36 (1852). *Campanula bononiensis* var. *simplex* (DC.) Steud., Nomencl. Bot.: 141 (1821).
   Campanula ruthenica M. Bieb., Fl. Taur.-Cauc. 1: 151 (1808). *Campanula bononiensis* var. *ruthenica* (M. Bieb.) A. DC. in DC., Prodr. 7: 470 (1839). *Campanula bononiensis* subvar. *ruthenica* (M. Bieb.) Nyman, Consp. Fl. Eur.: 478 (1879).
   Campanula obliquifolia Ten., Fl. Napol. 1: xv (1811). *Campanula bononiensis* var. *obliquifolia* (Ten.) A. DC. in DC., Prodr. 7: 470 (1839). *Campanula bononiensis* subsp. *obliquifolia* (Ten.) Arcang., Comp. Fl. Ital.: 455 (1882).
   Campanula lychnitis Hornem., Hort. Bot. Hafn.: 199 (1815). *Campanula bononiensis* var. *lychnitis* (Hornem.) A. DC. in DC., Prodr. 7: 470 (1839).
   Campanula thaliana Wallr., Sched. Crit.: 86 (1822).
   Campanula bononiensis var. *latifolia* Schur, Enum. Pl. Transsilv.: 437 (1866).
   Campanula tenuiflora Schur, Enum. Pl. Transsilv.: 438 (1866); non Ten., Fl. Napol. 3: 207 (1824). *Campanula bononiensis* var. *tenuiflora* Nyman, Consp. Fl. Eur.: 478 (1879). *Campanula bononiensis* f. *tenuiflora* (Nyman) Sóo, Acta Bot. Acad. Sci. Hung. 12: 366 (1966).
   Campanula cana Simonk., Enum. Fl. Transsilv.: 383 (1887); non Wall. in Roxb., Fl. Ind. 2: 101 (1824). *Campanula bononiensis* var. *cana* Nyman, Consp. Fl. Eur., Suppl. 2: 208 (1889).

Campanula bordesiana Maire, Bull. Soc. Hist. Nat. Afrique N. 20: 188 (1929). N. Africa. 20 ALG LBY 24 CHA. Hemicr. 2*n* = 80, 84.

Campanula bornmuelleri Nábělek, Spisy Přír. Fak. Masarykovy Univ. 70: 3 (1926), as 'bornmülleri'. Turkey. 34 TUR. Hemicr.

Campanula bravensis (Bolle) A. Chev., Rev. Int. Bot. Appl. Agric. Trop. 15: 889 (1935).
Cape Verde (Brava, Fogo, Santiago). 21 CVI. Hemicr. 2*n* = 54.
  * *Campanula jacobaea* var. *bravensis* Bolle, Bonplandia 9: 51 (1861).

Campanula buseri Damboldt, Notes Roy. Bot. Gard. Edinburgh 35: 47 (1976).
Turkey, Lebanon, Syria. 34 LBS TUR. Hemicr. 2*n* = 34.
  * *Trachelium tubulosum* Boiss., Diagn. Pl. Orient. (ser. 1) 11: 60 (1849). *Tracheliopsis tubulosa* (Boiss.) Buser, Bull. Herb. Boissier 2: 525 (1894). *Campanula tubulosa* (Boiss.) Engl. in Engl. & Prantl, Nat. Pflanzenfam., Nachtr. 1: 319 (1897); non Lam., Encycl. 1: 586 (1785). *Diosphaera tubulosa* (Boiss.) Bornm., Beih. Bot. Centralbl. 38(2): 338 (1921).
  Tracheliopsis tubulosa var. *berytensis* Buser, Bull. Herb. Boissier 2: 526 (1894).
  Tracheliopsis tubulosa var. *taurica* Buser, Bull. Herb. Boissier 2: 526 (1894).
  Tracheliopsis tubulosa var. *libanotica* Buser, Bull. Herb. Boissier 2: 527 (1894).

Campanula calamenthifolia Lam., Encycl. 1: 585 (1785).
S. Aegean Is. (Náxos, Ikaría). 13 GRC 34 EAI. Hemicr. or geophyte. 2*n* = 34.
  Campanula olivieri A. DC., Monogr. Campan.: 233 (1830).

Campanula calcarata Sommier & Levier, Nuovo Giorn. Bot. Ital. (n.s.) 2: 96 (1895).
Caucasus (Karacheyevo-Cherkessiya). 33 NCS. Hemicr. or geophyte.

Campanula calcicola W. W. Sm., Notes Roy. Bot. Gard. Edinburgh 12: 196 (1920).
SC. China (Sichuan, Yunnan). 36 CHC. Hemicr. or geophyte.

Campanula californica (Kellogg) A. Heller, Muhlenbergia 1: 46 (1904).
W. U.S.A. (N. & C. California). 76 CAL. Hemicr. or geophyte.
  * *Wahlenbergia californica* Kellogg, Proc. Calif. Acad. Sci. 2: 158 (1862).
  Campanula linnaeifolia A. Gray, Proc. Amer. Acad. Arts 7: 366 (1867).

Campanula calycialata V. Randjel. & Zlatković, Flora Medit. 8: 87 (1998).
Serbia. 13 YUG. Hemicr. or geophyte.

Campanula camptoclada Boiss., Diagn. Pl. Orient. (ser. 1) 11: 63 (1849).
Israel & Lebanon. 34 LBS PAL. Ther.

Campanula cana Wall. in Roxb., Fl. Ind. 2: 101 (1824).
Himalaya to SC. China (Sichuan, Yunnan) & Myanmar. 36 CHC CHT 40 EHM NEP WHM 41 MYA. Hemicr.
  Campanula pasumensis C. Marquand, J. Linn. Soc., Bot. 48: 196 (1929).
  Campanula aprica Nannf., Acta Horti Gothob. 5: 23 (1930).
  Campanula xylopoda Jeffrey, Notes Roy. Bot. Gard. Edinburgh 10: 17 (1917).
  Campanula tortuosa C. Y. Wu, Rep. Yunnan Trop. Subtrop. Fl. Res. Inst. 1: 62 (1965).

Campanula candida A. DC., Monogr. Campan.: 234 (1830).
Iran. 34 IRN. Hemicr.

Campanula cantabrica Feer, J. Bot. 28: 273 (1890).
N. Spain (Cordillera Cantábrica). 12 SPA. Hemicr. or geophyte. 2*n* = 34.
  Campanula cantabrica subsp. *occidentalis* M. Laínz, Mis Contrib. Conocim. Fl. Asturias: 65 (1982).

Campanula carnica Schiede ex Mert. & Koch in Röhl., Deutschl. Fl. (ed. 3) 2: 158 (1826).
Alps & Carpathians. 11 AUT 13 ITA ROM YUG. Hemicr. or geophyte. 2*n* = 34.
  * *Campanula linifolia* Scop., Annus Hist.-Nat. 2: 47 (1769); non L., Fl. Monsp.: 12 (1756).
  Campanula scheuchzeri var. *carnica* (Schiede ex Mert. & Koch) Posp., Fl. Oesterr. Küstenl.

2: 683 (1899). *Campanula rotundifolia* var. *linifolia* Bég. in Fiori & Paol., Fl. Italia 3: 185
(1903). *Campanula rotundifolia* subsp. *linifolia* Rouy, Fl. France 10: 81 (1908); non
Lapeyr., Hist. Pl. Pyrénées: 104 (1813); nec Arcang., Comp. Fl. Ital.: 454 (1882).

subsp. **carnica**
   Alps & Carpathians. 11 AUT 13 ITA ROM YUG. Hemicr. or geophyte. $2n = 34$.
   Campanula carnica var. *angustifolia* Schur, Enum. Pl. Transsilv.: 442 (1866).
   Campanula carnica var. *latifolia* Schur, Enum. Pl. Transsilv.: 442 (1866).
   Campanula scheuchzeri var. *kladniana* Schur, Enum. Pl. Transsilv.: 443 (1866).
      Campanula kladniana (Schur) Witasek, Abh. K. K. Zool.-Bot. Ges. Wien 1(3): 39
      (1902). *Campanula rotundifolia* subsp. *kladniana* (Schur) Tacik in Pawłowskiego &
      Jasiewicz, Fl. Polska 12: 76 (1971).
   Campanula carnica var. *hirta* Gelmi, Prosp. Fl. Trent: 109 (1893). *Campanula linifolia* var.
      hirta (Gelmi) Dalla Torre & Sarnth., Fl. Tirol 6(3): 442 (1911).
   Campanula linifolia var. *ciliata* Witasek ex Vacc., Cat. Pl. Vall. Aoste: 608 (1911).
   Campanula linifolia f. *latifrons* Hruby, Magyar Bot. Lapok 29: 225 (1930).
   Campanula linifolia subf. *integrifrons* Hruby, Magyar Bot. Lapok 29: 225 (1930).
   Campanula linifolia f. *angustifrons* Hruby, Magyar Bot. Lapok 29: 225 (1930).
   Campanula linifolia subf. *angustissima* Hruby, Magyar Bot. Lapok 29: 225 (1930).
   Campanula linifolia f. *umbrosa* Hruby, Magyar Bot. Lapok 29: 225 (1930).
   Campanula linifolia f. *vestina* Hruby, Magyar Bot. Lapok 29: 226 (1930).
   Campanula linifolia var. *portae* Hruby, Magyar Bot. Lapok 29: 226 (1930).

subsp. **puberula** Podlech, Feddes Repert. 71: 95 (1965).
   Alps. 13 ITA. Hemicr. or geophyte.

Campanula carpatha Halácsy, Consp. Fl. Graec. 2: 252 (1902). *Campanula tubulosa* var.
   carpatha (Halácsy) Hayek, Repert. Spec. Nov. Regni Veg. Beih. 30(2): 523 (1930).
   S. Aegean Is. (Kárpathos). 13 KRI. Hemicr. $2n = 34$.
   Campanula carpatha var. *oreophila* Phitos, Oesterr. Bot. Z. 112: 475 (1965).

Campanula carpatica Jacq., Hort. Vindob. 1: 22 (1770). *Campanula cordifolia* Vuk., Linnaea
26: 328 (1854); non K. Koch, Linnaea 19: 29 (1845). *Neocodon carpaticus* (Jacq.) Kolak. &
Serdyuk., Zametki Sist. Geogr. Rast. 40: 28 (1984).
   Carpathians; naturalized in NE. U.S.A. 11 CZE POL 13 ROM 14 UKR (75) cnt mic. Hemicr.
   $2n = 32, 34$.
   Campanula turbinata Schott in Schott, Nyman & Kotschy, Analect. Bot.: 14 (1854).
      Campanula carpatica subsp. *turbinata* (Schott) Nyman, Consp. Fl. Eur.: 482 (1879).
      Campanula carpatica var. *turbinata* (Schott) Anonymous, Garden (London) 45: 171
      (1893).
   Campanula carpatica var. *hemisphaerica* Schur, Enum. Pl. Transsilv.: 440 (1866).
   Campanula carpatica var. *subpilosa* Schur, Enum. Pl. Transsilv.: 440 (1866). *Campanula*
      carpatica f. *subpilosa* (Schur) Tacik in Pawłowskiego & Jasiewicz, Fl. Polska 12: 84 (1971).
   Campanula carpatica var. *dasycarpa* Schur, Enum. Pl. Transsilv.: 440 (1866). *Campanula*
      carpatica f. *dasycarpa* (Schur) Tacik in Pawłowskiego & Jasiewicz, Fl. Polska 12: 84
      (1971).
   Campanula carpatica var. *grandiflora* Schur, Enum. Pl. Transsilv.: 440 (1866), as
      'granpiflora'.
   Campanula carpatica var. *oreophila* Schur, Enum. Pl. Transsilv.: 440 (1866).
   Campanula carpatica var. *pelviformis* Anonymous, Rev. Hort. 1882: 509 (1882).
   Campanula hendersonii Anonymous, Gard. Chron. (ser. 2) 18: 502 (1882). *Campanula*
      carpatica var. *hendersonii* (Anonymous) W. T. Mill. in L. H. Bailey, Cycl. Amer. Hort.:
      231 (1900), as 'hendersoni'.
   Campanula turbinata f. *alba* Voss in Siebert & Voss, Vilm. Blumengärtn. (ed. 3) 1: 570
      (1894).
   Campanula turbinata f. *lilacina* Voss in Siebert & Voss, Vilm. Blumengärtn. (ed. 3) 1:
      570 (1894).

Campanula turbinata f. *pelviformis* Voss in Siebert & Voss, Vilm. Blumengärtn. (ed. 3) 1: 570 (1894).

Campanula fergusonii Ferguson, Rev. Hort. 1904: 557 (1904), as 'fergusoni'.

Campanula cashmeriana Benth. in Royle, Ill. Bot. Himal. Mts.: pl. 62 (1835).
Tadzhikistan to E. Himalyas; cult. 32 KGZ TZK 34 AFG 40 EHM PAK WHM. Hemicr. or geophyte. $2n = 28$.

Campanula evolvulacea Royle ex A. DC. in DC., Prodr. 7: 473 (1839). *Campanula cashmeriana* var. *evolvulacea* (Royle ex A. DC.) C. B. Clarke in Hook. f., Fl. Brit. India 3: 441 (1881).

Campanula ruderalis Aitch. & Hemsl., J. Linn. Soc., Bot. 19: 174 (1882).

Campanula incanescens var. *mollis* Korsh., Izv. Imp. Akad. Nauk (ser. 5) 4: 426 (1898).

Campanula incanescens var. *holosericea* Korsh., Izv. Imp. Akad. Nauk (ser. 5) 4: 433 (1898).

Campanula caucasica M. Bieb., Tabl. Prov. Mer Casp.: 112 (1798). *Campanula sibirica* var. *caucasica* (M. Bieb.) Trautv., Pl. Casp.-Caucas.: 96 (1877). *Campanula sibirica* subsp. *caucasica* (M. Bieb.) Viktorov, Novosti Sist. Vyssh. Rast. 32: 168 (2000).
Caucasus (Azerbaijan, Dagestan). 33 ALL. Hemicr. $2n = 102$.

Campanula caucasica var. *major* K. Koch, Linnaea 23: 637 (1850).

Campanula celsii A. DC., Monogr. Campan.: 217 (1830). *Campanula rupestris* subsp. *celsii* (A. DC.) Hayek, Repert. Spec. Nov. Regni Veg. Beih. 30(2): 525 (1930).
Greece. 13 GRC. Biennial. $2n = 34$.

subsp. **carystea** Phitos, Oesterr. Bot. Z. 112: 465 (1965).
Greece. 13 GRC. Biennial.

subsp. **celsii**
Greece. 13 GRC. Biennial. $2n = 34$.

subsp. **parnesia** Phitos, Oesterr. Bot. Z. 112: 464 (1965).
Greece. 13 GRC. Biennial.

subsp. **spathulifolia** (Turrill) Phitos, Oesterr. Bot. Z. 112: 464 (1965).
Greece. 13 GRC. Biennial.
* *Campanula rupestris* var. *spathulifolia* Turrill, Kew Bull. 10: 354 (1955).

Campanula cenisia L., Sp. Pl. (ed. 2): 1669. 1763. *Campanula rosulata* Vuk., Linnaea 26: 326 (1854).
Alps. 11 AUT SWI 12 FRA 13 ITA. Hemicr. $2n = 34$.

Campanula cervicaria L., Sp. Pl. 167 (1753). *Weitenwebera cervicaria* (L.) Opiz, Seznam: 105 (1852). *Campanula cephalaria* var. *macrophylla* Schloss. ex Vuk., Linnaea 26: 334 (1854). *Campanula echiifolia* Rupr., Fl. Ingr. 1: 655 (1856). *Campanula glomerata* var. *cervicaria* (L.) Kuntze, Taschen-Fl. Leipzig: 127 (1867).
Europe to S. Siberia & Kazakhstan; naturalized in NC. USA. 10 DEN FIN NOR SWE 11 AUT BGM CZE GER HUN POL SWI 12 FRA SPA 13 ALB BUL GRC ITA ROM YUG 14 BLR RUS RUE RUN UKR 30 ALT IRK KRA WSB 32 KAZ (74) min. Biennial. $2n = 24, 26, 34$.

Campanula dalmatica Tausch, Flora 10: 246 (1827). *Campanula cervicaria* var. *dalmatica* (Tausch) Nyman, Consp. Fl. Eur.: 477 (1879).

Campanula lingulata Rchb., Iconogr. Bot. Pl. Crit. 6: 12 (1828); non Waldst. & Kit., Descr. Icon. Pl. Hung. 1: 65 (1802). *Campanula cervicaria* var. *lingulata* Nyman, Consp. Fl. Eur.: 477 (1879).

Campanula capitata Schur, Verh. Mitth. Siebenbürg. Vereins Naturwiss. Hermannstadt 3: 88 (1852); non Sims, Bot. Mag. 21: tab. 811 (1805). *Campanula cervicaria* var. *capitata* Schur, Enum. Pl. Transsilv.: 435 (1866).

Campanula longifolia Schloss. & Vuk., Syll. Fl. Croat.: 72 (1857); non Lapeyr., Hist. Pl. Pyrénées: 107 (1813). *Campanula cervicaria* var. *longifolia* Nyman, Consp. Fl. Eur.: 477 (1879).

Campanula cervicaria var. *albiflora* Schur., Enum. Pl. Transsilv.: 435 (1866).

Campanula cervicaria var. *oblongifolia* Schur, Enum. Pl. Transsilv.: 435 (1866).

Campanula cervicaria var. *albiflora* Syr., Ill. Fl. Mosk. Gub. 3: 227 (1910); non Schur, Enum. Pl. Transsilv.: 435 (1866).

Campanula cespitosa Scop., Fl. Carniol. (ed. 2) 1: 143 (1772). *Campanula rotundifolia* var. *cespitosa* (Scop.) Willd., Sp. Pl. 1: 893 (1798), as 'caespitosa'. *Campanula rotundifolia* subsp. *cespitosa* (Scop.) Lapeyr., Hist. Pl. Pyrénées: 103 (1813), as 'caespitosa'. *Decaprisma cespitosa* (Scop.) Raf., Fl. Tellur. 2: 78 (1837).
E. Alps to Croatia. 11 AUT 13 ITA YUG. Hemicr. or geophyte. $2n = 34$.
 Campanula minor Honck., Verz. Gew. Teutschl.: 617 (1782); non Lam., Fl. Franç. 3: 339 (1779).
 Campanula cespitosa f. *austriaca* Beck, Fl. Nieder-Österreich: 1104 (1893). *Campanula cespitosa* subf. *austriaca* (Beck) Hruby, Magyar Bot. Lapok 29: 233 (1930).
 Campanula cespitosa var. *hirta* Dalla Torre & Sarnth., Fl. Tirol 6(3): 436 (1911).
 Campanula cespitosa f. *simplex* Hruby, Magyar Bot. Lapok 29: 233 (1930).
 Campanula cespitosa f. *ramosa* Hruby, Magyar Bot. Lapok 29: 233 (1930).
 Campanula cespitosa subf. *grandiflorens* Hruby, Magyar Bot. Lapok 29: 233 (1930).
 Campanula cespitosa subf. *normalis* Hruby, Magyar Bot. Lapok 29: 233 (1930).
 Campanula cespitosa subf. *hirta* Hruby, Magyar Bot. Lapok 29: 233 (1930).

Campanula × chevalieri Sennen, Bull. Soc. Bot. France 74: 386 (1927). C. rapunculoides × C. trachelium
Spain. 12 SPA. Hemicr. or geophyte.

Campanula chinensis D. Y. Hong, Acta Phytotax. Sin. 18: 247 (1980).
SC. China (Yunnan), Tibet. 36 CHC CHT. Hemicr.

Campanula choruhensis Kit Tan & Sorger, Notes Roy. Bot. Gard. Edinburgh 40: 333 (1982).
Turkey. 34 TUR. Hemicr. or geophyte.

Campanula chrysospleniifolia Franch., J. Bot. (Morot) 9: 364 (1895), as 'chrysosplenifolia'.
SC. China (Yunnan). 36 CHC. Hemicr. or geophyte.
 Campanula leucotricha C. Y. Wu, Rep. Yunnan Trop. Subtrop. Fl. Res. Inst. 1: 58 (1965).

Campanula ciliata Steven, Mém. Soc. Imp. Naturalistes Moscou 3: 256 (1812). *Campanula tridentata* var. *ciliata* (Steven) Trautv., Trudy Imp. S.-Peterburgsk. Bot. Sada 4: 387 (1876). *Hemisphaera ciliata* (Steven) Kolak., Okhrana Prir. Gruz. 12: 163 (1984).
Caucasus (Azerbaijan, Dagestan, Gruziya, Krasnodar). 33 ALL. Hemicr. or geophyte.
 Campanula ciliata var. *longifolia* Rupr., Bull. Acad. Imp. Sci. Saint-Pétersbourg 11: 204 (1867).
 Campanula ciliata var. *pontica* Albov, Bull. Herb. Boissier 2: 118 (1894).

Campanula circassica Fomin in N. M. Kusn., N. Bush & Fomin, Mater. Fl. Kavkaza 4(6): 52 (1905). *Hemisphaera circassica* (Fomin) Kolak. & Serdyuk., Okhrana Prir. Gruz. 12: 166 (1984).
Caucasus (Gruziya, Krasnodar). 33 ALL. Hemicr. or geophyte.
 Campanula anomala Fomin in N. M. Kuzn., N. Bush & Fomin, Mater. Fl. Kavkaza 4(6): 53 (1905). *Hemisphaera anomala* (Fomin) Kolak. & Serdyuk., Okhrana Prir. Gruz. 12: 162 (1984).

Campanula cochleariifolia Lam., Encycl. 1: 578 (1785), as 'cochlearifolia'. *Campanula rotundifolia* var. *cochleariifolia* (Lam.) Fiori, Nuov. Fl. Italia 2: 563 (1927).
Pyrénées to Balkans. 12 FRA SPA 13 ALB BUL ITA ROM YUG. Hemicr. or geophyte. $2n = 34, 68$.
 Campanula bellardii All., Fl. Pedem. 1: 108 (1785), as 'bellardi'. *Campanula pusilla* var. *bellardii* (All.) DC., Fl. Franç. 5: 432 (1815). *Campanula cespitosa* var. *bellardii* (All.)

Nyman, Consp. Fl. Eur.: 440 (1879), as 'bellardi'. *Campanula cochleariifolia* var.
bellardii (All.) Hayek & Hegi in Hegi, Ill. Fl. Mitt.-Eur. 6: 351 (1915).
Campanula leucanthemifolia Pourr., Hist. & Mém. Acad. Roy. Sci. Toulouse 3: 309
(1788). *Campanula pusilla* var. *pinguis* Gren. & Godr., Fl. Fr. 2: 417 (1853). *Campanula
pusilla* var. *leucanthemifolia* (Pourr.) Nyman, Consp. Fl. Eur.: 479 (1879).
Campanula pusilla Haenke in Jacq., Collect. 2: 79 (1789). *Campanula rotundifolia* var.
pusilla (Haenke) Willd., Sp. Pl. 1: 892 (1798). *Campanula rotundifolia* subsp. *pusilla*
(Haenke) Lapeyr., Hist. Pl. Pyrénées: 103 (1813).
Campanula pubescens F. W. Schmidt, Fl. Boëm. 2: 67 (1794). *Campanula cespitosa* var.
pubescens (F. W. Schmidt) A. DC., Monogr. Campan.: 284 (1830). *Campanula pusilla*
var. *pubescens* (F. W. Schmidt) Steud., Nomencl. Bot. (ed. 2) 1: 269 (1840); non DC.,
Fl. Franç. 5: 432 (1815). *Campanula cespitosa* subsp. *pubescens* (F. W. Schmidt)
Arcang., Comp. Fl. Ital.: 453 (1882). *Campanula bellardii* var. *pubescens* (F. W.
Schmidt) Bég. in Fiori & Paol., Fl. Italia 2: 183 (1900). *Campanula cochleariifolia* var.
pubescens (F. W. Schmidt) Dalla Torre & Sarnth., Fl. Tirol 6(3): 445 (1911).
Campanula pumila Sims, Bot. Mag. 15: tab. 512 (1801); non F. W. Schmidt, Fl. Boëm. 2:
71 (1794).
Campanula pusilla var. *pubescens* DC., Fl. Franç. 5: 432 (1815).
Campanula glacialis Shuttlew., Mag. Zool.-Bot. 2: 188 (1838).
Campanula compacta Hegetschw., Fl. Schweiz: 232 (1838-1839). *Campanula
cochleariifolia* subf. *compacta* (Hegetschw.) Hruby, Magyar Bot. Lapok 29: 258 (1930).
Campanula mixta Hegetschw., Fl. Schweiz: 231 (1838-39).
Campanula pubescens Hegetschw., Fl. Schweiz: 232 (1838-39); non F. W. Schmidt, Fl.
Boëm. 2: 67 (1794).
Campanula gracilis Jord., Mém. Acad. Roy. Sci. Lyon, Sect. Lett. (n.s.) 1: 333 (1851); non
G. Forst., Fl. Ins. Austr.: 15 (1786); nec Avé-Lall., Pl. Ital. Bor.: 10 (1829); nec (Boiss.
& Heldr.) Boiss. & Heldr. in Boiss., Diagn. Pl. Orient. (ser. 1) 11: 77 (1849).
Campanula pusilla var. *gracilis* Nyman, Consp. Fl. Eur.: 479 (1879). *Campanula
cochleariifolia* var. *gracilis* (Nyman) Vacc., Cat. Pl. Vall. Aoste: 619 (1911). *Campanula
cochleariifolia* subf. *gracilis* (Nyman) Hruby, Magyar Bot. Lapok 29: 258 (1930).
Campanula parvula Jord., Mém. Acad. Roy. Sci. Lyon, Sect. Lett. (n.s.) 1: 334 (1851).
Campanula pusilla var. *parvula* (Jord.) Nyman, Consp. Fl. Eur.: 479 (1879).
Campanula foudrasii Jord., Mém. Acad. Roy. Sci. Lyon, Sect. Lett. (n.s.) 1: 335 (1851), as
'foudrasi'. *Campanula pusilla* var. *foudrasii* (Jord.) Nyman, Consp. Fl. Eur.: 479 (1879),
as 'foudrasi'. *Campanula cochleariifolia* var. *foudrasii* (Jord.) Vacc., Cat. Pl. Vall. Aoste:
619 (1911).
Campanula tenella Jord., Mém. Acad. Roy. Sci. Lyon, Sect. Sci. (ser. 2) 1: 336 (1851).
Campanula pusilla var. *tenella* (Jord.) Nyman, Consp. Fl. Eur.: 479 (1879). *Campanula
pusilla* subsp. *tenella* (Jord.) Rouy, Fl. France 10: 76 (1908). *Campanula cochleariifolia*
var. *tenella* (Jord.) Vacc., Cat. Pl. Vall. Aoste: 616 (1911). *Campanula cochleariifolia* f.
tenella (Jord.) Hruby, Magyar Bot. Lapok 29: 259 (1930).
Campanula pulchella Jord., Mém. Acad. Roy. Sci. Lyon, Sect. Lett. (n.s.) 1: 337 (1851);
non Salisb., Prodr. Stirp. Chap. Allerton: 127 (1796); nec (Fisch. & C. A. Mey.) Boiss.,
Diagn. Pl. Orient. (ser. 1) 11: 76 (1849). *Campanula pusilla* var. *pulchella* Gren. &
Godr., Fl. Fr. 2: 417 (1853). *Campanula cochleariifolia* var. *pulchella* (Gren. & Godr.)
Vacc., Cat. Pl. Vall. Aoste: 616 (1911). *Campanula cochleariifolia* subf. *pulchella* (Gren.
& Godr.) Hruby, Magyar Bot. Lapok 29: 258 (1930).
Campanula pusilla f. *umbosa* J. Hofm., Oesterr. Bot. Wochenbl. 2: 193 (1852).
Campanula pusilla var. *umbrosa* (J. Hofm.) Beck, Fl. Nieder-Österreich: 1104 (1893).
Campanula cochleariifolia f. *umbrosa* (J. Hofm.) Vacc., Cat. Pl. Vall. Aoste: 618 (1911).
Campanula cochleariifolia var. *umbrosa* (J. Hofm.) Dalla Torre & Sarnth., Fl. Tirol 6(3):
444 (1911).
Campanula pusilla f. *vagans* J. Hofm., Oesterr. Bot. Wochenbl. 2: 194 (1852). *Campanula
cochleariifolia* var. *vagans* (J. Hofm.) Dalla Torre & Sarnth., Fl. Tirol 6(3): 445 (1911).
Campanula mathonetii Jord. ex Gren. & Godr., Fl. Fr. 2: 418 (1853), as 'mathoneti'.
Campanula pusilla var. *mathonetii* (Jord.) Nyman, Consp. Fl. Eur.: 479 (1879), as

'mathoneti'. *Campanula cochleariifolia* var. *mathonetii* (Jord. ex Gren. & Godr.) Vacc.,
Cat. Pl. Vall. Aoste: 617 (1911). *Campanula cochleariifolia* subf. *mathonetii* (Jord. ex
Gren. & Godr.) Hruby, Magyar Bot. Lapok 29: 258 (1930). *Campanula cochleariifolia* f.
mathonetii (Jord. ex Gren. & Godr.) Hruby, Magyar Bot. Lapok 29: 259 (1930).

Campanula subramulosa Jord. ex Gren. & Godr., Fl. Fr. 2: 418 (1853). *Campanula pusilla*
var. *subramulosa* (Jord. ex Gren. & Godr.) Nyman, Consp. Fl. Eur.: 479 (1879).
Campanula cochleariifolia var. *subramulosa* (Jord. ex Gren. & Godr.) Vacc., Cat. Pl.
Vall. Aoste: 618 (1911).

Campanula minutissima Schur, Sert. Fl. Transsilv.: 47 (1853).

Campanula reflexa Schur, Sert. Fl. Transsilv.: 47 (1853). *Campanula pusilla* var. *reflexa*
(Schur) Nyman, Consp. Fl. Eur.: 479 (1879). *Campanula cochleariifolia* subsp. *reflexa*
(Schur) Hruby, Magyar Bot. Lapok 29: 264 (1930).

Campanula hauryi Schott in Schott, Nyman & Kotschy, Analect. Bot.: 10 (1854).
Campanula rotundifolia var. *hauryi* (Schott) Nyman, Consp. Fl. Eur.: 479 (1879).
Campanula pusilla f. *hauryi* (Schott) Hayek, Abh. K. K. Zool.-Bot. Ges. Wien 4: 129
(1907). *Campanula cochleariifolia* var. *hauryi* (Schott) Hayek & Hegi in Hegi, Ill. Fl.
Mitt.-Eur. 6: 351 (1915). *Campanula cochleariifolia* f. *hauryi* (Schott) Hruby, Magyar
Bot. Lapok 29: 259 (1930).

Campanula perneglecta Schott in Schott, Nyman & Kotschy, Analect. Bot.: 11 (1854).

Campanula hochstetteri Schott in Schott, Nyman & Kotschy, Analect. Bot.: 12 (1854).
Campanula pusilla var. *hochstetteri* (Schott) Nyman, Consp. Fl. Eur.: 479 (1879).

Campanula tyrolensis Schott in Schott, Nyman & Kotschy, Analect. Bot.: 12 (1854).
Campanula pusilla var. *tyrolensis* (Schott) Nyman, Consp. Fl. Eur.: 479 (1879).
Campanula cochleariifolia f. *tyrolensis* (Schott) Vacc., Cat. Pl. Vall. Aoste: 618
(1911). *Campanula cochleariifolia* subf. *tyrolensis* (Schott) Hruby, Magyar Bot.
Lapok 29: 258 (1930).

Campanula modesta Schott in Schott, Nyman & Kotschy, Analect. Bot.: 13 (1854).
Campanula pusilla var. *modesta* (Schott) Nyman, Consp. Fl. Eur.: 479 (1879).
Campanula cochleariifolia f. *modesta* (Schott) Hruby, Magyar Bot. Lapok 29: 265 (1930).

Campanula notata Schott in Schott, Nyman & Kotschy, Analect. Bot.: 13 (1854).
Campanula pusilla subsp. *notata* (Schott) Nyman, Consp. Fl. Eur.: 479 (1879).
Campanula pusilla var. *notata* (Schott) Wohlf. in W. J. D. Koch, Syn. Deut. Schweiz.
Fl. (ed. 3) 2: 1266 (1893)

Campanula pusilla var. *hoppeana* Rupr. ex Rchb., Icon. Fl. Germ. Helv. 19: 117 (1859).
Campanula bellardii var. *hoppeana* (Rupr. ex Rchb.) Bég. in Fiori & Paol., Fl. Italia 2:
183 (1900). *Campanula cochleariifolia* var. *hoppeana* (Rupr. ex Rchb.) Vacc., Cat. Pl.
Vall. Aoste: 615 (1911). *Campanula cochleariifolia* f. *hoppeana* (Rupr. ex Rchb.) Hruby,
Magyar Bot. Lapok 29: 259 (1930).

Campanula pusilla var. *paniculata* Nägeli ex Rchb., Icon. Fl. Germ. Helv. 19: 117 (1859).
Campanula bellardii var. *paniculata* (Nageli ex Rchb.) Bég. in Fiori & Paol., Fl. Italia 2:
183 (1900). *Campanula cochleariifolia* var. *paniculata* (Nageli ex Rchb.) Dalla Torre &
Sarnth., Fl. Tirol 6(3): 445 (1911).

Campanula venusta Schur, Enum. Pl. Transsilv.: 442 (1866). *Campanula pusilla* var.
venusta (Schur) Nyman, Consp. Fl. Eur.: 479 (1879). *Campanula cochleariifolia* f.
venusta (Schur) Hruby, Magyar Bot. Lapok 29: 265 (1930).

Campanula pusilla var. *calycina* Willk. in Willk. & Lange, Prodr. Fl, Hispan. 2: 292 (1868).

Campanula pusilla f. *densa* Gsaller, Oesterr. Bot. Z. 20: 287 (1870). *Campanula
cochleariifolia* f. *densa* (Gsaller) Vacc., Cat. Pl. Vall. Aoste: 618 (1911).

Campanula pusilla f. *imbricata* Vuk., Rad Jugoslav. Akad. Znan. 57: 93 (1881).

Campanula pusilla f. *hirciana* Vuk., Rad Jugoslav. Akad. Znan. 57: 94 (1881).

Campanula pusilla var. *descensa* Beck, Fl. Nieder-Österreich: 1104 (1893). *Campanula
cochleariifolia* var. *descensa* (Beck) Săvul., Stud. Sp. Campanula: 48 (1916).

Campanula pusilla var. *foliosa* Krašan, Mitt. Naturwiss. Vereines Steiermark 30: 252
(1894). *Campanula cochleariifolia* var. *foliosa* (Krašan) Dalla Torre & Sarnth., Fl. Tirol
6(3): 445 (1911). *Campanula cochleariifolia* subf. *foliosa* (Krašan) Hruby, Magyar Bot.
Lapok 29: 258 (1930).

Campanula pusilla var. *delpontei* Chabert, Bull. Herb. Boissier 3: 148 (1895). *Campanula bellardii* var. *delpontei* (Chabert) Bég. in Fiori & Paol., Fl. Italia 2: 183 (1900). *Campanula cochleariifolia* f. *delpontei* (Chabert) Vacc., Cat. Pl. Vall. Aoste: 615 (1911).

Campanula pusilla var. *tubulosa* Chabert, Bull. Herb. Boissier 3: 147 (1895). *Campanula cochleariifolia* f. *tubulosa* (Chabert) Vacc., Cat. Pl. Vall. Aoste: 615 (1911).

Campanula pusilla var. *brachyantha* Murr, Deutsche Bot. Monatsschr. 17: 151 (1899). *Campanula cochleariifolia* var. *brachyantha* (Murr) Dalla Torre & Sarnth., Fl. Tirol 6(3): 445 (1911). *Campanula cochleariifolia* subf. *brachyantha* (Murr) Hruby, Magyar Bot. Lapok 29: 258 (1930).

Campanula pusilla var. *subacaulis* Murr, Deutsche Bot. Monatsschr. 17: 151 (1899). *Campanula cochleariifolia* var. *subacaulis* (Murr) Hruby, Magyar Bot. Lapok 29: 259 (1930).

Campanula pusilla var. *pirinica* Velen., Sitzungsber. Königl. Böhm. Ges. Wiss. Prag, Math.-Naturwiss. Kl. 8: 9 (1910). *Campanula cochleariifolia* f. *pirinica* (Velen.) Hayek, Repert. Spec. Nov. Regni Veg. Beih. 30(2): 536 (1930).

Campanula cochleariifolia f. *ramulosa* Witasek ex Vacc., Cat. Pl. Vall. Aoste: 620 (1911).

Campanula cochleariifolia subf. *grandiflorens* Hruby, Magyar Bot. Lapok 29: 258 (1930); non Magyar Bot. Lapok 29: 265 (1930).

Campanula cochleariifolia f. *veronicae* Hruby, Magyar Bot. Lapok 29: 258 (1930).

Campanula cochleariifolia subf. *aberrans* Hruby, Magyar Bot. Lapok 29: 259 (1930).

Campanula cochleariifolia subf. *reflexa* Hruby, Magyar Bot. Lapok 29: 259 (1930).

Campanula cochleariifolia subf. *gracilescens* Hruby, Magyar Bot. Lapok 29: 260 (1930).

Campanula cochleariifolia f. *subramulosa* Hruby, Magyar Bot. Lapok 29: 260 (1930).

Campanula cochleariifolia subsp. *croatica* Hruby, Magyar Bot. Lapok 29: 260 (1930).

Campanula cochleariifolia subf. *pubescens* Hruby, Magyar Bot. Lapok 29: 262 (1930).

Campanula cochleariifolia subf. *hoppeaniformis* Hruby, Magyar Bot. Lapok 29: 262 (1930).

Campanula cochleariifolia subf. *pygmaea* Hruby, Magyar Bot. Lapok 29: 262 (1930).

Campanula cochleariifolia subf. *multiflorens* Hruby, Magyar Bot. Lapok 29: 262 (1930).

Campanula cochleariifolia subf. *polyphylla* Hruby, Magyar Bot. Lapok 29: 262 (1930).

Campanula cochleariifolia f. *rosulata* Hruby, Magyar Bot. Lapok 29: 263 (1930).

Campanula cochleariifolia subf. *macedonica* Hruby, Magyar Bot. Lapok 29: 263 (1930).

Campanula cochleariifolia f. *incisiserrata* Hruby, Magyar Bot. Lapok 29: 263 (1930), as 'inciso-serrata'.

Campanula cochleariifolia f. *umbrosa* Hruby, Magyar Bot. Lapok 29: 265 (1930); non (J. Hofm.) Vacc., Cat. Pl. Vall. Aoste: 618 (1911).

Campanula cochleariifolia f. *multiflorens* Hruby, Magyar Bot. Lapok 29: 265 (1930).

Campanula cochleariifolia subf. *altior* Hruby, Magyar Bot. Lapok 29: 265 (1930).

Campanula cochleariifolia f. *pulchella* Hruby, Magyar Bot. Lapok 29: 265 (1930).

Campanula cochleariifolia f. *javorkae* Hruby, Magyar Bot. Lapok 29: 265 (1930).

Campanula cochleariifolia subf. *grandiflorens* Hruby, Magyar Bot. Lapok 29: 265 (1930); non Magyar Bot. Lapok 29: 258 (1930).

Campanula cochleariifolia f. *subcaespitosa* Hruby, Magyar Bot. Lapok 29: 268 (1930).

Campanula cochleariifolia f. *pseudomodesta* Hruby, Magyar Bot. Lapok 33: 156 (1934).

Campanula cochleariifolia f. *anagalloides* Hruby, Magyar Bot. Lapok 33: 156 (1934).

Campanula cochleariifolia f. *elongata* Hruby, Magyar Bot. Lapok 33: 158 (1934).

Campanula renatii Sennen, Diagn. Nouv.: 21 (1936), as 'renati'. *Campanula cochleariifolia* var. *renatii* (Sennen) P. Monts. in Losa & P. Monts., Aport. Conoc. Fl. Andorra. 120 (1951).

Campanula pusilla subsp. *veronicifolia* Sennen, Diagn. Nouv.: 21 (1936).

Campanula collina Sims, Bot. Mag. 24: tab. 927 (1806).
Turkey & Caucasus. 33 ALL 34 TUR. Hemicr. or geophyte. $2n = 48, 68$.

subsp. **collina**
Turkey & Caucasus (Armenia, Azerbaijan, Dagestan, Gruziya, Krasnodar). 33 ALL 34 TUR. Hemicr. or geophyte. $2n = 68$.
Campanula collina var. *major* M. Bieb., Fl. Taur.-Cauc. 1: 152 (1808).

Campanula collina var. *pumila* M. Bieb., Fl. Taur.-Cauc. 1: 152 (1808).
Campanula collina var. *dasycarpa* K. Koch, Linnaea 23: 643 (1850).
Campanula collina var. *leiocarpa* K. Koch, Linnaea 23: 643 (1850).
Campanula collina var. *grandiflora* Vatke, Linnaea 38: 708 (1874).
Campanula collina var. *eriocalyx* Trautv., Trudy Imp. S.-Peterburgsk. Bot. Sada 6: 65 (1879).
Campanula collina var. *leiocalyx* Trautv., Trudy Imp. S.-Peterburgsk. Bot. Sada 6: 65 (1879).
Campanula annae Kolak., Zametki Sist. Geogr. Rast. 16: 55 (1951).
Campanula collina subsp. *gapschimensis* Husseinov, Bot. Žurn. (Moscow & Leningrad) 55: 1500 (1970).
Campanula nefedovii Galushko, Novosti Sist. Vyssh. Rast. 8: 268 (1971).
Campanula galushkoi Prima, Bot. Žurn. (Moscow & Leningrad) 59: 369 (1974).

subsp. **fondervisii** (Albov) Ogan., Candollea 50: 289 (1995).
Caucasus (Gruziya). 33 TCS. Hemicr.
* *Campanula fondervisii* Albov, Bull. Herb. Boissier 2: 117 (1894). *Campanula collina* var. *fondervisii* (Albov) Albov, Prodr. Fl. Colchic.: 155 (1895).
Campanula migarica Schischk., Trudy Azerbajdžansk Otd. Zakavkazsk. Fil. Akad. Nauk SSSR 1: 49 (1933).
Campanula irinae A. Kuth., Zametki Sist. Geogr. Rast. 17: 96 (1953).

subsp. **sphaerocarpa** (Kolak.) Ogan., Candollea 50: 288 (1995).
Caucasus (Abkhaziya, Gruziya). 33 TCS. Hemicr. or geophyte. $2n = 48$.
Campanula collina var. *abchasica* Albov, Bull. Herb. Boissier 2: 118 (1894).
* *Campanula sphaerocarpa* Kolak., Soobšč. Akad. Nauk Gruzinsk. SSR 8: 237 (1947).
Campanula sphaerocarpa var. *grandiflora* Kolak., Soobšč. Akad. Nauk Gruzinsk. SSR 8: 239 (1947).
Campanula sphaerocarpa var. *multiflora* Kolak., Fl. Abkhas. 4: 171 (1949).
Campanula albovii Kolak., Zametki Sist. Geogr. Rast. 16: 54 (1951).
Campanula kluchorica Kolak., Zametki Sist. Geogr. Rast. 16: 56 (1951).
Campanula schistosa Kolak., Zametki Sist. Geogr. Rast. 16: 58 (1951).
Campanula schistosa var. *uniflora* Kolak., Zametki Sist. Geogr. Rast. 16: 58 (1951).

Campanula columnaris Contandr., Quézel & Zaffran, Bull. Soc. Bot. France 120: 333 (1974).
Greece. 13 GRC. Biennial. $2n = 16$.

Campanula conferta A. DC. in DC., Prodr. 7: 468 (1839).
Turkey & Iraq. 34 TUR IRQ. Hemicr.
Campanula conferta var. *glabra* A. DC. in DC., Prodr. 7: 468 (1839).

Campanula constantini Beauverd & Topali, Candollea 7: 266 (1937), as 'constantinii'.
Greece. 13 GRC. Biennial. $2n = 34$.

Campanula coriacea P. H. Davis, Notes Roy. Bot. Gard. Edinburgh 24: 29 (1962).
Turkey, Caucasus (Armenia, Nakhichevan), Iran. 33 TCS 34 IRN TUR. Hemicr. or geophyte. $2n = 34$.
Campanula radula var. *minor* Boiss., Fl. Orient. 3: 909 (1875).

Campanula × cottia Beyer, Verh. Bot. Vereins Prov. Brandenburg 58: 110 (1917). C. cochleariifolia × C. stenocodon.
Italy. 13 ITA. Hemicr. or geophyte.

Campanula crassipes Heuff., Oesterr. Bot. Z. 8: 27 (1858).
Serbia & Romania. 13 ROM YUG. Hemicr. or geophyte. $2n = 34$.

Campanula crenulata Franch., J. Bot. (Morot) 9: 365 (1895).
SC. China (Sichuan, Yunnan). 36 CHC. Hemicr.
Campanula nephrophylla C. Y. Wu, Rep. Yunnan Trop. Subtrop. Fl. Res. Inst. 1: 60 (1965).

Campanula cretica (A. DC.) D. Dietr., Syn. Pl. 1: 758 (1839).
Kriti. 13 KRI. Hemicr. or geophyte. *2n* = 34.
  **Symphyandra cretica* A. DC., Monogr. Campan.: 366 (1830).

Campanula creutzburgii Greuter, Boissiera 13: 133 (1967). *Campanula drabifolia* subsp.
creutzburgii (Greuter) Fed., Bot. J. Linn. Soc. 67: 281 (1973).
Kriti. 13 KRI. Ther. *2n* = 56.

Campanula crispa Lam., Encycl. 1: 581 (1785). *Quinquelocularia crispa* (Lam.) K. Koch,
Linnaea 23: 631 (1850).
Turkey, Caucasus (Armenia, Gruziya), Iran. 33 TCS 34 IRN TUR. Hemicr. or biennial.

Campanula cymaea Phitos, Oesterr. Bot. Z. 111: 212 (1964).
Greece. 13 GRC. Hemicr. or biennial. *2n* = 34.

Campanula cymbalaria Sm. in Sibth. & Sm., Fl. Graec. Prodr. 1: 139 (1806).
Turkey & Lebanon. 34 TUR LBS. Hemicr. *2n* = 34.
  Campanula billardierei A. DC., Monogr. Campan.: 303 (1830), as 'billardierii'.
  Campanula billardierei var. *major* A. DC. in DC., Prodr. 7: 474 (1839).

Campanula czerepanovii Fed., Novosti Sist. Vyssh. Rast. 1965: 237 (1965).
Caucasus (Dagestan). 33 NCS. Hemicr.

Campanula daghestanica Fomin in N. M. Kusn., N. Bush & Fomin, Mater. Fl. Kavkaza 4(6):
29 (1904). *Campanula sibirica* subsp. *daghestanica* (Fomin) Viktorov, Novosti Sist. Vyssh.
Rast. 32: 168 (2000).
Caucasus (Chechnya, Dagestan). 33 NCS. Hemicr.

Campanula damascena Labill., Icon. Pl. Syr. 5: 7 (1812).
Israel, Lebanon, Syria. 34 PAL LBS. Hemicr.
  Campanula aaronsohnii Evenari, Bull. Soc. Bot. Genève (ser. 2) 31: 403 (1941).

Campanula damboldtiana P. H. Davis & Sorger, Notes Roy. Bot. Gard. Edinburgh 37:
265 (1979).
Turkey. 34 TUR. Hemicr. or geophyte.

Campanula dasyantha M. Bieb., Fl. Taur.-Cauc. 3: 147 (1819). *Campanula pilosa* var.
dasyantha (M. Bieb.) Herder, Trudy Imp. S.-Peterburgsk. Bot. Sada 1: 288 (1873).
SW. Siberia to Aleutian Is. 30 ALT BRY CTA IRK KRA TVA YAK 31 KAM KHA KUR MAG
SAK 37 MON 38 JAP 70 ALU. Hemicr. or geophyte. *2n* =34, 68.

subsp. **dasyantha**
SW. Siberia to Mongolia & Russian Far East. 30 ALT BRY CTA IRK KRA TVA YAK 31
KAM KHA MAG 37 MON. Hemicr. or geophyte. *2n* = 34, 68.
Campanula pallasiana Vest ex Schult. in Roem. & Schult., Syst. Veg. 5: 138 (1819).
Campanula pilosa Pall. ex Schult. in Roem. & Schult., Syst. Veg. 5: 148 (1819).

subsp. **chamissonis** (Fed.) Viktorov, Novosti Sist. Vyssh. Rast. 34: 225 (2002).
Russian Far East to Japan (Hokkaido, Honshu) & Aleutian Is. 31 KAM KUR SAK 38 JAP
70 ALU. Hemicr. or geophyte. *2n* = 34.
  **Campanula chamissonis* Fed. in Kom., Fl. URSS 24: 279 (1957). *Campanula dasyantha*
  var. *chamissonis* (Fed.) Toyok. & Nosaka, Acta Phytotax. Geobot. 18: 193 (1960).
  Campanula dasyantha f. *albiflora* Miyabe & Tatew., Trans. Sapporo Nat. Hist. Soc. 13: 70
  (1933). *Campanula chamissonis* f. *albiflora* (Miyabe & Tatew.) T. Shimizu, New Alp.
  Fl. Japan 1: 311 (1982).

Campanula davisii Turrill, Bot. Mag. 171: pl. 283 (1956).
Turkey. 34 TUR. Hemicr. or biennial.

Campanula decumbens A. DC., Monogr. Campan.: 334 (1830).
Spain. 12 SPA. Hemicr.
Campanula dieckii Lange, Overs. Kongel. Danske Vidensk. Selsk. Forh. Medlemmers
Arbeider 1893: 195 (1893).

Campanula delavayi Franch., J. Bot. (Morot) 9: 364 (1895).
SC. China (Yunnan). 36 CHC. Hemicr. or geophyte.

Campanula delicatula Boiss., Diagn. Pl. Orient. (ser. 1) 11: 67 (1849).
E. Mediterranean. 13 GRC KRI 34 CYP EAI TUR. Ther. $2n = 16$.

Campanula dichotoma L. f., Cent. Pl. II: 10 (1756).
W. Mediterranean. 12 BAL 13 GRC ITA SIC 20 ALG MOR TUN. Ther. $2n = 24$.
Campanula decipiens Schult. in Roem. & Schult., Syst. Veg. 5: 142 (1819).
Campanula dichotoma var. *brachiata* A. DC., Monogr. Campan.: 237 (1830).
Campanula catinensis Tornab., Fl. Sicul.: 357 (1887).
Campanula dichotoma f. *alba* Voss in Siebert & Voss, Vilm. Blumengärtn. (ed. 3) 1:
569 (1894).

Campanula × digenea Wettst., Oesterr. Bot. Z. 70: 183 (1921). C. barbata × C. glomerata
Austria. 11 AUT. Hemicr. or geophyte.

Campanula dimorphantha Schweinf., Beitr. Fl. Aethiop.: 140 (1867).
Egypt to Ethiopia; Afghanistan to C. China (Shaanxi, Guangdong, Guizhou, Sichuan,
Yunnan), Taiwan, Vietnam & Sri Lanka. 20 EGY 24 ETH SUD 34 AFG 36 CHC CHN
CHS 38 TAI 40 ASS EHM IND NEP PAK SRL WHM 41 LAO MYA VIE. Ther.
Campanula canescens Wall. ex A. DC., Monogr. Campan. 292 (1830); non (Waldst. &
Kit.) Roth, Enum. Pl. Phaen. Germ. 1: 716 (1827). *Campanula benthamii* Wall. ex
Kitam., Fl. Afghan.: 377 (1960). *Campanula wallichii* Babu, J. Bombay Nat. Hist. Soc.
65: 808 (1969).
Cephalostigma spathulatum Thwaites, Enum. Pl. Zeyl. 422 (1864); non *Campanula
spatulata* Sm. in Sibth. & Sm., Fl. Graec. Prodr. 1: 137 (1806); nec Waldst. & Kit.,
Descr. Icon. Pl. Hung. 3: 286 (1809).
Campanula veronicifolia Hance, J. Bot. 9: 133 (1871).

Campanula divaricata Michx., Fl. Bor.-Amer. 1: 109 (1803).
SE. U.S.A. (Maryland to Tennessee & Alabama). 74 WVA 75 ALA GEO KTY MRY NCA SCA
TEN VRG. Hemicr. $2n = 34, 40$.
Campanula flexuosa Michx., Fl. Bor.-Amer. 1: 109 (1803).
Campanula divaricata f. *alba* Fosberg, Castanea 20: 60 (1955).

Campanula dolomitica E. A. Busch, Trudy Bot. Muz. 22: 215 (1930).
Caucasus (Severo-Osetiya). 33 NCS. Hemicr. or geophyte.

Campanula drabifolia Sm. in Sibth. & Sm., Fl. Graec. Prodr. 1: 142 (1806). *Roucela
drabifolia* (Sm.) Dumort., Comment. Bot.: 15 (1822), as 'drabaefolia'.
Greece. 13 GRC. Ther. $2n = 28, 34$.
Campanula attica Boiss. & Heldr. in Boiss., Diagn. Pl. Orient. (ser. 1) 11: 67 (1849).
Campanula drabifolia var. *major* Boiss., Fl. Orient. 3: 933 (1875). *Campanula drabifolia*
subsp. *attica* (Boiss. & Heldr.) Nyman, Consp. Fl. Eur.: 481 (1879).

Campanula dulcis Decne., Ann. Sci. Nat. Bot. (ser. 2) 2: 258 (1834).
Sinai, Jordan, Saudi Arabia. 34 PAL SIN 35 SAU. Hemicr.

Campanula dzaaku Albov, Bull. Herb. Boissier 2: 114 (1894). *Pseudocampanula dzaaku* (Albov) Kolak., Soobšč. Akad. Nauk Gruzinsk. SSR 97: 414 (1980).
Caucasus (Abkhaziya, Gruziya). 33 TCS. Hemicr. or geophyte.

Campanula dzyschrica Kolak., Zametki Sist. Geogr. Rast. 16: 57 (1951).
Caucasus (Abkhaziya). 33 TCS. Hemicr. or geophyte.

Campanula edulis Forssk., Fl. Aegypt.-Arab.: 44 (1775).
Sudan to Rwanda, Tanzania, Somalia & Arabian Pen. 23 RWA 24 ETH SOM SUD 25 KEN TAN UGA 35 SAU YEM. Hemicr. $2n = 28, 56, 70$.
  Campanula rigidipila Steud. & Hochst. ex A. Rich., Tent. Fl. Abyss. 2: 3 (1850). *Campanula schimperi* var. *rigidipila* (Steud. & Hochst. ex A. Rich.) Vatke, Linnaea 40: 201 (1876).
  Campanula esculenta A. Rich., Tent. Fl. Abyss. 2: 4 (1850); non Salisb., Prodr. Stirp. Chap. Allerton: 126 (1796). *Campanula rigidipila* var. *esculenta* (A. Rich.) Di Capua, Annuario Reale Ist. Bot. Roma 8: 236 (1904).
  Campanula sarmentosa Hochst. ex A. Rich., Tent. Fl. Abyss. 2: 4 (1850). *Campanula schimperi* var. *sarmentosa* (Hochst. ex A. Rich.) Vatke, Linnaea 40: 201 (1876). *Campanula rigidipila* var. *sarmentosa* (Hochst. ex A. Rich.) Engl., Hochgebirgsfl. Afrika: 410 (1892).
  Campanula quartiniana A. Rich., Tent. Fl. Abyss. 2: 5 (1850). *Campanula schimperi* var. *quartiniana* (A. Rich.) Vatke, Linnaea 40: 201 (1876). *Campanula rigidipila* var. *quartiniana* (A. Rich.) Engl., Hochgebirgsfl. Afrika: 410 (1892).
  Campanula schimperi Vatke, Linnaea 38: 712 (1874).

Campanula ekimiana Güner, Doğa Bilim Dergisi (ser. A$_2$) 9: 62 (1985).
Turkey. 34 TUR. Hemicr.

Campanula elatines L., Syst. Nat. (ed. 10) 2: 927 (1759). *Wahlenbergia elatines* (L.) Link, Handbuch 1: 631 (1829).
Italy. 13 ITA. Hemicr. $2n = 34$.
  Campanula elatina Hill, Syst. Veg. 8: 13 (1765).
  Campanula elatines var. *glaberrima* A. DC., Monogr. Campan.: 301 (1830). *Campanula subalpina* Delponte & Gras, Enum. Sem. Hort. Bot. Taurin. 1873: 35 (1873). *Campanula elatines* var. *subalpina* (Delponte & Gras) Nyman, Consp. Fl. Eur.: 481 (1879). *Campanula elatines* subsp. *glaberrima* (A. DC.) Arcang., Comp. Fl. Ital.: 453 (1882).
  Campanula barbeyi Feer, J. Bot. 28: 271 (1890).

Campanula elatinoides Moretti, Giorn. Fis. (ser. 2) 5: 110 (1822). *Campanula elatines* var. *elatinoides* (Moretti) Fiori, Nuov. Fl. Italia 2: 560 (1927).
Italy. 13 ITA. Hemicr. $2n = 34$.
  Campanula petraea Zanted. ex Moretti, Syll. Pl. Nov. 1: 154 (1824); *non* L., Syst. Nat. (ed. 10) 2: 926 (1759).

Campanula embergeri Litard. & Maire, Contrib. Étude Fl. Grand Atlas 2: 5 (1930). *Campanula filicaulis* subsp. *embergeri* (Litard. & Maire) Dobignard, J. Bot. Soc. Bot. France 20: 25 (2002).
Morocco. 20 MOR. Hemicr. $2n = 28$.

subsp. **embergeri**
Morocco. 20 MOR. Hemicr. $2n = 28$.

subsp. **schotteri** Quézel, Feddes Repert. Spec. Nov. Regni Veg. 56: 28 (1953). *Campanula filicaulis* var. *schotteri* (Quézel) Dobignard, J. Bot. Soc. Bot. France 20: 25 (2002).
Morocco. 20 MOR. Hemicr.

Campanula engurensis Kharadze, Zametki Sist. Geogr. Rast. 1: 7 (1938).
Caucasus (Gruziya). 33 TCS. Hemicr. or geophyte.

Campanula erinus L., Sp. Pl.: 169 (1753). *Roucela erinus* (L.) Dumort., Comment. Bot.: 14
(1822). *Wahlenbergia erinus* (L.) Link, Handbuch 1: 631 (1829). *Erinia campanula* Noulet, Fl.
Bass. Sous-Pyrén.: 404 (1837). *Campanula parviflora* St.-Lag., Ann. Soc. Bot. Lyon 7: 121
(1880); non Lam., Encycl. 1: 588 (1785); nec Salisb., Prodr. Stirp. Chap. Allerton: 126 (1796).
Macaronesia to W. Asia. 12 BAL COR FRA POR SAR SPA 13 ALB GRC ITA KRI SIC TUE
    YUG 14 KRY 20 ALG EGY LBY MOR TUN 21 AZO CNY MDR 33 TCS 34 CYP IRN IRQ
    PAL LBS TUR 35 SAU. Ther. $2n = 28, 56$.
    Campanula portensis L. in Loefl., Iter Hispan.: 111, 127 (1758).
    Campanula erinus f. *coerulea* Maire in Jahand. & Maire, Cat. Pl. Maroc: 735 (1934).
    Campanula erinus f. *albiflora* Maire in Jahand. & Maire, Cat. Pl. Maroc: 735 (1934).
    Campanula nanella P. A. Smirn., Bjull. Moskovsk. Obšč. Isp. Prir., Otd. Biol. (n.s.)
       70(3): 100 (1965).

Campanula escalerae Rech. f. & Schiman-Czeika in Rech. f., Fl. Iran. 13: 22 (1965).
Iran. 34 IRN. Hemicr.
   Campanula violifolia Pau, Trab. Mus. Nac. Ci. Nat., Ser. Bot. 14: 38 (1918); non Lam.,
    Encycl. 1: 587 (1785), as 'violaefolia'.

Campanula euboica Phitos, Oesterr. Bot. Z. 112: 467 (1965).
Greece. 13 GRC. Hemicr. or biennial. $2n = 34$.

Campanula euclasta Boiss., Diagn. Pl. Orient. (ser. 1) 11: 70 (1849).
Lebanon & Syria. 34 LBS. Hemicr. or geophyte.

Campanula eugeniae Fed. in Kom., Fl. URSS 24: 466 (1957).
Kirgizstan. 32 KGZ. Hemicr. or geophyte.

Campanula euxina (Velen.) Ančev in Velchev, Treta Nac. Konferenc. Bot. Sofia 1981:
242 (1983).
Bulgaria. 13 BUL. Hemicr. or geophyte. $2n = 34$.
   Campanula rotundifolia var. *euxina* Velen., Fl. Bulg., Suppl.: 185 (1898). *Campanula*
    rotundifolia subsp. *euxina* (Velen.) Hayek, Repert. Spec. Nov. Regni Veg. Beih. 30(2):
    538 (1930).

Campanula excisa Schleich. ex Murith, Guide Bot. Voy. Valais: 57 (1810). *Campanula*
grammosepala var. *leptosepala* Vuk., Linnaea 26: 325 (1854).
S. Alps. 11 SWI 13 ITA. Hemicr. $2n = 34$.

Campanula exigua Rattan, Bot. Gaz. (Crawfordsville) 11: 339 (1886).
W. U.S.A. (C. California). 76 CAL. Ther. $2n = 34$.

Campanula expansa Rudolph, Mém. Acad. Imp. Sci. St. Pétersbourg Hist. Acad. 4: 340
(1809). *Wahlenbergia expansa* (Rudolph) Steffen, Beih. Bot. Centralbl. 56 (abteil. B): 416
(1937). *Astrocodon expansus* (Rudolph) Fed. in Kom., Fl. URSS 24: 378 (1957).
Russian Far East (Khabarovsk, Magadan). 31 KHA MAG. Hemicr. $2n = 34, 40$.
   Campanula homallanthina Ledeb., Mém. Acad. Imp. Sci. St. Pétersbourg Hist. Acad. 5:
    524 (1811). *Platycodon homallanthinus* (Ledeb.) A. DC., Monogr. Campan.: 126
    (1830), as 'homallanthinum'. *Wahlenbergia homallanthina* (Ledeb.) A. DC. in DC.,
    Prodr. 7: 425 (1839). *Campanopsis homallanthina* (Ledeb.) Kuntze, Revis. Gen. Pl. 2:
    379 (1891).
   Platycodon homallanthinus var. *angustifolius* A. DC., Monogr. Campan.: 126 (1830), as
    'angustifolium'. *Wahlenbergia homallanthina* var. *angustifolia* A. DC. in DC., Prodr. 7:
    426 (1839).
   Campanula kruhseana Fisch. ex Regel & Tiling, Fl. Ajan.: 108 (1858). *Astrocodon*
    kruhseanus (Fisch. ex Regel & Tiling) Fed. in Kom., Fl. URSS 24: 377 (1957).

Campanula fastigiata Dufour ex Schult. in Roem. & Schult., Syst. Veg. 5: 136 (1819).
  Brachycodon fastigiatus (Dufour ex Schult.) Fed. in Kom., Fl. URSS 24: 342 (1957).
  Brachycodonia fastigiata (Dufour ex Schult.) Fed. in Karyagin, Fl. Azerbajdž. 8: 164 (1961).
  W. Mediterranean; Turkey to Caucasus (Gruziya, Azerbaijan), Tadzhikistan & Syria. 12 SPA
    20 ALG MOR 32 TZK 33 TCS 34 LBS TUR. Biennial or ther. $2n$ = 18.

Campanula fenestrellata Feer, J. Bot. 28: 272 (1890). *Campanula garganica* subsp.
  fenestrellata (Feer) Hayek, Repert. Spec. Nov. Regni Veg. Beih. 30(2): 533 (1930).
  Campanula elatines var. *fenestrellata* (Feer) L. H. Bailey in L. H. Bailey & E. Z. Bailey,
  Hortus Second: 135 (1941).
  Albania, Croatia, Macedonia. 13 ALB YUG. Hemicr. or geophyte. $2n$ = 34.

  subsp. **debarensis** (Rech. f.) Damboldt, Bot. Jahrb. Syst. 84: 358 (1965).
    Albania & Macedonia. 13 ALB YUG. Hemicr. or geophyte. $2n$ = 34.
    * *Campanula debarensis* Rech. f., Repert. Spec. Nov. Regni Veg. 47: 175 (1939).

  subsp. **fenestrellata**
    Croatia. 13 YUG. Hemicr. or geophyte. $2n$ = 34.
    Campanula lepida Feer, J. Bot. 28: 272 (1890). *Campanula garganica* var. *lepida* (Feer)
      Hayek, Repert. Spec. Nov. Regni Veg. Beih. 30(2): 533 (1930).

  subsp. **istriaca** (Feer) Damboldt, Bot. Jahrb. Syst. 84: 349 (1965).
    Croatia. 13 YUG. Hemicr. or geophyte. $2n$ = 34.
    * *Campanula istriaca* Feer, J. Bot. 28: 271 (1890). *Campanula elatines* var. *istriaca* (Feer)
      Fiori & Paol. in Fiori & Bég., Fl. Italia 3: 182 (1903). *Campanula garganica* subsp.
      istriaca (Feer) Hayek, Repert. Spec. Nov. Regni Veg. Beih. 30(2): 534 (1930).

Campanula ficarioides Timb.-Lagr., Mém. Acad. Sci. Toulouse (ser. 5) 6: 33 (1862).
  Campanula linifolia subsp. *ficarioides* (Timb.-Lagr.) Nyman, Consp. Fl. Eur.: 479 (1879).
  Campanula pusilla subsp. *ficarioides* (Timb.-Lagr.) Rouy, Fl. France 10: 76 (1908).
  Campanula scheuchzeri subsp. *ficarioides* (Timb.-Lagr.) O. Bolòs & Vigo, Collect. Bot.
  (Barcelona) 14: 101 (1983). *Campanula scheuchzeri* var. *ficarioides* (Timb.-Lagr.) O. Bolòs &
  Vigo, Collect. Bot. (Barcelona) 14: 101 (1983).
  Pyrénées. 12 FRA SPA. Hemicr. or geophyte. $2n$ = 102.
    Campanula gautieri Jeanb. & Timb.-Lagr., Bull. Soc. Sci. Phys. Nat. Toulouse 3: 621
      (1876). *Campanula pusilla* var. *gautieri* (Jeanb. & Timb.-Lagr.) Rouy, Fl. France 10: 77
      (1908). *Campanula ficarioides* subsp. *gautieri* (Jeanb. & Timb.-Lagr.) Podlech, Feddes
      Repert. 71: 146 (1965).

Campanula filicaulis Durieu, Expl. Sci. Algérie: pl. 62 (1849).
  N. Africa. 20 MOR ALG TUN. Hemicr. $2n$ = 16, 24, 26, 48, 50, 52, 72.
    Campanula reboudiana Pomel, Nouv. Mat. Fl. Atl.: 2 (1874). *Campanula filicaulis* var.
      reboudiana (Pomel) Maire, Bull. Soc. Hist. Nat. Afrique N. 19: 59 (1928).
    Campanula maroccana Ball, J. Linn. Soc., Bot. 16: 554 (1878). *Campanula atlantica* var.
      maroccana (Ball) Batt., Fl. Algérie, Suppl. Phan.: 62 (1910). *Campanula filicaulis* var.
      maroccana (Ball) Dobignard, J. Bot. Soc. Bot. France 20: 25 (2002); non Pau in Font
      Quer, Iter Marocc.: 634 (1927).
    Campanula maroccana var. *pseudoradicosa* Litard. & Maire, Mém. Soc. Sci. Nat. Maroc 6:
      14 (1924). *Campanula filicaulis* var. *pseudoradicosa* (Litard. & Maire) Emb. & Maire,
      Mém. Soc. Sci. Nat. Maroc 17: 51 (1927), as 'pseudo-radicosa'.
    Campanula filicaulis var. *cedretorum* H. Lindb., Acta Soc. Sci. Fenn., Ser. B, Opera Biol.
      1(2): 149 (1932).
    Campanula saxifragoides f. *elata* Faure & Maire, Bull. Soc. Hist. Nat. Afrique N. 29:
      431 (1938).
    Campanula saxifragoides var. *gattefossei* Maire & Weiller, Bull. Soc. Hist. Nat. Afrique N.
      29: 432 (1938).
    Campanula antiatlantica Maire, Weiller & Wilczek, Bull. Soc. Hist. Nat. Afrique N. 29:
      432 (1938).

Campanula filicaulis var. *parielii* Quézel, Feddes Repert. Spec. Nov. Regni Veg. 56: 26 (1953).
Campanula filicaulis var. *mairei* Quézel, Feddes Repert. Spec. Nov. Regni Veg. 56: 26 (1953).
Campanula filicaulis var. *pseudoantiatlantica* Quézel, Feddes Repert. Spec. Nov. Regni Veg. 56: 26 (1953).

Campanula flaccidula Vatke, Linnaea 38: 712 (1874).
Turkey to Syria & Iran. 34 IRQ IRN LBS TUR. Ther. $2n = 28$.
Campanula singarensis Boiss. & Hausskn. in Boiss., Pl. Orient. Nov. 1: 2 (1875).

Campanula floridana S. Watson ex A. Gray, Syn. Fl. N. Amer. 2(1): 13 (1878). *Rotantha floridana* (S. Watson ex A. Gray) Small, Man. S.E. Fl.: 1289 (1933).
SE. U.S.A. (Florida). 78 FLA. Hemicr.

Campanula foliosa Ten., Fl. Napol. 1: xvi (1811). *Campanula glomerata* var. *foliosa* (Ten.) Spreng., Syst. Veg. 1: 731 (1824).
Italy & Balkans. 13 ALB GRC ITA YUG. Hemicr. $2n = 34$.

Campanula formanekiana Degen & Dörfl., Denkschr. Kaiserl. Akad. Wiss., Math.-Naturwiss. Kl. 64: 728 (1897).
Greece & Macedonia. 13 GRC YUG. Biennial. $2n = 34$.
 * *Campanula cinerea* Formánek, Verh. Naturf. Vereines Brünn 32: 289 (1894); non L. f., Suppl. Pl.: 139 (1782); nec Hegetschw., Fl. Schweiz: 230 (1838-39).

Campanula forsythii (Arcang.) Podlech, Feddes Repert. 71: 81 (1965).
Sardegna. 12 SAR. Hemicr. or geophyte. $2n = 34$.
 * *Campanula rotundifolia* var. *forsythii* Arcang., Atti Soc. Tosc. Sci. Nat. Pisa Processi Verbali 5: 138 (1886).
Campanula macrorhiza var. *sardoa* Levier ex Tanfani in Parl., Fl. Ital. 8: 101 (1888).

Campanula fragilis Cirillo, Pl. Rar. Neapol. 1: 32 (1788).
Italy. 13 ITA. Hemicr. $2n = 32, 34$.

subsp. **cavolinii** (Ten.) Damboldt, Bot. Jahrb. Syst. 84: 331 (1965).
Italy. 13 ITA. Hemicr. $2n = 32$.
 * *Campanula cavolinii* Ten., Fl. Napol. 3: 201 (1824-29), as 'cavolini'.
Campanula cavolinii var. *hirta* Ten., Fl. Napol. 3: 201 (1824-1829).

subsp. **fragilis**
Italy. 13 ITA. Hemicr. $2n = 32, 34$.
Campanula diffusa Vahl, Symb. Bot. 1: 48 (1790). *Campanula fragilis* var. *diffusa* (Vahl) Steud., Nomencl. Bot. (ed. 2) 1: 267 (1840).
Campanula crassifolia Nees, Syll. Pl. Nov. 1: 6 (1824).
Campanula fragilis var. *hirsuta* Ten., Fl. Napol. 3: 201 (1824-29).
Campanula fragilis var. *hirta* Ten., Fl. Napol. 3: 201 (1824-29).
Campanula fragilis var. *hirsuta* A. DC., Monogr. Campan.: 307 (1830); non Ten., Fl. Napol. 3: 201 (1824-29). *Campanula fragilis* f. *hirsuta* Voss in Siebert & Voss, Vilm. Blumengärtn. (ed. 3) 1: 566 (1894).
Campanula barrelieri C. Presl, Symb. Bot. 1: 30 (1831). *Campanula fragilis* var. *barrelieri* (C. Presl) Nyman, Consp. Fl. Eur.: 480 (1879).
Campanula fragilis subsp. *canescens* Schousb. ex Arcang., Comp. Fl. Ital.: 453 (1882).
Campanula fragilis f. *grandiflora* Voss in Siebert & Voss, Vilm. Blumengärtn. (ed. 3) 1: 566 (1894).

Campanula fritschii Witasek, Abh. K. K. Zool.-Bot. Ges. Wien 1(3): 90 (1902). *Campanula rhomboidalis* var. *fritschii* (Witasek) Fiori, Nuov. Fl. Italia 2: 563 (1927).
SE. France (Alpes de Provence). 12 FRA. Hemicr. or geophyte. $2n = 68$.

Campanula fruticulosa (O. Schwarz & P. H. Davis) Damboldt, Notes Roy. Bot. Gard.
Edinburgh 35: 49 (1976).
Turkey. 34 TUR. Hemicr.
** Tracheliopsis fruticulosa* O. Schwarz & P. H. Davis, Hooker's Icon. Pl. 35: pl. 3498
(1950).

Campanula gansuensis L. Z. Wang & D. Y. Hong, Bot. Bull. Acad. Sin. 41: 159 (2000).
N. China (Gansu). 36 CHN. Ther.

Campanula garganica Ten., App. Ind. Sem. 1827: 3 (1827). *Campanula oxyphylla* var.
lasiopetala Vuk., Linnaea 26: 331 (1854). *Campanula elatines* var. *garganica* (Ten.) Fiori,
Nuov. Fl. Italia 2: 559 (1927).
SE. Italy (Monte Gargano) & Ionian Is. (Kefallinía). 13 GRC ITA. Hemicr. 2*n* = 34, 68.

subsp. **acarnanica** (Damboldt) Damboldt, Bot. Jahrb. Syst. 84: 358 (1965).
Ionian Is. (Kefallinía). 13 GRC. Hemicr. 2*n* = 34.
** Campanula acarnanica* Damboldt, Bot. Jahrb. Syst. 84: 341 (1965).

subsp. **cephallenica** (Feer) Hayek, Repert. Spec. Nov. Regni Veg. Beih. 30(2): 534 (1930).
Ionian Is. (Kefallinía). 13 GRC. Hemicr. 2*n* = 34.
** Campanula cephallenica* Feer, J. Bot. 28: 273 (1890).

subsp. **garganica**
SE. Italy (Monte Gargano). 13 ITA. Hemicr. 2*n* = 34, 68.
Wahlenbergia flaccida C. Presl, Symb. Bot. 1: 29 (1831); non A. DC., Monogr. Campan.:
138 (1830). *Wahlenbergia garganica* G. Don, Gen. Hist. 3: 738 (1834).
Campanula garganica var. *hirsuta* Anonymous, Garden (London) 46: 253 (1894).

Campanula gentilis Kovanda, Folia Geobot. Phytotax. 3: 407 (1968).
S. Germany & Czech Rep. 11 CZE GER. Hemicr. or geophyte. 2*n* = 34.
Campanula rotundifolia var. *conferta* Knaf in Opiz, Nomencl. Bot. 1: 95 (1831).

Campanula ghilanensis Pall. ex Schult. in Roem. & Schult., Syst. Veg. 5: 154 (1819).
Specularia ghilanensis (Pall. ex Schult.) A. DC., Monogr. Campan.: 350 (1830).
Iran. 34 IRN. Ther.

Campanula giesekiana Vest ex Schult. in Roem. & Schult., Syst. Veg. 5: 89 (1819).
Campanula uniflora var. *giesekiana* (Vest ex Schult.) A. DC., Monogr. Campan.: 339 (1830),
as 'gieseckiana'. *Campanula rotundifolia* var. *giesekiana* (Vest ex Schult.) Simmons, Ark. Bot.
6(17): 20 (1907), as 'gieseckiana'.
NE. Canada to Greenland, Iceland, and N. Europe. 10 FIN ICE NOR SWE 14 RUN 70 GNL
NUN 72 NFL LAB ONT QUE. Hemicr. or geophyte. 2*n* = 34, 68.
Campanula dubia A. DC., Monogr. Campan.: 286 (1830). *Campanula rotundifolia* var.
dubia (A. DC.) J. K. Henry, Fl. S. Brit. Columbia: 283 (1915). *Campanula rotundifolia*
f. *dubia* (A. DC.) Farw., Pap. Michigan Acad. Sci. 3: 105 (1924).
Campanula pratensis Bach. Pyl. ex A. DC., Monogr. Campan.: 287 (1830).
Campanula pratensis var. *angustifolia* A. DC., Monogr. Campan.: 287 (1830).
Campanula rotundifolia var. *arctica* Lange, Fl. Dan. 46: pl. 2711 (1867). *Campanula
arctica* (Lange) Hruby, Magyar Bot. Lapok 29: 183 (1930).
Campanula rotundifolia var. *uniflora* Lange, Meddel. Grønland 3: 93 (1880).
Campanula groenlandica Berlin, Öfvers. Förh. Kongl. Svenska Vetensk.-Akad. 41(7): 50
(1884). *Campanula gieseckiana* subsp. *groenlandica* (Berlin) Böcher, Biol. Skr. 11(4): 63
(1960). *Campanula rotundifolia* subsp. *groenlandica* (Berlin) A. Löve & D. Löve, Univ.
Colorado Stud., Ser. Biol. 24: 52 (1966).
Campanula rotundifolia f. *pygmaea* Hartz, Meddel. Grønland 18: 338 (1895).
Campanula rotundifolia f. *albiflora* Abrom., Biblioth. Bot. 42B: 63 (1899); non E. L. Rand
& Redfield, Fl. Mt. Desert Isl.: 124 (1894).

Campanula rotundifolia f. *lapponica* Witasek, Meddeland. Soc. Fauna Fl. Fenn. 29: 204
(1904). *Campanula rotundifolia* var. *lapponica* (Witasek) Simmons, Ark. Bot. 6(17): 21
(1907). *Campanula lapponica* (Witasek) C. Regel, Repert. Spec. Nov. Regni Veg. Beih.
82(1): 87 (1935).
Campanula rotundifolia f. *laciniata* J. Rousseau & Raymond, Naturaliste Canad. 79:
84 (1952).

Campanula gilliatii Milne-Redh. & Turrill, Bull. Misc. Inform. Kew 1930: 439 (1930).
Iran. 34 IRN. Hemicr.

Campanula × gisleri Brügger, Jahresber. Naturf. Ges. Graub. (ser. 2): 23-24: 104 (1880). C.
rhomboidalis × C. scheuchzeri
Switzerland. 11 SWI. Hemicr. or geophyte.

Campanula glomerata L., Sp. Pl.: 166 (1753). *Weitenwebera glomerata* (L.) Opiz, Seznam:
105 (1852).
Europe to Russian Far East, Korea & Japan (Kyushu); cult.; naturalized in E. Canada, N. &
W. U.S.A. 10 DEN FIN GRB SWE 11 AUT BGM CZE GER HUN NET POL SWI 12 FRA
SPA 13 ALB BUL GRC ITA ROM YUG 14 BLR BLT KRY RUC RUE RUN RUS RUW UKR
30 ALT BRY CTA IRK KRA TVA WSB 31 AMU KHA PRM 32 KAZ 33 ALL 34 IRN IRQ
TUR 36 CHI CHM CHX 37 MON 38 JAP (72) nsc ont que (74) ill min nda (75) mic mai
mas nwh nwy pen ver (76) uta. Hemicr. or geophyte. $2n$ = 30, 34, 60, 68, 90.

subsp. **caucasica** (Trautv.) Ogan., Candollea 50: 280 (1995).
Caucasus (Krasnodar, Karacheyevo-Cherkessiya, Kabardino-Balkariya, Severo-Osetiya,
Chechnya, Dagestan, Gruziya, Armenia, Azerbaijan). 33 ALL. Hemicr. or geophyte.
$2n$ = 90
* *Campanula glomerata* var. *caucasica* Trautv., Trudy Imp. S.-Peterburgsk. Bot. Sada 2:
564 (1873).
Campanula ruprechtii Grossh., Bull. Polytechn. Inst. Tiflis. 2: 201 (1926); non Boiss., Fl.
Orient. 3: 905 (1875), as 'ruprechti'.
Campanula trautvetteri Grossh. ex Fed. in Kom., Fl. URSS 24: 463 (1957).

subsp. **cervicarioides** (Schult.) Arcang., Comp. Fl. Ital.: 456 (1882).
S. Europe. 12 FRA 13 ITA YUG. Hemicr. or geophyte.
* *Campanula cervicarioides* Schult. in Roem. & Schult., Syst. Veg. 5: 128 (1819). *Campanula
glomerata* var. *cervicarioides* (Schult.) A. DC., Monogr. Campan.: 255 (1830).
Campanula glomerata race *cervicarioides* (Schult.) Rouy, Fl. France 10: 67 (1908).

subsp. **daqingshanica** D. Y. Hong & Y. Z. Zhao, Fl. Reipubl. Popul. Sin. 73(2): 184
(1983).
Inner Mongolia (Nei Mongol). 36 CHI. Hemicr. or geophyte.

subsp. **elliptica** (Kit. ex Schult.) O. Schwarz, Mitt. Thüring. Bot. Ges. 1: 118 (1949).
SE. Europe. 11 CZE HUN 13 BUL ITA ROM YUG 14 UKR. Hemicr. or geophyte. $2n$ = 30.
* *Campanula elliptica* Kit. ex Schult., Oestr. Fl. (ed. 2) 1: 410 (1814). *Campanula glomerata*
var. *elliptica* (Kit. ex Schult.) Spreng., Syst. Veg. 1: 731 (1824).
Campanula glomerata var. *ramosa* K. Koch, Linnaea 17: 280 (1843).
Campanula pulchra O. D. Wissjul. in M. Kotov, Fl. URSR 10: 474 (1961).

subsp. **farinosa** (Rochel ex Besser) Kirschl., Fl. Alsace 1: 378 (1851).
France & Italy to Kazakhstan. 11 AUT CZE GER HUN POL SWI 12 FRA 13 ITA ROM
YUG 14 BLR KRY RUC RUE RUS UKR 32 KAZ. Hemicr. or geophyte. $2n$ = 30.
* *Campanula glomerata* var. *farinosa* Rochel ex Besser, Cat. Hort. Cremeneci: 27 (1816).
Campanula farinosa (Rochel ex Besser) Andrz. ex Besser, Enum. Pl.: 10 (1821).
Campanula desertorum Weinm., Bull. Soc. Imp. Naturalistes Moscou 7: 58 (1837).
Campanula glomerata var. *desertorum* (Weinm.) Nyman, Consp. Fl. Eur., Suppl. 2:
207 (1889).

Campanula glaucophylla Schloss. & Vuk., Syll. Fl. Croat.: 72 (1857). *Campanula glomerata* var. *glaucophylla* (Schloss. & Vuk.) Nyman, Consp. Fl. Eur.: 478 (1879). *Campanula glomerata* f. *glaucophylla* (Schloss. & Vuk.) Hayek & Hegi in Hegi, Ill. Fl. Mitt.-Eur. 6: 340 (1915).

Campanula polessica O. D. Wissjul. in M. Kotov, Fl. URSR 10: 475 (1961). *Campanula glomerata* f. *polessica* (O. D. Wissjul.) Sóo, Acta Bot. Acad. Sci. Hung. 12: 365 (1966).

subsp. **glomerata**

Europe to Caucasus (Krasnodar, Kabardino-Balkariya, Dagestan, Adzhariya, Gruziya, Armenia, Nagorno Karabakh, Azerbaijan), Kazakhstan, Siberia & NW. China (Xinjiang); cult.; naturalized in E. Canada, N. & W. U.S.A. 10 DEN FIN GRB SWE 11 AUT BGM CZE GER HUN NET POL SWI 12 FRA SPA 13 ALB BUL GRC ITA ROM YUG 14 BLR BLT KRY RUC RUE RUN RUS RUW UKR 30 ALT CTA IRK KRA TVA WSB 32 KAZ 33 ALL 36 CHX (72) nsc ont que (74) ill min nda (75) mic mai mas nwh nwy pen ver (76) uta. Hemicr. or geophyte. $2n = 30, 34, 60, 68, 90$.

Campanula conglomerata Gueldenst., Reis. Russland 2: 343 (1791).

Gentiana collina With., Arr. Brit. Pl. (ed. 3) 2: 282 (1796).

Campanula glomerata var. *heterophylla* Lapeyr., Hist. Pl. Pyrénées: 106 (1813).

Campanula glomerata var. *ramosa* Lapeyr., Hist. Pl. Pyrénées: 106 (1813).

Campanula aggregata Willd., Enum. Pl., Suppl.: 10 (1814). *Marianthemum aggregatum* (Willd.) Schrank, Flora 7 (Beilage 2): 54 (1824). *Campanula glomerata* var. *aggregata* (Willd.) Spreng., Syst. Veg. 1: 731 (1824). *Campanula glomerata* subsp. *aggregata* (Willd.) Arcang., Comp. Fl. Ital.: 456 (1882). *Campanula glomerata* f. *aggregata* (Willd.) Voss in Siebert & Voss, Vilm. Blumengärtn. (ed. 3) 1: 569 (1894). *Campanula glomerata* race *aggregata* (Willd.) Rouy, Fl. France 10: 66 (1908).

Campanula congesta Vest ex Schult. in Roem. & Schult., Syst. Veg. 5: 125 (1819). *Campanula glomerata* var. *congesta* (Vest ex Schult.) Schur, Enum. Pl. Transsilv.: 434 (1866).

Campanula glomerata var. *minor* Gray, Nat. Arr. Brit. Pl. 2: 410 (1821).

Campanula glomerata var. *pusilla* A. DC., Monogr. Campan.: 255 (1830). *Campanula glomerata* f. *pusilla* (A. DC.) Hayek & Hegi in Hegi, Ill. Fl. Mitt.-Eur. 6: 340 (1915).

Campanula glomerata var. *sparsiflora* A. DC., Monogr. Campan.: 255 (1830).

Campanula glomerata var. *nana* Noulet, Fl. Bass. Sous-Pyrén.: 408 (1837).

Campanula glomerata var. *grandiflora* Kuntze, Taschen-Fl. Leipzig: 127 (1867).

Campanula glomerata var. *oblongata* Trautv., Trudy Imp. S.-Peterburgsk. Bot. Sada 6: 69 (1879).

Campanula glomerata var. *umbrosa* Trautv., Trudy Imp. S.-Peterburgsk. Bot. Sada 6: 69 (1879).

Campanula glomerata var. *ortleppii* H. Lév., Monde Pl. 10: 6 (1908), as 'ortleppi'.

Campanula glomerata var. *longifolia* Schloss. ex Dalla Torre & Sarnth., Fl. Tirol. 6(3): 453 (1911).

Campanula glomerata var. *semgensis* Husseinov, Bot. Žurn. (Moscow & Leningrad) 55: 1501 (1970).

subsp. **hispida** (Witasek) Hayek, Repert. Spec. Nov. Regni Veg. Beih. 30(2): 532 (1930).

Balkans, Caucasus (Krasnodar, Stavropol, Kabardino-Balkariya, Chechnya, Dagestan), W. Asia. 13 BUL YUG 33 NCS 34 IRN IRQ TUR. Hemicr. or geophyte. $2n = 60$.

* *Campanula glomerata* f. *hispida* Witasek, Ann. K. K. Naturhist. Hofmus. 20: 419 (1905).

Campanula glomerata var. *hispida* Fomin in N. M. Kusn., N. Bush & Fomin, Mater. Fl. Kavkaza 4(6): 104 (1906).

Campanula lamioides Witasek, Ann. K. K. Naturhist. Hofmus. 23: 191 (1909).

Campanula eocervicaria Nábělek, Spisy Pfiír. Fak. Masarykovy Univ. 70: 5 (1926), as 'eo-cervicaria'.

Campanula maleevii Fed. in Kom., Fl. URSS 24: 462 (1957). *Campanula glomerata* subsp. *maleevii* (Fed.) Ogan., Candollea 50: 280 (1995).

Campanula glomerata subsp. *glabriflora* Contandr., Quézel & Pamukç., Ann. Univ. Provence Sci. 46: 54 (1972), as 'grabriflora'.

subsp. **krylovii** Olonova, Turczaninowia 2(3): 8 (1999).
West Siberia. 30 WSB. Hemicr. or geophyte.

subsp. **oblongifolia** (K. Koch) Fed. in Takht. & Fed., Fl. Erevana: 267 (1972).
Turkey, Caucasus (Gruziya, Armenia, Nakhichevan, Azerbaijan, Nagorno Karabakh),
Iran. 33 TCS 34 IRN TUR. Hemicr. or geophyte. 2*n* = 90.
Campanula nicaeensis Schult. in Roem. & Schult., Syst. Veg. 5: 126 (1819). *Campanula
glomerata* var. *nicaeensis* (Schult.) Spreng., Syst. Veg. 1: 731 (1824).
* *Campanula glomerata* var. *oblongifolia* K. Koch, Linnaea 23: 639 (1850). *Campanula
oblongifolia* (K. Koch) Kharadze, Zametki Sist. Geogr. Rast. 15: 30 (1949).

subsp. **oblongifolioides** (Galushko) Ogan., Candollea 50: 280 (1995).
Caucasus (Kabardino-Balkariya, Karacheyevo-Cherkessiya). 33 NCS. Hemicr. or geophyte.
* *Campanula oblongifolioides* Galushko, Novosti Sist. Vyssh. Rast. 8: 267 (1971).

subsp. **panjutinii** (Kolak.) Viktorov, Novosti Sist. Vyssh. Rast. 34: 215 (2002). *Campanula
panjutinii* Kolak., Dokl. Akad. Nauk Armjansk. SSR 7: 223 (1947), as 'panjutini'.
Caucasus (Abkhaziya). 33 TCS. Hemicr. or geophyte.

subsp. **serotina** (Wettst.) O. Schwarz, Mitt. Thüring. Bot. Ges. 1: 118 (1949).
Alps. 13 ITA. Hemicr. or geophyte. 2*n* = 30.
* *Campanula serotina* Wettst., Denkschr. Kaiserl. Akad. Wiss., Math.-Naturwiss. Kl. 70:
337 (1900). *Campanula glomerata* var. *serotina* (Wettst.) Hayek & Hegi in Hegi, Ill. Fl.
Mitt.-Eur. 6: 341 (1915).

subsp. **speciosa** (Hornem. ex Spreng.) Domin, Preslia 13-15: 222 (1936).
SE. Siberia (Buryatiya, Chita) to Inner Mongolia (Nei Mongol), Manchuria, Russian Far
East, Korea & Japan (Kyushu); cult. 30 BRY CHT 31 AMU KHA PRM 36 CHI CHM 37
MON 38 JAP. Hemicr. or geophyte. 2*n* = 30.
* *Campanula speciosa* Hornem., Hort. Bot. Hafn.: 957 (1815); non Pourr., Hist. & Mém.
Acad. Roy. Sci. Toulouse 3: 309 (1788). *Campanula glomerata* var. *speciosa* Hornem. ex
Spreng., Syst. Veg. 1: 731 (1824). *Campanula glomerata* var. *grandiflora* Herder, Trudy
Imp. S.-Peterburgsk. Bot. Sada 1: 292 (1873); non Kuntze, Taschen-Fl. Leipzig: 127
(1867). *Campanula glomerata* f. *speciosa* (Hornem. ex Spreng.) Voss in Siebert & Voss,
Vilm. Blumengärtn. (ed. 3) 1: 569 (1894).
Campanula cephalotes Fisch. ex Schrank, Denkschr. Königl.-Baier. Bot. Ges. Regensburg
2: 32 (1822). *Campanula glomerata* subsp. *cephalotes* (Fisch. ex Schrank) D. Y. Hong,
Fl. Reipubl. Popul. Sin. 73(2): 84 (1983).
Campanula glomerata var. *dahurica* Fisch. ex Ker-Gawl., Bot. Reg. 8: pl. 620 (1822).
Campanula cephalotes f. *alba* Nakai, J. Jap. Bot. 20: 187 (1944). *Campanula glomerata* f.
alba (Nakai) U. C. La, Fl. Coreana 7: 36 (1999).
Campanula cephalotes var. *canescens* Maxim. ex Nakai, J. Jap. Bot. 20: 187 (1944).
Campanula glomerata f. *canescens* (Maxim. ex Nakai) Kitag., Neo-Lineam. Fl.
Manshur.: 606 (1979).

subsp. **subcapitata** (Popov) Fed., Bot. J. Linn. Soc. 67: 281 (1973).
Ukraine. 14 UKR. Hemicr. or geophyte.
* *Campanula subcapitata* Popov, Mater. Pozn. Fauny Fl. SSSR, Otd. Bot. (n.s.) 13(5): 295
(1949).

subsp. **symphytifolia** (Albov) Ogan., Candollea 50: 280 (1995).
Caucasus (Abkhaziya). 33 TCS. Hemicr. or geophyte.
* *Campanula glomerata* var. *symphytifolia* Albov, Prodr. Fl. Colchic.: 161 (1895).
Campanula symphytifolia (Albov) Kolak., Fl. Abkhas. 4: 180 (1949).

Campanula × **glomeratiformis** Murr ex Dalla Torre & Sarnth., Fl. Tirol. 6(3): 450 (1911). C.
glomerata × C. spicata
Austria. 11 AUT. Hemicr. or biennial.
Campanula × *pechlaneri* Murr ex Dalla Torre & Sarnth., Fl. Tirol. 6(3): 450 (1911).

Campanula glomeratoides D. Y. Hong, Acta Phytotax. Sin. 18: 247 (1980).
Tibet. 36 CHT. Hemicr.

Campanula goulimyi Turrill, Kew Bull. 1955: 354 (1955).
Greece. 13 GRC. Hemicr. $2n = 34$.

Campanula gracillima Podlech, Feddes Repert. 71: 78 (1965).
S. France (Mt. Lozère). 12 FRA. Hemicr. or geophyte. $2n = 34$.

Campanula grandis Fisch. & C. A. Mey., Index Sem. Hort. Petrop. 5: 34 (1839).
Turkey. 34 TUR. Hemicr. $2n = 16$.

subsp. **grandis**
Turkey 34 TUR. Hemicr. $2n = 16$.
Campanula latiloba A. DC. in DC., Prodr. 7: 478 (1839).

subsp. **rizeensis** (A. Güner) Lammers, Novon 9: 385 (1999).
Turkey. 34 TUR. Hemicr.
**Campanula latiloba* subsp. *rizeensis* A. Güner, Candollea 39: 348 (1984).

Campanula griffinii Morin, Madroño 27: 160 (1980).
W. U.S.A. (N. & C. California). 76 CAL. Ther. $2n = 34$.
**Campanula angustiflora* var. *exilis* J. T. Howell, Leafl. W. Bot. 2: 102 (1938); non
Campanula exilis (A. DC.) D. Dietr., Syn. Pl. 1: 754 (1839).

Campanula grossekii Heuff., Flora 16: 353 (1833).
Serbia, Romania, Bulgaria. 13 BUL ROM YUG. Hemicr. $2n = 34$.

Campanula guinochetii Quézel, Bull. Soc. Sci. Nat. Maroc 31: 260 (1953), as 'guinocheti'.
Morocco. 20 MOR. Hemicr. $2n = 28, 32$.

Campanula hagielia Boiss., Fl. Orient. 3: 899 (1875)
E. Aegean Is. (Chálki, Ródos) & Turkey. 34 EAI TUR. Hemicr. or biennial.
Campanula sporadum Feer, J. Bot. 28: 268 (1890). *Campanula tomentosa* var. *sporadum*
(Feer) Fiori, Annuario Reale Ist. Super. Agric. Forest. Naz. 9: 39 (1924).

Campanula haradjanii Rech. f., Ann. Naturhist. Mus. Wien 57: 86 (1950).
Turkey. 34 TUR. Ther.

Campanula hawkinsiana Hausskn. & Heldr. ex Hausskn., Mitt. Geogr. Ges. (Thüringen)
Jena 5: 87 (1887). *Campanula halacsyana* Baldacci, Malpighia 8: 280 (1894).
Albania & Greece. 13 ALB GRC. Hemicr. or geophyte. $2n = 22$.
Campanula hawkinsiana var. *scabriuscula* Hausskn., Mitth. Thüring. Bot. Vereins (ser. 2)
7: 62 (1895).

Campanula hedgei P. H. Davis, Notes Roy. Bot. Gard. Edinburgh 24: 30 (1962).
Turkey. 34 TUR. Hemicr.

Campanula hercegovina Degen & Fiala, Oesterr. Bot. Z. 44: 303 (1894).
Croatia & Hercegovina. 13 YUG. Hemicr. or geophyte. $2n = 34$.
Campanula tarana K. Malý, Glasn. Zemaljsk. Muz. Bosni Hercegovini 32: 126 (1920).
Campanula hercegovina f. *glaberrima* Witasek ex Hayek, Repert. Spec. Nov. Regni Veg.
Beih. 30(2): 535 (1930).
Campanula hercegovina f. *squarrosa* Degen & Fiala ex Hayek, Repert. Spec. Nov. Regni
Veg. Beih. 30(2): 535 (1930).
Campanula hercegovina f. *humilis* Hruby, Magyar Bot. Lapok 29: 253 (1930).
Campanula hercegovina f. *squarrosa* Hruby, Magyar Bot. Lapok 29: 253 (1930); non
Degen & Fiala ex Hayek, Repert. Spec. Nov. Regni Veg. Beih. 30(2): 535 (1930).

Campanula hercegovina f. *umbosa* Hruby, Magyar Bot. Lapok 29: 253 (1930).
Campanula hercegovina f. *ovalifolia* Hruby, Magyar Bot. Lapok 29: 253 (1930).
Campanula hercegovina subf. *brevicalycina* Hruby, Magyar Bot. Lapok 29: 254 (1930).
Campanula hercegovina subf. *longicalycina* Hruby, Magyar Bot. Lapok 29: 254 (1930).
Campanula hercegovina var. *fallens* Hruby, Magyar Bot. Lapok 29: 254 (1930).
Campanula hercegovina var. *pseudopusilla* Hruby, Magyar Bot. Lapok 29: 254 (1930).

Campanula hermanii Rech. f., Mitt. Bot. Staatssamml. München 16 (Beih.): 37 (1980).
Iran. 34 IRN. Hemicr.

Campanula herminii Hoffmanns. & Link, Fl. Portug. 2: 9 (1813).
Iberian Pen. 12 POR SPA. Hemicr. or geophyte. $2n = 32$.

Campanula heterophylla L., Sp. Pl.: 169, 1231 (1753).
S. Aegean Is. (Kykládes). 13 GRC. Hemicr. or geophyte.

Campanula hieracioides Kolak., Soobšč. Akad. Nauk Gruzinsk. SSR 8: 235 (1947). *Annaea hieracioides* (Kolak.) Kolak., Soobšč. Akad. Nauk Gruzinsk. SSR 94: 163 (1979).
Caucasus (Abkhaziya). 33 TCS. Hemicr. $2n = 34$.

Campanula hierapetrae Rech. f., Oesterr. Bot. Z. 84: 170 (1935).
Kriti. 13 KRI. Hemicr. $2n = 34$.

Campanula hierosolymitana Boiss., Diagn. Pl. Orient. (ser. 1) 11: 62 (1849).
Israel, Lebanon, Syria, Jordan. 34 PAL LBS. Ther.

Campanula hispanica Willk. in Willk. & Lange, Prodr. Fl. Hispan. 2: 291 (1868).
Campanula rotundifolia var. *hispanica* (Willk.) Fiori, Nuov. Fl. Italia 2: 561 (1927).
Campanula balcanica var. *hispanica* (Willk.) Hruby, Magyar Bot. Lapok 29: 244 (1930).
Campanula rotundifolia subsp. *hispanica* (Willk.) Rivas Goday & Borja ex O. Bolòs & Vigo, Fl. Països Catalans 3: 661 (1996).
W. Mediterranean. 12 FRA SPA 20 ALG. Hemicr. or geophyte. $2n = 34$.

subsp. **catalanica** Podlech, Feddes Repert. 71: 70 (1965). *Campanula rotundifolia* subsp. *catalanica* (Podlech) O. Bolòs & Vigo, Collect. Bot. (Barcelona) 14: 101 (1983). *Campanula rotundifolia* var. *catalanica* (Podlech) O. Bolòs & Vigo, Collect. Bot. (Barcelona) 14: 101 (1983). *Campanula macrorhiza* subsp. *catalanica* (Podlech) Rivas Mart., Itinera Geobot. 15: 699 (2002).
SW. Europe. 12 FRA SPA. Hemicr. or geophyte.
Campanula macrorhiza var. *gypsicola* Costa, Introd. Fl. Cataluña: 165 (1864). *Campanula gypsicola* (Costa) Witasek, Abh. K. K. Zool.-Bot. Ges. Wien 1(3): 70 (1902). *Campanula rotundifolia* var. *gypsicola* (Costa) O. Bolòs & Vigo, Collect. Bot. 14: 101 (1983). *Campanula macrorhiza* subsp. *gypsicola* (Costa) Rivas Mart., Itinera Geobot. 15: 699 (2002).

subsp. **hispanica**
W. Mediterranean. 12 SPA 20 ALG. Hemicr. or geophyte. $2n = 34$.
Campanula hispanica var. *glabra* Willk., Suppl. Prodr. Fl. Hispan.: 128 (1893).
Campanula granatensis Witasek, Abh. K. K. Zool.-Bot. Ges. Wien 1(3): 73 (1902).
Campanula rotundifolia var. *subramulosa* Batt., Fl. Algérie, Suppl. Phan.: 62 (1910).
Campanula burgalensis Font Quer, Trab. Mus. Ci. Nat., Ser. Bot. 5: 40 (1924).
Campanula paui Font Quer, Trab. Mus. Ci. Nat., Ser. Bot. 5: 40 (1924).
Campanula kahenae Maire, Bull. Soc. Hist. Nat. Afrique N. 19: 60 (1928).
Campanula guiraoi Sennen, Diagn. Nouv.: 92 (1936).
Campanula losae Sennen, Diagn. Nouv.: 222 (1936).
Campanula betetae Caball., Anales Jard. Bot. Madrid 2: 247 (1942).

Campanula hissarica Kamelin ex Rassulova in Ovcz., Fl. Tadzhik. SSR 9: 526 (1988).
Tadzhikistan. 32 TZK. Hemicr. or geophyte.

Campanula hofmannii (Pant.) Greuter & Burdet, Willdenowia 11: 40 (1981).
Bosnia. 13 YUG. Biennial. $2n = 34$.
 * *Symphyandra hofmannii* Pant., Oesterr. Bot. Z. 31: 347 (1881), as 'hofmanni'.

Campanula humillima A. DC. in DC., Prodr. 7: 474 (1839).
Iran. 34 IRN. Hemicr.

Campanula hypocrateriformis Dobignard, Saussurea 19: 112 (1989).
Morocco. 20 MOR. Ther.
 * *Campanula dichotoma* f. *pseudodichotoma* Maire ex Quézel, Feddes Repert. Spec. Nov.
   Regni Veg. 56: 16 (1953).
 Campanula dichotoma var. *pallida* Maire, Bull. Soc. Hist. Nat. Afrique N. 23: 199
   (1932).*Campanula afra* var. *pallida* (Maire) Dobignard, Candollea 47: 455 (1992).

Campanula hypopolia Trautv., Trudy Imp. S.-Peterburgsk. Bot. Sada 4: 389 (1876).
 Hemisphaera hypopolia (Trautv.) Kolak. & Serdyuk. in Kolak., Kolokol'ch. Kavkaz.: 124 (1991).
Caucasus (Severo-Osetiya, Gruziya). 33 ALL. Hemicr. $2n = 34$.

Campanula hystricula Pau, Trab. Mus. Nac. Ci. Nat., Ser. Bot. 14: 37 (1918).
Iran. 34 IRN. Hemicr.

Campanula iconia Phitos, Oesterr. Bot. Z. 112: 481 (1965).
Turkey. 34 TUR. Biennial.

Campanula imeretina Rupr., Bull. Acad. Imp. Sci. Saint-Pétersbourg 11: 217 (1867).
 Campanula sibirica var. *imeretina* (Rupr.) Trautv., Trudy Imp. S.-Peterburgsk. Bot. Sada 4:
 388 (1876).
Caucasus (Gruziya). 33 TCS. Hemicr. $2n = 34$.

Campanula immodesta Lammers, Novon 8: 34 (1998).
N. India (Uttar Pradesh) to Tibet & SC. China (Sichuan, Yunnan). 36 CHC CHT 40 EHM
   IND NEP. Hemicr.
 * *Campanula modesta* Hook. f. & Thomson, J. Proc. Linn. Soc., Bot. 2: 24 (1858); non
   Schott in Schott, Nyman & Kotschy, Analect. Bot.: 13 (1854).

Campanula incanescens Boiss., Diagn. Pl. Orient. (ser. 1) 7: 16 (1846).
Iran to Turkmenistan & Kirgizstan. 32 KGZ TKM TZK 34 IRN. Hemicr. or geophyte. $2n = 28$.
 Campanula fedtschenkiana Trautv., Trudy Imp. S.-Peterburgsk. Bot. Sada 6: 77 (1879).

Campanula incurva Aucher ex A. DC. in DC., Prodr. 7: 464 (1839).
Greece & E. Aegean Is. (Ikaria) 13 GRC 34 EAI. Biennial. $2n = 32$.
 Campanula leutweinii Heldr., Index Sem. Hort. Bot. Reg. Univ. Othonis Athenarum: 7
   (1860).

Campanula intercedens Witasek, Abh. K. K. Zool.-Bot. Ges. Wien 1(3): 43 (1902).
 Campanula rotundifolia var. *intercedens* (Witasek) Farw., Pap. Michigan Acad. Sci. 3: 105
 (1924). *Campanula rotundifolia* subsp. *intercedens* (Witasek) Á. Löve & D. Löve, Taxon 31:
 356 (1982).
E. North America (Nova Scotia to Ontario, Missouri & North Carolina). 72 NBR NSC ONT
   QUE 74 ILL IOW MIN MSO WIS 75 CNT INI MAI MAS MIC NWH NWJ NWY OHI PEN
   VER WVA 78 MRY NCA TEN VRG; cult. Hemicr. or geophyte. $2n = 34, 68$.
 Campanula virgata Raf. ex Schult. in Roem. & Schult., Syst. Veg. 5: 100 (1819); non
   Labill., Icon. Pl. Syr. 2: 11 (1791).

Campanula rotundifolia var. *alpina* Tuck., Amer. J. Sci. Arts 45: 27 (1843).
Campanula rotundifolia var. *dentata* Coleman, Kent Sci. Inst. Misc. Publ. 2: 24 (1874);
  non Schur, Enum. Pl. Transsilv.: 443 (1866).
Campanula rotundifolia var. *canescens* E. J. Hill, Amer. Naturalist 12: 818 (1878).
Campanula rotundifolia f. *albiflora* E. L. Rand & Redfield, Fl. Mt. Desert Isl.: 124 (1894).
Campanula rotundifolia f. *linifolia* Farw., Pap. Michigan Acad. Sci. 3: 105 (1924).
Campanula rotundifolia f. *cleistocodona* Lakela, Rhodora 47: 391 (1945).
Campanula latisepala f. *alba* J. Rousseau & Raymond, Naturaliste Canad. 79: 84
  (1952).

Campanula involucrata Aucher ex A. DC. in DC., Prodr. 7: 467 (1839).
  Turkey to Iran. 34 IRN IRQ TUR. Hemicr. $2n = 26$.

Campanula isaurica Contandr., Quézel & Pamukç., Ann. Univ. Provence Sci. 46: 53 (1972).
  Turkey. 34 TUR. Hemicr. $2n = 34$.
    Campanula ermenekensis Contandr. & Quézel, Bull. Soc. Bot. France 123: 431 (1976).

Campanula × iserana Kovanda, Thaiszia 9: 16 (1999). C. rhomboidalis × C. rotundifolia
  Czech Rep. 11 CZE. Hemicr. or geophyte.

Campanula isophylla Moretti, Giorn. Fis. (ser. 2) 7: 44, 98 (1824).
  NW. Italy; cult. 13 ITA. Hemicr. $2n = 32$.
    Campanula floribunda Viv., Fl. Libyc. Spec.: 67 (1824).
    Campanula isophylla f. *alba* Voss in Siebert & Voss, Vilm. Blumengärtn. (ed. 3) 1: 566
      (1894).
    Campanula × mayi Anonymous, Gard. Chron. (ser. 3) 26: 96 (1899).

Campanula jacobaea C. Sm. ex Webb, Icon. Pl. 8: pl. 762 (1848).
  Cape Verde (Santo Antão, São Vicente, São Nicolau, Santiago). 21 CVI. Hemicr. $2n = 54$.
    Campanula jacobaea var. *humilis* Bolle, Bonplandia 9: 50 (1861).
    Campanula jacobaea var. *hispida* Bolle, Bonplandia 9: 51 (1861).

Campanula jacquinii (Sieber) A. DC., Monogr. Campan.: 251 (1830), as 'jacquini'.
  Kriti. 13 KRI. Hemicr. $2n = 34$.
    Phyteuma jacquinii Sieber, Reise Kreta 1: 437 (1823), as 'jacquini'. *Trachelium jacquinii*
      (Sieber) Boiss., Fl. Orient. 3: 961 (1875), as 'jacquini'. *Diosphaera jacquinii* (Sieber)
      Buser, Bull. Herb. Boissier 2: 520 (1894), as 'jacquini'.

Campanula jadvigae Kolak., Bot. Mater. Gerb. Bot. Inst. Komarova Akad. Nauk SSSR 15:
  22 (1953).
  Caucasus (Abkhaziya). 33 TCS. Hemicr. or geophyte.

Campanula jaubertiana Timb.-Lagr., Bull. Soc. Bot. France 15: xcviii (1868). *Campanula
  linifolia* var. *jaubertiana* (Timb.-Lagr.) Nyman, Consp. Fl. Eur.: 479 (1879). *Campanula
  pusilla* var. *jaubertiana* (Timb.-Lagr.) Rouy, Fl. France 10: 76 (1908). *Campanula pusilla*
  subsp. *jaubertiana* (Timb.-Lagr.) Bonnier, Fl. Ill. France 7: 46 (1924). *Campanula
  cochleariifolia* var. *jaubertiana* (Timb.-Lagr.) O. Bolòs & Vigo, Collect. Bot. (Barcelona) 14:
  101 (1983). *Campanula cochlearifolia* subsp. *jaubertiana* (Timb.-Lagr.) Rivas Mart., Itinera
  Geobot. 15: 698 (2002).
  C. Pyrénées. 12 FRA SPA. Hemicr. or geophyte. $2n = 34$.
    Campanula andorrana Braun-Blanq., Commun. Stat. Int. Géobot. Médit. Montpellier
      87:230 (1945). *Campanula jaubertiana* subsp. *andorrana* (Braun-Blanq.) P. Monts. in
      Losa & P. Monts., Aport. Conoc. Fl. Andorra: 115 (1951). *Campanula cochleariifolia*
      subsp. *andorrana* (Braun-Blanq.) O. Bolòs & Vigo, Collect. Bot. (Barcelona) 14: 101
      (1983). *Campanula cochleariifolia* var. *andorrana* (Braun-Blanq.) O. Bolòs & Vigo,
      Collect. Bot. (Barcelona) 14: 101 (1983).

Campanula jordanovii Ančev & Kovanda, Preslia 61: 196 (1989).
Macedonia & Bulgaria. 13 BUL YUG. Hemicr. or geophyte. $2n = 34$.

Campanula jurjurensis Pomel, Nouv. Mat. Fl. Atl.: 257 (1875). *Campanula macrorhiza* var.
jurjurensis (Pomel) Chabert, Bull. Soc. Bot. France 36: 28 (1889). *Campanula rotundifolia*
var. *jurjurensis* (Pomel) Quézel, Feddes Repert. Spec. Nov. Regni Veg. 56: 47 (1953).
N. Algeria (Djebel Djurdjura Mts.). 20 ALG. Hemicr. or geophyte. $2n = 102$.
    Campanula macrorhiza var. *rotundata* Chabert, Bull. Soc. Bot. France 36: 28 (1889).

Campanula justiniana Witasek, Magyar Bot. Lapok. 5: 245 (1906). *Campanula linifolia* var.
justiniana (Witasek) Jáv., Magyar Fl.: 1077 (1925). *Campanula linifolia* subsp. *justiniana*
(Witasek) Hayek, Repert. Spec. Nov. Regni Veg. Beih. 30(2): 537 (1930). *Campanula*
balcanica subf. *justiniana* (Witasek) Hruby, Magyar Bot. Lapok 29: 242 (1930).
Slovenia & Croatia. 13 YUG. Hemicr. or geophyte. $2n = 34$.

Campanula kachethica Kantsch., Veštn. Tiflissk. Bot. Sada (n.s.) 4-5: 2 (1931).
Caucasus (Gruziyah). 33 TCS. Hemicr. or geophyte.

Campanula kadargavanica Amirkh. & Komzha in Naymenko, Bot. Issl. Zapov. 138 (1984).
Caucasus (Severo-Osetiya). 33 NCS. Hemicr. or geophyte.

Campanula kantschavelii Zagar., Bot. Žurn. SSSR 21: 285 (1936).
Caucasus (Gruziya). 33 TCS. Hemicr. or geophyte. $2n = 34$.

Campanula karadjana Bocquet, Compt. Rend. Séances Soc. Phys. Genève (n.s.) 3: 90 (1968).
Turkey. 34 TUR. Biennial.

Campanula kastellorizana A. Carlström, Willdenowia 15: 384 (1986).
E. Aegean Is. (Megísti). 34 EAI. Ther.

Campanula keniensis Thulin, Bot. Not. 128: 350 (1976).
Kenya. 25 KEN. Ther. $2n = 54$.

Campanula kermanica (Rech. f., Aellen & Esfand.) Rech. f., Fl. Iran. 13: 18 (1965).
Iran. 34 IRN. Hemicr.
    **Campanula incanescens* subsp. *kermanica* Rech. f., Aellen & Esfand., Österr. Akad. Wiss.,
    Math.-Naturwiss. Kl., Anz. 87: 196 (1950). *Campanula incanescens* var. *kermanica*
    (Rech. f., Aellen & Esfand.) Parsa, Fl. Iran 10: 220 (1980).

Campanula khorasanica (Rech. f. & Aellen) Rech. f., Fl. Iran. 13: 17 (1965).
Iran. 34 IRN. Hemicr. $2n = 28$.
    **Campanula incanescens* subsp. *khorasanica* Rech. f. & Aellen, Österr. Akad. Wiss., Math.-
    Naturwiss. Kl., Anz. 87: 197 (1950). *Campanula incanescens* f. *khorasanica* (Rech. f. &
    Aellen) Parsa, Fl. Iran 10: 221 (1980).

Campanula kiharae Kitam., Acta Phytotax. Geobot. 16: 131 (1956).
Afghanistan. 34 AFG. Hemicr. or geophyte.

Campanula kirikkaleensis Dönmez & Güner, Karaca Arbor. Mag. 2: 115 (1994).
Turkey. 34 TUR. Hemicr.

Campanula kolakovskyi Kharadze, Zametki Sist. Geogr. Rast. 13: 108 (1947).
Caucasus (Abkhaziya). 33 TCS. Hemicr. or geophyte.

Campanula kolenatiana C. A. Mey. ex Rupr., Bull. Acad. Imp. Sci. Saint-Pétersbourg 11:
216 (1867).
Caucasus (Krasnodar, Dagestan). 33 NCS. Hemicr. or geophyte.
    Campanula akhverdovii Kharadze, Zametki Sist. Geogr. Rast. 15: 18 (1949).

Campanula komarovii Maleev, Izv. Glavn. Bot. Sada SSSR 29: 424 (1930), as 'komarovi'.
Campanula sibirica subsp. *komarovii* (Maleev) Viktorov, Novosti Sist. Vyssh. Rast. 32:
166 (2000).
Caucasus (Krasnodar). 33 NCS. Hemicr. $2n = 34$.

Campanula kotschyana A. DC. in DC., Prodr. 7: 481 (1839).
Lebanon & Turkey. 34 LBS TUR. Ther.

Campanula koyuncui H. Duman, Edinburgh J. Bot. 56: 357 (1999).
Turkey. 34 TUR. Hemicr.

Campanula kremeri Boiss. & Reut., Pugill. Pl. Afr. Bor. Hispan.: 75 (1852). *Campanula
dichotoma* subsp. *kremeri* (Boiss. & Reut.) Nyman, Consp. Fl. Eur.: 477 (1879).
W. Mediterranean. 12 SPA 20 ALG MOR TUN. Ther. $2n = 24$.

Campanula kryophila Rupr., Bull. Acad. Imp. Sci. Saint-Pétersbourg 11: 211 (1867).
Campanula ardonensis var. *kryophila* (Rupr.) Boiss., Fl. Orient. 3: 904 (1875). *Hemisphaera
kryophila* (Rupr.) Kolak., Okhrana Prir. Gruz. 12: 166 (1984).
Caucasus (Severo-Osetiya). 33 NCS. Hemicr. or geophyte.

Campanula laciniata L., Sp. Pl.: 165 (1753).
Greece & Kriti. 13 GRC KRI. Hemicr. $2n = 34, 102$.
Campanula erucifolia Feer, J. Bot. 28: 268 (1890).

Campanula lactiflora M. Bieb., Fl. Taur.-Cauc. 1: 153 (1808). *Gadellia lactiflora* (M. Bieb.)
Schulkina, Novosti Sist. Vyssh. Rast. 16: 178 (1979).
Turkey to Caucasus (Krasnodar, Karacheyevo-Cherkessiya, Kabardino-Balkariya, Severo-
Osetiya, Chechnya, Dagestan, Abkhaziya, Gruziya) & Iran; naturalized in Great Britain
& North America. (10) grb 33 ALL 34 IRN TUR (71) brc (75) nwh. Hemicr. $2n = 34, 36$.
Campanula hispida Fisch. ex Hornem., Hort. Bot. Hafn.: 199 (1815); non Lej., Fl. Spa 2:
299 (1813).
Campanula volubilis Pall. ex Schult. in Roem. & Schult., Syst. Veg. 5: 98 (1819).
Campanula biserrata K. Koch, Linnaea 19: 29 (1845). *Campanula lactiflora* var.
glabriuscula K. Koch, Linnaea 23: 643 (1850). *Gadellia biserrata* (K. Koch) A. P.
Khokhr., Bjull. Moskovsk. Obšč. Isp. Prir., Otd. Biol. (n.s.) 96(6): 100 (1991).
Campanula lactiflora var. *olympica* Griseb., Spic. Fl. Rumel. 2: 284 (1846).
Campanula celtidifolia Boiss. & A. Huet in Boiss., Diagn. Pl. Orient. (ser. 2) 3: 111 (1856).
Campanula lactiflora var. *glabra* Trautv., Trudy Imp. S.-Peterburgsk. Bot. Sada 4: 390 (1876).
Campanula lactiflora var. *pilosa* Trautv., Pl. Casp.-Caucas.: 96 (1877).
Campanula lactiflora f. *coerulea* Voss in Siebert & Voss, Vilm. Blumengärtn. (ed. 3) 1:
569 (1894).

Campanula lamondiae Rech. f., Österr. Akad. Wiss., Math.-Naturwiss. Kl., Anz. 109: 173
(1973).
Iran. 34 IRN. Hemicr.

Campanula lanata Friv., Flora 19: 434 (1836).
Serbia & Bulgaria; naturalized in NE. U.S.A. 13 BUL YUG (75) mas. Biennial. $2n = 34$.
Campanula velutina Velen., Sitzungsber. Königl. Böhm. Ges. Wiss. Prag, Math.-
Naturwiss. Cl. 2: 54 (1890); non Desf., Fl. Atlant. 1: 180 (1798).

Campanula lasiocarpa Cham., Linnaea 4: 39 (1829).
Russian Far East & Japan (Hokkaido, Honshu) to NW. U.S.A. (Washington). 31 KAM KUR
MAG SAK 38 JAP 70 ALU ASK NWT YUK 71 ABT BRC 73 WAS. Hemicr. or geophyte.
$2n = 34$.
Campanula algida Fisch. ex A. DC., Monogr. Campan.: 338 (1830).

Campanula lasiocarpa f. *albiflora* Tatew., Trans. Sapporo Nat. Hist. Soc. 13: 382 (1934).
Campanula lasiocarpa var. *latisepala* Hultén, Fl. Aleutian Isl.: 313 (1937). *Campanula lasiocarpa* subsp. *latisepala* (Hultén) Hultén, Ark. Bot. (n.s.) 7: 127 (1968).

Campanula latifolia L., Sp. Pl.: 165 (1753). *Campanula serratifolia* var. *acutipetala* Vuk., Linnaea 26: 335 (1854).
Europe to Siberia, Kazakhstan & N. India (Uttar Pradesh); cult.; naturalized in U.S.A. 10 DEN FIN GRB NOR SWE 11 AUT CZE GER HUN POL SWI 12 FRA SPA 13 BUL ITA ROM YUG 14 BLR BLT KRY RUC RUE RUN RUS RUW UKR 31 ALT WSB 32 KAZ 33 ALL 34 IRN TUR 40 NEP PAK WHM (75) cnt mai pen ver (76) uta. Hemicr. 2*n* = 34.

subsp. **latifolia**
Europe to Siberia, Kazakhstan & Himalaya; cult.; naturalized in U.S.A. 10 DEN FIN GRB NOR SWE 11 AUT CZE GER HUN POL SWI 12 FRA SPA 13 BUL ITA ROM YUG 14 BLR BLT KRY RUC RUE RUN RUS RUW UKR 31 ALT WSB 32 KAZ 33 ALL 34 IRN TUR 40 NEP PAK WHM (75) cnt mai pen ver (76) uta. Hemicr. 2*n* = 34.
Campanula urticifolia All., Fl. Pedem. 1: 110 (1785); non Turra, Fl. Ital. Prodr.: 64 (1780).
Campanula eriocarpa M. Bieb., Fl. Taur.-Cauc. 1: 149 (1808). *Campanula latifolia* var. *eriocarpa* (M. Bieb.) Fisch. ex A. DC., Monogr. Campan.: 265 (1830).
Campanula macrantha Hornem., Suppl. Hort. Bot. Hafn.: 23 (1819). *Campanula latifolia* var. *macrantha* (Hornem.) Sims, Bot. Mag. 52: tab. 2553 (1825). *Campanula latifolia* f. *macrantha* (Hornem.) Voss in Siebert & Voss, Vilm. Blumengärtn. (ed. 3) 1: 567 (1894).
Campanula macrantha var. *polyantha* Hook., Bot. Mag. 61: tab. 3347 (1834).
Campanula latifolia var. *fimbriata* K. Koch, Linnaea 19: 29 (1845).
Campanula latifolia var. *leiocarpa* Trautv., Trudy Imp. S.-Peterburgsk. Bot. Sada 6: 72 (1879).
Campanula latifolia var. *intermedia* Trautv., Trudy Imp. S.-Peterburgsk. Bot. Sada 6: 72 (1879).
Campanula latifolia var. *canescens* Trautv., Trudy Imp. S.-Peterburgsk. Bot. Sada 6: 73 (1879).
Campanula latifolia f. *alba* Voss in Siebert & Voss, Vilm. Blumengärtn. (ed. 3) 1: 567 (1894).

subsp. **megrelica** (Manden. & Kuth.) Ogan., Candollea 50: 281 (1995).
Caucasus (Krasnodar, Abkhaziya, Gruziya). 33 ALL. Hemicr.
** Campanula megrelica* Manden. & Kuth., Zametki Sist. Geogr. Rast. 18: 62 (1955).

Campanula lavrensis (Tocl & Rohlena) Phitos, Oesterr. Bot. Z. 112: 460 (1965).
Greece. 13 GRC. Hemicr. 2*n* = 34.
** Campanula andrewsii* var. *lavrensis* Tocl & Rohlena, Sitzungsber. Königl. Böhm. Ges. Wiss. Prag, Math.-Naturwiss. Cl. 49: 2 (1902). *Campanula rupestris* f. *lavrensis* (Tocl & Rohlena) Hayek, Repert. Spec. Nov. Regni Veg. Beih. 30(2): 525 (1930).
Campanula andrewsii subsp. *breueri* Tocl & Rohlena, Sitzungsber. Königl. Böhm. Ges. Wiss. Prag, Math.-Naturwiss. Cl. 49: 2 (1902). *Campanula rupestris* f. *breueri* (Tocl & Rohlena) Hayek, Repert. Spec. Nov. Regni Veg. Beih. 30(2): 525 (1930). *Campanula lavrensis* var. *breueri* (Tocl & Rohlena) Phitos, Oesterr. Bot. Z. 112: 461 (1965).

Campanula lazica (Boiss. & Balansa) Kharadze, Zametki Sist. Geogr. Rast. 15: 17 (1949).
Turkey. 34 TUR. Hemicr.or geophyte.
** Symphyandra lazica* Boiss. & Balansa in Boiss., Fl. Orient. 3: 888 (1875).

Campanula ledebouriana Trautv., Trudy Imp. S.-Peterburgsk. Bot. Sada 2: 477 (1873). *Hemisphaera ledebouriana* (Trautv.) Kolak. & Serdyuk., Okhrana Prir. Gruz. 12: 166 (1984).
Turkey. 34 TUR. Hemicr. or geophyte.

Campanula lehmanniana Bunge, Beitr. Kenntn. Fl. Russlands: 211 (1852). *Hyssaria lehmanniana* (Bunge) Kolak., Soobšč. Akad. Nauk Gruzinsk. SSR 103: 151 (1981).
Kirgizstan & Tadzhikistan. 32 KGZ TZK. Hemicr. or geophyte.

subsp. **capusii** (Franch.) Viktorov, Novosti Sist. Vyssh. Rast. 34: 226 (2002).
Kirgizstan & Tadzhikistan. 32 KGZ TZK. Hemicr. or geophyte.

* *Campanula lehmanniana* var. *capusii* Franch., Mission Capus Pl. Turkest.: 115 (1883).
Campanula capusii (Franch.) Fed. in Kom., Fl. URSS 24: 289 (1957).

subsp. **integerrima** Kamelin ex Rassulova in Ovcz., Fl. Tadzhik. SSR 9: 526 (1988).
Tadzhikistan. 32 TZK. Hemicr. or geophyte.

subsp. **lehmanniana**
Kirgizstan & Tadzhikistan. 32 KGZ TZK. Hemicr. or geophyte.

subsp. **pseudohissarica** Kamelin ex Rassulova in Ovcz., Fl. Tadzhik. SSR 9: 525 (1988).
Kirgizstan & Tadzhikistan. 32 KGZ TZK. Hemicr. or geophyte.

Campanula leucantha Gilli, Feddes Repert. Spec. Nov. Regni Veg. 68: 87 (1963).
Afghanistan & Pakistan. 34 AFG 40 PAK. Hemicr.

Campanula leucoclada Boiss., Diagn. Pl. Orient. (ser. 2) 3: 109 (1856).
Afghanistan & Pakistan. 34 AFG 40 PAK. Hemicr. $2n = 30$.
Campanula griffithii Hook. f. & Thoms., J. Proc. Linn. Soc., Bot. 2: 22 (1858).

Campanula leucosiphon Boiss. & Heldr. in Boiss., Diagn. Pl. Orient. (ser. 1) 11: 64 (1849).
Turkey. 34 TUR. Hemicr. $2n = 34$.

Campanula lezgina (Alex.) Kolak. & Serdyuk., Zametki Sist. Geogr. Rast. 36: 46 (1980).
Caucasus (Azerbaijan). 33 TCS. Hemicr.
* *Symphyandra lezgina* Alex., Trudy Tiflissk. Bot. Sada 6(1): 68 (1902).

Campanula lingulata Waldst. & Kit., Descr. Icon. Pl. Hung. 1: 65 (1802). *Marianthemum lingulatum* (Waldst. & Kit.) Schur, Sert. Fl. Transsilv.: 48 (1853).
Italy to Turkey. 13 ALB BUL GRC ITA ROM YUG 34 TUR. Hemicr. $2n = 34$.
Campanula capitata Sims, Bot. Mag. 21: tab. 811 (1805). *Campanula lingulata* var.
capitata (Sims) Nyman, Consp. Fl. Eur.: 476 (1879).
Campanula cichoracea Sibth. & Sm., Fl. Graec. Prodr. 1: 140 (1806). *Campanula lingulata*
var. *cichoracea* (Sibth. & Sm.) Griseb., Spic. Fl. Rumel. 2: 288 (1846).
Campanula tenuiflora Ten., Fl. Napol. 3: 207 (1824). *Campanula lingulata* var. *tenuiflora*
(Ten.) Nyman, Consp. Fl. Eur.: 476 (1879).
Campanula lingulata var. *intybacea* Griseb., Spic. Fl. Rumel. 2: 288 (1846).

Campanula longisepala Podlech, Feddes Repert. 71: 97 (1965).
SE. France (Alpes de Provence). 12 FRA. Hemicr. or geophyte. $2n = 68$.

Campanula longistyla Fomin, Trudy Tiflissk. Bot. Sada 4(3): 37 (1904), as 'longestyla'.
Campanula sibirica subsp. *longistyla* (Fomin) Viktorov, Novosti Sist. Vyssh. Rast. 32:
165 (2000).
Caucasus (Krasnodar, Abkhaziya). 33 ALL. Hemicr. $2n = 34$.
Campanula schischkinii Kolak. & Sachokia, Spisok Rast. Gerb. Fl. SSSR Bot. Inst.
Vsesojuzn. Akad. Nauk 12: 113 (1953).
Campanula bzybica Jabr.-Kolak., Trudy Suhumsk. Bot. Sada 11: 457 (1958).

Campanula lourica Boiss., Diagn. Pl. Orient. (ser. 1) 11: 70 (1849).
Iran. 34 IRN. Hemicr.

Campanula × lundstroemii Fedde, Just's Bot. Jahresber. 42(2): 73 (1920), as 'lundströmii'.
C. rapunculoides × C. trachelium
Cult. Hemicr. or geophyte.

Campanula luristanica Freyn, Bull. Herb. Boissier 5: 791 (1897). *Campanula humillima* var.
luristanica (Freyn) Bornm., Beih. Bot. Centralbl. 28: 459 (1911).
Iran. 34 IRN. Hemicr.

Campanula lusitanica Loefl., Iter Hispan.: 111, 126 (1758).
Iberian Pen. & Morocco; cult.; naturalized on Madeira. 12 POR SPA 20 MOR (21) mdr. Ther.
2*n* = 18, 20.

subsp. **lusitanica**
Iberian Pen. & Morocco; cult.; naturalized on Madeira. 12 POR SPA 20 MOR 21 mdr. Ther.
2*n* = 18, 20.
Campanula loeflingii Brot., Fl. Lusit. 1: 287 (1804).
Campanula broussonetiana Schult. in Roem. & Schult., Syst. Veg. 5: 104 (1819).
Campanula durieui Boiss., Voy. Bot. Espagne: 402 (1841), as 'duriaei'. *Campanula loeflingii*
var. *durieui* (Boiss.) Nyman, Consp. Fl. Eur.: 482 (1879), as 'duriaei'.
Campanula vinciflora Pau, Bol. Soc. Esp. Hist. Nat. 21: 278 (1921), as 'vincaeflora;' non Vent.,
Jard. Malmaison: pl. 12 (1803), as 'vincaeflora'.
Campanula lusitanica var. *maura* Murb., Contr. Fl. Maroc 2: 50 (1923).
Campanula lusitanica var. *pseudophrygia* Lacaita, Cavanillesia 3: 22 (1930).
Campanula lusitanica f. *pallidiflora* Maire in Jahand. & Maire, Cat. Pl. Maroc: 735 (1934).
Campanula lusitanica f. *tenuis* Caball., Trab. Mus. Nac. Ci. Nat., Ser. Bot. 30: 5 (1935).
Campanula lusitanica var. *puberula* C. Vicioso, Anales Jard. Bot. Madrid 6(2): 78 (1946).

subsp. **matritensis** (A. DC.) Franco, Nova Fl. Portugal 2: 326, 569 (1984).
Iberian Pen. 12 POR SPA. Ther.
* *Campanula matritensis* A. DC., Monogr. Campan.: 332 (1830).
Campanula matritensis var. *nevadensis* A. DC. in DC., Prodr. 7: 481 (1839).

subsp. **specularioides** (Coss.) Aldasoro & L. Sáez, Anales Jard. Bot. Madrid 59: 173 (2001).
Spain. 12 SPA. Ther. 2*n* = 20.
* *Campanula specularioides* Coss., Notes Pl. Crit.: 41 (1849).

subsp. **transtagana** (Ros. Fernandes) Fed., Bot. J. Linn. Soc. 67: 281 (1973).
Portugal. 12 POR. Ther. 2*n* = 20.
* *Campanula transtagana* Ros. Fernandes, Bol. Soc. Brot. (ser. 2) 36: 121 (1962).

Campanula lycica Kit Tan & Sorger, Pl. Syst. Evol. 155: 96 (1987).
Turkey. 34 TUR. Ther.

Campanula lyrata Lam., Encycl. 1: 588 (1785).
E. Aegean Is. & Turkey. 13 TUE 34 EAI TUR. Hemicr. or biennial. 2*n* = 34.

subsp. **icarica** Runemark ex Phitos, Notes Roy. Bot. Gard. Edinburgh 35: 44 (1976).
E. Aegean Is. (Ikaría, Sámos, Kálymnos). 34 EAI. Hemicr. or biennial.

subsp. **lyrata**
E. Aegean Is. (Lésvos, Chíos, Ikaría, Sámos, Kálymnos) & Turkey. 13 TUE 34 EAI TUR.
Hemicr. or biennial. 2*n* = 34.
Campanula lyratella Feer, J. Bot. 28: 269 (1890).
Campanula esculenta Candargy, Bull. Soc. Bot. France 44: 148 (1897); non Salisb., Prodr.
Stirp. Chap. Allerton: 126 (1796); nec A. Rich., Tent. Fl. Abyss. 2: 4 (1850).
Campanula lyrata var. *albostrigosa* Rech. f., Ann. Naturhist. Mus. Wien 43: 321 (1929).

Campanula macrorhiza J. Gay ex A. DC., Monogr. Campan.: 302 (1830). *Campanula rotundifolia*
var. *macrorhiza* (J. Gay ex A. DC.) Bég. in Fiori & Paol, Fl. Italia 3: 185 (1903). *Campanula*
rotundifolia race *macrorhiza* (J. Gay ex A. DC.) Rouy, Fl. France 10: 80 (1908). *Campanula*
rotundifolia subsp. *macrorhiza* (J. Gay ex A. DC.) Guin., Bull. Soc. Bot. France 89: 74 (1942).
SE. Spain to NW. Italy. 12 FRA SPA 13 ITA. Hemicr. or geophyte. 2*n* = 34.
Campanula macrorhiza var. *pubescens* A. DC., Monogr. Campan.: 302 (1830).
Campanula nicaeensis Risso, Fl. Nice: 303 (1844); non Schult. in Roem. & Schult., Syst.
Veg. 5: 126 (1819). *Campanula macrorhiza* f. *nicaeensis* Hruby, Magyar Bot. Lapok 33:
150 (1934).

Campanula cuatrecasasii Pau, Trab. Mus. Ci. Nat. Barcelona 12: 442 (1929).
Campanula macrorhiza f. *squarrulosa* Hruby, Magyar Bot. Lapok 33: 149 (1934).
Campanula alpium-maritimarum Hruby, Magyar Bot. Lapok 33: 151 (1934).
Campanula hispanica f. *minor* Sennen, Diagn. Nouv.: 21 (1936).
Campanula suffruticosa Sennen, Diagn. Nouv.: 270 (1936).
Campanula linearifolia Sennen, Diagn. Nouv.: 270 (1936).
Campanula mariani Sennen, Diagn. Nouv.: 274 (1936).
Campanula sanctae-balmae Sennen, Diagn. Nouv.: 283 (1936).
Campanula rotundifolia subsp. *litardierei* Guin., Bull. Soc. Bot. France 89: 72 (1942).
Campanula rotundifolia subvar. *gracilis* Guin., Bull. Soc. Bot. France 89: 74 (1942).
Campanula rotundifolia subvar. *puberula* Guin., Bull. Soc. Bot. France 89: 74 (1942).
Campanula rotundifolia var. *ludoviciana* Guin., Bull. Soc. Bot. France 89: 74 (1942).
Campanula rotundifolia var. *eulitardierei* Guin., Bull. Soc. Bot. France 89: 74 (1942), as 'eu-litardierei'.
Campanula cuatrecasasii var. *gaditana* C. Vicioso, Anales Jard. Bot. Madrid 6(2): 78 (1946).
Campanula rotundifolia subsp. *aitanica* Pau ex O. Bolòs & Vigo, Collect. Bot. 14: 101 (1983).
   Campanula rotundifolia var. *aitanica* Pau ex O. Bolòs & Vigo, Collect. Bot. 14: 101 (1983).
   Campanula macrorhiza subsp. *aitanica* (Pau ex O. Bolòs & Vigo) Rivas Mart., Itinera Geobot. 15: 699 (2002).
Campanula rotundifolia var. *alcoiana* O. Bolòs & Vigo, Collect. Bot. 14: 101 (1983).

Campanula macrostachya Waldst. & Kit. ex Willd., Enum. Pl. 1: 213 (1809).
Campanula cervicaria subsp. *macrostachya* (Waldst. & Kit. ex Willd.) Tacik in Pawłowskiego & Jasiewicz, Fl. Polska 12: 65 (1971).
SE. Europe to Turkey. 11 CZE HUN 13 ALB BUL GRC ROM TUE YUG 14 UKR 34 TUR. Biennial. $2n = 16, 18, 32$.
Campanula multiflora Waldst. & Kit., Descr. Icon. Pl. Hung. 3: 292 (1811).
Campanula macedonica Boiss. & Orph. ex Freyn, Verh. K. K. Zool.-Bot. Ges. Wien 38: 618 (1888).

Campanula macrostyla Boiss. & Heldr. in Boiss., Diagn. Pl. Orient. (ser. 1) 11: 65 (1849).
Sicyocodon macrostylus (Boiss. & Heldr.) Feer, Bot. Jahrb. Syst. 12: 614 (1890).
Turkey. 34 TUR. Ther. $2n = 20$.
   Campanula macrostyla f. *rosea* Voss in Siebert & Voss, Vilm. Blumengärtn. (ed. 3) 1: 569 (1894).

Campanula mairei Pau ex Maire, Bull. Soc. Hist. Nat. Afrique N. 19: 59 (1928).
Morocco. 20 MOR. Hemicr. $2n = 16, 24, 34$.
   * *Campanula herminii* var. *atlantica* Jahand. & Maire, Bull. Soc. Hist. Nat. Afrique N. 14: 71 (1923); non *Campanula atlantica* Coss. & Durieu, Fl. Algérie, Dicotyl.: 573 (1889).
   Campanula mairei var. *anremerica* Litard. & Maire, Mém. Soc. Sci. Nat. Maroc 26: 29 (1931).
   Campanula mairei var. *tenera* Emb. & Maire, Bull. Soc. Hist. Nat. Afrique N. 23: 199 (1932).
   Campanula mairei f. *cordatifolia* Maire, Bull. Soc. Hist. Nat. Afrique N. 28: 367 (1937).
   Campanula mairei f. *ramosa* Quézel, Feddes Repert. Spec. Nov. Regni Veg. 56: 44 (1953).

Campanula marcenoi Brullo, Candollea 48: 494 (1993).
Sicilia. 13 SIC. Hemicr. or geophyte.

Campanula marchesettii Witasek, Abh. K. K. Zool.-Bot. Ges. Wien 1(3): 32 (1902). *Campanula rotundifolia* var. *marchesettii* (Witasek) Fiori, Nuov. Fl. Italia 2: 562 (1927). *Campanula rotundifolia* subsp. *marchesettii* (Witasek) D. E. Mey., Seznam: 257 (1952).
Italy & Austria to Croatia. 11 AUT 13 ITA YUG. Hemicr. or geophyte. $2n = 34, 68$.

Campanula mardinensis Bornm. & Sint., Mitt. Thüring. Bot. Vereins (n.s.) 20: 31 (1904).
Turkey & Iraq. 34 IRQ TUR. Hemicr.
   Campanula incanescens var. *exappendiculata* Bornm., Beih. Bot. Centralbl. 33: 165 (1915).

Campanula massalskyi Fomin in N. M. Kusn., N. Bush & Fomin, Mater. Fl. Kavkaza 4(6): 78 (1905).
Turkey & Caucasus (Armenia). 33 TCS 34 TUR. Hemicr. or geophyte.
Campanula ketzkhovelii Sosn. ex Grossh., Fl. Kavkaza 4: 69 (1934).

Campanula medium L., Sp. Pl.: 167 (1753). *Campanula florida* Salisb., Prodr. Stirp. Chap.
Allerton: 126 (1796). *Marianthemum medium* (L.) Schur, Sert. Fl. Transsilv.: 48 (1853). *Rapuntia medium* (L.) Chevall., Fl. Gén. Env. Paris (ed. 2) 2: 526 (1836), as 'media'.
France & Italy; cult.; naturalized in N. & W. Europe, North America & New Zealand. (10) grb (11) aut ger hun 12 FRA spa 13 ITA rom (51) nzn nzs (71) brc (72) nbr ont (73) mnt ore (75) cnt mic. Biennial. $2n = 34$.
Campanula grandiflora Lam., Fl. Franç. 3: 334 (1779); non Jacq., Hort. Bot. Vindob. 3: 4 (1776). *Medium grandiflorum* Spach, Hist. Nat. Vég. 9: 553 (1838).
Campanula bourdiniana Gand., Fl. Lyon.: 146 (1875). *Campanula medium* var. *bourdiniana* (Gand.) Nyman, Consp. Fl. Eur.: 476 (1879).
Campanula medium f. *alba* Voss in Siebert & Voss, Vilm. Blumengärtn. (ed. 3) 1: 564 (1894).
Campanula medium f. *caesia* Voss in Siebert & Voss, Vilm. Blumengärtn. (ed. 3) 1: 564 (1894).
Campanula medium f. *calycantha* Voss in Siebert & Voss, Vilm. Blumengärtn. (ed. 3) 1: 564 (1894).
Campanula medium f. *coerulea* Voss in Siebert & Voss, Vilm. Blumengärtn. (ed. 3) 1: 564 (1894).
Campanula medium f. *rosea* Voss in Siebert & Voss, Vilm. Blumengärtn. (ed. 3) 1: 564 (1894).
Campanula medium f. *striata* Voss in Siebert & Voss, Vilm. Blumengärtn. (ed. 3) 1: 564 (1894).
Campanula medium var. *calycanthema* Anonymous, Garden (London) 48: 295 (1895).

Campanula mekongensis Diels ex C. Y. Wu, Rep. Yunnan Trop. Subtrop. Fl. Res. Inst. 1: 58 (1965).
S. China (Guangxi, Yunnan). 36 CHC CHS. Hemicr. or geophyte.

Campanula merxmuelleri Phitos, Mitt. Bot. Staatssamml. München 5: 121 (1963).
C. Aegean Is. (Skyros). 13 GRC. Hemicr.

Campanula micrantha Bertol., Fl. Ital. 7: 623 (1851). *Campanula rotundifolia* var. *micrantha* (Bertol.) Tanfani in Parl., Fl. Ital. 8: 98 (1888). *Campanula scheuchzeri* subsp. *micrantha* (Bertol.) Arcang., Comp. Fl. Ital. (ed. 2): 637 (1894).
Italy (C. Apennines). 13 ITA. Hemicr. or geophyte. $2n = 34$.
Campanula marchesettii subsp. *apennina* Podlech, Feddes Repert. 71: 108 (1965). *Campanula apennina* (Podlech) Podlech, Mitt. Bot. Staatssamml. München 8: 216 (1970).

Campanula microdonta Koidz., Bot. Mag. (Tokyo) 33: 221 (1919). *Campanula punctata* subsp. *microdonta* (Koidz.) Kitam., Acta Phytotax. Geobot. 17: 7 (1957).
Japan (Izu Is.). 38 JAP. Hemicr. or geophyte.

Campanula mirabilis Albov, Bull. Herb. Boissier 3: 228 (1895).
Caucasus (Abkhaziya). 33 TCS. Hemicr. or biennial. $2n = 34, 102$.
Campanula regina Albov, Zap. Kavkazsk. Otd. Imp. Russk. Geogr. Obšč. 18: ? (1896).

Campanula moesiaca Velen., Sitzungsber. Königl. Böhm. Ges. Wiss. Prag, Math.-Naturwiss. Cl. 1892: 385 (1893).
Balkans. 13 ALB BUL GRC YUG. Biennial. $2n = 34$.
Campanula moesiaca var. *oblongifolia* K. Malý, Oesterr. Bot. Z. 57: 184 (1907).

Campanula mollis L., Sp. Pl. (ed. 2): 237 (1762).
W. Mediterranean. 12 SPA 20 ALG MOR. Hemicr. $2n = 24, 26, 46, 48, 50, 52$.

subsp. **mollis**
W. Mediterranean. 12 SPA 20 ALG MOR. Hemicr. $2n = 24, 26, 46, 48, 50, 52$.
Campanula velutina Desf., Fl. Atlant. 1: 180 (1798).
Campanula microphylla Cav., Anales Ci. Nat. 3: 19 (1800). *Campanula mollis* var. *microphylla* (Cav.) A. DC., Monogr. Campan.: 238 (1830). *Campanula velutina* var. *microphylla* (Cav.) G. López, Bol. Soc. Brot. (ser. 2) 53: 306 (1980), as 'microphyla'.

Campanula mollis var. *lasiantha* Lara, Anales Soc. Esp. Hist. Nat. 18: 38 (1889).
Campanula malacitana Degen & Hervier, Bull. Acad. Int. Géogr. Bot. 17: 56 (1907).
  Campanula velutina var. *malacitana* (Degen & Hervier) G. López, Fontqueria 21: 50 (1988).
Campanula malacitana var. *almeriensis* Degen & Hervier, Bull. Acad. Int. Géogr. Bot. 17: 58
  (1907). *Campanula velutina* var. *almeriensis* (Degen & Hervier) G. López, Bol. Soc. Brot.
  (ser. 2) 53: 305 (1980).
Campanula malacitana var. *gibraltarica* Degen & Hervier, Bull. Acad. Int. Géogr. Bot. 17: 58
  (1907). *Campanula velutina* var. *gibraltarica* (Degen & Hervier) G. López, Bol. Soc. Brot.
  (ser. 2) 53: 305 (1980).
Campanula malacitana var. *gienensis* Degen & Hervier, Bull. Acad. Int. Géogr. Bot. 17: 58 (1907).
Campanula mollis var. *oranensis* Maire, Mem. Soc. Sci. Nat. Maroc 17: 52 (1927).
Campanula mollis var. *rifana* Emb. & Maire, Mém. Soc. Sci. Nat. Maroc 17: 52 (1927). *Campanula*
  velutina var. *rifana* (Emb. & Maire) G. López, Bol. Soc. Brot. (ser. 2) 53: 306 (1980).
Campanula filicaulis var. *maroccana* Pau in Font Quer, Iter Marocc.: 634 (1927). *Campanula*
  mollis var. *maroccana* (Pau) Font Quer, Iter Marocc.: 634 (1927). *Campanula velutina* var.
  maroccana (Pau) G. López, Bol. Soc. Brot. (ser. 2) 53: 305 (1980).
Campanula velata var. *rifana* Maire, Cavanillesia 4: 16 (1931). *Campanula mollis* var.
  pseudovelata Maire, Bull. Soc. Hist. Nat. Afrique N. 28: 367 (1937); non *Campanula mollis*
  var. *rifana* Emb. & Maire, Mem. Soc. Sci. Nat. Maroc 17: 52 (1927).
Campanula mollis var. *longicornis* Maire in Jahand. & Maire, Cat. Pl. Maroc: 732 (1934).
Campanula mollis var. *tlemcenensis* Quézel, Feddes Repert. Spec. Nov. Regni Veg. 56: 32 (1953).
Campanula mollis var. *mesatlantica* Quézel, Feddes Repert. Spec. Nov. Regni Veg. 56: 32 (1953).
Campanula mollis var. *canescens* Quézel, Feddes Repert. Spec. Nov. Regni Veg. 56: 33 (1953).
Campanula mollis f. *transiens* Quézel, Feddes Repert. Spec. Nov. Regni Veg. 56: 33 (1953).
Campanula velutina var. *glaberrima* Escudero & Pajarón, Lazaroa 14: 200 (1994).

  subsp. **winkleri** (Lacaita) Greuter & Burdet, Willdenowia 13: 280 (1984).
    Spain. 12 SPA. Hemicr.
   * *Campanula mollis* var. *winkleri* Lacaita, Cavanillesia 3: 23 (1930). *Campanula velutina* subsp.
     winkleri (Lacaita) G. López, Bol. Soc. Brot. (ser. 2) 53: 304 (1980).

Campanula monodiana Maire, Bull. Soc. Hist. Nat. Afrique N. 34: 136 (1943).
  N. Chad (Mt. Tousside). 24 CHA. Hemicr.

Campanula moravica (Spitzn.) Kovanda, Folia Geobot. Phytotax. 3: 408 (1968).
  SE. Europe. 11 AUT CZE HUN ROM. Hemicr. or geophyte. $2n = 68, 102$.
   * *Campanula rotundifolia* var. *moravica* Spitzn., Verh. Naturf. Vereins Brünn 31: 193 (1893).

  subsp. **moravica**
    Czechoslovakia to Romania. 11 AUT CZE HUN ROM. Hemicr. or geophyte. $2n = 68$.

  subsp. **xylorrhiza** (O. Schwarz) Kovanda, Folia Geobot. Phytotax. 3: 409 (1968).
    Czechoslovakia & Austria. 11 AUT CZE. Hemicr. or geophyte. $2n = 102$.
   * *Campanula rotundifolia* subsp. *xylorrhiza* O. Schwarz, Mitt. Thüring. Bot. Ges. 1: 118 (1949).
     Campanula bertolae subsp. *xylorrhiza* (O. Schwarz) Podlech, Feddes Repert. 71: 107 (1965).

Campanula morettiana Rchb., Iconogr. Bot. Pl. Crit. 4: 18 (1826). *Campanula brachysepala* Vuk.,
Linnaea 26: 325 (1854).
  Italy. 13 ITA. Hemicr. or geophyte. $2n = 34, 68$.
   * *Campanula filiformis* Moretti, Giorn. Fis. (ser. 2) 9: 155 (1826); non Ruiz & Pav., Fl. Peruv. 2:
     55 (1799).

Campanula munzurensis P. H. Davis, Notes Roy. Bot. Gard. Edinburgh 24: 31 (1962).
  Turkey. 34 TUR. Hemicr.

Campanula × murrii Dalla Torre & Sarnth., Fl. Tirol. 6(3): 446 (1911). C. cochlearifolia × C.
scheuchzeri
  Austria & Switzerland. 11 AUT SWI. Hemicr. or geophyte.

Campanula × *semproniana* Beauverd, Bull. Soc. Bot. Genève (ser. 2) 4: 436 (1913).
Campanula × *pseudoscheuchzeri* Beyer, Verh. Bot. Vereins Prov. Brandenburg 58: 115 (1917).

Campanula myrtifolia Boiss. & Heldr. in Boiss., Diagn. Pl. Orient. (ser. 1) 11: 69 (1849).
Trachelium myrtifolium (Boiss. & Heldr.) Boiss., Fl. Orient. 3: 962 (1875). *Tracheliopsis myrtifolia* (Boiss. & Heldr.) O. Schwarz & P. H. Davis, Hooker's Icon. Pl. 35: pl. 3498 (1950).
Turkey. 34 TUR. Hemicr. 2*n* = 34.
    Tracheliopsis taurica Contandr., Quézel & Pamukç., Ann. Univ. Provence Sci. 46: 59 (1972), as 'tauricum'.

Campanula nakaoi Kitam., Acta Phytotax. Geobot. 15: 108 (1954).
Himalaya. 36 CHT 40 NEP. Hemicr. or geophyte.

Campanula nisyria Papatsou & Phitos, Notes Roy. Bot. Gard. Edinburgh 34: 204 (1975).
E. Aegean Is. (Nisyros). 34 EAI. Hemicr. or biennial. 2*n* =34.

Campanula numidica Durieu, Expl. Sci. Algérie: pl. 62 (1849).
Algeria. 20 ALG. Hemicr. 2*n* = 24.

Campanula nuristanica Rech. f. & Schiman-Czeika in Rech. f., Fl. Iran. 13: 25 (1965).
Afghanistan. 34 AFG. Hemicr.

Campanula occidentalis Y. Nyman, Willdenowia 20: 113 (1991).
Canary Is. & Morocco. 20 MOR 21 CNY. Ther. 2*n* = 24.

Campanula odontosepala Boiss., Diagn. Pl. Orient. (ser. 1) 11: 66 (1849). *Symphyandra odontosepala* (Boiss.) Esfand., Notes Roy. Bot. Gard. Edinburgh 38: 447 (1980).
Caucasus (Azerbaijan) & Iran. 33 TCS 34 IRN. Hemicr.
    Campanula faucium Ponert, Preslia 44: 50 (1972).
    Symphyandra repens Karpiss., Bjull. Moskovsk. Obšč. Isp. Prir., Otd. Biol. 84(6): 121 (1979).
    Campanula repens (Karpiss.) Czerep., Vasc. Pl. Russia: 155 (1995); non Lour., Fl. Cochinch. 1: 139 (1790).

Campanula oligosperma Damboldt in P. H. Davis, Fl. Turkey 6: 46 (1978).
Turkey. 34 TUR. Hemicr.
    * *Campanula crassifolia* Huber-Mor. & C. Simon, Bauhinia 2: 197 (1963); non Nees, Syll. Pl. Nov. 1: 6 (1824). *Campanula glaucophylla* Damboldt, Notes Roy. Bot. Gard. Edinburgh 35: 49 (1976); non Schloss. & Vuk., Syll. Fl. Croat.: 72 (1857) .

Campanula olympica Boiss., Diagn. Pl. Orient. (ser. 1) 4: 34 (1844).
Serbia to Turkey & Caucasus (Gruziya, Adzhariya). 13 BUL YUG 33 TCS 34 TUR. Hemicr. or biennial. 2*n* = 20.
    Campanula hemschinica K. Koch, Linnaea 23: 644 (1850). *Neocodon hemscinicus* (K. Koch) Kolak. & Serdyuk., Zametki Sist. Geogr. Rast. 40: 28 (1984).
    Campanula patula var. *confertiflora* Trautv., Pl. Messes: 69 (1874).

Campanula oreadum Boiss. & Heldr. in Boiss., Diagn. Pl. Orient. (ser. 2) 3: 107 (1856).
Greece. 13 GRC. Hemicr. 2*n* = 34.

Campanula orphanidea Boiss., Fl. Orient. 3: 897 (1875). *Petkovia orphanidea* (Boiss.) Stef., God. Sofiisk. Univ. Agron.-Lesoved. Fak. 14: 100 (1936).
Bulgaria & Greece. 13 BUL GRC. Biennial. 2*n* = 26.
    Symphyandra pangaea Heldr. & Charrel in Charrel, Géogr. Bot. Salonique: 29 (1892).

Campanula ossetica M. Bieb., Fl. Taur.-Cauc. 3: 145 (1819). *Symphyandra ossetica* (M. Bieb.) A. DC., Monogr. Campan.: 368 (1830).
Caucasus (Severo-Osetiya, Chechnya). 33 NCS. Hemicr. or geophyte. 2*n* = 34.

Campanula pallida Wall., Asiat. Res. 13: 375 (1820).
Afghanistan to SC. China (Guizhou, Sichuan, Yunnan) & Laos. 34 AFG 36 CHC CHT 40 ASS
EHM IND PAK NEP WHM 41 LAO MYA THA. Ther. $2n = 24, 28$.
Campanula ramulosa Wall. in Roxb., Fl. Ind. 2: 100 (1824). *Campanula colorata* var. *ramulosa*
(Wall.) Hook. f. & Thomson, J. Proc. Linn. Soc., Bot. 2: 23 (1858).
Campanula colorata Wall. in Roxb., Fl. Ind. 2: 101 (1824).
Campanula colorata var. *moorcroftiana* A. DC., Monogr. Campan.: 293 (1830).
Campanula colorata var. *tibetica* Hook. f. & Thomson, J. Proc. Linn. Soc., Bot. 2: 23 (1858).
Campanula pallida var. *tibetica* (Hook. f. & Thomson) H. Hara, J. Jap. Bot. 50: 270 (1975).
Campanula colorata var. *anomala* Hook. f. & Thomson, J. Proc. Linn. Soc., Bot. 2: 23 (1858).
Campanula himalayensis Klotzsch in Klotzsch & Garcke, Bot. Ergebn. Reise Waldemar: 73
(1862).
Campanula hoffmeisteri Klotzsch in Klotzsch & Garcke, Bot. Ergebn. Reise Waldemar: 74 (1862).
Campanula wightii Gamble, Fl. Madras: 740 (1921).
Campanula microcarpa C. Y. Wu, Rep. Yunnan Trop. Subtrop. Fl. Res. Inst. 1: 60 (1965).

Campanula pangea Hartvig, Willdenowia 28: 65 (1998).
Greece. 13 GRC. Biennial.

Campanula papillosa Halácsy in Maire & Petitm., Étud. Pl. Vasc. Grèce: 145 (1908).
Greece. 13 GRC. Hemicr. or geophyte.

Campanula paradoxa Kolak., Bjull. Glavn. Bot. Sada 102: 39 (1976).
Caucasus (Abkhaziya). 33 TCS. Hemicr.

Campanula parryi A. Gray, Syn. Fl. N. Amer. (ed. 2) 2(1): 395 (1886).
W. U.S.A. (Washington to Montana & New Mexico). 73 COL IDA MNT WAS WYO 76 ARI UTA
77 NWM. Hemicr.
Campanula planiflora Engelm., Bot. Gaz. (Crawfordsville) 7: 5 (1882); non Lam., Encycl. 1:
580 (1785); nec Willd., Enum. Pl.: 210 (1809).
Campanula parryi var. *idahoensis* McVaugh, Bull. Torrey Bot. Club 69: 241 (1942).
Campanula rentoniae Senior, Rhodora 51: 302 (1949), as 'rentonae'.

Campanula patula L., Sp. Pl.: 163 (1753). *Neocodon patulus* (L.) Kolak. & Serdyuk., Zametki Sist.
Geogr. Rast. 40: 29 (1984).
Europe to W. Siberia; naturalized in NE. U.S.A. 10 FIN GRB 11 AUT BGM CZE GER HUN NET
POL SWI 12 FRA SAR SPA 13 ALB BUL GRC ITA ROM YUG 14 BLR BLT KRY RUC RUE RUN
RUS RUW UKR 30 WSB (75) cnt nwh nwy. Hemicr. or biennial. $2n = 20, 40, 60, 68, 80$.

subsp. **abietina** (Griseb. & Schenk) Simonk., Enum. Fl. Transsilv.: 383 (1887).
SE. Europe. 11 CZE HUN 13 ALB BUL GRC ROM YUG 14 UKR. Hemicr. $2n = 20, 68, 80$.
Campanula patula var. *pauciflora* Rochel, Pl. Banat. Rar.: 42 (1826).
* *Campanula abietina* Griseb. & Schenk, Arch. Naturgesch. 18: 333 (1852). *Neocodon abietinus*
(Griseb. & Schenk) Kolak. & Serdyuk., Zametki Sist. Geogr. Rast. 40: 28 (1984).
Campanula stefanoffii Hermann, Izv. Bulg. Bot. Druž. 3: 43 (1929).
Campanula vajdae Pénzes, Bot. Közlem. 39: 92 (1942). *Campanula patula* var. *vajdae* (Pénzes)
Fed., Bot. J. Linn. Soc. 67: 281 (1973).

subsp. **alekovyi** Ančev, God. Sofiisk. Univ. Biol. Fak., 2 Bot. 85: 196 (1994).
Bulgaria. 13 BUL. Biennial.

subsp. **costae** (Willk.) Nyman, Consp. Fl. Eur.: 482 (1879).
Pyrénées. 12 FRA SPA . Biennial. $2n = 20$.
* *Campanula costae* Willk. in Willk. & Lange, Prodr. Fl. Hispan. 2: 294 (1868). *Campanula
patula* var. *costae* (Willk.) O. Bolòs & Vigo, Collect. Bot. (Barcelona) 14: 101 (1983).

subsp. **epigaea** (Janka ex Degen) Hayek, Repert. Spec. Nov. Regni Veg. Beih. 30(2): 546 (1930).
Serbia, Bulgaria, Greece. 13 BUL GRC YUG. Biennial. $2n = 20$.

Campanula epigaea Janka ex Degen, Oesterr. Bot. Z. 41: 194 (1891).

Campanula velenovskyi Adamovic, Oesterr. Bot. Z. 42: 408 (1892). *Campanula patula* f. *velenovskyi* (Adamovic) Hayek, Repert. Spec. Nov. Regni Veg. Beih. 30(2): 547 (1930).

subsp. **jahorinae** (K. Malý) Greuter & Burdet, Willdenowia 11: 40 (1981).
Italy & Bosnia-Hercegovina. 13 ITA YUG. Biennial. $2n = 40$.
* *Campanula patula* var. *jahorinae* K. Malý, Oesterr. Bot. Z. 57: 184 (1907).

subsp. **patula**
Europe to W. Siberia; naturalized in NE. U.S.A. 10 FIN GRB 11 AUT BGM CZE GER HUN NET POL SWI 12 FRA SAR SPA 13 ALB BUL GRC ITA ROM YUG 14 BLR BLT KRY RUC RUE RUN RUS RUW UKR 30 WSB (75) cnt nwh nwy. Biennial. $2n = 20, 40, 60$.
Campanula decurrens L., Sp. Pl.: 164 (1753). *Campanula paula* var. *decurrens* (L.) Schult. in Roem. & Schult., Syst. Veg. 5: 103 (1819).
Campanula patens Gueldenst., Reis. Russland 2: 343 (1791).
Campanula bellidifolia Lapeyr., Hist. Pl. Pyrénées, Suppl.: 36 (1818); non Adam in F. Weber & D. Mohr, Beitr. Naturk. 1: 47 (1805).
Campanula neglecta Schult. in Roem. & Schult., Syst. Veg. 5: 104 (1819); non Besser, Cat. Hort. Cremeneci: 28 (1816). *Campanula patula* var. *neglecta* A. DC., Monogr. Campan.: 329 (1830). *Campanula patula* race *neglecta* (A. DC.) Rouy, Fl. France 10: 69 (1908). *Campanula patula* subsp. *neglecta* (A. DC.) O. Schwarz, Mitt. Thüring. Bot. Ges. 1: 118 (1949).
Campanula patula var. *flaccida* Wallr., Sched. Crit.: 85 (1822). *Campanula flaccida* (Wallr.) Dalla Torre & Sarnth., Fl. Tirol 6(3): 456 (1911); non (A. DC.) D. Dietr., Syn. Pl. 1: 752 (1839). *Campanula patula* subsp. *flaccida* (Wallr.) Sóo, Acta Bot. Acad. Sci. Hung. 17: 124 (1972).
Campanula brachiata Seidl ex Opiz, Böhm. Phan. Crypt. Gew.: 168 (1823). *Campanula patula* var. *stricta* Opiz, Seznam: 25 (1852).
Campanula patula var. *latifolia* A. DC., Monogr. Campan.: 329 (1830).
Campanula patula var. *uniflora* Noulet, Fl. Bass. Sous-Pyrén.: 410 (1837).
Campanula patula var. *grandiflora* A. DC. in DC., Prodr. 7: 480 (1839).
Campanula patula var. *hirsuta* Schur, Enum. Pl. Transsilv.: 439 (1866).
Campanula patula var. *glauca* Kuntze, Taschen-Fl. Leipzig: 128 (1867).
Campanula patula var. *platyphylla* Kuntze, Taschen-Fl. Leipzig: 128 (1867).
Campanula patula var. *macrocalyx* Kuntze, Taschen-Fl. Leipzig: 128 (1867).
Campanula patula var. *calycina* Willk. in Willk. & Lange, Prodr. Fl. Hispan. 2: 294 (1868).
Campanula monanthos Pant., Verh. Vereins Natur-Heilk. Presburg (n.s.) 2: 54 (1871-1872). *Campanula allionii* subsp. *monanthos* (Pant.) Nyman, Consp. Fl. Eur., Suppl. 2: 207 (1889).
Campanula patula var. *thyrsiflora* Kuntze, Trudy Imp. S.-Peterburgsk. Bot. Sada 10: 207 (1887).
Campanula brodensis Formánek, Oesterr. Bot. Z. 40: 81 (1890).
Campanula patula var. *serratisepala* Murr, Deutsche Bot. Monatsschr. 20: 27 (1902).
Campanula patula var. *parva* Merino, Fl. Galicia 2: 300 (1906).
Campanula patula var. *albiflora* Syr., Ill. Fl. Mosk. Gub. 3: 225 (1910). *Campanula patula* var. *flaccida* Syr., Ill. Fl. Mosk. Gub. 3: 225 (1910).
Campanula patula var. *calyciserrata* Beyer, Verh. Bot. Vereins Prov. Brandenburg 58: 119 (1917).
Campanula patula [unranked] *peterfii* Sóo ex Jáv., Magyar Fl.: 1080 (1925). *Campanula patula* var. *peterfii* (Sóo ex Jáv.) Sóo, Bot. Közlem. 23: 153 (1926). *Campanula patula* subsp. *peterfii* (Sóo ex Jáv.) Sóo, Acta Bot. Acad. Sci. Hung. 17: 124 (1972).
Campanula patula subsp. *debilis* Kuvaev in Kuvaev, Shelgunova & Konstantinov, Fl. Okrestnost. Znamensk.: 150 (1992).

Campanula pelia (Halácsy) Hausskn. & Sint. ex Phitos, Phyton 10: 126 (1963).
Greece. 13 GRC. Hemicr. $2n = 34$.
* *Campanula andrewsii* var. *pelia* Halácsy, Magyar Bot. Lapok 11: 170 (1912).
Campanula thessala Maire, Bull. Soc. Bot. France 68: 376 (1921); non Gand., Contrib. Fl. Terr. Slav. Merid. 1: 21 (1883).
Campanula patenticalyx Bocquet, Compt. Rend. Séances Soc. Phys. Genève (n.s.) 3: 90 (1968).

Campanula pelviformis Lam., Encycl. 1: 586 (1785).
Kriti. 13 KRI. Biennial. $2n = 34$.
Campanula corymbosa Desf., Ann. Mus. Natl. Hist. Nat. 11: 139 (1808). *Campanula corymbifera* Desf. ex Poir. in Lam., Encycl., Suppl. 2: 64 (1811). *Campanula pelviformis* var. *corymbosa* (Desf.) Nyman, Consp. Fl. Eur.: 476 (1879).
Campanula pelviformis var. *micrantha* A. DC., Monogr. Campan.: 224 (1830).
Campanula corymbosa var. *parviflora* K. Koch, Linnaea 19: 28 (1845).

Campanula pendula M. Bieb., Fl. Taur.-Cauc. 1: 154 (1808). *Symphyandra pendula* (M. Bieb.) A. DC., Monogr. Campan.: 366 (1830). $2n = 34$.
Caucasus (Krasnodar, Karacheyevo-Cherkessiya, Kabardino-Balkariya, Severo-Oseriya, Chechnya, Dagestan, Abkhaziya, Gruziya, Azerbaijan). 33 ALL. Hemicr. or geophyte.
Symphyandra pendula var. *transcaucasica* Sommier & Levier, Enum. Pl. Cauc.: 313 (1900). *Symphyandra transcaucasica* (Sommier & Levier) Grossh., Opred. Rast. Kavkaz.: 424 (1949). *Campanula transcaucasica* (Sommier & Levier) Kolak. & Serdyuk., Zametki Sist. Geogr. Rast. 36: 46 (1980).
Symphyandra antiqua Kolak., Trudy Suhumsk. Bot. Sada 24: 99 (1978). *Campanula antiqua* (Kolak.) Kolak. & Serdyuk. in Kolak., Fl. Abchasii 1: 130 (1980).

Campanula peregrina L., Mant. Pl.: 204 (1771).
Cyprus, Turkey, Lebanon, Syria. 34 CYP LBS TUR. Biennial. $2n = 26$.
Campanula lanuginosa Lam., Encycl. 1: 584 (1785).
Campanula decurrens Zuccagni, Cent. Observ. Bot. 1: 16 (1806); non L., Sp. Pl.: 164 (1753); nec Thore, Essai Chloris: 64 (1803). *Campanula orientalis* Steud., Nomencl. Bot. (ed. 2) 1: 269 (1840).
Campanula alata Bové ex Vatke, Linnaea 38: 710 (1874); non Desf., Fl. Atlant. 1: 178 (1798).

Campanula perpusilla A. DC. in DC., Prodr. 7: 474 (1839).
Iraq & Iran. 34 IRN IRQ. Hemicr.

Campanula persepolitana Kotschy ex Boiss., Diagn. Pl. Orient. (ser. 1) 7: 19 (1846).
Iran. 34 IRN. Hemicr.

Campanula persicifolia L., Sp. Pl.: 164 (1753). *Campanula amygdalifolia* Salisb., Prodr. Stirp. Chap. Allerton: 126 (1796). *Campanula racemosa* var. *grandiflora* Vuk., Linnaea 26: 333 (1854). *Neocodon persicifolius* (L.) Kolak. & Serdyuk., Zametki Sist. Geogr. Rast. 40: 29 (1984).
Continental Europe to W. Siberia, Kazakhstan & Turkey; cult.; naturalized in Great Britain & North America. 10 DEN FIN grb NOR SWE 11 AUT BGM CZE GER HUN NET POL SWI 12 FRA SAR SPA 13 ALB BUL GRC ITA ROM TUE YUG 14 BLR BLT KRY RUC RUE RUN RUS RUW UKR 30 WSB 32 KAZ 34 TUR (71) brc (72) nbr ont que (73) ore (74) wis (75) cnt mas mic nwy pen ver (76) uta (78) nca vrg. Hemicr. $2n = 16, 18, 32$.

subsp. **persicifolia**
Continental Europe to W. Siberia, Kazakhstan & Turkey; cult.; naturalized in Great Britain & North America. 10 DEN FIN grb NOR SWE 11 AUT CZE BGM GER HUN NET POL SWI 12 FRA SAR SPA 13 ALB BUL GRC ITA ROM TUE YUG 14 BLR BLT KRY RUC RUE RUN RUS RUW UKR 30 WSB 32 KAZ 34 TUR (71) brc (72) nbr ont que (73) ore (74) wis (75) cnt mas mic nwy pen ver (76) uta (78) nca vrg. Hemicr. $2n = 16, 18, 32$.
Campanula linifolia L., Fl. Monsp.: 12 (1756).
Campanula vesula All., Fl. Pedem. 1: 108 (1785).
Campanula pumila F. W. Schmidt, Fl. Boëm. 2: 71 (1794). *Campanula persicifolia* var. *pumila* (F. W. Schmidt) Schult. in Roem. & Schult., Syst. Veg. 5: 106 (1819). *Campanula persicifolia* subvar. *pumila* (F. W. Schmidt) Nyman, Consp. Fl. Eur.: 481 (1879).
Campanula persicifolia var. *maxima* Sims, Bot. Mag. 12: tab. 397 (1798).
Campanula persicifolia var. *macrantha* A. DC., Monogr. Campan.: 322 (1830).
Campanula magellensis Ten., Fl. Napol. 1: xv (1811). *Campanula persicifolia* var. *pumila* Ten., Fl. Napol. 3: 205 (1824-29); non (F. W. Schmidt) Schult. in Roem. & Schult., Syst. Veg. 5: 106 (1819).

Campanula hispida Lej., Fl. Spa 2: 299 (1813). *Campanula persicifolia* var. *hispida* (Lej.) A. DC., Monogr. Campan.: 323 (1830). *Campanula persicifolia* subvar. *hispida* (Lej.) Rouy, Fl. France 10: 68 (1908).

Campanula dasycarpa Kit. in Schult., Oestr. Fl. (ed. 2) 1: 404 (1814). *Campanula persicifolia* var. *dasycarpa* (Kit.) A. DC., Monogr. Campan.: 323 (1830).

Campanula lanceolata J. Presl & C. Presl, Fl. Čech.: 49 (1819); non Lapeyr., Hist. Pl. Pyrénées: 105 (1815). *Campanula persicifolia* var. *lanceolata* Steud., Nomencl. Bot. (ed. 2) 1: 269 (1840).

Campanula persicifolia var. *angustifolia* A. DC., Monogr. Campan.: 322 (1830)

Campanula persicifolia var. *uniflora* Noulet, Fl. Bass. Sous-Pyrén.: 409 (1837).

Campanula persicifolia var. *parviflora* Peterm., Deutschl. Fl. ? (1846).

Campanula persicifolia var. *parviflora* Kirschl., Fl. Alsace 1: 276 (1851); non Peterm., Deutschl. Fl. ? (1846). *Campanula persicifolia* f. *kirschlegeri* Sóo, Acta Bot. Acad. Sci. Hung. 12: 366 (1966).

Campanula persicifolia var. *alpina* Schur, Enum. Pl. Transsilv.: 438 (1866).

Campanula persicifolia var. *eriocarpa* Schur, Enum. Pl. Transsilv.: 438 (1866).

Campanula persicifolia var. *glaberrima* Schur, Enum. Pl. Transsilv.: 438 (1866).

Campanula persicifolia var. *grandiflora* Schur, Enum. Pl. Transsilv.: 438 (1866).

Campanula persicifolia var. *humillima* Schur, Enum. Pl. Transsilv.: 438 (1866).

Campanula persicifolia var. *monstrosa* Schur, Enum. Pl. Transsilv.: 439 (1866).

Campanula persicifolia var. *hispidior* Trautv., Trudy Imp. S.-Peterburgsk. Bot. Sada 6: 87 (1879).

Campanula persicifolia var. *parviflora* Freyn, Verh. K. K. Zool.-Bot. Ges. Wien 38: 618 (1888); non Kirschl., Fl. Alsace 1: 276 (1851).

Campanula persicifolia f. *alba* Voss in Siebert & Voss, Vilm. Blumengärtn. (ed. 3)1: 570 (1894).

Campanula persicifolia f. *coronata* Voss in Siebert & Voss, Vilm. Blumengärtn. (ed. 3) 1: 570 (1894).

Campanula persicifolia var. *laevicaulis* Korsh., Tent. Fl. Ross. Orient.: 273 (1898).

Campanula persicifolia var. *lasiocarpa* Korsh., Tent. Fl. Ross. Orient. 273 (1898).

Campanula crystallocalyx Adamoviç, Denkschr. Kaiserl. Akad. Wiss.,Math.-Naturwiss. Kl. 74: 141 (1903). *Campanula persicifolia* var. *crystallocalyx* (Adamoviç) Hayek, Repert. Spec. Nov. Regni Veg. Beih. 30(2): 545 (1930).

Campanula persicifolia var. *suskalovicii* Adamoviç, Denkschr. Kaiserl. Akad. Wiss., Math.-Naturwiss. Kl. 74: 141 (1903).

Campanula persicifolia var. *eriocarpa* Syr., Ill. Fl. Mosk. Gub. 3: 226 (1910).

Campanula persicifolia var. *reflexa* Dalla Torre & Sarnth., Fl. Tirol. 6(3): 454 (1911).

subsp. **sessiliflora** (K. Koch ex Velen.) Fed. ex Greuter & Burdet, Willdenowia 12: 36 (1982). Slovenia, Serbia, Macedonia. 13 YUG. Hemicr. $2n = 18$.

* *Campanula sessiliflora* K. Koch, Linnaea 19: 30 (1845); non L. f., Suppl. Pl.: 139(182).
  Campanula persicifolia var. *sessiliflora* K. Koch ex Velen., Oesterr. Bot. Z. 52: 16 (1892).

Campanula persicifolia var. *latisepala* Degen & Dörfl., Denkschr. Kaiserl. Akad. Wiss., Math.-Naturwiss. Kl. 64: 728 (1897).

Campanula persicaster F. Herm., Fl. N.-Mitt.-Eur.: 977 (1956).

Campanula persicifolia var. *snjagovii* Velchev & P. Vassil., Izv. Bot. Inst. (Sofia) 24: 249 (1973), as 'snjagovi'.

subsp. **subpyrenaica** (Timb.-Lagr.) Fed., Bot. J. Linn. Soc. 67: 281 (1973). Pyrénées. 12 FRA SPA. Hemicr. $2n = 16$.

Campanula persicifolia var. *lasiocalyx* Gren. & Godr., Fl. Fr. 2: 420 (1853).

* *Campanula subpyrenaica* Timb.-Lagr., Mém. Acad. Sci. Toulouse (ser. 4) 5: 387 (1855).
  Campanula persicifolia var. *subpyrenaica* (Timb.-Lagr.) Nyman, Consp. Fl. Eur.: 481 (1879).

Campanula peshmenii Güner, Notes Roy. Bot. Gard. Edinburgh 41: 287 (1983).
Turkey. 34 TUR. Hemicr.

Campanula petiolata A. DC., Monogr. Campan.: 278 (1830). *Campanula rotundifolia* var. *petiolata* (A. DC.) J. K. Henry, Fl. S. Brit. Columbia: 283 (1915).

W. Canada to NE. Mexico (Coahuila, Nuevo León, Tamaulipas). 70 NUN NWT YUK 71 ABT
BRC MAN SAS 73 COL IDA MNT ORE WAS WYO 74 NEB NDA SDA 76 ARI CAL UTA 77
NWM TEX 79 MXE. Hemicr. or geophyte. $2n = 68$.
Campanula macdougalii Rydb., Bull. Torrey Bot. Club 28: 25 (1901).
Campanula stylocampa Eastw., Bull. Torrey Bot. Club 29: 525 (1902).
Campanula sacajaweana M. Peck, Proc. Biol. Soc. Wash. 50: 123 (1937).

Campanula petraea L., Syst. Nat. (ed. 10) 2: 926 (1759). *Campanula cephalaria* var. *gnaphalophylla*
Vuk., Linnaea 26: 335 (1854). *Tracheliopsis petraea* (L.) Buser, Bull. Herb. Boissier 2: 528 (1894).
S. Alps. 12 FRA 13 ITA. Hemicr. $2n = 34$.
Tracheliopsis albicans Buser, Bull. Herb. Boissier 2: 531 (1894). *Campanula albicans* (Buser)
Engl. in Engl. & Prantl, Nat. Pflanzenfam., Nachtr. 1: 319 (1897).

Campanula petrophila Rupr., Bull. Acad. Imp. Sci. Saint-Pétersbourg 11: 212 (1867). *Campanula*
tridentata var. *petrophila* (Rupr.) Trautv., Trudy Imp. S.-Peterburgsk. Bot. Sada 6: 58 (1879).
Hemisphaera petrophila (Rupr.) Kolak. & Serdyuk. in Kolak., Kolokol'ch. Kavkaz.: 123 (1991).
Caucasus (Kabardino-Balkariya, Severo-Osetiya, Chechnya, Dagestan, Gruziya, Azerbaijan). 33
ALL. Hemicr. or geophyte.
Campanula petrophila var. *linoides* Rupr., Bull. Acad. Imp. Sci. Saint-Pétersbourg 11: 214 (1867).
Campanula petrophila var. *borbalensis* Rupr., Bull. Acad. Imp. Sci. Saint-Pétersbourg 11: 214
(1867).
Campanula petrophila var. *longiflora* Rupr., Bull. Acad. Imp. Sci. Saint-Pétersbourg 11: 214
(1867).
Campanula petrophila var. *exappendiculata* Sommier & Levier, Enum. Pl. Cauc.: 320 (1900).
Campanula petrophila var. *angustiflora* Fomin in N. M. Kusn., N. Bush & Fomin, Mater. Fl.
Kavkaza 4(6): 71 (1905).

Campanula phrygia Jaub. & Spach, Ill. Pl. Orient. 3: 42 (1848).
Balkans to Israel & Syria. 13 ALB BUL GRC TUE YUG 34 LBS PAL TUR. Ther. $2n = 16$.
Campanula ramosissima var. *velutina* Griseb., Spic. Fl. Rumel. 2: 283 (1846). *Campanula*
phrygia f. *velutina* (Griseb.) Hayek, Repert. Spec. Nov. Regni Veg. Beih. 30(2): 548 (1930).
Campanula tillischii O. Schwarz, Repert. Spec. Nov. Regni Veg. 36: 142 (1934).

Campanula phyctidocalyx Boiss. & Noë in Boiss., Diagn. Pl. Orient. (ser. 2) 3: 114 (1856).
Turkey & Iran. 34 IRN TUR. Hemicr. $2n = 16$.
Campanula amabilis Leichtlin ex Spigolatore, Bull. Soc. Tosc. Ortic. 1904: 339 (1904).
Campanula phyctidocalyx var. *sulcata* Parsa, Fl. Iran 10: 226 (1980).

Campanula pinatzii Greuter & Phitos, Boissiera 13: 134 (1967). *Campanula drabifolia* subsp.
pinatzii (Greuter & Phitos) Fed., Bot. J. Linn. Soc. 67: 281 (1973).
S. Aegean Is. (Kárpathos, Kásos, Saría). 13 KRI. Ther. $2n = 20$.

Campanula pindicola Aldén, Bot. Not. 129: 308 (1976).
Greece. 13 GRC. Hemicr. or geophyte.

Campanula pinnatifida Huber-Mor., Bauhinia 2: 198 (1963).
Turkey. 34 TUR. Hemicr. or geophyte.
Campanula pinnatifida var. *germanicopolitana* Huber-Mor., Bauhinia 2: 199 (1963).
Campanula pinnatifida var. *robusta* Huber-Mor., Bauhinia 2: 199 (1963).

Campanula piperi Howell, Fl. N. W. Amer.: 409 (1901). *Astrocodon piperi* (Howell) A. P. Khokhr.,
Analiz Fl. Kolymsk. Nagor'ya: 122 (1989).
NW. U.S.A. (Washington). 73 WAS. Geophyte. $2n = 34$.
Campanula piperi f. *sovereigniana* E. English, Little Gardens 12(3): 13 (1940).

Campanula podocarpa Boiss., Diagn. Pl. Orient. (ser. 1) 11: 68 (1849).
  E. Aegean Is. (Ródos), Cyprus, Turkey. 34 CYP EAI TUR. Ther.
    Campanula cypria Rech. f., Ark. Bot. (n.s.) 1: 432 (1951).

Campanula pollinensis Podlech, Mitt. Bot. Staatssamml. München 8: 211 (1970). *Campanula scheuchzeri* subsp. *pollinensis* (Podlech) Bernardo, Gargano & Peruzzi, Inform. Bot. Ital. 36: 519 (2005).
  S. Italy. 13 ITA. Hemicr. or geophyte.

Campanula polyclada Rech. f. & Schiman-Czeika in Rech. f., Fl. Iran. 13: 19 (1965).
  Afghanistan & Pakistan. 34 AFG 40 PAK. Hemicr.

Campanula pontica Albov, Bull. Herb. Boissier 2: 116 (1894). *Neocodon ponticus* (Albov) Kolak. & Serdyuk., Zametki Sist. Geogr. Rast. 40: 29 (1984).
  Caucasus (Adzhariya, Gruziya) & Turkey. 33 TCS 34 TUR. Hemicr.

Campanula portenschlagiana Schult. in Roem. & Schult., Syst. Veg. 5: 93 (1819). *Campanula oxyphylla* var. *leiophylla* Vuk., Linnaea 26: 331 (1854).
  Croatia & Hercegovina; cult.; naturalized in Great Britain & Channel Is. (10) grb (12) fra 13
    YUG. Hemicr. 2*n* = 34, 80, 102.
    Campanula portenschlagiana var. *pubescens* A. DC. in DC., Prodr. 7: 476 (1839).
    Campanula portenschlagiana var. *pumila* D. Šoljan, Glasn. Zemaljsk. Muz. Bosne Hercegovine
      Sarajevu, Prir. Nauke 29: 45 (1990).
    Campanula portenschlagiana var. *hirsuta* D. Šoljan, Glasn. Zemaljsk. Muz. Bosne Hercegovine
      Sarajevu, Prir. Nauke 29: 46 (1990).
    Campanula portenschlagiana f. *grandiflora* D. Šoljan, Glasn. Zemaljsk. Muz. Bosne
      Hercegovine Sarajevu, Prir. Nauke 29: 46 (1990).

Campanula poscharskyana Degen, Magyar Bot. Lapok 7: 103 (1908).
  Croatia & Montenegro; cult.; naturalized in Great Britain & New Zealand. (10) grb 13 YUG
    (51) nzn nzs. Hemicr. 2*n* = 34.

Campanula postii (Boiss.) Engl. in Engl. & Prantl, Nat. Pflanzenfam., Nachtr. 1: 319 (1897).
  Turkey, Syria, Iraq. 34 IRQ LBS TUR. Hemicr. 2*n* = 34.
    * *Trachelium postii* Boiss., Fl. Orient., Suppl.: 336 (1888). *Tracheliopsis postii* (Boiss.) Buser, Bull.
      Herb. Boissier 2: 527 (1894). *Diosphaera postii* (Boiss.) Bornm., Beih. Bot. Centralbl. 38(2):
      338 (1921).
    Campanula shepardii Post, Bull. Herb. Boissier 1: 25 (1893), as 'shepardi'. *Tracheliopsis postii*
      subsp. *shepardii* (Post) P. H. Davis, Notes Roy. Bot. Gard. Edinburgh 22: 80 (1956), as
      'shepardi'.
    Campanula amana Rech. f., Ann. Naturhist. Mus. Wien 57: 85 (1950).
    Diosphaera hysterantha Rech. f. & Schiman-Czeika in Rech. f., Fl. Iran. 13: 39 (1965).

Campanula praesignis Beck, Fl. Nieder-Österreich: 1105 (1893). *Campanula rotundifolia* subsp. *praesignis* (Beck) Hayek in Hegi, Ill. Fl. Mitt.-Eur. 6: 353 (1916). *Campanula lanceolata* var. *praesignis* (Beck) Hruby, Magyar Bot. Lapok 29: 215 (1930).
  NE. Alps. 11 AUT. Hemicr. or geophyte. 2*n* = 34.
    Campanula praesignis var. *breynina* Beck, Fl. Nieder-Österreich: 1105 (1893). *Campanula
      rotundifolia* var. *breyniana* (Beck) Hayek in Hegi, Ill. Fl. Mitt.-Eur. 6: 354 (1916).
    Campanula breyniana (Beck) Hruby, Magyar Bot. Lapok 29: 231 (1930).

Campanula precatoria Timb.-Lagr., Mém. Acad. Sci. Toulouse (ser. 7) 5: 271 (1873). *Campanula lanceolata* var. *precatoria* (Timb.-Lagr.) Nyman, Consp. Fl. Eur.: 478 (1879).
  E. Pyrénées. 12 FRA SPA. Hemicr. or geophyte. 2*n* = 34.
    Campanula precatoria var. *hirsuta* Timb.-Lagr., Mém. Acad. Sci. Toulouse (ser. 7) 5: 272
      (1873). *Campanula lanceolata* var. *hirsuta* (Timb.-Lagr.) Rouy, Fl. France 10: 74 (1908).

Campanula precatoria var. *major* Timb.-Lagr., Mém. Acad. Sci. Toulouse (ser. 7) 5: 272 (1873).
Campanula lanceolata var. *major* (Timb.-Lagr.) Rouy, Fl. France 10: 74 (1908).
Campanula precatoria var. *tenuifolia* Timb.-Lagr., Mém. Acad. Sci. Toulouse (ser. 7) 5: 272
(1873). *Campanula lanceolata* var. *tenuifolia* (Timb.-Lagr.) Rouy, Fl. France 10: 74 (1908).
Campanula loretiana Witasek, Abh. K. K. Zool.-Bot. Ges. Wien 1(3): 88 (1902).

Campanula prenanthoides Durand, J. Acad. Nat. Sci. Philadelphica (ser. 2) 3: 93 (1855).
Asyneuma prenanthoides (Durand) McVaugh, Bartonia 23: 36 (1945). 2*n* = 32, 34.
W. U.S.A. (Oregon, California). 73 ORE 76 CAL.
Campanula filiflora Kellogg, Proc. Calif. Acad. Sci. 2: 5 (1861).
Campanula roezlii Regel, Gartenflora 21: 239 (1872), as 'roezli'.

Campanula propinqua Fisch. & C. A. Mey., Index Sem. Hort. Petrop. 2: 32 (1836). *Roucela
propinqua* (Fisch. & C. A. Mey.) Kharadze, Zametki Sist. Geogr. Rast. 32: 55 (1976).
Turkey to Caucasus (Armenia, Nakhichevan), Iran & Saudi Arabia. 33 TCS 34 IRN IRQ LBS TUR
35 SAU. Ther. 2*n* = 16.
Campanula hispidissima Hochst. in Lorent, Wander. Morgenlande: 331 (1845).
Campanula pestalozzae Boiss., Diagn. Pl. Orient. (ser. 1) 11: 62 (1849); non (Boiss.) Boiss.,
Diagn. Pl. Orient. (ser. 1) 11: 77 (1849).
Campanula propinqua var. *parviflora* Turrill, Bull. Misc. Inform. Kew 1929: 229 (1929).

Campanula pseudostenocodon Lacaita, Nuovo Giorn. Bot. Ital. (n.s.) 25: 22 (1918). *Campanula
rotundifolia* var. *pseudostenocodon* (Lacaita) Fiori, Nuov. Fl. Italia 2: 562 (1927). *Campanula
scheuchzeri* subsp. *pseudostenocodon* (Lacaita) Bernardo, Gargano & Peruzzi, Inform. Bot. Ital.
36: 519 (2005).
S. Italy (Apennines). 13 ITA. Hemicr. or geophyte. 2*n* = 102.

Campanula psilostachya Boiss. & Kotschy in Boiss., Diagn. Pl. Orient. (ser. 2) 3: 113 (1856).
Podanthum psilostachyum (Boiss. & Kotschy) Boiss., Fl. Orient. 3: 948 (1875). *Asyneuma
psilostachya* (Boiss. & Kotschy) Bornm., Beih. Biol. Centralbl. 38(2): 348 (1921).
Turkey. 34 TUR. Biennial. 2*n* = 16, 34.

Campanula ptarmicifolia Lam., Encycl. 1: 579 (1785), as 'ptarmicaefolia'.
Turkey. 34 TUR. Hemicr.
Campanula ptarmicifolia var. *capitellata* Damboldt, Notes Roy. Bot. Gard. Edinburgh 35:
49 (1976).

Campanula pterocaula Hausskn., Mitt. Thüring. Bot. Vereins (n.s.) 20: 34 (1904).
Turkey. 34 TUR. Biennial.

Campanula pubicalyx (P. H. Davis) Damboldt, Notes Roy. Bot. Gard. Edinburgh 35: 47 (1976).
Turkey. 34 TUR. Hemicr.
* *Tracheliopsis pubicalyx* P. H. Davis, Notes Roy. Bot. Gard. Edinburgh 22: 79 (1956).

Campanula pulla L., Sp. Pl.: 163 (1753).
NE. Alps. 11 AUT. Hemicr. or geophyte. 2*n* = 34.
Campanula pulla var. *ramosa* A. DC., Monogr. Campan.: 288 (1830).
Campanula pseudopulla Schur, Enum. Pl. Transsilv.: 441 (1866), as 'pseudo-pulla'. *Campanula
pulla* var. *pseudopulla* (Schur) Nyman, Consp. Fl. Eur.: 480 (1879), as 'pseudo-pulla'.

Campanula pulvinaris Hausskn. & Bornm., Mitt. Thüring. Bot. Vereins (n.s.) 20: 29 (1904).
Turkey. 34 TUR. Hemicr.

Campanula punctata Lam., Encycl. 1: 586 (1785).
SE. Siberia to Russian Far East, Japan (Hokkaido, Honshu, Shikoku, Kyushu), N. & C. China
(Nei Mongol, Gansu, Shaanxi, Shanxi, Hebei, Henan, Hubei, Sichuan); cult.; naturalized in

NE. U.S.A. 30 BRY CTA 31 AMU KHA PRM SAK 36 CHC CHI CHN CHS 38 JAP KOR (75)
nwh. Hemicr. or geophyte. $2n = 34$.
Campanula violifolia Lam., Encycl. 1: 587 (1785), as 'violaefolia'. *Campanula violae* Pers., Syn.
Pl. 1: 189 (1805).
Campanula nobilis Lindl., J. Hort. Soc. London 1: 232 (1846).
Campanula nobilis var. *alba* Van Houtte ex Planch., Fl. Serres Jard. Eur. 6: 95 (1850-51).
Campanula vanhouttei Carr., Rev. Hort. 1878: 420 (1878).
Campanula hybrida Rodigas, Ill. Hort. 30: 118, pl. 499 (1883); non L., Sp. Pl.: 168 (1753).
Campanula punctata f. *partita* Makino, Bot. Mag. (Tokyo) 19: 148 (1905).
Campanula punctata var. *rubriflora* Makino, Bot. Mag. (Tokyo) 22: 156 (1908). *Campanula
punctata* f. *rubriflora* (Makino) T. Shimizu in K. Iwats., T. Yamaz., Boufford & H. Ohba, Fl.
Japan 3a: 411 (1993).
Campanula hondoensis Kitam., Acta Phytotax. Geobot. 5: 212 (1936). *Campanula punctata*
subsp. hondoensis (Kitam.) Kitam., Acta Phytotax. Geobot. 17: 7 (1957). *Campanula
punctata* var. *hondoensis* (Kitam.) Ohwi ex T. Shimizu in K. Iwats., T. Yamaz., Boufford &
H. Ohba, Fl. Japan 3a: 411 (1993).
Campanula punctata f. *impunctata* N. Yonez., J. Phytogeogr. Taxon. 35: 41 (1987).
Campanula punctata f. *albiflora* T. Shimizu, J. Phytogeogr. Taxon. 37: 120 (1989).

Campanula pyramidalis L., Sp. Pl.: 164 (1753). *Campanula umbellulifera* Vuk., Linnaea 26:
333 (1854).
SE. Europe; cult.; naturalized in England & Channel Is. (Guernsey). (10) grb (12) fra 13 ALB
ITA YUG. Hemicr. $2n = 34$.
Campanula pyramidalis var. *calycina* A. DC., Monogr. Campan.: 310 (1830).
Campanula pyramidalis f. *alba* Voss in Siebert & Voss, Vilm. Blumengärtn. (ed. 3) 1: 568 (1894).
Campanula pyramidalis var. *compacta* Anonymous, Garden (London) 47: 86 (1894).

Campanula pyrenaica A. DC., Monogr. Campan.: 324 (1830).
Pyrénées. 12 FRA SPA. Ther. $2n = 16$.

Campanula quercetorum Huber-Mor. & C. Simon, Bauhinia 2: 200 (1963).
Turkey. 34 TUR. Hemicr.

Campanula radchensis Kharadze, Zametki Sist. Geogr. Rast. 13: 53 (1947). *Hemisphaera radchensis*
(Kharadze) Kolak., Okhrana Prir. Gruz. 12: 167 (1984).
Caucasus (Gruziya). 33 TCS. Hemicr. or geophyte.

Campanula raddeana Trautv., Bull. Acad. Imp. Sci. Saint-Pétersbourg 10: 395 (1866).
Caucasus (Gruziya). 33 TCS. Hemicr. or geophyte. $2n = 34, 102$.
Campanula brotheri Sommier & Levier, Nuovo Giorn. Bot. Ital. (n.s.) 4: 199 (1897).
Campanula kemulariae Fomin, Trudy Bot. Inst. Akad. Nauk SSSR (ser. 1) 3: 289 (1937).

Campanula radicosa Bory & Chaub., Nouv. Fl. Pélop.: 14 (1838).
Greece. 13 GRC. Hemicr. $2n = 34$.
Campanula colettae Beauverd & Topoli, Bull. Soc. Bot. Genève (ser. 2) 30: 283 (1940).

Campanula radula Fisch. ex Tchich., Asie Min., Bot. 2: 395 (1860).
Iraq. 34 IRQ. Hemicr.

Campanula raineri Perp., Biblioth. Ital. (Turin) 5: 134 (1817), as 'rainerii'. *Campanula sessiliflora*
Vuk., Linnaea 26: 327 (1854); non L. f., Suppl. Pl.: 139 (1782); nec K. Koch, Linnaea 19: 30 (1845).
SE. Alps. 13 ITA. Hemicr. $2n = 32, 34$.

Campanula ramosissima Sm. in Sibth. & Sm., Fl. Graec. Prodr. 1: 137 (1806).
SE. Europe. 13 ALB GRC ITA YUG. Ther. $2n = 20$.
Campanula lorei Pollini, Elem. Bot. 2: 148 (1811).

Campanula baldensis Balb., Cat. Pl. (ed. 1813): 20 (1813). *Loreia baldensis* (Balb.) Raf., Fl.
 Tellur. 3: 82 (1837).
Campanula ramosissima var. *glabrescens* Hausskn., Mitth. Thüring. Bot. Vereins (ser. 2) 7: 63
 (1895).

Campanula rapunculoides L., Sp. Pl.: 165 (1753). *Campanula morifolia* Salisb., Prodr. Stirp. Chap.
 Allerton: 126 (1796). *Campanula rigida* Stokes, Bot. Mat. Med. 1: 333 (1812), nom. illeg. *Cenekia
 rapunculoides* (L.) Opiz, Seznam: 36 (1852). *Campanula racemosa* var. *laxiflora* Vuk., Linnaea 26:
 332 (1854). *Campanula rapunculiformis* St.-Lag. in Cariot, Étude Fl. (ed. 8) 2: 548 (1889).
 Continental Europe to C. Siberia (Krasnoyarsk), Kazakhstan & Iran; cult.; naturalized in British
 Is., North America & New Zealand. 10 DEN FIN grb ire NOR SWE 11 ALL 12 FRA SPA 13
 ALB BUL GRC ITA ROM YUG 30 KRA 32 KAZ 33 ALL 34 IRN TUR (51) nzn nzs (71) abt brc
 man (72) nbr nfl nsc ont pei que (73) all (74) ill iow kan min mso nda neb sda wis (75) all
 (76) nev uta (77) all (78) del kty mry nca ten vrg. Hemicr. or geophyte. $2n = 68, 102$.
Campanula rhomboidalis Gorter, Fl. Ingr.: 34 (1761); non L., Sp. Pl.: 165 (1753).
Campanula nutans Lam., Fl. Franç. 3: 336 (1779).
Campanula urticifolia Turra, Fl. Ital. Prodr.: 64 (1780).
Campanula rhomboidea Falk, Topogr. Beitr. 2: 128 (1786).
Campanula secunda F. W. Schmidt, Fl. Boëm. 2: 74 (1794). *Campanula rapunculoides* f.
 secunda (F. W. Schmidt) Hayek & Hegi in Hegi, Ill. Fl. Mitt.-Eur. 6: 343 (1915).
Campanula hortensis Meerb., Pl. Select. Icon. Pict.: A2 (1798).
Campanula trachelioides M. Bieb., Fl. Taur.-Cauc. 1: 150 (1808). *Campanula rapunculoides* var.
 trachelioides (M. Bieb.) A. DC., Monogr. Campan.: 269 (1830). *Campanula rapunculoides* f.
 trachelioides (M. Bieb.) Hayek & Hegi in Hegi, Ill. Fl. Mitt.-Eur. 6: 343 (1915).
Campanula ucranica Besser, Cat. Jard. Bot. Krzemieniec Suppl. 2: 6 (1814). *Campanula
 neglecta* Besser, Cat. Hort. Cremeneci: 28 (1816). *Campanula crenata* Link, Enum. Hort.
 Berol. Alt. 1: 214 (1821). *Campanula rapunculoides* subvar. *neglecta* Nyman, Consp. Fl.
 Eur.: 478 (1879). *Campanula rapunculoides* var. *glabrata* Trautv., Trudy Imp. S.-Peterburgsk.
 Bot. Sada 6: 75 (1879). *Campanula rapunculoides* f. *crenata* Hayek & Hegi in Hegi, Ill. Fl.
 Mitt.-Eur. 6: 343 (1915). *Campanula rapunculoides* f. *ucranica* (Besser) Hayek & Hegi in
 Hegi, Ill. Fl. Mitt.-Eur. 6: 343 (1915).
Campanula lunariifolia Willd. ex Schult. in Roem. & Schult., Syst. Veg. 5: 92, 634 (1819), as
 'lunariaefolia'.
Campanula elegans Schult. in Roem. & Schult., Syst. Veg. 5: 105 (1819).
Campanula pyramidiflora Rchb., Cat. Sem. Hort. Dresd.: ? (1822).
Campanula speciosa Willd. ex Spreng., Syst. Veg. 1: 729 (1824); non Pourr., Hist. & Mém.
 Acad. Roy. Sci. Toulouse 3: 309 (1788); nec Hornem., Hort. Bot. Hafn.: 957 (1815).
Campanula infundibuliformis Sims, Bot. Mag. 53: tab. 2632 (1826).
Campanula rapunculoides var. *macrophylla* A. DC., Monogr. Campan.: 268 (1830).
Campanula rapunculoides var. *oenipontana* A. DC., Monogr. Campan.: 268 (1830).
Campanula rapunculoides var. *nana* A. DC., Monogr. Campan.: 269 (1830).
Campanula nemorosa A. DC., Monogr. Campan.: 274 (1830). *Campanula rapunculoides* var.
 nemorosa (A. DC.) Nyman, Consp. Fl. Eur.: 478 (1879).
Campanula rapunculoides var. *racemosa* Peterm. in ? *Cenekia rapunculoides* var. *racemosa*
 (Peterm.) Opiz, Seznam: 36 (1852).
Campanula rapunculoides var. *reflexa* Peterm. in ? *Cenekia rapunculoides* var. *reflexa* (Peterm.)
 Opiz, Seznam: 36 (1852).
Campanula rapunculoides var. *speciosa* Knaf in ? *Cenekia rapunculoides* var. *speciosa* (Knaf)
 Opiz, Seznam: 36 (1852).
Campanula cordifolia K. Koch, Linnaea 19: 29 (1845). *Campanula rapunculoides* f. *cordifolia* (K.
 Koch) Albov, Prodr. Fl. Colchic.: 160 (1895). *Campanula rapunculoides* var. *cordifolia* (K.
 Koch) Fomin in N. M. Kusn., N. Bush & Fomin, Mater. Fl. Kavkaza 4(6): 94 (1906).
 Campanula rapunculoides subsp. *cordifolia* (K. Koch) Damboldt, Notes Roy. Bot. Gard.,
 Edinburgh 35: 45 (1976).
Campanula rapunculoides var. *cordata* K. Koch, Linnaea 23: 641 (1850).
Campanula rapunculoides var. *grandiflora* K. Koch, Linnaea 23: 641 (1850).

Campanula rapunculoides var. *simplex* K. Koch, Linnaea 23: 641 (1850).
Campanula secundiflora Vis. & Pančić, Mem. Reale Ist. Veneto Sci. 10: 442 (1861).
Campanula rapunculoides var. *ramosissima* Schur, Enum. Pl. Transsilv.: 437 (1866).
Campanula rapunculoides var. *subsimplex* Schur, Enum. Pl. Transsilv.: 437 (1866).
Campanula grossheimii Kharadze, Zametki Sist. Geogr. Rast. 15: 29 (1949).
Campanula foliosa Galushko, Novosti Syst. Vyssh. Rast. 8: 266 (1971); non Ten., Fl. Napol.
   1: xvi (1811). *Campanula chysnysuensis* Czerep., Vasc. Pl. Russia: 154 (1995).

Campanula rapunculus L., Sp. Pl.: 164 (1753). *Campanula esculenta* Salisb., Prodr. Stirp. Chap.
   Allerton: 126 (1796). *Campanula racemosa* var. *paniculiformis* Vuk., Linnaea 26: 332 (1854), as
   'paniculaeformis'. *Campanula patula* var. *rapunculus* (L.) Kuntze, Taschen-Fl. Leipzig: 128
   (1867). *Campanula rapuncula* St.-Lag., Ann. Soc. Bot. Lyon 7: 121 (1880). *Neocodon rapunculus*
   (L.) Kolak. & Serdyuk., Zametki Sist. Geogr. Rast. 40: 27 (1984).
   Continental Europe to N. Africa & W. Asia; cult.; naturalized in Great Britain. (10) grb 11 AUT
   CZE BGM GER HUN NET POL SWI 12 COR FRA POR SPA 13 ALB BUL GRC ITA ROM TUE
   YUG 14 BLR KRY RUC RUS UKR 20 ALG MOR TUN 33 ALL 34 IRN LBS PAL TUR. Biennial.
   $2n = 20$.

subsp. **lambertiana** (A. DC.) Rech. f., Fl. Iran. 13: 34 (1965). $2n = 20$.
   Bulgaria to Krym, Caucasus & Iran. 13 BUL TUE 14 KRY 33 ALL 34 IRN LBS PAL TUR.
   Biennial.
    * *Campanula lambertiana* A. DC., Monogr. Campan.: 327 (1830). *Campanula rapunculus* var.
       lambertiana (A. DC.) Boiss., Fl. Orient. 3: 940 (1875). *Campanula rapunculus* f. *lambertiana*
       (A. DC.) Voss in Siebert & Voss, Vilm. Blumengärtn. (ed. 3) 1: 571 (1894). *Neocodon
       lambertianus* (A. DC.) Kolak. & Serdyuk., Zametki Sist. Geogr. Rast. 40: 29 (1984).
       Campanula rapunculus var. *spiciformis* Boiss., Fl. Orient. 3: 939 (1875). *Campanula rapunculus*
       f. *spiciformis* (Boiss.) Voss in Siebert & Voss, Vilm. Blumengärtn. (ed. 3) 1: 571 (1894).
       Campanula hyrcania Wettst., Denkschr. Kaiserl. Akad. Wiss., Math.-Naturwiss. Kl. 50: 69
          (1885).

subsp. **rapunculus**
   Continental Europe to N. Africa & Turkey; cult.; naturalized in Great Britain. (10) grb 11
   AUT BGM CZE GER HUN NET POL SWI 12 COR FRA POR SPA 13 ALB BUL GRC ITA
   ROM TUE YUG 14 BLR KRY RUC RUS UKR 20 ALG MOR TUN 34 TUR. Biennial.
   $2n = 20$.
   Campanula decurrens Thore, Essai Chloris: 64 (1803); non L., Sp. Pl.: 164 (1753).
   Campanula elatior Hoffmanns. & Link, Fl. Portug. 2: 11 (1813-20). *Campanula rapunculus*
      subvar. *elatior* (Hoffmanns. & Link) Nyman, Consp. Fl. Eur.: 482 (1879).
   Campanula verruculosa Hoffmanns. & Link, Fl. Portug. 2: 12 (1813-20). *Campanula
      rapunculus* var. *verruculosa* (Hoffmanns. & Link) Steud., Nomencl. Bot. (ed. 2) 1: 270
      (1840). *Campanula rapunculus* subsp. *verruculosa* (Hoffmanns. & Link) Nyman, Consp. Fl.
      Eur.: 482 (1879). *Campanula rapunculus* subvar. *verruculosa* (Hoffmanns. & Link) Emb. &
      Maire, Mém. Soc. Sci. Nat. Maroc 17: 52 (1927).
   Campanula calycina Boeber ex Schult. in Roem. & Schult., Syst. Veg. 5: 104 (1819). *Campanula
      rapunculus* var. *calycina* (Boeber ex Schult.) A. DC., Monogr. Campan.: 326 (1830).
   Campanula rapunculus var. *reclinata* Griseb., Spic. Fl. Rumel. 2: 284 (1846).
   Campanula rapunculus var. *hirsuta* Schur, Enum. Pl. Transsilv.: 439 (1866).
   Campanula rapunculus var. *racemosopaniculata* Willk. in Willk. & Lange, Prodr. Fl. Hispan. 2:
      294 (1868), as 'racemoso-paniculata'.
   Campanula rapunculus var. *cymosospicata* Willk. in Willk. & Lange, Prodr. Fl. Hispan. 2: 294
      (1868), as 'cymoso-spicata'.
   Campanula castellana Pau, Not. Bot. Fl. Españ. 1: 24 (1887).
   Campanula rapunculus var. *strigulosa* Batt., Fl. Algérie, Dicotyl.: 576 (1889).
   Campanula rapunculus var. *bracteosa* Willk., Suppl. Prodr. Fl. Hispan.: 130 (1893).
   Campanula rapunculus var. *hirta* Murr, Deutsche Bot. Monatsschr. 17: 151 (1899).
   Campanula rapunculus var. *micrantha* Beyer, Verh. Bot. Vereins Prov. Brandenburg 58: 119
      (1917).

Campanula rapunculus f. *hirsutissima* Faure, Bull. Soc. Hist. Nat. Afrique N. 14: 263 (1923).
  Campanula rapunculus subvar. *hirsutissima* (Faure) Maire in Jahand. & Mire, Cat. Pl.
    Maroc.: 735 (1934).
  Campanula rapunculus subvar. *hirtula* Maire, Bull. Soc. Hist. Nat. Afrique N. 23: 199 (1932).
  Campanula rapunculus "var. vel ssp". *losae* Sennen, Diagn. Nouv.: 85 (1936).
  Campanula rapunculus subvar. *santae* Quézel, Feddes Repert. Spec. Nov. Regni Veg. 56: 40
    (1953).

Campanula rashtiana Parsa, Fl. Iran 10: 226 (1980).
  Iran. 34 IRN. Hemicr. or geophyte.

Campanula raveyi Boiss., Diagn. Pl. Orient. (ser. 1) 4: 32 (1844).
  Turkey. 34 TUR. Ther.

Campanula reatina Lucchese, Fl. Medit. 3: 265 (1993).
  C. Italy. 13 ITA. Hemicr.

Campanula rechingeri Phitos, Oesterr. Bot. Z. 112: 470 (1965).
  C. Aegean Is. (Pipéri). 13 GRC. Hemicr. or biennial. $2n = 34$.

Campanula reiseri Halácsy, Oesterr. Bot. Z. 46: 15 (1896).
  W. Aegean Is. 13 GRC. Hemicr. $2n = 34$.
    Campanula reiseri var. *leonis* Halácsy, Oesterr. Bot. Z. 47: 96 (1897).
    Campanula reiseri var. *pelagia* Phitos, Oesterr. Bot. Z. 112: 469 (1965).

Campanula retrorsa Labill., Icon. Pl. Syr. 5: 5 (1812).
  Turkey to Iraq & Israel. 34 IRQ LBS PAL TUR. Ther.

Campanula reuteriana Boiss. & Balansa in Boiss., Diagn. Pl. Orient. (ser. 2) 3: 108 (1856).
  Turkey to Iran. 34 IRN IRQ LBS TUR. Ther. $2n = 20, 22$.
    Campanula propinqua [unranked] *grandiflora* Milne-Redh., Bot. Mag. 157: tab. 9349 (1934).
    Campanula cecilii Anonymous, Gard. Chron. (ser. 3) 89: 397 (1931).

Campanula reverchonii A. Gray, Syn. Fl. N. Amer. (ed. 2) 2(1): 396 (1886), as 'reverchoni'.
  SC. U.S.A. (C. Texas). 77 TEX. Ther.

Campanula rhodensis A. DC., Monogr. Campan.: 297 (1830). *Campanula drabifolia* var. *rhodensis*
  (A. DC.) Nyman, Consp. Fl. Eur.: 481 (1879).
  E. Aegean Is. (Ródos, Chálki). 34 EAI. Ther. $2n = 16$.

Campanula rhomboidalis L., Sp. Pl.: 165 (1753). *Campanula rotundifolia* subsp. *rhomboidalis* (L.)
  Bonnier, Fl. Ill. France 7: 47 (1924).
  Alps; naturalized in Great Britain & C. Europe. (10) grb 11 AUT bgm cze ger net SWI 12 FRA 13
    ITA. Hemicr. or geophyte. $2n = 34$.
    Campanula azurea Sol. ex Sims, Bot. Mag. 16: tab. 551 (1802).
    Campanula rhomboidalis var. *uniflora* Gaudin, Fl. Helv. 2: 150 (1828).
    Campanula rhomboidalis var. *reflexa* A. DC., Monogr. Campan.: 275 (1830). *Campanula
      rhomboidalis* f. *reflexa* (A. DC.) Voss in Siebert & Voss, Vilm. Blumengärtn. (ed. 3) 1: 568
      (1894).
    Campanula rhomboidalis var. *polypetala* A. DC., Monogr. Campan.: 275 (1830).
    Campanula rubra Anonymous, J. Fl. Jard. 1832: 120, pl. 120 (1832). *Campanula rhomboidalis*
      f. *rubra* (Anonymous) Voss in Siebert & Voss, Vilm. Blumengärtn. (ed. 3) 1:568 (1894).
    Campanula rhomboidalis var. *hispida* St.-Lag. in Cariot, Étude Fl. (ed. 8) 2: 549 (1889).
      Campanula rhomboidalis subvar. *hispida* (St.-Lag.) Rouy, Fl. France 10: 73 (1908).
    Campanula songeonii Chabert, Bull. Herb. Boissier 3: 146 (1895), as 'songeoni'. *Campanula
      rhomboidalis* var. *songeonii* (Chabert) Chabert, Bull. Soc. Bot. France 5: 308 (1908).

Campanula rhomboidalis race *songeonii* (Chabert) Rouy, Fl. France 10: 73 (1908), as 'songeoni'. *Campanula rotundifolia* var. *songeonii* (Chabert) Bonnier, Fl. Ill. France 7: 47 (1924).

Campanula rhomboidalis var. *villosa* Wohlf. in W. D. J. Koch, Syn. Deut. Schweiz. Fl. (ed. 3) 2: 1271 (1895).

Campanula rhomboidalis var. *calycina* Chabert, Bull. Soc. Bot. France 5: 307 (1908).

Campanula rhomboidalis var. *tuberosa* Chabert, Bull. Soc. Bot. France 5: 307 (1908).

Campanula rhomboidalis var. *glabrescens* Vacc., Cat. Pl. Vall. Aoste: 620 (1911).

Campanula rhomboidalis f. *paniculata* Vacc., Cat. Pl. Vall. Aoste: 621 (1911).

Campanula rhomboidalis f. *microphylla* Vacc., Cat. Pl. Vall. Aoste: 621 (1911).

Campanula rhomboidalis f. *macrophylla* Vacc., Cat. Pl. Vall. Aoste: 621 (1911).

Campanula rhomboidalis var. *pilosa* Vacc., Cat. Pl. Vall. Aoste: 621 (1911).

Campanula rhomboidalis f. *cordifolia* Witasek ex Vacc., Cat. Pl. Vall. Aoste: 622 (1911).

Campanula rhomboidalis var. *goudetiana* Beauverd, Bull. Soc. Bot. Genève (ser. 2) 12: 11 (1921).

Campanula rimarum Boiss., Fl. Orient. 3: 931 (1875).
Turkey. 34 TUR. Ther. 2*n* = 20.

Campanula robertsonii Gamble, Bull. Misc. Inform. Kew 1913: 187 (1913).
Myanmar. 41 MYA. Hemicr.

Campanula robinsiae Small, Torreya 26: 35 (1926). *Rotantha robinsiae* (Small) Small, Man. S.E. Fl.: 1289 (1933).
SE. U.S.A. (WC. Florida). 78 FLA. Ther.

Campanula romanica Săvul., Stud. Sp. Campanula: 60 (1916). *Campanula rotundifolia* subsp. *romanica* (Săvul.) Hayek, Repert. Spec. Nov. Regni Veg. Beih. 30(2): 538 (1930).
SE. Romania. 13 ROM. Hemicr. or geophyte. 2*n* = 34.

Campanula rosmarinifolia Kerr, Bull. Misc. Inform. Kew 1936: 35 (1936).
Thailand. 41 THA. Hemicr.

Campanula rotundifolia L., Sp. Pl.: 163 (1753). *Campanula variifolia* Salisb., Prodr. Stirp. Chap. Allerton: 126 (1796). *Campanula heterophylla* Gray, Nat. Arr. Brit. Pl. 2: 408 (1821). *Campanula diversifolia* Dumort., Fl. Belg.: 58 (1827). *Campanula grammosepala* var. *lobophylla* Vuk., Linnaea 26: 324 (1854).
Europe to S. Siberia, Manchuria (Heilongjiang) & Sakhalin; cult.; naturalized in New Zealand, Tierra del Fuego & Falkland Is. 10 DEN FIN FOR GRB IRE NOR SWE 11 AUT BGM CZE GER HUN NET POL SWI 12 FRA SPA 13 ALB BUL ITA ROM YUG 14 BLR BLT RUC RUE RUN RUW UKR 30 ALT BRY CTA IRK KRA TVA WSB YAK 31 AMU KHA MAG SAK 36 CHM (51) nzn nzs (85) ags (90) fal. Hemicr. or geophyte. 2*n* = 34, 68, 102.
Campanula minor Lam., Fl. Franç. 3: 339 (1779).
Campanula rapunculus O. F. Müll., Fl. Dan. 5: pl. 855 (1782); non L., Sp. Pl.: 164 (1753).
Campanula bocconei Vill., Hist. Pl. Dauphiné 1: 304 (1786), as 'bocconi'. *Campanula rotundifolia* var. *bocconei* (Vill.) Lapeyr., Hist. Pl. Pyrénées 104 (1813), as 'boccone'. *Campanula cespitosa* var. *bocconei* (Vill.) Steud., Nomencl. Bot. (ed. 2) 1: 266 (1840), as 'bocconi'. Campanula *rotundifolia* race *bocconei* (Vill.) Rouy, Fl. France 10: 79 (1908), as 'bocconi'.
Campanula tenuifolia Hoffm., Deutschl. Fl. 1: 100 (1791). *Campanula rotundifolia* var. *tenuifolia* (Hoffm.) Opiz, Seznam: 24 (1852).
Campanula linifolia Schrank, Prim. Fl. Salisb.: 70 (Apr-May 1792); *non* L., Fl. Monsp.: 12 56); *nec* Scop., Annus Hist.-Nat. 2: 47 (1769); *nec* Lam., Encycl. 1: 579 (1785).
Campanula angustifolia Lam., Tabl. Encycl. 2: 53 (1796). *Campanula rotundifolia* var. *angustifolia* (Lam.) Vacc., Cat. Pl. Vall. Aoste: 602 (1911).
Campanula rotundifolia var. *stricta* Schumach., Enum. Pl. 1: 69 (1801).
Campanula rotundifolia var. *glabra* Lapeyr., Hist. Pl. Pyrénées: 104 (1813).

Campanula lanceolata Lapeyr., Hist. Pl. Pyrénées: 105 (1813). *Campanula rhomboidalis* var. *lanceolata* (Lapeyr.) Loisel., Fl. Gall. (ed. 2) 1: 141 (1828).

Campanula antirrhina Schleich., Cat. Pl. Helv. (ed. 3): 10 (1815).

Campanula rotundifolia var. *velutina* DC., Fl. Franç. 6: 432 (1815).

Campanula heterodoxa Vest ex Schult. in Roem. & Schult., Syst. Veg. 5: 98 (1819). *Campanula linifolia* var. *heterodoxa* (Vest ex Schult.) Ledeb., Fl. Ross. 2: 888 (1846). *Campanula rotundifolia* var. *heterodoxa* (Vest ex Schult.) Trautv., Trudy Imp. S.-Peterburgsk. Bot. Sada 6: 79 (1879). *Campanula rotundifolia* subsp. *heterodoxa* (Vest ex Schult.) Tacik, Fragm. Florist. Geobot. 17: 233 (1971).

Campanula rotundifolia var. *hirta* Mert. & W. D. J. Koch in Röhl., Deutschl. Fl. (ed. 3) 2: 155 (1826). *Campanula rotundifolia* f. *hirta* (Mert. & W. D. J. Koch) Hruby, Magyar Bot. Lapok 29: 174 (1930); non Hruby, Magyar Bot. Lapok 29: 157 (1930).

Campanula rotundifolia var. *scabriuscula* Mert. & W. D. J. Koch in Röhl., Deutschl. Fl. (ed. 3) 2: 155 (1826). *Campanula rotundifolia* f. *scabriuscula* (Mert. & W. D. J. Koch) Hruby, Magyar Bot. Lapok 29: 174 (1930).

Campanula rotundifolia var. *parviflora* Lej., Comp. Fl. Belg. 1: 181 (1828).

Campanula rotundifolia var. *pubescens* Gaudin, Fl. Helv. 2: 144 (1828).

Campanula linifolia var. *langsdorffiana* A. DC., Monogr. Campan.: 279 (1830). *Campanula langsdorffiana* (A. DC.) Fisch. ex Trautv. & C. A. Mey. in Middend., Reise Sibir. 1 (2, Lfg.3): 60 (1856). *Campanula rotundifolia* var. *langsdorffiana* (A. DC.) Britt., Mem. Torrey Bot. Club 5: 309 (1894). *Campanula linifolia* f. *langsdorffiana* (A. DC.) Voss in Siebert & Voss, Vilm. Blumengärtn. (ed. 3) 1: 566 (1894).

Campanula rotundifolia var. *major* A. DC., Monogr. Campan.: 282 (1830).

Campanula cespitosa var. *imbricata* A. DC., Monogr. Campan.: 284 (1830).

Campanula lostrittii Ten., Stll. Pl. Fl. Neapol.: 96 (1831), as 'lostritti'.

Campanula rotundifolia var. *albiflora* G. Don, Gen. Hist. 3: 759 (1834). *Campanula rotundifolia* f. *albiflora* (G. Don) House, New York State Mus. Bull. 243-244: 21 (1921); non E. L. Rand & Redfield, Fl. Mt. Desert Isl.: 124 (1894); nec Abrom., Biblioth. Bot. 42B: 63 (1899).

Campanula cinerea Hegetschw., Fl. Schweiz: 230 (1838-1839); non L.f., Suppl. Pl.: 139 (1782).

Campanula rotundifolia var. *ovata* Peterm., Anal. Pfl.-Schlüss: 272 (1846). *Campanula rotundifolia* f. *ovata* (Peterm.) Hruby, Magyar Bot. Lapok 29: 174 (1930)

Campanula bielzii Schur, Sert. Fl. Transsilv.: 47 (1853).

Campanula pseudovaldensis Schur, Sert. Fl. Transsilv.: 47 (1853), as 'pseudo-valdensis'.

Campanula lanceolata Schur, Sert. Fl. Transsilv.: 47 (1853); non Lapeyr., Hist. Pl. Pyrénées: 105 (1815); nec J. Presl & C. Presl, Fl. Čech.: 49 (1819); nec Hegetschw., Fl. Schweiz: 230 (1838-39).

Campanula pennina Reut., Compt.-Rend. Trav. Soc. Hallér. 4: 115 (1856). *Campanula rotundifolia* var. *pennina* (Reut.) Nyman, Consp. Fl. Eur.: 479 (1879). *Campanula rotundifolia* subsp. *pennina* (Reut.) Witasek in Vacc., Cat. Pl. Vall. Aoste: 606 (1911).

Campanula rotundifolia var. *vulgaris* Neilr., Fl. Nied.-Oesterr.: 449 (1858).

Campanula rotundifolia var. *confertifolia* Reut., Cat. Pl. Vasc. Genève (ed. 2): 139 (1861).*Campanula confertifolia* (Reut.) Witasek, Abh. K. K. Zool.-Bot. Ges. Wien 1(3): 28 (1902). *Campanula rotundifolia* subsp. *confertifolia* (Reut.) Witasek ex Vacc., Cat. Pl. Vall. Aoste: 605 (1911).

Campanula rotundifolia var. *parviflora* Lange, Haandb. Danske Fl. (ed. 3): 188 (1864); non Lej., Comp. Fl. Belg. 1: 181 (1828).

Campanula rotundifolia var. *angustissima* Schur, Enum. Pl. Transsilv.: 444 (1866).

Campanula rotundifolia var. *bielziana* Schur, Enum. Pl. Transsilv.: 444 (1866).

Campanula arcuata var. *subrhomboidalis* Schur, Enum. Pl. Transsilv.: 445 (1866).

Campanula rotundifolia var. *montana* Syme in Sm., Eng. Bot. (ed. 3) 6: 13 (1866); non Lecoq & Lamotte, Cat. Pl. Plateau Central: 260 (1848). *Campanula rotundifolia* subsp. *montana* P. D. Sell in P. D. Sell & Murrell, Fl. Gr. Brit. Ireland 4: 530 (2006).

Campanula rotundifolia var. *delitschiana* Kuntze,Taschen-Fl. Leipzig: 128 (1867).

Campanula paenina Reut. ex Tissière, Guide Bot. Grand St. Bernard: 63 (1868). *Campanula rotundifolia* subsp. *paenina* (Reut. ex Tissière) Witasek ex Vacc., Cat. Pl. Vall. Aoste: 606 (1911).

Campanula lobata Schloss. & Vuk., Fl. Croat.: 949 (1869). *Campanula pusilla* f. *lobata* (Schloss. & Vuk.) Vuk., Rad Jugoslav. Akad. Znan. 57: 93 (1881).

Campanula rotundifolia f. *grandiflora* J. A. Knapp, Pfl. Galiz.: 173 (1872).

Campanula linifolia var. *major* Timb.-Lagr., Mém. Acad. Sci. Toulouse (ser. 7) 5: 269 (1873). *Campanula rotundifolia* var. *major* (Timb.-Lagr.) Rouy, Fl. France 10: 81 (1908); non A. DC., Monogr. Campan.: 282 (1830); nec Neilr., Fl. Wien: 298 (1846).

Campanula linifolia var. *tenuifolia* Timb.-Lagr., Mém. Acad. Sci. Toulouse (ser. 7) 5: 269 (1873). *Campanula rotundifolia* var. *stenophylla* Rouy, Fl. France 10: 81 (1908); non *Campanula rotundifolia* var. *tenuifolia* A. DC., Monogr. Campan.: 282 (1830); nec (Hoffm.) Schur, Enum. Pl. Transsilv.: 444 (1866).

Campanula pinifolia Uechtr. ex Pančić, Fl. Serbiae: 438 (1874). *Campanula rotundifolia* subvar. *pinifolia* (Uechtr. ex Pančić) Hruby, Magyar Bot. Lapok 29: 157 (1930).

Campanula pinifolia Vuk., Rad Jugoslav. Akad. Znan. 57: 92 (1881); non Uechtr. ex Pančić, Fl. Serbiae: 438 (1874).

Campanula rotundifolia var. *scopulicola* Lamotte, Prodr. Fl. Plat. Centr.: 503 (1881).

Campanula carnica var. *racemosa* Krašan, Mitt. Naturwiss. Vereines Steiermark 24: lxxxi (1888). *Campanula racemosa* (Krašan) Witasek, Abh. K. K. Zool.-Bot. Ges. Wien 1(3): 34 (1902); non Vuk., Linnaea 26: 332 (1854). *Campanula rotundifolia* subsp. *racemosa* (Krašan) Hayek & Hegi in Hegi, Ill. Fl. Mitt.-Eur. 6: 353 (1916). *Campanula balcanica* f. *racemosa* (Krašan) Hruby, Magyar Bot. Lapok 33: 241 (1934).

Campanula solstitialis A. Kern., Verh. K. K. Zool.-Bot. Ges. Wien 38: 669 (1888). *Campanula rotundifolia* var. *solstitialis* (A. Kern.) Beck, Fl. Nieder-Österreich: 1105 (1893). *Campanula rotundifolia* subsp. *solstitialis* (A. Kern.) Witasek ex Hayek & Hegi in Hegi, Ill. Fl. Mitt.-Eur. 6: 353 (1916).

Campanula rotundifolia var. *laxiflora* Beck, Fl. Nieder-Österreich: 1105 (1893).

Campanula rotundifolia f. *soldanelliflora* Voss in Siebert & Voss, Vilm. Blumengärtn. (ed. 3) 1: 567 (1894), as 'soldanellaeflora'. *Campanula rotundifolia* var. *soldanelliflora* (Voss) W. T. Mill. in L. H. Bailey, Cycl. Amer. Hort.: 232 (1900), as 'soldanellaeflora'.

Campanula rotundifolia var. *pygmaea* T. Wulff, Bot. Not. 1896: 53 (1896).

Campanula rotundifolia var. *calisii* Murr, Deutsche Bot. Monatsschr. 17: 84 (1899). *Campanula marchesettii* var. *calisii* (Murr) Dalla Torre & Sarnth., Fl. Tirol. 6(3): 440 (1911).

Campanula rotundifolia var. *bulgarica* Nejceff, God. Sofijsk. Univ. 2: 140 (1906); non *Campanula bulgarica* Witasek, Magyar Bot. Lapok 5: 244 (1906); nec Davidov, Trav. Soc. Bulg. Sci. Nat. 8: 91 (1915). *Campanula rotundifolia* f. *nejceffii* Hayek, Repert. Spec. Nov. Regni Veg. Beih. 30(2): 538 (1930).

Campanula rotundifolia var. *verlotii* Rouy, Fl. France 10: 79 (1908), as 'verloti'.

Campanula rotundifolia var. *reflexa* Syr., Ill. Fl. Mosk. Gub. 3: 221 (1910).

Campanula rotundifolia f. *calvescens* Witasek ex Vacc., Cat. Pl. Vall. Aoste: 603 (1911).

Campanula rotundifolia subsp. *pedemontana* Witasek in Vacc., Cat. Pl. Vall. Aoste: 604 (1911).

Campanula rotundifolia var. *minor* Witasek in Vacc., Cat. Pl. Vall. Aoste: 604 (1911).

Campanula rotundifolia var. *alpicola* Hayek & Hegi in Hegi, Ill. Fl. Mitt.-Eur. 6: 353 (1916). *Campanula rotundifolia* f. *alpicola* (Hayek & Hegi) Hruby, Magyar Bot. Lapok 33: 129 (1934).

Campanula rotundifolia var. *papillifera* Săvul., Stud. Sp. Campanula: 18 (1916).

Campanula rotundifolia var. *papillosa* Beyer, Verh. Bot. Vereins Prov. Brandenburg 58: 118 (1917).

Campanula legionensis Pau, Cavanillesia 1: 65 (1928). *Campanula rotundifolia* var. *legionensis* (Pau) Lacaita, Cavanillesia 3: 23 (1930). *Campanula rotundifolia* subsp. *legionensis* (Pau) Laínz, Bol. Inst. Estud. Asturianos, Supl. Ci. 16: 191 (1973).

Campanula rotundifolia f. *latifrons* Hruby, Magyar Bot. Lapok 29: 157 (1930); non Hruby, Magyar Bot. Lapok 29: 158 (1930).

Campanula rotundifolia f. *subcongesta* Hruby, Magyar Bot. Lapok 29: 157 (1930); non Hruby, Magyar Bot. Lapok 29: 174 (1930). *Campanula rotundifolia* f. *conferta* Sóo, Acta Bot. Acad. Sci. Hung. 12: 366 (1966).

Campanula rotundifolia f. *normalis* Hruby, Magyar Bot. Lapok 29: 157 (1930); non Hruby, Magyar Bot. Lapok 29: 174 (1930).

Campanula rotundifolia f. *graminifolia* Hruby, Magyar Bot. Lapok 29: 157 (1930).

Campanula rotundifolia f. *ovalifolia* Hruby, Magyar Bot. Lapok 29: 157 (1930).

Campanula rotundifolia f. *hirta* Hruby, Magyar Bot. Lapok 29: 157 (1930); non (Mert. & W. D. J. Koch) Hruby, Magyar Bot. Lapok 29: 174 (1930). *Campanula rotundifolia* f. *subhirta* Sóo, Acta Bot. Acad. Sci. Hung. 12: 365 (1966).

Campanula rotundifolia subf. *parviflora* Hruby, Magyar Bot. Lapok 29: 158 (1930). *Campanula rotundifolia* f. *micranthoides* Sóo, Acta Bot. Acad. Sci. Hung. 12: 365 (1966).

Campanula rotundifolia subf. *grandiflora* Hruby, Magyar Bot. Lapok 29: 158 (1930). *Campanula rotundifolia* f. *hrubyana* Sóo, Acta Bot. Acad. Sci. Hung. 12: 365 (1966).

Campanula rotundifolia f. *latifrons* Hruby, Magyar Bot. Lapok 29: 158 (1930); non Hruby, Magyar Bot. Lapok 29: 157 (1930). *Campanula rotundifolia* f. *borhidiana* Sóo, Acta Bot. Acad. Sci. Hung. 12: 366 (1966).

Campanula rotundifolia f. *elata* Hruby, Magyar Bot. Lapok 29: 165 (1930).

Campanula rotundifolia var. *pygmaea* Hruby, Magyar Bot. Lapok 29: 169 (1930); non T. Wulff, Bot. Not. 1896: 53 (1896).

Campanula rotundifolia var. *saxatilis* Hruby, Magyar Bot. Lapok 29: 170 (1930).

Campanula rotundifolia f. *umbrosa* Hruby, Magyar Bot. Lapok 29: 170 (1930).

Campanula rotundifolia subf. *ovalifolia* Hruby, Magyar Bot. Lapok 29: 171 (1930).

Campanula rotundifolia f. *tenerrima* Hruby, Magyar Bot. Lapok 29: 171 (1930).

Campanula rotundifolia subf. *reflexa* Hruby, Magyar Bot. Lapok 29: 171 (1930).

Campanula rotundifolia f. *silvicola* Hruby, Magyar Bot. Lapok 29: 171 (1930).

Campanula rotundifolia f. *luxurians* Hruby, Magyar Bot. Lapok 29: 171 (1930).

Campanula rotundifolia f. *subracemosa* Hruby, Magyar Bot. Lapok 29: 171 (1930).

Campanula rotundifolia subf. *bracteata* Hruby, Magyar Bot. Lapok 29: 171 (1930).

Campanula rotundifolia f. *normalis* Hruby, Magyar Bot. Lapok 29: 174 (1930); non Hruby, Magyar Bot. Lapok 29: 157 (1930).

Campanula rotundifolia f. *subcongesta* Hruby, Magyar Bot. Lapok 29: 174 (1930); non Hruby, Magyar Bot. Lapok 29: 157 (1930).

Campanula rotundifolia f. *frondosa* Hruby, Magyar Bot. Lapok 29: 174 (1930).

Campanula rotundifolia f. *glabrescens* Hruby, Magyar Bot. Lapok 29: 174 (1930).

Campanula rotundifolia subf. *perfoliosa* Hruby, Magyar Bot. Lapok 29: 175 (1930).

Campanula rotundifolia subvar. *tenuissima* Hruby, Magyar Bot. Lapok 29: 175 (1930).

Campanula arctica f. *simplex* Hruby, Magyar Bot. Lapok 29: 186 (1930).

Campanula rotundifolia subf. *brachyantha* Hruby, Magyar Bot. Lapok 33: 128 (1934).

Campanula rotundifolia f. *serpentini* Hruby, Magyar Bot. Lapok 33: 130 (1934).

Campanula tracheliifolia Losa ex Sennen, Diagn. Nouv.: 274 (1936).

Campanula delicatula Sennen & Losa in Sennen, Diagn. Nouv.: 275 (1936); non Boiss., Diagn. Pl. Orient. (ser. 1) 11: 67 (1849).

Campanula caballeroi Sennen & Losa in Sennen, Diagn. Nouv.: 276 (1936).

Campanula rotundifolia f. *laevis* Guin., Bull. Soc. Bot. France 89: 72 (1942).

Campanula rotundifolia f. *subverruculosa* Guin., Bull. Soc. Bot. France 89: 72 (1942).

Campanula rotundifolia var. *flexuosa* C. Vicioso, Anales Jard. Bot. Madrid 6(2): 79 (1946).

Campanula rotundifolia f. *albiflora* Sugaw. ex H. Hara, Enum. Sperm. Jap. 2: 96 (1952); non E. L. Rand & Redfield, Fl. Mt. Desert Isl.: 124 (1894); nec Abrom., Biblioth. Bot. 42B: 63 (1899); nec (G. Don) House, New York State Mus. Bull. 243-244: 21 (1921).

Campanula chinganensis A. I. Baranov, Phyton (Horn) 6: 57 (1955).

Campanula asturica Podlech, Mitt. Bot. Staatssamml. München 8: 213 (1970).

Campanula wiedmannii Podlech, Mitt. Bot. Staatssamml. München 8: 213 (1970).

Campanula rotundifolia var. *altitatrica* Tacik, Fragm. Florist. Geobot. 17: 231 (1971).

Campanula urbionensis Rivas Mart. & G. Navarro, Opusc. Bot. Pharm. Complut. 5: 65 (1989).

Campanula rumeliana (Hampe) Vatke, Linnaea 38: 705 (1874).
  Balkans. 13 BUL GRC TUE. Hemicr. 2n = 32, **34.**
  * *Trachelium rumelianum* Hampe, Flora 20: 234 (1837). *Phyteuma rumelianum* (Hampe) Griseb., Spic. Fl. Rumel. 2: 291 (1846), as 'rumelicum'. *Diosphaera dubia* Friv. ex Buser, Bull. Herb. Boissier 2: 521 (1894). *Diosphaera rumeliana* (Hampe) Bornm.,

Beih. Bot. Centralbl. 38(2): 338 (1921). *Trachelium jacquinii* subsp. *rumelianum*
(Hampe) Tutin, Bot. J. Linn. Soc. 71: 274 (1976). *Campanula jacquinii* subsp.
rumeliana (Hampe) Kit Tan, Endemic Pl. Greece, Peloponnese: 360 (2001).

subsp. **chalcidica** (Buser) Greuter & Burdet, Willdenowia 11: 40 (1981).
Greece. 13 GRC. Hemicr. 2n = 34.
 * *Diosphaera chalcidica* Buser, Bull. Herb. Boissier 2: 521 (1894). *Campanula chalcidica* (Buser)
Engl. in Engl. & Prantl, Nat. Pflanzenfam., Nachtr. 1: 319 (1897). *Diosphaera jacquinii*
var. *chalcidica* (Buser) Hayek, Repert. Spec. Nov. Regni Veg. Beih. 30(2): 553 (1930).
Diosphaera chalcidica var. *denudata* Buser, Bull. Herb. Boissier 2: 522 (1894). *Diosphaera*
jacquinii subf. *denudata* (Buser) Hayek, Repert. Spec. Nov. Regni Veg. Beih. 30(2):
553 (1930).

subsp. **rumeliana**
Balkans. 13 BUL GRC TUE. Hemicr. 2n = 32, 34.
Trachelium rumelianum var. *cinerascens* Vandas, Reliq. Formán.: 387 (1909). *Diosphaera*
rumeliana var. *cinerascens* (Vandas) Hayek, Repert. Spec. Nov. Regni Veg. Beih. 30(2):
552 (1930).
Trachelium jacquinii subsp. *dalgiciorum* Özhatay & Dane in Özhatay, Pl. Balkan Penins.
1: 157 (2001).

Campanula rupestris Sm. in Sibth. & Sm., Fl. Graec. Prodr. 1: 142 (1806). *Campanula*
tomentosa var. *brachyantha* Boiss., Fl. Orient. 3: 898 (1875). *Campanula tomentosa* subsp.
rupestris (Sm.) Nyman, Consp. Fl. Eur.: 476 (1879).
Greece. 13 GRC. Biennial. 2n = 34
Campanula lanuginosa Willd., Enum. Pl.: 213 (1809).

Campanula rupicola Boiss. & Spruner in Boiss., Diagn. Pl. Orient. (ser. 1) 7: 17 (1846).
Greece. 13 GRC. Hemicr. 2n = 32.

Campanula ruscinonensis Timb.-Lagr., Mém. Acad. Sci. Toulouse (ser. 7) 5: 275 (1873).
Campanula linifolia var. *ruscinonensis* (Timb.-Lagr.) Nyman, Consp. Fl. Eur.: 479 (1879).
Campanula rotundifolia race *ruscinonensis* (Timb.-Lagr.) Rouy, Fl. France 10: 79 (1908).
Campanula rotundifolia var. *ruscinonensis* (Timb.-Lagr.) O. Bolòs & Vigo, Collect. Bot.
(Barcelona) 14: 101 (1983).
E. Pyrénées. 12 FRA SPA. Hemicr. or geophyte. 2n = 34.

Campanula sabatia De Not., Prosp. Fl. Ligust.: 52 (1846). *Campanula macrorhiza* var. *sabatia*
(De Not.) Tanfani in Parl., Fl. Ital. 8: 101 (1888). *Campanula macrorhiza* subsp. *sabatia* (De
Not.) Arcang., Fl. Ital. (ed. 2) 637 (1894). *Campanula rotundifolia* var. *sabatia* (De Not.)
Fiori, Fl. Italia 2: 562 (1927).
NW. Italy. 13 ITA. Hemicr. or geophyte. 2n = 34.

Campanula samothracica (Degen) Greuter & Burdet, Willdenowia 11: 40 (1981).
N. Aegean Is. 13 GRC. Hemicr. 2n = 34.
 * *Symphyandra cretica* var. *samothracica* Degen, Oesterr. Bot. Z. 41: 333 (1891).
Symphyandra samothracica (Degen) Halácsy, Oesterr. Bot. Z. 45: 462 (1895).
Symphyandra cretica subsp. *samothracica* (Degen) Hayek, Repert. Spec. Nov. Regni
Veg. Beih. 30(2): 549 (1930).

subsp. **samothracica**
N. Aegean Is. (Samothráki). 13 GRC. Hemicr. 2n = 34.

subsp. **sporadum** (Halácsy) Greuter & Burdet, Willdenowia 11: 40 (1981).
N. Aegean Is. (Vóreioi Sporádes). 13 GRC. Hemicr. 2n = 34.
 * *Symphyandra sporadum* Halácsy, Oesterr. Bot. Z. 45: 461 (1895). *Symphyandra cretica* subsp.
sporadum (Halácsy) Hayek, Repert. Spec. Nov. Regni Veg. Beih. 30(2): 549 (1930).

Campanula sarmatica Ker Gawl., Bot. Reg. 3: pl. 237 (1817).
   Caucasus (Krasnodar, Karacheyevo-Cherkessiya, Kabardino-Balkariya, Severo-Osetiya, Chechnya, Dagestan, Abkhaziya, Gruziya, Azerbaijan). 33 ALL. 2n = 34.

   subsp. **calcarea** (Albov) Ogan., Candollea 50: 287 (1995).
      Caucasus (Abkhaziya). 33 TCS. Hemicr. or geophyte.
      * *Campanula sarmatica* var. *calcarea* Albov, Prodr. Fl. Colchic.: 155 (1895). *Campanula calcarea* (Albov) Albov ex Kharadze, Zametki Sist. Geogr. Rast. 13: 44 (1947).

   subsp. **sarmatica**
      Caucasus (Krasnodar, Kabardino-Balkariya, Severo-Osetiya, Chechnya, Dagestan, Gruziya, Azerbaijan). 33 ALL. Hemicr. or geophyte. 2n = 34.
      Campanula betonicifolia Biehler, Pl. Nov. Herb. Spreng.: 14 (1807), as 'betonicaefolia;' non Sm. in Sibth. & Sm., Fl. Graec. Prodr. 1: 141 (1806). *Campanula gummifera* Willd. ex Schult. in Roem. & Schult., Syst. Veg. 5: 144 (1819). *Marianthemum betonicifolium* Schrank, Flora 7 (Beilage 2): 54 (1824), as 'betonicaefolium'. *Medium gummiferum* (Willd. ex Schult.) Spach, Hist. Nat. Vég. 9: 556 (1838).
      Campanula commutata Schult. in Roem. & Schult., Syst. Veg. 5: 143 (1819).
      Campanula sarmatica var. *glabra* A. DC., Monogr. Campan.: 240 (1830). *Campanula sarmatica* f. *glabra* (A. DC.) Fomin in N. M. Kusn., N. Bush & Fomin, Mater. Fl. Kavkaza 4(6): 37 (1904).
      Campanula albiflora K. Koch, Linnaea 23: 634 (1850).
      Campanula sarmatica var. *subtomentosa* Trautv., Trudy Imp. S.-Peterburgsk. Bot. Sada 6: 67 (1879).
      Campanula brotherorum Feer, J. Bot. 28: 273 (1890).
      Campanula sarmatica var. *gracilis* Fomin in N. M. Kusn., N. Bush & Fomin, Mater. Fl. Kavkaza 4(6): 37 (1904).
      Campanula siegizmundii Fed. in Kom., Fl. URSS 24: 464 (1957), as 'siegizmundi'.

   subsp. **ramosissima** (Sommier & Levier) Ogan., Candollea 50: 287 (1995).
      Caucasus (Karacheyevo-Cherkessiya). 33 NCS. Hemicr.
      * *Campanula sarmatica* var. *ramosissima* Sommier & Levier, Enum. Pl. Cauc.: 316 (1900). *Campanula sarmatica* f. *ramosissima* (Sommier & Levier) Fomin in N. M. Kusn., N. Bush & Fomin, Mater. Fl. Kavkaza 4(6): 37 (1904). *Campanula sommieri* Kharadze, Zametki Sist. Geogr. Rast. 13: 46 (1947); non *Campanula ramosissima* Sm. in Sibth. & Sm., Fl. Graec. Prodr. 1: 137 (1806).
      Campanula sarmatica f. *tenuicaulis* Sommier & Levier, Enum. Pl. Cauc.: 317 (1900).

   subsp. **woronovii** (Kharadze) Ogan., Candollea 50: 287 (1995).
      Caucasus (Krasnodar, Karacheyevo-Cherkessiya). 33 NCS. Hemicr.
      * *Campanula woronovii* Kharadze, Zametki Sist. Geogr. Rast. 15: 22 (1949).

Campanula sartorii Boiss. & Heldr. in Boiss., Fl. Orient. 3: 919 (1875).
   S. Aegean Isl. (Ándros). 13 GRC. Hemicr. or geophyte. 2n = 34.

Campanula sauvagei Quézel, Feddes Repert. Spec. Nov. Regni Veg. 56: 37 (1953).
   Morocco. 20 MOR. Hemicr.

Campanula saxatilis L., Sp. Pl.: 167 (1753). Pentropis saxatilis (L.) Raf., Fl. Tellur. 2: 80 (1837).
   SW. Aegean Is. & Kriti. 13 GRC KRI. Hemicr. or geophyte. 2n = 34.

   subsp. **cytherea** Rech. f. & Phitos, Oesterr. Bot. Z. 112: 483 (1965).
      SW. Aegean Is. (Kýthira, Adikythira). 13 GRC. Hemicr. or geophyte.

   subsp. **saxatilis**
      Kriti. 13 KRI. Hemicr. or geophyte. 2n = 34.

Campanula saxifragoides Doum., Compt. Rend. Assoc. Franç. Avancem. Sci. 27(2): 455 (1898).
   N. Africa. 20 ALG MOR. Hemicr. 2n = 14, 16.

Campanula saxifragoides f. *latifolia* Litard. & Maire, Mém. Soc. Sci. Nat. Maroc 26: 29 (1931).

Campanula saxifragoides var. *guinetii* Quézel, Feddes Repert. Spec. Nov. Regni Veg. 56: 36 (1953).

Campanula saxonorum Gand., Bull. Soc. Bot. France 65: 54 (1918).
Turkey. 34 TUR. Ther.

Campanula scabrella Engelm., Bot. Gaz. (Crawfordsville) 6: 237 (1881).
W. U.S.A. (Washington to California & Montana). 73 IDA MNT ORE WAS 76 CAL. Hemicr.

Campanula scheuchzeri Vill., Prosp. Hist. Pl. Dauphiné: 22 (1779), as 'scheuczeri'.
Campanula valdensis All., Fl. Pedem. 1: 109 (1785). *Campanula linifolia* var. *scheuchzeri* (Vill.) Willd., Sp. Pl. 1: 893 (1798). *Campanula rotundifolia* subsp. *scheuchzeri* (Vill.) Lapeyr., Hist. Pl. Pyrénées: 103 (1813). *Campanula linifolia* var. *glabra* Schult. in Roem. & Schult., Syst. Veg. 5: 101 (1819). *Campanula valdensis* var. *scheuchzeri* (Vill.) Gaudin, Fl. Helv. 2: 147 (1828). *Campanula linifolia* var. *valdensis* A. DC., Monogr. Campan.: 279 (1830). *Campanula prostrata* Dulac, Fl. Hautes-Pyrénées: 457 (1867). *Campanula linifolia* subsp. *valdensis* (A. DC.) Nyman, Consp. Fl. Eur.: 479 (1879). *Campanula rotundifolia* var. *scheuchzeri* (Vill.) Fiek, Fl. Schlesien: 286 (1881). *Campanula linifolia* var. *humilis* St.-Lag. in Cariot, Étude Fl. (ed. 8) 2: 550 (1889). *Campanula linifolia* f. *scheuchzeri* (Vill.) Voss in Siebert & Voss, Vilm. Blumengärtn. (ed. 3) 1: 566 (1894). *Campanula linifolia* f. *valdensis* (A. DC.) Voss in Siebert & Voss, Vilm. Blumengärtn. (ed. 3) 1: 566 (1894). *Campanula scheuchzeri* subvar. *valdensis* (A. DC.) Rouy, Fl. France 10: 77 (1908). *Campanula scheuchzeri* var. *valdensis* (A. DC.) Vacc., Cat. Pl. Vall. Aoste: 611 (1911).
Pyrénées to S. Germany & Balkans. 11 GER SWI 12 FRA SPA 13 ALB BUL ITA SIC YUG. Hemicr. or geophyte. 2n = 68, 102.
Campanula schleicheri Suter, Helvet. Fl. 1: 124 (1802). *Campanula scheuchzeri* var. *schleicheri* (Suter) Beck, Fl. Nieder-Österreich: 1106 (1893).
Campanula linifolia var. *pubescens* Show, Giorn. Fis. (ser. 2) 7: 29 (1824).
Campanula valdensis var. *villosa* Gaudin, Fl. Helv. 2: 146 (1828).
Campanula valdensis var. *schleicheri* Gaudin, Fl. Helv. 2: 147 (1828).
Campanula scheuchzeri var. *glabra* W. D. J. Koch, Syn. Deut. Schweiz. Fl.: 468 (1837).
Campanula scheuchzeri var. *hirta* W. D. J. Koch, Syn. Deut. Schweiz. Fl.: 468 (1837).
Campanula schleicheri Hegetschw., Fl. Schweiz: 231 (1838-39); non Suter, Helvet. Fl. 1: 124 (1802). *Campanula hegetschweileri* Bech., Ber. Schweiz. Bot. Ges. 52: 531 (1942).
Campanula scheuchzeri var. *hirta* Gren. & Godr., Fl. Fr. 2: 416 (1853); non W. D. J. Koch, Syn. Deut. Schweiz. Fl.: 468 (1837).
Campanula monantha Schur, Sert. Fl. Transsilv.: 47 (1853). *Campanula scheuchzeri* var. *monantha* (Schur) Nyman, Consp. Fl. Eur.: 479 (1879).
Campanula consanguinea Schott in Schott, Nyman & Kotschy, Analect. Bot.: 8 (1854). *Campanula carnica* subsp. *consanguinea* (Schott) Nyman, Consp. Fl. Eur., Suppl. 2: 209 (1889). *Campanula scheuchzeri* var. *consanguinea* (Schott) Dalla Torre & Sarnth., Fl. Tirol 6(3): 448 (1911).
Campanula dilecta Schott in Schott, Nyman & Kotschy, Analect. Bot.: 8 (1854). *Campanula rotundifolia* subsp. *dilecta* (Schott) Nyman, Consp. Fl. Eur.: 479 (1879). *Campanula rotundifolia* var. *dilecta* (Schott) Wohlf. in W. D. J. Koch, Syn. Deut. Schweiz. Fl. (ed. 3) 2: 1269 (1893). *Campanula scheuchzeri* var. *dilecta* (Schott) Vacc., Cat. Pl. Vall. Aoste: 611 (1911). *Campanula rotundifolia* f. dilecta (Schott) Hruby, Magyar Bot. Lapok 29: 189 (1930).
Campanula styriaca Schott in Schott, Nyman & Kotschy, Analect. Bot.: 9 (1854). *Campanula scheuchzeri* var. *styriaca* (Schott) Nyman, Consp. Fl. Eur.: 479 (1879). *Campanula scheuchzeri* subf. *styriaca* (Schott) Hruby, Magyar Bot. Lapok 29: 193 (1930), as 'stiriaca'.

Campanula malyi Schott in Schott, Nyman & Kotschy, Analect. Bot.: 11 (1854).
 Campanula rotundifolia var. *malyi* (Schott) Nyman, Consp. Fl. Eur.: 479 (1879).
 Campanula scheuchzeri f. *malyi* (Schott) Dalla Torre & Sarnth., Fl. Tirol. 6(3): 448 (1911).
Campanula rotundifolia var. *grandiflora* Neilr., Fl. Nied.-Oesterr.: 448 (1858); non
 Wimm., Fl. Schles.: 241 (1840).
Campanula scheuchzeri var. *glabra* Schur, Verh. Mitth. Siebenbürg. Vereins Naturwiss.
 Hermanstadt 10: 138 (1859); non W. D. J. Koch, Syn. Deut. Schweiz. Fl.: 468 (1837).
Campanula scheuchzeri var. *minima* Schur, Verh. Mitth. Siebenbürg. Vereins Naturwiss.
 Hermanstadt 10: 138 (1859).
Campanula scheuchzeri var. *grandiflora* Schur, Enum. Pl. Transsilv.: 443 (1866).
Campanula scheuchzeri var. *hirsuta* Schur, Enum. Pl. Transsilv.: 443 (1866).
Campanula rotundifolia var. *balcanica* Adamović, Allg. Bot. Z. Syst. 5: 73 (1899).
 Campanula balcanica (Adamović) Hruby, Magyar Bot. Lapok 29: 234 (1930).
Campanula kerneri Witasek ex A. Kern. & Fritsch., Sched. Fl. Exc. Austro-Hung. 9: 38
 (1902). *Campanula scheuchzeri* var. *kerneri* (Witasek ex A. Kern. & Fritsch.) Dalla Torre
 & Sarnth., Fl. Tirol 6(3):448 (1911). *Campanula scheuchzeri* subsp. *kerneri* (Witasek ex
 A. Kern. & Fritsch.) Vacc., Cat. Pl. Vall. Aoste: 610 (1911). *Campanula scheuchzeri* subf.
 kerneri (Witasek ex A. Kern. & Fritsch.) Hruby, Magyar Bot. Lapok 29: 193 (1930)
Campanula scheuchzeri var. *glareosa* Vacc., Cat. Pl. Vall. Aoste: 610 (1911).
Campanula scheuchzeri var. *calycina* Witasek ex Vacc., Cat. Pl. Vall. Aoste: 611 (1911).
Campanula scheuchzeri f. *latifolia* Witasek ex Vacc., Cat. Pl. Vall. Aoste: 611 (1911).
Campanula scheuchzeri f. *elatior* Witasek ex Vacc., Cat. Pl. Vall. Aoste: 613 (1911).
Campanula scheuchzeri var. *villarsiana* Hayek, Fl. Steiermark 2: 455 (1912). *Campanula
 scheuchzeri* subvar. *villarsiana* (Hayek) Hruby, Magyar Bot. Lapok 29: 193 (1930).
Campanula scheuchzeri f. *oblongifolia* Beyer, Verh. Bot. Vereins Prov. Brandenburg 58:
 119 (1917).
Campanula rotundifolia subsp. *pseudarctica* Hruby, Magyar Bot. Lapok 29: 189 (1930).
Campanula scheuchzeri subf. *altior* Hruby, Magyar Bot. Lapok 29: 194 (1930).
Campanula scheuchzeri f. *ovalifolia* Hruby, Magyar Bot. Lapok 29: 194 (1930).
Campanula scheuchzeri f. *rhombifolia* Hruby, Magyar Bot. Lapok 29: 194 (1930).
Campanula scheuchzeri var. *luxurians* Hruby, Magyar Bot. Lapok 29: 195 (1930).
Campanula scheuchzeri f. *reflectans* Hruby, Magyar Bot. Lapok 29: 195 (1930).
Campanula scheuchzeri subf. *hirtescens* Hruby, Magyar Bot. Lapok 29: 195 (1930).
Campanula scheuchzeri var. *hirta* Hruby, Magyar Bot. Lapok 29: 196 (1930); non W. D. J.
 Koch, Syn. Deut. Schweiz. Fl.: 468 (1837); nec Gren. & Godr., Fl. Fr. 2: 416 (1853).
Campanula scheuchzeri f. *hirsutissima* Hruby, Magyar Bot. Lapok 29: 196 (1930).
Campanula scheuchzeri subsp. *rhaetica* Hruby, Magyar Bot. Lapok 29: 197 (1930).
Campanula scheuchzeri f. *humilior* Hruby, Magyar Bot. Lapok 29: 197 (1930).
Campanula scheuchzeri var. *intercedens* Hruby, Magyar Bot. Lapok 29: 197 (1930).
Campanula scheuchzeri f. *humilis* Hruby, Magyar Bot. Lapok 29: 197 (1930).
Campanula scheuchzeri f. *parvula* Hruby, Magyar Bot. Lapok 33: 134 (1934).
Campanula scheuchzeri subsp. *susplugasii* Braun-Blanq., Commun. Stat. Int. Géobot.
 Médit. Montpellier 87: 231 (1945), as 'susplugasi'. *Campanula scheuchzeri* var.
 susplugasii (Braun-Blanq.) O. Bolòs & Vigo, Collect. Bot. (Barcelona) 14: 102 (1983).

Campanula schimaniana Rech. f., Pl. Syst. Evol. 144: 149 (1984).
 Afghanistan & Pakistan. 34 AFG 40 PAK. Hemicr.

Campanula sciathia Phitos, Oesterr. Bot. Z. 111: 215 (1964).
 W. Aegean Is. (Skíathos). 13 GRC. Hemicr. or biennial. 2n = 34.

Campanula sclerophylla (Kolak.) Ogan., Candollea 50: 287 (1995).
 Caucasus (Krasnodar). 33 NCS.
 Mzymtella sclerophylla Kolak., Soobšč. Akad. Nauk Gruzinsk. SSR 103: 151 (1981).

Campanula sclerotricha Boiss., Diagn. Pl. Orient. (ser. 1) 11: 66 (1849).
Turkey & Caucasus (Nakhichevan) to Afghanistan. 33 TCS 34 AFG IRN IRQ TUR. Hemicr.
or biennial. 2n = 34.
Campanula sclerotricha subsp. *tchitounyi* Azn., Magyar Bot. Lapok. 16: 23 (1917).
Campanula striata Kitam., Acta Phytotax. Geobot. 16: 132 (1956).

Campanula scoparia (Boiss. & Hausskn.) Damboldt, Willdenowia 5: 241 (1969).
Turkey. 34 TUR. Hemicr.
* *Podanthum scoparium* Boiss. & Hausskn. in Boiss., Fl. Orient. 3: 953 (1875). *Asyneuma scoparium* (Boiss. & Hausskn.) Bornm., Beih. Bot. Centralbl. 38(2): 349 (1921).

Campanula scopelia Phitos, Oesterr. Bot. Z. 111: 214 (1964).
W. Aegean Isl. (Skópelos). 13 GRC. Hemicr. 2n = 34.

Campanula scouleri Hook. ex A. DC., Monogr. Campan.: 312 (1830).
NW. North America (SE. Alaska to California). 70 ASK 71 BRC 73 ORE WAS 76 CAL. Hemicr.
Campanula scouleri var. *glabra* Hook., Fl. Bor.-Amer. 2: 28 (1834).
Campanula scouleri var. *hirsutula* Hook., Fl. Bor.-Amer. 2: 28 (1834).

Campanula scutellata Griseb., Spic. Fl. Rumel. 2: 282 (1846).
Balkans. 13 BUL GRC YUG. Ther. 2n = 14.

Campanula secundiflora Vis. & Pančić, Mem. Inst. Veneto 10: 442 (1861).
Serbia. 13 YUG. Hemicr.

Campanula × segusina Beyer, Verh. Bot. Vereins Prov. Brandenburg 58: 117 (1917). C.
rotundifolia × C. scheuchzeri
W. Alps. 12 FRA 13 ITA. Hemicr. or geophyte.

Campanula semisecta Murb., Acta Univ. Lund. 33(12): 115 (1897). *Campanula dichotoma*
subsp. *semisecta* (Murb.) Rivas Mart., Itinera Geobot. 15: 698 (2002).
Spain. 12 SPA. Ther.
Campanula semisecta var. *basiclada* Murb., Acta Univ. Lund. 33(12): 116 (1897).

Campanula seraglio Kit Tan & Sorger, Notes Roy. Bot. Gard. Edinburgh 41: 528 (1984).
Turkey. 34 TUR. Hemicr.

Campanula serhouchensis Dobignard, J. Bot. Soc. Bot. France 20: 27 (2002).
Morocco. 20 MOR. Hemicr.

Campanula serrata (Kit.) Hendrych, Taxon 11: 123 (1962).
C. Europe. 11 CZE POL 12 FRA SPA 13 ROM YUG 14 UKR. Hemicr. or geophyte. 2n = 34.
* *Thesium serratum* Kit. in Schult., Oestr. Fl. (ed. 2) 1: 437 (1814).

subsp. **recta** (Dulac) Podlech, Feddes Repert. 71: 132 (1965).
Pyrénées to C. France. 12 FRA SPA. Hemicr. or geophyte. 2n = 34.
* *Campanula linifolia* Lam., Encycl. 1: 579 (1785); non L., Fl. Monsp.: 12 (1759); nec
Scop., Annus Hist.-Nat. 2: 47 (1769). *Campanula recta* Dulac, Fl. Hautes-Pyrénées:
458 (1867). *Campanula rotundifolia* subsp. *linifolia* Arcang., Comp. Fl. Ital.: 454
(1882); non Lapeyr., Hist. Pl. Pyrénées: 104 (1813).
Campanula linifolia var. *rohdii* Lecoq & Lamotte, Cat. Pl. Plateau Central: 260 (1848).
Campanula linifolia var. *ovalifolia* St.-Lag. in Cariot, Étude Fl. (ed. 8) 2: 550 (1889).
Campanula rotundifolia var. *ovalifolia* (St.-Lag.) Rouy, Fl. France 10: 81 (1908)

subsp. **serrata**
Carpathians. 11 CZE POL 13 BUL ROM YUG 14 UKR. Hemicr. or geophyte. 2n = 34.
Campanula microphylla Kit. in Schult., Oestr. Fl. (ed. 2) 1: 400 (1814); non Cav., Anales

Ci. Nat. 3: 19 (1800). *Campanula kitaibeliana* Schult. in Roem. & Schult., Syst. Veg. 5: 90 (1819).

Campanula redux Schott in Schott, Nyman & Kotschy, Analect. Bot.: 9 (1854). *Campanula scheuchzeri* subsp. *redux* (Schott) Nyman, Consp. Fl. Eur.: 479 (1879). *Campanula napuligera* var. *redux* (Schott) Hruby, Magyar Bot. Lapok 29: 219 (1930).

Campanula arcuata Schur, Verh. Mitth. Siebenbürg. Vereins Naturwiss. Hermanstadt 10: 138 (1859). *Campanula rotundifolia* var. *arcuata* (Schur) Nyman, Consp. Fl. Eur.: 479 (1879). *Campanula lanceolata* subsp. *arcuata* (Schur) Simonkai, Enum. Fl. Transsilv.: 385 (1887). *Campanula pseudolanceolata* var. *arcuata* (Schur) Porcius, Analele Acad. Române (ser. 2) 14: 196 (1893). *Campanula napuligera* f. *arcuata* (Schur) Hruby, Magyar Bot. Lapok 29: 217 (1930), as 'arenata'. *Campanula napuligera* var. *arcuata* (Schur) Morariu in Săvul., Fl. Reipubl. Popul. Roman. 9: 93 (1964).

Campanula rhomboidalis var. *angustifolia* Neilr., Aufz. Ungarn Slavon. Gefässpfl.: 145 (1865).

Campanula hornungiana Schur, Enum. Pl. Transsilv.: 442 (1866). *Campanula lanceolata* var. *hornungiana* (Schur) Simonkai, Enum. Fl. Transsilv.: 385 (1887). *Campanula pseudolanceolata* var. *hornungiana* (Schur) Porcius, Analele Acad. Române (ser. 2) 14: 202 (1893). *Campanula napuligera* var. *hornungiana* (Săvul.) Morariu in Săvul., Fl. Reipubl. Popul. Roman. 9: 97 (1964).

Campanula napuligera Schur, Enum. Pl. Transsilv.: 444 (1866).

Campanula rotundifolia var. *dentata* Schur, Enum. Pl. Transsilv.: 444 (1866).

Campanula rotundifolia var. *alpina* Schur, Enum. Pl. Transsilv.: 444 (1866); non Tuck., Amer. J. Sci. Arts 45: 27 (1843).

Campanula rotundifolia var. *grandiflora* J. A. Knapp, Pfl. Galiz.: 173 (1872); non Wimm., Fl. Schles.: 241 (1840); nec Neilr., Fl. Nied.-Oesterr.: 448 (1858).

Campanula pseudolanceolata Pant., Magyar Növényt. Lapok 6: 162 (1882). *Campanula rhomboidalis* subsp. *pseudolanceolata* (Pant.) Nyman, Consp. Fl. Eur., Suppl. 2: 208 (1889). *Campanula polymorpha* f. *pseudolanceolata* (Pant.) Hruby, Magyar Bot. Lapok 29: 203 (1930).

Campanula pseudolanceolata f. *elatior* Săvul., Stud. Sp. Campanula: 78 (1916). *Campanula napuligera* var. *elatior* (Săvul.) Morariu in Săvul., Fl. Reipubl. Popul. Roman. 9: 94 (1964).

Campanula pseudolanceolata f. *umbraticola* Săvul., Stud. Sp. Campanula: 78 (1916).

Campanula pseudolanceolata f. *transsilvanica* Săvul., Stud. Sp. Campanula: 78 (1916). *Campanula napuligera* f. *transsilvanica* (Săvul.) Morariu in Săvul., Fl. Reipubl. Popul. Roman. 9: 93 (1964).

Campanula pseudolanceolata f. *minima* Săvul., Stud. Sp. Campanula: 81 (1916). *Campanula napuligera* f. *minima* (Săvul.) Morariu in Săvul., Fl. Reipubl. Popul. Roman. 9: 97 (1964).

Campanula pseudolanceolata f. *albiflora* Săvul., Stud. Sp. Campanula: 81 (1916).

Campanula pseudolanceolata var. *porcii* Săvul., Stud. Sp. Campanula: 84 (1916).

Campanula pseudolanceolata subsp. *semiamplexicaulis* Săvul., Stud. Sp. Campanula: 86 (1916). *Campanula napuligera* f. *semiamplexicaulis* (Săvul.) Morariu in Săvul., Fl. Reipubl. Popul. Roman. 9: 98 (1964).

Campanula napuligera var. *stricta* Hruby, Magyar Bot. Lapok 29: 216 (1930).

Campanula napuligera f. *latifrons* Hruby, Magyar Bot. Lapok 29: 216 (1930).

Campanula napuligera f. *simplex* Hruby, Magyar Bot. Lapok 29: 216 (1930).

Campanula napuligera f. *angustifrons* Hruby, Magyar Bot. Lapok 29: 217 (1930).

Campanula napuligera f. *genuina* Hruby, Magyar Bot. Lapok 29: 217 (1930); non Hruby, Magyar Bot. Lapok 29: 220 (1930).

Campanula napuligera subf. *angustifrons* Hruby, Magyar Bot. Lapok 29: 217 (1930).

Campanula napuligera subf. *latifrons* Hruby, Magyar Bot. Lapok 29: 218 (1930).

Campanula napuligera f. *humilis* Hruby, Magyar Bot. Lapok 29: 218 (1930).

Campanula napuligera f. *glabrescens* Hruby, Magyar Bot. Lapok 29: 218 (1930); non Hruby, Magyar Bot. Lapok 29: 221 (1930).

Campanula napuligera var. *umbrosa* Hruby, Magyar Bot. Lapok 29: 218 (1930).

Campanula napuligera f. *intermedia* Hruby, Magyar Bot. Lapok 29: 218 (1930).

Campanula napuligera var. *hirsuta* Hruby, Magyar Bot. Lapok 29: 220 (1930).

Campanula napuligera f. *genuina* Hruby, Magyar Bot. Lapok 29: 220 (1930); non Hruby,
Magyar Bot. Lapok 29: 217 (1930).

Campanula napuligera subf. *brachyantha* Hruby, Magyar Bot. Lapok 29: 220 (1930).

Campanula napuligera subf. *tenella* Hruby, Magyar Bot. Lapok 29: 220 (1930).

Campanula napuligera f. *robusta* Hruby, Magyar Bot. Lapok 29: 220 (1930).

Campanula napuligera f. *glabrescens* Hruby, Magyar Bot. Lapok 29: 221 (1930); non
Hruby, Magyar Bot. Lapok 29: 218 (1930).

Campanula napuligera var. *longisepala* Nyár., Bul. Grăd. Bot. Univ. Cluj 14: 95 (1934).
Campanula napuligera f. *longisepala* (Nyár.) Morariu in Săvul., Fl. Reipubl. Popul.
Roman. 9: 101 (1964).

Campanula napuligera var. *scheuchzeriformis* Nyár., Bul. Grăd. Bot. Univ. Cluj 14: 95
(1934). *Campanula napuligera* f. *scheuchzeriformes* (Nyár.) Morariu in Săvul., Fl.
Reipubl. Popul. Roman. 9: 98 (1964).

Campanula napuligera f. *stenophylloides* Nyár., Bul. Grăd. Bot. Univ. Cluj 14: 96 (1934).
Campanula napuligera var. *stenophylloides* (Nyár.) Morariu in Săvul., Fl. Reipubl.
Popul. Roman. 9: 98 (1964).

Campanula napuligera f. *parvula* Morariu in Săvul., Fl. Reipubl. Popul. Roman. 9: 96 (1964).

Campanula napuligera var. *savulescui* Morariu in Săvul., Fl. Reipubl. Popul. Roman. 9: 96
(1964). *Campanula napuligera* f. *savulescui* Morariu in Săvul., Fl. Reipubl. Popul.
Roman. 9: 97 (1964).

Campanula napuligera var. *alpiniformis* Nyár. ex Morariu in Săvul., Fl. Reipubl. Popul.
Roman. 9: 96 (1964).

Campanula napuligera f. *setulosa* Morariu in Săvul., Fl. Reipubl. Popul. Roman. 9: 97
(1964).

Campanula sharsmithiae Morin, Madroño 27: 159 (1980).
W. U.S.A. (C. California). 76 CAL. Ther. 2n = 34.

Campanula shetleri Heckard, Madroño 20: 231 (1970).
W. U.S.A. (N. California). 76 CAL. Hemicr. 2n = 34.

Campanula sibirica L., Sp. Pl.: 167 (1753). *Campanula undulata* Moench, Suppl. Meth.: 189
(1802); non L. f., Suppl. Pl.: 142 (1782). *Campanula paniculata* Pohl, Tent. Fl. Bohem. 1:
207 (1809); non L. f., Suppl. Pl.: 139 (1782). *Medium sibiricum* (L.) Spach, Hist. Nat. Vég.
9: 554 (1838). *Nenningia paniculata* Opiz, Seznam: 68 (1852). *Marianthemum sibiricum* (L.)
Schur, Sert. Fl. Transsilv.: 48 (1853). *Campanula macrorrhiza* var. *chymatophylla* Vuk.,
Linnaea 26: 330 (1854).
Germany to Italy, Turkey, Iran, Kazakhstan, W. China (Xinjiang) & W. Siberia. 11 AUT
CZE GER HUN POL 13 ALB BUL ITA ROM YUG 14 BLR KRY RUC RUE RUS RUW UKR
30 WSB 32 KAZ 33 ALL 34 IRN TUR 36 CHX. Hemicr. or biennial. 2n = 26, 34, 102.

subsp. **brassicifolia** (Sommier & Levier) Ogan., Candollea 50: 284 (1995).
Caucasus (Gruziya). 33 TCS. Hemicr.
 * *Campanula brassicifolia* Sommier & Levier, Nuovo Giorn. Bot. Ital. (n.s.) 2: 94 (1895).

subsp. **charadzae** (Grossh.) Ogan., Candollea 50: 284 (1995).
Caucasus (Chechnya, Dagestan, Gruziya). 33 ALL. Hemicr.
 * *Campanula charadzae* Grossh., Dokl. Akad. Nauk Armjansk. SSR 7: 169 (1947).
Campanula longistyla f. *parviflora* Fomin in N. M. Kusn., N. Bush & Fomin, Mater. Fl.
Kavkaza 4(6): 21 (1904).

subsp. **charkeviczii** (Fed.) Fed., Bot. J. Linn. Soc. 67: 281 (1973).
Krym. 14 KRY. Hemicr. or biennial.
 * *Campanula charkeviczii* Fed. in Kom., Fl. URSS 24: 459 (1957).

subsp. **ciscaucasica** (Kharadze) Ogan., Candollea 50: 283 (1995).
Caucasus (Karacheyevo-Cherkessiya, Kabardino-Balkariya). 33 NCS.
* Campanula ciscaucasica Kharadze, Zametki Sist. Geogr. Rast. 17: 100 (1953).

subsp. **divergens** (Waldst. & Kit. ex Willd.) Nyman, Consp. Fl. Eur.: 476 (1879).
SE. Europe. 11 CZE HUN 13 ALB ITA YUG. Biennial. 2n = 34.
* *Campanula divergens* Waldst. & Kit. ex Willd., Enum. Pl.: 212 (1809). *Campanula
macrorrhiza* var. *trianthemos* Vuk., Linnaea 26: 330 (1854). *Marianthemum divergens*
(Waldst. & Kit.) Schur, Enum. Pl. Transsilv.: 433 (1866). *Campanula sibirica* var.
divergens (Waldst. & Kit. ex Willd.) Trautv., Pl. Messes: 68 (1874). *Campanula sibirica*
var. *major* Boiss., Fl. Orient. 3: 901 (1875).
Campanula spathulata Waldst. & Kit., Descr. Icon. Pl. Hung. 3: 286 (1812); non
Campanula spatulata Sm. in Sibth. & Sm., Fl. Graec. Prodr. 1: 137 (1806).
Marianthemum spathulatum Schur, Sert. Fl. Transsilv.: 48 (1853). *Campanula divergens*
var. *spathulata* (Schur) Schur, Enum. Pl. Transsilv.: 433 (1866). *Campanula sibirica* var.
spathulata (Schur) Nyman, Consp. Fl. Eur.: 476 (1879).
Campanula nutans Vahl ex Hornem., Hort. Bot. Hafn.: 201 (1815); non Lam., Fl. Franç.
3: 336 (1779).
Campanula divergens var. *cernua* Schult. in Roem. & Schult., Syst. Veg. 5: 146 (1819).
Marianthemum pannonicum Schrank, Denkschr. Königl.-Baier Bot. Ges. Regensburg 2: 34
(1822). *Campanula sibirica* var. *pannonica* (Schrank) Nyman, Consp. Fl. Eur.: 476 (1879).
Campanula sibirica var. *divergentiformis* Jáv., Magyar Fl.: 1073 (1925). *Campanula sibirica*
subsp. *divergentiformis* (Jáv.) Domin, Preslia 13-15: 222 (1936).

subsp. **elatior** (Fomin) Fed., Bot. J. Linn. Soc. 67: 281 (1973).
Ukraine to S. Russia & Caucasus (Krasnodar, Stavropol, Karacheyevo-Cherkessiya,
Kabardino-Balkariya, Severo-Osetiya, Chechnya). 14 RUS UKR 33 NCS. Biennial.
* *Campanula sibirica* f. *elatior* Fomin in N. M. Kusn., N. Bush & Fomin, Mater. Fl. Kavkaza
4(6): 23 (1904). *Campanula elatior* (Fomin) Grossh., Opred. Rast. Kavkaz. 418 (1949);
non Hoffmanns. & Link, Fl. Portug. 2: 11 (1813-20). *Campanula praealta* Galushko,
Fl. Severn. Kavkaza 3: 147 (1980).

subsp. **hohenackeri** (Fisch. & C. A. Mey.) Damboldt, Notes Roy. Bot. Gard. Edinburgh 35:
45 (1976).
Turkey, Caucasus (Kabardino-Balkariya, Severo-Osetiya, Chechnya, Dagestan, Gruziya,
Armenia, Azerbaijan, Nagorno Karabakh), Iran. 33 ALL 34 IRN TUR. Hemicr. or
biennial. 2n = 34.
Campanula parviflora Lam., Encycl. 1: 588 (1785). *Campanula sibirica* var. *parviflora*
(Lam.) Trautv., Trudy Imp. S.-Peterburgsk. Bot. Sada 6: 62 (1879). *Campanula sibirica*
f. *parviflora* (Lam.) Fomin in N. M. Kusn., N. Bush & Fomin, Mater. Fl. Kavkaza 4(6):
27 (1904).
* *Campanula hohenackeri* Fisch. & C. A. Mey., Index Sem. Hort. Petrop. 9 (suppl.): 9
(1844). *Campanula sibirica* var. *hohenackeri* (Fisch. & C. A. Mey.) Trautv., Trudy Imp.
S.-Peterburgsk. Bot. Sada 2: 564 (1873), as 'hohenackeriana'.
Campanula sibirica var. *ampliata* K. Koch, Linnaea 23: 636 (1850).
Campanula sibirica f. *corymbosa* Fomin in N. M. Kusn., N. Bush & Fomin, Mater. Fl.
Kavkaza 4(6): 26 (1904).
Campanula schelkownikowii Grossh., Dokl. Akad. Nauk Armyansk. SSSR 7: 170 (1947).
Campanula darialica Kharadze, Zametki Sist. Geogr. Rast. 17: 104 (1953).
Campanula hohenackeri var. *darialica* (Kharadze) Serdyuk., Zametki Sist. Geogr. Rast. 38:
20 (1982).
Campanula fedorovii Kharadze, Zametki Sist. Geogr. Rast. 19: 36 (1956).

subsp. **sibirica**
Germany to Italy, Balkans, Ukraine, C. & E. Russia, Kazakhstan, W. China (Xinjiang) &
W. Siberia. 11 AUT CZE GER HUN POL 13 ALB BUL ITA ROM YUG 14 BLR RUC RUE
RUW UKR 30 WSB 32 KAZ 36 CHX. Biennial. 2n = 34, 102.

Campanula sibirica var. *paniculata* A. DC., Monogr. Campan.: 244 (1830). *Campanula sibirica* subsp. *paniculata* (A. DC.) Arcang., Comp. Fl. Ital.: 457 (1882).
Campanula sibirica var. *abortiva* A. DC., Monogr. Campan.: 244 (1830). *Campanula sibirica* f. *abortiva* (A. DC.) Hayek & Hegi in Hegi, Ill. Fl. Mitt.-Eur. 6: 335 (1915).
Campanula sibirica var. *minor* Schur, Enum. Pl. Transsilv.: 433 (1866).

subsp. **talievii** (Juz.) Fed., Bot. J. Linn. Soc. 67: 281 (1973).
Krym. 14 KRY. Hemicr.
* *Campanula talievii* Juz., Bot. Mater. Gerb. Bot. Inst. Komarova Akad. Nauk SSSR 14: 39 (1951).

subsp. **taurica** (Juz.) Fed., Bot. J. Linn. Soc. 67: 281 (1973).
Krym. 14 KRY. Hemicr. or biennial. 2n = 26.
Campanula sibirica var. *taurica* Trautv., Pl. Messes: 68 (1874).
* *Campanula taurica* Juz., Bot. Mater. Gerb. Bot. Inst. Komarova Akad. Nauk SSSR14: 36 (1951).

Campanula sidonensis Boiss. & Blanche in Boiss., Diagn. Pl. Orient. (ser. 2) 3: 114 (1856).
Israel, Jordan, Lebanon. 34 LBS PAL. Ther.

Campanula simulans A. Carlström, Willdenowia 15: 381 (1986).
E. Aegean Is. (Kálymnos, Sámos, Tílos, Sými, Kos) & Turkey. 34 EAI TUR. Ther. 2n = 28.

Campanula sivasica Kit Tan & Yildiz, Notes Roy. Bot. Gard. Edinburgh 45: 447 (1988).
Turkey. 34 TUR. Ther.

Campanula × smithii T. Moore, Florist & Pomol. 1875: 209 (1875). C. cespitosa × C. fragilis
Cult. Hemicr.

Campanula songutica Amirkh. in Naymenko, Bot. Issl. Zapov.: 136 (1984).
Caucasus (Severo-Osetiya). 33 NCS. Hemicr. or geophyte.

Campanula sorgerae Phitos, Notes Roy. Bot. Gard. Edinburgh 35: 45 (1976).
Turkey. 34 TUR. Hemicr. or biennial.

Campanula sparsa Friv., Magyar Tud. Társ. Évk. 7: 201 (1840) .
Balkans. 13 ALB BUL GRC ROM TUE YUG. Ther. 2n = 20.
* *Campanula expansa* Friv., Flora 19: 434. 1836; non Rudolph, Mém. Acad. Imp. Sci. St. Pétersbourg Hist. Acad. 4: 340 (1809). *Campanula frivaldskyi* Steud., Nomencl. Bot. (ed. 2) 1: 267 (1840). *Neocodon sparsus* (Friv.) Kolak. & Serdyuk., Zametki Sist. Geogr. Rast. 40: 29 (1984).

subsp. **sparsa**
Balkans. 13 ALB BUL GRC ROM TUE YUG. Ther. 2n = 20.
Campanula welandii Heuff., Oesterr. Bot. Wochenbl. 7: 118 (1857). *Campanula expansa* var. *welandii* (Heuff.) Nyman, Consp. Fl. Eur.: 482 (1879).
Campanula crassa Formánek, Verh. Naturf. Vereines Brünn 32: 11 (1894). *Campanula expansa* subsp. *crassa* (Formánek) Formánek, Verh. Naturf. Vereines Brünn 32: 154 (1894).
Campanula expansa var. *macedonica* Formánek, Verh. Naturf. Vereines Brünn 35: 155 (1897).
Campanula revoluta Formánek, Verh. Naturf. Vereines Brünn 38: 190 (1900).

subsp. **sphaerothrix** (Griseb.) Hayek, Repert. Spec. Nov. Regni Veg. Beih. 30(2): 547 (1930).
Balkans. 13 ALB BUL GRC YUG. Ther. 2n = 20.
* *Campanula sphaerothrix* Griseb., Spic. Fl. Rumel. 2: 280 (1846). *Campanula expansa* var. *sphaerothrix* (Griseb.) Boiss., Fl. Orient. 3: 941 (1875).

Campanula spatulata Sm. in Sibth. & Sm., Fl. Graec. Prodr. 1: 137 (1806). *Campanula spruneriana* var. *alpina* Boiss., Fl. Orient. 3: 937 (1875). *Campanula spruneriana* subsp. *spatulata* (Sm.) Nyman, Consp. Fl. Eur.: 481 (1879), as 'spathulata'. *Neocodon spatulatus* (Sm.) Kolak. & Serdyuk., Zametki Sist. Geogr. Rast. 40: 29 (1984), as 'spathulatus'. Balkans & Kriti. 13 ALB BUL GRC KRI YUG. Hemicr. 2n = 20.

subsp. **filicaulis** (Halácsy) Phitos, Verh. Zool.-Bot. Ges. Wien 103-104: 228 (1964).
Kriti. 13 KRI. Hemicr.
Campanula pauciflora Desf., Ann. Mus. Natl. Hist. Nat. 11: 57 (1808).
* *Campanula sibthorpiana* var. *filicaulis* Halácsy, Consp. Fl. Graec. 2: 269 (1902).
Campanula spatulata var. *filicaulis* (Halácsy) Hayek, Repert. Spec. Nov. Regni Veg. Beih. 30(2): 545 (1930).

subsp. **spatulata**
Balkans. 13 ALB GRC YUG. Hemicr. 2n = 20.
Campanula sibthorpiana Halácsy, Consp. Fl. Graec. 2: 268 (1902); non (Schult.) Boiss., Diagn. Pl. Orient. (ser. 1) 11: 77 (1849). *Campanula spruneriana* subsp. *sibthorpiana* Maire & Petitm., Étud. Pl. Vasc. Grèce: 147 (1908). *Campanula spatulata* subsp. *sibthorpiana* (Maire & Petitm.) Hayek, Repert. Spec. Nov. Regni Veg. Beih. 30(2): 545 (1930).

subsp. **spruneriana** (Hampe) Hayek, Repert. Spec. Nov. Regni Veg. Beih. 30(2): 545 (1930).
Balkans. 13 ALB BUL GRC YUG. Hemicr. 2n = 20.
* *Campanula spruneriana* Hampe, Flora 25: 76 (1842).
Campanula spruneriana var. *hirsuta* Heldr. ex Halácsy, Consp. Fl. Graec. 2: 268 (1902). *Campanula spatulata* f. *hirsuta* (Heldr. ex Halácsy) Hayek, Repert. Spec. Nov. Regni Veg. Beih. 30(2): 545 (1930).
Campanula spruneriana var. *lepidota* Turrill, Bull. Misc. Inform. Kew 1918: 306 (1918). *Campanula spatulata* f. *lepidota* (Turrill) Hayek, Repert. Spec. Nov. Regni Veg. Beih. 30(2): 545 (1930).
Campanula sophiae Beauverd, Candollea 7: 271 (1937).

Campanula speciosa Pourr., Hist. & Mém. Acad. Roy. Sci. Toulouse 3: 309 (1788).
Campanula longifolia var. *pyramidalis* Lapeyr., Hist. Pl. Pyrénées: 107 (1813). *Campanula longifolia* var. *speciosa* (Pourr.) Steud., Nomencl. Bot.: 143 (1821).
Spain & France. 12 FRA SPA. Hemicr. or geophyte. 2n = 34, 68.

subsp. **affinis** (Schult.) Font Quer in Cadevall & Font Quer, Fl. Catalunya 4: 19 (1932).
Spain. 12 SPA. Hemicr. or geophyte.
* *Campanula affinis* Schult. in Roem. & Schult., Syst. Veg. 5: 140 (1819).
Campanula bolosii Vayr., Anales Soc. Esp. Hist. Nat. 8: 451 (1879). *Campanula affinis* subsp. *bolosii* (Vayr.) Fed., Bot. J. Linn. Soc. 67: 281 (1973).
Campanula vayredae Leresche, J. Bot. 17: 199 (1879).
Campanula affinis var. *ramosa* Marcet, Bol. Soc. Esp. Hist. Nat. 48: 92 (1950).
Campanula speciosa var. *brevifolia* Pau, Bol. Soc. Esp. Hist. Nat. 48: 92 (1950).
Campanula speciosa var. *stenophylla* Marcet, Bol. Soc. Esp. Hist. Nat. 48: 92 (1950), as 'stenophilla'.
Campanula × *martinezii* Marcet, Bol. Soc. Esp. Hist. Nat. 48: 92 (1950).

subsp. **speciosa**
Spain & France. 12 FRA SPA. Hemicr. or geophyte. 2n = 34, 68.
Campanula bicaulis Lapeyr., Fig. Fl. Pyrénées: 13 (1795-1801). *Campanula longifolia* var. *bicaulis* (Lapeyr.) Schult. in Roem. & Schult., Syst. Veg. 5: 139 (1819). *Campanula speciosa* var. *bicaulis* (Lapeyr.) A. DC., Monogr. Campan.: 248 (1830).
Campanula longifolia Lapeyr., Hist. Pl. Pyrénées: 107 (1813).
Marianthemum longifolium Schrank, Flora 7 (Beilage 2): 54 (1824).
Campanula oliveri Rouy & Gaut., Bull. Soc. Bot. France 41: 326 (1894). *Campanula speciosa* var. *oliveri* (Rouy & Gaut.) O. Bolòs & Vigo, Collect. Bot. (Barcelona) 14: 101 (1983).

Campanula corbariensis Rouy, Ill. Pl. Eur.: 160 (1905).
Campanula beltranii Sennen & Losa in Sennen, Diagn. Nouv.: 276 (1936).
Campanula speciosa f. *multicaulis* Marcet, Bol. Soc. Esp. Hist. Nat. 48: 92 (1950).

Campanula spicata L., Sp. Pl.: 166 (1753).
Alps. 11 AUT SWI 12 FRA 13 ITA YUG. Biennial. 2n = 34.
Campanula spicata var. *hornschuchii* A. DC., Monogr. Campan.: 261 (1830).
Campanula spicata var. *ramosa* A. DC. , Monogr. Campan.: 261 (1830).
Campanula spicata var. *canescens* St.-Lag. in Cariot, Étude Fl. (ed. 8) 2: 546 (1889).
Campanula spicata subvar. *canescens* (St.-Lag.) Rouy, Fl. France 10: 65 (1908).

Campanula × **spryginii** Saksonov & Tzvelev , Bot. Žurn. (Moscow & Leningrad) 79: 99 (1994). C. bononiensis × C. rapunculoides
E. Russia (Samara). 14 RUE. Hemicr. or geophyte.

Campanula staintonii Rech. f. & Schiman-Czeika in Rech. f., Fl. Iran. 13: 16 (1965).
Pakistan. 40 PAK. Hemicr.

Campanula stellaris Boiss., Diagn. Pl. Orient. (ser. 1) 11: 63 (1849).
Turkey to Israel & Jordan. 34 LBS PAL TUR. Ther.

Campanula stenocarpa Trautv. & C. A. Mey. in Middend., Reise Sibir. 1(2, Lfg. 3): 61 (1856).
Popoviocodonia stenocarpa (Trautv. & C. A. Mey.) Fed. in Kom., Fl. URSS 24: 374 (1957).
Russian Far East (Magadan). 31 MAG.

Campanula stenocodon Boiss. & Reut. in Boiss., Diagn. Pl. Orient. (ser. 2) 3: 112 (1856).
Campanula rotundifolia subsp. *stenocodon* (Boiss. & Reut.) Nyman, Consp. Fl. Eur.: 479 (1879). *Campanula scheuchzeri* subsp. *stenocodon* (Boiss. & Reut.) Arcang., Comp. Fl. Ital. (ed. 2): 637 (1894). *Campanula pusilla* var. *stenocodon* (Boiss. & Reut.) Rouy, Fl. France 10: 76 (1908). *Campanula rotundifolia* var. *stenocodon* (Boiss. & Reut.) Fiori, Nuov. Fl. Italia 2: 562 (1927).
SW. Alps. 12 FRA 13 ITA. Hemicr. or geophyte. 2n = 34.
Campanula stenocodon var. *latifolia* Beyer, Verh. Bot. Vereins Prov. Brandenburg 58: 114 (1917).

Campanula stenosiphon Boiss. & Heldr. in Boiss., Diagn. Pl. Orient. (ser. 1) 7: 18 (1846).
Campanula glomerata var. *stenosiphon* (Boiss. & Heldr.) Boiss., Fl. Orient. 3: 928 (1875).
Campanula glomerata subsp. *stenosiphon* (Boiss. & Heldr.) Nyman, Consp. Fl. Eur.: 478 (1879).
Greece. 13 GRC. Hemicr. 2n = 34.

Campanula stevenii M. Bieb., Fl. Taur.-Cauc. 3: 138 (1819), as 'steveni'.
E. Europe to Russian Far East. 13 ROM 14 RUE RUC RUS UKR 30 ALT BRY CTA IRK TVA WSB 31 AMU 32 KGZ 33 ALL 34 IRN TUR 36 CHX 37 MON. Hemicr. 2n = 20, 32, 34, 48, 58.
* *Campanula simplex* Steven, Mém. Soc. Imp. Naturalistes Moscou 3: 255 (1812); non DC. in Lam. & DC., Fl. Franç. (ed. 3) 3: 730 (1805). *Campanula steveniana* Schult. in Roem. & Schult., Syst. Veg. 5: 91 (1819). *Neocodon stevenii* (M. Bieb.) Kolak. & Serdyuk., Zametki Sist. Geogr. Rast. 40: 29 (1984).

subsp. **alberti** (Trautv.) Viktorov, Novosti Sist. Vyssh. Rast. 34: 231 (2002).
Kirgizstan & NW. China (Xinjiang). 32 KGZ 36 CHX. Hemicr.
* *Campanula alberti* Trautv., Trudy Imp. S.-Peterburgsk. Bot. Sada 6: 77 (1879). *Neocodon alberti* (Trautv.) Kolak. & Serdyuk., Zametki Sist. Geogr. Rast. 40: 28 (1984), as 'albertii'.

subsp. **altaica** (Ledeb.) Fed., Bot. J. Linn. Soc. 67: 281 (1973).
E. Europe to W. Siberia & Mongolia. 13 ROM 14 RUC RUS UKR 30 ALT WSB 37 MON. Hemicr. 2n = 32, 48.
Campanula sajanensis Fisch., Cat. Jard. Gorenki: 52 (1808).

* *Campanula altaica* Ledeb., Index Sem. Horti Dorpat.: 2 (1824). *Neocodon altaicus*
(Ledeb.) Kolak. & Serdyuk., Zametki Sist. Geogr. Rast. 40: 28 (1984).
Campanula rochelii Schur, Oesterr. Bot. Z. 11: 45 (1861).

subsp. **beauverdiana** (Fomin) Rech. f. & Schiman-Czeika in Rech. f., Fl. Iran. 13: 36 (1965).
Turkey, Caucasus (Gruziya, Armenia, Nakhichevan, Azerbaijan), Iran. 33 TCS 34 IRN
TUR. Hemicr.
* *Campanula beauverdiana* Fomin, Věstn. Tiflissk. Bot. Sada 1: 12 (1905). *Neocodon
beauverdianus* (Fomin) Kolak. & Serdyuk., Zametki Sist. Geogr. Rast. 40: 28 (1984).
Campanula stevenii var. *vesiculosa* Bornm., Bull. Herb. Boissier 7: 774 (1907).
Campanula stevenii f. *alpina* Bornm., Bull. Herb. Boissier 7: 774 (1907).
Campanula beauverdiana var. *aissori* Tamamsch., Repert. Spec. Nov. Regni Veg. 38:
171 (1935).

subsp. **stevenii**
Turkey, Caucasus (Krasnodar, Karacheyevo-Cherkessiya, Kabardino-Balkariya, Severo-
Osetiya, Dagestan, Abkhaziya, Gruziya, Armenia, Nakhichevan, Azerbaijan, Nagorno
Karabakh), Iran. 33 NCS TCS 34 IRN TUR. Hemicr. 2n = 20, 32.
Campanula vitinghoffiana Schult. in Roem. & Schult., Syst. Veg. 5: 102 (1819).

subsp. **turczaninovii** (Fed.) Viktorov, Novosti Sist. Vyssh. Rast. 34: 230 (2002).
S. Siberia, Mongolia, Russian Far East (Amur). 30 ALT BRY CTA IRK TVA 31 AMU 37
MON. Hemicr. 2n = 34, 58.
* *Campanula silenifolia* Fisch. ex A. DC., Monogr. Campan.: 320 (1830); non Host, Fl.
Austriac. 1: 268 (1827). *Campanula stevenii* var. *silenifolia* Regel, Bull. Soc. Imp.
Naturalistes Moscou 40(3): 187 (1868). *Campanula simplex* var. *silenifolia* (Regel)
Trautv., Trudy Imp. S.-Peterburgsk. Bot. Sada 5: 540 (1878). *Campanula turczaninovii*
Fed. in Kom., Fl. URSS 24: 304 (1957). *Neocodon turczaninovii* (Fed.) Kolak. &
Serdyuk., Zametki Sist. Geogr. Rast. 40: 29 (1984).
Campanula stevenii var. *integerrima* Regel, Bull. Soc. Imp. Naturalistes Moscou 40(3):
188 (1868).
Campanula stevenii var. *dasycarpa* Regel, Bull. Soc. Imp. Naturalistes Moscou 40(3): 188
(1868). *Campanula simplex* var. *dasycarpa* (Regel) Trautv., Trudy Imp. S.-Peterburgsk.
Bot. Sada 6: 85 (1879).

subsp. **wolgensis** (P. A. Smirn.) Fed., Bot. J. Linn. Soc. 67: 281 (1973).
E. Russia & W. Siberia. 14 RUE 30 WSB. Hemicr.
Campanula infundibulum Vest ex Schult. in Roem. & Schult., Syst. Veg. 5: 106 (1819).
Campanula stevenii var. *sibirica* A. DC., Monogr. Campan.: 321 (1830).
* *Campanula wolgensis* P. A. Smirn., Bjull. Moskovsk. Obšč. Isp. Prir., Otd. Biol. (n.s.)
52(3): 57 (1947). *Neocodon wolgensis* (P. A. Smirn.) Kolak. & Serdyuk., Zametki Sist.
Geogr. Rast. 40: 29 (1984).

Campanula stricta L., Sp. Pl. (ed. 2): 238 (1762).
Turkey to Iran. 34 IRN IRQ LBS TUR. Hemicr. or geophyte. 2n = 34.
Campanula jasionifolia A. DC. in DC., Prodr. 7: 463 (1839). *Campanula stricta* var.
jasionifolia (A. DC.) Boiss., Fl. Orient. 3: 924 (1875)
Campanula libanotica A. DC. in DC., Prodr. 7: 463 (1839). *Campanula stricta* var.
libanotica (A. DC.) Boiss., Fl. Orient. 3: 924 (1875).
Campanula syspirensis K. Koch, Linnaea 23: 639 (1850).
Campanula stricta var. *muricata* Trautv., Trudy Imp. S.-Peterburgsk. Bot. Sada 2: 563 (1873).
Campanula stricta f. *adpressa* Witasek, Ann. K. K. Naturhist. Hofmus. 20: 418 (1905).
Campanula stricta var. *alidagensis* Damboldt, Notes Roy. Bot. Gard. Edinburgh 35:
47 (1976).

Campanula strigillosa Boiss., Ann. Sci. Nat., Bot. (ser. 4) 2: 251 (1854). *Podanthum
strigillosum* (Boiss.) Boiss., Fl. Orient. 3: 957 (1875). *Asyneuma strigillosum* (Boiss.) Bornm.,
Beih. Bot. Centralbl. 38(2): 349 (1921).
Turkey. 34 TUR. Hemicr.

Campanula strigosa Banks & Sol. in Russell, Nat. Hist. Aleppo (ed. 2) 2: 246 (1794).
Campanula russelliana Schult. in Roem. & Schult., Syst. Veg. 5: 142 (1819).
Turkey to Israel & Iran. 34 IRN LBS PAL TUR. Ther. 2n = 20.
Campanula strigosa Vahl, Symb. Bot. 3: 34 (1794); non Banks & Sol. in Russell, Nat.
Hist. Aleppo (ed. 2) 2: 246 (1794).

Campanula suanetica Rupr., Bull. Acad. Imp. Sci. Saint-Pétersbourg 11: 215 (1867).
Caucasus (Gruziya). 34 TCS. Hemicr. or geophyte.
Campanula suanetica var. *appendiculata* Sommier & Levier, Enum. Pl. Cauc.: 323 (1900).
Campanula suanetica f. *appendiculata* (Sommier & Levier) Fomin in N. M. Kusn., N.
Bush & Fomin, Mater. Fl. Kavkaza 4(6): 74 (1905).

Campanula sulaimanii Nasir in Nasir & Ali, Fl. Pakistan 155: 13 (1984).
Pakistan. 40 PAK. Hemicr.

Campanula sulphurea Boiss., Diagn. Pl. Orient. (ser. 1) 11: 64 (1849).
Lebanon to Egypt. 20 EGY 34 LBS PAL SIN. Ther.
Campanula sulphurea f. *brachyantha* Evenari, Bull. Soc. Bot. Genève (ser. 2) 31: 407 (1940).

Campanula sylvatica Wall. in Roxb., Fl. Ind. 2: 97 (1824).
Himalaya to N. India (Uttar Pradesh). 40 EHM IND NEP WHM. Ther.
* *Campanula stricta* Wall., Asiat. Res. 13: 374 (1820); non L., Sp. Pl. (ed. 2): 238 (1762).
Campanula integerrima Buch.-Ham. ex D. Don, Prodr. Fl. Nepal.: 155 (1825).
Campanula caperonioides Klotzsch in Klotzsch & Garcke, Bot. Ergebn. Reise Waldemar:
73 (1862).

Campanula takesimana Nakai, Rep. Veg. Ooryongto: 42 (1919).
Korea (Ullung-do); cult. 38 KOR. Hemicr. or geophyte. 2n = 34.

Campanula tanfanii Podlech, Feddes Repert. 71: 95 (1965).
Italy (C. Apennines). 23 ITA. Hemicr. or geophyte. 2n = 34.
* *Campanula macrorhiza* var. *angustiflora* Tanfani in Parl., Fl. Ital. 8: 101 (1888); non
Campanula angustiflora Eastw., Proc. Calif. Acad. Sci. (ser. 3) 1: 132 (1898).
Campanula macrorhiza subsp. *angustiflora* (Tanfani) Arcang., Comp. Fl. Ital. (ed 2):
637 (1894). *Campanula rotundifolia* var. *angustiflora* (Tanfani) Guin., Bull. Soc. Bot.
France 89: 75 (1942).

Campanula tatrae Borbás, Magyar Bot. Lapok 1: 319 (1902).
Czechoslovakia & Poland. 11 CZE POL. Hemicr. or geophyte. 2n = 68.

subsp. **mentiens** (Witasek) Kovanda, Folia Geobot. Phytotax. 12: 66 (1977).
Czechoslovakia & Poland (W. Carpathians). 11 CZE POL. Hemicr. or geophyte. 2n = 68.
* *Campanula kladniana* subsp. *mentiens* Witasek, Magyar Bot. Lapok 5: 241 (1906).
Campanula mentiens (Witasek) Prain, Index Kew. suppl. 4: 35 (1913). *Campanula
kladniana* var. *mentiens* (Witasek) Pawł., Acta Soc. Bot. Poloniae 1: 5 (1923).
Campanula polymorpha f. *mentiens* (Witasek) Jáv., Magyar Fl.: 1079 (1925).
Campanula arctica subsp. *mentiens* (Witasek) Hruby in Domin & Podp., Klíč Úplné
Květ. Republ. Českoslov.: 534 (1928). *Campanula polymorpha* var. *mentiens* (Witasek)
Maroriu in Săvul., Fl. Reipubl. Popul. Roman. 9: 109 (1964). *Campanula rotundifolia*
var. *mentiens* (Witasek) Tacik, Monogr. Bot. 20: 254 (1966).

subsp. **sudetica** (Hruby) Kovanda, Folia Geobot. Phytotax. 12: 66 (1977).
Czechoslovakia (Sudeten Mts.). 11 CZE. Hemicr. or geophyte. 2n = 68.
* *Campanula rotundifolia* var. *sudetica* Hruby, Magyar Bot. Lapok 29: 165 (1930).
Campanula rotundifolia subsp. *sudetica* (Hruby) Soó, Acta Bot. Acad. Sci. Hung. 13:
306 (1967).

subsp. **tatrae**

Czechoslovakia & Poland (W. Carpathians). 11 CZE POL. Hemicr. or geophyte. 2n = 68.

Campanula scheuchzeri var. *stenophylla* Schur, Enum. Pl. Transsilv.: 443 (1866).
Campanula kladniana subsp. *stenophylla* (Schur) Witasek, Magyar Bot. Lapok 5: 238
(1906). *Campanula stenophylla* (Schur) Prain, Index Kew. suppl. 4: 35 (1913); non
Boiss. & Heldr. in Boiss., Diagn. Pl. Orient. (ser. 1) 11: 77 (1849). *Campanula
polymorpha* var. *stenophylla* (Schur) Hruby, Magyar Bot. Lapok 29: 205 (1930).
Campanula scheuchzeri var. *dacica* Porcius, Fl. Fanerogam. Naseud.: 98 (1881).
Campanula kladniana subsp. *polymorpha* Witasek, Magyar Bot. Lapok 5: 239(1906).
Campanula polymorpha (Witasek) Prain, Index Kew. suppl. 4: 35 (1913). *Campanula
kladniana* var. *polymorpha* (Witasek) Pawł., Acta Soc. Bot. Poloniae 1: 5 (1923).
Campanula rotundifolia subsp. *polymorpha* (Witasek) Tacik, Monogr. Bot. 20: 254 (1966).
Campanula scheuchzeriformis Hayek, Oesterr. Bot. Z. 70: 19 (1921). *Campanula
rotundifolia* var. *scheuchzeriformis* (Hayek) Hayek, Repert. Spec. Nov. Regni Veg. Beih.
30(2): 540 (1930). *Campanula balcanica* var. *scheuchzeriformis* (Hayek) Hruby, Magyar
Bot. Lapok 29: 251 (1930).
Campanula polymorpha var. *praticola* Hruby, Magyar Bot. Lapok 29: 198 (1930).
Campanula polymorpha var. *intercedens* Hruby, Magyar Bot. Lapok 29: 199 (1930).
Campanula polymorpha f. *latifolia* Hruby, Magyar Bot. Lapok 29: 199 (1930).
Campanula polymorpha f. *angustifolia* Hruby, Magyar Bot. Lapok 29: 200 (1930).
Campanula polymorpha f. *exigua* Hruby, Magyar Bot. Lapok 29: 200 (1930).
Campanula polymorpha f. *reflectans* Hruby, Magyar Bot. Lapok 29: 200 (1930).
Campanula polymorpha f. *umbrosa* Hruby, Magyar Bot. Lapok 29: 200 (1930).
Campanula polymorpha f. *kladnianioides* Nyár. ex Hruby, Magyar Bot. Lapok 29: 201(1930).
Campanula polymorpha subf. *umbrosa* Hruby, Magyar Bot. Lapok 29: 202 (1930).
Campanula polymorpha f. *lepida* Hruby, Magyar Bot. Lapok 29: 202 (1930).
Campanula polymorpha subf. *reflectans* Hruby, Magyar Bot. Lapok 29: 202 (1930).
Campanula polymorpha f. *sciaphila* Hruby, Magyar Bot. Lapok 29: 204 (1930).
Campanula polymorpha f. *brachyphylla* Hruby, Magyar Bot. Lapok 29: 206 (1930).
Campanula polymorpha f. *gracilis* Hruby, Magyar Bot. Lapok 29: 206 (1930).

Campanula telephioides Boiss. & Hausskn. in Boiss., Pl. Orient. Nov. 1: 1 (1875).
Turkey. 34 TUR. Hemicr.
Campanula telephioides subsp. *argutiserrata* Rech. f., Österr. Akad. Wiss., Math.-
Naturwiss. Kl., Anz. 87: 197 (1950).

Campanula telmessii Huber-Mor. & Phitos, Notes Roy. Bot. Gard. Edinburgh 35: 45 (1976),
as 'telmessi'.
Turkey. 34 TUR. Hemicr. or biennial.

Campanula tenuissima Dunn, Bull. Misc. Inform. Kew 1924: 385 (1924).
W. Himalaya. 40 WHM. Hemicr.

Campanula teucrioides Boiss., Diagn. Pl. Orient. (ser. 1) 4: 33 (1844).
Turkey. 34 TUR. Hemicr. 2n = 34.

Campanula thyrsoides L., Sp. Pl.: 167 (1753). *Campanula macrorrhiza* var. *thyrsoides* (L.)
Vuk., Linnaea 26: 330 (1854), as 'thyrsoidea'.
France to Germany, Balkans & Italy. 11 AUT GER SWI 12 FRA 13 BUL ITA YUG. Biennial.
2n = 34.

subsp. **carniolica** (Sünd.) Podlech, Ber. Bayer. Bot. Ges. 37: 111 (1964).
Alps. 11 AUT 13 ITA YUG. Biennial. 2n = 34.
* *Campanula thyrsoides* var. *carniolica* Sünd., Allg. Bot. Z. Syst. 26-27: 23 (1925).
Campanula carniolica (Sünd.) Crook, Campanulas: 56 (1951).

subsp. **thyrsoides**
> France to Germany, Balkans & Italy. 11 AUT GER SWI 12 FRA 13 BUL ITA YUG.
> Biennial. 2n = 34.
> *Campanula thyrsoides* var. *multicaulis* A. DC., Monogr. Campan.: 263 (1830).
> *Campanula thyrsoides* var. *glomerata* Saut., Flora 35: 622 (1852).

Campanula tokurii Ocak, Israel J. Pl. Sci. 51: 321 (2003). *Campanula pamphylica* subsp.
tokurii (Ocak) Akçiçek & Vural, Ann. Bot. Fenn. 42: 408 (2005), as 'tokuri'.
Turkey. 34 TUR. Biennial. 2n = 34.
> *Campanula argaea* subsp. *pamphylica* Contandr., Quézel & Pamukç., Ann. Univ.
> Provence Sci. 46: 54 (1972). *Campanula pamphylica* (Contandr., Quézel & Pamukç.)
> Akçiçek & Vural, Ann. Bot. Fenn. 42: 407 (2005).
> *Campanula pamphylica* subsp. *afyonica* Akçiçek & Vural, Ann. Bot. Fenn. 42: 408 (2005).

Campanula tomentosa Lam., Encycl. 1: 584 (1785).
> N. Aegean Is. (Chíos) & Turkey. 34 EAI TUR. Hemicr. or biennial. 2n = 34.
> *Campanula tomentosa* var. *ephesia* A. DC., Monogr. Campan.: 218 (1830).
> *Campanula appendiculata* A. DC. in DC., Prodr. 7: 462 (1839). *Campanula tomentosa* var.
> *appendiculata* (A. DC.) Nyman, Consp. Fl. Eur.: 476 (1879).
> *Campanula eriantha* Hampe, Flora 25: 76 (1842).
> *Campanula ephesia* Boiss., Fl. Orient. 3: 898 (1875).
> *Campanula mycalaea* Barbey & Major in Barbey, Lydie Lycie Carie: 80 (1890).
> *Campanula vardariana* Bocquet, Compt. Rend. Séances Soc. Phys. Genève (n.s.) 3: 90
> (1968).

Campanula tommasiniana K. Koch in F. W. Schultz, Arch. Fl. France Allem.: 229 (1850-54).
> *Campanula waldsteiniana* subsp. *tommasiniana* (K. Koch) Nyman, Consp. Fl. Eur.: 480 (1879).
> Croatia. 13 YUG. Hemicr. or geophyte. 2n = 34.

Campanula topaliana Beauverd, Candollea 7: 268 (1937).
> Greece. 13 GRC. Hemicr. 2n = 34.

> subsp. **cordifolia** Phitos, Oesterr. Bot. Z. 112: 458 (1965).
> > Greece. 13 GRC. Hemicr. 2n = 34.
> > *Campanula calavrytana* Beauverd & Topali, Candollea 7: 269 (1937).

> subsp. **delphica** Phitos, Oesterr. Bot. Z. 112: 459 (1965).
> > Greece. 13 GRC. Hemicr.

> subsp. **topaliana**
> > Greece. 13 GRC. Hemicr.

Campanula trachelium L., Sp. Pl.: 166 (1753). *Campanula urticifolia* Salisb., Prodr. Stirp.
Chap. Allerton: 127 (1796), as 'urticaefolia;' non Turra, Fl. Ital. Prodr.: 64 (1780); nec All.,
Fl. Pedem. 1: 110 (1785). *Campanula serratifolia* var. *ciliatosepala* Vuk., Linnaea 26: 335
(1854), as 'ciliato-sepala'.
> Europe to N. Africa, Lebanon, Iran, Kazakhstan & W. Siberia; cult.; naturalized in C.
> Canada & NE. U.S.A. 10 DEN FIN GRB IRE NOR SWE 11 AUT BGM CZE GER HUN NET
> POL SWI 12 FRA SPA 13 ALB BUL GRC ITA ROM SCI TUE YUG 14 BLR BLT KRY RUC
> RUE SUN RUS RUW UKR 20 ALG MOR TUN 30 ALT WSB 32 KAZ 34 IRN LBS TUR (71)
> man (72) ont que (74) wis (75) mai mas mic nwy ohi pen ver. Hemicr. 2n = 34.

> subsp. **athoa** (Boiss. & Heldr.) Nyman, Consp. Fl. Eur.: 478 (1879).
> > Balkans to Turkey. 13 ALB BUL GRC TUE YUG 34 TUR. Hemicr. 2n = 34.
> > *Campanula athoa* Boiss. & Heldr., Diagn. Pl. Orient. (ser. 2) 3: 110 (1856). *Campanula
> > trachelium* var. *orientalis* Boiss., Fl. Orient. 3: 922 (1875). *Campanula trachelium* var.
> > *athoa* (Boiss. & Heldr.) Degen, Fl. Veleb. 3: 94 (1938).

Campanula trachelium subsp. *dasycarpa* A. DC. ex Arcang., Comp. Fl. Ital.: 455 (1882).
Campanula trachelium var. *solitaria* Post, J. Linn. Soc., Bot. 24: 435 (1888).

subsp. **mauritanica** (Pomel) Quézel, Feddes Repert. Spec. Nov. Regni Veg. 56: 46, 65 (1953).
N. Africa. 20 ALG MOR TUN. 2n = 34.
* *Campanula trachelioides* Munby, Bull. Soc. Bot. France 2: 285 (1855); non M. Bieb., Fl.
Taur.-Cauc. 1: 150 (1808). *Campanula mauritanica* Pomel, Nouv. Mat. Fl. Atl.: 3 (1874).
Campanula trachelium var. *parviflora* Batt., Bull. Soc. Bot. France 44: 322 (1897).
Campanula trachelium var. *hirta* Quézel, Feddes Repert. Spec. Nov. Regni Veg. 56: 46
(1953).
Campanula trachelium var. *munbyana* Quézel, Feddes Repert. Spec. Nov. Regni Veg. 56:
47 (1953).
Campanula trachelium f. *numidica* Quézel, Feddes Repert. Spec. Nov. Regni Veg. 56:
47 (1953).
Campanula trachelium f. *maroccana* Quézel, Feddes Repert. Spec. Nov. Regni Veg. 56:
47 (1953).

subsp. **trachelium**
Europe to Lebanon, Iran, Kazakhstan & W. Siberia; cult.; naturalized in C. Canada &
NE. U.S.A. 10 DEN FIN GRB IRE NOR SWE 11 AUT BGM CZE GER HUN NET POL SWI
12 FRA SPA 13 ALB BUL GRC ITA ROM SIC YUG 14 BLR BLT KRY RUC RUE RUN RUS
RUW UKR 30 ALT WSB 32 KAZ 34 IRN LBS (71) man (72) ont que (74) wis (75) mai
mas mic nwy ohi pen ver. Hemicr. 2n = 34.
Campanula urticifolia F. W. Schmidt, Fl. Boëm. 2: 73 (1794); non Turra, Fl. Ital. Prodr.:
64 (1780); nec All., Fl. Pedem. 1: 110 (1785). *Campanula trachelium* var. *urticifolia*
Steud., Nomencl. Bot. (ed. 2) 1: 270 (1840), as 'urticaefolia'. *Campanula trachelium*
var. *dasycarpa* Gren. & Godr., Fl. Fr. 2: 411 (1853).
Campanula plicatula Dumort., Fl. Belg.: 58 (1827).
Campanula urticifolia var. *multiflora* Gaudin, Fl. Helv. 2: 158 (1828).
Campanula trachelium var. *monstrosa* A. DC., Monogr. Campan.: 267 (1830).
Campanula trachelium var. *cordifolia* Noulet, Fl. Bass. Sous-Pyrén.: 409 (1837).
Campanula trachelium var. *secundiflora* Noulet, Fl. Bass. Sous-Pyrén.: 409 (1837).
Campanula trachelium var. *subsessilis* Noulet, Fl. Bass. Sous-Pyrén.: 409 (1837).
Campanula trachelium var. *hispida* Peterm., Fl. Lips. Excurs.: ? (1838).
Campanula trachelium var. *albiflora* Peterm., Fl. Lips. Excurs.: ? (1838).
Campanula trachelium var. *virgata* Knaf in ?
Campanula trachelium var. *leucantha* Schur, Enum. Pl. Transsilv.: 436 (1866).
Campanula oligosantha Schur, Enum. Pl. Transsilv.: 436 (1866). *Campanula trachelium*
var. *oligosantha* (Schur) Nyman, Consp. Fl. Eur.: 478 (1879).
Campanula trachelium var. *nuda* Kuntze, Taschen-Fl. Leipzig: 127 (1867).
Campanula trachelium var. *gymnocarpa* Trautv., Trudy Imp. S.-Peterburgsk. Bot. Sada 6:
74 (1879).
Campanula trachelium f. *alba* Voss in Siebert & Voss, Vilm. Blumengärtn. (ed. 3) 1:
568 (1894).
Campanula trachelium var. *ramosissima* Marcet, Bol. Soc. Esp. Hist. Nat. 48: 94 (1950).

Campanula trachyphylla Schott & Kotschy ex Boiss., Fl. Orient. 3: 926 (1875).
Turkey. 34 TUR. Hemicr. 2n = 34.

Campanula transsilvanica Schur ex Andrae, Bot. Zeitung (Berlin) 13: 328 (1855).
Campanula cervicaria subsp. *transsilvanica* (Schur ex Andrae) Tacik in Pawłowskiego &
Jasiewicz, Fl. Polska 12: 67 (1971), as 'transilvanica'.
Carpathians. 13 BUL ROM. Biennial.
Campanula transsilvanica subsp. *davidovii* Urum., Magyar Bot. Lapok 19: 38 (1920).
Campanula transsilvanica var. *davidovii* (Urum.) Hayek, Repert. Spec. Nov. Regni Veg.
Beih. 30(2): 530 (1930).

Campanula trichopoda Boiss., Diagn. Pl. Orient. (ser. 1) 11: 68 (1849).
Lebanon & Syria. 34 LBS. Hemicr.

Campanula tridentata Schreb., Icon. Descr. Pl. 3: pl. 2 (1766). *Hemisphaera tridentata*
(Schreb.) Kolak., Kolokol'ch. Kavkaz.: 115 (1991).
Turkey & Caucasus (Krasnodar, Karacheyevo-Cherkessiya, Kabardino-Balkariya, Severo-
Osetiya, Chechnya, Dagestan, Abkhaziya, Adzhariya, Gruziya, Azerbaijan). 33 ALL 34
TUR. Hemicr. or geophyte. 2n = 34.

subsp. **biebersteiniana** (Schult.) Ogan., Bot. Žurn. (Moscow & Leningrad) 66: 403 (1981).
Caucasus (Krasnodar, Karacheyevo-Cherkessiya, Kabardino-Balkariya, Severo-Osetiya,
Chechnya, Dagestan, Abkhaziya, Gruziya, Azerbaijan). 33 ALL. Hemicr. or geophyte.
2n = 34.
* *Campanula rupestris* M. Bieb., Fl. Taur.-Cauc. 1: 154 (1808); non Sm. in Sibth. & Sm., Fl.
Graec. Prodr. 1: 142 (1806). *Campanula biebersteiniana* Schult. in Roem. & Schult.,
Syst. Veg. 5: 147 (1819). *Campanula tridentata* var. *rupestris* Trautv., Trudy Imp. S.-
Peterburgsk. Bot. Sada 2: 561 (1873).
Campanula tridens Rupr., Bull. Acad. Imp. Sci. Saint-Pétersbourg 11: 204 (1867).
Hemisphaera tridens (Rupr.) Kolak., Okhrana Prir. Gruz. 12: 168 (1984).
Campanula akuschensis Husseinov, Bot. Žurn. (Moscow & Leningrad) 55: 1500 (1970).
Campanula scapifoliosa A. P. Khokhr., Bjull. Moskovsk. Obšč. Isp. Prir., Otd. Biol. 96(4):
109 (1991).

subsp. **tridentata**
Turkey & Caucasus (Adzhariya, Gruziya, Armenia, Nakhichevan, Nagorno Karabakh).
33 TCS 34 TUR. Hemicr. or geophyte. 2n = 34.
Campanula biebersteiniana var. macrantha A. DC., Monogr. Campan.: 227 (1830).
Campanula bithynica A. DC. in DC., Prodr. 7: 460 (1839). *Campanula tridentata* var.
bithynica (A. DC.) Griseb., Spic. Fl. Rumel. 2: 287 (1846).
Campanula tridens var. *araratica* Rupr., Bull. Acad. Imp. Sci. Saint-Pétersbourg 11: 205
(1867).
Campanula tridens var. *ciliata* Rupr., Bull. Acad. Imp. Sci. Saint-Pétersbourg 11: 205 (1867).
Campanula tridens var. *crenatoserrata* Rupr., Bull. Acad. Imp. Sci. Saint-Pétersbourg 11:
205 (1867), as 'crenato-serrata'.
Campanula tridentata var. *stenophylla* Boiss., Fl. Orient. 3: 904 (1875).
Campanula tridentata var. *barbata* Fomin in N. M. Kusn., N. Bush & Fomin, Mater. Fl.
Kavkaza 4(6): 50 (1904). *Campanula tridens* var. *barbata* (Fomin) Kharadze, Zametki
Sist. Geogr. Rast. 15: 27 (1949).
Campanula tridentata var. *gracilis* Fomin in N. M. Kusn., N. Bush & Fomin, Mater. Fl.
Kavkaza 4(6): 51 (1905).

Campanula tristis Kitam., Acta Phytotax. Geobot. 16: 132 (1956), as 'trista'.
Afghanistan & Pakistan. 34 AFG 40 PAK. Hemicr.

Campanula troegerae Damboldt, Notes Roy. Bot. Gard. Edinburgh 35: 47 (1976).
Turkey. 34 TUR. Hemicr.

Campanula trojanensis Kovanda & Ančev, Preslia 61: 201 (1989).
Bulgaria. 13 BUL. Hemicr. or geophyte. 2n = 34, 68.

Campanula × truedingeri Murr, Neue Übers. Bl.-pfl. Vorarlberg: 304 (1924). C.
cochleariifolia × C. rotundifolia
Austria. 11 AUT. Hemicr. or geophyte.
Campanula × racemosa Beyer, Verh. Bot. Vereins Prov. Brandenburg 58: 115 (1917); non
Vuk., Linnaea 26: 332 (1854), pro sp.; nec (Krašan) Witasek, Abh. K. K. Zool.-Bot.
Ges. Wien 1(3): 34 (1902), pro sp.

Campanula tschuktschorum Jurtsev & Fed., Bot. Žurn. (Moscow & Leningrad) 59: 1452 (1974).
Russian Far East (Magadan). 31 MAG. Hemicr.

Campanula tubulosa Lam., Encycl. 1: 586 (1785).
Kriti. 13 KRI. Biennial. 2n = 34.
Campanula subidaea Gand., Fl. Crete: 68 (1916).

Campanula tymphaea Hausskn., Mitt. Geogr. Ges. (Thüringen) Jena 5: 87 (1887).
Albania, Macedonia, Greece. 13 ALB GRC YUG. Hemicr. 2n = 34.

Campanula uniflora L., Sp. Pl.: 163 (1753).
Circumboreal, extending south to Scandinavia, N. Russia, N. Siberia, Kamchatka, Alaska, SW. U.S.A. (New Mexico) & Labrador. 10 FIN ICE NOR SVA SWE 14 RUN 30 KRA WSB YAK 31 KAM MAG 70 ALU ASK GNL NWT NUN 71 ABT BRC MAN 72 LAB QUE 73 COL IDA MNT WYO 76 UTA 77 NWM. Hemicr. 2n = 34.

Campanula uyemurae (Kudô) Miyabe & Tatew., Trans. Sapporo Nat. Hist. Soc. 14: 81 (1935).
Russian Far East (Sakhalin). 31 SAK. Hemicr. or geophyte.
* *Adenophora uyemurae* Kudô, Bull. Kyûshû Univ. Forests 1: 88 (1931). *Popoviocodonia uyemurae* (Kudô) Fed. in Kom., Fl. URSS 24: 373 (1957).

Campanula vaillantii Quézel, Bull. Soc. Sci. Nat. Maroc 34: 310 (1955).
Morocco. 20 MOR. Hemicr. 2n = 14.

Campanula velata Pomel, Nouv. Mat. Fl. Atl.: 2 (1874).
N. Africa. 20 ALG MOR. Hemicr. 2n = 26.

subsp. **mesatlantica** (Litard. & Maire) Quézel, Feddes Repert. Spec. Nov. Regni Veg. 56: 19, 65 (1953).
Morocco. 20 MOR. Hemicr. 2n = 26.
* *Campanula velata* var. *mesatlantica* Litard. & Maire, Mém. Soc. Sci. Nat. Maroc 26: 27 (1931).

subsp. **serpylliformis** (Batt.) Quézel, Feddes Repert. Spec. Nov. Regni Veg. 56: 19, 65 (1953).
N. Africa. 20 ALG MOR. Hemicr.
* *Campanula serpylliformis* Batt. in Batt. & Trab., Atlas Fl. Alger: 11 (1886).

subsp. **velata**
Algeria. 20 ALG. Hemicr.

Campanula velebitica Borbás, Math. Természettud. Értes. 1: 81 (1883). *Campanula rotundifolia* subsp. *velebitica* (Borbás) Hayek, Repert. Spec. Nov. Regni Veg. Beih. 30(2): 541 (1930). *Campanula balcanica* var. *velebitica* (Borbás) Hruby, Magyar Bot. Lapok 29: 236 (1930). Balkans. 13 ALB BUL GRC YUG. Hemicr. or geophyte. 2n = 34, 68
Campanula farinulenta A. Kern. & Wettst., Oesterr. Bot. Z. 37: 80 (1887). *Campanula rotundifolia* var. *farinulenta* (A. Kern. & Wettst.) Hayek, Repert. Spec. Nov. Regni Veg. Beih. 30(2): 540 (1930). *Campanula balcanica* f. *farinulenta* (A. Kern. & Wettst.) Hruby, Magyar Bot. Lapok 29: 239 (1930).
Campanula velebitica f. *borbasiana* Witasek, Magyar Bot. Lapok. 5: 242 (1906). *Campanula rotundifolia* f. *borbasiana* (Witasek) Hayek, Repert. Spec. Nov. Regni Veg. Beih. 30(2): 539 (1930). *Campanula balcanica* f. *borbasiana* (Witasek) Hruby, Magyar Bot. Lapok 33: 146 (1934).
Campanula velebitica f. *farinulenta* Witasek, Magyar Bot. Lapok. 5: 242 (1906).
Campanula velebitica f. *incerta* Witasek, Magyar Bot. Lapok. 5: 242 (1906). *Campanula rotundifolia* f. *incerta* (Witasek) Hayek, Repert. Spec. Nov. Regni Veg. Beih. 30(2): 540 (1930). *Campanula balcanica* f. *incerta* (Witasek) Hruby, Magyar Bot. Lapok 33: 146 (1934).

Campanula velebitica f. *parviflora* Witasek, Magyar Bot. Lapok. 5: 242 (1906). *Campanula rotundifolia* f. *parviflora* (Witasek) Hayek, Repert. Spec. Nov. Regni Veg. Beih. 30(2): 540 (1930). *Campanula balcanica* f. *parviflora* (Witasek) Hruby, Magyar Bot. Lapok 29: 237 (1930).

Campanula velebitica f. *divaricata* Witasek, Magyar Bot. Lapok. 5: 242 (1906). *Campanula rotundifolia* f. *divaricata* (Witasek) Hayek, Repert. Spec. Nov. Regni Veg. Beih. 30(2): 539 (1930). *Campanula balcanica* f. *divaricata* (Witasek) Hruby, Magyar Bot. Lapok 33: 146 (1934).

Campanula bulgarica Witasek, Magyar Bot. Lapok. 5: 244 (1906). *Campanula rotundifolia* subsp. *bulgarica* (Witasek) Hayek, Repert. Spec. Nov. Regni Veg. Beih. 30(2): 538 (1930). *Campanula balcanica* subvar. *bulgarica* (Witasek) Hruby, Magyar Bot. Lapok 29: 250 (1930).

Campanula balcanica subf. *normalis* Hruby, Magyar Bot. Lapok 29: 239 (1930).

Campanula balcanica subf. *umbrosa* Hruby, Magyar Bot. Lapok 29: 240 (1930).

Campanula veneris Carlström, Willdenowia 15: 384 (1986).
Cyprus. 34 CYP. Ther.

Campanula versicolor Andrews, Bot. Repos. 6: pl. 396 (1804).
SE. Europe. 13 ALB BUL GRC ITA YUG. Hemicr. 2n = 34.
Campanula planiflora Willd., Enum. Pl.: 210 (1809); non Lam., Encycl. 1: 580 (1785). *Campanula willdenowiana* Schult. in Roem. & Schult., Syst. Veg. 5: 107 (1819).
Campanula corymbosa Ten., Fl. Napol. 1: xv (1811); non Desf., Ann. Mus. Natl. Hist. Nat. 11: 139 (1808). *Campanula tenorei* Moretti, Giorn. Fis. (ser. 2) 7: ? (1824), as 'tenorii'. *Campanula rosanii* Ten., Fl. Napol. 3: 205 (1824-29), as 'rosani'. *Campanula versicolor* subsp. *rosanii* Nyman, Consp. Fl. Eur.: 480 (1879), as 'rosani'.
Campanula versicolor var. *multiflora* A. DC., Monogr. Campan.: 308 (1830).
Campanula versicolor var. *thessala* Boiss., Fl. Orient. 3: 915 (1875).
Campanula mrkvickana Velen., Allg. Bot. Z. Syst. 11: 44 (1905). *Campanula versicolor* f. *mrkvickana* (Velen.) Hayek, Repert. Spec. Nov. Regni Veg. Beih. 30(2): 543 (1930).

Campanula waldsteiniana Schult. in Roem. & Schult., Syst. Veg. 5: 99 (1819).
Croatia & Hercegovina. 13 YUG. Hemicr. or geophyte. 2n = 34.
* *Campanula flexuosa* Waldst. & Kit., Descr. Icon. Pl. Hung. 2: 145 (1803-1805); non Michx., Fl. Bor.-Amer. 1: 109 (1803). *Campanula rupestris* Host, Fl. Austriac. 1: 263 (1827); non Sm. in Sibth. & Sm., Fl. Graec. Prodr. 1: 142 (1806); nec M. Bieb., Fl. Taur.-Cauc. 1: 154 (1808).

Campanula wanneri Rochel, Pl. Banat. Rar.: 41 (1828). *Symphyandra wanneri* (Rochel) Heuff., Verh. K. K. Zool.-Bot. Ges. Wien 8: 156 (1858).
Balkans. 13 BUL GRC ROM YUG. Hemicr. 2n = 34.

Campanula wattiana Nayar & Babu, J. Indian Bot. Soc. 49: 183 (1971).
W. Himalaya. 40 WHM. Hemicr.

Campanula wilkinsiana Greene, Pittonia 4: 38 (1899)
W. U.S.A. (N. California). 76 CAL. Hemicr. or geophyte.
Campanula baileyi Eastw., Bull. Torrey Bot. Club 29: 525 (1902).

Campanula willkommii Witasek, Abh. K. K. Zool.-Bot. Ges. Wien 1(3): 75 (1902).
S. Spain (Sierra Nevada). 12 SPA. Hemicr. or geophyte. 2n = 68.
Campanula nevadensis Pau, Bol. Soc. Aragonesa Ci. Nat. 8: 124, 130 (1909).

Campanula witasekiana Vierh., Mitt. Naturwiss. Vereins Univ. Wien 4: 72 (1906).
Campanula scheuchzeri f. *witasekiana* (Vierh.) Hayek, Abh. K. K. Zool.-Bot. Ges. Wien 4(2): 128 (1907). *Campanula hostii* var. *witasekiana* (Vierh.) K. Malý, Glasn. Zemaljsk. Muz.

Bosni Hercegovini 20: 555 (1908). *Campanula scheuchzeri* var. *witasekiana* (Vierh.) Dalla Torre & Sarnth., Fl. Tirol. 6(3): 448 (1911). *Campanula scheuchzeri* subsp. *witasekiana* (Vierh.) Hayek, Fl. Steiermark 2(1): 454 (1912).
Alps to Balkans. 11 AUT 13 BUL ITA YUG. Hemicr. or geophyte. 2n = 34.
  Campanula witasekiana var. *praticola* Hruby, Magyar Bot. Lapok 29: 207 (1930).

Campanula xylocarpa Kovanda, Folia Geobot. Phytotax. 1: 183 (1966).
  SE. Slovakia. 11 CZE. Hemicr. or geophyte. 2n = 34.

Campanula yaltirikii H. Duman, Edinburgh J. Bot. 56: 355 (1999).
  Turkey. 34 TUR. Hemicr.

Campanula yildirimlii Kit Tan & Sorger, Pl. Syst. Evol. 154: 124 (1986).
  Turkey. 34 TUR. Hemicr. or geophyte.

Campanula yunnanensis D. Y. Hong, Fl. Reipubl. Popul. Sin. 73(2): 184 (1983).
  SC. China (Yunnan). 36 CHC. Hemicr.

Campanula zangezura (Lipsky) Kolak. & Serdyuk., Zametki Sist. Geogr. Rast. 36: 46 (1980).
  Caucasus (Armenia, Nakhichevan) & Iran. 33 TCS 34 IRN. Hemicr. or geophyte.
  ** Symphyandra zangezura* Lipsky, Trudy Imp. S.-Peterburgsk. Bot. Sada 13: 317 (1894).

Campanula zeyensis Amirkh. & Tavasiev, Bjull. Moskovsk. Obšč. Isp. Prir., Otd. Biol. 84(6): 119 (1979).
  Caucasus (Severo-Osetiya). 33 NCS. Hemicr. or geophyte.

Synonyms:
Campanula sect. *Adenophora* (Fisch.) D. Dietr. === **Adenophora** Fisch.
Campanula sect. *Annaea* (Kolak.) Viktorov === **Campanula** L.
Campanula sect. *Azorina* (Feer) Kuntze === **Azorina** Feer
Campanula sect. *Campanopsis* R. Br. === **Wahlenbergia** Schrad. ex Roth
Campanula sect. *Codonia* Colla === **Wahlenbergia** Schrad. ex Roth
Campanula sect. *Codonopsis* (Wall.) D. Dietr. === **Codonopsis** Wall.
Campanula sect. *Diosphaera* (Buser) Kuntze === **Campanula** L.
Campanula sect. *Favratia* (Feer) Kuntze === **Favratia** Feer
Campanula sect. *Floerkeae* Colla === **Adenophora** Fisch.
Campanula sect. *Gadellia* (Kolak.) Viktorov === **Campanula** L.
Campanula sect. *Marianthemum* (Schrank) Schur === **Campanula** L.
Campanula sect. *Microcodon* (A. DC.) D. Dietr. === **Microcodon** A. DC.
Campanula sect. *Musschia* (Dumort.) D. Dietr. === **Musschia** Dumort.
Campanula sect. *Prismatocarpus* (L'Hér.) Schult. === **Prismatocarpus** L'Hér.
Campanula sect. *Pseudocampanula* (Kolak.) Viktorov === **Campanula** L.
Campanula sect. *Sicyocodon* (Feer) Kuntze === **Campanula** L.
Campanula sect. *Specularia* D. Dietr. === **Legousia** Durande
Campanula sect. *Theodorovia* (Kolak.) Viktorov === **Theodorovia** Kolak.
Campanula sect. *Tracheliopsis* (Buser) Kuntze === **Campanula** L.
Campanula sect. *Trachelium* (L.) Kuntze === **Trachelium** L.
Campanula sect. *Wahlenbergia* (Schrad. ex Roth) D. Dietr. === **Wahlenbergia** Schrad. ex Roth
Campanula subg. *Adenophora* (Fisch.) Borbás === **Adenophora** Fisch.
Campanula subg. *Annaea* (Kolak.) Ogan. === **Campanula** L.
Campanula subg. *Brachycodonia* (Fed.) Damboldt === **Campanula** L.
Campanula subg. *Legousia* (Durande) Raf. === **Legousia** Durande
Campanula subg. *Pseudocampanula* (Kolak.) Ogan. === **Campanula** L.
Campanula subg. *Roucela* (Dumort.) Damboldt === **Campanula** L.
Campanula subg. *Sicyocodon* (Feer) Damboldt === **Campanula** L.
Campanula subg. *Theodorovia* (Kolak.) Ogan. === **Theodorovia** Kolak.

Campanula [unranked] *Legousia* (Durande) Pers. === **Legousia** Durande

Campanula aaronsohnii Evenari === **Campanula damascena** Labill.

Campanula abietina Griseb. & Schenk === **Campanula patula** subsp. **abietina** (Griseb. & Schenk) Simonk.

Campanula acarnanica Damboldt === **Campanula garganica** subsp. **acarnanica** (Damboldt) Damboldt

Campanula acuminata Michx. === **Campanula americana** L.

Campanula acutangula Leresche & Levier === **Campanula arvatica** Lag. subsp. **arvatica**

Campanula adamii M. Bieb. === **Campanula bellidifolia** Adam subsp. **bellidifolia**

Campanula adpressa L. f. === **Wahlenbergia adpressa** (L. f.) Sond.

Campanula adscendens Vest ex Schult. === ?

Campanula adsurgens Levier & Leresche === **Campanula arvatica** subsp. **adsurgens** (Levier & Leresche) Damboldt

Campanula affinis Fisch. ex Rupr. === **Campanula bellidifolia** subsp. **aucheri** (A. DC.) Viktorov

Campanula affinis Schult. === **Campanula speciosa** subsp. **affinis** (Schult.) Font Quer

Campanula affinis subsp. *bolosii* (Vayr.) Fed. === **Campanula speciosa** subsp. **affinis** (Schult.) Font Quer

Campanula affinis var. *ramosa* Marcet === **Campanula speciosa** subsp. **affinis** (Schult.) Font Quer

Campanula afra var. *pallida* (Maire) Dobignard === **Campanula hypocrateriformis** Dobignard

Campanula aggregata Hegetschw. === ?

Campanula aggregata Willd. === **Campanula glomerata** L. subsp. **glomerata**

Campanula agrestis Wall. === **Wahlenbergia marginata** (Thunb. ex Murray) A. DC.

Campanula aizoon subsp. *aizoides* (Zaffran ex Greuter) Fed. === **Campanula aizoides** Zaffran ex Greuter

Campanula ajugifolia Sest. ex Schult. === ?

Campanula akhverdovii Kharadze === **Campanula kolenatiana** C. A. Mey. ex Rupr.

Campanula akuschensis Husseinov === **Campanula tridentata** subsp. **biebersteiniana** (Schult.) Ogan.

Campanula alata Bové ex Vatke === **Campanula peregrina** L.

Campanula alberti Trautv. === **Campanula stevenii** subsp. **alberti** (Trautv.) Viktorov

Campanula albicans (Buser) Engl. === **Campanula petraea** L.

Campanula albiflora K. Koch === **Campanula sarmatica** Ker Gawl. subsp. **sarmatica**

Campanula albovii Kolak. === **Campanula collina** subsp. **sphaerocarpa** (Kolak.) Ogan.

Campanula alburnica V. Brig. === **Asyneuma trichocalycinum** (Ten.) K. Malý

Campanula algida Fisch. ex A. DC. === **Campanula lasiocarpa** Cham.

Campanula alliariifolia subsp. *ochroleuca* Kem.-Nat. === **Campanula alliariifolia** Willd. subsp. **alliariifolia**

Campanula alliariifolia var. *alpestris* Kem.-Nat. === **Campanula alliariifolia** Willd. subsp. **alliariifolia**

Campanula alliariifolia var. *cordata* Trautv. === **Campanula alliariifolia** Willd. subsp. **alliariifolia**

Campanula alliariifolia var. *macrophylla* (Sims) A. DC. === **Campanula alliariifolia** Willd. subsp. **alliariifolia**

Campanula alliariifolia var. *rupestris* Kem.-Nat. === **Campanula alliariifolia** Willd. subsp. **alliariifolia**

Campanula alliariifolia var. *silvatica* Kem.-Nat. === **Campanula alliariifolia** Willd. subsp. **alliariifolia**

Campanula allionii Vill. === **Campanula alpestris** All.

Campanula allionii subsp. *monanthos* (Pant.) Nyman === **Campanula patula** L. subsp. **patula**

Campanula alpigena K. Koch === **Campanula bellidifolia** subsp. **aucheri** (A. DC.) Viktorov

Campanula alpina var. *albiiflora* Schur === **Campanula alpina** Jacq. subsp. **alpina**

Campanula alpina var. *calyculata* Schur === **Campanula alpina** Jacq. subsp. **alpina**

Campanula alpina var. *frigida* Schur === **Campanula alpina** Jacq. subsp. **alpina**

Campanula alpina var. *haynaldii* (Szontágh) Nyman === **Campanula alpina** Jacq. subsp. **alpina**
Campanula alpini L. === **Adenophora liliifolia** (L.) A. DC.
Campanula alpini var. *asperula* Borbás === **Adenophora liliifolia** (L.) A. DC.
Campanula alpini var. *botryantha* Borbás === **Adenophora liliifolia** (L.) A. DC.
Campanula alpini var. *hungarica* Borbás === **Adenophora liliifolia** (L.) A. DC.
Campanula alpini var. *villosula* Borbás === **Adenophora liliifolia** (L.) A. DC.
Campanula alpium-maritimarum Hruby === **Campanula macrorhiza** J. Gay ex A. DC.
Campanula altaica Ledeb. === **Campanula stevenii** subsp. **altaica** (Ledeb.) Fed.
Campanula altiflora (L'Hér.) Pers. === **Prismatocarpus altiflorus** L'Hér.
Campanula amabilis Leichtlin ex Spigolatore === **Campanula phyctidocalyx** Boiss. & Noë
Campanula amana Rech. f. === **Campanula postii** (Boiss.) Engl.
Campanula amasiae Post === ?
Campanula americana f. *albiflora* M. L. Grant === **Campanula americana** L.
Campanula americana f. *tubuliflora* Fernald === **Campanula americana** L.
Campanula americana var. *illinoensis* (Fresen.) Farw. === **Campanula americana** L.
Campanula americana var. *subulata* P. Beauv. ex A. DC. === **Campanula americana** L.
Campanula amoris Sennen === ?
Campanula amplexicaulis Michx. === **Triodanis perfoliata** (L.) Nieuwl.
Campanula amplexicaulis (Willd.) Boiss. === **Asyneuma amplexicaule** (Willd.) Hand.-Mazz.
Campanula amygdalifolia Salisb. === **Campanula persicifolia** L. subsp. **persicifolia**
Campanula andina var. *alexeenkoi* Fomin === **Campanula andina** Rupr.
Campanula andorrana Braun-Blanq. === **Campanula jaubertiana** Timb.-Lagr.
Campanula andrewsii subsp. *breueri* Tocl & Rohlena === **Campanula lavrensis** (Tocl & Rohlena) Phitos
Campanula andrewsii var. *lavrensis* Tocl & Rohlena === **Campanula lavrensis** (Tocl & Rohlena) Phitos
Campanula andrewsii var. *pelia* Halácsy === **Campanula pelia** (Halácsy) Hausskn. & Sint. ex Phitos
Campanula androsacea (A. DC.) D. Dietr. === **Wahlenbergia androsacea** A. DC.
Campanula angulata Raf. === **Triodanis perfoliata** subsp. **biflora** (Ruiz & Pav.) Lammers
Campanula angustiflora var. *exilis* J. T. Howell === **Campanula griffinii** Morin
Campanula angustifolia Lam. === **Campanula rotundifolia** L.
Campanula annae Kolak. === **Campanula collina** Sims subsp. **collina**
Campanula anomala Fomin === **Campanula circassica** Fomin
Campanula antiatlantica Maire, Weiller & Wilczek === **Campanula filicaulis** Durieu
Campanula antiqua (Kolak.) Kolak. & Serdyuk. === **Campanula pendula** M. Bieb.
Campanula antirrhina Schleich. === **Campanula rotundifolia** L.
Campanula aparinoides var. *erinoides* A. W. Wood === **Campanula aparinoides** Pursh
Campanula aparinoides var. *grandiflora* Holz. === **Campanula aparinoides** Pursh
Campanula aparinoides var. *rosea* Coleman === **Campanula aparinoides** Pursh
Campanula aparinoides var. *uliginosa* (Rydb.) Gleason === **Campanula aparinoides** Pursh
Campanula apennina (Podlech) Podlech === **Campanula micrantha** Bertol.
Campanula appendiculata A. DC. === **Campanula tomentosa** Lam.
Campanula aprica Nannf. === **Campanula cana** Wall.
Campanula arctica (Lange) Hruby === **Campanula giesekiana** Vest. ex Schult.
Campanula arctica f. *simplex* Hruby === **Campanula rotundifolia** L.
Campanula arctica subsp. *mentiens* (Witasek) Hruby === **Campanula tatrae** subsp. **mentiens** (Witasek) Kovanda
Campanula arctolinifolia Hruby === ?
Campanula arcuata Schur === **Campanula serrata** (Kit.) Hendrych subsp. **serrata**
Campanula arcuata var. *subrhomboidalis* Schur === **Campanula rotundifolia** L.
Campanula ardonensis var. *kryophila* (Rupr.) Boiss. === **Campanula kryophila** Rupr.
Campanula ardonensis var. *latifolia* Fomin === **Campanula ardonensis** Rupr.
Campanula arenaria Formánek === ?
Campanula argaea subsp. *pamphylica* Contandr., Quézel & Pamukç. === **Campanula tokurii** Ocak

Campanula argunensis Rupr. === **Campanula bellidifolia** subsp. **argunensis** (Rupr.) Viktorov
Campanula arida Kunth === **Wahlenbergia linarioides** (Lam.) A. DC.
Campanula aristata var. *longisepala* C. Marquand === **Campanula aristata** Wall.
Campanula armazica Kharadze === **Campanula bellidifolia** subsp. **aucheri** (A. DC.) Viktorov
Campanula aspera (Boiss.) Boiss. === **Asyneuma persicum** (A. DC.) Bornm.
Campanula aspera Moench === ?
Campanula asperrima Zuccagni   === ?
Campanula asteroides Lam. === **Campanula americana** L.
Campanula asturica Podlech === **Campanula rotundifolia** L.
Campanula athoa Boiss. & Heldr. === **Campanula trachelium** subsp. **athoa** (Boiss. & Heldr.) Nyman
Campanula atlantica Coss. & Durieu ex Batt. === **Campanula afganica** Pomel
Campanula atlantica var. *glabra* Bonnet === **Campanula afganica** Pomel
Campanula atlantica var. *guergourensis* Batt. === **Campanula afganica** Pomel
Campanula atlantica var. *maroccana* (Ball) Batt. === **Campanula filicaulis** Durieu
Campanula attica Boiss. & Heldr. === **Campanula drabifolia** Sm.
Campanula aucheri A. DC. in DC. === **Campanula bellidifolia** subsp. **aucheri** (A. DC.) Viktorov
Campanula aucheri var. *compacta* Fomin === **Campanula bellidifolia** subsp. **aucheri** (A. DC.) Viktorov
Campanula aurasiaca (Batt. & Trab.) Batt. & Trab. === **Asyneuma rigidum** subsp. **aurasiacum** (Batt. & Trab.) Damboldt
Campanula aurea L. f. === **Musschia aurea** (L. f.) Dumort.
Campanula aurea var. *angustifolia* Ker Gawl. === **Musschia aurea** (L. f.) Dumort.
Campanula axilliflora Borbás === **Adenophora stricta** Miq. subsp. **stricta**
Campanula azurea Sol. ex Sims === **Campanula rhomboidalis** L.
Campanula baileyi Eastw. === **Campanula wilkinsiana** Greene
Campanula balcanica (Adamović) Hruby === **Campanula scheuchzeri** Vill.
Campanula balcanica f. *borbasiana* (Witasek) Hruby === **Campanula velebitica** Borbás
Campanula balcanica f. *divaricata* (Witasek) Hruby === **Campanula velebitica** Borbás
Campanula balcanica f. *farinulenta* (A. Kern. & Wettst.) Hruby === **Campanula velebitica** Borbás
Campanula balcanica f. *incerta* (Witasek) Hruby === **Campanula velebitica** Borbás
Campanula balcanica f. *racemosa* (Krašan) Hruby === **Campanula rotundifolia** L.
Campanula balcanica subf. *justiniana* (Witasek) Hruby === **Campanula justiniana** Witasek
Campanula balcanica subf. *normalis* Hruby === **Campanula velebitica** Borbás
Campanula balcanica subf. *simplex* Hruby === **Campanula albanica** subsp. **sancta** (Hayek) Podlech
Campanula balcanica subf. *umbrosa* Hruby === **Campanula velebitica** Borbás
Campanula balcanica subvar. *bulgarica* (Witasek) Hruby === **Campanula velebitica** Borbás
Campanula balcanica var. *hispanica* (Willk.) Hruby === **Campanula hispanica** Willk.
Campanula balcanica var. *scheuchzeriformis* (Hayek) Hruby === **Campanula tatrae** Borbás subsp. **tatrae**
Campanula balcanica var. *velebitica* (Borbás) Hruby === **Campanula velebitica** Borbás
Campanula baldensis Balb. === **Campanula ramosissima** Sm.
Campanula banksiana (A. DC.) D. Dietr. === **Wahlenbergia banksiana** A. DC.
Campanula barbata f. *pusilla* (Gaudin) Hayek & Hegi === **Campanula barbata** L.
Campanula barbata var. *firmiana* (Vand.) Steud. === **Campanula barbata** L.
Campanula barbata var. *pusilla* Gaudin === **Campanula barbata** L.
Campanula barbata var. *strictopedunculata* Thomas ex Rchb. === **Campanula barbata** L.
Campanula barbata var. *uniflora* A. DC. === **Campanula barbata** L.
Campanula barbeyi Feer === **Campanula elatines** L.
Campanula barrelieri C. Presl === **Campanula fragilis** Cirillo subsp. **fragilis**
Campanula barrelieri Marnock === ?
Campanula baumgartenii subsp. *bohemica* (Hruby) Tacik === **Campanula bohemica** Hruby
Campanula bayerniana subsp. *choziatowskyi* (Fomin) Ogan. === **Campanula bayerniana** Rupr.

Campanula bayerniana var. *andina* (Rupr.) Trautv. === **Campanula andina** Rupr.

Campanula bayerniana var. *trautvetteri* Fomin === **Campanula bayerniana** Rupr.

Campanula beauverdiana Fomin === **Campanula stevenii** subsp. **beauverdiana** (Fomin) Rech. f. & Schiman-Czeika

Campanula beauverdiana var. *aissori* Tamamsch. === **Campanula stevenii** subsp. **beauverdiana** (Fomin) Rech. f. & Schima-Czeika

Campanula beckiana Hayek === **Campanula baumgartenii** subsp. **beckiana** (Hayek) Podlech

Campanula bellardii All. === **Campanula cochleariifolia** Lam.

Campanula bellardii var. *delpontei* (Chabert) Bég. === **Campanula cochleariifolia** Lam.

Campanula bellardii var. *hoppeana* (Rupr. ex Rchb.) Bég. === **Campanula cochleariifolia** Lam.

Campanula bellardii var. *paniculata* (Nageli ex Rchb.) Bég. === **Campanula cochleariifolia** Lam.

Campanula bellardii var. *pubescens* (F. W. Schmidt) Bég. === **Campanula cochleariifolia** Lam.

Campanula bellidifolia Lapeyr. === **Campanula patula** L. subsp. **patula**

Campanula bellidifolia var. *longisepala* Fomin === **Campanula bellidifolia** Adam subsp. **bellidifolia**

Campanula beltranii Sennen & Losa === **Campanula speciosa** Pourr. subsp. **speciosa**

Campanula benthamii Wall. ex Kitam. === **Campanula dimorphantha** Schweinf.

Campanula bergiana (Cham.) D. Dietr. === **Prismatocarpus fruticosus** (L.) L'Hér.

Campanula bertolae subsp. *xylorrhiza* (O. Schwarz) Podlech === **Campanula moravica** subsp. **xylorrhiza** (O. Schwarz) Kovanda

Campanula bertolae var. *re* (Colla) Gola === **Campanula bertolae** Colla

Campanula besenginica Fomin === **Campanula bellidifolia** subsp. **besenginica** (Fomin) Viktorov

Campanula betetae Caball. === **Campanula hispanica** Willk. subsp. **hispanica**

Campanula betonicifolia Biehler === **Campanula sarmatica** Ker Gawl. subsp. **sarmatica**

Campanula betonicifolia var. *byzantina* K. Koch === **Campanula betonicifolia** Sm.

Campanula betonicifolia var. *micrantha* K. Koch === **Campanula betonicifolia** Sm.

Campanula betonicifolia var. *multicaulis* K. Koch === **Campanula betonicifolia** Sm.

Campanula betulifolia var. *exappendiculata* Albov === **Campanula betulifolia** K. Koch

Campanula betulifolia var. *glaberrima* Parsa === **Campanula bayerniana** Rupr.

Campanula bicaulis Lapeyr. === **Campanula speciosa** Pourr. subsp. **speciosa**

Campanula biebersteiniana Schult. === **Campanula tridentata** subsp. **biebersteiniana** (Schult.) Ogan.

Campanula biebersteiniana var. *macrantha* A. DC. === **Campanula tridentata** Schreb. subsp. **tridentata**

Campanula bielzii Schur === **Campanula rotundifolia** L.

Campanula biflora Ruiz & Pav. === **Triodanis perfoliata** subsp. **biflora** (Ruiz & Pav.) Lammers

Campanula billardierei A. DC. === **Campanula cymbalaria** Sm.

Campanula billardierei var. *major* A. DC. === **Campanula cymbalaria** Sm.

Campanula biserrata K. Koch === **Campanula lactiflora** M. Bieb.

Campanula bithynica A. DC. === **Campanula tridentata** Schreb. subsp. **tridentata**

Campanula bocconei Vill. === **Campanula rotundifolia** L.

Campanula boissieri Vatke === **Asyneuma persicum** (A. DC.) Bornm.

Campanula bolosii Vayr. === **Campanula speciosa** subsp. **affinis** (Schult.) Font Quer

Campanula bononiensis f. *tenuiflora* (Nyman) Sóo === **Campanula bononiensis** L.

Campanula bononiensis subsp. *obliquifolia* (Ten.) Arcang. === **Campanula bononiensis** L.

Campanula bononiensis subvar. *ruthenica* (M. Bieb.) A. DC. === **Campanula bononiensis** L.

Campanula bononiensis var. *cana* Nyman === **Campanula bononiensis** L.

Campanula bononiensis var. *latifolia* Schur === **Campanula bononiensis** L.

Campanula bononiensis var. *lychnitis* (Hornem.) A. DC. === **Campanula bononiensis** L.

Campanula bononiensis var. *obliquifolia* (Ten.) A. DC. === **Campanula bononiensis** L.

Campanula bononiensis var. *ruthenica* (M. Bieb.) A. DC. === **Campanula bononiensis** L.

Campanula bononiensis var. *tenuiflora* Nyman === **Campanula bononiensis** L.

Campanula bononiensis var. *simplex* (DC.) Steud. === **Campanula bononiensis** L.

Campanula bourdiniana Gand. === **Campanula medium** L.

Campanula brachiata Seidl ex Opiz === **Campanula patula** L. subsp. **patula**

Campanula brachysepala Vuk. === **Campanula morettiana** Rchb.

Campanula bracteata Thunb. === **Roella spicata** L. f.

Campanula brassicifolia Sommier & Levier === **Campanula sibirica** subsp. **brassicifolia** (Sommier & Levier) Ogan.

Campanula brevibracteata (H. Buek) D. Dietr. === **Microcodon glomeratus** A. DC.

Campanula breyniana (Beck) Hruby === **Campanula praesignis** Beck

Campanula brodensis Formánek === **Campanula patula** L. subsp. **patula**

Campanula brotheri Sommier & Levier === **Campanula raddeana** Trautv.

Campanula brotherorum Feer === **Campanula sarmatica** Ker Gawl. subsp. **sarmatica**

Campanula broussonetiana Schult. === **Campanula lusitanica** Loefl. subsp. **lusitanica**

Campanula buekii D. Dietr. === **Wahlenbergia pilosa** H. Buek

Campanula bulgarica Davidov === ?

Campanula bulgarica Witasek === **Campanula velebitica** Borbás

Campanula burgalensis Font Quer === **Campanula hispanica** Willk. subsp. **hispanica**

Campanula bzybica Jabr.-Kolak. === **Campanula longistyla** Fomin

Campanula caballeroi Sennen & Losa === **Campanula rotundifolia** L.

Campanula calavrytana Beauverd & Topali === **Campanula topaliana** subsp. **cordifolia** Phitos

Campanula calcarea (Albov) Albov ex Kharadze === **Campanula sarmatica** subsp. **calcarea** (Albov) Ogan.

Campanula calycina Boeber ex Schult. === **Campanula rapunculus** L. subsp. **rapunculus**

Campanula camtschatica Pall. ex Schult. === ?

Campanula cana Simonk. === **Campanula bononiensis** L.

Campanula canariensis L. === **Canarina canariensis** (L.) Vatke

Campanula candolleana (Cham.) D. Dietr. === **Prismatocarpus candolleanus** Cham.

Campanula canescens (Waldst. & Kit.) Roth === **Asyneuma canescens** (Waldst. & Kit.) Griseb. & Schenk

Campanula canescens Wall. ex A. DC. === **Campanula dimorphantha** Schweinf.

Campanula cantabrica subsp. *occidentalis* M. Laínz === **Campanula cantabrica** Feer

Campanula capensis L. === **Wahlenbergia capensis** (L.) A. DC.

Campanula caperonioides Klotzsch === **Campanula sylvatica** Wall.

Campanula capillacea L. f. === **Wahlenbergia capillacea** (L. f.) A. DC.

Campanula capillaris Lodd. === **Wahlenbergia capillaris** (Lodd.) Sweet

Campanula capitata Schur === **Campanula cervicaria** L.

Campanula capitata Sims === **Campanula lingulata** Waldst. & Kit.

Campanula cappadocicum (Boiss.) Boiss. === **Asyneuma virgatum** (Labill.) Bornm. subsp. **virgatum**

Campanula capusii (Franch.) Fed. === **Campanula lehmanniana** subsp. **capusii** (Franch.) Viktorov.

Campanula cariensis A. DC. === **Campanula betonicifolia** Sm.

Campanula carnica subsp. *consanguinea* (Schott) Nyman === **Campanula scheuchzeri** Vill.

Campanula carnica var. *angustifolia* Schur === **Campanula carnica** Schiede ex Mert. & Koch subsp. **carnica**

Campanula carnica var. *hirta* Gelmi === **Campanula carnica** Schiede ex Mert. & Koch subsp. **carnica**

Campanula carnica var. *latifolia* Schur === **Campanula carnica** Schiede ex Mert. & Koch subsp. **carnica**

Campanula carnica var. *racemosa* Krašan === **Campanula rotundifolia** L.

Campanula carniolica (Sünd.) Crook === **Campanula thyrsoides** subsp. **carniolica** (Sünd.) Podlech

Campanula carnosa Wall. === **Peracarpa carnosa** (Wall.) Hook. f. & Thomson

Campanula carpatha var. *oreophila* Phitos === **Campanula carpatha** Halácsy

Campanula carpatica f. *dasycarpa* (Schur) Tacik === **Campanula carpatica** Jacq.

Campanula carpatica f. *subpilosa* (Schur) Tacik === **Campanula carpatica** Jacq.

Campanula carpatica subsp. *turbinata* (Schott) Nyman === **Campanula carpatica** Jacq.

Campanula carpatica var. *dasycarpa* Schur === **Campanula carpatica** Jacq.

Campanula carpatica var. *grandiflora* Schur === **Campanula carpatica** Jacq.

Campanula carpatica var. *hemisphaerica* Schur === **Campanula carpatica** Jacq.

Campanula carpatica var. *hendersonii* (Anonymous) W. T. Mill. === **Campanula carpatica** Jacq.

Campanula carpatica var. *oreophila* Schur === **Campanula carpatica** Jacq.

Campanula carpatica var. *pelviformis* Anonymous === **Campanula carpatica** Jacq.

Campanula carpatica var. *subpilosa* Schur === **Campanula carpatica** Jacq.

Campanula carpatica var. *turbinata* (Schott) Anonymous === **Campanula carpatica** Jacq.

Campanula cashmeriana var. *evolvulacea* (Royle ex A. DC.) C. B. Clarke === **Campanula cashmeriana** Benth.

Campanula castellana Pau === **Campanula rapunculus** L. subsp. **rapunculus**

Campanula catinensis Tornab. === **Campanula dichotoma** L. f.

Campanula caucasica var. *major* K. Koch === **Campanula caucasica** M. Bieb.

Campanula caudata Vis. === **Edraianthus dalmaticus** (A. DC.) A. DC.

Campanula cavolinii Ten. === **Campanula fragilis** subsp. **cavolinii** (Ten.) Damboldt

Campanula cavolinii var. *hirta* Ten. === **Campanula fragilis** subsp. **cavolinii** (Ten.) Damboldt

Campanula cecilii Anonymous === **Campanula reuteriana** Boiss. & Balansa

Campanula celebica (Blume) D. Dietr. === **Cyclocodon lancifolius** subsp. **celebicus** (Blume) K. E. Morris & Lammers

Campanula celtidifolia Boiss. & A. Huet === **Campanula lactiflora** M. Bieb.

Campanula cephalaria Vuk. === ?

Campanula cephalaria var. *cardiophylla* Vuk. === ?

Campanula cephalaria var. *gnaphalophylla* Vuk. === **Campanula petraea** L.

Campanula cephalaria var. *macrophylla* Schloss. ex Vuk. === **Campanula cervicaria** L.

Campanula cephalaria var. *ovaliphylla* Vuk. === ?

Campanula cephalaria var. *polyanthemos* Vuk. === ?

Campanula cephallenica Feer === **Campanula garganica** subsp. **cephallenica** (Feer) Hayek

Campanula cephalotes Fisch. ex Schrank === **Campanula glomerata** subsp. **speciosa** (Hornem. ex Spreng.) Domin

Campanula cephalotes f. *alba* Nakai === **Campanula glomerata** subsp. **speciosa** (Hornem. ex Spreng.) Domin

Campanula cephalotes var. *canescens* Maxim. ex Nakai === **Campanula glomerata** subsp. **speciosa** (Hornem. ex Spreng.) Domin

Campanula cernua Thunb. === **Wahlenbergia cernua** (Thunb.) A. DC.

Campanula cervicaria subsp. *macrostachya* (Waldst. & Kit. ex Willd.) Tacik === **Campanula macrostachya** Waldst. & Kit. ex Willd.

Campanula cervicaria subsp. *transsilvanica* (Schur ex Andrae) Tacik === **Campanula transsilvanica** Schur ex Andrae

Campanula cervicaria var. *albiflora* Schur === **Campanula cervicaria** L.

Campanula cervicaria var. *albiflora* Syr. === **Campanula cervicaria** L.

Campanula cervicaria var. *capitata* Schur === **Campanula cervicaria** L.

Campanula cervicaria var. *dalmatica* (Tausch) Nyman === **Campanula cervicaria** L.

Campanula cervicaria var. *lingulata* Nyman === **Campanula cervicaria** L.

Campanula cervicaria var. *longifolia* Nyman === **Campanula cervicaria** L.

Campanula cervicaria var. *oblongifolia* Schur === **Campanula cervicaria** L.

Campanula cervicarioides Schult. === **Campanula glomerata** subsp. **cervicarioides** (Schult.) Arcang.

Campanula cervicina D. Dietr. === **Wahlenbergia campanuloides** (Delile) Vatke

Campanula cespitosa f. *austriaca* Beck === **Campanula cespitosa** Scop.

Campanula cespitosa f. *ramosa* Hruby === **Campanula cespitosa** Scop.

Campanula cespitosa f. *simplex* Hruby === **Campanula cespitosa** Scop.

Campanula cespitosa subf. *austriaca* (Beck) Hruby === **Campanula cespitosa** Scop.

Campanula cespitosa subf. *grandiflorens* Hruby === **Campanula cespitosa** Scop.

Campanula cespitosa subf. *hirta* Hruby === **Campanula cespitosa** Scop.

Campanula cespitosa subf. *normalis* Hruby === **Campanula cespitosa** Scop.

Campanula cespitosa subsp. *pubescens* (F. W. Schmidt) Arcang. === **Campanula cochleariifolia** Lam.

Campanula cespitosa var. *bellardii* (All.) Nyman === **Campanula cochleariifolia** Lam.

Campanula cespitosa var. *bocconei* (Vill.) Steud. === **Campanula rotundifolia** L.

Campanula cespitosa var. *hirta* Dalla Torre & Sarnth. === **Campanula cespitosa** Scop.

Campanula cespitosa var. *imbricata* A. DC. === **Campanula rotundifolia** L.

Campanula cespitosa var. *pubescens* (F. W. Schmidt) A. DC. === **Campanula cochleariifolia** Lam.

Campanula chalcidica (Buser) Engl. === **Campanula rumeliana** subsp. **chalcidica** (Buser) Greuter & Burdet

Campanula chamissonis Fed. in Kom. === **Campanula dasyantha** subsp. **chamissonis** (Fed.) Viktorov

Campanula chamissonis f. *albiflora* (Miyabe & Tatew.) T. Shimizua === **Campanula dasyantha** subsp. **chamissonis** (Fed.) Viktorov

Campanula chanetii H. Lév. === **Adenophora polyantha** Nakai

Campanula charadzae Grossh. === **Campanula sibirica** subsp. **charadzae** (Grossh.) Ogan.

Campanula charkeviczii Fed. === **Campanula sibirica** subsp. **charkeviczii** (Fed.) Fed.

Campanula chilensis Mol. === **Wahlenbergia linarioides** (Lam.) A. DC.

Campanula chinganensis A. I. Baranov === **Campanula rotundifolia** L.

Campanula chondrophylla (H. Buek) D. Dietr. === **Monopsis simplex** (L.) E. Wimm.

Campanula choziatowskyi Fomin === **Campanula bayerniana** Rupr.

Campanula chrysogonii Sennen === ?

Campanula chysnysuensis Czerep. === **Campanula rapunculoides** L.

Campanula ciblesii Prodan === **Campanula alpina** Jacq. subsp. **alpina**

Campanula cichoracea Sm. === **Campanula lingulata** Waldst. & Kit.

Campanula cichoriiforme (Boiss.) Boiss. === **Asyneuma virgatum** subsp. **cichoriiforme** (Boiss.) Damboldt

Campanula ciliaris Salisb. === **Roella ciliata** L.

Campanula ciliata Thunb. === **Wahlenbergia thunbergii** (Schult.) B. Nord.

Campanula ciliata var. *longifolia* Rupr. === **Campanula ciliata** Steven

Campanula ciliata var. *pontica* Albov === **Campanula ciliata** Steven

Campanula cilicia Boiss. === **Asyneuma amplexicaule** (Willd.) Hand.-Mazz. subsp. **amplexicaule**

Campanula cinerea L. f. === **Wahlenbergia cinerea** (L. f.) Lammers

Campanula cinerea Formánek === **Campanula formanekiana** Degen & Dörfl.

Campanula cinerea Hegetschw. === **Campanula rotundifolia** L.

Campanula circaeoides Fr. Schmidt === **Peracarpa carnosa** (Wall.) Hook. f. & Thomson

Campanula ciscaucasica Kharadze === **Campanula sibirica** subsp. **ciscaucasica** (Kharadze) Ogan.

Campanula cissophylla Boiss. & Hausskn. === **Campanula acutiloba** Vatke

Campanula cissophylla var. *brachycalyx* Bornm. === **Campanula acutiloba** Vatke

Campanula clivosa (A. DC.) D. Dietr. === **Wahlenbergia angustifolia** (Roxb.) A. DC.

Campanula coa (A. DC.) D. Dietr. === **Legousia pentagonia** (L.) Thell.

Campanula cochleariifolia f. *anagalloides* Hruby === **Campanula cochleariifolia** Lam.

Campanula cochleariifolia f. *delpontei* (Chabert) Vacc. === **Campanula cochleariifolia** Lam.

Campanula cochleariifolia f. *densa* (Gsaller) Vacc. === **Campanula cochleariifolia** Lam.

Campanula cochleariifolia f. *elongata* Hruby === **Campanula cochleariifolia** Lam.

Campanula cochleariifolia f. *hauryi* (Schott) Hruby === **Campanula cochleariifolia** Lam.

Campanula cochleariifolia f *hoppeana* (Rupr. ex Rchb.) Hruby === **Campanula cochleariifolia** Lam.

Campanula cochleariifolia f. *incisiserrata* Hruby === **Campanula cochleariifolia** Lam.

Campanula cochleariifolia f. *javorkae* Hruby === **Campanula cochleariifolia** Lam.

Campanula cochleariifolia f. *mathonetii* (Jord. ex Gren. & Godr.) Hruby === **Campanula cochleariifolia** Lam.

Campanula cochleariifolia f. *modesta* (Schott) Hruby === **Campanula cochleariifolia** Lam.

Campanula cochleariifolia f. *multiflorens* Hruby === **Campanula cochleariifolia** Lam.

Campanula cochleariifolia f. *pirinica* (Velen.) Hayek === **Campanula cochleariifolia** Lam.

Campanula cochleariifolia f. *pseudomodesta* Hruby === **Campanula cochleariifolia** Lam.

Campanula cochleariifolia f. *pulchella* Hruby === **Campanula cochleariifolia** Lam.

Campanula cochleariifolia f. *ramulosa* Witasek ex Vacc. === **Campanula cochleariifolia** Lam.
Campanula cochleariifolia f. *rosulata* Hruby === **Campanula cochleariifolia** Lam.
Campanula cochleariifolia f. *subcaespitosa* Hruby === **Campanula cochleariifolia** Lam.
Campanula cochleariifolia f. *subramulosa* Hruby === **Campanula cochleariifolia** Lam.
Campanula cochleariifolia f. *tenella* (Jord.) Hruby === **Campanula cochleariifolia** Lam.
Campanula cochleariifolia f. *tubulosa* (Chabert) Vacc. === **Campanula cochleariifolia** Lam.
Campanula cochleariifolia f. *tyrolensis* (Schott) Vacc. === **Campanula cochleariifolia** Lam.
Campanula cochleariifolia f. *umbrosa* (J. Hofm.) Vacc. === **Campanula cochleariifolia** Lam.
Campanula cochleariifolia f. *umbrosa* Hruby === **Campanula cochleariifolia** Lam.
Campanula cochleariifolia f. *venusta* (Schur) Hruby === **Campanula cochleariifolia** Lam.
Campanula cochleariifolia f. *veronicae* Hruby === **Campanula cochleariifolia** Lam.
Campanula cochleariifolia subf. *aberrans* Hruby === **Campanula cochleariifolia** Lam.
Campanula cochleariifolia subf. *altior* Hruby === **Campanula cochleariifolia** Lam.
Campanula cochleariifolia subf. *brachyantha* (Murr) Hruby === **Campanula cochleariifolia** Lam.
Campanula cochleariifolia subf. *compacta* (Hegetschw.) Hruby === **Campanula cochleariifolia** Lam.
Campanula cochleariifolia subf. *foliosa* (Krašan) Hruby === **Campanula cochleariifolia** Lam.
Campanula cochleariifolia subf. *gracilescens* Hruby === **Campanula cochleariifolia** Lam.
Campanula cochleariifolia subf. *gracilis* (Nyman) Hruby === **Campanula cochleariifolia** Lam.
Campanula cochleariifolia subf. *grandiflorens* Hruby === **Campanula cochleariifolia** Lam.
Campanula cochleariifolia subf. *grandiflorens* Hruby === **Campanula cochleariifolia** Lam.
Campanula cochleariifolia subf. *hoppeaniformis* Hruby === **Campanula cochleariifolia** Lam.
Campanula cochleariifolia subf. *macedonica* Hruby === **Campanula cochleariifolia** Lam.
Campanula cochleariifolia subf. *multiflorens* Hruby === **Campanula cochleariifolia** Lam.
Campanula cochleariifolia subf. *polyphylla* Hruby === **Campanula cochleariifolia** Lam.
Campanula cochleariifolia subf. *pubescens* Hruby === **Campanula cochleariifolia** Lam.
Campanula cochleariifolia subf. *pulchella* (Gren. & Godr.) Hruby === **Campanula cochleariifolia** Lam.
Campanula cochleariifolia subf. *pygmaea* Hruby === **Campanula cochleariifolia** Lam.
Campanula cochleariifolia subf. *reflexa* Hruby === **Campanula cochleariifolia** Lam.
Campanula cochleariifolia subf. *tyrolensis* (Schott) Hruby === **Campanula cochleariifolia** Lam.
Campanula cochleariifolia subsp. *andorrana* (Braun-Blanq.) O. Bolòs & Vigo === **Campanula jaubertiana** Timb.-Lagr.
Campanula cochleariifolia subsp. *croatica* Hruby === **Campanula cochleariifolia** Lam.
Campanula cochlearifolia subsp. *jaubertiana* (Timb.-Lagr.) Rivas Mart. === **Campanula jaubertiana** Timb.-Lagr.
Campanula cochleariifolia subsp. *reflexa* (Schur) Hruby === **Campanula cochleariifolia** Lam.
Campanula cochleariifolia subsp. *septentrionalis* Hruby === ?
Campanula cochleariifolia var. *andorrana* (Braun-Blanq.) O. Bolòs & Vigo === **Campanula jaubertiana** Timb.-Lagr.
Campanula cochleariifolia var. *bellardii* (All.) Hayek & Hegi === **Campanula cochleariifolia** Lam.
Campanula cochleariifolia var. *brachyantha* (Murr) Dalla Torre & Sarnth. === **Campanula cochleariifolia** Lam.
Campanula cochleariifolia var. *descensa* (Beck) Săvul. === **Campanula cochleariifolia** Lam.
Campanula cochleariifolia var. *foliosa* (Murr) Dalla Torre & Sarnth. === **Campanula cochleariifolia** Lam.
Campanula cochleariifolia var. *foudrasii* (Jord.) Vacc. === **Campanula cochleariifolia** Lam.
Campanula cochleariifolia var. *gracilis* (Nyman) Vacc. === **Campanula cochleariifolia** Lam.
Campanula cochleariifolia var. *hauryi* (Schott) Hayek & Hegi === **Campanula cochleariifolia** Lam.
Campanula cochleariifolia var. *hoppeana* (Rupr. ex Rchb.) Vacc. === **Campanula cochleariifolia** Lam.
Campanula cochleariifolia var. *jaubertiana* (Timb.-Lagr.) O. Bolòs & Vigo === **Campanula jaubertiana** Timb.-Lagr.

Campanula cochleariifolia var. *mathonetii* (Jord. ex Gren. & Godr.) Vacc. === **Campanula cochleariifolia** Lam.

Campanula cochleariifolia var. *paniculata* (Nageli ex Rchb.) Dalla Torre & Sarnth. === **Campanula cochleariifolia** Lam.

Campanula cochleariifolia var. *pubescens* (F. W. Schmidt) Dalla Torre & Sarnth. === **Campanula cochleariifolia** Lam.

Campanula cochleariifolia var. *pulchella* (Gren. & Godr.) Vacc. === **Campanula cochleariifolia** Lam.

Campanula cochleariifolia var. *renatii* (Sennen) P. Monts. === **Campanula cochleariifolia** Lam.

Campanula cochleariifolia var. *subacaulis* (Murr) Hruby === **Campanula cochleariifolia** Lam.

Campanula cochleariifolia var. *subramulosa* (Jord. ex Gren. & Godr.)Vacc. === **Campanula cochleariifolia** Lam.

Campanula cochleariifolia var. *tenella* (Jord.) Vacc. === **Campanula cochleariifolia** Lam.

Campanula cochleariifolia var. *umbrosa* (J. Hofm.) Dalla Torre & Sarnth. === **Campanula cochleariifolia** Lam.

Campanula cochleariifolia var. *vagans* (J. Hofm.) Dalla Torre & Sarnth. === **Campanula cochleariifolia** Lam.

Campanula colettae Beauverd & Topoli === **Campanula radicosa** Bory & Chaub.

Campanula collina subsp. *gapschimensis* Husseinov === **Campanula collina** subsp. **collina**

Campanula collina var. *abchasica* Albov === **Campanula collina** subsp. **sphaerocarpa** (Kolak.) Ogan.

Campanula collina var. *dasycarpa* K. Koch === **Campanula collina** subsp. **collina**

Campanula collina var. *eriocalyx* Trautv. === **Campanula collina** Sims subsp. **collina**

Campanula collina var. *fondervisii* (Albov) Albov === **Campanula collina** subsp. **fondervisii** (Albov) Ogan.

Campanula collina var. *grandiflora* Vatke === **Campanula collina** Sims subsp. **collina**

Campanula collina var. *leiocalyx* Trautv. === **Campanula collina** Sims subsp. **collina**

Campanula collina var. *leiocarpa* K. Koch === **Campanula collina** Sims subsp. **collina**

Campanula collina var. *major* M. Bieb. === **Campanula collina** Sims subsp. **collina**

Campanula collina var. *pumila* M. Bieb. === **Campanula collina** Sims subsp. **collina**

Campanula coloradoensis Buckley === **Triodanis coloradoensis** (Buckley) McVaugh

Campanula colorata Wall. === **Campanula pallida** Wall.

Campanula colorata var. *anomala* Hook. f. & Thomson === **Campanula pallida** Wall.

Campanula colorata var. *moorcroftiana* A. DC. === **Campanula pallida** Wall.

Campanula colorata var. *ramulosa* (Wall.) Hook. f. & Thomsen === **Campanula pallida** Wall.

Campanula colorata var. *tibetica* Hook. f. & Thomson === **Campanula pallida** Wall.

Campanula compacta Boiss. & Heldr. === **Asyneuma compactum** Damboldt

Campanula compacta Hegetschw. === **Campanula cochleariifolia** Lam.

Campanula commutata Schult. === **Campanula sarmatica** Ker Gawl. subsp. **sarmatica**

Campanula conferta var. *glabra* A. DC. === **Campanula conferta** A. DC.

Campanula confertifolia (Reut.) Witasek === **Campanula rotundifolia** L.

Campanula congesta Vest ex Schult. === **Campanula glomerata** L. subsp. **glomerata**

Campanula conglomerata Gueldenst. === **Campanula glomerata** L. subsp. **glomerata**

Campanula consanguinea Schott === **Campanula scheuchzeri** Vill.

Campanula controversa Boiss. === **Asyneuma rigidum** (Willd.) Grossh. subsp. **rigidum**

Campanula corbariensis Rouy === **Campanula speciosa** Pourr. subsp. **speciosa**

Campanula corcontica Šourek === **Campanula bohemica** Hruby subsp. **bohemica**

Campanula cordata Peterm. === ?

Campanula cordata Tausch === **Adenophora liliifolia** (L.) A. DC.

Campanula cordata Vis. === **Legousia speculum-veneris** (L.) Durande ex Vill.

Campanula cordifolia K. Koch === **Campanula rapunculoides** L.

Campanula cordifolia Vuk. === **Campanula carpatica** Jacq.

Campanula coronata Ker Gawl. === **Adenophora stenanthina** (Ledeb.) Kitag. subsp. **stenanthina**

Campanula coronopifolia Fisch. ex Schult. === **Adenophora gmelinii** (Biehler) Fisch. subsp. **gmelinii**

Campanula coronopifolia var. *odontosepala* Borbás === **Adenophora gmelinii** (Biehler) Fisch. subsp. **gmelinii**

Campanula corymbifera Desf. ex Poir. === **Campanula pelviformis** Lam.

Campanula corymbosa Desf. === **Campanula pelviformis** Lam.

Campanula corymbosa Ten. === **Campanula versicolor** Andrews

Campanula corymbosa var. *parviflora* K. Koch === **Campanula pelviformis** Lam.

Campanula costae Willk. === **Campanula patula** subsp. **costae** (Willk.) Nyman

Campanula crassa Formánek = **Campanula sparsa** Friv. subsp. **sparsa**

Campanula crassifolia Huber-Mor. & C. Simon === **Campanula oligosperma** Damboldt

Campanula crassifolia Nees === **Campanula fragilis** Cirillo subsp. **fragilis**

Campanula crenata Link === **Campanula rapunculoides** L.

Campanula crystallocalyx Adamović === **Campanula persicifolia** L. subsp. **persicifolia**

Campanula cuatrecasasii Pau === **Campanula macrorhiza** J. Gay ex A. DC.

Campanula cuatrecasasii Sennen & Losa === ?

Campanula cuatrecasasii var. *gaditana* C. Vicioso === **Campanula macrorhiza** J. Gay ex A. DC.

Campanula cylindrica (Pax & K. Hoffm.) Nannf. === **Campanula aristata** Wall.

Campanula cypria Rech. f. === **Campanula podocarpa** Boiss.

Campanula dalmatica (A. DC.) D. Dietr. === **Edraianthus dalmaticus** (A. DC.) A. DC.

Campanula dalmatica Tausch === **Campanula cervicaria** L.

Campanula daralaghezica (Grossh.) Kolak. & Serdyuk. === **Campanula armena** Steven

Campanula darialica Kharadze === **Campanula sibirica** subsp. **hohenackeri** (Fisch. & C. A. Mey.) Damboldt

Campanula dasyantha f. *albiflora* Miyabe & Tatew. === **Campanula dasyantha** subsp. **chamissonis** (Fed.) Viktorov

Campanula dasyantha var. *chamissonis* (Fed.) Toyok. & Nosaka === **Campanula dasyantha** subsp. **chamissonis** (Fed.) Viktorov

Campanula dasycarpa Kit. === **Campanula persicifolia** L. subsp. **persicifolia**

Campanula davurica Siev. === ?

Campanula debarensis Rech. f. === **Campanula fenestrellata** subsp. **debarensis** (Rech. f.) Damboldt

Campanula debilis (H. Buek) D. Dietr. === **Wahlenbergia debilis** H. Buek

Campanula declinata Moench === **Campanula americana** L.

Campanula decloetiana Ortmann === ?

Campanula decumbens var. *pseudospecularioides* G. Lopéz === **Campanula lusitanica** subsp. **specularioides** (Coss.) Aldasoro & L. Sáez × **Campanula decumbens** A. DC.

Campanula decurrens L. === **Campanula patula** L. subsp. **patula**

Campanula decurrens Thore === **Campanula rapunculus** L. subsp. **rapunculus**

Campanula decurrens Zuccagni === **Campanula peregrina** L.

Campanula dehiscens Roxb. === **Wahlenbergia marginata** (Thunb. ex Murray) A. DC.

Campanula delicatula Sennen & Losa === **Campanula rotundifolia** L.

Campanula densifoliata Sennen & Losa === ?

Campanula denticulata Burch. === **Wahlenbergia denticulata** (Burch.) A. DC.

Campanula denticulata (Fisch.) Spreng. === **Adenophora tricuspidata** (Fisch. ex Schult.) A. DC.

Campanula denticulata Boiss. & A. Huet === **Campanula betulifolia** K. Koch

Campanula denudata (A. DC.) D. Dietr. === **Wahlenbergia denudata** A. DC.

Campanula desertorum Weinm. === **Campanula glomerata** subsp. **farinosa** (Rochel ex Besser) Kirschl.

Campanula dichotoma f. *alba* Voss === **Campanula dichotoma** L. f.

Campanula dichotoma f. *pseudodichotoma* Maire ex Quézel === **Campanula hypocrateriformis** Dobignard

Campanula dichotoma subsp. *afra* (Cav.) Maire === **Campanula afra** Cav.

Campanula dichotoma subsp. *kremeri* (Boiss. & Reut.) Nyman === **Campanula kremeri** Boiss. & Reut.

Campanula dichotoma subsp. *semisecta* (Murb.) Rivas Mart. === **Campanula semisecta** Murb.

Campanula dichotoma var. *brachiata* A. DC. === **Campanula dichotoma** L. f.

Campanula dichotoma var. *pallida* Maire === **Campanula hypocrateriformis** Dobignard

Campanula diffusa (A. DC.) D. Dietr. === **Microcodon linearis** (L. f.) H. Buek
Campanula diffusa (L. f.) D. Dietr. === **Prismatocarpus diffusus** (L. f.) A. DC.
Campanula diffusa Vahl === **Campanula fragilis** Cirillo subsp. **fragilis**
Campanula dilecta Schott === **Campanula scheuchzeri** Vill.
Campanula dinarica A. Kern. === **Edraianthus dinaricus** (A. Kern.) Wettst.
Campanula divaricata f. *alba* Fosberg === **Campanula divaricata** Michx.
Campanula divergens Waldst. & Kit. ex Willd. === **Campanula sibirica** subsp. **divergens** (Waldst. & Kit. ex Willd.) Nyman
Campanula divergens var. *cernua* Schult. === **Campanula sibirica** subsp. **divergens** (Waldst. & Kit. ex Willd.) Nyman
Campanula divergens var. *spathulata* (Schur) Schur === **Campanula sibirica** subsp. **divergens** (Waldst. & Kit. ex Willd.) Nyman
Campanula diversifolia (A. DC.) D. Dietr. === **Wahlenbergia procumbens** (L. f.) A. DC.
Campanula diversifolia Dumort. === **Campanula rotundifolia** L.
Campanula doluchanovii Kharadze === **Campanula bellidifolia** subsp. **argunensis** (Rupr.) Viktorov
Campanula draba Burm. f. === ?
Campanula drabifolia subsp. *attica* (Boiss. & Heldr.) Nyman === **Campanula drabifolia** Sm.
Campanula drabifolia subsp. *creutzburgii* (Greuter) Fed. === **Campanula creutzburgii** Greuter
Campanula drabifolia subsp. *pinatzii* (Greuter & Phitos) Fed. === **Campanula pinatzii** Greuter & Phitos
Campanula drabifolia var. *major* Boiss. === **Campanula drabifolia** Sm.
Campanula drabifolia var. *rhodensis* (A. DC.) Nyman === **Campanula rhodensis** A. DC.
Campanula dubia A. DC. === **Campanula giesekiana** Vest ex Schult.
Campanula dunantii (A. DC.) D. Dietr. === **Wahlenbergia dunantii** A. DC.
Campanula durieui Boiss. === **Campanula lusitanica** Loefl. subsp. **lusitanica**
Campanula echiifolia Rupr. === **Campanula cervicaria** L.
Campanula ecklonii (H. Buek) D. Dietr. === **Wahlenbergia ecklonii** H. Buek
Campanula elata Colla === ?
Campanula elatina Hill === **Campanula elatines** L.
Campanula elatines subsp. *glaberrima* (A. DC.) Arcang. === **Campanula elatines** L.
Campanula elatines var. *elatinoides* (Moretti) Fiori === **Campanula elatinoides** Moretti
Campanula elatines var. *fenestrellata* (Feer) L. H. Bailey === **Campanula fenestrellata** Feer
Campanula elatines var. *garganica* (Ten.) Fiori === **Campanula garganica** Ten.
Campanula elatines var. *glaberrima* A. DC. === **Campanula elatines** L.
Campanula elatines var. *istriaca* (Feer) Fiori & Paol. === **Campanula fenestrellata** subsp. **istriaca** (Feer) Damboldt
Campanula elatines var. *subalpina* (Delponte & Gras) Nyman === **Campanula elatines** L.
Campanula elatior (Fomin) Grossh. === **Campanula sibirica** subsp. **elatior** (Fomin) Fed.
Campanula elatior Hoffmanns. & Link === **Campanula rapunculus** L. subsp. **rapunculus**
Campanula elegans Schult. === **Campanula rapunculoides** L.
Campanula elegantissima Grossh. === **Campanula bayerniana** Rupr.
Campanula elliptica Kit. ex Schult. === **Campanula glomerata** subsp. **elliptica** (Kit. ex Schult.) O. Schwarz
Campanula elongata Willd. === **Wahlenbergia campanuloides** (Delile) Vatke
Campanula ensifolia Lam. === **Heterochaenia ensifolia** (Lam.) A. DC.
Campanula eocervicaria Nábĕlek === **Campanula glomerata** subsp. **hispida** (Witasek) Hayek
Campanula ephesia Boiss. === **Campanula tomentosa** Lam.
Campanula epigaea Janka ex Degen === **Campanula patula** subsp. **epigaea** (Janka ex Degen) Hayek
Campanula eriantha Hampe === **Campanula tomentosa** Lam.
Campanula erinoides L. === ?
Campanula erinus f. *albiflora* Maire === **Campanula erinus** L.
Campanula erinus f. *coerulea* Maire === **Campanula erinus** L.

Campanula eriocarpa M. Bieb. === **Campanula latifolia** L. subsp. **latifolia**

Campanula ermenekensis Contandr. & Quézel === **Campanula isaurica** Contandr., Quézel & Pamukç.

Campanula erucifolia Feer === **Campanula laciniata** L.

Campanula erysimoides Vest ex Schult. === **Adenophora gmelinii** (Biehler) Fisch. subsp. **gmelinii**

Campanula esculenta Salisb. === **Campanula rapunculus** L. subsp. **rapunculus**

Campanula esculenta A. Rich. === **Campanula edulis** Forssk.

Campanula esculenta Candargy === **Campanula lyrata** Lam. subsp. **lyrata**

Campanula evolvulacea Royle ex A. DC. === **Campanula cashmeriana** Benth.

Campanula exigua Formánek === ?

Campanula exilis (A. DC.) D. Dietr. === **Wahlenbergia exilis** A. DC.

Campanula expansa Friv. === **Campanula sparsa** Friv.

Campanula expansa subsp. *crassa* (Formánek) Formánek === **Campanula sparsa** Friv. subsp. **sparsa**

Campanula expansa var. *macedonica* Formánek === **Campanula sparsa** Friv. subsp. **sparsa**

Campanula expansa var. *sphaerothrix* (Griseb.) Boiss. === **Campanula sparsa** subsp. **sphaerothrix** (Griseb.) Hayek

Campanula expansa var. *welandii* (Heuff.) Nyman === **Campanula sparsa** Friv. subsp. **sparsa**

Campanula exul Schott === ?

Campanula falcata (Ten.) Schult. === **Legousia falcata** (Ten.) Fritsch ex Janch.

Campanula fallax Rupr. === **Campanula bellidifolia** subsp. **aucheri** (A. DC.) Viktorov

Campanula farinosa (Rochel ex Besser) Andrz. ex Besser === **Campanula glomerata** subsp. **farinosa** (Rochel ex Besser) Kirschl.

Campanula farinulenta A. Kern. & Wettst. === **Campanula velebitica** Borbás

Campanula fasciculata L. f. === **Wahlenbergia desmantha** Lammers

Campanula faucium Ponert === **Campanula odontosepala** Boiss.

Campanula fedorovii Kharadze === **Campanula sibirica** subsp. **hohenackeri** (Fisch. & C. A. Mey.) Damboldt

Campanula fedtschenkiana Trautv. === **Campanula incanescens** Boiss.

Campanula fergusonii Ferguson === **Campanula carpatica** Jacq.

Campanula fernandeziana (A. DC.) D. Dietr. === **Wahlenbergia fernandeziana** A. DC.

Campanula ficarioides subsp. *gautieri* (Jeanb. & Timb.-Lagr.) Podlech === **Campanula ficarioides** Timb.-Lagr.

Campanula ficarioides var. *major* Timb.-Lagr. === ?

Campanula filicaulis subsp. *embergeri* (Litard. & Maire) Dobignard === **Campanula embergeri** Litard. & Maire

Campanula filicaulis var. *cedretorum* H. Lindb. === **Campanula filicaulis** Durieu

Campanula filicaulis var. *mairei* Quézel === **Campanula filicaulis** Durieu

Campanula filicaulis var. *maroccana* (Ball) Dobignard === **Campanula filicaulis** Durieu

Campanula filicaulis var. *maroccana* Pau === **Campanula mollis** L. subsp. **mollis**

Campanula filicaulis var. *parielii* Quézel === **Campanula filicaulis** Durieu

Campanula filicaulis var. *pseudoantiatlantica* Quézel === **Campanula filicaulis** Durieu

Campanula filicaulis var. *pseudoradicosa* (Litard. & Maire) Emb. & Maire === **Campanula filicaulis** Durieu

Campanula filicaulis var. *reboudiana* (Pomel) Maire === **Campanula filicaulis** Durieu

Campanula filicaulis var. *schotteri* (Quézel) Dobignard === **Campanula embergeri** subsp. **schotteri** Quézel

Campanula filiflora Kellogg === **Campanula prenanthoides** Durand

Campanula filiformis Moretti === **Campanula morettiana** Rchb.

Campanula filiformis Ruiz & Pav. === **Wahlenbergia linarioides** (Lam.) A. DC.

Campanula finitima Fomin === **Campanula betulifolia** K. Koch

Campanula firma Boiss. === **Asyneuma amplexicaule** subsp. **aucheri** (A. DC.) Bornm.

Campanula firmiana Vand. === **Campanula barbata** L.

Campanula fischeri Schult. in Roem. & Schult. === **Adenophora liliifolia** (L.) A. DC.

Campanula fischeriana Spreng. === **Adenophora gmelinii** (Biehler) Fisch. subsp. **gmelinii**

Campanula flaccida (A. DC.) D. Dietr. === ?
Campanula flaccida (Wallr.) Dalla Torre & Sarnth. === **Campanula patula** L. subsp. **patula**
Campanula flagellaris Halácsy === ?
Campanula flagellaris Kunth === **Triodanis perfoliata** (L.) Nieuwl. subsp. **perfoliata**
Campanula flexuosa Michx. === **Campanula divaricata** Michx.
Campanula flexuosa Waldst. & Kit. === **Campanula waldsteiniana** Schult.
Campanula floribunda Viv. === **Campanula isophylla** Moretti
Campanula florida Salisb. === **Campanula medium** L.
Campanula foliosa Galushko === **Campanula rapunculoides** L.
Campanula fominii Grossh. === **Campanula bellidifolia** subsp. **meyerana** (Rupr.) Viktorov
Campanula fondervisii Albov === **Campanula collina** subsp. **fondervisii** (Albov) Ogan.
Campanula fontanesiana Boiss. === **Asyneuma rigidum** (Willd.) Grossh. subsp. **rigidum**
Campanula foudrasii Jord. === **Campanula cochleariifolia** Lam.
Campanula fragilis f. *grandiflora* Voss === **Campanula fragilis** Cirillo subsp. **fragilis**
Campanula fragilis f. *hirsuta* (A. DC.) Voss === **Campanula fragilis** Cirillo subsp. **fragilis**
Campanula fragilis subsp. *canescens* Schousb. ex Arcang. === **Campanula fragilis** Cirillo
    subsp. **fragilis**
Campanula fragilis var. *barrelieri* (C. Presl) Nyman === **Campanula fragilis** Cirillo subsp.
    fragilis
Campanula fragilis var. *diffusa* (Vahl) Steud. === **Campanula fragilis** Cirillo subsp. **fragilis**
Campanula fragilis var. *hirsuta* A. DC. === **Campanula fragilis** Cirillo subsp. **fragilis**
Campanula fragilis var. *hirsuta* Ten. === **Campanula fragilis** Cirillo subsp. **fragilis**
Campanula fragilis var. *hirta* Ten. === **Campanula fragilis** Cirillo subsp. **fragilis**
Campanula frivaldskyi Steud. === **Campanula sparsa** Friv.
Campanula froedinii Rech. f. === **Campanula bellidifolia** subsp. **aucheri** (A. DC.) Viktorov
Campanula fruticosa Hill === **Prismatocarpus pedunculatus** (P. J. Bergius) A. DC.
Campanula fruticosa L. === **Prismatocarpus fruticosus** (L.) L'Hér.
Campanula fulgens Wall. === **Asyneuma fulgens** (Wall.) Briq.
Campanula galushkoi Prima === **Campanula collina** Sims subsp. **collina**
Campanula garganica subsp. *fenestrellata* (Feer) Hayek === **Campanula fenestrellata** Feer
Campanula garganica subsp. *istriaca* (Feer) Hayek === **Campanula fenestrellata** subsp.
    istriaca (Feer) Damboldt
Campanula garganica var. *acarnanica* (Damboldt) Damboldt === **Campanula garganica**
    subsp. **acarnanica** (Damboldt) Damboldt
Campanula garganica var. *hirsuta* Anonymous === **Campanula garganica** Ten. subsp.
    garganica
Campanula garganica var. *lepida* (Feer) Hayek === **Campanula fenestrellata** Feer subsp.
    fenestrellata
Campanula gautieri Jeanb. & Timb.-Lagr. === **Campanula ficarioides** Timb.-Lagr.
Campanula gelida Kovanda === **Campanula bohemica** subsp. **gelida** (Kovanda)
    Kovanda
Campanula gentianoides Lam. === **Platycodon grandiflorus** (Jacq.) A. DC.
Campanula gieseckiana subsp. *groenlandica* (Berlin) Böcher === **Campanula giesekiana** Vest
    ex Schult.
Campanula gilanica Rupr. === **Campanula bellidifolia** subsp. **aucheri** (A. DC.) Viktorov
Campanula glacialis Shuttlew. === **Campanula cochleariifolia** Lam.
Campanula glauca Thunb. ex Murray === **Platycodon grandiflorus** (Jacq.) A. DC.
Campanula glaucophylla Damboldt === **Campanula oligosperma** Damboldt
Campanula glaucophylla Schloss. & Vuk. === **Campanula glomerata** subsp. **farinosa** (Rochel
    ex Besser) Kirschl.
Campanula glomerata Hegetschw. === ?
Campanula glomerata f. *aggregata* (Willd.) Voss === **Campanula glomerata** L. subsp. **glomerata**
Campanula glomerata f. *alba* (Nakai) U. C. La === **Campanula glomerata** subsp. **speciosa**
    (Hornem. ex Spreng.) Domin
Campanula glomerata f. *canescens* (Maxim. ex Nakai) Kitag. === **Campanula glomerata**
    subsp. **speciosa** (Hornem. ex Spreng.) Domin

Campanula glomerata f. *glaucophylla* (Schloss. & Vuk.) Hayek & Hegi === **Campanula glomerata** subsp. **farinosa** (Rochel ex Besser) Kirschl.

Campanula glomerata f. *hispida* Witasek === **Campanula glomerata** subsp. **hispida** (Witasek) Hayek

Campanula glomerata f. *polessica* (O. D. Wissjul.) Sóo === **Campanula glomerata** subsp. **farinosa** (Rochel ex Besser) Kirschl.

Campanula glomerata f. *pusilla* (A. DC.) Hayek & Hegi === **Campanula glomerata** L. subsp. **glomerata**

Campanula glomerata f. *speciosa* (Hornem. ex Spreng.) Voss === **Campanula glomerata** subsp. **speciosa** (Hornem. ex Spreng.) Domin

Campanula glomerata race *aggregata* (Willd.) Rouy === **Campanula glomerata** L. subsp. **glomerata**

Campanula glomerata race *cervicarioides* (Schult.) Rouy === **Campanula glomerata** subsp. **cervicarioides** (Schult.) Arcang.

Campanula glomerata subsp. *aggregata* (Willd.) Arcang. === **Campanula glomerata** L. subsp. **glomerata**

Campanula glomerata subsp. *cephalotes* (Fisch. ex Schrank) D. Y. Hong === **Campanula glomerata** subsp. **speciosa** (Hornem. ex Spreng.) Domin

Campanula glomerata subsp. *fatrae* (Borbás) Sóo === ?

Campanula glomerata subsp. *glabriflora* Contandr., Quézel & Pamukç. === **Campanula glomerata** subsp. **hispida** (Witasek) Hayek

Campanula glomerata subsp. *maleevii* (Fed.) Ogan. === **Campanula glomerata** subsp. **hispida** (Witasek) Hayek

Campanula glomerata subsp. *salviifolia* (Wallr.) O. Schwarz === ?

Campanula glomerata subsp. *stenosiphon* (Boiss. & Heldr.) Nyman === **Campanula stenosiphon** Boiss. & Heldr.

Campanula glomerata var. *aggregata* (Willd.) Spreng. === **Campanula glomerata** L. subsp. **glomerata**

Campanula glomerata var. *artvinense* Kuntze === **Sachokiella macrochlamys** (Boiss. & A. Huet) Kolak.

Campanula glomerata var. *capituliformis* K. Koch === ?

Campanula glomerata var. *caucasica* Trautv. === **Campanula glomerata** subsp. **caucasica** (Trautv.) Ogan.

Campanula glomerata var. *cervicaria* (L.) Kuntze === **Campanula cervicaria** L.

Campanula glomerata var. *cervicarioides* (Schult.) A. DC. === **Campanula glomerata** subsp. **cervicarioides** (Schult.) Arcang.

Campanula glomerata var. *congesta* (Vest ex Schult.) Schur === **Campanula glomerata** L. subsp. **glomerata**

Campanula glomerata var. *cordifolia* Rohlena === ?

Campanula glomerata var. *coriacea* Fomin === ?

Campanula glomerata var. *dahurica* Fisch. ex Ker-Gawl. === **Campanula glomerata** subsp. **speciosa** (Hornem. ex Spreng.) Domin

Campanula glomerata var. *desertorum* (Weinm.) Nyman === **Campanula glomerata** subsp. **farinosa** (Rochel ex Besser) Kirschl.

Campanula glomerata var. *elliptica* (Kit. ex Schult.) Spreng. === **Campanula glomerata** subsp. **elliptica** (Kit. ex Schult.) O. Schwarz

Campanula glomerata var. *farinosa* Rochel ex Besser === **Campanula glomerata** subsp. **farinosa** (Rochel ex Besser) Kirschl.

Campanula glomerata var. *fatrae* Borbás === ?

Campanula glomerata var. *foliosa* (Ten.) Spreng. === **Campanula foliosa** Ten.

Campanula glomerata var. *glaucophylla* (Schloss. & Vuk.) Nyman === **Campanula glomerata** subsp. **farinosa** (Rochel ex Besser) Kirschl.

Campanula glomerata var. *grandiflora* Herder === **Campanula glomerata** subsp. **speciosa** (Hornem. ex Spreng.) Domin

Campanula glomerata var. *grandiflora* Kuntze === **Campanula glomerata** L. subsp. **glomerata**

Campanula glomerata var. *heterophylla* Lapeyr. === **Campanula glomerata** L. subsp. **glomerata**

Campanula glomerata var. *hispida* Fomin === **Campanula glomerata** subsp. **hispida** (Witasek) Hayek

Campanula glomerata var. *latibracteata* Sommier & Levier === ?

Campanula glomerata var. *longifolia* Schloss. ex Dalla Torre & Sarnth. === **Campanula glomerata** L. subsp. **glomerata**

Campanula glomerata var. *minor* Gray === **Campanula glomerata** L. subsp. **glomerata**

Campanula glomerata var. *mollis* Tausch === ?

Campanula glomerata var. *nana* Peterm. === ?

Campanula glomerata var. *nana* Noulet === **Campanula glomerata** L. subsp. **glomerata**

Campanula glomerata var. *nicaeensis* (Schult.) Spreng. === **Campanula glomerata** subsp. **oblongifolia** (K. Koch) Fed.

Campanula glomerata var. *oblongata* Trautv. === **Campanula glomerata** L. subsp. **glomerata**

Campanula glomerata var. *oblongifolia* K. Koch === **Campanula glomerata** subsp. **oblongifolia** (K. Koch) Fed.

Campanula glomerata var. *ortleppii* H. Lév. === **Campanula glomerata** L. subsp. **glomerata**

Campanula glomerata var. *pusilla* A. DC. === **Campanula glomerata** L. subsp. **glomerata**

Campanula glomerata var. *ramosa* K. Koch === **Campanula glomerata** subsp. **elliptica** (Kit. ex Schult.) O. Schwarz

Campanula glomerata var. *ramosa* Lapeyr. === **Campanula glomerata** L. subsp. **glomerata**

Campanula glomerata var. *salviifolia* Wallr. === ?

Campanula glomerata var. *semgensis* Husseinov === **Campanula glomerata** L. subsp. **glomerata**

Campanula glomerata var. *serotina* (Wettst.) Hayek & Hegi === **Campanula glomerata** subsp. **serotina** (Wettst.) O. Schwarz

Campanula glomerata var. *sparsiflora* A. DC. === **Campanula glomerata** subsp. **glomerata**

Campanula glomerata var. *speciosa* Hornem. ex Spreng. === **Campanula glomerata** L. subsp. **speciosa** (Hornem. ex Spreng.) Domin

Campanula glomerata var. *stenosiphon* (Boiss. & Heldr.) Boiss. === **Campanula stenosiphon** Boiss. & Heldr.

Campanula glomerata var. *symphytifolia* Albov === **Campanula glomerata** subsp. **symphytifolia** (Albov) Ogan.

Campanula glomerata var. *umbrosa* Trautv. === **Campanula glomerata** L. subsp. **glomerata**

Campanula gmelinii Biehler === **Adenophora gmelinii** (Biehler) Fisch.

Campanula gmelinii var. *pomponiifolia* Fisch. ex Borbás === **Adenophora gmelinii** (Biehler) Fisch. subsp. **gmelinii**

Campanula gracilis Avé-Lall. === ?

Campanula gracilis (Boiss. & Heldr.) Boiss. & Heldr. === **Asyneuma lobelioides** (Willd.) Hand.-Mazz.

Campanula gracilis G. Forst. === **Wahlenbergia gracilis** (G. Forst.) A. DC.

Campanula gracilis Jord. === **Campanula cochleariifolia** Lam.

Campanula gracilis f. *stricta* (R. Br.) A. Voss === **Wahlenbergia stricta** (R. Br.) Sweet

Campanula gracilis var. *capillaris* R. Br. === **Wahlenbergia capillaris** (Lodd.) Sweet

Campanula gracilis var. *littoralis* (Labill.) R. Br. === **Wahlenbergia littoralis** (Labill.) Sweet

Campanula gracilis var. *revoluta* Colla === **Wahlenbergia berteroi** Hook. & Arn.

Campanula gracilis var. *stricta* R. Br. === **Wahlenbergia stricta** (R. Br.) Sweet

Campanula graminifolia L. === **Edraianthus graminifolius** (L.) A. DC.

Campanula graminifolia var. *linearifolia* Vuk. === **Edraianthus pumilio** (Port. ex Schult.) A. DC.

Campanula graminifolia var. *setifolia* Vuk. === **Edraianthus graminifolius** (L.) A. DC. subsp. **graminifolius**

Campanula graminifolia var. *sordidifolia* Vuk. === **Edraianthus tenuifolius** (Waldst. & Kit.) A. DC.

Campanula grammosepala Vuk. === ?

Campanula grammosepala var. *cardiophylla* Vuk. === ?

Campanula grammosepala var. *leptosepala* Vuk. === **Campanula excisa** Schleich. ex Murith

Campanula grammosepala var. *lobophylla* Vuk. === **Campanula rotundifolia** L.

Campanula grammosepala var. *spathiphylla* Vuk. === ?

Campanula granatensis Witasek === **Campanula hispanica** Willk. subsp. **hispanica**

Campanula grandiflora Jacq. === **Platycodon grandiflorus** (Jacq.) A. DC.

Campanula grandiflora Lam. === **Campanula medium** L.

Campanula griffithii Hook. f. & Thoms. === **Campanula leucoclada** Boiss.

Campanula grisebachiana Gand. === ?

Campanula groenlandica Berlin === **Campanula giesekiana** Vest ex Schult.

Campanula grossheimii Kharadze === **Campanula rapunculoides** L.

Campanula guiraoi Sennen === **Campanula hispanica** Willk. subsp. **hispanica**

Campanula gumbetica Boiss. === **Campanula andina** Rupr.

Campanula gummifera Willd. ex Schult. === **Campanula sarmatica** Ker Gawl. subsp. **sarmatica**

Campanula gundelia K. Koch === **Campanula alliariifolia** Willd. subsp. **alliariifolia**

Campanula gypsicola (Costa) Witasek === **Campanula hispanica** subsp. **catalanica** Podlech

Campanula hakkiarica P. H. Davis === **Theodorovia karakuschensis** (Grossh.) Kolak.

Campanula halacsyana Baldacci === **Campanula hawkinsiana** Hausskn. & Heldr. ex Hausskn.

Campanula hastifolia Salisb. === **Canarina canariensis** (L.) Vatke

Campanula hauryi Schott === **Campanula cochleariifolia** Lam.

Campanula hausmannii Rchb. f. === **Campanula barbata** L. × **Phyteuma hemisphaericum** L.

Campanula hawkinsiana var. *scabriuscula* Hausskn. === **Campanula hawkinsiana** Hausskn. & Heldr. ex Hausskn.

Campanula haynaldii Szontágh === **Campanula alpina** Jacq. subsp. **alpina**

Campanula hederacea L. === **Wahlenbergia hederacea** (L.) Rchb.

Campanula hederifolia Salisb. === **Wahlenbergia hederacea** (L.) Rchb.

Campanula hegetschweileri Bech. === **Campanula scheuchzeri** Vill.

Campanula hellenica (Hayek) Podlech === **Campanula albanica** Witasek subsp. **albanica**

Campanula hemschinica K. Koch === **Campanula olympica** Boiss.

Campanula hendersonii Anonymous === **Campanula carpatica** Jacq.

Campanula hercegovina f. *glaberrima* Witasek ex Hayek === **Campanula hercegovina** Degen & Fiala

Campanula hercegovina f. *humilis* Hruby === **Campanula hercegovina** Degen & Fiala

Campanula hercegovina f. *ovalifolia* Hruby === **Campanula hercegovina** Degen & Fiala

Campanula hercegovina f. *squarrosa* Degen & Fiala ex Hayek === **Campanula hercegovina** Degen & Fiala

Campanula hercegovina f. *squarrosa* Hruby === **Campanula hercegovina** Degen & Fiala

Campanula hercegovina f. *umbosa* Hruby === **Campanula hercegovina** Degen & Fiala

Campanula hercegovina subf. *brevicalycina* Hruby === **Campanula hercegovina** Degen & Fiala

Campanula hercegovina subf. *longicalycina* Hruby === **Campanula hercegovina** Degen & Fiala

Campanula hercegovina var. *fallens* Hruby === **Campanula hercegovina** Degen & Fiala

Campanula hercegovina var. *pseudopusilla* Hruby === **Campanula hercegovina** Degen & Fiala

Campanula herminii var. *atlantica* Jahand. & Maire === **Campanula mairei** Pau ex Maire

Campanula hesperidifolia Boiss. === **Asyneuma lobelioides** (Willd.) Hand.-Mazz.

Campanula heterodoxa Vest ex Schult. === **Campanula rotundifolia** L.

Campanula heterophylla Gray === **Campanula rotundifolia** L.

Campanula himalayensis Klotzsch === **Campanula pallida** Wall.

Campanula hirsuta Pant. === ?

Campanula hirsutissima Sennen & Mauricio === **Campanula afra** Cav.

Campanula hirta Hegetschw. === ?

Campanula hirta (Ten.) Schult. === **Legousia speculum-veneris** (L.) Durande ex Vill.

Campanula hispanica f. *minor* Sennen === **Campanula macrorhiza** J. Gay ex A. DC.

Campanula hispanica var. *glabra* Willk. === **Campanula hispanica** Willk. subsp. **hispanica**

Campanula hispida Fisch. ex Hornem. === **Campanula lactiflora** M. Bieb.

Campanula hispida Lej. === **Campanula persicifolia** L. subsp. **persicifolia**

Campanula hispidissima Hochst. === **Campanula propinqua** Fisch. & C. A. Mey.

Campanula hispidula L. f. === **Microcodon hispidulus** (L. f.) Sond.

Campanula hochstetteri Schott === **Campanula cochleariifolia** Lam.

Campanula hoffmeisteri Klotzsch === **Campanula pallida** Wall.

Campanula hohenackeri Fisch. & C. A. Mey. === **Campanula sibirica** subsp. **hohenackeri** (Fisch. & C. A. Mey.) Damboldt

Campanula hohenackeri var. *darialica* (Kharadze) Serdyuk. === **Campanula sibirica** subsp. **hohenackeri** (Fisch. & C. A. Mey.) Damboldt

Campanula homallanthina Ledeb. === **Campanula expansa** Rudolph

Campanula hondoensis Kitam. === **Campanula punctata** Lam.

Campanula hornungiana Schur === **Campanula serrata** (Kit.) Hendrych subsp. **serrata**

Campanula hortensis Meerb. === **Campanula rapunculoides** L.

Campanula hostii Baumg. === ?

Campanula hostii var. *uniflora* A. DC. === ?

Campanula hostii var. *witasekiana* (Vierh.) K. Malý === **Campanula witasekiana** Vierh.

Campanula humilis (A. DC.) D. Dietr. === **Wahlenbergia lobelioides** subsp. **riparia** (A. DC.) Thulin

Campanula humillima var. *luristanica* (Freyn.) Bornm. === **Campanula luristanica** Freyn

Campanula hybrida L. === **Legousia hybrida** (L.) Delarbre

Campanula hybrida Rodigas === **Campanula punctata** Lam.

Campanula hygrophila Rupr. === **Campanula bellidifolia** subsp. **aucheri** (A. DC.) Viktorov

Campanula hygrophila var. *obovata* Rupr. === **Campanula bellidifolia** subsp. **aucheri** (A. DC.) Viktorov

Campanula hyrcania Wettst. === **Campanula rapunculus** sunsp. **lambertiana** (A. DC.) Rech. f.

Campanula illinoensis Fresen. === **Campanula americana** L.

Campanula incanescens f. *khorasanica* (Rech. f. & Aellen) Parsa === **Campanula khorasanica** (Rech. f. & Aellen) Rech. f.

Campanula incanescens subsp. *kermanica* Rech. f., Aellen & Esfand. === **Campanula kermanica** (Rech. f., Aellen & Esfand.) Rech. f.

Campanula incanescens subsp. *khorasanica* Rech. f. & Aellen === **Campanula khorasanica** (Rech. f. & Aellen) Rech. f.

Campanula incanescens var. *exappendiculata* Bornm. === **Campanula mardinensis** Bornm. & Sint.

Campanula incanescens var. *holosericea* Korsh. === **Campanula cashmeriana** Benth.

Campanula incanescens var. *kermanica* (Rech. f., Aellen & Esfand.) Parsa === **Campanula kermanica** (Rech. f., Aellen & Esfand.) Rech. f.

Campanula incanescens var. *mollis* Korsh. === **Campanula cashmeriana** Benth.

Campanula inconcessa Schott === ?

Campanula indica (A. DC.) D. Dietr. === **Wahlenbergia marginata** (Thunb. ex Murray) A. DC.

Campanula infundibuliformis Sims === **Campanula rapunculoides** L.

Campanula infundibulum Vest ex Schult. === **Campanula stevenii** subsp. **wolgensis** (P. A. Smirn.) Fed.

Campanula integerrima Buch.-Ham. ex D. Don === **Campanula sylvatica** Wall.

Campanula intermedia Gand. === ?

Campanula intermedia Schult. === **Adenophora liliifolia** (L.) A. DC.

Campanula interrupta (L'Hér.) Pers. === **Prismatocarpus pedunculatus** (P. J. Bergius) A. DC.

Campanula irinae A. Kuth. === **Campanula collina** subsp. **fondervisii** (Albov) Ogan.

Campanula isophylla f. *alba* Voss === **Campanula isophylla** Moretti

Campanula istriaca Feer === **Campanula fenestrellata** subsp. **istriaca** (Feer) Damboldt

Campanula italica Arv.-Touv. === ?

Campanula jacobaea var. *bravensis* Bolle === **Campanula bravensis** (Bolle) A. Chev.

Campanula jacobaea var. *hispida* Bolle === **Campanula jacobaea** C. Sm. ex Webb

Campanula jacobaea var. *humilis* Bolle === **Campanula jacobaea** C. Sm. ex Webb

Campanula jacquinii subsp. *rumeliana* (Hampe) Kit Tan === **Campanula rumeliana** (Hampe) Vatke

Campanula japonica (Miq.) Vatke === **Asyneuma japonicum** (Miq.) Briq.

Campanula jasionifolia A. DC. === **Campanula stricta** L.

Campanula javanica (Blume) D. Dietr. === **Codonopsis javanica** (Blume) Miq.

Campanula juncea (H. Buek) D. Dietr. === **Wahlenbergia juncea** (H. Buek) Lammers

Campanula juncea Hill === ?

Campanula juncea Wettst. === **Asyneuma junceum** Parolly

Campanula kahenae Maire === **Campanula hispanica** Willk. subsp. **hispanica**

Campanula karakuschensis Grossh. === **Theodorovia karakuschensis** (Grossh.)Kolak.

Campanula kemulariae Fomin === **Campanula raddeana** Trautv.

Campanula kerneri Witasek ex A. Kern. & Fritsch. === **Campanula scheuchzeri** Vill.

Campanula ketzkhovelii Sosn. === **Campanula massalskyi** Fomin

Campanula khasiana Hook. f. & Thomson === **Adenophora khasiana** (Hook. f. & Thomson) Feer

Campanula kirpicznikovii Fed. === **Campanula alliariifolia** Willd. subsp. **alliariifolia**

Campanula kitaibeliana Schult. === **Campanula serrata** (Kit.) Hendrych subsp. **serrata**

Campanula kladniana (Schur) Witasek === **Campanula carnica** Schiede ex Mert. & Koch subsp. **carnica**

Campanula kladniana subsp. *mentiens* Witasek === **Campanula tatrae** subsp. **mentiens** (Witasek) Kovanda

Campanula kladniana subsp. *polymorpha* Witasek === **Campanula tatrae** Borbás subsp. **tatrae**

Campanula kladniana subsp. *stenophylla* (Schur) Witasek === **Campanula tatrae** Borbás subsp. **tatrae**

Campanula kladniana var. *mentiens* (Witasek) Pawł. === **Campanula tatrae** subsp. **mentiens** (Witasek) Kovanda

Campanula kluchorica Kolak. === **Campanula collina** subsp. **sphaerocarpa** (Kolak.) Ogan.

Campanula krebsii (Cham.) D. Dietr. === **Wahlenbergia krebsii** Cham.

Campanula kruhseana Fisch. ex Regel & Tiling === **Campanula expansa** Rudolph

Campanula lactiflora f. *coerulea* Voss === **Campanula lactiflora** M. Bieb.

Campanula lactiflora var. *glabra* Trautv. === **Campanula lactiflora** M. Bieb.

Campanula lactiflora var. *glabriuscula* K. Koch === **Campanula lactiflora** M. Bieb.

Campanula lactiflora var. *olympica* Griseb. === **Campanula lactiflora** M. Bieb.

Campanula lactiflora var. *pilosa* Trautv. === **Campanula lactiflora** M. Bieb.

Campanula lagranensis Sennen & Losa === ?

Campanula lamarckii (Fisch.) D. Dietr. === **Adenophora lamarckii** Fisch.

Campanula lamarckii var. *botryantha* Borbás === **Adenophora lamarckii** Fisch.

Campanula lambertiana A. DC. === **Campanula rapunculus** subsp. **lambertiana** (A. DC.) Rech. f.

Campanula lamiifolia Adam === **Campanula alliariifolia** Willd. subsp. **alliariifolia**

Campanula lamiifolia var. *macrophylla* (Sims) A. DC. === **Campanula alliariifolia** Willd. subsp. **alliariifolia**

Campanula lamioides Witasek === **Campanula glomerata** subsp. **hispida** (Witasek) Hayek

Campanula lanceolata Hegetschw. === ?

Campanula lanceolata J. Presl & C. Presl === **Campanula persicifolia** L. subsp. **persicifolia**

Campanula lanceolata Lapeyr. === **Campanula rotundifolia** L.

Campanula lanceolata Schur === **Campanula rotundifolia** L.

Campanula lanceolata subsp. *arcuata* (Schur) Simonkai === **Campanula serrata** (Kit.) Hendrych subsp. **serrata**

Campanula lanceolata var. *hirsuta* (Timb.-Lagr.) Rouy === **Campanula precatoria** Timb.-Lagr.

Campanula lanceolata var. *hornungiana* (Schur) Simonkai === **Campanula serrata** (Kit.) Hendrych subsp. **serrata**

Campanula lanceolata var. *major* (Timb.-Lagr.) Rouy === **Campanula precatoria** Timb.-Lagr.

Campanula lanceolata var. *praesignis* (Beck) Hruby === **Campanula praesignis** Beck

Campanula lanceolata var. *precatoria* (Timb.-Lagr.) Nyman === **Campanula precatoria** Timb.-Lagr.

Campanula lanceolata var. *tenuifolia* (Timb.-Lagr.) Rouy === **Campanula precatoria** Timb.-Lagr.

Campanula lancifolia Witasek === **Campanula baumgartenii** Becker subsp. **baumgartenii**

Campanula lancifolia Roxb. === **Cyclocodon lancifolius** (Roxb.) Kurz

Campanula langsdorffiana (Fisch. ex A. DC.) Trautv. & C. A. Mey. === **Campanula rotundifolia** L.

Campanula lanuginosa Lam. === **Campanula peregrina** L.

Campanula lanuginosa Willd. === **Campanula rupestris** Sm.

Campanula lapponica (Witasek) C. Regel === **Campanula giesekiana** Vest ex Schult.
Campanula larrainii Bertero ex Colla === **Wahlenbergia fernandeziana** A. DC.
Campanula lasiocarpa f. *albiflora* Tatew. === **Campanula lasiocarpa** Cham.
Campanula lasiocarpa subsp. *latisepala* (Hultén) Hultén === **Campanula lasiocarpa** Cham.
Campanula lasiocarpa var. *latisepala* Hultén === **Campanula lasiocarpa** Cham.
Campanula latifolia f. *alba* Voss === **Campanula latifolia** L. subsp. **latifolia**
Campanula latifolia f. *macrantha* (Fisch. ex Sims) Voss === **Campanula latifolia** L. subsp. **latifolia**
Campanula latifolia var. *canescens* Trautv. === **Campanula latifolia** L. subsp. **latifolia**
Campanula latifolia var. *eriocarpa* (M. Bieb.) Fisch. ex A. DC. === **Campanula latifolia** L. subsp. **latifolia**
Campanula latifolia var. *fimbriata* K. Koch === **Campanula latifolia** L. subsp. **latifolia**
Campanula latifolia var. *intermedia* Trautv. === **Campanula latifolia** L. subsp. **latifolia**
Campanula latifolia var. *leiocarpa* Trautv. === **Campanula latifolia** L. subsp. **latifolia**
Campanula latifolia var. *macrantha* Fisch. ex Sims === **Campanula latifolia** L. subsp. **latifolia**
Campanula latiloba A. DC. === **Campanula grandis** Fisch. & C. A. Mey. subsp. **grandis**
Campanula latiloba subsp. *rizeensis* A. Güner === **Campanula grandis** subsp. **rizeensis** (A. Güner) Lammers
Campanula latisepala Hultén === **Campanula alaskana** (A. Gray) Wight ex J. P. Anderson
Campanula latisepala f. *alba* J. Rousseau & Raymond === **Campanula intercedens** Witasek
Campanula latisepala var. *dubia* Hultén === **Campanula alaskana** (A. Gray) Wight ex J. P. Anderson
Campanula lavandulifolia Reinw. ex Blume === **Wahlenbergia marginata** (Thunb. ex Murray) A. DC.
Campanula lavrensis var. *breueri* (Tocl & Rohlena) Phitos === **Campanula lavrensis** (Tocl & Rohlena) Phitos
Campanula legionensis Pau === **Campanula rotundifolia** L.
Campanula lehmanniana var. *capusii* Franch. === **Campanula lehmanniana** subsp. **capusii** (Franch.) Viktorov.
Campanula lepida Feer === **Campanula fenestrellata** Feer subsp. **fenestrellata**
Campanula leptosiphon Pau & Sennen === **Campanula afra** Cav.
Campanula leskovii Fed. === **Campanula alliariifolia** Willd. subsp. **alliariifolia**
Campanula letschchumensis Kem.-Nat.=== **Campanula alliariifolia** subsp. **letschchumensis** (Kem.-Nat.) Ogan.
Campanula leucanthemifolia Pourr. === **Campanula cochleariifolia** Lam.
Campanula leucotricha C. Y. Wu === **Campanula chrysosplenifolia** Franch.
Campanula leutweinii Heldr. === **Campanula incurva** Aucher ex A. DC.
Campanula libanotica A. DC. === **Campanula stricta** L.
Campanula ligularis Lam. === ?
Campanula liliifolia L. === **Adenophora liliifolia** (L.) A. DC.
Campanula liliifolia var. *angustifolia* M. Bieb. === **Adenophora liliifolia** (L.) A. DC.
Campanula liliifolia var. *hirtula* Borbás === **Adenophora liliifolia** (L.) A. DC.
Campanula liliifolia var. *hybrida* Schult. === **Adenophora liliifolia** (L.) A. DC.
Campanula liliifolia var. *pycnodonta* Borbás === **Adenophora gmelinii** (Biehler) Fisch. subsp. **gmelinii**
Campanula liliifolia var. *stylosa* (Lam.) Besser === **Adenophora liliifolia** (L.) A. DC.
Campanula liliifolia var. *suaveolens* (Schrad. ex Hornem.) Schult. === **Adenophora liliifolia** (L.) A. DC.
Campanula limonifolium L. === **Asyneuma limonifolium** (L.) Janch.
Campanula linariifolia (A. DC.) D. Dietr. === **Prismatocarpus campanuloides** (L. f.) Sond.
Campanula linarioides Lam. === **Wahlenbergia linarioides** (Lam.) A. DC.
Campanula linearifolia Sennen === **Campanula macrorhiza** J. Gay ex A. DC.
Campanula linearis L. f. === **Microcodon linearis** (L. f.) H. Buek
Campanula lingulata Rchb. === **Campanula cervicaria** L.
Campanula lingulata var. *capitata* (Sims) Nyman === **Campanula lingulata** Waldst. & Kit.
Campanula lingulata var. *cichoracea* (Sm.) Griseb. === **Campanula lingulata** Waldst. & Kit.

Campanula lingulata var. *intybacea* Griseb. === **Campanula lingulata** Waldst. & Kit.

Campanula lingulata var. *tenuiflora* (Ten.) Nyman === **Campanula lingulata** Waldst. & Kit.

Campanula linifolia Hegetschw. === ?

Campanula linifolia L. === **Campanula persicifolia** L. subsp. **persicifolia**

Campanula linifolia Lam. === **Campanula serrata** subsp. **recta** (Dulac) Podlech

Campanula linifolia Schrank === **Campanula rotundifolia** L.

Campanula linifolia Scop. === **Campanula carnica** Schiede ex Mert. & Koch subsp. **carnica**

Campanula linifolia f. *angustifrons* Hruby === **Campanula carnica** Schiede ex Mert. & Koch subsp. **carnica**

Campanula linifolia f. *langsdorffiana* (Fisch. ex A. DC.) Voss === **Campanula rotundifolia** L.

Campanula linifolia f. *latifrons* Hruby === **Campanula carnica** Schiede ex Mert. & Koch subsp. **carnica**

Campanula linifolia f. *ramulosa* Hruby === **Campanula albanica** Witasek subsp. **albanica**

Campanula linifolia f. *scheuchzeri* (Vill.) Voss === **Campanula scheuchzeri** Vill.

Campanula linifolia f. *simplex* Hruby === **Campanula albanica** Witasek subsp. **albanica**

Campanula linifolia f. *umbrosa* Hruby === **Campanula carnica** Schiede ex Mert. & Koch subsp. **carnica**

Campanula linifolia f. *valdensis* (A. DC.) Voss === **Campanula scheuchzeri** Vill.

Campanula linifolia f. *vestina* Hruby === **Campanula carnica** Schiede ex Mert. & Koch subsp. **carnica**

Campanula linifolia subf. *angustissima* Hruby === **Campanula carnica** Schiede ex Mert. & Koch subsp. **carnica**

Campanula linifolia subf. *integrifrons* Hruby === **Campanula carnica** Schiede ex Mert. & Koch subsp. **carnica**

Campanula linifolia subsp. *albanica* (Witasek) Hayek === **Campanula albanica** Witasek

Campanula linifolia subsp. *ficarioides* (Timb.-Lagr.) Nyman === **Campanula ficarioides** Timb.-Lagr.

Campanula linifolia subsp. *justiniana* (Witasek) Hayek === **Campanula justiniana** Witasek

Campanula linifolia subsp. *valdensis* (A. DC.) Nyman === **Campanula scheuchzeri** Vill.

Campanula linifolia var. *albanica* (Witasek) Hruby === **Campanula albanica** Witasek

Campanula linifolia var. *ciliata* Witasek ex Vacc. === **Campanula carnica** Schiede ex Mert. & Koch subsp. **carnica**

Campanula linifolia var. *glabra* Schult. === **Campanula scheuchzeri** Vill.

Campanula linifolia var. *heterodoxa* Ledeb. === **Campanula rotundifolia** L.

Campanula linifolia var. *hirta* (Gelmi) Dalla Torre & Sarnth. === **Campanula carnica** Schiede ex Mert. & Koch subsp. **carnica**

Campanula linifolia var. *humilis* St.-Lag. === **Campanula scheuchzeri** Vill.

Campanula linifolia var. *jaubertiana* (Timb.-Lagr.) Nyman === **Campanula jaubertiana** Timb.-Lagr.

Campanula linifolia var. *justiniana* (Witasek) Jáv. === **Campanula justiniana** Witasek

Campanula linifolia var. *langsdorffiana* Fisch. ex A. DC. === **Campanula rotundifolia** L.

Campanula linifolia var. *major* Timb.-Lagr. === **Campanula rotundifolia** L.

Campanula linifolia var. *ovalifolia* St.-Lag. === **Campanula serrata** subsp. **recta** (Dulac) Podlech

Campanula linifolia var. *portae* Hruby === **Campanula carnica** Schiede ex Mert. & Koch subsp. **carnica**

Campanula linifolia var. *pubescens* Show === **Campanula scheuchzeri** Vill.

Campanula linifolia var. *rohdii* Lecoq & Lamotte === **Campanula serrata** subsp. **recta** (Dulac) Podlech

Campanula linifolia var. *ruscinonensis* (Timb.-Lagr.) Nyman === **Campanula ruscinonensis** Timb.-Lagr.

Campanula linifolia var. *scheuchzeri* (Vill.) Willd. === **Campanula scheuchzeri** Vill.

Campanula linifolia var. *tenuifolia* Timb.-Lagr. === **Campanula rotundifolia** L.

Campanula linifolia var. *valdensis* A. DC. === **Campanula scheuchzeri** Vill.

Campanula linnaeifolia A. Gray === **Campanula californica** (Kellogg) A. Heller

Campanula littoralis Labill. === **Wahlenbergia littoralis** (Labill.) Sweet

Campanula lobata Schloss. & Vuk. === **Campanula rotundifolia** L.

Campanula lobelioides L. f. === **Wahlenbergia lobelioides** (L. f.) Schrad. ex Link
Campanula lobelioides (Willd.) Vatke === **Asyneuma lobelioides** (Willd.) Hand.-Mazz.
Campanula loeflingii Brot. === **Campanula lusitanica** Loefl. subsp. **lusitanica**
Campanula loeflingii var. *durieui* (Boiss.) Nyman === **Campanula lusitanica** Loefl. subsp.
　lusitanica
Campanula longibracteata (H. Buek) D. Dietr. === **Treichelia longibracteata** (H. Buek) Vatke
Campanula longifolia Lapeyr. === **Campanula speciosa** Pourr. subsp. **speciosa**
Campanula longifolia Schloss. & Vuk. === **Campanula cervicaria** L.
Campanula longifolia var. *bicaulis* (Lapeyr.) Schult. === **Campanula speciosa** Pourr. subsp.
　speciosa
Campanula longifolia var. *pyramidalis* Lapeyr. === **Campanula speciosa** Pourr.
Campanula longifolia var. *speciosa* (Pourr.) Steud. === **Campanula speciosa** Pourr.
Campanula longirostris Banks ex D. Dietr. === **Prismatocarpus crispus** L'Hér.
Campanula longistyla f. *parviflora* Fomin === **Campanula sibirica** subsp. **charadzae**
　(Grossh.) Ogan.
Campanula lorei Pollini === **Campanula ramosissima** Sm.
Campanula loretiana Witasek === **Campanula precatoria** Timb.-Lagr.
Campanula losae Sennen === **Campanula hispanica** Willk. subsp. **hispanica**
Campanula lostrittii Ten. === **Campanula rotundifolia** L.
Campanula ludoviciana Riddell === **Triodanis perfoliata** subsp. **biflora** (Ruiz & Pav.) Lammers
Campanula lunariifolia Willd. ex Schult. === **Campanula rapunculoides** L.
Campanula lusitanica f. *pallidiflora* Maire === **Campanula lusitanica** Loefl. subsp. **lusitanica**
Campanula lusitanica f. *tenuis* Caball. === **Campanula lusitanica** Loefl. subsp. **lusitanica**
Campanula lusitanica var. *maura* Murb. === **Campanula lusitanica** Loefl. subsp. **lusitanica**
Campanula lusitanica var. *pseudophrygia* Lacaita === **Campanula lusitanica** Loefl. subsp.
　lusitanica
Campanula lusitanica var. *puberula* C. Vicioso === **Campanula lusitanica** Loefl. subsp.
　lusitanica
Campanula lychnitis Hornem. === **Campanula bononiensis** L.
Campanula lycia (Boiss.) Boiss. === **Asyneuma lycium** (Boiss.) Bornm.
Campanula lyrata var. *albostrigosa* Rech. f. === **Campanula lyrata** Lam. subsp. **lyrata**
Campanula lyratella Feer === **Campanula lyrata** Lam. subsp. **lyrata**
Campanula lyrifolia Salisb. === **Michauxia campanuloides** L'Hér.
Campanula macdougalii Rydb. === **Campanula petiolata** A. DC.
Campanula macedonica Boiss. & Orph. ex Freyn === **Campanula macrostachya** Waldst. &
　Kit. ex Willd.
Campanula macrantha (Fisch. ex Sims) Hook. === **Campanula latifolia** L. subsp. **latifolia**
Campanula macrantha var. *polyantha* Hook. === **Campanula latifolia** L. subsp. **latifolia**
Campanula macrochlamys Boiss. & A. Huet === **Sachokiella macrochlamys** (Boiss. & A.
　Huet) Kolak.
Campanula macrophylla Sims === **Campanula alliariifolia** Willd. subsp. **alliariifolia**
Campanula macrorhiza f. *nicaeensis* (Risso) Hruby === **Campanula macrorhiza** J. Gay ex A. DC.
Campanula macrorhiza f. *squarrulosa* Hruby === **Campanula macrorhiza** J. Gay ex A. DC.
Campanula macrorhiza subsp. *aitanica* (Pau ex O. Bolòs & Vigo) Rivas Mart. === **Campanula**
　macrorhiza J. Gay ex A. DC.
Campanula macrorhiza subsp. *angustiflora* (Tanfani) Arcang. === **Campanula tanfanii** Podlech
Campanula macrorhiza subsp. *catalanica* (Podlech) Rivas Mart. === **Campanula hispanica**
　subsp. **catalanica** Podlech
Campanula macrorhiza subsp. *gypsicola* (Costa) Rivas Mart. === **Campanula hispanica**
　subsp. **catalanica** Podlech
Campanula macrorhiza subsp. *sabatia* (De Not.) Arcang. === **Campanula sabatia** De Not.
Campanula macrorhiza subvar. *re* (Colla) Hruby === **Campanula bertolae** Colla
Campanula macrorhiza var. *angustiflora* Tanfani === **Campanula tanfanii** Podlech
Campanula macrorhiza var. *bertolae* (Colla) Hruby === **Campanula bertolae** Colla
Campanula macrorhiza var. *gypsicola* Costa === **Campanula hispanica** subsp. **catalanica**
　Podlech

Campanula macrorhiza var. *jurjurensis* (Pomel) Quézel === **Campanula jurjurensis** Pomel
Campanula macrorhiza var. *pubescens* A. DC. === **Campanula macrorhiza** J. Gay ex A. DC.
Campanula macrorhiza var. *rotundata* (Pomel) Chabert === **Campanula jurjurensis** Pomel
Campanula macrorhiza var. *sabatia* (De Not.) Tanfani === **Campanula sabatia** De Not.
Campanula macrorhiza var. *sardoa* Levier ex Tanfani === **Campanula forsythii** (Arcang.)
    Podlech
Campanula macrorrhiza Vuk. === ?
Campanula macrorrhiza var. *thyrsoides* (L.) Vuk. === **Campanula thyrsoides** L.
Campanula macrorrhiza var. *chymatophylla* Vuk. === **Campanula sibirica** L.
Campanula macrorrhiza var. *pogonopetala* Vuk. === **Campanula barbata** L.
Campanula macrorrhiza var. *polycaulis* Vuk. === **Campanula alpina** Jacq.
Campanula macrorrhiza var. *trianthemos* Vuk.=== **Campanula sibirica** subsp. **divergens**
    (Waldst. & Kit. ex Willd.) Nyman
Campanula macrostyla f. rosea Voss === **Campanula macrostyla** Boiss. & Heldr.
Campanula madagascariensis (A. DC.) D. Dietr. === **Wahlenbergia madagascariensis**
    A. DC.
Campanula magellensis Ten. === **Campanula persicifolia** L. subsp. **persicifolia**
Campanula mairei f. *cordatifolia* Maire === **Campanula mairei** Pau ex Maire
Campanula mairei f. *ramosa* Quézel === **Campanula mairei** Pau ex Maire
Campanula mairei var. *anremerica* Litard. & Maire === **Campanula mairei** Pau ex Maire
Campanula makaschvilii E. A. Busch === **Campanula alliariifolia** Willd. subsp. **alliariifolia**
Campanula malacitana Degen & Hervier === **Campanula mollis** L. subsp. **mollis**
Campanula malacitana var. *almeriensis* Degen & Hervier === **Campanula mollis** L. subsp.
    mollis
Campanula maleevii Fed. === **Campanula glomerata** subsp. **hispida** (Witasek) Hayek
Campanula malyi Schott === **Campanula scheuchzeri** Vill.
Campanula marchesettii subsp. *apennina* Podlech === **Campanula micrantha** Bertol.
Campanula marchesettii var. *calisii* (Murr) Dalla Torre & Sarnth. === **Campanula
    marchesettii** Witasek
Campanula marginata Thunb. ex Murray === **Wahlenbergia marginata** (Thunb. ex Murray)
    A. DC.
Campanula mariani Sennen === **Campanula macrorhiza** J. Gay ex A. DC.
Campanula maroccana Ball === **Campanula filicaulis** Durieu
Campanula maroccana var. *pseudoradicosa* Litard. & Maire === **Campanula filicaulis** Durieu
Campanula marschalliana Boiss. === **Asyneuma campanuloides** (M. Bieb. ex Sims) Bornm.
Campanula marsupiiflora Schult. === **Adenophora stenanthina** (Ledeb.) Kitag. subsp.
    stenanthina
Campanula × *martinezii* Marcet === **Campanula speciosa** subsp. **affinis** (Schult.) Font Quer
Campanula massonii (A. DC.) D. Dietr. === **Wahlenbergia massonii** A. DC.
Campanula mathionetii Jord. ex Gren. & Godr. === **Campanula cochleariifolia** Lam.
Campanula matritensis A. DC. === **Campanula lusitanica** subsp. **matritensis** (A. DC.)
    Franco
Campanula matritensis var. *nevadensis* A. DC. === **Campanula lusitanica** subsp. **matritensis**
    (A. DC.) Franco
Campanula mauritanica Pomel === **Campanula trachelium** subsp. **mauritanica** (Pomel)
    Quézel
Campanula × *mayi* Anonymous === **Campanula isophylla** Moretti
Campanula medium f. *alba* Voss === **Campanula medium** L.
Campanula medium f. *caesia* Voss === **Campanula medium** L.
Campanula medium f. *calycantha* Voss === **Campanula medium** L.
Campanula medium f. *coerulea* Voss === **Campanula medium** L.
Campanula medium f. *rosea* Voss === **Campanula medium** L.
Campanula medium f. *striata* Voss === **Campanula medium** L.
Campanula medium var. *bourdiniana* (Gand.) Nyman === **Campanula medium** L.
Campanula medium var. *calycanthema* Anonymous === **Campanula medium** L.
Campanula megaphylla Gand. === ?

Campanula megrelica Manden. & Kuth. === **Campanula latifolia** subsp. **megrelica** (Manden. & Kuth.) Ogan.

Campanula mentiens (Witasek) Prain === **Campanula tatrae** subsp. **mentiens** (Witasek) Kovanda

Campanula meyerana Rupr. === **Campanula bellidifolia** subsp. **meyerana** (Rupr.) Viktorov

Campanula michauxioides Boiss. === **Asyneuma michauxioides** (Boiss.) Damboldt

Campanula michauxioides var. *dilacerata* Bornm. === **Asyneuma michauxioides** (Boiss.) Damboldt

Campanula michauxioides var. *hirtifolia* Huber-Mor. === **Asyneuma michauxioides** (Boiss.) Damboldt

Campanula microcarpa C. Y. Wu === **Campanula pallida** Wall.

Campanula microcodon D. Dietr. === **Microcodon glomeratus** A. DC.

Campanula microphylla Cav. === **Campanula mollis** L. subsp. **mollis**

Campanula microphylla Kit. === === **Campanula serrata** (Kit.) Hendrych subsp. **serrata**

Campanula migarica Schischk. === **Campanula collina** subsp. **fondervisii** (Albov) Ogan.

Campanula mikoi Borbás === **Adenophora liliifolia** (L.) A. DC.

Campanula minae Strobl === **Asyneuma trichocalycinum** (Ten.) K. Malý

Campanula minor Honck. === **Campanula cespitosa** Scop.

Campanula minor Lam. === **Campanula rotundifolia** L.

Campanula minsteriana Grossh. === **Theodorovia karakuschensis** (Grossh.) Kolak.

Campanula minuta Savi === ?

Campanula minutissima Schur === **Campanula cochleariifolia** Lam.

Campanula mixta Hegetschw. === **Campanula cochleariifolia** Lam.

Campanula modesta Hook. f. & Thomson === **Campanula immodesta** Lammers

Campanula modesta Schott === **Campanula cochleariifolia** Lam.

Campanula moesiaca var. *oblongifolia* K. Malý === **Campanula moesiaca** Velen.

Campanula moldavica Gand. === ?

Campanula molineri Colla === ?

Campanula mollis f. *transiens* Quézel === **Campanula mollis** L. subsp. **mollis**

Campanula mollis var. *canescens* Quézel === **Campanula mollis** L. subsp. **mollis**

Campanula mollis var. *lasiantha* Lara === **Campanula mollis** L. subsp. **mollis**

Campanula mollis var. *longicornis* Maire === **Campanula mollis** L. subsp. **mollis**

Campanula mollis var. *maroccana* (Pau) Font Quer === **Campanula mollis** L. subsp. **mollis**

Campanula mollis var. *mesatlantica* Quézel === **Campanula mollis** L. subsp. **mollis**

Campanula mollis var. *microphylla* (Cav.) A. DC. === **Campanula mollis** L. subsp. **mollis**

Campanula mollis var. *oranensis* Maire === **Campanula mollis** L. subsp. **mollis**

Campanula mollis var. *pseudovelata* Maire === **Campanula mollis** L. subsp. **mollis**

Campanula mollis var. *rifana* Emb. & Maire === **Campanula mollis** L. subsp. **mollis**

Campanula mollis var. *tlemcenensis* Quézel === **Campanula mollis** L. subsp. **mollis**

Campanula monantha Schur === **Campanula scheuchzeri** Vill.

Campanula monanthos Pant. === **Campanula patula** L. subsp. **patula**

Campanula monocephala Trautv. === **Cryptocodon monocephalus** (Trautv.) Fed.

Campanula montana Delarbre === ?

Campanula montevidensis Spreng. === **Triodanis perfoliata** subsp. **biflora** (Ruiz & Pav.) Lammers

Campanula morifolia Salisb. === **Campanula rapunculoides** L.

Campanula mrkvckiana Velen. === **Campanula versicolor** Andrews

Campanula multicaulis Boiss. === **Asyneuma persicum** (A. DC.) Bornm.

Campanula multicaulis Witasek === ?

Campanula multiflora Waldst. & Kit. === **Campanula macrostachya** Waldst. & Kit. ex Willd.

Campanula murrayana Karsch === ?

Campanula mycalaea Barbey & Major === **Campanula tomentosa** Lam.

Campanula myosotidifolia Boiss. === **Campanula argentea** Lam.

Campanula nana Lam. === **Campanula alpestris** All.

Campanula nanella P. A. Smirn. === **Campanula erinus** L.

Campanula napuligera Schur === **Campanula serrata** (Kit.) Hendrych subsp. **serrata**

Campanula napuligera f. *angustifrons* Hruby === **Campanula serrata** (Kit.) Hendrych subsp. **serrata**

Campanula napuligera f. *arcuata* (Schur) Hruby === **Campanula serrata** (Kit.) Hendrych subsp. **serrata**

Campanula napuligera f. *genuina* Hruby === **Campanula serrata** (Kit.) Hendrych subsp. **serrata**

Campanula napuligera f. *genuina* Hruby === **Campanula serrata** (Kit.) Hendrych subsp. **serrata**

Campanula napuligera f. *glabrescens* Hruby === **Campanula serrata** (Kit.) Hendrych subsp. **serrata**

Campanula napuligera f. *glabrescens* Hruby === **Campanula serrata** (Kit.) Hendrych subsp. **serrata**

Campanula napuligera f. *humilis* Hruby === **Campanula serrata** (Kit.) Hendrych subsp. **serrata**

Campanula napuligera f. *intermedia* Hruby === **Campanula serrata** (Kit.) Hendrych subsp. **serrata**

Campanula napuligera f. *latifrons* Hruby === **Campanula serrata** (Kit.) Hendrych subsp. **serrata**

Campanula napuligera f. *longisepala* (Nyár.) Morariu === **Campanula serrata** (Kit.) Hendrych subsp. **serrata**

Campanula napuligera f. *minima* (Săvul.) Morariu === **Campanula serrata** (Kit.) Hendrych subsp. **serrata**

Campanula napuligera f. *robusta* Hruby === **Campanula serrata** (Kit.) Hendrych subsp. **serrata**

Campanula napuligera f. *savulescui* Morariu === **Campanula serrata** (Kit.) Hendrych subsp. **serrata**

Campanula napuligera f. *scheuchzeriformes* (Nyár.) Morariu === **Campanula serrata** (Kit.) Hendrych subsp. **serrata**

Campanula napuligera f. *semiamplexicaulis* (Săvul.) Morariu === **Campanula serrata** (Kit.) Hendrych subsp. **serrata**

Campanula napuligera f. *setulosa* Morariu === **Campanula serrata** (Kit.) Hendrych subsp. **serrata**

Campanula napuligera f. *simplex* Hruby === **Campanula serrata** (Kit.) Hendrych subsp. **serrata**

Campanula napuligera f. *stenophylloides* Nyár. === **Campanula serrata** (Kit.) Hendrych subsp. **serrata**

Campanula napuligera f. *transsilvanica* (Săvul.) Morariu === **Campanula serrata** (Kit.) Hendrych subsp. **serrata**

Campanula napuligera subf. *angustifrons* Hruby === **Campanula serrata** (Kit.) Hendrych subsp. **serrata**

Campanula napuligera subf. *brachyantha* Hruby === **Campanula serrata** (Kit.) Hendrych subsp. **serrata**

Campanula napuligera subf. *latifrons* Hruby === **Campanula serrata** (Kit.) Hendrych subsp. **serrata**

Campanula napuligera subf. *tenella* Hruby === **Campanula serrata** (Kit.) Hendrych subsp. **serrata**

Campanula napuligera var. *alpiniformis* Nyár. ex Morariu === **Campanula serrata** (Kit.) Hendrych subsp. **serrata**

Campanula napuligera var. *arcuata* (Schur) Morariu === **Campanula serrata** (Kit.) Hendrych subsp. **serrata**

Campanula napuligera var. *elatior* (Săvul.) Morariu === **Campanula serrata** (Kit.) Hendrych subsp. **serrata**

Campanula napuligera var. *hirsuta* Hruby === **Campanula serrata** (Kit.) Hendrych subsp. **serrata**

Campanula napuligera var. *hornungiana* (Săvul.) Morariu === **Campanula serrata** (Kit.) Hendrych subsp. **serrata**

Campanula napuligera var. *longisepala* Nyár. === **Campanula serrata** (Kit.) Hendrych subsp. **serrata**

Campanula napuligera var. *redux* (Schott) Hruby === **Campanula serrata** (Kit.) Hendrych subsp. **serrata**

Campanula napuligera var. *savulescui* Morariu === **Campanula serrata** (Kit.) Hendrych subsp. **serrata**

Campanula napuligera var. *scheuchzeriformis* Nyár. === **Campanula serrata** (Kit.) Hendrych subsp. **serrata**

Campanula napuligera var. *stenophylloides* (Nyár.) Morariu === **Campanula serrata** (Kit.) Hendrych subsp. **serrata**

Campanula napuligera var. *stricta* Hruby === **Campanula serrata** (Kit.) Hendrych subsp. **serrata**

Campanula napuligera var. *umbrosa* Hruby === **Campanula serrata** (Kit.) Hendrych subsp. **serrata**

Campanula nefedovii Galushko === **Campanula collina** Sims subsp. **collina**

Campanula neglecta Schult. === **Campanula patula** L. subsp. **patula**

Campanula neglecta Besser === **Campanula rapunculoides** L.

Campanula nemorosa A. DC. === **Campanula rapunculoides** L.

Campanula nephrophylla C. Y. Wu === **Campanula crenulata** Franch.

Campanula nevadensis Pau === **Campanula willkommii** Witasek

Campanula nicaeensis Schult. === **Campanula glomerata** subsp. **oblongifolia** (K. Koch) Fed.

Campanula nicaeensis Risso === **Campanula macrorhiza** J. Gay ex A. DC.

Campanula nitida Aiton === **Campanula americana** L.

Campanula nobilis Lindl. === **Campanula punctata** Lam.

Campanula notata Schott, Nyman & Kotschy === **Campanula cochleariifolia** Lam.

Campanula nudicaulis (A. DC.) D. Dietr. === **Wahlenbergia androsacea** A. DC.

Campanula nutabunda Guss. === **Wahlenbergia lobelioides** subsp. **nutabunda** (Guss.) Murb.

Campanula nutans Lam. === **Campanula rapunculoides** L.

Campanula nutans Vahl ex Hornem. === **Campanula sibirica** subsp. **divergens** (Waldst. & Kit. ex Willd.) Nyman

Campanula obliqua Jacq. === **Campanula americana** L.

Campanula obliquifolia Ten. === **Campanula bononiensis** L.

Campanula oblongifolia (K. Koch) Kharadze === **Campanula glomerata** subsp. **oblongifolia** (K. Koch) Fed.

Campanula oblongifolioides Galushko === **Campanula glomerata** subsp. **oblongifolioides** (Galushko) Ogan.

Campanula ochroleuca (Kem.-Nat.) Kem.-Nat. === **Campanula alliariifolia** Willd. subsp. **alliariifolia**

Campanula ochroleuca var. *alpestris* (Kem.-Nat.) Kem.-Nat. === **Campanula alliariifolia** Willd. subsp. **alliariifolia**

Campanula ochroleuca var. *rupestris* (Kem.-Nat.) Kem.-Nat. === **Campanula alliariifolia** Willd. subsp. **alliariifolia**

Campanula ochroleuca var. *silvatica* (Kem.-Nat.) Kem.-Nat. === **Campanula alliariifolia** Willd. subsp. **alliariifolia**

Campanula oligosantha Schur === **Campanula trachelium** L. subsp. **trachelium**

Campanula oliveri Rouy & Gaut. === **Campanula speciosa** Pourr. subsp. **speciosa**

Campanula olivieri A. DC. === **Campanula calamenthifolia** Lam.

Campanula orbelica Pančić === **Campanula alpina** subsp. **orbelica** (Pančić) Urum.

Campanula orientalis Steud. === **Campanula peregrina** L.

Campanula ottoniana Schult. === **Wahlenbergia parvifolia** (P. J. Bergius) Lammers

Campanula ovata (D. Don) Spreng. === **Percarpa carnosa** (Wall.) Hook. f. & Thomson

Campanula oxyphylla Vuk. === ?

Campanula oxyphylla var. *lasiopetala* Vuk. === **Campanula garganica** Ten.

Campanula oxyphylla var. *leiophylla* Vuk. === **Campanula portenschlagiana** Schult.

Campanula paenina Reut. ex Tissière === **Campanula rotundifolia** L.

Campanula pallasiana Vest ex Schult. === **Campanula dasyantha** M. Bieb. subsp. **dasyantha**

Campanula pallida var. *tibetica* (Hook. f. & Thomson) H. Hara === **Campanula pallida** Wall.

Campanula pallidiflora Rupr. === **Campanula bellidifolia** subsp. **aucheri** (A. DC.) Viktorov

Campanula pamphylica (Contandr., Quézel & Pamukç.) Akçiçek & Vural === **Campanula tokurii** Ocak

Campanula pamphylica subsp. *afyonica* Akçiçek & Vural === **Campanula tokurii** Ocak

Campanula pamphylica subsp. *tokurii* (Ocak) Akçiçek & Vural === **Campanula tokurii** Ocak

Campanula paniculata L. f. === **Wahlenbergia paniculata** (L. f.) A. DC.

Campanula paniculata Pohl === **Campanula sibirica** L.

Campanula paniculata Turra ex Sacc. === ?

Campanula panjutinii Kolak. === **Campanula glomerata** subsp. **panjutinii** (Kolak.) Viktorov

Campanula parnassica Boiss. & Spruner === **Edraianthus parnassicus** (Boiss. & Spruner) Halácsy

Campanula parryi var. *idahoensis* McVaugh === **Campanula parryi** A. Gray

Campanula parviflora Lam. === **Campanula sibirica** subsp. **hohenackeri** (Fisch. & C. A. Mey.) Damboldt

Campanula parviflora Salisb. === **Wahlenbergia lobelioides** (L. f.) Schrad. ex Link

Campanula parviflora St.-Lag. === **Campanula erinus** L.

Campanula parvula Jord. === **Campanula cochleariifolia** Lam.

Campanula pasumensis C. Marquand === **Campanula cana** Wall.

Campanula patens Gueldenst. === **Campanula patula** L. subsp. **patula**

Campanula patenticalyx Bocquet === **Campanula pelia** (Halácsy) Hausskn. & Sint. ex Phitos

Campanula patula f. *velenovskyi* (Adamović) Hayek === **Campanula patula** subsp. **epigaea** (Janka ex Degen) Hayek

Campanula patula race *neglecta* (A. DC.) Rouy === **Campanula patula** L. subsp. **patula**

Campanula patula subsp. *flaccida* (Wallr.) Sóo === **Campanula patula** L. subsp. **patula**

Campanula patula subsp. *neglecta* (A. DC.) O. Schwarz === **Campanula patula** L. subsp. **patula**

Campanula patula subsp. *peterfii* (Sóo ex Jáv.) Sóo === **Campanula patula** L. subsp. **patula**

Campanula patula [unranked] *peterfii* Sóo ex Jáv. === **Campanula patula** L. subsp. **patula**

Campanula patula var. *albiflora* Syr. === **Campanula patula** L. subsp. **patula**

Campanula patula var. *calycina* Willk. === **Campanula patula** L. subsp. **patula**

Campanula patula var. *calyciserrata* Beyer === **Campanula patula** L. subsp. **patula**

Campanula patula var. *confertiflora* Trautv. === **Campanula olympica** Boiss.

Campanula patula var. *costae* (Willk.) O. Bolòs & Vigo === **Campanula patula** subsp. **costae** (Willk.) Nyman

Campanula patula var. *decumbens* (A. DC.) Cuatrec. === **Campanula decumbens** A. DC.

Campanula patula var. *decurrens* (L.) Schult. === **Campanula patula** L. subsp. **patula**

Campanula patula var. *flaccida* Syr. === **Campanula patula** L. subsp. **patula**

Campanula patula var. *flaccida* Wallr. === **Campanula patula** L. subsp. **patula**

Campanula patula var. *glauca* Kuntze === **Campanula patula** L. subsp. **patula**

Campanula patula var. *grandiflora* A. DC. === **Campanula patula** L. subsp. **patula**

Campanula patula var. *hirsuta* Schur === **Campanula patula** L. subsp. **patula**

Campanula patula var. *jahorinae* K. Malý === **Campanula patula** subsp. **jahorinae** (K. Malý) Greuter & Burdet

Campanula patula var. *latifolia* A. DC. === **Campanula patula** L. subsp. **patula**

Campanula patula var. *macrocalyx* Kuntze === **Campanula patula** L. subsp. **patula**

Campanula patula var. *neglecta* A. DC. === **Campanula patula** L. subsp. **patula**

Campanula patula var. *parva* Merino === **Campanula patula** L. subsp. **patula**

Campanula patula var. *pauciflora* Rochel === **Campanula patula** subsp. **abietina** Griseb. & Schenk

Campanula patula var. *peterfii* (Sóo ex Jáv.) Sóo === **Campanula patula** L. subsp. **patula**

Campanula patula var. *platyphylla* Kuntze === **Campanula patula** L. subsp. **patula**

Campanula patula var. *rapunculus* (L.) Kuntze === **Campanula rapunculus** L.

Campanula patula var. *serratisepala* Murr === **Campanula patula** L. subsp. **patula**

Campanula patula var. *stricta* Opiz === **Campanula patula** L. subsp. **patula**

Campanula patula var. *thyrsiflora* Kuntze === **Campanula patula** L. subsp. **patula**

Campanula patula var. *uniflora* Noulet === **Campanula patula** L. subsp. **patula**

Campanula patula var. *vajdae* (Pénzes) Fed. === **Campanula patula** subsp. **abietina** (Griseb. & Schenk) Simonk.

Campanula pauciflora Desf. === **Campanula spatulata** subsp. **filicaulis** (Halácsy) Phitos

Campanula paui Font Quer === **Campanula hispanica** Willk. subsp. **hispanica**

Campanula × *pechlaneri* Murr ex Dalla Torre & Sarnth. === **Campanula × glomeratiformis** Murr ex Dalla Torre & Sarnth.

Campanula peduncularis Wall. === **Wahlenbergia peduncularis** (Wall. ex A. DC.) Hook. f. & Thomson

Campanula pedunculata J. F. Gmel. === **Prismatocarpus pedunculatus** (P. J. Bergius) A. DC.

Campanula pelviformis var. *corymbosa* (Desf.) Nyman === **Campanula pelviformis** Lam.

Campanula pelviformis var. *micrantha* A. DC. === **Campanula pelviformis** Lam.

Campanula pennina Reut. === **Campanula rotundifolia** L.

Campanula pentagonia L. === **Legousia pentagonia** (L.) Thell.

Campanula pentagonophylla Vuk. === **Wahlenbergia hederacea** (L.) Rchb.

Campanula pereskiifolia (Fisch. ex Schult.) Schult. === **Adenophora pereskiifolia** (Fisch. ex Schult.) Fisch. ex G. Don

Campanula perfoliata L. === **Triodanis perfoliata** (L.) Nieuwl.

Campanula periplocifolia Lam. === **Adenophora liliifolia** (L.) A. DC.

Campanula perneglecta Schott === **Campanula cochleariifolia** Lam.

Campanula perpallens Borbás === **Adenophora liliifolia** (L.) A. DC.

Campanula persica A. DC. === **Asyneuma persicum** (A. DC.) Bornm.

Campanula persicaster F. Herm. === **Campanula persicifolia** subsp. **sessiliflora** (K. Koch ex Velen.) Fed. ex Greuter & Burdet

Campanula persicifolia f. *alba* Voss === **Campanula persicifolia** L. subsp. **persicifolia**

Campanula persicifolia f. *coronata* Voss === **Campanula persicifolia** L. subsp. **persicifolia**

Campanula persicifolia f. *kirschlegeri* Sóo === **Campanula persicifolia** L. subsp. **persicifolia**

Campanula persicifolia subvar. *hispida* (Lej.) Rouy === **Campanula persicifolia** L. subsp. **persicifolia**

Campanula persicifolia subvar. *pumila* (F. W. Schmidt) Nyman === **Campanula persicifolia** L. subsp. **persicifolia**

Campanula persicifolia var. *alpina* Schur === **Campanula persicifolia** L. subsp. **persicifolia**

Campanula persicifolia var. *angustifolia* A. DC. === **Campanula persicifolia** L. subsp. **persicifolia**

Campanula persicifolia var. *crystallocalyx* (Adamović) Hayek === **Campanula persicifolia** L. subsp. **persicifolia**

Campanula persicifolia var. *dasycarpa* (Kit.) A. DC. === **Campanula persicifolia** L. subsp. **persicifolia**

Campanula persicifolia var. *eriocarpa* Schur === **Campanula persicifolia** L. subsp. **persicifolia**

Campanula persicifolia var. *eriocarpa* Syr. === **Campanula persicifolia** L. subsp. **persicifolia**

Campanula persicifolia var. *glaberrima* Schur === **Campanula persicifolia** L. subsp. **persicifolia**

Campanula persicifolia var. *grandiflora* Schur === **Campanula persicifolia** L. subsp. **persicifolia**

Campanula persicifolia var. *hispida* (Lej.) A. DC. === **Campanula persicifolia** L. subsp. **persicifolia**

Campanula persicifolia var. *hispidior* Trautv. === **Campanula persicifolia** L. subsp. **persicifolia**

Campanula persicifolia var. *humillima* Schur === **Campanula persicifolia** L. subsp. **persicifolia**

Campanula persicifolia var. *laevicaulis* Korsh. === **Campanula persicifolia** L. subsp. **persicifolia**

Campanula persicifolia var. *lanceolata* Steud. === **Campanula persicifolia** L. subsp. **persicifolia**

Campanula persicifolia var. *lasiocalyx* Gren. & Godr. === **Campanula persicifolia** subsp. **subpyrenaica** (Timb.-Lagr.) Fed.

Campanula persicifolia var. *lasiocarpa* Korsh. === **Campanula persicifolia** L. subsp. **persicifolia**

Campanula persicifolia var. *latisepala* Degen & Dörfl. === **Campanula persicifolia** subsp. **sessiliflora** (K. Koch ex Velen.) Fed. ex Greuter & Burdet

Campanula persicifolia var. *macrantha* A. DC. === **Campanula persicifolia** L. subsp. **persicifolia**

Campanula persicifolia var. *maxima* Sims === **Campanula persicifolia** L. subsp. **persicifolia**

Campanula persicifolia var. *monstrosa* Schur === **Campanula persicifolia** L. subsp. **persicifolia**

Campanula persicifolia var. *parviflora* Freyn === **Campanula persicifolia** L. subsp. **persicifolia**

Campanula persicifolia var. *parviflora* Kirschl. === **Campanula persicifolia** L. subsp. **persicifolia**

Campanula persicifolia var. *parviflora* Peterm. === **Campanula persicifolia** L. subsp. **persicifolia**

Campanula persicifolia var. *pumila* (F. W. Schmidt) Schult. === **Campanula persicifolia** L. subsp. **persicifolia**

Campanula persicifolia var. *pumila* Ten. === **Campanula persicifolia** L. subsp. **persicifolia**
Campanula persicifolia var. *reflexa* Dalla Torre & Sarnth. === **Campanula persicifolia** L. subsp. **persicifolia**
Campanula persicifolia var. *sessiliflora* K. Koch ex Velen. === **Campanula persicifolia** subsp. **sessiliflora** (K. Koch ex Velen.) Fed. ex Greuter & Burdet
Campanula persicifolia var. *snjagovii* Velchev & P. Vassil. === **Campanula persicifolia** subsp. **sessiliflora** (K. Koch ex Velen.) Fed. ex Greuter & Burdet
Campanula persicifolia var. *subpyrenaica* (Timb.-Lagr.) Nyman === **Campanula persicifolia** subsp. **subpyrenaica** (Timb.-Lagr.) Fed.
Campanula persicifolia var. *suskalovicii* Adamović === **Campanula persicifolia** L. subsp. **persicifolia**
Campanula persicifolia var. *uniflora* Noulet === **Campanula persicifolia** L. subsp. **persicifolia**
Campanula pestalozzae (Boiss.) Boiss. === **Asyneuma limonifolium** subsp. **pestalozzae** (Boiss.) Damboldt
Campanula pestalozzae Boiss. === **Campanula propinqua** Fisch. & C. A. Mey.
Campanula petraea Zanted. ex Moretti === **Campanula elatinoides** Moretti
Campanula petrophila var. *angustiflora* Fomin === **Campanula petrophila** Rupr.
Campanula petrophila var. *borbalensis* Rupr. === **Campanula petrophila** Rupr.
Campanula petrophila var. *exappendiculata* Sommier & Levier === **Campanula petrophila** Rupr.
Campanula petrophila var. *linoides* Rupr. === **Campanula petrophila** Rupr.
Campanula petrophila var. *longiflora* Rupr. === **Campanula petrophila** Rupr.
Campanula phrygia f. *velutina* (Griseb.) Hayek === **Campanula phrygia** Jaub. & Spach
Campanula phyctidocalyx var. *sulcata* Parsa === **Campanula phyctidocalyx** Boiss. & Noë
Campanula phyteumoides Sibth. ex Zuccagni === **Asyneuma limonifolium** (L.) Janch. subsp. **limonifolium**
Campanula pichleri Vis. === **Asyneuma pichleri** (Vis.) D. Lakušić & F. Conti
Campanula pilosa Pall. ex Schult. === **Campanula dasyantha** M. Bieb. subsp. **dasyantha**
Campanula pilosa var. *dasyantha* (M. Bieb.) Herder === **Campanula dasyantha** M. Bieb.
Campanula pilosa var. *pontica* K. Koch === **Campanula bellidifolia** subsp. **aucheri** (A. DC.) Viktorov
Campanula pinifolia Uechtr. ex Pančić === **Campanula rotundifolia** L.
Campanula pinifolia Vuk. === **Campanula rotundifolia** L.
Campanula pinnatifida var. *germanicopolitana* Huber-Mor. === **Campanula pinnatifida** Huber-Mor.
Campanula pinnatifida var. *robusta* Huber-Mor. === **Campanula pinnatifida** Huber-Mor.
Campanula piperi f. *sovereigniana* E. English === **Campanula piperi** Howell
Campanula planiflora Engelm. === **Campanula parryi** A. Gray
Campanula planiflora Lam. === **Campanula americana** L.
Campanula planiflora Willd. === **Campanula versicolor** Andrews
Campanula plasonii Formánek === ?
Campanula plicata Pers. === **Prismatocarpus crispus** L'Hér.
Campanula plicatula Dumort. === **Campanula trachelium** L.
Campanula podanthoides Boiss. & Hausskn. === **Asyneuma linifolium** subsp. **podanthoides** (Boiss. & Hausskn.) Damboldt
Campanula polessica O. D. Wissjul. === **Campanula glomerata** subsp. **farinosa** (Rochel ex Besser) Kirschl.
Campanula polyantha Schult. === ?
Campanula polymorpha (Witasek) Prain === **Campanula tatrae** Borbás subsp. **tatrae**
Campanula polymorpha f. *angustifolia* Hruby === **Campanula tatrae** Borbás subsp. **tatrae**
Campanula polymorpha f. *brachyphylla* Hruby === **Campanula tatrae** Borbás subsp. **tatrae**
Campanula polymorpha f. *exigua* Hruby === **Campanula tatrae** Borbás subsp. **tatrae**
Campanula polymorpha f. *gracilis* Hruby === **Campanula tatrae** Borbás subsp. **tatrae**
Campanula polymorpha f. *kladnianioides* Nyár. ex Hruby === **Campanula tatrae** Borbás subsp. **tatrae**
Campanula polymorpha f. *latifolia* Hruby === **Campanula tatrae** Borbás subsp. **tatrae**
Campanula polymorpha f. *lepida* Hruby === **Campanula tatrae** Borbás subsp. **tatrae**

Campanula polymorpha f. *mentiens* (Witasek) Jáv. === **Campanula tatrae** subsp. **mentiens** (Witasek) Kovanda

Campanula polymorpha f. *pseudolanceolata* (Pant.) Hruby === **Campanula serrata** (Kit.) Hendrych subsp. **serrata**

Campanula polymorpha f. *reflectans* === **Campanula tatrae** Borbás subsp. **tatrae**

Campanula polymorpha f. *sciaphila* Hruby === **Campanula tatrae** Borbás subsp. **tatrae**

Campanula polymorpha f. *umbrosa* === **Campanula tatrae** Borbás subsp. **tatrae**

Campanula polymorpha subf. *reflectans* Hruby === **Campanula tatrae** Borbás subsp. **tatrae**

Campanula polymorpha subf. *umbrosa* Hruby === **Campanula tatrae** Borbás subsp. **tatrae**

Campanula polymorpha var. *intercedens* Hruby === **Campanula tatrae** Borbás subsp. **tatrae**

Campanula polymorpha var. *mentiens* (Witasek) Maroriu === **Campanula tatrae** subsp. **mentiens** (Witasek) Kovanda

Campanula polymorpha var. *praticola* Hruby === **Campanula tatrae** Borbás subsp. **tatrae**

Campanula polymorpha var. *stenophylla* (Schur) Hruby === **Campanula tatrae** Borbás subsp. **tatrae**

Campanula porosa L. f. === **Samolus valerandi** L. (Primulaceae)

Campanula portenschlagiana f. *grandiflora* D. Šoljan === **Campanula portenschlagiana** Schult.

Campanula portenschlagiana var. *hirsuta* D. Šoljan === **Campanula portenschlagiana** Schult.

Campanula portenschlagiana var. *pubescens* A. DC. === **Campanula portenschlagiana** Schult.

Campanula portenschlagiana var. *pumila* D. Šoljan === **Campanula portenschlagiana** Schult.

Campanula portensis L. === **Campanula erinus** L.

Campanula pourretii Jeanb. & Timb.-Lagr. === ?

Campanula praealta Galushko === **Campanula sibirica** subsp. **elatior** (Fomin) Fed.

Campanula praecox Miégev. === ?

Campanula praesignis var. *breynina* Beck === **Campanula praesignis** Beck

Campanula pratensis Bach. Pyl. ex A. DC. === **Campanula giesekiana** Vest. ex Schult.

Campanula pratensis var. *angustifolia* A. DC. === **Campanula giesekiana** Vest. ex Schult.

Campanula precatoria var. *hirsuta* Timb.-Lagr. === **Campanula precatoria** Timb.-Lagr.

Campanula precatoria var. *major* Timb.-Lagr. === **Campanula precatoria** Timb.-Lagr.

Campanula precatoria var. *tenuifolia* Timb.-Lagr. === **Campanula precatoria** Timb.-Lagr.

Campanula preissii de Vriese === ?

Campanula primulifolia Brot. === **Campanula alata** Desf.

Campanula prismatocarpus Aiton === **Prismatocarpus nitidus** L'Hér.

Campanula procumbens L. f. === **Wahlenbergia procumbens** (L. f.) A. DC.

Campanula propinqua [unranked] *grandiflora* Milne-Redh. === **Campanula reuteriana** Boiss. & Balansa

Campanula propinqua var. *parviflora* Turrill === **Campanula propinqua** Fisch. & C. A. Mey.

Campanula prostrata Dulac === **Campanula scheuchzeri** Vill.

Campanula pseudolanceolata Pant. === **Campanula serrata** (Kit.) Hendrych subsp. **serrata**

Campanula pseudolanceolata f. *albiflora* Săvul. === **Campanula serrata** (Kit.) Hendrych subsp. **serrata**

Campanula pseudolanceolata f. *elatior* Săvul. === **Campanula serrata** (Kit.) Hendrych subsp. **serrata**

Campanula pseudolanceolata f. *minima* Săvul. === **Campanula serrata** (Kit.) Hendrych subsp. **serrata**

Campanula pseudolanceolata f. *transsilvanica* Săvul. === **Campanula serrata** (Kit.) Hendrych subsp. **serrata**

Campanula pseudolanceolata f. *umbraticola* Săvul. === **Campanula serrata** (Kit.) Hendrych subsp. **serrata**

Campanula pseudolanceolata subsp. *semiamplexicaulis* Săvul. === **Campanula serrata** (Kit.) Hendrych subsp. **serrata**

Campanula pseudolanceolata var. *arcuata* (Schur) Porcius === **Campanula serrata** (Kit.) Hendrych subsp. **serrata**

Campanula pseudolanceolata var. *hornungiana* (Schur) Porcius === **Campanula serrata** (Kit.) Hendrych subsp. **serrata**

Campanula pseudolanceolata var. *porcii* Săvul. === **Campanula serrata** (Kit.) Hendrych subsp. **serrata**

Campanula pseudopulla Schur === **Campanula pulla** L.

Campanula × *pseudoscheuchzeri* Beyer === **Campanula** × **murrii** Dalla Torre & Sarnth.

Campanula pseudovaldensis Schur === **Campanula rotundifolia** L.

Campanula ptarmicifolia var. *capitellata* Damboldt === **Campanula ptarmicifolia** Lam.

Campanula pubescens F. W. Schmidt === **Campanula cochleariifolia** Lam.

Campanula pubescens Hegetschw. === **Campanula cochleariifolia** Lam.

Campanula pubiflora Rupr. === **Campanula bellidifolia** subsp. **argunensis** (Rupr.) Viktorov

Campanula pulchella (Fisch. & C. A. Mey.) Boiss. === **Asyneuma pulchellum** (Fisch. & C. A. Mey.) Bornm.

Campanula pulchella Jord. === **Campanula cochleariifolia** Lam.

Campanula pulchella Salisb. === **Legousia speculum-veneris** (L.) Durande ex Vill.

Campanula pulchra O. D. Wissjul. === **Campanula glomerata** subsp. **elliptica** (Kit. ex Schultz) O. Schwarz

Campanula pulla var. *pseudopulla* (Schur) Nyman === **Campanula pulla** L.

Campanula pulla var. *ramosa* A. DC. === **Campanula pulla** L.

Campanula pulliformis Rouy === ?

Campanula pumila F. W. Schmidt === **Campanula persicifolia** L. subsp. **persicifolia**

Campanula pumila Sims === **Campanula cochleariifolia** Lam.

Campanula pumilio Port. ex Schult. === **Edraianthus pumilio** (Port. ex Schult.) A. DC.

Campanula pumilio var. *major* Vis. === **Edraianthus dinaricus** (A. Kern.) Wettst.

Campanula punctata f. *albiflora* T. Shimizu === **Campanula punctata** Lam.

Campanula punctata f. *impunctata* N. Yonez. === **Campanula punctata** Lam.

Campanula punctata f. *partita* Makino === **Campanula punctata** Lam.

Campanula punctata f. *rubriflora* (Makino) T. Shimizu === **Campanula punctata** Lam.

Campanula punctata subsp. *hondoensis* (Kitam.) Kitam. === **Campanula punctata** Lam.

Campanula punctata subsp. *microdonta* (Koidz.) Kitam. === **Campanula microdonta** Koidz.

Campanula punctata var. *hondoensis* (Kitam.) Ohwi ex T. Shimizu === **Campanula punctata** Lam.

Campanula punctata var. *rubriflora* Makino === **Campanula punctata** Lam.

Campanula punduana D. Dietr. === **Cyclocodon parviflorus** (Wall. ex A. DC.) Hook. f. & Thomson

Campanula purpurea (Wall.) Spreng. === **Codonopsis purpurea** Wall.

Campanula pusilla Haenke === **Campanula cochleariifolia** Lam.

Campanula pusilla Hegetschw. === ?

Campanula pusilla f. *densa* Gsaller === **Campanula cochleariifolia** Lam.

Campanula pusilla f. *hauryi* (Schott) Hayek === **Campanula cochleariifolia** Lam.

Campanula pusilla f. *hirciana* Vuk. === **Campanula cochleariifolia** Lam.

Campanula pusilla f. *imbricata* Vuk. === **Campanula cochleariifolia** Lam.

Campanula pusilla f. *lobata* (Schloss. & Vuk.) Vuk. === **Campanula rotundifolia** L.

Campanula pusilla f. *umbrosa* J. Hofm. === **Campanula cochleariifolia** Lam.

Campanula pusilla f. *vagans* J. Hofm. === **Campanula cochleariifolia** Lam.

Campanula pusilla subsp. *ficarioides* (Timb.-Lagr.) Rouy === **Campanula ficarioides** Timb.-Lagr.

Campanula pusilla subsp. *jaubertiana* (Timb.-Lagr.) Bonnier === **Campanula jaubertiana** Timb.-Lagr.

Campanula pusilla subsp. *notata* (Schott) Nyman === **Campanula cochleariifolia** Lam.

Campanula pusilla subsp. *tenella* (Jord.) Rouy === **Campanula cochleariifolia** Lam.

Campanula pusilla subsp. *veronicifolia* Sennen === **Campanula cochleariifolia** Lam.

Campanula pusilla var. *bellardii* (All.) DC. === **Campanula cochleariifolia** Lam.

Campanula pusilla var. *brachyantha* Murr === **Campanula cochleariifolia** Lam.

Campanula pusilla var. *calycina* Willk. === **Campanula cochleariifolia** Lam.

Campanula pusilla var. *delpontei* Chabert === **Campanula cochleariifolia** Lam.

Campanula pusilla var. *descensa* Beck === **Campanula cochleariifolia** Lam.

Campanula pusilla var. *foliosa* Krašan === **Campanula cochleariifolia** Lam.

Campanula pusilla var. *foudrasii* (Jord.) Nyman === **Campanula cochleariifolia** Lam.

Campanula pusilla var. *gautieri* (Jeanb. & Timb.-Lagr.) Rouy === **Campanula ficarioides** Timb.-Lagr.

Campanula pusilla var. *gracilis* (Jord.) Nyman === **Campanula cochleariifolia** Lam.

Campanula pusilla var. *hochstetteri* (Schott) Nyman === **Campanula cochleariifolia** Lam.

Campanula pusilla var. *hoppeana* Rupr. ex Rchb. === **Campanula cochleariifolia** Lam.

Campanula pusilla var. *jaubertiana* (Timb.-Lagr.) Rouy === **Campanula jaubertiana** Timb.-Lagr.

Campanula pusilla var. *leucanthemifolia* (Pourr.) Nyman === **Campanula cochleariifolia** Lam.

Campanula pusilla var. *mathonetii* (Jord.) Nyman === **Campanula cochleariifolia** Lam.

Campanula pusilla var. *modesta* (Schott) Nyman === **Campanula cochleariifolia** Lam.

Campanula pusilla var. *notata* (Schott) Wohlf. === **Campanula cochleariifolia** Lam.

Campanula pusilla var. *paniculata* Nageli ex Rchb. === **Campanula cochleariifolia** Lam.

Campanula pusilla var. *parvula* (Jord.) Nyman === **Campanula cochleariifolia** Lam.

Campanula pusilla var. *pinguis* Gren. & Godr. === **Campanula cochleariifolia** Lam.

Campanula pusilla var. *pirinica* Velen. === **Campanula cochleariifolia** Lam.

Campanula pusilla var. *pubescens* (F. W. Schmidt) Steud. === **Campanula cochleariifolia** Lam.

Campanula pusilla var. *pubescens* DC. === **Campanula cochleariifolia** Lam.

Campanula pusilla var. *pulchella* Gren. & Godr. === **Campanula cochleariifolia** Lam.

Campanula pusilla var. *reflexa* (Schur) Nyman === **Campanula cochleariifolia** Lam.

Campanula pusilla var. *reniformis* (Pers.) Steud. === **Campanula baumgartenii** Becker subsp. **baumgartenii**

Campanula pusilla var. *stenocodon* (Boiss. & Reut.) Rouy === **Campanula stenocodon** Boiss. & Reut.

Campanula pusilla var. *subacaulis* Murr === **Campanula cochleariifolia** Lam.

Campanula pusilla var. *subramulosa* (Jord. ex Gren. & Godr.) Nyman === **Campanula cochleariifolia** Lam.

Campanula pusilla var. *tenella* (Jord.) Nyman === **Campanula cochleariifolia** Lam.

Campanula pusilla var. *tubulosa* Chabert === **Campanula cochleariifolia** Lam.

Campanula pusilla var. *tyrolensis* (Schott) Nyman === **Campanula cochleariifolia** Lam.

Campanula pusilla var. *umbrosa* (J. Hofm.) Beck === **Campanula cochleariifolia** Lam.

Campanula pusilla var. *venusta* (Schur) Nyman === **Campanula cochleariifolia** Lam.

Campanula pygmaea DC. === **Borago laxiflora** DC. (**Boraginaceae**)

Campanula pygmaea (H. Buek) D. Dietr. === **Microcodon glomeratus** A. DC.

Campanula pyramidalis f. *alba* Voss === **Campanula pyramidalis** L.

Campanula pyramidalis var. *calycina* A. DC. === **Campanula pyramidalis** L.

Campanula pyramidalis var. *compacta* Anonymous === **Campanula pyramidalis** L.

Campanula pyramidata Gilib. === **Campanula bononiensis** L.

Campanula pyramidiflora Rchb. === **Campanula rapunculoides** L.

Campanula quadrifida R. Br. === **Wahlenbergia quadrifida** (R. Br.) A. DC.

Campanula quartiniana A. Rich. === **Campanula edulis** Forssk.

Campanula rabelaisiana Schult. === **Adenophora gmelinii** (Biehler) Fisch. subsp. **gmelinii**

Campanula × *racemosa* Beyer === **Campanula** × **truedingeri** Murr

Campanula racemosa (Krašan) Witasek === **Campanula rotundifolia** L.

Campanula racemosa Vuk. === **?**

Campanula racemosa var. *arctiflora* Vuk. === **Campanula bononiensis** L.

Campanula racemosa var. *grandiflora* Vuk. === **Campanula persicifolia** L.

Campanula racemosa var. *laxiflora* Vuk. === **Campanula rapunculoides** L.

Campanula racemosa var. *paniculiformis* Vuk. === **Campanula rapunculus** L.

Campanula radula var. *minor* Boiss. === **Campanula coriacea** P. H. Davis

Campanula ramosissima var. *glabrescens* Hausskn. === **Campanula ramosissima** Sm.

Campanula ramosissima var. *velutina* Griseb. === **Campanula phrygia** Jaub. & Spach

Campanula ramulosa Wall. === **Campanula pallida** Wall.

Campanula rapunculiformis St.-Lag. === **Campanula rapunculoides** L.

Campanula rapunculoides f. *cordifolia* (K. Koch) Albov === **Campanula rapunculoides** L.

Campanula rapunculoides f. *crenata* Hayek & Hegi === **Campanula rapunculoides** L.

Campanula rapunculoides f. *secunda* (F. W. Schmidt) Hayek & Hegi === **Campanula rapunculoides** L.

Campanula rapunculoides f. *trachelioides* (M. Bieb.) Hayek & Hegi === **Campanula rapunculoides** L.

Campanula rapunculoides f. *ucranica* (Besser) Hayek & Hegi === **Campanula rapunculoides** L.

Campanula rapunculoides subvar. *neglecta* Nyman === **Campanula rapunculoides** L.

Campanula rapunculoides var. *cordata* K. Koch === **Campanula rapunculoides** L.

Campanula rapunculoides var. *cordifolia* (K. Koch) Fomin === **Campanula rapunculoides** L.

Campanula rapunculoides var. *glabrata* Trautv. === **Campanula rapunculoides** L.

Campanula rapunculoides var. *grandiflora* K. Koch === **Campanula rapunculoides** L.

Campanula rapunculoides var. *macrophylla* A. DC. === **Campanula rapunculoides** L.

Campanula rapunculoides var. *nana* A. DC. === **Campanula rapunculoides** L.

Campanula rapunculoides var. *nemorosa* (A. DC.) Nyman === **Campanula rapunculoides** L.

Campanula rapunculoides var. *oenipontana* A. DC. === **Campanula rapunculoides** L.

Campanula rapunculoides var. *racemosa* Peterm. === **Campanula rapunculoides** L.

Campanula rapunculoides var. *ramosissima* Schur === **Campanula rapunculoides** L.

Campanula rapunculoides var. *reflexa* Peterm. === **Campanula rapunculoides** L.

Campanula rapunculoides var. *simplex* K. Koch === **Campanula rapunculoides** L.

Campanula rapunculoides var. *speciosa* Knaf === **Campanula rapunculoides** L.

Campanula rapunculoides var. *subsimplex* Schur === **Campanula rapunculoides** L.

Campanula rapunculoides var. *trachelioides* (M. Bieb.) A. DC. === **Campanula rapunculoides** L.

Campanula rapunculus O. F. Müll. === **Campanula rotundifolia** L.

Campanula rapunculus f. *hirsutissima* Faure === **Campanula rapunculus** L. subsp. **rapunculus**

Campanula rapunculus f. *lambertiana* (A. DC.) Voss === **Campanula rapunculus** subsp. **lambertiana** (A. DC.) Rech. f.

Campanula rapunculus f. *spiciformis* (Boiss.) Voss === **Campanula rapunculus** subsp. **lambertiana** (A. DC.) Rech. f.

Campanula rapunculus subsp. *verruculosa* (Hoffmanns. & Link) Nyman === **Campanula rapunculus** L. subsp. **rapunculus**

Campanula rapunculus subvar. *elatior* (Hoffmanns. & Link) Nyman === **Campanula rapunculus** L. subsp. **rapunculus**

Campanula rapunculus subvar. *hirsutissima* (Faure) Maire === **Campanula rapunculus** L. subsp. **rapunculus**

Campanula rapunculus subvar. *hirtula* Maire === **Campanula rapunculus** L. subsp. **rapunculus**

Campanula rapunculus subvar. *santae* Quézel === **Campanula rapunculus** L. subsp. **rapunculus**

Campanula rapunculus subvar. *verruculosa* (Hoffmanns. & Link) Emb. & Maire === **Campanula rapunculus** L. subsp. **rapunculus**

Campanula rapunculus var. *bracteosa* Willk. === **Campanula rapunculus** L. subsp. **rapunculus**

Campanula rapunculus var. *calycina* (Boeber ex Schult.) A. DC. === **Campanula rapunculus** L. subsp. **rapunculus**

Campanula rapunculus var. *cymosospicata* Willk. === **Campanula rapunculus** L. subsp. **rapunculus**

Campanula rapunculus var. *hirsuta* Schur === **Campanula rapunculus** L. subsp. **rapunculus**

Campanula rapunculus var. *hirta* Murr === **Campanula rapunculus** L. subsp. **rapunculus**

Campanula rapunculus var. *lambertiana* (A. DC.) Boiss. === **Campanula rapunculus** subsp. **lambertiana** (A. DC.) Rech. f.

Campanula rapunculus var. *micrantha* Beyer === **Campanula rapunculus** L. subsp. **rapunculus**

Campanula rapunculus var. *racemosopaniculata* Willk. === **Campanula rapunculus** L. subsp. **rapunculus**

Campanula rapunculus var. *reclinata* Griseb. === **Campanula rapunculus** L. subsp. **rapunculus**

Campanula rapunculus var. *spiciformis* Boiss. === **Campanula rapunculus** subsp. **lambertiana** (A. DC.) Rech. f.

Campanula rapunculus var. *strigulosa* Batt. === **Campanula rapunculus** L. subsp. **rapunculus**

Campanula rapunculus var. *verruculosa* (Hoffmanns. & Link) Steud. === **Campanula rapunculus** L. subsp. **rapunculus**

Campanula rapunculus "var. vel ssp". *losae* Sennen === **Campanula rapunculus** L. subsp. **rapunculus**

Campanula re Colla === **Campanula bertolae** Colla
Campanula reboudiana Pomel === **Campanula filicaulis** Durieu
Campanula recta Dulac === **Campanula serrata** subsp. **recta** (Dulac) Podlech
Campanula redux Schott === **Campanula serrata** (Kit.) Hendrych subsp. **serrata**
Campanula reflexa Schur === **Campanula cochleariifolia** Lam.
Campanula regina Albov === **Campanula mirabilis** Albov
Campanula reiseri var. *leonis* Halácsy === **Campanula reiseri** Halácsy
Campanula reiseri var. *pelagia* Phitos === **Campanula reiseri** Halácsy
Campanula remotiflora Siebold & Zucc. === **Adenophora remotiflora** (Siebold & Zucc.) Miq.
Campanula renatii Sennen === **Campanula cochleariifolia** Lam.
Campanula reniformis Schur === ?
Campanula rentoniae Senior === **Campanula parryi** A. Gray
Campanula repanda (Sm.) Boiss. === **Asyneuma limonifolium** (L.) Janch. subsp. **limonifolium**
Campanula repens (Karpiss.) Czerep. === **Campanula odontosepala** Boiss.
Campanula repens Lour. === **Dentella repens** (L.) J. R. Forst. & G. Forst. (**Rubiaceae**)
Campanula revoluta Formánek === **Campanula sparsa** Friv. subsp. **sparsa**
Campanula rhomboidalis Gorter === **Campanula rapunculoides** L.
Campanula rhomboidalis f. *cordifolia* Witasek ex Vacc. === **Campanula rhomboidalis** L.
Campanula rhomboidalis f. *macrophylla* Vacc. === **Campanula rhomboidalis** L.
Campanula rhomboidalis f. *microphylla* Vacc. === **Campanula rhomboidalis** L.
Campanula rhomboidalis f. *paniculata* Vacc. === **Campanula rhomboidalis** L.
Campanula rhomboidalis f. *reflexa* (A. DC.) Voss === **Campanula rhomboidalis** L.
Campanula rhomboidalis f. *rubra* (Anonymous) Voss === **Campanula rhomboidalis** L.
Campanula rhomboidalis race *songeonii* (Chabert) Rouy === **Campanula rhomboidalis** L.
Campanula rhomboidalis subsp. *pseudolanceolata* (Pant.) Nyman === **Campanula serrata** (Kit.) Hendrych subsp. **serrata**
Campanula rhomboidalis subvar. *hispida* (St.-Lag.) Rouy === **Campanula rhomboidalis** L.
Campanula rhomboidalis var. *alpini* (L.) L. === **Adenophora liliifolia** (L.) A. DC.
Campanula rhomboidalis var. *angustifolia* Neilr. === **Campanula serrata** (Kit.) Hendrych subsp. **serrata**
Campanula rhomboidalis var. *calycina* Chabert === **Campanula rhomboidalis** L.
Campanula rhomboidalis var. *fritschii* (Witasek) Fiori === **Campanula fritschii** Witasek
Campanula rhomboidalis var. *glabrescens* Vacc. === **Campanula rhomboidalis** L.
Campanula rhomboidalis var. *goudetiana* Beauverd === **Campanula rhomboidalis** L.
Campanula rhomboidalis var. *hispida* St.-Lag. === **Campanula rhomboidalis** L.
Campanula rhomboidalis var. *lanceolata* (Lapeyr.) Loisel. === **Campanula rotundifolia** L.
Campanula rhomboidalis var. *pilosa* Vacc. === **Campanula rhomboidalis** L.
Campanula rhomboidalis var. *polypetala* A. DC. === **Campanula rhomboidalis** L.
Campanula rhomboidalis var. *reflexa* A. DC. === **Campanula rhomboidalis** L.
Campanula rhomboidalis var. *songeonii* (Chabert) Chabert === **Campanula rhomboidalis** L.
Campanula rhomboidalis var. *tuberosa* Chabert === **Campanula rhomboidalis** L.
Campanula rhomboidalis var. *uniflora* Gaudin === **Campanula rhomboidalis** L.
Campanula rhomboidalis var. *villosa* Wohlf. === **Campanula rhomboidalis** L.
Campanula rhomboidea Falk === **Campanula rapunculoides** L.
Campanula rhomboidea Borbás === **Adenophora pereskiifolia** (Fisch. ex Schult.) G. Don
Campanula richteri (Borbás) Borbás === **Adenophora tricuspidata** (Fisch. ex Schult.) A. DC.
Campanula rigida Stokes === **Campanula rapunculoides** L.
Campanula rigida (Willd.) Boiss. === **Asyneuma rigidum** (Willd.) Grossh.
Campanula rigescens Pall. ex Schult. === ?
Campanula rigidipila Steud. & Hochst. ex A. Rich. === **Campanula edulis** Forssk.
Campanula rigidipila var. *esculenta* (A. Rich.) Di Capua === **Campanula edulis** Forssk.
Campanula rigidipila var. *quartiniana* (A. Rich.) Engl. === **Campanula edulis** Forssk.
Campanula rigidipila var. *sarmentosa* (Hochst. ex A. Rich.) Engl. === **Campanula edulis** Forssk.
Campanula riparia (A. DC.) Leprieur & Perottet ex D. Dietr. === **Wahlenbergia lobelioides** subsp. **riparia** (A. DC.) Thulin
Campanula rochelii Schur === **Campanula stevenii** subsp. **altaica** (Ledeb.) Fed.

Campanula roezlii Regel === **Campanula prenanthoides** Durand

Campanula rohdii Loisel. === **Campanula ficarioides** Timb.-Lagr.

Campanula rosanii Ten. === **Campanula versicolor** Andrews

Campanula rosea Noronha === ?

Campanula rosulata Vuk. === **Campanula cenisia** L.

Campanula rotundifolia f. *albiflora* Abrom. === **Campanula giesekiana** Vest ex Schult.

Campanula rotundifolia f. *albiflora* E. L. Rand & Redfield === **Campanula intercedens** Witasek

Campanula rotundifolia f. *albiflora* (G. Don) House === **Campanula rotundifolia** L.

Campanula rotundifolia f. *albiflora* Sugaw. ex H. Hara === **Campanula rotundifolia** L.

Campanula rotundifolia f. *alpicola* (Hayek & Hegi) Hruby === **Campanula rotundifolia** L.

Campanula rotundifolia f. *borbasiana* (Witasek) Hayek === **Campanula velebitica** Borbás?

Campanula rotundifolia f. *borhidiana* Só === **Campanula rotundifolia** L.

Campanula rotundifolia f. *calvescens* Witasek ex Vacc. === **Campanula rotundifolia** L.

Campanula rotundifolia f. *cleistocodona* Lakela === **Campanula intercedens** Witasek

Campanula rotundifolia f. *conferta* Só === **Campanula rotundifolia** L.

Campanula rotundifolia f. *dilecta* (Schott) Hruby === **Campanula rotundifolia** Vill.

Campanula rotundifolia f. *divaricata* (Witasek) Hayek === **Campanula velebitica** Borbás

Campanula rotundifolia f. *dubia* (A. DC.) Farw. === **Campanula giesekiana** Vest ex Schult.

Campanula rotundifolia f. *elata* Hruby === **Campanula rotundifolia** L.

Campanula rotundifolia f. *frondosa* Hruby === **Campanula rotundifolia** L.

Campanula rotundifolia f. *glabrescens* Hruby === **Campanula rotundifolia** L.

Campanula rotundifolia f. *graminifolia* Hruby === **Campanula rotundifolia** L.

Campanula rotundifolia f. *grandiflora* J. A. Knapp === **Campanula rotundifolia** L.

Campanula rotundifolia f. *hirta* (Mert. & W. D. J. Koch) Hruby === **Campanula rotundifolia** L.

Campanula rotundifolia f. *hirta* Hruby === **Campanula rotundifolia** L.

Campanula rotundifolia f. *hrubyana* Só === **Campanula rotundifolia** L.

Campanula rotundifolia f. *incerta* (Witasek) Hayek === **Campanula velebitica** Borbás

Campanula rotundifolia f. *laciniata* J. Rousseau & Raymond === **Campanula giesekiana** Vest ex Schult.

Campanula rotundifolia f. *laevis* Guin. === **Campanula rotundifolia** L.

Campanula rotundifolia f. *lapponica* Witasek === **Campanula giesekiana** Vest ex Schult.

Campanula rotundifolia f. *latifrons* Hruby === **Campanula rotundifolia** L.

Campanula rotundifolia f. *latifrons* Hruby === **Campanula rotundifolia** L.

Campanula rotundifolia f. *linifolia* Farw. === **Campanula intercedens** Witasek

Campanula rotundifolia f. *luxurians* Hruby === **Campanula rotundifolia** L.

Campanula rotundifolia f. *micranthoides* Só === **Campanula rotundifolia** L.

Campanula rotundifolia f. *nejceffii* Hayek === **Campanula rotundifolia** L.

Campanula rotundifolia f. *normalis* Hruby === **Campanula rotundifolia** L.

Campanula rotundifolia f. *normalis* Hruby === **Campanula rotundifolia** L.

Campanula rotundifolia f. *ovalifolia* Hruby === **Campanula rotundifolia** L.

Campanula rotundifolia f. *ovata* (Peterm.) Hruby === **Campanula rotundifolia** L.

Campanula rotundifolia f. *parviflora* (Witasek) Hayek === **Campanula velebitica** Borbás

Campanula rotundifolia f. *pygmaea* Hartz === **Campanula giesekiana** Vest ex Schult.

Campanula rotundifolia f. *scabriuscula* (Mert. & W. D. J. Koch) Hruby === **Campanula rotundifolia** L.

Campanula rotundifolia f. *serpentini* Hruby === **Campanula rotundifolia** L.

Campanula rotundifolia f. *silvicola* Hruby === **Campanula rotundifolia** L.

Campanula rotundifolia f. *soldanelliflora* Voss === **Campanula rotundifolia** L.

Campanula rotundifolia f. *subcongesta* Hruby === **Campanula rotundifolia** L.

Campanula rotundifolia f. *subcongesta* Hruby === **Campanula rotundifolia** L.

Campanula rotundifolia f. *subhirta* Só === **Campanula rotundifolia** L.

Campanula rotundifolia f. *subracemosa* Hruby === **Campanula rotundifolia** L.

Campanula rotundifolia f. *subverruculosa* Guin. === **Campanula rotundifolia** L.

Campanula rotundifolia f. *tenerrima* Hruby === **Campanula rotundifolia** L.

Campanula rotundifolia f. *umbrosa* Hruby === **Campanula rotundifolia** L.

Campanula rotundifolia race *macrorhiza* (J. Gay ex A. DC.) Rouy === **Campanula macrorhiza** J. Gay ex A. DC.

Campanula rotundifolia race *ruscinonensis* (Timb.-Lagr.) Rouy === **Campanula ruscinonensis** Timb.-Lagr.

Campanula rotundifolia subf. *bracteata* Hruby === **Campanula rotundifolia** L.

Campanula rotundifolia subf. *grandiflora* Hruby === **Campanula rotundifolia** L.

Campanula rotundifolia subf. *ovalifolia* Hruby === **Campanula rotundifolia** L.

Campanula rotundifolia subf. *parviflora* Hruby === **Campanula rotundifolia** L.

Campanula rotundifolia subf. *perfoliosa* Hruby === **Campanula rotundifolia** L.

Campanula rotundifolia subf. *reflexa* Hruby === **Campanula rotundifolia** L.

Campanula rotundifolia subsp. *aitanica* Pau ex O. Bolòs & Vigo === **Campanula macrorhiza** J. Gay ex A. DC.

Campanula rotundifolia subsp. *baumgartenii* (Becker) Rouy === **Campanula baumgartenii** Becker

Campanula rotundifolia subsp. *bertolae* (Colla) Vacc. === **Campanula bertolae** Colla

Campanula rotundifolia subsp. *catalanica* (Podlech) O. Bolòs & Vigo === **Campanula macrorhiza** J. Gay ex A. DC.

Campanula rotundifolia subsp. *cespitosa* (Scop.) Lapeyr. === **Campanula cespitosa** Scop.

Campanula rotundifolia subsp. *confertifolia* (Reut.) Witasek ex Vacc. === **Campanula rotundifolia** L.

Campanula rotundifolia subsp. *dilecta* (Schott) Nyman === **Campanula scheuchzeri** Vill.

Campanula rotundifolia subsp. *euxina* (Velen.) Hayek === **Campanula euxina** (Velen.) Ančev

Campanula rotundifolia subsp. *groenlandica* (Berlin) A. Löve & D. Löve === **Campanula giesekiana** Vest ex Schult.

Campanula rotundifolia subsp. *hellenica* Hayek === **Campanula albanica** Witasek subsp. **albanica**

Campanula rotundifolia subsp. *heterodoxa* (Vest ex Schult.) Tacik === **Campanula rotundifolia** L.

Campanula rotundifolia subsp. *hispanica* (Willk.) Rivas Goday & Borja ex O. Bolòs & Vigo === **Campanula macrorhiza** J. Gay ex A. DC.

Campanula rotundifolia subsp. *intercedens* (Witasek) Á. Löve & D. Löve === **Campanula intercedens** Witasek

Campanula rotundifolia subsp. *kladniana* (Schur) Tacik === **Campanula carnica** Schiede ex Mert. & Koch subsp. **carnica**

Campanula rotundifolia subsp. *legionensis* (Pau) Laínz === **Campanula rotundifolia** L.

Campanula rotundifolia subsp. *linifolia* Arcang. === **Campanula serrata** subsp. **recta** (Dulac) Podlech

Campanula rotundifolia subsp. *linifolia* Lapeyr. === ?

Campanula rotundifolia subsp. *linifolia* Rouy === **Campanula carnica** Schiede ex Mert. & Koch subsp. **carnica**

Campanula rotundifolia subsp. *litardierei* Guin. === **Campanula macrorhiza** J. Gay ex A. DC.

Campanula rotundifolia subsp. *macrorhiza* (J. Gay ex A. DC.) Guin. === **Campanula macrorhiza** J. Gay ex A. DC.

Campanula rotundifolia subsp. *marchesettii* (Witasek) D. E. Mey. === **Campanula marchesettii** Witasek

Campanula rotundifolia subsp. *montana* P. D. Sell === **Campanula rotundifolia** L.

Campanula rotundifolia subsp. *paenina* (Reut. ex Tissière) Witasek ex Vacc. === **Campanula rotundifolia** L.

Campanula rotundifolia subsp. *pedemontana* Witasek === **Campanula rotundifolia** L.

Campanula rotundifolia subsp. *pennina* (Reut.) Witasek === **Campanula rotundifolia** L.

Campanula rotundifolia subsp. *polymorpha* (Witasek) Tacik === **Campanula tatrae** Borbás subsp. **tatrae**

Campanula rotundifolia subsp. *praesignis* (Beck) Hayek === **Campanula praesignis** Beck

Campanula rotundifolia subsp. *pseudarctica* Hruby === **Campanula scheuchzeri** Vill.

Campanula rotundifolia subsp. *pusilla* (Haenke) Lapeyr. === **Campanula cochleariifolia** Lam.

Campanula rotundifolia subsp. *racemosa* (Krašan) Hayek & Hegi === **Campanula rotundifolia** L.

Campanula rotundifolia subsp. *rhomboidalis* (L.) Bonnier === **Campanula rhomboidalis** L.

Campanula rotundifolia subsp. *romanica* (Savul.) Hayek === **Campanula romanica** Savul.

Campanula rotundifolia subsp. *sancta* Hayek === **Campanula albanica** subsp. **sancta** (Hayek) Podlech

Campanula rotundifolia subsp. *scheuchzeri* (Vill.) Lapeyr. === **Campanula scheuchzeri** Vill.

Campanula rotundifolia subsp. *solstitialis* (A. Kern.) Witasek ex Hayek & Hegi === **Campanula rotundifolia** L.

Campanula rotundifolia subsp. *stenocodon* (Boiss. & Reut.) Nyman === **Campanula stenocodon** Boiss. & Reut.

Campanula rotundifolia subsp. *sudetica* (Hruby) Soó === **Campanula tatrae** subsp. **sudetica** (Hruby) Kovanda

Campanula rotundifolia subsp. *velebitica* (Borbás) Hayek === **Campanula velebitica** Borbás

Campanula rotundifolia subsp. *xylorrhiza* O. Schwarz === **Campanula moravica** subsp. **xylorrhiza** (O. Schwarz) Kovanda

Campanula rotundifolia subvar. *gracilis* Guin. === **Campanula macrorhiza** J. Gay ex A. DC.

Campanula rotundifolia subvar. *pinifolia* (Uechtr. ex Pančić) Hruby === **Campanula rotundifolia** L.

Campanula rotundifolia subvar. *puberula* Guin. === **Campanula macrorhiza** J. Gay ex A. DC.

Campanula rotundifolia subvar. *re* (Colla) Hruby === **Campanula bertolae** Colla

Campanula rotundifolia subvar. *tenuissima* Hruby === **Campanula rotundifolia** L.

Campanula rotundifolia var. *aitanica* Pau ex O. Bolòs & Vigo === **Campanula macrorhiza** J. Gay ex A. DC.

Campanula rotundifolia var. *alaskana* A. Gray === **Campanula alaskana** (A. Gray) Wight ex J. P. Anderson

Campanula rotundifolia var. *albiflora* G. Don === **Campanula rotundifolia** L.

Campanula rotundifolia var. *alcoiana* O. Bolòs & Vigo === **Campanula macrorhiza** J. Gay ex A. DC.

Campanula rotundifolia var. *alpicola* Hayek & Hegi === **Campanula rotundifolia** L.

Campanula rotundifolia var. *alpina* Schur === **Campanula serrata** (Kit.) Hendrych subsp. **serrata**

Campanula rotundifolia var. *alpina* Tuck. === **Campanula intercedens** Witasek

Campanula rotundifolia var. *altitatrica* Tacik === **Campanula rotundifolia** L.

Campanula rotundifolia var. *angustiflora* (Tanfani) Guin. === **Campanula tanfanii** Podlech

Campanula rotundifolia var. *angustifolia* (Lam.) Vacc. === **Campanula rotundifolia** L.

Campanula rotundifolia var. *angustissima* Schur === **Campanula rotundifolia** L.

Campanula rotundifolia var. *arctica* Lange === **Campanula giesekiana** Vest. ex Schult.

Campanula rotundifolia var. *arcuata* (Schur) Nyman === **Campanula serrata** (Kit.) Hendrych subsp. **serrata**

Campanula rotundifolia var. *balcanica* Adamović === **Campanula scheuchzeri** Vill.

Campanula rotundifolia var. *baumgartenii* (Becker) Nyman === **Campanula baumgartenii** Becker

Campanula rotundifolia var. *bertolae* (Colla) Nyman === **Campanula bertolae** Colla

Campanula rotundifolia var. *bielziana* Schur === **Campanula rotundifolia** L.

Campanula rotundifolia var. *bocconei* (Vill.) Lapeyr. === **Campanula rotundifolia** L.

Campanula rotundifolia var. *breyniana* (Beck) Hayek === **Campanula praesignis** Beck

Campanula rotundifolia var. *bulgarica* Nejceff === **Campanula rotundifolia** L.

Campanula rotundifolia var. *calisii* Murr === **Campanula marchesettii** Witasek

Campanula rotundifolia var. *canescens* E. J. Hill === **Campanula intercedens** Witasek

Campanula rotundifolia var. *catalanica* (Podlech) O. Bolòs & Vigo === **Campanula hispanica** subsp. **catalanica** Podlech

Campanula rotundifolia var. *cespitosa* (Scop.) Willd. === **Campanula cespitosa** Scop.

Campanula rotundifolia var. *cochleariifolia* (Lam.) Fiori === **Campanula cochlearifolia** Lam.

Campanula rotundifolia var. *conferta* Knaf === **Campanula gentilis** Kovanda

Campanula rotundifolia var. *confertifolia* Reut. === **Campanula rotundifolia** L.

Campanula rotundifolia var. *decloetiana* (Ortmann) Nyman === ?

Campanula rotundifolia var. *delitschiana* Kuntze === **Campanula rotundifolia** L.

Campanula rotundifolia var. *dentata* Coleman === **Campanula intercedens** Witasek

Campanula rotundifolia var. *dentata* Schur === **Campanula serrata** (Kit.) Hendrych subsp. **serrata**

Campanula rotundifolia var. *dilecta* (Schott) Wohlf. === **Campanula scheuchzeri** Vill.

Campanula rotundifolia var. *dubia* (A. DC.) J. K. Henry === **Campanula giesekiana** Vest ex Schult.

Campanula rotundifolia var. *eulitardierei* Guin. === **Campanula macrorhiza** J. Gay ex A. DC.

Campanula rotundifolia var. *euxina* Velen. === **Campanula euxina** (Velen.) Ančev

Campanula rotundifolia var. *farinulenta* (A. Kern. & Wettst.) Hayek === **Campanula velebitica** Borbás

Campanula rotundifolia var. *flexuosa* C. Vicioso === **Campanula rotundifolia** L.

Campanula rotundifolia var. *forsythii* Arcangeli === **Campanula forsythii** (Arcangeli) Podlech

Campanula rotundifolia var. *giesekiana* (Vest ex Schult.) Simmons === **Campanula giesekiana** Vest ex Schult.

Campanula rotundifolia var. *glabra* Lapeyr. === **Campanula rotundifolia** L.

Campanula rotundifolia var. *grandiflora* J. A. Knapp === **Campanula serrata** (Kit.) Hendrych subsp. **serrata**

Campanula rotundifolia var. *grandiflora* Neilr. === **Campanula scheuchzeri** Vill.

Campanula rotundifolia var. *grandiflora* Wimm.=== **Campanula bohemica** Hruby subsp. **bohemica**

Campanula rotundifolia var. *gypsicola* (Costa) O. Bolòs & Vigo === **Campanula hispanica** subsp. **catalanica** Podlech

Campanula rotundifolia var. *hauryi* (Schott) Nyman === **Campanula cochleariifolia** Lam.

Campanula rotundifolia var. *heterodoxa* (Vest ex Schult.) Trautv. === **Campanula rotundifolia** L.

Campanula rotundifolia var. *hirsuta* Macoun === **Campanula alaskana** (A. Gray) Wight ex J. P. Anderson

Campanula rotundifolia var. *hirta* Mert. & W. D. J. Koch === **Campanula rotundifolia** L.

Campanula rotundifolia var. *hispanica* (Willk.) Fiori === **Campanula hispanica** Willk.

Campanula rotundifolia var. *hostii* (Baumg.) Nyman === ?

Campanula rotundifolia var. *intercedens* (Witasek) Farw. === **Campanula intercedens** Witasek

Campanula rotundifolia var. *jurjurensis* (Pomel) Quézel === **Campanula jurjurensis** Pomel

Campanula rotundifolia var. *lancifolia* Mert. & W. D. J. Koch === **Campanula baumgartenii** Becker subsp. **baumgartenii**

Campanula rotundifolia var. *langsdorffiana* (Fisch. ex A. DC.) Britt. === **Campanula rotundifolia** L.

Campanula rotundifolia var. *lapponica* (Witasek) Simmons === **Campanula giesekiana** Vest ex Schult.

Campanula rotundifolia var. *laxiflora* Beck === **Campanula rotundifolia** L.

Campanula rotundifolia var. *legionensis* (Pau) Lacaita === **Campanula rotundifolia** L.

Campanula rotundifolia var. *linifolia* Bég. === **Campanula carnica** Schiede ex Mert. & Koch

Campanula rotundifolia var. *ludoviciana* Guin. === **Campanula macrorhiza** J. Gay ex A. DC.

Campanula rotundifolia var. *macrorhiza* (J. Gay ex A. DC.) Bég. === **Campanula macrorhiza** J. Gay ex A. DC.

Campanula rotundifolia var. *major* A. DC. === **Campanula rotundifolia** L.

Campanula rotundifolia var. *major* Neilr. === **Campanula baumgartenii** subsp. **beckiana** (Hayek) Podlech

Campanula rotundifolia var. *major* (Timb.-Lagr.) Rouy === **Campanula rotundifolia** L.

Campanula rotundifolia var. *malyi* (Schott) Nyman === **Campanula scheuchzeri** Vill.

Campanula rotundifolia var. *marchesettii* (Witasek) Fiori === **Campanula marchesettii** Witasek

Campanula rotundifolia var. *mentiens* (Witasek) Tacik === **Campanula tatrae** subsp. **mentiens** (Witasek) Kovanda

Campanula rotundifolia var. *micrantha* (Bertol.) Tanfani === **Campanula micrantha** Bertol.

Campanula rotundifolia var. *minor* Witasek === **Campanula rotundifolia** L.

Campanula rotundifolia var. *montana* Lecoq & Lamotte === ?

Campanula rotundifolia var. *montana* Syme === **Campanula rotundifolia** L.

Campanula rotundifolia var. *moravica* Spitzner === **Campanula moravica** (Spitzner) Kovanda

Campanula rotundifolia var. *multiflora* Neilr. === **Campanula baumgartenii** subsp. **beckiana** (Hayek) Podlech

Campanula rotundifolia var. *ovalifolia* (St.-Lag.) Rouy === **Campanula serrata** subsp. **recta** (Dulac) Podlech

Campanula rotundifolia var. *ovata* Peterm. === **Campanula rotundifolia** L.

Campanula rotundifolia var. *papillifera* Săvul. === **Campanula rotundifolia** L.

Campanula rotundifolia var. *papillosa* Beyer === **Campanula rotundifolia** L.

Campanula rotundifolia var. *parviflora* Lange === **Campanula rotundifolia** L.

Campanula rotundifolia var. *parviflora* Lej. === **Campanula rotundifolia** L.

Campanula rotundifolia var. *petiolata* (A. DC.) J. K. Henry === **Campanula petiolata** A. DC.

Campanula rotundifolia var. *pseudostenocodon* (Lacaita) Fiori === **Campanula pseudostenocodon** Lacaita

Campanula rotundifolia var. *pubescens* Gaudin === **Campanula rotundifolia** L.

Campanula rotundifolia var. *pusilla* (Haenke) Willd. === **Campanula cochleariifolia** Lam.

Campanula rotundifolia var. *pygmaea* Hruby === **Campanula rotundifolia** L.

Campanula rotundifolia var. *pygmaea* T. Wulff === **Campanula rotundifolia** L.

Campanula rotundifolia var. *re* (Colla) Nyman === **Campanula bertolae** Colla

Campanula rotundifolia var. *reflexa* Syr. === **Campanula rotundifolia** L.

Campanula rotundifolia var. *reniformis* Pers. === **Campanula baumgartenii** Becker subsp. **baumgartenii**

Campanula rotundifolia var. *rotundata* Chabert === **Campanula jurjurensis** Pomel

Campanula rotundifolia var. *ruscinonensis* (Timb.-Lagr.) O. Bolòs & Vigo === **Campanula ruscinonensis** Timb.-Lagr.

Campanula rotundifolia var. *sabatia* (De Not.) Fiori === **Campanula sabatia** De Not.

Campanula rotundifolia var. *saxatilis* Hruby === **Campanula rotundifolia** L.

Campanula rotundifolia var. *scabriuscula* Mert. & W. D. J. Koch === **Campanula rotundifolia** L.

Campanula rotundifolia var. *scheuchzeri* (Vill.) Fiek === **Campanula scheuchzeri** Vill.

Campanula rotundifolia var. *scheuchzeriformis* (Hayek) Hayek === **Campanula tatrae** Borbás subsp. **tatrae**

Campanula rotundifolia var. *scopulicola* Lamotte === **Campanula rotundifolia** L.

Campanula rotundifolia var. *soldanelliflora* (Voss) W. T. Mill. === **Campanula rotundifolia** L.

Campanula rotundifolia var. *solstitialis* (A. Kern.) Beck === **Campanula rotundifolia** L.

Campanula rotundifolia var. *songeonii* (Chabert) Bonnier === **Campanula rhomboidalis** L.

Campanula rotundifolia var. *stenocodon* (Boiss. & Reut.) Fiori === **Campanula stenocodon** Boiss. & Reut.

Campanula rotundifolia var. *stenophylla* Rouy === **Campanula rotundifolia** L.

Campanula rotundifolia var. *stricta* Schumach. === **Campanula rotundifolia** L.

Campanula rotundifolia var. *subramulosa* Batt. === **Campanula hispanica** Willk. subsp. **hispanica**

Campanula rotundifolia var. *sudetica* Hruby === **Campanula tatrae** subsp. **sudetica** (Hruby) Kovanda

Campanula rotundifolia var. *tenuifolia* (Hoffm.) Opiz === **Campanula rotundifolia** L.

Campanula rotundifolia var. *uniflora* Lange === **Campanula giesekiana** Vest ex Schult.

Campanula rotundifolia var. *velutina* DC. === **Campanula rotundifolia** L.

Campanula rotundifolia var. *verlotii* Rouy === **Campanula rotundifolia** L.

Campanula rotundifolia var. *vulgaris* Neilr.=== **Campanula rotundifolia** L.

Campanula rubra Anonymous === **Campanula rhomboidalis** L.

Campanula ruderalis Aitch. & Hemsl. === **Campanula cashmeriana** Benth.

Campanula rumenica Gand. === ?

Campanula rupestris Host === **Campanula waldsteiniana** Schult.

Campanula rupestris M. Bieb. === **Campanula tridentata** subsp. **biebersteiniana** (Schult.) Ogan.

Campanula rupestris f. *breueri* (Tocl & Rohlena) Hayek === **Campanula lavrensis** (Tocl & Rohlena) Phitos

Campanula rupestris f. *lavrensis* (Tocl & Rohlena) Hayek === **Campanula lavrensis** (Tocl & Rohlena) Phitos

Campanula rupestris subsp. *anchusiflora* (Salisb. ex Sm.) Hayek === **Campanula anchusiflora** Salisb. ex Sm.

Campanula rupestris subsp. *andrewsii* (A. DC.) Hayek === **Campanula andrewsii** A. DC.

Campanula rupestris subsp. celsii (A. DC.) Hayek === **Campanula celsii** A. DC. subsp. **celsii**

Campanula rupestris var. *spathulifolia* Turrill === **Campanula celsii** subsp. **spathulifolia** (Turrill) Phitos

Campanula ruprechtii Boiss. === **Campanula bellidifolia** subsp. **aucheri** (A. DC.) Viktorov

Campanula ruprechtii Grossh. === **Campanula glomerata** subsp. **caucasica** (Trautv.) Ogan.

Campanula russelliana Schult. === **Campanula strigosa** Banks & Sol.

Campanula ruthenica M. Bieb. === **Campanula bononiensis** L.

Campanula sacajaweana M. Peck === **Campanula petiolata** A. DC.

Campanula sajanensis Fisch. === **Campanula stevenii** subsp. **altaica** (Ledeb.) Fed.

Campanula sajanensis Pall. ex Schultes === **Adenophora stenanthina** (Ledeb.) Kitag. subsp. **stenanthina**

Campanula salicifolia Boiss. === **Asyneuma canescens** (Waldst. & Kit.) Griseb. & Schenk subsp. **canescens**

Campanula samothracica subsp. *sporadum* (Halácsy) Greuter & Burdet === **Campanula samothracica** (Degen) Greuter & Burdet

Campanula sanctae-balmae Sennen === **Campanula macrorhiza** J. Gay ex A. DC.

Campanula sarmatica f. *glabra* (A. DC.) Fomin === **Campanula sarmatica** Ker Gawl. subsp. **sarmatica**

Campanula sarmatica f. *ramosissima* (Sommier & Levier) Fomin === **Campanula sarmatica** subsp. **ramosissima** (Sommier & Levier) Ogan.

Campanula sarmatica f. *tenuicaulis* Sommier & Levier === **Campanula sarmatica** subsp. **ramosissima** (Sommier & Levier) Ogan.

Campanula sarmatica var. *calcarea* Albov === **Campanula sarmatica** subsp. **calcarea** (Albov) Ogan.

Campanula sarmatica var. *glabra* A. DC. === **Campanula sarmatica** Ker Gawl. subsp. **sarmatica**

Campanula sarmatica var. *gracilis* Fomin === **Campanula sarmatica** Ker Gawl. subsp. **sarmatica**

Campanula sarmatica var. *ramosissima* Sommier & Levier === **Campanula sarmatica** subsp. **ramosissima** (Sommier & Levier) Ogan.

Campanula sarmatica var. *subtomentosa* Trautv. === **Campanula sarmatica** Ker Gawl. subsp. **sarmatica**

Campanula sarmentosa Hochst. ex A. Rich. === **Campanula edulis** Forssk.

Campanula savalanica Fed. === **Campanula bayerniana** Rupr.

Campanula saxicola R. Br. === **Wahlenbergia saxicola** (R. Br.) A. DC.

Campanula saxifraga M. Bieb. === **Campanula bellidifolia** subsp. **saxifraga** (M. Bieb.) Viktorov

Campanula saxifraga f. *stenophylla* Fomin === **Campanula bellidifolia** subsp. **saxifraga** (M. Bieb.) Viktorov

Campanula saxifraga subsp. *argunensis* (Rupr.) Ogan. === **Campanula bellidifolia** subsp. **argunensis** (Rupr.) Viktorov

Campanula saxifraga subsp. *aucheri* (A. DC.) Ogan. === **Campanula bellidifolia** subsp. **aucheri** (A. DC.) Viktorov

Campanula saxifraga subsp. *meyerana* (Rupr.) Ogan. === **Campanula bellidifolia** subsp. **meyerana** (Rupr.) Viktorov

Campanula saxifraga var. *leptorhiza* Sommier & Levier === **Campanula bellidifolia** subsp. **saxifraga** (M. Bieb.) Viktorov

Campanula saxifraga var. *transcaucasica* Rupr. === **Campanula bellidifolia** subsp. **aucheri** (A. DC.)Viktorov

Campanula saxifragoides f. *elata* Faure & Maire === **Campanula filicaulis** Durieu

Campanula saxifragoides f. *latifolia* Litard. & Maire === **Campanula saxifragoides** Doum.

Campanula saxifragoides var. *gattefossei* Maire & Weiller === **Campanula filicaulis** Durieu

Campanula saxifragoides var. *guinetii* Quézel === **Campanula saxifragoides** Doum.

Campanula scandens Pall. ex Schult. === ?

Campanula scapifoliosa A. P. Khokhr. === **Campanula tridentata** subsp. **biebersteiniana** (Schult.) Ogan.

Campanula schelkownikowii Grossh. === **Campanula sibirica** subsp. **hohenackeri** (Fisch. & C. A. Mey.) Damboldt

Campanula scheuchzeri f. *elatior* Witasek ex Vacc. === **Campanula scheuchzeri** Vill.

Campanula scheuchzeri f. *hirsutissima* Hruby === **Campanula scheuchzeri** Vill.

Campanula scheuchzeri f. *humilior* Hruby === **Campanula scheuchzeri** Vill.

Campanula scheuchzeri f. *humilis* Hruby === **Campanula scheuchzeri** Vill.

Campanula scheuchzeri f. *latifolia* Witasek ex Vacc. === **Campanula scheuchzeri** Vill.

Campanula scheuchzeri f. *malyi* (Schott) Dalla Torre & Sarnth. === **Campanula scheuchzeri** Vill.

Campanula scheuchzeri f. *oblongifolia* Beyer === **Campanula scheuchzeri** Vill.

Campanula scheuchzeri f. *ovalifolia* Hruby === **Campanula scheuchzeri** Vill.

Campanula scheuchzeri f. *parvula* Hruby === **Campanula scheuchzeri** Vill.

Campanula scheuchzeri f. *reflectans* Hruby === **Campanula scheuchzeri** Vill.

Campanula scheuchzeri f. *rhombifolia* Hruby === **Campanula scheuchzeri** Vill.

Campanula scheuchzeri f. *witasekiana* (Vierh.) Hayek === **Campanula witasekiana** Vierh.

Campanula scheuchzeri race *pourretii* (Jeanb. & Timb.-Lagr.) Rouy === ?

Campanula scheuchzeri subf. *altior* Hruby === **Campanula scheuchzeri** Vill.

Campanula scheuchzeri subf. *hirtescens* Hruby === **Campanula scheuchzeri** Vill.

Campanula scheuchzeri subf. *kerneri* (Witasek ex A. Kern. & Fritsch.) Hruby === **Campanula scheuchzeri** Vill.

Campanula scheuchzeri subf. *styriaca* (Schott) Hruby === **Campanula scheuchzeri** Vill.

Campanula scheuchzeri subsp. *ficarioides* (Timb.-Lagr.) O. Bolòs & Vigo === **Campanula ficarioides** Timb.-Lagr.

Campanula scheuchzeri subsp. *kerneri* (Witasek ex A. Kern. & Fritsch.) Vacc. === **Campanula scheuchzeri** Vill.

Campanula scheuchzeri subsp. *micrantha* (Bertol.) Arcang. === **Campanula micrantha** Bertol.

Campanula scheuchzeri subsp. *pollinensis* (Podlech) Bernardo, Gargano & Peruzzi === **Campanula pollinensis** Podlech

Campanula scheuchzeri subsp. *pseudostenocodon* (Lacaita) Bernardo, Gargano & Peruzzi === **Campanula pseudostenocodon** Lacaita

Campanula scheuchzeri subsp. *redux* (Schott) Nyman === **Campanula serrata** (Kit.) Hendrych subsp. **serrata**

Campanula scheuchzeri subsp. *rhaetica* Hruby === **Campanula scheuchzeri** Vill.

Campanula scheuchzeri subsp. *stenocodon* (Boiss. & Reut.) Arcang. === **Campanula stenocodon** Boiss. & Reut.

Campanula scheuchzeri subsp. *susplugasii* Braun-Blanq. === **Campanula scheuchzeri** Vill.

Campanula scheuchzeri subsp. *witasekiana* (Vierh.) Hayek === **Campanula witasekiana** Vierh.

Campanula scheuchzeri subvar. *valdensis* (A. DC.) Rouy === **Campanula scheuchzeri** Vill.

Campanula scheuchzeri subvar. *villarsiana* (Hayek) Hruby === **Campanula scheuchzeri** Vill.

Campanula scheuchzeri var. *calycina* Witasek ex Vacc. === **Campanula scheuchzeri** Vill.

Campanula scheuchzeri var. *carnica* (Schiede ex Mert. & Koch) Posp. === **Campanula carnica** Schiede ex Mert. & Koch

Campanula scheuchzeri var. *consanguinea* (Schott) Dalla Torre & Sarnth. === **Campanula scheuchzeri** Vill.

Campanula scheuchzeri var. *dacica* Porcius === **Campanula tatrae** Borbás subsp. **tatrae**

Campanula scheuchzeri var. *dilecta* (Schott) Vacc. === **Campanula scheuchzeri** Vill.

Campanula scheuchzeri var. *ficarioides* (Timb.-Lagr.) O. Bolòs & Vigo === **Campanula ficarioides** Timb.-Lagr.

Campanula scheuchzeri var. *glabra* Schur === **Campanula scheuchzeri** Vill.

Campanula scheuchzeri var. *glabra* W. D. J. Koch === **Campanula scheuchzeri** Vill.

Campanula scheuchzeri var. *glareosa* Vacc. === **Campanula scheuchzeri** Vill.

Campanula scheuchzeri var. *grandiflora* Schur === **Campanula scheuchzeri** Vill.
Campanula scheuchzeri var. *hirsuta* Schur === **Campanula scheuchzeri** Vill.
Campanula scheuchzeri var. *hirta* Gren. & Godr. === **Campanula scheuchzeri** Vill.
Campanula scheuchzeri var. *hirta* Hruby === **Campanula scheuchzeri** Vill.
Campanula scheuchzeri var. *hirta* W. D. J. Koch === **Campanula scheuchzeri** Vill.
Campanula scheuchzeri var. *inconcessa* (Schott) Nyman === ?
Campanula scheuchzeri var. *intercedens* Hruby === **Campanula scheuchzeri** Vill.
Campanula scheuchzeri var. *kerneri* (Witasek ex A. Kern. & Fritsch.) Dalla Torre & Sarnth. === **Campanula scheuchzeri** Vill.
Campanula scheuchzeri var. *kladniana* Schur === **Campanula carnica** Schiede ex Mert. & Koch subsp. **carnica**
Campanula scheuchzeri var. *luxurians* Hruby === **Campanula scheuchzeri** Vill.
Campanula scheuchzeri var. *minima* Schur === **Campanula scheuchzeri** Vill.
Campanula scheuchzeri var. *monantha* (Schur) Nyman === **Campanula scheuchzeri** Vill.
Campanula scheuchzeri var. *rohdii* (Loisel.) Rouy === **Campanula ficarioides** Timb.-Lagr.
Campanula scheuchzeri var. *schleicheri* (Suter) Beck === **Campanula scheuchzeri** Vill.
Campanula scheuchzeri var. *stenophylla* Schur === **Campanula tatrae** Borbás subsp. **tatrae**
Campanula scheuchzeri var. *styriaca* (Schott) Nyman === **Campanula scheuchzeri** Vill.
Campanula scheuchzeri var. *susplugasii* (Braun-Blanq.) O. Bolòs & Vigo === **Campanula scheuchzeri** Vill.
Campanula scheuchzeri var. *valdensis* (A. DC.) Vacc. === **Campanula scheuchzeri** Vill.
Campanula scheuchzeri var. *villarsiana* Hayek === **Campanula scheuchzeri** Vill.
Campanula scheuchzeri var. *witasekiana* (Vierh.) Dalla Torre & Sarnth. === **Campanula witasekiana** Vierh.
Campanula scheuchzeriformis Hayek === **Campanula tatrae** Borbás subsp. **tatrae**
Campanula schimperi Vatke === **Campanula edulis** Forssk.
Campanula schimperi var. *quartiniana* (A. Rich.) Vatke === **Campanula edulis** Forssk.
Campanula schimperi var. *rigidipila* (Steud. & Hochst. ex A. Rich.) Vatke === **Campanula edulis** Forssk.
Campanula schimperi var. *sarmentosa* (Hochst. ex A. Rich.) Vatke === **Campanula edulis** Forssk.
Campanula schischkinii Kolak. & Sachokia === **Campanula longistyla** Fomin
Campanula schistosa Kolak. === **Campanula collina** subsp. **sphaerocarpa** (Kolak.) Ogan.
Campanula schleicheri Hegetschw. === **Campanula scheuchzeri** Vill.
Campanula schleicheri Suter === **Campanula scheuchzeri** Vill.
Campanula sclerotricha subsp. *tchitounyi* Azn. === **Campanula sclerotricha** Boiss.
Campanula scouleri var. *glabra* Hook. === **Campanula scouleri** Hook. ex A. DC.
Campanula scouleri var. *hirsutula* Hook. === **Campanula scouleri** Hook. ex A. DC.
Campanula secunda F. W. Schmidt === **Campanula rapunculoides** L.
Campanula secundiflora Vis. & Pančić === **Campanula rapunculoides** L.
Campanula semisecta var. *basiclada* Murb. === **Campanula semisecta** Murb.
Campanula serotina Wettst. === **Campanula glomerata** subsp. **serotina** (Wettst.) O. Schwarz
Campanula serpyllifolia Vis. === **Edraianthus serpyllifolius** (Vis.) A. DC.
Campanula serpylliformis Batt. === **Campanula velata** subsp. **serpylliformis** (Batt.) Quézel
Campanula serrata subsp. *recta* (Dulac) Podlech === **Campanula rotundifolia** L.
Campanula serratifolia Vuk. === ?
Campanula serratifolia var. *acutipetala* Vuk. === **Campanula latifolia** L.
Campanula serratifolia var. *ciliatosepala* Vuk. === **Campanula trachelium** L.
Campanula sessiliflora K. Koch === **Campanula persicifolia** subsp. **sessiliflora** (K. Koch ex Velen.) Fed. ex Greuter & Burdet
Campanula sessiliflora L. f. === **Wahlenbergia subulata** (L'Hér.) Lammers
Campanula sessiliflora Vuk. === **Campanula raineri** Perp.
Campanula sessilis (Eckl. ex A. DC.) D. Dietr. === **Prismatocarpus sessilis** Eckl. ex A. DC.
Campanula setacea D. Dietr. === **Wahlenbergia capillacea** (L. f.) A. DC.
Campanula seticalyx Gand. === ?
Campanula setosa J. C. Wendl. === ?

Campanula sewerzowii Regel === **Sergia sewerzowii** (Regel) Fed.

Campanula shepardii Post === **Campanula postii** (Boiss.) Engl.

Campanula sibirica f. *abortiva* (A. DC.) Hayek & Hegi === **Campanula sibirica** L. subsp. **sibirica**

Campanula sibirica f. *corymbosa* Fomin === **Campanula sibirica** subsp. **hohenackeri** (Fisch. & C. A. Mey.) Damboldt

Campanula sibirica f. *elatior* Fomin === **Campanula sibirica** subsp. **elatior** (Fomin) Fed.

Campanula sibirica f. *nana* Fomin === ?

Campanula sibirica f. *parviflora* Fomin === **Campanula sibirica** subsp. **hohenackeri** (Fisch. & C. A. Mey.) Damboldt

Campanula sibirica subsp. *caucasica* (M. Bieb.) Viktorov === **Campanula caucasica** M. Bieb.

Campanula sibirica subsp. *daghestanica* (Fomin) Viktorov === **Campanula daghestanica** Fomin

Campanula sibirica subsp. *komarovii* (Maleev) Viktorov === **Campanula komarovii** Maleev

Campanula sibirica subsp. *longistyla* (Fomin) Viktorov === **Campanula longistyla** Fomin

Campanula sibirica subsp. *paniculata* (A. DC.) Arcang. === **Campanula sibirica** L. subsp. **sibirica**

Campanula sibirica var. *ampliata* K. Koch === **Campanula sibirica** subsp. **hohenackeri** (Fisch. & C. A. Mey.) Damboldt

Campanula sibirica var. *abortiva* A. DC. === **Campanula sibirica** L. subsp. **sibirica**

Campanula sibirica var. *caucasica* (M. Bieb.) Trautv. === **Campanula caucasica** M. Bieb.

Campanula sibirica var. *divergens* (Waldst. & Kit. ex Willd.) Trautv. === **Campanula sibirica** subsp. **divergens** (Waldst. & Kit. ex Willd.) Nyman

Campanula sibirica var. *divergentiformis* Jáv. === **Campanula sibirica** subsp. **divergens** (Waldst. & Kit. ex Willd.) Nyman

Campanula sibirica var. *hohenackeri* (Fisch. & Mey.) Trautv. === **Campanula sibirica** subsp. **hohenackeri** (Fisch. & C. A. Mey.) Damboldt

Campanula sibirica var. *imeretina* (Rupr.) Trautv. === **Campanula imeretina** Rupr.

Campanula sibirica var. *major* Boiss. === **Campanula sibirica** subsp. **divergens** (Waldst. & Kit. ex Willd.) Nyman

Campanula sibirica var. *minor* Schur === **Campanula sibirica** L. subsp. **sibirica**

Campanula sibirica var. *paniculata* A. DC. === **Campanula sibirica** L. subsp. **sibirica**

Campanula sibirica var. *pannonica* (Schrank) Nyman === **Campanula sibirica** subsp. **divergens** (Waldst. & Kit. ex Willd.) Nyman

Campanula sibirica var. *parviflora* (Lam.) Trautv. === **Campanula sibirica** subsp. **hohenackeri** (Fisch. & C. A. Mey.) Damboldt

Campanula sibirica var. *saxicola* K. Koch === ?

Campanula sibirica var. *spathulata* (Schur) Nyman === **Campanula sibirica** subsp. **divergens** (Waldst. & Kit. ex Willd.) Nyman

Campanula sibirica var. *taurica* Trautv. === **Campanula sibirica** subsp. **taurica** (Juz.) Fed.

Campanula sibthorpiana (Schult.) Boiss. === **Asyneuma rigidum** subsp. **sibthorpianum** (Schult.) Damboldt

Campanula sibthorpiana Halácsy === **Campanula spatulata** Sm. subsp. **spatulata**

Campanula sibthorpiana var. *filicaulis* Halácsy === **Campanula spatulata** subsp. **filicaulis** (Halácsy) Phitos

Campanula sieberi (A. DC.) D. Dietr. === **Wahlenbergia quadrifida** (R. Br.) A. DC.

Campanula siegizmundii Fed. === **Campanula sarmatica** Ker Gawl. subsp. **sarmatica**

Campanula silenifolia Fisch. ex A. DC. === **Campanula stevenii** subsp. **turczaninovii** (Fed.) Viktorov

Campanula silenifolia Host == **Edraianthus pumilio** (Port. ex Schult.) A. DC.

Campanula simplex DC. === **Campanula bononiensis** L.

Campanula simplex Steven === **Campanula stevenii** M. Bieb.

Campanula simplex var. *dasycarpa* (Regel) Trautv. === **Campanula stevenii** subsp. **turczaninovii** (Fed.) Viktorov

Campanula simplex var. *silenifolia* (Regel) Trautv. === **Campanula stevenii** subsp. **turczaninovii** (Fed.) Viktorov

Campanula sinai (A. DC.) Boiss. === **Asyneuma rigidum** subsp. **sinai** (A. DC.) Damboldt
Campanula sinensis (A. DC.) D. Dietr. === **Adenophora sinensis** A. DC.
Campanula singarensis Boiss. & Hausskn. === **Campanula flaccidula** Vatke
Campanula solstitialis A. Kern. === **Campanula rotundifolia** L.
Campanula sommieri Kharadze === **Campanula sarmatica** subsp. **ramosissima** (Sommier &
  Levier) Ogan.
Campanula songeonii Chabert === **Campanula rhomboidalis** L.
Campanula sophiae Beauverd === **Campanula spatulata** subsp. **spruneriana** (Hampe) Hayek
Campanula sosnowskyi Kharadze === **Campanula bellidifolia** Adam subsp. **bellidifolia**
Campanula sparsiflora (A. DC.) D. Dietr. === **Microcodon sparsiflorus** A. DC.
Campanula spathulata Waldst. & Kit. === **Campanula sibirica** subsp. **divergens** (Waldst. &
  Kit. ex Willd.) Nyman
Campanula spatulata f. *hirsuta* (Heldr. ex Halácsy) Hayek === **Campanula spatulata** subsp.
  spruneriana (Hampe) Hayek
Campanula spatulata f. *lepidota* (Turrill) Hayek === **Campanula spatulata** subsp.
  spruneriana (Hampe) Hayek
Campanula spatulata subsp. *sibthorpiana* (Maire & Petitm.) Hayek === **Campanula spatulata**
  Sm. subsp. **spatulata**
Campanula spatulata var. *filicaulis* (Halácsy) Hayek === **Campanula spatulata** subsp.
  filicaulis (Halácsy) Phitos
Campanula speciosa Hornem. === **Campanula glomerata** subsp. **speciosa** (Hornem. ex
  Spreng.) Domin
Campanula speciosa Willd. ex Spreng. === **Campanula rapunculoides** L.
Campanula speciosa f. *multicaulis* Marcet === **Campanula speciosa** Pourr. subsp. **speciosa**
Campanula speciosa var. *bicaulis* (Lapeyr.) A. DC. === **Campanula speciosa** Pourr.
  subsp. **speciosa**
Campanula speciosa var. *brevifolia* Pau === **Campanula speciosa** subsp. **affinis** (Schult.)
  Font Quer
Campanula speciosa var. *oliveri* (Rouy & Gaut.) O. Bolòs & Vigo === **Campanula speciosa**
  Pourr. subsp. **speciosa**
Campanula speciosa var. *stenophylla* Marcet === **Campanula speciosa** subsp. **affinis** (Schult.)
  Font Quer
Campanula speculum-veneris L. === **Legousia speculum-veneris** (L.) Durande ex Vill.
Campanula sphaerocarpa Kolak. === **Campanula collina** subsp. **sphaerocarpa** (Kolak.)
  Ogan.
Campanula sphaerocarpa var. *grandiflora* Kolak. === **Campanula collina** subsp. **sphaerocarpa**
  (Kolak.) Ogan.
Campanula sphaerocarpa var. *multiflora* Kolak. === **Campanula collina** subsp. **sphaerocarpa**
  (Kolak.) Ogan.
Campanula sphaerothrix Griseb. === **Campanula sparsa** subsp. **sphaerothrix** (Griseb.) Hayek
Campanula spicata subvar. *canescens* (St.-Lag.) Rouy === **Campanula spicata** L.
Campanula spicata var. *canescens* St.-Lag. === **Campanula spicata** L.
Campanula spicata var. *hornschuchii* A. DC. === **Campanula spicata** L.
Campanula spicata var. *ramosa* A. DC. === **Campanula spicata** L.
Campanula spinulosa (A. DC.) D. Dietr. === ?
Campanula sporadum Feer === **Campanula hagielia** Boiss.
Campanula spreta Schult. === **Adenophora liliifolia** (L.) A. DC.
Campanula spruneriana Hampe === **Campanula spatulata** subsp. **spruneriana** (Hampe) Hayek
Campanula spruneriana subsp. *sibthorpiana* Maire & Petitm. === **Campanula spatulata** Sm.
  subsp. **spatulata**
Campanula spruneriana subsp. *spatulata* (Sm.) Nyman === **Campanula spatulata** Sm.
Campanula spruneriana var. *alpina* Boiss. === **Campanula spatulata** Sm.
Campanula spruneriana var. *hirsuta* Heldr. ex Halácsy === **Campanula spatulata** subsp.
  spruneriana (Hampe) Hayek
Campanula spruneriana var. *lepidota* Turrill === **Campanula spatulata** subsp. **spruneriana**
  (Hampe) Hayek

Campanula spuria Pallas ex Schult. === **Legousia hybrida** (L.) Delarbre

Campanula stefanoffii Hermann === **Campanula patula** subsp. **abietina** (Griseb. & Schenk) Simonk.

Campanula stellarioides (Cham.) D. Dietr. === **Wahlenbergia stellarioides** Cham.

Campanula stellata Thunb. === **Viola decumbens** L. (Violaceae)

Campanula stenanthina Ledeb. === **Adenophora stenanthina** (Ledeb.) Kitag.

Campanula stenocodon var. *latifolia* Beyer === **Campanula stenocodon** Boiss. & Reut.

Campanula stenophylla Boiss. & Heldr. === **Asyneuma linifolium** (Boiss. & Heldr.) Bornm. subsp. **linifolium**

Campanula stenophylla (Schur) Prain === **Campanula tatrae** Borbás subsp. **tatrae**

Campanula stenopoda Gand. === ?

Campanula steveniana Schult. === **Campanula stevenii** M. Bieb.

Campanula stevenii f. *alpina* Bornm. === **Campanula stevenii** subsp. **beauverdiana** (Fomin) Rech. f. & Schiman-Czeika

Campanula stevenii var. *dasycarpa* Regel === **Campanula stevenii** subsp. **turczaninovii** (Fed.) Viktorov

Campanula stevenii var. *integerrima* Regel === **Campanula stevenii** subsp. **turczaninovii** (Fed.) Viktorov

Campanula stevenii var. *sibirica* A. DC. === **Campanula stevenii** subsp. **wolgensis** (P. A. Smirn.) Fed.

Campanula stevenii var. *silenifolia* Regel === **Campanula stevenii** subsp. **turczaninovii** (Fed.) Viktorov

Campanula stevenii var. *vesiculosa* Bornm. === **Campanula stevenii** subsp. **beauverdiana** (Fomin) Rech. f. & Schiman-Czeika

Campanula st.-helenae D. Dietr. === **Wahlenbergia linifolia** (Roxb.) A. DC.

Campanula stolonifera Mign. === ?

Campanula striata Kitam. === **Campanula sclerotricha** Boiss.

Campanula stricta Sm. === **Wahlenbergia stricta** (R. Br.) Sweet

Campanula stricta Wall. === **Campanula sylvatica** Wall.

Campanula stricta f. *adpressa* Witasek === **Campanula stricta** L.

Campanula stricta var. *alidagensis* Damboldt === **Campanula stricta** L.

Campanula stricta var. *jasionifolia* (A. DC.) Boiss. === **Campanula stricta** L.

Campanula stricta var. *libanotica* (A. DC.) Boiss. === **Campanula stricta** L.

Campanula stricta var. *muricata* Trautv. === **Campanula stricta** L.

Campanula strigosa Vahl === **Campanula strigosa** Banks & Sol.

Campanula stylidioides (Boiss.) Boiss. === **Asyneuma limonifolium** (L.) Janch. subsp. **limonifolium**

Campanula stylocampa Eastw. === **Campanula petiolata** A. DC.

Campanula stylosa Lam. === **Adenophora liliifolia** (L.) A. DC.

Campanula stylosa var. *periplocifolia* (Lam.) Pers. === **Adenophora liliifolia** (L.) A. DC.

Campanula styriaca Schott === **Campanula scheuchzeri** Vill.

Campanula suanetica f. *appendiculata* (Sommier & Levier) Fomin === **Campanula suanetica** Rupr.

Campanula suanetica var. *appendiculata* Sommier & Levier === **Campanula suanetica** Rupr.

Campanula suaveolens Schrad. ex Hornem. === **Adenophora liliifolia** (L.) A. DC.

Campanula subalpina Delponte & Gras === **Campanula elatines** L.

Campanula subcapitata Popov === **Campanula glomerata** subsp. **subcapitata** (Popov) Fed.

Campanula subidaea Gand. === **Campanula tubulosa** Lam.

Campanula subramulosa Jord. ex Gren. & Godr. === **Campanula cochleariifolia** Lam.

Campanula subpyrenaica Timb.-Lagr. === **Campanula persicifolia** subsp. **subpyrenaica** (Timb.-Lagr.) Fed.

Campanula subulata Thunb. === **Prismatocarpus fruticosus** (L.) L'Hér.

Campanula subuniflora Lam. === ?

Campanula suffruticosa Sennen === **Campanula macrorhiza** J. Gay ex A. DC.

Campanula sulphurea f. *brachyantha* Evenari === **Campanula sulphurea** Boiss.

Campanula swellendamensis (H. Buek) D. Dietr. === **Wahlenbergia ecklonii** H. Buek

Campanula symphytifolia (Albov) Kolak. === **Campanula glomerata** subsp. **symphytifolia** (Albov) Ogan.

Campanula syriaca Willd. ex Schult. === **Legousia falcata** (Ten.) Fritsch ex Janch.

Campanula syspirensis K. Koch === **Campanula stricta** L.

Campanula takhtadzhianii Fed. === **Campanula bayerniana** Rupr.

Campanula talievii Juz. === **Campanula sibirica** subsp. **talievii** (Juz.) Fed.

Campanula tarana K. Malý === **Campanula hercegovina** Degen & Fiala

Campanula taurica Juz. === **Campanula sibirica** subsp. **taurica** (Juz.) Fed.

Campanula tauricola Boiss. & Balansa === **Asyneuma rigidum** (Willd.) Grossh. subsp. **rigidum**

Campanula telephioides subsp. *argutiserrata* Rech. f. === **Campanula telephioides** Boiss. & Hausskn.

Campanula tenella Jord. === **Campanula cochleariifolia** Lam.

Campanula tenella L. f. === **Wahlenbergia tenella** (L. f.) Lammers

Campanula tenerrima (H. Buek) D. Dietr. === **Prismatocarpus tenerrimus** H. Buek

Campanula tenorei Moretti === **Campanula versicolor** Andrews

Campanula tenuiflora Schur === **Campanula bononiensis** L.

Campanula tenuiflora Ten. === **Campanula lingulata** Waldst. & Kit.

Campanula tenuifolia Waldst. & Kit. === **Edraianthus tenuifolius** (Waldst. & Kit.) A. DC.

Campanula tenuifolia Hoffm. === **Campanula rotundifolia** L.

Campanula tetraphylla Thunb. === **Adenophora triphylla** (Thunb.) A. DC.

Campanula thaliana Wallr. === **Campanula bononiensis** L.

Campanula thalictrifolia (Wall.) Spreng. === **Codonopsis thalictrifolia** Wall.

Campanula thessala Gand. === ?

Campanula thessala Maire === **Campanula pelia** (Halácsy) Hausskn. & Sint. ex Phitos

Campanula thomsonii Hook. f. === **Asyneuma thomsonii** (Hook. f.) Bornm.

Campanula thunbergii Schult. === **Wahlenbergia thunbergii** (Schult.) B. Nord.

Campanula thyrsoides var. *carniolica* Sünd. === **Campanula thyrsoides** subsp. **carniolica** (Sünd.) Podlech

Campanula thyrsoides var. *glomerata* Saut. === **Campanula thyrsoides** L. subsp. **thyrsoides**

Campanula thyrsoides var. *multicaulis* A. DC. === **Campanula thyrsoides** L. subsp. **thyrsoides**

Campanula tillischii O. Schwarz === **Campanula phrygia** Jaub. & Spach

Campanula toletana Sennen === ?

Campanula tomentosa subsp. *rupestris* (Sm.) Nyman === **Campanula rupestris** Sm.

Campanula tomentosa var. *appendiculata* (A. DC.) Nyman === **Campanula tomentosa** Lam.

Campanula tomentosa var. *brachyantha* Boiss. === **Campanula rupestris** Sm.

Campanula tomentosa var. *ephesia* A. DC. === **Campanula tomentosa** Lam.

Campanula tomentosa var. *sporadum* (Feer) Fiori === **Campanula hagielia** Boiss.

Campanula tortuosa C. Y. Wu === **Campanula cana** Wall.

Campanula tracheliifolia Losa ex Sennen === **Campanula rotundifolia** L.

Campanula trachelioides M. Bieb. === **Campanula rapunculoides** L.

Campanula trachelioides Munby === **Campanula trachelium** subsp. **mauritanica** (Pomel) Quézel

Campanula trachelium f. *alba* Voss === **Campanula trachelium** L. subsp. **trachelium**

Campanula trachelium f. *maroccana* Quézel === **Campanula trachelium** subsp. **mauritanica** (Pomel) Quézel

Campanula trachelium f. *numidica* Quézel === **Campanula trachelium** subsp. **mauritanica** (Pomel) Quézel.

Campanula trachelium subsp. *dasycarpa* A. DC. ex Arcang. === **Campanula trachelium** subsp. **athoa** (Boiss. & Heldr.) Nyman

Campanula trachelium var. *albiflora* Peterm. === **Campanula trachelium** L. subsp. **trachelium**

Campanula trachelium var. *athoa* (Boiss. & Heldr.) Degen === **Campanula trachelium** subsp. **athoa** (Boiss. & Heldr.) Nyman

Campanula trachelium var. *dasycarpa* Gren. & Godr. === **Campanula trachelium** L. subsp. **trachelium**

Campanula trachelium var. *gymnocarpa* Trautv. === **Campanula trachelium** L. subsp. **trachelium**

Campanula trachelium var. *hirta* Quézel === **Campanula trachelium** subsp. **mauritanica** (Pomel) Quézel

Campanula trachelium var. *hispida* Peterm. === **Campanula trachelium** L. subsp. **trachelium**

Campanula trachelium var. *leucantha* Schur === **Campanula trachelium** L. subsp. **trachelium**

Campanula trachelium var. *monstrosa* A. DC. === **Campanula trachelium** L. subsp. **trachelium**

Campanula trachelium var. *munbyana* Quézel === **Campanula trachelium** subsp. **mauritanica** (Pomel) Quézel

Campanula trachelium var. *nuda* Kuntze === **Campanula trachelium** L. subsp. **trachelium**

Campanula trachelium var. *oligosantha* (Schur) Nyman === **Campanula trachelium** L. subsp. **trachelium**

Campanula trachelium var. *orientalis* Boiss. === **Campanula trachelium** subsp. **athoa** (Boiss. & Heldr.) Nyman

Campanula trachelium var. *parviflora* Battand. === **Campanula trachelium** subsp. **mauritanica** (Pomel) Quézel

Campanula trachelium var. *ramosissima* Marcet === **Campanula trachelium** L. subsp. **trachelium**

Campanula trachelium var. *secundiflora* Noulet === **Campanula trachelium** L. subsp. **trachelium**

Campanula trachelium var. *solitaria* Post === **Campanula trachelium** subsp. **athoa** (Boiss. & Heldr.) Nyman

Campanula trachelium var. *subsessilis* Noulet === **Campanula trachelium** L. subsp. **trachelium**

Campanula trachelium var. *urticifolia* Steud. === **Campanula trachelium** L. subsp. **trachelium**

Campanula trachelium var. *virgata* Knaf === **Campanula trachelium** L. subsp. **trachelium**

Campanula transcaucasica (Sommier & Levier) Kolak. & Serdyuk. === **Campanula pendula** M. Bieb.

Campanula transsilvanica subsp. *davidovii* Urum. === **Campanula transsilvanica** Schur ex Andrae

Campanula transsilvanica var. *davidovii* (Urum.) Hayek === **Campanula transsilvanica** Schur ex Andrae

Campanula transtagana Ros. Fernandes === **Campanula lusitanica** subsp. **transtagana** (Ros. Fernandes) Fed.

Campanula trautvetteri Grossh. ex Fed. === **Campanula glomerata** subsp. **caucasica** (Trautv.) Ogan.

Campanula triangularis Parsa === ?

Campanula trichocalycinum Ten. === **Asyneuma trichocalycinum** (Ten.) K. Malý

Campanula tricuspidata Fisch. ex Schult. === **Adenophora tricuspidata** (Fisch. ex Schult.) A. DC.

Campanula tridens Rupr. === **Campanula tridentata** subsp. **biebersteiniana** (Schult.) Ogan.

Campanula tridens var. *araratica* Rupr. === **Campanula tridentata** Schreb. subsp. **tridentata**

Campanula tridens var. *barbata* (Fomin) Kharadze === **Campanula tridentata** Schreb. subsp. **tridentata**

Campanula tridens var. *ciliata* Rupr. === **Campanula tridentata** Schreb. subsp. **tridentata**

Campanula tridens var. *crenatoserrata* Rupr. === **Campanula tridentata** Schreb. subsp. **tridentata**

Campanula tridentata var. *stenophylla* Boiss. === **Campanula tridentata** Schreb. subsp. **tridentata**

Campanula tridentata var. *ardonensis* (Rupr.) Trautv. === **Campanula ardonensis** Rupr.

Campanula tridentata var. *argunensis* (Rupr.) Trautv. === **Campanula bellidifolia** subsp. **argunensis** (Rupr.) Viktorov

Campanula tridentata var. *barbata* Fomin === **Campanula tridentata** Schreb. subsp. **tridentata**

Campanula tridentata var. *bellidifolia* (Adam) Trautv. === **Campanula bellidifolia** Adam

Campanula tridentata var. *bithynica* (A. DC.) Griseb. === **Campanula tridentata** Schreb. subsp. **tridentata**

Campanula tridentata var. *ciliata* (Stev.) Trautv. === **Campanula ciliata** Steven

Campanula tridentata var. *gilanica* (Rupr.) Trautv. === **Campanula bellidifolia** subsp. **aucheri** (A. DC.) Viktorov

Campanula tridentata var. *gracilis* Fomin === **Campanula tridentata** Schreb. subsp. **tridentata**

Campanula tridentata var. *petrophila* Trautv. === **Campanula bellidifolia** subsp. **argunensis** (Rupr.) Viktorov

Campanula tridentata var. *petrophila* (Rupr.) Trautv. === **Campanula petrophila** Rupr.

Campanula tridentata var. *pubiflora* (Rupr.) Trautv. === **Campanula bellidifolia** subsp. **argunensis** (Rupr.) Viktorov

Campanula tridentata var. *rupestris* Trautv. === **Campanula tridentata** subsp. **biebersteiniana** (Schult.) Ogan.

Campanula tridentata var. *saxifraga* Trautv. === **Campanula bellidifolia** subsp. **saxifraga** (M. Bieb.) Viktorov

Campanula trilocularis Turra === **Campanula alpestris** All.

Campanula triphylla Thunb. === **Adenophora triphylla** (Thunb.) A. DC.

Campanula truncata (Wall. ex A. DC.) D. Dietr. === **Cyclocodon lancifolius** (Roxb.) Kurz subsp. **lancifolius**

Campanula tubulosa (Boiss.) Engl. === **Campanula buseri** Damboldt

Campanula tubulosa var. *carpatha* (Halácsy) Hayek === **Campanula carpatha** Halácsy

Campanula turbinata Schott === **Campanula carpatica** Jacq.

Campanula turbinata f. *alba* Voss === **Campanula carpatica** Jacq.

Campanula turbinata f. *lilacina* Voss === **Campanula carpatica** Jacq.

Campanula turbinata f. *pelviformis* Voss === **Campanula carpatica** Jacq.

Campanula turczaninovii Fed. === **Campanula stevenii** subsp. **turczaninovii** (Fed.) Viktorov

Campanula tyrolensis Schott === ?

Campanula ucranica Besser. === **Campanula rapunculoides** L.

Campanula uliginosa Rydb. === **Campanula aparinoides** Pursh

Campanula umbellulifera Vuk. === **Campanula pyramidalis** L.

Campanula umbrosa F. Dietr. === **Adenophora liliifolia** (L.) A. DC.

Campanula undulata L. f. === **Wahlenbergia undulata** (L. f.) A. DC.

Campanula undulata Moench === **Campanula sibirica** L.

Campanula unidentata L. f. === **Wahlenbergia unidentata** (L. f.) Lammers

Campanula uniflora var. *giesekiana* (Vest ex Schult.) A. DC. === **Campanula giesekiana** Vest ex Schult.

Campanula urbionensis Rivas Mart. & G. Navarro === **Campanula rotundifolia** L.

Campanula urticifolia All. === **Campanula latifolia** L. subsp. **latifolia**

Campanula urticifolia F. W. Schmidt === **Campanula trachelium** L. subsp. **trachelium**

Campanula urticifolia Hegetschw. === ?

Campanula urticifolia Salisb. === **Campanula trachelium** L.

Campanula urticifolia Turra === **Campanula rapunculoides** L.

Campanula vajdae Pénzes === **Campanula patula** subsp. **abietina** (Griseb. & Schenk) Simonk.

Campanula valdensis All. === **Campanula scheuchzeri** Vill.

Campanula valdensis var. *scheuchzeri* (Vill.) Gaudin === **Campanula scheuchzeri** Vill.

Campanula valdensis var. *schleicheri* Gaudin === **Campanula scheuchzeri** Vill.

Campanula valdensis var. *villosa* Gaudin === **Campanula scheuchzeri** Vill.

Campanula vanhouttei Carr. === **Campanula punctata** Lam.

Campanula vardariana Bocquet === **Campanula tomentosa** Lam.

Campanula variifolia Salisb. === **Campanula rotundifolia** L.

Campanula vayredae Leresche === **Campanula speciosa** subsp. **affinis** (Schult.) Font Quer

Campanula velata var. *mesatlantica* Litard. & Maire === **Campanula velata** subsp. **mesatlantica** (Litard.& Maire) Quézel

Campanula velata var. *rifana* Maire === **Campanula mollis** L. subsp. **mollis**

Campanula velebitica f. *borbasiana* Witasek === **Campanula velebitica** Borbás

Campanula velebitica f. *divaricata* Witasek === **Campanula velebitica** Borbás

Campanula velebitica f. *farinulenta* Witasek === **Campanula velebitica** Borbás

Campanula velebitica f. *incerta* Witasek === **Campanula velebitica** Borbás

Campanula velebitica f. *parviflora* Witasek === **Campanula velebitica** Borbás

Campanula velenovskyi Adamović === **Campanula patula** subsp. **epigaea** (Janka ex Degen) Hayek

Campanula velutina Desf. === **Campanula mollis** L. subsp. **mollis**

Campanula velutina Velen. === **Campanula lanata** Friv.

Campanula velutina var. *almeriensis* (Degen & Hervier) G. López === **Campanula mollis** L. subsp. **mollis**

Campanula velutina var. *gibraltarica* (Degen & Hervier) G. López === **Campanula mollis** L. subsp. **mollis**

Campanula velutina var. *glaberrima* Escudero & Pajarón === **Campanula mollis** L. subsp. **mollis**

Campanula velutina var. *malacitana* (Degen & Hervier) G. López === **Campanula mollis** L. subsp. **mollis**

Campanula velutina var. *maroccana* (Pau) G. López === **Campanula mollis** L. subsp. **mollis**

Campanula velutina var. *microphylla* (Cav.) G. López === **Campanula mollis** L. subsp. **mollis**

Campanula velutina var. *rifana* (Emb. & Maire) G. López === **Campanula mollis** L. subsp. **mollis**

Campanula venusta Schur === **Campanula cochleariifolia** Lam.

Campanula verbenifolia Sieber === **?**

Campanula veronicifolia Hance === **Campanula dimorphantha** Schweinf.

Campanula verruculosa Hoffmanns. & Link === **Campanula rapunculus** L. subsp. **rapunculus**

Campanula versicolor f. *mrkvickana* (Velen.) Hayek === **Campanula versicolor** Andrews

Campanula versicolor subsp. *rosanii* Nyman === **Campanula versicolor** Andrews

Campanula versicolor var. *multiflora* A. DC. === **Campanula versicolor** Andrews

Campanula versicolor var. *thessala* Boiss. === **Campanula versicolor** Andrews

Campanula verticillata Hill === **Triodanis perfoliata** (L.) Nieuwl. subsp. **perfoliata**

Campanula verticillata Pall. === **Adenophora triphylla** (Thunb.) A. DC.

Campanula vesula All. === **Campanula persicifolia** L. subsp. **persicifolia**

Campanula vidalii H. C. Watson === **Azorina vidalii** (H. C. Watson) Feer

Campanula vinciflora Pau === **Campanula lusitanica** Loefl. subsp. **lusitanica**

Campanula vinciflora Vent. === **Wahlenbergia gracilis** (G. Forst.) A. DC.

Campanula violae Pers. === **Campanula punctata** Lam.

Campanula violifolia Lam. === **Campanula punctata** Lam.

Campanula violifolia Pau === **Campanula escalerae** Rech. f. & Schiman-Czeika

Campanula virgata Labill. === **Asyneuma virgatum** (Labill.) Bornm. subsp. **virgatum**

Campanula virgata Raf. ex Schult. === **Campanula intercedens** Witasek

Campanula viridis (Wall.) Spreng. === **Codonopsis viridis** Wall.

Campanula vitinghoffiana Schult. === **Campanula stevenii** M. Bieb. subsp. **stevenii**

Campanula volubilis Pall. ex Schult. === **Campanula lactiflora** M. Bieb.

Campanula × *vrtacensis* Ravnik === **Campanula cochleariifiolia** Lam. × **Favratia zoysii** (Jacq.) Feer

Campanula waldsteiniana subsp. *tommasiniana* (K. Koch) Nyman === **Campanula tommasiniana** K. Koch

Campanula wallichii Babu === **Campanula dimorphantha** Schweinf.

Campanula welandii Heuff. === **Campanula sparsa** Friv. subsp. **sparsa**

Campanula wiedmannii Podlech === **Campanula rotundifolia** L.

Campanula wightii Gamble === **Campanula pallida** Wall.

Campanula willdenowiana Boiss. === **Asyneuma limonifolium** subsp. **pestalozzae** (Boiss.) Damboldt

Campanula willdenowiana Schult. === **Campanula versicolor** Andrews

Campanula witasekiana var. *praticola* Hruby === **Campanula witasekiana** Vierh.

Campanula wolgensis P. A. Smirn. === **Campanula stevenii** subsp. **wolgensis** (P. A. Smirn.) Fed.

Campanula woronovii Kharadze === **Campanula sarmatica** subsp. **woronovii** (Kharadze) Ogan.

Campanula xylopoda Jeffrey === **Campanula cana** Wall.

Campanula zeyheri (H. Buek) D. Dietr. === **Wahlenbergia krebsii** Cham. subsp. **krebsii**

Campanula zoysii Jacq. === **Favratia zoysii** (Jacq.) Feer

Campanulastrum

The single North American species that has been assigned here is rather distinct from all other species of *Campanula* on the basis of morphology, palynology, and chromosome number. As a result, some workers have recommended the resurrection of *Campanulastrum*. However, a preliminary molecular phylogeny (Eddie et al. 2003, under **General**) suggests that *C. americana* belongs to a clade (within the larger "Rapunculus clade") that otherwise comprises *C. divaricata* and all sampled species of *Legousia* and *Triodanis*. Until such time as the relationships of North American Campanuloideae have been resolved adequately, it seems best to leave this species in an admittedly unnatural *Campanula*. Type [designated by the author]: *Campanula americana* L.

> Small, J. K. (1903) *Campanulastrum* Small. In Flora of the southeastern United States: 1141, 1338. New York: published privately. En. — Establishment of genus as distinct from *Campanula*.
> Shetler, S. G. & J. F. Matthews (1967). Generic position of *Campanula americana* L. [abstract]. ASB Bull. 14(2): 40. En. — Advocate recognition of *Campanulastrum*.
> Shetler, S. G. & N. R. Morin (1986). Seed morphology in North American Campanulaceae. Ann. Missouri Bot. Gard. 73: 653-688, illus. En. — Seed coat morphology supports recognition of genus as distinct from *Campanula*.
> Nowicke, J. W., S. G. Shetler & N. Morin (1992). Exine structure of pantoporate *Campanula* (Campanulaceae) species. Ann. Missouri Bot. Gard. 79: 65-80, illus. En. — Pollen very similar to that *C. exigua*, *C. californica*, *C. griffinii*, and *C. sharsmithiae* of California.

Synonyms:
Campanulastrum Small === **Campanula** L.
Campanulastrum americanum (L.) Small === **Campanula americana** L.

Campanumoea

Although recognized by many authors, this genus in its traditional circumscription is clearly biphyletic (Murthy 1983, Morris & Lammers 1997, Hong & Pan 1998). While segregation of sect. *Cyclocodon* (Griff. ex Hook. f. & Thomson) C. B. Clarke as a distinct genus alleviates the biphyly, it does not obviate all problems. Morphology (Moeliono 1960), palynology (Murthy 1983, Morris & Lammers 1997), and DNA data (Eddie et al. 2003, under **General**) indicate that the type species of *Campanumoea* is best included within *Codonopsis*. Type [designated by C. B. Clarke in Hook. f., Fl. Brit. Ind. 3: 436 (1881)]: *Campanumoea javanica* Blume.

> Blume, C. L. (1826). *Campanumoea*, Bl. In Bijdragen tot de flora van Nederlandsch Indië: 726-727. Batavia: Lands Drukkerij. La. — Establishment of genus.
> Moeliono, B. (1960). *Codonopsis*. In C. G. G. J. van Steenis (ed.), Flora Malesiana (ser. I) 6(1): 118-121, illus. Djakarta: Noordhoff-Kolff. En. — Morphological data support inclusion of *Campaumoea* s.l. in *Codonopsis*.
> Murthy, G. V. S. (1983). Pollen morphology of Indian *Campanumoea* Bl. – a revision of the genus. J. Palynol. 18: 55-59, illus. En. — Pollen of sect. *Campanumoea* indistinguishable from *Codonopsis,* that of sect. *Cyclocodon* distinct.
> Hong, D. Y. (1983). *Campanumoea* Bl. In Flora Reipublicae Popularis Sinicae 73(2): 70-74, illus. Beijing: Science Press. Ch. — Flora with keys, descriptions, and full nomenclature.
> Morris, K. E. & T. G. Lammers (1997). Circumscription of *Codonopsis* and the allied genera *Campanumoea* and *Leptocodon* (Campanulaceae: Campanuloideae). I. Palynological data. Bot. Bull. Acad. Sin. 38: 277-284, illus. En. — Pollen data support merging sect. *Campanumoea* with *Codonopsis* and recognition of sect. *Cyclocodon* at generic rank.

Hong, D. Y. & K. Y. Pan (1998). The restoration of the genus *Cyclocodon* (Campanulaceae) and its evidence from pollen and seed-coat. Acta Phytotax. Sin. 36: 106-110, illus. — Further evidence of biphyletic nature of *Campanumoea* s. l.

Synonyms:
Campanumoea Blume === **Codonopsis** Wall.
Campanumoea [unranked] *Codonopsis* (Wall.) Endl. === **Codonopsis** Wall.
Campanumoea sect. *Cyclocodon* (Griff. ex Hook. f. & Thomson) C. B. Clarke === **Cyclocodon** Griff. ex Hook. f. & Thomson
Campanumoea axillaris Oliv. === **Cyclocodon lancifolius** (Roxb.) Kurz subsp. **lancifolius**
Campanumoea celebica Blume === **Cyclocodon lancifolius** subsp. **celebicus** (Blume) K. E. Morris & Lammers
Campanumoea cordata Miq. === **Codonopsis javanica** (Blume) Miq. subsp. **javanica**
Campanumoea gracilis (Hook. f.) G. Nichols. === **Codonopsis gracilis** Hook. f.
Campanumoea inflata (Hook. f.) C. B. Clarke === **Codonopsis inflata** Hook. f.
Campanumoea japonica Maxim. === **Codonopsis javanica** subsp. **japonica** (Makino) Lammers
Campanumoea japonica Siebold ex ex E. Morren === **Codonopsis lanceolata** (Siebold & Zucc.) Trautv.
Campanumoea javanica Blume === **Codonopsis javanica** (Blume) Miq.
Campanumoea javanica subsp. *japonica* (Makino) D. Y. Hong === **Codonopsis javanica** subsp. **japonica** (Makino) Lammers
Campanumoea javanica var. *japonica* Makino === **Codonopsis javanica** subsp. **japonica** (Makino) Lammers
Campanumoea labordei H. Lév. === **Codonopsis javanica** (Blume) Miq. subsp. **javanica**
Campanumoea lanceolata Siebold & Zucc. === **Codonopsis lanceolata** (Siebold & Zucc.) Trautv.
Campanumoea lancifolia (Roxb.) Merrill === **Cyclocodon lancifolius** (Roxb.) Kurz
Campanumoea maximowiczii Honda === **Codonopsis javanica** (Blume) Miq. subsp. **japonica** (Makino) Lammers
Campanumoea parviflora (Wall. ex A. DC.) Benth. ex C. B. Clarke === **Cyclocodon parviflorus** (Wall. ex A. DC.) Hook. f. & Thomson
Campanumoea pilosula Franch. === **Codonopsis pilosula** (Franch.) Nannf.
Campanumoea truncata (Wall. ex A. DC.) Diels === **Cyclocodon lancifolius** (Roxb.) Kurz subsp. **lancifolius**
Campanumoea violifolia H. Lév. === **Codonopsis micrantha** Chipp

Campylocera

Nuttall (1842) segregated this genus from *Triodanis* (for which he used the name *Dysmicodon*) on the basis of the unicellular capsule of its cleistogamous flowers, which often splits longitudinally from its apex. Type [by monotypy]: *Campylocera leptocarpa* Nutt.

Nuttall, T. (1842). Description and notices of new or rare plants in the natural orders Lobeliaceae, Campanulaceae, Vaccinieae, Ericaceae, collected in a journey over the continent of North America, and during a visit to the Sandwich Islands, and Upper California. Trans. Amer. Philos. Soc. (n.s.) 8: 251-272. En. — Establishment of genus.
Gray, A. (1876). Miscellaneous botanical contributions, *Specularia* Heister. Proc. Amer. Acad. Arts 11: 81-82. En. — Reduction of genus to section of European *Legousia* (under the name *Specularia*).

Synonyms:
Campylocera Nutt. === **Triodanis** Raf.
Campylocera leptocarpa Nutt. === **Triodanis leptocarpa** (Nutt.) Nieuwl.
Campylocera leptocarpa var. *glabella* Nutt. === **Triodanis leptocarpa** (Nutt.) Nieuwl.

Campylosiphon

This name was created as part of an ill-advised reform of nomenclature. Although Saint-Lager (1880) may have intended it as an avowed replacement for *Siphocampylus,* he validated only a single combination at specific rank, and it was not for the type species. As such, he created a unispecific segregate genus rather than an illegitimate replacement name. Type [by monotypy]: *Campylosiphon lycioides* (Cham.) St.-Lag.

> Saint-Lager, J. B. (1880). Réforme del la nomenclature botanique. Ann. Soc. Bot. Lyon 7: 1-154. Fr. — Establishment of genus.

Synonyms:
Campylosiphon St.-Lag. === **Siphocampylus** Pohl
Campylosiphon lycioides (Cham.) St.-Lag. === **Siphocampylus lycioides** (Cham.) G. Don.

Canarina

Campanuloideae, 3 species, Canary Islands and eastern Africa. In a preliminary molecular phylogeny (Eddie et al. 2003, under **General**), the type species was one of four branches of the "platycodonoid clade", which otherwise included *Cyananthus, Codonopsis,* and *Platycodon.* The treatment here follows Hedberg (1961). *2n* = 34. Type [by monotypy]: *Canarina campanula* L.

> Linnaeus, C. (1771). *Canaria* [sic]. In Mantissa plantarum altera: 148, 225, 588. Stockholm: Laurentius Salvius. La. — Establishment of genus, misspelled at first but corrected in an addendum.
>
> Webb, P. B. & S. Berthelot (1844). Trib. I. Canarineae. Nob. In Histoire naturelle des Iles Canaries 2(3): 1-3. Paris: Béthune. La. — Floristic treatment for Canary Islands, with description and full nomenclature; establishment of genus as type of tribe within Campanulaceae.
>
> • Burtt, B. L. (1938). *Canarina eminii.* Bot. Mag. 161: tab. 9531 + 4 pp., illus. En. — Portrait with description and discussion of biogeography and infraspecific variation.
>
> Bailey, L. H. (1953). *Canarina.* In The garden of bellflowers in North America: 149-150, illus. New York: MacMillan. En. — Account of *C. canariensis* in cultivation, with description.
>
> • Hedberg, O. (1961). Monograph of the genus *Canarina* L. (Campanulaceae). Svensk Bot. Tidskr. 55: 17-62, illus., maps. En. — Monograph with key, descriptions, full nomenclature, and specimen citations.
>
> Meikle, R. D. (1972). Proposal to conserve the generic name 8656 *Canarina* L. (1771) (Campanulaceae) against *Mindium* Adanson (1763). Taxon 21: 542-543. En. — Formal motion to protect name.
>
> McVaugh, R. (1974). Report of the Committee for Spermatophyta. Conservation of generic names, XVII. Taxon 23: 819-824. En. — Recommendation to reject *Mindium* so that it threatens neither *Canarina* or *Michauxia.*
>
> Stafleu, F. A. & E. G. Voss (1975). Synopsis of proposals on botanical nomenclature Leningrad 1975. Taxon 24: 201-251. En. — General Committee recommends conservation.
>
> • Thulin, M. (1976). *Canarina.* In Flora of tropical East Africa, Campanulaceae: 2-4, illus. Rotterdam: Balkema. En. — Flora with key, descriptions, full nomenclature, and specimen citations.
>
> • Thulin, M. (1977). *Canarina* L. In Flore d'Afrique centrale (Zaire-Rwanda-Burundi), spermatophytes, Campanulaceae: 2-5, illus. Meise: Jardin Botanique National de Belgique. Fr. — Flora with description, full nomenclature, and specimen citations.
>
> Kunkel, M. A. & G. Kunkel (1978). Bicácaro (o bicacarea). In Flora de Gran Canaria 2: 110 + pl. 98, illus. Las Palmas: Ediciones del Excelentísimo Cabildo Insular. Sp. — Portrait of *C. canariensis,* with description.

- Thulin, M. (1983). *Canarina* L. In E. Launert (ed.), Flora Zambesiaca 7(1): 88-89, illus. Kew: Royal Botanic Garden. En. — Flora of Mozambique, Malawi, Zambia, Zimbabwe and Botswana, with description, full nomenclature & specimen citations.

  Vogel, S., C. Westerkamp, B. Thiel & K. Gessner (1984). Ornithophilie auf den Canarischen Inseln. Pl. Syst. Evol. 146: 225-248. Ge. — Pollination of *C. canariensis*.

  Olesen, J. M. (1985). The Macaronesian bird-flower element and its relation to bird and bee opportunists. Bot. J. Linn. Soc. 91: 395-414. En. — Pollination of *C. canariensis*.

  Kolakovskii, A. A. (1990). [New data on the morphology of the flower and fruit in the family Campanulaceae.] Soobšč. Akad. Nauk Gruzinsk. SSR 139: 381-384, illus. Ru. — Fruit anatomy of *C. canariensis* unique within family.

  Westerkamp, C. (1990). Blumenvögel auf den Islas Canarias. Trochilus 11: 79-81, illus. Ge. — Notes on ornithophily in flora of Canary Islands, including *C. canariensis*.

  Caballero, M., M. C. Cid & A. Gonzalez (1992). Adaptacion al cultivo como planta ornamental de *Canarina canariensis* (L.) Vatke. I: Comparacion de parametros de crecimiento y desarrollo de diversas poblaciones. Bot. Macarones. 19-20: 15-26, illus. Sp. — Conditions required for culture as pot plant.

  Caballero, M. & M. C. Cid (1992). Adaptacion al cultivo como planta ornamental de *Canarina canariensis* (L.) Vatke. II: Estudio sobre la germinacion de semillas. Bot. Macarones. 19-20: 27-38, illus. Sp. — Seed germination requirements in cultivation.

  Caballero, M., M. C. Cid & A. Gonzalez (1992). Adaptacion al cultivo como planta ornamental de *Canarina canariensis* (L.) Vatke. III: Respuestas al fotoperiod y regimen termico. Bot. Macarones. 19-20: 39-44. Sp. — Effects of day length and temperatures on growth and flowering in cultivation.

- Huxley, A. J., ed. (1992). *Canarina*. In The new Royal Horticultural Society dictionary of gardening 1: 500. London: MacMillan. En. — Account of all three species in cultivation.

Canarina L., Mant. Pl.: 148, as 'Canaria', 588 (1771), nom. cons.
  Macaronesia & trop. E. Africa. 21 23 24 25 26.
  Mindium Adans., Fam. Pl. 2: 134, 578 (1763), as 'Mindion', nom. rej.
  Pernetya Scop., Intr. Hist. Nat.: 150 (1777); non *Pernettya* Gaudich., Ann. Sci. Nat., Paris 5: 102 (1825), nom. et orth. cons.

Canarina abyssinica Engl., Bot. Jahrb. Syst. 32: 116 (1902).
  Sudan to Tanzania. 24 ETH SUD 25 KEN TAN UGA. Cl. tuber geophyte. $2n = 34$.
  Canarina abyssinica var. *umbrosa* Engl., Bot. Jahrb. Syst. 32: 116 (1902).

Canarina canariensis (L.) Vatke, Linnaea 38: 700 (1874).
  Canary Is. (Palma, Gomera, Tenerife, Gran Canaria); cult. 21 CNY. Cl. tuber geophyte. $2n = 34$.
  ** Campanula canariensis* L., Sp. Pl.: 168 (1753). *Canarina campanula* L., Mant. Pl.: 225, 588 (1771). *Campanula hastifolia* Salisb., Prodr. Stirp. Chap. Allerton: 127 (1796), as 'hastaefolia'. *Mindium canariense* (L.) Raf., Fl. Tellur. 2: 79 (1837).
  Canarina canariensis var. *angustifolia* Kunkel, Cuad. Bot. Canar. 28: 56 (1977).

Canarina eminii Asch. ex Schweinf., Sitzungsber. Ges. Naturf. Freunde Berlin 1892: 173 (1893).
  Ethiopia to Zaïre & Malawi. 23 BUR RWA ZAI 24 ETH SUD 25 KEN TAN UGA 26 MLW. Cl. tuber geophyte. $2n = 34$.
  Canarina elegantissima T. C. E. Fr., Notizbl. Bot. Gart. Berlin-Dahlem 8: 392 (1923).
  Canarina eminii var. *elgonensis* T. C. E. Fr., Notizbl. Bot. Gart. Berlin-Dahlem 8: 395 (1923).

Synonyms:
Canarina abyssinica var. *umbrosa* Engl. === **Canarina abyssinica** Engl.
Canarina campanula L. === **Canarina canariensis** (L.) Vatke
Canarina canariensis var. *angustifolia* Kunkel === **Canarina canariensis** (L.) Vatke

Canarina elegantissima T. C. E. Fr. === **Canarina eminii** Asch. ex Schweinf.
Canarina eminii var. *elgonensis* T. C. E. Fr. === **Canarina eminii** Asch. ex Schweinf.
Canarina moluccana Roxb. === **Cyclocodon lancifolius** (Roxb.) Kurz ssp. **lancifolius**
Canarina zanguebar Lour. === ?

Canonanthus

The name *Canonanthus,* like *Byrsanthes* (q.v.), applies to the small group of species treated by McVaugh (1949) as *Siphocampylus* subsect. *Nivei.* Type [by monotypy]: *Canonanthus campanulatus* G. Don.

> Don, G. (1834). *Canonanthus.* In A general history of the dichlamydeous plants 3: 698, 718-719. London: J. G. & F. Rivington. En. — Establishment of genus as distinct from *Lobelia.*
>
> McVaugh, R. (1949). Studies in Studies in South American Lobelioideae (Campanulaceae) with special reference to Colombian species. II. A new subsection of *Siphocampylus* sect. *Eusiphocampylus.* Brittonia 6: 453-458. En. — Treatment of group at infrageneric rank, with key, descriptions, full nomenclature, and specimen citations.

Synonyms:
Canonanthus G. Don === **Siphocampylus** Pohl
Canonanthus campanulatus G. Don === **Siphocampylus affinis** (Mirb.) McVaugh

Cardinalis

Like *Dortmanna* and *Rapuntium* (q.v.), this name represents a revival of a pre-Linnaean name for *Lobelia,* because of the view that that name was best applied to a species of Goodeniaceae. Based on: *Lobelia* L.

> Fabricius, P. C. (1759). Enumeratio methodica plantarum horti medici Helmstadiensis. Helmstadt: Johann Drimbiorn. La. — Establishment of name.

Synonyms:
Cardinalis Riv. ex Fabr. === **Lobelia** L.

Cenekia

This genus was segregated from *Campanula* to accomodate species with long reflexed calyx lobes, campanulate corolla, and three-chambered 10-angled capsules dehiscent toward the base. Type [designated here]: *Cenekia rapunculoides* (L.) Opiz.

> Berchtold, F. (1839). Oekonomisch-technische flora Böhmens, band 2, abteil. 2. Prague: Thomas Thabor. Ge. — Segregation of genus from *Campanula.*

Synonyms:
Cenekia Opiz === **Campanula** L.
Cenekia rapunculoides (L.) Opiz === **Campanula rapunculoides** L.
Cenekia rapunculoides var. *racemosa* (Peterm.) Opiz === **Campanula rapunculoides** L.
Cenekia rapunculoides var. *reflexa* (Peterm.) Opiz === **Campanula rapunculoides** L.
Cenekia rapunculoides var. *speciosa* (Knaf) Opiz === **Campanula rapunculoides** L.
Cenekia simplex (DC.) Opiz === **Campanula bononiensis** L.

Centropogon

Lobelioideae, 212 species, Neotropics, from southern Mexico to Bolivia and Brazil, with two species in the Lesser Antilles; the greatest concentration of species is in the northern Andes. The genus is divided into subg. *Centropogon* and subg. *Siphocampyloides* (Benth.) Wilbur (Wilbur 1976). The former subgenus was divided into four sections by Stein (1987), but those names have not been validated yet. The latter subgenus is divided into four sections: sect. *Burmeisteroides* Gleason, sect. *Niveopsis* Lammers, sect. *Siphocampyloides* Benth., and sect. *Wimmeriopsis* McVaugh; the last two are divided into two subsections each (McVaugh 1949, Lammers 1998). Morphological data support a close relationship with *Burmeistera* and especially *Siphocampylus* (Lammers 1998, 2002; Buss et al. 2001; Batterman & Lammers 2004). This relationship is supported by preliminary analyses of molecular data (E. Knox, pers. comm.; A. Antonelli, pers. comm.); in fact, once more data are available, it will undoubtedly prove necessary to redraw the boundaries of *Centropogon* and *Siphocampylus* or to merge them entirely. The most recent monograph is that by Wimmer (1943, 1953, 1968). $2n = 28$. Type [designated by Pfeiff., Nomencl. Bot. 1: 650 (1873)]: *Lobelia surinamensis* L.

Presl, C. (1836). *Centropogon*. In Prodromus monographiae Lobeliacearum: 48-49. Prague: Theophilus Haase. La. — Establishment of genus as distinct from *Lobelia*.

Hooker, W. J. (1845). *Siphocampylos* [sic] *coccineus*. Bot. Mag. 71: tab. 4178 + 2 pp. text, illus. En. — Portrait of *C. coccineus* with description; erroneously said to be from Brazil.

Hooker, W. J. (1847). *Siphocampylos* [sic] *glandulosa*. Bot. Mag. 71: tab. 4331 + 2 pp. text, illus. En. — Portrait of *C. glandulosus*, with description.

Decaisne, J. (1848). *Centropogon glandulosus* Dne. Rev. Hort. (ser. 3) 2: 241-422 + fig. 22. illus. Fr. — Portrait with description.

Houllet, R. J. B. (1868). *Centropogon* hybridus *lucyanus*. Rev. Hort. 40: 291 + 1 pl., illus. Fr. — Portrait and description of alleged artificial hybrid between *C. cornutus* and *Siphocampylus betulifolius*.

Gleason, H. A. (1921). A rearrangement of the Bolivian species of *Centropogon* and *Siphocampylus*. Bull. Torrey Bot. Club 48: 189-201, illus. En. — Floristic treatment.

• Gleason, H. A. (1924). Studies on the flora of northern South America. I. *Centropogon*, section *Burmeisteroides*. Bull. Torrey Bot. Club 51: 443-448, illus. En. — Establishment of new section, with key to its species.

• Gleason, H. A. (1925). Studies on the flora of northern South America. II. The stellate-tomentose species of *Centropogon*. Bull. Torrey Bot. Club 51: 443-448, illus. En. — Taxonomic treatment of a major group of species, with key, full nomenclature, and specimen citations.

Standley, P. C. (1938). Flora of Costa Rica, *Centropogon* Presl. Field Mus. Nat. Hist., Bot. Ser. 18: 1409-1414. En. — Floristic treatment with descriptions.

• McVaugh, R. (1943). *Centropogon* Presl. In North American Flora 32A: 114-127. New York: New York Botanical Garden. En. — Flora with key, descriptions, infrageneric classification, and full nomenclature.

• Wimmer, F. E. (1943). *Centropogon* Presl. In A. Engler & L. Diels, Das Pflanzenreich IV. 276b: 161-260 + 4 plates, illus. Leipzig: Wilhelm Engelmann. Ge., La. — Monograph with keys, descriptions, infrageneric classification, full nomenclature, and specimen citations.

• McVaugh, R. (1949). Studies in South American Lobelioideae (Campanulaceae) with special reference to Colombian species. III. A suggested reclassification of *Centropogon*, including revisions of the sections *Wimmeriopsis* and *Burmeisteroides*. Brittonia 6: 458-492. En. — Revision of infrageneric classification, including keys to species of subsect. *Colombiani* McVaugh and sect. *Burmeisteroides*, full nomenclature, and specimen citations.

• Wimmer, F. E. (1953). *Centropogon*. In A. Engler & L. Diels, Das Pflanzenreich IV. 276b: 768-774. Berlin: Akademie-Verlag. Ge., La. — Supplement to Wimmer (1943).

Esson, J. G. (1955). *Centropogon* hybridus *lucyanus*. Addisonia 23(2): 3-4 + pl. 746, illus. En. — Portrait, description, and notes on cultivation techniques of alleged artificial hybrid between *C. cornutus* and *Siphocampylus betulifolius*.

- Wimmer, F. E. (1968). *Centropogon* Presl. In A. Engler & L. Diels, Das Pflanzenreich IV. 276c: 838-843 + Taf. 14-15, illus. Berlin: Akademie-Verlag. Ge., La. — Second supplement to Wimmer (1943).

Colwell, R. K., B. J. Betts, P. Bunnell, F. L. Carpenter & P. Feinsinger (1974). Competition for the nectar of *Centropogon valerii* by the hummingbird *Colibri thalassinus* and the flower-piercer *Diglossa plumbea,* and its evolutionary implications. Condor 76: 447-452. En. — Species is major food source in Costa Rica for nectarivorous birds and nectar mite *Rhinoseius colwelli.*

Belem, C. I. F. (1976). Descrição palinológica de espécies de Campanulaceae dos gêneros *Centropogon* e *Siphocampylus.* Revista Brasil. Biol. 36: 861-870, illus. Por. — Palynological survey of Brazilian species.

- Nash, D. L. (1976). Flora of Guatemala, *Centropogon* Presl. Fieldiana Bot. 24: 403-408, illus. En. — Floristic treatment with key and descriptions.

- Wilbur, R. L. (1976). A synopsis of the Costa Rican species of the genus *Centropogon* Presl (Campanulaceae, Lobelioideae). Brenesia 8: 59-59-84. En. — Floristic treatment with key, descriptions, full nomenclature, and specimen citations.

- Wilbur, R. L. (1977). Flora of Panama, *Centropogon.* Ann. Missouri Bot. Gard. 63: 612-632, illus. En. — Floristic treatment with key and descriptions.

Fournet, J. (1978). *Centropogon* Presl. In Flore illustrée des phanérogames de Guadeloupe et de Martinique: 595-596. Paris: Institut National de la Recherche Agronomique. Fr. — Flora with key and descriptions.

- Jeppesen, S. (1981). *Centropogon* Presl. In G. Harling & B. Sparre (eds.), Flora of Ecuador 14: 48-122, illus. Stockholm: Swedish Natural Science Research Council. En. — Flora with keys, descriptions, full nomenclature, and specimen citations.

Schultes, R. E. (1983). De plantis toxicariis e mundo novo tropicale commentationes XXXII. Bot. Mus. Leafl. 29: 251-272. En. — Extract of *C. ferrugineus* used to treat dysentery.

Colwell, R. K. (1985). Stowaways on the hummingbird express. Nat. Hist. 94(7): 56-63, illus. En. — Nectar mite *Rhinoseius colwelli* dispersed among *Centropogon* flowers by pollinating hummingbirds.

- Stein, B. A. (1987) Systematics and evolution of *Centropogon* subgenus *Centropogon* (Campanulaceae: Lobelioideae). viii + 465 pp., illus., maps. St. Louis: Washington University. En. — Unpublished Ph.D. dissertation with keys, descriptions, full nomenclature, and specimen citations; several new taxa described here are not yet validly published.

Howard, R. A. (1989). *Centropogon* Presl. In Flora of the Lesser Antilles 6: 497-500, illus. Jamaica Plain: Arnold Arboretum. En. — Flora with key and descriptions.

Koptur, S., E. N. Dávila, D. R. Gordon, B. J. Davis McPhail, C. G. Murphy & J. B. Slowinski (1991). The effect of pollen removal on the duration of the staminate phase of *Centropogon talamancensis.* Brenesia 33: 15-18, illus. En. — Experimental removal of pollen reduces duration of male phase of flower and hastens onset of female phase.

Harvey, Y. (1992). *Centropogon cornutus.* Kew Mag. 9: 3-7 + pl. 187, illus. En. — Portrait with description, full nomenclature, and notes on cultivation.

Huxley, A. J., ed. (1992). *Centropogon.* In The new Royal Horticultural Society dictionary of gardening 1: 556. London: MacMillan. En. — Account of species in cultivation.

Stein, B. A. (1992). Sicklebill hummingbirds, ants, and flowers: plant animal interactions and evolutionary relationships in Andean Lobeliaceae. BioScience 42: 27-33, illus. En. — Impact of mutualisms with animals on phylogeny of genus.

Bernardello, L. M., L. Galetto, J. Jaramillo & E. Grijalba (1994). Floral nectar composition of some species from Reserva Río Guajalito, Ecuador. Biotropica 26: 113-116. En. — Nectar of *C. solanifolius* is sucrose-dominant, consistent with hummingbird pollination.

Weiss, M. R. (1996). Pollen-feeding fly alters floral phenotypic gender in *Centropogon solanifolius* (Campanulaceae). Biotropica 28: 770-773, illus. En. — Larvae of drosophilid *Zygothrica neolinea* infest anther tubes, eating pollen; this reduces duration of male phase of flower and hastens onset of female phase.

- Lammers, T. G. (1998). Review of the neotropical endemics *Burmeistera, Centropogon,* and *Siphocampylus* (Campanulaceae: Lobeliodeae), with descriptionr of 18 new species and a new section. Brittonia 50: 233-262, illus. En. — Summary of current knowledge of genus and each of its infrageneric taxa.
- Buss, C. C., T. G. Lammers & R. R. Wise (2001). Seed coat morphology and its systematic implications in *Cyanea* and other genera of Lobelioideae (Campanulaceae). Amer. J. Bot. 88: 1301-1308, illus. En. — Testal pattern supports close relationship of genus to *Siphocampylus.*
- Wilbur, R. L. (2001). *Centropogon* C. Presl. In W. D. Stevens, C. Ulloa U., A. Pool & O. M. Montiel (eds.), Flora de Nicaragua 1: 559-560. St. Louis: Missouri Botanical Garden Press. Sp. — Flora with key and descriptions.

    Lammers, T. G. (2002). Seventeen new species of Lobelioideae (Campanulaceae) from South America. Novon 12: 206-233, illus. En. — Update of Lammers (1998).
- Batterman, M. R. W., & T. G. Lammers (2004). Branched foliar trichomes of Lobelioideae (Campanulaceae) and the infrageneric classification of *Centropogon.* Syst. Bot. 29: 448-458, illus. En. — Study via scanning electron microscopy of branched hairs supports infrageneric classification.

Centropogon C. Presl, Prodr. Monogr. Lobel.: 48 (1836). *Lobelia* [unranked] *Centropogon* (C. Presl) Heynh., Nom. Bot. Hort. 1: 470 (1840).
Trop. America. 79 80 81 82 83 84.

Centropogon acrodentatus E. Wimm., Repert. Spec. Nov. Regni Veg. 19: 241 (1924).
Venezuela. 82 VEN. Nanophan.

Centropogon aequatorialis E. Wimm., Repert. Spec. Nov. Regni Veg. 19: 241 (1924).
Ecuador. 83 ECU. Nanophan. or cham.

Centropogon alatus Gleason, Bull. Torrey Bot. Club 52: 12 (1925).
Venezuela. 82 VEN. Nanophan.

Centropogon albertinus E. Wimm., Repert. Spec. Nov. Regni Veg. 38: 15 (1935).
Colombia. 83 CLM. Nanophan.

Centropogon albolimbatus E. Wimm., Repert. Spec. Nov. Regni Veg. 19: 242 (1924).
Colombia. 83 CLM. Nanophan.
Centropogon serratus Gleason, Bull. Torrey Bot. Club 52: 54 (1925). *Centropogon albolimbatus* var. *serratus* (Gleason) E. Wimm., Pflanzenr. IV.276b: 187 (1943).
Centropogon albolimbatus var. *concolor* E. Wimm., Repert. Spec. Nov. Regni Veg. 29: 63 (1931).
Centropogon albolimbatus var. *villosus* E. Wimm., Pflanzenr. IV.276b: 187 (1943).

Centropogon albostellatus Jeppesen in Harling & Sparre, Fl. Ecuador 14: 56 (1981).
Ecuador. 83 ECU. Nanophan.

Centropogon alens E. Wimm., Repert. Spec. Nov. Regni Veg. 29: 65 (1931).
Colombia. 83 CLM. Nanophan.

Centropogon alsophilus E. Wimm. in J. F. Macbr., Fl. Peru 6: 402 (1937).
Ecuador & Peru. 83 ECU PER. Nanophan. or cham.

Centropogon altus E. Wimm. in J. F. Macbr., Fl. Peru 6: 402 (1937).
Peru. 83 PER. Nanophan.

Centropogon amplicorollinus (E. Wimm.) B. A. Stein in Brako & Zarucchi, Cat. Fl. Pl. Gymnosp. Peru: 1254 (1993).
Colombia & Peru. 83 CLM PER. Nanophan. or cham.
* *Centropogon planchonis* var. *amplicorollinus* E. Wimm., Pflanzenr. IV.276b: 184 (1943).

Centropogon aquilinus E. Wimm., Pflanzenr. IV.276b: 259 (1943).
Bolivia. 83 BOL. Nanophan.
Centropogon aquilinus var. *campestris* E. Wimm., Pflanzenr. IV.276b: 260 (1943).
Centropogon aquilinus var. *integer* E. Wimm., Pflanzenr. IV.276b: 260 (1943).

Centropogon arachnocalyx Lammers, Brittonia 50: 246 (1998).
Colombia. 83 CLM. Nanophan.

Centropogon arcuatus E. Wimm., Repert. Spec. Nov. Regni Veg. 19: 242 (1924).
Ecuador. 83 ECU. Nanophan. or cham.

Centropogon argutus E. Wimm., Repert. Spec. Nov. Regni Veg. 29: 77 (1931).
Ecuador & Peru. 83 ECU PER. Cl. phan.

Centropogon asclepiadeus (Willd. ex Schult.) E. Wimm., Repert. Spec. Nov. Regni Veg. 22: 198 (1926).
Colombia. 83 CLM. Nanophan.
* *Lobelia asclepiadea* Willd. ex Schult. in Roem. & Schult., Syst. Veg. 5: 57 (1819).

Centropogon astrotrichus E. Wimm., Notizbl. Bot. Gart. Berlin-Dahlem 10: 735 (1929).
Peru. 83 PER. Nanophan.

Centropogon aurostellatus E. Wimm., Pflanzenr. IV.276b: 249 (1943).
Colombia. 83 CLM. Nanophan.

Centropogon australis (E. Wimm.) Gleason, Bull. Torrey Bot. Club 52: 12 (1925).
Venezuela. 82 VEN. Nanophan.
* *Centropogon grandidentatus* var. *australis* E. Wimm., Repert. Spec. Nov. Regni Veg. 19: 45 (1924).
Centropogon australis var. *taraxacoides* E. Wimm., Pflanzenr. IV.276b: 232 (1943).

Centropogon ayavacensis (Willd. ex Schult.) Lammers, Novon 9: 386 (1999).
Colombia. 83 CLM. Nanophan.
* *Lobelia ayavacensis* Willd. ex Schult. in Roem. & Schult., Syst. Veg. 5: 57 (1819). *Lobelia willdenowiana* C. Presl, Prodr. Monogr. Lobel.: 39 (1836); non Schult. in Roem. & Schult., Syst. Veg. 5: 634 (1819). *Siphocampylus umbellatus* var. *willdenowianus* A. DC. in DC., Prodr. 7: 406 (1839). *Siphocampylus umbellatus* var. *ayavacensis* (Willd. ex Schult.) Steud., Nomencl. Bot. (ed. 2) 2: 592 (1841). *Centropogon willdenowianus* (A. DC.) E. Wimm., Repert. Spec. Nov. Regni Veg. 22: 204 (1926).

subsp. **ayavacensis**
Colombia. 83 CLM. Nanophan.
Siphocampylus stellatus Gleason, Bull. Torrey Bot. Club 52: 67 (1925).

subsp. **cylindricus** (Gleason) Lammers, Novon 9: 386 (1999).
Colombia. 83 CLM. Nanophan.
* *Siphocampylus cylindricus* Gleason, Bull. Torrey Bot. Club 52: 66 (1925). *Centropogon cylindricus* (Gleason) E. Wimm., Pflanzenr. IV.276b: 257 (1943). *Centropogon willdenowianus* subsp. *cylindricus* (Gleason) McVaugh, Brittonia 6: 480 (1949).

Centropogon azuayensis Jeppesen in Harling & Sparre, Fl. Ecuador 14: 58 (1981).
Ecuador. 83 ECU. (Cl.) nanophan.

Centropogon baezanus Jeppesen in Harling & Sparre, Fl. Ecuador 14: 59 (1981).
Ecuador. 83 ECU. Cl. phan.

Centropogon balslevii Jeppesen in Harling & Sparre, Fl. Ecuador 14: 60 (1981).
Ecuador. 83 ECU. Nanophan. or cham.

Centropogon bangii Zahlbr., Bull. Torrey Bot. Club 24: 372 (1897), as 'bangi'.
Bolivia. 83 BOL. Cham.

Centropogon belliflorus E. Wimm., Repert. Spec. Nov. Regni Veg. 38: 13 (1935).
Colombia. 83 CLM. Nanophan.

Centropogon beniteziae Lammers, Novon 12: 215 (2002).
Venezuela. 82 VEN. Nanophan. or cham.

Centropogon berteroanus (Spreng.) A. DC. in DC., Prodr. 7: 345 (1839), as 'berterianus'.
Lesser Antilles (Guadeloupe, Dominica, St. Lucia). 81 LEE WIN. Nanophan.
* *Lobelia berteroana* Spreng., Syst. Veg. 1: 712 (1824), as 'berteriana'. *Siphocampylus berteroanus* (Spreng.) G. Don, Gen. Hist. 3: 703 (1834), as 'berterianus'.

Centropogon beslerioides (Kunth) A. DC. in DC., Prodr. 7: 346 (1839).
Colombia. 83 CLM. Nanophan.
* *Lobelia beslerioides* Kunth in Humb., Bonpl. & Kunth, Nov. Gen. Sp. 3: 306 (quarto), 238 (folio) (1819). *Siphocampylus beslerioides* (Kunth) G. Don, Gen. Hist. 3: 702 (1834).

Centropogon brachysiphoniatus Zahlbr., Repert. Spec. Nov. Regni Veg. 14: 139 (1915).
Ecuador. 83 ECU. Nanophan.
Centropogon mojandensis Benoist, Bull. Soc. Bot. France 84: 635 (1938).

Centropogon brittonianus Zahlbr., Bull. Torrey Bot. Club 24: 373 (1897).
Bolivia. 83 BOL. Nanophan.
* *Siphocampylus giganteus* var. *latifolius* Britton, Bull. Torrey Bot. Club 19: 373 (1892); non Vatke, Linnaea 38: 734 (1874).

Centropogon bruneotomentosus E. Wimm. in J. F. Macbr., Fl. Peru 6: 404 (1937), as 'bruneo-tomentosus'.
Peru. 83 PER. Nanophan.

Centropogon burmeisteroides McVaugh, Brittonia 6: 490 (1949).
Colombia. 83 CLM. Nanophan.

Centropogon calycinus Benth., Pl. Hartweg.: 212 (1845).
Ecuador. 83 ECU. Nanophan. or cham.
Centropogon tubulosus Zahlbr., Repert. Spec. Nov. Regni Veg. 14: 139 (1915). *Centropogon calycinus* var. *tubulosus* (Zahlbr.) E. Wimm., Biblioth. Bot. 116: 156 (1937).
Centropogon caligatus E. Wimm., Repert. Spec. Nov. Regni Veg. 22: 195 (1926).

Centropogon candidatus Lammers, Novon 12: 217 (2002).
Colombia. 83 CLM. Nanophan.

Centropogon caoutchouc (Kunth) E. Wimm., Repert. Spec. Nov. Regni Veg. 19: 391 (1924).
Columbia to Peru. 83 CLM ECU PER. (Cl.) nanophan.
* *Lobelia caoutchouc* Kunth in Humb., Bonpl. & Kunth, Nov. Gen. Sp. 3: 304 (quarto), 237 (folio) (1819). *Siphocampylus caoutchouc* (Kunth) G. Don, Gen. Hist. 3: 701 (1834).
Lobelia cautschuk Willd. ex Schult. in Roem. & Schult., Syst. Veg. 5: 58 (1819).

Centropogon capitatus Drake, J. Bot. (Morot) 3: 238 (1889).
Ecuador & Peru. 83 ECU PER. Nanophan. or cham.
Centropogon capitatus f. *hirtus* Zahlbr., Ann. K. K. Naturhist. Hofmus. 6: 436 (1891), as 'hirta'.
Centropogon capitatus var. *fieldii* E. Wimm. in J. F. Macbr., Fl. Peru 6: 406 (1937).
Centropogon capitatus var. *trichandrus* E. Wimm. in J. F. Macbr., Fl. Peru 6: 406 (1937).

Centropogon carnosus Zahlbr., Repert. Spec. Nov. Regni Veg. 13: 536 (1915).
Colombia. 83 CLM. Cham.
Centropogon carpinoides Gleason, Bull. Torrey Bot. Club 52: 62 (1925).

Centropogon carolinae E. Wimm., Pflanzenr. IV.276b: 228 (1943).
Colombia. 83 CLM. Nanophan.

Centropogon cazaletii Jeppesen in Harling & Sparre, Fl. Ecuador 14: 67 (1981).
Ecuador. 83 ECU. Nanophan. or cham.

Centropogon chiltasonensis Jeppesen in Harling & Sparre, Fl. Ecuador 14: 67 (1981).
Ecuador. 83 ECU. Cl. phan.

Centropogon chontalensis Jeppesen in Harling & Sparre, Fl. Ecuador 14: 69 (1981).
Ecuador. 83 ECU. Cl. nanophan.

Centropogon cinnabarinus E. Wimm., Repert. Spec. Nov. Regni Veg. 26: 9 (1929).
Colombia. 83 CLM. Nanophan.

Centropogon coccineus (Hook.) Regel ex B. D. Jacks., Index Kew. 1: 479 (1893).
Nicaragua to Colombia. 80 COS NIC PAN 83 CLM. Cham.
* *Siphocampylus coccineus* Hook., Bot. Mag. 71: tab. 4178 (1845).
Siphocampylus rugosus Poit. & Vilm., Rev. Hort. (ser. 2) 1: 186 (1842); non (C. Presl) A. DC. in DC., Prodr. 7: 399 (1839).
Siphocampylus coccineus var. *leucostoma* Van Houtte, Fl. Serres Jard. Eur. 7: 5 (1851). *Centropogon coccineus* var. *leucostoma* (Van Houtte) E. Wimm., Pflanzenr. IV.276b: 217 (1943).
Siphocampylus radicans Kuntze, Revis. Gen. Pl. 2: 381 (1891). *Centropogon radicans* (Kuntze) McVaugh, Ann. Missouri Bot. Gard. 27: 353 (1940).
Siphocampylus roseus Donn. Sm., Bot. Gaz. (Crawfordsville) 23: 249 (1897).

Centropogon comosus Gleason, Bull. Torrey Bot. Club 52: 13 (1925).
Ecuador. 83 ECU. (Cl.) nanophan.

Centropogon congestus Gleason, Bull. Torrey Bot. Club 52: 52 (1925). *Centropogon macrophyllus* var. *congestus* (Gleason) McVaugh, Ann. Missouri Bot. Gard. 27: 352 (1940).
Costa Rica to Colombia. 80 COS PAN 83 CLM. Nanophan. or cham.
Centropogon ovalifolius var. *asperatulus* Zahlbr., Repert. Spec. Nov. Regni Veg. 14: 137 (1915).
Centropogon gesnerioides Gleason, Bull. Torrey Bot. Club 52: 53 (1925).
Centropogon diocleus E. Wimm., Ann. Missouri Bot. Gard. 24: 209 (1937).
Centropogon ovalifolius var. *sneidernii* E. Wimm., Pflanzenr. IV.276b: 187 (1943).
Centropogon caninus var. *hirsutulus* E. Wimm., Pflanzenr. IV.276b: 769 (1953).

Centropogon connatilobatus Lammers, Brittonia 50: 248 (1998).
Venezuela. 82 VEN. Nanophan.

Centropogon cordifolius Benth., Pl. Hartweg.: 77 (1841). *Siphocampylus guatemalensis* Vatke, Linnaea 38: 730 (1874); non *Siphocampylus cordifolius* Otto & A. Dietr., Allg. Gartenzeitung 12: 371 (1844).
S. Mexico to Nicaragua. 79 MXS MXT 80 GUA HON NIC. Nanophan. or cham.
Centropogon cordatus M. Martens & Galeotti, Bull. Acad. Roy. Sci. Bruxelles 9(2): 40 (1842).

Centropogon cornutus (L.) Druce, Bot. Exch. Club Soc. Brit. Isles 3: 416 (1914).
Panama to Bolivia & SE. Brazil; Lesser Antilles (Antigua, Guadeloupe, Martinique, Grenada, Trinidad-Tobago); cult. 80 PAN 81 LEE TRT WIN 82 FRG GUY SUR VEN 83 BOL CLM ECU PER 84 BZC BZE BZL BZN. (Cl.) nanophan. or cham. $2n = 28$.

* *Lobelia cornuta* L., Sp. Pl.: 930 (1753).

Lobelia surinamensis L., Sp. Pl. (ed. 2): 1320 (1763). *Siphocampylus surinamensis* (L.) G. Don, Gen. Hist. 3: 702 (1834). *Centropogon surinamensis* (L.) C. Presl, Prodr. Monogr. Lobel.: 48 (1836).

Lobelia obscura L., Pl. Surin.: 14 (1775).

Lobelia laevigata L. f., Suppl. Pl.: 392 (1782). *Centropogon laevigatus* (L. f.) A. DC. in DC., Prodr. 7: 344 (1839). *Centropogon cornutus* var. *laevigatus* (L. f.) E. Wimm. in J. F. Macbr., Fl. Peru 6: 407 (1937).

Lobelia andropogon Cav., Anales Hist. Nat. 2: 106 (1800). *Siphocampylus andropogon* (Cav.) G. Don, Gen. Hist. 3: 703 (1834). *Centropogon andropogon* (Cav.) A. DC. in DC., Prodr. 7: 345 (1839).

Lobelia spectabilis Kunth in Humb., Bonpl. & Kunth, Nov. Gen. Sp. 3: 306 (quarto), 239 (folio) (1819). *Siphocampylus spectabilis* (Kunth) G. Don, Gen. Hist. 3: 702 (1834).

Lobelia bonplandiana Schult. in Roem. & Schult., Syst. Veg. 5: 57 (1819). *Centropogon bonplandianus* (Schult.) C. Presl, Prodr. Monogr. Lobel.: 48 (1836).

Lobelia purpurea Vell., Fl. Flumin., Icon. 8: pl. 156 (1831); non Breiter, Hort. Breiter.: 249 (1817); nec Lindl., Edward's Bot. Reg. 16: pl. 1325 (1830). *Centropogon cornutus* f. *vellozianus* E. Wimm., Pflanzenr. IV.276b: 200 (1943).

Siphocampylus macranthus Pohl, Pl. Bras. Icon. Descr. 2: 105 (1831).

Centropogon fastuosus Scheidw., Allg. Gartenzeitung 9: 396 (1841).

Centropogon oblongus Benth., Pl. Hartweg.: 214 (1845).

Centropogon surinamensis var. *angustifolius* Zahlbr., Ann. K. K. Naturhist. Hofmus. 6: 437 (1891). *Centropogon cornutus* var. *angustifolius* (Zahlbr.) E. Wimm., Pflanzenr. IV.276b: 201 (1943).

Centropogon surinamensis var. *vestitus* Pilger, Bot. Jahrb. Syst. 30: 200 (1901), as 'vestita'.

Centropogon intermedius Zahlbr., Repert. Spec. Nov. Regni Veg. 14: 135 (1915). *Centropogon bonplandianus* var. *intermedius* (Zahlbr.) E. Wimm., Pflanzenr. IV.276b: 197 (1943).

Centropogon puerilis E. Wimm., Repert. Spec. Nov. Regni Veg. 29: 68 (1931). *Centropogon cornutus* f. *leucostomus* E. Wimm. in J. F. Macbr., Fl. Peru 6: 407 (1937).

Centropogon bonplandiaus f. *glabrescens* E. Wimm., Pflanzenr. IV.276b: 197 (1943).

Centropogon cornutus f. *ynesae* E. Wimm., Pflanzenr. IV.276b: 200 (1943).

Centropogon cornutus f. *leucanthus* E. Wimm., Pflanzenr. IV.276b: 200 (1943).

Centropogon costaricae (Vatke) McVaugh, N. Amer. Fl. 32A: 121 (1943).
Costa Rica & Panama. 80 COS PAN. Nanophan. or cham.

* *Siphocampylus costaricae* Vatke, Linnaea 38: 730 (1874).

Centropogon porphyrodontus Donn. Sm., Bot. Gaz. (Crawfordsville) 44: 114 (1907).

Centropogon brumalis Standl., Fl. of Costa Rica: 1410 (1938).

Centropogon cordifolius var. dentatus E. Wimm. ex Standl., Fl. of Costa Rica: 1410 (1938). *Centropogon costaricae* f. *dentatus* (E. Wimm. ex Standl.) McVaugh, N. Amer. Fl. 32A: 121 (1943).

Centropogon cupreus E. Wimm., Repert. Spec. Nov. Regni Veg. 29: 70 (1931).
Colombia. 83 CLM. Nanophan.

Centropogon curvatus Gleason, Bull. Torrey Bot. Club 52: 55 (1925).
Colombia & Ecuador. 83 CLM ECU. Nanophan. or cham.

Centropogon colombiensis E. Wimm., Repert. Spec. Nov. Regni Veg. 22: 196 (1926).

Centropogon curvatus var. *pearcei* E. Wimm., Repert. Spec. Nov. Regni Veg. 29: 62 (1931).

Centropogon curvatus f. *minor* E. Wimm., Repert. Spec. Nov. Regni Veg. 29: 62 (1931).

Centropogon darienensis Wilbur, Ann. Missouri Bot. Gard. 63: 620 (1977).
Panama. 80 PAN. Cham.

Centropogon david-smithii Lammers, Brittonia 50: 255 (1998).
Peru. 83 PER. Nanophan.

Centropogon delicatulus McVaugh, Brittonia 9: 30 (1957).
Costa Rica. 80 COS. Nanophan. or cham.
Centropogon oaxacanus var. *berterioides* E. Wimm., Pflanzenr. IV.276c: 841 (1968).

Centropogon densiflorus Benth., Pl. Hartweg.: 138 (1845).
Ecuador. 83 ECU. Cl. phan.
Centropogon gracilis Drake, J. Bot. (Morot) 3: 238 (1889).

Centropogon dianae Lammers, Brittonia 50: 257 (1998).
Peru. 83 PER. Cham.

Centropogon dillonii Lammers, Brittonia 50: 243 (1998).
Peru. 83 PER. Nanophan. or cham.

Centropogon dissectus E. Wimm., Repert. Spec. Nov. Regni Veg. 22: 197 (1926).
Ecuador. 83 ECU. Nanophan. or cham.

Centropogon dombeyanus (C. Presl) E. Wimm. in J. F. Macbr., Fl. Peru 6: 408 (1937).
Peru. 83 PER. Nanophan.
* *Lobelia dombeyana* C. Presl, Prodr. Monogr. Lobel.: 39 (1836). *Siphocampylus dombeyanus* (C. Presl) A. DC. in DC., Prodr. 7: 406 (1839).

Centropogon eborinus E. Wimm., Repert. Spec. Nov. Regni Veg. 19: 244 (1924).
Colombia. 83 CLM. Nanophan.

Centropogon eilersii Lammers & M. O. Dillon, Novon 12: 218 (2002).
Peru. 83 PER. Nanophan.

Centropogon ellipticus Gleason, Bull. Torrey Bot. Club 52: 8 (1925).
Colombia. 83 CLM. Nanophan. or cham.

Centropogon elmanus E. Wimm., Repert. Spec. Nov. Regni Veg. 19: 389 (1924).
Venezuela. 82 VEN. Nanophan.

Centropogon erianthus (Benth.) Benth. & Hook. f. ex Drake, J. Bot. (Morot) 3: 237 (1889).
Ecuador & Peru. 83 ECU PER. Cl. phan.
* *Siphocampylus erianthus* Benth., Pl. Hartweg.: 139 (1845).
Centropogon erianthus var. *brachysepalus* E. Wimm. in J. F. Macbr., Fl. Peru 6: 409 (1937).
Centropogon erianthus var. *diversipilis* E. Wimm., Pflanzenr. IV.276c: 842 (1968).

Centropogon erythraeus Drake, J. Bot. (Morot) 3: 237 (1889).
Ecuador. 83 ECU. Nanophan.

Centropogon eurystomus E. Wimm., Repert. Spec. Nov. Regni Veg. 29: 60 (1931).
Ecuador. 83 ECU. Cham.

Centropogon ewanii E. Wimm., Brittonia 8: 110 (1955).
Venezuela. 82 VEN. Nanophan.

Centropogon exsertus Lammers, Brittonia 50: 248 (1998).
Colombia. 83 CLM. Nanophan.

Centropogon featherstonei Gleason, Bull. Torrey Bot. Club 52: 17 (1925).
Peru. 83 PER. Cl. phan.

Centropogon ferrugineus (L. f.) Gleason, Bull. Torrey Bot. Club 52: 11 (1925).
Guatemala to Peru. 80 COS ELS GUA PAN 82 VEN 83 CLM ECU PER. Cl. nanophan.

Lobelia ferruginea L. f., Suppl. Pl.: 394 (1782). *Siphocampylus ferrugineus* (L. f.) G. Don, Gen. Hist. 3: 701 (1834).

Lobelia barbata Cav., Icon. 6: 12 (1800). *Siphocampylus barbatus* (Cav.) G. Don, Gen. Hist. 3: 701 (1834). *Centropogon barbatus* (Cav.) Planch., Fl. Serres Jard. Eur. 6: 16 (1850).

Centropogon costaricanus Planch. & Oerst., Vidensk. Meddel. Dansk Naturhist. Foren. Kjøbenhavn 1857: 156 (1857). *Centropogon affinis* var. *costaricanus* (Planch. & Oerst.) Zahlbr., Ann. K. K. Naturhist. Hofmus. 6: 437 (1891). *Centropogon ferrugineus* var. *costaricanus* (Planch. & Oerst.) McVaugh, Ann. Missouri Bot. Gard. 27: 352 (1940).

Siphocampylus regelii Vatke, Linnaea 38: 732 (1874).

Siphocampylus regelii var. *umbrosus* Vatke, Linnaea 38: 733 (1874).

Centropogon affinis var. *venezuelanus* E. Wimm., Repert. Spec. Nov. Regni Veg. 19: 242 (1924). *Centropogon ferrugineus* var. *venezuelanus* (E. Wimm.) McVaugh, Ann. Missouri Bot. Gard. 27: 351 (1940).

Centropogon poasensis Gleason, Torreya 25: 92 (1925).

Centropogon jahnii Gleason, Bull. Torrey Bot. Club 52: 10 (1925). *Centropogon ferrugineus* var. *jahnii* (Gleason) E. Wimm., Repert. Spec. Nov. Regni Veg. 22: 199 (1926).

Centropogon ferrugineus var. *glabrifilaris* E. Wimm., Repert. Spec. Nov. Regni Veg. 22: 199 (1926).

Centropogon costaricanus var. *tomentellus* E. Wimm., Ann. Naturhist. Mus. Wien 46: 240 (1933). *Centropogon ferrugineus* var. *tomentellus* (E. Wimm.) McVaugh, N. Amer. Fl. 32A: 126 (1943).

Centropogon costaricanus var. *cufodontidis* E. Wimm., Ann. Naturhist. Mus. Wien 46: 240 (1933).

Centropogon ferrugineus var. *nitens* E. Wimm., Pflanzenr. IV.276b: 225 (1943).

Centropogon fimbriatulus McVaugh, Brittonia 6: 471 (1949).
Ecuador. 83 ECU. Cham.

Centropogon floccosus Planch., Fl. Serres Jard. Eur. 6: 16 (1850).
Colombia. 83 CLM. Nanophan.

Centropogon floricomus McVaugh, J. Wash. Acad. Sci. 39: 160 (1949).
Panama. 80 PAN. Cham.

Centropogon flos-mutisii E. Wimm., Trab. Mus. Nac. Ci. Nat., Ser. Bot. 26: 27 (1933).
Colombia. 83 CLM. Nanophan.
Centropogon flos-mutisii f. *fuscoviridis* E. Wimm., Trab. Mus. Nac. Ci. Nat., Ser. Bot. 26: 28 (1933), as 'fusco-viridis'.
Centropogon flos-mutisii f. *haticensis* E. Wimm., Repert. Spec. Nov. Regni Veg. 38: 15 (1935).

Centropogon foetidus (Kunth) Zahlbr., Repert. Spec. Nov. Regni Veg. 13: 536 (1915).
Colombia. 83 CLM. Nanophan.
Lobelia foetida Kunth in Humb., Bonpl. & Kunth, Nov. Gen. Sp. 3: 305 (quarto), 238 (folio) (1819). *Siphocampylus foetidus* (Kunth) G. Don, Gen. Hist. 3: 702 (1834).
Centropogon rex E. Wimm., Repert. Spec. Nov. Regni Veg. 26: 8 (1929).

Centropogon foliosus Rusby, Descr. S. Amer. Pl.: 146 (1920), as 'foliosum'.
Colombia. 83 CLM. Nanophan.

Centropogon fulvus Gleason, Bull. Torrey Bot. Club 52: 7 (1925).
Colombia. 83 CLM. Cl. nanophan.

Centropogon fuscus (G. Don) E. Wimm., Repert. Spec. Nov. Regni Veg. 22: 201 (1926).
Peru. 83 PER. Nanophan.
Siphocampylus fuscus G. Don, Gen. Hist. 3: 704 (1834). *Lobelia fusca* (G. Don) C. Presl, Prodr. Monogr. Lobel.: 40 (1836).

Centropogon gamosepalus Zahlbr., Ann. K. K. Naturhist. Hofmus. 6: 434 (1891).
Ecuador & Peru. 83 ECU PER. Nanophan. or cham.
Centropogon grandicephalus Zahlbr., Bot. Jahrb. Syst. 37: 454 (1906).

Centropogon gesneriformis Drake, J. Bot. (Morot) 3: 239 (1889), as 'gesneraeformis'.
Ecuador & Peru. 83 ECU PER. Nanophan. or cham. $2n$ = 28.

Centropogon glabrifilis (E. Wimm.) Jeppesen in Harling & Sparre, Fl. Ecuador 14: 83 (1981).
Colombia & Ecuador. 83 CLM ECU. Nanophan.
* *Centropogon floccosus* f. *glabrifilis* E. Wimm., Pflanzenr. IV.276c: 842 (1968).

Centropogon glandulosus (Hook.) Decne., Rev. Hort. (ser. 3) 2: 421 (1848).
Colombia. 83 CLM. Cham.
* *Siphocampylus glandulosus* Hook., Bot. Mag. 73: tab. 4331 (1847), as 'glandulosa'.
Siphocampylus glandulosus var. *ovatus* E. Wimm., Pflanzenr. IV.276b: 212 (1943).

Centropogon glaucotomentosus E. Wimm., Repert. Spec. Nov. Regni Veg. 29: 73 (1931).
Colombia. 83 CLM. (Cl.) nanophan.
Centropogon erianthus var. *stella* E. Wimm., Repert. Spec. Nov. Regni Veg. 29: 76 (1931).
Centropogon glaucotomentosus var. *correana* E. Wimm., J. Wash. Acad. Sci. 40: 354 (1950).

Centropogon gleasonii (E. Wimm.) E. Wimm., Pflanzenr. IV.276b: 243 (1943).
Peru. 83 PER. Nanophan.
* *Centropogon fuscus* var. *gleasonii* E. Wimm., Repert. Spec. Nov. Regni Veg. 38: 16 (1935).

Centropogon gloriosus (Britton) Zahlbr., Bull. Torrey Bot. Club 24: 373 (1897).
Bolivia. 83 BOL. Nanophan.
* *Siphocampylus gloriosus* Britton, Bull. Torrey Bot. Club 19: 373 (1892).

Centropogon grandidentatus (Schltdl.) Zahlbr., Ann. K. K. Naturhist. Hofmus. 6: 439 (1891).
C. Mexico to Venezuela. 79 MXC MXG MXS MXT 80 COS ELS GUA HON PAN 82 VEN 83 CLM. Nanophan. or cham.
* *Lobelia grandidentata* Schltdl., Linnaea 9: 262 (1834). *Siphocampylus grandidentatus* (Schltdl.) D. Dietr., Syn. Pl. 1: 737 (1839), as 'quadridentatus'.
Centropogon affinis M. Martens & Galeotti, Bull. Acad. Roy. Sci. Bruxelles 9(2): 40 (1842), as 'affine'.
Laurentia insignis Brandegee, Univ. Calif. Publ. Bot. 4: 388 (1913).
Centropogon grandidentatus var. *diversidens* E. Wimm., Repert. Spec. Nov. Regni Veg. 19: 244 (1924).
Centropogon grandidentatus f. *incisus* E. Wimm., Repert. Spec. Nov. Regni Veg. 19: 244 (1924), as 'incisa'.
Siphocampylus oaxacanus E. Wimm., Repert. Spec. Nov. Regni Veg. 19: 260 (1924).
Centropogon oaxacanus (E. Wimm.) E. Wimm., Pflanzenr. IV.276b: 217 (1943).
Centropogon grandidentatus f. *denticulatus* E. Wimm., Pflanzenr. IV.276b: 220 (1943).
Centropogon grandidentatus f. *vulcanicus* E. Wimm., Pflanzenr. IV.276c: 841 (1968).

Centropogon grandis (L. f.) C. Presl, Prodr. Monogr. Lobel.: 48 (1836).
Colombia. 83 CLM. Nanophan. or cham.
* *Lobelia grandis* L. f., Suppl. Pl.: 394 (1782). *Siphocampylus grandis* (L. f.) G. Don, Gen. Hist. 3: 702 (1834).
Centropogon grandis var. *hirtellus* E. Wimm., Repert. Spec. Nov. Regni Veg. 29: 62 (1931).
Centropogon rostellatus E. Wimm., Repert. Spec. Nov. Regni Veg. 29: 63 (1931).
Centropogon aristi E. Wimm., Repert. Spec. Nov. Regni Veg. 38: 10 (1935).
Centropogon aristi var. *subterpilis* E. Wimm., Brittonia 8: 110 (1955).

Centropogon granulosus C. Presl, Prodr. Monogr. Lobel.: 49 (1836).
Nicaragua to Bolivia. 80 COS NIC PAN 82 VEN 83 BOL CLM ECU PER. Cl. nanophan.
$2n = 28$.
Centropogon cuspidatus A. DC. in DC., Prodr. 7: 346 (1839). *Centropogon granulosus* var.
cuspidatus (A. DC.) E. Wimm., Repert. Spec. Nov. Regni Veg. 29: 67 (1931).
Centropogon nutans Planch. & Oerst., Vidensk. Meddel. Dansk Naturhist. Foren.
Kjøbenhavn 1857: 156 (1857).
Centropogon warscewiczii Vatke, Linnaea 38: 716 (1874); non Van Houtte ex Regel,
Gartenflora 7: 374 (1858).
Siphocampylus aggregatus Rusby, Bull. New York Bot. Gard. 8: 122 (1912), as
'aggregata'. *Centropogon aggregatus* (Rusby) Gleason, Bull. Torrey Bot. Club 48: 199
(1921). *Centropogon granulosus* var. *aggregatus* (Rusby) E. Wimm., Pflanzenr.
IV.276b: 170 (1943).
Centropogon cardinalis Zahlbr. & Rech., Meded. Rijks-Herb. 19: 51 (1913). *Centropogon
aggregatus* var. *cardinalis* (Zahlbr. & Rech.) E. Wimm. in J. F. Macbr., Fl. Peru 6: 402
(1937). *Centropogon granulosus* f. *cardinalis* (Zahlbr. & Rech.) E. Wimm., Pflanzenr.
IV.276b: 170 (1943).
Centropogon parvulus Gleason, Bull. Torrey Bot. Club 52: 56 (1925).
Centropogon densiflorus var. *lugens* E. Wimm., Repert. Spec. Nov. Regni Veg. 22: 196 (1926).
Centropogon granulosus var. *rutilus* E. Wimm., Repert. Spec. Nov. Regni Veg. 22: 201 (1926).
Centropogon tortilis E. Wimm., Repert. Spec. Nov. Regni Veg. 22: 217 (1926).
Centropogon casapiensis E. Wimm., Repert. Spec. Nov. Regni Veg. 29: 66 (1931).
Centropogon holtonis E. Wimm., Repert. Spec. Nov. Regni Veg. 29: 66 (1931).
Centropogon erastus E. Wimm., Repert. Spec. Nov. Regni Veg. 38: 8 (1935).
Centropogon holtonis var. *albanensis* E. Wimm., Repert. Spec. Nov. Regni Veg. 38: 8
(1935).
Centropogon augostanus E. Wimm., Repert. Spec. Nov. Regni Veg. 38: 9 (1935).
Centropogon cumulatus E. Wimm., Pflanzenr. IV.276b: 178 (1943).
Centropogon vinosus E. Wimm., Pflanzenr. IV.276b: 179 (1943).
Centropogon planchonis var. *heteranthus* E. Wimm., Pflanzenr. IV.276b: 184 (1943).
Centropogon lateriflorus E. Wimm., Pflanzenr. IV.276b: 193 (1943).
Centropogon panamensis Wilbur, Ann. Missouri Bot. Gard. 63: 628 (1977).

Centropogon gutierrezii (Planch. & Oerst.) E. Wimm., Repert. Spec. Nov. Regni Veg. 22:
202 (1926).
Costa Rica & Panama. 80 COS PAN. Cham.
* *Siphocampylus gutierrezii* Planch. & Oerst., Vidensk. Meddel. Dansk Naturhist. Foren.
Kjøbenhavn 1857: 155 (1857).
Siphocampylus thysanopetalus Vatke, Linnaea 38: 731 (1874).

Centropogon hartwegii (Benth.) Benth. & Hook. f. ex B. D. Jacks., Index Kew. 2: 1274
(1895), as 'hartwegi'.
Colombia & Ecuador. 83 CLM ECU. Nanophan.
* *Siphocampylus hartwegii* Benth., Pl. Hartweg.: 139 (1845), as 'hartwegi'.
Centropogon hartwegii var. *parvus* E. Wimm., Repert. Spec. Nov. Regni Veg. 29: 76 (1931).

Centropogon hazenii (Gleason) E. Wimm., Repert. Spec. Nov. Regni Veg. 22: 202 (1926).
Colombia. 83 CLM. Nanophan.
* *Siphocampylus hazenii* Gleason, Bull. Torrey Bot. Club 52: 68 (1925). *Centropogon
nigricans* var. *hazenii* (Gleason) E. Wimm., Pflanzenr. IV.276b: 773 (1953).

Centropogon herzogii Zahlbr. & Rech., Meded. Rijks-Herb. 19: 49 (1913), as 'herzogi'.
Bolivia. 83 BOL. Nanophan.

Centropogon heteropilis E. Wimm., Repert. Spec. Nov. Regni Veg. 29: 72 (1931).
Ecuador. 83 ECU. Nanophan. or cham.

Centropogon hirsutus Gleason, Bull. Torrey Bot. Club 52: 14 (1925).
Colombia. 83 CLM. Nanophan.

Centropogon hirtiflorus Drake, J. Bot. (Morot) 3: 239 (1889).
Ecuador. 83 ECU. Nanophan. or cham.

Centropogon hirtus (Cav.) C. Presl, Prodr. Monogr. Lobel.: 48 (1836).
Peru. 83 PER. Cham.
* *Lobelia hirta* Cav., Anales Hist. Nat. 2: 108 (1800). *Siphocampylus hirtus* (Cav.) G. Don, Gen. Hist. 3: 702 (1834).
Centropogon exasperatus C. Presl, Prodr. Monogr. Lobel.: 48 (1836).
Centropogon angustus Gleason, Bull. Torrey Bot. Club 52: 54 (1925).

Centropogon hyalinus McVaugh, Brittonia 6: 489 (1949).
Venezuela & Colombia. 82 VEN 83 CLM. Cham.

Centropogon hylephilus E. Wimm., Repert. Spec. Nov. Regni Veg. 29: 74 (1931).
Colombia. 83 CLM. Cham.

Centropogon hypotrichus E. Wimm. in J. F. Macbr., Fl. Peru 6: 415 (1937).
Peru. 83 CLM. Nanophan. or cham.

Centropogon incanus (Britton) Zahlbr., Bull. Torrey Bot. Club 24: 374 (1897).
Peru & Bolivia. 83 BOL PER. Nanophan.
* *Siphocampylus incanus* Britton, Bull. Torrey Bot. Club 19: 373 (1892).

Centropogon intonsus Gleason, Bull. Torrey Bot. Club 52: 9 (1925).
Ecuador & Peru. 83 ECU PER. Nanophan.
Centropogon ferrugineus var. *vulgatus* E. Wimm., Repert. Spec. Nov. Regni Veg. 22: 198 (1926).

Centropogon irazuensis Wilbur, Brittonia 24: 422 (1972).
Costa Rica. 80 COS. Nanophan. or cham.

Centropogon isabellinus E. Wimm., Repert. Spec. Nov. Regni Veg. 19: 245 (1924).
Peru. 83 PER. Nanophan.

Centropogon jeppesenii Lammers, Brittonia 50: 252 (1998).
Ecuador. 83 ECU. Cham.

Centropogon joergensenii Lammers, Novon 12: 219 (2002).
Ecuador. 83 ECU. (Cl.) nanophan.

Centropogon karstenii Zahlbr., Ann. K. K. Naturhist. Hofmus. 6: 436 (1891).
Colombia. 83 CLM. Nanophan.

Centropogon knoxii Lammers, Brittonia 50: 249 (1998).
Peru. 83 PER. Nanophan.

Centropogon lagotis E. Wimm., Pflanzenr. IV.276b: 248 (1943).
Colombia. 83 CLM. Nanophan.
Centropogon lagotis var. *gracilior* E. Wimm., Pflanzenr. IV.276b: 772 (1953).

Centropogon lanceolatus E. Wimm., Repert. Spec. Nov. Regni Veg. 19: 245 (1924).
Venezuela. 82 VEN. Nanophan.

Centropogon latifolius E. Wimm., Repert. Spec. Nov. Regni Veg. 38: 7 (1935).
Peru. 83 PER. Nanophan. or cham.

Centropogon latisepalus Gleason, Bull. Torrey Bot. Club 52: 6 (1925).
Colombia. 83 CLM. Nanophan.

Centropogon laxus Zahlbr., Repert. Spec. Nov. Regni Veg. 14: 134 (1915).
Colombia. 83 CLM. Nanophan.
  Centropogon brachyandrus E. Wimm., Repert. Spec. Nov. Regni Veg. 29: 68 (1931).

Centropogon lehmannii Zahlbr., Repert. Spec. Nov. Regni Veg. 13: 536 (1915), as 'lehmanni'.
Colombia. 83 CLM. Nanophan.

  subsp. **asservatus** (Gleason) McVaugh, Brittonia 6: 488 (1949).
    Colombia. 83 CLM. Nanophan.
   * *Centropogon asservatus* Gleason, Bull. Torrey Bot. Club 51: 445 (1924). *Centropogon
      lehmannii* var. *asservatus* (Gleason) E. Wimm., Pflanzenr. IV.276b: 255 (1943).

  subsp. **lehmanii**
    Colombia. 83 CLM. Nanophan.
    Centropogon altisilvanus E. Wimm., Repert. Spec. Nov. Regni Veg. 29: 71 (1931).
    Centropogon altisilvanus var. *minor* E. Wimm., Repert. Spec. Nov. Regni Veg. 29: 72 (1931).

Centropogon leucocarpus McVaugh, J. Wash. Acad. Sci. 39: 161 (1949).
Panama. 80 PAN. Cham.

Centropogon leucophyllus Gleason, Bull. Torrey Bot. Club 52: 61 (1925).
Colombia. 83 CLM. Cham.
  Centropogon leucophyllus var. *truncatus* E. Wimm., Repert. Spec. Nov. Regni Veg. 29: 70
    (1931). *Centropogon leucophyllus* f. *truncatus* (E. Wimm.) E. Wimm., Pflanzenr.
    IV.276b: 215 (1943), as 'truncata'.
  Centropogon minimus McVaugh, Brittonia 6: 467 (1949). *Centropogon leucophyllus* var.
    minimus (McVaugh) E. Wimm., Pflanzenr. IV.276b: 770 (1953).

Centropogon lianeus E. Wimm., Repert. Spec. Nov. Regni Veg. 38: 12 (1935).
Colombia. 83 CLM. Cl. phan.

Centropogon licayensis Gleason, Bull. Torrey Bot. Club 52: 16 (1925).
Ecuador. 83 ECU. Nanophan.
  Centropogon quitensis E. Wimm., Repert. Spec. Nov. Regni Veg. 29: 75 (1931).
  Centropogon paraxylochus E. Wimm., Biblioth. Bot. 116: 157 (1937).
  Centropogon erianthus f. *chagalensis* E. Wimm., Pflanzenr. IV.276b: 250 (1943).

Centropogon lindenianus E. Wimm., Repert. Spec. Nov. Regni Veg. 19: 246 (1924).
Colombia. 83 CLM. (Cl.) nanophan.

Centropogon linnaeanus E. Wimm., Repert. Spec. Nov. Regni Veg. 29: 61 (1931).
Colombia. 83 CLM. Nanophan.

Centropogon llanganatensis Jeppesen in Harling & Sparre, Fl. Ecuador 14: 92 (1981).
Ecuador. 83 ECU. Nanophan.

Centropogon longifolius E. Wimm., Repert. Spec. Nov. Regni Veg. 19: 246 (1924).
Peru. 83 PER. Nanophan.

Centropogon longipetiolatus E. Wimm., Repert. Spec. Nov. Regni Veg. 26: 5 (1929).
Peru. 83 PER. Nanophan.
  Centropogon flexuosus var. *ethnicus* E. Wimm., Repert. Spec. Nov. Regni Veg. 29: 65 (1931).

Centropogon loretensis E. Wimm. in J. F. Macbr., Fl. Peru 6: 417 (1937).
Colombia to Peru. 83 CLM ECU PER. Nanophan. or cham.

Centropogon luteus E. Wimm., Repert. Spec. Nov. Regni Veg. 19: 247 (1924).
Colombia to Peru. 83 CLM ECU PER. (Cl.) nanophan.
Centropogon aurantiacus Gleason, Bull. Torrey Bot. Club 52: 7 (1925).

Centropogon luteynii Wilbur, Ann. Missouri Bot. Gard. 63: 627 (1977).
Panama. 80 PAN. Cham.

Centropogon macbridei Gleason, Bull. Torrey Bot. Club 52: 18 (1925).
Peru. 83 PER. Cl. nanophan.

Centropogon macrocarpus Zahlbr., Bot. Jahrb. Syst. 37: 452 (1906). *Burmeistera macrocarpa*
(Zahlbr.) E. Wimm., Repert. Spec. Nov. Regni Veg. 30: 41 (1932).
Peru. 83 PER. Nanophan.

Centropogon macrophyllus (G. Don) E. Wimm., Notizbl. Bot. Gart. Berlin-Dahlem 10:
733 (1929).
Peru & Bolivia. 83 BOL PER. Nanophan. or cham.
* *Siphocampylus macrophyllus* G. Don, Gen. Hist. 3: 704 (1834). *Lobelia macrophylla* (G.
Don) C. Presl, Prodr. Monogr. Lobel. 39 (1836).
Centropogon amplifolius Vatke, Linnaea 38: 716 (1874).
Centropogon macrophyllus f. *minoratus* E. Wimm. in J. F. Macbr., Fl. Peru 6: 419 (1937).

Centropogon magnificus Zahlbr. & Rech., Meded. Rijks-Herb. 19: 50 (1913).
Bolivia. 83 BOL. Phan. or nanophan.

Centropogon mandonis Zahlbr., Ann. K. K. Naturhist. Hofmus. 6: 438 (1891).
Peru & Bolivia. 83 BOL PER. Nanophan.

Centropogon medusa E. Wimm., Repert. Spec. Nov. Regni Veg. 22: 203 (1926).
Ecuador. 83 ECU. (Cl.) nanophan.

Centropogon mellitus E. Wimm., Repert. Spec. Nov. Regni Veg. 29: 73 (1931).
Colombia & Peru. 83 CLM PER. Nanophan.
Centropogon mellitus var. *peruvianus* E. Wimm., Pflanzenr. IV.276b: 248 (1943).

Centropogon microcalyx E. Wimm., Repert. Spec. Nov. Regni Veg. 38: 14 (1935).
Colombia. 83 CLM. Nanophan.

Centropogon monagensis McVaugh, Brittonia 6: 474 (1949).
Venezuela. 82 VEN. Cham.

Centropogon nervosus E. Wimm. ex Gleason, Bull. Torrey Bot. Club 52: 13 (1925).
Peru. 83 PER. Nanophan.

Centropogon nigricans Zahlbr., Repert. Spec. Nov. Regni Veg. 14: 141 (1915).
Ecuador. 83 ECU. Nanophan.
Centropogon andreanus Gleason, Bull. Torrey Bot. Club 51: 444 (1924).

Centropogon occultus Gleason, Bull. Torrey Bot. Club 52: 16 (1925).
Colombia & Ecuador. 83 CLM ECU. Nanophan.
Centropogon occultus f. *denudatus* E. Wimm., Repert. Spec. Nov. Regni Veg. 29: 75 (1931).
Centropogon occultus var. *phaeus* E. Wimm., Pflanzenr. IV.276b: 239 (1943).
Centropogon espinosanus E. Wimm., Pflanzenr. IV.276b: 249 (1943).

Centropogon oligotrichus E. Wimm., Repert. Spec. Nov. Regni Veg. 29: 72 (1931), as
'oligitrichus'.
Ecuador. 83 ECU. Nanophan.

Centropogon ozotrichus E. Wimm., Repert. Spec. Nov. Regni Veg. 26: 10 (1929).
Colombia. 83 CLM. Cham.

Centropogon palmanus (Donn. Sm.) E. Wimm., Repert. Spec. Nov. Regni Veg. 38: 7 (1935).
Costa Rica. 80 COS. Cham.
   Centropogon nematosepalus var. *palmanus* Donn. Sm., Bot. Gaz. (Crawfordsville) 44:
      115 (1907).

Centropogon pamplonensis E. Wimm., Repert. Spec. Nov. Regni Veg. 19: 248 (1924).
Colombia. 83 CLM. Nanophan.
   Centropogon majalis E. Wimm., Repert. Spec. Nov. Regni Veg. 22: 202 (1931).
   Centropogon pamplonensis var. *fratellus* E. Wimm., Repert. Spec. Nov. Regni Veg. 38:
      11 (1935).

Centropogon papillosus E. Wimm., Pflanzenr. IV.276b: 177 (1943).
Ecuador. 83 ECU. Nanophan. or cham.

Centropogon parviflorus (Zahlbr.) Jeppesen in Harling & Sparre, Fl. Ecuador 14: 98 (1981).
Ecuador. 83 ECU. (Cl.) nanophan.
   Centropogon barbatus var. *parviflorus* Zahlbr., Ann. K. K. Naturhist. Hofmus. 6: 436
      (1891). *Centropogon ferrugineus* var. *parviflorus* (Zahlbr.) Gleason, Bull. Torrey Bot.
      Club 52: 11 (1925).
   Centropogon saltuum var. *ferrugineoides* E. Wimm., Pflanzenr. IV.276b: 235 (1943).

Centropogon pedicellaris Gleason, Bull. Torrey Bot. Club 52: 57 (1925).
Colombia. 83 CLM. Nanophan. or cham.
   Centropogon pedicellaris var. *gallerensis* Gleason, Bull. Torrey Bot. Club 52: 58 (1925).
   Centropogon pedicellaris var. *breviflorus* E. Wimm., Pflanzenr. IV.276b: 769 (1953).

Centropogon perlongus Gleason, Bull. Torrey Bot. Club 52: 19 (1925).
Peru. 83 PER. Nanophan.

Centropogon peruvianus (E. Wimm.) McVaugh, Brittonia 6: 462 (1949).
Peru. 83 PER. Nanophan.
   Burmeistera peruviana E. Wimm., Repert. Spec. Nov. Regni Veg. 38: 5 (1935).
   Burmeistera asteriscus E. Wimm., Repert. Spec. Nov. Regni Veg. 38: 5 (1935). *Centropogon
      asteriscus* (E. Wimm.) McVaugh, Brittonia 6: 462 (1949).

Centropogon phoeniceus Jeppesen in Harling & Sparre, Fl. Ecuador 14: 99 (1981).
Ecuador. 83 ECU. Nanophan. or cham.

Centropogon pichinchensis Zahlbr., Repert. Spec. Nov. Regni Veg. 14: 180 (1915).
Colombia & Ecuador. 83 CLM ECU. Nanophan.
   Centropogon minensis Benoist, Bull. Soc. Bot. France 84: 635 (1938).
   Centropogon nebularum Benoist, Bull. Soc. Bot. France 84: 636 (1938).

Centropogon pilalensis Jeppesen in Harling & Sparre, Fl. Ecuador 14: 102 (1981).
Ecuador. 83 ECU. Nanophan.

Centropogon pinguis E. Wimm., Repert. Spec. Nov. Regni Veg. 29: 70 (1931).
Colombia. 83 CLM. Nanophan. or cham.

Centropogon polytrichus E. Wimm., Repert. Spec. Nov. Regni Veg. 26: 11 (1929).
Colombia. 83 CLM. Nanophan.

Centropogon preslii E. Wimm., Repert. Spec. Nov. Regni Veg. 19: 248 (1924).
Colombia to Peru. 83 CLM ECU PER. Nanophan.

Centropogon pulcher Zahlbr., Bot. Jahrb. Syst. 37: 451 (1906).
Peru. 83 PER. Cl. nanophan. *2n* = 28.

Centropogon pyropus E. Wimm., Repert. Spec. Nov. Regni Veg. 38: 14 (1935).
Colombia. 83 CLM. Nanophan.

Centropogon quebradanus E. Wimm., Repert. Spec. Nov. Regni Veg. 29: 61 (1931).
Ecuador. 83 ECU. Cham.

Centropogon reflexus C. Presl, Prodr. Monogr. Lobel.: 49 (1836).
Peru. 83 PER. Cl. phan.
  Centropogon auratus E. Wimm., Repert. Spec. Nov. Regni Veg. 19: 243 (1924).
  Centropogon rubrovenosus Gleason, Torreya 25: 93 (1925).

Centropogon reticulatus Drake, J. Bot. (Morot) 3: 238 (1889).
Ecuador & Peru. 83 ECU PER. Nanophan. or cham.

Centropogon rimbachii E. Wimm., Repert. Spec. Nov. Regni Veg. 26: 5 (1929).
Ecuador. 83 ECU. Nanophan. or cham.

Centropogon roraimanus E. Wimm., Repert Spec. Nov. Regni Veg. 26: 6 (1929).
Venezuela & Guyana. 82 GUY VEN. Cham.

Centropogon roseus Rusby, Bull. New York Bot Gard. 8: 123 (1912).
Peru & Bolivia. 83 BOL PER. Cham.
  Centropogon inflatus E. Wimm., Repert. Spec. Nov. Regni Veg. 29: 60 (1931).

Centropogon rubiginosus E. Wimm., Repert. Spec. Nov. Regni Veg. 19: 249 (1924).
Ecuador. 83 ECU. (Cl.) nanophan.
  Centropogon ferrugineus f. *subaequus* E. Wimm., Pflanzenr. IV.276c: 841 (1968).

Centropogon rubroaureus E. Wimm., Repert. Spec. Nov. Regni Veg. 38: 13 (1935), as
'rubro-aureus'.
Colombia. 83 CLM. Nanophan.

Centropogon rubrodentatus Jeppesen in Harling & Sparre, Fl. Ecuador 14: 107 (1981).
Ecuador. 83 ECU. Cham.

Centropogon rufus E. Wimm., Repert. Spec. Nov. Regni Veg. 19: 249 (1924).
Peru. 83 PER. Nanophan.

Centropogon saltuum E. Wimm., Repert. Spec. Nov. Regni Veg. 22: 203 (1926).
Ecuador. 83 ECU. Nanophan.
  Centropogon flaviceps E. Wimm., Repert. Spec. Nov. Regni Veg. 38: 12 (1935).

Centropogon salviiformis Zahlbr., Repert. Spec. Nov. Regni Veg. 14: 138 (1915), as
'salviaeformis'.
Colombia & Ecuador. 83 CLM ECU. Nanophan.
  Centropogon milleanus E. Wimm., Repert. Spec. Nov. Regni Veg. 19: 390 (1924).
  Centropogon barbatellus Gleason, Bull. Torr. Bot. Club 52: 11 (1925).

Centropogon scabellus E. Wimm., Repert. Spec. Nov. Regni Veg. 26: 6 (1929).
Colombia. 83 CLM. Nanophan.
Centropogon altibracteolatus E. Wimm., Repert. Spec. Nov. Regni Veg. 38: 9 (1935).

Centropogon scabiosus E. Wimm., Repert. Spec. Nov. Regni Veg. 19: 388 (1924).
Peru. 83 PER. Nanophan.

Centropogon sciaphilus Zahlbr., Ann. K. K. Naturhist. Hofmus. 6: 435 (1891).
Peru. 83 PER. Nanophan. or cham.
Centropogon ciliatus Gleason, Bull. Torrey Bot Club 52: 58 (1925).

Centropogon silvaticus E. Wimm., Repert. Spec. Nov. Regni Veg. 38: 6 (1935).
Peru. 83 PER. Cham.

Centropogon simulans Lammers, Brittonia 50: 251 (1998).
Peru. 83 PER. Nanophan.

Centropogon smithii E. Wimm., Ann. Naturhist. Mus. Wien 46: 240 (1933).
Costa Rica & Panama. 80 COS PAN. Cham.
* *Siphocampylus discolor* Donn. Sm., Bot. Gaz. (Crawfordsville) 23: 248 (1897); non
Centropogon discolor Kunth & Bouché in Kunth, Sp. Nov. Hort. Berol. 1847: 13 (1848).

Centropogon sodiroanus Zahlbr., Repert. Spec. Nov. Regni Veg. 14: 181 (1915).
Ecuador. 83 ECU. Nanophan.

Centropogon solanifolius Benth., Pl. Hartweg.: 139 (1845).
Costa Rica to Ecuador. 80 COS PAN 82 VEN 83 CLM ECU. Nanophan. or cham. $2n = 28$.
Centropogon prostratus Benth., Pl. Hartweg.: 212 (1845).
Centropogon discolor Kunth & Bouché in Kunth, Sp. Nov. Hort. Berol. 1847: 13 (1848).
Centropogon speciosus Planch., Fl. Serres Jard. Eur. 6: 16 (1850). *Centropogon solanifolius*
var. *speciosus* (Planch.) E. Wimm., Pflanzenr. IV.276b: 190 (1943).
Centropogon longipes Regel, Gartenflora 3: 3 (1854). *Siphocampylus longipes* (Regel) Vatke,
Linnaea 38: 733 (1874).
Centropogon planchonis Zahlbr., Repert. Spec. Nov. Regni Veg. 14: 133 (1915).
Centropogon subfalcatus Zahlbr., Repert. Spec. Nov. Regni Veg. 14: 136 (1915).
Centropogon ovalifolius Zahlbr., Repert. Spec. Nov. Regni Veg. 14: 136 (1915).
Centropogon montanus E. Wimm., Repert. Spec. Nov. Regni Veg. 19: 247 (1924).
Centropogon riparius E. Wimm., Repert. Spec. Nov. Regni Veg. 19: 248 (1924).
Centropogon semperflorens E. Wimm., Repert. Spec. Nov. Regni Veg. 19: 250 (1924).
Centropogon flexuosus E. Wimm., Repert. Spec. Nov. Regni Veg. 22: 200 (1926).
Centropogon psilandrus E. Wimm., Repert. Spec. Nov. Regni Veg. 26: 7 (1929).
Centropogon xestus E. Wimm., Repert. Spec. Nov. Regni Veg. 29: 64 (1931).
Centropogon solanifolius var. *hirtellus* E. Wimm., Ann. Naturhist. Mus. Wien 46: 241 (1933).
Centropogon solanifolius var. *inconstans* E. Wimm., Biblioth. Bot. 116: 156 (1937).
Centropogon austin-smithii Standl., Fl. of Costa Rica: 1409 (1938).
Centropogon montanus f. *subpubescens* E. Wimm., Pflanzenr. IV.276b: 190 (1943).
Centropogon nubicola Gómez-Laur. & L. D. Gómez, Phytologia 51: 477 (1982).

Centropogon solisii Jeppesen in Harling & Sparre, Fl. Ecuador 14: 114 (1981).
Ecuador. 83 ECU. Nanophan.

Centropogon steinii Lammers, Brittonia 50: 245 (1998).
Ecuador. 83 ECU. Cl. nanophan.

Centropogon steyermarkii Jeppesen in Harling & Sparre, Fl. Ecuador 14: 115 (1981).
Ecuador. 83 ECU. Nanophan.

Centropogon subandinus Zahlbr., Repert. Spec. Nov. Regni Veg. 14: 141 (1915).
Ecuador. 83 ECU. Nanophan. or cham.
Centropogon subandinus var. *ventricosus* E. Wimm., Repert. Spec. Nov. Regni Veg. 26: 8 (1929).

Centropogon subcordatus Zahlbr., Repert. Spec. Nov. Regni Veg. 14: 140 (1915).
Colombia to Peru. 83 CLM ECU PER. Nanophan.

Centropogon suberianthus Zahlbr., Repert. Spec. Nov. Regni Veg. 14: 134 (1915).
Colombia. 83 CLM. Nanophan.

Centropogon talamancensis Wilbur, Brittonia 21: 355 (1970).
Costa Rica. 80 COS. Cham.

Centropogon tenuifolius E. Wimm., Repert. Spec. Nov. Regni Veg. 38: 13 (1935).
Colombia. 83 CLM. Cham.

Centropogon tessmannii E. Wimm., Repert. Spec. Nov. Regni Veg. 26: 8 (1929).
Ecuador & Peru. 83 ECU PER. Cham.
Centropogon tessmannii var. *tenuiflorus* E. Wimm., Repert. Spec. Nov. Regni Veg. 29: 69 (1931).

Centropogon tovarensis Planch. & Linden ex Planch., Fl. Serres Jard. Eur. 8: 145 (1853).
Venezuela. 82 VEN. Cham.

Centropogon trachyanthus E. Wimm., Repert. Spec. Nov. Regni Veg. 29: 65 (1931).
Ecuador. 83 ECU. Cl. nanophan. $2n = 28$.

Centropogon trianae Zahlbr., Repert. Spec. Nov. Regni Veg. 14: 137 (1915).
Colombia. 83 CLM. Nanophan.
Centropogon trianae var. *cuspidata* Zahlbr., Repert. Spec. Nov. Regni Veg. 14: 138 (1915).
Centropogon decemlobus Gleason, Bull. Torrey Bot. Club 52: 59 (1925). *Centropogon trianae* var. *decemlobus* (Gleason) E. Wimm., Repert. Spec. Nov. Regni Veg. 29: 68 (1931).
Centropogon purdieanus Gleason, Bull. Torrey Bot. Club 52: 60 (1925). *Centropogon trianae* var. *purdieanus* (Gleason) E. Wimm., Repert. Spec. Nov. Regni Veg. 29: 68 (1931).

Centropogon trichodes E. Wimm., Repert. Spec. Nov. Regni Veg. 29: 69 (1931).
Ecuador. 83 ECU. Nanophan. or cham.

Centropogon ulloae Lammers, Novon 12: 221 (2002).
Ecuador. 83 ECU. Nanophan.

Centropogon umbrosus E. Wimm., Repert. Spec. Nov. Regni Veg. 19: 250 (1924).
Peru. 83 PER. Nanophan.
Centropogon caninus E. Wimm., Notizbl. Bot. Gart. Berlin-Dahlem 10: 734 (1929).
Centropogon gesnerioides var. *zelans* E. Wimm., Notizbl. Bot. Gart. Berlin-Dahlem 10: 733 (1929).
Centropogon gesnerioides var. *viperinus* E. Wimm., Repert. Spec. Nov. Regni Veg. 38: 6 (1935).

Centropogon uncialis McVaugh, J. Wash. Acad. Sci. 39: 159 (1949).
Colombia. 83 CLM. Nanophan. or cham.

Centropogon uncinatus Zahlbr., Beih. Bot. Centralbl. 13: 84 (1903).
Ecuador. 83 ECU. Nanophan.

Centropogon unduavensis (Britton) Zahlbr., Bull. Torrey Bot. Club 24: 374 (1897).
Bolivia. 83 BOL. Nanophan.
* *Siphocampylus unduavensis* Britton, Bull. Torrey Bot. Club 19: 373 (1892).
Centropogon unduavensis var. *aduanus* E. Wimm., Pflanzenr. IV.276b: 230 (1943).

Centropogon unicolor E. Wimm., Repert. Spec. Nov. Regni Veg. 38: 11 (1935).
Colombia. 83 CLM. Nanophan.

Centropogon urceolatus E. Wimm., Pflanzenr. IV.276b: 253 (1943).
Colombia. 83 CLM. Nanophan. or cham.

Centropogon ursinus Jeppesen in Harling & Sparre, Fl. Ecuador 14: 120 (1981).
Ecuador. 83 ECU. Nanophan.

Centropogon urubambae E. Wimm., Repert. Spec. Nov. Regni Veg. 38: 7 (1935).
Peru. 83 PER. Cl. phan.
Centropogon urubambae var. *estrellanus* E. Wimm., Repert. Spec. Nov. Regni Veg. 38: 8
(1935). *Centropogon estrellanus* (E. Wimm.) E. Wimm., Pflanzenr. 1V.276b: 192 (1943).

Centropogon valerii Standl., Fl. of Costa Rica: 1413 (1938). *Centropogon grandidentatus* var.
valerii (Standl.) McVaugh, N. Amer. Fl. 32A: 125 (1943).
Costa Rica. 80 COS. Cham.
Centropogon valerii var. *kupperi* Suesseng., Bot. Jahrb. Syst. 72: 287 (1942).
Centropogon stenophyllus E. Wimm., Pflanzenr. IV.276c: 841 (1968).

Centropogon varelae E. Wimm., Pflanzenr. IV.276c: 842 (1968).
Colombia. 83 CLM. Cham.

Centropogon varicus McVaugh, Ann. Missouri Bot. Gard. 52: 401 (1965).
Peru. 83 PER. Cl. nanophan.

Centropogon vaughianus E. Wimm., Brittonia 8: 110 (1955).
Colombia. 83 CLM. Cham.

Centropogon verbascifolius (C. Presl) Gleason, Bull. Torrey Bot. Club 52: 18 (1925).
Colombia & Peru. 83 CLM PER. Nanophan.
* *Lobelia verbascifolia* C. Presl, Prodr. Monogr. Lobel.: 38 (1836). *Siphocampylus*
verbascifolius (C. Presl) A. DC. in DC., Prodr. 7: 402 (1839).
Centropogon cinereus Gleason, Bull. Torrey Bot. Club 52: 15 (1925).
Centropogon cinereus f. *odontosepalus* E. Wimm., Repert. Spec. Nov. Regni Veg. 29: 75
(1931).

Centropogon vernicosus Zahlbr., Ann. K. K. Naturhist. Hofmus. 6: 440 (1891).
Peru. 83 PER. Nanophan.
Centropogon pilosulus E. Wimm., Notizbl. Bot. Gart. Berlin-Dahlem 10: 732 (1929).
Centropogon pamplonensis var. *peruvianus* E. Wimm., Notizbl. Bot. Gart. Berlin-Dahlem
10: 734 (1929).
Centropogon pamplonensis var. *glabriflorus* E. Wimm., Notizbl. Bot. Gart. Berlin-Dahlem
10: 735 (1929).

Centropogon viriduliflorus E. Wimm., Repert. Spec. Nov. Regni Veg. 38: 16 (1935).
Peru. 83 PER. Nanophan.

Centropogon vitifolius Lammers, Novon 12: 222 (2002).
Peru. 83 PER. Cham.

Centropogon vittariifolius McVaugh, J. Wash. Acad. Sci. 39: 159 (1949), as 'vittariaefolius'.
Colombia. 83 CLM. Nanophan.

Centropogon vulpinus E. Wimm., Repert. Spec. Nov. Regni Veg. 29: 76 (1931).
Colombia. 83 CLM. Nanophan.
Centropogon vulpinus var. *sonsonensis* E. Wimm., J. Wash. Acad. Sci. 40: 354 (1950).

Centropogon warscewiczii Van Houtte ex Regel, Gartenflora 7: 374 (1858).
Colombia. 83 CLM. Nanophan.
Centropogon erythranthus Zahlbr., Repert. Spec. Nov. Regni Veg. 13: 535 (1915), as
'erythanthrus'.
Centropogon vimineus E. Wimm., Repert. Spec. Nov. Regni Veg. 22: 204 (1926). *Centropogon
warscewiczii* var. *vimineus* (E. Wimm.) E. Wimm., Pflanzenr. IV.276b: 217 (1943).
Centropogon warscewiczii var. *jervisaei* E. Wimm., Repert. Spec. Nov. Regni Veg. 38:
10 (1935).
Centropogon warscewiczii f. *brevilobus* E. Wimm., Repert. Spec. Nov. Regni Veg. 38:
11 (1935).

Centropogon weberbaueri Zahlbr., Bot. Jahrb. Syst. 37: 453 (1906).
Peru. 83 PER. Nanophan.

Centropogon wilburii Lammers, Brittonia 50: 253 (1998).
S. Mexico (Oaxaca). 79 MXS. Cham.

Centropogon wimmeri Standl., Fl. of Costa Rica: 1414 (1938), as 'wimmerii'.
Costa Rica. 80 COS. Cham.

Centropogon yarumalensis E. Wimm., Repert. Spec. Nov. Regni Veg. 29: 74 (1931).
Colombia. 83 CLM. Nanophan.
Centropogon yarumalensis var. *puniceus* E. Wimm., Pflanzenr. IV.276c: 842 (1968).

Centropogon yungasensis Britton, Bull. Torrey Bot. Club 19: 371 (1892), as 'yungasense'.
Peru & Bolivia. 83 BOL PER. Cl. phan.
Centropogon yungasensis var. *angustior* Zahlbr., Bot. Jahrb. Syst. 37: 452 (1906).
Centropogon ostrinus E. Wimm., Repert. Spec. Nov. Regni Veg. 29: 67 (1931).

Centropogon zamorensis Jeppesen in Harling & Sparre, Fl. Ecuador 14: 121 (1981).
Ecuador. 83 ECU. Nanophan.

Synonyms:
Centropogon sect. *Burmeistera* (Triana) Schönl. === **Burmeistera** Triana
Centropogon affinis M. Martens & Galeotti === **Centropogon grandidentatus** (Schltdl.) Zahlbr.
Centropogon affinis var. *costaricanus* (Planch. & Oerst.) Zahlbr. === **Centropogon ferrugineus**
(L. f.) Gleason
Centropogon affinis var. *venezuelanus* E. Wimm. === **Centropogon ferrugineus** (L. f.) Gleason
Centropogon aggregatus (Rusby) Gleason === **Centropogon granulosus** C. Presl
Centropogon aggregatus var. *cardinalis* (Zahlbr. & Rech.) E. Wimm. === **Centropogon
granulosus** C. Presl
Centropogon albolimbatus var. *concolor* E. Wimm. === **Centropogon albolimbatus** E. Wimm.
Centropogon albolimbatus var. *serratus* (Gleason) E. Wimm. === **Centropogon albolimbatus**
E. Wimm.
Centropogon albolimbatus var. *villosus* E. Wimm. === **Centropogon albolimbatus** E. Wimm.
Centropogon altibracteolatus E. Wimm. === **Centropogon scabellus** E. Wimm.
Centropogon altisilvanus E. Wimm. === **Centropogon lehmannii** Zahlbr. subsp. **lehmannii**
Centropogon altisilvanus var. *minor* E. Wimm. === **Centropogon lehmannii** Zahlbr.
subsp. **lehmannii**

Centropogon amplifolius Vatke === **Centropogon macrophyllus** (G. Don) E. Wimm.

Centropogon andreanus Gleason === **Centropogon nigricans** Zahlbr.

Centropogon andropogon (Cav.) A. DC. in DC. === **Centropogon cornutus** (L.) Druce

Centropogon angustus Gleason === **Centropogon hirtus** (Cav.) C. Presl

Centropogon aquilinus var. *campestris* E. Wimm. === **Centropogon aquilinus** E. Wimm.

Centropogon aquilinus var. *integer* E. Wimm. === **Centropogon aquilinus** E. Wimm.

Centropogon argentinus Griseb. === **Siphocampylus argentinus** (Griseb.) Hieron. ex E. Wimm.

Centropogon aristi E. Wimm. === **Centropogon grandis** (L. f.) C. Presl

Centropogon aristi var. *subterpilis* E. Wimm. === **Centropogon grandis** (L. f.) C. Presl

Centropogon asservatus Gleason === **Centropogon lehmannii** subsp. **asservatus** (Gleason) McVaugh

Centropogon asteriscus (E. Wimm.) McVaugh === **Centropogon peruvianus** (E. Wimm.) McVaugh

Centropogon augostanus E. Wimm. === **Centropogon granulosus** C. Presl

Centropogon aurantiacus Gleason === **Centropogon luteus** E. Wimm.

Centropogon auratus E. Wimm. === **Centropogon reflexus** C. Presl

Centropogon aurobarbatus E. Wimm. === **Burmeistera microphylla** Donn. Sm.

Centropogon austin-smithii Standl. === **Centropogon solanifolius** Benth.

Centropogon australis var. *taraxacoides* E. Wimm. === **Centropogon australis** (E. Wimm.) Gleason

Centropogon barbatellus Gleason === **Centropogon salviiformis** Zahlbr.

Centropogon barbatus (Cav.) Planch. === **Centropogon ferrugineus** (L. f.) Gleason

Centropogon barbatus var. *parviflorus* Zahlbr. === **Centropogon parviflorus** (Zahlbr.) Jeppesen

Centropogon bonplandianus (Schult.) C. Presl === **Centropogon cornutus** (L.) Druce

Centropogon bonplandiaus f. *glabrescens* E. Wimm. === **Centropogon cornutus** (L.) Druce

Centropogon bonplandianus var. *intermedius* (Zahlbr.) E. Wimm.

Centropogon brachyandrus E. Wimm. === **Centropogon laxus** Zahlbr.

Centropogon breviflorus Gleason === **Burmeistera breviflora** (Gleason) E. Wimm.

Centropogon brittonianus var. *brevidentatus* Zahlbr. & Rech. === **Siphocampylus radiatus** Rusby

Centropogon brumalis Standl. === **Centropogon costaricae** (Vatke) McVaugh

Centropogon caldasensis Gleason === **Burmeistera caldasensis** (Gleason) E. Wimm.

Centropogon caligatus E. Wimm. === **Centropogon calycinus** Benth.

Centropogon calochlamys Donn. Sm. === **Lobelia calochlamys** (Donn. Sm.) Wilbur

Centropogon calycinus var. *tubulosus* (Zahlbr.) E. Wimm. === **Centropogon calycinus** Benth.

Centropogon caninus E. Wimm. === **Centropogon umbrosus** E. Wimm.

Centropogon caninus var. *hirsutulus* E. Wimm. === **Centropogon congestus** Gleason

Centropogon capitatus f. *hirtus* Zahlbr. === **Centropogon capitatus** Drake

Centropogon capitatus var. *fieldii* E. Wimm. === **Centropogon capitatus** Drake

Centropogon capitatus var. *trichandrus* E. Wimm. === **Centropogon capitatus** Drake

Centropogon cardinalis Zahlbr. & Rech. === **Centropogon granulosus** C. Presl

Centropogon carpinoides Gleason === **Centropogon carnosus** Zahlbr.

Centropogon casapiensis E. Wimm. === **Centropogon granulosus** C. Presl

Centropogon chamissonianus (A. DC.) Kanitz === **Siphocampylus umbellatus** (Kunth) G. Don

Centropogon ciliatus Gleason === **Centropogon sciaphilus** Zahlbr.

Centropogon cinereus Gleason === **Centropogon verbascifolius** (C. Presl) Gleason

Centropogon cinereus f. *odontosepalus* E. Wimm. === **Centropogon verbascifolius** (C. Presl) Gleason

Centropogon coccineus var. *leucostomus* (Van Houtte) E. Wimm. === **Centropogon coccineus** (Hook.) Regel ex B. D. Jacks.

Centropogon coleoides (Vatke) Zahlbr. === **Burmeistera coleoides** (Vatke) E. Wimm.

Centropogon colombiensis E. Wimm. === **Centropogon curvatus** Gleason

Centropogon cordatus M. Martens & Galeotti === **Centropogon cordifolius** Benth.

Centropogon cordifolius var. *dentatus* E. Wimm. ex Standl. === **Centropogon costaricae** (Vatke) McVaugh

Centropogon cornutus f. *leucanthus* E. Wimm. === **Centropogon cornutus** (L.) Druce

Centropogon cornutus f. *leucostomus* E. Wimm. === **Centropogon cornutus** (L.) Druce

Centropogon cornutus f. *vellozianus* E. Wimm. === **Centropogon cornutus** (L.) Druce
Centropogon cornutus f. *ynesae* E. Wimm. === **Centropogon cornutus** (L.) Druce
Centropogon cornutus var. *angustifolius* (Zahlbr.) E. Wimm. === **Centropogon cornutus** (L.) Druce
Centropogon cornutus var. *laevigatus* (L. f.) E. Wimm. === **Centropogon cornutus** (L.) Druce
Centropogon costaricae f. *dentatus* (E. Wimm. ex Standl.) McVaugh === **Centropogon costaricae** (Vatke) McVaugh
Centropogon costaricanus Planch. & Oerst. === **Centropogon ferrugineus** (L. f.) Gleason
Centropogon costaricanus var. *cufodontidis* E. Wimm. === **Centropogon ferrugineus** (L. f.) Gleason
Centropogon costaricanus var. *tomentellus* E. Wimm. === **Centropogon ferrugineus** (L. f.) Gleason
Centropogon crassifolius E. Wimm. === **Burmeistera crassifolia** (E. Wimm.) E. Wimm.
Centropogon crassifolius var. *ovatifolius* E. Wimm. === **Burmeistera crassifolia** (E. Wimm.) E. Wimm.
Centropogon cumulatus E. Wimm. === **Centropogon granulosus** C. Presl
Centropogon curvatus f. *minor* E. Wimm. === **Centropogon curvatus** Gleason
Centropogon curvatus var. *pearcei* E. Wimm. === **Centropogon curvatus** Gleason
Centropogon cuspidatus A. DC. === **Centropogon granulosus** C. Presl
Centropogon cylindricus (Gleason) E. Wimm. === **Centropogon ayavacensis** subsp. **cylindricus** (Gleason) Lammers
Centropogon decemlobus Gleason === **Centropogon trianae** Zahlbr.
Centropogon declinatus (Rusby) E. Wimm. === **Siphocampylus declinatus** Rusby
Centropogon densiflorus var. *lugens* E. Wimm. === **Centropogon granulosus** C. Presl
Centropogon diocleus E. Wimm. === **Centropogon congestus** Gleason
Centropogon discolor Kunth & Bouché === **Centropogon solanifolius** Benth.
Centropogon dubius (Zahlbr.) E. Wimm. === **Siphocampylus dubius** Zahlbr.
Centropogon erastus E. Wimm. === **Centropogon granulosus** C. Presl
Centropogon erianthus f. *chagalensis* E. Wimm. === **Centropogon licayensis** Gleason
Centropogon erianthus var. *brachysepalus* E. Wimm. === **Centropogon erianthus** (Benth.) Benth. & Hook. f. ex Drake
Centropogon erianthus var. *diversipilis* E. Wimm. === **Centropogon erianthus** (Benth.) Benth. & Hook. f. ex Drake
Centropogon erianthus var. *stella* E. Wimm. === **Centropogon glaucotomentosus** E. Wimm.
Centropogon erythranthus Zahlbr. === **Centropogon warscewiczii** Van Houtte ex Regel
Centropogon espinosanus E. Wimm. === **Centropogon occultus** Gleason
Centropogon estrellanus (E. Wimm.) E. Wimm. === **Centropogon urubambae** E. Wimm.
Centropogon exasperatus C. Presl === **Centropogon hirtus** (Cav.) C. Presl
Centropogon fastuosus Scheidw. === **Centropogon cornutus** (L.) Druce
Centropogon ferrugineus f. *subaequus* E. Wimm. === **Centropogon rubiginosus** E. Wimm.
Centropogon ferrugineus var. *costaricanus* (Planch. & Oerst.) McVaugh === **Centropogon ferrugineus** (L. f.) Gleason
Centropogon ferrugineus var. *glabrifilaris* E. Wimm. === **Centropogon ferrugineus** (L. f.) Gleason
Centropogon ferrugineus var. *jahnii* (Gleason) E. Wimm. === **Centropogon ferrugineus** (L. f.) Gleason
Centropogon ferrugineus var. *nitens* E. Wimm. === **Centropogon ferrugineus** (L. f.) Gleason
Centropogon ferrugineus var. *parviflorus* (Zahlbr.) Gleason === **Centropogon parviflorus** (Zahlbr.) Jeppesen
Centropogon ferrugineus var. *tomentellus* (E. Wimm.) McVaugh === **Centropogon ferrugineus** (L. f.) Gleason
Centropogon ferrugineus var. *venezuelanus* (E. Wimm.) McVaugh === **Centropogon ferrugineus** (L. f.) Gleason
Centropogon ferrugineus var. *vulgatus* E. Wimm. === **Centropogon intonsus** Gleason
Centropogon flaviceps E. Wimm. === **Centropogon saltuum** E. Wimm.
Centropogon flexuosus E. Wimm. === **Centropogon solanifolius** Benth.

Centropogon flexuosus var. *ethnicus* E. Wimm. === **Centropogon longipetiolatus** E. Wimm.

Centropogon floccosus f. *glabrifilis* E. Wimm. === **Centropogon glabrifilis** (E. Wimm.) Jeppesen

Centropogon flos-mutisii f. *fusco-viridis* E. Wimm. === **Centropogon flos-mutisii** E. Wimm.

Centropogon flos-mutisii f. *haticensis* E. Wimm. === **Centropogon flos-mutisii** E. Wimm.

Centropogon formosus E. Wimm. === **Burmeistera formosa** (E. Wimm.) Jeppesen

Centropogon fuscus var. *gleasonii* E. Wimm. === **Centropogon gleasonii** (E. Wimm.) E. Wimm.

Centropogon gesnerioides Gleason === **Centropogon congestus** Gleason

Centropogon gesnerioides var. *viperinus* E. Wimm. === **Centropogon umbrosus** E. Wimm.

Centropogon gesnerioides var. *zelans* E. Wimm. === **Centropogon umbrosus** E. Wimm.

Centropogon glabratus (Kunth) Planch. & Oerst. === **Burmeistera glabrata** (Kunth) Benth. & Hook. f. ex B. D. Jacks.

Centropogon glaucotomentosus var. *correana* E. Wimm. === **Centropogon glaucotomentosus** E. Wimm.

Centropogon glaucus E. Wimm. === **Burmeistera glauca** (E. Wimm.) Gleason

Centropogon gracilis Drake === **Centropogon densiflorus** Benth.

Centropogon grandicephalus Zahlbr. === **Centropogon gamosepalus** Zahlbr.

Centropogon grandidentatus f. *denticulatus* E. Wimm. === **Centropogon grandidentatus** (Schltdl.) Zahlbr.

Centropogon grandidentatus f. *incisus* E. Wimm. === **Centropogon grandidentatus** (Schltdl.) Zahlbr.

Centropogon grandidentatus f. *vulcanicus* E. Wimm. === **Centropogon grandidentatus** (Schltdl.) Zahlbr.

Centropogon grandidentatus var. *australis* E. Wimm. === **Centropogon australis** (E. Wimm.) Gleason

Centropogon grandidentatus var. *diversidens* E. Wimm. === **Centropogon grandidentatus** (Schltdl.) Zahlbr.

Centropogon grandidentatus var. *valerii* (Standl.) McVaugh === **Centropogon valerii** Standl.

Centropogon grandis var. *hirtellus* E. Wimm. === **Centropogon grandis** (L. f.) C. Presl

Centropogon granulosus f. *cardinalis* (Zahlbr. & Rech.) E. Wimm. === **Centropogon granulosus** C. Presl

Centropogon granulosus var. *aggregatus* (Rusby) E. Wimm. === **Centropogon granulosus** C. Presl

Centropogon granulosus var. *cuspidatus* (A. DC.) E. Wimm. === **Centropogon granulosus** C. Presl

Centropogon granulosus var. *rutilus* E. Wimm. === **Centropogon granulosus** C. Presl

Centropogon gravidus Gleason === **Burmeistera variabilis** (Gleason) E. Wimm.

Centropogon griseus Gleason === **Siphocampylus humboldtianus** A. DC.

Centropogon guatemalensis B. L. Rob. === **Lobelia guatemalensis** (B. L. Rob.) Wilbur

Centropogon hartwegii var. *parvus* E. Wimm. === **Centropogon hartwegii** (Benth.) Benth. & Hook. f. ex B. D. Jacks.

Centropogon hitchcockii Gleason === **Siphocampylus sanguineus** Zahlbr.

Centropogon holtonis E. Wimm. === **Centropogon granulosus** C. Presl

Centropogon holtonis var. *albanensis* E. Wimm. === **Centropogon granulosus** C. Presl

Centropogon inflatus E. Wimm. === **Centropogon roseus** Rusby

Centropogon intermedius Zahlbr. === **Centropogon cornutus** (L.) Druce

Centropogon jahnii Gleason === **Centropogon ferrugineus** (L. f.) Gleason

Centropogon laciniatus (G. Don) Zahlbr. === ?

Centropogon laevigatus (L. f.) A. DC. === **Centropogon cornutus** (L.) Druce

Centropogon lagotis var. *gracilior* E. Wimm. === **Centropogon lagotis** E. Wimm.

Centropogon lateriflorus E. Wimm. === **Centropogon granulosus** C. Presl

Centropogon lehmannii var. *asservatus* (Gleason) E. Wimm. === **Centropogon lehmannii** subsp. **asservatus** (Gleason) McVaugh

Centropogon leucophyllus var. *minimus* (McVaugh) E. Wimm. === **Centropogon leucophyllus** Gleason

Centropogon leucophyllus f. *truncatus* (E. Wimm.) E. Wimm. === **Centropogon leucophyllus** Gleason

Centropogon leucophyllus var. *truncatus* E. Wimm. === **Centropogon leucophyllus** Gleason

Centropogon lignescens E. Wimm. === **Burmeistera virescens** (Benth.) Benth. & Hook. ex Hemsl.

Centropogon longipes Regel === **Centropogon solanifolius** Benth.

Centropogon × *lucyanus* Houllet === **Centropogon cornutus** (L.) Druce × **Siphocampylus betulifolius** (Cham.) G. Don

Centropogon macrophyllus f. *minoratus* E. Wimm. === **Centropogon macrophyllus** (G. Don) E. Wimm.

Centropogon macrophyllus var. *congestus* (Gleason) McVaugh === **Centropogon congestus** Gleason

Centropogon macrostemon (C. Presl) Zahlbr. === **Siphocampylus macrostemon** (C. Presl) A. DC.

Centropogon majalis E. Wimm. === **Centropogon pamplonensis** E. Wimm.

Centropogon mellitus var. *peruvianus* E. Wimm. === **Centropogon mellitus** E. Wimm.

Centropogon milleanus E. Wimm. === **Centropogon salviiformis** Zahlbr.

Centropogon minensis Benoist === **Centropogon pichinchensis** Zahlbr.

Centropogon minimus McVaugh === **Centropogon leucophyllus** Gleason

Centropogon mojandensis Benoist === **Centropogon brachysiphoniatus** Zahlbr.

Centropogon montanus E. Wimm. === **Centropogon solanifolius** Benth.

Centropogon montanus f. *subpubescens* E. Wimm. === **Centropogon solanifolius** Benth.

Centropogon mutisianus (Kunth) Gleason === **Burmeistera mutisiana** (Kunth) E. Wimm.

Centropogon nebularum Benoist === **Centropogon pichinchensis** Zahlbr.

Centropogon nematosepalus Donn. Sm. === **Siphocampylus nematosepalus** (Donn. Sm.) E. Wimm.

Centropogon nematosepalus var. *palmanus* Donn. Sm. === **Centropogon palmanus** (Donn. Sm.) E. Wimm.

Centropogon nigricans var. *hazenii* (Gleason) E. Wimm. === **Centropogon hazenii** (Gleason) E. Wimm.

Centropogon nubicola Gómez-Laur. & L. D. Gómez === **Centropogon solanifolius** Benth.

Centropogon nutans Planch. & Oerst. === **Centropogon granulosus** C. Presl

Centropogon oaxacanus (E. Wimm.) E. Wimm. === **Centropogon grandidentatus** (Schltdl.) Zahlbr.

Centropogon oaxacanus var. *berterioides* E. Wimm. === **Centropogon delicatulus** McVaugh

Centropogon oblongus Benth. === **Centropogon cornutus** (L.) Druce

Centropogon occultus f. *denudatus* E. Wimm. === **Centropogon occultus** Gleason

Centropogon occultus var. *phaeus* E. Wimm. === **Centropogon occultus** Gleason

Centropogon ostrinus E. Wimm. === **Centropogon yungasensis** Britton

Centropogon ovalifolius Zahlbr. === **Centropogon solanifolius** Benth.

Centropogon ovalifolius var. *asperatulus* Zahlbr. === **Centropogon congestus** Gleason

Centropogon ovalifolius var. *glabristamineus* E. Wimm. === ?

Centropogon ovalifolius var. *sneidernii* E. Wimm. === **Centropogon congestus** Gleason

Centropogon pallidus Drake === **Burmeistera pallida** (Drake) E. Wimm.

Centropogon pamplonensis var. *fratellus* E. Wimm. === **Centropogon pamplonensis** E. Wimm.

Centropogon pamplonensis var. *glabriflorus* E. Wimm. === **Centropogon vernicosus** Zahlbr.

Centropogon pamplonensis var. *peruvianus* E. Wimm. === **Centropogon vernicosus** Zahlbr.

Centropogon panamensis Wilbur === **Centropogon granulosus** C. Presl

Centropogon paraxylochus E. Wimm. === **Centropogon licayensis** Gleason

Centropogon parvulus Gleason === **Centropogon granulosus** C. Presl

Centropogon pedicellaris var. *breviflorus* E. Wimm. === **Centropogon pedicellaris** Gleason

Centropogon pedicellaris var. *gallerensis* Gleason === **Centropogon pedicellaris** Gleason

Centropogon pilosulus E. Wimm. === **Centropogon vernicosus** Zahlbr.

Centropogon pilosulus var. *quindiuensis* E. Wimm. === ?

Centropogon planchonis Zahlbr. === **Centropogon solanifolius** Benth.

Centropogon planchonis var. amplicorollinus E. Wimm. === **Centropogon amplicorollinus** (E. Wimm.) B. A. Stein

Centropogon planchonis var. *heteranthus* E. Wimm. === **Centropogon granulosus** C. Presl

Centropogon poasensis Gleason === **Centropogon ferrugineus** (L. f.) Gleason

Centropogon popayanensis (Benth.) Benth. & Hook. f. ex B. D. Jacks. === **Siphocampylus popayanensis** Benth.

Centropogon porphyrodontus Donn. Sm. === **Centropogon costaricae** (Vatke) McVaugh

Centropogon prostratus Benth. === **Centropogon solanifolius** Benth.

Centropogon psilandrus E. Wimm. === **Centropogon solanifolius** Benth.

Centropogon puerilis E. Wimm. === **Centropogon cornutus** (L.) Druce

Centropogon purdieanus Gleason === **Centropogon trianae** Zahlbr.

Centropogon quitensis E. Wimm. === **Centropogon licayensis** Gleason

Centropogon radicans (Kuntze) McVaugh === **Centropogon coccineus** (Hook.) Regel ex B. D. Jacks.

Centropogon rex E. Wimm. === **Centropogon foetidus** (Kunth) Zahlbr.

Centropogon riparius E. Wimm. === **Centropogon solanifolius** Benth.

Centropogon rostellatus E. Wimm. === **Centropogon grandis** (L. f.) C. Presl

Centropogon rubrosepalus E. Wimm. === **Burmeistera rubrosepala** (E. Wimm.) E. Wimm.

Centropogon rubrovenosus Gleason === **Centropogon reflexus** C. Presl

Centropogon saltuum var. *ferrugineoides* E. Wimm. === **Centropogon parviflorus** (Zahlbr.) Jeppesen

Centropogon scandens Planch. & Oerst. === ?

Centropogon semperflorens E. Wimm. === **Centropogon solanifolius** Benth.

Centropogon serratus Gleason === **Centropogon albolimbatus** E. Wimm.

Centropogon solanifolius var. *hirtellus* E. Wimm. === **Centropogon solanifolius** Benth.

Centropogon solanifolius var. *inconstans* E. Wimm. === **Centropogon solanifolius** Benth.

Centropogon solanifolius var. *speciosus* (Planch.) E. Wimm. === **Centropogon solanifolius** Benth.

Centropogon speciosus Planch. === **Centropogon solanifolius** Benth.

Centropogon stenophyllus E. Wimm. === **Centropogon valerii** Standl.

Centropogon subandinus var. *ventricosus* E. Wimm. === **Centropogon subandinus** Zahlbr.

Centropogon subfalcatus Zahlbr. === **Centropogon solanifolius** Benth.

Centropogon surinamensis (L.) C. Presl === **Centropogon cornutus** (L.) Druce

Centropogon surinamensis var. *angustifolius* Zahlbr. === **Centropogon cornutus** (L.) Druce

Centropogon surinamensis var. *vestitus* Pilger === **Centropogon cornutus** (L.) Druce

Centropogon tessmannii var. *tenuiflorus* E. Wimm. === **Centropogon tessmannii** E. Wimm.

Centropogon tortilis E. Wimm. === **Centropogon granulosus** C. Presl

Centropogon trianae var. *cuspidata* Zahlbr. === **Centropogon trianae** Zahlbr.

Centropogon trianae var. *decemlobus* (Gleason) E. Wimm. === **Centropogon trianae** Zahlbr.

Centropogon trianae var. *purdieanus* (Gleason) E. Wimm. === **Centropogon trianae** Zahlbr.

Centropogon tubulosus Zahlbr. === **Centropogon calycinus** Benth.

Centropogon unduavensis var. *aduanus* E. Wimm. === **Centropogon unduavensis** (Britton) Zahlbr.

Centropogon urubambae var. *estrellanus* E. Wimm. === **Centropogon urubambae** E. Wimm.

Centropogon valerii var. *kupperi* Suesseng. === **Centropogon valerii** Standl.

Centropogon variabilis Gleason === **Burmeistera variabilis** (Gleason) E. Wimm.

Centropogon vimineus E. Wimm. === **Centropogon warscewiczii** Van Houtte ex Regel

Centropogon vinosus E. Wimm. === **Centropogon granulosus** C. Presl

Centropogon virescens (Benth.) Planch. & Oerst. === **Burmeistera virescens** (Benth.) Benth. & Hook. ex Hemsl.

Centropogon vulpinus var. *sonsonensis* E. Wimm. === **Centropogon vulpinus** E. Wimm.

Centropogon warscewiczii Vatke === **Centropogon granulosus** C. Presl

Centropogon warscewiczii f. *brevilobus* E. Wimm. === **Centropogon warscewiczii** Van Houtte ex Regel

Centropogon warscewiczii var. *jervisaei* E. Wimm. === **Centropogon warscewiczii** Van Houtte ex Regel

Centropogon warscewiczii var. *vimineus* (E. Wimm.) E. Wimm. === **Centropogon warscewiczii** Van Houtte ex Regel

Centropogon weddellii E. Wimm. === **Siphocampylus rusbyanus** Britton

Centropogon willdenowianus (A. DC.) E. Wimm. === **Centropogon ayavacensis** (Willd. ex Schult.) Lammers subsp. **ayavacensis**

Centropogon willdenowianus subsp. *cylindricus* (Gleason) McVaugh === **Centropogon ayavacensis** subsp. **cylindricus** (Gleason) Lammers

Centropogon xestus E. Wimm. === **Centropogon solanifolius** Benth.

Centropogon yarumalensis var. *puniceus* E. Wimm. === **Centropogon yarumalensis** E. Wimm.

Centropogon yungasensis var. *angustior* Zahlbr. === **Centropogon yungasensis** Britton

Cephalostigma

Although this genus has been recognized by many authors, it cannot be seperated from *Wahlenbergia* in any meaningful fashion (Tuyn 1960, Thulin 1975). Type [designated by E. Phillips, Gen. S. Afr. Fl. Pl. (ed. 2) 755 (1951)]: *Cephalostigma paniculatum* A. DC.

de Candolle, A. (1830). *Cephalostigma*. In Monographie des Campanulées: 117-118. Paris: Veuve Desray. La. — Establishment of genus as distinct from *Campanula* and *Wahlenbergia*.

Tuyn, P. (1960). *Wahlenbergia*. In C. G. G. J. van Steenis (ed.), Flora Malesiana (ser. I) 6(1): 110-118, illus., map. Djakarta: Noordhoff-Kolff. En. — Evidence for merger of genus with *Wahlenbergia*.

Thulin, M. (1975). The genus *Wahlenbergia* s. lat. (Campanulaceae) in tropical Africa and Madagascar. Symb. Bot. Upsal. 21: 1-223, illus., maps. En. — Data favor assigning species of *Cephalostigma* to *Wahlenbergia*.

Synonyms:

Cephalostigma A. DC. === **Wahlenbergia** Schrad. ex Roth

Cephalostigma bahiense A. DC. === **Wahlenbergia perrottetii** (A. DC.) Thulin

Cephalostigma bahiense var. *major* A. DC. === **Wahlenbergia perrottetii** (A. DC.) Thulin

Cephalostigma candolleanum Hiern. === **Wahlenbergia candolleana** (Hiern) Thulin

Cephalostigma erectum (Roth ex Schult.) Vatke === **Wahlenbergia erecta** (Roth ex Schult.) Tuyn

Cephalostigma erectum var. *coeruleum* Chiov. === **Wahlenbergia hirsuta** (Edgew.) Tuyn

Cephalostigma erectum var. *luteum* Chiov. === **Wahlenbergia flexuosa** (Hook. f. & Thomson) Thulin

Cephalostigma flexuosa Hook. f. & Thomson === **Wahlenbergia flexuosa** (Hook. f. & Thomson) Thulin

Cephalostigma fluminale J. M. Black === **Wahlenbergia fluminalis** (J. M. Black) E. Wimm. ex H. Eichler

Cephalostigma fockeanum Schinz === **Wahlenbergia androsacea** A. DC.

Cephalostigma hirsutum Edgew. === **Wahlenbergia hirsuta** (Edgew.) Tuyn

Cephalostigma hookeri C. B. Clarke === **Wahlenbergia hookeri** (C. B. Clarke) Tuyn

Cephalostigma nanellum R. E. Fr. === **Monopsis zeyheri** (Sond.) Thulin

Cephalostigma paniculatum A. DC. === **Wahlenbergia candollei** Tuyn

Cephalostigma perotifolia (Willd. ex Schult.) Hutch. & Dalziel === **Wahlenbergia erecta** (Roth ex Schult.) Tuyn

Cephalostigma perrottetii A. DC. === **Wahlenbergia perrottetii** (A. DC.) Thulin

Cephalostigma prieurei A. DC. === **Wahlenbergia perrottetii** (A. DC.) Thulin

Cephalostigma pyramidale Schinz === **Wahlenbergia ramosissima** subsp. **lateralis** (Brehmer) Thulin

Cephalostigma ramosissima Hemsl. === **Wahlenbergia ramosissima** (Hemsl.) Thulin

Cephalostigma schimperi Hochst. ex A. Rich. === **Wahlenbergia erecta** (Roth ex Schult.) Tuyn

Cephalostigma spathulatum Thwaites === **Campanula dimorphantha** Schweinf.

Cervicina

Although this name has priority, it was formally rejected in 1906 in favor of *Wahlenbergia*. Type [by monotypy]: *Cervicina campanuloides* Delile.

> Delile, A. (1813-14). Flore d' Égypte, avec explication des planches. In Description de l' Éygypte, histoire naturelle 2: 145-320, illus. Paris: Imprimerie Impériale. Fr. — Establishment of genus.
>
> Fairchild, D. (1911). Seeds and plants imported during the period from April 1 to June 30, 1910: inventory no. 23; nos. 27481 to 28324. USDA Bur. Pl. Industr. Bull. 208: 1-88. En. — H. C. Skeels argues for use of *Cervicina* over *Wahlenbergia* on basis of priority.

Synonyms:
Cervicina Delile === **Wahlenbergia** Schrad. ex Roth
Cervicina campanuloides Delile === **Wahlenbergia campanuloides** (Delile) Vatke
Cervicina denudata (A. DC.) S. Moore === **Wahlenbergia denudata** A. DC.
Cervicina depressa (J. M. Wood & M. Evans) S. Moore === **Wahlenbergia depressa** J. M. Wood & M. Evans
Cervicina gracilis (G. Forst.) Britten === **Wahlenbergia gracilis** (G. Forst.) A. DC.
Cervicina hederacea (L.) Druce === **Wahlenbergia hederacea** (L.) Rchb.
Cervicina huillana (A. DC.) Hiern === **Gunillaea emirnensis** (A. DC.) Thulin
Cervicina lobelioides (L. f.) Druce === **Wahlenbergia lobelioides** (L. f.) Schrad. ex Link
Cervicina pendula Druce === **Wahlenbergia lobelioides** (L. f.) Schrad. ex Link
Cervicina undulata (L. f.) Skeels === **Wahlenbergia undulata** (L. f.) A. DC.

Chrysangia

Link (1829) apparently was unaware of the earlier name *Musschia* when he created this genus to house *Campanula aurea* L. f. of Madeira. Type [by monotypy]: *Chrysangia aurea* (L.) Link.

> Link, T. H. F. (1829). *Chrysangia*. In Handbuch zur Erkennung der nutzbarsten und am häufigsten vorkommenden Gewächse 1: 632. Berlin: Haude & Spenerschen. Ge. — Establishment of genus.

Synonyms:
Chrysangia Link === **Musschia** Dumort.
Chrysangia aurea (L. f.) Link === **Musschia aurea** (L. f.) Dumort.

Clermontia

Lobelioideae, 22 species, Hawaiian Islands, with a pronounced concentration of taxa on the geologically younger islands of Maui and Hawai'i. The genus is divided into two sections, *Clermontia* and *Clermontioideae* (Hillebr.) Rock, each divided in turn into three series (Lammers 1991). *Clermontia* is the sister-group to *Cyanea;* this clade is in turn sister to a clade comprising *Brighamia* and *Delissea* (Givnish et al. 1995, under **Cyanea**; E. Knox, pers. comm.; A. Antonelli, pers. comm.). The treatment here follows Lammers (1991). $2n = 28$. Type [designated by Rock, Monogr. Stud. Haw. Lobelioid. 285 (1919)]: *Clermontia oblongifolia* Gaudich.

> Gaudichaud, C. (1826). *Clermontia*. In Voyage autour du monde, entrepris par order du roi, ... exécuté sur les corvettes de S. M. *l'Uranie* et *la Physicienne*, pendant les années 1817, 1818, 1819 et 1820. Botanique: 459 + pl. 71-73, illus. Paris: Pillet-ainé. La., Fr. — Establishment of genus.
>
> Gray, A. (1861). Notes on Lobeliaceae, Goodeniaceae, &c. of the collections of the U.S. South Pacific Exploring Expedition. Proc. Amer. Acad. Arts 5: 146-152. En. — Synopsis of genus, reducing it to two species.

• Hillebrand, W. (1888). *Clermontia,* Gaud. In Flora of the Hawaiian Islands: 239-244. London: Williams & Norgate. En. — Flora with key, descriptions, infrageneric classification, full nomenclature, and specimen citations.

Rock, J. F. (1913). *Clermontia* Gaud. In The indigenous trees of the Hawaiian Islands: 471-489, 511-512, illus. Honolulu: privately published. En. — Treatment of arborescent species, with key, descriptions, full nomenclature, and specimen citations.

• Rock, J. F. (1919). A monographic study of the Hawaiian species of the tribe Lobelioideae family Campanulaceae, *Clermontia* Gaudichaud. Mem. Bernice Pauahi Bishop Mus. 7(2): 283-340, illus. En. — Monograph with key, descriptions, infrageneric classification, full nomenclature, and specimen citations.

Ka'aiakamanu, D. M. & J. K. Akina (1922). Ohawainui. In A. Akana (transl.), Hawaiian herbs of medicinal value: 30. Honolulu: Territrial Board of Health. En. — Fruit of *C. arborescens* edible and used in treating asthma; latex used to treat deep wounds and as a galactagogue.

St. John, H. (1939). New Hawaiian species of *Clermontia,* including a revision of the *Clermontia grandiflora* group. Hawaiian plant studies 6. Occas. Pap. Bernice Pauahi Bishop Mus. 15: 1-19, illus. En. — Key, descriptions, and specimen citations for small group of taxa.

• Wimmer, F. E. (1943). *Clermontia* Gaudich. In A. Engler & L. Diels, Das Pflanzenreich IV. 276b: 79-96, illus. Leipzig: Wilhelm Engelmann. Ge., La. — Monograph with key, descriptions, infrageneric classification, full nomenclature, and specimen citations.

Selling, O. H. (1947). Studies in Hawaiian pollen statistics II. The pollens of the Hawaiian phanerogams. Lobeliaceae. Spec. Publ. Bernice Pauahi Bishop Mus. 38: 318-320 + pl. 49, illus. En. — Descriptive palynology of two species.

• Wimmer, F. E. (1953). *Clermontia.* In A. Engler & L. Diels, Das Pflanzenreich IV. 276b: 761-762. Berlin: Akademie-Verlag. Ge., La. — Supplement to Wimmer (1943).

Spieth, H. T. (1966). Hawaiian honeycreeper, *Vestiaria coccinea* (Forster), feeding on lobeliad flowers, *Clermontia arborescens* (Mann) Hillebr. Amer. Naturalist 100: 470-473, illus. En. — Pollination by endemic bird.

Stone, B. C. (1967). A review of the endemic genera of Hawaiian plants. *Clermontia* Gaud. Bot. Rev. (Lancaster) 33: 245. En. — Brief descriptive treatment with full nomenclature.

• Wimmer, F. E. (1968). *Clermontia* Gaudich. In A. Engler & L. Diels, Das Pflanzenreich IV. 276c: 825-829 + Taf. 8-11, illus. Berlin: Akademie-Verlag. Ge., La. — Second supplement to Wimmer (1943).

St. John, H. (1969). Types of sections in *Clermontia, Cyanea,* and *Delissea* (Lobeliaceae). Taxon 18: 483. En. — Formal nomenclature of infrageneric taxa.

• Cory, C. (1984). Pollination biology of two species of Hawaiian Lobeliaceae (*Clermontia kakeana* and *Cyanea angustifolia*) and their presumed coevolved relationship with native honeycreepers (Drepanididae), illus. Fullerton: California State University. En. — Unpublished M.A. thesis on reproductive biology, documenting self-fertilization in nature.

Lammers, T. G. & C. E. Freeman (1986). Ornithophily among the Hawaiian Lobelioideae (Campanulaceae): evidence from floral nectar sugar compositions. Amer. J. Bot. 73: 1613-1619. En. — Ornithophilous pollination inferred from hexose-dominant nectars.

Lammers, T. G. (1988). Chromosome numbers and their systematic implications in Hawaiian Lobelioideae (Campanulaceae). Amer. J. Bot. 75:1130-1134. En. — Summary of known chromosome numbers in genus.

Lammers, T. G., S. G. Weller & A. K. Sakai (1988). Japanese White-eye, an introduced passerine, visits the flowers of *Clermontia arborescens,* an endemic Hawaiian lobelioid. Pacific Sci. 41: 74-78. En. — Apparent replacement pollinator for extinct native birds.

Lammers, T. G. (1990). *Clermontia* Gaud. In W. L. Wagner, D. R. Herbst & S. H. Sohmer, Manual of the flowering plants of Hawai'i: 423-437, illus. Honolulu: University of Hawaii Press. En. — Flora with key and descriptions.

Duvall, F. II (1991). Cultivation of *Clermontia, Cyanea, Lobelia,* and *Trematolobelia* from seed. Hawaii's Forests & Wildlife 6(3): 12-13, illus. En. — Practical methods for cultivation of endangered rainforest species.

- Lammers, T. G. (1991). Systematics of *Clermontia* (Campanulaceae-Lobelioideae). Syst. Bot. Monogr. 32: 1-97, illus., maps. — Monograph with keys, descriptions, infrageneric classification, full nomenclature, and specimen citations.

  Lammers, T. G. (1994). Typification of the names of Hawaiian Lobelioideae (Campanulaceae) published by Wilhelm Hillebrand or based upon his specimens. Taxon 43: 545-572. En. — Formal nomenclature of numerous species.

- Lammers, T. G. (1995). Patterns of speciation and biogeography in *Clermontia* (Campanulaceae, Lobelioideae). In W. L. Wagner & V. A. Funk (eds.), Hawaiian biogeography: evolution on a hot spot archipelago: 338-362, illus., maps. Washington: Smithsonian Institution Press. En. — Morphology-based phylogeny of genus used to infer patterns of diversification within archipelago.

  DiLaurenzio, L., B. Johansen & V. A. Albert (1998). The homeotic double-corolla phenotype in Hawaiian *Clermontia* (Lobelioideae: Campanulaceae) is caused by overexpression of a B-function MADS box gene. Amer. J. Bot. 85 (6, suppl.): 124 [abstract]. En. — Genetic basis of unique floral morphology of sect. *Clermontia*.

  Small, M. F. (1999). Floral arrangements. Nat. Hist. 108(4): 46-47, illus. En. — Popular account of information presented by DiLaurenzio et al. (1998).

Clermontia Gaudich., Voy. Uranie: 459 (1829). *Delissea* sect. *Clermontia* (Gaudich.) Baill., Hist. Pl. 8: 364 (1885).
Hawaiian Is. 63.

Clermontia arborescens (H. Mann) Hillebr., Fl. Hawaiian Isl.: 242 (1888).
Hawaiian Is. (Moloka'i, Lāna'i, Maui). 63 HAW. Phan. or nanophan.
  * *Cyanea arborescens* H. Mann, Proc. Amer. Acad. Arts 7: 183 (1867). *Clermontia mannii* H. St. John, Nord. J. Bot. 3: 544 (1983).

subsp. **arborescens**
Hawaiian Is. (W. Maui). 63 HAW. Phan. or nanophan.
Clermontia furcata E. Wimm. in O. Deg. & I. Deg., Fl. Hawaiiensis, fam. 339 (1956).

subsp. **waihiae** (Wawra) Lammers, Syst. Bot. 13: 497 (1988).
Hawaiian Is. (Maui). 63 HAW. Phan. or nanophan.
  * *Delissea waihiae* Wawra, Flora 56: 8 (1873).

subsp. **waikoluensis** (H. St. John) Lammers, Syst. Bot. Monogr. 32: 25 (1991).
Hawaiian Is. (Moloka'i, Lāna'i). 63 HAW. Phan. or nanophan.
  * *Clermontia waikoluensis* H. St. John, Phytologia 63: 352 (1987).

Clermontia calophylla E. Wimm., Pflanzenr. IV.276b: 85 (1943).
Hawaiian Is. (Hawai'i). 63 HAW. Phan. or nanophan. $2n = 28$.
  Clermontia parviflora var. *grandis* Rock, Monogr. Stud. Haw. Lobelioid.: 339 (1919).
  Clermontia montis-loa var. *tenuifolia* Skottsb., Acta Horti Gothob. 2: 269 (1926).
  Clermontia parviflora var. *intermedia* Skottsb., Acta Horti Gothob. 15: 489 (1944).

Clermontia clermontioides (Gaudich.) A. Heller, Minnesota Bot. Stud. 1: 906 (1897).
Hawaiian Is. (Hawai'i). 63 HAW. Phan. or nanophan. $2n = 28$.
  * *Delissea clermontioides* Gaudich., Voy. Bonite, Bot.: pl. 47 (1842). *Clermontia gaudichaudii* Hillebr., Fl. Hawaiian Isl.: 243 (1888).

subsp. **clermontioides**
Hawaiian Is. (Ka'u & S. Kona Distr. of Hawai'i) 63 HAW. Phan. or nanophan.
Clermontia coerulea Hillebr., Fl. Hawaiian Isl.: 243 (1888).
Clermontia coerulea var. *brevidens* Skottsb., Acta Horti Gothob. 2: 268 (1926). *Clermontia coerulea* subsp. *brevidens* (Skottsb.) H. St. John, List & Summary Fl. Pl. Hawaiian Isl.: 337 (1973).
Clermontia coerulea var. *degeneri* Skottsb., Acta Horti Gothob. 15: 495 (1944).
Clermontia coerulea f. *flavescens* E. Wimm., Pflanzenr. IV.276b: 762 (1953).

Clermontia coerulea var. *parvifolia* Rock, Occas. Pap. Bernice Pauahi Bishop Mus. 22: 39 (1957).
Clermontia konaensis H. St. John, Pacific Sci. 30: 37 (1976).

subsp. **rockiana** (E. Wimm.) Lammers, Syst. Bot. 13: 498 (1988).
Hawaiian Is. (N. Kona Distr. of Hawai'i) 63 HAW. Phan. or nanophan. *2n* = 28.
* *Clermontia rockiana* E. Wimm., Repert. Spec. Nov. Regni Veg. 26: 4 (1929).
Clermontia coerulea var. *greenwelliana* E. Wimm., Pflanzenr. IV.276b: 762 (1953).
Clermontia loyana Rock, Occas. Pap. Bernice Pauahi Bishop Mus. 22: 36 (1957).
Clermontia hualalaiensis H. St. John, Phytologia 63: 351 (1987).

Clermontia drepanomorpha Rock, Indig. Trees Haw. Isl.: 473 (1913).
Hawaiian Is. (Hawai'i). 63 HAW. Phan. or nanophan. *2n* = 28.

Clermontia fauriei H. Lév., Repert. Spec. Nov. Regni Veg. 10: 156 (1911).
Hawaiian Is. (Kaua'i, O'ahu). 63 HAW. Phan. or nanophan.
Clermontia clermontioides var. *epiphytica* Hochr., Candollea 5: 294 (1934).
Clermontia clermontioides var. *hirsutiflora* Rock, Occas. Pap. Bernice Pauahi Bishop Mus. 22: 39 (1957).

Clermontia grandiflora Gaudich., Voy. Uranie: 459 (1829). *Lobelia grandiflora* (Gaudich.) Endl., Ann. Wiener Mus. Naturgesch. 1: 170 (1836).
Hawaiian Is. (Moloka'i, Lāna'i, Maui). 63 HAW. Phan. or nanophan. *2n* = 28.
Delissea filigera Wawra, Flora 56: 31 (1873).

subsp. **grandiflora**
Hawaiian Is. (W. Maui). 63 HAW. Phan. or nanophan. *2n* = 28.
Clermontia reticulata f. *pilifera* H. St. John, Occas. Pap. Bernice Pauahi Bishop Mus. 15: 8 (1939).
Clermontia grandiflora f. *hamata* E. Wimm., Pflanzenr. IV.276b: 89 (1943).
Clermontia earina H. St. John, Nord. J. Bot. 3: 543 (1983).
Clermontia spatulata H. St. John, Phytologia 63: 352 (1987).

subsp. **maxima** Lammers, Syst. Bot. Monogr. 32: 77 (1991).
Hawaiian Is. (E. Maui). 63 HAW. Phan. or nanophan.

subsp. **munroi** (H. St. John) Lammers, Syst. Bot. 13: 499 (1988).
Hawaiian Is. (Moloka'i, Lāna'i, Maui). 63 HAW. Phan. or nanophan.
Clermontia forbesii H. St. John, Occas. Pap. Bernice Pauahi Bishop Mus. 15: 4 (1939).
Clermontia grandiflora var. *forbesii* (H. St. John) E. Wimm., Pflanzenr. IV.276b: 89 (1943).
Clermontia hirsutinervis H. St. John, Occas. Pap. Bernice Pauahi Bishop Mus. 15: 6 (1939). *Clermontia grandiflora* f. *hirsutinervis* (H. St. John) E. Wimm., Pflanzenr. IV.276b: 89 (1943).
Clermontia reticulata H. St. John, Occas. Pap. Bernice Pauahi Bishop Mus. 15: 7 (1939).
Clermontia grandiflora var. *vulgata* E. Wimm., Pflanzenr. IV.276b: 89 (1943).
* *Clermontia munroi* H. St. John, Occas. Pap. Bernice Pauahi Bishop Mus. 15: 8 (1939).
Clermontia molokaiensis H. St. John, Occas. Pap. Bernice Pauahi Bishop Mus. 15: 9 (1939).
Clermontia subpetiolata H. St. John, Occas. Pap. Bernice Pauahi Bishop Mus. 15: 10 (1939). *Clermontia grandiflora* var. *subpetiolata* (H. St. John) E. Wimm., Pflanzenr. IV.276b: 90 (1943).
Clermontia wailauensis H. St. John, Occas. Pap. Bernice Pauahi Bishop Mus. 15: 11 (1939).
Clermontia grandiflora f. *nitida* E. Wimm., Pflanzenr. IV.276c: 828 (1968).

Clermontia hawaiiensis (Hillebr.) Rock, Indig. Trees Haw. Isl.: 477 (1913).
Hawaiian Is. (Hawai'i). 63 HAW. Phan. or nanophan.
* *Clermontia macrocarpa* var. *hawaiiensis* Hillebr., Fl. Hawaiian Isl. 241 (1888).
Clermontia kohalae var. *hiloensis* E. Wimm., Pflanzenr. IV.276b: 84 (1943).
Clermontia subteralbula H. St. John, Phytologia 63: 352 (1987), as 'subteralbulus'.

Clermontia kakeana Meyen, Reise 2: 139 (1834). *Lobelia kakeana* (Meyen) Endl., Ann. Wiener Mus. Naturgesch. 1: 170 (1836).

Hawaiian Is. (O'ahu, Moloka'i, Maui). 63 HAW. Phan. or nanophan. $2n = 28$.

    Clermontia macrocarpa Gaudich., Voy. Bonite, Bot.: pl. 49 (1842). *Clermontia grandiflora* var. *longifolia* A. Gray, Proc. Amer. Acad. Arts 5: 150 (1861).

    Clermontia macrophylla Nutt., Trans. Amer. Philos. Soc. (ser. 2) 8: 251 (1842).

    Clermontia macrocarpa var. *cymosa* Hillebr., Fl. Hawaiian Isl.: 241 (1888).

    Clermontia macrocarpa var. *rosea* Hillebr., Fl. Hawaiian Isl.: 241 (1888). *Clermontia kakeana* f. *rosea* (Hillebr.) O. Deg. & I. Deg., Fl. Hawaiiensis, fam. 339 (1958). *Clermontia kakeana* var. *rosea* (Hillebr.) H. St. John, Phytologia 63: 351 (1987).

    Clermontia kakeana var. *forbesii* O. Deg. & I. Deg., Fl. Hawaiiensis, fam. 339 (1958).

    Clermontia kakeana f. *gracilis* E. Wimm., Pflanzenr. IV.276c: 826 (1968).

    Clermontia montis-loa f. *molokaiensis* E. Wimm., Pflanzenr. IV.276c: 827 (1968).

    Clermontia kakeana var. *orientalis* H. St. John, Pacific Sci. 25: 60 (1971).

    Clermontia glabra H. St. John, Phytologia 63: 350 (1987).

    Clermontia mauiensis H. St. John, Phytologia 63: 351 (1987).

Clermontia kohalae Rock, Indig. Trees Haw. Isl.: 476 (1913).

Hawaiian Is. (Hawai'i). 63 HAW. Phan. or nanophan.

    Clermontia kohalae var. *robusta* Rock, Coll. Hawaii Publ. Bull. 2: 41 (1913).

    Clermontia leptoclada var. *holopsila* E. Wimm., Pflanzenr. IV.276b: 82 (1943).

    Clermontia leptoclada var. *urceolata* Rock, Occas. Pap. Bernice Pauahi Bishop Mus. 22: 39 (1957).

    Clermontia convallis E. Wimm., Pflanzenr. IV.276c: 827 (1968).

    Clermontia epilosa H. St. John, Phytologia 63: 350 (1987).

Clermontia × leptoclada Rock, Indig. Trees Haw. Isl.: 477 (1913), pro sp. C. drepanomorpha × C. kohalae

Hawaiian Is. (Hawai'i). 63 HAW. Phan. or nanophan.

Clermontia lindseyana Rock, Occas. Pap. Bernice Pauahi Bishop Mus. 23: 67 (1962).

Hawaiian Is. (Hawai'i). 63 HAW. Phan. or nanophan.

    * *Clermontia hawaiiensis* var. *grandis* Rock, Occas. Pap. Bernice Pauahi Bishop Mus. 22: 43 (1957).

    Clermontia lindseyana var. *livida* Rock, Occas. Pap. Bernice Pauahi Bishop Mus. 23: 69 (1962).

    Clermontia albimontis H. St. John, Phytologia 63: 350 (1987).

    Clermontia viridis H. St. John, Phytologia 63: 352 (1987).

Clermontia micrantha (Hillebr.) Rock, Monogr. Stud. Haw. Lobelioid.: 334 (1919).

Hawaiian Is. (Lāna'i, W. Maui). 63 HAW. Nanophan.

    * *Clermontia multiflora* var. *micrantha* Hillebr., Fl. Hawaiian Isl.: 242 (1888).

    Clermontia multiflora f. *montana* Rock, Indig. Trees Haw. Isl.: 511 (1913).

Clermontia montis-loa Rock, Coll. Hawaii Publ. Bull. 2: 40 (1913).

Hawaiian Is. (Hawai'i). 63 HAW. Phan. or nanophan. $2n = 28$.

    Clermontia montis-loa f. *globosa* Rock, Monogr. Stud. Haw. Lobelioid.: 337 (1919).

Clermontia multiflora Hillebr., Fl. Hawaiian Isl.: 242 (1888).

Hawaiian Is. (O'ahu, W. Maui). 63 HAW. Nanophan.

Clermontia oblongifolia Gaudich., Voy. Uranie: 459 (1829). *Lobelia oblongifolia* (Gaudich.) Endl., Ann. Wiener Mus. Naturgesch. 1: 170 (1836). *Clermontia grandiflora* var. *oblongifolia* (Gaudich.) A. Gray, Proc. Amer. Acad. Arts 5: 150 (1861).

Hawaiian Is. (O'ahu, Moloka'i, Lāna'i, Maui). 63 HAW. Phan. or nanophan. $2n = 28$.

subsp. **brevipes** (E. Wimm.) Lammers, Syst. Bot. 13: 500 (1988).
Hawaiian Is. (Moloka'i). 63 HAW. Phan. or nanophan.
   * *Clermontia oblongifolia* f. *brevipes* E. Wimm., Pflanzenr. IV.276b: 88 (1943).

subsp. **mauiensis** (Rock) Lammers, Syst. Bot. 13: 500 (1988).
Hawaiian Is. (Lāna'i, Maui). 63 HAW. Phan. or nanophan.
   * *Clermontia oblongifolia* var. *mauiensis* Rock, Indig. Trees Haw. Isl.: 476 (1913). *Clermontia oblongifolia* f. *mauiensis* (Rock) O. Deg., Fl. Hawaiiensis, fam. 339 (1937).

subsp. **oblongifolia**
Hawaiian Is. (O'ahu). 63 HAW. Phan. or nanophan. $2n = 28$.
Clermontia oblongifolia f. *kaalae* O. Deg., Fl. Hawaiiensis, fam. 339 (1937).
Clermontia aspera E. Wimm., Pflanzenr. IV.276c: 827 (1968).

Clermontia pallida Hillebr., Fl. Hawaiian Isl.: 241 (1888).
Hawaiian Is. (Moloka'i). 63 HAW. Phan. or nanophan.
   Clermontia pallida var. *ramosissima* Rock, Monogr. Stud. Haw. Lobelioid.: 319 (1919).
   Clermontia oblongifolia f. *glabra* H. St. John, Nord. J. Bot. 3: 545 (1983).

Clermontia × paradisia E. Wimm., Pflanzenr. IV.276b: 761 (1953), pro sp. C. kohalae × C. parviflora
Hawaiian Is. (Hawai'i). 63 HAW. Nanophan.

Clermontia parviflora Gaudich. ex A. Gray, Proc. Amer. Acad. Arts 5: 150 (1861).
Hawaiian Is. (Hawai'i). 63 HAW. Nanophan. $2n = 28$.
   Clermontia parviflora var. *pleiantha* Hillebr., Fl. Hawaiian Isl.: 242 (1888).
   Cyanea blinii H. Lév., Repert. Spec. Nov. Regni Veg. 10: 156 (1911).
   Clermontia parviflora var. *umbraticola* Skottsb., Acta Horti Gothob. 2: 270 (1926).

Clermontia peleana Rock, Indig. Trees Haw. Isl.: 483 (1913).
Hawaiian Is. (E. Maui, Hawai'i). 63 HAW. Phan. or nanophan.

subsp. **peleana**
Hawaiian Is. (Puna to N. Hilo Distr. of Hawai'i). 63 HAW. Phan. or nanophan.

subsp. **singuliflora** (Rock) Lammers, Syst. Bot. Monogr. 32: 32 (1991).
Hawaiian Is. (E. Maui, Hamakua Distr. of Hawai'i). 63 HAW. Phan. or nanophan.
   * *Clermontia gaudichaudii* var. *singuliflora* Rock, Indig. Trees Haw. Isl.: 512 (1913).
     Clermontia singuliflora (Rock) Rock, Monogr. Stud. Haw. Lobelioid.: 293 (1919).
     Clermontia clermontioides var. *singuliflora* (Rock) Hochr., Candollea 5: 294 (1934).
   Clermontia gaudichaudii var. *barbata* Rock, Monogr. Stud. Haw. Lobelioid.: 293 (1919).
     Clermontia clermontioides var. *mauiensis* Hochr., Candollea 5: 294 (1934). *Clermontia clermontioides* var. *barbata* (Rock) H. St. John, List & Summary Fl. Pl. Hawaiian Isl.: 337 (1973).

Clermontia persicifolia Gaudich., Voy. Uranie: 459 (1829). *Lobelia persicifolia* (Gaudich.) Endl., Ann. Wiener Mus. Naturgesch. 1: 170 (1836); non Lam., Encycl. 3: 584 (1792); nec Cav., Icon. 6: 12 (1800). *Lobelia clermontiana* Gaudich. ex D. Dietr., Syn. Pl. 1: 728 (1839).
Hawaiian Is. (O'ahu). 63 HAW. Phan. or nanophan.
   Clermontia epiphytica H. St. John, Occas. Pap. Bernice Pauahi Bishop Mus. 15: 21 (1939).

Clermontia pyrularia Hillebr., Fl. Hawaiian Isl.: 243 (1888).
Hawaiian Is. (Hawai'i). 63 HAW. Phan.
   Delissea obtusa var. *mollis* A. Gray, Proc. Amer. Acad. Arts 5: 148 (1861).

Clermontia samuelii C. N. Forbes, Occas. Pap. Bernice Pauahi Bishop Mus. 7: 37 (1920).
Hawaiian Is. (E. Maui). 63 HAW. Phan. or nanophan.

subsp. **hanaensis** (H. St. John) Lammers, Syst. Bot. 13: 500 (1988).
   Hawaiian Is. (E. Maui). 63 HAW. Phan. or nanophan.
   ** Clermontia hanaensis* H. St. John, Occas. Pap. Bernice Pauahi Bishop Mus. 15: 12 (1939).
   Clermontia kipahuluensis H. St. John, Phytologia 63: 351 (1987).

subsp. **samuelii**
   Hawaiian Is. (E. Maui). 63 HAW. Phan. or nanophan.
   Clermontia rosacea H. St. John, Pacific Sci. 25: 61 (1971).
   Clermontia gracilis H. St. John, Phytologia 63: 350 (1987).

Clermontia tuberculata C. N. Forbes, Occas. Pap. Bernice Pauahi Bishop Mus. 5: 8 (1912).
   Hawaiian Is. (East Maui). 63 HAW. Phan.
   Clermontia tuberculata var. *subtuberculata* H. St. John, Phytologia 40: 97 (1978).

Clermontia waimeae Rock, Coll. Hawaii Publ. Bull. 2: 40 (1913).
   Hawaiian Is. (Hawai‘i). 63 HAW. Phan. or nanophan.
   Clermontia parviflora var. *calycina* Rock, Indig. Trees Haw. Isl.: 512 (1913).
   Clermontia waimeae var. *obovata* Rock, Occas. Pap. Bernice Pauahi Bishop Mus. 22: 41 (1957).
   Clermontia waimeae var. *longisepala* Rock, Occas. Pap. Bernice Pauahi Bishop Mus. 22: 42 (1957).
   Clermontia waimeae f. *lanceolata* E. Wimm., Pflanzenr. IV.276c: 831 (1968).
   Clermontia bicolorata H. St. John, Phytologia 63: 350 (1987).
   Clermontia kahuaensis H. St. John, Phytologia 63: 351 (1987).

Synonyms:
Clermontia albimontis H. St. John === **Clermontia lindseyana** Rock
Clermontia aspera E. Wimm. === **Clermontia oblongifolia** Gaudich. subsp. **oblongifolia**
Clermontia bicolorata H. St. John === **Clermontia waimeae** Rock
Clermontia carinifera H. Lév. === ?
Clermontia clermontioides var. *barbata* (Rock) H. St. John === **Clermontia peleana** subsp. **singuliflora** (Rock) Lammers
Clermontia clermontioides var. *epiphytica* Hochr. === **Clermontia fauriei** H. Lév.
Clermontia clermontioides var. *hirsutiflora* Rock === **Clermontia fauriei** H. Lév.
Clermontia clermontioides var. *mauiensis* Hochr. === **Clermontia peleana** subsp. **singuliflora** (Rock) Lammers
Clermontia clermontioides var. *singuliflora* (Rock) Hochr. === **Clermontia peleana** subsp. **singuliflora** (Rock) Lammers
Clermontia coerulea Hillebr. === **Clermontia clermontioides** (Gaudich.) A. Heller subsp. **clermontioides**
Clermontia coerulea f. *flavescens* E. Wimm. === **Clermontia clermontioides** (Gaudich.) A. Heller subsp. **clermontioides**
Clermontia coerulea subsp. *brevidens* (Skottsb.) H. St. John === **Clermontia clermontioides** (Gaudich.) A. Heller subsp. **clermontioides**
Clermontia coerulea var. *brevidens* Skottsb. === **Clermontia clermontioides** (Gaudich.) A. Heller subsp. **clermontioides**
Clermontia coerulea var. *degeneri* Skottsb. === **Clermontia clermontioides** (Gaudich.) A. Heller subsp. **clermontioides**
Clermontia coerulea var. *greenwelliana* E. Wimm. === **Clermontia clermontioides** subsp. **rockiana** (E. Wimm.) Lammers
Clermontia coerulea var. *parvifolia* Rock === **Clermontia clermontioides** (Gaudich.) A. Heller subsp. **clermontioides**
Clermontia convallis E. Wimm. === **Clermontia kohalae** Rock
Clermontia earina H. St. John === **Clermontia grandiflora** Gaudich. subsp. **grandiflora**
Clermontia epilosa H. St. John === **Clermontia kohalae** Rock
Clermontia epiphytica H. St. John === **Clermontia persicifolia** Gaudich.

Clermontia forbesii H. St. John === **Clermontia grandiflora** subsp. **munroi** (H. St. John) Lammers

Clermontia fulva H. Lév. === ?

Clermontia furcata E. Wimm. === **Clermontia arborescens** (H. Mann) Hillebr. subsp. **arborescens**

Clermontia gaudichaudii Hillebr. === **Clermontia clermontioides** (Gaudich.) A. Heller subsp. **clermontioides**

Clermontia gaudichaudii var. *barbata* Rock === **Clermontia peleana** subsp. **singuliflora** (Rock) Lammers

Clermontia gaudichaudii var. *singuliflora* Rock === **Clermontia peleana** subsp. **singuliflora** (Rock) Lammers

Clermontia glabra H. St. John === **Clermontia kakeana** Meyen

Clermontia gracilis H. St. John === **Clermontia samuelii** C. N. Forbes subsp. **samuelii**

Clermontia grandiflora f. *hamata* E. Wimm. === **Clermontia grandiflora** Gaudich. subsp. **grandiflora**

Clermontia grandiflora f. *hirsutinervis* (H. St. John) E. Wimm. === **Clermontia grandiflora** subsp. **munroi** (H. St. John) Lammers

Clermontia grandiflora f. *nitida* E. Wimm. === **Clermontia grandiflora** subsp. **munroi** (H. St. John) Lammers

Clermontia grandiflora var. *forbesii* (H. St. John) E. Wimm. === **Clermontia grandiflora** subsp. **munroi** (H. St. John) Lammers

Clermontia grandiflora var. *longifolia* A. Gray === **Clermontia kakeana** Meyen

Clermontia grandiflora var. *oblongifolia* (Gaudich.) A. Gray === **Clermontia oblongifolia** Gaudich.

Clermontia grandiflora var. *subpetiolata* (H. St. John) E. Wimm. === **Clermontia grandiflora** subsp. **munroi** (H. St. John) Lammers

Clermontia grandiflora var. *vulgata* E. Wimm. === **Clermontia grandiflora** subsp. **munroi** (H. St. John) Lammers

Clermontia haleakalensis Rock === **Cyanea pohaku** Lammers

Clermontia hanaensis H. St. John === **Clermontia samuelii** subsp. **hanaensis** (H. St. John) Lammers

Clermontia hawaiiensis var. *grandis* Rock === **Clermontia lindseyana** Rock

Clermontia hirsutinervis H. St. John === **Clermontia grandiflora** subsp. **munroi** (H. St. John) Lammers

Clermontia hualalaiensis H. St. John === **Clermontia clermontioides** subsp. **rockiana** (E. Wimm.) Lammers

Clermontia kahuaensis H. St. John === **Clermontia waimeae** Rock

Clermontia kakeana f. *gracilis* E. Wimm. === **Clermontia kakeana** Meyen

Clermontia kakeana f. *rosea* (Hillebr.) O. Deg. & I. Deg. === **Clermontia kakeana** Meyen

Clermontia kakeana var. *forbesii* O. Deg. & I. Deg. === **Clermontia kakeana** Meyen

Clermontia kakeana var. *orientalis* H. St. John === **Clermontia kakeana** Meyen

Clermontia kakeana var. *rosea* (Hillebr.) H. St. John === **Clermontia kakeana** Meyen

Clermontia kipahuluensis H. St. John === **Clermontia samuelii** subsp. **hanaensis** (H. St. John) Lammers

Clermontia kohalae var. *hiloensis* E. Wimm. === **Clermontia hawaiiensis** (Hillebr.) Rock

Clermontia kohalae var. *robusta* Rock === **Clermontia kohalae** Rock

Clermontia konaensis H. St. John === **Clermontia clermontioides** (Gaudich.) A. Heller subsp. **clermontioides**

Clermontia leptoclada var. *holopsila* E. Wimm. === **Clermontia kohalae** Rock

Clermontia leptoclada var. *urceolata* Rock === **Clermontia kohalae** Rock

Clermontia lindseyana var. *livida* Rock === **Clermontia lindseyana** Rock

Clermontia loyana Rock === **Clermontia clermontioides** subsp. **rockiana** (E. Wimm.) Lammers

Clermontia macrocarpa Gaudich. === **Clermontia kakeana** Meyen

Clermontia macrocarpa var. *cymosa* Hillebr. === **Clermontia kakeana** Meyen

Clermontia macrocarpa var. *hawaiiensis* Hillebr. === **Clermontia hawaiiensis** (Hillebr.) Rock

Clermontia macrocarpa var. *rosea* Hillebr. === **Clermontia kakeana** Meyen

Clermontia macrophylla Nutt. === **Clermontia kakeana** Meyen

Clermontia mannii H. St. John === **Clermontia arborescens** (H. Mann) Hillebr.

Clermontia mauiensis H. St. John === **Clermontia kakeana** Meyen

Clermontia molokaiensis H. St. John === **Clermontia grandiflora** subsp. **munroi**
(H. St. John) Lammers

Clermontia montis-loa f. *globosa* Rock === **Clermontia montis-loa** Rock

Clermontia montis-loa f. *molokaiensis* E. Wimm. === **Clermontia kakeana** Meyen

Clermontia montis-loa var. *tenuifolia* Skottsb. === **Clermontia calophylla** E. Wimm.

Clermontia multiflora f. *montana* Rock === **Clermontia micrantha** (Hillebr.) Rock

Clermontia multiflora var. *micrantha* Hillebr. === **Clermontia micrantha** (Hillebr.) Rock

Clermontia munroi H. St. John === **Clermontia grandiflora** subsp. **munroi** (H. St. John)
Lammers

Clermontia oblongifolia f. *mauiensis* (Rock) O. Deg. === **Clermontia oblongifolia** subsp.
mauiensis (Rock) Lammers

Clermontia oblongifolia f. *glabra* H. St. John === **Clermontia pallida** Hillebr.

Clermontia oblongifolia f. *kaalae* O. Deg. === **Clermontia oblongifolia** Gaudich. subsp.
oblongifolia

Clermontia oblongifolia f. *brevipes* E. Wimm. === **Clermontia oblongifolia** subsp. **brevipes**
(E. Wimm.) Lammers

Clermontia oblongifolia var. *mauiensis* Rock === **Clermontia oblongifolia** subsp. **mauiensis**
(Rock) Lammers

Clermontia pallida var. *ramosissima* Rock === **Clermontia pallida** Hillebr.

Clermontia parviflora var. *calycina* Rock === **Clermontia waimeae** Rock

Clermontia parviflora var. *grandis* Rock === **Clermontia calophylla** E. Wimm.

Clermontia parviflora var. *intermedia* Skottsb. === **Clermontia calophylla** E. Wimm.

Clermontia parviflora var. *pleiantha* Hillebr. === **Clermontia parviflora** Gaudich. ex
A. Gray

Clermontia parviflora var. *umbraticola* Skottsb. === **Clermontia parviflora** Gaudich. ex
A. Gray

Clermontia pluriflora H. St. John === **Cyanea hirtella** (H. Mann) Hillebr.

Clermontia reticulata H. St. John === **Clermontia grandiflora** subsp. **munroi** (H. St. John)
Lammers

Clermontia reticulata f. *pilifera* H. St. John === **Clermontia grandiflora** Gaudich. subsp.
grandiflora

Clermontia rockiana E. Wimm. === **Clermontia clermontioides** subsp. **rockiana** (E.
Wimm.) Lammers

Clermontia rosacea H. St. John === **Clermontia samuelii** subsp. **samuelii**

Clermontia singuliflora (Rock) Rock === **Clermontia peleana** subsp. **singuliflora** (Rock)
Lammers

Clermontia spatulata H. St. John === **Clermontia grandiflora** Gaudich. subsp. **grandiflora**

Clermontia subpetiolata H. St. John === **Clermontia grandiflora** subsp. **munroi**
(H. St. John) Lammers

Clermontia subteralbula H. St. John === **Clermontia hawaiiensis** (Hillebr.) Rock

Clermontia tuberculata var. *subtuberculata* H. St. John === **Clermontia tuberculata**
C. N. Forbes

Clermontia viridis H. St. John === **Clermontia lindseyana** Rock

Clermontia waikoluensis H. St. John === **Clermontia arborescens** subsp. **waikoluensis** (H. St.
John) Lammers

Clermontia wailauensis H. St. John === **Clermontia grandiflora** subsp. **munroi** (H. St. John)
Lammers

Clermontia waimeae f. *lanceolata* E. Wimm. === **Clermontia waimeae** Rock

Clermontia waimeae var. *longisepala* Rock === **Clermontia waimeae** Rock

Clermontia waimeae var. *obovata* Rock === **Clermontia waimeae** Rock

Clintonia

This name cannot be used because it is a later homonym of *Clintonia* Raf. (Convallariaceae). Type [by monotypy]: *Clintonia elegans* Dougl. ex Lindl.

> Lindley, J. (1829). *Clintonia elegans.* Elegant clintonia. Edwards' Bot. Reg. 15: fol. 1241 + 2 pp., illus. En. — Establishment of genus.
> Rafinesque, C. S. (1832). On 3 n. sp. of *Clintonia.* Atlantic J.: 120. En. — Validation of name *Bolelia* (q.v.) to replace illegitimate *Clintonia.*

Synonyms:
Clintonia Dougl. ex Lindl. === **Downingia** Torr.
Clintonia bergiana (Cham.) G. Don === **Grammatotheca bergiana** (Cham.) C. Presl
Clintonia corymbosa A. DC. === **Downingia elegans** (Dougl. ex Lindl.) Torr.
Clintonia elegans Dougl. ex Lindl. === **Downingia elegans** (Dougl. ex Lindl.) Torr.
Clintonia pulchella Lindl. === **Downingia pulchella** (Lindl.) Torr.
Clintonia pulchella var. *pumila* Benth. === **Downingia pulchella** (Lindl.) Torr.
Clintonia pusilla A. DC. === **Downingia pusilla** (A. DC.) Torr.

Codiphus

Rafinesque's (1837) segregation of this from *Prismatocarpus* was largely based on the mistaken impression that the corolla was tubular and the anthers connate. Type [by monotypy]: *Codiphus nitidus* (L'Hér.) Raf.

> Rafinesque, C. S. (1837). Flora Telluriana, pars secunda. Philadelphia: published privately. En. — Establishment of genus as distinct from *Campanula* and *Prismatocarpus.*

Synonyms:
Codiphus Raf. === **Prismatocarpus** L'Hér.
Codiphus nitidus (L'Hér.) Raf. === **Prismatocarpus nitidus** L'Hér.

Codonopsis

Campanuloideae, 59 species, southern and eastern Asia, from Afghanistan to Kamchatka and as far south as Java, with the greatest concentration of species in China. As circumscribed here, the genus includes *Leptocodon* and *Campanumoea* sect. *Campanumoea.* Shen & Hong (1983), who treated *Campanumoea* and *Leptocodon* as distinct, divided *Codonopsis* into subg. *Codonopsis* (two sections), subg. *Obconicapsula* D. Y. Hong, and subg. *Pseudocodonopsis* Kom. In a preliminary phylogenetic analysis based on nucleic acid data (Eddie et al. 2003, under **General**), the seven species of subg. *Codonopsis* sampled formed a clade that also included the type species of *Campanumoea, Leptocodon,* and *Platycodon.* This clade formed a tetrachotomy with (1) the type of *Codonopsis* subg. *Obconicapsula,* (2) the type of *Canarina,* and (3) the single included species of *Cyananthus.* This larger four-branched clade (the "platycodonoid clade") was then sister to the remainder of the Campanuloideae. Until additional species (particularly representatives of subg. *Pseudocodonopsis*) can be analyzed, *Platycodon* is here maintained as distinct and the sole species of subg. *Obconicapsula* is retained in *Codonopsis.* Although no modern monograph or revision is available for the genus, the floristic accounts for China (Shen & Hong 1983, Hong 1983) provide keys and descriptions for about three-fourths of the species. Grey-Wilson's (1990) account of species in cultivation is also useful, as it provides keys and descriptions for over half the members of the genus, including the commonest and most widely distributed. $2n = 16$. Type [designated by Fed. in Kom., Fl. URSS 24: 433 (1957)]: *Codonopsis viridis* Wall.

Roxburgh, W. (1824). *Codonopsis*. In Flora Indica 2: 103-107. Serampore: Mission Press. En. — Establishment of genus.

Lindley, J. (1842). *Glossocomia* [sic] *ovata*. Ovate Pouchbell. Edwards' Bot. Reg. 28: pl. 3 + 2 pp., illus. En. — Portrait and description of *C. ovata,* with discussion of recognition of *Glosocomia* as distinct and the correct spelling of that name.

Hooker, W. J. (1856). *Codonopsis rotundifolia*. Round-leaved codonopsis. Bot. Mag. 82: tab. 4942 + 2 pp., illus. En. — Portrait with description and brief comments on generic boundaries.

Hooker, W. J. (1857). *Codonopsis rotundifolia;* var. *grandiflora*. Round-leaved codonopsis; large-flowered var. Bot. Mag. 83: tab. 5018 + 1 pg., illus. En. — Portrait and description of *C. rotundifolia*.

Hooker, W. J. (1863). *Codonopsis cordata*. Heart-leaved codonopsis. Bot. Mag. 89: tab. 5372 + 2 pp., illus. En. — Portrait and description of *C. javanica* subsp. *javanica*.

Oliver, D. (1891a). *Codonopsis tangshen,* Oliv. Hooker's Icon. Pl. 20: pl. 1966 + 1 pg., illus. En. — Portrait with description.

Oliver, D. (1891b). *Codonopsis henryi,* Oliv. Hooker's Icon. Pl. 20: pl. 1967 + 1 pg., illus. En. — Portrait with description.

Oliver, D. (1895). *Codonopsis convolvulacea,* Kurz. Hooker's Icon. Pl. 24: pl. 2385 + 1 pg., illus. En. — Portrait with description.

Skan, S. A. (1906). *Codonopsis tangshen*. Bot. Mag. 132: tab. 8090 + 2 pp., illus. En. — Portrait with description and information on pharmaceutical usage; 600 tons of roots exported from Hankow to rest of China annually.

Wilson, E. H. (1907). T'ang-shên. (*Codonopsis tangshen,* Oliv.). Bull. Misc. Inform. Kew 1907: 9, illus. En. — Details on use of roots as substitute for ginseng; 500 tons of roots exported from Hankow to rest of China annually.

• Chipp, T. F. (1908). A revision of the genus *Codonopsis,* Wall. J. Linn. Soc., Bot. 38: 374-391, illus. En., La. — Monograph with key, descriptions, full nomenclature, and specimen citations, as well as discussion of unique non-apical calyx lobe insertion found in some species; note on pg. 378 indicates this paper appeared after Komarov (1908).

• Komarov, V. L. (1908). Prolegomena ad floras chinae nec non Mongoliae. VII. Revisio critica specierum generis *Codonopsis* Wall. Trudy Imp. S.-Peterburgsk. Bot. Sada 29: 98-124 + pl. II, illus. La. — Floristic treatment with key, descriptions, infrageneric classification, full nomenclature, and specimen citations.

Prain, D. (1908). *Codonopsis convolvulacea*. Bot. Mag. 134: tab. 8178 + 3 pp., illus. En. — Portrait with description and notes on cultivation.

• Anthony, J. (1926). A key to the genus *Codonopsis,* Wall., with an account of two undescribed species. Notes Roy. Bot. Gard. Edinburgh 15: 173-190 + pl. ccxvi –ccxvii, illus. En. — Update of Chipp (1908), incorporating Komarov's (1908) infrageneric classification, with revised key, some descriptions, full nomenclature, and some specimen citations.

Stapf, O. (1930). *Codonopsis ovata*. Bot. Mag. 154: tab. 9087 + 4 pp., illus. En. — Portrait with description.

Anonymous (1931). The genus *Codonopsis*. Gard. Chron. (ser. 3) 90: 112, illus. En. — Brief summary of genus from perspective of horticulture.

• Nannfeldt, J. A. (1931). Some notes on the genus *Codonopsis* Wall. Notes Roy. Bot. Gard. Edinburgh 16: 149-160 + pl. 227-230, illus. En. — Update of Chipp (1908) and Anthony (1926).

Stapf, O. (1931). *Codonopsis meleagris*. Bot. Mag. 154: tab. 9237 + 3 pp., illus. En. — Portrait.

Cox, E. H. M. (1936). *Codonopsis* in cultivation. New Flora & Silva 9: 25-30 + fig. vi-xii, illus. En. — Popular account of species in cultivation.

Fletcher, H. R. (1937). The genus *Codonopsis* in cultivation. Gard. Chron. (ser. 3) 102: 80-83,120-121, illus. En. — Account of species in cultivation, with key and descriptions.

Wilkie, D. (1937). New and interesting plants. *Codonopsis vinciflora* (Kom). New Flora & Silva 10: 58-59 + fig. xxi, illus. En. — Introduction to horticulture.

Marquand, C. V. B. & F. Ballard (1938). *Codonopsis subglobosa*. Bot. Mag. 147: tab. 8879 + 2 pp., illus. En. — Portrait with description.

Ballard, F. (1939). *Codonopsis convolvulacea* var. *forrestii*. Bot. Mag. 162: tab. 9581 + 4 pp., illus. En. — Portrait of *C. forrestii*, with description.

Nannfeldt, J. A. (1940). Bemerkungen über einige morphologische Eigentümlichkeiten in der Gattung *Codonopsis* Wall., nebst beschreibung drei neuer Arten. Svensk Bot. Tidskr. 34: 381-388 + pl. IV-V, illus. Ge. — Discussion of unique calyx lobe insertion in some species.

Fletcher, H. R. (1946). *Codonopsis mollis*. Bot. Mag. 164: tab. 9677 + 4 pp., illus. En. — Portrait and description of *C. thalictrifolia*.

Fletcher, H. R. (1949). *Codonopsis vinciflora*. Bot. Mag. 166: tab. 59 + 4 pp., illus. En. — Portrait with description and notes on cultivation.

Fletcher, H. R. (1950a). *Codonopsis macrocalyx*. Bot. Mag. 167: tab. 94 + 4 pp., illus. En. — Portrait with description.

Fletcher, H. R. (1950b). *Codonopsis rotundifolia* var. *angustifolia*. Bot. Mag. 167: tab. 131 + 4 pp., illus. En. — Portrait of *C. rotundifolia* with description and notes on cultivation.

Bailey, L. H. (1953). *Codonopsis*. In The garden of bellflowers in North America: 142-150, illus.New York: MacMillan. En. — Account of species cultivated in North America, with key and descriptions.

Fedorov, A. A. (1957). *Codonopsis* Wall. In V. L. Komarov, Flora URSS 24: 433-440, illus. Leningrad: Academia Scientiae URSS. Ru. — Flora for former Soviet Union, with key, descriptions, infrageneric classification, and full nomenclature.

• Moeliono, B. (1960). *Codonopsis*. In C. G. G. J. van Steenis (ed.), Flora Malesiana (ser. I) 6(1): 118-121, illus. Djakarta: Noordhoff-Kolff. En. — Flora with key and descriptions; argues that *Campanumoea* (including *Cyclocodon*, q.v.) only differs in fruit type; describes unique calyx lobe insertion of some species.

Rechinger, K. H. & H. Schiman-Czeika (1965). *Codonopsis*. In K. H. Rechinger (ed.), Flora Iranica 13: 3-5 + tab. 1, illus. Graz: Akademische Druck- und Verlagsanstalt. Ge., La. — Flora with key and descriptions.

Thuan, N. V. (1969). *Codonopsis* Wall. In A. Aubréville (ed.), Flore du Camdodge du Laos et du Vietnam 9: 7-12, illus. Paris: Muséum National d'Histoire Naturelle. Fr. — Flora with key and descriptions.

Finlay, M. K. (1972). On *Codonopsis* which grow at Keillour. J. Roy. Hort. Soc. 97: 82-87 + figs. 33-34, illus. En. — Popular account of species cultivated at garden in Scotland.

Shen, L. D., S. Y. Tang, S. G. Yuen & C. K. Hsieh (1975). [The genus *Codonopsis* in Szechuan.] Acta Phytotax. Sin. 13(3): 51-68. Ch. — Floristic treatment with key, full nomenclature, and specimen citations.

Matthews, Y. S. (1980). The genus *Codonopsis*. Bull. Alpine Gard. Soc. Gr. Brit. 48: 96-108, illus. En. — Popular account of species in cultivation.

• Murthy, G. V. S. (1983). Pollen morphology of Indian *Campanumoea* Bl. – a revision of the genus. J. Palynol. 18: 55-59, illus. En. — Pollen data support merger of *Campanumoea* sect. *Campanumoea* with *Codonoopsis*.

• Hong, D. Y. (1983). [*Campanumoea* Bl.] Sect. *Campanumoea; Leptocodon* Hook. f. et Thoms. In Flora Reipublicae Popularis Sinicae 73(2): 69-72, 74-76, illus. Beijing: Science Press. Ch. — Flora of China, with key, descriptions, infrageneric classification, and full nomenclature.

• Shen, L. D. & D. Y. Hong (1983). *Codonopsis* Wall. In Flora Reipublicae Popularis Sinicae 73(2): 32-69, illus. Beijing: Science Press. Ch. — Flora of China, with key, descriptions, infrageneric classification, and full nomenclature.

Nasir, E. (1984). *Codonopsis*. In Flora of Pakistan 155: 1-5, illus. Islamabad: Pakistan Agricultural Research Council. En. — Flora with key and descriptions.

Yoo, K. O., & W. T. Lee (1989). A taxonomic study of the genus *Codonopsis* in Korea. Korean J. Pl. Taxon. 19: 81-102, illus. Ko. — Floristic treatment with key and full nomenclature; includes detailed studies of morphology, anatomy, palynology, and cytology.

• Grey-Wilson, C. (1990). A survey of *Codonopsis* in cultivation. Plantsman 12(2): 65-99, illus. En.— Account of species in cultivation, with key and descriptions.

Huxley, A. J., ed. (1992). *Codonopsis* Wallich. In The new Royal Horticultural Society dictionary of gardening 1: 666-668, illus. London: MacMillan. En. — Account of species in cultivation, with descriptions.

Shimizu, T. (1993). *Codonopsis* Wall.; *Campanumoea* Bl. sect. *Campanumoea*. In K. Iwatsuki, T. Yamazaki, D. E. Boufford & H. Ohba, Flora of Japan 3a: 413-414. Tokyo: Kodansha. En. — Flora with key and descriptions.

• Grey-Wilson, C. (1995). *Codonopsis convolvulacea* and its allies. New Plantsman 2: 213-225, illus. En. — Synopsis of subg. *Pseudocodonopsis,* with key, descriptions, and full nomenclature.

• Wang, Z. T., G. Y. Ma, P. F. Tu, G. J. Xu & T. B. Ng (1995). Chemotaxonomic study of *Codonopsis* (family Campanulaceae) and its related genera. Biochem. Syst. Ecol. 23: 809-812. En. — Secondary chemistry supports Shen & Hong's (1983) division into three subgenera.

• Morris, K. E. & T. G. Lammers (1997). Circumscription of *Codonopsis* and the allied genera *Campanumoea* and *Leptocodon* (Campanulaceae: Campanuloideae). I. Palynological data. Bot. Bull. Acad. Sin. 38: 277-284. En. — Pollen data support subgeneric classification of Shen & Hong (1983) and inclusion of *Leptocodon* and *Campanumoea* sect. *Campanumoea* in subg. *Codonopsis.*

Lammers, T. G. (1998). *Codonopsis* Wall. In T. C. Huang (ed.), Flora of Taiwan (ed. 2) 4: 782-784, illus. Taipei: National Taiwan University. En. — Flora with key and descriptions.

Zhang, Y. B., F. N. Ngan, Z. T. Wang, T. B. Ng, P. P. But, P. C. Shaw & J. Wang (1999). Random primed polymerase chain reaction differentiates *Codonopsis pilosula* from different localities. Pl. Med. (Stuttgart) 65: 157-160. En. — DNA fingerprinting techniques identify geographic sources of a widely used pharmaceutical.

Wei, Z. X. (2001). Pollen morphology of some *Codonopsis* and relafed [sic] species of Campanulaceae. Acta Bot. Yunnan. 23: 335-338 + pl. I-II, illus. Ch. — Comparative pollen morphology.

Codonopsis Wall. in Roxb., Fl. Ind. 2: 103 (1824). *Campanula* sect. *Codonopsis* (Wall.) D. Dietr., Syn Pl. 1: 757 (1839). *Campanumoea* [unranked] *Codonopsis* (Wall.) Endl., Gen. Pl.: 1392 (1841).
Temp. & trop. Asia. 31 32 34 36 37 38 40 41 42.
Glosocomia D. Don, Prodr. Fl. Nepal.: 158 (1825).
Campanumoea Blume, Bijdr.: 726 (1826).
Codonopsis subg. *Leptocodon* Hook. f., Ill. Himal. Pl.: pl. 16A (1855). *Leptocodon* (Hook. f.) Lem., Ill. Hort. 3 (Misc.): 49 (1856).

Codonopsis affinis Hook. f. & Thomson, J. Proc. Linn. Soc., Bot. 2: 12 (1858).
Tibet to Myanmar. 36 CHT 40 ASS EHM NEP WHM 41 MYA. Cl. tuber geophyte.
Codonopsis affinis var. *birmanica* C. B. Clarke in Hook. f., Fl. Brit. India 3: 431 (1881).

Codonopsis alpina Nannf., Notes Roy. Bot. Gard. Edinburgh 16: 154 (1931).
SW. China (Yunnan, Tibet). 36 CHC CHT. Geophyte.
Codonopsis foetens var. *major* Hand.-Mazz., Akad. Wiss. Wien, Math.-Naturwiss. Kl., Anz. 61: 169 (1924).

Codonopsis argentea P. C. Tsoong, Contr. Inst. Bot. Natl. Acad. Peiping 3: 92 (1935).
S. China (Guizhou). 36 CHC. Geophyte.

Codonopsis benthamii Hook. f. & Thomson, J. Proc. Linn. Soc., Bot. 2: 14 (1858), as 'benthami'.
Himalaya to Myanmar. 40 ASS EHM NEP 41 MYA. Cl. tuber geophyte.

Codonopsis bhutanica Ludlow, J. Roy. Hort. Soc. 97: 127 (1972).
Himalaya. 40 EHM NEP. Geophyte.

Codonopsis bicolor Nannf., Acta Horti Gothob. 5: 24 (1930).
C. & SW. China (Gansu, Qinghai, Sichuan, Yunnan, Tibet). 36 CHC CHN CHQ CHT.
Geophyte.

Codonopsis bragaensis Grey-Wilson, Plantsman 12: 99 (1990).
Nepal. 40 NEP. Cl. tuber geophyte.

Codonopsis bulleyana Forrest ex Diels, Notes Roy. Bot. Gard. Edinburgh 5: 171 (1912).
SW. China (Sichuan, Yunnan, Tibet). 36 CHC CHT. Geophyte.
Cyananthus mairei H. Lév., Cat. Pl. Yun-nan: 25 (1916).

Codonopsis canescens Nannf., Svensk Bot. Tidskr. 34: 386 (1940).
C. & SW. China (Qinghai, Sichuan, Tibet). 36 CHC CHQ CHT. Geophyte.

Codonopsis cardiophylla Diels ex Kom., Trudy Imp. S.-Peterburgsk. Bot. Sada 29: 117 (1908).
C. China (Shanxi, Shaanxi, Hubei). 36 CHC CHN. Geophyte.

Codonopsis chimiliensis J. Anthony, Notes Roy. Bot. Gard. Edinburgh 15: 184 (1926).
SC. China (Yunnan) & Myanmar. 36 CHC YN 41 MYA. Geophyte.

Codonopsis chlorocodon C. Y. Wu, Rep. Yunnan Trop. Subtrop. Fl. Res. Inst. 1: 82 (1965).
Codonopsis viridiflora var. *chlorocodon* (C. Y. Wu) S. H. Huang in C. Y. Wu & C. Chen, Fl.
Yunnan. 5: 490 (1991).
SC. China (Sichuan, Yunnan). 36 CHC. Geophyte.

Codonopsis clematidea (Schrenk) C. B. Clarke in Hook. f., Fl. Brit. India 3: 433 (1881).
E. Kazakhstan to Pakistan & W. China (Tibet, Xinjiang); cult. 32 KAZ KGZ TZK 34 AFG 36
CHT CHX 40 PAK WHM. Geophyte. 2*n* = 16.
  * *Wahlenbergia clematidea* Schrenk in Fisch. & C. A. Mey., Enum. Pl. Nov. 1: 38 (1841).
    Glosocomia clematidea (Schrenk) Fisch. ex Regel, Gartenflora 5: 226 (1856).
  Codonopsis ovata var. *ramosissima* Hook. f. & Thomson, J. Proc. Linn. Soc., Bot. 2: 15
    (1858).
  Codonopsis ovata var. *cuspidata* Chipp, J. Linn. Soc., Bot. 38: 385 (1908).
  Codonopsis ovata var. *obtusa* Chipp, J. Linn. Soc., Bot. 38: 385 (1908). *Codonopsis obtusa*
    (Chipp) Nannf., Acta Horti Gothob. 5: 28 (1930). *Codonopsis clematidea* var. *obtusa*
    (Chipp) Kitam., Fl. Afghan.: 383 (1960).

Codonopsis convolvulacea Kurz, J. Bot. 11: 195 (1873).
SC. China (Sichuan, Yunnan) & Myanmar. 36 CHC 41 MYA. Cl. tuber geophyte.

Codonopsis cordifolioidea P. C. Tsoong, Contr. Inst. Bot. Natl. Acad. Peiping 3: 93 (1935).
SC. China (Yunnan). 36 CHC. Cl. tuber geophyte.

Codonopsis deltoidea Chipp, J. Linn. Soc., Bot. 38: 387 (1908).
C. China (Gansu, Sichuan). 36 CHC CHN. Cl. tuber geophyte.

Codonopsis dicentrifolia (C. B. Clarke) W. W. Sm., Rec. Bot. Surv. India 4: 388 (1913).
Tibet to West Bengal & Assam. 36 CHT 40 ASS EHM IND NEP. Geophyte.
  * *Wahlenbergia dicentrifolia* C. B. Clarke in Hook. f., Fl. Brit. India 3: 430 (1881).
    Campanopsis dicentrifolia (C. B. Clarke) Kuntze, Revis. Gen. Pl. 2: 379 (1891).

Codonopsis efilamentosa W. W. Sm., Notes Roy. Bot. Gard. Edinburgh 8: 107 (1913).
Codonopsis convolvulacea var. *efilamentosa* (W. W. Sm.) L. T. Shen, Fl. Reipubl. Popul. Sin.
73(2): 69 (1983).
SC. China (Yunnan) & Myanmar. 36 CHC, 41 MYA. Cl. tuber geophyte.

Codonopsis farreri J. Anthony, Notes Roy. Bot. Gard. Edinburgh 15: 181 (1926).
SC. China (Yunnan) & Myanmar. 36 CHC, 41 MYA. Cl. tuber geophyte.
Codonopsis farreri var. *grandiflora* S. H. Huang, Acta Bot. Yunnan. 6: 394 (1984).

Codonopsis foetens Hook. f. & Thomson, J. Proc. Linn. Soc., Bot. 2: 16 (1858).
E. Himalaya. 40 EHM. Geophyte.

Codonopsis forrestii Diels, Notes Roy. Bot. Gard. Edinburgh 5: 171 (1912). *Codonopsis convolvulacea* var. *forrestii* (Diels) Ballard, Bot. Mag. 162: tab. 9581 (1939). *Codonopsis convolvulacea* subsp. *forrestii* (Diels) D. Y. Hong & L. M. Ma in H. W. Li & Z. Y. Zhu, Fl. Sichuanica 10: 546 (1992).
SC. China (Guizhou, Sichuan, Yunnan). 36 CHC. Cl. tuber geophyte.
Codonopsis mairei H. Lév., Cat. Pl. Yun-nan: 24 (1916).
Codonopsis forrestii var. *heterophylla* C. Y. Wu, Rep. Yunnan Trop. Subtrop. Fl. Res. Inst. 1: 80 (1965).
Codonopsis forrestii var. *hirsuta* P. C. Tsoong & L. T. Shen, Acta Phytotax. Sin. 13(3): 55 (1975).

Codonopsis gombalana C. Y. Wu, Rep. Yunnan Trop. Subtrop. Fl. Res. Inst. 1: 81 (1965).
SC. China (Yunnan). 36 CHC. Geophyte.

Codonopsis gracilis Hook. f., Ill. Himal. Pl.: pl. 16A (1855). *Leptocodon gracilis* (Hook. f.) Lem., Ill. Hort. 3 (Misc.): 49 (1856). *Campanumoea gracilis* (Hook. f.) G. Nichols., Ill. Dict. Gard. 1: 259 (1884).
Nepal to SC. China (Sichuan, Yunnan) & Myanmar. 36 CHC 40 EHM NEP 41 MYA. Geophyte.

Codonopsis grey-wilsonii J. M. H. Shaw, New Plantsman 3: 93 (1996).
Himalaya. 40 EHM NEP. Cl. tuber geophyte.
* *Codonopsis nepalensis* Grey-Wilson, Plantsman 12: 99 (1990); non H. Hara, J. Jap. Bot. 53: 139 (1978).

Codonopsis henryi Oliv., Hooker's Icon. Pl. 20: pl. 1967 (1891).
C. China (Hubei, Sichuan). 36 CHC. Cl. tuber geophyte.

Codonopsis hongii Lammers, Novon 11: 67 (2001).
SW. China (Yunnan, Tibet). 36 CHC CHT. Geophyte.
* *Leptocodon hirsutus* D. Y. Hong, Acta Phytotax. Sin. 18: 246 (1980). *Codonopsis hirsuta* (D. Y. Hong) K. E. Morris & Lammers, Novon 9: 387 (1999); non (Hand.-Mazz.) D. Y. Hong & L. M. Ma in H. W. Li & Z. Y. Zhu, Fl. Sichuanica 10: 546 (1992).

Codonopsis inflata Hook. f., Ill. Himal. Pl.: pl. 16C (1855). *Campanumoea inflata* (Hook. f.) C. B. Clarke in Hook. f., Fl. Brit. India 3: 436 (1881).
Tibet to E. Himalaya. 36 CHT 40 EHM NEP. Cl. tuber geophyte.

Codonopsis javanica (Blume) Miq., Fl. Ned. Ind. 2: 566 (1857).
S. China to E. India, Thailand, Jawa, Taiwan & Japan. 36 CHC CHS 38 JAP TAI 40 ASS EHM IND 41 LAO MYA THA VIE 42 JAW SUM. Cl. tuber geophyte.
* *Campanumoea javanica* Blume, Bijdr. 727 (1826). *Campanula javanica* (Blume) D. Dietr., Syn. Pl. 1: 758 (1839).

subsp. **japonica** (Makino) Lammers, Bot. Bull. Acad. Sin. 33: 285 (1992).
S. China (Anhui, Zhejiang, Jiangxi, Fujian, Guangdong, Guangxi, Hunan, Hubei, Guizhou, Sichuan), Taiwan, Japan (Honshu, Shikoku, Kyushu). 36 CHC CHS 38 JAP TAI. Cl. tuber geophyte.

* *Campanumoea japonica* Maxim., Bull. Acad. Imp. Sci. Saint-Pétersbourg 12: 67(1867); non Siebold ex E. Morren, Belgique Hort. 14: 337 (1863). *Campanumoea javanica* var. *japonica* Makino, Bot. Mag. (Tokyo) 22: 155 (1908). *Campanumoea maximowiczii* Honda, Bot. Mag. (Tokyo) 50: 389 (1936). *Campanumoea javanica* subsp. *japonica* (Makino) D. Y. Hong, Fl. Reipubl. Popul. Sin. 73(2): 71 (1983).

subsp. **javanica**
  S. China (Guangdong, Guangxi, Guizhou, Yunnan) to West Bengal, Thailand & Jawa. 36 CHC CHS 38 JAP TAI 40 ASS EHM IND 41 LAO MYA THA VIE 42 JAW SUM. Cl. tuber geophyte.
  Codonopsis cordata Hassk., Retzia 1: 9 (1855). *Campanumoea cordata* (Hassk.) Miq., Fl. Ned. Ind., Eerste Bijv.: 234 (1861).
  Campanumoea labordei H. Lév., Bull. Soc. Agric. Sarthe 39: 323 (1904).
  Codonopsis cordifolia Kom., Trudy Imp. S.-Peterburgsk. Bot. Sada 29: 108 (1908).

Codonopsis kawakamii Hayata, Mat. Fl. Formosa: 165 (1911).
  Taiwan. 38 TAI. Cl. tuber geophyte.

Codonopsis lanceolata (Siebold & Zucc.) Trautv., Trudy Imp. S.-Peterburgsk. Bot. Sada 6: 46 (1879).
  Russian Far East to E. China (Hebei, Shanxi, Shandong, Jiangsu, Anhui, Zhejiang, Fujian, Henan, Hubei, Hunan), Korea & Japan (Hokkaido, Honshu, Shikoku, Kyushu). 31 KHA PRM 36 CHC CHI CHM CHN CHS 38 JAP KOR. Cl. tuber geophyte. $2n = 16$.
  * *Campanumoea lanceolata* Siebold & Zucc., Fl. Jap. 1: 174 (1841). *Glosocomia lanceolata* (Siebold & Zucc.) Maxim., Bull. Acad. Imp. Sci. Saint-Pétersbourg 31: 62 (1886).
  Glosocomia hortensis Rupr., Bull. Cl. Phys.-Math. Acad. Imp. Sci. Saint-Pétersbourg 15: 209 (1857) .
  Campanumoea japonica Siebold ex ex E. Morren, Belgique Hort. 14: 337 (1863).
  Codonopsis bodinieri H. Lév., Fl. Kouy-Tchéou: 57 (1915).
  Codonopsis yesoensis Nakai, Rep. Veg. Daisetsu Mts.: 56, 71 (1930).
  Codonopsis lanceolata var. *emaculata* Honda, Bot. Mag. Tokyo (50): 436 (1936).
    Codonopsis lanceolata f. *emaculata* (Honda) H. Hara, Enum. Sperm. Jap. 2: 98 (1952).
  Codonopsis lanceolata var. *omurae* T. Koyama, J. Jap. Bot. 32: 61 (1957).

Codonopsis levicalyx L. T. Shen, Acta Phytotax. Sin. 13(3): 55 (1975).
  SW. China (Sichuan, Tibet). 36 CHC CHT. Cl. tuber geophyte.
    Codonopsis levicalyx var. *hirsuticalyx* L. T. Shen, Fl. Reipubl. Popul. Sin. 73(2): 183 (1983).

Codonopsis limprichtii Lingelsh. & Borza, Repert. Spec. Nov. Regni Veg. 13: 391 (1914).
  Codonopsis convolvulacea var. *limprichtii* (Lingel & Borza) Anthony, Notes Roy. Bot. Gard. Edinburgh 15: 178 (1926).
  SC. China (Guizhou, Sichuan, Yunnan). 36 CHC. (Cl.) tuber geophyte.
    Codonopsis graminifolia H. Lév., Cat. Pl. Yun-nan: 24 (1916).
    Codonopsis limprichtii var. *hirsuta* Hand.-Mazz., Akad. Wiss. Wien, Math.-Naturwiss. Kl., Anz. 61: 169 (1924). *Codonopsis convolvulacea* var. *hirsuta* (Hand.-Mazz.) Nannf. in Hand.-Mazz., Symb. Sin. 7: 1077 (1936). *Codonopsis hirsuta* (Hand.-Mazz.) D. Y. Hong & L. M. Ma in H. W. Li & Z. Y. Zhu, Fl. Sichuanica 10: 546 (1992).
    Codonopsis limprichtii var. *pinifolia* Hand.-Mazz., Akad. Wiss. Wien, Math.-Naturwiss. Kl., Anz. 61: 170 (1924). *Codonopsis convolvulacea* var. *pinifolia* (Hand.-Mazz.) Nannf. in Hand.-Mazz., Symb. Sin. 7: 1077 (1936).

Codonopsis longifolia D. Y. Hong, Acta Phytotax. Sin. 18: 245 (1980).
  Tibet. 36 CHT. Cl. tuber geophyte.

Codonopsis macrocalyx Diels, Notes Roy. Bot. Gard. Edinburgh 5: 170 (1912).
  SW. China (Sichuan, Yunnan, Tibet). 36 CHC CHT. Cl. tuber geophyte.
    Codonopsis macrocalyx var. *parviloba* J. Anthony, Notes Roy. Bot. Gard. Edinburgh 15: 183 (1926).

Codonopsis meleagris Diels, Notes Roy. Bot. Gard. Edinburgh 5: 172 (1912).
  SC China (Yunnan). 36 CHC. Geophyte.

Codonopsis micrantha Chipp, J. Linn. Soc., Bot. 38: 382 (1908).
  SC. China (Sichuan, Yunnan). 36 CHC. Cl. tuber geophyte.
    Campanumoea violifolia H. Lév., Cat. Pl. Yun-Nan: 24 (1916).

Codonopsis microtubulosa Z. T. Wang & G. J. Xu, Acta Phytotax. Sin. 31: 184 (1993).
  SC. China (Sichuan). 36 CHC. Cl. tuber geophyte.

Codonopsis minima Nakai, Bot. Mag. (Tokyo) 29: 7 (1915), as 'minina'.
  Korea. 38 KOR. Cl. tuber geophyte. $2n = 16$.

Codonopsis nepalensis H. Hara, J. Jap. Bot. 53: 139 (1978).
  Nepal. 40 NEP. Cl. tuber geophyte.

Codonopsis nervosa (Chipp) Nannf., Acta Horti Gothob. 5: 26 (1930).
  SW. China (Sichuan, Tibet). 36 CHC CHT. Geophyte.
    * *Codonopsis ovata* var. *nervosa* Chipp, J. Linn. Soc., Bot. 38: 385 (1908).
    Codonopsis macrantha Nannf., Notes Roy. Bot. Gard. Edinburgh 16: 157 (1931).
      Codonopsis nervosa var. *macrantha* (Nannf.) L. T. Shen, Fl. Reipubl. Popul. Sin. 73(2):
      57 (1983). *Codonopsis nervosa* subsp. *macrantha* (Nannf.) D. Y. Hong & L. M. Ma in H.
      W. Li & Z. Y. Zhu, Fl. Sichuanica 10: 541 (1992).

Codonopsis ovata Benth. in Royle, Ill. Bot. Himal. Mts.: pl. 69 (1835). *Wahlenbergia roylei* A.
  DC. in DC., Prodr. 7: 425 (1839); non *Wahlenbergia ovata* D. Don, Prodr. Fl. Nepal.: 156
  (1825). *Glosocomia ovata* (Benth.) Lindl., Edwards' Bot. Reg. 28: pl. 3 (1842).
  Pakistan & W. Himalaya. 40 PAK WHM. Geophyte. $2n = 16$

Codonopsis pianmaensis S. H. Huang, Acta Bot. Yunnan. 6: 393 (1984).
  SC. China (Yunnan). 36 CHC. Geophyte.

Codonopsis pilosula (Franch.) Nannf., Acta Horti Gothob. 5: 29 (1930).
  Russian Far East (Khabarovsk, Primorye) to Mongolia, Tibet, S. China & Korea. 31 KHA
    PRM 36 CHC CHI CHM CHN CHQ CHS 37 MON 38 KOR. Cl. tuber geophyte. $2n = 16$.
    * *Campanumoea pilosula* Franch., Pl. David. 1: 192 (1883).
    Codonopsis silvestris Kom., Trudy Imp. S.-Petersburgsk. Bot. Sada 18: 425 (1901).
    Codonopsis modesta Nannf., Acta Horti Gothob. 5: 29 (1930). *Codonopsis pilosula* var.
      modesta (Nannf.) L. T. Shen, Fl. Reipubl. Popul. Sin. 73(2): 41 (1983).
    Codonopsis handeliana Nannf. in Hand.-Mazz., Symb. Sin. 7: 1078 (1936). *Codonopsis
      pilosula* var. *handeliana* (Nannf.) L. T. Shen, Fl. Reipubl. Popul. Sin. 73(2): 41 (1983).
      Codonopsis pilosula subsp. *handeliana* (Nannf.) D. Y. Hong & L. M. Ma in H. W. Li &
      Z. Y. Zhu, Fl. Sichuanica 10: 532 (1992).
    Codonopsis glaberrima Nannf., Svensk Bot. Tidskr. 34: 387 (1940). *Codonopsis pilosula*
      var. *glaberrima* (Nannf.) P. C. Tsoong, Acta Phytotax. Sin. 13(3): 58 (1975).
    Codonopsis volubilis Nannf., Svensk Bot. Tidskr. 34: 388 (1940). *Codonopsis pilosula* var.
      volubilis (Nannf.) L. T. Shen, Fl. Reipubl. Popul. Sin. 73(2): 41 (1983).

Codonopsis purpurea Wall. in Roxb., Fl. Ind. 2: 105 (1824). *Campanula purpurea* (Wall.)
  Spreng., Syst. Veg. 4(2): 78 (1827). *Wahlenbergia purpurea* (Wall.) A. DC. in DC., Prodr. 7:
  425 (1839). *Glosocomia purpurea* (Wall.) Rupr., Bull. Cl. Phys.-Math. Acad. Imp. Sci. Saint-
  Pétersbourg 15: 210 (1857).
  Himalaya to SC. China (Yunnan). 36 CHC CHT 40 ASS EHM NEP WHM. Geophyte.

Codonopsis retroserrata Z. T. Wang & G. J. Xu, Acta Phytotax. Sin. 31: 186 (1993).
  SC. China (Sichuan). 36 CHC. Cl. tuber geophyte.

Codonopsis rosulata W. W. Sm., Notes Roy. Bot. Gard. Edinburgh 13: 157 (1921).
SC. China (Sichuan). 36 CHC. Geophyte.

Codonopsis rotundifolia Benth. in Royle, Ill. Bot. Himal. Mts.: pl. 62 (1835). *Wahlenbergia rotundifolia* (Benth.) A. DC. in DC., Prodr. 7: 425 (1839). *Glosocomia rotundifolia* (Benth.) Rupr., Bull. Cl. Phys.-Math. Acad. Imp. Sci. Saint-Pétersbourg 15: 210 (1857).
Pakistan to N. India (Punjab, Uttar Pradesh) & E. Himalaya. 40 EHM IND NEP PAK WHM. Cl. tuber geophyte. $2n = 16$.
  Codonopsis lurida Lindl., Edwards' Bot. Reg. 25: 82 (1839). *Glosocomia lurida* (Lindl.) Lindl., Edwards' Bot. Reg. 28: pl. 3 (1842). *Wahlenbergia lurida* (Lindl.) G. Manetti, Cat. Pl. Hort. Modic.: 106 (1842).
  Codonopsis rotundifolia var. *grandiflora* Hook., Bot. Mag. 83: tab. 5018 (1857).
  Codonopsis rotundifolia var. *angustifolia* Nannf., Bot. Mag. 167: tab. 131 (1950).

Codonopsis subglobosa W. W. Sm., Notes Roy. Bot. Gard. Edinburgh 8: 108 (1913).
SC. China (Sichuan, Yunnan). 36 CHC. Geophyte.

Codonopsis subscaposa Kom., Trudy Imp. S.-Peterburgsk. Bot. Sada 29: 114 (1908).
SC. China (Sichuan, Yunnan). 36 CHC. Geophyte.

Codonopsis subsimplex Hook. f. & Thomson, J. Proc. Linn. Soc., Bot. 2: 16 (1858).
Tibet to West Bengal & Assam. 36 CHT 40 ASS EHM NEP. Geophyte. $2n = 16$.

Codonopsis tangshen Oliv., Hooker's Icon. Pl. 20: pl. 1966 (1891).
C. China (Shaanxi, Hubei, Hunan, Guizhou, Sichuan). 36 CHC CHN CHS. Cl. tuber geophyte.

Codonopsis thalictrifolia Wall. in Roxb., Fl. Ind. 2: 106 (1824). *Glosocomia tenera* D. Don, Prodr. Fl. Nepal.: 158 (1825). *Campanula thalictrifolia* (Wall.) Spreng., Syst. Veg. 4(2): 77 (1827). *Wahlenbergia thalictrifolia* (Wall.) A. DC. in DC., Prodr. 7: 425 (1839).
Tibet to Assam. 36 CHT 40 ASS EHM NEP. Geophyte.
  Codonopsis mollis Chipp, J. Linn. Soc., Bot. 38: 381 (1908). *Codonopsis thalictrifolia* var. *mollis* (Chipp) L. T. Shen, Fl. Reipubl. Popul. Sin. 73(2): 55 (1983).

Codonopsis tsinlingensis Pax & K. Hoffm., Repert. Spec. Nov. Regni Veg. Beih. 12: 500 (1922).
C. China (Gansu, Shaanxi, Sichuan). 36 CHN CHC. Geophyte.

Codonopsis tubulosa Kom., Trudy Imp. S.-Peterburgsk. Bot. Sada 29: 112 (1908).
SC. China (Guizhou, Sichuan, Yunnan). 36 CHC. Cl. tuber geophyte.
  Codonopsis pilosa Chipp, J. Linn. Soc., Bot. 38: 388 (1908).
  Codonopsis accrescenticalyx H. Lév., Cat. Pl. Yun-Nan: 24 (1916).
  Codonopsis macrocalyx var. *coerulescens* Hand.-Mazz., Akad. Wiss. Wien, Math.-Naturwiss. Kl., Anz. 61: 169 (1924).

Codonopsis ussuriensis (Rupr. & Maxim.) Hemsl., J. Linn. Soc., Bot. 26: 6 (1889).
Russian Far East to Manchuria (Heilongjiang, Jilin), Korea & Japan (Hokkaido, Honshu, Shikoku, Kyushu). 31 KHA PRM 36 CHM 38 JAP KOR. Cl. tuber geophyte. $2n = 16$.
  * *Glosocomia ussuriensis* Rupr. & Maxim., Bull. Cl. Phys.-Math. Acad. Imp. Sci. Saint-Pétersbourg 15: 223 (1857). *Codonopsis lanceolata* var. *ussuriensis* (Rupr. & Maxim.) Trautv., Trudy Imp. S.-Peterburgsk. Bot. Sada 6: 47 (1879). *Glosocomia lanceolata* var. *ussuriensis* (Rupr. & Maxim.) Regel, Index Hort. Sem. Petrop. 1886: 92 (1886).
  Glosocomia lanceolata var. *obtusa* Regel, Bull. Cl. Phys.-Math. Acad. Imp. Sci. Saint-Pétersbourg 15: 223 (1857) .
  Codonopsis ussuriensis f. *viridiflora* J. Ohara, J. Phytogeogr. Taxon 33: 72 (1985).

Codonopsis vinciflora Kom., Trudy Imp. S.-Peterburgsk. Bot. Sada 29: 103 (1908).
Codonopsis convolvulacea var. *vinciflora* (Kom.) L. T. Shen, Fl. Reipubl. Popul. Sin. 73(2): 68 (1983). *Codonopsis convolvulacea* subsp. *vinciflora* (Kom.) D. Y. Hong in C. Y. Wu, Fl. Xizang. 4: 582 (1985).
SW. China (Sichuan, Yunnan, Tibet) to Bhutan. 36 CHC CHT 40 EHM. Cl. tuber geophyte.

Codonopsis viridiflora Maxim., Bull. Acad. Imp. Sci. Saint-Pétersbourg 27: 496 (1882).
C. China (Ningxia, Gansu, Shaanxi, Qinghai, Sichuan). 36 CHC CHI CHN CHQ. Cl. tuber geophyte.

Codonopsis viridis Wall. in Roxb., Fl. Ind. 2: 103 (1824). *Campanula viridis* (Wall.) Spreng., Syst. Veg. 4(2): 78 (1827). *Wahlenbergia viridis* (Wall.) A. DC. in DC., Prodr. 7: 424 (1839). *Glosocomia viridis* (Wall.) Rupr., Bull. Cl. Phys.-Math. Acad. Imp. Sci. Saint-Pétersbourg 15: 210 (1857)
Pakistan to N. India (Uttar Pradesh) & Assam. 40 ASS EHM IND NEP PAK WHM. Cl. tuber geophyte.
Codonopsis griffithii C. B. Clarke in Hook. f., Fl. Brit. India 3: 431 (1881). *Codonopsis viridis* var. *hirsuta* Chipp, J. Linn. Soc., Bot. 38: 386 (1908).

Codonopsis xizangensis D. Y. Hong, Acta Phytotax. Sin. 18: 246 (1980).
Tibet. 36 CHT. Geophyte.

Synonyms:
Codonopsis subg. *Leptocodon* Hook. f. === **Codonopsis** Wall.
Codonopsis accrescenticalyx H. Lév. === **Codonopsis tubulosa** Kom.
Codonopsis affinis var. *birmanica* C. B. Clarke === **Codonopsis affinis** Hook. f. & Thomson
Codonopsis albiflora Griff. === **Cyclocodon lancifolius** (Roxb.) Kurz subsp. **lancifolius**
Codonopsis bodinieri H. Lév. === **Codonopsis lanceolata** (Siebold & Zucc.) Trautv.
Codonopsis celebica (Blume) Miq. === **Cyclocodon lancifolius** subsp. **celebicus** (Blume) K. E. Morris & Lammers
Codonopsis clematidea var. *obtusa* (Chipp) Kitam. === **Codonopsis clematidea** (Schrenk) C. B. Clarke
Codonopsis convolvulacea subsp. *forrestii* (Diels) D. Y. Hong & L. M. Ma === **Codonopsis forrestii** Diels
Codonopsis convolvulacea subsp. *vinciflora* (Kom.) D. Y. Hong === **Codonopsis vinciflora** Kom.
Codonopsis convolvulacea var. *efilamentosa* (W. W. Sm.) L. T. Shen === **Codonopsis efilamentosa** W. W. Sm.
Codonopsis convolvulacea var. *forrestii* (Diels) Ballard === **Codonopsis forrestii** Diels
Codonopsis convolvulacea var. *hirsuta* (Hand.-Mazz.) Nannf. === **Codonopsis limprichtii** Lingelsh. & Borza
Codonopsis convolvulacea var. *limprichtii* (Lingel & Borza) J. Anthony === **Codonopsis limprichtii** Lingelsh. & Borza
Codonopsis convolvulacea var. *pinifolia* (Hand.-Mazz.) Nannf. === **Codonopsis limprichtii** Lingelsh. & Borza
Codonopsis convolvulacea var. *vinciflora* (Kom.) L. T. Shen === **Codonopsis vinciflora** Kom.
Codonopsis cordata Hassk. === **Codonopsis javanica** (Blume) Miq. subsp. **javanica**
Codonopsis cordifolia Kom. === **Codonopsis javanica** (Blume) Miq. subsp. **javanica**
Codonopsis draco Pampan. === ?
Codonopsis farreri var. *grandiflora* S. H. Huang === **Codonopsis farreri** J. Anthony
Codonopsis foetens var. *major* Hand.-Mazz. === **Codonopsis alpina** Nannf.
Codonopsis forrestii var. *heterophylla* C. Y. Wu === **Codonopsis forrestii** Diels
Codonopsis forrestii var. *hirsuta* P. C. Tsoong & L. T. Shen === **Codonopsis forrestii** Diels
Codonopsis glaberrima Nannf. === **Codonopsis pilosula** (Franch.) Nannf.
Codonopsis graminifolia H. Lév. === **Codonopsis limprichtii** Lingelsh. & Borza
Codonopsis griffithii C. B. Clarke === **Codonopsis viridis** Wall.

Codonopsis handeliana Nannf. === **Codonopsis pilosula** (Franch.) Nannf.

Codonopsis hirsuta (Hand.-Mazz.) D. Y. Hong & L. M. Ma === **Codonopsis limprichtii** Lingelsh. & Borza

Codonopsis hirsuta (D. Y. Hong) K. E. Morris & Lammers === **Codonopsis hongii** Lammers

Codonopsis japonica Miq. === ?

Codonopsis lanceolata f. *emaculata* (Honda) H. Hara === **Codonopsis lanceolata** (Siebold & Zucc.) Trautv.

Codonopsis lanceolata var. *emaculata* Honda === **Codonopsis lanceolata** (Siebold & Zucc.) Trautv.

Codonopsis lanceolata var. *omurae* T. Koyama === **Codonopsis lanceolata** (Siebold & Zucc.) Trautv.

Codonopsis lanceolata var. *ussuriensis* (Rupr. & Maxim.) Trautv. === **Codonopsis ussuriensis** (Rupr. & Maxim.) Hemsl.

Codonopsis lancifolia (Roxb.) Moeliono === **Cyclocodon lancifolius** (Roxb.) Kurz

Codonopsis lancifolia subsp. *celebica* (Blume) Moeliono === **Cyclocodon lancifolius** subsp. **celebicus** (Blume) K. E. Morris & Lammers

Codonopsis leucocarpa Miq. === **Cyclocodon lancifolius** subsp. **celebicus** (Blume) K. E. Morris & Lammers

Codonopsis levicalyx var. *hirsuticalyx* L. T. Shen === **Codonopsis levicalyx** L. T. Shen

Codonopsis limprichtii var. *hirsuta* Hand.-Mazz. === **Codonopsis limprichtii** Lingelsh. & Borza

Codonopsis limprichtii var. *pinifolia* Hand.-Mazz. === **Codonopsis limprichtii** Lingelsh. & Borza

Codonopsis lurida Lindl. === **Codonopsis rotundifolia** Benth.

Codonopsis macrantha Nannf. === **Codonopsis nervosa** (Chipp) Nannf.

Codonopsis macrocalyx var. *coerulescens* Hand.-Mazz. === **Codonopsis tubulosa** Kom.

Codonopsis macrocalyx var. *parviloba* J. Anthony === **Codonopsis macrocalyx** Diels

Codonopsis mairei H. Lév. === **Codonopsis forrestii** Diels

Codonopsis modesta Nannf. === **Codonopsis pilosula** (Franch.) Nannf.

Codonopsis mollis Chipp === **Codonopsis thalictrifolia** Wall.

Codonopsis nepalensis Grey-Wilson === **Codonopsis grey-wilsonii** J. M. H. Shaw

Codonopsis nervosa subsp. *macrantha* (Nannf.) D. Y. Hong & L. M. Ma === **Codonopsis nervosa** (Chipp) Nannf.

Codonopsis nervosa var. *macrantha* (Nannf.) L. T. Shen === **Codonopsis nervosa** (Chipp) Nannf.

Codonopsis obtusa (Chipp) Nannf. === **Codonopsis clematidea** (Schrenk) C. B. Clarke

Codonopsis ovata var. *cuspidata* Chipp === **Codonopsis clematidea** (Schrenk) C. B. Clarke

Codonopsis ovata var. *nervosa* Chipp === **Codonopsis nervosa** (Chipp) Nannf.

Codonopsis ovata var. *obtusa* Chipp === **Codonopsis clematidea** (Schrenk) C. B. Clarke

Codonopsis ovata var. *ramosissima* Hook. f. & Thomson === **Codonopsis clematidea** (Schrenk) C. B. Clarke

Codonopsis parviflora Wall. ex A. DC. === **Cyclocodon parviflorus** (Wall. ex A. DC.) Hook. f. & Thomson

Codonopsis pilosa Chipp === **Codonopsis tubulosa** Kom.

Codonopsis pilosula subsp. *handeliana* (Nannf.) D. Y. Hong & L. M. Ma === **Codonopsis pilosula** (Franch.) Nannf.

Codonopsis pilosula var. *glaberrima* (Nannf.) P. C. Tsoong === **Codonopsis pilosula** (Franch.) Nannf.

Codonopsis pilosula var. *handeliana* (Nannf.) L. T. Shen === **Codonopsis pilosula** (Franch.) Nannf.

Codonopsis pilosula var. *modesta* (Nannf.) L. T. Shen === **Codonopsis pilosula** (Franch.) Nannf.

Codonopsis pilosula var. *volubilis* (Nannf.) L. T. Shen === **Codonopsis pilosula** (Franch.) Nannf.

Codonopsis rotundifolia var. *angustifolia* Nannf. === **Codonopsis rotundifolia** Benth.

Codonopsis rotundifolia var. *grandiflora* Hook. === **Codonopsis rotundifolia** Benth.

Codonopsis silvestris Kom. === **Codonopsis pilosula** (Franch.) Nannf.

Codonopsis thalictrifolia var. *mollis* (Chipp) L. T. Shen === **Codonopsis thalictrifolia** Wall.

Codonopsis truncata Wall. ex A. DC. === **Cyclocodon lancifolius** (Roxb.) Kurz subsp. **lancifolius**

Codonopsis ussuriensis f. *viridiflora* J. Ohara === **Codonopsis ussuriensis** (Rupr. & Maxim.) Hemsl.

Codonopsis vinciflora Kom. === **Codonopsis convolvulacea** Kurz

Codonopsis viridiflora var. *chlorocodon* (C. Y. Wu) S. H. Huang === **Codonopsis chlorocodon** C. Y. Wu

Codonopsis viridis var. *hirsuta* Chipp === **Codonopsis viridis** Wall.

Codonopsis volubilis Nannf. === **Codonopsis pilosula** (Franch.) Nannf.

Codonopsis yesoensis Nakai === **Codonopsis lanceolata** (Siebold & Zucc.) Trautv.

Colensoa

Hooker (1852) segregated this genus from *Lobelia* on the basis of its baccate fruit. Recognition is not supported by molecular data (E. Knox, pers. comm.); its type is part of a clade that also includes numerous species of *Lobelia* sect. *Delostemon* (E. Wimm.) J. Murata and sect. *Pratia* (Gaudich.) J. Murata, plus the types of *Dielsantha* and *Isotoma*. Type [by monotypy]: *Colensoa physaloides* (A. Cunn.) Hook. f.

> Hooker, J. D. (1852). *Colensoa*, Hook. fil. In The botany of the antarctic voyage of H.M. discovery ships *Erebus* and *Terror* in the years 1839-1843. II. Flora Novae-Zelandiae 1: 156-157. London: Reeve Bros. En. — Establishment of genus as distinct from *Lobelia*.
>
> Baillon, H. (1885). *Pratia* Gaudich. In Histoire des plantes 8: 366. Paris: L. Hachette. Fr. — Included in *Pratia* as one of four sections.
>
> Hemsley, W. B. (1886). *Pratia borneensis*, Hemsl. Hooker's Icon. Pl. 16: pl. 1532 + 1 pg., illus. En. — Included in *Pratia* as one of three sections.
>
> Hooker, J. D. (1886). *Colensoa physaloides*. Bot. Mag. 112: tab. 6864 + 2 pp., illus. En. — Comments on relationship of genus to *Pratia*.

Synonyms:
Colensoa Hook. f. === **Lobelia** L.
Colensoa physaloides (A. Cunn.) Hook. f. === **Lobelia physaloides** A. Cunn.

Concilium

This segregate from *Campanula* was founded on the cylindric capsule, subrotate corolla, and the mistaken impression that the ovary was five-loculed. Type [designated by the author]: *Concilium pedunculare* Raf.

> Rafinesque, C. S. (1837). Flora Telluriana, pars secunda. Philadelphia: published privately. En. — Estsablishment of genus as distinct from *Campanula*.

Synonyms:
Concilium Raf. === **Prismatocarpus** L'Hér.
Concilium pedunculare Raf. === **Prismatocarpus fruticosus** (L.) L'Hér.

Craterocapsa

Campanuloideae, 5 species, southern Africa. In a molecular phylogeny (Eddie et al. 2003, under **General**), the sole included species formed a clade (the "wahlenbergioid clade") with the type of *Roella*; this was sister to the much larger "campanuloid" clade. Type [designated by the authors]: *Craterocapsa tarsodes* Hilliard & B. L. Burtt.

- Hilliard, O. M., & B. L. Burtt (1973). Notes on some plants of southern Africa chiefly from Natal: III. Campanulaceae. Notes Roy. Bot. Gard. Edinburgh 32: 314-327, illus., map. En. — Establishment of genus as distinct from *Roella* and *Wahlenbergia*, with key, descriptions, full nomenclature, and specimen citations.

Dyer, R. A. (1975). *Craterocapsa* Hilliard & Burtt. In The genera of southern African flowering plants 1: 641. Pretoria: Department of Agricultural Technical Services. En. — Generic description.

Thulin, M. (1983b). *Craterocapsa* Hilliard & Burtt. In E. Launert (ed.), Flora Zambesiaca 7(1): 114-115, illus. Kew: Royal Botanic Garden. En. — Flora of Mozambique, Malawi, Zambia, Zimbabwe and Botswana, with keys, descriptions, full nomenclature & specimen citations.

Welman, W. G. (2000). *Craterocapsa* Hilliard & B. L. Burtt. In O. A. Leistner (ed.), Seed plants of southern Africa: families and genera. Strelitzia 10: 200. En. — Generic description.

van Heerden, F. R., A. M. Viljoen & S. Mohoto (2002). A phytochemical investigation of *Craterocapsa tarsodes*, a plant used for the treatment of epilepsy by the Northern Sotho people of South Africa. S. African J. Bot. 68: 77-79. En. — Contains acetoside and a pinocembrin neohesperidoside that may account for its traditional use in treating epilepsy.

Craterocapsa Hilliard & B. L. Burtt, Notes Roy. Bot. Gard. Edinburgh 32: 314 (1973).
S. Africa. 26 27.

Craterocapsa alfredica D. Y. Hong, Taxon 51: 733 (2003).
KwaZulu-Natal. 27 NAT. Cham. or hemicr.

Craterocapsa congesta Hilliard & B. L. Burtt, Notes Roy. Bot. Gard. Edinburgh 32: 320 (1973).
S. Africa. 27 CPP LES NAT. Cham. or hemicr.

Craterocapsa insizwae (Zahlbr.) Hilliard & B. L. Burtt, Notes Roy. Bot. Gard. Edinburgh 32: 323 (1973).
KwaZulu-Natal. 27 NAT. Cham. or hemicr.
  * *Roella insizwae* Zahlbr., Ann. K. K. Naturhist. Hofmus. 18: 401 (1903).
  Wahlenbergia ovalis Brehmer, Bot. Jahrb. Syst. 53: 130 (1915).

Craterocapsa montana (A. DC.) Hilliard & B. L. Burtt, Notes Roy. Bot. Gard. Edinburgh 32: 323 (1973).
S. Africa. 27 CPP LES OFS. Cham. or hemicr.
  * *Wahlenbergia montana* A. DC. in DC., Prodr. 7: 430 (1839). *Campanopsis montana* (A. DC.) Kuntze, Revis. Gen. Pl. 2: 379 (1891).
  Wahlenbergia basutica E. Phillips, Ann. S. African Mus. 16: 181 (1917).

Craterocapsa tarsodes Hilliard & B. L. Burtt, Notes Roy. Bot. Gard. Edinburgh 32: 324 (1973).
Zimbabwe to S. Africa. 26 ZIM 27 CPP LES NAT OFS SWZ TVL. Cham. or hemicr.
  Wahlenbergia montana var. *glabrata* Sond. in Harv. & Sond., Fl. Cap. 3: 573 (1865).
  Wahlenbergia montana var. *angustisepala* Brehmer, Bot. Jahrb. Syst. 53: 131 (1915).

Cremochilus

This genus was distinguished from *Siphocampylus* on the basis of the four-lobed upper lip of its corolla. Its type is currently assigned to *Siphocampylus* subsect. *Isochilus* E. Wimm., along with *S. isochilus*. Type [by monotypy]: *Cremochilus meridensis* Turcz.

Turczaninow, N. (1852). Decas septima. Generum plantarum non descriptorum adjectis descriptionis nonnullarum specierum. Bull. Soc. Imp. Naturalistes Moscou 25(3): 138-181. La. — Establishment of genus as distinct from *Siphocampylus*.

Synonyms:
Cremochilus Turcz. === **Siphocampylus** Pohl
Cremochilus meridensis Turcz. === **Siphocampylus sceptrum** Decne.

Cryptocodon

Campanuloideae, 1 species, Tadzhikistan. The genus has not been included yet in a molecular phylogeny, but on the basis of morphology and serology (Gudkova & Borshchenko 1991), one would predict that it is related to *Asyneuma, Petromarula, Physoplexis,* and *Phyteuma.* The exemplars of those four genera formed a clade within the "Rapunculus clade" in a recent study (Eddie et al. 2003, under **General**). Type [designated by the author]: *Cryptocodon monocephalus* (Trautv.) Fed.

> Fedorov, A. A. (1957). *Cryptocodon* Fed., gen. nov. In V. L. Komarov, Flora URSS 24: 422-426, 474-475, illus. Leningrad: Academia Scientiae URSS. Ru. — Establishment of genus as distinct from *Asyneuma* and *Campanula.*
> Gudkova, I. Y. & G. P. Borshchenko (1991). [The serological study of the Campanulaceae. The phylogenetic relations in the tribe Phyteumateae.] Bot. Žurn. 76(6): 809-817. Ru. — Protein serology supports close relationship to *Asyneuma, Petromarula, Phyteuma,* and *Sergia,* and a somewhat more distant relationship to *Cylindrocarpa.*

Cryptocodon Fed. in Kom., Fl. URSS 24: 474 (1957).
Middle Asia. 32.

Cryptocodon monocephalus (Trautv.) Fed. in Kom., Fl. URSS 24: 425 (1957).
Tadzhikistan. 32 TZK. Hemicr.
    * *Campanula monocephala* Trautv., Trudy Imp. S.-Peterburgsk. Bot. Sada 6: 64 (1879).
      Phyteuma monocephalum (Trautv.) Pavlov, Bjull. Moskovsk. Obšč. Isp. Prir., Otd. Biol. (n.s.) 42: 127, 131 (1933).
      Phyteuma occultans Popov & Vved., Trudy Turkestansk. Naučn. Obšč. Sredne-Aziatsk. Gosud. Universitete 2: 30 (1925).

Cyananthus

Campanuloideae, 23 species, easterrn Asia, from western China to Himalaya and Myanmar. The genus is divided (Shrestha 1997) into subg. *Cyananthus* and subg. *Micranthus* K. Shrestha; the former comprises four sections: *Annui* (Y. S. Lian) D. Y. Hong & L. M. Ma (two subsections), *Cyananthus, Stenolobi* (Franch.) Y. S. Lian (two subsections), and *Suffruticulosi* K. Shrestha. Preliminary phylogenetic analyses of molecular data (Eddie et al. 2003, under **General**) show that the genus belongs to a clade comprising *Canarina, Codonopsis, and Platycodon;* this "platycodonoid clade" is sister to the remainder of the Campanuloideae. The treatment here follows Shrestha (1997). $2n$ = 14. Type [designated by Reveal & Hoogland, Taxon 40: 650 (1991)]: *Cyananthus lobatus* Wall. ex Benth.

> Royle, J. F. (1836). Illustrations of the botany and other branches of the natural history of the Himalayan Mountains and of the flora of Cashmere, 2 vols., illus. London: Wm. H. Alland. En. — Establishment of genus as member of Polemoniaceae.
> Hooker, J. D. (1880). *Cyananthus lobatus.* Bot. Mag. 106: tab. 6485 + 2 pp., illus. En. — Portrait with description.
> • Franchet, M. A. (1887). Le genre *Cyananthus.* J. Bot (Morot) 1: 241-247, 257-260, 279-282. Fr. — Treatment with non-dichotomous key, descriptions, infrageneric classification, full nomenclature, and specimen citations.
> • Marquand, C. V. B. (1924). Revision of the genus *Cyananthus.* Bull. Misc. Inform. Kew 1924: 241-255, illus. En. — Treatment with key, full nomenclature, and specimen citations; argues against infrageneric classification of Franchet (1887).
> Marquand, C. V. B. (1935). *Cyananthus longiflorus.* Bot. Mag. 158: tab. 9387 + 2 pp., illus. En. — Portrait with description.
> • Cowan, J. M. (1938). Concerning the genus *Cyananthus.* New Fl. & Silva 10: 108-115, 181-190 + figs. xxxiv-xxxvii, lvii-lxii, illus. En. — Overview of genus from horticultural perspective.

Marquand, C. V. B. (1938). *Cyananthus neglectus*. Bot. Mag. 147: tab. 8909 + 2 pp., illus. En. — Portrait of *C. incanus* subsp. *petiolatus,* with description.

Ballard, F. (1939). *Cyananthus macrocalyx*. Bot. Mag. 162: tab. 9586 + 2 pp., illus. En. — Portrait of *C. macrocalyx* subsp. *macrocalyx,* with description.

Grahame, R. E. (1940). Cyananthuses. Gard. Chron. (ser. 3) 107: 10. En. — Popular account of genus for alpine gardeners.

Ballard, F. (1940). *Cyananthus microphyllus*. Bot. Mag. 162: tab. 9598 + 3 pp., illus. En. — Portrait of *C. microphyllus* subsp. *microphyllus,* with description.

Cowan, J. M. (1943). *Cyananthus sherriffii*. Bot. Mag. 164: tab. 9655 + 2 pp., illus. En. — Portrait with description.

Bailey, L. H. (1953). *Cyananthus*. In The garden of bellflowers in North America: 128-132, illus. New York: MacMillan. En. — Account of species cultivated in North America, with key and descriptions.

Kumar, V. & K. P. S. Chauhan (1975). Cytology of *Cyananthus linifolius* Wall. Cat. Proc. Indian Sci. Congr. Assoc. 62 (3) Group B: 130. En. — First chromosome number report for genus.

• Lian, Y. S. (1983). *Cyananthus* Wall. ex Benth. In Flora Reipublicae Popularis Sinicae 73(2): 5-28, illus. Beijing: Science Press. Ch. — Flora of China, with key, descriptions, infrageneric classification, and full nomenclature.

Ducháčová, O. (1985). *Cyananthus microphyllus* Edgew. (Campanulaceae). Skalničky 2: 65-66. Cz. — Brief descriptive account.

• Hong, D. Y., and L. M. Ma (1991). Systematics of the genus *Cyananthus* Wall. ex Royle. Acta Phytotax. Sin. 29: 25-51 + pl. 1, illus., maps. Ch. — Treatment with key, descriptions, infrageneric classification, full nomenclature, and specimen citations; designated *C. integer* as type seven months before Reveal & Hoogland's (1991) choice of *C. lobatus,* but this action was ignored when the name was conserved.

Reveal, J. L. & R. D. Hoogland (1991). Proposal to conserve 8660 *Cyananthus* Wallich ex Bentham (Campanulaceae) over *Cyananthus* Rafinesque (Asteraceae). Taxon 40: 650-651. En. — Formal proposal to conserve the name; designation of a different lectotype than that chosen earlier by Hong & Ma (1991).

Huxley, A. J., ed. (1992). *Cyananthus*. In The new Royal Horticultural Society dictionary of gardening 1: 788. London: MacMillan. En. — Account of species in cultivation.

Shrestha, K. K. & T. I. Kravtsova (1992). [Seed coat anatomy and ultrasculpture in the genus *Cyananthus* (Campanulaceae) and its relation to its systematics.] Bot. Žurn. 77(6): 18-29 + pl. I-III, illus. Ru. — Little variation noted within genus.

Shrestha, K. K. (1992). The taxonomic significance of the petiole anatomy in the genus *Cyananthus* (Campanulaceae). Bot. Žurn. 77(8): 83-89, illus. En. — Some correlation of anatomical data with other features, *e.g.,* phloem lactifers most consistently present in sect. *Stenolobi*.

Shrestha, K. K. & V. F. Tarasevich (1992). [Comparative pollen morphology of genus *Cyananthus* in relation to its systematics and its position within the family Campanulaceae.] Bot. Žurn. 77(10): 1-13 + pl. I-IV, illus. Ru. — Application of palynological data to taxonomy of genus; suggests relationship to *Codonopsis* and *Ostrowskia*.

Brummitt, R. K. (1994). Report of the Committee for Spermatophyta: 40. Taxon 43: 113-126. En. — Conservation of generic name recommended.

Chadwell, C. & H. Hay (1995). Himalayan Cyananthuses. Sino-Himalayan Pl. Assoc. Newsl. 11: 20-27, illus. En. — Descriptive account from horticultural perspective, with emphasis on exploration and discovery in natural habitats.

• Shrestha, K. K. (1997). Taxonomic revision of the Sino-Himalayan genus *Cyananthus* (Campanulaceae). Acta Phytotax. Sin. 35: 396-433, illus. En. — Treatment with keys, descriptions, infrageneric classification, full nomenclature, and specimen citations.

Cyananthus Wall. ex Benth. in Royle, Ill. Bot. Himal. Mts.: 309 (1836), nom. cons.; non Raf., Anal. Bot.: 192 (1815), nom. rejic.

E. Asia. 36 40 41.

Cyananthus cordifolius Duthie, Bull. Misc. Inform. Kew 1912: 37 (1912).
Tibet to Himalaya. 36 CHT 40 NEP WHM. Hemicr.

Cyananthus cronquistii K. Shrestha, Kew Bull. 49: 143 (1994).
SC. China (Yunnan). 36 CHC. Ther.
Cyananthus hookeri var. *grandiflorus* C. Marquand, Bull. Misc. Inform. Kew 1924:
249 (1924).
Cyananthus hookeri var. *levicalyx* Y. S. Lian, Fl. Reipubl. Popul. Sin. 73(2): 183 (1983).

Cyananthus delavayi Franch., J. Bot. (Morot) 1: 280 (1887).
SC. China (Sichuan, Yunnan). 36 CHC. Hemicr.
* *Cyananthus barbatus* Franch., Bull. Soc. Bot. France 32: 9 (1885); non Edgew., Trans.
Linn. Soc. London 20: 82 (1846), as 'barbata'.
Cyananthus microrhombeus C. Y. Wu, Rep. Yunnan Trop. Subtrop. Fl. Res. Inst. 1:
85 (1965).

Cyananthus dolichosceles C. Marquand, Bull. Misc. Inform. Kew 1924: 250 (1924).
SW. China (Sichuan, Tibet). 36 CHC CHT. Hemicr.

Cyananthus fasciculatus C. Marquand, Bull. Misc. Inform. Kew 1924: 247 (1924).
SC. China (Sichuan, Yunnan). 36 CHC. Ther.

Cyananthus flavus C. Marquand, Bull. Misc. Inform. Kew 1924: 247 (1924).
SC. China (Sichuan, Yunnan). 36 CHC. Hemicr.

subsp. **flavus**
SC. China (Yunnan). 36 CHC. Hemicr.
Cyananthus flavus var. *glaber* C. Y. Wu, Rep. Yunnan Trop. Subtrop. Fl. Res. Inst. 1:
84 (1965).

subsp. **montanus** (C. Y. Wu) D. Y. Hong & L. M. Ma, Acta Phytotax. Sin. 29: 46 (1991).
SC. China (Sichuan, Yunnan). 36 CHC. Hemicr.
* *Wahlenbergia mairei* H. Lév., Repert. Spec. Nov. Regni Veg. 12: 285 (1913). *Atropanthe
mairei* (H. Lév.) H. Lév., Bull. Acad. Int. Géogr. Bot. 25: 37 (1915). *Cyananthus mairei*
(H. Lév.) Cowan, New Fl. & Silva 10: 188 (1938); non H. Lév., Cat. Pl. Yun-Nan: 25
(1916). *Cyananthus montanus* C. Y. Wu, Rep. Yunnan Trop. Subtrop. Fl. Res. Inst. 1:
89 (1965). *Cyananthus albiflorus* D. F. Chamb., Notes Roy. Bot. Gard. Edinburgh 35:
252 (1977).

Cyananthus formosus Diels, Notes Roy. Bot. Gard. Edinburgh 5: 172 (1912).
SC. China (Sichuan, Yunnan). 36 CHC. Hemicr.
Cyananthus chungdienensis C. Y. Wu, Rep. Yunnan Trop. Subtrop. Fl. Res. Inst. 1: 83 (1965).

Cyananthus hayanus C. Marquand, New Fl. & Silva 8: 207 (1936), as 'hayana'.
Nepal. 40 NEP. Hemicr.

Cyananthus himalaicus K. Shrestha, Brittonia 44: 253 (1992).
Nepal. 40 NEP. Hemicr.

Cyananthus hookeri C. B. Clarke in Hook. f., Fl. Brit. India 3: 435 (1881).
Himalaya to W. China (Gansu, Qinghai, Sichuan, Yunnan). 36 CHC CHN CHQ CHT 40
EHM NEP. Ther.
Cyananthus hookeri var. *hispidus* Franch., J. Bot. (Morot) 1: 281 (1887).
Cyananthus hookeri var. *densus* C. Marquand, Bull. Misc. Inform. Kew 1924: 249 (1924).

Cyananthus incanus Hook. f. & Thomson, J. Proc. Linn. Soc., Bot. 2: 20 (1858).
Himalaya to W. China. 36 CHC CHQ CHT 40 EHM NEP. Hemicr.

subsp. **incanus**
  Himalaya to W. China (Qinghai, Sichuan). 36 CHC CHQ CHT 40 EHM NEP. Hemicr.

subsp. **orientalis** K. Shrestha, Acta Phytotax. Sin. 35: 407 (1997).
  Himalaya to Tibet. 36 CHT 40 EHM NEP. Hemicr.
  Cyananthus incanus var. *parvus* C. Marquand, Bull. Misc. Inform. Kew 1924: 252 (1924).
  Cyananthus incanus var. *decumbens* Y. S. Lian, Acta Phytotax. Sin. 17: 122 (1979).

subsp. **petiolatus** (Franch.) D. Y. Hong & L. M. Ma, Acta Phytotax. Sin. 29: 44 (1991).
  SC. China (Sichuan). 36 CHC. Hemicr.
  * *Cyananthus petiolatus* Franch., Bull. Soc. Philom. Paris (ser. 8) 3: 147 (1891).
  Cyananthus neglectus C. Marquand, Bot. Mag. 147: tab. 8909 (1938).
  Cyananthus pilifolius C. Y. Wu, Rep. Yunnan Trop. Subtrop. Fl. Res. Inst. 1: 87 (1965).
    Cyananthus petiolatus var. *pilifolius* (C. Y. Wu) Y. S. Lian, Fl. Reipubl. Popul. Sin. 73(2):
    21 (1983).
  Cyananthus pilifolius f. *leiocalyx* C. Y. Wu, Rep. Yunnan Trop. Subtrop. Fl. Res. Inst. 1: 88
    (1965); non C. Y. Wu, Rep. Yunnan Trop. Subtrop. Fl. Res. Inst. 1: 89 (1965).
  Cyananthus pilifolius f. *leiocalyx* C. Y. Wu, Rep. Yunnan Trop. Subtrop. Fl. Res. Inst. 1: 89
    (1965); non C. Y. Wu, Rep. Yunnan Trop. Subtrop. Fl. Res. Inst. 1: 88 (1965).
  Cyananthus pilifolius var. *minor* C. Y. Wu, Rep. Yunnan Trop. Subtrop. Fl. Res. Inst. 1:
    89 (1965).
  Cyananthus pilifolius var. *pallidocoeruleus* C. Y. Wu, Rep. Yunnan Trop. Subtrop. Fl. Res.
    Inst. 1: 89 (1965).

Cyananthus inflatus Hook. f. & Thomson, J. Proc. Linn. Soc., Bot. 2: 21 (1858).
  Himalaya to Myanmar & SW. China (Sichuan, Tibet, Yunnan). 36 CHC CHT 40 ASS EHM
    IND NEP 41 MYA. Ther. $2n = 14$.
  Cyananthus forrestii Diels, Notes Roy. Bot. Gard. Edinburgh 5: 173 (1912).
  Cyananthus pseudoinflatus P. C. Tsoong, Contr. Inst. Bot. Natl. Acad. Peiping 3: 109
    (1935), as 'pseudo-inflatus'.

Cyananthus integer Wall. ex Benth. in Royle, Ill. Bot. Himal. Mts.: 309 (1836), as 'integra'.
  W. Himalaya. 40 WHM. Hemicr.
  Cyananthus barbatus Edgew., Trans. Linn. Soc. London 20: 82 (1846), as 'barbata'.

Cyananthus leiocalyx (Franch.) Cowan, New Fl. & Silva 10: 187 (1938).
  E. Himalaya & SW. China (Yunnan, Tibet). 36 CHC CHT 40 EHM. Hemicr.
  * *Cyananthus incanus* var. *leiocalyx* Franch., J. Bot. (Morot) 1: 279 (1887).

subsp. **leiocalyx**
  E. Himalaya & SW. China (Yunnan, Tibet). 36 CHC CHT 40 EHM. Hemicr.

subsp. **lucidus** K. Shrestha, Acta Phytotax. Sin. 35: 409 (1997).
  Tibet. 36 CHT. Hemicr.

Cyananthus lichiangensis W. W. Sm., Notes Roy. Bot. Gard. Edinburgh 8: 109 (1913).
  SW. China (Sichuan, Yunnan, Tibet). 36 CHC CHT. Ther.

Cyananthus lobatus Wall. ex Benth. in Royle, Ill. Bot. Himal. Mts.: 309 (1836), as 'lobata'.
  N. India (Punjab, Uttar Pradesh) to SW. China (Tibet, Yunnan) & Myanmar. 36 CHC CHT
    40 ASS EHM IND NEP WHM 41 MYA. Hemicr.
  Cyananthus lobatus var. *farreri* C. Marquand, Bull. Misc. Inform. Kew 1924: 247 (1924).

Cyananthus longiflorus Franch., J. Bot. (Morot) 1: 280 (1887).
  SC. China (Yunnan). 36 CHC. Hemicr.
  Cyananthus argenteus C. Marquand, Bull. Misc. Inform. Kew 1924: 253 (1924).
  Cyananthus obtusilobus C. Marquand, Bull. Misc. Inform. Kew 1924: 254 (1924).

Cyananthus macrocalyx Franch., J. Bot. (Morot) 1: 279 (1887).
W. China to Myanmar. 36 CHC CHN CHQ CHT 40 ASS EHM NEP 41 MYA. Hemicr.

subsp. **macrocalyx**
W. China (Gansu, Qinghai, Sichuan, Tibet, Yunnan). 36 CHC CHN CHQ CHT. Hemicr.
Cyananthus macrocalyx var. *flavopurpureus* C. Marquand, Bull. Misc. Inform. Kew 1924:
252 (1924), as 'flavo-purpureus'.
Cyananthus neurocalyx C. Y. Wu, Rep. Yunnan Trop. Subtrop. Fl. Res. Inst. 1: 86 (1965).

subsp. **spathulifolius** (Nannf.) K. Shrestha, Acta Phytotax. Sin. 35: 412 (1997).
SW. China (Sichuan, Tibet, Yunnan) to Myanmar. 36 CHC CHT 40 ASS EHM NEP 41
MYA. Hemicr.
* *Cyananthus spathulifolius* Nannf., Acta Horti Gothob. 5: 30 (1930).

Cyananthus microphyllus Edgew., Trans. Linn. Soc. London 20: 81 (1846), as
'microphylla'. *Cyananthus linifolius* Wall. ex Hook. f. & Thomson, J. Proc. Linn. Soc., Bot.
2: 20 (1858).
N. India (Punjab, Uttar Pradesh) to Tibet. 36 CHT 40 IND NEP WHM. Hemicr. $2n = 14$.

subsp. **microphyllus**
N. India (Punjab, Uttar Pradesh) to Tibet. 36 CHT 40 IND NEP WHM. Hemicr. $2n = 14$.
Cyananthus nepalensis Kitam., Acta Phytotax. Geobot. 15: 109 (1954).

subsp. **williamsii** K. Shrestha, Acta Phytotax. Sin. 35: 419 (1997), as 'williamsonii'.
Nepal. 40 NEP. Hemicr.

Cyananthus pedunculatus C. B. Clarke in Hook. f., Fl. Brit. India 3: 434 (1881).
Tibet to Himalaya. 36 CHT 40 EHM NEP. Hemicr.
Cyananthus sericeus Y. S. Lian, Acta Phytotax. Sin. 17: 122 (1979).

Cyananthus pilosus (C. Marquand) K. Shrestha, Acta Phytotax. Sin. 35: 405 (1997).
SW. China (Yunnan, Tibet). 36 CHC CHT. Hemicr.
* *Cyananthus macrocalyx* var. *pilosus* C. Marquand, Bull. Misc. Inform. Kew 1924: 251
(1924).

Cyananthus sherriffii Cowan, New Fl. & Silva 10: 181 (1938).
Tibet. 36 CHT. Hemicr.

Cyananthus wardii C. Marquand, J. Linn. Soc., Bot. 48: 196 (1929).
Tibet. 36 CHT. Hemicr.

Synonyms:
Cyananthus albiflorus D. F. Chamb. === **Cyananthus flavus** subsp. **montanus** (C. Y. Wu) D.
Y. Hong & L. M. Ma
Cyananthus argenteus C. Marquand === **Cyananthus longiflorus** Franch.
Cyananthus barbatus Edgew. === **Cyananthus integer** Wall. ex Benth.
Cyananthus barbatus Franch. === **Cyananthus delavayi** Franch.
Cyananthus chungdienensis C. Y. Wu === **Cyananthus formosus** Diels
Cyananthus flavus var. *glaber* C. Y. Wu === **Cyananthus flavus** C. Marquand var. **flavus**
Cyananthus forrestii Diels === **Cyananthus inflatus** Hook. f. & Thomson
Cyananthus hookeri var. *densus* C. Marquand === **Cyananthus hookeri** C. B. Clarke
Cyananthus hookeri var. *hispidus* Franch. === **Cyananthus hookeri** C. B. Clarke
Cyananthus hookeri var. *levicalyx* Y. S. Lian === **Cyananthus cronquistii** K. Shrestha
Cyananthus incanus var. *bicolor* Cowan === ?
Cyananthus incanus var. *decumbens* Y. S. Lian === **Cyananthus incanus** subsp. **orientalis**
K. Shrestha
Cyananthus incanus var. *leiocalyx* Franch. === **Cyananthus leiocalyx** (Franch.) Cowan

Cyananthus incanus var. *nudicalyx* Cowan === ?

Cyananthus incanus var. *parvus* C. Marquand === **Cyananthus incanus** subsp. **orientalis** K. Shrestha

Cyananthus inflatus var. *rufus* Franch. === ?

Cyananthus inflatus var. *sylvestris* C. Marquand === ?

Cyananthus linifolius Wall. ex Hook. f. & Thomson === **Cyananthus microphyllus** Edgew.

Cyananthus lobatus var. *farreri* C. Marquand === **Cyananthus lobatus** Wall. ex Benth.

Cyananthus macrocalyx var. *flavopurpureus* C. Marquand === **Cyananthus macrocalyx** Franch. subsp. **macrocalyx**

Cyananthus macrocalyx var. *pilosus* C. Marquand === **Cyananthus pilosus** (C. Marquand) K. Shrestha

Cyananthus mairei H. Lév. === **Codonopsis bulleyana** Forrest ex Diels

Cyananthus mairei (H. Lév.) Cowan === **Cyananthus flavus** subsp. **montanus** (C. Y. Wu) D. Y. Hong & L. M. Ma

Cyananthus microrhombeus C. Y. Wu === **Cyananthus delavayi** Franch.

Cyananthus microrhombeus var. *leiocalyx* C. Y. Wu === ?

Cyananthus montanus C. Y. Wu === **Cyananthus flavus** subsp. **montanus** (C. Y. Wu) D. Y. Hong & L. M. Ma

Cyananthus neglectus C. Marquand === **Cyananthus incanus** subsp. **petiolatus** (Franch.) D. Y. Hong & L. M. Ma

Cyananthus nepalensis Kitam. === **Cyananthus microphyllus** Edgew. subsp. **microphyllus**

Cyananthus neurocalyx C. Y. Wu === **Cyananthus macrocalyx** Franch. subsp. **macrocalyx**

Cyananthus obtusilobus C. Marquand === **Cyananthus longiflorus** Franch.

Cyananthus pedunculatus var. *crenatus* C. Marquand === ?

Cyananthus petiolatus Franch. === **Cyananthus incanus** subsp. **petiolatus** (Franch.) D. Y. Hong & L. M. Ma

Cyananthus petiolatus var. *pilifolius* (C. Y. Wu) Y. S. Lian === **Cyananthus incanus** subsp. **petiolatus** (Franch.) D. Y. Hong & L. M. Ma

Cyananthus pilifolius C. Y. Wu === **Cyananthus incanus** subsp. **petiolatus** (Franch.) D. Y. Hong & L. M. Ma

Cyananthus pilifolius f. *leiocalyx* C. Y. Wu === **Cyananthus incanus** subsp. **petiolatus** (Franch.) D. Y. Hong & L. M. Ma

Cyananthus pilifolius f. *leiocalyx* C. Y. Wu === **Cyananthus incanus** subsp. **petiolatus** (Franch.) D. Y. Hong & L. M. Ma

Cyananthus pilifolius var. *pallidocoeruleus* C. Y. Wu === **Cyananthus incanus** subsp. **petiolatus** (Franch.) D. Y. Hong & L. M. Ma

Cyananthus pilifolius var. *minor* C. Y. Wu === **Cyananthus incanus** subsp. **petiolatus** (Franch.) D. Y. Hong & L. M. Ma

Cyananthus pseudoinflatus P. C. Tsoong === **Cyananthus inflatus** Hook. f. & Thomson

Cyananthus sericeus Y. S. Lian === **Cyananthus pedunculatus** C. B. Clarke

Cyananthus spathulifolius Nannf. === **Cyananthus macrocalyx** subsp. **spathulifolius** (Nannf.) K. Shrestha

Cyanea

Lobelioideae, 78 species, Hawaiian Islands. Although the genus has been divided into as many as five sections (Rock 1919), recent analyses of molecular data (Givnish et al. 1995) and seed morphology (Buss et al. 2001) suggest that the genus might best be divided into just two: sect. *Cyanea* and sect. *Delisseoideae* (Hillebr.) Rock. Phylogenetic analyses of DNA data (Givnish et al. 1995; E. Knox, pers. comm.; A. Antonelli, pers. comm.) indicate that *Cyanea* is sister to *Clermontia* (q.v.). The treatment here follows Lammers (1990), as modified subsequently (Lammers 1992, 1996, 1998, 1999, 2004, 2005; Lammers & Lorence 1993; Lammers et al. 1993). 2n = 28. Type [by monotypy]: *Cyanea grimesiana* Gaudich.

Gaudichaud, C. (1826). *Cyanea*. In Voyage autour du monde, entrepris par order du roi ... exécuté sur les corvettes de S. M. *l'Uranie* et *la Physicienne,* pendant les années 1817, 1818, 1819 et 1820. Botanique: 457-458 + pl. 74, illus. Paris: Pillet-ainé. La., Fr. — Establishment of genus.

Gray, A. (1861). Notes on Lobeliaceae, Goodeniaceae, &c. of the collections of the U.S. South Pacific Exploring Expedition. Proc. Amer. Acad. Arts 5: 146-152. En. — Circumscription altered to encompass all Hawaiian Lobelioideae with persistent foliaceous or enlarged (but non-petaloid) calyx lobes, thus encompassing some species of *Rollandia*.

- Hillebrand, W. (1888). *Rollandia,* Gaud.; *Cyanea,* Gaud. In Flora of the Hawaiian Islands: 244-248, 251-264. London: Williams & Norgate. En. — Flora with keys, descriptions, infrageneric classification, full nomenclature, and specimen citations.

Rock, J. F. (1913). *Cyanea* Gaud. In The indigenous trees of the Hawaiian Islands: 489-494, 508-511, illus. Honolulu: published privately. En. — Floristic treatment of arborescent species, with key, descriptions, full nomenclature, and specimen citations.

- Rock, J. F. (1919). A monographic study of the Hawaiian species of the tribe Lobelioideae family Campanulaceae, *Cyanea* Gaudichaud, *Rollandia* Gaudichaud. Mem. Bernice Pauahi Bishop Mus. 7(2):153-282, 363-383, illus. En. — Monograph with keys, descriptions, infrageneric classification, full nomenclature, and specimen citations.

Skottsberg, C. (1927). Iakttagelser över blomningen hos *Cyanea hirtella* (H. Mann) Rock. Acta Horti Gothob. 3: 43-55, illus. Sw. — Observations of flowering and pollen discharge in individual grown from seed in Göteborg greenhouse.

- Wimmer, F. E. (1943). *Cyanea* Gaudich., *Rollandia* Gaudich. In A. Engler & L. Diels, Das Pflanzenreich IV. 276b: 46-79, 96-104, illus. Leipzig: Wilhelm Engelmann. Ge., La. — Monograph with keys, descriptions, infrageneric classification, full nomenclature, and specimen citations.

Selling, O. H. (1947). Studies in Hawaiian pollen statistics II. The pollens of the Hawaiian phanerogams. Lobeliaceae. Spec. Publ. Bernice Pauahi Bishop Mus. 38: 318-320 + pl. 49, illus. En. — Descriptive palynology of two species.

- Wimmer, F. E. (1953). *Cyanea, Rollandia*. In A. Engler & L. Diels, Das Pflanzenreich IV. 276b: 760-763. Berlin: Akademie-Verlag. Ge., La. — Supplement to Wimmer (1943).

Carlquist, S. (1962). Ontogeny and comparative anatomy of thorns of Hawaiian Lobeliaceae. Amer. J. Bot. 49: 413-419, illus. En. — Anatomy of epidermal prickles.

Stone, B. C. (1967). A review of the endemic genera of Hawaiian plants. *Cyanea* Gaud., *Rollandia* Gaud. Bot. Rev. (Lancaster) 33: 243-246. En. — Brief descriptive treatment with full nomenclature.

- Wimmer, F. E. (1968). *Cyanea* Gaudich., *Rollandia* Gaudich. In A. Engler & L. Diels, Das Pflanzenreich IV. 276c: 817-825 + Taf. 1-8, illus. Berlin: Akademie-Verlag. Ge., La. — Second supplement to Wimmer (1943).

Degener, O., I. Degener & H. Hörmann (1969). *Cyanea carlsonii* Rock and the unnatural distribution of *Sphagnum palustre* L. Phytologia 19: 1-4, illus. En. — Conservation status of *C. hamatiflora* subsp. *carlsonii,* with notes on its discovery and naming.

St. John, H. (1969). Types of sections in *Clermontia, Cyanea,* and *Delissea* (Lobeliaceae). Taxon 18: 483. En. — Formal nomenclature of infrageneric taxa.

- Cory, C. (1984). Pollination biology of two species of Hawaiian Lobeliaceae (*Clermontia kakeana* and *Cyanea angustifolia*) and their presumed coevolved relationship with native honeycreepers (Drepanididae), illus. Fullerton: California State University. En. — Unpublished M.A. thesis on reproductive biology, documenting self-fertilization.

Lammers, T. G. & C. E. Freeman (1986). Ornithophily among the Hawaiian Lobelioideae (Campanulaceae): evidence from floral nectar sugar compositions. Amer. J. Bot. 73: 1613-1619. En. — Ornithophily inferred from hexose-dominant nectars.

- St. John (1987). Enlargement of *Delissea* (Lobeliaceae). Hawaiian plant studies 138. Phytologia 63: 79-90. En. — Transfer of all taxa in *Cyanea* into *Delissea,* in light of St. John & Takeuchi (1987).

- St. John, H. & W. Takeuchi (1987). Are the distinctions of *Delissea* valid? Hawaiian plant studies 137. Phytologia 63: 129-130. En. — *Delissea* indistinguishable from *Cyanea.*

Lammers, T. G. (1988). Chromosome numbers and their systematic implications in Hawaiian Lobelioideae (Campanulaceae). Amer. J. Bot. 75:1130-1134. En. — Summary of known chromosome numbers in genus.

Lammers, T. G. (1990a). Sequential paedomorphosis among the endemic Hawaiian Lobelioideae (Campanulaceae). Taxon 39: 206-211, illus., maps. En. — Evolution within *C. solanacea* complex involved increasing juvenilization of adult morphology.

• Lammers, T. G. (1990b). *Cyanea* Gaud., *Rollandia* Gaud. In W. L. Wagner, D. R. Herbst & S. H. Sohmer, Manual of the flowering plants of Hawai'i: 437-467, 480-485, illus. Honolulu: University of Hawaii Press. En. — Flora with key and descriptions; argues against St. John's (1987) merger with *Delissea*.

Hobdy, R. W., A. C. Medeiros & L. L. Loope (1990). *Cyanea obtusa* and *Cyanea lobata* (Lobeliaceae): recent apparent extinctions of two Maui endemics. Newsl. Hawaiian Bot. Soc. 29: 3-6. En. — Species long thought extinct were rediscovered in 1980s but subsequently extirpated.

Duvall, F. II (1991). Cultivation of *Clermontia, Cyanea, Lobelia,* and *Trematolobelia* from seed. Hawaii's Forests & Wildlife 6(3): 12-13, illus. En. — Practical methods for cultivation of endangered rainforest species.

Lammers, T. G. (1992). Two new combinations in the endemic Hawaiian genus *Cyanea* (Campanulaceae: Lobelioideae). Novon 2: 129-131. En. — Update of Lammers (1990).

Wichman, C. (1992). *Cyanea linearifolia*. Exciting rediscovery of "extinct" plant. Bull. Pacific Trop Bot. Gard. 22: 39-41 + cover, illus. En. — Description of events surrounding discovery of *C. kuhihewa* (originally mistaken for *C. linearifolia*).

Lammers, T. G., T. J. Givnish & K. J. Sytsma (1993). Merger of the endemic Hawaiian genera *Cyanea* and *Rollandia* (Campanulaceae: Lobelioideae). Novon 3: 437-441. En. — *Rollandia* subsumed into *Cyanea* on basis of data from molecular phylogenetics.

Lammers, T. G. & D. H. Lorence (1993). A new species of *Cyanea* (Campanulaceae: Lobelioideae) from Kaua'i, and the resurrection of *C. remyi*. Novon 3: 431-436, illus. En. — Update of Lammers (1990).

• Givnish, T. J., K. J. Sytsma, J. F. Smith & W. J. Hahn (1994). Thorn-like prickles and heterophylly in *Cyanea*: adaptations to extinct avian browsers on Hawaii? Proc. Natl. Acad. Sci. USA 91: 2810-2814, illus., maps. En. — Epidermal prickles and dissected juvenile foliage adaptation to predation by flightless geese and ducks known only from fossils.

Lammers, T. G. (1994). Typification of the names of Hawaiian Lobelioideae (Campanulaceae) published by Wilhelm Hillebrand or based upon his specimens. Taxon 43: 545-572. En. — Formal nomenclature of numerous species.

• Givnish, T. J., K. J. Sytsma, J. F. Smith & W. J. Hahn (1995). Molecular evolution, adaptive radiation, and geographic speciation in *Cyanea* (Campanulaceae, Lobelioideae). In W. L. Wagner & V. A. Funk (eds.), Hawaiian biogeography: evolution on a hot spot archipelago: 288-337, illus., maps. Washington: Smithsonian Institution Press. En. — Molecular phylogeny of genus used to infer patterns of diversification within archipelago.

Lammers, T. G. (1996). A new linear-leaved *Cyanea* (Campanulaceae: Lobelioideae) from Kaua'i, and the "rediscovery" of *Cyanea linearifolia*. Brittonia 48: 237-240, illus. En. — Update of Lammers (1990); Wichman's (1992) rediscovery actually an undescribed species.

Wichman, C. (1997). Limahuli battles Koster's curse: rare lobelia returns. News of the [National Tropical Botanical] Garden 2(3): 7, illus. En. — Conservation efforts to save *C. kuhihewa*, threatened by introduced *Clidemia hirta* (Melastomataceae).

Lammers, T. G. (1998). New names and new combinations in Campanulaceae. Novon 8: 31-35. En. — Update of Lammers (1990).

Lammers, T. G. (1999). A new *Lobelia* from Mexico, with additional new combinations in world Campanulaceae. Novon 9: 381-389, illus. En. — Update of Lammers (1990).

• Buss, C. C., T. G. Lammers & R. R. Wise (2001). Seed coat morphology and its systematic implications in *Cyanea* and other genera of Lobelioideae (Campanulaceae). Amer. J. Bot. 88: 1301-1308, illus. En. — Scanning electron microscopy showed unique testal pattern for sect. *Delisseoideae*.

Lammers, T. G. (2004). Five new species of the endemic Hawaiian genus *Cyanea* (Campanulaceae: Lobelioideae). Novon 14: 84-101. — Update of Lammers (1990); includes keys to all taxa in *C. grimesiana* complex, *C. scabra* complex, and sect. *Delisseoideae*.

Lammers, T. G. (2005). Revision of *Delissea* (Campanulaceae-Lobelioideae). Syst. Bot. Monogr. 73: 1-75, illus., maps. En. — *Delissea rivularis* transferred to *Cyanea*.

Cyanea Gaudich., Voy. Uranie: pl. 75 (1828). *Kittelia* Rchb., Handb. Nat. Pfl.-Syst.: 186 (1837). *Delissea* sect. *Cyanea* (Gaudich.) Baill., Hist. Pl. 8: 364 (1885).
Hawaiian Is. 63 HAW.
  Rollandia Gaudich., Voy. Uranie: pl. 74 (1828).
  Macrochilus C. Presl, Prodr. Monogr. Lobel.: 47 (1836).

Cyanea aculeatiflora Rock, Indig. Trees Haw. Isl.: 509 (1913). *Delissea aculeatiflora* (Rock) H. St. John, Phytologia 63: 79 (1987).
Hawaiian Is. (E. Maui). 63 HAW. Phan.
  Cyanea solenocalyx var. *schizocalyx* Hillebr., Fl. Hawaiian Isl.: 258 (1888).

Cyanea acuminata (Gaudich.) Hillebr., Fl. Hawaiian Isl.: 254 (1888).
Hawaiian Is. (O'ahu). 63 HAW. Nanophan.
  Delissea acuminata Gaudich., Voy. Uranie: pl. 76 (1828). *Lobelia acuminata* (Gaudich.) Hook. & Arn., Bot. Beechey Voy.: 88 (1832); non Sw., Prodr.: 117 (1787). *Lobelia delisseana* Gaudich. ex D. Dietr., Syn. Pl. 1: 728 (1839).
  Delissea acuminata var. *latifolia* Wawra, Flora 56: 7 (1873). *Cyanea acuminata* f. *latifolia* (Wawra) E. Wimm., Pflanzenr. IV.276b: 73 (1943). *Delissea acuminata* f. *latifolia* (Wawra) H. St. John, Phytologia 63: 79 (1987).
  Cyanea acuminata var. *calycina* Hosaka, Occas. Pap. Bernice Pauahi Bishop Mus. 14: 29 (1938). *Delissea acuminata* var. *calycina* (Hosaka) H. St. John, Phytologia 63: 80 (1987).
  Cyanea occultans H. St. John, Phytologia 48: 143 (1981). *Delissea occultans* (H. St. John) H. St. John, Phytologia 63: 86 (1987).

Cyanea angustifolia (Cham.) Hillebr., Fl. Hawaiian Isl.: 253 (1888).
Hawaiian Is. (O'ahu, Moloka'i, Lāna'i, W. Maui). 63 HAW. Nanophan. or phan. 2n = 28.
  Lobelia angustifolia Cham., Linnaea 8: 219 (1833). *Delissea angustifolia* (Cham.) G. Don, Gen. Hist. 3: 699 (1834). *Delissea acuminata* var. *angustifolia* (Cham.) A. Gray, Proc. Amer. Acad. Arts 5: 148 (1861).
  Delissea honoluluensis Wawra, Flora 56: 11 (1873).
  Cyanea angustifolia var. *racemosa* Hillebr., Fl. Hawaiian Isl.: 253 (1888). *Delissea angustifolia* var. *racemosa* (Hillebr.) H. St. John, Phytologia 63: 80 (1987).

Cyanea arborea Hillebr., Fl. Hawaiian Isl.: 261 (1888).
Hawaiian Is. (E. Maui). 63 HAW. Phan.
  Delissea arborea H. Mann, Proc. Amer. Acad. Arts 7: 180 (1867); non (G. Forst.) C. Presl, Prodr. Monogr. Lobel.: 47 (1836). *Cyanea longifolia* A. Heller, Minnesota Bot. Stud. 1: 909 (1897).

Cyanea asarifolia H. St. John, Bot. Mag. (Tokyo) 88: 61 (1975). *Delissea asarifolia* (H. St. John) H. St. John, Phytologia 63: 81 (1987).
Hawaiian Is. (Kaua'i). 63 HAW. Nanophan.

Cyanea aspleniifolia (H. Mann) Hillebr., Fl. Hawaiian Isl.: 260 (1888), as 'asplenifolia'.
Hawaiian Is. (Maui). 63 HAW. Nanophan.
  Delissea aspleniifolia H. Mann, Proc. Amer. Acad. Arts 7: 182 (1867), as 'asplenifolia'.

Cyanea calycina (Cham.) Lammers, Novon 8: 32 (1998).
Hawaiian Is. (O'ahu). 63 HAW. Nanophan. 2n = 28.

* *Lobelia calycina* Cham., Linnaea 8: 222 (1833). *Rollandia calycina* (Cham.) G. Don, Gen.
    Hist. 3: 699 (1834). *Delissea calycina* (Cham.) C. Presl, Prodr. Monogr. Lobel.: 47
    (1836). *Rollandia lanceolata* subsp. *calycina* (Cham.) Lammers, Syst. Bot. 13: 507
    (1988). *Cyanea lanceolata* subsp. *calycina* (Cham.) Lammers, Givnish & Sytsma,
    Novon 3: 439 (1993).
  Cyanea aspera A. Gray, Proc. Amer. Acad. Arts 5: 148 (1861).
  Rollandia humboldtiana var. *tomentella* Wawra, Flora 56: 32 (1873). *Rollandia lanceolata*
    var. *tomentella* (Wawra) E. Wimm., Pflanzenr. IV.276b: 102 (1943).
  Rollandia kaalae Wawra, Flora 56: 45 (1873). *Rollandia calycina* var. *kaalae* (Wawra) E.
    Wimm., Pflanzenr. IV.276b: 101 (1943).
  Rollandia scabra Wawra, Flora 56: 46 (1873).
  Rollandia lanceolata [unranked] *tomentosa* Hillebr., Fl. Hawaiian Isl.: 247 (1888), as
    'Tomentosae'. *Rollandia lanceolata* var. *tomentosa* (Hillebr.) Drake, Ill. Fl. Ins. Pacif.:
    218 (1892).
  Rollandia bidentata H. St. John, Occas. Pap. Bernice Pauahi Bishop Mus. 15: 25 (1939).
  Rollandia waianaeensis H. St. John, Occas. Pap. Bernice Pauahi Bishop Mus. 15: 27 (1939).
  Rollandia obatae H. St. John, Phytologia 63: 367 (1987).

Cyanea comata Hillebr., Fl. Hawaiian Isl.: 256 (1888). *Delissea comata* (Hillebr.) H. St. John,
  Phytologia 63: 81 (1987).
  Hawaiian Is. (E. Maui). 63 HAW. Nanophan.

Cyanea copelandii Rock, Bull. Torrey Bot. Club 44: 231 (1917). *Delissea copelandii* (Rock) H.
  St. John, Phytologia 63: 81 (1987).
  Hawaiian Is. (E. Maui, Hawai'i). 63 HAW. Cl. nanophan.

  subsp. **copelandii**
    Hawaiian Is. (Hawai'i). 63 HAW. Cl. nanophan.

  subsp. **haleakalaensis** (H. St. John) Lammers, Syst. Bot. 13: 501 (1988).
    Hawaiian Is. (E. Maui). 63 HAW. Cl. nanophan.
    Cyanea haleakalaensis H. St. John, Pacific Sci. 25: 65 (1971). *Delissea haleakalaensis* (H.
      St. John) H. St. John, Phytologia 63: 84 (1987).

Cyanea coriacea (A. Gray) Hillebr., Fl. Hawaiian Isl.: 254 (1888).
  Hawaiian Is. (Kaua'i). 63 HAW. Phan. or nanophan.
  * *Delissea coriacea* A. Gray, Proc. Amer. Acad. Arts 5: 147 (1861).
  Cyanea fauriei H. Lév., Repert. Spec. Nov. Regni Veg. 10: 156 (1911). *Cyanea coriacea* var.
    fauriei (H. Lév.) E. Wimm., Pflanzenr. IV.276b: 72 (1943). *Delissea fauriei* (H. Lév.) H.
    St. John, Phytologia 63: 342 (1987); non H. Lév., Repert. Spec. Nov. Regni Veg. 12:
    505 (1913).
  Cyanea coriacea var. *degeneriana* E. Wimm., Pflanzenr. IV.276b: 761 (1953). *Delissea*
    coriacea var. *degeneriana* (E. Wimm.) H. St. John, Phytologia 63: 82 (1987).
  Delissea coriacea var. *deltoidea* H. St. John, Phytologia 63: 340 (1987). *Delissea perlmanii*
    H. St. John, Phytologia 63: 346 (1987), as 'perlmannii'.
  Delissea coriacea var. *haupuensis* H. St. John, Phytologia 63: 340 (1987).
  Delissea coriacea var. *lumahaiensis* H. St. John, Phytologia 63: 340 (1987).

Cyanea crispa (Gaudich.) Lammers, Givnish & Sytsma, Novon 3: 439. 1993.
  Hawaiian Is. (O'ahu). 63 HAW. Nanophan.
  * *Rollandia crispa* Gaudich., Voy. Uranie: 459 (1829). *Lobelia crispa* (Gaudich.) Endl., Ann.
    Wiener Mus. Naturgesch. 1: 170 (1836); non Graham, Edinburgh New Philos. J. 1:
    173 (1826). *Cyanea rollandia* A. Gray, Proc. Amer. Acad. Arts 5: 149 (1861).
  Rollandia crispa var. *muricata* Rock, Monogr. Stud. Haw. Lobelioid.: 381 (1919).

Cyanea cylindrocalyx (Rock) Lammers, Novon 8: 31 (1998).
  Hawaiian Is. (Hawai'i). 63 HAW. Nanophan.

Cyanea grimesiana var. *cylindrocalyx* Rock, Bull. Torrey Bot. Club 44: 235 (1917). *Delissea grimesiana* var. *cylindrocalyx* (Rock) H. St. John, Phytologia 63: 83 (1987). *Cyanea grimesiana* subsp. *cylindrocalyx* (Rock) Lammers, Syst. Bot. 13: 502 (1988).

Cyanea dolichopoda Lammers & Lorence, Novon 3: 432 (1993).
Hawaiian Is. (Kaua'i). 63 HAW. Nanophan.

Cyanea dunbariae Rock, Monogr. Stud. Haw. Lobelioid.: 265 (1919), as 'dunbarii'. *Delissea dunbariae* (Rock) H. St. John, Phytologia 63: 82 (1987).
Hawaiian Is. (Moloka'i). 63 HAW. Nanophan.

Cyanea duvalliorum Lammers & H. Oppenh., Novon 14: 89 (2004).
Hawaiian Is. (E. Maui). 63 HAW. Phan.

Cyanea eleeleensis (H. St. John) Lammers, Novon 2: 129 (1992).
Hawaiian Is. (Kaua'i). 63 HAW. Nanophan.
* *Delissea eleeleensis* H. St. John, Phytologia 63: 341 (1987).

Cyanea elliptica (Rock) Lammers, Syst. Bot. 13: 501 (1988).
Hawaiian Is. (Moloka'i, Lāna'i, Maui). 63 HAW. Nanophan.
Delissea undulata var. *serrulata* Wawra, Flora 56: 8 (1873).
Cyanea angustifolia var. *hillebrandii* Rock, Bull. Torrey Bot. Club 44: 234 (1917). *Delissea angustifolia* var. *hillebrandii* (Rock) H. St. John, Phytologia 63: 80 (1987).
Cyanea angustifolia var. *lanaiensis* Rock, Bull. Torrey Bot. Club 44: 235 (1917). *Delissea angustifolia* var. *lanaiensis* (Rock) H. St. John, Phytologia 63: 80 (1987). *Delissea angustifolia* f. *lanaiensis* (Rock) H. St. John, Phytologia 63: 80 (1987).
* *Cyanea angustifolia* f. *elliptica* Rock, Occas. Pap. Bernice Pauahi Bishop Mus. 22: 64 (1957). *Cyanea angustifolia* var. *elliptica* (Rock) E. Wimm., Pflanzenr. IV.276c: 821 (1968). *Delissea angustifolia* f. *elliptica* (Rock) H. St. John, Phytologia 63: 80 (1987).
Cyanea angustifolia var. *isabella* E. Wimm., Pflanzenr. IV.276c: 821 (1968). *Delissea angustifolia* var. *isabella* (E. Wimm.) H. St. John, Phytologia 63: 80 (1987).
Delissea molokaiensis H. St. John, Phytologia 64: 166 (1988).

Cyanea fissa (H. Mann) Hillebr., Fl. Hawaiian Isl.: 255 (1888).
Hawaiian Is. (Kaua'i). 63 HAW. Phan. or nanophan.
* *Delissea fissa* H. Mann, Proc. Amer. Acad. Arts 7: 182 (1867).
Cyanea humilis Wawra, Flora 56: 47 (1873). *Delissea humilis* (Wawra) H. St. John, Phytologia 63: 342 (1987).
Cyanea multispicata H. Lév., Repert. Spec. Nov. Regni Veg. 10: 157 (1911). *Delissea multispicata* (H. Lév.) H. St. John, Phytologia 63: 86 (1987).
Cyanea gayana Rock, Indig. Trees Haw. Isl.: 510 (1913). *Delissea gayana* (Rock) H. St. John, Phytologia 63: 83 (1987). *Cyanea fissa* subsp. *gayana* (Rock) Lammers, Syst. Bot. 13: 502 (1988).
Cyanea gayana var. *duvelii* Rock, Occas. Pap. Bernice Pauahi Bishop Mus. 22: 50 (1957). *Delissea gayana* var. *duvelii* (Rock) H. St. John, Phytologia 63: 83 (1987).
Cyanea gayana var. *wainihaensis* Rock, Occas. Pap. Bernice Pauahi Bishop Mus. 22: 56 (1957). *Delissea gayana* var. *wainihaensis* (Rock) H. St. John, Phytologia 63: 83 (1987).
Delissea albilineata H. St. John, Phytologia 63: 339 (1987).
Delissea cataracta H. St. John, Phytologia 63: 339 (1987).
Delissea decumbens H. St. John, Phytologia 63: 340 (1987).
Delissea denticulata H. St. John, Phytologia 63: 341 (1987).
Delissea duploserrata H. St. John, Phytologia 63: 341 (1987).
Delissea glabrifolia H. St. John, Phytologia 63: 342 (1987).
Delissea keaensis H. St. John, Phytologia 63: 343 (1987).
Delissea latibasilaris H. St. John, Phytologia 63: 344 (1987).
Delissea limahuliensis H. St. John, Phytologia 63: 344 (1987).

Delissea longicalyx H. St. John, Phytologia 63: 345 (1987).
Delissea lumahaiensis H. St. John, Phytologia 63: 345 (1987).
Delissea multiramosa H. St. John, Phytologia 63: 345 (1987).
Delissea napaliensis H. St. John, Phytologia 63: 345 (1987).
Delissea parva H. St. John, Phytologia 63: 346 (1987).
Delissea pluriflora H. St. John, Phytologia 63: 346 (1987).
Delissea ramosa H. St. John, Phytologia 63: 347 (1987).
Delissea subintegra H. St. John, Phytologia 63: 348 (1987).
Delissea subsessilis H. St. John, Phytologia 63: 348 (1987).
Delissea waipaensis H. St. John, Phytologia 63: 349 (1987).

Cyanea floribunda E. Wimm., Pflanzenr. IV.276b: 761 (1953). *Delissea floribunda* (E. Wimm.) H. St. John, Phytologia 63: 83 (1987).
Hawaiian Is. (Hawai'i). 63 HAW. Nanophan. $2n = 28$.
Cyanea pilosa var. *bondiana* Rock, Indig. Trees Haw. Isl.: 508 (1913). *Cyanea bondiana* (Rock) Rock, Occas. Pap. Bernice Pauahi Bishop Mus. 23: 72 (1962). *Delissea bondiana* (Rock) H. St. John, Phytologia 63: 81 (1987).
Cyanea pilosa var. *densiflora* Rock, Indig. Trees Haw. Isl.: 508 (1913). *Cyanea densiflora* (Rock) Rock, Occas. Pap. Bernice Pauahi Bishop Mus. 23: 72 (1962). *Delissea densiflora* (Rock) H. St. John, Phytologia 63: 82 (1987).
Cyanea pilosa var. *glabrifolia* Rock, Indig. Trees Haw. Isl.: 508 (1913).
Delissea nemorosa H. St. John, Phytologia 64: 167 (1988).
Delissea subobtusa H. St. John, Phytologia 64: 167 (1988).
Delissea aurantiaca H. St. John, Phytologia 64: 172 (1988).

Cyanea gibsonii Hillebr., Fl. Hawaiian Isl.: 263 (1888). *Delissea gibsonii* (Hillebr.) H. St. John, Phytologia 63: 83 (1987). *Cyanea macrostegia* subsp. *gibsonii* (Hillebr.) Lammers, Syst. Bot. 13: 503 (1988).
Hawaiian Is. (Lāna'i). 63 HAW. Phan. or nanophan.

Cyanea giffardii Rock, Bull. Torrey Bot. Club 45: 133 (1918). *Delissea giffardii* (Rock) H. St. John, Phytologia 63: 83 (1987).
Hawaiian Is. (Hawai'i). 63 HAW. Phan.

Cyanea glabra (E. Wimm.) H. St. John, Phytologia 48: 144 (1981).
Hawaiian Is. (E. Maui). 63 HAW. Nanophan.
* *Cyanea knudsenii* var. *glabra* E. Wimm., Pflanzenr. IV.276b: 75 (1943). *Delissea glabra* (E. Wimm.) H. St. John, Phytologia 63: 83 (1987).
Delissea lyonii H. St. John, Phytologia 64: 166 (1988).

Cyanea grimesiana Gaudich., Voy. Uranie: pl. 75 (1828). *Lobelia grimesiana* (Gaudich.) Hook. & Arn., Bot. Beechey Voy.: 88 (1832). *Delissea grimesiana* (Gaudich.) H. St. John, Phytologia 63: 83 (1987).
Hawaiian Is. (O'ahu, Moloka'i). 63 HAW. Nanophan.

subsp. **grimesiana**
Hawaiian Is. (Ko'olau Mts. & N. Wai'anae Mts. of O'ahu, Moloka'i). 63 HAW. Nanophan.

subsp. **obatae** (H. St. John) Lammers, Syst. Bot. 13: 502 (1988).
Hawaiian Is. (S. Wai'anae Mts. of O'ahu). 63 HAW. Nanophan.
Cyanea grimesiana var. *hirsutifolia* Rock, Occas. Pap. Bernice Pauahi Bishop Mus. 22: 64 (1957). *Delissea grimesiana* var. *hirsutifolia* (Rock) H. St. John, Phytologia 63: 83 (1987).
* *Cyanea grimesiana* var. *obatae* H. St. John, Phytologia 40: 97 (1978). *Delissea grimesiana* var. *obatae* (H. St. John) H. St. John, Phytologia 63: 83 (1987).

Cyanea habenata (H. St. John) Lammers, Novon 8: 33 (1998).
Hawaiian Is. (Kaua'i). 63 HAW. Nanophan.
   * *Delissea habenata* H. St. John, Phytologia 63: 342 (1987).

Cyanea hamatiflora Rock, Indig. Trees Haw. Isl.: 510 (1913). *Delissea hamatiflora* (Rock) H.
St. John, Phytologia 63: 84 (1987).
Hawaiian Is. (E. Maui, Hawai'i). 63 HAW. Phan.

subsp. **carlsonii** (Rock) Lammers, Syst. Bot. 13: 502 (1988).
   Hawaiian Is. (Hawai'i). 63 HAW. Phan.
   * *Cyanea carlsonii* Rock, Occas. Pap. Bernice Pauahi Bishop Mus. 22: 60 (1957). *Delissea
      carlsonii* (Rock) H. St. John, Phytologia 63: 81 (1987).

subsp. **hamatiflora**
   Hawaiian Is. (E. Maui). 63 HAW. Phan.
   Delissea sessilis H. St. John, Phytologia 64: 167 (1988).

Cyanea hardyi Rock, Bull. Torrey Bot. Club 44: 236 (1917). *Cyanea coriacea* var. *hardyi*
(Rock) E. Wimm., Pflanzenr. IV.276b: 72 (1943). *Delissea coriacea* var. *hardyi* (Rock) H. St.
John, Phytologia 63: 82 (1987).
Hawaiian Is. (Kaua'i). 63 HAW. Phan. or nanophan.
   Delissea nigra H. St. John, Phytologia 63: 346 (1987).

Cyanea hirtella (H. Mann) Hillebr., Fl. Hawaiian Isl.: 255 (1888).
Hawaiian Is. (Kaua'i). 63 HAW. Phan. or nanophan. $2n$ = 28.
   * *Delissea hirtella* H. Mann, Proc. Amer. Acad. Arts 7: 179 (1867).
   Cyanea sylvestris A. Heller, Minnesota Bot. Stud. 1: 909 (1897). *Delissea sylvestris* (A.
      Heller) H. St. John, Phytologia 63: 89 (1987).
   Cyanea feddei H. Lév., Repert. Spec. Nov. Regni Veg. 10: 156 (1911).
   Cyanea communis Rock, Coll. Hawaii Publ. Bull. 2: 41 (1913).
   Cyanea knudsenii Rock, Monogr. Stud. Haw. Lobelioid.: 213 (1919). *Delissea knudsenii*
      (Rock) H. St. John, Phytologia 63: 84 (1987).
   Cyanea eriantha Skottsb., Acta Horti Gothob. 2: 266 (1926). *Cyanea sylvestris* var.
      eriantha (Skottsb.) E. Wimm., Pflanzenr. IV.276b: 61 (1943). *Delissea sylvestris* var.
      eriantha (Skottsb.) H. St. John, Phytologia 63: 89 (1987).
   Cyanea degeneriana E. Wimm., Pflanzenr. IV.276b: 69 (1943). *Delissea degeneriana* (E.
      Wimm.) H. St. John, Phytologia 63: 82 (1987).
   Cyanea chockii Rock, Occas. Pap. Bernice Pauahi Bishop Mus. 22: 58 (1957). *Delissea
      chockii* (Rock) H. St. John, Phytologia 63: 81 (1987).
   Cyanea hirtella var. *striata* E. Wimm., Pflanzenr. IV.276c: 824 (1968). *Delissea hirtella* var.
      striata (E. Wimm.) H. St. John, Phytologia 63: 84 (1987).
   Cyanea hirtella var. *subglabra* E. Wimm., Pflanzenr. IV.276c: 824 (1968). *Delissea hirtella*
      var. *subglabra* (E. Wimm.) H. St. John, Phytologia 63: 84 (1987).
   Delissea alba H. St. John, Phytologia 63: 339 (1987).
   Delissea brevipedicellata H. St. John, Phytologia 63: 339 (1987).
   Delissea chartacea H. St. John, Phytologia 63: 340 (1987).
   Delissea christensenii H. St. John, Phytologia 63: 340 (1987).
   Delissea deltoidea H. St. John, Phytologia 63: 340 (1987).
   Delissea divergens H. St. John, Phytologia 63: 341 (1987).
   Delissea excurrens H. St. John, Phytologia 63: 342 (1987).
   Delissea fruticosa H. St. John, Phytologia 63: 342 (1987).
   Delissea hypoleuca H. St. John, Phytologia 63: 342 (1987).
   Delissea iliahiatilis H. St. John, Phytologia 63: 343 (1987).
   Delissea impedicellata H. St. John, Phytologia 63: 343 (1987).
   Delissea kealawelaensis H. St. John, Phytologia 63: 343 (1987).
   Delissea leiophylla H. St. John, Phytologia 63: 344 (1987).
   Delissea ligulata H. St. John, Phytologia 63: 344 (1987).

Delissea longiantherae H. St. John, Phytologia 63: 345 (1987).
Delissea paliensis H. St. John, Phytologia 63: 346 (1987).
Delissea purpurea H. St. John, Phytologia 63: 347 (1987).
Delissea robinsonii H. St. John, Phytologia 63: 347 (1987).
Delissea scopuli H. St. John, Phytologia 63: 347 (1987).
Delissea simplex H. St. John, Phytologia 63: 348 (1987), as 'simples'.
Delissea simplex f. *maculata* H. St. John, Phytologia 63: 348 (1987).
Delissea subacuminata H. St. John, Phytologia 63: 348 (1987).
Delissea umbrosa H. St. John, Phytologia 63: 348 (1987).
Delissea virgata H. St. John, Phytologia 63: 349 (1987).
Delissea waioliensis H. St. John, Phytologia 63: 349 (1987).
Clermontia pluriflora H. St. John, Phytologia 63: 351 (1987).

Cyanea horrida (Rock) O. Deg. & Hosaka, Fl. Hawaiiensis, fam. 339 (1940).
Hawaiian Is. (E. Maui). 63 HAW. Phan.
* **Cyanea ferox* var. *horrida* Rock, Bull. Torrey Bot. Club 44: 235 (1917). *Delissea horrida*
  (Rock) H. St. John, Phytologia 63: 84 (1987).
* *Cyanea lobata* var. *hamakuae* Rock, Monogr. Stud. Haw. Lobelioid.: 247 (1919). *Delissea*
  lobata var. *hamakuae* (Rock) H. St. John, Phytologia 63: 85 (1987).
* *Delissea kipahuluensis* H. St. John, Phytologia 64: 166 (1988).

Cyanea humboldtiana (Gaudich.) Lammers, Givnish & Sytsma, Novon 3: 439 (1993).
Hawaiian Is. (O'ahu). 63 HAW. Nanophan.
* **Rollandia humboldtiana* Gaudich., Voy. Bonite, Bot.: pl. 76 (1844).
* *Delissea* racemosa H. Mann, Proc. Amer. Acad. Arts 7: 181 (1867). *Rollandia racemosa*
  (H. Mann) Hillebr., Fl. Hawaiian Isl.: 246 (1888).
* *Rollandia pedunculosa* Wawra, Flora 56: 46 (1873).
* *Rollandia humboldtiana* f. *albida* H. St. John, Occas. Pap. Bernice Pauahi Bishop Mus.
  15: 357 (1940).

Cyanea kahiliensis (H. St. John) Lammers, Novon 8: 33 (1998).
Hawaiian Is. (Kaua'i). 63 HAW. Nanophan.
* **Delissea kahiliensis* H. St. John, Phytologia 63: 343 (1987).
* *Delissea inermis* H. St. John, Phytologia 63: 343 (1987).
* *Delissea inermis* H. St. John, Phytologia 64: 172 (1988); non H. St. John, Phytologia 63:
  343 (1987).
* *Cyanea spathulata* subsp. *longipetiolata* Lammers, Syst. Bot. 13: 504 (1988).

Cyanea kolekoleensis (H. St. John) Lammers, Novon 2: 130 (1992).
Hawaiian Is. (Kaua'i). 63 HAW. Phan. or nanophan.
* **Delissea kolekoleensis* H. St. John, Phytologia 63: 344 (1987).

Cyanea koolauensis Lammers, Givnish & Sytsma, Novon 3: 439 (1993).
Hawaiian Is. (O'ahu). 63 HAW. Nanophan.
* **Rollandia longiflora* var. *angustifolia* Hillebr., Fl. Hawaiian Isl.: 246 (1888). *Rollandia*
  angustifolia (Hillebr.) Rock, Bull. Torrey Bot. Club 45: 136 (1918); non *Cyanea*
  angustifolia (Cham.) Hillebr., Fl. Hawaiian Isl.: 253 (1888).
* *Rollandia angustifolia* var. *ochreata* E. Wimm., Pflanzenr. IV.276b: 99 (1943).

Cyanea kuhihewa Lammers, Brittonia 48: 238 (1996).
Hawaiian Is. (Kaua'i). 63 HAW. Nanophan.

Cyanea kunthiana Hillebr., Fl. Hawaiian Isl.: 264 (1888). *Cyanea bishopii* Rock, Indig. Trees
Haw. Isl.: 509 (1913). *Delissea bishopii* H. St. John, Phytologia 63: 81 (1987); non *Delissea*
kunthiana Gaudich., Voy. Bonite: pl. 77 (1844).
Hawaiian Is. (Maui). 63 HAW. Nanophan.
* *Delissea waikamoiensis* H. St. John, Phytologia 64: 167 (1988).

Cyanea lanceolata (Gaudich.) Lammers, Givnish & Sytsma, Novon 3: 439 (1993).
Hawaiian Is. (O'ahu). 63 HAW. Nanophan.
  * *Rollandia lanceolata* Gaudich., Voy. Uranie: 458 (1829). *Lobelia lanceolata* (Gaudich.)
    Hook. & Arn., Bot. Beechey Voy.: 88 (1832). *Lobelia rollandia* Gaudich. ex D. Dietr.,
    Syn. Pl. 1: 728 (1839). *Delissea lanceolata* (Gaudich.) A. Gray, Proc. Amer. Acad. Arts
    5: 147 (1861).
  Lobelia ambigua Cham., Linnaea 8: 221 (1833). *Rollandia ambigua* (Cham.) G. Don,
    Gen. Hist. 3: 698 (1834). *Delissea ambigua* (Cham.) C. Presl, Prodr. Monogr. Lobel.:
    47 (1836).
  Rollandia lanceolata var. *grandifolia* A. DC. in DC., Prodr. 7: 344 (1839). *Rollandia
    grandifolia* (A. DC.) Hillebr., Fl. Hawaiian Isl.: 245 (1888).
  Rollandia delessertiana Gaudich., Voy. Bonite, Bot.: pl. 75 (1844). *Delissea delessertiana*
    (Gaudich.) A. Gray, Proc. Amer. Acad. Arts 5: 147 (1861).
  Rollandia lanceolata var. *viridiflora* Rock, Monogr. Stud. Haw. Lobelioid.: 373 (1919).
  Rollandia lanceolata f. *rockii* E. Wimm., Pflanzenr. IV.276b: 102 (1943).
  Rollandia lanceolata var. *kipapaensis* Hosaka ex Fosberg, Occas. Pap. Bernice Pauahi
    Bishop Mus. 12(15): 10 (1936).
  Cyanea coronata E. Wimm., Pflanzenr. IV.276b: 59 (1943). *Delissea coronata* (E. Wimm.)
    H. St. John, Phytologia 63: 82 (1987).
  Rollandia lanceolata var. *glaberrima* E. Wimm., Pflanzenr. IV.276b: 763 (1953).

Cyanea leptostegia A. Gray, Proc. Amer. Acad. Arts 5: 149 (1861). *Delissea leptostegia* (A.
  Gray) H. St. John, Phytologia 63: 84 (1987).
  Hawaiian Is. (Kaua'i). 63 HAW. Phan. $2n = 28$.
  Delissea coriacea var. *pinnatiloba* A. Gray ex H. Mann, Proc. Amer. Acad. Arts 7: 178 (1867).
  Cyanea leptostegia var. *velutina* Skottsb., Acta Horti Gothob. 2: 264 (1926). *Delissea
    leptostegia* var. *velutina* (Skottsb.) H. St. John, Phytologia 63: 85 (1987).

Cyanea linearifolia Rock, Occas. Pap. Bernice Pauahi Bishop Mus. 22: 60 (1957). *Delissea
  linearifolia* (Rock) H. St. John, Phytologia 63: 85 (1987).
  Hawaiian Is. (Kaua'i). 63 HAW. Nanophan.

Cyanea lobata H. Mann, Proc. Amer. Acad. Arts 7: 183 (1867). *Delissea lobata* (H. Mann) H.
  St. John, Phytologia 63: 85 (1987).
  Hawaiian Is. (Lāna'i, W. Maui). 63 HAW. Nanophan.

  subsp. **baldwinii** (C. N. Forbes & G. C. Munro) Lammers, Novon 9: 387 (1999).
    Hawaiian Is. (Lāna'i). 63 HAW. Nanophan.
    * *Cyanea baldwinii* C. N. Forbes & G. C. Munro, Occas. Pap. Bernice Pauahi Bishop Mus.
      7: 43 (1920). *Delissea baldwinii* (C. N. Forbes & G. C. Munro) H. St. John, Phytologia
      63: 81 (1987).

  subsp. **lobata**
    Hawaiian Is. (W. Maui). 63 HAW. Nanophan.

Cyanea longiflora (Wawra) Lammers, Givnish & Sytsma, Novon 3: 439 (1993).
  Hawaiian Is. (O'ahu). 63 HAW. Nanophan.
  * *Rollandia longiflora* Wawra, Flora 56: 44 (1873).
  Rollandia lanceolata var. *brevipes* E. Wimm., Pflanzenr. IV.276b: 763 (1953).
  Rollandia degeneriana E. Wimm. in O. Deg. & I. Deg., Fl. Hawaiiensis, fam. 339 (1956).

Cyanea longissima (Rock) H. St. John, Phytologia 40: 98 (1978).
  Hawaiian Is. (E. Maui). 63 HAW. Nanophan.
  * *Cyanea scabra* var. *longissima* Rock, Monogr. Stud. Haw. Lobelioid.: 259 (1919). *Delissea
    longissima* (Rock) H. St. John, Phytologia 63: 85 (1987).
  Cyanea scabra var. *variabilis* Rock, Monogr. Stud. Haw. Lobelioid.: 255 (1919). *Delissea
    scabra* var. *variabilis* (Rock) H. St. John, Phytologia 63: 88 (1987).

Cyanea macrostegia Hillebr., Fl. Hawaiian Isl.: 263 (1888). *Delissea macrostegia* (Hillebr.) H. St. John, Phytologia 63: 85 (1987).
Hawaiian Is. (Maui). 63 HAW. Phan. or nanophan.
Cyanea atra Hillebr., Fl. Hawaiian Isl.: 263 (1888). *Delissea atra* (Hillebr.) H. St. John, Phytologia 63: 81 (1987).
Cyanea macrostegia var. *parvibracteata* Rock, Coll. Hawaii Publ. Bull. 2: 43 (1913). *Delissea macrostegia* var. *parvibracteata* (Rock) H. St. John, Phytologia 63: 85 (1987).
Cyanea mariana E. Wimm., Pflanzenr. IV.276b: 57 (1943). *Delissea mariana* (E. Wimm.) H. St. John, Phytologia 63: 85 (1987).
Cyanea bicolor H. St. John, Pacific Sci. 25: 63 (1971). *Delissea bicolor* (H. St. John) H. St. John, Phytologia 63: 81 (1987).
Cyanea hanaensis H. St. John, Phytologia 62: 433 (1987). *Delissea hanaensis* (H. St. John) H. St. John, Phytologia 64: 165 (1988).
Delissea globosa H. St. John, Phytologia 64: 165 (1988).
Delissea latior H. St. John, Phytologia 64: 166 (1988).

Cyanea magnicalyx Lammers, Novon 14: 85 (2004).
Hawaiian Is. (W. Maui). 63 HAW. Nanophan.

Cyanea mannii (Brigham ex H. Mann) Hillebr., Fl. Hawaiian Isl.: 253 (1888).
Hawaiian Is. (Moloka'i). 63 HAW. Phan. or nanophan.
* *Delissea mannii* Brigham ex H. Mann, Proc. Amer. Acad. Arts 7: 182 (1867).
Delissea kawelaensis H. St. John, Phytologia 64: 166 (1988).

Cyanea maritae Lammers & H. Oppenh., Novon 14: 92 (2004).
Hawaiian Is. (E. Maui). 63 HAW. Nanophan.

Cyanea marksii Rock, Occas. Pap. Bernice Pauahi Bishop Mus. 22: 52 (1957). *Delissea marksii* (Rock) H. St. John, Phytologia 63: 85 (1987).
Hawaiian Is. (Hawai'i). 63 HAW. Nanophan.
Cyanea tritomantha var. *lydgatei* Hillebr. ex Rock, Monogr. Stud. Haw. Lobelioid.: 103 (1919). *Delissea tritomantha* var. *lydgatei* (Hillebr. ex Rock) H. St. John, Phytologia 63: 89 (1987).

Cyanea mauiensis (Rock) Lammers, Novon 8: 31 (1998).
Hawaiian Is. (Maui). 63 HAW. Nanophan.
* *Cyanea grimesiana* var. *mauiensis* Rock, Monogr. Stud. Haw. Lobelioid.: 251 (1919). *Delissea grimesiana* var. *mauiensis* (Rock) H. St. John, Phytologia 63: 84 (1987).
Cyanea grimesiana var. *lydgatei* Rock, Monogr. Stud. Haw. Lobelioid.: 251 (1919). *Delissea grimesiana* var. *lydgatei* (Rock) H. St. John, Phytologia 63: 83 (1987).

Cyanea mceldowneyi Rock, Occas. Pap. Bernice Pauahi Bishop Mus. 22: 43 (1957). *Delissea mceldowneyi* (Rock) H. St. John, Phytologia 63: 85 (1987), as 'mceldownei'.
Hawaiian Is. (E. Maui). 63 HAW. Nanophan.

Cyanea membranacea Rock, Occas. Pap. Bernice Pauahi Bishop Mus. 22: 63 (1957). *Delissea membranacea* (Rock) H. St. John, Phytologia 63: 86 (1987).
Hawaiian Is. (O'ahu). 63 HAW. Nanophan.
Cyanea angustifolia var. *tomentella* Hillebr., Fl. Hawaiian Isl.: 253 (1888). *Delissea angustifolia* var. *tomentella* (Hillebr.) H. St. John, Phytologia 63: 80 (1987).
Cyanea angustifolia f. *subpubescens* E. Wimm., Pflanzenr. IV.276b: 71 (1943). *Delissea angustifolia* f. *subpubescens* (E. Wimm.) H. St. John, Phytologia 63: 80 (1987).
Cyanea coriacea var. *serratifolia* Rock, Occas. Pap. Bernice Pauahi Bishop Mus. 22: 65 (1957). *Delissea coriacea* var. *serratifolia* (Rock) H. St. John, Phytologia 63: 82 (1987). *Delissea serratifolia* (Rock) H. St. John, Phytologia 63: 347 (1987).
Cyanea dentata E. Wimm., Pflanzenr. IV.276c: 821 (1968). *Delissea dentata* (E. Wimm.) H. St. John, Phytologia 63: 82 (1987).

Cyanea minutiflora Lammers, Novon 14: 98 (2004).
Hawaiian Is. (Kaua'i). 63 HAW. Nanophan.

Cyanea munroi (Hosaka) Lammers, Novon 8: 31 (1998).
Hawaiian Is. (Moloka'i, Lāna'i). 63 HAW. Nanophan.
* *Cyanea grimesiana* var. *munroi* Hosaka, Occas. Pap. Bernice Pauahi Bishop Mus. 14: 30 (1938). *Delissea grimesiana* var. *munroi* (Hosaka) H. St. John, Phytologia 63: 84 (1987).

Cyanea obtusa (A. Gray) Hillebr., Fl. Hawaiian Isl.: 254 (1888).
Hawaiian Is. (Maui). 63 HAW. Phan. or nanophan.
* *Delissea obtusa* A. Gray, Proc. Amer. Acad. Arts 5: 148 (1861).
Delissea angusta H. St. John, Phytologia 64: 165 (1988).
Delissea puberula H. St. John, Phytologia 64: 167 (1988).

Cyanea parvifolia (C. N. Forbes) Lammers, Givnish & Sytsma, Novon 3: 439 (1993).
Hawaiian Is. (Kaua'i). 63 HAW. Nanophan.
* *Rollandia parvifolia* C. N. Forbes, Occas. Pap. Bernice Pauahi Bishop Mus. 5: 10 (1912).

Cyanea pilosa A. Gray, Proc. Amer. Acad. Arts 5: 149 (1861). *Delissea pilosa* (A. Gray) H. Mann, Proc. Amer. Acad. Arts 7: 182 (1867).
Hawaiian Is. (Hawai'i). 63 HAW. Nanophan.

subsp. **longipedunculata** (Rock) Lammers, Syst. Bot. 13: 503 (1988).
Hawaiian Is. (N. Hilo to Ka'u Distr. of Hawai'i). 63 HAW. Nanophan.
* *Cyanea longipedunculata* Rock, Occas. Pap. Bernice Pauahi Mus. 22: 54 (1957). *Delissea longipedunculata* (Rock) H. St. John, Phytologia 63: 85 (1987).

subsp. **pilosa**
Hawaiian Is. (Kohala to Hamakua Distr. of Hawai'i). 63 HAW. Nanophan.
Cyanea pilosa var. *megacarpa* Rock, Indig. Trees Haw. Isl.: 509 (1913). *Cyanea megacarpa* (Rock) Rock, Occas. Pap. Bernice Pauahi Bishop Mus. 23: 73 (1962). *Delissea megacarpa* (Rock) H. St. John, Phytologia 63: 86 (1987).

Cyanea pinnatifida (Cham.) E. Wimm., Pflanzenr. IV.276b: 63 (1943).
Hawaiian Is. (O'ahu). 63 HAW. Nanophan.
* *Lobelia pinnatifida* Cham., Linnaea 8: 220 (1833). *Rollandia pinnatifida* (Cham.) G. Don, Gen. Hist. 3: 698 (1834). *Delissea pinnatifida* (Cham.) C. Presl, Prodr. Monogr. Lobel.: 47 (1836).
Cyanea selachicauda O. Deg., Fl. Hawaiiensis, fam. 339 (1932).
Rollandia alba H. St. John & W. N. Takeuchi, Phytologia 63: 367 (1987).

Cyanea platyphylla (A. Gray) Hillebr., Fl. Hawaiian Isl.: 264 (1888).
Hawaiian Is. (Hawai'i). 63 HAW. Nanophan.
* *Delissea platyphylla* A. Gray, Proc. Amer. Acad. Arts 5: 148 (1861).
Cyanea noli-me-tangere Rock, Bull. Torrey Bot. Club 44: 229 (1917). *Delissea noli-me-tangere* (Rock) H. St. John, Phytologia 63: 86 (1987).
Cyanea fernaldii Rock, Bull. Torrey Bot. Club 44: 231 (1917). *Delissea fernaldii* (Rock) H. St. John, Phytologia 63: 82 (1987).
Cyanea rollandioides Rock, Bull. Torrey Bot. Club 45: 135 (1918). *Delissea rollandioides* (Rock) H. St. John, Phytologia 63: 87 (1987).
Cyanea bryanii Rock, Occas. Pap. Bernice Pauahi Bishop Mus. 22: 47 (1957). *Delissea bryanii* (Rock) H. St. John, Phytololgia 63: 81 (1987).
Cyanea pulchra Rock, Occas. Pap. Bernice Pauahi Bishop Mus. 22: 58 (1957). *Delissea pulchra* (Rock) H. St. John, Phytologia 63: 87 (1987).
Cyanea crispihirta E. Wimm., Pflanzenr. IV.276c: 823 (1968), as 'crispohirta'. *Delissea crispihirta* (E. Wimm.) H. St. John, Phytologia 63: 82 (1987).

Delissea angustior H. St. John, Phytologia 64: 165 (1988).
Delissea inflatispinosa H. St. John, Phytologia 64: 165 (1988).
Delissea kohalaensis H. St. John, Phytologia 64: 166 (1988).

Cyanea pohaku Lammers, Syst. Bot. 13: 503 (1988).
Hawaiian Is. (E. Maui). 63 HAW. Phan.
* *Clermontia haleakalensis* Rock, Indig. Trees Haw. Isl.: 489 (1913); non *Cyanea haleakalaensis* H. St. John, Pacific Sci. 25: 65 (1971).

Cyanea procera Hillebr., Fl. Hawaiian Isl.: 262 (1888). *Delissea procera* (Hillebr.) H. St. John, Phytologia 63: 87 (1987).
Hawaiian Is. (Moloka'i). 63 HAW. Phan.

Cyanea profuga C. N. Forbes, Occas. Pap. Bernice Pauahi Bishop Mus. 6: 186 (1916). *Delissea profuga* (C. N. Forbes) H. St. John, Phytologia 63: 87 (1987).
Hawaiian Is. (Moloka'i). 63 HAW. Nanophan.

Cyanea pseudofauriei Lammers, Novon 14: 96 (2004).
Hawaiian Is. (Kaua'i). 63 HAW. Nanophan.
Cyanea coriacea f. *gratiosa* E. Wimm., Pflanzenr. IV.276b: 760 (1953). *Delissea coriacea* var. *gratiosa* (E. Wimm.) H. St. John, Phytologia 63: 82 (1987).

Cyanea purpurellifolia (Rock) Lammers, Givnish & Sytsma, Novon 3: 439 (1993).
Hawaiian Is. (O'ahu). 63 HAW. Nanophan.
* *Rollandia purpurellifolia* Rock, Coll. Hawaii Publ. Bull. 2: 44 (1913).

Cyanea pycnocarpa (Hillebr.) E. Wimm., Pflanzenr. IV.276b: 53 (1943).
Hawaiian Is. (Hawai'i). 63 HAW. Phan.
* *Cyanea arborea* var. *pycnocarpa* Hillebr., Fl. Hawaiian Isl.: 262 (1888). *Delissea pycnocarpa* (Hillebr.) H. St. John, Phytologia 63: 87 (1987).

Cyanea quercifolia (Hillebr.) E. Wimm., Pflanzenr. IV.276b: 64 (1943).
Hawaiian Is. (E. Maui). 63 HAW. Phan.
* *Cyanea solanacea* var. *quercifolia* Hillebr., Fl. Hawaiian Isl.: 259 (1888). *Delissea quercifolia* (Hillebr.) H. St. John, Phytologia 63: 87 (1987).

Cyanea recta (Wawra) Hillebr., Fl. Hawaiian Isl.: 255 (1888).
Hawaiian Is. (Kaua'i). 63 HAW. Nanophan.
* *Delissea recta* Wawra, Flora 56: 30 (1873).

Cyanea remyi Rock, Bull. Torrey Bot. Club 44: 233 (1917). *Delissea remyi* (Rock) H. St. John, Phytologia 63: 87 (1987).
Hawaiian Is. (Kaua'i). 63 HAW. Nanophan. 2*n* = 28.

Cyanea rivularis Rock, Indig. Trees Haw. Isl.: 511 (1913). *Delissea rivularis* (Rock) E. Wimm., Pflanzenr. IV.276b: 43 (1943).
Hawaiian Is. (Kaua'i). 63 HAW. Phan.

Cyanea salicina H. Lév., Repert. Spec. Nov. Regni Veg. 12: 505 (1913).
Hawaiian Is. (Kaua'i). 63 HAW. Nanophan.
Cyanea larrisonii Rock, Bull. Torrey Bot. Club 42: 77 (1915). *Delissea larrisonii* (Rock) H. St. John, Phytologia 63: 84 (1987).
Cyanea rockii E. Wimm., Pflanzenr. IV.276c: 823 (1968). *Delissea rockii* (E. Wimm.) H. St. John, Phytologia 63: 87 (1987).

Cyanea scabra Hillebr., Fl. Hawaiian Isl.: 256 (1888). *Delissea scabra* (Hillebr.) H. St. John, Phytologia 63: 88 (1987).
  Hawaiian Is. (W. Maui). 63 HAW. Nanophan.
    Cyanea holophylla Hillebr., Fl. Hawaiian Isl.: 257 (1888). *Delissea holophylla* (Hillebr.) H. St. John, Phytologia 63: 84 (1987).

Cyanea sessilifolia (O. Deg.) Lammers, Novon 8: 32 (1998).
  Hawaiian Is. (O'ahu). 63 HAW. Nanophan.
    * *Rollandia sessilifolia* O. Deg., Fl. Hawaiiensis, fam. 339 (1932).

Cyanea shipmanii Rock, Occas. Pap. Bernice Pauahi Bishop Mus. 22: 44 (1957). *Delissea shipmanii* (Rock) H. St. John, Phytologia 63: 88 (1987).
  Hawaiian Is. (Hawai'i). 63 HAW. Phan. or nanophan.
    Cyanea grimesiana var. *citrullifolia* A. Gray, Proc. Amer. Acad. Arts 5: 148 (1861).

Cyanea solanacea Hillebr., Fl. Hawaiian Isl.: 259 (1888). *Delissea solanacea* (Hillebr.) H. St. John, Phytologia 63: 88 (1987).
  Hawaiian Is. (Moloka'i). 63 HAW. Phan. or nanophan.
    Cyanea ferox Hillebr., Fl. Hawaiian Isl.: 259 (1888). *Delissea ferox* (Hillebr.) H. St. John, Phytologia 63: 82 (1987).
    Cyanea ferox var. *laevicalyx* Skottsb., Acta Horti Gothob. 15: 487 (1944). *Delissea ferox* var. *laevicalyx* (Skottsb.) H. St. John, Phytologia 63: 82 (1987).

Cyanea solenocalyx Hillebr., Fl. Hawaiian Isl.: 258 (1888). *Delissea solenocalyx* (Hillebr.) H. St. John, Phytologia 63: 88 (1987).
  Hawaiian Is. (Moloka'i). 63 HAW. Phan. or nanophan.
    Cyanea wailauensis Rock, Coll. Hawaii Publ. Bull. 2: 43 (1913). *Delissea wailauensis* (Rock) H. St. John, Phytologia 63: 89 (1987).
    Cyanea solenocalyx var. *glabrata* Rock, Monogr. Stud. Haw. Lobelioid.: 171 (1919). *Cyanea solenocalyx* f. *glabrata* (Rock) O. Deg., Fl. Hawaiiensis, fam. 339 (1934).
    Cyanea ovatisepala E. Wimm., Pflanzenr. IV.276b: 57 (1943). *Delissea ovatisepala* (E. Wimm.) H. St. John, Phytologia 63: 86 (1987).
    Cyanea solenocalyx var. *latifolia* E. Wimm., Pflanzenr. IV.276b: 69 (1943). *Delissea solenocalyx* var. *latifolia* (E. Wimm.) H. St. John, Phytologia 63: 88 (1987).
    Delissea olokuiensis H. St. John, Phytologia 64: 167 (1988).
    Delissea waikoluensis H. St. John, Phytologia 64: 168 (1988).

Cyanea spathulata (Hillebr.) A. Heller, Minnesota Bot. Stud. 1: 909 (1897).
  Hawaiian Is. (Kaua'i). 63 HAW. Nanophan.
    * *Cyanea coriacea* var. *spathulata* Hillebr., Fl. Hawaiian Isl.: 254 (1888). *Delissea spathulata* (Hillebr.) H. St. John, Phytologia 63: 88 (1987).
    Rollandia fauriei H. Lév., Repert. Spec. Nov. Regni Veg. 12: 506 (1913).

Cyanea stictophylla Rock, Indig. Trees Haw. Isl.: 509 (1913). *Delissea stictophylla* (Rock) H. St. John, Phytologia 63: 88 (1987).
  Hawaiian Is. (Hawai'i). 63 HAW. Phan. or nanophan.
    Cyanea palakea C. N. Forbes, Occas. Pap. Bernice Pauahi Bishop Mus. 6: 188 (1916). *Delissea palakea* (C. N. Forbes) H. St. John, Phytologia 63: 86 (1987).
    Cyanea quercifolia var. *atropurpurea* E. Wimm., Pflanzenr. IV.276b: 64 (1943). *Delissea quercifolia* var. *atropurpurea* (E. Wimm.) H. St. John, Phytologia 63: 87 (1987).
    Cyanea stictophylla var. *inermis* Rock, Occas. Pap. Bernice Pauahi Bishop Mus. 22: 63 (1957). *Delissea stictophylla* var. *inermis* (Rock) H. St. John, Phytologia 63: 88 (1987).
    Cyanea nelsonii H. St. John, Pacific Sci. 30: 37 (1976). *Delissea nelsonii* (H. St. John) H. St. John, Phytologia 63: 86 (1987).

Cyanea st.-johnii (Hosaka) Lammers, Givnish & Sytsma, Novon 3: 440 (1993).
Hawaiian Is. (O'ahu). 63 HAW. Nanophan.
* *Rollandia st.-johnii* Hosaka, Occas. Pap. Bernice Pauahi Bishop Mus. 11(13): 15 (1935).
Rollandia st.-johnii var. *obtusisepala* E. Wimm., Pflanzenr. IV.276b: 762 (1953).

Cyanea superba (Cham.) A. Gray, Proc. Amer. Acad. Arts 5: 149 (1861).
Hawaiian Is. (O'ahu). 63 HAW. Phan.
* *Lobelia superba* Cham., Linnaea 8: 223 (1833). *Macrochilus superbus* (Cham.) C. Presl,
Prodr. Monogr. Lobel.: 47 (1836). *Delissea superba* (Cham.) H. St. John, Phytologia
63: 89 (1987).

subsp. **regina** (Wawra) Lammers, Syst. Bot. 13: 504 (1988).
Hawaiian Is. (S. Ko'olau Mts. of O'ahu). 63 HAW. Phan.
* *Delissea regina* Wawra, Flora 56: 9 (1873). *Cyanea superba* var. *regina* (Wawra) Hillebr.,
Fl. Hawaiian Isl.: 261 (1888), as 'reginae'. *Cyanea regina* (Wawra) Rock, Monogr.
Stud. Haw. Lobelioid.: 159 (1919).

subsp. **superba**
Hawaiian Is. (N. Wai'anae Mts. of O'ahu). 63 HAW. Phan.
Cyanea superba var. *velutina* Rock, Monogr. Stud. Haw. Lobelioid.: 157 (1919). *Delissea
superba* var. *velutina* (Rock) H. St. John, Phytologia 63: 89 (1987).

Cyanea tritomantha A. Gray, Proc. Amer. Acad. Arts 5: 149 (1861). *Delissea tritomantha* (A.
Gray) H. St. John, Phytologia 63: 89 (1987).
Hawaiian Is. (Hawai'i). 63 HAW. Phan.
Cyanea submuricata E. Wimm., Pflanzenr. IV.276b: 760 (1953). *Delissea submuricata* (E.
Wimm.) H. St. John, Phytologia 63: 89 (1987).
Cyanea magnifica E. Wimm., Pflanzenr. IV.276c: 817 (1968). *Delissea magnifica* (E.
Wimm.) H. St. John, Phytologia 63: 85 (1987).

Cyanea truncata (Rock) Rock, Bull. Torrey Bot. Club 44: 234 (1917).
Hawaiian Is. (O'ahu). 63 HAW. Nanophan.
* *Rollandia truncata* Rock, Coll. Hawaii Publ. Bull. 2: 44 (1913). *Delissea truncata* (Rock) H.
St. John, Phytologia 63: 89 (1987).
Cyanea juddii C. N. Forbes, Occas. Pap. Bernice Pauahi Bishop Mus. 6: 184 (1916).
Cyanea truncata var. *juddii* (C. N. Forbes) H. St. John, Occas. Pap. Bernice Pauahi
Bishop Mus. 15: 22 (1939). *Delissea truncata* var. *juddii* (C. N. Forbes) H. St. John,
Phytologia 63: 89 (1987).

Cyanea undulata C. N. Forbes, Occas. Pap. Bernice Pauahi Bishop Mus. 5: 12 (1912).
Delissea forbesii H. St. John, Phytologia 63: 83 (1987); non *Delissea undulata* Gaudich.,
Voy. Uranie: pl. 78 (1826). *Delissea lydgatei* H. St. John, Phytologia 63: 345 (1987).
Hawaiian Is. (Kaua'i). 63 HAW. Nanophan.

Synonyms:
Cyanea acuminata f. *latifolia* (Wawra) E. Wimm. === **Cyanea acuminata** (Gaudich.)
Hillebr.
Cyanea acuminata var. *calycina* Hosaka === **Cyanea acuminata** (Gaudich.) Hillebr.
Cyanea angustifolia f. *elliptica* Rock === **Cyanea elliptica** (Rock) Lammers
Cyanea angustifolia f. *subpubescens* E. Wimm. === **Cyanea membranacea** Rock
Cyanea angustifolia var. *elliptica* (Rock) E. Wimm. === **Cyanea elliptica** (Rock) Lammers
Cyanea angustifolia var. *hillebrandii* Rock === **Cyanea elliptica** (Rock) Lammers
Cyanea angustifolia var. *isabella* E. Wimm. === **Cyanea elliptica** (Rock) Lammers
Cyanea angustifolia var. *lanaiensis* Rock === **Cyanea elliptica** (Rock) Lammers
Cyanea angustifolia var. *racemosa* Hillebr. === **Cyanea angustifolia** (Cham.) Hillebr.
Cyanea angustifolia var. *tomentella* Hillebr. === **Cyanea membranacea** Rock
Cyanea arborea var. *pycnocarpa* Hillebr. === **Cyanea pycnocarpa** (Hillebr.) E. Wimm.

Cyanea arborescens H. Mann === **Clermontia arborescens** (H. Mann) Hillebr.

Cyanea argutidentata E. Wimm. === **Delissea argutidentata** (E. Wimm.) H. St. John

Cyanea aspera A. Gray === **Cyanea calycina** (Cham.) Lammers

Cyanea atra Hillebr. === **Cyanea macrostegia** Hillebr.

Cyanea atra var. *lobata* Rock === **Cyanea horrida** (Rock) O. Deg. & Hosaka — **Cyanea macrostegia** Hillebr.

Cyanea baldwinii C. N. Forbes & G. C. Munro === **Cyanea lobata** subsp. **baldwinii** (C. N. Forbes & G. C. Munro) Lammers

Cyanea bicolor H. St. John === **Cyanea macrostegia** Hillebr.

Cyanea bishopii Rock === **Cyanea kunthiana** Hillebr.

Cyanea blinii H. Lév. === **Clermontia parviflora** Gaudich. ex A. Gray

Cyanea bondiana (Rock) Rock === **Cyanea floribunda** E. Wimm.

Cyanea bonita Rock === ?

Cyanea bryanii Rock === **Cyanea platyphylla** (A. Gray) Hillebr.

Cyanea carlsonii Rock === **Cyanea hamatiflora** subsp. **carlsonii** (Rock) Lammers

Cyanea chockii Rock === **Cyanea hirtella** (H. Mann) Hillebr.

Cyanea communis Rock === **Cyanea hirtella** (H. Mann) Hillebr.

Cyanea coriacea f. *gratiosa* E. Wimm. === **Cyanea pseudofauriei** Lammers

Cyanea coriacea var. *degeneriana* E. Wimm. === **Cyanea coriacea** (A. Gray) Hillebr.

Cyanea coriacea var. *fauriei* (H. Lév.) E. Wimm. === **Cyanea coriacea** (A. Gray) Hillebr.

Cyanea coriacea var. *hardyi* (Rock) E. Wimm. === **Cyanea hardyi** Rock

Cyanea coriacea var. *serratifolia* Rock === **Cyanea membranacea** Rock

Cyanea coriacea var. *spathulata* Hillebr. === **Cyanea spathulata** (Hillebr.) A. Heller

Cyanea coronata E. Wimm. === **Cyanea lanceolata** (Gaudich.) Lammers, Givnish & Sytsma

Cyanea crispihirta E. Wimm. === **Cyanea platyphylla** (A. Gray) Hillebr.

Cyanea degeneriana E. Wimm. === **Cyanea hirtella** (H. Mann) Hillebr.

Cyanea densiflora (Rock) Rock === **Cyanea floribunda** E. Wimm.

Cyanea dentata E. Wimm. === **Cyanea membranacea** Rock

Cyanea eriantha Skottsb. === **Cyanea hirtella** (H. Mann) Hillebr.

Cyanea fauriei H. Lév. === **Cyanea coriacea** (A. Gray) Hillebr.

Cyanea feddei H. Lév. === **Cyanea hirtella** (H. Mann) Hillebr.

Cyanea fernaldii Rock === **Cyanea platyphylla** (A. Gray) Hillebr.

Cyanea ferox Hillebr. === **Cyanea solanacea** Hillebr.

Cyanea ferox var. *horrida* Rock === **Cyanea horrida** (Rock) O. Deg. & Hosaka

Cyanea ferox var. *laevicalyx* Skottsb. === **Cyanea solanacea** Hillebr.

Cyanea fissa subsp. *gayana* (Rock) Lammers === **Cyanea fissa** (H. Mann) Hillebr.

Cyanea gayana Rock === **Cyanea fissa** (H. Mann) Hillebr.

Cyanea gayana var. *duvelii* Rock === **Cyanea fissa** (H. Mann) Hillebr.

Cyanea gayana var. *wainihaensis* Rock === **Cyanea fissa** (H. Mann) Hillebr.

Cyanea grimesiana subsp. *cylindrocalyx* (Rock) Lammers === **Cyanea cylindrocalyx** (Rock) Lammers

Cyanea grimesiana var. *citrullifolia* A. Gray === **Cyanea shipmanii** Rock

Cyanea grimesiana var. *cylindrocalyx* Rock === **Cyanea cylindrocalyx** (Rock) Lammers

Cyanea grimesiana var. *hirsutifolia* Rock === **Cyanea grimesiana** subsp. **obatae** (H. St. John) Lammers

Cyanea grimesiana var. *lydgatei* Rock === **Cyanea mauiensis** (Rock) Lammers

Cyanea grimesiana var. *mauiensis* Rock === **Cyanea mauiensis** (Rock) Lammers

Cyanea grimesiana var. *munroi* Hosaka === **Cyanea munroi** (Hosaka) Lammers

Cyanea grimesiana var. *obatae* H. St. John === **Cyanea grimesiana** subsp. **obatae** (H. St. John) Lammers

Cyanea haleakalaensis H. St. John === **Cyanea copelandii** subsp. **haleakalaensis** (H. St. John) Lammers

Cyanea hanaensis H. St. John === **Cyanea macrostegia** Hillebr.

Cyanea hirtella var. *striata* E. Wimm. === **Cyanea hirtella** (H. Mann) Hillebr.

Cyanea hirtella var. *subglabra* E. Wimm. === **Cyanea hirtella** (H. Mann) Hillebr.

Cyanea holophylla Hillebr. === **Cyanea scabra** Hillebr.

Cyanea holophylla var. *obovata* Rock === ?

Cyanea humilis Wawra === **Cyanea fissa** (H. Mann) Hillebr.

Cyanea juddii C. N. Forbes === **Cyanea truncata** (Rock) Rock

Cyanea knudsenii Rock === **Cyanea hirtella** (H. Mann) Hillebr.

Cyanea knudsenii var. *glabra* E. Wimm. === **Cyanea glabra** (E. Wimm.) H. St. John

Cyanea lanceolata subsp. *calycina* (Cham.) Lammers, Givnish & Sytsma === **Cyanea calycina** (Cham.) Lammers

Cyanea larrisonii Rock === **Cyanea salicina** H. Lév.

Cyanea leptostegia var. *velutina* Skottsb. === **Cyanea leptostegia** A. Gray

Cyanea lobata var. *hamakuae* Rock === **Cyanea horrida** (Rock) O. Deg. & Hosaka

Cyanea longifolia A. Heller === **Cyanea arborea** Hillebr.

Cyanea longipedunculata Rock === **Cyanea pilosa** subsp. **longipedunculata** (Rock) Lammers

Cyanea macrostegia subsp. *gibsonii* (Hillebr.) Lammers === **Cyanea gibsonii** Hillebr.

Cyanea macrostegia var. *parvibracteata* Rock === **Cyanea macrostegia** Hillebr.

Cyanea macrostegia var. *viscosa* Rock === **Cyanea aculeatiflora** Rock × **Cyanea macrostegia** Hillebr.

Cyanea magnifica E. Wimm. === **Cyanea tritomantha** A. Gray

Cyanea mariana E. Wimm. === **Cyanea macrostegia** Hillebr.

Cyanea megacarpa (Rock) Rock === **Cyanea pilosa** A. Gray subsp. **pilosa**

Cyanea multispicata H. Lév. === **Cyanea fissa** (H. Mann) Hillebr.

Cyanea nelsonii H. St. John === **Cyanea stictophylla** Rock

Cyanea noli-me-tangere Rock === **Cyanea platyphylla** (A. Gray) Hillebr.

Cyanea occultans H. St. John === **Cyanea acuminata** (Gaudich.) Hillebr.

Cyanea ovatisepala E. Wimm. === **Cyanea solenocalyx** Hillebr.

Cyanea palakea C. N. Forbes === **Cyanea stictophylla** Rock

Cyanea pilosa var. *bondiana* Rock === **Cyanea floribunda** E. Wimm.

Cyanea pilosa var. *densiflora* Rock === **Cyanea floribunda** E. Wimm.

Cyanea pilosa var. *glabrifolia* Rock === **Cyanea floribunda** E. Wimm.

Cyanea pilosa var. *megacarpa* Rock === **Cyanea pilosa** A. Gray subsp. **pilosa**

Cyanea pulchra Rock === **Cyanea platyphylla** (A. Gray) Hillebr.

Cyanea quercifolia var. *atropurpurea* E. Wimm. === **Cyanea stictophylla** Rock

Cyanea regina (Wawra) Rock === **Cyanea superba** subsp. **regina** (Wawra) Lammmers

Cyanea rockii E. Wimm. === **Cyanea salicina** H. Lév.

Cyanea rollandia A. Gray === **Cyanea crispa** (Gaudich.) Lammers, Givnish & Sytsma

Cyanea rollandioides Rock === **Cyanea platyphylla** (A. Gray) Hillebr.

Cyanea scabra f. *sinuata* (Rock) E. Wimm. === ?

Cyanea scabra var. *longissima* Rock === **Cyanea longissima** (Rock) H. St. John

Cyanea scabra var. *sinuata* Rock === ?

Cyanea scabra var. *variabilis* Rock === **Cyanea longissima** (Rock) H. St. John

Cyanea selachicauda O. Deg. === **Cyanea pinnatifida** (Cham.) E. Wimm.

Cyanea solanacea var. *quercifolia* Hillebr. === **Cyanea quercifolia** (Hillebr.) E. Wimm.

Cyanea solenocalyx f. *glabrata* (Rock) O. Deg. === **Cyanea solenocalyx** Hillebr.

Cyanea solenocalyx var. *glabrata* Rock === **Cyanea solenocalyx** Hillebr.

Cyanea solenocalyx var. *latifolia* E. Wimm. === **Cyanea solenocalyx** Hillebr.

Cyanea solenocalyx var. *schizocalyx* Hillebr. === **Cyanea aculeatiflora** Rock

Cyanea spathulata subsp. *longipetiolata* Lammers === **Cyanea kahiliensis** (H. St. John) Lammers

Cyanea stictophylla var. *inermis* Rock === **Cyanea stictophylla** Rock

Cyanea submuricata E. Wimm. === **Cyanea tritomantha** A. Gray

Cyanea superba var. *regina* (Wawra) Hillebr. === **Cyanea superba** var. **regina** (Wawra) Lammers

Cyanea superba var. *velutina* Rock === **Cyanea superba** (Cham.) A. Gray subsp. **superba**

Cyanea sylvestris A. Heller === **Cyanea hirtella** (H. Mann) Hillebr.

Cyanea sylvestris var. *eriantha* (Skottsb.) E. Wimm. === **Cyanea hirtella** (H. Mann) Hillebr.
Cyanea tritomantha var. *lydgatei* Hillebr. ex Rock === **Cyanea marksii** Rock
Cyanea truncata var. *juddii* (C. N. Forbes) H. St. John === **Cyanea truncata** (Rock) Rock
Cyanea wailauensis Rock === **Cyanea solenocalyx** Hillebr.

Cyclocodon

Campanuloideae, 2 species, southern and eastern Asia, from Himalaya to Japan, south to
New Guinea. These species have been included in *Codonopsis* or its segregate *Campanumoea*
in the past, but palynological and other data (Morris & Lammers 1997, Hong & Pan 1998)
support recognition at generic rank. Type [designated by Pfeiff., Nomencl. Bot. 1: 964
(1874)]: *Codonopsis parviflora* Wall. ex A. DC.

M'Clelland, J., ed. (1854). Notulae ad plantas Asiaticas. Part IV. Dicotyledonous plants. By
the late William Griffith, Esq., F.L.S. Calcutta: C. A. Serrao. En. — First appearance of
the name, but not validly published because the inclusion of two species without a
seperate generic description violates ICBN Art. 42.1(a).

Hooker, J. D. & T. Thomson (1858). Praecursores ad floram Indicam: being sketches of the
natural families of Indian plants, with remarks on their distribution, structure, and
affinities. Series I. Stylidieae, Goodenovieae, et Campanulaceae (including
Lobeliaceae). J. Proc. Linn. Soc., Bot. 2: 1-29. En. — Validation of the generic name.

Oliver, D. (1888). *Campanumoea axillaris*, Oliv. Hooker's Icon. Pl. 18: pl. 1775 + 1 pg.,
illus. En. — Portrait and description of *C. lancifolius* subsp. *lancifolius*.

• Hong, D. Y. (1983). [*Campanumoea* Bl.] Sect. *Cyclocodon* (Griff.) C. B. Cl. In Flora
Reipublicae Popularis Sinicae 73(2): 72-74, illus. Beijing: Science Press. Ch. — Flora of
China, with key, descriptions, and full nomenclature.

• Murthy, G. V. S. (1983). Pollen morphology of Indian *Campanumoea* Bl. – a revision of the
genus. J. Palynol. 18: 55-59, illus. En. — Pollen morphology suggests *Campanumoea* is
biphyletic; first modern support for recognition of *Cyclocodon*.

Shimizu, T. (1993). *Campanumoea* Bl. sect. *Cyclocodon* (Griff.) C. B. Clarke. In K. Iwatsuki,
T. Yamazaki, D. E. Boufford & H. Ohba, Flora of Japan 3a: 415. Tokyo: Kodansha. En. —
Flora with description.

• Morris, K. E. & T. G. Lammers (1997). Circumscription of *Codonopsis* and the allied genera
Campanumoea and *Leptocodon* (Campanulaceae: Campanuloideae). I. Palynological
data. Bot. Bull. Acad. Sin. 38: 277-284, illus. En. — Pollen data support recognition of
Cyclocodon as distinct.

• Hong, D. Y. & K. Y. Pan (1998). The restoration of the genus *Cyclocodon* (Campanulaceae)
and its evidence from pollen and seed-coat. Acta Phytotax. Sin. 36: 106-110. Ch. —
Additional support for reinstatement of genus.

Lammers, T. G. (1998). *Cyclocodon* Griff. Flora of Taiwan (ed. 2) 4: 784-787, illus. Taipei:
National Taiwan University. En. — Flora with description.

Cyclocodon Griff. ex Hook. f. & Thomson, J. Proc. Linn. Soc., Bot. 2: 17 (1858).
Campanumoea sect. *Cyclocodon* (Griff. ex Hook. f. & Thomson) C. B. Clarke in Hook. f.,
Fl. Brit. Ind. 3: 436 (1881).
Temp. & trop. Asia. 36 38 40 41 42 43.

Cyclocodon lancifolius (Roxb.) Kurz, Flora 55: 303 (1872), as 'lancifolium'.
Tibet to Japan, Nansei-shoto, Taiwan, Philippines & New Guinea. 36 CHC CHS CHT 38
JAP NNS TAI 40 ASS BAN EHM IND 41 THA VIE 42 BOR JAW MLY MOL PHI SUL SUM
43 NWG. Geophyte.
Campanula lancifolia Roxb., Fl. Ind. 2: 96 (1824). *Campanumoea lancifolia* (Roxb.) Merr.,
Enum. Philipp. Fl. Pl. 3: 587 (1923). *Codonopsis lancifolia* (Roxb.) Moeliono in
Steenis, Fl. Males. (ser. 1) 6: 120 (1960).

subsp. **celebicus** (Blume) K. E. Morris & Lammers, Novon 9: 387 (1999).

Tibet to SC. China (Yunnan), Vietnam, Philippines & New Guinea. 36 CHC CHT 41 THA VIE 42 BOR JAW MLY MOL PHI SUL SUM 43 NWG. Geophyte.

** Campanumoea celebica* Blume, Bijdr.: 727 (1826). *Campanula celebica* (Blume) D. Dietr., Syn. Pl. 1: 758 (1839). *Codonopsis celebica* (Blume) Miq., Fl. Ned. Ind. 2: 566 (1857). *Codonopsis lancifolia* subsp. *celebica* (Blume) Moeliono in Steenis, Fl. Males. (ser. 1) 6: 121 (1960). *Cyclocodon celebicus* (Blume) D. Y. Hong, Acta Phytotax. Sin. 26: 109 (1998).

Codonopsis leucocarpa Miq., Fl. Ned. Ind. 2: 565 (1857).

subsp. **lancifolius**

E. India (West Bengal) to S. China (Hubei, Hunan, Fujian, Guangdong, Guangxi, Guizhou, Sichuan, Yunnan), Japan (Kyushu), Nansei-shoto (Iriomote, Okinawa), Taiwan, Philippines & Maluku. 36 CHC CHS 38 JAP NNS TAI 40 ASS BAN EHM IND 42 JAW MOL PHI SUM. Geophyte.

Codonopsis truncata Wall. ex A. DC., Monogr. Campan.: 122 (1830). *Campanula truncata* (Wall. ex A. DC.) D. Dietr., Syn. Pl. 1: 757 (1839). *Cyclocodon truncatus* (Wall. ex A. DC.) Hook. f. & Thomson, J. Proc. Linn. Soc., Bot. 2: 18 (1858), as 'truncatum'. *Campanumoea truncata* (Wall. ex A. DC.) Diels, Bot. Jahrb. Syst. 29: 606 (1901).

Canarina moluccana Roxb., Fl. Ind. (ed. 1832) 2: 173 (1832).

Codonopsis albiflora Griff., Not. Pl. Asiat. 4: 279 (1854).

Campanumoea axillaris Oliv., Hooker's Icon. Pl. (ser. 3) 8: pl. 1775 (1888).

Cyclocodon parviflorus (Wall. ex A. DC.) Hook. f. & Thomson, J. Proc. Linn. Soc., Bot. 2: 18 (1858), as 'parviflorum'.

E. India (West Bengal) to SC. China (Yunnan) & Laos. 36 CHC 40 ASS BAN EHM IND 41 LAO. Geophyte.

** Codonopsis parviflora* Wall. ex A. DC., Monogr. Campan.: 123 (1830). *Campanula punduana* D. Dietr., Syn. Pl. 1: 757 (1839); non *Campanula parviflora* Lam., Encycl. 1: 588 (1785). *Campanumoea parviflora* (Wall. ex A. DC.) Benth. ex C. B. Clarke in Hook. f., Fl. Brit. India 3: 426 (1881).

Synonyms:

Cyclocodon celebicus (Blume) D. Y. Hong === **Cyclocodon lancifolius** subsp. **celebicus** (Blume) K. E. Morris & Lammers

Cyclocodon truncatus (Wall. ex A. DC.) Hook. f. & Thomson === **Cyclocodon lancifolius** (Roxb.) Kurz subsp. **lancifolius**

Cylindrocarpa

Campanuloideae, 1 species, Kirghizstan. Like *Cryptocodon*, one would predict on the basis of morphology and serology (Gudkova & Borshchenko 1991) that *Cylindrocarpa* is related to *Asyneuma, Petromarula, Physoplexis,* and *Phyteuma*. Type [by monotypy]: *Cylindrocarpa sewerzowii* (Regel) Regel.

Regel, E. (1877). Descriptiones plantarum novarum et minus cognitarum, fasciculus V. Trudy Imp. S.-Peterburgsk. Bot. Sada 5: 217-272. La. — Establishment of genus.

Fedorov, A. A. (1957). *Cylindrocarpa* Rgl. In V. L. Komarov, Flora URSS 24: 426-427, illus. Leningrad: Academia Scientiae URSS. Ru. — Floristic account for former Soviet Union, with description and full nomenclature.

Gudkova, I. Y. & G. P. Borshchenko (1991). [The serological study of the Campanulaceae. The phylogenetic relations in the tribe Phyteumateae.] Bot. Žurn. 76(6): 809-817. Ru. — Protein serology supports relationship to *Asyneuma, Cryptocodon, Petromarula, Phyteuma,* and *Sergia*.

Cylindrocarpa Regel, Trudy Imp. S.-Peterburgsk. Bot. Sada 5: 258 (1877). *Euregelia* Kuntze, Revis. Gen. Pl. 3(3): 403 (1898). *Phyteuma* sect. *Cylindrocarpa* (Regel) Kuntze in T. Post & Kuntze, Lex. Gen. Phan. 437 (1904).
Middle Asia. 32.

Cylindrocarpa sewerzowii (Regel) Regel, Trudy Imp. S.-Peterburgsk. Bot. Sada 5: 258 (1877), as 'sewerzowi'.
Kirgizstan. 32 KGZ. Hemicr.
   * *Phyteuma sewerzowii* Regel, Bull. Soc. Imp. Naturalistes Moscou 40(3): 184 (1867), as 'sewerzowi'. *Euregelia sewerzowii* (Regel) Kuntze, Revis. Gen. Pl. 3(3): 403 (1898).

Cyphia

Cyphioideae, 64 species, tropical and southern Africa. Although the genus has been divided into two sections and six subsections (Wimmer 1968), that classification has been criticized as unnatural by Thulin (1978). One preliminary phylogenetic analysis of molecular data (Cosner et al. 1994, under **General**) indicated that *Cyphia* is the sister group of the remainder of the family, while another (Lundberg & Bremer 2003, under **General**) placed it as the sister group of just the Campanuloideae. The most recent monograph is that of Wimmer (1968). $2n = 18$. Type [by monotypy]: *Cyphia bulbosa* (L.) P. J. Bergius.

Bergius, P. J. (1767). Descriptiones plantarum ex Capite Bonae Spei. Stockholm: Laurentius Salvius. La. — Establishment of genus as distinct from *Lobelia*.
Ker Gawler, J. B. (1822). *Cyphia phyteuma*. Rampion-flowered Cyphia. Bot. Reg. 8: tab. 625 + 3 pp., illus. En., La. — Portrait with description.
* Schönland, S. (1890). Notes on *Cyphia volubilis*, Willd. Trans. S. Afr. Philos. Soc. 6: 44-51, illus. En. — First detailed description of unique stigmatic cavity.
* Marloth, R. (1932). Campanulaceae. In The flora of South Africa 3: 208-215 + pl. 54, illus. Capetown: Darter Bros. En. — Stigmatic cavity illustrated and described; tubers contain inulin.
Adamson, R. S. (1950). *Cyphia* Berg. In R. S. Adamson & T. M. Salter, Flora of the Cape Peninsula: 750-753. Cape Town: Juta. En. — Flora with key and descriptions.
* Wimmer, F. E. (1968). *Cyphia* Berg. In A. Engler & L. Diels, Das Pflanzenreich IV. 276c: 935-1014 + taf. 32-69, illus. Berlin: Akademie-Verlag. Ge., La. — Monograph with key, descriptions, infrageneric classification, full nomenclature, and specimen citations.
Dyer, R. A. (1975). *Cyphia* Berg. In The genera of southern African flowering plants 1: 644. Pretoria: Department of Agricultural Technical Services. En. — Generic description.
Dunbar, A. & H. G. Wallentinus (1976). On pollen of Campanulaceae III. A numerical taxonomic investigation. Bot. Not. 129: 69-72. En. — Cluster analysis of palynological data suggests Cyphioideae s. l. not a natural group.
* Thulin, M. (1978). *Cyphia* (Lobeliaceae) in tropical Africa. Bot. Not. 131: 455-471, illus., maps. En. — Monograph of species found outside S. Africa, with key, synonymy, specimen citations, and comments on infrageneric classification; report for Cape Verde based on specimen from Angola.
Roessler, H. (1981). Die Gattung *Cyphia* Berg. neu für Südwestafrika. Mitt. Bot. Staatssamml. München 17: 243-244, map. Ge. — First report of genus from Namibia.
* Thulin, M. (1983). *Cyphia* Berg. In E. Launert (ed.), Flora Zambesiaca 7(1): 148-157, illus. Kew: Royal Botanic Garden. En. — Flora of Mozambique, Malawi, Zambia, Zimbabwe and Botswana, with keys, descriptions, full nomenclature & specimen citations.
* Thulin, M. (1984). *Cyphia*. Flora of tropical East Africa, Lobeliaceae: 49-56, illus. Rotterdam: Balkema. En. — Flora of Kenya, Tanzania & Uganda, with key and descriptions.
* Thulin, M. (1985). *Cyphia* Berg. Flore d'Afrique centrale (Zaire-Rwanda-Burundi), Lobeliaceae: 52-63, illus. Meise: Jardin Botanique National de Belgique. Fr. — Flora with key and descriptions.

Figueiredo, E. (1995). Flora de Cabo Verde. Plantas vasculares. 86. Campanulaceae. 11 pp., illus. Lisbon: Praia. Port. — Genus does not occur here despite earlier reports.

Goldblatt, P. & J. Manning (2000). Cape plants. A conspectus of the Cape flora of South Africa. *Cyphia*. Strelitzia 9: 388-390. En. — Checklist for Cape of Good Hope, with brief species desciptions.

Welman, W. G. (2000). *Cyphia* P. J. Bergius. In O. A. Leistner (ed.), Seed plants of southern Africa: families and genera. Strelitzia 10: 339. En. — Generic description.

Cupido, C. N. & F. Conrad (2003). *Cyphia*. A pretty perennial herb. Veld & Flora 89: 62-63, illus. En. — Brief descriptive account of genus and some common species.

Leins, P. & C. Erbar (2003). The pollen box in Cyphiaceae (Campanulales). Int. J. Pl. Sci. 164 (5, suppl.): S321-S328, illus. En. — Structure of the stamens, style and stigma and their function in secondary pollen presentation.

Plisko, M. A. (2003). [Fruit and seed structure of two species of the genus *Cyphia* (Cyphiaceae).] Bot. Žurn. 88(11): 107-113, illus. Ru. — Anatomical details of *C. bulbosa* and *C. sylvatica*.

Leins, P. & C. Erbar (2005). Floral morphological studies in the South African *Cyphia stenopetala* Diels (Cyphiaceae). Int. J. Pl. Sci. 166: 207-217, illus. En. — Flowering is proterogynous and pollination autogamous.

Cyphia P. J. Bergius, Descr. Pl. Cap.: 172 (1767). *Cyphopsis* Kuntze, Revis. Gen. Pl. 3(2): 186 (1898).
Trop. & S. Africa. 23 24 25 26 27.
    Cyphia [unranked] *Cyphiella* C. Presl in Eckl. & Zeyh., Enum. Pl. Afric. Austral.: 392 (1837). *Cyphia* subg. *Cyphiella* (C. Presl) C. Presl in E. Mey., Comm. Pl. Afr. Austr.: 293 (1838). *Cyphiella* (C. Presl) Spach, Hist. Nat. Vég. 9: 584 (1838). *Cyphia* sect. *Cyphiella* (C. Presl) E. Wimm., Pflanzenr. IV.276c: 936 (1968).

Cyphia alba N. E. Br., Bull. Misc. Inform. Kew 1906: 165 (1906).
Zimbabwe. 26 ZIM. Geophyte.
    Cyphia alba f. *purpurea* E. Wimm., Pflanzenr. IV.276c: 973 (1968).

Cyphia alicedalensis E. Wimm., Pflanzenr. IV.276c: 1005 (1968).
Cape Provinces. 27 CPP. Cl. tuber geophyte.

Cyphia angustifolia Eckl. & Zeyh. ex C. Presl in Eckl. & Zeyh., Enum. Pl. Afric. Austral.: 390 (1837).
Cape Provinces. 27 CPP. Cl. tuber geophyte.

Cyphia aspergilloides E. Wimm., Kew Bull. 1952: 146 (1952).
S. Africa. 27 NAT OFS. Cl. tuber geophyte.
    Cyphia aspergilloides var. *brevipes* E. Wimm., Pflanzenr. IV.276c: 1012 (1968).

Cyphia basiloba E. Wimm., Pflanzenr. IV.276c: 1013 (1968).
Cape Provinces. 27 CPP. Cl. tuber geophyte.

Cyphia belfastica E. Wimm., Pflanzenr. IV.276c: 940 (1968).
Transvaal. 27 TVL. Geophyte.

Cyphia bolusii E. Phillips, Ann. S. African Mus. 9: 459 (1917).
Swaziland. 27 SWZ. Geophyte.

Cyphia brachyandra Thulin, Bot. Not. 131: 466 (1978).
Tanzania & Malawi. 25 TAN 26 MLW. Cl. tuber geophyte. $2n = 18$.

Cyphia brevifolia Thulin, Bot. Not. 131: 459 (1978).
Angola. 26 ANG. (Cl.) geophyte.

Cyphia brummittii Thulin in Launert, Fl. Zambes. 7: 151 (1983).
  Malawi. 26 MLW. Cl. tuber geophyte.

Cyphia bulbosa (L.) P. J. Bergius, Descr. Pl. Cap.: 172 (1767).
  Cape Provinces. 27 CPP. (Cl.) geophyte.
   * *Lobelia bulbosa* L., Sp. Pl.: 933 (1753). *Lobelia cyphia* J. F. Gmel., Syst. Nat.: 357 (1791).
      Cyphia capensis J. F. Gmel., Syst. Nat.: 370 (1791), as 'Cyphium capense'. *Lobelia digitata* Thunb., Fl. Cap. 2: 50 (1818).
   Lobelia tuberosa Burm. f., Prodr. Fl. Cap.: 29 (1768).
   Cyphia botrys Willd. ex Schult. in Roem. & Schult., Syst. Veg. 5: 477 (1819).
   Cyphia bulbosa var. *pinnatifida* C. Presl in E. Mey., Comm. Pl. Afr. Austr.: 297 (1838).
   Cyphia bulbosa var. *angustiloba* A. DC. in DC., Prodr. 7: 499 (1839).
   Cyphia bulbosa var. *orientalis* E. Phillips, Ann. S. African Mus. 9: 456 (1917).
   Cyphia bulbosa f. *lobata* E. Wimm., Pflanzenr. IV.276c: 965 (1968).
   Cyphia bulbosa var. *hafstroemii* E. Wimm., Pflanzenr. IV.276c: 965 (1968).
   Cyphia bulbosa var. *acocksii* E. Wimm., Pflanzenr. IV.276c: 965 (1968).
   Cyphia bulbosa var. *leiandra* E. Wimm., Pflanzenr. IV.276c: 966 (1968).
   Cyphia kerastes E. Wimm., Pflanzenr. IV.276c: 980 (1968).
   Cyphia stephensii E. Wimm., Pflanzenr. IV.276c: 1013 (1968).

Cyphia comptonii Bond, J. S. African Bot. 6: 62 (1940).
  Cape Provinces. 27 CPP. Geophyte.

Cyphia corylifolia Harv., Thes. Cap. 2: 39 (1862).
  KwaZulu-Natal. 27 NAT. Cl. tuber geophyte.
   Cyphia corylifolia f. *singuliflora* E. Wimm., Pflanzenr. IV.276c: 1011 (1968).

Cyphia couroublei Bamps & Malaisse, Bull. Jard. Bot. Belg. 56: 488 (1986).
  Zaïre. 23 ZAI. Geophyte.

Cyphia crenata (Thunb.) C. Presl, Prodr. Monogr. Lobel.: 51 (1836).
  Cape Provinces. 27 CPP. Cl. tuber geophyte.
   * *Lobelia crenata* Thunb., Prodr. Fl. Cap.: 39 (1794).
   Cyphia atriplicifolia C. Presl in Eckl. & Zeyh., Enum. Pl. Afric. Austral.: 389 (1837).
   Cyphia dregeana C. Presl in E. Mey., Comm. Pl. Afr. Austr.: 294 (1838).
   Cyphia smithiae L. Bolus, J. Bot. 68: 78 (1930). *Cyphia crenata* var. *smithiae* (L. Bolus) E. Wimm., Kew Bull. 1952: 145 (1952).
   Cyphia crenata var. *angustifolia* E. Wimm., Pflanzenr. IV.276c: 1009 (1968).

Cyphia decora Thulin, Bot. Not. 131: 462 (1978).
  Malawi. 26 MLW. (Cl.) geophyte. $2n = 18$.

Cyphia digitata Willd., Sp. Pl. 1: 953 (1798).
  Namibia & Cape Provinces. 27 CPP NAM. Cl. geophyte.
   Cyphia tomentosa C. Presl in E. Mey., Comm. Pl. Afr. Austr.: 295 (1838). *Cyphia digitata* var. *tomentosa* (C. Presl) Sond. in Harv. & Sond., Fl. Cap. 3: 604 (1865).
   Cyphia tomentosa var. *minor* C. Presl in E. Mey., Comm. Pl. Afr. Austr.: 295 (1838).
   Cyphia dentariifolia C. Presl in E. Mey., Comm. Pl. Afr. Austr.: 295 (1838), as 'dentariaefolia'.
   Cyphia polydactyla C. Presl in E. Mey., Comm. Pl. Afr. Austr.: 296 (1838).
   Cyphia tomentosa var. *glabra* A. DC. in DC., Prodr. 7: 499 (1839).
   Cyphia dentariifolia var. *luttigii* E. Wimm., Pflanzenr. IV.276c: 996 (1968).
   Cyphia dentariifolia var. *psilandra* E. Wimm., Pflanzenr. IV.276c: 996 (1968).
   Cyphia digitata f. *glabrifilis* E. Wimm., Pflanzenr. IV.276c: 998 (1968).
   Cyphia digitata var. *trimera* E. Wimm., Pflanzenr. IV.276c: 999 (1968).
   Cyphia digitata subsp. *gracilis* E. Wimm., Pflanzenr. IV.276c: 999 (1968).

Cyphia elata Harv., Thes. Cap. 2: 39 (1862). *Cyphopsis elata* (Harv.) Kuntze, Revis. Gen. Pl. 3(2): 186 (1898).
S. Africa. 27 CPP LES NAT OFS SWZ TVL. Geophyte.
Cyphia elata var. *glabra* Harv. in Harv. & Sond., Fl. Cap. 3: 601 (1865).
Cyphia oblongifolia Harv. & Sond., Fl. Cap. 3: 601 (1865). *Cyphia elata* var. *oblongifolia* (Harv. & Sond.) E. Phillips, Ann. S. African Mus. 9: 457 (1917).
Cyphia gerrardii Harv. in Harv. & Sond., Fl. Cap. 3: 601 (1865), as 'gerrardi'. *Cyphopsis gerrardii* (Harv.) Kuntze, Rev. Gen. Pl. 3(2): 186 (1898). *Cyphia elata* var. *gerrardii* (Harv.) E. Wimm., Pflanzenr. IV.276c: 969 (1968).
Cyphia wilmsiana Diels, Bot. Jahrb. Syst. 26: 113 (1898).
Cyphia elata var. *globularis* E. Wimm., Pflanzenr. IV.276c: 968 (1968).
Cyphia elata f. *flanagana* E. Wimm., Pflanzenr. IV.276c: 969 (1968).
Cyphia elata f. *truncata* E. Wimm., Pflanzenr. IV.276c: 969 (1968).
Cyphia elata f. *decurrens* E. Wimm., Pflanzenr. IV.276c: 969 (1968).
Cyphia elata f. *thodeana* E. Wimm., Pflanzenr. IV.276c: 970 (1968).

Cyphia erecta De Wild., Ann. Mus. Congo, Bot. (ser. 4) 1: 162 (1903).
Zaïre to Tanzania & Malawi. 23 ZAI 25 TAN 26 MLW ZAM. Geophyte.
Cyphia scandens De Wild., Ann. Mus. Congo, Bot. (ser. 4) 1: 163 (1903).
Cyphia regularis E. Wimm., Kew Bull. 1952: 143 (1952).
Cyphia rhodesiaca E. Wimm., Pflanzenr. IV.276c: 975 (1968).
Cyphia erecta var. *wetteana* E. Wimm., Pflanzenr. IV.276c: 975 (1968).

Cyphia eritreana E. Wimm., Pflanzenr. IV.276c: 988 (1968).
Ethiopia & Eritrea. 24 ETH ERI. Cl. tuber geophyte.

Cyphia galpinii E. Wimm., Pflanzenr. IV.276c: 1003 (1968).
Cape Provinces. 27 CPP. Cl. tuber geophyte.

Cyphia gamopetala P. A. Duvign. & Denaeyer, Bull. Soc. Roy. Bot. Belgique 96: 133 (1963).
Zaïre. 23 ZAI. Geophyte.

Cyphia georgica E. Wimm., Pflanzenr. IV.276c: 980 (1968).
Cape Provinces. 27 CPP. Geophyte.

Cyphia glabra E. Wimm., Kew Bull. 1952: 146 (1952).
Transvaal. 27 TVL. Cl. tuber geophyte.

Cyphia glandulifera Hochst. ex A. Rich., Tent. Fl. Abyss. 2: 8 (1850).
Eritrea to Somalia & Tanzania. 24 ERI ETH SOM 25 KEN TAN. Geophyte.
Cyphia glandulifera f. *obovatifolia* E. Wimm., Pfanzenr. IV.276c: 956 (1968).
Cyphia glandulifera f. *subcuspidata* E. Wimm., Pfanzenr. IV.276c: 958 (1968).

Cyphia heterophylla C. Presl in Eckl. & Zeyh., Enum. Pl. Afric. Austral.: 389 (1837).
Cape Provinces. 27 CPP. Cl. geophyte.

Cyphia incisa (Thunb.) Willd., Sp. Pl. 1: 953 (1798).
Cape Provinces. 27 CPP. Geophyte.
* *Lobelia incisa* Thunb., Prodr. Fl. Cap.: 39 (1794).
Lobelia cardamines Thunb., Prodr. Fl. Cap.: 39 (1794). *Cyphia cardamines* (Thunb.) Willd., Sp. Pl. 1: 953 (1798). *Cyphia incisa* var. *cardamines* (Thunb.) E. Phillips, Ann. S. African Mus. 9: 454 (1917).
Cyphia incisa var. *bracteata* E. Phillips, Ann. S. African Mus. 9: 455 (1917).
Cyphia incisa var. *sinuata* E. Wimm., Pflanzenr. IV.276c: 955 (1968).
Cyphia incisa var. *lyrata* E. Wimm., Pflanzenr. IV.276c: 955 (1968).

Cyphia lasiandra Diels, Bot. Jahrb. Syst. 26: 111 (1898).
Zaïre to Tanzania, Mozambique & Angola. 23 BUR ZAI 25 TAN 26 ANG MLW MOZ.
(Cl.) geophyte.
Cyphia nyasica Baker, Bull. Misc. Inform. Kew 1898: 157 (1898).
Cyphia antunesii Engl., Bot. Jahrb. Syst. 32: 147 (1902).
Cyphia cacondensis R. D. Good, J. Bot. 65 (Suppl. 2): 68 (1927).
Cyphia peteriana E. Wimm., Kew Bull. 1952: 147 (1952).
Cyphia zernyana E. Wimm., Pflanzenr. IV.276c: 1006 (1968).
Cyphia exelliana E. Wimm., Pflanzenr. IV.276c: 1010 (1968).
Cyphia floribunda E. Wimm., Pflanzenr. IV.276c: 1013 (1968).

Cyphia linarioides C. Presl in Eckl. & Zeyh., Enum. Pl. Afric. Austral.: 391 (1837). *Cyphia undulata* var. *linarioides* (C. Presl) E. Wimm., Pflanzenr. IV.276c: 979 (1968).
Cape Provinces. 27 CPP. Geophyte.
Cyphia volubilis var. *racemosa* Griess., Linnaea 5: 420 (1830).
Cyphia campestris Eckl. & Zeyh. ex C. Presl in Eckl. & Zeyh., Enum. Pl. Afric. Austral.: 391 (1837).
Cyphia tenuifolia A. DC. in DC., Prodr. 7: 498 (1839).
Cyphia campestris var. *nudiuscula* E. Wimm., Pflanzenr. IV.276c: 981 (1968).

Cyphia longiflora Schltr., Bot. Jahrb. Syst. 27: 194 (1899).
Cape Provinces. 27 CPP. Cl. geophyte.

Cyphia longifolia N. E. Br., Bull. Misc. Inform. Kew 1908: 435 (1908).
S. Africa. 27 NAT CPP. Geophyte.
Cyphia longifolia var. *baurii* E. Phillips, Ann. S. African Mus. 9: 460 (1917).

Cyphia longilobata E. Phillips, Ann. S. African Mus. 9: 465 (1917).
Cape Provinces. 27 CPP. Cl. geophyte.

Cyphia longipedicellata E. Wimm., Kew Bull. 1952: 144 (1952).
Cape Provinces. 27 CPP. Geophyte.

Cyphia maculosa E. Phillips, Ann. S. African Mus. 9: 463 (1917).
Cape Provinces. 27 CPP. Cl. tuber geophyte.

Cyphia mafingensis Thulin in Launert, Fl. Zambes. 7: 150 (1983).
Malawi. 26 MLW. Geophyte.

Cyphia mazoensis S. Moore, J. Bot. 45: 46 (1907).
Zambia to Mozambique. 26 MLW MOZ ZAM ZIM. (Cl.) geophyte.
Cyphia mazoensis var. *stellaris* E. Wimm., Kew Bull. 1952: 144 (1952).
Cyphia rivularis E. Wimm., Kew Bull. 1952: 145 (1952).
Cyphia subscandens E. Wimm., Pflanzenr. IV.276c: 949 (1968).

Cyphia natalensis E. Phillips, Ann. S. African Mus. 9: 464 (1917).
S. Africa. 27 CPP NAT. Cl. tuber geophyte.

Cyphia nyikensis Thulin, Bot. Not. 131: 461 (1978).
Malawi. 26 MLW. Geophyte.

Cyphia oligotricha Schltr., Bot. Jahrb. Syst. 27: 195 (1899).
Cape Provinces. 27 CPP. Geophyte.

Cyphia pectinata E. Wimm., Pflanzenr. IV.276c: 970 (1968).
Swaziland. 27 SWZ. Geophyte.

Cyphia persicifolia C. Presl in E. Mey., Comm. Pl. Afr. Austr.: 296 (1838).
S. Africa. 27 CPP TVL. Geophyte.
Cyphia linarioides var. *major* C. Presl in Eckl. & Zeyh., Enum. Pl. Afric. Austral.: 391
(1837). *Cyphia assimilis* Sond. in Harv. & Sond., Fl. Cap. 3: 600 (1865).

Cyphia phillipsii E. Wimm., Pflanzenr. IV.276c: 938 (1968).
Cape Provinces. 27 CPP. Geophyte.
* *Cyphia assimilis* var. *latifolia* E. Phillips, Ann. S. African Mus. 9: 461 (1917).

Cyphia phyteuma (L.) Willd., Sp. Pl. 1: 953 (1798).
Cape Provinces. 27 CPP. Geophyte.
* *Lobelia phyteuma* L., Sp. Pl.: 930 (1753). *Cyphia phyteumifolia* St.-Lag., Ann. Soc. Bot.
Lyon 7: 124 (1880).
Lobelia nudicaulis Lam., Encycl. 3: 590 (1792).
Cyphia serrata Spreng., Neue Entd. 1: 274 (1820).
Cyphia phyteuma var. *grandidentata* E. Wimm., Pflanzer. IV.276c: 959 (1968).
Cyphia phyteuma var. *ciliata* E. Wimm., Pflanzer. IV.276c: 960 (1968).

Cyphia reducta E. Wimm., Kew Bull. 1952: 144 (1952).
Mozambique & Zimbabwe. 26 MOZ ZIM. Cl. tuber geophyte.

Cyphia revoluta E. Wimm., Pflanzenr. IV.276c: 972 (1968).
Cape Provinces. 27 CPP. Geophyte.

Cyphia richardsiae E. Wimm., Pflanzenr. IV.276c: 973 (1968).
Zaïre, Tanzania & Malawi. 23 ZAI 25 TAN 26 MLW. Geophyte.

Cyphia rogersii S. Moore, J. Bot. 56: 9 (1918).
S. Africa. 27 NAT SWZ TVL. Cl. geophyte.

subsp. **rogersii**
Swaziland & Transvaal. 27 SWZ TVL. Cl. geophyte.

subsp. **winteri** E. Wimm., Pflanzenr. IV.276c: 948 (1968).
Transvaal & KwaZulu-Natal. 27 NAT SWZ TVL. Cl. geophyte.

Cyphia rupestris E. Wimm., Pflanzenr. IV.276c: 972 (1968).
Tanzania. 25 TAN. Geophyte.

Cyphia salteri E. Wimm., Pflanzenr. IV.276c: 1003 (1968).
Cape Provinces. 27 CPP. Cl. tuber geophyte.

Cyphia schlechteri E. Phillips, Ann. S. African Mus. 9: 466 (1917).
Cape Provinces. 27 CPP. Cl. geophyte.

Cyphia smutsii E. Wimm., Pflanzenr. IV.276c: 971 (1968).
Transvaal. 27 TVL. Geophyte.

Cyphia stenopetala Diels, Bot. Jahrb. Syst. 26: 112 (1898).
S. Africa. 27 BOT CPP TVL. Cl. geophyte.
Cyphia bechuanensis Bremek. & Oberm., Ann. Transvaal Mus. 16: 437 (1935).
Cyphia stenopetala var. *johannesburgensis* E. Wimm., Pflanzenr. IV.276c: 946 (1968).

Cyphia stenophylla (E. Wimm.) E. Wimm., Pflanzenr. IV.276c: 971 (1968).
S. Africa. 27 NAT OFS TVL. Geophyte.
* *Cyphia elata* var. *stenophylla* E. Wimm., Kew Bull. 1952: 143 (1952).

Cyphia stheno Webb. in Hook., Niger Fl.: 148 (1849).
  Angola. 26 ANG. Geophyte.
    Cyphia lobelioides Welw. ex Hiern., Cat. Afr. Pl. 1: 628 (1898).

Cyphia subtubulata E. Wimm., Pflanzenr. IV.276c: 950 (1968).
  Cape Provinces. 27 CPP. Cl. geophyte.

Cyphia sylvatica Eckl. ex C. Presl in Eckl. & Zeyh., Enum. Pl. Afric. Austral.: 388 (1837).
  S. Africa. 27 CPP NAM NAT. Cl. geophyte.
    Cyphia salicifolia C. Presl in Eckl. & Zeyh., Enum. Pl. Afric. Austral.: 390 (1837). *Cyphia
      sylvatica* var. *salicifolia* (C. Presl) E. Wimm., Pflanzenr. IV.276c: 1004 (1968).
    Cyphia tortilis N. E. Br., Bull. Misc. Inform. Kew 1894: 356 (1894).
    Cyphia sylvatica f. *sublobata* E. Wimm., Pflanzenr. IV.276c: 1004 (1968).
    Cyphia sylvatica var. *graminea* E. Wimm., Pflanzenr. IV.276c: 1005 (1968).

Cyphia tenera Diels, Bot. Jahrb. Syst. 26: 112 (1898).
  Cape Provinces. 27 CPP. Cl. tuber geophyte.

Cyphia transvaalensis E. Phillips, Ann. S. African Mus. 9: 462 (1917).
  Transvaal. 27 TVL. Cl. geophyte.

Cyphia triphylla E. Phillips, Ann. S. African Mus. 9: 466 (1917).
  S. Africa. 27 CPP LES OFS. Geophyte.
    Cyphia volubilis var. *parviflora* Cham., Linnaea 8: 225 (1833). *Cyphopsis volubilis* var.
      parviflora (Cham.) Kuntze, Revis. Gen. Pl. 3(2): 186 (1898).
    Cyphia linarioides var. *micrantha* C. Presl in E. Mey., Comm. Pl. Afr. Austr.: 294 (1838).

Cyphia tysonii E. Phillips, Ann. S. African Mus. 9: 470 (1917).
  S. Africa. 27 CPP NAT. Cl. geophyte.

Cyphia ubenensis Engl., Bot. Jahrb. Syst. 30: 419 (1901).
  Tanzania. 25 TAN. Geophyte.

Cyphia undulata Eckl. ex C. Presl in Eckl. & Zeyh., Enum. Pl. Afric. Austral.: 390 (1837).
  Cape Provinces. 27 CPP. Geophyte.
    Cyphia anomala Eckl. & Zeyh. ex C. Presl in Eckl. & Zeyh., Enum. Pl. Afric. Austral.:
      390 (1837).
    Cyphia undulata f. *longifolia* E. Wimm., Pflanzenr. IV.276c: 979 (1968).

Cyphia volubilis (Burm. f.) Willd., Sp. Pl. 1: 952 (1798).
  Cape Provinces. 27 CPP. Cl. geophyte.
    * *Lobelia volubilis* Burm. f., Prodr. Fl. Cap.: 29 (1768). *Cyphopsis volubilis* (Burm. f.)
      Kuntze, Revis. Gen. Pl. 3(2): 186 (1898).
    Cyphia volubilis var. *vulgarior* Cham., Linnaea 8: 224 (1833).
    Cyphia volubilis var. *intermedia* Cham., Linnaea 8: 224 (1833).
    Cyphia angustiloba C. Presl in Eckl. & Zeyh., Enum. Pl. Afric. Austral.: 389 (1837).
    Cyphia longipetala C. Presl in Eckl. & Zeyh., Enum. Pl. Afric. Austral.: 389 (1837).
      Cyphia volubilis var. *intermedia* E. Wimm., Pflanzenr. IV.276c: 987 (1968); non
      Cham., Linnaea 8: 224 (1833).
    Cyphia latipetala C. Presl in Eckl. & Zeyh., Enum. Pl. Afric. Austral.: 391 (1837). *Cyphia
      volubilis* var. *latipetala* (C. Presl) E. Wimm., Pflanzenr. IV:276c: 986 (1968).
    Cyphia digitata var. *major* C. Presl in Eckl. & Zeyh., Enum. Pl. Afric. Austral.: 391 (1837).
      Cyphia volubilis f. *major* (C. Presl) E. Wimm., Pflanzenr. IV.276c: 987 (1968).
    Cyphia dentata E. Wimm., Pflanzenr. IV.276c: 983 (1968).
    Cyphia volubilis var. *longipes* E. Wimm., Planzenr. IV.276c: 985 (1968).
    Cyphia volubilis var. *banksiana* E. Wimm., Planzenr. IV.276c: 986 (1968).
    Cyphia tricuspis E. Wimm., Pflanzenr. IV.276c: 987 (1968).

Cyphia zeyheriana C. Presl in Eckl. & Zeyh., Enum. Pl. Afric. Austral.: 392 (1837).
   Cape Provinces. 27 CPP. Cl. geophyte.
      Cyphia eckloniana C. Presl in Eckl. & Zeyh., Enum. Pl. Afric. Austral.: 392 (1837). *Cyphia*
         zeyheriana var. *eckloniana* (C. Presl) Sond. in Harv. & Sond., Fl. Cap. 3: 604 (1865).
      Cyphia ranunculifolia E. Wimm., Pflanzenr. IV.276c: 945 (1968).

Synonyms:

Cyphia sect. *Cyphiella* (C. Presl) E. Wimm. === **Cyphia** P. J. Bergius

Cyphia subg. *Cyphiella* (C. Presl) C. Presl === **Cyphia** P. J. Bergius

Cyphia [unranked] Cyphiella C. Presl === **Cyphia** P. J. Bergius

Cyphia alba f. *purpurea* E. Wimm. === **Cyphia alba** N. E. Br.

Cyphia angustiloba C. Presl === **Cyphia volubilis** (Burm. f.) Willd.

Cyphia anomala Eckl. & Zeyh. ex C. Presl === **Cyphia undulata** Eckl. ex C. Presl

Cyphia antunesii Engl. === **Cyphia lasiandra** Diels

Cyphia aspergilloides var. *brevipes* E. Wimm. === **Cyphia aspergilloides** E. Wimm.

Cyphia assimilis Sond. === **Cyphia persicifolia** C. Presl

Cyphia assimilis var. *latifolia* E. Phillips === **Cyphia phillipsii** E. Wimm.

Cyphia atriplicifolia C. Presl === **Cyphia crenata** (Thunb.) C. Presl

Cyphia bechuanensis Bremek. & Oberm. === **Cyphia stenopetala** Diels

Cyphia botrys Willd. ex Schult. === **Cyphia bulbosa** (L.) P. J. Bergius

Cyphia bulbosa f. *lobata* E. Wimm. === **Cyphia bulbosa** (L.) P. J. Bergius

Cyphia bulbosa var. *acocksii* E. Wimm. === **Cyphia bulbosa** (L.) P. J. Bergius

Cyphia bulbosa var. *angustiloba* A. DC. === **Cyphia bulbosa** (L.) P. J. Bergius

Cyphia bulbosa var. *hafstroemii* E. Wimm. === **Cyphia bulbosa** (L.) P. J. Bergius

Cyphia bulbosa var. *leiandra* E. Wimm. === **Cyphia bulbosa** (L.) P. J. Bergius

Cyphia bulbosa var. *orientalis* E. Phillips === **Cyphia bulbosa** (L.) P. J. Bergius

Cyphia bulbosa var. *pinnatifida* C. Presl === **Cyphia bulbosa** (L.) P. J. Bergius

Cyphia cacondensis R. D. Good === **Cyphia lasiandra** Diels

Cyphia campestris Eckl. & Zeyh. ex C. Presl === **Cyphia linarioides** C. Presl

Cyphia campestris var. *nudiuscula* E. Wimm. === **Cyphia linarioides** C. Presl

Cyphia capensis J. F. Gmel. === **Cyphia bulbosa** (L.) P. J. Bergius

Cyphia cardamines (Thunb.) Willd. === **Cyphia incisa** (Thunb.) Willd.

Cyphia corylifolia f. *singuliflora* E. Wimm. === **Cyphia corylifolia** Harv.

Cyphia crenata var. *angustifolia* E. Wimm. === **Cyphia crenata** (Thunb.) C. Presl

Cyphia crenata var. *smithiae* (L. Bolus) E. Wimm. === **Cyphia crenata** (Thunb.) C. Presl

Cyphia dentariifolia C. Presl === **Cyphia digitata** Willd.

Cyphia dentariifolia var. *luttigii* E. Wimm. === **Cyphia digitata** Willd.

Cyphia dentariifolia var. *psilandra* E. Wimm. === **Cyphia digitata** Willd.

Cyphia dentata E. Wimm. === **Cyphia volubilis** (Burm. f.) Willd.

Cyphia digitata f. *glabrifilis* E. Wimm. === **Cyphia digitata** Willd.

Cyphia digitata subsp. *gracilis* E. Wimm. === **Cyphia digitata** Willd.

Cyphia digitata var. *major* C. Presl === **Cyphia volubilis** (Burm. f.) Willd.

Cyphia digitata var. *tomentosa* (C. Presl) Sond. === **Cyphia digitata** Willd.

Cyphia digitata var. *trimera* E. Wimm. === **Cyphia digitata** Willd.

Cyphia dregeana C. Presl === **Cyphia crenata** (Thunb.) C. Presl

Cyphia eckloniana C. Presl === **Cyphia zeyheriana** C. Presl

Cyphia elata f. *decurrens* E. Wimm. === **Cyphia elata** Harv.

Cyphia elata f. *flanagana* E. Wimm. === **Cyphia elata** Harv.

Cyphia elata f. *thodeana* E. Wimm. === **Cyphia elata** Harv.

Cyphia elata f. *truncata* E. Wimm. === **Cyphia elata** Harv.

Cyphia elata var. *gerrardii* (Harv.) E. Wimm. === **Cyphia elata** Harv.

Cyphia elata var. *glabra* Harv. === **Cyphia elata** Harv.

Cyphia elata var. *globularis* E. Wimm. === **Cyphia elata** Harv.

Cyphia elata var. *oblongifolia* (Sond. & Harv.) E. Phillips === **Cyphia elata** Harv.

Cyphia elata var. *stenophylla* E. Wimm. === **Cyphia stenophylla** (E. Wimm.) E. Wimm.

Cyphia exelliana E. Wimm. === **Cyphia lasiandra** Diels

Cyphia erecta var. *wetteana* E. Wimm. === **Cyphia erecta** De Wild.
Cyphia floribunda E. Wimm. === **Cyphia lasiandra** Diels
Cyphia gerrardii Harv. === **Cyphia elata** Harv.
Cyphia glandulifera f. *obovatifolia* E. Wimm. === **Cyphia glandulifera** Hochst. ex A. Rich.
Cyphia glandulifera f. *subcuspidata* E. Wimm. === **Cyphia glandulifera** Hochst. ex A. Rich.
Cyphia incisa var. *bracteata* E. Phillips === **Cyphia incisa** (Thunb.) Willd.
Cyphia incisa var. *cardamines* (Thunb.) E. Phillips === **Cyphia incisa** (Thunb.) Willd.
Cyphia incisa var. *lyrata* E. Wimm. === **Cyphia incisa** (Thunb.) Willd.
Cyphia incisa var. *sinuata* E. Wimm. === **Cyphia incisa** (Thunb.) Willd.
Cyphia kerastes E. Wimm. === **Cyphia bulbosa** (L.) P. J. Bergius
Cyphia latipetala C. Presl === **Cyphia volubilis** (Burm. f.) Willd.
Cyphia linarioides var. *major* C. Presl === **Cyphia persicifolia** C. Presl
Cyphia linarioides var. *micrantha* C. Presl === **Cyphia triphylla** E. Phillips
Cyphia lobelioides Welw. ex Hiern. === **Cyphia stheno** Webb.
Cyphia longifolia var. *baurii* E. Phillips === **Cyphia longifolia** N. E. Br.
Cyphia longipetala C. Presl === **Cyphia volubilis** (Burm. f.) Willd.
Cyphia mazoensis var. *stellaris* E. Wimm. === **Cyphia mazoensis** S. Moore
Cyphia nyasica Baker f. === **Cyphia lasiandra** Diels
Cyphia oblongifolia Sond. & Harv. === **Cyphia elata** Harv.
Cyphia peteriana E. Wimm. === **Cyphia lasiandra** Diels
Cyphia phyteuma var. *ciliata* E. Wimm. === **Cyphia phyteuma** (L.) Willd.
Cyphia phyteuma var. *grandidentata* E. Wimm. === **Cyphia phyteuma** (L.) Willd.
Cyphia phyteumifolia St.-Lag. === **Cyphia phyteuma** (L.) Willd.
Cyphia pinnata (Lam.) Schult. === ?
Cyphia polydactyla C. Presl === **Cyphia digitata** Willd.
Cyphia ranunculifolia E. Wimm. === **Cyphia zeyheriana** C. Presl
Cyphia regularis E. Wimm. === **Cyphia erecta** De Wild.
Cyphia rhodesiaca E. Wimm. === **Cyphia erecta** De Wild.
Cyphia rivularis E. Wimm. === **Cyphia mazoensis** S. Moore
Cyphia salicifolia C. Presl === **Cyphia sylvatica** Eckl. ex C. Presl
Cyphia scandens De Wild. === **Cyphia erecta** De Wild.
Cyphia serrata Spreng. === **Cyphia phyteuma** (L.) Willd.
Cyphia smithiae L. Bolus === **Cyphia crenata** (Thunb.) C. Presl
Cyphia stenopetala var. *johannesburgensis* E. Wimm. === **Cyphia stenopetala** Diels
Cyphia stephensii E. Wimm. === **Cyphia bulbosa** (L.) P. J. Bergius
Cyphia subscandens E. Wimm. === **Cyphia mazoensis** S. Moore
Cyphia sylvatica f. *sublobata* E. Wimm. === **Cyphia sylvatica** Eckl. ex C. Presl
Cyphia sylvatica var. *salicifolia* (C. Presl) E. Wimm. === **Cyphia sylvatica** Eckl. ex C. Presl
Cyphia sylvatica var. *graminea* E. Wimm. === **Cyphia sylvatica** Eckl. ex C. Presl
Cyphia tenuifolia A. DC. === **Cyphia linarioides** C. Presl
Cyphia tomentosa C. Presl === **Cyphia digitata** Willd.
Cyphia tomentosa var. *glabra* A. DC. === **Cyphia digitata** Willd.
Cyphia tomentosa var. *minor* C. Presl === **Cyphia digitata** Willd.
Cyphia tortilis N. E. Br. === **Cyphia sylvatica** Eckl. ex C. Presl
Cyphia tricuspis E. Wimm. === **Cyphia volubilis** (Burm. f.) Willd.
Cyphia undulata f. *longifolia* E. Wimm. === **Cyphia undulata** Eckl. ex C. Presl
Cyphia undulata var. *linarioides* (C. Presl) E. Wimm. === **Cyphia linarioides** C. Presl
Cyphia volubilis f. *major* (C. Presl) E. Wimm. === **Cyphia volubilis** (Burm. f.) Willd.
Cyphia volubilis var. *banksiana* E. Wimm. === **Cyphia volubilis** (Burm. f.) Willd.
Cyphia volubilis var. *intermedia* Cham. === **Cyphia volubilis** (Burm. f.) Willd.
Cyphia volubilis var. *intermedia* E. Wimm. === **Cyphia volubilis** (Burm. f.) Willd.
Cyphia volubilis var. *latipetala* (C. Presl) E. Wimm. === **Cyphia volubilis** (Burm. f.) Willd.
Cyphia volubilis var. *longipes* E. Wimm. === **Cyphia volubilis** (Burm. f.) Willd.
Cyphia volubilis var. *parviflora* Cham. === **Cyphia triphylla** E. Phillips
Cyphia volubilis var. *racemosa* Griess. === **Cyphia linarioides** C. Presl
Cyphia volubilis var. *vulgarior* Cham. === **Cyphia volubilis** (Burm. f.) Willd.

Cyphia wilmsiana Diels === **Cyphia elata** Harv.
Cyphia zeyheriana var. *eckloniana* (C. Presl) Sond. === **Cyphia zeyheriana** C. Presl
Cyphia zernyana E. Wimm. === **Cyphia lasiandra** Diels

Cyphiella

This genus was segregated from *Cyphia* to accomodate the species with a nearly regular sympetalous corolla. Based on: *Cyphia* [unranked] *Cyphiella* C. Presl. Type [designated here]: *Cyphia zeyheriana* C. Presl.

> Spach, E. (1838). Histoire naturelle des végétaux. Phanerogames, vol. 9. Paris: Librairie Encyclopédique de Roret. La. — Establishment of genus as distinct from *Cyphia* s. s.

Synonyms:
Cyphiella (C. Presl) Spach === **Cyphia** P. J. Bergius

Cyphocarpus

Cyphocarpoideae, 3 species, northern Chile. A preliminary phylogenetic analysis of molecular data (Cosner et al. 1994, under **General**) indicated that *Cyphocarpus* is sister to a major clade comprising Campanuloideae, Lobelioideae, and Nemacladoideae. Although all three species had been published at the time of Wimmer's death in 1961, his monograph of the genus (Wimmer 1968) accounted for only the type species. The treatment here follows Ricardi (1959), who provided a key by which the species may be distinguished. Type [by monotypy]: *Cyphocarpus rigescens* Miers.

> Miers, J. (1848). Contributions to the botany of South America. *Cyphocarpus*. London J. Bot. 7: 59-64. En. — Establishment of genus (and subfamily Cyphocarpoideae), with detailed discussion of its relationships.
>
> Gay, C. (1849). Cyphocarpo. - *Cyphocarpus*. In Historia fisica y politica de Chile, Botanica 4: 335-337. Santiago: Museo Historia Natural. Sp. — Flora with description.
>
> Miers, J. (1850). *Cyphocarpus rigescens*. Illustrations of South American plants: pl. XXVII + 1 pg., illus. London: H. Bailliere. En. — Portrait with dissections.
>
> Reiche, K. (1905). Estudios criticos sobre la flora de Chile, *Cyphocarpus*. - Miers. Anales Univ. Chile 117: 453-454. Sp. — Floristic account with description and full nomenclature.
>
> • Ricardi, M. (1959). Un *Cyphocarpus* nuevo para Chile. Bol. Soc. Argent. Bot. 7: 247-250, illus. Sp. — Includes key to all three species.
>
> • Wimmer, F. E. (1968). *Cyphocarpus* Miers. In A. Engler & L. Diels, Das Pflanzenreich IV. 276c: 922-923 + taf. 30, illus. Berlin: Akademie Verlag. Ge., La. — Monograph with description, full nomenclature, and specimen citations, but only for type species.
>
> Dunbar, A. & H. G. Wallentinus (1976). On pollen of Campanulaceae III. A numerical taxonomic investigation. Bot. Not. 129: 69-72. En. — Cluster analysis of palynological data suggests Cyphioideae s. l. is not a natural group.
>
> Dunbar, A. (1984). Pollen morphology in Campanulaceae IV. Nord. J. Bot. 4: 1-19, illus. En. — Grains similar to those of Lobelioideae: 3-colporate with reticulate sculpturing and divided footlayer

Cyphocarpus Miers, London J. Bot. 7: 62 (1848).
 N. Chile. 85.

Cyphocarpus innocuus Sandwith, Bull. Misc. Inform. Kew 1931: 102 (1931).
 N. Chile (Limarí). 85 CLC. Ther.

Cyphocarpus psammophilus Ricardi, Bol. Soc. Argent. Bot. 7: 247 (1959).
 N. Chile (Huasco). 85 CLN. Ther.

Cyphocarpus rigescens Miers, London J. Bot. 7: 63 (1848).
   N. Chile (Copiapó, Huasco, Elqui). 85 CLC CLN. Ther.
      Cyphocarpus rigescens f. *glabratus* Phil. ex E. Wimm., Pflanzenr. IV.276c: 923 (1968).

Synonyms:
Cyphocarpus rigescens f. *glabratus* Phil. ex E. Wimm. === **Cyphocarpus rigescens** Miers.

Cyphopsis

Kuntze (1898) published this name as an avowed substitute for *Cyphia,* which he regarded
as a homonym of *Cuphea* P. Browne (Lythraceae). Under current provisions of the ICBN
(Art. 53.3 Ex. 10), it is not. Based on: *Cyphia* P. J. Bergius.

   Kuntze, O. (1898). *Cyphopsis.* In Revisio generum plantarum vascularium omnium atque
      cellularium multarum secundum leges nomenclaturae internationales 3(2): 186.
      Leipzig: Arthur Felix. Ge. — Validation of name.

Synonyms:
Cyphopsis Kuntze === **Cyphia** P. J. Bergius
Cyphopsis bulbosa (L.) Kuntze === **Cyphia bulbosa** (L.) P. J. Bergius
Cyphopsis elata (Harv.) Kuntze === **Cyphia elata** Harv.
Cyphopsis gerrardii (Harv.) Kuntze === **Cyphia elata** Harv.
Cyphopsis volubilis (Burm. f.) Kuntze === **Cyphia volubilis** (Burm. f.) Willd.
Cyphopsis volubilis var. *parviflora* (Cham.) Kuntze === **Cyphia triphylla** E. Phillips

Cyrtandroidea

This was described as a new genus of woody Lobelioideae from the Marquesas, but its type is
in fact a species of Gesneriaceae (Burtt 1967, Gillett 1973). Type [by monotypy]:
Cyrtandroidea jonesii F. Br.

   Brown, F. B. H. (1935). Flora of southeastern Polynesia III. Dicotyledons. *Cyrtandroidea* F.
      Brown, new genus. Bernice P. Bishop Mus. Bull. 130: 323-325, illus. En. —
      Establishment of genus in Lobelioideae, with comments on similarities to Gesneriaceae.
   Wimmer, F. E. (1943). *Cyrtandroidea* F. Brown. In A. Engler & L. Diels, Das Pflanzenreich
      IV. 276b: 104, illus. Leipzig: Wilhelm Engelmann. Ge., La. — Monograph with
      description and specimen citations; comments: "An genus ad Lobelioideas revera
      pertinet, mihi dubium est".
   Burtt, B. L. (1967). Studies in the Gesneeriaceae of the Old World XXVIII: the acquisition
      of *Cyrtandroidea.* Notes Roy. Bot. Gard. Edinburgh 28: 217-218. En. — Genus removed
      from Campanulaceae and classified under Gesneriaceae.
   Gillett, G. W. (1973). The genus *Cyrtandra* (Gesneriaceae) in the South Pacific. Univ. Calif.
      Publ. Bot. 66: 1-59, illus., maps. En. — Type species transferred to *Cyrtandra* J. R. Forst.
      & G. Forst.

Synonyms:
Cyrtandroidea F. Br. === **Cyrtandra** (Gesneriaceae)
Cyrtandroidea jonesii F. Br. === **Cyrtandra jonesii** (F.Br.) G. W. Gillett (Gesneriaceae)

Decaprisma

In Rafinesque's (1837) view, *Campanula* consisted solely of species with exappendiculate
calyx, campanulate corolla, and smooth three-locular poricidal capsules. He segregated this
species from it because of its five-angled capsule, which he thought was five-locular, though
it is in fact three-locular. Type [designated by the author]: *Decaprisma cespitosa* (Scop.) Raf.

Rafinesque, C. S. (1837). Flora Telluriana, pars secunda. Philadelphia: published privately. En. — Establishment of genus as distinct from *Campanula*.

Synonyms:
Decaprisma Raf. === **Campanula** L.
Decaprisma cespitosa (Scop.) Raf. === **Campanula cespitosa** Scop.

Delissea

Lobelioideae, 15 species, Hawaiian Islands. The genus is divided (Lammers 2005) into sect. *Delissea,* sect. *Macranthae* (Hillebr.) Rock, and sect. *Rhytidospermae* Lammers. Although St. John and Takeuchi (1987) felt that the genus could not be distinguished from *Cyanea,* recent analyses of molecular data (Givnish et al. 1995; E. Knox, pers. comm.; A. Antonelli, pers. comm.) and seed morphology (Buss et al. 2001) indicate that *Delissea* is the sister-group of *Brighamia,* while *Cyanea* is sister to *Clermontia.* The treatment here follows Lammers (2005). *2n* = 28. *Type* [by monotypy]: *Delissea undulata* Gaudich.

Gaudichaud, C. (1826). *Delissea.* In Voyage autour du monde, entrepris par order du roi ... exécuté sur les corvettes de S. M. *l'Uranie* et *la Physicienne,* pendant les années 1817, 1818, 1819 et 1820. Botanique: 457 + pl. 76-78, illus. Paris: Pillet-ainé. La., Fr. — Establishment of genus.

Gray, A. (1861). Notes on Lobeliaceae, Goodeniaceae, &c. of the collections of the U.S. South Pacific Exploring Expedition. Proc. Amer. Acad. Arts 5: 146-152. En. — Circumscription expanded to include all baccate Hawaiian species with minute tooth-like calyx lobes, bringing some species of *Rollandia* to *Delissea.*

Baillon, H. (1885). *Delissea* Gaudich. In Histoire des plantes 8: 364. Paris: L. Hachette. Fr. — Circumscription expanded to include *Clermontia, Cyanea,* and *Rollandia* as sections.

• Hillebrand, W. (1888). *Delissea,* Gaud. In Flora of the Hawaiian Islands: 248-251. London: Williams & Norgate. En. — Flora with keys, descriptions, infrageneric classification, full nomenclature, and specimen citations; circumscription narrowed to that employed here.

von Post, T. E. (1903). *Delissea* Gaud. 1826. In Lexikon generum phanerogamarum: 166. Stuttgart: Deutsche Verlags-Anstalt. Ge., La. — *Clermontia* and *Cyanea* included as sections, but not *Rollandia.*

• Rock, J. F. (1919). A monographic study of the Hawaiian species of the tribe Lobelioideae family Campanulaceae, *Delissea* Gaudichaud. Mem. Bernice Pauahi Bishop Mus. 7(2): 341-360, illus. En. — Monograph with key, descriptions, infrageneric classification, full nomenclature, and specimen citations.

• Wimmer, F. E. (1943). *Delissea* Gaudich. In A. Engler & L. Diels, Das Pflanzenreich IV. 276b: 42-46, illus. Leipzig: Wilhelm Engelmann. Ge., La. — Monograph with key, descriptions, infrageneric classification, full nomenclature, and specimen citations.

Stone, B. C. (1967). A review of the endemic genera of Hawaiian plants. *Delissea* Gaud. Bot. Rev. (Lancaster) 33: 245. En. — Brief descriptive treatment with full nomenclature.

• Wimmer, F. E. (1968). *Delissea* Gaudich. In A. Engler & L. Diels, Das Pflanzenreich IV. 276c: 817. Berlin: Akademie-Verlag. Ge., La. — Supplement to Wimmer (1943).

St. John, H. (1969). Types of sections in *Clermontia, Cyanea,* and *Delissea* (Lobeliaceae). Taxon 18: 483. En. — Formal nomenclature of infrageneric taxa.

St. John, H. (1977). The variations of *Delissea subcordata* Gaud. Lobeliaceae. Hawaiian plant studies 62. Phytologia 37: 417-419. En. — Examination of geographically correlated morphological variation among conspecific populations.

St. John, H. (1985). Earlier dates of valid publication of some genera and species in Gaudichaud's botany of the *Uranie* voyage. Taxon 34: 663-665. En. — Establishment of correct publication date and typification.

St. John (1987). Enlargement of *Delissea* (Lobeliaceae). Hawaiian plant studies 138. Phytologia 63: 79-90. En. — Transfer of all taxa of *Cyanea* into *Delissea,* following St. John & Takeuchi (1987).

St. John, H. & W. Takeuchi (1987). Are the distinctions of *Delissea* valid? Hawaiian plant studies 137. Phytologia 63: 129-130. En. — *Delissea* indistinguishable from *Cyanea*.
- Lammers, T. G. (1990). *Delissea* Gaud. In W. L. Wagner, D. R. Herbst & S. H. Sohmer, Manual of the flowering plants of Hawai'i: 467-472, illus. Honolulu: University of Hawaii Press. En. — Treatment with key, descriptions, and synonymy; circumscribed following Hillebrand (1888) and Rock (1919).

  Lammers, T. G. (1994). Typification of the names of Hawaiian Lobelioideae (Campanulaceae) published by Wilhelm Hillebrand or based upon his specimens. Taxon 43: 545-572. En. — Formal nomenclature of several species.
- Givnish, T. J., K. J. Sytsma, J. F. Smith & W. J. Hahn (1995). Molecular evolution, adaptive radiation, and geographic speciation in *Cyanea* (Campanulaceae, Lobelioideae). In W. L. Wagner & V. A. Funk (eds.), Hawaiian biogeography: evolution on a hot spot archipelago: 288-337, illus., maps. Washington: Smithsonian Institution Press. En. — Molecular phylogeny refutes merger with *Cyanea*.
- Buss, C. C., T. G. Lammers & R. R. Wise (2001). Seed coat morphology and its systematic implications in *Cyanea* and other genera of Lobelioideae (Campanulaceae). Amer. J. Bot. 88: 1301-1308, illus. En. — Seed surface morphology supports distinctness from *Cyanea* and close relationship to *Brighamia*.
- Lammers, T. G. (2005). Revision of *Delissea* (Campanulaceae-Lobelioideae). Syst. Bot. Monogr. 73: 1-75, illus., maps. En. — Monograph with key, descriptions, infrageneric classification, full nomenclature, and specimen citations.

Delissea Gaudich., Voy. Uranie: pl. 78 (1826).
Hawaiian Is. 63.

Delissea argutidentata (E. Wimm.) H. St. John, Pacific Sci. 13: 181 (1959).
Hawaiian Is. (Hawai'i). 63 HAW. Phan. or nanophan.
  * *Cyanea argutidentata* E. Wimm., Pflanzenr. IV.276b: 75 (1943). *Delissea undulata* var. *argutidentata* (E. Wimm.) E. Wimm., Pflanzenr. IV.276c: 817 (1968).
  Delissea konaensis H. St. John, Phytologia 59: 315 (1986).

Delissea fallax Hillebr., Fl. Hawaiian Isl.: 251 (1888).
Hawaiian Is. (Hawai'i). 63 HAW. Nanophan.

Delissea fauriei H. Lév., Repert. Spec. Nov. Regni Veg. 12: 505 (1913).
Hawaiian Is. (Moloka'i). 63 HAW. Nanophan.

Delissea kauaiensis (Lammers) Lammers, Syst. Bot. Monogr. 73: 19 (2005).
Hawaiian Is. (Kaua'i). 63 HAW. Phan. or nanophan.
  * *Delissea undulata* subsp. *kauaiensis* Lammers, Syst. Bot. 13: 505 (1988). *Delissea niihauensis* subsp. *kauaiensis* (Lammers) Lammers, Novon 9: 388 (1999).

Delissea laciniata Hillebr., Fl. Hawaiian Isl.: 249 (1888).
Hawaiian Is. (O'ahu). 63 HAW. Nanophan.

Delissea lanaiensis (Rock) Lammers, Novon 8: 34 (1998).
Hawaiian Is. (Lāna'i). 63 HAW. Nanophan.
  * *Delissea sinuata* var. *lanaiensis* Rock, Monogr. Stud. Haw. Lobelioid.: 353 (1919).
  Delissea sinuata subsp. *lanaiensis* (Rock) Lammers, Syst. Bot. 13: 505 (1988).

Delissea lauliiana Lammers, Syst. Bot. 13: 505 (1988).
Hawaiian Is. (O'ahu). 63 HAW. Nanophan.
  * *Delissea laciniata* var. *parvifolia* Rock, Monogr. Stud. Haw. Lobelioid.: 349 (1919).

Delissea niihauensis H. St. John, Pacific Sci. 13: 177 (1959). *Delissea undulata* subsp. *niihauensis* (H. St. John) Lammers, Syst. Bot. 13: 505 (1988).
Hawaiian Is. (Ni'ihau). 63 HAW. Phan.

Delissea parviflora Hillebr., Fl. Hawaiian Isl.: 251 (1888).
Hawaiian Is. (Hawai'i). 63 HAW. Nanophan.

Delissea rhytidosperma H. Mann, Proc. Amer. Acad. Arts 7: 180 (1867).
Hawaiian Is. (Kaua'i). 63 HAW. Nanophan. *2n* = 28.
Delissea kealiae Wawra, Flora 56: 10 (1873).

Delissea sinuata Hillebr., Fl. Hawaiian Isl.: 250 (1888).
Hawaiian Is. (O'ahu). 63 HAW. Nanophan.

Delissea subcordata Gaudich., Voy. Uranie: pl. 77 (1828). *Lobelia subcordata* (Gaudich.)
Endl., Ann. Wiener Mus. Naturgesch. 1: 170 (1836).
Hawaiian Is. (O'ahu). 63 HAW. Nanophan.

subsp. **obtusifolia** (Wawra) Lammers, Syst. Bot. Monogr. 73: 34 (2005).
Hawaiian Is. (C. Ko'olau Mts. of O'ahu). 63 HAW. Nanophan.
Delissea subcordata var. *obtusifolia* Wawra, Flora 56: 7 (1873).

subsp. **subcordata**
Hawaiian Is. (S. & C. Ko'olau Mts. of O'ahu). 63 HAW. Nanophan.
Delissea subcordata var. *waialaeensis* H. St. John, Phytologia 37: 419 (1977).
Delissea subcordata var. *waikaneensis* H. St. John, Phytologia 37: 419 (1977).

Delissea takeuchii Lammers, Syst. Bot. Monogr. 73: 38 (2005).
Hawaiian Is. (O'ahu). 63 HAW. Nanophan.

Delissea undulata Gaudich., Voy. Uranie: pl. 78 (1826). *Lobelia undulata* (Gaudich.) Endl.,
Ann. Wiener Mus. Naturgesch. 1: 170 (1836).
Hawaiian Is. (W. Maui). 63 HAW. Phan.

Delissea waianaeensis Lammers, Syst. Bot. Monogr. 73: 34 (2005).
Hawaiian Is. (O'ahu). 63 HAW. Nanophan.
Delissea subcordata var. *kauaiensis* H. St. John, Phytologia 37: 418 (1977).

Synonyms:
Delissea sect. *Clermontia* (Gaudich.) Baill. === **Clermontia** Gaudich.
Delissea sect. *Cyanea* (Gaudich.) Baill. === **Cyanea** Gaudich.
Delissea aculeatiflora (Rock) H. St. John === **Cyanea aculeatiflora** Rock
Delissea acuminata Gaudich. === **Cyanea acuminata** (Gaudich.) Hillebr.
Delissea acuminata f. *latifolia* (Wawra) H. St. John === **Cyanea acuminata** (Gaudich.) Hillebr.
Delissea acuminata var. *angustifolia* (Cham.) A. Gray === **Cyanea angustifolia** (Cham.) Hillebr.
Delissea acuminata var. *calycina* (Hosaka) H. St. John === **Cyanea acuminata** (Gaudich.) Hillebr.
Delissea acuminata var. *latifolia* Wawra === **Cyanea acuminata** (Gaudich.) Hillebr.
Delissea acuta H. St. John === ?
Delissea alba H. St. John === **Cyanea hirtella** (H. Mann) Hillebr.
Delissea albilineata H. St. John === **Cyanea fissa** (H. Mann) Hillebr.
Delissea ambigua (Cham.) C. Presl === **Cyanea lanceolata** (Gaudich.) Lammers, Givnish
& Sytsma
Delissea angusta H. St. John === **Cyanea obtusa** (A. Gray) Hillebr.
Delissea angustifolia (Cham.) G. Don === **Cyanea angustifolia** (Cham.) Hillebr.
Delissea angustifolia f. *elliptica* (Rock) H. St. John === **Cyanea elliptica** (Rock) Lammers
Delissea angustifolia f. *lanaiensis* (Rock) H. St. John === **Cyanea elliptica** (Rock) Lammers
Delissea angustifolia f. *subpubescens* (E. Wimm.) H. St. John === **Cyanea membranacea** Rock
Delissea angustifolia var. *hillebrandii* (Rock) H. St. John === **Cyanea elliptica** (Rock) Lammers
Delissea angustifolia var. *isabella* (E. Wimm.) H. St. John === **Cyanea elliptica** (Rock) Lammers
Delissea angustifolia var. *lanaiensis* (Rock) H. St. John === **Cyanea elliptica** (Rock) Lammers
Delissea angustifolia var. *racemosa* (Hillebr.) H. St. John === **Cyanea angustifolia** (Cham.) Hillebr.

Delissea angustifolia var. *tomentella* (Hillebr.) H. St. John === **Cyanea membranacea** Rock

Delissea angustior H. St. John === **Cyanea platyphylla** (A. Gray) Hillebr.

Delissea arborea (G. Forst.) C. Presl === **Sclerotheca arborea** (G. Forst.) A. DC.

Delissea arborea H. Mann === **Cyanea arborea** Hillebr.

Delissea asarifolia (H. St. John) H. St. John === **Cyanea asarifolia** H. St. John

Delissea aspleniifolia H. Mann === **Cyanea aspleniifolia** (H. Mann) Hillebr.

Delissea atra (Hillebr.) H. St. John === **Cyanea macrostegia** Hillebr.

Delissea atra var. *lobata* (Rock) H. St. John === **Cyanea horrida** (Rock) O. Deg. & Hosaka —
Cyanea macrostegia Hillebr.

Delissea aurantiaca H. St. John === **Cyanea floribunda** E. Wimm.

Delissea baldwinii (C. N. Forbes & G. C. Munro) H. St. John === **Cyanea lobata** H. Mann

Delissea bicolor (H. St. John) H. St. John === **Cyanea macrostegia** Hillebr.

Delissea bishopii H. St. John === **Cyanea kunthiana** Hillebr.

Delissea bondiana (Rock) H. St. John === **Cyanea floribunda** E. Wimm.

Delissea brevipedicellata H. St. John === **Cyanea hirtella** (H. Mann) Hillebr.

Delissea bryanii (Rock) H. St. John === **Cyanea platyphylla** (A. Gray) Hillebr.

Delissea calycina (Cham.) C. Presl === **Cyanea calycina** (Cham.) Lammers

Delissea carlsonii (Rock) H. St. John === **Cyanea hamatiflora** subsp. **carlsonii** (Rock)
Lammers

Delissea cataracta H. St. John === **Cyanea fissa** (H. Mann) Hillebr.

Delissea chartacea H. St. John === **Cyanea hirtella** (H. Mann) Hillebr.

Delissea chockii (Rock) H. St. John === **Cyanea hirtella** (H. Mann) Hillebr.

Delissea christensenii H. St. John === **Cyanea hirtella** (H. Mann) Hillebr.

Delissea clermontioides Gaudich. === **Clermontia clermontioides** (Gaudich.) A. Heller

Delissea comata (Hillebr.) H. St. John === **Cyanea comata** Hillebr.

Delissea copelandii (Rock) H. St. John === **Cyanea copelandii** Rock

Delissea coriacea A. Gray === **Cyanea coriacea** (A. Gray) Hillebr.

Delissea coriacea var. *degeneriana* (E. Wimm.) H. St. John === **Cyanea coriacea** (A. Gray)
Hillebr.

Delissea coriacea var. *deltoidea* H. St. John === **Cyanea coriacea** (A. Gray) Hillebr.

Delissea coriacea var. *gratiosa* (E. Wimm.) H. St. John === **Cyanea pseudofauriei** Lammers

Delissea coriacea var. *hardyi* (Rock) H. St. John === **Cyanea hardyi** Rock

Delissea coriacea var. *haupuensis* H. St. John === **Cyanea coriacea** (A. Gray) Hillebr.

Delissea coriacea var. *lumahaiensis* H. St. John === **Cyanea coriacea** (A. Gray) Hillebr.

Delissea coriacea var. *pinnatiloba* A. Gray ex H. Mann === **Cyanea leptostegia** A. Gray

Delissea coriacea var. *serratifolia* (Rock) H. St. John === **Cyanea membranacea** Rock

Delissea coronata (E. Wimm.) H. St. John === **Cyanea lanceolata** (Gaudich.) Lammers,
Givnish & Sytsma

Delissea crispihirta (E. Wimm.) H. St. John === **Cyanea platyphylla** (A. Gray) Hillebr.

Delissea decumbens H. St. John === **Cyanea fissa** (H. Mann) Hillebr.

Delissea degeneriana (E. Wimm.) H. St. John === **Cyanea hirtella** (H. Mann) Hillebr.

Delissea delessertiana (Gaudich.) A. Gray === **Cyanea lanceolata** (Gaudich.) Lammers,
Givnish & Sytsma

Delissea deltoidea H. St. John === **Cyanea hirtella** (H. Mann) Hillebr.

Delissea dentata (E. Wimm.) H. St. John === **Cyanea membranacea** Rock

Delissea denticulata H. St. John === **Cyanea fissa** (H. Mann) Hillebr.

Delissea densiflora (Rock) H. St. John === **Cyanea floribunda** E. Wimm.

Delissea divergens H. St. John === **Cyanea hirtella** (H. Mann) Hillebr.

Delissea dunbariae (Rock) H. St. John === **Cyanea dunbariae** Rock

Delissea duploserrata H. St. John === **Cyanea fissa** (H. Mann) Hillebr.

Delissea eleeleensis H. St. John === **Cyanea eleeleensis** (H. St. John) Lammers

Delissea excurrens H. St. John === **Cyanea hirtella** (H. Mann) Hillebr.

Delissea fauriei (H. Lév.) H. St. John === **Cyanea coriacea** (A. Gray) Hillebr.

Delissea fernaldii (Rock) H. St. John === **Cyanea platyphylla** (A. Gray) Hillebr.

Delissea ferox (Hillebr.) H. St. John === **Cyanea solanacea** Hillebr.

Delissea ferox var. *laevicalyx* (Skottsb.) H. St. John === **Cyanea solanacea** Hillebr.

Delissea filigera Wawra === **Clermontia grandiflora** Gaudich.
Delissea fissa H. Mann === **Cyanea fissa** (H. Mann) Hillebr.
Delissea floribunda (E. Wimm.) H. St. John === **Cyanea floribunda** E. Wimm.
Delissea forbesii H. St. John === **Cyanea undulata** C. N. Forbes
Delissea fruticosa H. St. John === **Cyanea hirtella** (H. Mann) Hillebr.
Delissea gayana (Rock) H. St. John === **Cyanea fissa** (H. Mann) Hillebr.
Delissea gayana var. *duvelii* (Rock) H. St. John === **Cyanea fissa** (H. Mann) Hillebr.
Delissea gayana var. *wainihaensis* (Rock) H. St. John === **Cyanea fissa** (H. Mann) Hillebr.
Delissea gibsonii (Hillebr.) H. St. John === **Cyanea gibsonii** Hillebr.
Delissea giffardii (Rock) H. St. John === **Cyanea giffardii** Rock
Delissea glabra (E. Wimm.) H. St. John === **Cyanea glabra** (E. Wimm.) H. St. John
Delissea glabrifolia H. St. John === **Cyanea fissa** (H. Mann) Hillebr.
Delissea globosa H. St. John === **Cyanea macrostegia** Hillebr.
Delissea grimesiana (Gaudich.) H. St. John === **Cyanea grimesiana** Gaudich.
Delissea grimesiana var. *cylindrocalyx* (Rock) H. St. John === **Cyanea cylindrocalyx** (Rock)
    Lammers
Delissea grimesiana var. *hirsutifolia* (Rock) H. St. John === **Cyanea grimesiana** subsp. **obatae**
    (H. St. John) Lammers
Delissea grimesiana var. *lydgatei* (Rock) H. St. John === **Cyanea mauiensis** (Rock) Lammers
Delissea grimesiana var. *mauiensis* (Rock) H. St. John === **Cyanea mauiensis** (Rock) Lammers
Delissea grimesiana var. *munroi* (Hosaka) H. St. John === **Cyanea munroi** (Hosaka) Lammers
Delissea grimesiana var. *obatae* (H. St. John) H. St. John === **Cyanea grimesiana** subsp.
    obatae (H. St. John) Lammers
Delissea habenata H. St. John === **Cyanea habenata** (H. St. John) Lammers
Delissea haleakalaensis (H. St. John) H. St. John === **Cyanea copelandii** subsp.
    haleakalaensis (H. St. John) Lammers
Delissea hamatiflora (Rock) H. St. John === **Cyanea hamatiflora** Rock
Delissea hanaensis (H. St. John) H. St. John === **Cyanea macrostegia** Hillebr.
Delissea hirtella H. Mann === **Cyanea hirtella** (H. Mann) Hillebr.
Delissea hirtella var. *striata* (E. Wimm.) H. St. John === **Cyanea hirtella** (H. Mann) Hillebr.
Delissea hirtella var. *subglabra* (E. Wimm.) H. St. John === **Cyanea hirtella** (H. Mann) Hillebr.
Delissea holophylla (Hillebr.) H. St. John === **Cyanea scabra** Hillebr.
Delissea holophylla var. *obovata* (Rock) H. St. John === ?
Delissea honoluluensis Wawra === **Cyanea angustifolia** (Cham.) Hillebr.
Delissea horrida (Rock) H. St. John === **Cyanea horrida** (Rock) O. Deg. & Hosaka
Delissea humilis (Wawra) H. St. John === **Cyanea fissa** (H. Mann) Hillebr.
Delissea hypoleuca H. St. John === **Cyanea hirtella** (H. Mann) Hillebr.
Delissea iliahiatilis H. St. John === **Cyanea hirtella** (H. Mann) Hillebr.
Delissea impedicellata H. St. John === **Cyanea hirtella** (H. Mann) Hillebr.
Delissea inermis H. St. John === **Cyanea kahiliensis** (H. St. John) Lammers
Delissea inermis H. St. John === **Cyanea kahiliensis** (H. St. John) Lammers
Delissea inflatispinosa H. St. John === **Cyanea platyphylla** (A. Gray) Hillebr.
Delissea kahiliensis H. St. John === **Cyanea kahiliensis** (H. St. John) Lammers
Delissea kawelaensis H. St. John === **Cyanea mannii** (Brigham ex H. Mann) Hillebr.
Delissea keaensis H. St. John === **Cyanea fissa** (H. Mann) Hillebr.
Delissea kealawelaensis H. St. John === **Cyanea hirtella** (H. Mann) Hillebr.
Delissea kealiae Wawra === **Delissea rhytidosperma** H. Mann
Delissea kipahuluensis H. St. John === **Cyanea horrida** (Rock) O. Deg. & Hosaka
Delissea knudsenii (Rock) H. St. John === **Cyanea hirtella** (H. Mann) Hillebr.
Delissea kohalaensis H. St. John === **Cyanea platyphylla** (A. Gray) Hillebr.
Delissea kolekoleensis H. St. John === **Cyanea kolekoleensis** (H. St. John) Lammers
Delissea konaensis H. St. John === **Delissea argutidentata** (E. Wimm.) H. St. John
Delissea kunthiana Gaudich. === ?
Delissea laciniata var. *parvifolia* Rock === **Delissea lauliiana** Lammers
Delissea lanceolata (Gaudich.) A. Gray === **Cyanea lanceolata** (Gaudich.) Lammers, Givnish
    & Sytsma

Delissea larrisonii (Rock) H. St. John === **Cyanea salicina** H. Lév.

Delissea latibasilaris H. St. John === **Cyanea fissa** (H. Mann) Hillebr.

Delissea latior H. St. John === **Cyanea macrostegia** Hillebr.

Delissea leiophylla H. St. John === **Cyanea hirtella** (H. Mann) Hillebr.

Delissea leptostegia (A. Gray) H. St. John === **Cyanea leptostegia** A. Gray

Delissea leptostegia var. *velutina* (Skottsb.) H. St. John === **Cyanea leptostegia** A. Gray

Delissea ligulata H. St. John === **Cyanea hirtella** (H. Mann) Hillebr.

Delissea limahuliensis H. St. John === **Cyanea fissa** (H. Mann) Hillebr.

Delissea linearifolia (Rock) H. St. John === **Cyanea linearifolia** Rock

Delissea lobata (H. Mann) H. St. John === **Cyanea lobata** H. Mann

Delissea lobata var. *hamakuae* (Rock) H. St. John === **Cyanea horrida** (Rock) O. Deg. & Hosaka

Delissea longiantherae H. St. John === **Cyanea hirtella** (H. Mann) Hillebr.

Delissea longicalyx H. St. John === **Cyanea fissa** (H. Mann) Hillebr.

Delissea longipedunculata (Rock) H. St. John === **Cyanea pilosa** subsp. **longipedunculata** (Rock) Lammers

Delissea longissima (Rock) H. St. John === **Cyanea longissima** (Rock) H. St. John

Delissea lumahaiensis H. St. John === **Cyanea fissa** (H. Mann) Hillebr.

Delissea lydgatei H. St. John === **Cyanea undulata** C. N. Forbes

Delissea lyonii H. St. John === **Cyanea glabra** (E. Wimm.) H. St. John

Delissea macrostachys (Hook. & Arn.) C. Presl === **Trematolobelia macrostachys** (Hook. & Arn.) Zahlbr. ex Rock

Delissea macrostegia (Hillebr.) H. St. John === **Cyanea macrostegia** Hillebr.

Delissea macrostegia var. *parvibracteata* (Rock) H. St. John === **Cyanea macrostegia** Hillebr.

Delissea macrostegia var. *viscosa* (Rock) H. St. John === **Cyanea aculeatiflora** Rock × **Cyanea macrostegia** Hillebr.

Delissea magnifica (E. Wimm.) H. St. John === **Cyanea tritomantha** A. Gray

Delissea mannii Brigham ex H. Mann === **Cyanea mannii** (Brigham ex H. Mann) Hillebr.

Delissea mariana (E. Wimm.) H. St. John === **Cyanea macrostegia** Hillebr.

Delissea marksii (Rock) H. St. John === **Cyanea marksii** Rock

Delissea mceldowneyi (Rock) H. St. John === **Cyanea mceldowneyi** Rock

Delissea megacarpa (Rock) H. St. John === **Cyanea pilosa** A. Gray subsp. **pilosa**

Delissea membranacea (Rock) H. St. John === **Cyanea membranacea** Rock

Delissea molokaiensis H. St. John === **Cyanea elliptica** (Rock) Lammers

Delissea multiramosa H. St. John === **Cyanea fissa** (H. Mann) Hillebr.

Delissea multispicata (H. Lév.) H. St. John === **Cyanea fissa** (H. Mann) Hillebr.

Delissea muriculata H. St. John === ?

Delissea napaliensis H. St. John === **Cyanea fissa** (H. Mann) Hillebr.

Delissea nelsonii (H. St. John) H. St. John === **Cyanea stictophylla** Rock

Delissea nemorosa H. St. John === **Cyanea floribunda** E. Wimm.

Delissea nigra H. St. John === **Cyanea hardyi** Rock

Delissea niihauensis subsp. *kauaiensis* (Lammers) Lammers === **Delissea kauaiensis** (Lammers) Lammers

Delissea noli-me-tangere (Rock) H. St. John === **Cyanea platyphylla** (A. Gray) Hillebr.

Delissea obtusa A. Gray === **Cyanea obtusa** (A. Gray) Hillebr.

Delissea obtusa var. *mollis* A. Gray === **Clermontia pyrularia** Hillebr.

Delissea occultans (H. St. John) H. St. John === **Cyanea acuminata** (Gaudich.) Hillebr.

Delissea olokuiensis H. St. John === **Cyanea solenocalyx** Hillebr.

Delissea ovatisepala (E. Wimm.) H. St. John === **Cyanea solenocalyx** Hillebr.

Delissea palakea (C. N. Forbes) H. St. John === **Cyanea stictophylla** Rock

Delissea paliensis H. St. John === **Cyanea hirtella** (H. Mann) Hillebr.

Delissea parva H. St. John === **Cyanea fissa** (H. Mann) Hillebr.

Delissea perlmanii H. St. John === **Cyanea coriacea** (A. Gray) Hillebr.

Delissea pilosa (A. Gray) H. Mann === **Cyanea pilosa** A. Gray

Delissea pinnatifida (Cham.) C. Presl === **Cyanea pinnatifida** (Cham.) E. Wimm.

Delissea platyphylla A. Gray === **Cyanea platyphylla** (A. Gray) Hillebr.

Delissea pluriflora H. St. John === **Cyanea fissa** (H. Mann) Hillebr.

Delissea procera (Hillebr.) H. St. John === **Cyanea procera** Hillebr.

Delissea profuga (C. N. Forbes) H. St. John === **Cyanea profuga** C. N. Forbes

Delissea puberula H. St. John === **Cyanea obtusa** (A. Gray) Hillebr.

Delissea pulchra (Rock) H. St. John === **Cyanea platyphylla** (A. Gray) Hillebr.

Delissea purpurea H. St. John === **Cyanea hirtella** (H. Mann) Hillebr.

Delissea pycnocarpa (Hillebr.) H. St. John === **Cyanea pycnocarpa** (Hillebr.) E. Wimm.

Delissea quercifolia (Hillebr.) H. St. John === **Cyanea quercifolia** (Hillebr.) E. Wimm.

Delissea quercifolia var. *atropurpurea* (E. Wimm.) H. St. John === **Cyanea stictophylla** Rock

Delissea racemosa H. Mann === **Cyanea humboldtiana** (Gaudich.) Lammers, Givnish & Sytsma

Delissea ramosa H. St. John === **Cyanea fissa** (H. Mann) Hillebr.

Delissea recta Wawra === **Cyanea recta** (Wawra) Hillebr.

Delissea regina Wawra === **Cyanea superba** subsp. **regina** (Wawra) Lammers

Delissea remyi (Rock) H. St. John === **Cyanea remyi** Rock

Delissea rivularis (Rock) E. Wimm. === **Cyanea rivularis** Rock

Delissea robinsonii H. St. John === **Cyanea hirtella** (H. Mann) Hillebr.

Delissea rockii (E. Wimm.) H. St. John === **Cyanea salicina** H. Lév.

Delissea rollandioides (Rock) H. St. John === **Cyanea platyphylla** (A. Gray) Hillebr.

Delissea scabra (Hillebr.) H. St. John === **Cyanea scabra** Hillebr.

Delissea scabra var. *variabilis* (Rock) H. St. John === **Cyanea longissima** (Rock) H. St. John

Delissea scopuli H. St. John === **Cyanea hirtella** (H. Mann) Hillebr.

Delissea serratifolia (Rock) H. St. John === **Cyanea membranacea** Rock

Delissea sessilis H. St. John === **Cyanea hamatiflora** Rock subsp. **hamatiflora**

Delissea shipmanii (Rock) H. St. John === **Cyanea shipmanii** Rock

Delissea simplex H. St. John === **Cyanea hirtella** (H. Mann) Hillebr.

Delissea simplex f. *maculata* H. St. John === **Cyanea hirtella** (H. Mann) Hillebr.

Delissea sinuata subsp. *lanaiensis* (Rock) Lammers === **Delissea lanaiensis** (Rock) Lammers

Delissea sinuata var. *lanaiensis* Rock === **Delissea lanaiensis** (Rock) Lammers

Delissea solanacea (Hillebr.) H. St. John === **Cyanea solanacea** Hillebr.

Delissea solenocalyx (Hillebr.) H. St. John === **Cyanea solenocalyx** Hillebr.

Delissea solenocalyx var. latifolia (E. Wimm.) H. St. John === **Cyanea solenocalyx** Hillebr.

Delissea spathulata (Hillebr.) H. St. John === **Cyanea spathulata** (Hillebr.) A. Heller

Delissea stictophylla (Rock) H. St. John === **Cyanea stictophylla** Rock

Delissea stictophylla var. *inermis* (Rock) H. St. John === **Cyanea stictophylla** Rock

Delissea subacuminata H. St. John === **Cyanea hirtella** (H. Mann) Hillebr.

Delissea subcordata var. *kauaiensis* H. St. John === **Delissea waianaeensis** Lammers

Delissea subcordata var. *obtusifolia* Wawra === **Delissea subcordata** subsp. **obtusifolia** (Wawra) Lammers

Delissea subcordata var. *waialaeensis* H. St. John === **Delissea subcordata** Gaudich. subsp. **subcordata**

Delissea subcordata var. *waikaneensis* H. St. John === **Delissea subcordata** Gaudich. subsp. **subcordata**

Delissea subintegra H. St. John === **Cyanea fissa** (H. Mann) Hillebr.

Delissea submuricata (E. Wimm.) H. St. John === **Cyanea tritomantha** A. Gray

Delissea subobtusa H. St. John === **Cyanea floribunda** E. Wimm.

Delissea subsessilis H. St. John === **Cyanea fissa** (H. Mann) Hillebr.

Delissea superba (Cham.) H. St. John === **Cyanea superba** (Cham.) A. Gray

Delissea superba var. *velutina* (Rock) H. St. John === **Cyanea superba** (Cham.) A. Gray subsp. **superba**

Delissea sylvestris (A. Heller) H. St. John === **Cyanea hirtella** (H. Mann) Hillebr.

Delissea sylvestris var. *eriantha* (Skottsb.) H. St. John === **Cyanea hirtella** (H. Mann) Hillebr.

Delissea tritomantha (A. Gray) H. St. John === **Cyanea tritomantha** A. Gray
Delissea tritomantha var. *lydgatei* (Hillebr. ex Rock) H. St. John === **Cyanea marksii** Rock
Delissea truncata (Rock) H. St. John === **Cyanea truncata** (Rock) Rock
Delissea truncata var. *juddii* (C. N. Forbes) H. St. John === **Cyanea truncata** (Rock) Rock
Delissea umbrosa H. St. John === **Cyanea hirtella** (H. Mann) Hillebr.
Delissea undulata subsp. *kauaiensis* Lammers === **Delissea kauaiensis** (Lammers) Lammers
Delissea undulata subsp. *niihauensis* (H. St. John) Lammers === **Delissea niihauensis**
    H. St. John
Delissea undulata var. *argutidentata* (E. Wimm.) E. Wimm. === **Delissea argutidentata**
    (E. Wimm.) H. St. John
Delissea undulata var. *serrulata* Wawra === **Cyanea elliptica** (Rock) Lammers
Delissea virgata H. St. John === **Cyanea hirtella** (H. Mann) Hillebr.
Delissea waihiae Wawra === **Clermontia arborescens** subsp. **waihiae** (Wawra) Lammers
Delissea waikamoiensis H. St. John === **Cyanea kunthiana** Hillebr.
Delissea waikoluensis H. St. John === **Cyanea solenocalyx** Hillebr.
Delissea wailauensis (Rock) H. St. John === **Cyanea solenocalyx** Hillebr.
Delissea waioliensis H. St. John === **Cyanea hirtella** (H. Mann) Hillebr.
Delissea waipaensis H. St. John === **Cyanea fissa** (H. Mann) Hillebr.

Dialypetalum

Lobelioideae, 5 species, Madagascar. Although this genus seems taxonomically isolated within Lobelioideae on the basis of morphology, a preliminary molecular phylogeny (E. Knox, pers. comm.) shows that it is embedded within a major clade of woody tetraploids and hexaploids that also includes the *Centropogon-Burmeistera-Siphocampylus* clade, the Hawaiian clade, and the East African giant *Lobelia* clade. The most recent monograph is that of Wimmer (1953b). Type [by monotypy]: *Dialypetalum floribundum* Benth.

    Bentham, G. (1876). *Dialypetalum,* Benth. gen. nov. In G. Bentham & J. D. Hooker,
        Genera plantarum, 2: 553-554. London: L. Reeve. La. — Establishment of genus.
    Hooker, J. D. (1876). *Dialypetalum floribundum,* Benth. Hooker's Icon. Pl. 12: 68-69 + pl.
        1178, illus. En. — Portrait and description, published almost simultaneously with
        Bentham (1876).
    • Wimmer, F. E. (1953a). *Dialypetalum* Benth. In Flore de Madagascar et des Comores
        (plantes vasculaires). 186. Lobéliacées (Lobeliaceae): 30-41, illus. Paris: Firmin-Didot.
        Fr. — Flora covering all species, with key, descriptions, full nomenclature, and
        specimen citations.
    • Wimmer, F. E. (1953b). *Dialypetalum* Benth. In A. Engler & L. Diels, Das Pflanzenreich IV.
        276b: 720-724, illus. Berlin: Akademie-Verlag. Ge., La. — Monograph with key,
        descriptions, full nomenclature, and specimen citations.

Dialypetalum Benth. in Benth. & Hook. f., Gen. Pl. 2: 553 (1876).
    Madagascar. 29.

Dialypetalum compactum Zahlbr., Repert. Spec. Nov. Regni Veg. 4: 7 (1907).
    Madagascar. 29 MDG. Cham. or hemicr.
        Dialypetalum compactum var. *minoriflorum* E. Wimm., Pflanzenr. IV.276b: 721 (1953).
        Dialypetalum compactum f. *subpubescens* E. Wimm., Pflanzenr. IV.276b: 721 (1953).

Dialypetalum floribundum Benth. in Benth. & Hook. f., Gen. Pl. 2: 553 (1876), as 'floribunda'.
    Madagascar. 29 MDG. Cham. or hemicr.
        Dialypetalum floribundum var. *clavifolium* E. Wimm., Pflanzenr. IV.276b: 722 (1953),
            as 'clavaefolium'.
        Dialypetalum floribundum var. *glabrifolium* E. Wimm., Pflanzenr. IV.276b: 722 (1953).
        Dialypetalum floribundum var. *naniflorum* E. Wimm., Pflanzenr. IV.276b: 722 (1953).

Dialypetalum humbertianum E. Wimm., Pflanzenr. IV.276b: 723 (1953).
Madagascar. 29 MDG. Cham. or hemicr.

Dialypetalum × **hybridum** E. Wimm., Pflanzenr. IV.276b: 724 (1953). D. compactum × D.
floribundum
Madagascar. 29 MDG. Cham. or hemicr.

Dialypetalum montanum E. Wimm., Pflanzenr. IV.276b: 722 (1953).
Madagascar. 29 MDG. Cham. or hemicr.

Dialypetalum stenopetalum E. Wimm., Pflanzenr. IV.276b: 723 (1953).
Madagascar. 29 MDG. Cham. or hemicr.

Synonyms:
Dialypetalum compactum f. *subpubescens* E. Wimm. === **Dialypetalum compactum** Zahlbr.
Dialypetalum compactum var. *minoriflorum* E. Wimm. === **Dialypetalum compactum** Zahlbr.
Dialypetalum floribundum var. *clavifolium* E. Wimm. === **Dialypetalum floribundum** Benth.
Dialypetalum floribundum var. *glabrifolium* E. Wimm. === **Dialypetalum floribundum** Benth.
Dialypetalum floribundum var. *naniflorum* E. Wimm. === **Dialypetalum floribundum** Benth.

Diastatea

Lobelioideae, 5 species, Neotropics, from central Mexico to Panama, with one species
extending south to Bolivia. The genus has sometimes been included with *Hippobroma,
Isotoma, Solenopsis, Palmerella, Porterella,* and *Wimmerella* in the broadly circumscribed genus
Laurentia (q.v.), but it seems more closely related to certain species of *Lobelia*, e.g., *L.
xalapensis.* The most recent monograph is that of Wimmer (1953, 1968). Type [by
monotypy]: *Diastatea virgata* Scheidw.

Scheidweiler, M. J. F. (1841). Beschreibung einiger neuen Pflanzen. Allg. Gartenzeitung 9:
396-397. Ge. — Establishment of genus.
- McVaugh, R. (1940). A revision of "Laurentia" and allied genera in North America. Bull.
Torrey Bot. Club 67: 778-798. En. — Taxonomic treatment with key, descriptions, full
nomenclature, and specimen citations.
- McVaugh, R. (1943). *Diastatea* Scheidw. In North American Flora 32A: 26-30. New York:
New York Botanical Garden. En. — Flora with key, descriptions, and full nomenclature.
- Wimmer, F. E. (1953). *Diastatea* Scheidw. In A. Engler & L. Diels, Das Pflanzenreich IV.
276b: 380-386, illus. Berlin: Akademie-Verlag. Ge., La. — Monograph with key and
descriptions.
- Wimmer, F. E. (1968). *Diastatea* Scheidw. In A. Engler & L. Diels, Das Pflanzenreich IV.
276c: 852-853. Berlin: Akademie-Verlag. Ge., La. — Supplement to Wimmer (1953).
- Nash, D. L. (1976). Flora of Guatemala, *Diastatea* Scheidweiler. Fieldiana Bot. 24: 408-411,
illus. En. — Floristic treatment with key and descriptions.
Wilbur, R. L. (1977). Flora of Panama, *Diastatea*. Ann. Missouri Bot. Gard. 63: 632-635,
illus. En. — Floristic treatment with description.
Jeppesen, S. (1981). *Diastatea* Scheidw. In G. Harling & B. Sparre (eds.), Flora of Ecuador
14: 122-124, illus. Stockholm: Swedish Natural Science Research Council. En. —
Floristic treatment with description.
Serra, L. & M. B. Crespo (1997). An outline revision of the subtribe Siphocampylinae
(Lobeliaceae). Lagascalia 19: 881-888, illus., maps. En. — Key to distinguish genus from
alleged allies.
Wilbur, R. L. (2001). *Diastatea* Scheidw. In W. D. Stevens, C. Ulloa Ulloa, A. Pool & O. M.
Montiel (eds.), Flora de Nicaragua 1: 560-561. St. Louis: Missouri Botanical Garden
Press. Sp. — Flora with key and descriptions.

Diastatea Scheidw., Allg. Gartenzeitung 9: 396 (1841).
Neotropics. 79 80 82 83.

Diastatea costaricensis McVaugh, Bull. Torrey Bot. Club 67: 789 (1940).
Guatemala to Costa Rica. 80 COS GUA HON NIC. Ther.

Diastatea expansa McVaugh, Bull. Torrey Bot. Club 67: 787 (1940).
C. Mexico (México State). 79 MXC. Ther.

Diastatea micrantha (Kunth) McVaugh, Bull. Torrey Bot. Club 67: 143 (1940).
C. Mexico to Bolivia. 79 MXC MXE MXG MXS MXT 80 COS ELS GUA HON NIC PAN 82 VEN 83 BOL CLM ECU PER. Ther.
* *Lobelia micrantha* Kunth in Humb., Bonpl. & Kunth, Nov. Gen. Sp. 3: 316 (quarto), 247 (folio) (1819). *Rapuntium micranthum* (Kunth) C. Presl, Prodr. Monogr. Lobel.: 25 (1836). *Dortmanna micrantha* (Kunth) Kuntze, Revis. Gen. Pl. 2: 972 (1891). *Laurentia micrantha* (Kunth) Zahlbr., Bull. Torrey Bot. Club 24: 386 (1897); non (E. Mey.) A. DC. in DC., Prodr. 7: 411 (1839).
  Lobelia subtilis Kunth in Humb., Bonpl. & Kunth, Nov. Gen. Sp. 3: 317 (quarto), 247 (folio) (1819). *Rapuntium subtile* (Kunth) C. Presl, Prodr. Monogr. Lobel.: 25 (1836). *Lobelia micrantha* var. *subtilis* (Kunth) Vatke, Linnaea 38: 721 (1874).
  Lobelia ruderalis Willd. ex Schult. in Roem. & Schult., Syst. Veg. 5: 56 (1819).
  Lobelia draba Willd. ex Schult. in Roem. & Schult., Syst. Veg. 5: 67 (1819).
  Lobelia parviflora M. Martens & Galeotti, Bull. Acad. Roy. Sci. Bruxelles 9(2): 41 (1842). *Dortmanna parviflora* (M. Martens & Galeotti) Kuntze, Revis. Gen. Pl. 2: 973 (1891).
  Lobelia minutiflora Kunze, Linnaea 16: 318 (1842), as 'minutiflorum'. *Dortmanna minutiflora* (Kunze) Kuntze, Revis. Gen. Pl. 2: 972 (1891).
  Laurentia ovatifolia B. L. Rob., Proc. Amer. Acad. Arts 26: 166 (1891). *Laurentia micrantha* var. *ovatifolia* (B. L. Rob.) E. Wimm. in J. F. Macbr., Fl. Peru 6: 476 (1937). *Diastatea micrantha* var. *ovatifolia* (B. L. Rob.) E. Wimm., Ann. Naturhist. Mus. Wien 56: 332 (1948).
  Laurentia pedunculata Brandegee, Univ. Calif. Publ. Bot. 6: 73 (1914).
  Laurentia micrantha var. *longibracteata* E. Wimm., Revista Sudamer. Bot. 2: 104 (1935). *Diastatea micrantha* var. *longibracteata* (E. Wimm.) E. Wimm., Ann. Naturhist. Mus. Wien 56: 332 (1948).
  Laurentia maximiliana E. Wimm., Repert. Spec. Nov. Regni Veg. 38: 78 (1935). *Diastatea maximiliana* (E. Wimm.) E. Wimm., Ann. Naturhist. Mus. Wien 56: 332 (1948).
  Diastatea serrata Standl. & L. O. Williams, Ceiba 1: 91 (1950).

Diastatea tenera (A. Gray) McVaugh, Bull. Torrey Bot. Club 67: 143 (1940).
C. Mexico to Guatemala. 79 MXC MXS 80 GUA. Ther.
* *Palmerella tenera* A. Gray, Proc. Amer. Acad. Arts 22: 433 (1887). *Lobelia palmeri* Greene, Pittonia 1: 297 (1889); non *Lobelia tenera* Kunth in Humb., Bonpl. & Kunth, Nov. Gen. Sp. 3: 314 (1819).
  Laurentia pinetorum Brandegee, Univ. Calif. Publ. Bot. 4: 92 (1910).

Diastatea virgata Scheidw., Allg. Gartenzeitung 9: 396 (1841).
C. & S. Mexico. 79 MXC MXC MXS. Ther.

subsp. **virgata**
S. Mexico (Vera Cruz, Oaxaca). 79 MXG MXS. Ther.
  Lobelia ramosissima M. Martens & Galeotti, Bull. Acad. Roy. Sci. Bruxelles 9(2):42 (1842). *Laurentia ramosissima* (M. Martens & Galeotti) Benth. & Hook. f. ex Hemsl., Biol. Centr.-Amer., Bot. 2: 265 (1881).

subsp. **ciliata** (McVaugh) Lammers, Novon 16: 69 (2006).
C. & SW. Mexico (Michoacán, México State, Morelos, Guerrero). 79 MXC MXS. Ther.
* *Diastatea virgata* var. *ciliata* McVaugh, Bull. Torrey Bot. Club 67: 793 (1940).

Synonyms:
Diastatea ghiesbreghtii (Kuntze) E. Wimm. === ?
Diastatea irasuensis (Planch. & Oerst.) E. Wimm. === **Lobelia irasuensis** Planch. & Oerst.
Diastatea lemairei E. Wimm. === ?
Diastatea maximiliana (E. Wimm.) E. Wimm. === **Diastatea micrantha** (Kunth) McVaugh
Diastatea micrantha var. *longibracteata* (E. Wimm.) E. Wimm. === **Diastatea micrantha**
  (Kunth) McVaugh
Diastatea micrantha var. *ovatifolia* (B. L. Rob.) E. Wimm. === **Diastatea micrantha** (Kunth)
  McVaugh
Diastatea serrata Standl. & L. O. Williams === **Diastatea micrantha** (Kunth) McVaugh
Diastatea virgata var. *ciliata* McVaugh === **Diastatea virgata** Scheidw.

Dielsantha

Lobelioideae, 1 species, western Africa. Wimmer (1953) grouped this genus with the western
North American genera *Downingia* and *Howellia* (q.v.) as subtribe Howelliinae E. Wimm.,
due to similarities in their capsular fruits. However, preliminary analysis of DNA sequence
data (E. Knox, pers. comm.) places *Dielsantha* as part of a clade comprised largely of African
species of *Lobelia* sect. *Delostemon* (E. Wimm.) J. Murata. The genus was monographed by
Wimmer (1953). Type [by monotypy]: *Dielsantha galeopsoides* (Engl. & Diels) E. Wimm.

> Wimmer, F. E. (1948). Vorarbeiten zur Monographie der Campanulaceae-Lobelioideae: II.
>   Trib. Lobelieae. *Dielsantha* E. Wimm. nov. gen. Ann. Naturhist. Mus. Wien 56: 372-373.
>   Ge., La. — Establishment of genus as distinct from *Lobelia*.
> • Wimmer, F. E. (1953). *Dielsantha* E. Wimm. In A. Engler & L. Diels, Das Pflanzenreich IV.
>   276b: 743-745, illus. Berlin: Akademie-Verlag. Ge., La. — Monograph with description,
>   full nomenclature, and specimen citations.
> Wimmer, F. E. (1963). *Dielsantha* E. Wimm. In J. Hutchinson & J. M. Dalziel, Flora of west
>   tropical Africa (ed. 2) 2: 315. London: Crown Agents for Overseas Governments. En. —
>   Flora with description, full nomenclature, and specimen citations.

Dielsantha E. Wimm., Ann. Naturhist. Mus. Wien 56: 372 (1948).
  Trop. W. Africa. 22 23.

Dielsantha galeopsoides (Engl. & Diels) E. Wimm., Ann. Naturhist. Mus. Wien 56: 373 (1948).
  Nigeria to Bioko & Congo. 22 NGA 23 CMN CON EQG GAB GGI. Cham. or hemicr.
    * *Lobelia galeopsoides* Engl. & Diels, Bot. Jahrb. Syst. 26: 118 (1898).
    Lobelia sylvicola Lejoly & Lisowski, Acta Bot. Gallica 147: 120 (2000).

Diosphaera

Although this genus has been recognized by many authors, others (e.g., Davis 1950, 1972;
Damboldt 1976) have considered it to be too ill-defined to justify segregation from
Campanula. However, results of a preliminary molecular phylogeny (Eddie et al. 2003, under
General) suggest that it may eventually prove desirable to recognize the genus; the sole
species sampled fell out near the base of the "Campanulaceae s. str. clade". Type [designated
here]: *Diosphaera jacquinii* (Sieber) Buser.

> Buser, R. (1894). Contributions a la connaissance des Campanulacées. I. *Trachelium* L.
>   revisum. Bull. Herb. Boissier 2: 501-532 + pl. XV-XIX, illus. Fr. — Establishment of
>   genus as distinct from *Phyteuma* and *Trachelium*.
> Davis, P. H. (1950). *Tracheliopsis antilibanotica* P. H. Davis. Hooker's Icon. Pl. 35: tab. 3498
>   + 5 pp., illus. En. — Doubts merit of recognizing genus.
> Davis, P. H. (1972). *Campanula myrtifolia*. Bot. Mag. 179: tab. 637 + 3 pp., illus. En. —
>   Update of Davis (1950), advocating inclusion of species of *Diosphaera* in *Campanula*.

- Damboldt, J. (1976). Materials for a flora of Turkey XXXII: Campanulaceae. Notes Roy. Bot. Gard. Edinburgh 35: 39-52, illus. En. — Genus included in synonymy of *Campanula* sect. *Tracheliopsis*.

Synonyms:
Diosphaera Buser === **Campanula** L.
Diosphaera asperuloides (Boiss. & Orph.) Buser === **Campanula asperuloides** (Boiss. & Orph.) Engl.
Diosphaera asperuloides var. *denudata* Buser === **Campanula asperuloides** (Boiss. & Orph.) Engl. subsp. **asperuloides**
Diosphaera asperuloides var. *vestita* Buser === **Campanula asperuloides** (Boiss. & Orph.) Engl. subsp. **asperuloides**
Diosphaera chalcidica Buser === **Campanula rumeliana** subsp. **chalcidica** (Buser) Greuter & Burdet
Diosphaera chalcidica var. *denudata* Buser === **Campanula rumeliana** subsp. **chalcidica** (Buser) Greuter & Burdet
Diosphaera dubia Friv. ex Buser === **Campanula rumeliana** (Hampe) Vatke subsp. **rumeliana**
Diosphaera hysterantha Rech. f. & Schiman-Czeika === **Campanula postii** (Boiss.) Engl.
Diosphaera jacquinii (Sieber) Buser === **Campanula jacquinii** (Sieber) A. DC.
Diosphaera jacquinii subf. *denudata* (Buser) Hayek === **Campanula rumeliana** subsp. **chalcidica** (Buser) Greuter & Burdet
Diosphaera jacquinii var. *chalcidica* (Buser) Hayek === **Campanula rumeliana** subsp. **chalcidica** (Buser) Greuter & Burdet
Diosphaera postii (Boiss.) Bornm. === **Campanula postii** (Boiss.) Engl.
Diosphaera rumeliana (Hampe) Bornm. === **Campanula rumeliana** (Hampe) Vatke
Diosphaera rumeliana var. *cinerascens* (Vandas) Hayek === **Campanula rumeliana** (Hampe) Vatke subsp. **rumeliana**
Diosphaera tubulosa (Boiss.) Bornm. === **Campanula buseri** Damboldt

Dobrowskya

This genus was distinguished largely on the basis of its linear revolute stigmatic lobes; this feature now defines the genus *Monopsis* (q.v.), and *Dobrowskya* is treated (Wimmer 1953) as one of three sections in that genus. Type [designated by Pfeiff., Nomencl. Bot. 1: 1117 (1874)]: *Lobelia unidentata* W. T. Aiton.

Presl, C. (1836). *Dobrowskya*. In Prodromus monographiae Lobeliacearum: 8-10. Prague: Theophilus Haase. La. — Establishment of genus as distinct from *Lobelia*.
Urban, I. (1881). Die Bestäubungseinrichtungen bei den Lobeliaceen nebst einer Monographie der afrikanischen Lobeliaceen-Gattung *Monopsis*. Jahrb. Königl. Bot. Gart. Berlin 1: 260-277, illus. Ge., La. — Reduced to sectional rank under *Monopsis* and its circumscription expanded to include *Parastranthus* (q.v.).
Wimmer, F. E. (1953). *Monopsis* Salisb. In A. Engler & L. Diels, Das Pflanzenreich IV. 276b: 698-713, 783-784, illus. Berlin: Akademie-Verlag. Ge., La. — Removal of *Parastranthus* from sect. *Dobrowskya* to form a third section.

Synonyms:
Dobrowskya C. Presl === **Monopsis** Salisb.
Dobrowskya anceps C. Presl === **Lobelia anceps** L. f.
Dobrowskya aspera (Spreng.) A. DC. === **Monopsis simplex** (L.) E. Wimm.
Dobrowskya dregeana C. Presl === **Monopsis scabra** (Thunb.) Urb.
Dobrowskya eckloniana C. Presl ex Eckl. & Zeyh. === **Monopsis simplex** (L.) E. Wimm.
Dobrowskya laevicaulis C. Presl === **Monopsis unidentata** subsp. **laevicaulis** (C. Presl) Phillipson
Dobrowskya massoniana C. Presl === **Monopsis unidentata** (W. T. Aiton) E. Wimm.

Dobrowskya polyphylla C. Presl === **Monopsis scabra** (Thunb.) Urb.
Dobrowskya scabra (Thunb.) A. DC. === **Monopsis scabra** (Thunb.) Urb.
Dobrowskya scabra var. *dregeana* (C. Presl) Sond. === **Monopsis scabra** (Thunb.) Urb.
Dobrowskya scabra var. *glabrata* Sond. === **Monopsis unidentata** subsp. **laevicaulis** (C. Presl)
   Phillipson
Dobrowskya serratifolia A. DC. in DC. === **Monopsis debilis** (L. f.) C. Presl
Dobrowskya stellarioides C. Presl === **Monopsis stellarioides** (C. Presl) Urb.
Dobrowskya stricta C. Presl ex Eckl. & Zeyh. === **Monopsis unidentata** (W. T. Aiton) E.
   Wimm. subsp. **unidentata**
Dobrowskya tenella (L.) Sond. === **Wahlenbergia parvifolia** (P. J. Bergius) Lammers
Dobrowskya thunbergiana C. Presl === **Monopsis scabra** (Thunb.) Urb.
Dobrowskya unidentata (W. T. Aiton) A. DC. === **Monopsis unidentata** (W. T. Aiton) E. Wimm.

Dominella

This genus was established to accomodate a species that resembles *Lysipomia* in all features
but its dorsally cleft corolla. Type [by monotypy]: *Dominella crassimarginata* E. Wimm.

   Wimmer, F. E. (1953). *Dominella* E. Wimm. In A. Engler & L. Diels, Das Pflanzenreich IV.
      276b: 754, illus. Berlin: Akademie-Verlag. Ge., La. — Establishment of genus.
   • Jeppesen, S. (1981). *Lysipomia* H. B. K. In G. Harling & B. Sparre (eds.), Flora of Ecuador
      14: 136-151, illus. Stockholm: Swedish Natural Science Research Council. En. — Type
      of *Dominella* transferred to *Lysipomia,* with comment that some species of latter also
      have cleft corolla tube.

Synonyms:
Dominella E. Wimm. === **Lysipomia** Kunth
Dominella crassimarginata E. Wimm. === **Lysipomia crassimarginata** (E. Wimm.)
   Jeppesen

Dortmanna

The revival of this early name and its application to the genus *Lobelia* was prompted by
Kuntze's (1891, 1898) conviction that, based on absolute priority, the type of the latter was
L. plumieri L., a species of Goodeniaceae now called *Scaevola plumieri* (L.) Vahl. With the
subsequent selection of *L. cardinalis* as the type of *Lobelia*, that interpretation was no longer
possible. Type [by monotypy]: *Lobelia dortmanna* L.

   Hill, J. (1756). Water gladiole. *Dortmanna.* In The British herbal: 126. London: T. Osborne
      & J. Shipton. En. — Establishment of genus in post-Linnaean times.
   Kuntze, O. (1891). *Dortmannia* [sic]. In Revisio generum plantarum vascularium omnium
      atque cellularium multarum secundum leges nomenclaturae internationales 2: 379-
      380, 971-973. Leipzig: Arthur Felix. Ge. — Substitution of name for *Lobelia,* and
      alteration of spelling to "Dortmannia".
   Kuntze, O. (1898). *Dortmannia* [sic]. In Revisio generum plantarum vascularium omnium
      atque cellularium multarum secundum leges nomenclaturae internationales 3(2): 186-
      188. Leipzig: Arthur Felix. Ge. — Addendum to Kuntze (1891).

Synonyms:
Dortmanna O. O. Rudbeck ex Hill === **Lobelia** L.
Dortmanna acuminata (Sw.) Kuntze === **Lobelia acuminata** Sw.
Dortmanna acuminata var. *pubescens* Kuntze === **Lobelia robusta** Graham
Dortmanna acutidens (Hook. f.) Kuntze === **Lobelia acutidens** Hook. f.
Dortmanna alpina (Vell.) Kuntze === **?**
Dortmanna alsinoides (Lam.) Kuntze === **Lobelia alsinoides** Lam.

Dortmanna amoenum (Michx.) Kuntze === **Lobelia amoena** Michx.

Dortmanna amygdalina (Willd. ex Schult.) Kuntze === **Lobelia laxiflora** Kunth subsp. **laxiflora**

Dortmanna anceps (L. f.) Kuntze === **Lobelia anceps** L. f.

Dortmanna appendiculata (A. DC.) Kuntze === **Lobelia appendiculata** A. DC.

Dortmanna aquatica (Cham.) Kuntze === **Lobelia aquatica** Cham.

Dortmanna arguta (Lindl.) Kuntze === **Lobelia excelsa** Bonpl.

Dortmanna assurgens (L.) Kuntze === **Lobelia assurgens** L.

Dortmanna bergiana (Cham.) Kuntze === **Grammatotheca bergiana** (Cham.) C. Presl

Dortmanna berlandieri (A. DC.) Kuntze === **Lobelia berlandieri** A. DC.

Dortmanna berterii (A. DC.) Kuntze === **Lobelia tupa** L.

Dortmanna besseriana (C. Presl) Kuntze === **Lobelia polyphylla** Hook. & Arn.

Dortmanna bicalcarata Kuntze === **Lobelia tupa** L.

Dortmanna blanda (D. Don) Kuntze === **Lobelia bridgesii** Hook. & Arn.

Dortmanna boivinii (Sond.) Kuntze === **Lobelia boivinii** Sond.

Dortmanna boykinii (Torr. & A. Gray ex A. DC.) Kuntze === **Lobelia boykinii** Torr. & A. Gray ex A. DC.

Dortmanna bracteosa (C. Presl) Kuntze === **Lobelia polyphylla** Hook. & Arn.

Dortmanna brevifolia (Nutt. ex A. DC.) Kuntze === **Lobelia brevifolia** Nutt. ex A. DC.

Dortmanna breynii (Lam.) Kuntze === **Monopsis lutea** (L.) Urb.

Dortmanna bridgesii (Hook. & Arn.) Kuntze === **Lobelia bridgesii** Hook. & Arn.

Dortmanna browniana (Schult.) Kuntze === **Lobelia gibbosa** Labill.

Dortmanna campanuloides (Thunb.) Kuntze === **Lobelia chinensis** Lour.

Dortmanna camporum (Pohl) Kuntze === **Lobelia camporum** Pohl

Dortmanna canbyi (A. Gray) Kuntze === **Lobelia canbyi** A. Gray

Dortmanna capillifolia (C. Presl) Kuntze === **Lobelia capillifolia** (C. Presl) A. DC.

Dortmanna cardinalis (L.) Kuntze === **Lobelia cardinalis** L.

Dortmanna caudata (Griseb.) Kuntze === **Lobelia caudata** (Griseb.) Urb.

Dortmanna chamaedryifolia (C. Presl) Kuntze === **Lobelia chamaedryifolia** (C. Presl) A. DC.

Dortmanna chinensis (Lour.) Kuntze === **Lobelia chinensis** Lour.

Dortmanna chireensis (A. Rich.) Kuntze === **Lobelia chireensis** A. Rich.

Dortmanna circaeoides (C. Presl) Kuntze === **Lobelia circaeoides** (C. Presl) A. DC.

Dortmanna cirsiifolia (Lam.) Kuntze === **Lobelia cirsiifolia** Lam.

Dortmanna cliffortiana (L.) Kuntze === **Lobelia cliffortiana** L.

Dortmanna cliffortiana var. *xalapensis* (Kunth) Kuntze === **Lobelia xalapensis** Kunth

Dortmanna collina (Kunth) Kuntze === **Lobelia collina** Kunth

Dortmanna colorata (Wall.) Kuntze === **Lobelia colorata** Wall.

Dortmanna columnaris (Hook. f.) Kuntze === **Lobelia columaris** Hook. f.

Dortmanna concolor Kuntze === **Lobelia laxiflora** Kunth subsp. **laxiflora**

Dortmanna conglobata (Lam.) Kuntze === **Lobelia conglobata** Lam.

Dortmanna cordata Kuntze === **Lobelia preslii** A. DC.

Dortmanna cordigera (Cav.) Kuntze === **Lobelia cardinalis** L.

Dortmanna cornopifolia (L.) Kuntze === **Lobelia coronopifolia** L.

Dortmanna coronopifolia f. *azurea* Kuntze === **Lobelia caerulea** Sims

Dortmanna coronopifolia f. *azurea* Kuntze === **Lobelia caerulea** Sims

Dortmanna coronopifolia f. *azurea* Kuntze === **Lobelia tomentosa** L. f.

Dortmanna coronopifolia f. *bicolor* Kuntze === **Lobelia caerulea** Sims

Dortmanna coronopifolia f. *rosea* Kuntze === **Lobelia caerulea** Sims

Dortmanna coronopifolia f. *rosea* Kuntze === **Lobelia caerulea** Sims

Dortmanna coronopifolia f. *rosea* Kuntze === **Lobelia coronopifolia** L.

Dortmanna coronopifolia var. *caerulea* (Sims) Kuntze === **Lobelia caerulea** Sims

Dortmanna coronopifolia var. *ceratophylla* (C. Presl) Kuntze === **Lobelia chamaepitys** Lam.

Dortmanna coronopifolia var. *glabrescens* (C. Presl) Kuntze === **Lobelia caerulea** Sims

Dortmanna coronopifolia var. *macularis* (C. Presl) Kuntze === **Lobelia caerulea** Sims

Dortmanna coronopifolia var. *multiflora* (A. DC.) Kuntze === **Lobelia tomentosa** L. f.

Dortmanna coronopifolia var. *normalis* Kuntze === **Lobelia coronopifolia** L.

Dortmanna coronopifolia var. *paucidentata* (C. Presl) Kuntze === **Lobelia tomentosa** L. f.

Dortmanna coronopifolia var. *tomentosa* (L. f.) Kuntze === **Lobelia tomentosa** L. f.
Dortmanna corymbosa (G. Don) Kuntze === **Lobelia jasionoides** (A. DC.) E. Wimm.
Dortmanna cymbalaria (Griseb.) Kuntze === **Lobelia nana** Kunth
Dortmanna cyphioides (Harv.) Kuntze === **Lobelia cyphioides** Harv.
Dortmanna debilis (L. f.) Kuntze === **Monopsis debilis** (L. f.) C. Presl
Dortmanna debilis var. *natalensis* Kuntze === **Lobelia erinus** L.
Dortmanna decipiens (Sond.) Kuntze === **Monopsis decipiens** (Sond.) Thulin
Dortmanna deckenii (Asch.) Kuntze === **Lobelia deckenii** (Asch.) Hemsl.
Dortmanna decurrens (Cav.) Kuntze === **Lobelia decurrens** Cav.
Dortmanna decurrentifolia Kuntze === **Lobelia decurrentifolia** (Kuntze) K. Schum.
Dortmanna dentata (Cav.) Kuntze === **Lobelia dentata** Cav.
Dortmanna depressa (L. f.) Kuntze === **Monopsis debilis** (L. f.) C. Presl
Dortmanna digitalifolia (Griseb.) Kuntze === **Lobelia digitalifolia** (Griseb.) Urb.
Dortmanna dioica (R. Br.) Kuntze === **Lobelia dioica** R. Br.
Dortmanna divaricata (Hook. & Arn.) Kuntze === **Lobelia divaricata** Hook. & Arn.
Dortmanna diversifolia (Willd. ex Schult.) Kuntze === ?
Dortmanna doniana Kuntze === **Siphocampylus pavonis** E. Wimm.
Dortmanna dregeana (C. Presl) Kuntze === **Lobelia dregeana** (C. Presl) A. DC.
Dortmanna eckloniana (C. Presl) Kuntze === **Lobelia filicaulis** (C. Presl) Schönl.
Dortmanna ehrenbergii (Vatke) Kuntze === **Lobelia ehrenbergii** Vatke
Dortmanna engelmanniana Kuntze === **Lobelia cardinalis** L.
Dortmanna ensifolia (A. DC.) Kuntze === **Lobelia salicina** Lam.
Dortmanna ensiformis (Vell.) Kuntze === ?
Dortmanna erecta Kuntze === **Lobelia erectiuscula** H. Hara
Dortmanna erinoides (L.) Kuntze === **Lobelia erinus** L.
Dortmanna erinus (L.) Kuntze === **Lobelia erinus** L.
Dortmanna erinus var. *bellidifolia* (L. f.) Kuntze === **Lobelia erinus** L.
Dortmanna exaltata (Pohl) Kuntze === **Lobelia exaltata** Pohl
Dortmanna excelsa (Bonpl.) Kuntze === **Lobelia excelsa** Bonpl.
Dortmanna fastigiata (Kunth) Kuntze === **Lobelia fastigiata** Kunth
Dortmanna feayana (A. Gray) Kuntze === **Lobelia feayana** A. Gray
Dortmanna fenestralis (Cav.) Kuntze === **Lobelia fenestralis** Cav.
Dortmanna fervens (Thunb.) Kuntze === **Lobelia fervens** Thunb.
Dortmanna filiformis (Lam.) Kuntze === **Lobelia filiformis** Lam.
Dortmanna fistulosa (Vell.) Kuntze === **Lobelia fistulosa** Vell.
Dortmanna flavescens (A. DC.) Kuntze === **Lobelia stricta** Sw.
Dortmanna flexuosa Kuntze === **Lobelia erinus** L.
Dortmanna fulgens (Humb. & Bonpl. ex Willd.) Kuntze === **Lobelia cardinalis** L.
Dortmanna gardneriana (Kanitz) Kuntze === **Lobelia fastigiata** Kunth
Dortmanna gattingeri (A. Gray) Kuntze === **Lobelia gattingeri** A. Gray
Dortmanna gaudichaudii (A. DC.) Kuntze === **Lobelia gaudichaudii** A. DC.
Dortmanna ghiesbreghtii Kuntze === ?
Dortmanna gibbosa (Labill.) Kuntze === **Lobelia gibbosa** Labill.
Dortmanna giberroa (Hemsl.) Kuntze === **Lobelia giberroa** Hemsl.
Dortmanna glandulosa (Walt.) Kuntze === **Lobelia glandulosa** Walt.
Dortmanna gracilis (C. Presl) Kuntze === **Lobelia andrewsii** Lammers
Dortmanna graminea (Lam.) Kuntze === **Lobelia cardinalis** L.
Dortmanna griffithii (Hook. f. & Thomson) Kuntze === **Lobelia griffithii** Hook. f. & Thomson
Dortmanna gruina (Cav.) Kuntze === **Lobelia gruina** Cav.
Dortmanna haenkeana (C. Presl) Kuntze === **Lobelia laxiflora** Kunth subsp. **laxiflora**
Dortmanna hartwegii (A. DC. ex Benth.) Kuntze === **Lobelia hartwegii** A. DC. ex Benth.
Dortmanna heterophylla (Labill.) Kuntze === **Lobelia heterophylla** Labill.
Dortmanna hilaireana (Kanitz) Kuntze === **Lobelia hilaireana** (Kanitz) E. Wimm.
Dortmanna hirsuta (L.) Kuntze === **Gnidia hirsuta** (L.) Thulin (Thymelaeaceae)
Dortmanna humifusa (A. DC.) Kuntze === **Unigenes humifusa** (A. DC.) E. Wimm.
Dortmanna hyssopifolia (C. Presl) Kuntze === **Lobelia polyphylla** Hook. & Arn.

Dortmanna imberbis (Griseb.) Kuntze === **Lobelia imberbis** (Griseb.) Urb.

Dortmanna inconspicua (A. Rich.) Kuntze === **Lobelia inconspicua** A. Rich.

Dortmanna infesta (Griseb.) Kuntze === **Lobelia cirsiifolia** Lam.

Dortmanna inflata (L.) Kuntze === **Lobelia inflata** L.

Dortmanna kalmii (L.) Kuntze === **Lobelia kalmii** L.

Dortmanna lacustris O. O. Rudbeck ex G. Don === **Lobelia dortmanna** L.

Dortmanna lasiantha (C. Presl) Kuntze === **Lobelia linearis** Thunb.

Dortmanna lavendulcea (Klotzsch) Kuntze === **Lobelia erinus** L.

Dortmanna laxiflora (Kunth) Kuntze === **Lobelia laxiflora** Kunth

Dortmanna leptostachys (A. DC.) Kuntze === **Lobelia spicata** Lam.

Dortmanna leschenaultiana (C. Presl) Kuntze === **Lobelia leschenaultiana** (C. Presl) Skottsb.

Dortmanna linarioides (C. Presl) Kuntze === **Lobelia linarioides** (C. Presl) A. DC.

Dortmanna linearis (Thunb.) Kuntze === **Lobelia linearis** Thunb.

Dortmanna longifolia (C. Presl) Kuntze === **Lobelia cardinalis** L.

Dortmanna loxensis (Willd. ex Schult.) Kuntze === **Siphocampylus loxensis** (Willd. ex Schult.) Vatke

Dortmanna lucaeanum (C. Presl) Kuntze === **Lobelia bridgesii** Hook. & Arn.

Dortmanna ludoviciana Kuntze === **Lobelia flaccidifolia** Small

Dortmanna lutea (L.) Kuntze === **Monopsis lutea** (L.) Urb.

Dortmanna lutea var. *euphrasioides* (Sond.) Kuntze === **Monopsis lutea** (L.) Urb.

Dortmanna macrostachys (Hook. & Arn.) Kuntze === **Trematolobelia macrostachys** (Hook. & Arn.) Zahlbr. ex Rock

Dortmanna madagascariensis (Schult.) Kuntze === **Lobelia fervens** Thunb. subsp. **fervens**

Dortmanna martagon (Griseb.) Kuntze === **Lobelia martagon** (Griseb.) Hitchc.

Dortmanna melleri (Hemsl.) Kuntze === **Lobelia trullifolia** Hemsl. subsp. **trullifolia**

Dortmanna membranacea (R. Br.) Kuntze === **Lobelia membranacea** R. Br.

Dortmanna mexicana Kuntze === **Lobelia flexuosa** (C. Presl) A. DC.

Dortmanna micrantha (Kunth) Kuntze === **Diastatea micrantha** (Kunth) McVaugh

Dortmanna microsperma (F. Muell.) Kuntze === **Lobelia gibbosa** Labill.

Dortmanna minutiflora (Kunze) Kuntze === **Diastatea micrantha** (Kunth) McVaugh

Dortmanna mishmica (C. B. Clarke) Kuntze === **Lobelia mishmica** C. B. Clarke

Dortmanna modesta (Wedd.) Kuntze === **Lobelia modesta** Wedd.

Dortmanna mollis (Graham) Kuntze === **Lobelia xalapensis** Kunth

Dortmanna montana (Fresen.) Kuntze === **Lobelia rhynchopetalum** Hemsl.

Dortmanna monticola (Kunth) Kuntze === **Lobelia xalapensis** Kunth

Dortmanna mucronata (Cav.) Kuntze === **Lobelia tupa** L.

Dortmanna muscoides (Cham.) Kuntze === **Lobelia muscoides** Cham.

Dortmanna nana (Kunth) Kuntze === **Lobelia nana** Kunth

Dortmanna neglecta Kuntze === **Lobelia longicaulis** Brandegee

Dortmanna neriifolia Kuntze === **Lobelia grayana** E. Wimm.

Dortmanna nicotianifolia (Roth ex Schult.) Kuntze === **Lobelia nicotianifolia** Roth ex Schult.

Dortmanna nummularioides (Cham.) Kuntze === **Lobelia nummularioides** Cham.

Dortmanna nuttallii (Schult.) Kuntze === **Lobelia nuttallii** Schult.

Dortmanna obovata (G. Don) Kuntze === **Siphocampylus obovatus** (G. Don) E. Wimm.

Dortmanna ocimoides (Kunze) Kuntze === **Lobelia xalapensis** Kunth

Dortmanna organensis (Gardner) Kuntze === **Lobelia organensis** Gardner

Dortmanna orizabae (M. Martens & Galeotti) Kuntze === **Lobelia gruina** Cav.

Dortmanna ovata (G. Don) Kuntze === **Siphocampylus ovatus** (G. Don) E. Wimm.

Dortmanna paludosa (Nutt.) G. Don === **Lobelia paludosa** Nutt.

Dortmanna parviflora (M. Martens & Galeotti) Kuntze === **Diastatea micrantha** (Kunth) McVaugh

Dortmanna patula (L. f.) Kuntze === **Lobelia patula** L. f.

Dortmanna pauciflora (Kunth) Kuntze === **Lobelia gruina** Cav.

Dortmanna persicifolia (Lam.) Kuntze === **Lobelia persicifolia** Lam.

Dortmanna philippiana Kuntze === **Lobelia tupa** L.

Dortmanna phyllostachya (Engelm.) Kuntze === **Lobelia cardinalis** L.

Dortmanna pinifolia (L.) Kuntze === **Lobelia pinifolia** L.
Dortmanna pinifolia var. *comosa* Kuntze === **Lobelia pinifolia** L.
Dortmanna pinifolia var. *laxa* Kuntze === **Lobelia pinifolia** L.
Dortmanna polyphylla (Hook. & Arn.) Kuntze === **Lobelia polyphylla** Hook. & Arn.
Dortmanna pratioides (Benth.) Kuntze === **Lobelia pratioides** Benth.
Dortmanna puberula (Michx.) Kuntze === **Lobelia puberula** Michx.
Dortmanna pubescens (Aiton) Kuntze === **Lobelia pubescens** Aiton
Dortmanna pulchella (Vatke) Kuntze === **Lobelia pulchella** Vatke
Dortmanna purpurascens (R. Br.) Kuntze === **Lobelia purpurascens** R. Br.
Dortmanna purpurea (G. Don) Kuntze === **Lobelia polyphylla** Hook. & Arn.
Dortmanna pusilla (C. Presl) Kuntze === **Lobelia divaricata** Hook. & Arn.
Dortmanna pyramidalis (Wall.) Kuntze === **Lobelia pyramidalis** Wall.
Dortmanna quadrangularis (R. Br.) Kuntze === **Lobelia quadrangularis** R. Br.
Dortmanna racemosa (C. Presl) Kuntze === **Lobelia cirsiifolia** Lam.
Dortmanna radicans (Thunb.) Kuntze === **Lobelia chinensis** Lour.
Dortmanna rapunculoides (Kunth) Kuntze === **Lobelia gruina** Cav.
Dortmanna reinwardtiana (C. Presl) Kuntze === **Lobelia heyneana** Schult.
Dortmanna reticulata (Willd. ex Schult.) Kuntze === **Siphocampylus reticulatus** (Willd. ex
    Schult.) Klotzsch & H. Karst. ex Vatke
Dortmanna retrorsa (Willd. ex Schult.) Kuntze === **Siphocampylus retrorsus** (Willd. ex
    Schult.) Vatke
Dortmanna rhombifolia (de Vriese) Kuntze === **Lobelia rhombifolia** de Vriese
Dortmanna rhytidosperma (Benth.) Kuntze === **Lobelia rhytidosperma** Benth.
Dortmanna robusta (Graham) Kuntze === **Lobelia robusta** Graham
Dortmanna rosea (Wall.) Kuntze === **Lobelia rosea** Wall.
Dortmanna rotundifolia (Juss. ex A. DC.) Kuntze === **Lobelia rotundifolia** Juss. ex A. DC.
Dortmanna roughii (Hook. f.) Kuntze === **Lobelia roughii** Hook. f.
Dortmanna rupestris (Kunth) Kuntze === **Lobelia tenera** Kunth
Dortmanna sartorii (Vatke) Kuntze === **Lobelia sartorii** Vatke
Dortmanna scabra (Thunb.) Kuntze === **Monopsis scabra** (Thunb.) Urb.
Dortmanna scabra var. *albiflora* Kuntze === **Monopsis unidentata** (W. T. Aiton) E. Wimm.
    subsp. **unidentata**
Dortmanna scabra var. *bicolor* Kuntze === **Monopsis scabra** (Thunb.) Urb.
Dortmanna scabra var. *violacea* Kuntze === **Monopsis unidentata** (W. T. Aiton) E. Wimm.
    subsp. **unidentata**
Dortmanna scaevolifolia (Roxb.) Kuntze === **Lobelia scaevolifolia** Roxb.
Dortmanna schimperi (Hochst. ex A. Rich.) Kuntze === **Lobelia schimperi** Hochst. ex
    A. Rich.
Dortmanna senegalensis (A. DC.) Kuntze === **Lobelia erinus** L.
Dortmanna sessilifolia (Lamb.) Kuntze === **Lobelia sessilifolia** Lamb.
Dortmanna setacea (Thunb.) Kuntze === **Lobelia setacea** Thunb.
Dortmanna siphilitica (L.) Kuntze === **Lobelia siphilitica** L.
Dortmanna sonchifolia (Sw.) Kuntze === **Siphocampylus sonchifolius** (Sw.) McVaugh
Dortmanna sonderiana Kuntze === **Lobelia sonderiana** (Kuntze) Lammers
Dortmanna spartioides (C. Presl) Kuntze === **Lobelia linearis** Thunb.
Dortmanna spicata (Lam.) Kuntze === **Lobelia spicata** Lam.
Dortmanna splendens (Humb. & Bonpl. ex Willd.) Kuntze === **Lobelia cardinalis** L.
Dortmanna stellarioides (C. Presl) Kuntze === **Monopsis stellarioides** (C. Presl) Urb.
Dortmanna stenophylla (Benth.) Kuntze === **Lobelia stenophylla** Benth.
Dortmanna stricta (Sw.) Kuntze === **Lobelia stricta** Sw.
Dortmanna subcuneata (Miq.) Kuntze === **Lobelia zeylanica** L.
Dortmanna subdentata (C. Presl) Kuntze === **Lobelia polyphylla** Hook. & Arn.
Dortmanna subnuda (Benth.) Kuntze === **Lobelia subnuda** Benth.
Dortmanna subpubera (Wedd.) Kuntze === **Lobelia subpubera** Wedd.
Dortmanna succulenta (Blume) Kuntze === **Lobelia zeylanica** L.
Dortmanna surrepens (Hook. f.) Kuntze === **Lobelia surrepens** Hook. f.

Dortmanna tenera (Kunth) Kuntze === **Lobelia tenera** Kunth
Dortmanna tenuior (R. Br.) Kuntze === **Lobelia tenuior** R. Br.
Dortmanna thapsoidea (Schott ex Pohl) Kuntze === **Lobelia thapsoidea** Schott ex Pohl
Dortmanna thermalis (Thunb.) Kuntze === **Lobelia thermalis** Thunb.
Dortmanna tomentosa (L. f.) Kuntze === **Lobelia tomentosa** L. f.
Dortmanna trialata (Buch.-Ham. ex D. Don) Kuntze === **Lobelia heyneana** Schult.
Dortmanna trigona (Roxb.) Kuntze === **Lobelia alsinoides** Lam.
Dortmanna trigona var. *affinis* (C. Presl) Kuntze === **Lobelia zeylanica** L.
Dortmanna trigona var. *intermedia* Kuntze === ?
Dortmanna trigona var. *microcarpa* (C. B. Clarke) Kuntze === **Lobelia microcarpa** C. B. Clarke
Dortmanna trigona var. *nummulariifolia* Kuntze === **Lobelia macraeana** E. Wimm.
Dortmanna trigona var. *terminalis* (C. B. Clarke) Kuntze === **Lobelia terminalis** C. B. Clarke
Dortmanna trigonocaulis (F. Muell.) Kuntze === **Lobelia trigonocaulis** F. Muell.
Dortmanna trinitensis (Kunth) Kuntze === **Lobelia fastigiata** Kunth
Dortmanna triquetrum (L.) Kuntze === **Lobelia comosa** L.
Dortmanna trullifolia (Hemsl.) Kuntze === **Lobelia trullifolia** Hemsl.
Dortmanna tupa (L.) Kuntze === **Lobelia tupa** L.
Dortmanna uranocoma (Cham.) Kuntze === **Lobelia fistulosa** Vell.
Dortmanna urens (L.) Kuntze === **Lobelia urens** L.
Dortmanna vagans (Balf. f.) Kuntze === **Lobelia vagans** Balf. f.
Dortmanna valdiviana (Phil.) Kuntze === **Legenere valdiviana** (Phil.) E. Wimm.
Dortmanna vanreenensis Kuntze === **Lobelia vanreenensis** (Kuntze) K. Schum.
Dortmanna warscewiczii (Vatke) Kuntze === **Lobelia irasuensis** Planch. & Oerst. subsp **irasuensis**
Dortmanna xalapensis (Kunth) Kuntze === **Lobelia xalapensis** Kunth
Dortmanna yuccoides (Hillebr.) Kuntze === **Lobelia yuccoides** Hillebr.
Dortmanna zeylanica (L.) Kuntze === **Lobelia zeylanica** L.

Downingia

Lobelioideae, 13 species, western North America, from British Columbia and Saskatchewan south to Baja California, with a single species also occurring in Chile and Argentina; California is the center of diversity, with all 13 native there. Phylogenetic analyses of molecular data (Schultheis 2001; E. Knox, pers. comm.) support the monophyly of the genus and show that it belongs to a clade comprising *Howellia*, *Legenere*, *Palmerella*, and *Porterella*. The treatment here follows Weiler (1962), as modified by Lammers (2006, under **Special**). 2*n* = 12, 16, 18, 20, 22, 24. Based on: *Clintonia* Douglas ex Lindl., non Raf.

Lindley, J. (1829). *Clintonia elegans*. Elegant clintonia. Edwards' Bot. Reg. 15: fol. 1241 + 2 pp., illus. En. — Establishment of genus under the illegitimate name *Clintonia* (q.v.).

Rafinesque, C. S. (1832). On 3 n. sp. of *Clintonia*. Atlantic J.: 120. En. — Validation of *Bolelia* (q.v.) to replace illegitimate *Clintonia*.

Lindley, J. (1836). *Clintonia pulchella*. Edwards' Bot. Reg. 22: fol. 1909 + 1 pg., illus. En. — Portrait of *D. pulchella*.

Torrey, J. (1857). Report on the botany of the expedition. In War Department, Explorations and surveys for a railroad route from the Mississippi River to the Pacific Ocean. Route near the thirty-fifth parallel, explored by Lieutenant A. W. Whipple, topographical engineers, in 1853 and 1854. Washington: Government Printing Office. En. — Validation of *Downingia* to replace illegitimate *Clintonia*; Torrey explicitly declined to take up *Wittea*, but apparently was unaware of *Bolelia*.

Hooker, J. D. (1876). *Laurentia carnosula* [sic]. Bot. Mag. 102: tab. 6257 + 2 pp., illus. En. — Portrait of *D. pulchella*, with description; misidentification noted in index to volume.

Greene, E. L. (1890). Some genera of Rafinesque. *Bolelia*. Pittonia 2: 124-127. En. — Brief synopsis of genus, arguing for priority of *Bolelia* over *Downingia*.

Greene, E. L. (1906). Doctor Torrey and *Downingia*. Torreya 6: 145-147. En. — Follow-up to Greene (1890), advocating use of *Bolelia* for genus.

Greene, E. L. (1910). The genus *Downingia*. Leafl. Bot. Observ. Crit. 2: 43-45. En. — Relents on use of *Bolelia* (Greene 1890, 1906) and advocates using *Downingia*.

Jepson, W. L. (1922). Revision of the California species of the genus *Downingia* Torr. Madroño 1: 98-102, illus. En. — Floristic treatment with key, descriptions, full nomenclature, and specimen citations.

Hoover, R. F. (1937). A provisional key to the species of *Downingia* known in California. Leafl. W. Bot. 2: 33-35. En. — Brief floristic summary with key.

Hoover, R. F. (1940). Observations on Californian plants – I. Leafl. W. Bot. 2: 273-278. En. — Realization that *D. pusilla* of Chile is conspecific with *D. humilis* of California.

• McVaugh, R. (1941). A monograph on the genus *Downingia*. Mem. Torrey Bot. Club 19(4): 1-57, illus., maps. En. — Monograph with keys, descriptions, full nomenclature, and specimen citations.

McVaugh, R. (1943). *Downingia* Torr. In North American Flora 32A: 15-25. New York: New York Botanical Garden. En. — Flora with key, descriptions, and full nomenclature.

• Wimmer, F. E. (1953). *Downingia* Torr. In A. Engler & L. Diels, Das Pflanzenreich IV. 276b: 733-743, illus. Berlin: Akademie-Verlag. Ge., La. — Monograph with key, descriptions, full nomenclature, and specimen citations.

• Wood, C. E., Jr. (1961). A study of hybridization in *Downingia* (Campanulaceae). J. Arnold Arbor. 42: 219-262, illus. En. — Study of species' relationships based on crossability data.

• Weiler, J. H. (1962). A biosystematic study of the genus *Downingia*. xxiv + 189 pp., illus., maps. Berkeley: University of California. — Unpublished Ph.D. dissertation with keys, descriptions, full nomenclature, and specimen citations.

Raven, P. H. (1963). Amphitropical relationships in the floras of North and South America. Quart. Rev. Biol. 38: 151-177, maps. En. — Discussion of the disjunct distribution of *D. pusilla*.

Kaplan, D. R. (1967). Floral morphology, organogenesis, and interpretation of the inferior ovary in *Downingia bacigalupii*. Amer. J. Bot. 54: 1274-1290, illus. En. — Anatomical study documenting appendicular origin of hypanthium.

Kaplan, D. R. (1968a). Structure and development of the perianth in *Downingia bacigalupii*. Amer. J. Bot. 55: 406-420, illus. En. — Organogenesis of the sympetalous corolla.

Kaplan, D. R. (1968b). Histogenesis of the androecium and gynoecium in *Downingia bacigalupii*. Amer. J. Bot. 55: 933-950, illus. En. — Organogenesis of the stamens and carpels.

Kaplan, D. R. (1969a). Sporogenesis and gametogenesis in *Downingia* (Campanulaceae; Lobelioideae). Bull. Torrey Bot. Club 96: 418-434, illus. En. — First part of a detailed embryological study

Kaplan, D. R. (1969b). Seed development in *Downingia*. Phytomorphology 19: 253-278, illus. En. — Conclusion of embryological study.

Munz, P. A. & D. D. Keck (1970). *Downingia* Torr. In A California flora: 1069-1072. Berkeley: University of California Press. En. — Floristic treatment with key, descriptions, and synonymy.

Foster, R. I. (1972). Explosive chromosome evolution in *Downingia yina*. viii + 85 pp., illus., maps. Davis: University of California. — Unpublished Ph.D. dissertation documents extensive aneuploid series correlated with geography.

Cronquist, A., A. H. Holmgren, N. H. Holmgren, J. L. Reveal & P. K. Holmgren (1984). *Downingia* Torr. nom. conserv. In Intermountain flora 4: 521-525, illus. New York: New York Botanical Garden. En. — Flora with key, descriptions, and full nomenclature.

Briggs, M. & R. Fitzgerald (1986). Aliens and adventives. *Downingia elegans* refound in East Sussex. Bot. Soc. Brit. Isles News 44: 20-21. En. — Naturalization in England.

Lenski, H. (1986). *Downingia elegans* (Dougl.) Torr.: Eine mit Grassaatgut eingeschleppte Adventivpflanze. Göttinger Florist. Rundbr. 19: 75-77, illus. Ge. — Introduction to Germany in grass seed.

Martin, B. D. & E. W. Lathrop (1986). Niche partitioning in *Downingia bella* and *D. cuspidata* (Campanulaceae) in the vernal pools of the Santa Rosa Plateau Preserve, California. Madroño 33: 284-299, illus. En. — Study of isolating mechanisms in a pair of closely allied species.

Lenski, H. (1987). Nachtrag zu *Downingia elegans* (Douglas) Torr. Florist. Rundbr. 21: 24. Ge. — Update on adventive populations reported by Lenski (1986).

Huxley, A. J., ed. (1992). *Downingia*. In The new Royal Horticultural Society dictionary of gardening 2: 93. London: MacMillan. En. — Account of species in cultivation.

• Ayers, T. (1993). *Downingia*. In J. C. Hickman (ed.), The Jepson manual, higher plants of California: 460-462, illus. Berkeley: University of California Press. En. — Flora with keys, descriptions, and synonymy.

• Schultheis, L. M. (2001). Systematics of *Downingia* (Campanulaceae) based on molecular sequence data: implications for floral and chromosome evolution. Syst. Bot. 26: 603-621. En. — Reveals two cytologically-correlated clades; indicates multiple origins for certain floral morphologies; and calls into question the circumscription of some species.

Downingia Torr., Pacific Railr. Rep. 4(5): 116 (1857), nom. cons.
W. North America & S. South America; naturalized in Europe. (10) (11) 71 73 76 79 85.
* *Clintonia* Douglas ex Lindl., Edwards' Bot. Reg. 15: pl. 1241 (1829); non Raf., Amer. Monthly Mag. Crit. Rev. 2: 266 (1818). *Bolelia* Raf., Atlantic J. 1: 120 (1832), nom. rejic. *Gynampsis* Raf., Herb. Raf.: 48 (1833). *Wittea* Kunth, Abh. Königl. Akad. Wiss. Berlin 1848: 32 (1850).

Downingia bacigalupii Weiler, Madroño 16: 256 (1962).
N. Oregon to SW. Idaho & N. California. 73 IDA ORE 76 CAL NEV. Ther. 2*n* = 24.

Downingia bella Hoover, Leafl. W. Bot. 2: 2 (1937). *Downingia concolor* var. *bella* (Hoover) E. Wimm., Pflanzenr. IV.276b: 741 (1953).
C. California. 76 CAL. Ther. 2*n* = 22.

Downingia bicornuta A. Gray, Syn. Fl. N. Amer. (ed. 2) 2(1): 395 (1886). *Bolelia bicornuta* (A. Gray) Greene, Pittonia 2: 127 (1890).
S. Oregon to SW. Idaho & C. California. 73 IDA ORE 76 CAL NEV. Ther. 2*n* = 22.
Downingia sikota Applegate, Contr. Dudley Herb. 1: 97 (1929).
Downingia bicornuta var. *picta* Hoover, Leafl. W. Bot. 2: 4 (1937).

Downingia concolor Greene, Bull. Calif. Acad. Sci. 2: 153 (1886). *Bolelia concolor* (Greene) Greene, Pittonia 2: 127 (1890).
C. & S. California. 76 CAL. Ther. 2*n* = 16, 18.

subsp. **concolor**
C. California. 76 CAL. Ther. 2*n* = 16.
Downingia tricolor Greene, Pittonia 2: 79 (1890). *Bolelia tricolor* (Greene) Greene, Pittonia 2: 127 (1890). *Bolelia concolor* var. *tricolor* (Greene) Jeps., Fl. W. Calif.: 481 (1901). *Downingia concolor* var. *tricolor* (Greene) Jeps., Fl. W. Calif. (ed. 2): 402 (1911).

subsp. **brevior** (McVaugh) R. M. Beauch., Phytologia 59: 438 (1986).
S. California (Cuyamaca Lake in San Diego Co.). 76 CAL. Ther. 2*n* = 18.
* *Downingia concolor* var. *brevior* McVaugh, Mem. Torrey Bot. Club 19(4): 20 (1941).

Downingia cuspidata (Greene) Rattan, Anal. Key West Coast Bot. (ed. 3): 48 (1898).
N. California to N. Baja California. 76 CAL 79 MXN. Ther. 2*n* = 22.
* *Bolelia cuspidata* Greene, Erythea 3: 101 (1895).
Downingia pulchella var. *arcana* Jeps., Madroño 1: 100 (1922).
Downingia immaculata Munz & I. M. Johnst., Bull. Torrey Bot. Club 51: 300 (1924).
Downingia pallida Hoover, Leafl. W. Bot. 2: 1 (1937).

Downingia elegans (Douglas ex Lindl.) Torr. in Wilkes, U. S. Expl. Exped. 17: 375 (1874).
S. British Columbia to N. California; naturalized in N. Europe. (10) grb (11) ger 71 BRC 73 IDA ORE WAS 76 CAL. Ther. 2*n* = 20.

Clintonia elegans Douglas ex Lindl., Edwards' Bot. Reg. 15: pl. 1241 (1829). *Gynampsis flexuosa* Raf., Herb. Raf.: 48 (1833). *Bolelia flexuosa* Raf., Herb. Raf.: 48 (1833). *Lobelia douglasii* A. W. Wood, Classbook Bot.: 478 (1861); non *Lobelia elegans* de Vriese in Lehm., Pl. Preiss.: 396 (1845). *Bolelia elegans* (Dougl. ex Lindl.) Greene, Pittonia 2: 126 (1890).
  Clintonia corymbosa A. DC. in DC., Prodr. 7: 347 (1839). *Downingia elegans* var. *corymbosa* (A. DC.) A. Gray, Proc. Amer. Acad. Arts 8: 393 (1873). *Downingia corymbosa* (A. DC.) A. Nelson & J. F. Macbr., Bot. Gaz. (Crawfordsville) 55: 382 (1913).
  Downingia brachypetala Gand., Bull. Soc. Bot. France 65: 55 (1918). *Downingia elegans* var. *brachypetala* (Gand.) McVaugh, Mem. Torrey Bot. Club 19(4): 55 (1941).
  Downingia elegans f. *rosea* H. St. John, Res. Stud. State Coll. Wash. 1: 105 (1929).

Downingia insignis Greene, Pittonia 2: 80 (1890). *Bolelia insignis* (Greene) Greene, Pittonia 2: 126 (1890).
  SE. Oregon to C. California. 73 ORE 76 CAL NEV. Ther. $2n = 22$.

Downingia laeta (Greene) Greene, Leafl. Bot. Observ. Crit. 2: 45 (1910).
  S. Alberta & S. Saskatchewan to SW. Wyoming, N. Utah & N. California. 71 ABT SAS 73 IDA MNT ORE WYO 76 CAL NEV UTA. Ther. $2n = 22$.
  Bolelia laeta Greene, Erythea 1: 238 (1893).
  Bolelia brachyantha Rydb., Mem. New York Bot. Gard. 1: 483 (1900). *Downingia brachyantha* (Rydb.) A. Nelson & J. F. Macbr., Bot. Gaz. (Crawfordsville) 55: 382 (1913).

Downingia montana Greene, Pittonia 2: 104 (1890). *Bolelia montana* (Greene) Greene, Pittonia 2: 127 (1890). *Downingia bicornuta* var. montana (Greene) Jeps., Madroño 1: 102 (1922).
  S. Oregon to C. California. 73 ORE 76 CAL. Ther. $2n = 22$.

Downingia ornatissima Greene, Pittonia 2: 80 (1890). *Bolelia ornatissima* (Greene) Greene, Pittonia 2: 127 (1890).
  N. & C. California. 76 CAL. Ther. $2n = 24$.
  Downingia mirabilis J. T. Howell, Leafl. W. Bot. 1: 221 (1936).
  Downingia mirabilis var. *eximia* Hoover, Leafl. W. Bot. 2: 6 (1937). *Downingia ornatissima* var. *eximia* (Hoover) McVaugh, Mem. Torrey Bot. Club 19(4): 24 (1941).

Downingia pulchella (Lindl.) Torr., Rep. Explor. RR Pacif. Ocean 4(5): 116 (1857).
  C. California. 76 CAL. Ther. $2n = 22$.
  Clintonia pulchella Lindl., Edwards' Bot. Reg. 22: pl. 1909 (1836). *Bolelia pulchella* (Lindl.) Greene, Pittonia 2: 126 (1890).
  Clintonia pulchella var. *pumila* Benth., Pl. Hartweg.: 321 (1849).
  Bolelia pulchella f. *alba* Voss in Siebert & Voss, Vilm. Blumengärtn. (ed. 3) 1: 577 (1894).
  Bolelia pulchella f. *atrocinerea* Voss in Siebert & Voss, Vilm. Blumengärtn. (ed. 3) 1: 577 (1894).
  Bolelia pulchella f. *atropurpurea* Voss in Siebert & Voss, Vilm. Blumengärtn. (ed. 3) 1: 577 (1894).

Downingia pusilla (G. Don ex A. DC.) Torr. in Wilkes, U. S. Expl. Exped. 17: 375 (1874).
  C. California; C. Chile to S. Argentina. 76 CAL 85 AGS CLC CLS. Ther. $2n = 22$.
  Clintonia pusilla G. Don ex A. DC. in DC., Prodr. 7: 347 (1839). *Bolelia pusilla* (G. Don ex A. DC.) Greene, Pittonia 2: 126 (1890).

subsp. **humilis** (Greene) Lammers, Novon 16: 70 (2006).
  C. California; NC. Chile (Valparaíso, Colchagua, Ñuble, Concepción). 76 CAL 85 CLC. Ther. $2n = 22$.
  Bolelia humilis Greene, Pittonia 2: 226 (1892). *Downingia humilis* (Greene) Rattan, Anal. Key West Coast Bot. (ed. 3): 47 (1898).

subsp. **pusilla**
>SC. Chile (Bío Bío, Valdivia) to S. Argentina (Neuquen, Chubut, Río Negro, Santa Cruz). 85 AGS CLC CLS. Ther. *2n* = 22.

Downingia yina Applegate, Contr. Dudley Herb. 1: 97 (1929).
>Washington to N. California. 73 ORE WAS 76 CAL. Ther. *2n* = 12, 16, 20, 24.
>>*Downingia willamettensis* M. Peck, Proc. Biol. Soc. Wash. 47: 187 (1934). *Downingia yina* var. *major* McVaugh, N. Amer. Fl. 32A: 24 (1943).
>>*Downingia pulcherrima* M. Peck, Proc. Biol. Soc. Wash. 50: 94 (1937).

Synonyms:

Downingia bicornuta var. *montana* (Greene) Jeps. === **Downingia montana** Greene

Downingia bicornuta var. *picta* Hoover === **Downingia bicornuta** A. Gray

Downingia brachyantha (Rydb.) A. Nelson & J. F. Macbr. === **Downingia laeta** (Greene) Greene

Downingia brachypetala Gand. === **Downingia elegans** (Dougl. ex Lindl.) Torr.

Downingia concolor var. *bella* (Hoover) E. Wimm. === **Downingia bella** Hoover

Downingia concolor var. *brevior* McVaugh === **Downingia concolor** subsp. **brevior** (McVaugh) R. M. Beauch.

Downingia concolor var. *tricolor* (Greene) Jeps. === **Downingia concolor** Greene subsp. **concolor**

Downingia corymbosa (A. DC.) A. Nelson & J. F. Macbr. === **Downingia elegans** (Dougl. ex Lindl.) Torr.

Downingia elegans f. *rosea* H. St. John === **Downingia elegans** (Dougl. ex Lindl.) Torr.

Downingia elegans var. *brachypetala* (Gand.) McVaugh === **Downingia elegans** (Dougl. ex Lindl.) Torr.

Downingia elegans var. *corymbosa* (A. DC.) A. Gray === **Downingia elegans** (Dougl. ex Lindl.) Torr.

Downingia humilis (Greene) Rattan === **Downingia pusilla** (A. DC.) Torr.

Downingia immaculata Munz & I. M. Johnst. === **Downingia cuspidata** (Greene) Rattan

Downingia mirabilis J. T. Howell === **Downingia ornatissima** Greene

Downingia mirabilis var. *eximia* Hoover === **Downingia ornatissima** Greene

Downingia ornatissima var. *eximia* (Hoover) McVaugh === **Downingia ornatissima** Greene

Downingia pallida Hoover === **Downingia cuspidata** (Greene) Rattan

Downingia pulchella var. *arcana* Jeps. === **Downingia cuspidata** (Greene) Rattan

Downingia pulcherrima M. Peck === **Downingia yina** Applegate

Downingia sikota Applegate === **Downingia bicornuta** A. Gray

Downingia tricolor Greene === **Downingia concolor** Greene subsp. **concolor**

Downingia willamettensis M. Peck === **Downingia yina** Applegate

Downingia yina var. *major* McVaugh === **Downingia yina** Applegate

Dysmicodon

Nuttall (1842) adopted this name for *Triodanis* because its basionym at infrageneric rank had several months priority over Rafinesque's name (which he misspelled "Triodallus"). Under the ICBN, however, priority only operates within a given rank (Art. 11.2). Based on: *Specularia* [unranked] *Dysmicodon* Endl. Type [designated by Peng & Lammers, Bot. Bull. Acad. Sin. 39: 213 (1998)]: *Campanula flagellaris* Kunth.

>Nuttall, T. (1842). Description and notices of new or rare plants in the natural orders Lobeliaceae, Campanulaceae, Vaccinieae, Ericaceae, collected in a journey over the continent of North America, and during a visit to the Sandwich Islands, and Upper California. Trans. Amer. Philos. Soc. (n.s.) 8: 251-272. En. — Establishment of genus as distinct from *Legousia* (under the illegitimate name *Specularia*).
>
>Gray, A. (1876). Miscellaneous botanical contributions, *Specularia* Heister. Proc. Amer. Acad. Arts 11: 81-82. En. — Reduction of genus to section of European *Legousia* (under the illegitimate name *Specularia*).

Synonyms:
Dysmicodon (Endl.) Nutt. === **Triodanis** Raf.
Dysmicodon californicus Nutt. === **Triodanis perfoliata** subsp. **biflora** (Ruiz & Pav.) Lammers
Dysmicodon ovatus Nutt. === **Triodanis perfoliata** subsp. **biflora** (Ruiz & Pav.) Lammers
Dysmicodon perfoliatus (L.) Nutt. === **Triodanis perfoliata** (L.) Nieuwl.

Echinocodon

Campanuloideae, 1 species, central China. Hong (1984) regarded it as closely allied to *Codonopsis* and *Platycodon* on the basis of its colpate pollen and chromosome number. $2n =$ 16. Type [designated by the author]: *Echinocodon lobophyllus* D. Y. Hong.

> Hong, D. Y. (1984). *Echinocodon* Hong, a new genus of Campanulaceae and its systematic position. Acta Phytotax. Sin. 22: 183-184, illus. En. — Establishment of genus.
> • Ying, T. S., Y. L. Zhang & D. E. Boufford (1993). *Echinocodon* Hong. In The endemic genera of seed plants in China: 297-299, illus., map. Beijing: Science Press. En. — Descriptive account, including details of pollen morphology.

Echinocodon D. Y. Hong, Acta Phytotax. Sin. 22: 183 (1984), as 'Echicocodon'.
  E. Asia. 36.

Echinocodon lobophyllus D. Y. Hong, Acta Phytotax. Sin. 22: 183 (1984).
  C. China (NW. Hubei). 36 CHC. Hemicr. $2n = 16$.

Echinocodon

Kolakovskii (1986) segregated this species from *Campanula* on the basis of its supposedly different mode of capsular dehiscence. Unfortunately, he was unaware of the prior use of the same name for a different species of Campanulaceae. Type [designated by the author]: *Echinocodon primulifolius* (Brot.) Kolak.

> Kolakovskii, A. A. (1986). [*Echinocodon:* a new Mediterranean genus of the bluebell family.] Soobšč. Akad. Nauk Gruzinsk. SSR 121: 385-388, illus. Ru. — Establishment of genus as distinct from *Campanula*.

Synonyms:
Echinocodon Kolak. === **Campanula** L.
Echinocodon primulifolius (Brot.) Kolak. === **Campanula alata** Desf.

Echinocodonia

This name was created as an avowed substitute for illegitimate *Echinocodon* Kolak.; note that the requisite combination for its type species has never been validated. In a preliminary molecular phylogeny (Eddie et al. 2003, under **General**), its type was the sister-species of *Campanula peregrina*, within a small clade that otherwise comprised the types of *Gadellia* and *Musschia,* and which was regarded as "transitional" between the "Campanulaceae s. str. clade" and the "Rapunculus clade". Based on: *Ecinocodon* Kolak.

> Kolakovskii, A. A. (1994). [The conspectus of the system of the Old World Campanulaceae.] Bot. Žurn. 79(1): 109-124. Ru. — Validation of name as substitute for illegitimate *Echinocodon* Kolak., non D. Y. Hong.

Synonyms:
Echinocodonia Kolak. === **Campanula** L.

Edraianthus

Campanuloideae, 13 species, the Balkans, with one extending west to Italy and Sicily. The genus is divided (Mayer & Blečić 1969) into sect. *Edraianthus,* sect. *Spathulati* Janch., and sect. *Uniflori* Wettst. Some authors (e.g., Hooker 1875, 1880) have opined that *Edraianthus* is related to (or even congeneric with) the austral genus *Wahlenbergia.* However, recent analyses of DNA data (Eddie et al. 2003, under **General**) support a close relationship to *Campanula;* the type of *Edraianthus* and a member of sect. *Uniflori* formed a clade with the type of *Campanula,* within the "Campanulaceae s. str. clade" and at considerable distance from the "Wahlenbergioid" exemplars. The name has been rendered as "Hedraeanthus" by some authors (e.g., von Wettstein 1887, Beck 1893), but the spelling used here is now conserved. The last monograph was that of von Wettstein (1887). $2n = 32$. Type [designated by Lammers, Taxon 45: 563 (1996)]: *Edraianthus graminifolius* (L.) A. DC.

Meisner, C. F. (1839). *Edraianthus.* In Plantarum vascularium genera secundum ordines naturales digesta 2: 149. Leipzig: Weidmann. La. — Establishment of genus as distinct from *Campanula* and *Wahlenbergia.*

Hooker, J. D. (1875). *Wahlenbergia kitaibelii.* Bot. Mag. 101: tab. 6188 + 2 pp., illus. En. — Portrait of *E. graminifolius* subsp. *graminifolius,* with comments on advisability of subsuming *Edraianthus* into *Wahlenbergia.*

Hooker, J. D. (1880). *Wahlenbergia tenuifolia.* Bot. Mag. 106: tab. 6482 + 2 pp., illus. En. — Portrait of *E. tenuifolius,* with additional comments on merger of *Edraianthus* with *Wahlenbergia.*

• von Wettstein, R. (1887). Monographie der Gattung *Hedraeanthus.* Denkschr. Kaiserl. Akad. Wiss., Math.-Naturwiss. Kl. 53: 185-212, illus., map. Ge. — Monograph with key, descriptions, infrageneric classification, full nomenclature, and specimen citations.

• Beck, G. (1893). Die Gattung *Hedraeanthus.* Wiener Ill. Gart.-Zeitung 18: 287-299, illus. Ge. — Synopsis of genus with brief descriptions, revised infrageneric classification, and full nomenclature.

• Janchen, E. (1910). Die *Edraianthus*-Arten der Balkanländer. Mitt. Naturwiss. Vereines Univ. Wien (n.s.) 8: 1-40 + Taf. I-IV, illus. Ge. — Revision of Wettstein (1887), with key, descriptions, revised infrageneric classification, and full nomenclature.

Stefanoff, B. (1936). [Über die systematische Stellung einiger Arten der Familie Campanulaceae.] God. Sofiisk. Univ. Agron.-Lesoved. Fak. 14: 93-104. Bu. — Argues for treating *Halacsyella* (q.v.) as a section of *Edraianthus.*

Ingwersen, W. E. T. (1944). *Edraianthus stenocalyx.* Gard. Chron. (ser. 3) 115: 106-107, illus. En. — Descriptive account from horticultutral perspective.

Bailey, L. H. (1953). *Edraianthus.* In The garden of bellflowers in North America: 139-143, illus. New York: MacMillan. En. — Account of species cultivated in North America, with key and descriptions.

Mayer, E., and V. Blečić (1969). Zur Taxonomie und Chorologie von *Edraianthis* sectio *Uniflori.* Phyton (Horn) 18: 241-247, map. Ge. — Synopsis of section.

Maloş, C. (1972). Considérations sur l'espèce *Edraianthus kitaibelii* (L.) DC. Rev. Roumaine Biol., Sér. Bot. 17: 63-68, illus. Fr. — Argues for recognition of *E. kitaibelii.*

• Lakušić, R. (1973). Prirodni sistem populacija I vrsta roda *Edraianthus* DC. God. Biol. Inst. u Sarajevu 26 (Suppl.): 1-130, illus. Se. — Floristic treatment with key, descriptions, revised infrageneric classification, full nomenclature, and specimen citations.

• Kuzmanov, B. (1976). *Edraianthus* A. DC. In T. G. Tutin, V. H. Heywood, N., A. Burges, D. M. Moore, D. H. Valentine, S. M. Walters & D. A. Webb (eds.), Flora Europaea 4: 99-100. Cambridge: Cambridge University Press. En. — Floristic treatment with key and descriptions.

Mededović, S. (1980). [Some characteristics of the chromosome compliments, pollen, and seed coat in *Edraianthus dalmaticus* DC. and *Edraianthus tenuifolius* (W. K.) DC.] God. Biol. Inst. u Sarajevu 33: 113-128, illus. Se. — Cytological and micromorphological data.

Kolakovskii, A. A. (1982). [The biological 'mechanism' ensuring dissemination in *Edraianthus.*] Soobšč. Akad. Nauk Gruzinsk. SSR 105: 361-364, illus. Ru. — Anatomy of capsular dehiscence and seed dispersal.

Lakušić, R. (1988). *Protoedraianthus* Lakušić status nov. (syn.: *Edraianthus* DC. subgen. *Protoedraianthus* Lakušić subgen. nov.). Zbornik Referata Naučnog skupa 'Minerali, Stijene, Izumrli' i živi svijet Bosne i Hercegovine: 263-272, illus. Se. — Establishment of segregate genus and its classification.

Huxley, A. J., ed. (1992). *Edraianthus*. The new Royal Horticultural Society dictionary of gardening 2: 143. London: MacMillan. En. — Account of species in cultivation.

Kress, C. H. (1994). Buschelglocken des Balkans. Gartenpraxis 20(3): 8-11, illus. Ge. — Descriptive account from gardening perspective.

Lammers, T. G. (1996). Proposal to conserve the name *Edraianthus* against *Pilorea* (Campanulaceae). Taxon 45: 563-564. — Formal proposal to permit continued use of name and to stabilize spelling.

Brummitt, R. K. (1999). Report of the Committee for Spermatophyta: 48. Taxon 48: 359-371. En. — Committee recommends conservation.

Apostolova, I. & A. Ganeva (2000). New data on *Edraianthus serbicus* (Kern.) Petrovic in Bulgaria. Phytologia Balcan. 6: 65-71, maps. En. — Low competitiveness and poor seed germination blamed for species' rarity.

Edraianthus A. DC. in Meisn., Pl. Vasc. Gen. 2: 149 (1839), nom. et orth. cons.
SE. Europe. 13.
* *Wahlenbergia* sect. *Edraiantha* A. DC., Monogr. Campan.: 129 (1830). *Campanopsis* sect. *Edraiantha* (A. DC.) Kuntze in T. Post & Kuntze, Lex. Gen. Phan.: 95 (1904), as 'Hedraeanthus'.
Pilorea Raf., Fl. Tellur. 2: 80 (1837), nom. rejic.
Halacsyella Janch., Mitt. Naturwiss. Vereins Wiss. Univ. Wien (ser. 2) 8: 39 (1910). *Edraianthus* sect. *Halacsyella* (Janch.) Stef., God. Sofiisk. Univ. Agron.-Lesoved. Fak. 14: 104 (1936).
Protoedraianthus Lakušić in Zemaljski Muzej Bosne i Hercegovine, Zbornik Referata Naučnog: 264 (1988).

Edraianthus dalmaticus (A. DC.) A. DC. in Meisn., Pl. Vasc. Gen. 2: 149 (1839).
Croatia. 13 YUG. Hemicr. or geophyte.
* *Wahlenbergia dalmatica* A. DC., Monogr. Campan.: 134 (1830). *Campanula dalmatica* (A. DC.) D. Dietr., Syn. Pl. 1: 752 (1839); non Tausch, Flora 10: 246 (1827). *Campanopsis dalmatica* (A. DC.) Kuntze, Revis. Gen. Pl. 2: 379 (1891).
Campanula caudata Vis., Fl. Dalmat. 2: 136 (1847). *Edraianthus caudatus* (Vis.) Rchb., Icon. Fl. Germ. Helv. 19: 109 (1859). *Wahlenbergia caudata* (Vis.) Vatke, Linnaea 38: 702 (1874).

Edraianthus dinaricus (A. Kern.) Wettst., Denkschr. Kaiserl. Akad. Wiss., Math.-Naturwiss. Kl. 53: 192 (1887).
Croatia. 13 YUG. Hemicr. or geophyte.
* *Campanula pumilio* var. *major* Vis., Fl. Dalmat., Suppl.: 106 (1872). *Campanula dinarica* A. Kern., Ber. Naturwiss.-Med. Vereins Innsbruck 3: lxxi (1872). *Edraianthus serpyllifolius* subsp. *dinaricus* (A. Kern.) Nyman, Consp. Fl. Eur., Suppl. 2: 212 (1889).

Edraianthus glisicii Černjavski & Soška, Bull. Inst. Jard. Bot. Univ. Belgrade 4: 88 (1937), as 'glišićii'.
Montenegro. 13 YUG. Hemicr. or geophyte.
Protoedraianthus majae Lakušić in Zemaljski Muzej Bosne i Hercegovine, Zbornik Referata Naučnog: 264 (1988).

Edraianthus graminifolius (L.) A. DC. in Meisn., Pl. Vasc. Gen. 2: 149 (1839).
Sicilia, Italy & Balkans. 13 ALB GRC ITA ROM SIC YUG. Hemicr. or geophyte. $2n = 32$.
* *Campanula graminifolia* L., Sp. Pl.: 166 (1753). *Wahlenbergia graminifolia* (L.) A. DC., Monogr. Campan.: 130 (1830). *Pilorea graminifolia* (L.) Raf., Fl. Tellur. 2: 80 (1837), as 'graminif'. *Campanula graminifolia* var. *setifolia* Vuk., Linnaea 26: 327 (1854). *Campanopsis graminifolia* (A. DC.) Kuntze, Revis. Gen. Pl. 2: 379 (1891).

subsp. **graminifolius**

Italy & Balkans. 13 ALB GRC ITA ROM YUG. Hemicr. or geophyte. $2n = 32$.

Wahlenbergia kitaibelii A. DC., Monogr. Campan.: 131 (1830). *Edraianthus kitaibelii* (A. DC.) A. DC. in Meisn., Pl. Vasc. Gen. 2: 149 (1839). *Campanopsis kitaibelii* (A. DC.) Kuntze, Revis. Gen. Pl. 2: 379 (1891).

Edraianthus croaticus A. Kern., Österr. Bot. Z. 22: 390 (1872). *Wahlenbergia croaticus* (A. Kern.) Vatke, Linnaea 38: 702 (1874). *Edraianthus kitaibelii* subsp. *croaticus* (A. Kern.) Nyman, Consp. Fl. Eur.: 486 (1879).

Edraianthus kitaibelii var. *alpinus* Wettst., Denkschr. Kaiserl. Akad. Wiss., Math.-Naturwiss. Kl. 53: 196 (1887). *Edraianthus graminifolius* f. *alpinus* (Wettst.) Janch., Mitt. Naturwiss. Vereines Univ. Wien (ser. 2) 8: 27 (1910).

Edraianthus kitaibelii var. *mediterraneus* Wettst., Denkschr. Kaiserl. Akad. Wiss., Math.-Naturwiss. Kl. 53: 196 (1887). *Edraianthus graminifolius* f. *mediterraneus* (Wettst.) Hayek, Repert. Spec. Nov. Regni Veg. Beih. 30(2): 562 (1930).

Edraianthus kitaibelii var. *subalpinus* Wettst., Denkschr. Kaiserl. Akad. Wiss.,Math.-Naturwiss. Kl. 53: 196 (1887). *Edraianthus graminifolius* f. *subalpinus* (Wettst.) Janch., Mitt. Naturwiss. Vereines Univ. Wien (ser. 2) 8: 27 (1910).

Edraianthus graminifolius var. *elatus* Wettst., Denkschr. Kaiserl. Akad. Wiss., Math.-Naturwiss. Kl. 53: 201 (1887).

Edraianthus graminifolius var. *pusilla* Wettst., Denkschr. Kaiserl. Akad. Wiss., Math.-Naturwiss. Kl. 53: 201 (1887).

Edraianthus graminifolius var. *australis* Wettst., Denkschr. Kaiserl. Akad. Wiss., Math.-Naturwiss. Kl. 53: 201 (1887). *Edraianthus graminifolius* f. *australis* (Wettst.) Janch., Mitt. Naturwiss. Vereines Univ. Wien (ser. 2) 8: 29 (1910).

Edraianthus montenegrinus Horák, Österr. Bot. Z. 50: 163 (1900). *Edraianthus graminifolius* f. *montenegrinus* (Horák) Hayek, Repert. Spec. Nov. Regni Veg. Beih. 30(2): 562 (1930).

Edraianthus graminifolius f. *ginzbergeri* H. Lindb., Finska Vetensk.-Soc. Förh. 48(13): 105 (1906). *Edraianthus graminifolius* var. *ginzbergeri* (H. Lindb.) Hayek, Repert. Spec. Nov. Regni Veg. Beih. 30(2): 562 (1930).

Edraianthus graminifolius f. *baldaccii* Janch., Mitt. Naturwiss. Vereines Univ. Wien (ser. 2) 8: 28 (1910).

Edraianthus graminifolius subsp. *albanicus* Degen & Kümm., Bot. Közlem. 19: 28 (1920). *Edraianthus graminifolius* subvar. *albanicus* (Degen & Kümm.) Hayek, Repert. Spec. Nov. Regni Veg. Beih. 30(2): 562 (1930).

Edraianthus horvatii Lakusić, God. Biol. Inst. u Sarajevu 26 (Suppl.): 44 (1974).

Edraianthus vesovicii Lakusić, God. Biol. Inst. u Sarajevu 26 (Suppl.): 79 (1974), as 'vešovićii'.

Edraianthus zogovicii Lakusić, God. Biol. Inst. u Sarajevu 26 (Suppl.): 81 (1974), as 'zogovićii'.

subsp. **siculus** (Strobl) Lakusić ex Greuter & Burdet, Willdenowia 13: 280 (1984). Hemicr. or geophyte.

Sicilia. 13 SIC.

* *Edraianthus siculus* Strobl, Flora 66: 551 (1883). *Edraianthus graminifolius* var. *siculus* (Strobl) Nyman, Consp. Fl. Eur., Suppl. 2: 212 (1889).

Edraianthus hercegovinus K. Malý, Glasn. Zemaljsk. Muz. Bosni Hercegovini 18: 277 (1906). *Wahlenbergia hercegovina* (K. Malý) K. Malý, Wiss. Mitt. Bosnien & Herzegovina 10: 674 (1907). *Edraianthus tenuifolius* subsp. *hercegovinus* (K. Malý) Hayek, Repert. Spec. Nov. Regni Veg. Beih. 30(2): 561 (1930).

Bosnia-Hercegovina. 13 YUG. Hemicr. or geophyte.

Edraianthus × intermedius Degen, Fl. Veleb. 3: 110 (1938). E. graminifolius × E. tenuifolius

Croatia. 13 YUG. Hemicr. or geophyte.

Edraianthus × linifolius Gusmus, Möller's Deutsche Gärtn.-Zeitung 19: 152 (1904).
E. pumilio × E. serpyllifolius
Croatia. 13 YUG. Hemicr. or geophyte.

Edraianthus × murbeckii Wettst., Acta Univ. Lund. 27: 93 (1891). E. graminifolius ×
E. serpyllifolius
Bosnia-Hercegovina. 13 YUG. Hemicr. or geophyte.

Edraianthus niveus Beck, Wiener Ill. Gart.-Zeitung 18: 296 (1893). *Edraianthus graminifolius*
subsp. *niveus* (Beck) Janch., Mitt. Naturwiss. Vereines Univ. Wien (n.s.) 8: 29 (1910).
Bosnia-Hercegovina. 13 YUG. Hemicr. or geophyte.

Edraianthus parnassicus (Boiss. & Spruner) Halácsy, Denkschr. Kaiserl. Akad. Wiss., Math.-
Naturwiss. Kl. 61: 247 (1894).
Greece. 13 GRC. Hemicr. or geophyte. $2n = 32$.
* *Campanula parnassica* Boiss. & Spruner in Boiss., Diagn. Pl. Orient. (ser. 1) 7: 17 (1846).
Halacsyella parnassica (Boiss. & Spruner) Janch., Mitt. Naturwiss. Vereins Wiss. Univ.
Wien (ser. 2) 8: 39 (1910).

Edraianthus pumilio (Port. ex Schult.) A. DC. in Meisn., Pl. Vasc. Gen. 2: 149 (1839).
Croatia. 13 YUG. Hemicr. or geophyte.
* *Campanula pumilio* Port. ex Schult. in Roem. & Schult., Syst. Veg. 5: 136 (1819).
Campanula silenifolia Host, Fl. Austriac. 1: 268 (1827). *Wahlenbergia pumilio* (Port. ex
Schult.) A. DC., Monogr. Campan.: 134 (1830). *Campanula graminifolia* var.
linearifolia Vuk., Linnaea 26: 327 (1854). *Campanopsis pumilio* (Port. ex Schult.)
Kuntze, Revis. Gen. Pl. 2: 379 (1891).

Edraianthus serbicus Petroviã, Fl. Agr. Nyss.: 549 (1882) .
E. Serbia & W. Bulgaria. 13 BUL YUG. Hemicr. or geophyte. $2n = 32$.
Edraianthus serbicus subsp. *stankovicii* Lakusić, God. Biol. Inst. u Sarajevu 26 (Suppl.): 27
(1974), as 'stankovićii'.

Edraianthus serpyllifolius (Vis.) A. DC. in Meisn., Pl. Vasc. Gen. 2: 149 (1839).
Croatia to Bosnia-Hercegovina & Albania. 13 ALB YUG. Hemicr. or geophyte.
* *Campanula serpyllifolia* Vis., Flora 12 (Ergänzungsbl. 1): 6 (1829). *Wahlenbergia
serpyllifolia* (Vis.) Vatke, Linnaea 38: 702 (1874). *Campanopsis serpyllifolia* (Vis.)
Kuntze, Revis. Gen. Pl. 2: 379 (1891).

Edraianthus sutjeskae Lakušić, God. Biol. Inst. u Sarajevu 26 (Suppl.): 94 (1974).
NW. Serbia. 13 YUG. Hemicr. or geophyte.

subsp. **sutjeskae**
NW. Serbia. 13 YUG. Hemicr. or geophyte.

subsp. **maslesae** Lakušić, God. Biol. Inst. u Sarajevu 26 (Suppl.): 98 (1974).
NW. Serbia. 13 YUG. Hemicr. or geophyte.

Edraianthus tenuifolius (Waldst. & Kit.) A. DC. in Meisn., Pl. Vasc. Gen. 2: 149 (1839).
Croatia to Bosnia-Hercegovina & Greece. 13 ALB GRC YUG. Hemicr. or geophyte. $2n = 32$.
* *Campanula tenuifolia* Waldst. & Kit., Descr. Icon. Pl. Hung. 2: 168 (1805). *Wahlenbergia
tenuifolia* (Waldst. & Kit.) A. DC., Monogr. Campan.: 133 (1830). *Campanula
graminifolia* var. *sordidifolia* Vuk., Linnaea 26: 327 (1854). *Campanopsis tenuifolia*
(Waldst. & Kit.) Kuntze, Revis. Gen. Pl. 2: 379 (1891).
Edraianthus caricinus Schott in Schott, Nyman & Kotschy, Analect. Bot.: 6 (1854).
Edraianthus tenuifolius subsp. *caricinus* (Schott) Nyman, Consp. Fl. Eur.: 486 (1879).

Edraianthus wettsteinii Halácsy & Bald., Oesterr. Bot. Z. 41: 371 (1891).
Montenegro. 13 YUG. Hemicr. or geophyte.

subsp. **lovcenicus** E. Mayer & Blečić, Phyton (Horn) 13: 246 (1969).
Montenegro. 13 YUG. Hemicr. or geophyte.

subsp. **wettsteinii**
Montenegro. 13 YUG. Hemicr. or geophyte.

Synonyms:
Edraianthus sect. *Halacsyella* (Janch.) Stef. === **Edraianthus** A. DC.
Edraianthus caricinus Schott === **Edraianthus tenuifolius** (Waldst. & Kit.) A. DC.
Edraianthus caudatus (Vis.) Rchb. === **Edraianthus dalmaticus** (A. DC.) A. DC.
Edraianthus croaticus A. Kern. === **Edraianthus graminifolius** (L.) A. DC. subsp. **graminifolius**
Edraianthus graminifolius f. *alpinus* (Wettst.) Janch. === **Edraianthus graminifolius** (L.) A. DC. subsp. **graminifolius**
Edraianthus graminifolius f. *australis* (Wettst.) Janch. === **Edraianthus graminifolius** (L.) A. DC. subsp. **graminifolius**
Edraianthus graminifolius f. *baldaccii* Janch. === **Edraianthus graminifolius** (L.) A. DC. subsp. **graminifolius**
Edraianthus graminifolius f. *ginzbergeri* H. Lindb. === **Edraianthus graminifolius** (L.) A. DC. subsp. **graminifolius**
Edraianthus graminifolius f. *mediterraneus* (Wettst.) Hayek === **Edraianthus graminifolius** (L.) A. DC. subsp. **graminifolius**
Edraianthus graminifolius f. *montenegrinus* (Horák) Hayek === **Edraianthus graminifolius** (L.) A. DC. subsp. **graminifolius**
Edraianthus graminifolius f. *subalpinus* (Wettst.) Janch. === **Edraianthus graminifolius** (L.) A. DC. subsp. **graminifolius**
Edraianthus graminifolius subsp. *albanicus* Degen & Kümm. === **Edraianthus graminifolius** (L.) A. DC. subsp. **graminifolius**
Edraianthus graminifolius subsp. *niveus* (Beck) Janch. === **Edraianthus niveus** Beck
Edraianthus graminifolius subvar. *albanicus* (Degen & Kümm.) Hayek === **Edraianthus graminifolius** (L.) A. DC. subsp. **graminifolius**
Edraianthus graminifolius var. *australis* Wettst. === **Edraianthus graminifolius** (L.) A. DC. subsp. **graminifolius**
Edraianthus graminifolius var. *elatus* Wettst. === **Edraianthus graminifolius** (L.) A. DC. subsp. **graminifolius**
Edraianthus graminifolius var. *ginzbergeri* (H. Lindb.) Hayek === **Edraianthus graminifolius** (L.) A. DC. subsp. **graminifolius**
Edraianthus graminifolius var. *pusilla* Wettst. === **Edraianthus graminifolius** (L.) A. DC. subsp. **graminifolius**
Edraianthus graminifolius var. *siculus* (Strobl) Nyman === **Edraianthus graminifolius** subsp. **siculus** (Strobl) Lakusić ex Greuter & Burdet
Edraianthus horvatii Lakusić === **Edraianthus graminifolius** (L.) A. DC. subsp. **graminifolius**
Edraianthus kitaibelii (A. DC.) A. DC. === **Edraianthus graminifolius** (L.) A. DC. subsp. **graminifolius**
Edraianthus kitaibelii subsp. *croaticus* (A. Kern.) Nyman === **Edraianthus graminifolius** (L.) A. DC. subsp. **graminifolius**
Edraianthus kitaibelii var. *alpinus* Wettst. === **Edraianthus graminifolius** (L.) A. DC. subsp. **graminifolius**
Edraianthus kitaibelii var. *mediterraneus* Wettst. === **Edraianthus graminifolius** (L.) A. DC. subsp. **graminifolius**
Edraianthus kitaibelii var. *subalpinus* Wettst. === **Edraianthus graminifolius** (L.) A. DC. subsp. **graminifolius**
Edraianthus montenegrinus Horák === **Edraianthus graminifolius** (L.) A. DC. subsp. **graminifolius**
Edraianthus owerinianus Rupr. === **Muehlbergella oweriniana** (Rupr.) Feer
Edraianthus serbicus subsp. *stankovicii* Lakusić === **Edraianthus serbicus** Petroviã
Edraianthus serpyllifolius subsp. *dinaricus* (A. Kern.) Nyman === **Edraianthus dinaricus** (A. Kern.) Wettst.

Edraianthus siculus Strobl === **Edraianthus graminifolius** subsp. **siculus** (Strobl) Lakusić ex
Greuter & Burdet
Edraianthus tenuifolius subsp. *caricinus* (Schott) Nyman === **Edraianthus tenuifolius**
(Waldst. & Kit.) A. DC.
Edraianthus tenuifolius subsp. *hercegovinus* (K. Malý) Hayek === **Edraianthus hercegovinus**
K. Malý
Edraianthus vesovicii Lakusić === **Edraianthus graminifolius** (L.) A. DC. subsp. **graminifolius**
Edraianthus zogovicii Lakusić === **Edraianthus graminifolius** (L.) A. DC. subsp. **graminifolius**

Enchysia

This genus was segregated from *Lobelia* largely on the basis of its entire corolla tube and
subequal corolla lobes. For many years, the name was treated as a synonym of *Laurentia* s.l.
(e.g., Wimmer 1953, under **General**). This led to a recent attempt (Serra and Crespo 1997,
Crespo et al. 1998) to resurrect the name from synonymy and apply it to the South African
species of *Laurentia* (q.v.). However, the type species of *Enchysia* is a synonym of *Lobelia
erinus* (Thulin et al. 1986), and as such, this name is properly a synonym of *Lobelia*. The
name *Wimmerella* (q.v.) was created subsequently (Serra et al. 1999) to accomodate the
South African species of *Laurentia*. Type [designated by Pfeiff., Nomencl. Bot. 1: 1199
(1874)]: *Lobelia erinoides* L.

> Presl, C. (1836). *Enchysia*. In Prodromus monographiae Lobeliacearum: 40. Prague:
> Theophilus Haase. La. — Establishment of genus as distinct from *Lobelia*.
> Thulin, M., P. B. Phillipson & D. O. Wijnands (1986). Typification of *Lobelia erinus* L. and
> *Lobelia erinoides* L. Taxon 35: 729-729. En. — Species designated as lectotype of
> *Enchysia* was based on a specimen of *Lobelia erinus*.
> Serra, L. & M. B. Crespo (1997). An outline revision of the subtribe Siphocampylinae
> (Lobeliaceae). Lagascalia 19: 881-888, illus., map. En. — Name misapplied to South
> African species of *Laurentia*.
> Crespo, M. B., L. Serra & A. Juan (1998). *Solenopsis* (Lobeliaceae): a genus endemic in the
> Mediterranean Region. Pl. Syst. Evol. 210: 211-229, illus., maps. En. — Name
> misapplied to South African species of *Laurentia*.
> Serra, L., M. B. Crespo & T. G. Lammers (1999). *Wimmerella,* a new South African genus of
> Lobelioideae (Campanulaceae). Novon 9: 414-418. En. — Establishment of new genus
> for species erroneously called *Enchysia*.

Synonyms:
Enchysia C. Presl === **Lobelia** L.
Enchysia baueri C. Presl === **Isotoma fluviatilis** (R. Br.) F. Muell. ex Benth.
Enchysia baueri var. *major* C. Presl === **Isotoma fluviatilis** (R. Br.) F. Muell. ex Benth.
Enchysia dentata A. DC. === **Wimmerella secunda** (L. f.) Serra, M. B. Crespo & Lammers
Enchysia erecta A. DC. in DC. === **Wimmerella secunda** (L. f.) Serra, M. B. Crespo
& Lammers
Enchysia erinoides (L.) C. Presl === **Lobelia erinus** L.
Enchysia gaudichaudii C. Presl === **Isotoma fluviatilis** (R. Br.) F. Muell. ex Benth.
Enchysia lessonii C. Presl === **Isotoma fluviatilis** (R. Br.) F. Muell. ex Benth.
subsp. **fluviatilis**
Enchysia repens (Thunb.) C. Presl === **Lobelia anceps** L. f.
Enchysia repens var. *reflexa* Kunze === **Wimmerella secunda** (L. f.) Serra, M. B. Crespo
& Lammers
Enchysia scapigera (R. Br.) C. Presl === **Isotoma scapigera** (R. Br.) G. Don
Enchysia scapigera var. *pusilla* (R. Br.) A. DC. === **Isotoma scapigera** (R. Br.) G. Don
Enchysia secunda (L. f.) Sond. === **Wimmerella secunda** (L. f.) Serra, M. B. Crespo
& Lammers
Enchysia secunda var. *dentata* (A. DC.) Sond. === **Wimmerella secunda** (L. f.) Serra,
M. B. Crespo & Lammers

Enchysia secunda var. *glabrata* Sond. === **Wimmerella secunda** (L. f.) Serra, M. B. Crespo &
Lammers
Enchysia secunda var. *reflexa* (Kunze) Sond. === **Wimmerella secunda** (L. f.) Serra, M. B.
Crespo & Lammers

Erinia

In segregating *Campanula erinus* as a distinct genus, Noulet (1837) apparently was unaware
of the priority of the name *Roucela* (q.v.). Type [by monotypy]: *Erinia campanula* Noulet.

Noulet, J.-B. (1837). Érinie. In Flore du Bassin Sous-Pyrénéen: 404. Toulouse: J.-B. Paya. Fr.
— Establishment of genus as distinct from *Campanula*.

Synonyms:
Erinia Noulet === **Campanula** L.
Erinia campanula Noulet === **Campanula erinus** L.

Euhaynaldia

This name was an avowed substitute for *Haynaldia* Kanitz (q.v.), which is a later homonym
of *Haynaldia* Schur (Poaceae) and *Haynaldia* Schulzer, a fungus. Based on: *Haynaldia* Kanitz.

Borbás, V. 1880. Növénytani apróságok V. Triticumok és ágas *Anthoxanthum*. Földmüv.
Érdek. 8: 331 (1880). Hu. — Validation of name as substitute for illegitimate
Haynaldia Kanitz.

Synonyms:
Euhaynaldia Borbás === **Lobelia** L.
Euhaynaldia exaltata (Pohl) Borbás === **Lobelia exaltata** Pohl
Euhaynaldia organensis (Gardner) Borbás === **Lobelia organensis** Gardner
Euhaynaldia thapsoidea (Schott ex Pohl) Borbás === **Lobelia thapsoidea** Schott ex Pohl

Euregelia

This name was provided as an avowed substitute for *Cylindrocarpa* (q.v.), which Kuntze felt
was a later homonym of the brown algal genus *Cylindrocarpus* H. Crouan & P. Crouan.
However, that interpretation is not supported by the current ICBN (Art. 53.5 Ex. 10). Based
on: *Cylindrocarpa* Regel.

Kuntze, O. (1898). *Cylindrocarpus*. In Revisio generum plantarum vascularium omnium
atque cellularium multarum secundum leges nomenclaturae internationales 3(3): 403.
Leipzig: Arthur Felix. Ge. — Validation of name as substitute for *Cylindrocarpa*.

Synonyms:
Euregelia Kuntze === **Cylindrocarpa** Regel
Euregelia sewerzowii (Regel) Kuntze === **Cylindrocarpa sewerzowii** (Regel) Regel

Favratia

Campanuloideae, 1 species, Europe. The species typically has been included in *Campanula*
but is judged distinct by virtue of its urceolate appendiculate corolla. $2n$ = 34. Type [by
monotypy]: *Favratia zoysii* (Jacq.) Feer.

Feer, H. (1890). Beiträge zur Systematik und Morphologie der Campanulaceen. I. *Favratia*,
gen. nov. Bot. Jahrb. Syst. 12: 608-610 + Taf. VI, illus. Ge. — Establishment of genus as
distinct from *Campanula*.

Turrill, W. B. (1916). *Campanula zoysii*. Bot. Mag. 142: tab. 8666 + 2 pp., illus. En. —
Portrait with description and notes on pollination.
Ravnik, V. (1967). *Campanula cochlearifolia* × *C. zoysii* = *Campanula* × *vrtacensis* Ravnik
hybr. nov. Phyton (Horn) 12: 169-172, illus. Ge. — Description of intergeneric hybrid.

Favratia Feer, Bot. Jahrb. Syst. 12: 610 (1890). *Campanula* sect. *Favratia* (Feer) Kuntze in T.
Post & Kuntze, Lex. Gen. Phan.: 95 (1904).
Europe. 11 13.

Favratia zoysii (Jacq.) Feer, Bot. Jahrb. Syst. 12: 610 (1890).
SE. Alps; cult. 11 AUT 13 ITA YUG. Hemicr. 2*n* = 34.
 * *Campanula zoysii* Jacq., Collect. Bot. 2: 122 (1788).

Fedorovia

Kolakovskii (1980) segregated this species from *Campanula* on the basis of its unique
schizocarpic fruit, but gave it a name that was a later homonym of *Fedorovia* Yakovlev
(Fabaceae). It was subsequently renamed *Theodorovia* (q.v.). Type [designated by the author]:
Fedorovia karakuschensis (Grossh.) Kolak.

Kolakovskii, A. A. (1980). [*Fedorovia* - a new monotypical genus from limestone of South
Transcaucasia.] Soobšč. Akad. Nauk Gruzinsk. SSR 97: 685-688, illus. Ru. —
Establishment of genus as distinct from *Campanula*.
Oganesian, M. (1995). Proposal to conserve *Campanula karakuschensis* Grossh. against
Campanula minsteriana Grossh. (Campanulaceae). Taxon 44: 241. En. — Formal
proposal to protect name of type species from competitor with priority.
Brummitt, R. K. (1997). Report of the Committee for Spermatophyta: 45. Taxon 46: 323-
328. En. — Proposal for conservation not recommended.
Nicolson, D. H. (1999). Report of the General Committee: 8. Taxon 48: 373-378. En. —
Proposal for conservation "on hold".

Synonyms:
Fedorovia Kolak. === **Theodorovia** Kolak.
Fedorovia karakuschensis (Grossh.) Kolak. === **Theodorovia karakuschensis** (Grossh.) Kolak.

Feeria

Campanuloideae, 1 species, Morocco. The genus has been allied to or included in *Trachelium*
by some authors, but a preliminary phylogenetic analysis based on molecular data (Eddie et
al. 2003, under **General**) placed it in a small clade comprising the type species of *Azorina*
and *Roucela* plus *Campanula mollis*, at some distance from the type of *Trachelium*, which was
part of the "Campanulaceae s. str. clade". 2*n* = 34. Type [by monotypy]: *Feeria angustifolia*
(Schousb.) Buser.

Buser, R. (1894). Contributions a la connaissance des Campanulacées. I. Genus *Trachelium*
revisum. Bull. Herb. Boissier 2: 501-532 + pl. XV-XIX, illus. Fr. — Establishment of
genus as distinct from *Trachelium*.
Quézel, P. (1953). Les Campanulacées d'Afrique du Nord, *Feeria* Buser. Feddes Repert.
Spec. Nov. Regni Veg. 56: 9. Fr. — Brief synopsis.

Feeria Buser, Bull. Herb. Boissier 2: 517 (1894).
NW. Africa. 20.

Feeria angustifolia (Schousb.) Buser, Bull. Herb. Boissier 2: 518 (1894).
E. & C. Morocco. 20 MOR. Hemicr. 2*n* = 34.
 * *Trachelium angustifolium* Schousb., Iagttag. Vextrig. Marokko: 85 (1800).

Floerkea

Although this name has priority, it cannot be used for *Adenophora* because it is a later homonym of *Floerkea* Willd. (Limnanthaceae). Type [by monotypy]: *Campanula marsupiiflora* Schult.

> Sprengel, K. P. J. (1818). Campanuleen. In Anleitung zur Kenntnis der Gewächse (ed. 2): 522-525. Halle: C. A. Kummel. Ge. — Establishment of genus as distinct from *Campanula*.

Synonyms:
Floerkea Spreng. === **Adenophora** Fisch.

Gadellia

This genus is often recognized in current literature (e.g., Czerepanov 1995, under **Asia**), and does differ in life-form from most perennial species of *Campanula*. A preliminary phylogeny based on molecular data (Eddie et al. 2003, under **General**) placed its type species as part of a small "transitional" clade comprising the types of *Musschia* and *Echinocodonia*, plus *Campanula peregrina*. Type [designated by the author]: *Gadellia lactiflora* (M. Bieb.) Shulkina.

> Schulkina, T. V. (1979). De positione systematica *Campanula lactiflora* Bieb. Novosti Sist. Vyssh. Rast. 16: 175-179+ Fig. 1, illus. Ru. — Establishment of genus as distinct from *Campanula*.

Synonyms:
Gadellia Shulkina === **Campanula** L.
Gadellia biserrata (K. Koch) A. P. Khokhr. === **Campanula lactiflora** M. Bieb.
Gadellia lactiflora (M. Bieb.) Shulkina === **Campanula lactiflora** M. Bieb.

Galeatella

This section of *Lobelia* was accorded generic rank because of its "large pollen, greater chromosome number and other unusual features" (Degener & Degener 1962). Two of its species were included in a preliminary phylogenetic analysis of DNA data (E. Knox, pers. comm.), where they formed the sister-group of *Trematolobelia* within the Hawaiian clade. Based on: *Lobelia* sect. *Galeatella* E. Wimm. Type [designated by O. Deg. & I. Deg. Fl. Hawaiiensis, fam. 339 (1962)]: *Lobelia villosa* (Rock) H. St. John & Hosaka.

> Degener, O. & I. Degener (1962). Flora Hawaiiensis, fam. 339, *Galeatella*. Honolulu: published privately. En. — Establishment of genus, with key to species.
> Stone, B. C. (1967). A review of the endemic genera of Hawaiian plants. *Galeatella* Deg. & Deg. Bot. Rev. (Lancaster) 33: 246-247. En. — Discusses merits of recognizing genus as distinct from *Lobelia*.
> Degener, O. & I. Degener (1974). Prodromus of *Galeatella* and *Neowimmeria*. Honolulu: published privately. En. — Synopsis with full nomenclature.

Synonyms:
Galeatella (E. Wimm.) O. Deg. & I. Deg. === **Lobelia** L.
Galeatella gaudichaudii (A. DC.) O. Deg. & I. Deg. === **Lobelia gaudichaudii** A. DC.
Galeatella gaudichaudii var. *koolauensis* (Hosaka & Fosberg) O. Deg. & I. Deg. === **Lobelia gaudichaudii** subsp. **koolauensis** (Hosaka & Fosberg) Lammers
Galeatella gloria-montis (Rock) O. Deg. & I. Deg. === **Lobelia gloria-montis** Rock

Galeatella gloria-montis f. *sanguinea* (H. St. John & Hosaka) O. Deg. & I. Deg. === **Lobelia gloria-montis** Rock

Galeatella gloria-montis var. *bryanii* (H. St. John & Hosaka) O. Deg. & I. Deg. === **Lobelia gloria-montis** Rock

Galeatella gloria-montis var. *kukuiensis* (H. St. John & Hosaka) O. Deg. & I. Deg. === **Lobelia gloria-montis** Rock

Galeatella gloria-montis var. *longibracteata* (Rock) O. Deg. & I. Deg. === **Lobelia gloria-montis** Rock

Galeatella gloria-montis var. *molokaiensis* (O. Deg.) O. Deg. & I. Deg. === **Lobelia gloria-montis** Rock

Galeatella kauaiensis (A. Gray) O. Deg. & I. Deg. === **Lobelia kauaiensis** (A. Gray) A. Heller

Galeatella kauaiensis var. *hirsuta* (H. St. John & Hosaka) O. Deg. & I. Deg. === **Lobelia kauaiensis** (A. Gray) A. Heller

Galeatella longibracteata (Rock) O. Deg. & I. Deg. === **Lobelia gloria-montis** Rock

Galeatella villosa (Rock) O. Deg. & I. Deg. === **Lobelia villosa** (Rock) H. St. John & Hosaka

Githopsis

Campanuloideae, 4 species, western North America, from southern Vancouver Island to northern Baja California and Isla de Guadeloupe. In a preliminary phylogeny based on molecular data (Eddie et al. 2003, under **General**), *G. diffusa* formed a clade with *Heterocodon rariflorus;* this small clade then formed the basal branch of the "Rapunculus clade". The treatment here follows Morin (1983), as modified by Lammers (2006, under **Special**). $2n$ = 18, 20, 36, 38, 40. Type [by monotypy]: *Githopsis specularioides* Nutt.

> Nuttall, T. (1842). Description and notices of new or rare plants in the natural orders Lobeliaceae, Campanulaceae, Vaccinieae, Ericaceae, collected in a journey over the continent of North America, and during a visit to the Sandwich Islands, and Upper California. Trans. Amer. Philos. Soc. (n.s.) 8: 251-272. En. — Establishment of genus.
>
> Ewan, J. (1939). A review of the genus *Githopsis*. Rhodora 41: 302-313, illus. En. — Synopsis with key, full nomenclature, and specimen citations.
>
> Munz, P. A. & D. D. Keck (1970). *Githopsis* Nutt. In A California flora: 1064-1065. Berkeley: University of California Press. En. — Flora with key, descriptions, and synonymy.
>
> • Morin, N. (1983). Systematics of *Githopsis* (Campanulaceae). Syst. Bot. 8: 436-468, illus., maps. En. — Monograph with key, descriptions, full nomenclature, and specimen citations.
>
> • Morin, N. (1993). *Githopsis*. In J. C. Hickman (ed.), The Jepson manual, higher plants of California: 462-464, illus. Berkeley: University of California Press. En. — Floristic treatment with keys, descriptions, and synonymy; covers all species in genus.

Githopsis Nutt., Trans. Amer. Philos. Soc. (n.s.) 8: 258 (1842).
W. North America. 71 73 76 79.

Githopsis diffusa A. Gray, Proc. Amer. Acad. Arts 17: 221 (1882).
N. California to N. Baja California; Guadeloupe I. 76 CAL 79 MXN MXI. Ther. $2n$ = 20, 40.

subsp. **candida** (Ewan) Morin, Syst. Bot. 8: 464 (1983).
S. California (San Diego Co.). 76 CAL. Ther. $2n$ = 20.
* *Githopsis specularioides* subsp. *candida* Ewan, Rhodora 41: 308 (1939). *Githopsis diffusa* var. *candida* (Ewan) Morin, Syst. Bot. 8: 464 (1983).

subsp. **diffusa**
C. California to N. Baja California. 76 CAL 79 MXN. Ther. $2n$ = 20.
Githopsis gilioides Ewan, Rhodora 41: 311 (1939).

subsp. **filicaulis** (Ewan) Morin, Syst. Bot. 8: 465 (1983).
S. California (Riverside & San Diego Co.). 76 CAL. Ther. $2n$ = 20.
* *Githopsis filicaulis* Ewan, Rhodora 41: 312 (1939).

subsp. **guadalupensis** (Morin) Lammers, Novon 16: 70 (2006).
Guadelupe I. 79 MXI. Ther.
* *Githopsis diffusa* var. *guadalupensis* Morin, Syst. Bot. 8: 464 (1983).

subsp. **robusta** Morin, Syst. Bot. 8: 463 (1983).
N. & C. California. 76 CAL. Ther. $2n = 40$.

Githopsis pulchella Vatke, Linnaea 38: 714 (1874).
California. 76 CAL. Ther. $2n = 18, 20$.

subsp. **campestris** Morin, Syst. Bot. 8: 455 (1983).
N. California. 76 CAL. Ther.

subsp. **pulchella**
NC. California. 76 CAL. Ther. $2n = 18, 20$.
Githopsis specularioides var. *glabra* Jeps., Man. Fl. Pl. Calif.: 974 (1925). *Githopsis pulchella* var. *glabra* (Jeps.) Morin, Syst. Bot. 8: 453 (1983).

subsp. **serpentinicola** Morin, Syst. Bot. 8: 456 (1983).
SC. California. 76 CAL. Ther. $2n = 20$.

Githopsis specularioides Nutt., Trans. Amer. Philos. Soc. (n.s.) 8: 258 (1842).
S. Vancouver I. to S. California. 71 BRC 73 ORE WAS 76 CAL. Ther. $2n = 36, 38, 40$.
Githopsis specularioides var. *hirsuta* Nutt., Trans. Amer. Philos. Soc. (n.s.) 8: 258 (1842).
Githopsis calycina Benth., Pl. Hartweg.: 321 (1849).

Githopsis tenella Morin, Syst. Bot. 8: 465 (1983).
SC. California. 76 CAL. Ther. $2n = 18$.

Synonyms:
Githopsis calycina Benth. === **Githopsis specularioides** Nutt.
Githopsis calycina var. *hirsuta* Benth. === **Githopsis specularioides** Nutt.
Githopsis diffusa var. *candida* (Ewan) Morin === **Githopsis diffusa** subsp. **candida** (Ewan) Morin
Githopsis diffusa var. *guadaloupensis* Morin === **Githopsis diffusa** subsp. **guadalupensis** (Morin) Lammers
Githopsis filicaulis Ewan === **Githopsis diffusa** subsp. **filicaulis** (Ewan) Morin
Githopsis gilioides Ewan === **Githopsis diffusa** A. Gray subsp. **diffusa**
Githopsis latifolia Eastw. === **Legousia speculum-veneris** (L.) Durande ex Vill.
Githopsis pulchella var. *glabra* (Jeps.) Morin === **Githopsis pulchella** Vatke subsp. **pulchella**
Githopsis specularioides subsp. *candida* Ewan === **Githopsis diffusa** subsp. **candida** (Ewan) Morin
Githopsis specularioides var. *glabra* Jeps. === **Githopsis pulchella** Vatke subsp. **pulchella**
Githopsis specularioides var. *hirsuta* Nutt. === **Githopsis specularioides** Nutt.

Glosocomia

This name has been used for those species of *Codonopsis* (q.v.) in which the calyx lobes are apical, as in almost all other Campanulaceae, rather than appearing to arise in a lateral or basal position on the hypanthium as in many *Codonopsis*. Although Lindley (1842) argued that the name should be spelled with a double "s" on etymological grounds, the original spelling is maintained here. Type [by monotypy]: *Glosocomia tenera* D. Don.

Don, D. (1825). *Glosocomia*. In Prodromus florae Nepalensis: 158. London: J. Gale. La. — Establishment of genus.
Lindley, J. (1842). *Glossocomia* [sic] *ovata*. Ovate Pouchbell. Edwards' Bot. Reg. 28: pl. 3 + 2 pp., illus. En. — Supports distinctness of *Glosocomia* from *Codonopsis;* insists correct spelling is "Glossocomia".

Chipp, T. F. (1908). A revision of the genus *Codonopsis,* Wall. J. Linn. Soc., Bot. 38: 374-391, illus. En., La. — Advocates merger of *Glosocomia* and *Codonopsis.*

Komarov, V. L. (1908). Prolegomena ad floras chinae nec non Mongoliae. VII. Revisio critica specierum generis *Codonopsis* Wall. Trudy Imp. S.-Peterburgsk. Bot. Sada 29: 98-124 + pl. II, illus. La. — Advocates merger of *Glosocomia* and *Codonopsis.*

Synonyms:

Glosocomia D. Don === **Codonopsis** Wall.

Glosocomia clematidea (Schrenk) Fisch. ex Regel === **Codonopsis clematidea** (Schrenk) C. B. Clarke

Glosocomia hortensis Rupr. === **Codonopsis lanceolata** (Siebold & Zucc.) Trautv.

Glosocomia lanceolata (Siebold & Zucc.) Maxim. === **Codonopsis lanceolata** (Siebold & Zucc.) Trautv.

Glosocomia lanceiolata var. *obtusa* Regel === **Codonopsis ussuriensis** (Rupr. & Maxim.) Hemsl.

Glosocomia lanceolata var. *ussuriensis* (Rupr. & Maxim.) Regel === **Codonopsis ussuriensis** (Rupr. & Maxim.) Hemsl.

Glosocomia lurida (Lindl.) Lindl. === **Codonopsis rotundifolia** Benth.

Glosocomia ovata (Benth.) Lindl. === **Codonopsis ovata** Benth.

Glosocomia purpurea (Wall.) Rupr. === **Codonopsis purpurea** Wall.

Glosocomia rotundifolia (Benth.) Rupr. === **Codonopsis rotundifolia** Benth.

Glosocomia tenera D. Don === **Codonopsis thalictrifolia** Wall.

Glosocomia ussuriensis Rupr. & Maxim. === **Codonopsis ussuriensis** (Rupr. & Maxim.) Hemsl.

Glosocomia viridis (Wall.) Rupr. === **Codonopsis viridis** Wall.

Grammatotheca

Lobelioideae, 1 species, South Africa, and naturalized in Australia. In preliminary molecular analyses (E. Knox, pers. comm.; A. Antonelli, pers. comm.), it belonged to a clade comprising some of the sampled species of *Lobelia* sect. *Delostemon* (E. Wimm.) J. Murata. Type [designated by Pfeiff., Nomencl. Bot. 1: 1493 (1874)]: *Clintonia bergiana* (Cham.) G. Don.

Presl, C. (1836). *Grammatotheca.* In Prodromus monographiae Lobeliacearum: 43-45. Prague: Theophilus Haase. La. — Establishment of genus as distinct from *Downingia.*

Bentham, G. & F. Mueller (1868). *Lobelia,* Linn. In Flora Australiensis: a description of the plants of the Australian Territory 4: 122-131. London: L. Reeve. En. — Argues that *Grammatotheca* is not disctinct from *Lobelia.*

Adamson, R. S. (1950). *Grammatotheca* Presl. In R. S. Adamson & T. M. Salter, Flora of the Cape Peninsula: 759. Cape Town: Juta. En. — Flora with description.

• Wimmer, F. E. (1953). *Grammatotheca* Presl. In A. Engler & L. Diels, Das Pflanzenreich IV. 276b: 695-698, illus. Berlin: Akademie-Verlag. Ge., La. — Monograph with key and descriptions.

Dyer, R. A. (1975). *Grammatotheca* C. Presl. In The genera of southern African flowering plants 1: 645-646. Pretoria: Department of Agricultural Technical Services. En. — Generic description.

Elliot, W. R. & D. L. Jones (1990). *Grammatotheca.* Encyclopedia of Australian plants suitable for cultivation 5: 3, illus. Melbourne: Lothian. En. — Descriptive account from horticultural perspective.

van der Vlugt, P. (1992). Zwei amphibische Lobeliengewächse aus Südafrika und Australien. Aqua Planta 17(3): 83-89, illus. Ge. — Cultivation of *Grammatotheca bergiana* in aquaria and water gardens.

Goldblatt, P. & J. Manning (2000). Cape plants. A conspectus of the Cape flora of South Africa. *Grammatotheca.* Strelitzia 9: 390. En. — Synopsis of genus with brief descriptions of this species and an additional unnamed one.

Welman, W. G. (2000). *Grammatotheca* C. Presl. In O. A. Leistner (ed.), Seed plants of southern Africa: families and genera. Strelitzia 10: 339. En. — Generic description.

Grammatotheca C. Presl, Prodr. Monogr. Lobel.: 43 (1836). *Rapuntium* sect. *Grammatotheca* (C. Presl) Kuntze in T. Post & Kuntze, Lex. Gen. Phan.: 478 (1903).
S. Africa; naturalized in Australia. 27 (50).

Grammatotheca bergiana (Cham.) C. Presl, Prodr. Monogr. Lobel.: 44 (1836).
S. Africa; naturalized in W. Australia. 27 CPP NAT (50) wau. Hemicr.
* *Lobelia bergiana* Cham., Linnaea 8: 217 (1833). *Clintonia bergiana* (Cham.) G. Don, Gen. Hist. 3: 718 (1834). *Grammatotheca erinoides* var. *thunbergiana* Sond. in Harv. & Sond., Fl. Cap. 3: 532 (1865). *Dortmanna bergiana* (Cham.) Kuntze, Revis. Gen. Pl. 2: 972 (1891).
Grammatotheca dregeana C. Presl, Prodr. Monogr. Lobel.: 44 (1836). *Grammatotheca erinoides* var. *dregeana* (C. Presl) Sond. in Harv. & Sond., Fl. Cap. 3: 532 (1865).
Grammatotheca eckloniana C. Presl, Prodr. Monogr. Lobel.: 44 (1836). *Grammatotheca erinoides* var. *eckloniana* (C. Presl) E. Wimm., Ann. Naturhist. Mus. Wien 56: 370 (1948). *Grammatotheca bergiana* var. *eckloniana* (C. Presl) E. Wimm., Pflanzenr. IV.276b: 697 (1953).
Grammatotheca meyeriana C. Presl, Prodr. Monogr. Lobel.: 44 (1836).
Grammatotheca mundtiana C. Presl, Prodr. Monogr. Lobel.: 44 (1836).
Lobelia macrocarpa de Vriese in Lehm., Pl. Preiss.: 396 (1845).
Lobelia macrocarpa var. *genistoides* de Vriese in Lehm., Pl. Preiss.: 397 (1845).
Lobelia amplexicaulis de Vriese in Lehm., Pl. Preiss.: 397 (1845).
Lobelia stenotheca F. Muell., Fragm. 2: 20 (1860).
Grammatotheca erinoides var. *pedunculata* E. Wimm., Ann. Naturhist. Mus. Wien 56: 370 (1948). *Grammatotheca bergiana* var. *pedunculata* (E. Wimm.) E. Wimm., Pflanzenr. IV.276b: 697 (1953).
Grammatotheca bergiana var. *foliosa* E. Wimm., Pflanzenr. IV.276b: 697 (1953).

Synonyms:
Grammatotheca bergiana var. *eckloniana* (C. Presl) E. Wimm. === **Grammatotheca bergiana** (Cham.) C. Presl
Grammatotheca bergiana var. *foliosa* E. Wimm. === **Grammatotheca bergiana** (Cham.) C. Presl
Grammatotheca bergiana var. *pedunculata* (E. Wimm.) E. Wimm. === **Grammatotheca bergiana** (Cham.) C. Presl
Grammatotheca dregeana C. Presl === **Grammatotheca bergiana** (Cham.) C. Presl
Grammatotheca eckloniana C. Presl === **Grammatotheca bergiana** (Cham.) C. Presl
Grammatotheca erinoides (L.) Sond. === **Lobelia erinus** L.
Grammatotheca erinoides var. *dregeana* (C. Presl) Sond. === **Grammatotheca bergiana** (Cham.) C. Presl
Grammatotheca erinoides var. *eckloniana* (C. Presl) E. Wimm. === **Grammatotheca bergiana** (Cham.) C. Presl
Grammatotheca erinoides var. *pedunculata* E. Wimm. === **Grammatotheca bergiana** (Cham.) C. Presl
Grammatotheca erinoides var. *thunbergiana* Sond. === **Grammatotheca bergiana** (Cham.) C. Presl
Grammatotheca meyeriana C. Presl === **Grammatotheca bergiana** (Cham.) C. Presl
Grammatotheca mundtiana C. Presl === **Grammatotheca bergiana** (Cham.) C. Presl

Gunillaea

Campanuloideae, 2 species, tropical Africa. On the basis of morphology, it is related to *Prismatocarpus* and *Wahlenbergia*. The treatment here follows Thulin (1983). $2n = 18$. Type [designated by the author]: *Gunillaea emirnensis* (A. DC.) Thulin.

Thulin, M. (1974). *Gunillaea* and *Namacodon*. Two new genera of Campanulaceae in Africa. Bot. Notiser 127: 165-182, illus., maps. En. — Establishment of genus as distinct from *Wahlenbergia*.

- Thulin, M. (1983). *Gunillaea* Thulin. In E. Launert (ed.), Flora Zambesiaca 7(1): 111-114, illus. En. — Key to both species, descriptions, full nomenclature & specimen citations.

  Welman, W. G. (2000). *Gunillaea* Thulin. In O. A. Leistner (ed.), Seed plants of southern Africa: families and genera. Strelitzia 10: 200-201. En. — Generic description.

Gunillaea Thulin, Bot. Not. 127: 166 (1974).
  Trop. Africa. 23 25 26 27 29.

Gunillaea emirnensis (A. DC.) Thulin, Bot. Not. 127: 166 (1974).
  Zaïre to Tanzania, Zimbabwe & Angola; Madagascar. 23 ZAI 25 TAN 26 ANG MLW ZAM ZIM 29 MDG. Ther.
  * *Wahlenbergia emirnensis* A. DC. in DC., Prodr. 7: 432 (1839). *Campanopsis emirnensis* (A. DC.) Kuntze, Revis. Gen. Pl. 2: 379 (1891).
    Wahlenbergia huillana A. DC., Ann. Sci. Nat., Bot. (ser. 5) 6: 333 (1866). *Campanopsis huillana* (A. DC.) Kuntze, Revis. Gen. Pl. 2: 379 (1891). *Cervicina huillana* (A. DC.) Hiern, Cat. Afr. Pl. 1: 631 (1898).
    Wahlenbergia huillana var. *pusilla* A. DC., Ann. Sci. Nat., Bot. (ser. 5) 6: 333 (1866).

Gunillaea rhodesica (Adamson) Thulin, Bot. Not. 127: 168 (1974).
  Zaïre to Mozambique, Botswana & Namibia. 23 ZAI 26 ANG MOZ ZAM ZIM 27 BOT NAM. Ther. $2n = 18$.
  * *Prismatocarpus rhodesicus* Adamson, J. S. African Bot. 17: 123 (1952).

Gynampsis

Like his earlier *Bolelia* (q.v.), this name was proposed by Rafinesque (1833) as an avowed substitute for the illegitimate name *Clintonia*. He did not explain why he felt the need to proffer a second replacement. Based on: *Clintonia* Dougl. ex Lindl., non Raf.

  Rafinesque, C. S. (1833). Atlantic Journal. Extra of N. 8. Herbarium rafinesquianum. Prodromus. Philadelphia: published privately. En. — Validation of name to replace illegitimate *Clintonia*.

Synonyms:
Gynampsis Raf. === **Downingia** Torr.
Gynampsis flexuosa Raf. === **Downingia elegans** (Dougl. ex Lindl.) Torr.

Halacsyella

Because this species more closely resembles in its habit various species of *Campanula* (e.g., *C. glomerata*, *C. tymphaea*) than other members of *Edraianthus*, Janchen (1910) proposed segregating it as the genus *Halacsyella*. Type [by monotypy]: *Halacsyella parnassica* (Boiss. & Spruner) Janch.

  Janchen, E. (1910). Die *Edraianthus*-Arten der Balkanländer. Mitt. Naturwiss. Vereines Univ. Wien (n.s.) 8: 1-40 + Taf. I-IV, illus. Ge. — Establishment of genus as distinct from *Edraianthus*.

  Stefanoff, B. (1936). [Über die systematische Stellung einiger Arten der Familie Campanulaceae.] God. Sofiisk. Univ. Agron.-Lesoved. Fak. 14: 93-104. Bu. — Argues for retention of genus in *Edraianthus*, but at sectional rank.

Synonyms:
Halacsyella Janch. === **Edraianthus** A. DC.
Halacsyella parnassica (Boiss. & Spruner) Janch. === **Edraianthus parnassicus** (Boiss. & Spruner) Halácsy

Hanabusaya

Campanuloideae, 1 species, Korea. Many authors have considered this genus to be related to or congeneric with *Symphyandra* (q.v.), due to its connate anthers. However, recent molecular phylogenies (Kim et al. 1999; Eddie et al. 2003, under **General**) indicate that its relationships are with *Adenophora;* both genera are characterized by flowers with a very prominent nectary. 2*n* = 34. Type [by monotypy]: *Symphyandra asiatica* Nakai.

Nakai, T. (1911). [A preliminary note on a new genus of Campanulaceae found in Korea.] Bot. Mag. (Tokyo) 25 (Jap.): 160-162. Ja. — Establishment of genus as distinct from *Symphyandra.*

Nakai, T. (1911). *Hanabusaya* Nakai. J. Coll. Sci. Imp. Univ. Tokyo 31: 62-63 + tab. XIII, illus. La. — Reiteration of generic protologue, and validation of species' name.

Turrill, W. B. (1920). *Symphyandra asiatica.* Bot. Mag. 146: tab. 8837 + 2 pp., illus. En. — Portrait of *H. asiatica,* with description; argues that genus does not merit recognition

Nakai, T. (1921). Notulae ad plantas Japoniae et Koreae XXV. Bot. Mag (Tokyo) 35: 139-153. En. — Rebuttal to Turrill's (1920) synonymization.

Kobayashi, Y. (1979). [*Hanabusaya asiatica* and Yoshikata Hanabusa.] J. Jap. Bot. 54: 189-192, illus. Ja. — Historical account of man commemorated by generic name.

• Lee, S. T., Y. M. An & K. R. Park (1986). [Palynological relationship of *Hanabusaya asiatica* Nakai within the Campanulaceae.] Korean J. Pl. Taxon. 16: 25-37, illus. Ko. — Palynological data suggest relationship to *Campanula* (especially *C. punctata*).

Lee, S. T. (1988). [Palynological evaluation of some endemic genera in Korea.] Korean J. Pl. Taxon. 18: 19-31. Ko. — Palynological data suggest relationship to *Campanula* (especially *C. punctata*).

Lee, S. T., Y. J. Chung & J. K. Lee (1988). A palynotaxonomic study of the Korean Campanulaceae. Korean J. Pl. Taxon. 18: 115-131, illus. Ko. — Although overall pollen morphology shows relationship to *Campanula,* greater pore number and size of surface features support distinctness of genus.

Lee, T. B. (1989). *Hanabusaya asiatica* Nakai, *Hanabusaya latisepala* Nakai. In [Illustrated Flora of Korea:] 723, illus. Hyangmunsa: Soul T'ukpylosi. Ko. — Flora with descriptions.

Huxley, A. J., ed. (1992). *Symphyandra.* In The new Royal Horticultural Society dictionary of gardening 1: 414. London: MacMillan. En. — Account of *H. asiatica* in cultivation.

Yoo, K. O. & W. T. Lee (1995). Seed morphology of Korean Campanulaceae. Korean J. Pl. Taxon. 25: 61-70, illus. Ko. — Seed surface pattern unique among Korean species.

• Yoo, K. O., W. T. Lee, N. S. Kim & H. T. Lim (1997). Comparative studies on the *Hanabusaya asiatica* and its allied groups by RAPD analysis. Pl. Spec. Biol. 12: 49-54. En. — Preliminary molecular data emphasize distinctness of genus.

• Im, R. J. (1999). *Keumkangsania.* In Flora Coreana 7: 13-15, illus. Pyongyang: Science & Technology Publishing House. Ko. — Flora, with key and descriptions.

• Kim, Y. D., J. K. Lee, Y. B. Suh, S. T. Lee, S. H. Kim & R. K. Jansen (1999). Molecular evidence for the phylogenetic position of *Hanabusaya asiatica* Nakai (Campanulaceae), an endemic species in Korea. J. Pl. Biol. 42: 168-173. En. — *Hanabusaya* is sister to clade comprising four species of *Adenophora.*

Hanabusaya Nakai, Bot. Mag. (Tokyo) 25(Jap.): 161 (1911).
E. Asia. 38.
Keumkangsania Kim, Fl. Coreana 6: 94 (1976).

Hanabusaya asiatica (Nakai) Nakai, J. Coll. Sci. Imp. Univ. Tokyo 31: 62 (1911).
Korea. 38 KOR. Hemicr. or geophyte. 2*n* = 34.
* *Symphyandra asiatica* Nakai, Bot. Mag. (Tokyo) 23: 188 (1909). *Keumkangsania asiatica* (Nakai) Kim, Fl. Coreana 6: 94 (1976).
Hanabusaya latisepala Nakai, Bot. Mag. (Tokyo) 35: 147 (1921). *Keumkangsania latisepala* (Nakai) Kim, Fl. Coreana 6: 95 (1976). *Hanabusaya asiatica* var. *latisepala* (Nakai) W. Lee, Lineam. Fl. Koreana: 1071 (1996).
Hanabusaya asiatica f. *alba* T. B. Lee, J. Korean Pl. Taxon. 6: 17 (1975).

Synonyms:
Hanabusaya asiatica f. *alba* T. B. Lee === **Hanabusaya asiatica** (Nakai) Nakai
Hanabusaya asiatica var. *latisepala* (Nakai) W. Lee === **Hanabusaya asiatica** (Nakai) Nakai
Hanabusaya latisepala Nakai === **Hanabusaya asiatica** (Nakai) Nakai

Haynaldia

Kanitz (1877) segregated these Brazilian species from *Lobelia* on the basis of their robust habit, large bracts, and winged seeds. Unfortunately, the name is a later homonym of *Haynaldia* Schur (Poaceae) and *Haynaldia* Schulzer, a fungus. In recent molecular phylogenies (E. Knox, pers. comm.; A. Antonelli, pers. comm.), these species belonged to the clade that comprised giant African species of *Lobelia* sect. *Rhynchopetalum*. Type [designated here]: *Haynaldia exaltata* (Pohl) Kanitz.

> Kanitz, A. (1877). *Haynaldia* novum genus Lobeliacearum. Magyar Növényt. Lapok 1: 3. La. — Establishment of genus as distinct from *Lobelia*.

Synonyms:
Haynaldia Kanitz === **Lobelia** L.
Haynaldia exaltata (Pohl) Kanitz === **Lobelia exaltata** Pohl
Haynaldia exaltata var. *ramosa* Kanitz === **Lobelia exaltata** Pohl
Haynaldia hilaireana Kanitz === **Lobelia hilaireana** (Kanitz) E. Wimm.
Haynaldia organensis (Gardner) Kanitz === **Lobelia organensis** Gardner
Haynaldia organensis var. *insignis* Kanitz === **Lobelia organensis** Gardner
Haynaldia thapsoidea (Schott ex Pohl) Kanitz === **Lobelia thapsoidea** Schott ex Pohl
Haynaldia uranocoma (Cham.) Kanitz === **Lobelia fistulosa** Vell.

Hecale

Rafinesque (1837) segregated this genus from *Campanula* on the basis of its 3-5-parted flowers and bilocular ovary, features perfectly at home in *Wahlenbergia*. Type [designated by the author]: *Hecale lobelioides* (L. f.) Raf.

> Rafinesque, C. S. (1837). Flora Telluriana, pars secunda. Philadelphia: published privately. En. — Establishment of genus as distinct from *Campanula,* and allied to his genus *Codiphus* (q.v.).

Synonyms:
Hecale Raf. === **Wahlenbergia** Schrad. ex Roth
Hecale lobelioides (L. f.) Raf. === **Wahlenbergia lobelioides** (L. f.) Schrad. ex Link

Hemisphaera

This genus was segregated from *Campanula* on the basis of perceived differences in capsule structure. Its species are treated by most authors as members of *Campanula* subg. *Scapiflorae* (Boiss.) Ogan. or sect. *Rupestres* (Boiss.) Kharadze. Type [designated by the author]: *Hemisphaera doluchanovii* (Kharadze) Kolak.

> Kolakovskii, A. A. (1984). [Semisfera novyi vysokogno-Kavkazskii rod kolokol'chikovykh.] Okhr. Prir. Gruzii 12: 157-171, illus. Ru. — Establishment of genus as distinct from *Campanula*.

Synonyms:
Hemisphaera Kolak. === **Campanula** L.
Hemisphaera alpigena (K. Koch) Kolak. === **Campanula bellidifolia** subsp. **aucheri** (A. DC.) Viktorov

Hemisphaera anomala (Fomin) Kolak. & Serdyuk. === **Campanula circassica** Fomin

Hemisphaera ardonensis (Rupr.) Kolak. & Serdyuk. === **Campanula ardonensis** Rupr.

Hemisphaera argunensis (Rupr.) Kolak. & Serdyuk. === **Campanula bellidifolia** subsp. **argunensis** (Rupr.) Viktorov

Hemisphaera armazica (Kharadze) Kolak. === **Campanula bellidifolia** subsp. **aucheri** (A. DC.) Viktorov

Hemisphaera aucheri (A. DC.) Kolak. === **Campanula bellidifolia** subsp. **aucheri** (A. DC.) Viktorov

Hemisphaera bellidifolia (Adam) Kolak. & Serdyuk. === **Campanula bellidifolia** Adam

Hemisphaera besengica (Fomin) Kolak. & Serdyuk. === **Campanula bellidifolia** subsp. **besenginica** (Fomin) Viktorov

Hemisphaera ciliata (Steven) Kolak. === **Campanula ciliata** Steven

Hemisphaera circassica (Fomin) Kolak. & Serdyuk. === **Campanula circassica** Fomin

Hemisphaera doluchanovii (Kharadze) Kolak. === **Campanula bellidifolia** subsp. **argunensis** (Rupr.) Viktorov

Hemisphaera fominii (Grossh.) Kolak. & Serdyuk. === **Campanula bellidifolia** subsp. **meyerana** (Rupr.) Viktorov

Hemisphaera hypopolia (Trautv.) Kolak. & Serdyuk. === **Campanula hypopolia** Trautv.

Hemisphaera kryophila (Rupr.) Kolak. === **Campanula kryophila** Rupr.

Hemisphaera ledebouriana (Trautv.) Kolak. & Serdyuk. === **Campanula ledebouriana** Trautv.

Hemisphaera meyerana (Rupr.) Kolak. & Serdyuk. === **Campanula bellidifolia** subsp. **meyerana** (Rupr.) Viktorov

Hemisphaera petrophila (Rupr.) Kolak. & Serdyuk. === **Campanula petrophila** Rupr.

Hemisphaera radchensis (Kharadze) Kolak. === **Campanula radchensis** Kharadze

Hemisphaera ruprechtii Kolak. & Serdyuk. === **Campanula bellidifolia** subsp. **aucheri** (A. DC.) Viktorov

Hemisphaera saxifraga (M. Bieb.) Kolak. === **Campanula bellidifolia** subsp. **saxifraga** (M. Bieb.) Viktorov

Hemisphaera sosnowskyi (Kharadze) Kolak. === **Campanula bellidifolia** Adam subsp. **bellidifolia**

Hemisphaera tridens (Rupr.) Kolak. === **Campanula tridentata** subsp. **biebersteiniana** (Schult.) Ogan.

Hemisphaera tridentata (Schreb.) Kolak. === **Campanula tridentata** Schreb.

Heterochaenia

Campanuloideae, 3 species, Réunion in the western Indian Ocean. On the basis of morphology and biogeography, this genus is believed to be related to *Nesocodon* (q.v.). Type [by monotypy]: *Heterochaenia ensifolia* (Lam.) A. DC.

Meisner, C. F. (1839). *Heterochaenia*. Alph. DC., in litt. In Plantarum vascularium genera 2: 149. Leipzig: Weidmann. La. — Establishment of genus as distinct from *Wahlenbergia*.

* Badré, F., T. Cadet & M. Malplanche (1972). Étude systématique et palynologique du genre *Heterochaenia* (Campanulaceae) endémique des Mascareignes. Adansonia (n.s.) 12: 267-278, illus. Fr. — Monograph with key, descriptions, and full nomenclature.
* Badré, F. (1976). *Heterochaenia* DC. In Flore des Mascareignes. 111. Campanulacées: 12-16, illus. Paris: ORSTOM. Fr. — Flora with key, descriptions, and full nomenclature.

Heterochaenia A. DC. in Meisn., Pl. Vasc. Gen. 2: 149 (1839). *Wahlenbergia* sect. *Heterochaenia* (A. DC.) Cordem., Fl. Réunion: 498 (1895).
Mascarenes. 29.

Heterochaenia borbonica Badré & Cadet, Adansonia (n.s.) 12: 270 (1972).
Réunion. 29 REU. Nanophan.

Heterochaenia ensifolia (Lam.) A. DC. in Meisn., Pl. Vasc. Gen. 2: 149 (1839).
   Réunion. 29 REU. Nanophan.
   ** Campanula ensifolia* Lam., Encycl. 1: 582 (1785). *Wahlenbergia ensifolia* (Lam.) A. DC.,
      Monogr. Campan.: 162 (1830).

Heterochaenia rivalsii Badré & Cadet, Adansonia (n.s.) 12: 268 (1972).
   Réunion. 29 REU. Nanophan.

Heterocodon

Campanuloideae, 1 species, western North America. The genus was included with the
species of *Triodanis* in *Specularia* s.l. by McVaugh (1941), but that classification is not
supported by a recent phylogenetic analysis based on molecular data (Eddie et al. 2003,
under **General**). In that study, *Heterocodon* was the sister-group of the sole species of
Githopsis sampled; this pair was then sister to the remainder of the "Rapunculus clade".
Type [by monotypy]: *Heterocodon rariflorus* Nutt.

   Nuttall, T. (1842). Description and notices of new or rare plants in the natural orders
      Lobeliaceae, Campanulaceae, Vaccinieae, Ericaceae, collected in a journey over the
      continent of North America, and during a visit to the Sandwich Islands, and Upper
      California. Trans. Amer. Philos. Soc. (n.s.) 8: 251-272. En. — Establishment of genus
      as intermediate between *Triodanis* (under the name *Dysmicodon*, q.v.) and
      Campanula.
   McVaugh, R. (1941). A new name for *Heterocodon rariflorum* [sic] Nutt. Leafl. W. Bot. 3: 48.
      En. — Transfer of type species to *Specularia* s.l.
 • McVaugh, R. (1945). The genus *Triodanis* Rafinesque, and its relationships to *Specularia*
      and *Campanula*. Wrightia 1: 13-52, maps. En. — Reverses earlier opinion (McVaugh
      1941) and recognizes *Heterocodon* as distinct from *Campanula*, *Legousia* (as *Specularia*
      s.s.), and *Triodanis*.
   Munz, P. A. & D. D. Keck (1970). *Heterocodon* Nutt. In A California flora: 1064. Berkeley:
      University of California Press. En. — Flora with description and synonymy.
   Cronquist, A., A. H. Holmgren, N. H. Holmgren, J. L. Reveal & P. K. Holmgren (1984).
      Heterocodon Nutt. In Intermountain flora 4: 518-519, illus. New York: New York
      Botanical Garden. En. — Flora with description and full nomenclature.
 • Shetler, S. G. & N. R. Morin (1986). Seed morphology in North American Campanulaceae.
      Ann. Missouri Bot. Gard. 73: 653-688, illus. En. — Seed coat morphology supports a
      relationship to *Legousia* and *Triodanis*.
 • Morin, N. (1993). *Heterocodon*. In J. C. Hickman (ed.), The Jepson manual, higher
      plants of California: 464, illus. Berkeley: University of California Press. En. — Flora
      with description.

Heterocodon Nutt., Trans. Amer. Philos. Soc. (n.s.) 8: 255 (1842).
   W. North America. 71 73 76.

Heterocodon rariflorus Nutt., Trans. Amer. Philos. Soc. (n.s.) 8: 255 (1842), as 'rariflorum'.
   Specularia rariflora (Nutt.) McVaugh, Leafl. W. Bot. 3: 48 (1941).
   S. British Columbia to NW. Colorado & S. California. 71 BRC 73 COL IDA MNT ORE WAS
      WYO 76 CAL NEV. Ther.

Synonyms:
Heterocodon brevipes (Hemsl.) Hand.-Mazz. & Nannf. === **Homocodon brevipes** (Hemsl. D.
   Y. Hong
Heterocodon minimus Kellogg === **Triodanis perfoliata** subsp. **biflora** (Ruiz & Pav.)
   Lammers

Heterotoma

Lobelioideae, 1 species, Mexico and Central America. This genus was for many years circumscribed broadly, encompassing all Lobelioideae with spurred corollas. Ayers (1990) narrowed the circumscription radically, transferring all species except the type to *Lobelia*. However, a very preliminary phylogenetic analysis based on DNA sequence data (Koopman & Ayers 2005) recently indicated that the type of *Heterotoma* is part of a clade largely comprised of spurred species of *Lobelia*. $2n = 14$. Type [by monotypy]: *Heterotoma lobelioides* Zucc.

Zuccarini, J. G. (1832). Plantarum novarum vel minus cognitarum, quae in horto botanico herbarioque regio nonacensi cervantur, fasciculus primus. Flora 15 (2. Beibl.): 57-102. La. — Establishment of genus.

Standley, P. C. (1938). Flora of Costa Rica, *Heterotoma* Zucc. Field Mus. Nat. Hist., Bot. Ser. 18: 1414-1415. En. — Floristic treatment with descriptions.

McVaugh, R. (1943). *Heterotoma* Zucc. In North American Flora 32A: 30-35. New York: New York Botanical Garden. En. — Flora with key, descriptions, and full nomenclature.

• Wimmer, F. E. (1953). *Heterotoma* Zucc. In A. Engler & L. Diels, Das Pflanzenreich IV. 276b: 713-720, illus. Berlin: Akademie-Verlag. Ge., La. — Monograph with key, descriptions, infrageneric classification, full nomenclature, and specimen citations.

Nash, D. L. (1976). Flora of Guatemala, *Heterotoma* Zuccarini. Fieldiana Bot. 24: 411-413, illus. En. — Floristic treatment with key and descriptions.

• Ayers, T. J. (1990). Systematics of *Heterotoma* (Campanulaceae) and the evolution of nectar spurs in the New World Lobelioideae. Syst. Bot. 15: 296-327, illus., maps. En. — Establishment of new generic boundaries, based on preliminary cladistic analysis.

Wilbur, R. L. (2001). *Heterotoma* Zucc. In W. D. Stevens, C. Ulloa U., A. Pool & O. M. Montiel (eds.), Flora de Nicaragua 1: 561-562. St. Louis: Missouri Botanical Garden Press. Sp. — Flora with description.

• Koopman, M. M. & T. J. Ayers (2005). Nectar spur evolution in the Mexican lobelias (Campanulaceae: Lobelioideae). Amer. J. Bot. 92: 558-562. En. — Preliminary molecular phylogeny suggests *Heterotoma* embedded within *Lobelia,* forming clade with spurred species of *Lobelia*.

Heterotoma Zucc., Flora 15(2, Beibl.): 100 (1832).
Mexico & Central America. 79 80.
Myopsia C. Presl, Prodr. Monogr. Lobel.: 8 (1836).

Heterotoma lobelioides Zucc., Flora 15 (2, Beibl.): 101 (1832). *Lobelia lobelioides* (Zucc.)
Koopman & T. J. Ayers, Amer. J. Bot. 92: 561 (2005).
C. Mexico to Costa Rica. 79 MXC MXE MXS MXT 80 COS ELS GUA HON NIC. Cham. or hemicr. $2n = 14$.

subsp. **glabra** (T. J. Ayers) Lammers, Novon 16: 70 (2006).
C. Mexico (San Luis Potosí, México State, Jalisco, Guerrero). 79 MXE MXC MXS. Cham. or hemicr.
* *Heterotoma lobelioides* var. *glabra* T. J. Ayers, Syst. Bot. 15: 311 (1990).

subsp. **lobelioides**
S. Mexico (Michoacan, Guerrero, Oaxaca, Chiapas) to Costa Rica. 79 MXS MXT 80 COS ELS GUA HON NIC. Cham. or hemicr. $2n = 14$.
Myopsia mexicana C. Presl, Prodr. Monogr. Lobel.: 8 (1836).
Lobelia calcarata Bertol., Novi Comment. Acad. Sci. Inst. Bononiensis 4: 409 (1840).
Heterotoma tonelii E. Ortiges, Gartenflora 12: 50 (1863). *Heterotoma lobelioides* var. *tonelii* (E. Ortiges) E. Wimm., Ann. Naturhist. Mus. Wien 56: 371 (1948).

Synonyms:

Heterotoma arabidoides (Hook. & Arn.) Benth. & Hook. ex Hemsl. === **Lobelia flexuosa** (C. Presl) A. DC. subsp. **flexuosa**

Heterotoma aurita Brandegee === **Lobelia aurita** (Brandegee) T. J. Ayers

Heterotoma cordifolia (Hook. & Arn.) McVaugh === **Lobelia cordifolia** Hook. & Arn.

Heterotoma cordifolia var. *intermedia* (Hemsl.) E. Wimm. === **Lobelia flexuosa** subsp. **intermedia** (Hemsl.) Lammers

Heterotoma cordifolia var. *tenella* (Turcz.) E. Wimm. === **Lobelia volcanica** T. J. Ayers

Heterotoma endlichii E. Wimm. === **Lobelia endlichii** (E. Wimm.) T. J. Ayers

Heterotoma flexuosa (C. Presl) McVaugh === **Lobelia flexuosa** (C. Presl) A. DC.

Heterotoma flexuosa var. *liebmanniana* E. Wimm. === **Lobelia flexuosa** (C. Presl) A. DC. subsp. **flexuousa**

Heterotoma gibbosa S. Watson === **Lobelia endlichii** (E. Wimm.) T. J. Ayers

Heterotoma goldmanii Fernald === **Lobelia goldmanii** (Fernald) T. J. Ayers

Heterotoma intermedia Hemsl. === **Lobelia flexuosa** subsp. **intermedia** (Hemsl.) Lammers

Heterotoma lobelioides var. *glabra* T. J. Ayers === **Heterotoma lobelioides** subsp. **glabra** (T. J. Ayers) Lammers

Heterotoma lobelioides var. *tonelii* (E. Ortiges) E. Wimm. === **Heterotoma lobelioides** Zucc. subsp. **lobelioides**

Heterotoma macrocentron Benth. === **Lobelia macrocentron** (Benth.) T. J. Ayers

Heterotoma pringlei B. L. Rob. === **Lobelia gypsophila** T. J. Ayers

Heterotoma riparia E. Wimm. === ?

Heterotoma salvadorensis E. Wimm. === **Lobelia cordifolia** Hook. & Arn.

Heterotoma stenodonta Fernald === **Lobelia stenodonta** (Fernald) McVaugh

Heterotoma tenella Turcz. === **Lobelia volcanica** T. J. Ayers

Heterotoma tonelii E. Ortiges === **Heterotoma lobelioides** Zucc. subsp. **lobelioides**

Hippobroma

Lobelioideae, 1 species, Jamaica, but now naturalized throughout much of the tropics. Although it has been allied by many authors (e.g., Gleason 1924, Standley 1938) to Australian *Isotoma* (sometimes under the illegitimate name *Laurentia;* e.g., Wimmer 1953, Tuyn 1960), a preliminary molecular phylogeny (E. Knox, pers. comm.) places it in a clade otherwise comprised of various New World species of *Lobelia*. Specifically, it is sister within that clade to *L. portoricensis*, the only sampled species of *Lobelia* sect. *Tylomium* (C. Presl) Benth., a group endemic to the Caribbean. This supports data from seed morphology (Buss et al. 2001), which suggested a close relationship to that section. $2n$ = 28. Type [by monotypy]: *Hippobroma longiflora* (L.) G. Don

Don, G. (1834). *Hippobroma*. In A general history of the dichlamydeous plants: 698, 717. London: J. G. & F. Rivington. En. — Establishment of genus as distinct from *Lobelia*.

Gleason, H. A. (1924). *Isotoma longiflora*. Addisonia 9: 21-22 + pl. 299, illus. En. — Portrait of *H. longiflora*, with description; argues that species not generically distinct from *Isotoma* of Australia.

Standley, P. C. (1938). Flora of Costa Rica, *Isotoma* Lindl. Field Mus. Nat. Hist., Bot. Ser. 18: 1415. En. — Floristic treatment with description.

• McVaugh, R. (1940). A revision of "Laurentia" and allied genera in North America. Bull. Torrey Bot. Club 67: 778-798. En. — Monograph with description and full nomenclature.

• McVaugh, R. (1943). *Hippobroma* G. Don. In North American Flora 32A: 99-100. New York: New York Botanical Garden. En. — Flora with description and full nomenclature.

Kausik, S. B. & K. Subramanyam (1945). An embryological study of *Isotoma longiflora* Presl. Proc. Indian Acad. Sci., B 21: 269-278, illus. En. — Genus stated to be more similar embryologically to Campanuloideae than to other Lobelioideae.

Kausik, S. B. & K. Subramanyam (1947). Embryogeny of *Isotoma longiflora* Presl. Proc. Indian Acad. Sci., B 26: 164-167, illus. En. — Supplement to Kausik & Subramanyam (1945).

- Wimmer, F. E. (1953). [*Laurentia* (Mich.) Adans.] Sectio III. *Isotoma* (R. Br.) Endl. In A. Engler & L. Diels, Das Pflanzenreich IV. 276b: 398-407, illus. Berlin: Akademie-Verlag. Ge., La. — Monograph with key, descriptions, full nomenclature, and specimen citations.

  Melville, R. (1960). The pollination mechanism of *Isotoma axillaris* Lindl. and the generic status of *Isotoma* Lindl. Kew Bull. 14: 277-279, illus. En. — Argues that *Hippobroma* not related to Australian *Isotoma*.

  Thanikaimoni, G. (1965). *Laurentia longiflora* (Linn.) Endl. in Pondicherry. J. Bombay Nat. Hist. Soc. 62: 323-324. En. — Spread of *H. longiflora* in India as naturalized species.

  Adams, C. D. (1972). *Hippobroma*. In Flowering plants of Jamaica: 737. Mona: University of the West Indies. En. — Flora with description.

  Nash, D. L. (1976). Flora of Guatemala, *Hippobroma* G. Don. Fieldiana Bot. 24: 412-415, illus. En. — Floristic treatment with description and full nomenclature.

  Wilbur, R. L. (1977). Flora of Panama, *Hippobroma*. Ann. Missouri Bot. Gard. 63: 635, illus. En. — Floristic treatment with description, full nomenclature, and specimen citations.

  Ferrari, J. M., T. S. M. Grandi & T. H. Viana (1980). Estudo botânico de *Isotoma longiflora* (Willd.) Presl. Oreades 7(12-13): 29-40, illus. Por. — Anatomy and palynology of *H. longiflora*.

  Panigrahi, G., P. Daniel & M. V. Viswanathan (1981). A note on *Isotoma longiflora* (L.) Presl (Campanulaceae) in India. Indian J. Forest. 4: 151-152. En. — Brief account of naturalized *H. longiflora* in India, with description.

  Jeppesen, S. (1981). *Hippobroma* G. Don. In G. Harling & B. Sparre (eds.), Flora of Ecuador 14: 124-125. Stockholm: Swedish Natural Science Research Council. En. — Flora with description, full nomenclature, and specimen citations.

  Lammers, T. G. (1990). *Hippobroma* G. Don. In W. L. Wagner, D. R. Herbst & S. H. Sohmer, Manual of the flowering plants of Hawai'i: 472, illus. Honolulu: University of Hawaii Press. En. — Flora with description.

  Smith, A. C. (1991). *Hippobroma* G. Don. In Flora Vitiensis nova 5: 244-246, illus. Lāwa'i: National Tropical Botanical Garden. En. — Flora with description, full nomenclature, and specimen citations.

  Serra, L. & M. B. Crespo (1997). An outline revision of the subtribe Siphocampylinae (Lobeliaceae). Lagascalia 19: 881-888, illus., maps. En. — Key to distinguish genus from alleged allies.

- Buss, C. C., T. G. Lammers & R. R. Wise (2001). Seed coat morphology and its systematic implications in *Cyanea* and other genera of Lobelioideae (Campanulaceae). Amer. J. Bot. 88: 1301-1308, illus. En. — Related to *Lobelia* sect. *Tylomium* of Caribbean.

  Wilbur, R. L. (2001). *Hippobroma* G. Don. In W. D. Stevens, C. Ulloa U., A. Pool & O. M. Montiel (eds.), Flora de Nicaragua 1: 562. St. Louis: Missouri Botanical Garden Press. Sp. — Flora with description and full nomenclature.

Hippobroma G. Don, Gen. Hist. 3: 717 (1834). *Laurentia* [unranked] *Hippobroma* (G. Don) Endl., Gen. Pl.: 512 (1838).
Jamaica; widely naturalized in tropics. (29) (40) (42) (60) (63) (78) (79) (80) 81 (82) (83) (84).

Hippobroma longiflora (L.) G. Don, Gen. Hist. 3: 717 (1834).
Jamaica; cult.; widely naturalized in tropics. (29) mau mdg reu rod (40) ind srl (42) all (60) fij (63) haw (78) fla (79) all (80) all 81 cub dom hai JAM lee pue win (82) ven (83) clm ecu per (84) all. Ther., biennial, or hemicr. $2n = 28$.
  * *Lobelia longiflora* L., Sp. Pl.: 930 (1753). *Rapuntium longiflorum* (L.) Mill., Gard. Dict. (ed. 8) (1768). *Isotoma longiflora* (L.) C. Presl, Prodr. Monogr. Lobel.: 42 (1836). *Laurentia longiflora* (L.) Peterm., Pflanzenreich: 444 (1845). *Solenopsis longiflora* (L.) M. R. Almeida, Fl. Maharashtra 3A: 155 (2001).
  Isotoma runcinata Hassk., Bonplandia 7: 181 (1859). *Laurentia longiflora* var. *runcinata* (Hassk.) E. Wimm., Ann. Naturhist. Mus. Wien 56: 337 (1948). *Isotoma longiflora* var. *runcinata* (Hassk.) Panigrahi, P. Daniel & M. V. Viswan., Indian J. Forest. 4: 151 (1981).

Holostigma

This genus was erected to accomodate the dioecious Australian species treated by recent authors as *Lobelia* sect. *Dioica* (E. Wimm.) J. Murata. Type [by monotypy]: *Holostigma dioicum* (R. Br.) G. Don.

> Don, G. (1834). *Hippobroma*. In A general history of the dichlamydeous plants: 698, 716. London: J. G. & F. Rivington. En. — Establishment of genus as distinct from *Lobelia*.

Synonyms:
Holostigma G. Don === **Lobelia** L.
Holostigma dioicum (R. Br.) G. Don === **Lobelia dioica** R. Br.

Holostigmateia

Reichenbach (1841) proposed this name as an avowed substitute for *Holostigma* G. Don (q. v.) to avoid confusion with *Holostigma* Spach (Onagraceae), even though the former had priority. The requisite combination for the type species was never validated. Based on: *Holostigma* G. Don.

> Reichenbach, H. G. L. (1841). Der Deutsche Botaniker. Erster Band. Das Herbarienbuch. Dresden: Arnold. Ge. — Validation of name as substitute for *Holostigma*.

Synonyms:
Holostigmateia Rchb. === **Lobelia** L.

Homocodon

Campanuloideae, 2 species, southern China. On the basis of morphology, a relationship to *Peracarpa* is likely. Type [by monotypy]: *Homocodon brevipes* (Hemsl.) D. Y. Hong.

> Hemsley, W. B. (1903). *Wahlenbergia brevipes,* Hemsl. Hooker's Icon. Pl. 28: pl. 2768, illus. En. — Portrait of *H. brevipes,* with original description.
> Handel-Mazzetti, H. & J. A. Nannfeldt (1936). *Heterocodon* Nutt. In H. Handel-Mazzetti, Symbolae Sinicae 7: 1075-1076. Wien: Julius Springer. Ge. — Transfer to *Heterocodon*.
> Hong, D. Y. (1980). *Homocodon* Hong – a new genus of Campanulaceae from China. Acta Phytotax. Sin. 18: 473-475, illus. Ch. — Establishment of genus as distinct from *Heterocodon* and *Peracarpa*.
> Hong, D. Y. (1983). *Homocodon* Hong. In Flora Reipublicae Popularis Sinicae 73(2): 143-144. Beijing: Science Press. Ch. — Flora of China, with description and full nomenclature.
> • Ying, T. S., Y. L. Zhang & D. E. Boufford (1993). *Homocodon* Hong. In The endemic genera of seed plants in China: 300-303, illus., map. Beijing: Science Press. En. — Descriptive account, including details of pollen morphology.

Homocodon D. Y. Hong, Acta Phytotax. Sin. 18: 473 (1980).
  E. Asia. 36.

Homocodon brevipes (Hemsl.) D. Y. Hong, Acta Phytotax. Sin. 18: 474 (1980).
  SC. China (Guizhou, Sichuan, Yunnan). 36 CHC. Ther.
  * *Wahlenbergia brevipes* Hemsl., Hooker's Icon. Pl. 28: pl. 2768 (1903). *Heterocodon brevipes* (Hemsl.) Hand.Mazz. & Nannf. in Hand.-Mazz., Symb. Sin. 7: 1075 (1936).
  Wahlenbergia monantha H. J. P. Winkl. ex H. Limpr., Repert. Spec. Nov. Regni Veg. Beih. 12: 501 (1922).

Homocodon pedicellatus D. Y. Hong & L. M. Ma, Acta Phytotax. Sin. 29: 268 (1991), as 'pedicellatum'.
  SC. China (Sichuan). 36 CHC. Ther.

Howellia

Lobelioideae, 1 species, western U.S.A. Phylogenetic analysis of molecular data (Schultheis 2001; E. Knox, pers. comm.) indicates that this genus belongs to a clade that also includes *Downingia, Palmerella, Porterella,* and *Legenere. 2n* = 22. Type [by monotypy]: *Howellia aquatilis* A. Gray.

Gray, A. (1879). *Howellia*, nov. gen. Lobeliacearum. Proc. Amer. Acad. Arts 15: 43-44. En. — Establishment of genus.
- McVaugh, R. (1943). *Howellia* A. Gray. In North American Flora 32A: 14. New York: New York Botanical Garden. En. — Flora with description.
- Wimmer, F. E. (1953). *Howellia* A. Gray. In A. Engler & L. Diels, Das Pflanzenreich IV. 276b: 732, illus. Berlin: Akademie-Verlag. Ge., La. — Monograph with description, full nomenclature, and specimen citations.
Hitchcock, C. L., A. Cronquist & M. Ownbey (1959). *Howellia* Gray. In Vascular plants of the Pacific Northwest 4: 482-495, illus. Seattle: University of Washington Press. En. — Flora with description and full nomenclature.
Hitchcock, C. L. & A. Cronquist (1973). *Howellia* Gray. In Flora of the Pacific Northwest: 460, illus. Seattle: University of Washington Press. En. — Flora with description.
McCune, B. (1982). Noteworthy collections. *Howellia aquatilis* Gray (Campanulaceae). Madroño 29: 123-124. En. — Range extension eastward into Montana.
- Lesica, P., R. F. Leary, F. W. Allendorf & D. E. Bilderback (1988). Lack of genic diversity within and among populations of an endangered plant, *Howellia aquatilis*. Conserv. Biol. 2: 275-282, map. En. — Enzyme electrophoresis shows no variation in 18 loci examined in four widely distributed populations; includes brief descriptive account and information on conservation status.
Isle, D. W. (1997). Rediscovery of water howellia for California. Fremontia 25: 29-32, illus. En. — Narrative account of rediscovery of extant populations in N. California, with overview of conservation biology.

Howellia A. Gray, Proc. Amer. Acad. Arts 15: 43 (1879).
　W. U.S.A. 73 76.

Howellia aquatilis A. Gray, Proc. Amer. Acad. Arts 15: 43 (1879).
　Washington to W. Montana & N. California. 73 IDA MNT ORE WAS 76 CAL. Hydrother.
　2n = 22.

Synonyms:
Howellia limosa Greene === **Legenere valdiviana** (Phil.) E. Wimm.
Howellia valdiviana (Phil.) E. Wimm. === **Legenere valdiviana** (Phil.) E. Wimm.

Hypsela

As defined by Wimmer (1943), *Hypsela* accomodated species that resembled *Pratia* (q.v.) except for their entire (vs. dorsally slit) corolla tube. After Chiapella (1997) suggested that the type of *Hypsela* was conspecific with the type of *Pratia,* the three remaining species in the genus were transferred to *Isotoma* (Lammers 1999). This action has been supported by a preliminary analysis of DNA sequences (E. Knox, pers. comm.), which placed *H. reniformis* and *H. tridens* in different parts of a clade that also included numerous Australasian species of *Isotoma* (though not the type) and *Lobelia* (including *Pratia*). Type [by monotypy]: *Hypsela reniformis* (Kunth) C. Presl.

Presl, C. (1836). *Hypsela*. In Prodromus monographiae Lobeliacearum: 45. Prague: Theophilus Haase. La. — Establishment of genus as distinct from *Lysipomia*.
Baillon, H. (1885). *Pratia* Gaudich. In Histoire des plantes 8: 366. Paris: L. Hachette. Fr. — *Hypsela* treated as one of four sections of *Pratia*.

Wimmer, F. E. (1943). *Hypsela* Presl. In A. Engler & L. Diels, Das Pflanzenreich IV. 276b: 119-122, illus. Leipzig: Wilhelm Engelmann. Ge., La. — Monograph with key, descriptions, full nomenclature, and specimen citations.

Wimmer, F. E. (1968). *Hypsela* Presl. In A. Engler & L. Diels, Das Pflanzenreich IV. 276c: 834. Berlin: Akademie-Verlag. La. — Supplement to Wimmer (1943).

Chiapella, J. (1997). Nota sobre la identidad de *Pratia repens* (Campanulaceae: Lobelioideae). Bol. Soc. Argent. Bot. 32: 123-135, illus. Sp. — Merger of type species of *Hypsela* with that of *Pratia*.

Lammers, T. G. (1999). Nomenclatural consequences of the synonymization of *Hypsela reniformis* (Campanulaceae: Lobelioideae). Novon 9: 73-76. En. — Transfer of remaining three species to *Isotoma*.

Synonyms:
Hypsela C. Presl === **Lobelia** L.
Hypsela atacamensis (Phil.) F. Phil. === **Lobelia oligophylla** (Wedd.) Lammers
Hypsela longiflora (Hook. f.) F. Phil. === **Lobelia oligophylla** (Wedd.) Lammers
Hypsela oligophylla (Wedd.) Benth. & Hook. f. ex Zahlbr. === **Lobelia oligophylla** (Wedd.) Lammers
Hypsela reniformis (Kunth) C. Presl === **Lobelia oligophylla** (Wedd.) Lammers
Hypsela rivalis E. Wimm. === **Isotoma rivalis** (E. Wimm.) Lammers
Hypsela sessiliflora E. Wimm. === **Isotoma sessiliflora** (E. Wimm.) Lammers
Hypsela subsessilis (Wedd.) Benth. & Hook. f. ex Zahlbr. === **Lobelia oligophylla** (Wedd.) Lammers
Hypsela tridens E. Wimm. === **Isotoma tridens** (E. Wimm.) Lammers

Hyssaria

Kolakovskii (1981) segregated this genus on the basis of minor differences in fruit structure. Type [designated by the author]: *Hyssaria lehmanniana* (Bunge) Kolak.

Kolakovskii, A. A. (1981). [Another two new monotypical genera of Campanulaceae for the flora of the USSR.] Soobšč. Akad. Nauk Gruzinsk. SSR 103: 149-152, illus. Ru. — Establishment of genus as distinct from *Campanula*.

Synonyms:
Hyssaria Kolak. === **Campanula** L.
Hyssaria lehmanniana (Bunge) Kolak. === **Campanula lehmanniana** Bunge

Ireon

This name has priority over *Wahlenbergia*, but is a later homonym of *Ireon* Burm. f. (Droseraceae). Type [by monotypy]: *Ireon parvifolia* (P. J. Bergius) Scop.

Scopoli, J. A. (1777). Introductio ad historiam naturalem. Prague: Wolfgang Gerle. La. — Establishment of genus.

Synonyms:
Ireon Scop. === **Wahlenbergia** Schrad. ex Roth
Ireon ciliatum Raf. === **Prismatocarpus fruticosus** (L.) L'Hér.
Ireon parvifolia (P. J. Bergius) Scop. === **Wahlenbergia parvifolia** (P. J. Bergius) Lammers

Isolobus

This genus (which current authors treat as *Lobelia* subg. *Isolobus* (A. DC.) Y. S. Lian or *Lobelia* sect. *Isolobus* (A. DC.) C. B. Clarke) was created for herbaceous species with unilabiate corollas. Type [designated by J. Murata, J. Fac. Sci. Univ. Tokyo, Sect. 3, Bot. 15: 357 (1995)]: *Isolobus radicans* (Thunb.) A. DC.

de Candolle, A. (1839). *Isolobus*. In A. P. de Candolle, Prodromus systematis naturalis regni vegetabilis 7: 352-354. Paris: Treuttel & Würtz. La. — Establishment of genus as distinct from *Lobelia, Monopsis,* and *Pratia.*

Adamson, R. S. (1950). *Isolobus* A. DC. In R. S. Adamson & T. M. Salter, Flora of the Cape Peninsula: 759. Cape Town: Juta. En. — One of the few 20th Century floras to recognize the genus.

Synonyms:

Isolobus A. DC. === **Lobelia** L.

Isolobus caespitosus (Blume) Hassk. === **Lobelia chinensis** Lour.

Isolobus campanuloides (Thunb.) A. DC. === **Lobelia chinensis** Lour.

Isolobus concolor (R. Br.) A. DC. === **Lobelia concolor** R. Br.

Isolobus corymbosus (G. Don) A. DC. === **Lobelia jasionoides** (A. DC.) E. Wimm.

Isolobus corymbosus var. *jasionoides* (A. DC.) Sond. === **Lobelia jasionoides** (A. DC.) E. Wimm.

Isolobus corymbosus var. *sparsiflorus* Sond. === **Lobelia jasionoides** (A. DC.) E. Wimm.

Isolobus cunninghamii A. DC. === **Lobelia concolor** R. Br.

Isolobus ecklonianus (C. Presl) Sond. === **Lobelia filicaulis** (C. Presl) Schönl.

Isolobus ecklonianus var. *spathulatus* (R. D. Good) Adamson === **Lobelia jasionoides** (A. DC.) E. Wimm.

Isolobus jasionoides A. DC. === **Lobelia jasionoides** (A. DC.) E. Wimm.

Isolobus keri A. DC. === **Lobelia chinensis** Lour.

Isolobus radicans (Thunb.) A. DC. === **Lobelia chinensis** Lour.

Isolobus roxburghianus A. DC. === **Lobelia chinensis** Lour.

Isolobus sparsiflorus A. DC. === **Lobelia filicaulis** (C. Presl) Schönl.

Isotoma

Lobelioideae, 14 species, Australasia. Early authors (e.g., Sims 1826, Hooker 1858) suggested that most of the species assigned here were not truly congeners of the type species, *I. hypocrateriformis,* a view supported by a preliminary molecular phylogeny (E. Knox, pers. comm.). In that analysis, *I. hypocrateriformis* was part of a clade comprised in large part of Australasian species of *Lobelia* sect. *Delostemon* (E. Wimm.) J. Murata, while the remaining sampled species of *Isotoma* were scattered throughout a second Australasian clade, among species referable to *Lobelia* sect. *Dioica* (E. Wimm.) J. Murata, sect. *Isolobus* (A. DC.) C. B. Clarke, sect. *Paramezleria* E. Wimm., and sect. *Pratia* (Gaudich.) J. Murata. At least the molecular phylogeny supports the view embraced here that none of these species is at all related to *Hippobroma, Palmerella, Solenopsis,* or *Wimmerella;* some earlier authors had treated these five genera as a single genus under the illegitimate name *Laurentia* (q.v.). The most recent monograph was that by Wimmer (1953). $2n$ = 14, 28, 56. Based on: *Lobelia* [unranked] *Isotoma* R. Br. Type [by monotypy]: *Lobelia hypocrateriformis* R. Br.

Brown, R. (1810). Lobelia. In Prodromus florae Novae Hollaniae et Insulae van-Diemen: 562-565. London: J. Johnson. La. — Establishment of taxon upon which genus was based.

Lindley, J. (1826). *Isotoma axillaris*. Bot. Reg. 12: pl. 964+ 2 pp., illus. En. — Establishment of genus as distinct from *Lobelia.*

Sims, J. (1826). *Lobelia senecioides*. Blue pedunculated lobelia. Bot. Mag. 53: tab. 2702 + 2 pp., illus. En. — Portrait of *I. axillaris,* with description; argues that this species does not accord well with *Isotoma* as originally described on basis of *I. hypocrateriformis.*

Hooker, W. J. (1858). *Isotoma senceioides;* var. *subpinnatifida.* Groundsel-leaved isotoma; subpinnatifid var. Bot. Mag. 84: tab. 5073 + 2 pp., illus. En. — Portrait of *I. axillaris,* with description; reiterates Sims' (1826) argument that this species does not accord well with *I. hypocrateriformis.*

Bentham, G. & F. Mueller (1868). *Isotoma,* Lindl. In Flora Australiensis: a description of the plants of the Australian Territory 4: 134-137. London: L. Reeve. En. — Flora with key, descriptions, full nomenclature, and specimen citations; suggests genus not truly distinct from *Lobelia.*

Subramanyam, K. (1951). Flower structure and seed development in *Isotoma fluviatilis*. Proc. Natl. Inst. Sci. India, Pt. B, Biol. Sci. 17: 275-285, illus. En. — Detailed embryological study, including chromosome count.

• Wimmer, F. E. (1953). [*Laurentia* (Mich.) Adans.] sectio III. *Isotoma* (R. Br.) Endl. In A. Engler & L. Diels, Das Pflanzenreich IV. 276b: 398-407, illus. Berlin: Akademie-Verlag. Ge., La. — Monograph with key, descriptions, full nomenclature, and specimen citations.

• Melville, R. (1960). The pollination mechanism of *Isotoma axillaris* Lindl. and the generic status of *Isotoma* Lindl. Kew Bull. 14: 277-279, illus. En. — Description of mechanism of pollen discharge; supports exclusion of *Hippobroma* and *Palmerella* from genus.

James, S. H. (1965). Complex hybridity in *Isotoma petraea*. I. The occurrence of interchange heterozygosity, autogamy and a balanced lethal system. Heredity (London) 20: 341-353 + pl. I, illus., map. En. — Cytogenetic studies document occurence of structural homozygosity and interchange heterozygosity; mechanism for autogamy described.

McComb, J. A. (1968). The occurrence of unisexuality and polyploidy in *Isotoma fluviatilis*. Austral. J. Bot. 16: 525-537, maps. En. — Documentation of diploid and tetraploid hermaphrodites, diploid dioecious populations, and one gynodioecious population.

McComb, J. A. (1969). The genetic basis of unisexuality in *Isotoma fluviatilis*. Austral. J. Bot. 17: 515-526. En. — Genetic study of dioecious populations.

Beltran, I. C. & S. H. James (1970). Complex hybridity in *Isotoma petraea:* III. Lethal system in ☉ 12 Bencubbin. Austral. J. Bot. 18: 223-232. En. — Evidence of transposition hybridity in a structurally heterozygous population.

James, S. H. (1970). Complex hybridity in *Isotoma petraea*. II. Components and operation of a possible evolutionary mechanism. Heredity (London) 25: 53-78. En. — Although structural heterozygosity usually overcomes sterility, this species shows 90% sterility in its structural heterozygotes and full fertility in structural homozygotes; infers pronounced advantage to heterozygosity *per se.*

Beltran, I. C. (1970). Embryology of *Isotoma petraea*. Austral. J. Bot. 18: 213-221, illus. En. — Descriptive study of ovule development, embryo sac formation, and embryogeny.

McComb, J. A. (1970). A revision of the species *Isotoma fluviatilis*. Contr. New South Wales Natl. Herb. 4: 106-11, illus. En. — Biosystematic revision, dividing species into three subspecies; key, descriptions, full nomenclature, and specimen citations.

Beltran, I. C. & S. H. James (1974). Complex hybridity in *Isotoma petraea:* 4. Heterosis in interpopulational hybrids. Austral. J. Bot. 22: 251-264, map. En. — Biosystematic studies among structural heterozygotes demonstrate hybrid vigor.

Moore, L. B. (1974). *Hypsela rivalis* and other pygmy plants. J. Canterbury Bot. Soc. 7: 12-15, illus. En. — Brief popular account of *I. rivalis* and its similarity to *I. fluviatilis* and *Lobelia perpusilla.*

James, S. H. (1982). The relevance of genetic systems in *Isotoma petraea* to conservation practice. In R. H. Groves & W. D. L. Ride (eds.), Species at risk: research in Australia: 63-71. Berlin: Springer Verlag. En. — Application of knowledge of population genetics to endangered species preservation.

• Brantjes, N. B. M. (1983). Regulated pollen issue in *Isotoma,* Campanulaceae, and evolution of secondary pollen presentation. Acta Bot. Neerl. 32: 213-222, illus. En. — Structure and function of anther tube and its relation to pollination.

James, S. H., A. P. Wylie, M. S. Johnson, S. A. Carstairs & G. A. Simpson (1983). Complex hybridity in *Isotoma petraea* V. Allozyme variation and the pursuit of hybridity. Heredity (London) 51: 653-663. En. — Electrophoretic estimates of genetic variation show far greater allozymic variability in fixed heterozygotes than in structural homozygotes.

James, S. H. (1984). The pursuit of hybridity and population divergence in *Isotoma petraea*. In W. F. Grant (ed.), Plant Biosystematics: 159-168. Toronto: Academic Press. En. — Summary of cytogenetic and biosystematic studies of structurally heterozygous populations.

Pate, J. S., N. E. Carson, J. Rullo & J. Kuo (1985). Biology of fire ephemerals of the sandplains of the Kwongan of south-western Australia. Austral. J. Pl. Physiol. 12: 641-655, illus. En. — *I. hypocrateriformis* adapted to fire regime.
• Elliot, W. R. & D. L. Jones (1990). *Isotoma* (R. Br.) Lindley. Encyclopedia of Australian plants suitable for cultivation 5: 450-453, illus. Melbourne: Lothian. En. — Synopsis of genus from horticultural perspective, with descriptions.
Serra, L. & M. B. Crespo (1997). An outline revision of the subtribe Siphocampylinae (Lobeliaceae). Lagascalia 19: 881-888, illus., maps. En. — Key to distinguish genus from alleged allies.
Lammers, T. G. (1999). Nomenclatural consequences of the synonymization of *Hypsela reniformis* (Campanulaceae: Lobelioideae). Novon 9: 73-76. En. — Addition of three species of *Hypsela* (but not its type).

Isotoma (R. Br.) Lindl., Bot. Reg. 12: pl. 964 (1826).
Australasia. 50 51.
  * *Lobelia* [unranked] *Isotoma* R. Br., Prodr.: 565 (1810). *Laurentia* [unranked] *Isotoma* (R. Br.) Endl., Gen. Pl.: 512 (1838). *Laurentia* subg. *Isotoma* (R. Br.) Peterm., Pflanzenreich: 444 (1845). *Laurentia* sect. *Isotoma* (R. Br.) E. Wimm., Ann. Naturhist. Mus. Wien 56: 335 (1948).

Isotoma anethifolia Summerh., Bull. Misc. Inform. Kew 1932: 318 (1932). *Laurentia anethifolia* (Summerh.) E. Wimm., Ann. Naturhist. Mus. Wien 56: 336 (1948).
Queensland & New South Wales. 50 NSW QLD. Hemicr. $2n = 14$.

Isotoma armstrongii E. Wimm., Repert. Spec. Nov. Regni Veg. 38: 77 (1935). *Laurentia armstrongii* (E. Wimm.) E. Wimm., Ann. Naturhist. Mus. Wien 56: 336 (1948).
N. Australia. 50 NSW NTA QLD WAU. Hemicr. $2n = 14$.

Isotoma axillaris Lindl., Bot. Reg. 12: pl. 964 (1826). *Lobelia senecionis* Spreng., Syst. Veg. 4(2): 75 (1827). *Laurentia axillaris* (Lindl.) E. Wimm., Ann. Naturhist. Mus. Wien 56: 336 (1948).
E. Australia; cult. 50 NSW QLD VIC. Hemicr. $2n = 14$.
  Lobelia senecioides A. Cunn. ex Sims, Bot. Mag. 53: tab. 2702 (1826). *Isotoma senecioides* (A. Cunn. ex Sims) A. DC. in DC., Prodr. 7: 412 (1839).
  Isotoma senecioides var. *subpinnatifida* Hook., Bot. Mag. 84: tab. 5073 (1858).

Isotoma baueri C. Presl, Prodr. Monogr. Lobel.: 42 (1836). *Laurentia ferdinandii* E. Wimm., Ann. Naturhist. Mus. Wien 56: 336 (1948), as 'ferdinandi;' non *Laurentia baueri* (C. Presl) A. DC. in DC., Prodr. 7: 411 (1839).
Queensland. 50 QLD. Hemicr.

Isotoma fluviatilis (R. Br.) F. Muell. ex Benth., Fl. Austral. 4: 136 (1867).
E. Australia & Tasmania. 50 NSW QLD SOA TAS VIC. Hemicr. $2n = 14, 28$.
  * *Lobelia fluviatilis* R. Br., Prodr.: 563 (1810). *Rapuntium fluviatile* (R. Br.) C. Presl, Prodr. Monogr. Lobel.: 30 (1836). *Laurentia fluviatilis* (R. Br.) E. Wimm., Ann. Naturhist. Mus. Wien 56: 335 (1948).
  Enchysia baueri C. Presl, Prodr. Monogr. Lobel.: 40 (1836). *Laurentia baueri* (C. Presl) A. DC. in DC., Prodr. 7: 411 (1839).
  Enchysia baueri var. *major* C. Presl, Prodr. Monogr. Lobel.: 40 (1836). *Laurentia baueri* var. *major* (C. Presl) A. DC. in DC., Prodr. 7: 411 (1839).
  Enchysia gaudichaudii C. Presl, Prodr. Monogr. Lobel.: 41 (1836). *Laurentia gaudichaudii* (C. Presl) A. DC. in DC., Prodr. 7: 411 (1839).

  subsp. **australis** McComb, Contr. New South Wales Natl. Herb. 4: 109 (1970).
    SE. Australia (South Australia to New South Wales) & Tasmania. 50 NSW SOA TAS VIC. Hemicr. $2n = 28$.

  subsp. **borealis** McComb, Contr. New South Wales Natl. Herb. 4: 108 (1970).
    E. Australia (New South Wales, Queensland). 50 NSW QLD. Hemicr. $2n = 14$.

subsp. **fluviatilis**
> SE. Australia (New South Wales). 50 NSW. Hemicr. $2n = 14$.
> *Lobelia inundata* R. Br., Prodr.: 563 (1810). *Rapuntium inundatum* (R. Br.) C. Presl, Prodr. Monogr. Lobel.: 30 (1836). *Isotoma fluviatilis* var. *inundata* (R. Br.) Benth., Fl. Austral. 4: 137 (1867). *Laurentia fluviatilis* var. *inundata* (R. Br.) E. Wimm., Ann. Naturhist. Mus. Wien 56: 335 (1948).
> *Enchysia lessonii* C. Presl, Prodr. Monogr. Lobel.: 40 (1836). *Laurentia fluviatilis* var. *lessonii* (C. Presl) E. Wimm., Ann. Naturhist. Mus. Wien 56: 335 (1948).

Isotoma gulliveri F. Muell., Fragm. 10: 39 (1876), as 'gulliverii'. *Laurentia gulliveri* (F. Muell.) E. Wimm., Ann. Naturhist. Mus. Wien 56: 336 (1948), as 'gulliverii'.
> Queensland. 50 QLD. Ther.

Isotoma hypocrateriformis (R. Br.) Druce, Bot. Soc. Exch. Club Brit. Isles 4: 629 (1917).
> Western Australia. 50 WAU. Ther. $2n = 28$
> * *Lobelia hypocrateriformis* R. Br., Prodr.: 565 (1810). *Isotoma brownii* G. Don, Gen. Hist. 3: 716 (1834). *Laurentia hypocrateriformis* (R. Br.) E. Wimm., Pflanzenr. IV.276b: 13 (1943).
> *Isotoma brevifolia* C. Presl, Prodr. Monogr. Lobel.: 43 (1836).
> *Lobelia lehmannii* de Vriese in Lehm., Pl. Preiss.: 394 (1845), as 'lehmanni'.

Isotoma luticola Carolin, Telopea 2: 63 (1980).
> Northern Territory. 50 NTA. Hemicr.

Isotoma petraea F. Muell., Linnaea 25: 420 (1852). *Laurentia petraea* (F. Muell.) E. Wimm., Ann. Naturhist. Mus. Wien 56: 336 (1948).
> Australia. 50 NSW NTA QLD SOA VIC WAU. Hemicr. $2n = 14$.
> *Isotoma petraea* f. *coerulea* Voss in Siebert & Voss, Vilm. Blumengärtn. (ed. 3) 1: 578 (1894).

Isotoma pusilla Benth. in Endl., Enum. Pl.: 75 (1837). *Laurentia pusilla* (Benth.) A. DC. in DC., Prodr. 7: 411 (1839).
> Western Australia. 50 WAU. Ther.
> *Lobelia elegans* de Vriese in Lehm., Pl. Preiss.: 396 (1845).

Isotoma rivalis (E. Wimm.) Lammers, Novon 9: 75 (1999).
> New Zealand. 51 NZN NZS. Hemicr. $2n = 14$.
> * *Hypsela rivalis* E. Wimm., Pflanzenr. IV.276b: 121 (1943).

Isotoma scapigera (R. Br.) G. Don, Gen. Hist. 3: 716 (1834).
> SW. Australia. 50 SOA WAU. Ther. $2n = 56$.
> * *Lobelia scapigera* R. Br., Prodr.: 565 (1810). *Enchysia scapigera* (R. Br.) C. Presl, Prodr. Monogr. Lobel.: 41 (1836). *Laurentia scapigera* (R. Br.) Endl. ex E. Wimm., Ann. Naturhist. Mus. Wien 56: 336 (1948).
> *Lobelia scapigera* var. *biuncialis* R. Br., Prodr.: 565 (1810). *Isotoma scapigera* var. *biuncialis* (R. Br.) G. Don, Gen. Hist. 3: 716 (1834).
> *Lobelia scapigera* var. *pusilla* R. Br., Prodr.: 565 (1810). *Isotoma scapigera* var. *pusilla* (R. Br.) G. Don, Gen. Hist. 3: 716 (1834). *Enchysia scapigera* var. *pusilla* (R. Br.) A. DC. in DC., Prodr. 7: 409 (1839).
> *Lobelia ophiocephala* de Vriese in Lehm., Pl. Preiss.: 397 (1845).
> *Lobelia monanthos* de Vriese in Lehm., Pl. Preiss.: 398 (1845), as 'monanthus'.

Isotoma sessiliflora (E. Wimm.) Lammers, Novon 9: 75 (1999).
> New South Wales. 50 NSW. Ther.
> * *Hypsela sessiliflora* E. Wimm., Pflanzenr. IV.276b: 121 (1943).

Isotoma tridens (E. Wimm.) Lammers, Novon 9: 75 (1999).
> New South Wales & Victoria. 50 NSW VIC. Ther.
> * *Hypsela tridens* E. Wimm., Pflanzenr. IV.276b: 121 (1943).

Synonyms:

Isotoma brevifolia C. Presl === **Isotoma hypocrateriformis** (R. Br.) Druce

Isotoma brownii G. Don === **Isotoma hypocrateriformis** (R. Br.) Druce

Isotoma fluviatilis var. *inundata* (R. Br.) Benth. === **Isotoma fluviatilis** (R. Br.) F. Muell. ex Benth. subsp. **fluviatilis**

Isotoma longiflora (L.) C. Presl === **Hippobroma longiflora** (L.) G. Don

Isotoma longiflora var. *runcinata* (Hassk.) Panigrahi, P. Daniel & M. V. Viswan. === **Hippobroma longiflora** (L.) G. Don

Isotoma petraea f. *coerulea* Voss === **Isotoma petraea** F. Muell.

Isotoma runcinata Hassk. === **Hippobroma longiflora** (L.) G. Don

Isotoma scapigera var. *biuncialis* (R. Br.) G. Don === **Isotoma scapigera** (R. Br.) G. Don

Isotoma scapigera var. *pusilla* (R. Br.) G. Don === **Isotoma scapigera** (R. Br.) G. Don

Isotoma senecioides (A. Cunn. ex Sims) A. DC. === **Isotoma axillaris** Lindl.

Isotoma senecioides var. *subpinnatifida* Hook. === **Isotoma axillaris** Lindl.

Jasione

Campanuloideae, 15 species, Mediterranean. Phylogenetic analysis of molecular data (Eddie et al. 2003, under **General**) supports the monophyly of this well-marked genus; all five species sampled formed a well supported clade, one of the three "transitional taxa" that form a pentachotomy with the "Campanulaceae s. str. clade" and the "Rapunculus clade" within the campanuloid clade. The most recent monograph was that of Schmeja (1931). *2n* = 12, 14, 18, 24, 36, 48, 60. Type [by monotypy]: *Jasione montana* L.

Linnaeus, C. (1753). *Jasione*. In Species plantarum: 928-929. Stockholm: Laurentius Salvius. La. — Establishment of genus in Syngenesia Monogamia.

Linnaeus, C. (1754). *Jasione*. In Genera plantarum: 400. Stockholm: Laurentius Salvius. La. — Description of genus to accompany preceding.

Sims, J. (1820). *Jasione perennis*. Perennial sheeps-bit. Bot. Mag. 48: tab. 2198 + 2 pp., illus. En. — Portrait of *J. laevis* subsp. *laevis*, with description.

Stojanow, N. (1926). Über den Formenkreis von *Jasione supina* (Sieb.) DC. Notizbl. Bot. Gart. Berlin-Dahlem 9: 545-561, illus., map. Ge. — Treatment of small group of species, with key, descriptions, full nomenclature, and specimen citations.

- Schmeja, O. (1931). Beitrag zur Kenntnis der Gattung *Jasione*. Beih. Bot. Centralbl. 48(2): 1-51 + Taf. I-II. Ge. — Monograph with key, descriptions, infrageneric classification, full nomenclature, and specimen citations.

Stefanoff, B. (1936). [Über die systematische Stellung einiger Arten der Familie Campanulaceae.] God. Sofiisk. Univ. Agron.-Lesoved. Fak. 14: 93-104. Bu. — Segregation of *J. bulgarica* and *J. foliosa* as segregate genera *Jasionella* and *Urumovia*, respectively.

- Fedorov, A. A. (1957). *Jasione* L. In V. L. Komarov, Flora URSS 24: 446-448, illus. Ru. — Leningrad: Academia Scientiae URSS. Flora for former Soviet Union, with key, descriptions, infrageneric classification, and full nomenclature.

Kovanda, M. (1968). Cytotaxonomic studies in the genus *Jasione* L. I. *Jasione montana* L. Folia Geobot. Phytotax. 3: 193-199. En. — Widespread sampling showed no variation in chromosome number.

Tutin, T. G. (1973). Notes on *Jasione* L. in Europe. Bot. J. Linn. Soc. 67: 276-279. En. — Brief notes preparatory to Tutin (1976).

- Tutin, T. G. (1976). *Jasione* L. In T. G. Tutin, V. H. Heywood, N., A. Burges, D. M. Moore, D. H. Valentine, S. M. Walters & D. A. Webb (eds.), Flora Europaea 4: 100-102. Cambridge: Cambridge University Press. En. — Flora with key and descriptions.

Damboldt, J. (1978). *Jasione* L. In P. H. Davis (ed.), Flora of Turkey and the East Aegean Islands 6: 86-89, maps. Edinburgh: University Press. En. — Flora treatment with key and descriptions.

Parnell, J. (1982a). Cytotaxonomy of *Jasione montana* L. in the British Isles. Watsonia 14: 147-151, illus. En. — Chromosome number variation.

Parnell, J. (1982b). Some observations on the breeding biology of *Jasione montana* L. J. Life Sci. Roy. Dublin Soc. 4: 1-7, illus. En. — Reproductive biology.

Parnell, J. (1985). Biological flora of the British Isles – *Jasione montana* L. J. Ecol. 73: 241-258, illus. En. — Descriptive and ecological account.

• Parnell, J. (1987). Variation in *Jasione montana* L. (Campanulaceae) and related species in Europe and North Africa. Watsonia 16: 249-267, illus., maps. En. — Based on biometrical study of morphological variation, species divided into two subspecies, one with six varieties; supports recognition of *J. corymbosa*, *J. heldreichii*, and *J. penicillata*.

Leitao, M. T. & J. Paiva (1988). El endemismo lusitano de *Jasione* L. (Campanulaceae). Lagascalia 15 (Suppl.): 341-344, illus., maps. Sp. — Description of taxa endemic to Portugal.

Parnell, J. (1988). Numerical analysis - its potential application to *Jasione* L., *Jovibarba* Opiz and *Sempervivum* L. In C. Moriarty (ed.), Taxonomy: putting plants and animals in their place: 31-50. Dublin: Royal Irish Academy. En. — Use of morphometrics to resolve classification of annual and biennial members.

Perez C., J. L. (1988). Las subespecies de *Jasione crispa* (Pourret) Sampaio (Campanulaceae) en la provincia corologica Luso-Extremadurense. Stud. Bot. (Salamanca) 6: 53-66, map. Sp. — Infraspecific variation in Iberian peninsula.

Huxley, A. J., ed. (1992). *Jasione*. The new Royal Horticultural Society dictionary of gardening 2: 712. London: MacMillan. En. — Account of species in cultivation.

Bokhari, M. H. & F. Sales (2001). *Jasione* (Campanulaceae) anatomy in the Iberian peninsula and its taxonomic significance. Edinburgh J. Bot. 58: 405-422, illus. En. — Various features of stem and leaf anatomy are species specific; anatomical features correlate with ecology.

Sales, F. & I. C. Hedge (2001a). Notulae taxinomicae, chorologicae, nomenclaturales, bibliographicae aut philologicae in opus "Flora Iberica" intendentes. Anales Jard. Bot. Madrid 59: 163-172. En. — Preparatory notes for next.

• Sales, F. & I. C. Hedge (2001b). *Jasione* L. In S. Castroviejo (ed.), Flora Iberica 14: 153-170, illus. Madrid: Real Jardín Botánico. Sp. — Flora of Spain with keys, descriptions, and full nomenclature.

Jasione L., Sp. Pl.: 928 (1753). *Ovilla* Adans., Fam. Pl. 2: 134, 586 (1763).
Europe & N. Africa; naturalized in Macaronesia and E. North America. 10 11 12 13 14 20 (21) 34 (75) (78).
Jasionella Stoj. & Stef., Fl. Bulg. (ed. 2): 986 (1933).
Urumovia Stef., God. Sofiisk. Univ. Agron.-Lesoved. Fak. 14: 103 (1936).

Jasione bulgarica Stoj. & Stef., Oesterr. Bot. Z. 70: 105 (1921). *Jasionella bulgarica* (Stoj. & Stef.) Stoj. & Stef., Fl. Bulg. (ed. 2): 986 (1933).
Bulgaria. 13 BUL. Hemicr. $2n = 12$.

Jasione cavanillesii C. Vicioso, Anales Jard. Bot. Madrid 6(2): 80 (1946). *Jasione amethystina* subsp. *cavanillesii* (C. Vicioso) C. Vicioso & Lainz ex Lainz, Bol. Inst. Estud. Asturianos 5: 30 (1962). *Jasione humilis* subsp. *cavanillesii* (C. Vicioso) Rivas Mart., Anales Inst. Bot. Cavanilles 21: 270 (1963). *Jasione crispa* subsp. *cavanillesii* (C. Vicioso) Tutin, Bot. J. Linn. Soc. 67: 278 (1973).
NW. Spain. 12 SPA. Hemicr. $2n = 12$.

Jasione corymbosa Poir. ex Schult. in Roem. & Schult., Syst. Veg. 5: 474 (1819). *Jasione montana* var. *corymbosa* (Poir. ex Schult.) Schmeja, Beih. Bot. Centralbl. 48(2): 31 (1931). *Jasione montana* subsp. *corymbosa* (Poir. ex Schult.) Greuter & Burdet, Willdenowia 11: 40 (1981). $2n = 12$.
W. Mediterranean. 12 POR SPA 20 ALG MOR. Ther. $2n = 12$.

subsp. **corymbosa**
W. Mediterranean. 12 POR SPA 20 ALG MOR. Ther. $2n = 12$.
Jasione corymbosa var. *battandieri* Maire, Bull. Soc. Hist. Nat. Afrique N. 28: 367 (1937).

subsp. **glabra** (Durieu ex Boiss. & Reut.) Batt. & Trab., Fl. Algérie, Dicotyl.: 571 (1889).
NW. Africa. 20 ALG MOR. Ther.
* *Jasione glabra* Durieu ex Boiss. & Reut., Pugill. Pl. Afr. Bor. Hispan.: 72 (1852). *Jasione montana* subsp. *glabra* (Durieu ex Boiss. & Reut.) Greuter & Burdet, Willdenowia 11: 40 (1981). *Jasione corymbosa* var. *glabra* (Durieu ex Boiss. & Reut.) J. Parn., Watsonia 16: 265 (1987).

Jasione crispa (Pourr.) Samp., Ann. Sci. Acad. Polytecn. Porto 14: 161 (1921).
W. Mediterranean. 12 FRA POR SPA 20 ALG MOR TUN. Hemicr. $2n$ = 12, 18, 24, 36, 48.
* *Phyteuma crispum* Pourr., Hist. & Mém. Acad. Roy. Sci. Toulouse 3: 324 (1788). *Jasione montana* var. *humilis* Pers., Syn. Pl. 2: 215 (1806). *Jasione humilis* (Pers.) Loisel., Not. Fl. France: 42 (1810). *Jasione perennis* subsp. *humilis* (Pers.) Bonnier & Layens, Tabl. Syn. Pl. Vasc. France: 196 (1894). *Ovilla humilis* (Pers.) Bubani, Fl. Pyren. 2: 22 (1899).

subsp. **arvernensis** Tutin, Bot. J. Linn. Soc. 67: 278 (1973).
SC. France. 12 FRA. Hemicr. $2n$ = 36.

subsp. **crispa**
W. Mediterranean. 12 FRA SPA 20 ALG MOR TUN. Hemicr. $2n$ = 12, 24, 36, 48.
Jasione fallax Willk., Flora 35: 198 (1852). *Jasione humilis* var. *fallax* (Willk.) Nyman, Consp. Fl. Eur.: 487 (1879).
Jasione bovei Boiss. & Reut., Pugill. Pl. Afr. Bor. Hispan.: 74 (1852). *Jasione sessiliflora* var. *bovei* (Boiss. & Reut.) Batt. & Trab., Fl. Algérie, Dicotyl.: 571 (1889).
Jasione humilis var. *pygmaea* Willk. in Willk. & Lange, Prodr. Fl. Hispan. 2: 283 (1868). *Jasione sessiliflora* f. *pygmaea* (Willk.) Pau, Anales Soc. Esp. Hist. Nat. 29: 288 (1900). *Jasione humilis* subsp. *centralis* Rivas Mart., Anales Inst. Bot. Cavanilles 21: 270 (1963). *Jasione amethystina* subsp. *centralis* (Rivas Mart.) Rivas Mart., Publ. Inst. Biol. Aplicada 42: 121 (1967). *Jasione crispa* subsp. *centralis* (Rivas Mart.) Tutin, Bot. J. Linn. Soc. 67: 278 (1973).
Jasione atlantica Ball, J. Bot. 11: 373 (1873). *Jasione amethystina* var. *atlantica* (Ball) Litard. & Maire, Mém. Soc. Sci. Nat. Maroc 26: 30 (1931). *Jasione humilis* var. *atlantica* (Ball) Maire in Jahand. & Maire, Cat. Pl. Maroc: 737 (1934). *Jasione humilis* f. *atlantica* (Ball) Quézel, Feddes Repert. Spec. Nov. Regni Veg. 56: 6, 62 (1953). *Jasione crispa* var. *atlantica* (Ball) Küpfer, Natural. Monspel., Sér. Bot. 29: 41 (1980).
Jasione megalocalyx Pomel, Nouv. Mat. Fl. Atl.: 258 (1875). *Jasione humilis* var. *megalocalyx* (Pomel) Jahand. & Maire, Cat. Pl. Maroc: 737 (1934). *Jasione sessiliflora* subsp. *megalocalyx* (Pomel) Batt. & Trab., Fl. Algérie, Dicotyl.: 571 (1889). *Jasione humilis* f. *megalocalyx* (Pomel) Quézel, Feddes Repert. Spec. Nov. Regni Veg. 56: 6, 62 (1953).
Jasione coespitans Pomel, Nouv. Mat. Fl. Atl.: 258 (1875).
Jasione nuriensis Sennen, Bol. Soc. Ibér. Ci. Nat. 28: 175 (1930).
Jasione brevisepala Rothm., Cavanillesia 7: 121 (1935). *Jasione crispa* subsp. *brevisepala* (Rothm.) Rivas Mart., Candollea 31: 112 (1976).

subsp. **lanuginella** (Litard. & Maire) J. Lambinon & J. Lewalle, Bull., Soc. Échange Pl. Vasc. Eur. Occid. Bassin Médit. 21: 59 (1986).
Morocco. 20 MOR. Hemicr.
* *Jasione amethystina* var. *lanuginella* Litard. & Maire, Mém. Soc. Sci. Nat. Maroc 26: 30 (1931). *Jasione humilis* var. *lanuginella* (Litard. & Maire) Maire, Bull. Soc. Hist. Nat. Afrique N. 23: 199 (1932).
Jasione amethystina var. *longidentata* Litard. & Maire, Mém. Soc. Sci. Nat. Maroc 26: 30 (1931). *Jasione humilis* var. *longidentata* (Litard. & Maire) Maire, Bull. Soc. Hist. Nat. Afrique N. 23: 199 (1932).

subsp. **mariana** (Willk.) Rivas Mart., Anales Inst. Bot. Cavanilles 18: 45 (1972).
C. Spain. 12 SPA. Hemicr. $2n$ = 12.

Jasione mariana Willk. in Willk. & Lange, Prodr. Fl. Hispan. 2: 284 (1868). *Jasione humilis* subsp. *mariana* (Willk.) Rivas Mart., Anales Inst. Bot. Cavanilles 21: 270 (1963). *Jasione amethystina* subsp. *mariana* (Willk.) Rivas Mart., Publ. Inst. Biol. Aplicada 42: 121 (1967).

subsp. **varduliensis** Uribe-Ech., Claves Ilustr. Fl. País Vasco y Territor. Limítr.: 767 (1999).
N. Spain. 12 SPA. Hemicr.

subsp. **tomentosa** (A. DC.) Rivas Mart., Anales Inst. Bot. Cavanilles 18: 45 (1972).
WC. & S. Spain. 12 SPA. Hemicr. $2n = 24, 48$.
Jasione humilis var. *tomentosa* A. DC., Monogr. Campan.: 105 (1830). *Jasione sessiliflora* var. *eriantha* Boiss. & Reut., Diagn. Pl. Nov. Hisp.: 21 (1842). *Jasione humilis* subsp. *tomentosa* (A. DC.) Rivas Mart., Anales Inst. Bot. Cavanilles 21: 270 (1963). *Jasione amethystina* subsp. *tomentosa* (A. DC.) Rivas Mart., Publ. Inst. Biol. Aplicada 42: 121 (1967). *Jasione tomentosa* (A. DC.) Rivas. Mart., Lagascalia 15 (extra): 117 (1988). *Jasione sessiliflora* subsp. *tomentosa* (A. DC.) Rivas Mart., Itinera Geobot. 15: 703 (2002).
Jasione diapensifolia Gand., Bull. Soc. Bot. France 59: 61 (1912).
Jasione crispa subsp. *segurensis* Mota, C. Díaz, Gómez-Merc. & F. Valle, Lagascalia 15 (extra): 479 (1988).

subsp. **tristis** (Bory) G. Lopéz, Anales Jard. Bot. Madrid 56: 166 (1998).
S. Spain. 12 SPA. Hemicr. $2n = 18, 36$.
Jasione amethystina Lag. & Rodr., Anales Ci. Nat. 5: 271 (1802). *Ovilla amethystina* (Lag. & Rodr.) Bubani, Fl. Pyren. 2: 22 (1899). *Jasione crispa* raca *amethystina* (Lag. & Rodr.) Samp., Ann. Sci. Acad. Polytecn. Porto 14: 161 (1921). *Jasione humilis* subsp. *amethystina* (Lag. & Rodr.) Rivas Mart., Anales Inst. Bot. Cavanilles 21: 270 (1963). *Jasione crispa* subsp. *amethystina* (Lag. & Rodr.) Tutin, Bot. J. Linn. Soc. 67: 278 (1973).
Jasione tristis Bory, Ann. Gén. Sci. Phys. 3: 3 (1820). *Jasione amethystina* var. *tristis* (Bory) A. DC. in DC., Prodr. 7: 416 (1839).

Jasione foliosa Cav., Icon. 2: 38 (1793). *Urumovia foliosa* (Cav.) Stef., God. Sofiisk. Univ. Agron.-Lesoved. Fak. 14: 103 (1936).
S. & C. Spain. 12 SPA. Hemicr. $2n = 12$.

subsp. **foliosa**
S. Spain. 12 SPA. Hemicr. $2n = 12$.
Phyteuma minutum Schult. in Roem. & Schult., Syst. Veg. 5: 86 (1819). *Jasione minuta* (Schult.) Pau, Anales Soc. Esp. Hist. Nat. 27: 90 (1898). *Jasione foliosa* var. *minuta* (Schult.) Cuatrec., Trab. Mus. Ci. Nat. Barcelona 12: 443 (1929). *Jasione foliosa* subsp. *minuta* (Schult.) Font Quer, Cavanillesia 7: 79 (1935).
Phyteuma rigidifolium Dufour ex Schult. in Roem. & Schult., Syst. Veg. 5: 87 (1819).
Jasione nummulariifolia Pau, Anales Soc. Esp. Hist. Nat. 27: 90 (1898), as 'nummulariaefolia'.

subsp. **mansanetiana** (Roselló & Peris) Rivas Mart., Itinera Geobot. 15: 703 (2002).
C. Spain. 12 SPA. Hemicr.
Jasione mansanetiana Roselló & Peris, Bol. Soc. Castellonense Cultura 68: 209 (1992).

Jasione heldreichii Boiss. & Orph. in Boiss., Diagn. Pl. Orient. (ser. 2) 6: 120 (1859). *Jasione montana* var. *heldreichii* (Boiss. & Orph.) Nyman, Consp. Fl. Eur.: 486 (1879). *Jasione montana* f. *heldreichii* (Boiss. & Orph.) Schmeja, Beih. Bot. Centralbl. 48(2): 33 (1931).
Balkans, E. Aegean Is. (Lésvos) & W. Turkey. 13 ALB BUL GRC ROM TUE YUG 34 EAI TUR. Hemicr. or biennial. $2n = 12$.
Jasione montana var. *dentata* A. DC. in DC., Prodr. 7: 415 (1839). *Jasione dentata* (A. DC.) Halácsy, Consp. Fl. Graec. 2: 280 (1902).
Jasione jankae Neilr., Aufz. Ungarn Slavon. Gefässpfl., Nachtr.: 43 (1870). *Jasione montana* f. *jankae* (Neilr.) Schmeja, Beih. Bot. Centralbl. 48(2): 33 (1931).
Jasione glabra Velen., Oesterr. Bot. Z. 34: 424 (1884); non Durieu ex Boiss. & Reut., Pugill. Pl. Afr. Bor. Hispan.: 72 (1852). *Jasione heldreichii* var. *glabra* Nyman, Consp. Fl. Eur., Suppl. 2: 212 (1889). *Jasione heldreichii* var. *microcephala* Velen., Fl. Bulg.: 374 (1891).

Jasione dentata f. *canescens* Gajiç, Glasn. Prir. Mus. Beogradu, Ser. B, Biol. 28: 64 (1973).
Jasione dentata f. *dubravkiana* Gajiç, Glasn. Prir. Mus. Beogradu, Ser. B, Biol. 28: 64 (1973).
Jasione dentata f. *integrifolia* Gajiç, Glasn. Prir. Mus. Beogradu, Ser. B, Biol. 28: 65 (1973).
Jasione heldreichii var. *papillosa* J. Parn., Watsonia 16: 266 (1987).

Jasione idaea Stoj., Notizbl. Bot. Gart. Berlin-Dahlem 9: 556 (1926).
W. Turkey. 34 TUR. Hemicr. $2n = 12$.

Jasione laevis Lam., Fl. Franç. 2: 3 (1779).
Spain to Luxembourg; naturalized in Finland. (10) fin 11 BGM 12 FRA SPA. Hemicr. $2n = 12, 24, 60$.

subsp. **carpetana** (Boiss. & Reut.) Rivas Mart., Publ. Inst. Biol. Aplicada 42: 122 (1967).
C. Spain. 12 SPA. Hemicr. $2n = 12$.
**Jasione carpetana* Boiss. & Reut. in Boiss., Voy. Bot. Espagne: 745 (1845). *Jasione perennis* var. *carpetana* (Boiss. & Reut.) Willk. in Willk. & Lange, Prodr. Fl. Hispan. 2: 284 (1868). *Jasione perennis* subsp. *carpetana* (Boiss. & Reut.) Nyman, Consp. Fl. Eur.: 487 (1879).

subsp. **laevis**
Spain to Luxembourg; naturalized in Finland. (10) fin 11 BGM 12 FRA SPA. Hemicr. $2n = 12, 24, 60$.
Jasione montana var. *perennis* L. f., Suppl. Pl.: 392 (1782). *Jasione perennis* (L. f.) Lam., Encycl. 3: 216 (1789). *Ovilla perennis* (L. f.) Bubani, Fl. Pyren. 2: 21 (1899).
Jasione perennis var. *pygmaea* Gren. & Godr., Fl. Fr. 2: 399 (1853). *Jasione pygmaea* (Gren. & Godr.) Timb.-Lagr., Bull. Soc. Sci. Phys. Nat. Toulouse 6: 149 (1883-84). *Jasione perennis* race *pygmaea* (Gren. & Godr.) Sennen, Bull. Soc. Bot. France 74: 387 (1927). *Jasione laevis* var. *pygmaea* (Gren. & Godr.) O. Bolòs & Vigo, Collect. Bot. (Barcelona) 14: 102 (1983).
Jasione perennis proles *pyrenaica* Sennen, Bol. Soc. Aragonesa Ci. Nat. 15: 251 (1916).
Jasione perennis race *pyrenaea* Sennen, Bull. Soc. Bot. France 74: 387 (1927). *Jasione perennis* subsp. *pyrenaea* (Sennen) Sennen, Butl. Inst. Catalana Hist. Nat. 32: 115 (1932).
Jasione tajae Sennen, Bol. Soc. Ibér. Ci. Nat. 28: 173 (1930).
Jasione perennis var. *gracilis* Sennen, Butl. Inst. Catalana Hist. Nat. 32: 115 (1932).
Jasione laevis subsp. *gredensis* Rivas Mart. & Sancho, Lazaroa 6: 181 (1985).

Jasione maritima (Duby) Dufour ex Merino, Fl. Galicia 2: 291 (1906).
NW. Portugal to SW. France. 12 FRA POR SPA. Hemicr. $2n = 12$.
**Jasione montana* var. *maritima* Duby, Bot. Gall. 1: 311 (1828). *Jasione humilis* var. *maritima* (Duby) Willk. in Willk. & Lange, Prodr. Fl. Hispan. 2: 283 (1868). *Jasione montana* race *maritima* (Duby) Rouy, Fl. France 10: 92 (1908). *Jasione montana* subsp. *maritima* (Duby) C. Vicioso, Anales Jard. Bot. Madrid 6: 80 (1946). *Jasione crispa* subsp. *maritima* (Duby) Tutin, Bot. J. Linn. Soc. 67: 278 (1973).
Jasione montana var. *sabularia* Cout., Fl. Portugal: 603 (1913). *Jasione maritima* var. *sabularia* (Cout.) Sales & Hedge, Anales Jard. Bot. Madrid 59: 168 (2001). *Jasione sessiliflora* subsp. *sabularia* (Cout.) Rivas Mart., Itinera Geobot. 15: 703 (2002).
Jasione montana var. *imbricans* J. Parn., Watsonia 16: 264 (1987).

Jasione montana L., Sp. Pl.: 928. 1753. *Jasione vulgaris* Gaterau, Descr. Pl. Montauban: 153 (1789). *Ovilla globulariiflora* Rupr., Fl. Ingr. 1: 653 (1856), as 'globulariaeflora'.
Europe & N. Africa; cult.; naturalized in Macaronesia and E. U.S.A. 10 DEN FIN GRB IRE NOR SWE 11 CZE GER HUN POL 12 COR FRA POR SAR SPA 13 ALB BUL ITA SIC TUE YUG 14 ALL 20 ALG MOR TUN (21) azo mdr (75) cnt mas nwj nwy pen rho (78) mry nca. Biennial or ther. $2n = 12, 14, 24$.

subsp. **cornuta** (Ball) Greuter & Burdet, Willdenowia 11: 40 (1981).
Morocco. 20 MOR. Ther.
**Jasione cornuta* Ball, J. Bot. 11: 373 (1873). *Jasione corymbosa* subsp. *cornuta* (Ball) Murb., Contr. Fl. Maroc 2: 50 (1923).

subsp. **montana**

Europe & N. Africa; cult.; naturalized in Macaronesia & E. U.S.A. 10 DEN FIN GRB IRE NOR SWE 11 CZE GER HUN POL 12 COR FRA POR SAR SPA 13 ALB BUL ITA SIC TUE YUG 14 ALL 20 ALG MOR TUN (21) azo mdr (75) cnt mas nwj nwy pen rho (78) mry nca. Biennial or ther. $2n = 12, 14, 24$.

Jasione undulata Lam., Fl. Franç. 2: 3 (1779).

Jasione montana var. *prolifera* DC. in Lam. & DC., Fl. Franç. (ed. 3) 3: 717 (1805). *Jasione prolifera* Bellardi ex Colla, Herb. Pedem. 4: 35 (1835).

Jasione montana var. *major* Mert. & W. D. J. Koch in Röhl., Deutschl. Fl. (ed. 3) 2: 147 (1826).

Jasione glabrata Dumort., Fl. Belg.: 58 (1827).

Jasione montana var. *litoralis* Fr., Nov. Fl. Suec. Alt.: 269 (1828).

Jasione montana var. *laevis* Duby, Bot. Gall. 1: 311 (1828).

Jasione montana var. *hirsuta* Duby, Bot. Gall. 1: 311 (1828).

Jasione lusitanica A. DC., Monogr. Campan.: 105 (1830). *Jasione montana* var. *lusitanica* (A. DC.) Schmeja, Beih. Bot. Centralbl. 48(2): 33 (1931).

Jasione humilis Jan, Elench. Pl.: 3 (1831); non (Pers.) Loisel., Not. Fl. France: 42 (1810). *Jasione montana* subvar. *humilis* Nyman, Consp. Fl. Eur.: 486 (1879).

Jasione montana var. *maritima* Bréb., Fl. Normandie: 179 (1835); non Duby, Bot. Gall. 1: 311 (1828)

Jasione montana var. *littoralis* W. D. J. Koch, Syn. Fl. Germ. Helv.: 463 (1837); non *Jasione montana* var. *litoralis* Fr., Nov. Fl. Suec. Alt.: 269 (1828).

Jasione montana var. *glabra* Peterm., Fl. Lips. Excurs.: 168 (1838).

Jasione montana var. *tenella* Peterm., Fl. Lips. Excurs.: 168 (1838).

Jasione montana var. *stolonifera* A. DC. in DC., Prodr. 7: 415 (1839). *Jasione montana* subsp. *stolonifera* (A. DC.) Arcang., Comp. Fl. Ital.: 450 (1882).

Jasione montana var. *major* Boreau, Fl. Centre France 2: 286 (1840); non Mert. & W. D. J. Koch in Röhl., Deutschl. Fl. (ed. 3) 2: 147 (1826).

Jasione montana var. *nana* Boreau, Fl. Centre France 2: 296 (1840).

Jasione perennis var. *carionii* Boreau, Fl. Centre France 2: 286 (1840), as 'carioni'. *Jasione carionii* (Boreau) Boreau, Fl. Centre France (ed. 3) 2: 425 (1857), as 'carioni'. *Jasione montana* var. *carionii* (Boreau) Nyman, Consp. Fl. Eur.: 486 (1879), as 'carioni'. *Jasione perennis* var. *prostrata* St.-Lag. in Cariot, Étude Fl. (ed. 8) 2: 541 (1889).

Jasione montana var. *littoralis* Boiss., Voy. Bot. Espagne: 396 (1841); non W. D. J. Koch, Syn. Fl. Germ. Helv.: 463 (1837); nec *Jasione montana* var. *litoralis* Fr., Nov. Fl. Suec. Alt.: 269 (1828). *Jasione corymbosa* var. *littoralis* Pau, Mem. Real Soc. Esp. Hist. Nat. 12: 208 (1924).

Jasione montana var. *bracteosa* Willk., Bot. Zeitung (Berlin) 5: 863 (1847).

Jasione blepharodon Boiss. & Reut., Pugill. Pl. Afr. Bor. Hispan.: 72 (1852). *Jasione corymbosa* subsp. *blepharodon* (Boiss. & Reut.) Batt. & Trab., Fl. Algérie, Dicotyl.: 571 (1889). *Jasione corymbosa* var. *blepharodon* (Boiss. & Reut.) Cout., Notas Fl. Portugal 2: 15 (1915). *Jasione montana* subsp. *blepharodon* (Boiss. & Reut) Rivas Mart., Candollea 31: 113 (1976).

Jasione echinata Boiss. & Reut., Pugill. Pl. Afr. Bor. Hispan.: 73 (1852). *Jasione montana* var. *echinata* (Boiss. & Reut.) Willk. in Willk. & Lange, Prodr. Fl. Hispan. 2: 282 (1868). *Jasione montana* subsp. *echinata* (Boiss. & Reut.) Nyman, Consp. Fl. Eur.: 486 (1879). *Jasione montana* f. *echinata* (Boiss. & Reut.) Schmeja, Beih. Bot. Centralbl. 48(2): 33 (1931).

Jasione rosularis Boiss. & Reut., Pugill. Pl. Afr. Bor. Hispan.: 74 (1852). *Jasione laevis* subsp. *rosularis* (Boiss. & Reut.) Tutin, Bot. J. Linn. Soc. 67: 277 (1973).

Jasione montana var. *nana* Gren. & Godr., Fl. Fr. 2: 399 (1853); non Boreau, Fl. Centre France 2: 296 (1840). *Jasione montana* subvar. *nana* Rouy, Fl. France 10: 91 (1908).

Jasione montana var. *gracilis* Lange, Vidensk. Meddel. Dansk Naturhist. Foren. Kjøbenhavn 1861: 105 (1861). *Jasione montana* subsp. *gracilis* (Lange) Rivas Mart., Itinera Geobot. 15: 703 (2002).

Jasione stricta Pomel, Nouv. Mat. Fl. Atl.: 1 (1874).

Jasione montana var. *gracilis* Timb.-Lagr., Bull. Soc. Sci. Phys. Nat. Toulouse 3: 419
(1875); non Lange, Vidensk. Meddel. Dansk Naturhist. Foren. Kjøbenhavn 1861: 105
(1861). *Jasione montana* var. *timbalii* Rouy, Fl. France 10: 91 (1908), as 'timbali'.
Jasione montana var. *umbellata* Trautv., Trudy Imp. S.-Petersburgsk. Bot. Sada 6: 44
(1879); non Lapeyr., Hist. Pl. Pyrénées: 102 (1813).
Jasione montana subsp. *depressa* Huet de Pav. ex Arcang., Comp. Fl. Ital.: 450 (1882).
Jasione montana f. *depressa* (Huet de Pav. ex Arcang.) Schmeja, Beih. Bot. Centralbl.
48(2): 34 (1831).
Jasione espadanae Pau, Not. Bot. Fl. Españ. 1: 19 (1887).
Jasione montana var. *hispida* Beck, Fl. Nieder-Österreich: 1110 (1893).
Jasione euphrasiifolia Porta, Atti Imp. Regia Accad. Rovereto (ser. 3) 2: 214 (1896), as
'euphrasioefolia'.
Jasione montana var. *aetnensis* Lojac., Fl. Sicul. 2(1): 229 (1903).
Jasione montana var. *boreaui* Rouy, Fl. France 10: 91 (1908), as 'boraei'.
Jasione montana race *mediterranea* Rouy, Fl. France 10: 92 (1908). *Jasione montana* subsp.
mediterranea (Rouy) Gamisans, Cat. Pl. Vasc. Corse: 99 (1985).
Jasione ambigua Merino, Fl. Galicia 3: 593 (1909).
Jasione macrocalyx Gand., Bull. Soc. Bot. France 59: 59 (1912).
Jasione montana var. *latifolia* Pugsley, J. Bot. 59: 215 (1921).
Jasione macrocephala Sennen, Bull. Soc. Bot. France 74: 387 (1927).
Jasione montana var. *glaberrima* Podp. in Podp. & Domin, Klíč Květ. Českoslov.: 542
(1928); non Merino, Brotéria, Sér. Bot. 12: 114 (1914).
Jasione cartilaginea Sennen, Bol. Soc. Ibér. Ci. Nat. 28: 176 (1930). *Jasione montana* race
cartilaginea (Sennen) Sennen, Bull. Soc. Bot. France 74: 387 (1927).
Jasione montana var. *megaphylla* C. Vicioso, Anales Jard. Bot. Madrid 6(2): 79 (1946).
Jasione montana var. *latifolia* C. Vicioso, Anales Jard. Bot. Madrid 6(2): 80 (1946); non
Pugsley, J. Bot. 59: 215 (1921).
Jasione montana monstr. *pedicellata* De Langhe, Bull. Soc. Roy. Bot. Belgique 106: 71
(1973).

Jasione orbiculata Griseb. ex Velen., Fl. Bulg.: 375 (1891). *Jasione laevis* subsp. *orbiculata*
(Griseb. ex Velen.) Tutin, Bot. J. Linn. Soc. 70: 18 (1975).
SE. Europe. 13 ALB BUL GRC ITA ROM YUG. Hemicr. $2n = 12$.
Jasione supina var. *hirsuta* Wettst., Biblioth. Bot. 26: 75 (1892). *Jasione orbiculata* f.
hirsuta (Wettst.) Stoj., Notizbl. Bot. Gart. Berlin-Dahlem 9: 553 (1926).
Jasione orbiculata var. *balcanica* Urum., Oesterr. Bot. Z. 49: 56 (1899).
Jasione orbiculata var. *bosniaca* Stoj., Notizbl. Bot. Gart. Berlin-Dahlem 9: 553 (1926).
Jasione orbiculata var. *italica* Stoj., Notizbl. Bot. Gart. Berlin-Dahlem 9: 554 (1926).
Jasione orbiculata var. *supinoides* Stoj., Notizbl. Bot. Gart. Berlin-Dahlem 9: 554 (1926).

Jasione penicillata Boiss., Elench. Pl. Nov.: 63 (1838). *Jasione blepharodon* subsp. *penicillata*
(Boiss.) Rivas Goday, Bol. Soc. Brot. 47 (Suppl.): 168 (1974). *Jasione montana* subsp.
penicillata (Boiss.) Rivas Mart., Candollea 31: 113 (1976).
S. Spain. 12 SPA. Ther.

Jasione sessiliflora Boiss. & Reut., Diagn. Pl. Nov. Hisp.: 21 (1842). *Jasione humilis* var. *montana*
Willk. in Willk. & Lange, Prodr. Fl. Hispan. 2: 282 (1868). *Jasione humilis* subsp. *sessiliflora*
(Boiss. & Reut.) Nyman, Consp. Fl. Eur.: 487 (1879). *Jasione maritima* var. *sessiliflora* (Boiss. &
Reut.) Pau & Merino in Merino, Fl. Galicia 2: 292 (1906). *Jasione amethystina* subsp.
sessiliflora (Boiss. & Reut.) Rivas Mart., Publ. Inst. Biol. Aplicada 42: 121 (1967). *Jasione crispa*
subsp. *sessiliflora* (Boiss. & Reut.) Rivas Mart., Anales Inst. Bot. Cavanilles 27: 154 (1973).
W. Mediterranean. 12 POR SPA 20 MOR. Hemicr. $2n = 12, 24$.
Jasione sessiliflora var. *canescens* Boiss. & Reut., Diagn. Pl. Nov. Hisp.: 21 (1842). *Jasione
humilis* var. *campestris* Willk. in Willk. & Lange, Prodr. Fl. Hispan. 2: 282 (1868).
Jasione maritima var. *campestris* (Willk.) Merino, Fl. Galicia 2: 292 (1906). *Jasione
amethystina* f. *microcephala* Rivas Mart., Publ. Inst. Biol. Aplicada 42: 121 (1967).

Jasione appressifolia Pau, Not. Bot. Fl. Españ. 1: 19 (1887).

Jasione castellana Sennen, Butl. Inst. Catalana Hist. Nat. 32: 101 (1932). *Jasione tristis*
subsp. *castellana* (Sennen) Sennen, Butl. Inst. Catalana Hist. Nat. 32: 101 (1932).

Jasione tristis subsp. *ateridae* Sennen, Butl. Inst. Catalana Hist. Nat. 32: 101 (1932).
Jasione ateridae (Sennen) Sennen, Diagn. Nouv.: 88 (1936).

Jasione humilis subvar. *mesatlantica* Emb. & Maire, Bull. Soc. Hist. Nat. Afrique N. 23:
198 (1932).

Jasione perennis var. *cedretorum* Pau & Font Quer in Font Quer, Iter Marocc.: 643 (1927).
Jasione humilis var. *cedretorum* (Pau & Font Quer) Maire, Bull. Soc. Hist. Nat. Afrique
N. 23: 198 (1932). *Jasione crispa* var. *cedretorum* (Pau & Font Quer) Küpfer, Natural.
Monspel., Sér. Bot. 29: 41 (1980). *Jasione crispa* subsp. *cedretorum* (Pau & Font Quer)
Dobignard, Candollea 52: 139 (1997).

Jasione humilis subvar. *tazzekana* Emb. & Maire, Bull. Soc. Hist. Nat. Afrique N. 23:
198(1932).

Jasione crispa subsp. *serpentinica* P. Silva, Agron. Lusit. 30: 225 (1970). *Jasione sessiliflora*
subsp. *serpentinica* (P. Silva) Rivas Mart. & Sánchez Mata, Itinera Geobot. 15: 703 (2002).

Jasione crispa var. *praelittoralis* Mascl., Collect. Bot. (Barcelona) 8: 124 (1972).

Jasione sphaerocephala Brullo, Marcenò & Pavone, Nordic J. Bot. 1: 137 (1981).
SW. Italy. 13 ITA. Hemicr. $2n = 14$.

Jasione supina Sieber ex Spreng., Syst. Veg. 1: 810 (1824). *Phyteuma supina* (Sieber ex
Spreng.) G. Don, Gen. Hist. 3: 749 (1834).
N. & W. Turkey. 34 TUR. Hemicr. $2n = 12$

subsp. **akmanii** Damboldt, Notes Roy. Bot. Gard. Edinburgh 35: 51 (1976).
N. Turkey. 34 TUR. Hemicr.

subsp. **pontica** (Boiss.) Damboldt, Notes Roy. Bot. Gard. Edinburgh 35: 51 (1976).
N. Turkey. 34 TUR. Hemicr.
**Jasione supina* var. *pontica* Boiss., Fl. Orient. 3: 886 (1875). *Jasione pontica* (Boiss.) Hand.-
Mazz., Ann. K. K. Naturhist. Hofmus. 23: 192 (1909).
Jasione pontica var. *microcephala* Freyn ex Stoj., Notizbl. Bot. Gart. Berlin-Dahlem 9:
551 (1926).

subsp. **supina**
W. Turkey. 34 TUR. Hemicr. $2n = 12$.
Jasione supina var. *hirtula* Stoj., Notizbl. Bot. Gart. Berlin-Dahlem 9: 554 (1926).

subsp. **tmolea** (Stoy.) Damboldt, Notes Roy. Bot. Gard. Edinburgh 35: 51 (1976).
W. Turkey. 34 TUR. Hemicr.
**Jasione tmolea* Stoy., Notizbl. Bot. Gart. Berlin-Dahlem 9: 550 (1926).

Synonyms:
Jasione ambigua Merino === **Jasione montana** L. subsp. **montana**
Jasione amethystina Lag. & Rodr. === **Jasione crispa** subsp. **tristis** (Bory) G. Lopéz
Jasione amethystina f. *glaberrima* H. Lindb. === ?
Jasione amethystina f. *glabra* Schmeja === ?
Jasione amethystina f. *glabra* Schmeja === ?
Jasione amethystina f. *hirsuta* Schmeja === ?
Jasione amethystina f. *hirsuta* Schmeja === ?
Jasione amethystina f. *microcephala* Rivas Mart. === **Jasione sessiliflora** Boiss. & Reut.
Jasione amethystina subsp. *cavanillesii* (C. Vicioso) C. Vicioso & Lainz ex Lainz === **Jasione
cavanillesii** C. Vicioso
Jasione amethystina subsp. *centralis* (Rivas Mart.) Rivas Mart. === **Jasione crispa** (Pourr.)
Samp. subsp. **crispa**
Jasione amethystina subsp. *sessiliflora* (Boiss. & Reut.) Rivas Mart. === **Jasione sessiliflora**
Boiss. & Reut.

Jasione amethystina subsp. *mariana* (Willk.) Rivas Mart. === **Jasione crispa** subsp. **mariana** (Willk.) Rivas Mart.

Jasione amethystina subsp. *tomentosa* (A. DC.) Rivas Mart. === — **Jasione crispa** subsp. **tomentosa** (A. DC.) Rivas Mart.

Jasione amethystina var. *atlantica* (Ball) Litard. & Maire === **Jasione crispa** (Pourr.) Samp. subsp. **crispa**

Jasione amethystina var. *intermedia* Willk. === ?

Jasione amethystina var. *lanuginella* Litard. & Maire === **Jasione crispa** subsp. **lanuginella** (Litard. & Maire) J. Lambinon & J. Lewalle

Jasione amethystina var. *longidentata* Litard. & Maire === **Jasione crispa** subsp. **lanuginella** (Litard. & Maire) J. Lambinon & J. Lewalle

Jasione amethystina var. *tristis* (Bory) A. DC. === **Jasione crispa** subsp. **tristis** (Bory) G. Lopéz

Jasione appressifolia Pau === **Jasione sessiliflora** Boiss. & Reut.

Jasione ateridae (Sennen) Sennen === **Jasione sessiliflora** Boiss. & Reut.

Jasione atlantica Ball === **Jasione crispa** (Pourr.) Samp. subsp. **crispa**

Jasione ausfeldii Regel === **Brunonia australis** Sm. ex R. Br. (Goodeniaceae)

Jasione blepharodon Boiss. & Reut. === **Jasione montana** L. subsp. **montana**

Jasione blepharodon subsp. *penicillata* (Boiss.) Rivas Goday === **Jasione penicillata** Boiss.

Jasione bovei Boiss. & Reut. === **Jasione crispa** (Pourr.) Samp. subsp. **crispa**

Jasione brevisepala Rothm. === **Jasione crispa** (Pourr.) Samp. subsp. **crispa**

Jasione caespitosa Roth === ?

Jasione capensis P. J. Bergius === **Alpidea capensis** (P. J. Bergius) R. A. Dyer (Apiaceae).

Jasione carionii (Boreau) Boreau === **Jasione montana** L. subsp. **montana**

Jasione carpetana Boiss. & Reut. === **Jasione laevis** subsp. **carpetana** (Boiss. & Reut.) Rivas Mart.

Jasione cartilaginea Sennen === **Jasione montana** L. var. **montana**

Jasione castellana Sennen === **Jasione sessiliflora** Boiss. & Reut.

Jasione coespitans Pomel === **Jasione crispa** (Pourr.) Samp. subsp. **crispa**

Jasione cornuta Ball === **Jasione montana** subsp. **cornuta** (Ball) Greuter & Burdet

Jasione corymbosa subsp. *blepharodon* (Boiss. & Reut.) Batt. & Trab. === **Jasione montana** L. subsp. **montana**

Jasione corymbosa subsp. *cornuta* (Ball) Murb. === **Jasione montana** subsp. **cornuta** (Ball) Greuter & Burdet

Jasione corymbosa var. *battandieri* Maire === **Jasione corymbosa** Poir. ex Schult. subsp. **corymbosa**

Jasione corymbosa var. *blepharodon* (Boiss. & Reut.) Cout. === **Jasione montana** L. subsp. **montana**

Jasione corymbosa var. *glabra* (Durieu ex Boiss. & Reut.) J. Parn. === **Jasione corymosa** subsp. **glabra** (Durieu ex Boiss. & Reut.) Batt. & Trab.

Jasione corymbosa var. *littoralis* Pau === **Jasione montana** L. subsp. **montana**

Jasione crispa raca *amethystina* (Lag. & Rodr.) Samp. === **Jasione crispa** subsp. **tristis** (Bory) G. Lopéz

Jasione crispa subsp. *amethystina* (Lag. & Rodr.) Tutin === **Jasione crispa** subsp. **tristis** (Bory) G. Lopéz

Jasione crispa subsp. *brevisepala* (Rothm.) Rivas Mart. === **Jasione crispa** (Pourr.) Samp. subsp. **crispa**

Jasione crispa subsp. *cavanillesii* (C. Vicioso) Tutin === **Jasione cavanillesii** C. Vicioso

Jasione crispa subsp. *cedretorum* (Pau & Font Quer) Dobignard === **Jasione sessiliflora** Boiss. & Reut.

Jasione crispa subsp. *centralis* (Rivas Mart.) Tutin === **Jasione crispa** (Pourr.) Samp. subsp. **crispa**

Jasione crispa subsp. *maritima* (Duby) Tutin === **Jasione maritima** (Duby) Dufour ex Merino

Jasione crispa subsp. *segurensis* Mota, C. Díaz, Gómez-Merc. & F. Valle === **Jasione crispa** subsp. **tomentosa** (A. DC.) Rivas Mart.

Jasione crispa subsp. *serpentinica* P. Silva === **Jasione sessiliflora** Boiss. & Reut.

Jasione crispa subsp. *sessiliflora* (Boiss. & Reut.) Rivas Mart. === **Jasione sessiliflora** Boiss. & Reut.

Jasione crispa subsp. *tristis* (Bory) G. Lopéz === **Jasione crispa** subsp. **amethystina** (Bory) G. Lopéz

Jasione crispa var. *atlantica* (Ball) Küpfer === **Jasione crispa** (Pourr.) Samp. subsp. **crispa**

Jasione crispa var. *cedretorum* (Pau & Font Quer) Küpfer === **Jasione sessiliflora** Boiss. & Reut.

Jasione crispa var. *praelittoralis* Mascl. === **Jasione sessiliflora** Boiss. & Reut.

Jasione dentata (A. DC.) Halácsy === **Jasione heldreichii** Boiss. & Orph.

Jasione dentata f. *canescens* Gajiç === **Jasione heldreichii** Boiss. & Orph.

Jasione dentata f. *dubravkiana* Gajiç === **Jasione heldreichii** Boiss. & Orph.

Jasione dentata f. *integrifolia* Gajiç === **Jasione heldreichii** Boiss. & Orph.

Jasione diapensifolia Gand. === **Jasione crispa** subsp. **tomentosa** (A. DC.) Rivas Mart.

Jasione echinata Boiss. & Reut. === **Jasione montana** L. subsp. **montana**

Jasione espadanae Pau === **Jasione montana** L. subsp. **montana**

Jasione euphrasiifolia Porta === **Jasione montana** L. subsp. **montana**

Jasione fallax Willk. === **Jasione crispa** (Pourr.) Samp. subsp. **crispa**

Jasione foliosa subsp. *minuta* (Schult.) Font Quer === **Jasione foliosa** Cav. subsp. **foliosa**

Jasione foliosa var. *minuta* (Schult.) Cuatrec. === **Jasione foliosa** Cav. subsp. **foliosa**

Jasione genisticola Gand. === ?

Jasione glabra Durieu ex Boiss. & Reut. === **Jasione corymosa** subsp. **glabra** (Durieu ex Boiss. & Reut.) Batt. & Trab.

Jasione glabra Velen. === **Jasione heldreichii** Boiss. & Orph.

Jasione glabrata Dumort. === **Jasione montana** L. subsp. **montana**

Jasione heldreichii var. *glabra* Nyman === **Jasione heldreichii** Boiss. & Orph.

Jasione heldreichii var. *microcephala* Velen. === **Jasione heldreichii** Boiss. & Orph.

Jasione heldreichii var. *papillosa* J. Parn. === **Jasione heldreichii** Boiss. & Orph.

Jasione humilis Jan === **Jasione montana** L. subsp. **montana**

Jasione humilis (Pers.) Loisel. === **Jasione crispa** (Pourr.) Samp.

Jasione humilis f. *atlantica* (Ball) Quézel === **Jasione crispa** (Pourr.) Samp. subsp. **crispa**

Jasione humilis f. *megalocalyx* (Pomel) Quézel === **Jasione crispa** (Pourr.) Samp. subsp. **crispa**

Jasione humilis subsp. *amethystina* (Lag. & Rodr.) Rivas Mart. === **Jasione crispa** subsp. **tristis** (Bory) G. Lopéz

Jasione humilis subsp. *cavanillesii* (C. Vicioso) Rivas Mart. === **Jasione cavanillesii** C. Vicioso

Jasione humilis subsp. *centralis* Rivas Mart. === **Jasione crispa** (Pourr.) Samp. subsp. **crispa**

Jasione humilis subsp. *mariana* (Willk.) Rivas Mart. === **Jasione crispa** subsp. **mariana** (Willk.) Rivas Mart.

Jasione humilis subsp. *sessiliflora* (Boiss. & Reut.) Nyman === **Jasione sessiliflora** Boiss. & Reut.

Jasione humilis subsp. *tomentosa* (A. DC.) Rivas Mart. === **Jasione crispa** subsp. **tomentosa** (A. DC.) Rivas Mart.

Jasione humilis subvar. *mesatlantica* Emb. & Maire === **Jasione sessiliflora** Boiss. & Reut.

Jasione humilis subvar. *tazzekana* Emb. & Maire === **Jasione sessiliflora** Boiss. & Reut.

Jasione humilis var. *atlantica* (Ball) Maire === **Jasione crispa** (Pourr.) Samp. subsp. **crispa**

Jasione humilis var. *campestris* Willk. === **Jasione sessiliflora** Boiss. & Reut.

Jasione humilis var. *cedretorum* Pau & Font Quer === **Jasione sessiliflora** Boiss. & Reut.

Jasione humilis var. *fallax* (Willk.) Nyman === **Jasione crispa** (Pourr.) Samp. subsp. **crispa**

Jasione humilis var. *lanuginella* (Litard. & Maire) Maire === **Jasione crispa** subsp. **lanuginella** (Litard. & Maire) J. Lambinon & J. Lewalle

Jasione humilis var. *longidentata* (Litard. & Maire) Maire === **Jasione crispa** subsp. **lanuginella** (Litard. & Maire) J. Lambinon & J. Lewalle

Jasione humilis var. *maritima* (Duby) Willk. === **Jasione maritima** (Duby) Dufour ex Merino

Jasione humilis var. *megalocalyx* (Pomel) Jahand. & Maire === **Jasione crispa** (Pourr.) Samp. subsp. **crispa**

Jasione humilis var. *montana* Willk. === **Jasione sessiliflora** Boiss. & Reut.

Jasione humilis var. *pygmaea* Willk. === **Jasione crispa** (Pourr.) Samp. subsp. **crispa**

Jasione humilis var. *tomentosa* A. DC. === **Jasione crispa** subsp. **tomentosa** (A. DC.) Tutin

Jasione jankae Neilr. === **Jasione heldreichii** Boiss. & Orph.

Jasione laevis subsp. *gredensis* Rivas Mart. & Sancho === **Jasione laevis** Lam. subsp. **laevis**

Jasione laevis subsp. *orbiculata* (Griseb. ex Velen.) Tutin === **Jasione orbiculata** Griseb. ex Velen.

Jasione laevis subsp. *rosularis* (Boiss. & Reut.) Tutin === **Jasione montana** L. subsp. **montana**

Jasione laevis var. *pygmaea* (Gren. & Godr.) O. Bolòs & Vigo === **Jasione laevis** Lam. subsp. **laevis**

Jasione lusitanica A. DC. === **Jasione montana** L. subsp. **montana**

Jasione macrocalyx Gand. === **Jasione montana** L. subsp. **montana**

Jasione macrocephala Sennen === **Jasione montana** L. subsp. **montana**

Jasione mansanetiana Roselló & Peris === **Jasione foliosa** subsp. **mansanetiana** (Roselló & Peris) Rivas Mart.

Jasione mariana Willk. === **Jasione crispa** subsp. **mariana** (Willk.) Rivas Mart.

Jasione maritima (Duby) Dufour ex Merino === **Jasione crispa** subsp. **maritima** (Duby) Tutin

Jasione maritima var. *campestris* (Willk.) Merino === **Jasione sessiliflora** Boiss. & Reut.

Jasione maritima var. *sabularia* (Cout.) Sales & Hedge === **Jasione maritima** (Duby) Dufour ex Merino

Jasione maritima var. *sessiliflora* (Boiss. & Reut.) Pau & Merino === **Jasione sessiliflora** Boiss. & Reut.

Jasione megalocalyx Pomel === **Jasione crispa** (Pourr.) Samp. subsp. **crispa**

Jasione minuta (Schult.) Pau === **Jasione foliosa** Cav. subsp. **foliosa**

Jasione montana f. *depressa* (Huet de Pav. ex Arcang.) Schmeja === **Jasione montana** L. subsp. **montana**

Jasione montana f. *echinata* (Boiss. & Reut.) Schmeja === **Jasione montana** L. subsp. **montana**

Jasione montana f. *glabra* Schmeja === ?

Jasione montana f. *glabra* Schmeja === ?

Jasione montana f. *glabra* Schmeja === ?

Jasione montana f. *heldreichii* (Boiss. & Orph.) Schmeja === **Jasione heldreichii** Boiss. & Orph.

Jasione montana f. *hirsuta* Schmeja === ?

Jasione montana f. *hirsuta* Schmeja === ?

Jasione montana f. *hirsuta* Schmeja === ?

Jasione montana f. *jankae* (Neilr.) Schmeja === **Jasione heldreichii** Boiss. & Orph.

Jasione montana f. *rossularis* Schmeja === ?

Jasione montana monstr. *pedicellata* De Langhe === **Jasione montana** L. subsp. **montana**

Jasione montana race *cartilaginea* (Sennen) Sennen === **Jasione montana** L. subsp. **montana**

Jasione montana race *maritima* (Duby) Rouy === **Jasione maritima** (Duby) Dufour ex Merino

Jasione montana race *mediterranea* Rouy === **Jasione montana** L. subsp. **montana**

Jasione montana subsp. *blepharodon* (Boiss. & Reut) Rivas Mart. === **Jasione montana** L. subsp. **montana**

Jasione montana subsp. *corymbosa* (Poir. ex Schult.) Greuter & Burdet === **Jasione corymbosa** Poir. ex Schult.

Jasione montana subsp. *depressa* Huet de Pav. ex Arcang. === **Jasione montana** L. subsp. **montana**

Jasione montana subsp. *echinata* (Boiss. & Reut.) Nyman === **Jasione montana** L. subsp. **montana**

Jasione montana subsp. *glabra* (Durieu ex Boiss. & Reut.) Greuter & Burdet === **Jasione corymbosa** subsp. **glabra** (Durieu ex Boiss. & Reut.) Batt. & Trab.

Jasione montana subsp. *gracilis* (Lange) Rivas Mart. === **Jasione montana** L. subsp. **montana**

Jasione montana subsp. *maritima* (Duby) C. Vicioso === **Jasione maritima** (Duby) Dufour ex Merino

Jasione montana subsp. *mediterranea* (Rouy) Gamisans === **Jasione montana** L. subsp. **montana**

Jasione montana subsp. *stolonifera* (A. DC.) Arcang. === **Jasione montana** L. subsp. **montana**

Jasione montana subvar. *humilis* Nyman === **Jasione montana** L. subsp. **montana**

Jasione montana subvar. *nana* Rouy === **Jasione montana** L. subsp. **montana**

Jasione montana var. *boreaui* Rouy === **Jasione montana** L. subsp. **montana**

Jasione montana var. *bracteosa* Willk. === **Jasione montana** L. subsp. **montana**

Jasione montana var. *carionii* (Boreau) Nyman === **Jasione montana** L. subsp. **montana**

Jasione montana var. *corymbosa* (Poir. ex Schult.) Schmeja === **Jasione corymbosa** Poir. ex Schult.

Jasione montana var. *dentata* A. DC. === **Jasione heldreichii** Boiss. & Orph.

Jasione montana var. *echinata* (Boiss. & Reut.) Willk. === **Jasione montana** L. subsp. **montana**

Jasione montana var. *glaberrima* Merino === ?

Jasione montana var. *glaberrima* Podp. === **Jasione montana** L. subsp. **montana**

Jasione montana var. *glabra* Peterm. === **Jasione montana** L. subsp. **montana**

Jasione montana var. *gracilis* Lange === **Jasione montana** L. subsp. **montana**

Jasione montana var. *gracilis* Timb.-Lagr. === **Jasione montana** L. subsp. **montana**

Jasione montana var. *heldreichii* (Boiss. & Orph.) Nyman === **Jasione heldreichii** Boiss. & Orph.

Jasione montana var. *hirsuta* Duby === **Jasione montana** L. subsp. **montana**

Jasione montana var. *hispida* Beck === **Jasione montana** L. subsp. **montana**

Jasione montana var. *humilis* Pers. === **Jasione crispa** (Pourr.) Samp.

Jasione montana var. *imbricans* J. Parn. === **Jasione maritima** (Duby) Dufour ex Merino

Jasione montana var. *laevis* Duby === **Jasione montana** L. subsp. **montana**

Jasione montana var. *latifolia* C. Vicioso === **Jasione montana** L. subsp. **montana**

Jasione montana var. *latifolia* Pugsley === **Jasione montana** L. subsp. **montana**

Jasione montana var. *litoralis* Fr. === **Jasione montana** L. subsp. **montana**

Jasione montana var. *littoralis* Boiss. === **Jasione montana** L. subsp. **montana**

Jasione montana var. *littoralis* W. D. J. Koch === **Jasione montana** L. subsp. **montana**

Jasione montana var. *lusitanica* (A. DC.) Schmeja === **Jasione montana** L. subsp. **montana**

Jasione montana var. *major* Boreau === **Jasione montana** L. subsp. **montana**

Jasione montana var. *major* Mert. & W. D. J. Koch === **Jasione montana** L. subsp. **montana**

Jasione montana var. *maritima* Bréb. === **Jasione montana** L. subsp. **montana**

Jasione montana var. *maritima* Duby === **Jasione maritima** (Duby) Dufour ex Merino

Jasione montana var. *megaphylla* C. Vicioso === **Jasione montana** L. subsp. **montana**

Jasione montana var. *nana* Boreau === **Jasione montana** L. subsp. **montana**

Jasione montana var. *nana* Gren. & Godr. === **Jasione montana** L. subsp. **montana**

Jasione montana var. *penicillata* (Boiss.) Rivas Mart. === **Jasione penicillata** Boiss.

Jasione montana var. *perennis* L. f. === **Jasione laevis** Lam. subsp. **laevis**

Jasione montana var. *prolifera* DC. === **Jasione montana** L. subsp. **montana**

Jasione montana var. *sabularia* Cout. === **Jasione maritima** (Duby) Dufour ex Merino

Jasione montana var. *stolonifera* A. DC. === **Jasione montana** L. subsp. **montana**

Jasione montana var. *tenella* Peterm. === **Jasione montana** L. subsp. **montana**

Jasione montana var. *timbalii* Rouy === **Jasione montana** L. subsp. **montana**

Jasione montana var. *umbellata* Lapeyr. === ?

Jasione montana var. *umbellata* Trautv. === **Jasione montana** L. subsp. **montana**

Jasione nummulariifolia Pau === **Jasione foliosa** Cav. subsp. **foliosa**

Jasione nuriensis Sennen === **Jasione crispa** (Pourr.) Samp. subsp. **crispa**

Jasione orbiculata f. *hirsuta* (Wettst.) Stoj. === **Jasione orbiculata** Griseb. ex Velen.

Jasione orbiculata var. *balcanica* Urum. === **Jasione orbiculata** Griseb. ex Velen.

Jasione orbiculata var. *bosniaca* Stoj. === **Jasione orbiculata** Griseb. ex Velen.

Jasione orbiculata var. *italica* Stoj. === **Jasione orbiculata** Griseb. ex Velen.

Jasione orbiculata var. *supinoides* Stoj. === **Jasione orbiculata** Griseb. ex Velen.

Jasione perennis (L. f.) Lam. === **Jasione laevis** Lam. subsp. **laevis**

Jasione perennis f. *carpetana* Schmeja === ?

Jasione perennis f. *megacarpaea* Coustur. & Gand. === ?

Jasione perennis f. *rosularis* Schmeja === ?

Jasione perennis proles *pyrenaica* Sennen === **Jasione laevis** Lam. subsp. **laevis**

Jasione perennis race *pygmaea* (Gren. & Godr.) Sennen === **Jasione laevis** Lam. subsp. **laevis**

Jasione perennis race *pyrenaea* Sennen === **Jasione laevis** Lam. subsp. **laevis**

Jasione perennis subsp. *carpetana* (Boiss. & Reut.) Nyman === **Jasione laevis** subsp. **carpetana** (Boiss. & Reut.) Rivas Mart.

Jasione perennis subsp. *humilis* (Pers.) Bonnier & Layens === **Jasione crispa** (Pourr.) Samp.

Jasione perennis subsp. *pyrenaea* (Sennen) Sennen === **Jasione laevis** Lam. subsp. **laevis**

Jasione perennis var. *alpestris* Willk. === ?

Jasione perennis var. *carionii* Boreau === **Jasione montana** L. subsp. **montana**

Jasione perennis var. *carpetana* (Boiss. & Reut.) Willk. === **Jasione laevis** subsp. **carpetana** (Boiss. & Reut.) Rivas Mart.

Jasione perennis var. *cedretorum* Pau & Font Quer === **Jasione sessiliflora** Boiss. & Reut.

Jasione perennis var. *gracilis* Sennen === **Jasione laevis** Lam. subsp. **laevis**

Jasione perennis var. *intermedia* Coss. === ?

Jasione perennis var. *prostrata* St.-Lag. === **Jasione montana** L. subsp. **montana**

Jasione perennis var. *pygmaea* Gren. & Godr. === **Jasione laevis** Lam. subsp. **laevis**

Jasione pontica (Boiss.) Hand.-Mazz. === **Jasione supina** subsp. **pontica** (Boiss.) Damboldt

Jasione pontica var. *microcephala* Freyn ex Stoj. === **Jasione supina** subsp. **pontica** (Boiss.) Damboldt

Jasione prolifera Bellardi ex Colla === **Jasione montana** L. subsp. **montana**

Jasione pygmaea (Gren. & Godr.) Timb.-Lagr. === **Jasione laevis** Lam. subsp. **laevis**

Jasione rosularis Boiss. & Reut. === **Jasione montana** L. subsp. **montana**

Jasione sessiliflora f. *pygmaea* (Willk.) Pau === **Jasione crispa** (Pourr.) Samp. subsp. **crispa**

Jasione sessiliflora subsp. *megalocalyx* (Pomel) Batt. & Trab. === **Jasione crispa** (Pourr.) Samp. subsp. **crispa**

Jasione sessiliflora subsp. *sabularia* (Cout.) Rivas Mart. === **Jasione maritima** (Duby) Dufour ex Merino

Jasione sessiliflora subsp. *serpentinica* (P. Silva) Rivas Mart. & Sánchez Mata === **Jasione sessiliflora** Boiss. & Reut.

Jasione sessiliflora subsp. *tomentosa* (A. DC.) Rivas Mart.=== **Jasione crispa** subsp. **tomentosa** (A. DC.) Rivas Mart.

Jasione sessiliflora var. *bovei* (Boiss. & Reut.) Batt. & Trab. === **Jasione crispa** (Pourr.) Samp. subsp. **crispa**

Jasione sessiliflora var. *canescens* Boiss. & Reut. === **Jasione sessiliflora** Boiss. & Reut.

Jasione sessiliflora var. *eriantha* Boiss. & Reut. === **Jasione crispa** subsp. **tomentosa** (A. DC.) Rivas Mart.

Jasione stricta Pomel === **Jasione montana** L. subsp. **montana**

Jasione supina var. *hirsuta* Wettst. === **Jasione orbiculata** Griseb. ex Velen.

Jasione supina var. *hirtula* Stoj. === **Jasione supina** Sieber ex Spreng.

Jasione supina var. *pontica* Boiss. === **Jasione supina** subsp. **pontica** (Boiss.) Damboldt

Jasione tajae Sennen === **Jasione laevis** Lam. subsp. **laevis**

Jasione tmolea Stoy. === **Jasione supina** subsp. **tmolea** (Stoy.) Damboldt

Jasione tomentosa (A. DC.) Rivas. Mart. === **Jasione crispa** subsp. **tomentosa** (A. DC.) Rivas Mart.

Jasione tristis Bory === **Jasione crispa** subsp. **tristis** (Bory) G. Lopéz

Jasione tristis subsp. *ateridae* Sennen === **Jasione sessiliflora** Boiss. & Reut.

Jasione tristis subsp. *castellana* (Sennen) Sennen === **Jasione sessiliflora** Boiss. & Reut.

Jasione undulata Lam. === **Jasione montana** L. subsp. **montana**

Jasione vulgaris Gaterau === **Jasione montana** L. subsp. **montana**

Jasionella

This genus was segregated from *Jasione* on the basis of rather vague differences in morphology. Type [by monotypy]: *Jasionella bulgarica* (Stoj. & Stef.) Stoj. & Stef.

> Stefanoff, B. (1936). [Über die systematische Stellung einiger Arten der Familie Campanulaceae.] God. Sofiisk. Univ. Agron.-Lesoved. Fak. 14: 93-104. Bu. — Argues for recognition of the genus.

Synonyms:

Jasionella Stoj. & Stef. === **Jasione** L.

Jasionella bulgarica (Stoj. & Stef.) Stoj. & Stef. === **Jasione bulgarica** Stoj. & Stef.

Juchia

Post and Kuntze (1903) believed this name had priority over *Siphocampylus*. However, Necker's book subsequently was declared *opus utique oppressus* by the ICBN and the name was not validly published there (Art. 32.7). Instead, it was validated by Kuntze's acceptance of it (in Post & Kuntze 1903) together with his indirect citation of Necker's *effectively* published work as the source of its description (Art. 32.1). This later date of publication makes it a later homonym of *Juchia* Roem. (Cucurbitaceae). Type [designated by the author]: *Lobelia columneae* L. f.

> Necker, N. J. de (1790). *Juchia.* In Elementa botanica 1: 133. Paris: Bossange. La. — Original description of genus, but not validly published.
> von Post, T. E., & O. Kuntze (1903). Lexicon generum phanerogamarum. La. — Validation of the name as a substitute for *Siphocampylus,* via indirect reference to preceding.

Synonyms:
Juchia Neck. ex Kuntze === **Siphocampylus** Pohl

Keumkangsania

This name was intentionally substituted for *Hanabusaya* on nationalistic grounds; it was felt inappropriate that a uniquely Korean plant be named in honor of a Japanese man. Whatever the merits of such an argument, this is not permitted under the ICBN (Art 51.1). Based on: *Hanabusaya* Nakai.

Synonyms:
Keumkangsania Kim === **Hanabusaya** Nakai
Keumkangsania asiatica (Nakai) Kim === **Hanabusaya asiatica** (Nakai) Nakai
Keumkangsania latisepala (Nakai) Kim === **Hanabusaya asiatica** (Nakai) Nakai

Kittelia

This name was proposed as an avowed replacement for *Cyanea* because of that name's homonymy with *Cyanea* Péron & Lesuer, a genus of Scyphozoa (Animalia: Cnidaria). However, due to the independence of the ICBN from zoological nomenclature (Principle I), this is not a consideration. Note that none of the species were formally transferred to the new name. Based on: *Cyanea* Gaudich.

> von Reichenbach, C. L. (1837). Handbuch des natürlichen Pflanzensystems. Dresden: Arnold. Ge. — Validation of name as replacement for *Cyanea.*

Synonyms:
Kittelia Rchb. === **Cyanea** Gaudich.

Laurentia

The prevailing opinion (Meikle 1979, Brummitt 2000) is that this name was an illegitimate renaming of *Lobelia*, despite Lammers' (1997) argument that Adanson (1763) cited *Lobelia laurentia* as the type of *Laurentia*. Had the correctness of this position been acknowledged, *Laurentia* would be the correct name for *Solenopsis*. Wimmer (1953, 1968) not only employed the name in that sense, but circumscribed it far more broadly, also encompassing the genera recognized here as *Hippobroma, Isotoma, Palmerella, Porterella,* and *Wimmerella*. The dismemberment of *Laurentia* s.l. embraced here is supported by a preliminary molecular phylogeny (E. Knox. pers. comm.), in which the species assigned to that genus by Wimmer (1953, 1968) were dispersed among six disparate clades: (1) all sampled species of

Wimmerella plus *Lobelia anceps;* (2) both sampled species of *Solenopsis;* (3) a clade that included *Palmerella* and *Porterella* among species of *Downingia, Howellia,* and *Legenere;* (4) a clade comprised of *Hippobroma* and the sole sampled species of *Lobelia* sect. *Tylomium* (C. Presl) Benth.; (5) a clade that included the type of *Isotoma* among Australasian species of *Lobelia* sect. *Delostemon* (E. Wimm.) J. Murata; and (6) a clade that comprised the remainder of *Isotoma* scattered among species referable to *Lobelia* sect. *Dioica* (E. Wimm.), sect. *Isolobus* (A. DC.) C. B. Clarke, sect. *Paramezleria* E. Wimm., and sect. *Pratia* (Gaudich.) J. Murata. Based on: *Lobelia* L.

Adanson, M. (1763). Les campanules. *Campanulae.* In Familles des plantes 2: 132-134. Paris: Vincent. Fr. — Validation of name as replacement for *Lobelia,* because that name was restricted to *L. plumieri* L., a species of Goodeniaceae now called *Scaevola plumieri* (L.) Vahl.

Wimmer, F. E. (1953). *Laurentia* (Mich.) Adans. In A. Engler & L. Diels, Das Pflanzenreich IV. 276b: 386-407, illus. Berlin: Akademie-Verlag. Ge., La. — Monograph with key, descriptions, infrageneric classification, full nomenclature, and specimen citations.

Wimmer, F. E. (1968). *Laurentia* (Mich.) Adams. [sic]. Das Pflanzenreich IV.276c: 853-855. Berlin: Akademie-Verlag. Ge., La. — Supplement to Wimmer (1953).

Meikle, R. D. (1979). Some notes on *Laurentia* Adanson (Campanulaceae). Kew Bull. 34: 373-375. En. — Name superfluous when published because *Lobelia* cited as synonym in protologue.

Lammers, T. G. (1997). Proposal to conserve the name *Laurentia* (Campanulaceae) with a conserved type. Taxon 46: 795-799. En. — Argues that name was not superfluous when published.

Brummitt, R. K. (2000). Report of the Committee for Spermatophyta: 49. Taxon 49: 261-278. En. — Proposal to conserve *Laurentia* not recommended.

Synonyms:

Laurentia Adans. === **Lobelia** L.

Laurentia sect. *Isotoma* (R. Br.) E. Wimm. === **Isotoma** (R. Br.) Lindl.

Laurentia sect. *Palmerella* (A. Gray) E. Wimm. === **Palmerella** A. Gray

Laurentia subg. *Enchysia* (C. Presl) Peterm. === **Lobelia** L.

Laurentia subg. *Isotoma* (R. Br.) Peterm. === **Isotoma** (R. Br.) Lindl.

Laurentia subg. *Solenopsis* (C. Presl) Peterm. === **Solenopsis** C. Presl

Laurentia [unranked] *Enchysia* (C. Presl) Endl. === **Lobelia** L.

Laurentia [unranked] *Hippobroma* (C. Presl) Endl. === **Hippobroma** G. Don

Laurentia [unranked] *Isotoma* (R. Br.) Endl. === **Isotoma** (R. Br.) Lindl.

Laurentia [unranked] *Solenopsis* (C. Presl) Endl. === **Solenopsis** C. Presl

Laurentia anethifolia (Summerh.) E. Wimm. === **Isotoma anethifolia** Summerh.

Laurentia arabidea (C. Presl) A. DC. === **Wimmerella arabidea** (C. Presl) Serra, M. B. Crespo & Lammers

Laurentia armstrongii (E. Wimm.) E. Wimm. === **Isotoma armstrongii** E. Wimm.

Laurentia axillaris (Lindl.) E. Wimm. === **Isotoma axillaris** Lindl.

Laurentia baueri (C. Presl) A. DC. === **Isotoma fluviatilis** (R. Br.) F. Muell. ex Benth.

Laurentia baueri var. *major* (C. Presl) A. DC. === **Isotoma fluviatilis** (R. Br.) F. Muell. ex Benth.

Laurentia bicolor (Batt.) Maire & T. Stephenson === **Solenopsis bicolor** (Batt.) Greuter & Burdet

Laurentia bifida (Thunb.) Sond. === **Wimmerella bifida** (Thunb.) Serra, M. B. Crespo & Lammers

Laurentia bivonae Tineo === **Solenopsis bivonae** (Tineo) M. B. Crespo, Serra & Juan

Laurentia canariensis (C. Presl) A. DC. === **Solenopsis laurentia** (L.) C. Presl

Laurentia carnosula (Hook. & Arn.) Benth. ex A. Gray === **Porterella carnosula** (Hook. & Arn.) Torr.

Laurentia commutata Tod. === ?

Laurentia debilis (A. Gray) McVaugh === **Palmerella debilis** A. Gray

Laurentia debilis var. *serrata* (A. Gray) McVaugh === **Palmerella debilis** subsp. **serrata** (A. Gray) Lammers

Laurentia dregeana (C. Presl) A. DC. === **Wimmerella bifida** (Thunb.) Serra, M. B. Crespo & Lammers

Laurentia erecta (A. DC.) Schönl. === **Wimmerella secunda** (L. f.) Serra, M. B. Crespo & Lammers

Laurentia erinoides (L.) G. Nichols. === **Lobelia erinus** L.

Laurentia erinoides f. *erecta* (A. DC.) E. Wimm. === **Wimmerella secunda** (L. f.) Serra, M. B. Crespo & Lammers

Laurentia erinoides f. *dentata* (A. DC.) E. Wimm. === **Wimmerella secunda** (L. f.) Serra, M. B. Crespo & Lammers

Laurentia etbaica Schweinf. === **Wahlenbergia lobelioides** subsp. **nutabunda** (Guss.) Murb.

Laurentia eximia (A. Nelson) A. Nelson === **Porterella carnosula** (Hook. & Arn.) Torr.

Laurentia ferdinandii E. Wimm. === **Isotoma baueri** C. Presl

Laurentia fluviatilis (R. Br.) E. Wimm. === **Isotoma fluviatilis** (R. Br.) F. Muell. ex Benth.

Laurentia fluviatilis var. *inundata* (R. Br.) E. Wimm. === **Isotoma fluviatilis** (R. Br.) F. Muell. ex Benth. subsp. **fluviatilis**

Laurentia fluviatilis var. *lessonii* (C. Presl) E. Wimm. === **Isotoma fluviatilis** (R. Br.) F. Muell. ex Benth. subsp. **fluviatilis**

Laurentia frontidentata E. Wimm. === **Wimmerella frontidentata** (E. Wimm.) Serra, M. B. Crespo & Lammers

Laurentia gasparrinii f. *albiflora* (Merino) E. Wimm. === **Solenopsis laurentia** (L.) C. Presl

Laurentia gasparrinii subsp. *tenella* Bolòs & Vigo === **Solenopsis bivonae** (Tineo) M. B. Crespo, Serra & Juan

Laurentia gasparrinii var. *subacaulis* (Pomel ex Batt. E. Wimm. === **Solenopsis laurentia** (L.) C. Presl

Laurentia gaudichaudii (C. Presl) A. DC. === **Isotoma fluviatilis** (R. Br.) F. Muell. ex Benth.

Laurentia giftbergensis (E. Phillips) E. Wimm. === **Wimmerella giftbergensis** (E. Phillips) Serra, M. B. Crespo & Lammers

Laurentia gulliveri (F. Muell.) E. Wimm. === **Isotoma gulliveri** F. Muell.

Laurentia hederacea Sond. === **Wimmerella hederacea** (Sond.) Serra, M. B. Crespo & Lammers

Laurentia hedyotidea Schltr. === **Wimmerella hedyotidea** (Schltr.) Serra, M. B. Crespo & Lammers

Laurentia hedyotidea var. *major* E. Wimm. === **Wimmerella hedyotidea** (Schltr.) Serra, M. B. Crespo & Lammers

Laurentia hypocrateriformis (R. Br.) E. Wimm. === **Isotoma hypocrateriformis** (R. Br.) Druce

Laurentia insignis Brandegee === **Centropogon grandidentatus** (Schlecht.) Zahlbr.

Laurentia irasuensis (Planch. & Oerst.) E. Wimm. === **Lobelia irasuensis** Planch. & Oerst.

Laurentia longiflora (L.) Peterm. === **Hippobroma longiflora** (L.) G. Don

Laurentia longiflora Schltr. === **Wimmerella longitubus** (E. Wimm.) Serra, M. B. Crespo & Lammers

Laurentia longiflora var. *runcinata* (Hassk.) E. Wimm. === **Hippobroma longiflora** (L.) G. Don

Laurentia longitubus E. Wimm. === **Wimmerella longitubus** (E. Wimm.) Serra, M. B. Crespo & Lammers

Laurentia mariae E. Wimm. === **Wimmerella mariae** (E. Wimm.) Serra, M. B. Crespo & Lammers

Laurentia maximiliana E. Wimm. === **Diastatea micrantha** (Kunth) McVaugh

Laurentia michelii A. DC. === **Solenopsis laurentia** (L.) C. Presl

Laurentia michelii f. *albiflora* Merino === **Solenopsis laurentia** (L.) C. Presl

Laurentia michelii var. *bicolor* Batt. === **Solenopsis bicolor** (Batt.) Greuter & Burdet

Laurentia michelii var. *integrifolia* Lange === **Solenopsis laurentia** (L.) C. Presl

Laurentia michelii var. *perpusilla* Maire === **Solenopsis laurentia** (L.) C. Presl

Laurentia michelii var. *subacaulis* Pomel ex Batt. === **Solenopsis laurentia** (L.) C. Presl

Laurentia micrantha (E. Mey.) A. DC. === **Wimmerella pygmaea** (Thunb.) Serra, M. B. Crespo & Lammers

Laurentia micrantha (Kunth) A. Zahlbr. === **Diastatea micrantha** (Kunth) McVaugh

Laurentia micrantha var. *longibracteata* E. Wimm. === **Diastatea micrantha** (Kunth) McVaugh

Laurentia micrantha var. *ovatifolia* (B. L. Rob.) E. Wimm. === **Diastatea micrantha** (Kunth) McVaugh

Laurentia minuta f. *balearica* E. Wimm. === **Solenopsis balearica** (E. Wimm.) Aldasoro, Castrov., Sales & Hedge

Laurentia minuta f. *nobilis* E. Wimm. === **Solenopsis bivonae** (Tineo) M. B. Crespo, Serra & Juan

Laurentia minuta var. *minima* (Sims) A. DC. === ?

Laurentia ovatifolia B. L. Rob. === **Diastatea micrantha** (Kunth) McVaugh

Laurentia pedunculata Brandegee === **Diastatea micrantha** (Kunth) McVaugh

Laurentia petraea (F. Muell.) E. Wimm. === **Isotoma petraea** F. Muell.

Laurentia pinetorum Brandegee === **Diastatea tenera** (A. Gray) McVaugh

Laurentia platycalyx F. Muell. === **Lobelia platycalyx** (F. Muell.) F. Muell.

Laurentia pusilla (Benth.) A. DC. === **Isotoma pusilla** Benth.

Laurentia pygmaea (Thunb.) Sond. === **Wimmerella pygmaea** (Thunb.) Serra, M. B. Crespo & Lammers

Laurentia pygmaea var. *glabra* Sond. === **Wimmerella pygmaea** (Thunb.) Serra, M. B. Crespo & Lammers

Laurentia pygmaea var. *obtusiloba* Sond. === **Wimmerella pygmaea** (Thunb.) Serra, M. B. Crespo & Lammers

Laurentia radicans Schönl. === **Lobelia dodiana** E. Wimm.

Laurentia ramosissima (M. Martens & Galeotti) Benth. & Hook. f. ex Hemsl. === **Diastatea virgata** Scheidw.

Laurentia repens (Thunb.) Benth. === **Lobelia anceps** L. f.

Laurentia scapigera (R. Br.) Endl. ex E. Wimm. === **Isotoma scapigera** (R. Br.) G. Don

Laurentia secunda (L. f.) Kuntze === **Wimmerella secunda** (L. f.) Serra, M. B. Crespo & Lammers

Laurentia secunda f. *dentata* (A. DC.) E. Wimm. === **Wimmerella secunda** (L. f.) Serra, M. B. Crespo & Lammers

Laurentia secunda f. *erecta* (A. DC.) E. Wimm. === **Wimmerella secunda** (L. f.) Serra, M. B. Crespo & Lammers

Laurentia tenella A. DC. === **Solenopsis bivonae** (Tineo) M. B. Crespo, Serra & Juan

Legenere

Lobelioideae, 1 species, western North America and southern South America. Preliminary phylogenetic analyses of molecular data (Schultheis 2001; E. Knox, pers. comm.) show that this genus belongs to a clade that otherwise comprises *Downingia*, *Howellia*, *Palmerella*, and *Porterella*. Type [designated by the author]: *Howellia limosa* Greene.

McVaugh, R. (1943). *Legenere* McVaugh, gen. nov. In North American Flora 32A: 13-14. New York: New York Botanical Garden. En. — Establishment of genus as distinct from *Howellia*.

Wimmer, F. E. (1953). *Legenere* McVaugh. In A. Engler & L. Diels, Das Pflanzenreich IV. 276b: 725-726, illus. Berlin: Akademie-Verlag. Ge., La. — Monograph with key, descriptions, full nomenclature, and specimen citations.

Raven, P. H. (1963). Amphitropical relationships in the floras of North and South America. Quart. Rev. Biol. 38: 151-177, maps. En. — Biogeographic discussion of disjunct distribution.

Munz, P. A. & D. D. Keck (1970). *Legenere* McVaugh. In A California flora: 1072. Berkeley: University of California Press. En. — Flora with description and synonymy.

- Ruiz de Ciolfi, E. N. (1976). *Legenere* McVaugh (Campanulaceae) nueva cita para la Argentina. Bol. Soc. Argent. Bot. 17: 176-178, illus. Sp. — Floristic treatment with description, full nomenclature, and specimen citations; North and South American populations conspecific.
  Morin, N. (1993). *Legenere.* In J. C. Hickman (ed.), The Jepson manual, higher plants of California: 464, illus. Berkeley: University of California Press. En. — Flora with description.
  Chiapella, J. (1999). *Legenere* McVaugh. In M. N. Correa (ed.), Flora Patagonica 6: 483-484, illus. Buenos Aires: Instituto Nacional de Tecnologia Agropecuaria. Sp. — Flora for southern Argentina, with description, full nomenclature, and specimen citations.

Legenere McVaugh, N. Amer. Fl. 32A: 13 (1943).
  W. North America & S. South America. 76 85.

Legenere valdiviana (Phil.) E. Wimm., Pflanzenr. IV.276b: 726 (1953).
  C. California; SC. Chile (Valdivia, Osorno) & S. Argentina (Chubut). 76 CAL 85 AGS CLS. Therophyte or hydrotherophyte.
  ** Mezleria valdiviana* Phil., Bot. Zeitung (Berlin) 22: 217 (1864). *Lobelia valdiviana* (Phil.) Vatke, Linnaea 38: 725 (1874). *Dortmanna valdiviana* (Phil.) Kuntze, Revis. Gen. Pl. 2: 973 (1891). *Howellia valdiviana* (Phil.) E. Wimm., Ann. Naturhist. Mus. Wien 56: 373 (1948).
  Howellia limosa Greene, Pittonia 2: 81 (1890). *Legenere limosa* (Greene) McVaugh, N. Amer. Fl. 32A: 13 (1943).

Synonyms:
Legenere limosa (Greene) McVaugh === **Legenere valdiviana** (Phil.) E. Wimm.

Legousia

Campanuloideae, 7 species, Europe to Macaronesia, North Africa and Middle Asia. This genus has sometimes been called by the pre-Linnaean name *Specularia,* which not only was validly published later but also included *Legousia* as a synonym. Under that name, *Legousia* has sometimes been circumscribed to include the New World genera *Heterocodon* and *Triodanis.* The former action is contradicted by the results of a preliminary molecular phylogeny (Eddie et al. 2003, under **General**), which places *Heterocodon* as the sister-group of *Githopsis.* That same study does support a close relationship between *Legousia* and *Triodanis,* but also indicates that a single genus comprising both would have to include *Campanula americana* and *C. divaricata* as well. 2*n* = 14, 16, 20, 26. Type [by monotypy]: *Legousia arvensis* Durande.

  Durande, J. F. (1782). Flore de Bourgogne. Dijon: Frantin. Fr. — Establishment of genus as distinct from *Campanula.*
  Ker Gawler, J. B. (1815). *Campanula pentagonia.* Five-angled bell-flower. Bot. Reg. 1: tab. 56 +2 pp., illus. En. — Portrait of *L. pentagonia,* with description; advocates retaining it and similar species in *Campanula.*
  de Candolle, A. (1830). *Specularia.* In Monographie des Campanulées: 344-352. Paris:Veuve Desray. La. — Reverts to pre-Linnaean name *Specularia;* followed by most writers for over a century.
  Bailey, L. H. (1953). *Specularia.* In The garden of bellflowers in North America: 112-114, illus. New York: MacMillan. En. — Account of species cultivated in North America, with descriptions.
  Fedorov, A. A. (1957). *Legouzia* Durand [sic]. In V. L. Komarov, Flora URSS 24: 427-432, illus. Leningrad: Academia Scientiae URSS. Ru. — Flora of former Soviet Union, with key, descriptions, and full nomenclature.
  Rechinger, K. H. & H. Schiman-Czeika (1965). *Legousia.* In K. H. Rechinger (ed.), Flora Iranica 13: 5-7. Graz: Akademische Druck- und Verlagsanstalt. Ge., La. — Flora with key and descriptions.

Longevialle, M. (1970). Sur la présence en Corse du *Legousia castellana* (Lange) Sampaio. Bull. Soc. Bot. France 117: 321-324. Fr. — Presence of *L. scabra* on Corse used to support hypothesis of geological connection between island and Iberian mainland.
• Tutin, T. G. (1976). *Legousia* Durande. In T. G. Tutin, V. H. Heywood, N., A. Burges, D. M. Moore, D. H. Valentine, S. M. Walters & D. A. Webb (eds.), Flora Europaea 4: 94. Cambridge: Cambridge University Press. En. — Flora with key and descriptions.
Damboldt, J. (1978). *Legousia* Durande. In P. H. Davis (ed.), Flora of Turkey and the East Aegean Islands 6: 83-85. Edinburgh: University Press. En. — Flora with key and descriptions.
Feinbrun-Dothan, N. (1978). *Legousia* Durande. Flora Palaestina, part 3 (text): 283-285. Jerusalem: Israel Academy of Sciences & Humanities. En. — Flora with key and descriptions.
Brewis, A. (1994). *Legousia speculum-veneris*. Biol. Soc. Brit. Isles News 66: 38-39, illus. En. — Report of species in Britain.
Huxley, A. J., ed. (1992). *Legousia*. The new Royal Horticultural Society dictionary of gardening 3: 37. London: MacMillan. En. — Account of species in cultivation.

Legousia Durande, Fl. Bourgogne 1: 37 (1782). *Campanula* [unranked] *Legousia* (Durande) Pers., Syn. Pl. 1: 192 (1805). *Campanula* subg. *Legousia* (Durande) Raf., Fl. Ludov.: 55 (1817), as 'Legouzia'. *Specularia* Heist. ex A. DC., Monogr. Campan.: 344 (1830). *Campanula* sect. *Specularia* D. Dietr., Syn. Pl. 1: 757 (1839). *Pentagonia* Moehring ex Kuntze, Revis. Gen. Pl. 2: 381 (1891); non Heist. ex Fabr., Enum. (ed. 2): 336 (1763), nom. rejic.; nec Benth., Bot. Voy. Sulphur: pl. 39 (1844), nom. cons.
Europe to Macaronesia, N. Africa & C. Asia. 10 11 12 13 14 20 21 32 33 34 35.

Legousia falcata (Ten.) Fritsch ex Janch., Mitt. Naturwiss. Vereines Univ. Wien (n.s.) 5: 100 (1907).
Macaronesia to Caucasus (Abkhaziya, Armenia), Turkmenistan & Iran. 12 BAL COR FRA POR SAR SPA 13 GRC ITA KRI SIC YUG 20 ALG LBY MOR TUN 21 CNY MDR 32 TKM 33 TCS 34 CYP EAI IRN IRQ LBS PAL TUR. Ther. $2n = 20, 26$.
Prismatocarpus falcatus Ten., Fl. Napol. 1: xvi (1811). *Campanula falcata* (Ten.) Schult. in Roem. & Schult., Syst. Veg. 5: 154 (1819). *Specularia falcata* (Ten.) A. DC., Monogr. Campan.: 345 (1830). *Pentagonia falcata* (Ten.) Kuntze, Revis. Gen. Pl. 2: 381 (1891). *Triodanis falcata* (Ten.) McVaugh, Wrightia 1: 28 (1945).
Campanula syriaca Willd. ex Schult. in Roem. & Schult., Syst. Veg. 5: 133 (1819).
Specularia falcata var. *pusilla* Boiss., Fl. Orient. 3: 960 (1875).

Legousia hybrida (L.) Delarbre, Fl. Auvergne (ed. 2): 47 (1800).
W. Europe to Macaronesia, N. Africa, Iran & Caucasus (Krasnodar, Azerbaijan); cult. 10 GRB 11 BGM GER NET SWI 12 BAL COR FRA POR SAR SPA 13 ALB GRC ITA KRI ROM SIC TUE YUG 14 KRY UKR 20 ALG LBY NOR TUN 21 MDR 33 NCS TCS 34 CYP IRN LBS PAL TUR. Ther. $2n = 20$.
Campanula hybrida L., Sp. Pl.: 168 (1753). *Prismatocarpus hybridus* (L.) L'Hér., Sert. Angl.: 3 (1789). *Prismatocarpus confertus* Moench, Methodus: 496 (1794). *Legousia parviflora* Gray, Nat. Arr. Brit. Pl. 2: 410 (1821). *Specularia hybrida* (L.) A. DC., Monogr. Campan.: 348 (1830). *Specularia parviflora* St.-Lag. in Cariot, Étude Fl. (ed. 8) 2: 551 (1889). *Pentagonia hybrida* (L.) Kuntze, Revis. Gen. Pl. 2: 381 (1891). *Specularia conferta* E. H. L. Krause in J. Sturm, Deutschl. Fl. (ed. 2) 12: 271 (1904). *Legousia conferta* Samp., Lista Esp. Herb. Portug.: 127 (1913).
Campanula spuria Pall. ex Schult. in Roem. & Schult., Syst. Veg. 5: 154 (1819).
Specularia calycina Lojac., Fl. Sicul. 2(1): 229 (1903).

Legousia julianii (Batt.) Briq., Candollea 4: 332 (1931), as 'juliani'.
Algeria. 20 ALG. Ther.
Specularia julianii Batt., Fl. Algérie, Dicotyl., appendix 2: xiv (1890), as 'juliani'.

Legousia pentagonia (L.) Thell., Vierteljahrsschr. Naturf. Ges. Zürich 52: 465 (1907).
Greece to Caucasus (Adzhariya) & Iran. 13 BUL GRC KRI TUE 33 TCS 34 EAI IRN IRQ LBS
PAL TUR. Ther. 2*n* =20.
* * *Campanula pentagonia* L., Sp. Pl.: 169 (1753). *Prismatocarpus pentagonia* (L.) L'Hér., Sert.
Angl.: 3 (1789). *Specularia pentagonia* (L.) A. DC., Monogr. Campan.: 344 (1830).
Specularia coa A. DC., Monogr. Campan.: 350 (1830). *Campanula coa* (A. DC.) D. Dietr.,
Syn. Pl. 1: 758 (1839). *Specularia pentagonia* var. *coa* (A. DC.) Nyman, Consp. Fl. Eur.:
483 (1879). *Pentagonia coa* (A. DC.) Kuntze, Revis. Gen. Pl. 2: 381 (1891).
Specularia pentagonia var. *pubescens* A. DC. in DC., Prodr. 7: 489 (1839).

Legousia scabra (Lowe) Gamisans, Cat. Pl. Vasc. Corse: 100 (1985).
Macaronesia & W. Mediterranean. 12 FRA COR POR SPA 20 MOR 21 AZO. Ther.
* * *Prismatocarpus scaber* Lowe, Trans. Cambridge Philos. Soc. 6: 538 (1838). *Specularia
falcata* var. *scabra* (Lowe) A. DC. in DC., Prodr. 7: 490 (1839). *Specularia falcata* f.
scabra (Lowe) H. Lindb., Acta Soc. Sci. Fenn., Ser. B, Opera Biol. 1(2): 150 (1932).
Legousia falcata var. *scabra* (Lowe) Jahand. & Maire, Cat. Pl. Maroc: 735 (1934).
Specularia castellana Lange, Ann. Sci. Nat. Bot. (ser. 4) 2: 372 (1854). *Specularia falcata*
subsp. *castellana* (Lange) Nyman, Consp. Fl. Eur.: 483 (1879). *Legousia castellana*
(Lange) Samp., Lista Esp. Herb. Portug.: 127 (1913).
Specularia castellana var. *grandiflora* Willk. in Willk. & Lange, Prodr. Fl. Hispan. 2: 297
(1868). *Legousia scabra* var. *grandiflora* (Willk.) O. Bolòs & Vigo, Fl. Països Catalans 3:
666 (1996).

Legousia skvortsovii Prosk., Bjull. Moskovsk. Obšč. Isp. Prir., Otd. Biol. 85(4): 95 (1980).
Turkmenistan. 32 TKM. Ther.

Legousia speculum-veneris (L.) Durande ex Vill., Hist. Pl. Dauphiné 1: 338 (1786), as
'speculum veneris'.
W. Europe to Egypt, Saudi Arabia & Iran; cult. 10 GRB 11 AUT BGM GER HUN NET SWI
12 COR FRA POR SAR SPA 13 ALB BUL GRC ITA KRI ROM SIC TUE YUG 20 EGY 34
CYP EAI IRN IRQ LBS PAL TUR 35 SAU. Ther. 2*n* = 14, 16, 20.
* * *Campanula speculum-veneris* L., Sp. Pl.: 168 (1753), as 'speculum. ♀'. *Legousia arvensis*
Durande, Fl. Bourgogne 1: 37 (1782). *Prismatocarpus speculum-veneris* (L.) L'Hér., Sert.
Angl.: 3 (1789), as 'speculum'. *Campanula pulchella* Salisb., Prodr. Stirp. Chap.
Allerton: 127 (1796). *Legousia durandei* Delarbre, Fl. Auvergne (ed. 2): 45 (1800).
Specularia speculum-veneris (L.) A. DC., Monogr. Campan.: 346 (1830), as 'speculum'.
Specularia vulgaris Kitt., Taschenb. Fl. Deutschl. (ed. 2): 483 (1843).
Prismatocarpus hirtus Ten., Fl. Napol. 1: xvi (1811). *Campanula hirta* (Ten.) Schult. in
Roem. & Schult., Syst. Veg. 5: 153 (1819). *Specularia speculum-veneris* var. *hirta* (Ten.)
Nyman, Consp. Fl. Eur.: 483 (1879). *Specularia speculum-veneris* subsp. *hirta* (Ten.)
Arcang., Comp. Fl. Ital.: 450 (1882). *Specularia hirta* (Ten.) Gand., Fl. Cret.: 69 (1916).
Campanula cordata Vis., Stirp. Dalmat. Spec.: 5 (1826). *Prismatocarpus cordatus* (Vis.)
Rchb., Fl. Germ. Excurs.: 858 (1832). *Specularia cordata* (Vis.) Heynh., Alph. Aufz.
Gew.: 689 (1847). *Specularia speculum-veneris* var. *cordata* (Vis.) Nyman, Consp. Fl.
Eur.: 483 (1879).
Specularia speculum-veneris var. *calycina* A. DC., Monogr. Campan.: 347 (1830). *Legousia
speculum-veneris* f. *calycina* (A. DC.) Hayek, Repert. Spec. Nov. Regni Veg. Beih. 30(2):
551 (1930).
Specularia speculum-veneris var. *pubescens* A. DC., Monogr. Campan.: 347 (1830).
Specularia speculum-veneris subvar. *pubescens* (A. DC.) Rouy, Fl. France 10: 57 (1908).
Specularia speculum-veneris var. *libanotica* A. DC., Monogr. Campan.: 347, 791 (1830).
Prismatocarpus speculum-veneris var. *hirtus* K. Koch, Linnaea 19: 30 (1845).
Specularia speculum-veneris var. *stricta* Griseb., Spic. Fl. Rumel. 2: 279 (1846). *Legousia
speculum-veneris* subf. *stricta* (Griseb.) Hayek, Repert. Spec. Nov. Regni Veg. Beih.
30(2): 551 (1930).
Specularia arvensis Montandon in Friche-Joset, Guide Bot. Sundgau: 156 (1868).

Specularia speculum-veneris var. *racemosa* Boiss., Fl. Orient. 3: 959 (1875).

Specularia speculum-veneris f. *plena* Voss in Siebert & Voss, Vilm. Blumengärtn. (ed. 3) 1: 563 (1894).

Specularia speculum-veneris f. *procumbens* Voss in Siebert & Voss, Vilm. Blumengärtn. (ed. 3) 1: 563 (1894).

Specularia speculum-veneris var. *procumbens* Anonymous, Rev. Hort. 1897: 254 (1897). *Specularia polypiflora* David., Trav. Soc. Bulg. Sci. Nat. 8: 92 (1915). *Legousia speculum-veneris* f. *polypiflora* (David.) Hayek, Repert Spec. Nov. Regni Veg. Beih. 30(2): 551 (1930).

Githopsis latifolia Eastw., Proc. Calif. Acad. Sci. (ser. 4) 20: 154 (1931).

Synonyms:

Legousia sect. *Campylocera* (Nutt.) Hayek & Hegi === **Triodanis** Raf.

Legousia arvensis Durande === **Legousia speculum-veneris** (L.) Durande ex Vill.

Legousia balearica Sennen === ?

Legousia biflora (Ruiz & Pav.) Britt. === **Triodanis perfoliata** subsp. **biflora** (Ruiz & Pav.) Lammers

Legousia castellana (Lange) Samp. === **Legousia scabra** (Lowe) Gamisans

Legousia coloradoensis (Buckl.) A. Heller === **Triodanis coloradoensis** (Buckley) McVaugh

Legousia conferta Samp. === **Legousia hybrida** (L.) Delarbre

Legousia durandei Delarbre === **Legousia speculum-veneris** (L.) Durande ex Vill.

Legousia falcata var. *scabra* (Lowe) Jahand. & Maire === **Legousia scabra** (Lowe) Gamisans

Legousia leptocarpa (Nutt.) Britt. === **Triodanis leptocarpa** (Nutt.) Nieuwl.

Legousia parviflora Gray === **Legousia hybrida** (L.) Delarbre

Legousia perfoliata (L.) Britt. === **Triodanis perfoliata** (L.) Nieuwl.

Legousia scabra var. *grandiflora* (Willk.) O. Bolòs & Vigo === **Legousia scabra** (Lowe) Gamisans

Legousia speculum-veneris f. *calycina* (A. DC.) Hayek === **Legousia speculum-veneris** (L.) Durande ex Vill.

Legousia speculum-veneris f. *polypiflora* (David.) Hayek === **Legousia speculum-veneris** (L.) Durande ex Vill.

Legousia speculum-veneris subf. *stricta* (Griseb.) Hayek === **Legousia speculum-veneris** (L.) Durande ex Vill.

Legousia speculum-veneris var. *maroccana* Pau & Font Quer === ?

Leptocodon

Leptocodon (Hook. f.) Lem. of Asia was segregated from *Codonopsis* (q.v.) on the basis of bearing five glands that alternate with the stamens atop the ovary. Aside from this single chatacter, the two genera are indistinguishable (Grey-Wilson 1990, Morris & Lammers 1997). Their merger has been supported by a preliminary molecular phylogeny (Eddie et al. 2003, under **General**) in which the type species formed a clade with all sampled species of *Codonopsis* subg. *Codonopsis,* plus the type species of *Campanumoea* and *Platycodon*. Based on: *Codonopsis* subg. *Leptocodon* Hook. Type [by monotypy]: *Codonopsis gracilis* Hook. f.

Hooker, J. D. (1855). Illustrations of Himalayan plants. London: published privately. En. — Establishment of taxon upon which genus based.

Lemaire, C. (1856). Plantes recommandées. (Espèces nouvelles). Ill. Hort. 3 (Misc.): 48-50. Fr. — Establishment of genus as distinct from *Codonopsis*.

Hong, D. Y. (1983). *Leptocodon* Hook. f. et Thoms. Flora Reipublicae Popularis Sinicae 73(2): 74-76, illus. Beijing: Science Press. Ch. — Floristic treatment with keys and descriptions.

Grey-Wilson, C. (1990). A survey of *Codonopsis* in cultivation. Plantsman 12(2): 65-99, illus. En. — Advocates subsuming *Leptocodon* into *Codonopsis* on basis of overall morphology.

Morris, K. E. & T. G. Lammers (1997). Circumscription of *Codonopsis* and the allied genera *Campanumoea* and *Leptocodon* (Campanulaceae: Campanuloideae). I. Palynological data. Bot. Bull. Acad. Sin. 38: 277-284. En. — Pollen morphology indistinguishable from that in *Codonopsis* subg. *Codonopsis*.

Synonyms:
Leptocodon (Hook. f.) Lem. === **Codonopsis** Wall.
Leptocodon gracilis (Hook. f.) Lem. === **Codonopsis gracilis** Hook. f.
Leptocodon hirsutus D. Y. Hong === **Codonopsis hongii** Lammers

Leptocodon

This name, which pertains to plants of South Africa, is a later homonym of *Leptocodon* (Hook. f.) Lem. of Asia; it is now known by the replacement name *Treichelia* (q.v.). Type [by monotypy]: *Leptocodon longibracteatus* (H. Buek) Sond.

> Sonder, O. W. (1865). *Leptocodon*, Sond. In W. H. Harvey & O. W. Sonder, Flora Capensis 3: 584-585. Dublin: Hodges Smith. En. — Establishment of genus as distinct from *Microcodon*.

Synonyms:
Leptocodon Sond. === **Treichelia** Vatke
Leptocodon longibracteatus (H. Buek) Sond. === **Treichelia longibracteata** (H. Buek) Vatke

Lightfootia

Lightfootia was recognized for many years (e.g., Candolle 1830, 1839, under **General**; Adamson 1955) as an African genus related to but distinct from *Wahlenbergia*. Tuyn (1960) questioned their distinctness and merged their Asian representatives, a move completed by Thulin (1975) and Lammers (1995). Even if it proved possible and desireable to recognize these species at generic rank, the name *Lightfootia* could not be used because it is a later homonym of *Lightfootia* Sw. (Flacourtiaceae). Type [designated by E. Phillips, Gen. S. Afr. Fl. Pl. (ed. 2): 756 (1951)]: *Lightfootia oxycoccoides* L'Hér.

> L'Héritier de Brutelle, C.-L. (1789). Sertum anglicum. Paris: Didot. La. — Establishment of genus as distinct from *Campanula*.
>
> Adamson, R. S. (1953). Notes on nomenclature in *Lightfootia*. J. S. African Bot. 19: 157-159. En. — Clarification of application of questionable names.
>
> Adamson, R. S. (1955). The South African species of *Lightfootia*. J. S. African Bot. 21: 155-218. En. — Monograph, lacking only the tropical representatives.
>
> Tuyn, P. (1960). *Wahlenbergia*. In C. G. G. J. van Steenis (ed.), Flora Malesiana (ser. I) 6: 110-118, illus., map. Djakarta: Noordhoff-Kolff. En. — Advocates subsuming *Lightfootia* into *Wahlenbergia*.
>
> Lambinon, J. & P. Duvigneaud (1961). Étude systématique et phytogéographique d'un groupe de *Lightfootia* à inflorescence contractée. Bull. Soc. Roy. Bot. Belgique 93: 41-53, illus., map. Fr. — Monograph of group of four species in tropical Africa, with key.
>
> Thulin, M. (1975). The genus *Wahlenbergia* s. lat. (Campanulaceae) in tropical Africa and Madagascar. Symb. Bot. Upsal. 21: 1-223. En. — Transfer of tropical species of *Lightfootia* to *Wahlenbergia*.
>
> Lammers, T. G. (1995). Transfer of the southern African species of *Lightfootia*, nom. illeg., to *Wahlenbergia* (Campanulaceae, Campanuloideae). Taxon 44: 333-339. En. — Transfer of South African species of *Lightfootia* to *Wahlenbergia*.

Synonyms:
Lightfootia L'Hér. === **Wahlenbergia** Schrad. ex Roth
Lightfootia abyssinica Hochst. ex A. Rich. === **Wahlenbergia abyssinica** (Hochst. ex A. Rich.) Thulin
Lightfootia abyssinica var. *cinerea* Engl. & Gilg. === **Wahlenbergia napiformis** (A. DC.) Thulin
Lightfootia abyssinica var. *glaberrima* Engl. === **Wahlenbergia napiformis** (A. DC.) Thulin

Lightfootia abyssinica var. *tenuis* Oliv. === **Wahlenbergia abyssinica** (Hochst. ex A. Rich.) Thulin subsp. **abyssinica**

Lightfootia adpressa (L. f.) A. DC. === **Wahlenbergia adpressa** (L. f.) Sond.

Lightfootia albanensis Sond. === **Wahlenbergia cinerea** (L. f.) Lammers

Lightfootia albens Spreng. ex A. DC. === **Wahlenbergia albens** (Spreng. ex A. DC.) Lammers

Lightfootia albicaulis Sond. === **Wahlenbergia albicaulis** (Sond.) Lammers

Lightfootia angustifolia A. DC. === **Wahlenbergia rubens** (H. Buek) Lammers

Lightfootia annua A. DC. === **Wahlenbergia annua** (A. DC.) Thulin

Lightfootia anomala A. DC. === **Wahlenbergia thunbergiana** (H. Buek) Lammers

Lightfootia arabidifolia Engl. === **Wahlenbergia krebsii** subsp. **arguta** (Hook. f.) Thulin

Lightfootia arenaria A. DC. === **Wahlenbergia erecta** (Roth ex Schult.) Tuyn

Lightfootia arenicola Meikle === **Wahlenbergia abyssinica** (Hochst. ex A. Rich.) Thulin subsp. **abyssinica**

Lightfootia asparagoides Adamson === **Wahlenbergia asparagoides** (Adamson) Lammers

Lightfootia axillaris Sond. === **Wahlenbergia axillaris** (Sond.) Lammers

Lightfootia bequaertii DeWild. & Ledoux === **Wahlenbergia capitata** (Baker) Thulin

Lightfootia brachiata Adamson === **Wahlenbergia brachiata** (Adamson) Lammers

Lightfootia brachyphylla Adamson === **Wahlenbergia brachyphylla** (Adamson) Lammers

Lightfootia buekii Sond. === **Wahlenbergia thunbergii** (Schult.) B. Nord.

Lightfootia calcarea Adamson === **Wahlenbergia calcarea** (Adamson) Lammers

Lightfootia caledonica Sond. === **Wahlenbergia dieterlenii** (E. Phillips) Lammers

Lightfootia campestris Engl. === **Wahlenbergia napiformis** (A. DC.) Thulin

Lightfootia capillaris H. Buek === **Wahlenbergia thulinii** Lammers

Lightfootia capitata Baker === **Wahlenbergia capitata** (Baker) Thulin

Lightfootia cartilaginea M. B. Scott === **Wahlenbergia scottii** Thulin

Lightfootia ciliata Sond. === **Wahlenbergia thunbergii** (Schult.) B. Nord.

Lightfootia ciliata Spreng. === **Prismatocarpus fruticosus** (L.) L'Hér.

Lightfootia ciliata var. *debilis* Sond. === **Wahlenbergia thunbergii** (Schult.) B. Nord.

Lightfootia ciliata var. *major* Sond.=== **Wahlenbergia thunbergii** (Schult.) B. Nord.

Lightfootia ciliata var. *pubescens* (A. DC.) Sond. === **Wahlenbergia thunbergii** (Schult.) B. Nord.

Lightfootia cinerea (L. f.) Sond. === **Wahlenbergia cinerea** (L. f.) Lammers

Lightfootia collomioides A. DC. === **Wahlenbergia collomoides** (A. DC.) Thulin

Lightfootia collomioides subsp. *katangensis* Lambinon === **Wahlenbergia collomoides** (A. DC.) Thulin

Lightfootia cordata Adamson === **Wahlenbergia cordata** (Adamson) Lammers

Lightfootia corymbosa Kuntze === **Wahlenbergia huttonii** (Sond.) Thulin

Lightfootia debilis A. DC. === **Wahlenbergia ramosissima** subsp. **lateralis** (Brehmer) Thulin

Lightfootia denticulata (Burch.) Sond. === **Wahlenbergia denticulata** (Burch.) A. DC.

Lightfootia denticulata var. *podanthoides* Markgr. === **Wahlenbergia denticulata** (Burch.) A. DC.

Lightfootia denticulata var. *transvaalensis* Adamson === **Wahlenbergia denticulata** (Burch.) A. DC.

Lightfootia dieterlenii E. Phillips === **Wahlenbergia dieterlenii** (E. Phillips) Lammers

Lightfootia diffusa H. Buek === **Wahlenbergia tenella** (L. f.) Lammers

Lightfootia diffusa var. *palustris* Adamson === **Wahlenbergia tenella** (L. f.) Lammers

Lightfootia diffusa var. *stokoei* Adamson === **Wahlenbergia tenella** (L. f.) Lammers

Lightfootia dinteri Engl. ex Dinter === **Wahlenbergia denticulata** (Burch.) A. DC.

Lightfootia divaricata Engl. === **Wahlenbergia abyssinica** (Hochst. ex A. Rich.) Thulin subsp. **abyssinica**

Lightfootia divaricata H. Buek === **Wahlenbergia thunbergii** (Schult.) B. Nord.

Lightfootia divaricata var. *debilis* (Sond.) Adamson === **Wahlenbergia thunbergii** (Schult.) B. Nord.

Lightfootia divaricata var. *filifolia* Adamson === **Wahlenbergia thunbergii** (Schult.) B. Nord.

Lightfootia effusa Adamson === **Wahlenbergia effusa** (Adamson) Lammers

Lightfootia elata Chiov. === **Wahlenbergia abyssinica** (Hochst. ex A. Rich.) Thulin
subsp. **abyssinica**
Lightfootia elegans Gilli === **Wahlenbergia capitata** (Baker) Thulin
Lightfootia ellenbeckii Engl. === **Wahlenbergia abyssinica** (Hochst. ex A. Rich.) Thulin
subsp. **abyssinica**
Lightfootia erecta R. D. Good === **Wahlenbergia longifolia** (A. DC.) Lammers
Lightfootia ericoidella P. A. Duvign. & Denaeyer === **Wahlenbergia ericoidella** (P. A. Duvign.
& Denaeyer) Thulin
Lightfootia exilis A. DC. === **Wahlenbergia ramosissima** subsp. **lateralis** (Brehmer) Thulin
Lightfootia fasciculata (L. f.) Spreng. === **Wahlenbergia desmantha** Lammers
Lightfootia fruticosa (L.) Druce === **Prismatocarpus fruticosus** (L.) L'Hér.
Lightfootia glomerata Engl. === **Wahlenbergia napiformis** (A. DC.) Thulin
Lightfootia glomerata var. *capitata* (Baker) Lambinon === **Wahlenbergia capitata** (Baker)
Thulin
Lightfootia glomerata var. *subspicata* Engl. === **Wahlenbergia capitata** (Baker) Thulin
Lightfootia goetzeana Engl. === **Wahlenbergia denticulata** (Burch.) A. DC.
Lightfootia gracilis A. DC. === **Wahlenbergia arcta** Thulin
Lightfootia gracilis (G. Forst.) Miq. === **Wahlenbergia gracilis** (G. Forst.) A. DC.
Lightfootia gracilis var. *lavandulifolia* (Reinw. ex Blume) Miq. === **Wahlenbergia marginata**
(Thunb. ex Murray) A. DC.
Lightfootia gracillima R. E. Fr. === **Wahlenbergia paludicola** Thulin
Lightfootia graminicola M. B. Scott === **Wahlenbergia napiformis** (A. DC.) Thulin
Lightfootia grandifolia Engl. === **Wahlenbergia abyssinica** (Hochst. ex A. Rich.) Thulin
subsp. **abyssinica**
Lightfootia grisea H. Buek === **Wahlenbergia cinerea** (L. f.) Lammers
Lightfootia hirsuta (Edgew.) E. Wimm. ex Hepper === **Wahlenbergia hirsuta** (Edgew.) Tuyn
Lightfootia huttonii Sond. === **Wahlenbergia huttonii** (Sond.) Thulin
Lightfootia intermedia H. Buek === **Wahlenbergia thunbergii** (Schult.) B. Nord.
Lightfootia intricata Dinter & Markgr. === **Wahlenbergia denticulata** (Burch.) A. DC.
Lightfootia juncea (H. Buek) Sond. === **Wahlenbergia juncea** (H. Buek) Lammers
Lightfootia kagerensis S. Moore === **Wahlenbergia napiformis** (A. DC.) Thulin
Lightfootia laricifolia Engl. & Gilg === **Wahlenbergia denticulata** (Burch.) A. DC.
Lightfootia laricina H. Buek === **Wahlenbergia albens** (Spreng. ex A. DC.) Lammers
Lightfootia laricina H. Buek === **Wahlenbergia thunbergii** (Schult.) B. Nord.
Lightfootia laxiflora Sond. === **Wahlenbergia laxiflora** (Sond.) Lammers
Lightfootia leptophylla C. H. Wright === **Wahlenbergia collomoides** (A. DC.) Thulin
Lightfootia loddigesii A. DC. === **Wahlenbergia tenerrima** (H. Buek) Lammers
Lightfootia longifolia A. DC. === **Wahlenbergia longifolia** (A. DC.) Lammers
Lightfootia longifolia var. *corymbosa* Adamson === **Wahlenbergia longifolia** (A. DC.)
Lammers
Lightfootia longifolia var. *lanuginosa* Cham. === **Wahlenbergia longifolia** (A. DC.) Lammers
Lightfootia longifolia var. *oppositifolia* Sond. === **Wahlenbergia longifolia** (A. DC.) Lammers
Lightfootia lycopodioides A. DC. === **Wahlenbergia unidentata** (L. f.) Lammers
Lightfootia lycopodioides Mildbr. === **Wahlenbergia huttonii** (Sond.) Thulin
Lightfootia macrostachys A. DC. === **Wahlenbergia macrostachys** (A. DC.) Lammers
Lightfootia madagascariensis A. DC. === **Wahlenbergia abyssinica** (Hochst. ex A. Rich.)
Thulin subsp. **abyssinica**
Lightfootia madagascariensis var. *glabra* Engl. === **Wahlenbergia abyssinica** (Hochst. ex A.
Rich.) Thulin subsp. **abyssinica**
Lightfootia marginata A. DC. === **Wahlenbergia napiformis** (A. DC.) Thulin
Lightfootia marginata var. *lucens* Lambinon === **Wahlenbergia napiformis** (A. DC.) Thulin
Lightfootia microphylla Adamson === **Wahlenbergia microphylla** (Adamson) Lammers
Lightfootia mucronulata H. Buek === **Wahlenbergia thunbergii** (Schult.) B. Nord.
Lightfootia multicaulis Adamson === **Wahlenbergia adamsonii** Lammers
Lightfootia multiflora Adamson === **Wahlenbergia polyantha** Lammers
Lightfootia namaquana Sond. === **Wahlenbergia sonderi** Lammers

Lightfootia napiformis A. DC. === **Wahlenbergia napiformis** (A. DC.) Thulin

Lightfootia nodosa H. Buek === **Wahlenbergia nodosa** (H. Buek) Lammers

Lightfootia oppositifolia A. DC. === **Wahlenbergia thulinii** Lammers

Lightfootia oxycoccoides L'Hér. === **Wahlenbergia parvifolia** (P. J. Bergius) Lammers

Lightfootia paniculata A. DC. === **Wahlenbergia candolleana** (Hiern) Thulin

Lightfootia paniculata Sond. === **Wahlenbergia magaliesbergensis** Lammers

Lightfootia parvifolia (P. J. Bergius) Adamson === **Wahlenbergia parvifolia** (P. J. Bergius) Lammers

Lightfootia pauciflora Adamson === **Wahlenbergia oligantha** Lammers

Lightfootia perotifolia (Willd. ex Schult) E. Wimm. ex Agnew === **Wahlenbergia erecta** (Roth ex Schult.) Tuyn

Lightfootia planifolia Adamson === **Wahlenbergia riversdalensis** Lammers

Lightfootia polycephala Mildbr. === **Wahlenbergia polycephala** (Mildbr.) Thulin

Lightfootia pubescens A. DC. === **Wahlenbergia thunbergii** (Schult.) B. Nord.

Lightfootia ramosissima (Hemsl.) E. Wimm. ex Hepper === **Wahlenbergia ramosissima** (Hemsl.) Thulin

Lightfootia rigida Adamson === **Wahlenbergia neorigida** Lammers

Lightfootia robusta A. DC. === **Wahlenbergia robusta** (A. DC.) Sond.

Lightfootia rubens H. Buek === **Wahlenbergia rubens** (H. Buek) Lammers

Lightfootia rubens var. *brachyphylla* Adamson === **Wahlenbergia rubens** (H. Buek) Lammers

Lightfootia rubioides A. DC. === **Wahlenbergia rubioides** (A. DC.) Lammers

Lightfootia rubioides var. *stokoei* Adamson === **Wahlenbergia rubioides** (A. DC.) Lammers

Lightfootia rupestris Engl. === **Wahlenbergia abyssinica** (Hochst. ex A. Rich.) Thulin subsp. **abyssinica**

Lightfootia scoparia Wild === **Wahlenbergia subaphylla** subsp. **scoparia** (Wild) Thulin

Lightfootia sessiliflora (L. f.) Spreng. === **Wahlenbergia subulata** (L'Hér.) Lammers

Lightfootia sodenii Engl. === **Wahlenbergia abyssinica** (Hochst. ex A. Rich.) Thulin subsp. **abyssinica**

Lightfootia spicata H. Buek === **Wahlenbergia macrostachys** (A. DC.) Lammers

Lightfootia squarrosa Adamson === **Wahlenbergia levynsiae** Lammers

Lightfootia stricta Adamson === **Wahlenbergia neostricta** Lammers

Lightfootia subaphylla Baker === **Wahlenbergia subaphylla** (Baker) Thulin

Lightfootia subulata Engl. === **Wahlenbergia abyssinica** (Hochst. ex A. Rich.) Thulin subsp. **abyssinica**

Lightfootia subulata L'Hér. === **Wahlenbergia subulata** (L'Hér.) Lammers

Lightfootia subulata var. *congesta* Adamson === **Wahlenbergia subulata** (L'Hér.) Lammers

Lightfootia subulata var. *tenuifolia* Adamson === **Wahlenbergia subulata** (L'Hér.) Lammers

Lightfootia tenella (L. f.) A. DC. === **Wahlenbergia tenella** (L. f.) Lammers

Lightfootia tenella Lodd. === **Wahlenbergia tenerrima** (H. Buek) Lammers

Lightfootia tenella f. *coerulescens* Kuntze === **Wahlenbergia tenerrima** (H. Buek) Lammers

Lightfootia tenella f. *flaviflora* Kuntze === **Wahlenbergia tenella** (L. f.) Lammers

Lightfootia tenella var. *brevivalvis* A. DC. === **Wahlenbergia tenella** (L. f.) Lammers

Lightfootia tenella var. *diffusa* (Buek) Zahlbr. === **Wahlenbergia tenella** (L. f.) Lammers

Lightfootia tenella var. *fasciculata* (L. f.) Sond. === **Wahlenbergia desmantha** Lammers

Lightfootia tenella var. *longivalvis* A. DC. === **Wahlenbergia nodosa** (H. Buek) Lammers

Lightfootia tenella var. *microphylla* Sond. === **Wahlenbergia nodosa** (H. Buek) Lammers

Lightfootia tenella var. *montana* Adamson === **Wahlenbergia tenerrima** (H. Buek) Lammers

Lightfootia tenella var. *rigida* Sond. === **Wahlenbergia nodosa** (H. Buek) Lammers

Lightfootia tenella var. *tenerrima* (H. Buek) Sond. === **Wahlenbergia tenerrima** (H. Buek) Lammers

Lightfootia tenerrima H. Buek === **Wahlenbergia tenerrima** (H. Buek) Lammers

Lightfootia tenuifolia A. DC. === **Wahlenbergia denticulata** (Burch.) A. DC.

Lightfootia tenuis Adamson === **Wahlenbergia pyrophila** Lammers

Lightfootia thunbergiana H. Buek === **Wahlenbergia thunbergiana** (H. Buek) Lammers

Lightfootia thymifolia H. Buek === **Wahlenbergia thunbergii** (Schult.) B. Nord.

Lightfootia uitenhagensis H. Buek === **Wahlenbergia thunbergii** (Schult.) B. Nord.

Lightfootia umbellata Adamson === **Wahlenbergia umbellata** (Adamson) Lammers
Lightfootia unidentata (L. f.) A. DC. === **Wahlenbergia unidentata** (L. f.) Lammers
Lightfootia unindentata var. *lycopodioides* (A. DC.) Sond. === **Wahlenbergia unidentata**
  (L. f.) Lammers
Lightfootia unindentata var. *pubescens* Sond. === **Wahlenbergia unidentata** (L. f.) Lammers
Lightfootia welwitschii A. DC. === **Wahlenbergia welwitschii** (A. DC.) Thulin

Lobelia

Lobelioideae, 405 species, cosmopolitan in distribution. Nearly 37% of the species are African and another 22% North American. Asia has 12% of the species, Australasia about 10%, South America and the Caribbean 8% each, and the Hawaiian Islands 3%. Two species (one North American, one African) extend into Europe. The genus is divided into three subgenera, each comprising several sections (Murata 1995; as modified by Lammers 1999, 2004). The autonymic subgenus comprises sect. *Delostemon* (E. Wimm.) J. Murata, sect. *Heyneana* J. Murata, sect. *Cryptostemon* (E. Wimm.) J. Murata, and sect. *Lobelia;* subg. *Isolobus* (A. DC.) Y. S. Lian includes sect. *Dioica* (E. Wimm.) J. Murata, sect. *Pratia* (Gaudich.) J. Murata, sect. *Paramezleria* E. Wimm., and sect. *Isolobus* (A. DC.) C. B. Clarke; and subg. *Tupa* (G. Don) E. Wimm. includes sect. *Tupa* (G. Don) Benth., sect. *Colensoa* (Hook. f.) J. Murata, sect. *Homochilus* A. DC., sect. *Tylomium* (C. Presl) Benth., sect. *Rhynchopetalum* (Fresen.) Benth., sect. *Revolutella* E. Wimm., and sect. *Galeatella* E. Wimm. Preliminary analyses of molecular data (Knox & Palmer 1998; E. Knox, pers. comm.; A. Antonelli, pers. comm.) make it clear that this genus will have to be dismembered in the future. Species assigned here to *Lobelia* are found throughout the phylogeny of the subfamily, from the first branch to the last; *all* of the other genera thus far sampled are embedded among species of *Lobelia.* Once such a remodeling of generic classification has been completed, it is likely that *Lobelia* would only include the 21 species in eastern North America, and the remaining 384 species would be assigned to new or resurrected genera. Unfortunately, the only rational way to avoid this massive restructuring of the subfamily would be to treat the entire subfamily as a single genus. The last monograph of the genus in its entirety was that of Wimmer (1953, 1968). $2n$ = 12, 14, 18, 24, 26, 28, 38, 42, 70, 140. Type [designated by Hitchc. & M. L. Green in Anon., Nomencl. Prop. Brit. Bot.: 184 (1929)]: *Lobelia cardinalis* L. Under ICBN Art. 10.5(b), this supercedes the earlier choice of *Lobelia dortmanna* L. [Britton & A. Br., Illus. Fl. N. US 3: 299 (1913)]; cf. McNeill et al. (1987).

Linnaeus, C. (1753). *Lobelia.* In Species plantarum: 929-933. Stockholm: Laurentius Salvius. La. — Establishment of genus in Syngenesia Monogamia.
Linnaeus, C. (1754). *Lobelia.* In Genera plantarum: 401. Stockholm: Laurentius Salvius. La. — Description of genus to accompany preceding.
Miller, P. (1754). The gardeners dictionary ... abridged from the last folio edition (ed. 4). London: J. & J. Rivington. En. — Restriction of *Lobelia* to species of Goodeniaceae; resurrected pre-Linnaean name *Rapuntium* (q.v.) for species here treated as *Lobelia.*
Koelreuter, I. T. (1777). Lobeliae hybridae. Acta Acad. Sci. Imp. Petrop. 1: 185-192. La. — Earliest reference to experimental hybrids of *L. cardinalis* and *L. siphilitica.*
Moench, C. (1794). *Rapuntium.* In Methodus plantas horti botanici et agri Marburgensis, a staminum situ describendi: 655-656. Marburg: Nova Libraria Academiae. La. — Like Miller (1754), used *Rapuntium* for *Lobelia* and applied *Lobelia* to genus of Goodeniaceae.
Sims, J. (1801). *Lobelia bicolor.* Spotted lobelia. Bot. Mag. 15: tab. 514 + 1 pg., illus. En. — Portrait of *L. erinus.*
Andrews, H. (1803a). *Lobelia coronopifolia.* Buck's-horn-leaved lobelia. Bot. Repos. 5: pl. 339 + 1 pg., illus. En. — Portrait with description.
Andrews, H. (1803b). *Lobelia gracilis.* Slender-stemmed [sic] lobelia. Bot. Repos. 5: pl. 340 + 1pg., illus. En. — Portrait of *L. andrewsii,* with description.
Sims, J. (1810). *Lobelia gigantea.* Gigantic lobelia. Bot. Mag. 32: tab. 1325 + 2 pp., illus. En. — Portrait of *L. excelsa,* with description.

Ker Gawler, J. B. (1815). *Lobelia splendens*. Shining lobelia. Bot. Reg. 1: fol. 60 + 2 pp., illus. En. — Portrait of *L. cardinalis*, with description.

Sims, J. (1817). *Lobelia ilicifolia*. Holly-leaved lobelia. Bot. Mag. 44: tab. 1896 + 1 pg., illus. En. — Portrait of *L. purpurascens*, with description.

Sims, J. (1820). *Lobelia racemosa*. Green-flowered lobelia. Bot. Mag. 47: tab. 2137 + 1 pg., illus. En. — Portrait of *L. cirsiifolia*, with description.

Sims, J. (1821a). *Lobelia kalmii*. Kalm's lobelia. Bot. Mag. 48: tab. 2238 + 2 pp., illus. En. — Portrait with description.

Sims, J. (1821b). *Lobelia pedunculata*. Long-stalked lobelia. Bot. Mag. 48: tab. 2251 + 1 pg., illus. En. — Portrait of *L. coronopifolia*, with description.

Sims, J. (1821c). *Lobelia decumbens*. Decumbent lobelia. Bot. Mag. 49: tab. 2277 + 2 pp., illus. En. — Portrait of *L. anceps,* with description.

Hooker, W. J. (1823). *Lobelia micrantha*. Small-flowered lobelia. In Exotic Flora 1: pl. 44 + 2 pp., illus. En. — Portrait of *L. heyneana,* with description.

Sims, J. (1823). *Lobelia pyramidalis*. Branchy lobelia. Bot. Mag. 50: tab. 2387 + 2 pp., illus. En. — Portrait with description.

Sweet, R. (1823-1825). *Lobelia inflata*. Brit. Fl. Gard. 1: pl. 99 + 2 pp., illus. En. — Portrait with description.

Sims, J. (1825). *Lobelia tupa*. Mullein-leaved lobelia. Bot. Mag. 52: tab. 2550 + 2 pp., illus. En. — Portrait with description.

Lindley, J. (1826). *Lobelia arguta*. Fine-toothed lobelia. Bot. Reg. 12: fol. 973 + 2 pp., illus. En. — Portrait of *L. excelsa,* with description.

Sims, J. (1826a). *Lobelia corymbosa*. Corymbose African lobelia. Bot. Mag. 53: tab. 2693 + 2 pp., illus. En. — Portrait of *L. jasionoides,* with description.

Sims, J. (1826b). *Lobelia caerulea*. Blue-flowered lobelia. Bot. Mag. 53: tab. 2701 + 2 pp., illus. En. — Portrait with description.

Sweet, R. (1827-1829). *Lobelia tupa*. Mullein-leaved lobelia. Brit. Fl. Gard. 3: pl. 284 + 2 pp., illus. En. — Portrait with description.

Rishel, J. (1828). *Lobelia inflata*. In The Indian physician: 26-27. New Berlin: Joseph Miller. En. — Influential early account of its pharmaceutical use.

Lindley, J. (1830a). *Lobelia purpurea*. Purple lobelia. Bot. Reg. 16: fol. 1325 + 2 pp., illus. En. — Portrait of *L. polyphylla,* with description.

Lindley, J. (1830b). *Pratia begonifolia* [sic]. Begonia-leaved pratia. Bot. Reg. 16: fol. 1373 + 2 pp., illus. En. — Portrait of *L. nummularia,* with description.

von Martius, K. F. (1830). *Lobelia cavanillesii*. In Auswahl merkwürdiger Pflanzen des königlichen botanischen Gartens zu München: 12-14 + tab. 9, illus. Frankfurt: Brönner. Fr., Ge. — Portrait of *L. laxiflora* subsp. *angustifolia*, with description.

Rafinesque, C. S. (1830). *Lobelia inflata*. In Medical flora; or manual of the medical botany of the United States of North America: 22-26 + pl. 60, illus. Philadelphia: Atkinson & Alexander. En. — Early account of use as pharmaceutical, with comments on use of *L. cardinalis* and *L. siphilitica* to treat syphilis.

Lindley, J. (1831). Low's purple lobelia. Bot. Reg. 17: fol. 1445 + 1 pg., illus. En. — Portrait of *L. × speciosa*, with description.

Sweet, R. (1831). *Lobelia decurrens*. Winged-stemmed lobelia. Brit. Fl. Gard. (ser. 2) 1: pl. 86 + 2 pp., illus. En. — Portrait of subsp. *decurrens*, with description.

Hooker, W. J. (1833). *Lobelia mucronata*. Sharp-pointed lobelia. Bot. Mag. 60: tab. 3207 + 2 pp., illus. En. — Portrait of *L. tupa,* with description.

Lindley, J. (1833). *Lobelia tupa*. The tupa-poison plant. Bot. Reg. 19: fol. 1612 + 2 pp., illus. En. — Portrait with description.

Sweet, R. (1833). *Lobelia colorata*. Red-leaved lobelia. Brit. Fl. Gard. (ser. 2) 2: pl. 180 + 2 pp., illus. En. — Portrait of *L. amoena,* with description.

Hooker, W. J. (1834). *Lobelia puberula,* β. Blue downy lobelia variety. Bot. Mag. 61: tab. 3292 + 2 pp., illus. En. — Portrait.

Don, D. (1835a). *Lobelia polyphylla*. Leafy lobelia. Brit. Fl. Gard. (ser. 2) 3: pl. 242 + 2 pp., illus. En. — Portrait with description.

Don, D. (1835b). *Tupa blanda*. Blush-flowered tupa. Brit. Fl. Gard. (ser. 2) 4: pl. 308 + 2 pp., illus. En. — Portrait of *L. bridgesii,* with description.

Griffith, R. E. (1836). *Lobelia syphilitica*. Amer. Pharm. J. 8: 191-192. En. — Description of use to treat syphilis.

Lindley, J. (1836). *Lobelia decurrens*. Winged-stemmed lobelia. Bot. Reg. 22: fol. 1842 + 1 pg., illus. En. — Portrait of subsp. *decurrens,* with description.

Paxton, J. (1836). *Lobelia fulgens* (var. *propinqua*). Paxton's Mag. Bot. 2: 52 + plate, illus. En. — Portrait of *L. cardinalis,* with description.

• Presl, C. (1836). *Rapuntium*. Tourn. Mill. Gaertn. Moench. In Prodromus monographiae Lobeliacearum: 11-31. Prague: Theophilus Haase. La. — Uses this name for species here treated as *Lobelia* (following Miller 1754 and Moench 1794), but inexplicably applies *Lobelia* to species here called *Siphocampylus*.

Hayne, F. G., J. F. Brandt & J. T. C. Ratzeburg (1837). *Lobelia antisyphilitica*. In Getreue Darstellung und Beschreibung der in der Arzneykunde Gewächase 13: pl. 9 + 2 pp., illus. Berlin: published privately. Ge. — Portrait of *L. siphilitica,* with description and discussion of use in treating syphilis.

Hooker, W. J. (1837a). *Lobelia polyphylla*. Many-leaved lobelia. Bot. Mag. 64: tab. 3550 + 2 pp., illus. En. — Portrait with description.

Hooker, W. J. (1837b). *Lobelia cavanillesii*. Cavanilles' lobelia. Bot. Mag. 64: tab. 3600 + 2 pp., illus. En. — Portrait of *L. laxiflora* subsp. *angustifolia,* with description.

Hooker, W. J. (1837c). *Lobelia siphilitica; hybrida*. Hybrid var. of the blue American lobelia. Bot. Mag. 64: tab. 3604 + 1 pg., illus. En. — Portrait of *L.* × *speciosa,* with description; equates it with "Low's purple lobelia" of Lindley (1831).

Don, D. (1838a). *Lobelia cardinalis;* var. *milleri*. Miller's lobelia. Brit. Fl. Gard. (ser. 2) 4: pl. 372 + 2 pp., illus. En. — Portrait of *L.* × *speciosa,* with description.

Don, D. (1838b). *Siphocampylus bicolor*. Two-coloured siphocampylus. Brit. Fl. Gard. (ser. 2) 4: pl. 389 + 2 pp., illus. En. — Portrait of *L. laxiflora* subsp. *laxiflora,* with description.

Knowles, G. B. & F. Westcott (1838). *Siphocampylus bicolor*. (Two-coloured siphocampylus.) Fl. Cab. 2: 97-98 + pl. 69, illus. En. — Portrait of *L. laxiflora* subsp. *laxiflora,* with description.

Maund, B. & J. S. Henslow (1838). *Lobelia ramosa*. Branching lobelia. In The Botanist: pl. 93 + 2 pp., illus. En. — Portrait of *L. tenuior,* with description.

• de Candolle, A. (1839). *Holostigma, Isolobus, Lobelia, Tupa, Rhynchopetalum, Enchysia*. In A. P. de Candolle, Prodromus systematis naturalis regni vegetabilis 7: 352-354, 357-396, 408-409, 784-786. Paris: Treuttel & Würtz. La. — Monograph with descriptions, infrageneric classifications, full nomenclature, and specimen citations; circumscription of genus relatively narrow.

Hooker, W. J. (1839). *Lobelia bridgesii*. Mr. Bridges' lobelia. Bot. Mag. 65: tab. 3671 + 2 pp., illus. En. — Portrait with description.

Otto, F. & A. Dietrich (1839). Ueber die großen, dunkelroth blühenden Lobelien unserer Gärten. Allg. Gartenzeitung 7: 297-300. Ge. — Description of various strains of *L. cardinalis* in cultivation.

Paxton, J. (1839a). *Lobelia heterophylla*. Paxton's Mag. Bot. 6: 197-198 + plate, illus. En. — Portrait with description and notes on cultivation.

Paxton, J. (1839b). *Lobelia ignea*. Paxton's Mag. Bot. 6: 247-248 + plate, illus. En. — Portrait of *L. cardinalis,* with description and notes on cultivation.

Hooker, W. J. (1840). *Lobelia heterophylla*. Various-leaved lobelia. Bot. Mag. 66: tab. 3784 + 2 pp., illus. En. — Portrait with description.

Arnott, G. A. W. (1841). *Lobelia trigona,* Roxb. Icon. Pl. 4: tab. 358 + 1 pg., illus. En. — Portrait of *L. alsinoides,* with description.

Paxton, J. (1842). *Lobelia heterophylla;* var. *major*. Paxton's Mag. Bot. 9: 101-102 + plate, illus. En. — Portrait of *L. heterophylla,* with description and notes on cultivation.

Hooker, W. J. (1843a). *Lobelia splendens;* var. β., *atro-sanguinea*. Shining lobelia; dark purple-leaved var. Bot. Mag. 69: tab. 4002 + 2 pp., illus. En. — Portrait of *L. cardinalis,* with description.

Hooker, W. J. (1843b). *Lobelia physaloides*. A. Cunn. Icon. Pl. 6: tab. 555-556 + 1 pg., illus. En. — Portrait and description.

Lemaire, C. (1843). Tupa élégant. *Tupa blanda*. Hort. Universel 4: 259-262 + plate, illus. Fr. — Portrait of *L. bridgesii*, with description.

Paxton, J. (1843). *Lobelia erinus;* var. *grandiflora*. Paxton's Mag. Bot. 10: 75-76 + plate, illus. En. — Portrait of *L. erinus*, with description and notes on cultivation.

Hooker, W. J. (1845). *Lobelia thapsoidea*. Mullein-like lobelia. Bot. Mag. 71: tab. 4150 + 2 pp., illus. En. — Portrait with description.

de Jonghe, J. (1845). Du genre *Lobelia* et de sa culture. Rev. Hort. (ser. 2) 4: 37-43. Fr. — Discussion of species and variants in cultivation.

Paxton, J. (1848). Floricultural notices. New, rare, or interesting plants in flower at the principal suburban nurseries and gardens. Paxton's Mag. Bot. 14: 262-263. En. — Announcement of several new cultivars of *L. cardinalis*.

• Paxton, J. (1849a). *Lobelia fulgens* var. Paxton's Mag. Bot. 15: 7-12 + plate, illus. En. — Overview of genus and synopsis of entire subfamily, with emphasis on genera in cultivation; portrait of new cultivars described by Paxton (1848).

Paxton, J. (1849b). *Lobelia coelestis*. Paxton's Mag. Bot. 15: 103 + plate, illus. En. — Portrait of *L. siphilitica*, with description.

Hooker, W. J. (1850). *Tupa crassicaulis*. Thick-stemmed tupa. Bot. Mag. 76: tab. 4505 + 2 pp., illus. En. — Portrait of *L. ghiesbreghtii*, with description.

Hooker, W. J. (1857a). *Lobelia splendens:* var. *ignea*. Shining lobelia: blood-red-leaved var. Bot. Mag. 83: tab. 4960 + 2 pp., illus. En. — Portrait of *L. cardinalis*, with description.

Hooker, W. J. (1857b). *Lobelia texensis*. Texas lobelia. Bot. Mag. 83: tab. 4964 + 2 pp., illus. En. — Portrait of *L. cardinalis*, with description.

Hooker, W. J. (1858). *Lobelia trigonocaulis*. Triangular-stemmed lobelia. Bot. Mag. 84: tab. 5088 + 2 pp., illus. En. — Portrait with description.

Hooker, W. J. (1866). *Lobelia nicotianaefolia*. Tobacco-leaved lobelia. Bot. Mag. 92: tab. 5587 + 2 pp., illus. En. — Portrait with description.

Lemaire, C. (1866). *Lobelia coronopifolia* (?). Ill. Hort. 13: pl. 485 + 2 pp., illus. Fr. — Portrait with description.

Bentham, G. & F. Mueller (1868). *Lobelia*, Linn.; *Pratia*, Gaudich. In Flora Australiensis: a description of the plants of the Australian Territory 4: 122-133. London: L.. Reeve. En. — Flora with keys, descriptions, infrageneric classification, full nomenclature, and specimen citations.

• Bentham, G. (1876). *Hypsela*, Presl; *Pratia*, Gaudich.; *Colensoa* Hook. f.; *Lobelia*, Linn. In G. Bentham & J. D. Hooker, Genera plantarum 2: 550-553. London: L. Reeve. La. — Generic monograph with descriptions; circumscription broader than Candolle (1839), with many former genera reduced to sectional rank (e.g., *Rhynchopetalum, Trimeris, Tupa, Tylomium*).

Kanitz, A. (1877). *Haynaldia* novum genus Lobeliacearum. Magyar Növényt. Lapok 1: 3. La. — Segregation of robust wing-seeded Brazilian species.

Schneck, J. (1878). More about lobelias. Bot. Gaz. (Crawfordsville) 3: 35-36. En. — Discovery of naturally occurring *L.* × *speciosa,* as well as plant that conforms in every way to *L. siphilitica* except for its red corolla.

Bower, F. O. (1883). On the structure of the stem of *Rhynchopetalum montanum* (Fresen.). Bot. J. Linn. Soc. 20: 440-445, illus. En. — Anatomical study of Afro-alpine pachycaul.

Hemsley, W. B. (1886). *Pratia borneensis,* Hemsl. Hooker's Icon. Pl. 16: pl. 1532 + 1 pg., illus. En. — Portrait of *L. borneensis,* with description.

Hooker, J. D. (1886). *Colensoa physaloides*. Bot. Mag. 112: tab. 6864 + 2 pp., illus. En. — Portrait of *L. physaloides,* with description.

• Schönland, S. (1889). *Lobelia* L., *Pratia* Gaud., *Hypsela* Presl In A. Engler & K. Prantl, Die natürlichen Pflanzenfamilien IV. 5: 66-69, illus. Leipzig: Wilhelm Engelmann. Ge. — Generic monographic with keys, descriptions, and infrageneric classification; broad circumscription similar to Bentham (1876).

- Kuntze, O. (1891). *Dortmannia* [sic]. In Revisio generum plantarum vascularium omnium atque cellularium multarum secundum leges nomenclaturae internationales 2: 379-380, 971-973. Leipzig: Arthur Felix. Ge. — Substitution of pre-Linnaean name *Dortmanna* (q.v.) for *Lobelia*, and application of the latter to genus of Goodeniaceae.

Sauvageau, C. (1893). *Lobelia gerardi.* Rev. Hort. 65: 519-520. Fr. — Discussion of *L. × speciosa* in cultivation.

Meehan, T. (1897). *Lobelia syphilitica.* Blue cardinal. Meehan's Monthly 7: 61-62 + pl. 4, illus. Portrait with description.

Anonymous (1898). *Lobelia rivoirei.* Gard. Chron. (ser. 3) 25: 232-233, illus. En. — Description of new cultivar of *L. × speciosa.*

Hooker, J. D. (1898). *Lobelia intertexta,* Baker. Bot. Mag. 124: tab. 7615 + 2 pp., illus. En. — Portrait of *L. trullifolia* subsp. *trullifolia,* with description.

- Kuntze, O. (1898). *Dortmannia* [sic]. In Revisio generum plantarum vascularium omnium atque cellularium multarum secundum leges nomenclaturae internationales 3(2): 186-188. Leipzig: Arthur Felix. Ge. — Supplement to Kuntze (1891).

Ames, O. (1901). *Lobelia inflata × cardinalis.* Rhodiora 3: 296-298. En. — Description of putative hybrid.

Hemsley, W. B. & P. H. E. Powell-Cotton (1901). The tree lobelias of tropical Africa. Gard. Chron. (ser. 3) 29: 417-418, illus. En. — Popular account of Afro-montane giants.

Ames, O. (1903). *Lobelia × syphilitico-cardinalis.* Rhodora 5: 284-286 + pl. 49, illus. En. — Description of artificially produced *L. × speciosa.*

Holm, H. T. (1907). Medicinal plants of North America 10. Merck's Rep. 16: 341-343. En. — Use of *L. inflata* as pharmaceutical.

Saunders, E. R. (1912). Double flowers. J. Roy. Hort. Soc. 38: 469-482, illus. En. — Description of *L. erinus* with doubled corolla.

Anonymous (1915). *Lobelia laxiflora* var. *angustifolia.* Gard. Chron. (ser. 3) 57: 262-263, illus. En. — Portrait of *L. laxiflora* subsp. *angustifolia,* with notes on cultivation.

Turrill, W. B. (1916). *Lobelia holstii.* Bot. Mag. 142: tab. 8648 + 2 pp., illus. En. — Portrait with description.

- Rock, J. F. (1919). A monographic study of the Hawaiian species of the tribe Lobelioideae family Campanulaceae, *Lobelia* Linn. Mem. Bernice Pauahi Bishop Mus. 7(2): 111-138, illus. En. — Regional monograph with keys, descriptions, full nomenclature, and specimen citations.

- Fries, R. E. & T. C. E. Fries (1922). Die riesen-lobelien Afrikas. Svensk. Bot. Tidsskr. 16: 383-416, illus. Ge. — Monograph of sect. *Rhynchopetalum,* with key, descriptions, full nomenclature, and specimen citations.

Anonymous (1923). Perennial lobelias at Wisley, 1921. J. Roy. Hort. Soc. 48: 239-240. En. — Results of garden trials of cultivars of *L. cardinalis, L. siphilitica, L. × speciosa,* and *L. tupa.*

de Vilmorin, R & M. Simonet (1927). Nombre des chromosomes dans les genres *Lobelia, Linum* et chez quelques autres espèces végétales. Compt.-Rend. Hebd. Séances Mém. Soc. Biol. 96: 166-168, illus. Fr. — Early cytological study, showing most species examimed with $x = 7$.

- Skottsberg, C. (1928). On some arborescent species of *Lobelia* from tropical Asia. Acta Horti Gothob. 4: 1-26, illus. En. — Discussion of *L. nicotianifolia* and its allies in southen Asia, including possible relationship to Hawaiian species.

Boynton, K. R. (1930). *Lobelia sessilifolia.* Violet lobelia. Addisonia 14: 59-60 + pl. 478, illus. En. — Portrait with description.

Braun, E. L. (1931). A hybrid *Lobelia.* Bot. Gaz. (Crawfordsville) 91: 462-463. En. — Spontaneous appearance in common garden of *L. puberula × L. siphilitica.*

Fraser, L. (1931). An investigation of *Lobelia gibbosa* and *Lobelia dentata.* I. Mycorhiza, latex system and general biology. Proc. Linn. Soc. New South Wales 56: 497-525, illus. En. — Anatomy and ecology of two Australian species.

Saunders, E. R. (1931). A study of the relations of the single and double forms of *Lobelia erinus* L. Z. Pflanzenzücht. 17: 136-146, illus. — Morphology of flowers with doubled corolla.

True, R. H. (1932). The introduction of *Lobelia siphilitica* into medicine. Amer. J. Pharm. 104: 279-281. En. — History of use to treat syphilis.

- Bruce, E. A. (1934). The giant lobelias of East Africa. Bull. Misc. Inform. Kew 1934: 61-88, 274, illus. En. — Monograph of sect. *Rhynchopetalum* (dividing them among five informal series), with keys, descriptions, full nomenclature, and specimen citations.

Cotton, A. D. (1934). The tree senecios and tree lobelias of high African mountains. New Fl. & Silva 7: 6-10 + figs. I-V, illus. En. — Popular account of Afro-montane giants.

Hauman, L. (1934a). Notes sur les lobeliae geants du Congo Belge. Rev. Zool. Bot. Africaines 25 (Suppl.): 13-20 + 3 pl., illus., map Fr. — Floristic treatment of sect. *Rhynchopetalum* in Zaire, with key.

- Hauman, L. (1934b). Les *Lobelia* geants montagnes du Congo Belge. Mém. Inst. Roy. Colon. Belge, Sect. Sci. Nat. (8°) 2: 1-52 + pl. 1-7, illus. Fr. — Expanded version of Hauman (1934a), with key, descriptions, full nomenclature, and specimen citations.

- McVaugh, R. (1936). Studies in the taxonomy and distribution of the eastern North American species of *Lobelia*. Rhodora 38: 241-263, 276-298, 305-329, 346-362 + pl. 435-436, illus., maps. En. — Monograph of sect. *Lobelia,* with key, descriptions, full nomenclature, and specimen citations.

Sharpe, M. R. (1936). Color variations in the cardinal flower. Horticulture 14: 479. En. — Unusual corolla colors in naturally occurring plants of *L. cardinalis*.

Fernald, M. L. & L. Griscom (1937). The identity of *Lobelia glandulosa* Walt. Rhodora 39: 497. En. — Clarification of problematic identity.

Okuno, S. (1937). Karyological studies on some species of *Lobelia*. Cytologia, Fuji Jubilee Vol.: 897-902, illus. En. — Determinations in several species support $x = 7$ for genus.

McVaugh, R. (1938). Notes on the lobelias in the herbarium of Stephen Elliott. Rhodora 40: 175-177. En. — Complete list of *Lobelia* specimens therein, and an attempt to clarify the application of some early names in SE. U.S.A.

- St. John, H. & E. Y. Hosaka (1938). Notes on Hawaiian species of *Lobelia*. Hawaiian plant studies 5. Occas. Pap. Bernice Pauahi Bishop Mus. 14: 117-126. En. — Monograph of sect. *Galeatella,* with key, descriptions, full nomenclature, and specimen citations.

Creasey, L. B. (1939). *Lobelia linearis*. Gard. Chron. (ser. 3) 105: 6. En. — Introduction of this South African species to cultivation.

- McVaugh, R. (1940). A key to the North American species of *Lobelia* (sect. *Hemipogon*). Amer. Midl. Naturalist 24: 681-702, illus. En. — Overview (including key) of all sections represented in North America; synopsis of North American representatives of sect. *Hemipogon* Benth., with key, full nomenclature, and specimen citations.

- McVaugh, R. (1943). *Lobelia* L., *Pratia* Gaud. In North American Flora 32A: 35-99, 110-114. New York: New York Botanical Garden. En. — Flora with keys, descriptions, infrageneric classification, and full nomenclature.

- Wimmer, F. E. (1943). *Pratia* Gaudich., *Hypsela* Presl. In A. Engler & L. Diels, Das Pflanzenreich IV. 276b: 104-122, illus. Leipzig: Wilhelm Engelmann. Ge., La. — Monograph with keys, descriptions, infrageneric classification, full nomenclature, and specimen citations.

Fernald, M. L. (1947). The geographic varieties of *Lobelia puberula*. Rhodora 49: 182-186, illus. En. — Revision of McVaugh's (1936) informal treatment of infraspecific variation.

Warner, M. F. (1949). Cardinalis barberini. Natl. Hort. Mag. 28: 130-135, illus. En. — Discovery and introduction to horticulture of *L. cardinalis* in the early 17th Century.

Adamson, R. S. (1950). *Mezleria* Presl, *Lobelia* L., *Isolobus* A. DC. In R. S. Adamson & T. M. Salter, Flora of the Cape Peninsula: 753-759. En. — Flora with keys and descriptions.

Turrill, W. B. (1951). *Pratia angulata*. Bot. Mag. 168: tab. 171B + 3 pp., illus. En. — Portrait of *L. angulata,* with description and notes on cultivation.

Woodhead, N. (1951). Biological flora of the British Isles. *Lobelia dortmanna* L. J. Ecol. 39: 458-464, illus. En. — Natural history of this hydrophytic species.

Leinfellner, W. (1952). Pseudodimere Gynözeen bei *Lobelia cardinalis*. Oesterr. Bot. Z. 99: 220-227, illus. Ge. — Gynoecium anatomy.

- Wimmer, F. E. (1953a). *Lobelia* L. In Flore de Madagascar et des Comores (plantes vasculaires). 186. Lobéliacées (Lobeliaceae): 4-28, illus. Paris: Firmin-Didot. Fr. — Flora for Madagascar and Comoros, with key, descriptions, infrageneric classification, full nomenclature, and specimen citations.
- Wimmer, F. E. (1953b). *Lobelia* L. In A. Engler & L. Diels, Das Pflanzenreich IV. 276b: 408-695 + 4 pl., illus. Berlin: Akademie-Verlag. Ge., La. — Monograph with keys, descriptions, infrageneric classification, full nomenclature, and specimen citations.
- Wimmer, F. E. (1953c). *Pratia*. In A. Engler & L. Diels, Das Pflanzenreich IV. 276b: 763-767. Berlin: Akademie-Verlag. Ge., La. — Supplement to Wimmer (1943).

Anonymous (1954). Scientific Committee. October 20, 1953. J. Roy. Hort. Soc. 79 (Proc.): 1. En. — Discussion of nomenclature of perennial hybrids.

Pugsley, H. C. (1955). *Lobelia cardinalis* × *L. syphilitica* hybrids. J. Roy. Hort. Soc. 80: 282-283. En. — Results of a program of artificial hybridization for horticulture.

Fedorov, A. A. (1957). *Lobelia* L. In V. L. Komarov, Flora URSS 24: 451-453. Leningrad: Academia Scientiae URSS. Ru. — Flora for former Soviet Union, with key, descriptions, infrageneric classification, and full nomenclature.

Hedberg, O. (1957). Afroalpine vascular plants, a taxonomic revision. Symb. Bot. Upsal. 15: 1-411, illus. En. — Floristic treatment of high elevation species.

Pugsley, H. C. (1957). Breeding herbaceous *Lobelia*. J. Roy. Hort. Soc. 82: 23-24. En. — Supplement to Pugsley (1955).

Anonymous (1959). Scientific Committee. August 11, 1959. September 1, 1959. September 15, 1959. J. Roy. Hort. Soc. 84 (Proc.): 52-53. En. — Discussion of nomenclature of perennial hybrids, especially "L. vedrariensis".

- Bowden, W. M. (1959a). Phylogenetic relationships of twenty-one species of *Lobelia* L. section *Lobelia*. Bull. Torrey Bot. Club 86: 94-108. En. — Evolutionary hypothesis based on biosystematic data and morphology; includes informal classification and evolutionary tree.
- Bowden, W. M. (1959b). Cytotaxonomy of *Lobelia* L. section *Lobelia*. I. Three diverse species and seven small flowered species. Canad. J. Genet. Cytol. 1: 49-64. En. — Biosystematic and cytogenetic approaches used to study intra- and interspecific relationships.

McVaugh, R. (1959). *Lobelia splendens* Humb. & Bonpl. ex Willd., a poorly understood member of the *Lobelia cardinals* [sic] complex. Bol. Soc. Bot. México 23: 48-54. En. — Variation within widespread polymorphic species.

- Bowden, W. M. (1960a). Cytotaxonomy of *Lobelia* L. section *Lobelia*. II. Four narrow-leaved species and five medium flowered species. Canad. J. Genet. Cytol. 2: 11-27. En. — Continuation of Bowden (1959b).
- Bowden, W. M. (1960b). Cytotaxonomy of *Lobelia* L. section *Lobelia*. III. *Lobelia siphilitica* L. and *L. cardinalis* L. Canad. J. Genet. Cytol. 2: 234-251. En. — Continuation of Bowden (1959b, 1960a).
- Moeliono, B. (1960). *Lobelia*. In C. G. G. J. van Steenis (ed.), Flora Malesiana (ser. I) 6(1): 121-136, illus. Djakarta: Noordhoff-Kolff. En. — Flora with key and descriptions; first to argue for inclusion of *Pratia,* on grounds that fruit characters inconstant..
- Allan, H. H. (1961). *Pratia* Gaud., *Lobelia* L. In Flora of New Zealand 1: 796-801, 1031-1032. Wellington: R. E. Owen. En. — Flora with keys, descriptions and full nomenclature.

Bowden, W. M. (1961a). Interspecific hybridization in *Lobelia* L. section *Lobelia*. Canad. J. Bot. 39: 1679-1693. En. — Overview of results of 1400 experimental crosses involving all species.

Bowden, W. M. (1961b). Nineteen artificial bispecific *Lobelia* hybrids. Canad. J. Genet. Cytol. 3: 403-423. En. — Cytotaxonomic details for 19 of the 33 bispecific hybrids reported by Bowden (1961a).

Turrill, W. B. (1962). *Hypsela reniformis*. Bot. Mag. 173: tab. 395B + 3 pp., illus. En. — Portrait of *L. oligophylla,* with description and notes on cultivation.

Alexander, E. J. (1963). *Lobelia paludosa*. Addisonia 24: 45-46 + pl. 791, illus. En. — Portrait with description.

Bowden, W. M. (1964a). Cytogenetics of *Lobelia* × *speciosa* Sweet (*L. cardinalis* L. × *L. siphilitica* L.). Canad. J. Genet. Cytol.. 6: 121-139. En. — Crossing relationships of pair of closely related species.

• Bowden, W. M. (1964b). The phylogenetic significance of meiotic chromosome behavior in some polyspecific tetraploid *Lobelia* hybrids. Canad. J. Genet. Cytol.. 6: 364-369. En. — Phylogenetic inferences in sect. *Lobelia* based on results from program of experimental hybridizations described by Bowden (1961a).

• Braga, R. E. (1964-1965). Lobelias do Brasil. Tribuna Farm. 32: 1-8, 41-54; 33: 3-24, illus. Por. — Floristic treatment, with keys (including all genera of subfamily in Brazil), descriptions, and full nomenclature.

Finnis, V. (1966). *Lobelia tupa*. J. Roy. Hort. Soc. 91: 132 + fig. 48, illus. En. — Portrait with information on cultivation.

Caudle, C. & J. M. Baskin (1968). Germination and dormancy in cedar glade plants. III. *Lobelia gattingeri*. J. Tennessee Acad. Sci. 43: 116-117. En. — Seed ecology of highly localized spring annual.

• Wimmer, F. E. (1968). *Pratia* Gaudich., *Hypsela* Presl, *Lobelia* L. In A. Engler & L. Diels, Das Pflanzenreich IV. 276c: 832-834, 855-889 + Taf. 12, 20-30, illus. Berlin: Akademie-Verlag. Ge., La. — Supplement to Wimmer (1943, 1953).

Van Thuan, N. (1969). *Pratia* Gaud., *Lobelia* L. In A. Aubréville (ed.), Flore de Cambodge, du Laos et du Vietnam 9: 25-39, illus. Paris: Muséum National d'Histoire Naturelle. Fr. — Flora of Indochina, with keys, descriptions, and full nomenclature; inexplicably designates a non-Linnaean species as generic type.

Krochmal, A., L. Wilken & M. Chien (1970). Lobeline content of *Lobelia inflata*: structural, environmental and developmental effects. USDA Forest Serv. Res. Pap. NE-178: 1-13, illus. En. — Variation in alkaloid content in commercially harvested species.

Caprotti, E. (1971). Le lobelie giganti del Monte Kenia e del Kilimangiaro. Natura (Italy) 62: 187-196, illus. It. — Natural history of Afro-montane giants.

Adams, C. D. (1972). *Lobelia* L. In Flowering plants of Jamaica: 734-736. Mona: University of the West Indies. En. — Flora with keys, descriptions, full nomenclature, and specimen citations; argues for inclusion of *Pratia* on grounds that all Jamaican species very similar, irrespective of fruit type.

Badré, F. & T. Cadet (1972). Lobeliaceae des Mascareignes, *Lobelia* Linn. Adansonia (n.s.) 11: 670-683, illus. Fr. — Floristic treatment for Mauritius, Réunion, and Rodrigues, with keys, descriptions, full nomenclature, and specimen citations.

Degener, O. & I. Degener (1974). Prodromus of *Galeatella* and *Neowimmeria*, illus. Honolulu: published privately. En. — Synopsis of sect. *Galeatella* and sect. *Revolutella* (as genera), with full nomenclature.

Killick, D. J. B. (1974). *Lobelia aberdarica*. Flow. Pl. Africa 43: pl. 1692, illus. En. — Portrait with description.

• Mabberley, D. J. (1974). The pachycaul lobelias of Africa and St. Helena. Kew Bull. 29: 535-584, illus. En. — Monograph of sect. *Rhynchopetalum*, with keys, descriptions, infrasectional classification, full nomenclature, and specimen citations.

Moore, L. B. (1974). *Hypsela rivalis* and other pygmy plants. J. Canterbury Bot. Soc. 7: 12-15, illus. En. — Brief popular account of *L. perpusilla* and its similarity to *Isotoma rivalis*.

Witherspoon, J. T. (1974). Rediscovery of a hybrid *Lobelia* (Campanulaceae) in Missouri. Southw. Naturalist 19: 329. En. — Natural population of *L.* × *speciosa*, one of few ever reported.

Dyer, R. A. (1975). *Lobelia* L. In The genera of southern African flowering plants 1: 645. Pretoria: Department of Agricultural Technical Services. En. — Generic description.

• Mabberley, D. J. (1975a). The giant lobelias: toxicity, inflorescence and tree-building in the Campanulaceae. New Phytol. 75: 289-295, illus. En. — Discussion of diverse aspects of evolution and phylogeny in sect. *Rhynchopetalum*, building on Mabberley (1974); considers woodiness primitive within subfamily.

• Mabberley, D. J. (1975b). The giant lobelias: pachycauly, biogeography, ornithophily and continental drift. New Phytol. 74: 365-374, illus., maps. En. — Additional aspects of evolution and phylogeny in sect. *Rhynchopetalum*, building on Mabberley (1974).

- Badré, F. (1976). *Lobelia* L. In Flore des Mascareignes. 111. Campanulacées: 2-12, illus. Fr. Paris: ORSTOM. — Flora for Mauritius, Réunion, and Rodrigues, with keys, descriptions, and full nomenclature.
- Nash, D. L. (1976). Flora of Guatemala; *Lobelia* Linnaeus, *Pratia* Gaudichaud. Fieldiana Bot. 24: 414-431, illus. En. — Floristic treatment with keys, descriptions, and full nomenclature.

  Rourke, J. P. (1976). Some observations on *Lobelia valida* - the *galjoenbloem*. Veld Fl. 62: 10-11, illus. En. — Popular account of South African species.
- Tutin, T. G. (1976). *Lobelia* L. In T. G. Tutin, V. H. Heywood, N., A. Burges, D. M. Moore, D. H. Valentine, S. M. Walters & D. A. Webb (eds.), Flora Europaea 4: 102. Cambridge: Cambridge University Press. En. — Flora of Europe with key, descriptions, and full nomenclature.
- Mabberley, D. J. (1977). The origin of the afroalpine pachycaul flora and its implications. Gard. Bull. Singapore 29: 41-55, illus. En. — Further elaboration of hypothesis (Mabberley 1975a, 1975b) that woodiness is primitive among Lobelioideae.
- Wilbur, R. L. (1977). Flora of Panama, *Lobelia*. Ann. Missouri Botanical Gard. 63: 637-646, illus. En. — Floristic treatment with keys, descriptions, full nomenclature, and specimen citations.

  Baskin, J. M., & C. C. Baskin (1979). The ecological life cycle of the cedar glade endemic *Lobelia gattingeri*. Bull. Torrey Bot. Club 106: 176-181, illus. En. — Growth and reproduction in highly localized spring annual.

  Moreira, E. A. (1979). Contribuição para o estudo fitoquímico de *Lobelia hassleri* A. Zahlb e *Lobelia stellfeldii* R. Braga. Campanulaceae: 1. Tribuna Farm. 47: 13-39, illus. Por. — Chemotaxonomic study of two Brazilian species.
- Jeppesen, S. (1981). *Hypsela* Presl, *Lobelia* L. In G. Harling & B. Sparre (eds.), Flora of Ecuador 14: 125-136, illus. Stockholm: Swedish Natural Science Research Council. En. — Flora with keys, descriptions, full nomenclature, and specimen citations.

  Schultes, R. E. (1981). Iconography of New World plant hallucinogens. Arnoldia 41: 80-125, illus. En. — Use of *L. tupa* as hallucinogen by indigenous peoples of Chile.

  Bowden, W. M. (1982). The taxonomy of *Lobelia* × *speciosa* s.l. and its parental species, *L. siphilitica* and *L. cardinalis* s.l. (Lobeliaceae). Canad. J. Bot. 60: 2054-2070. En. — Taxonomic account with keys, infraspecific classifications, and evolutionary tree.
- Lian, Y. S. (1983). Lobelioideae Schonland [sic]. In Flora Reipublicae Popularis Sinicae 73(2): 144-173, illus. Beijing: Science Press. Ch. — Flora of China, with keys, descriptions, infrageneric classifications, and full nomenclature.
- Thulin, M. (1983). *Lobelia* L. In E. Launert (ed.), Flora Zambesiaca 7(1): 118-145, illus. Kew: Royal Botanic Garden. En. — Flora of Mozambique, Malawi, Zambia, Zimbabwe and Botswana, with keys, descriptions, full nomenclature and specimen citations.

  Bowden, W. M. (1984). Perennial tetraploid lobelia hybrids. The Garden 109(2): 55-57, illus. En. — Summary of horticultural development program involving induced polyploid cultivars of *L.* × *speciosa*.

  Devlin, B. & A. G. Stephenson (1984). Factors that influence the duration of the staminate and pistillate phases of *Lobelia cardinalis*. Bot. Gaz. (Crawfordsville) 145: 323-328. En. — Environmental impacts on dichogamy.
- Thulin, M. (1984). *Lobelia*. In Flora of tropical east Africa, Lobeliaceae: 2-46, illus. Rotterdam: Balkema. En. — Flora with keys, descriptions, full nomenclature, and specimen citations.

  Ebinger, J. E. (1985). *Lobelia cardinalis* × *L. siphilitica* in Illinois. Trans. Illinois Acad. Sci. 78: 29-31. En. — One of few known spontaneous occurrences of *L.* × *speciosa,* including backcrosses to *L. siphilitica*.

  McGregor, R. L. (1985). Studies on the validity of *Lobelia spicata* infraspecific taxa in the prairies and plains of central North America with notes on *Lobelia appendiculata*. Contr. Univ. Kansas Herb. 16: 1-10. En. — Revision of infraspecific classification in widespread polymorphic species.

  Sastre, C. (1985a). Nomenclature de deux especes de *Lobelia* L. des Petites Antilles. Phytologia 58: 167-168. Fr. — Attempt to clarify application of names in sect. *Tylomium*.

Sastre, C. (1985b). Endemovicariance et speciation: application a la systematique des *Lobelia* L. des petites Antilles. Compt. Rend. Acad. Sci. Paris, Sér. 3, Sci. Vie 300: 161-164. Fr. — Comments on evolution and biogeography in sect. *Tylomium.*

Sivarajan, V. V. (1985). *Lobelia zeylanica* Linn., a misunderstood species and related Indian taxa. J. Econ. Taxon. Bot. 7: 221-223, illus. En. — Attempt to clarify application of name; includes description and key to all species of sect. *Holopogon* in India.

• Thulin, M. (1985). *Lobelia* L. In Flora d'Afrique Centrale (Zaire-Rwanda-Burundi), spermatophytes. Lobeliaceae: 2-48, illus. Meise: Jardin Botanique National de Belgique. Fr. — Flora with keys, descriptions, full nomenclature, and specimen citations.

Ono, M., S. Kobayashi & N. Kawakubo (1986). Present situation of endangered plant species in the Bonin (Ogasawara) Islands. Ogasawara Res. 12: 1-32 + pl. 1-4, illus. En. — Conservation status of endemic *L. boninensis.*

Thulin, M. (1986). The identity of *Lobelia hirsuta* L. Taxon 35: 721-725, illus. En. — Linnaean name from South Africa actually applies to species of Thymeleaceae.

Thulin, M., P. B. Phillipson & D. O. Wijnands (1986). Typification of *Lobelia erinus* L. and *L. erinoides* L. Taxon 35: 725-729, illus. En. — Clarification of two names from South Africa.

• Toelken, H. R. (1986). *Lobelia* L., *Pratia* Gaudich. In Flora of South Australia 3: 1369-1376, illus. Adelaide: Government Printing Division. En. — Flora with keys, descriptions, and full nomenclature.

Devlin, B. & A. G. Stephenson (1987). Sexual variations among plants of a perfect flowered species. Amer. Naturalist 130: 119-218. En. — Occurrence of gynodioecy in *L. cardinalis.*

Evans, M. & J. D. Sole (1987). The first Kentucky record of *Lobelia appendiculata* A. DC. var. *gattingeri* (Gray) McVaugh. Castanea 52: 226-227. En. — Discovery of cedar glade endemic *L. gattingeri* outside Tennessee and Alabama.

Farmer, A. M. (1987). Terrestrial and aquatic variants of *Lobelia dortmanna*. Watsonia 16: 432-435. En. — Correlation of environmental factors with growth form in a hydrophyte.

McNeill, J., E. A. Odell, L. L. Consaul & D. S. Katz (1987). American Code and later lectotypifications of Linnaean generic names dating from 1753: a case study in discrepancies. Taxon 36: 350-401. En. — Conflicting lectotypifications of genus.

Devlin, B. (1988). The effects of stress on reproductive characters of *Lobelia cardinalis*. Ecology 69: 1716-1720, illus. En. — Ecological impacts on floral morphology.

• Haridasan, V. K. & P. K. Mukherjee (1988). Lobeliaceae; *Lobelia, Pratia*. In P. K. Hajra & M. Sanjappa (eds.), Fascicles of flora of India 19: 41-63, illus. Calcutta: Botanical Survey of India. En. — Flora with keys, descriptions, and full nomenclature.

Matthews, V. (1988). *Lobelia tupa.* Kew Mag. 5: 157-161 + pl. 112, illus. En. — Portrait with description and notes on cultivation; account of poisoning via accidental inhalation of dried latex.

Pringle, J. S. (1988). Nomenclature of the white cardinal flower, *Lobelia cardinalis* f. *alba* (Campanulaceae). Pl. Press (Mississauga) 5: 4-5, illus. En. — Correct name of corolla color variant.

Nelson, E. C. (1988). *Lobelia cardinalis* f. *alba* (McNab) St John - a correction. Notes Roy. Bot. Gard. Edinburgh 45: 375-376. En. — Repetition of Pringle (1988).

Mason, D. (1989). Perennial lobelias at Longstock Park Gardens. The Garden 114: 221-224, illus. En. — Cultivars and cultivation methods for *L. cardinalis, L. siphilitica,* and their hybrid.

Farmer, A. M. (1989). Biological flora of the British Isles, no. 165. *Lobelia dortmanna* L. J. Ecol. 77: 1161-1173, illus. En. — Natural history of hydrophytic species.

• Ayers, T. J. (1990). Systematics of *Heterotoma* (Campanulaceae) and the evolution of nectar spurs in the New World Lobelioideae. Syst. Bot. 15: 296-327, illus., maps En. — Nearly all species of *Heterotoma* except type transferred to *Lobelia;* includes key for all species of *Lobelia* with nectar spurs.

• Lammers, T. G. (1990). *Lobelia.* In W. L. Wagner, D. R. Herbst & S. H. Sohmer, Manual of the flowering plants of Hawai'i: 473-480, illus. Honolulu: University of Hawaii Press. En. — Flora with keys, descriptions, and synonymy.

Murray, B. G. & E. K. Cameron (1990). An update on the cytogeography of *Pratia*. New Zealand Bot. Soc. Newslett. 22: 7-8. En. — Species in New Zealand show extreme variation in ploidy levels.

Lancaster, R. (1991). Plant profile. *Lobelia laxiflora* var *angustifolia*. The Garden 116: 474-475, illus. En. — Use of in horticulture.

• Wilbur, R. L. (1991). Synopsis of the Mexican and Central American representatives of *Lobelia* section *Tylomium* (Campanulaceae: Lobelioideae). Sida 14: 555-567. En. — Regional monograph, including key, descriptions, full nomenclature, and specimen citations.

Harvey, Y. (1992). *Lobelia organensis*. Kew Mag. 9: 120-124 + pl. 200, illus. En. — Portrait and description, with notes on cultivation.

• Hong, D. Y. & T. J. Zhang (1992). A revision of *Lobelia* subgen. *Tupa* (Campanulaceae) in China. Acta Phytotax. Sin. 30: 146-162, illus. Ch. — Regional synopsis with key, full nomenclature, and specimen citations.

Huxley, A. J., ed. (1992). *Lobelia*. In The new Royal Horticultural Society dictionary of gardening 3: 104-106, illus. London: MacMillan. En. — Account of species in cultivation.

Kindscher, K. (1992). *Lobelia inflata*. In Medicinal wild plants of the prairie: 146-150. Lawrence: University Press of Kansas. En. — Ethnobotany of species used commercially.

Lammers, T. G. & N. Hensold (1992). Chromosome numbers in Campanulaceae. II. The *Lobelia tupa* complex of Chile. Amer. J. Bot. 79: 585-586, illus. En. — Species of sect. *Tupa* all hexaploid.

• Murata, J. (1992a). Systematic implication of seed coat morphology in *Lobelia* (Campanulaceae: Lobelioideae). J. Fac. Sci. Univ. Tokyo, Sect. 3, Bot. 15: 155-172, illus. En. — Detailed survey of seed coat morphology and internal wall structure.

Murata, J. (1992b). Morphology and chromosome number of *Lobelia loochooensis* Koidz. J. Jap. Bot. 67: 282-285, illus. En. — Descriptive account of rare island endemic.

Murray, B. G., E. K. Cameron & L. S. Standring (1992). Chromosome numbers, karyotypes, and nuclear DNA variation in *Pratia* Gaudin [sic] (Lobeliaceae). New Zealand J. Bot. 30: 181-187. En. — Follow-up to Murray & Cameron (1990); $2n = 14$ to 140.

• Smith, P. J. (1992). *Lobelia, Pratia*. In G. J. Harden (ed.), Flora of New South Wales 3: 124-128, illus. Sydney: New South Wales University Press. En. — Flora with keys and descriptions.

Zhang, M. Z. & J. C. Wang (1992). The chemical components of *Lobelia davidii* Franch. Acta Bot. Sin. 34: 58-61. Ch. — Sterols, triterpenes, and several alkaloids isolated and identified.

Elliot, W. R. & D. L. Jones (1993). *Lobelia* L. Encyclopedia of Australian plants suitable for cultivation 6: 208-213, illus. Melbourne: Lothian. En. — Synopsis of Australian species from horticultural perspective, with descriptions.

Knox, E. B. (1993a). The species of giant *Senecio* (Compositae) and giant *Lobelia* (Lobeliaceae) in eastern Africa. Contr. Univ. Michigan Herb. 19: 241-257, illus. En. — Checklist of species of sect. *Rhynchopetalum*.

Knox, E. B. (1993b). The conservation status of the giant senecios and giant lobelias in eastern Africa. Opera Bot. 121: 195-216, maps. En. — Positive conservation outlook for species of sect. *Rhynchopetalum*.

• Knox, E. B., S. R. Downie & J. D. Palmer (1993). Chloroplast genome rearrangements and the evolution of giant lobelias from herbaceous ancestors. Mol. Biol. Evol. 10: 414-430. En. — First application of the methods of molecular biology to systematic questions in subfamily; supports herbaceous ancestry of pachycaul taxa.

Knox, E. B. & R. R. Kowal (1993). Chromosome numbers of the East African giant senecios and giant lobelias and their evolutiuionary significance. Amer. J. Bot. 80: 847-853. En. — Cytological data and phylogeny of sect. *Rhynchopetalum;* all taxa sampled tetraploid.

Lammers, T. G. (1993). *Lobelia bridgesii*. Kew Mag. 10: 70-75 + pl. 220, illus. En. — Portrait with description and notes on cultivation.

- Shimizu, T. (1993). *Lobelia* L. In K. Iwatsuki, T. Yamazaki, D. E. Boufford & H. Ohba (eds.), Flora of Japan 3a: 416-418. Tokyo: Kodansha. En. — Flora with keys, descriptions, infrageneric classification, and full nomenclature.

  Howard, R. A. (1994). Eighteenth century West Indian pharmaceuticals. Harvard Pap. Bot. 5: 69-91. En. — North American *L. siphilitica* cultivated in Caribbean to treat veneral disease; native *L. cirsiifolia* similarly used.

- Liogier, H. (1995). *Lobelia* L. In La flora de la Española 7: 442-450, illus. San Pedro de Macorís: Ediciones de la Universidad Central del Este. Sp. — Flora of Hispaniola with keys, descriptions, and full nomenclature.

- Murata, J. (1995). A revision of infrageneric classification of *Lobelia* (Campanulaceae-Lobelioideae) with special reference to seed coat morphology. J. Fac. Sci. Univ. Tokyo, Sect. 3, Bot. 15: 349-371, illus. En. — Revision of classification of genus, dividing it into three subgenera and 14 sections.

- Chiapella, J. (1996). Nota sobre la identidad de *Pratia repens* (Campanulaceae: Lobelioideae). Bol. Soc. Argent. Bot. 32: 123-135, illus. Sp. — Morphometric analysis supports conspecificity of *Pratia repens* and *Hypsela reniformis*.

  Crespo, M. B., L. Serra & N. Turland (1996). Lectotypification of four names in *Lobelia* (Lobeliaceae). Taxon 45: 117-120. En. — Nomenclatural housekeeping.

  Ferguson, K. (1996). *Lobelia vagans*. Bot. Mag. 13: 200-203 + pl. 304, illus. En. — Portrait with description and notes on cultivation.

- Chiapella, J. & S. G. Tressens (1997). *Lobelia* (Campanulaceae – Lobelioideae): nueva citas y clave para las especies Argentinas. Bonplandia 9: 245-250, illus. Sp. — Key to species in Argentina.

- Liogier, H. (1997). *Lobelia* L. In Descriptive flora of Puerto Rico and adjacent islands 5: 211-216, illus. San Juan: Editorial de la Universidad de Puerto Rico. En. — Flora with keys, descriptions, and full nomenclature.

- Thompson, S. W. & T. G. Lammers (1997). Phenetic analysis of morphological variation in the *Lobelia cardinalis* complex (Campanulaceae: Lobelioideae). Syst. Bot. 22: 315-331, map. En. — Morphometric analysis of widespread polymorphic species shows it cannot be subdivided meaningfully.

  Wilbur, R. L. (1997). *Calcaratolobelia* (Campanulaceae): a new genus of spurred lobelioids from Mexico and Central America. Sida 17: 555-564. En. — Establishment of new genus for spurred species of *Lobelia*, with key to species groups within it.

  Fetene, M., M. Gashaw & P. Nauke (1998). Microclimate and ecophysiological significance of the tree-like life-form of *Lobelia rhynchopetalum* in a tropical alpine environment. Oecologia 113: 332-340, illus. En. — Tree-like life-form hypothesized to be mechanism to mitigate strong diurnal temperature fluctuations near ground level.

- Knox, E. B. & J. D. Palmer (1998). Chloroplast DNA evidence on the origin and radiation of the giant lobelias in Eastern Africa. Syst. Bot. 23: 109-149, illus., maps. En. — Phylogeny of sect. *Rhynchopetalum* based on molecular data.

  Paviour, M. (1998). Lobelias chileñas - 'las olvidadas'. Cornish Garden 41: 40-47, illus. En. — Recent re-introduction of species of sect. *Tupa* to horticulture

  Knox, E. B. & J. D. Palmer (1999). The chloroplast genome arrangement of *Lobelia thuliniana* (Lobeliaceae): expansion of the inverted repeat in an ancestor of the Campanulales. Pl. Syst. Evol. 214: 49-64. En. — Mutations in DNA and their systematic utility.

  Lammers, T. G. (1999). A new *Lobelia* from Mexico, with additional new combinations in world Campanulaceae. Novon 9: 381-389, illus. En. — Notes that *Mezleria* is referable to *Monopsis* not *Lobelia,* and correct name for subg. *Mezleria* of Wimmer (1953) and Murata (1995).

  Vanzela, A. L. L., A. Cuadrado, A. O. S. Vieira & N. Jouve (1999). Genome characterization and relationships between two species of the genus *Lobelia* (Campanulaceae) determined by repeated DNA sequences. Pl. Syst. Evol. 214: 211-218. En. — Mutations in DNA and their systematic utility.

  Goldblatt, P. & J. Manning (2000). Cape plants. A conspectus of the Cape flora of South Africa. *Lobelia*. Strelitzia 9: 390-391. En. — Checklist for Cape of Good Hope, with brief species descriptions and informal infrageneric classification.

- Lammers, T. G. (2000). Revision of *Lobelia* sect. *Tupa* (Campanulaceae: Lobelioideae). Sida 19: 87-110. En. — Monograph, with key, descriptions, full nomenclature, and specimen citations.
  Welman, W. G. (2000). *Lobelia* L. In O. A. Leistner (ed.), Seed plants of southern Africa: families and genera. Strelitzia 10: 339-340. En. — Generic description.
  Ruas, P. M., A. L. L. Vanzela, A. O. Vieira, C. Bernini & C. F. Ruas (2001). Karyotype studies in Brazilian species of *Lobelia* L., subgenus *Tupa* (Campanulaceae). Revista Brasil. Bot. 24: 249-254, illus., map. En. — All species sampled were tetraploid.
  Wilbur, R. L. (2001). *Calcaratolobelia* Wilbur, *Lobelia* L. In W. D. Stevens, C. U. Ulloa, A. Pool & O. M. Montiel (eds.), Flora de Nicaragua 1: 558-559, 562-564. St. Louis: Missouri Botanical Garden Press. Sp. — Flora with keys, descriptions, and full nomenclature.
- Lammers, T. G. (2004). Revision of *Lobelia* sect. *Homochilus* (Campanulaceae: Lobelioideae).Sida 21: 591-623, illus. En. — Monograph, with key, descriptions, full nomenclature, and specimen citations; advocates recognition of sect. *Tylomium* within the infrageneric classification of Murata (1995).
  Murray, B. G., P. M. Datson, E. L. Y. Lai, K. M. Sheath & E. K. Cameron (2004). Polyploidy, hybridization and evolution in *Pratia* (Campanulaceae). New Zealand J. Bot. 42: 905-920, maps. En. — Change in ploidy level played major role in evolution of group of species in New Zealand.

Lobelia L., Sp. Pl.: 929 (1753). *Rapuntium* Mill., Gard. Dict. (abr. ed. 4) (1754). *Cardinalis* Riv. ex Fabr., Enum.: 122 (1759). *Laurentia* Adans., Fam. Pl. 2: 134, 568 (1763). *Mecoschistum* Dulac, Fl. Hautes-Pyrénées: 459 (1867).
Cosmopolitan. 10 11 12 14 20 21 22 23 24 25 26 27 28 29 30 31 35 36 38 40 41 42 43 50 51 (60) 61 63 70 71 72 73 74 75 76 77 78 79 80 81 82 83 84 85 90.
Dortmanna O. O. Rudbeck ex Hill, Brit. Herb.: 126 (1756).
Pratia Gaudich., Ann. Sci. Nat. 5: 103 (1825). *Lobelia* [unranked] *Pratia* (Gaudich.) Heynh., Nom. Bot. Hort. 1: 473 (1840).
Tupa G. Don, Gen. Hist. 3: 700 (1834). *Lobelia* [unranked] *Tupa* (G. Don) Heynh., Nom. Bot. Hort. 1: 473 (1840). *Lobelia* sect. *Tupa* (G. Don) Benth. in Benth. & Hook. f., Gen. Pl. 2: 552 (1876). *Rapuntium* sect. *Tupa* (G. Don) Kuntze in T. Post & Kuntze, Lex. Gen. Phan.: 478 (1903). *Lobelia* subg. *Tupa* E. Wimm., Ann. Naturhist. Mus. Wien 56: 364 (1948).
Holostigma G. Don, Gen. Hist. 3: 716 (1834). *Monopsis* [unranked] *Holostigma* (G. Don)Endl., Gen. Pl.: 511 (1838). *Holostigmateia* Rchb., Deut. Bot. Herb.-Buch, Syn. Reduct.: 57 (1841).
Tylomium C. Presl, Prodr. Monogr. Lobel.: 31 (1836). *Tupa* sect. *Tylomium* (C. Presl) A. DC. in DC., Prodr. 7: 394 (1839). *Lobelia* sect. *Tylomium* (C. Presl) Benth. in Benth. & Hook. f., Gen. Pl. 2: 552 (1876). *Rapuntium* sect. *Tylomium* (C. Presl) Kuntze in T. Post & Kuntze, Lex. Gen. Phan.: 478 (1903).
Enchysia C. Presl, Prodr. Monogr. Lobel.: 40 (1836). *Laurentia* [unranked] *Enchysia* (C. Presl) Endl., Gen. Pl.: 512 (1838). *Lobelia* [unranked] *Enchysia* (C. Presl) Heynh., Nom. Bot. Hort. 1: 470 (1840). *Laurentia* subg. *Enchysia* (C. Presl) Peterm., Pflanzenreich: 444 (1845).
Hypsela C. Presl, Prodr. Monogr. Lobel.: 45 (1836). *Lysipomia* sect. *Hypsela* (C. Presl) A.DC. in DC., Prodr. 7: 350 (1839). *Pratia* sect. *Hypsela* (J. D. Hook.) Baill., Hist. Pl. 8: 366 (1885).
Trimeris C. Presl, Prodr. Monogr. Lobel.: 46 (1836). *Lobelia* sect. *Trimeris* (C. Presl) A. DC. in DC., Prodr. 7: 357 (1839). *Rapuntium* sect. *Trimeris* (C. Presl) Kuntze in T. Post & Kuntze, Lex. Gen. Phan.: 478 (1903).
Rhynchopetalum Fresen., Flora 21: 603 (1838). *Tupa* sect. *Rhynchopetalum* (Fresen.) A. Rich., Tent. Fl. Abyss. 2: 9 (1850). *Lobelia* sect. *Rhynchopetalum* (Fresen.) Benth. in Benth. & Hook. f., Gen. Pl. 2: 552 (1876).
Pratia [unranked] *Bernonia* Endl., Gen. Pl.: 512 (1838). *Piddingtonia* A. DC. in DC., Prodr. 7: 341 (1839). *Lobelia* [unranked] *Piddingtonia* (A. DC.) Heynh., Nom. Bot. Hort. 1: 473 (1840).

Isolobus A. DC. in DC., Prodr. 7: 352 (1839). *Lobelia* [unranked] *Isolobus* (A. DC.)
Heynh., Nom. Bot. Hort. 1: 471 (1840). *Monopsis* [unranked] *Isolobus* (A. DC.) Endl.,
Gen Pl.: 1391 (1841). *Lobelia* sect. *Isolobus* (A. DC.) C. B. Clarke in Hook. f., Fl. Brit.
India 3: 425 (1881). *Rapuntium* sect. *Isolobus* (A. DC.) Kuntze in T. Post & Kuntze,
Lex. Gen. Phan.: 478 (1903). *Lobelia* subg. *Isolobus* (A. DC.) Y. S. Lian, Fl. Reipubl.
Popul. Sin. 73(2): 154 (1983).
Colensoa Hook. f., Fl. Nov.-Zel. 1: 156 (1852). *Pratia* sect. *Colensoa* (Hook. f.) Baill., Hist.
Pl. 8: 366 (1885). *Lobelia* sect. *Colensoa* (Hook. f.) Murata, J. Fac. Sci. Univ. Tokyo,
Sect. 3, Bot. 15: 358 (1995).
Speirema Hook. f. & Thomson, J. Proc. Linn. Soc., Bot. 2: 27 (1858). *Pratia* sect. *Speirema*
(Hook. f.) Baill., Hist. Pl. 8: 366 (1885).
Haynaldia Kanitz, Magyar. Növényt. Lapok 1: 3 (1877); non Schur, Enum. Pl. Transsilv.:
807 (1866); nec Schulzer, Verh. K. K. Zool.-Bot. Ges. Wien 16: 37 (1866). *Euhaynaldia*
Borbás, Földmüv. Érdek. 8: 331 (1880). *Lobelia* sect. *Euhaynaldia* (Borbás) Zahlbr.,
Vidensk. Meddel. Dansk Naturhist. Foren. Kjøbenhavn 1895: 69 (1895). *Rapuntium*
sect. *Haynaldia* Kuntze in T. Post & Kuntze, Lex. Gen. Phan.: 478 (1903).
Petromarula Belli ex Nieuwl. & Lunell, Amer. Midl. Naturalist 5: 13 (1917); non Vent. ex
R. Hedw., Gen. Pl.: 139 (1806).
Lobelia sect. *Galeatella* E. Wimm., Ann. Naturhist. Mus. Wien 56: 369 (1948). *Galeatella*
(E. Wimm.) O. Deg. & I. Deg., Fl. Hawaiiensis, fam. 339 (1962).
Lobelia sect. *Revolutella* E. Wimm., Ann. Naturhist. Mus. Wien 56: 369 (1948).
Neowimmeria O. Deg. & I. Deg., Phytologia 12: 73 (1965).
Calcaratolobelia Wilbur, Sida 17: 561 (1997).

Lobelia aberdarica R. E. Fr. & T. C. E. Fr., Svensk Bot. Tidskr. 16: 403 (1922).
Kenya & Uganda. 25 KEN UGA. Nanophan. or herb. phan. 2*n* = 28.

Lobelia acrochilus (E. Wimm.) E. B. Knox, Contr. Univ. Michigan Herb. 19: 247 (1993),
as 'acrochila'.
Ethiopia. 24 ETH. Nanophan or herb. phan. 2*n* = 28.
  * *Lobelia rhynchopetalum* var. *acrochilus* E. Wimm., Ann. Naturhist. Mus. Wien 56: 368
    (1948).

Lobelia acuminata Sw., Prodr.: 117 (1788). *Rapuntium acuminatum* (Sw.) C. Presl, Prodr.
Monogr. Lobel.: 24 (1836). *Siphocampylus acuminatus* (Sw.) G. Don in Sweet, Hort. Brit.
(ed. 3): 424 (1839). *Tupa acuminata* (Sw.) A. DC. in DC., Prodr. 7: 396 (1839). *Dortmanna
acuminata* (Sw.) Kuntze, Revis. Gen. Pl. 2: 379 (1891). *Pratia acuminata* (Sw.) McVaugh, N.
Amer. Fl. 32A: 111 (1943).
Jamaica. 81 JAM. Nanophan. or cham.
  Lobelia alexia E. Wimm., Repert. Spec. Nov. Regni Veg. 38: 86 (1935).

Lobelia acutidens Hook. f., J. Proc. Linn. Soc., Bot. 7: 204 (1864). *Dortmanna acutidens*
(Hook. f.) Kuntze, Revis, Gen. Pl. 2: 972 (1891).
Equitorial Guinea. 23 EQG. Hemicr. or ther.

Lobelia adnexa E. Wimm., Ann. Naturhist. Mus. Wien 56: 352 (1948).
Sierra Leone to Zimbabwe. 22 NGA SIE 23 CMN 25 TAN 26 MLW ZIM. Ther.

Lobelia agrestis E. Wimm., Pflanzenr. IV.276b: 510 (1953).
Madagascar. 29 MDG. Ther.

Lobelia aguana E. Wimm., Repert. Spec. Nov. Regni Veg. 38: 86 (1935).
S. Mexico (Guerrero, Oaxaca, Chiapas) & W. Guatemala. 79 MXS MXT 80 GUA HON.
Nanophan. or cham.
  Lobelia laxiflora var. *insignis* Donn. Sm., Bot. Gaz. (Crawfordsville) 16: 12 (1891).

Lobelia alsinoides Lam., Encycl. 3: 588 (1792). *Rapuntium alsinoides* (Lam.) C. Presl, Prodr.
Monogr. Lobel.: 22 (1836). *Dortmanna alsinoides* (Lam.) Kuntze, Revis. Gen. Pl. 2: 972 (1891).
Himalaya to S. China (Guangdong, Yunnan), Malesia & New Guinea. 36 CHC CHS CHT
40 ASS BAN EHM IND NEP SRL WHM 41 LAO MYA THA VIE 42 JAW SUL 43 NWG.
Ther. *2n* = 14.
Lobelia triangulata Roxb., Hort. Bengal.: 16 (1814).
Lobelia stipularis Roth ex Schult. in Roem. & Schult., Syst. Veg. 5: 67 (1819).
Lobelia trigona Roxb., Fl. Ind. 2: 111 (1824). *Dortmanna trigona* (Roxb.) Kuntze, Revis.
Gen. Pl. 2: 380 (1891).
Lobelia chinensis var. *cantonensis* E. Wimm. ex Danguy in Lecomte & Humbert, Fl. Indo-
Chine 3: 681 (1930). *Lobelia alsinoides* var. *cantonensis* (E. Wimm. ex Danguy) E.
Wimm., Ann. Naturhist. Mus. Wien 56: 360 (1948).
Lobelia chinensis var. *hirta* E. Wimm. ex Danguy in Lecomte & Humbert, Fl. Indo-Chine
3: 681 (1930). *Lobelia alsinoides* var. *hirta* (E. Wimm. ex Danguy) E. Wimm., Ann.
Naturhist. Mus. Wien 56: 360 (1948).
Lobelia chinensis f. *elongata* Danguy in Lecomte & Humbert, Fl. Indo-Chine 3: 681 (1930).
Lobelia alsinoides f. *elongata* (Danguy) E. Wimm., Pflanzenr. IV.276b: 573 (1953).

Lobelia alticaulis Proctor, Bull. Inst. Jamaica, Sci. Ser. 16: 69 (1967).
Jamaica. 81 JAM. Nanophan. or cham.

Lobelia amoena Michx., Fl. Bor.-Amer. 2: 153 (1803). *Rapuntium amoenum* (Michx.) C. Presl,
Prodr. Monogr. Lobel.: 23 (1836). *Dortmanna amoenum* (Michx.) Kuntze, Revis. Gen. Pl. 2:
972 (1891).
SE. U.S.A. (Virginia to Florida & Alabama). 78 ALA FLA GEO NCA SCA TEN VRG. Hemicr.
2n = 28.
Lobelia colorata Sweet, Brit. Fl. Gard. 5: pl. 180 (1833); non Wall., Pl. Asiat. Rar. 2: 42
(1831). *Rapuntium coloratum* C. Presl, Prodr. Monogr. Lobel.: 30 (1836); non (Wall.)
C. Presl, Prodr. Monogr. Lobel.: 24 (1836). *Lobelia hortensis* A. DC. in DC., Prodr. 7:
377 (1839).

Lobelia anatina E. Wimm., Repert. Spec. Nov. Regni Veg. 19: 385 (1924).
SW. U.S.A. (Arizona, New Mexico) & N. Mexico. 76 ARI 77 NWM 79 MXE MXN. Hemicr.
2n = 14.
Lobelia anatina var. *riskindii* M. C. Johnst., Nordic J. Bot. 2: 3 (1982).

Lobelia anceps L. f., Suppl. Pl.: 395 (1782). *Rapuntium anceps* (L. f.) C. Presl, Prodr. Monogr.
Lobel.: 14 (1836). *Dortmanna anceps* (L. f.) Kuntze, Revis. Gen. Pl. 2: 972 (1891).
Circumaustral, extending north to Mozambique, Australia, Rapa, Pitcairn, Juan Fernández
Is. & N. Chile. 26 MOZ 27 CPP NAT 29 MDG 50 NFK NSW QLD SOA TAS VIC WAU 51
CTM KER NZN NZS 61 PIT TUB 85 CLC CLN CLS JNF. Hemicr. *2n* = 14.
Lobelia repens Thunb., Prodr. Pl. Cap.: 40 (1794). *Enchysia repens* (Thunb.) C. Presl,
Prodr. Monogr. Lobel.: 40 (1836). *Lobelia anceps* var. *minor* Sond. in Harv. & Sond.,
Fl. Cap. 3: 547 (1865). *Laurentia repens* (Thunb.) Benth. in Benth. & Hook. f., Gen.
Pl. 2: 549 (1876). *Lobelia alata* var. *minor* (Sond.) E. Wimm., Ann. Naturhist. Mus.
Wien 56: 343 (1948).
Lobelia alata Labill., Nov. Holl. Pl. 1: 51 (1805). *Rapuntium alatum* (Labill.) C. Presl,
Prodr. Monogr. Lobel.: 14 (1836). *Lobelia anceps* var. *alata* (Labill.) Heynh., Nom. Bot.
Hort. 1: 471 (1840).
Lobelia cuneiformis Labill., Nov. Holl. Pl. 1: 51 (1805). *Lobelia alata* var. *cuneiformis*
(Labill.) R. Br., Prodr.: 562 (1810). *Rapuntium cuneiforme* (Labill.) C. Presl, Prodr.
Monogr. Lobel.: 14 (1836). *Lobelia anceps* var. *cuneiformis* (Labill.) A. DC. in DC.,
Prodr. 7: 376 (1839).
Lobelia rhizophyta Spreng., Novi Provent.: 25 (1818).
Lobelia decumbens Sims, Bot. Mag. 49: tab. 2277 (1821); non A. Rich. ex Schult. in
Roem. & Schult., Syst. Veg. 5: 67 (1819).

Lobelia alata var. *stolonifera* Endl., Prodr. Fl. Norfolk.: 50 (1833).

Lobelia rupincola Bertero ex Colla, Herb. Pedem. 4: 41 (1835).

Dobrowskya anceps C. Presl, Prodr. Monogr. Lobel.: 10 (1836).

Rapuntium ottonianum C. Presl, Prodr. Monogr. Lobel.: 14 (1836). *Lobelia ottoniana* (C. Presl) A. DC. in DC., Prodr. 7: 370 (1839). *Lobelia alata* var. *ottoniana* (C. Presl) E. Wimm., Ann. Naturhist. Mus. Wien 56: 344 (1948).

Lobelia angustifolia Benth. in Endl., Fenzl, Benth. & Schott, Enum. Pl.: 74 (1837); non Cham., Linnaea 8: 219 (1833).

Lobelia erecta de Vriese in Lehm., Pl. Preiss.: 395 (1845).

Lobelia stricta de Vriese in Lehm., Pl. Preiss.: 396 (1845); non Sw., Prodr.: 117 (1787); nec R. Br., Prodr.: 564 (1810); nec M. Martens & Galeotti, Bull. Acad. Roy. Sci. Bruxelles 9(2): 43 (1842).

Lobelia uncinata de Vriese in Lehm., Pl. Preiss.: 398 (1845).

Lobelia saxicola de Vriese in Lehm., Pl. Preiss.: 398 (1845).

Lobelia rudatisii Schltr., Bot. Jahrb. Syst. 57: 621 (1922). *Lobelia alata* var. *rudatisii* (Schltr.) E. Wimm., Ann. Naturhist. Mus. Wien 56: 344 (1948).

Lobelia longiracemosa R. D. Good, J. Bot. 62: 49 (1924).

Lobelia anceps var. *vagans* R. D. Good, J. Bot. 62: 49 (1924).

Lobelia alata var. *longisepala* E. Wimm., Kew Bull. 1952: 138 (1952).

Lobelia andrewsii Lammers, Novon 11: 67 (2001).

E. Australia (Queensland, New South Wales). 50 NSW QLD. Ther. 2*n* = 16.

* *Lobelia gracilis* Andrews, Bot. Repos. 5: pl. 340 (1803); non Salisb., Prodr. Stirp. Chap. Allerton: 129 (1796). *Rapuntium gracile* C. Presl, Prodr. Monogr. Lobel.: 21 (1836). *Dortmanna gracilis* (C. Presl) Kuntze, Revis. Gen. Pl. 2: 972 (1891).

Lobelia gracilis var. *rosea* J. W. Loudon, Ladies' Flower-gard. Ornam. Annuals 1: 164 (1840). *Lobelia gracilis* f. *rosea* (J. W. Loudon) E. Wimm., Pflanzenr. IV.276c: 563 (1953).

Lobelia gracilis var. *major* Benth., Fl. Austral. 4: 125 (1868).

Lobelia angulata G. Forst., Fl. Ins. Austr.: 58 (1786). *Lobelia repanda* Martyn in Mill., Gard. Dict. (ed. 9) (1797). *Rapuntium angulatum* (G. Forst.) C. Presl, Prodr. Monogr. Lobel.: 30 (1836). *Pratia angulata* (G. Forst.) Hook. f., Fl. Antarct.: 43 (1844).

New Zealand, Antipodean Is., Chatham Is.; naturalized in Great Britain. (10) grb 51 ATP CTM NZN NZS. Hemicr. 2*n* = 14, 28, 70, 140.

Lobelia rugulosa Graham, Edinburgh New Philos. J. 8: 186 (1829). *Rapuntium rugulosum* (Graham) C. Presl, Prodr. Monogr. Lobel.: 30 (1836). *Pratia rugulosa* (Graham) G. Don in Sweet, Hort. Brit. (ed. 3): 425 (1839).

Lobelia littoralis R. Cunn. ex A. Cunn., Ann. Nat. Hist. 2: 50 (1839).

Pratia arenaria Hook. f., Fl. Antarct.: 41 (1844). *Pratia angulata* var. *arenaria* (Hook. f.) Hook. f., Fl. Nov.-Zel. 1: 157 (1852).

Piddingtonia palliardii Lehm., Neue Allg. Deutsche Garten- Blumenzeitung 7: 337 (1851).

Pratia linnaeoides Hook. f., Handb. New Zealand Fl.: 172 (1867). *Lobelia linnaeoides* (Hook. f.) Petrie, Trans. & Proc. New Zealand Inst. 23: 405 (1891).

Pratia angulata var. *minor* Carse, Trans. & Proc. New Zealand Inst. 60: 307 (1929).

Pratia angulata var. *obovata* E. Wimm., Pflanzenr. IV.276b: 111 (1943).

Lobelia appendiculata A. DC. in DC., Prodr. 7: 376 (1839). *Dortmanna appendiculata* (A. DC.) Kuntze, Revis. Gen. Pl. 2: 972 (1891).

SC. U.S.A. (Kansas to Texas & Alabama). 74 KAN OKL 77 TEX 78 ALA ARK LOU MSI. Ther. 2*n* = 14.

Lobelia aquaemontis E. Wimm., Pflanzenr. IV.276b: 553 (1953).

Transvaal. 27 TVL. Hemicr. or ther.

Lobelia aquatica Cham., Linnaea 8: 211 (1833). *Rapuntium aquaticum* (Cham.) C. Presl,
Prodr. Monogr. Lobel.: 21 (1836). *Dortmanna aquatica* (Cham.) Kuntze, Revis. Gen. Pl. 2:
972 (1891).
Hispaniola; Colombia to French Guiana, Brazil & NE. Argentina (Corrientes). 81 DOM
HAI 82 FRG GUY SUR VEN 83 BOL CLM PER 84 ALL 85 AGE PAR. Hydrother.
Lobelia domingensis A. DC. in DC., Prodr. 7: 359 (1839).
Lobelia bracteolata Vatke, Linnaea 38: 721 (1874); non A. DC. in DC., Prodr. 7: 371
(1839).
Lobelia aquatica var. *gracilis* E. Wimm., Pflanzenr. IV.276b: 570 (1953).

Lobelia archboldiana (Merr. & L. M. Perry) Moeliono in Steenis, Fl. Males. (ser. 1) 6:
131 (1960).
New Guinea. 43 NWG. Hemicr. or ther.
* *Pratia archboldiana* Merr. & L. M. Perry, J. Arnold Arbor. 30: 59 (1949).

Lobelia ardisiandroides Schltr., Bot. Jahrb. Syst. 57: 623 (1922).
Cape Provinces. 27 CPP. Hemicr.

Lobelia arnhemiaca E. Wimm., Ann. Naturhist. Mus. Wien 56: 346 (1948).
Australia (Northern Territory). 50 NTA. Hemicr. or ther.

Lobelia assurgens L., Syst. Nat. (ed. 10): 1237 (1759). *Tylomium assurgens* (L.) C. Presl, Prodr.
Monogr. Lobel.: 32 (1836). *Tupa assurgens* (L.) A. DC. in DC., Prodr. 7: 394 (1839).
Dortmanna assurgens (L.) Kuntze, Revis. Gen. Pl. 2: 972 (1891).
Cuba, Haiti, Jamaica. 81 CUB HAI JAM. Nanophan. $2n = 28$.

subsp. **assurgens**
Jamaica. 81 JAM. Nanophan.
Lobelia assurgens var. *jamaicensis* Urb., Symb. Antill. 1: 453 (1899). *Lobelia jamaicensis*
(Urb.) Urb., Ark. Bot. 23A(5): 104 (1930).

subsp. **santa-clarae** (McVaugh) Lammers, Novon 16: 70 (2006).
Cuba & Haiti. 81 CUB HAI. Nanophan. $2n = 28$.
Siphocampylus cubensis A. Rich., Hist. Phys. Cuba, Pl. Vasc.: 68 (1841) .
Lobelia assurgens var. *santa-clarae* McVaugh, N. Amer. Fl. 32A: 84 (1943).

Lobelia aurita (Brandegee) T. J. Ayers, Syst. Bot. 15: 324 (1990).
NW. Mexico (Baja California Sur). 79 MXN. Ther.
* *Heterotoma aurita* Brandegee, Proc. Calif. Acad. Sci. (ser. 2) 3: 149 (1891).
Calcaratolobelia aurita (Brandegee) Wilbur, Sida 17: 563 (1997).
Lobelia cotensis M. E. Jones, Contr. W. Bot. 15: 152 (1929).
Lobelia amabilis M. E. Jones, Contr. W. Bot. 18: 68 (1933).

Lobelia australiensis Lammers, Novon 8: 34 (1998).
Australia (?). 50+. Hemicr. or ther.
* *Pratia prostrata* E. Wimm., Repert. Spec. Nov. Regni Veg. 38: 3 (1935); non *Lobelia
prostrata* Zahlbr., Bull. Herb. Boissier (ser. 2) 7: 447 (1907).

Lobelia bambusetii R. E. Fr. & T. C. E. Fr., Svensk Bot. Tidskr. 16: 401 (1922), as 'bambuseti'.
Kenya. 25 KEN. Phan. $2n = 28$.

Lobelia barkerae E. Wimm., Pflanzenr. IV.276b: 781 (1953), as 'barkeri'.
Cape Provinces. 27 CPP. Cham.

Lobelia barnsii Exell, Cat. Vasc. Pl. S. Tomé 229 (1944).
São Tomé. 23 GGI. Nanophan or herb. phan.

Lobelia baumannii Engl., Bot. Jahrb. Syst. 19 (Beibl. 47): 51 (1894), as 'baumanni'.
Zaïre to Kenya & Mozambique. 23 BUR ZAI 25 KEN TAN 26 MLW MOZ ZIM ZAM.
Hemicr. $2n = 28$.
Lobelia baumannii var. *cuneata* E. Wimm., Notizbl. Bot. Gart. Berlin-Dahlem 12: 104
(1934).
Lobelia baumannii f. *alba* E. Wimm., Notizbl. Bot. Gart. Berlin-Dahlem 12: 104 (1934).
Lobelia kummeriae Engl. ex T. C. E. Fr., Notizbl. Bot. Gart. Berlin-Dahlem 8: 398 (1923).
Lobelia baumannii var. *kummeriae* (Engl. ex T. C. E. Fr.) E. Wimm., Notizbl. Bot. Gart.
Berlin-Dahlem 12: 104 (1934).

Lobelia beaugleholei Albr., Muelleria 7: 525 (1992).
SE. Australia (South Australia, Victoria). 50 SOA VIC. Hemicr. or geophyte.

Lobelia beddomeana E. Wimm., Pflanzenr. IV.276b: 645 (1953).
S. India (Tamil Nadu). 40 IND. Ther.

Lobelia benthamii F. Muell., Syst. Census Austral. Pl.: 85 (1882), as 'benthami'.
E. & SC. Australia. 50 NSW QLD SOA TAS VIC. Hemicr. $2n = 14, 28, 42$.
* *Pratia puberula* Benth., Fl. Austral. 4: 133 (1868); non *Lobelia puberula* Michx., Fl. Bor.-
Amer. 2: 152 (1803).
Pratia puberula f. *arguta* E. Wimm., Pflazenr. IV.276b: 110 (1943).

Lobelia bequaertii De Wild., Rev. Zool. Afr. 8 (Suppl. Bot.): 31 (1920), as 'bequaerti'. *Lobelia
deckenii* subsp. *bequaertii* (De Wild.) Mabb., Kew Bull. 29: 574 (1974).
Zaïre & Uganda. 23 ZAI 25 UGA. Nanophan. or herb. phan. $2n = 28$.

Lobelia berlandieri A. DC. in DC., Prodr. 7: 367 (1839). *Dortmanna berlandieri* (A. DC.)
Kuntze, Revis. Gen. Pl. 2: 972 (1891).
SC. U.S.A. (Texas) to E. Mexico. 77 TEX 79 MXE MXG. Ther. $2n = 14$.

subsp. **berlandieri**
SC. U.S.A. (SE. Texas) & NE. Mexico (San Luis Potosí, Nuevo León, Tamaulipas &
Veracruz). 77 TEX 79 MXE MXG. Ther. $2n = 14$.
Lobelia calcarea E. Wimm., Repert. Spec. Nov. Regni. Veg. 38: 83 (1935).

subsp. **brachypoda** (A. Gray) Lammers, Novon 16: 70 (2006).
SC. U.S.A. (S. Texas) & N. Mexico (Chihuahua, Coahuila, Nuevo León & Tamaulipas).
77 TEX 79 MXE. Ther.
**Lobelia cliffortiana* var. *brachypoda* A. Gray, Syn. Fl. N. Amer. 2(1): 7 (1878). *Lobelia
brachypoda* (A. Gray) Small, Fl. S.E. U.S.: 1147 (1903). *Lobelia berlandieri* var.
brachypoda (A. Gray) McVaugh, Bartonia 23: 40 (1945).

Lobelia blantyrensis E. Wimm., Ann. Naturhist. Mus. Wien 56: 363 (1948).
Malawi & Mozambique. 26 MLW MOZ. Hemicr.

Lobelia boivinii Sond. in Harv. & Sond., Fl. Cap. 3: 546 (1865), as 'boivini'. *Dortmanna
boivinii* (Sond.) Kuntze, Revis. Gen. Pl. 2: 972 (1891).
Cape Provinces. 27 CPP. Hemicr.

Lobelia boninensis Koidz. in Matsum., Icon. Pl. Koisik. 2: 19 (1914).
Ogasawara-shoto (Keetaajima, Chichijima, Hahajima, Mukohjima, Anejima, Imohtojima,
Meijima). 38 OGA. Nanophan. or herb. phan. $2n = 28$.
Lobelia boninensis f. *viridis* Nakai, Bot. Mag. (Tokyo) 34 (Jap.): 324 (1920).
Lobelia boninensis f. *rubra* Nakai, Bot. Mag. (Tokyo) 34 (Jap.): 324 (1920).

Lobelia borneensis (Hemsl.) Moeliono in Steenis, Fl. Males. (ser. 1) 6: 133 (1960).
Borneo, Sulawesi, Lesser Sunda Is. (Flores). 42 BOR LSI SUL. Cham. or hemicr.

Pratia borneensis Hemsl., Hooker's Icon. Pl. 16: pl. 1532 (1886).
Pratia borneensis var. *grandiflora* Stapf, Trans. Linn. Soc. London, Bot. 4: 188 (1894).

Lobelia boykinii Torr. & A. Gray ex A. DC. in DC., Prodr. 7: 374 (1839). *Dortmanna boykinii* (Torr. & A. Gray ex A. DC.) Kuntze, Revis. Gen. Pl. 2: 972 (1891).
E. U.S.A. (New Jersey to Florida & Mississippi). 75 DEL NWJ 78 ALA FLA GEO MSI NCA SCA. Hel. $2n = 14$.

Lobelia brachyantha Merr. & L. M. Perry, J. Arnold Arbor. 22: 385 (1941). *Pratia brachyantha* (Merr. & L. M. Perry) E. Wimm., Pflanzenr. IV.276c: 833 (1968).
New Guinea. 43 NWG. Hemicr. or ther.

Lobelia brasiliensis A. O. S. Vieira & G. J. Sheph., Novon 8: 457 (1998).
E. Brazil (Distríto Federal). 84 BZC. Nanophan. or herb. phan. $2n = 28$.

Lobelia brevifolia Nutt. ex A. DC. in DC., Prodr. 7: 377 (1839). *Dortmanna brevifolia* (Nutt. ex A. DC.) Kuntze, Revis. Gen. Pl. 2: 972 (1891).
SE. U.S.A. (Louisiana to Florida). 78 ALA FLA LOU MSI. Hemicr. $2n = 14$.
Lobelia ludoviciana A. W. Wood, Class-book Bot.: 476 (1861).

Lobelia brevisepala (Y. S. Lian) Lammers, Novon 8: 34 (1998).
SC. China (Yunnan). 36 CHC. Cham. or hemicr.
** Pratia brevisepala* Y. S. Lian, Fl. Reipubl. Popul. Sin. 73(2): 189 (1983).

Lobelia bridgesii Hook. & Arn., J. Bot. (Hooker) 1: 278 (1834). *Rapuntium bridgesii* (Hook. & Arn.) C. Presl, Prodr. Monogr. Lobel.: 28 (1836). *Tupa bridgesii* (Hook. & Arn.) A. DC. in DC., Prodr. 7: 394 (1839). *Dortmanna bridgesii* (Hook. & Arn.) Kuntze, Revis. Gen. Pl. 2: 972 (1891).
SC. Chile (Cautín to Valdivia). 85 CLS. Cham. or hemicr. $2n = 42$.
Tupa blanda D. Don in Sweet, Brit. Fl. Gard. 7: pl. 308 (1835). *Rapuntium blandum* (D. Don) C. Presl, Prodr. Monogr. Lobel.: 27 (1836). *Lobelia blanda* (D. Don) Heynh., Nom. Bot. Hort. 1: 473 (1840). *Dortmanna blanda* (D. Don) Kuntze, Revis. Gen. Pl. 2: 972 (1891).
Rapuntium lucaeanum C. Presl, Prodr. Monogr. Lobel.: 27 (1836). *Lobelia lucaeanum* (C. Presl) A. DC. in DC., Prodr. 7: 383 (1839). *Dortmanna lucaeanum* (C. Presl) Kuntze, Revis. Gen. Pl. 2: 972 (1891).

Lobelia brigittalis E. H. L. Krause, Beih. Bot. Centralbl. 32(2): 337 (1914).
Windward Is. (St. Vincent). 81 WIN. Nanophan. or cham.

Lobelia burttii E. A. Bruce, Bull. Misc. Inform. Kew 1933: 473 (1933). *Lobelia deckenii* subsp. *burttii* (E. A. Bruce) Mabb., Kew Bull. 29: 575 (1974).
Tanzania. 25 TAN. Nanophan. or herb. phan. $2n = 28$.

subsp. **burttii**
Tanzania. 25 TAN. Nanophan. or herb. phan.

subsp. **meruensis** E. B. Knox, Contr. Univ. Michigan Herb. 19: 247 (1993).
Tanzania. 25 TAN. Nanophan. or herb. phan. $2n = 28$.

subsp. **telmaticola** E. B. Knox, Contr. Univ. Michigan Herb. 19: 248 (1993).
Tanzania. 25 TAN. Nanophan. or herb. phan.

Lobelia cacuminis Britton & P. Wilson, Bull. Torrey Bot. Club 50: 50 (1923). *Lobelia oxyphylla* var. *cacuminis* (Britton & P. Wilson) E. Wimm., Pflanzenr. IV.276b: 621 (1953). *Lobelia oxyphylla* subsp. *cacuminis* (Britton & P. Wilson) Borhidi, Acta Bot. Acad. Sci. Hung. 25: 36 (1979).
Cuba. 81 CUB. Nanophan.

Lobelia caeciliae E. Wimm., Ann. Naturhist. Mus. Wien 56: 348 (1948).
S. Mexico (Chiapas). 79 MXT. Cham. or hemicr.

Lobelia caerulea Sims, Bot. Mag. 53: tab. 2701 (1826). *Rapuntium caeruleum* (Sims) C. Presl,
Prodr. Monogr. Lobel.: 20 (1836). *Lobelia coronopifolia* var. *caerulea* (Sims) Sond. in Harv. &
Sond., Fl. Cap. 3: 543 (1865), as 'coerulea'. *Dortmanna coronopifolia* var. *caerulea* (Sims)
Kuntze, Revis. Gen. Pl. 3(2): 187 (1898), as 'coerulea'.
S. Africa. 26 MOZ 27 CPP NAT. Cham. or hemicr.
  Rapuntium maculare C. Presl, Prodr. Monogr. Lobel.: 20 (1836). *Lobelia macularis* (C.
    Presl) A. DC. in DC., Prodr. 7: 364 (1839). *Lobelia coronopifolia* var. *macularis* (C.
    Presl) Sond. in Harv. & Sond., Fl. Cap. 3: 543 (1865). *Dortmanna coronopifolia* var.
    macularis (C. Presl) Kuntze, Revis. Gen. Pl. 3(2): 187 (1898). *Lobelia caerulea* var.
    macularis (C. Presl) E. Wimm., Pflanzenr. IV.276b: 593 (1953).
  Rapuntium maculare var. *procerum* C. Presl, Prodr. Monogr. Lobel.: 21 (1836). *Lobelia
    macularis* var. *procera* (C. Presl) A. DC. in DC., Prodr. 7: 364 (1839). *Lobelia caerulea* f.
    procera (C. Presl) E. Wimm., Pflanzenr. IV.276b: 594 (1953).
  Rapuntium caeruleum var. *glabrescens* C. Presl in Eckl. & Zeyh., Enum. Pl. Afric. Austral.:
    399 (1837). *Lobelia caerulea* var. *glabrescens* (C. Presl) A. DC. in DC., Prodr. 7: 363
    (1839). *Lobelia coronopifolia* var. *glabrescens* (C. Presl) Sond. in Harv. & Sond., Fl. Cap.
    3: 543 (1865). *Dortmanna coronopifolia* var. *glabrescens* (C. Presl) Kuntze, Revis. Gen.
    Pl. 3(2): 187 (1898).
  Rapuntium caeruleum var. *latidens* C. Presl in Eckl. & Zeyh., Enum. Pl. Afric. Austral.: 399
    (1837). *Lobelia caerulea* var. *latidens* (C. Presl) A. DC. in DC., Prodr. 7: 363 (1839).
  Lobelia backhouseana Lem., Ill. Hort. 13: pl. 485 (1866). *Lobelia caerulea* var.
    backhouseana (Lem.) E. Wimm., Pflanzenr. IV.276b: 593 (1953).
  Dortmanna coronopifolia f. *azurea* Kuntze, Revis. Gen. Pl. 3(2): 187 (1898); non Kuntze,
    Revis. Gen. Pl. 3(2): 187 (1898); nec Kuntze, Revis. Gen. Pl. 3(2): 187 (1898).
  Dortmanna coronopifolia f. *azurea* Kuntze, Revis. Gen. Pl. 3(2): 187 (1898); non Kuntze,
    Revis. Gen. Pl. 3(2): 187 (1898); nec Kuntze, Revis. Gen. Pl. 3(2): 187 (1898).
  Dortmanna coronopifolia f. *bicolor* Kuntze, Revis. Gen. Pl. 3(2): 187 (1898).
  Dortmanna coronopifolia f. *rosea* Kuntze, Revis. Gen. Pl. 3(2): 187 (1898); non Kuntze,
    Revis. Gen. Pl. 3(2): 187 (1898).
  Dortmanna coronopifolia f. *rosea* Kuntze, Revis. Gen. Pl. 3(2): 187 (1898); non Kuntze,
    Revis. Gen. Pl. 3(2): 187 (1898).

Lobelia caledoniana C. D. Adams, Phytologia 21: 67 (1971).
Jamaica. 81 JAM. Nanophan.

Lobelia calochlamys (Donn. Sm.) Wilbur, Sida 14: 562 (1991).
Guatemala. 80 GUA. Cham. or hemicr.
  * *Centropogon calochlamys* Donn. Sm., Bot. Gaz. (Crawfordsville) 46: 112 (1908). *Pratia
    calochlamys* (Donn. Sm.) E. Wimm., Repert. Spec. Nov. Regni Veg. 29: 50 (1931).

Lobelia camporum Pohl, Pl. Bras. Icon. Descr. 2: 100 (1831). *Rapuntium camporum* (Pohl) C.
Presl, Prodr. Monogr. Lobel.: 15 (1836). *Dortmanna camporum* (Pohl) Kuntze, Revis. Gen.
Pl. 2: 972 (1891).
Brazil (Minas Gerais, Rio de Janeiro, São Paulo, Paraná, Rio Grande do Sul) & NE.
  Argentina (Misiones). 84 BZL BZS 85 AGE. Ther.
  Lobelia camporum var. *lundiana* A. DC. in DC., Prodr. 7: 375 (1839).
  Lobelia camporum f. *minor* E. Wimm., Revista Sudamer. Bot. 2: 105 (1935).
  Lobelia camporum f. *angustifolia* E. Wimm., Pflanzenr. IV.276b: 454 (1953).
  Lobelia paulista E. Wimm., Pflanzenr. IV.276c: 857 (1968).

Lobelia canbyi A. Gray, Manual (ed. 5): 284 (1867). *Dortmanna canbyi* (A. Gray) Kuntze,
Revis. Gen. Pl. 2: 972 (1891). Hemicr. $2n = 14$.
E. U.S.A. (New Jersey to Georgia). 75 DEL MRY NWJ 79 GEO NCA SCA TEN.

Lobelia capillifolia (C. Presl) A. DC. in DC., Prodr. 7: 362 (1839). *Rapuntium capillifolium* C. Presl, Prodr. Monogr. Lobel.: 19 (1836). *Dortmanna capillifolia* (C. Presl) Kuntze, Revis. Gen. Pl. 2: 972 (1891).
Cape Provinces. 27 CPP. Cham.
    Lobelia capillifolia var. *velutina* A. DC. in DC., Prodr. 7: 362 (1839).

Lobelia cardinalis L., Sp. Pl.: 930 (1753). *Rapuntium cardinale* (L.) Mill., Gard. Dict. (ed. 8) (1768), as 'cardinalis'. *Rapuntium coccineum* Moench, Suppl. Meth.: 277 (1802). *Lobelia coccinea* Stokes, Bot. Mat. Med. 1: 343 (1812). *Dortmanna cardinalis* (L.) Kuntze, Revis. Gen. Pl. 2: 380 (1891).
S. Canada to N. Colombia; cult. 72 NBR ONT QUE 73 COL 74 ILL IOW KAN MIN MSO NEB OKL WIS 75 ALL 76 ALL 77 ALL 78 ALL 79 MXC MXE MXG MXN MXS MXT 80 BLZ COS GUA HON NIC PAN 83 CLM. Hemicr. $2n = 14$.
    Lobelia graminea Lam., Encycl. 3: 583 (1792). *Rapuntium gramineum* (Lam.) C. Presl, Prodr. Monogr. Lobel.: 26 (1836). *Dortmanna graminea* (Lam.) Kuntze, Revis. Gen. Pl. 2: 972 (1891). *Lobelia cardinalis* subsp. *graminea* (Lam.) McVaugh, Ann. Missouri Bot. Gard. 27: 347 (1940). *Lobelia cardinalis* var. *graminea* (Lam.) McVaugh, Ann. Missouri Bot. Gard. 27: 348 (1940).
    Lobelia cordigera Cav., Icon. 6: 14 (1800). *Rapuntium cordigerum* (Cav.) C. Presl, Prodr. Monogr. Lobel.: 26 (1836). *Dortmanna cordigera* (Cav.) Kuntze, Revis. Gen. Pl. 2: 972 (1891). *Lobelia cardinalis* f. *cordigera* (Cav.) Bowden, Canad. J. Bot. 60: 2060 (1982).
    Lobelia fulgens Humb. & Bonpl. ex Willd., Hort. Berol. 2: pl. 85 (1809). *Rapuntium fulgens* (Humb. & Bonpl. ex Willd.) C. Presl, Prodr. Monogr. Lobel.: 26 (1836). *Lobelia splendens* var. *fulgens* (Humb. & Bonpl. ex Willd.) S. Watson, Proc. Amer. Acad. Arts 18: 110 (1883). *Dortmanna fulgens* (Humb. & Bonpl. ex Willd.) Kuntze, Revis. Gen. Pl. 2: 972 (1891).
    Lobelia splendens Humb. & Bonpl. ex Willd., Hort. Berol. 2: pl. 86 (1809). *Rapuntium splendens* (Humb. & Bonpl. ex Willd.) C. Presl, Prodr. Monogr. Lobel.: 26 (1836). *Lobelia fulgens* var. *glabriuscula* Klotzsch ex Walp., Repert. Bot. Syst. 2: 707 (1843). *Dortmanna splendens* (Humb. & Bonpl. ex Willd.) Kuntze, Revis. Gen. Pl. 2: 973 (1891). *Lobelia cardinalis* f. *splendens* (Humb. & Bonpl. ex Willd.) Bowden, Canad. J. Bot. 60: 2060 (1982).
    Lobelia texensis Raf., Atlantic J. 176 (1833). *Lobelia cardinalis* var. *texensis* (Raf.) Rothr., Rep. U.S. Geogr. Surv., Wheeler: 182 (1879).
    Lobelia cardinalis var. *alba* J. McNab, Edinburgh New Philos. J. 19: 61 (1835). *Lobelia cardinalis* f. *alba* (J. McNab) H. St. John, Rhodora 21: 218 (1919).
    Rapuntium longifolium C. Presl, Prodr. Monogr. Lobel.: 26 (1836). *Lobelia longifolia* (C. Presl) A. DC. in DC., Prodr. 7: 382 (1839). *Dortmanna longifolia* (C. Presl) Kuntze, Revis. Gen. Pl. 2: 972 (1891).
    Lobelia fulgens var. *propinqua* Paxton, Paxton's Mag. Bot. 2: 52 (1836). *Lobelia cardinalis* var. *propinqua* (Paxton) Bowden, Canad. J. Bot. 60: 2060 (1982).
    Lobelia ignea Paxton, Paxton's Mag. Bot. 6: 165 (1839); non Vell., Fl. Flumin., Icon. 8: pl. 1287 (1831). *Lobelia splendens* var. *ignea* Hook., Bot. Mag. 61: tab. 4960 (1857). *Lobelia splendens* f. *ignea* (Hook.) Voss in Siebert & Voss, Vilm. Blumengärtn. (ed. 3) 1: 575 (1894).
    Lobelia princeps Otto & A. Dietr., Allg. Gartenzeitung 7: 298 (1839).
    Lobelia punicea Otto & A. Dietr., Allg. Gartenzeitung 7: 299 (1839).
    Lobelia splendens var. *atrosanguinea* Hook., Bot. Mag. 69: tab. 4002 (1843). *Lobelia fulgens* f. *atrosanguinea* (Hook.) Voss in Siebert & Voss, Vilm. Blumengärtn. (ed. 3) 1: 575 (1894).
    Lobelia coccinea de Jonghe, Rev. Hort. (ser. 2) 4: 39 (1845); non Stokes, Bot. Mat. Med. 1: 343 (1812).
    Lobelia fulgens var. *multiflora* Paxton, Paxton's Mag. Bot. 14: 263 (1848). *Lobelia cardinalis* var. *multiflora* (Paxton) McVaugh, Ann. Missouri Bot. Gard. 27: 349 (1940). *Lobelia cordigera* var. *multiflora* (Paxton) E. Wimm., Pflanzenr. IV.276b: 418 (1953).

Lobelia fulgens var. *pyramidalis* Paxton, Paxton's Mag. Bot. 14: 263 (1848). *Lobelia cordigera* var. *pyramidalis* (Paxton) E. Wimm., Pflanzenr. IV.276b: 418 (1953).

Lobelia marryattiae Paxton, Paxton's Mag. Bot. 14: 263 (1848), as 'marryatti'. *Lobelia fulgens* var. *marryattiae* (Paxton) Paxton, Paxton's Mag. Bot. 15: 7 (1849), as 'marryattae'. *Lobelia cordigera* var. *marryattiae* (Paxton) E. Wimm., Pflanzenr. IV.276b: 418 (1953), as 'marryattae'.

Lobelia mucronata Engelm. in Wislizenus, Mem. Tour N. Mexico: 108 (1848); non Cav., Icon. 6: 11 (1800). *Dortmanna engelmanniana* Kuntze, Revis. Gen. Pl. 2: 972 (1891); non *Dortmanna mucronata* (Cav.) Kuntze, Revis. Gen. Pl. 2: 972 (1891).

Lobelia phyllostachya Engelm. in Wislizenus, Mem. Tour N. Mexico: 108 (1848). *Dortmanna phyllostachya* (Engelm.) Kuntze, Revis. Gen. Pl. 2: 973 (1891). *Lobelia cardinalis* var. *phyllostachya* (Engelm.) McVaugh, Ann. Missouri Bot. Gard. 27: 349 (1940). *Lobelia graminea* var. *phyllostachya* (Engelm.) E. Wimm., Pflanzenr. IV.276b: 414 (1953).

Lobelia porphyrantha Decne. ex Groenland, Rev. Hort. (ser. 4) 9: 120 (1860).

Lobelia cardinalis var. *candida* A. W. Wood, Amer. Bot. Fl.: 195 (1870).

Lobelia cardinalis var. *integerrima* A. W. Wood, Amer. Bot. Fl.: 195 (1870).

Lobelia cardinalis var. *angustifolia* Vatke, Linnaea 38: 722 (1874).

Lobelia cardinalis var. *glandulosa* N. Coleman, Kent Sci. Inst. Misc. Publ. 2: 24 (1874).

Lobelia ramosa Burb., Gard. Chron. (ser. 2) 15: 105 (1881); non Benth. in Maund, Botanist: pl. 93 (1838).

Lobelia kerneri L. Nagy, Gartenflora 38: 302 (1889). *Lobelia punicea* var. *kerneri* (L. Nagy) E. Wimm., Ann. Naturhist. Mus. Wien. 56: 339 (1948). *Lobelia graminea* f. *kerneri* (L. Nagy) E. Wimm., Pflanzenr. IV.276b: 415 (1953).

Lobelia fulgens f. *rosea* Voss in Siebert & Voss, Vilm. Blumengärtn. (ed. 3) 1: 575 (1894).

Lobelia fulgens f. *maculata* Voss in Siebert & Voss, Vilm. Blumengärtn. (ed. 3) 1: 575 (1894).

Lobelia cardinalis f. *rosea* H. St. John, Rhodora 21: 217 (1919).

Lobelia cardinalis var. *pseudosplendens* McVaugh, Ann. Misouri Bot. Gard. 27: 348 (1940). *Lobelia graminea* var. *pseudosplendens* (McVaugh) E. Wimm., Pflanzenr. IV.276b: 414 (1953).

Lobelia cardinalis var. *hispidula* E. Wimm., Ann. Naturhist. Mus. Wien 56: 338 (1948). *Lobelia cardinalis* f. *hispidula* (E. Wimm.) Bowden, Canad. J. Bot. 60: 2057 (1982).

Lobelia graminea var. *intermedia* E. Wimm., Ann. Naturhist. Mus. Wien 56: 339 (1948).

Lobelia cordigera var. *fatalis* E. Wimm., Ann. Naturhist. Mus. Wien 56: 340 (1948).

Lobelia cardinalis var. *meridionalis* Bowden, Canad. J. Genet. Cytol. 2: 241 (1960).

Lobelia caudata (Griseb.) Urb., Symb. Antill. 1: 455 (1899).

Jamaica. 81 Jam. Cham. or hemicr.

* *Tupa caudata* Griseb., Fl. Brit. W.I.: 386 (1861). *Dortmanna caudata* (Griseb.) Kuntze, Revis. Gen. Pl. 2: 972 (1891).

Lobelia chamaedryifolia (C. Presl) A. DC. in DC., Prodr. 7: 364 (1839), as 'chamaedryfolia'.

S. Africa. 27 CPP NAT. Cham. or hemicr.

* *Rapuntium chamaedryifolium* C. Presl, Prodr. Monogr. Lobel.: 21 (1836), as 'chamaedrifolium'. *Dortmanna chamaedryifolia* (C. Presl) Kuntze, Revis. Gen. Pl. 2: 972 (1891), as 'chamaedrifolia'.

Lobelia chamaepitys Lam., Encycl. 3: 590 (1792), as 'chamaepithys'. *Rapuntium chamaepitys* (Lam.) C. Presl, Prodr. Monogr. Lobel.: 20 (1836).

Cape Provinces. 27 CPP. Cham.

Rapuntium ceratophyllum C. Presl in Eckl. & Zeyh., Enum. Pl. Afric. Austral.: 399 (1837). *Lobelia ceratophylla* (C. Presl) D. Dietr., Syn. Pl. 1: 728 (1839). *Lobelia tomentosa* var. *ceratophylla* (C. Presl) Sond. in Harv. & Sond., Fl. Cap. 3: 543 (1865). *Dortmanna coronopifolia* var. *ceratophylla* (C. Presl) Kuntze, Revis. Gen. Pl. 3(2): 187 (1898). *Lobelia chamaepitys* var. *ceratophylla* (C. Presl) E. Wimm., Pflanzenr. IV.276b: 591 (1953).

Rapuntium ceratophyllum var. *glabratum* C. Presl in E. Mey., Comm. Pl. Afr. Austr.: 289 (1838). *Lobelia ceratophylla* var. *glabrata* (C. Presl) A. DC. in DC., Prodr. 7: 362 (1839).

Lobelia cheranganiensis Thulin, Nordic J. Bot. 3: 379 (1983).
Kenya. 25 KEN. Hemicr. $2n = 26$.

Lobelia chevalieri Danguy, Bull. Mus. Hist. Nat. (Paris) (ser. 2) 1: 263 (1929).
S. China (Guangdong, Guangxi, Yunnan) to Japan (Kyushu), Nansei-shoto & Taiwan. 36 CHC CHS 38 JAP NNS TAI. Ther.
Lobelia hancei H. Hara, J. Jap. Bot. 17: 23 (1941). *Lobelia alsinoides* subsp. *hancei* (H. Hara) Lammers, Bot. Bull. Acad. Sin. 33: 286 (1992).

Lobelia chinensis Lour., Fl. Cochinchin.: 514 (1790). *Rapuntium chinense* (Lour.) C. Presl, Prodr. Monogr. Lobel.: 13 (1836). *Dortmanna chinensis* (Lour.) Kuntze, Revis. Gen. Pl. 2: 972 (1891).
C. Himalaya to Japan (Hokkaido, Honshu, Shikoku, Kyushu), Nansei-shoto (Kume, Okinawa), Taiwan, Jawa & Sri Lanka. 36 CHS 38 JAP KOR NNS TAI 40 ASS BAN IND NEP SRL 41 CBD LAO THA VIE 42 JAW MLY. Hemicr. or geophyte. $2n = 14, 42$.
Lobelia radicans Thunb., Trans. Linn. Soc. London 2: 330 (1794). *Pratia thunbergii* G. Don, Gen. Hist. 3: 700 (1834); non *Pratia radicans* G. Don, Gen. Hist. 3: 700 (1834). *Rapuntium radicans* (Thunb.) C. Presl, Prodr. Monogr. Lobel.: 14 (1836). *Isolobus radicans* (Thunb.) A. DC. in DC., Prodr. 7: 353 (1839). *Dortmanna radicans* (Thunb.) Kuntze, Revis. Gen. Pl. 2: 380 (1891).
Lobelia campanuloides Thunb., Trans. Linn. Soc. London 2: 331 (1794). *Lobelia japonica* F. Dietr., Vollst. Lex. Gärtn. 5: 552 (1805). *Rapuntium campanuloides* (Thunb.) C. Presl, Prodr. Monogr. Lobel.: 14 (1836). *Isolobus campanuloides* (Thunb.) A. DC. in DC., Prodr. 7: 353 (1839). *Dortmanna campanuloides* (Thunb.) Kuntze, Revis. Gen. Pl. 2: 972 (1891), as 'campanulodes'.
Lobelia caespitosa Blume, Bijdr.: 729 (1826). *Rapuntium caespitosum* (Blume) C. Presl, Prodr. Monogr. Lobel.: 13 (1836). *Isolobus caespitosus* (Blume) Hassk., Bonplandia 7: 180 (1859).
Pratia radicans G. Don, Gen. Hist. 3: 700 (1834). *Isolobus roxburghianus* A. DC. in DC., Prodr. 7: 353 (1839); non *Isolobus radicans* (Thunb.) A. DC. in DC., Prodr. 7: 353 (1839). *Lobelia roxburgiana* (A. DC.) Heynh., Nom. Bot. Hort. 1: 471 (1840); non *Lobelia radicans* Thunb., Trans. Linn. Soc. London 2: 330 (1794).
Isolobus keri A. DC. in DC., Prodr. 7: 353 (1839), as 'kerii'. *Lobelia keri* (A. DC.) Heynh., Nom. Bot. Hort. 1: 471 (1840), as 'kerii'.
Lobelia radicans f. *plena* Makino in Iinuma, Sômoku-Dzusetsu (ed. 3) 4: 1170 (1912). *Lobelia chinensis* f. *plena* (Makino) H. Hara, J. Jap. Bot. 17: 23 (1941).
Lobelia radicans f. *lactiflora* Hisauti, J. Jap. Bot. 16: 565 (1940). *Lobelia chinensis* f. *lactiflora* (Hisauti) H. Hara, J. Jap. Bot. 17: 23 (1941).
Lobelia radicans var. *albiflora* E. Wimm., Akad. Wiss. Wien, Math.-Naturwiss. Kl., Anz. 61: 112 (1925). *Lobelia chinensis* var. *albiflora* (E. Wimm.) E. Wimm., Pflanzenr. IV.276b: 611 (1953).

Lobelia chireensis A. Rich., Tent. Fl. Abyss. 2: 6 (1850). *Dortmanna chireensis* (A. Rich.) Kuntze, Revis. Gen. Pl. 2: 972 (1891).
Ethiopia to Zimbabwe. 23 BUR 24 ETH 25 TAN 26 ZIM ZAM. Ther.
Lobelia victorialis E. Wimm., Pflanzenr. IV.276b: 776 (1953).

Lobelia christii Urb., Symb. Antill. 7: 420 (1912).
Haiti. 81 HAI. Cham. or hemicr.

Lobelia circaeoides (C. Presl) A. DC. in DC., Prodr. 7: 379 (1839).
Mexico (?). 79 +. Hemicr.?
   * *Rapuntium circaeoides* C. Presl, Prodr. Monogr. Lobel.: 25 (1836). *Dortmanna circaeoides* (C. Presl) Kuntze, Revis. Gen. Pl. 2: 972 (1891).

Lobelia cirsiifolia Lam., Encycl. 3: 584 (1792), as 'cirsifolia'. *Siphocampylus cirsiifolius* (Lam.) G. Don, Gen. Hist. 3: 703 (1834), as 'circiifolius'. *Rapuntium cirsiifolium* (Lam.) C. Presl, Prodr. Monogr. Lobel.: 27 (1836). *Tupa cirsiifolia* (Lam.) A. DC. in DC., Prodr. 7: 395 (1839). *Dortmanna cirsiifolia* (Lam.) Kuntze, Revis. Gen. Pl. 2: 972 (1891).
Lesser Antilles (St. Kitts, Dominica, St. Lucia, St. Vincent, Grenada). 81 LEE WIN. Cham. or hemicr.
    Lobelia racemosa Sims, Bot. Mag. 47: tab. 2137 (1820); non Vest ex Schult. in Roem. & Schult., Syst. Veg. 5: 58 (1819). *Rapuntium racemosum* C. Presl, Prodr. Monogr. Lobel.: 25 (1836). *Tupa racemosa* (C. Presl) A. DC. in DC., Prodr. 7: 394 (1839). *Dortmanna racemosa* (C. Presl) Kuntze, Revis. Gen. Pl. 2: 973 (1891).
    Tupa infesta Griseb., Fl. Brit. W.I.: 387 (1861). *Dortmanna infesta* (Griseb.) Kuntze, Revis. Gen. Pl. 2: 972 (1891). *Lobelia infesta* (Griseb.) Urb., Symb. Antill. 1: 455 (1899).
    Lobelia heterodonta Sprague, Gard. Chron. (ser. 3) 36: 252 (1904).
    Lobelia ryanii Rendle, J. Bot. 73: 277 (1935). *Lobelia racemosa* subsp. *ryanii* (Rendle) Sastre, Biogeographica 73: 33 (1997).
    Lobelia cirsiifolia var. *opilionis* E. Wimm., Ann. Naturhist. Mus. Wien 56: 365 (1948).

Lobelia clavata E. Wimm., Repert. Spec. Nov. Regni Veg. 38: 78 (1935).
SC. China (Guizhou, Yunnan). 36 CHC. Hemicr.

Lobelia cliffortiana L., Sp. Pl.: 931 (1753). *Rapuntium cliffortianum* (L.) Mill., Gard. Dict. (ed. 8) (1768). *Rapuntium cordifolium* Moench, Suppl. Meth.: 277 (1802). *Dortmanna cliffortiana* (L.) Kuntze, Revis. Gen. Pl. 2: 380 (1891).
Cuba to Lesser Antilles (St. Lucia); naturalized in the Mascarenes. (29) mau reu rod 81 CUB DOM HAI JAM PUE WIN. Ther. $2n = 14, 28$.
    Lobelia tetragona Blume, Bijdr.: 729 (1826).
    Lobelia chenopodiifolia Wall. ex G. Don, Gen. Hist. 3: 709 (1834). *Rapuntium chenopodiifolium* (Wall. ex G. Don) C. Presl, Prodr. Monogr. Lobel.: 31 (1836), as 'chenopodifolium'.
    Lobelia bisserrata Sessé & Moç., Fl. Mexic. (ed. 2): 198 (1894).

Lobelia cobaltica S. Moore, J. Linn. Soc., Bot. 40: 124 (1911).
Mozambique & Zimbabwe. 26 MOZ ZAM. Cham.
    Lobelia longipilosa E. Wimm., Kew Bull. 1952: 142 (1952).

Lobelia cochlearifolia Diels, Bot. Jahrb. Syst. 26: 117 (1898).
KwaZulu-Natal. 27 NAT. Hemicr.

Lobelia collina Kunth in Humb., Bonpl. & Kunth, Nov. Gen. Sp. 3: 312 (quarto), 244 (folio) (1819). *Rapuntium collinum* (Kunth) C. Presl, Prodr. Monogr. Lobel.: 23 (1836). *Dortmanna collina* (Kunth) Kuntze, Revis. Gen. Pl. 2: 972 (1891).
Ecuador. 83 ECU. Hemicr.
    Lobelia linifolia Willd. ex Schult. in Roem. & Schult., Syst. Veg. 5: 67 (1819).
    Lobelia phyteumoides Willd. ex Schult. in Roem. & Schult., Syst. Veg. 5: 68 (1819).

Lobelia colorata Wall., Pl. Asiat. Rar. 2: 42 (1831). *Rapuntium coloratum* (Wall.) C. Presl, Prodr. Monogr. Lobel.: 24 (1836); non (Sweet) C. Presl, Prodr. Monogr. Lobel.: 30 (1836). *Dortmanna colorata* (Wall.) Kuntze, Revis. Gen. Pl. 2: 972 (1891).
E. Himalaya to SC. China (Guizhou, Yunnan) & Thailand. 36 CHC 40 ASS EHM 41 THA. Hemicr.

subsp. **colorata**
    E. Himalaya to SC. China (Guizhou, Yunnan) & Thailand. 36 CHC 40 ASS EHM 41 THA. Hemicr.
    Lobelia colorata var. *dsolinhoensis* E. Wimm., Akad. Wiss. Wien, Math.-Naturwiss.Kl., Anz. 61: 110 (1925).

Lobelia colorata var. *baculus* E. Wimm., Akad. Wiss. Wien, Math.-Naturwiss. Kl., Anz. 61: 110 (1925).
Lobelia palustris Kerr, Bull. Misc. Inform. Kew 1936: 35 (1936).
Lobelia leucanthera Kerr, Bull. Misc. Inform. Kew 1936: 34 (1936).

subsp. **guizhouensis** T. J. Zhang & D. Y. Hong, Acta Phytotax. Sin. 30: 161 (1992).
SC. China (Guizhou). 36 CHC. Hemicr.

subsp. **taliensis** (Diels) T. J. Zhang & D. Y. Hong, Acta Phytotax. Sin. 30: 161 (1992).
SC. China (Yunnan). 36 CHC. Hemicr.
* *Lobelia taliensis* Diels, Notes Roy. Bot. Gard. Edinburgh 5: 170 (1912).
Lobelia fossarum E. Wimm., Akad. Wiss. Wien, Math.-Naturwiss. Kl., Anz. 61: 109 (1925).
Lobelia hybrida C. Y. Wu, Rep. Yunnan Trop. Subtrop. Fl. Res. Inst. 1: 92 (1965); non Voss in Siebert & Voss, Vilm. Blumengärtn. (ed. 3) 1: 576 (1894).

Lobelia columaris Hook. f., J. Proc. Linn. Soc., Bot. 6: 14 (1862). *Tupa columnaris* (Hook. f.) Vatke, Linnaea 38: 725 (1874). *Dortmanna columnaris* (Hook. f.) Kuntze, Revis. Gen. Pl. 2: 972 (1891).
Nigeria to Equitorial Guinea. 22 NGA 23 CMN EQG. Nanophan. or herb. phan.
Lobelia conraui R. E. Fr. & T. C. E. Fr., Svensk Bot. Tidskr. 16: 395 (1922).

Lobelia comosa L., Sp. Pl.: 933 (1753). *Lobelia triquetra* var. *comosa* (L.) Schult. in Roem. & Schult., Syst. Veg. 5: 44 (1819).
Cape Provinces. 27 CPP. Cham. $2n = 14, 42$.
Lobelia corymbosa P. J. Bergius, Descr. Pl. Cap.: 344 (1767).
Lobelia triquetra L., Mant. Pl.: 120 (1767). *Rapuntium triquetrum* (L.) C. Presl, Prodr. Monogr. Lobel.: 16 (1836). *Dortmanna triquetrum* (L.) Kuntze, Revis. Gen. Pl. 2: 973 (1891).
Lobelia capitata Burm. f., Prodr. Pl. Cap.: 29 (1768).
Lobelia heteromalla Schrad., Cat. Götting. (1827). *Rapuntium heteromallum* (Schrad.) C. Presl, Prodr. Monogr. Lobel.: 15 (1836).
Lobelia triquetra var. *alba* G. Don, Gen. Hist. 3: 710 (1834).
Rapuntium microdon C. Presl, Prodr. Monogr. Lobel.: 15 (1836). *Lobelia microdon* (C. Presl) A. DC. in DC., Prodr. 7: 369 (1839). *Lobelia erinus* var. *microdon* (C. Presl) Sond. in Harv. & Sond, Fl. Cap. 3: 545 (1865). *Lobelia triquetra* var. *microdon* (C. Presl) E. Wimm., Ann. Naturhist. Mus. Wien 56: 356 (1948). *Lobelia comosa* var. *microdon* (C. Presl) E. Wimm., Pflanzenr. IV.276b: 550 (1953).
Lobelia triquetra var. *secundata* Sond. in Harv. & Sond., Fl. Cap. 3: 546 (1865). *Lobelia comosa* var. *secundata* (Sond.) E. Wimm., Pflanzenr. IV.276b: 550 (1953).
Lobelia triquetra var. *foliosa* E. Wimm., Ann. Naturhist. Mus. Wien 56: 357 (1948). *Lobelia comosa* var. *foliosa* (E. Wimm.) E. Wimm., Pflanzenr. IV.276b: 550 (1953).

Lobelia comptonii E. Wimm., Pflanzenr. IV.276b: 782 (1953).
Cape Provinces. 27 CPP. Hemicr.
Lobelia esterhuyseniae E. Wimm., Pflanzenr. IV.276c: 872 (1968).

Lobelia concolor R. Br., Prodr.: 563 (1810). *Isolobus concolor* (R. Br.) A. DC. in DC., Prodr. 7: 354 (1839). *Pratia concolor* (R. Br.) Druce, Bot. Soc. Exch. Club Brit. Isles 4: 641 (1917).
E. & SC. Australia. 50 NSW QLD SOA VIC. Hemicr. $2n = 28$.
Pratia erecta Gaudich., Voy. Uranie: 456 (1829).
Isolobus cunninghamii A. DC. in DC., Prodr. 7: 354 (1839). *Pratia cunninghamii* (A. DC.) Hook. f., Fl. Antarct.: 42 (1844).
Pratia cunninghamii var. *longipes* Hook. f., Fl. Antarct.: 42 (1844). *Pratia concolor* var. *longipes* (Hook. f.) E. Wimm., Pflanzenr. IV.276b: 109 (1943).

Lobelia conferta Merr. & L. M. Perry, J. Arnold Arbor. 30: 59 (1949). *Pratia conferta* (Merr. &
L. M. Perry) E. Wimm., Pflanzenr. IV.276b: 764 (1953).
New Guinea. 43 NWG. Hemicr.

Lobelia conglobata Lam., Encycl. 3: 585 (1792). *Tylomium conglobatum* (Lam.) C. Presl,
Prodr. Monogr. Lobel.: 32 (1836). *Tupa conglobata* (Lam.) A. DC. in DC., Prodr. 7: 395
(1839). *Dortmanna conglobata* (Lam.) Kuntze, Revis. Gen. Pl. 2: 972 (1891).
Windward Is. (Martinique). 81 WIN. Cham. or hemicr.

Lobelia cordifolia Hook. & Arn., Bot. Beechey Voy.: 301 (1838). *Heterotoma cordifolia* (Hook.
& Arn.) McVaugh, Bull. Torrey Bot Club 67: 143 (1940). *Calcaratolobelia cordifolia* (Hook.
& Arn.) Wilbur, Sida 17: 562 (1997).
NW. Mexico to Costa Rica. 79 MXC MXN MXS MXT 80 COS ELS HON NIC. Ther. 2n = 14.
    Heterotoma salvadorensis E. Wimm., Ann. Naturhist. Mus. Wien 56: 371 (1948).

Lobelia corniculata Thulin, Nordic J. Bot. 3: 380 (1983).
S. Africa. 27 NAT SWZ. Cham. or hemicr.

Lobelia coronopifolia L., Sp. Pl.: 933 (1753). *Rapuntium cornopifolium* (L.) C. Presl, Prodr.
Monogr. Lobel.: 20 (1836). *Dortmanna cornopifolia* (L.) Kuntze, Revis. Gen. Pl. 2: 972 (1891).
Cape Provinces. 27 CPP. Cham. 2n = 12.
    Lobelia pedunculata Sims, Bot. Mag. 48: tab. 2251 (1821); non R. Br., Prodr.: 563 (1810).
        Lobelia simsii Sweet, Hort. Brit.: 247 (1826). *Rapuntium simsii* (Sweet) C. Presl, Prodr.
        Monogr. Lobel.: 20 (1836). *Lobelia recurvifolia* Spreng., Syst. Veg. 5: 420 (1828).
        Lobelia coronopifolia var. *simsii* (Sweet) E. Wimm., Pflanzenr. IV.276b: 592 (1953).
    Lobelia thunbergii Sweet, Hort. Brit.: 247 (1826). *Rapuntium thunbergii* (Sweet) C. Presl,
        Prodr. Monogr. Lobel.: 20 (1836).
    Rapuntium coronopifolium var. *latifolium* C. Presl in Eckl. & Zeyh., Enum. Pl.
        Afric.Austral.: 399 (1837).
    Lobelia coronopifolia var. *uniflora* A. DC. in DC., Prodr. 7: 363 (1839).
    Dortmanna coronopifolia var. *normalis* Kuntze, Revis. Gen. Pl. 3(2): 187 (1898).
    Dortmanna coronopifolia f. *rosea* Kuntze, Revis. Gen. Pl. 3(2): 187 (1898); non Kuntze,
        Revis. Gen. Pl. 3(2): 187 (1898).
    Lobelia coronopifolia f. *albiflora* E. Wimm., Pflanzenr. IV.276b: 592 (1953).

Lobelia cubana Urb., Symb. Antill. 1: 455 (1899).
Cuba. 81 CUB. Nanophan.
   Tupa montana C. Wright ex Griseb., Cat. Pl. Cub.: 159 (1866); non *Lobelia montana*
        Reinw. ex Blume, Bijdr.: 728 (1826).

Lobelia cuneifolia Link & Otto, Icon. Pl. Select.: pl. 39 (1825). *Rapuntium ovatum* C. Presl,
Prodr. Monogr. Lobel.: 16 (1836); non (G. Don) C. Presl, Prodr. Monogr. Lobel.: 29 (1836).
Cape Provinces. 27 CPP. Hemicr. 2n = 42.
    Rapuntium ovatum var. *hirsutum* C. Presl in Eckl. & Zeyh., Enum. Pl. Afric. Austral.:
        397 (1837). *Lobelia cuneifolia* var. *hirsutum* (C. Presl) A. DC. in DC., Prodr. 7:
        785 (1839).
    Lobelia hederacea A. DC. in DC., Prodr. 7: 370 (1839); non Cham., Linnaea 8: 212
        (1833).
    Lobelia pilosa Schltr., Bot. Jahrb. Syst. 57: 620 (1922); non (Benth.) Kuntze, Revis. Gen.
        Pl. 2: 378 (1891).
    Lobelia sylvatica Fourc., Trans. Roy. Soc. South Africa 21: 82 (1932).
    Lobelia cuneifolia var. *ananda* E. Wimm., Pflanzenr. IV.276c: 865 (1968).

Lobelia cyanea E. Wimm., Pflanzenr. IV.276b: 779 (1953).
Madagascar. 29 MDG. Hemicr.

Lobelia cymbalarioides Engl., Bot. Jahrb. Syst. 19 (Beibl. 47): 50 (1894).
Tanzania. 25 TAN. Hemicr. $2n = 38$.
Lobelia afromontana T. C. E. Fr., Notizbl. Bot. Gart. Berlin-Dahlem 8: 401 (1923).

Lobelia cyphioides Harv., Thes. Cap. 2: 40 (1862). *Dortmanna cyphioides* (Harv.) Kuntze, Revis. Gen. Pl. 2: 972 (1891), as 'cyphiodes'.
Cape Provinces. 27 CPP. Hemicr.

Lobelia darlingensis (E. Wimm.) Albr., Telopea 5: 791 (1994).
E. Australia. 50 NSW QLD VIC. Hemicr.
**Pratia darlingensis* E. Wimm., Pflanzenr. IV.276b: 108 (1943).

Lobelia dasyphylla E. Wimm., Pflanzenr. IV.276b: 782 (1953).
Cape Provinces. 27 CPP. Hemicr.

Lobelia davidii Franch., Pl. David. 1: 191 (1883), as 'davidi'.
S. & E. China (Anhui, Zhejiang, Jiangxi, Fujian, Guangdong, Guangxi, Hunan, Hubei, Guizhou, Sichuan, Yunnan). 36 CHC CHS. Cham. or hemicr.
Lobelia dolichothyrsa Diels, Bot. Jahrb. Syst. 29: 607 (1901). *Lobelia davidii* var. *dolichothyrsa* (Diels) E. Wimm., Pflanzenr. IV.276b: 658 (1953).
Lobelia kwangsiensis E. Wimm., Ann. Naturhist. Mus. Wien 56: 367 (1948). *Lobelia davidii* var. *kwangsiensis* (E. Wimm.) Y. S. Lian, Fl. Reipubl. Popul. Sin. 73(2): 166 (1983).
Lobelia davidii var. *glaberrima* E. Wimm., Ann. Naturhist. Mus. Wien 56: 367 (1948).
Lobelia oligantha C. Y. Wu, Rep. Yunnan Trop. Subtrop. Fl. Res. Inst. 1: 96 (1965).
Lobelia davidii var. *sichuanensis* Y. S. Lian, Fl. Reipubl. Popul. Sin. 73(2): 188 (1983).

Lobelia deckenii (Asch.) Hemsl. in Oliv., Fl. Trop. Afr. 3: 466 (1877).
Tanzania. 25 TAN. Nanophan. or herb. phan. $2n = 28$.
**Tupa deckenii* Asch., Sitzungsber. Ges. Naturf. Freunde Berlin 1868: 23 (1869). *Tupa kerstenii* Vatke, Linnaea 38: 725 (1874). *Dortmanna deckenii* (Asch.) Kuntze, Revis. Gen. Pl. 2: 972 (1891).

subsp. **deckenii**
Tanzania. 25 TAN. Nanophan. or herb. phan. $2n = 28$.
Lobelia tayloriana Baker f., J. Bot. 32: 67 (1894). *Lobelia deckenii* var. *tayloriana* (Baker f.) E. Wimm., Pflanzenr. IV.276c: 883 (1968).
Lobelia deckenii var. *cacuminum* R. E. Fr. & T. C. E. Fr., Svensk Bot. Tidskr. 16: 411 (1922).

subsp. **incipiens** E. B. Knox, Contr. Univ. Michigan Herb. 19: 249 (1993).
Tanzania. 25 TAN. Nanophan. or herb. phan.

Lobelia decurrens Cav., Icon. 6: 13 (1800). *Tupa decurrens* (Cav.) G. Don in Sweet, Hort. Brit. (ed. 3): 424 (1839). *Rapuntium decurrens* (Cav.) C. Presl, Prodr. Monogr. Lobel.: 24 (1836). *Dortmanna decurrens* (Cav.) Kuntze, Rev. Gen. Pl. 2: 972 (1891).
Peru. 83 PER. Cham. or hemicr. $2n = 14$.

subsp. **decurrens**
S. Peru (Lima to Arequipa). 83 PER. Cham. or hemicr. $2n = 14$.
Lobelia foliosa Kunth in Humb., Bonpl. & Kunth, Nov. Gen. Sp. 3: 310 (quarto), 242 (folio) (1819). *Rapuntium foliosum* (Kunth) C. Presl, Prodr. Monogr. Lobel.: 24 (1836). *Lobelia decurrens* var. *foliosa* (Kunth) Heynh., Nom. Bot. Hort. 1: 471 (1840).
Lobelia decurrens var. *jaensis* E. Wimm. in J. F. Macbr., Fl. Peru 6: 478 (1937).

subsp. **parviflora** Lammers, Sida 21: 616 (2004). Cham. or hemicr.
N. Peru (Piura to Ancash and Huánuco). 83 PER.

Lobelia decurrentifolia (Kuntze) K. Schum., Just's Bot. Jahresber. 26(1): 373 (1900).
Cape Provinces. 27 CPP. Cham. or hemicr.
**Dortmanna decurrentifolia* Kuntze, Revis. Gen. Pl. 3(2): 187 (1898).

Lobelia dentata Cav., Icon. 6: 14 (1800). *Rapuntium dentatum* (Cav.) C. Presl, Prodr. Monogr.
Lobel.: 21 (1836). *Dortmanna dentata* (Cav.) Kuntze, Revis. Gen. Pl. 2: 972 (1891).
SE. Australia (New South Wales). 50 NSW. Ther. *2n* = 20.

Lobelia diastateoides McVaugh, Amer. Midl. Naturalist 24: 695 (1940).
S. Mexico (Oaxaca, Chiapas). 79 MXS MXT. Hemicr.

Lobelia diazlunae Rzed. & Calderón, Acta Bot. Méx. 40: 59 (1997).
W. Mexico (Nayarit). 79 MXS. Geophyte.

Lobelia dichroma Schltr., Bot. Jahrb. Syst. 57: 623 (1922).
Cape Provinces. 27 CPP. Cham.

Lobelia dielsiana E. Wimm., Repert. Spec. Nov. Regni Veg. 22: 194 (1926).
SW. Mexico (Guerrero). 79 MXS. Hemicr.

Lobelia digitalifolia (Griseb.) Urb., Symb. Antill. 1: 455 (1899).
Lesser Antilles (Montserrat, Guadeloupe, Dominica). 81 LEE WIN. Cham. or hemicr.
 **Tupa digitalifolia* Griseb., Fl. Brit. W.I.: 387 (1861). *Dortmanna digitalifolia* (Griseb.)
 Kuntze, Revis. Gen. Pl. 2: 972 (1891).
 Lobelia guadeloupensis Urb., Symb. Antill. 1: 454 (1899). *Lobelia digitalifolia* var.
 guadeloupensis (Urb.) McVaugh, N. Amer. Fl. 32A: 90 (1943).

Lobelia dioica R. Br., Prodr.: 565 (1810). *Holostigma dioicum* (R. Br.) G. Don, Gen. Hist.
3: 716 (1834), as 'dioica'. *Monopsis dioica* (R. Br.) C. Presl, Prodr. Monogr. Lobel.: 11
(1836), as 'dioeca'. *Dortmanna dioica* (R. Br.) Kuntze, Revis. Gen. Pl. 2: 972 (1891),
as 'dioeca'.
NW. Australia. 50 NTA WAU. Hemicr.

Lobelia dissecta M. B. Moss, Bull. Misc. Inform. Kew 1929: 197 (1929).
Sudan to Burundi. 23 BUR 24 SUD 25 UGA. Hemicr. or ther.
 ' *Lobelia dissecta* subsp. *humidulorum* Friis & Vollesen, Kew Bull. 37: 471 (1982).

Lobelia divaricata Hook. & Arn., Bot. Beechey Voy.: 301 (1838). *Dortmanna divaricata*
(Hook. & Arn.) Kuntze, Revis. Gen. Pl. 2: 972 (1891). Ther.
C. Mexico (Coahuila to Nayarit, Michoacán & México State). 79 MXC MXE MXS.
 Rapuntium pusillum C. Presl, Prodr. Monogr. Lobel.: 22 (1836). *Lobelia pusilla* (C. Presl)
 A. DC. in DC., Prodr. 7: 379 (1839); non G. Don, Gen. Hist. 3: 711 (1834).
 Dortmanna pusilla (C. Presl) Kuntze, Revis. Gen. Pl. 2: 973 (1891).
 Lobelia schmitzii E. Wimm., Repert. Spec. Nov. Regni Veg. 26: 2 (1929).
 Lobelia seleriana E. Wimm., Repert. Spec. Nov. Regni Veg. 38: 82 (1935). *Lobelia*
 berlandieri var. *seleriana* (E. Wimm.) E. Wimm., Pflanzenr. IV.276b: 465 (1953).
 Lobelia trivialis E. Wimm., Repert. Spec. Nov. Regni Veg. 38: 83 (1935).

Lobelia djurensis Engl. & Diels, Bot. Jahrb. Syst. 26: 116 (1898).
Senegal to Sudan. 22 GHA GUI IVO MLI NGA SEN TOG 23 CAF 24 SUD. Ther.
 Lobelia baoulensis A. Chev., Mem. Soc. Bot. France 58 (Mém. 8d): 180 (1912).

Lobelia dodiana E. Wimm., Pflanzenr. IV.276c: 879 (1968).
Cape Provinces. 27 CPP. Ther.
 Laurentia radicans Schönland, Trans. Roy. Soc. South Africa 1: 446 (1910); non *Lobelia*
 radicans Thunb., Trans. Linn. Soc. London 2: 330 (1794). *Lobelia dodiana* var.
 radicans (Schönland) E. Wimm., Pflanzenr. IV.276c: 879 (1968).

Lobelia donanensis P. Royen, Kew Bull. 20: 305 (1966).
New Guinea. 43 NWG. Hemicr.

Lobelia doniana Skottsb., Acta Horti Gothob. 4: 19 (1928). *Lobelia seguinii* var. *doniana* (Skottsb.) E. Wimm., Ann. Naturhist. Mus. Wien 56: 367 (1948).
Himalaya to SC. China (Yunnan) & Myanmar. 36 CHC CHT 40 ASS EHM IND NEP 41 MYA. Cham. or hemicr.

Lobelia dortmanna L., Sp. Pl.: 929 (1753). *Lobelia lacustris* Salisb., Prodr. Stirp. Chap. Allerton: 128 (1796). *Dortmanna lacustris* O. O. Rudbeck ex G. Don, Gen. Hist. 3: 715 (1834). *Rapuntium dortmanna* (L.) C. Presl, Prodr. Monogr. Lobel.: 18 (1836).
North America (British Columbia to Oregon, Maryland & Newfoundland); Føroyar; British Is.; France to Belarus & N. Russia. 10 DEN FIN FOR GRB IRE NOR SWE 11 BGM GER NET POL 12 FRA 14 BLR BLT RUW RUN 71 BRC MAN SAS 72 ALL 73 ORE WAS 74 MIN WIS 75 CNT MAI MAS NWH NWJ NWY PEN RHO VER 78 MRY. Hydrohemicr. $2n = 14$.
Lobelia dortmanna f. *terrestris* Sylvèn, Bot. Tidsskr. 21: 99 (1897) .

Lobelia dregeana (C. Presl) A. DC. in DC., Prodr. 7: 371 (1839).
S. Africa. 27 CPP LES NAM OFS TVL. Cham. or hemicr.
  * *Rapuntium dregeanum* C. Presl, Prodr. Monogr. Lobel.: 22 (1836). *Dortmanna dregeana* (C. Presl) Kuntze, Revis. Gen. Pl. 2: 972 (1891).
  Lobelia omphalodoides Schltr., Bot. Jahrb. Syst. 57: 616 (1922).

Lobelia dressleri Wilbur, Ann. Missouri Bot. Gard. 61: 889 (1974).
Panama. 80 PAN. Hemicr.

Lobelia dunbariae Rock, Monogr. Stud. Haw. Lobelioid.: 133 (1919), as 'dunbarii'.
Neowimmeria dunbariae (Rock) O. Deg. & I. Deg., Phytologia 12: 73 (1965).
Hawaiian Is. (Moloka'i). 63 HAW. Nanophan.

subsp. **dunbariae**
Hawaiian Is. (Moloka'i). 63 HAW. Nanophan.

subsp. **paniculata** (Rock) Lammers, Syst. Bot. 13: 506 (1988).
Hawaiian Is. (Moloka'i). 63 HAW. Nanophan.
  * *Lobelia hillebrandii* var. *paniculata* Rock, Monogr. Stud. Haw. Lobelioid.: 135 (1919). *Neowimmeria paniculata* (Rock) O. Deg. & I. Deg., Prodr. Galeatella Neowimmeria: 10 (1974).
  Lobelia costata E. Wimm., Ann. Naturhist. Mus. Wien 56: 369 (1948). *Neowimmeria costata* (E. Wimm.) O. Deg. & I. Deg., Phytologia 12: 73 (1965).

Lobelia duriprati T. C. E. Fr., Notizbl. Bot. Gart. Berlin-Dahlem 8: 407 (1923).
Kenya, Uganda, Tanzania. 25 ALL. Geophyte. $2n = 38$.

Lobelia ehrenbergii Vatke, Linnaea 38: 719 (1874). *Dortmanna ehrenbergii* (Vatke) Kuntze, Revis. Gen. Pl. 2: 972 (1891).
N. Mexico (Chihuahua to Tamaulipas & Hidalgo). 79 MXE. Hemicr.

subsp. **ehrenbergii**
NE. Mexico (Nuevo León, Tamaulipas, San Luis Potosí, Hidalgo). 79 MXE. Hemicr.

subsp. **gracilens** (A. Gray) Lammers, Novon 16: 71 (2006).
NW. Mexico (Chihuahua, Durango). 79 MXE. Hemicr.
  * *Lobelia gracilens* A. Gray, Proc. Amer. Acad. Arts 21: 393 (1886). *Lobelia ehrenbergii* var. *gracilens* (A. Gray) McVaugh, Amer. Midl. Naturalist 24: 695 (1940).

Lobelia elongata Small, Fl. S.E. U.S.: 1144 (1903).
Louisiana to Georgia & Delaware. 78 ALA DEL GEO LOU MRY MSI NCA SCA VRG. Hemicr. $2n = 14, 28$.
  Lobelia glandulosa var. *glabra* A. DC. in DC., Prodr. 7: 378 (1839).
  Lobelia georgiana var. *azurea* E. Wimm., Pflanzenr. IV.276b: 429 (1953).

Lobelia endlichii (E. Wimm.) T. J. Ayers, Syst. Bot. 15: 319 (1990).
NW. Mexico (Chihuahua). 79 MXE. Ther. $2n = 14$.
Heterotoma gibbosa S. Watson, Proc. Amer. Acad. Arts 23: 280 (1888); non *Lobelia gibbosa* Labill., Nov. Holl. Pl. 1: 50 (1805). *Calcaratolobelia gibbosa* (S. Watson) Wilbur, Sida 17: 563 (1997).
* *Heterotoma endlichii* E. Wimm., Repert. Spec. Nov. Regni Veg. 26: 1 (1929).

Lobelia erectiuscula H. Hara, J. Jap. Bot. 40: 328 (1965).
Nepal to Myanmar. 36 CHT 40 ASS EHM IND NEP 41 MYA. Nanophan. or cham.
* *Lobelia erecta* Hook. f. & Thomson, J. Proc. Linn. Soc., Bot. 2: 28 (1858); non de Vriese in Lehm., Pl. Preiss.: 395 (1845). *Dortmanna erecta* Kuntze, Revis. Gen. Pl. 2: 972 (1891).

Lobelia erinus L., Sp. Pl.: 932 (1753). *Rapuntium erinus* (L.) Mill., Gard. Dict. (ed. 8) (1768), as 'erinum'. *Dortmanna erinus* (L.) Kuntze, Revis. Gen. Pl. 2: 972 (1891).
Senegal to Somalia & S. Africa; cult.; naturalized in Macaronesia, Australia & Hawaiian Is. (21) azo mdr 22 NGR SEN 23 ZAI 24 SOM SUD 25 TAN 26 ALL 27 ALL (50) soa (63) haw. Hemicr. or ther. $2n = 14, 28, 42$.
Lobelia erinoides L., Sp. Pl.: 932 (1753). *Rapuntium erinoides* (L.) Mill., Gard. Dict. (ed. 8) (1768). *Lobelia erinifolia* Salisb., Prodr. Stirp. Chap. Allerton: 129 (1796). *Monopsis conspicua* Salisb., Trans. London Hort. Soc. 2: 37 (1817). *Enchysia erinoides* (L.) C. Presl, Prodr. Monogr. Lobel.: 40 (1836). *Grammatotheca erinoides* (L.) Sond. in Harv. & Sond., Fl. Cap. 3: 532 (1865). *Monopsis debilis* var. *conspicua* Sond. in Harv. & Sond., Fl. Cap. 3: 534 (1865). *Laurentia erinoides* (L.) G. Nichols., Ill. Dict. Gard. 2: 238 (1885). *Dortmanna erinoides* (L.) Kuntze, Revis. Gen. Pl. 2: 972 (1891). *Monopsis simplex* var. *conspicua* (Sond.) E. Wimm., Ann. Naturhist. Mus. Wien 56: 370. 1948.
Lobelia bellidifolia L. f., Suppl. Pl.: 396 (1782), as 'bellidiflora'. *Rapuntium bellidifolium* (L. f.) C. Presl, Prodr. Monogr. Lobel.: 15 (1836). *Lobelia erinus* var. *bellidifolia* (L. f.) Sond. in Harv. & Sond., Fl. Cap. 3: 544 (1865). *Dortmanna erinus* var. *bellidifolia* (L. f.) Kuntze, Revis. Gen. Pl. 3(2): 187 (1898).
Lobelia bicolor Sims, Bot. Mag. 15: tab. 514 (1801). *Rapuntium bicolor* (Sims) C. Presl, Prodr. Monogr. Lobel.: 17 (1836). *Lobelia erinus* f. *bicolor* (Sims) Zahlbr., Ann. K. K. Naturhist. Hofmus. 18: 405 (1903).
Lobelia schrankii Sweet, Hort. Brit. (ed. 2): 323 (1830). *Lobelia erinus* var. *schrankii* (Sweet) E. Wimm., Ann. Naturhist. Mus. Wien 56: 351 (1948).
Lobelia procumbens J. Forbes, Hort. Woburn.: 33 (1833).
Rapuntium flexuosum C. Presl, Prodr. Monogr. Lobel.: 16 (1836); non C. Presl, Prodr. Monogr. Lobel.: 23 (1836). *Lobelia natalensis* A. DC. in DC., Prodr. 7: 369 (1839); non *Lobelia flexuosa* (C. Presl) A. DC. in DC., Prodr. 7: 378 (1839). *Dortmanna flexuosa* Kuntze, Revis. Gen. Pl. 2: 972 (1891). *Lobelia filiformis* var. *natalensis* (A. DC.) E. Wimm., Ann. Naturhist. Mus. Wien 56: 354 (1948).
Rapuntium acutangulum C. Presl, Prodr. Monogr. Lobel.: 16 (1836). *Lobelia acutangula* (C. Presl) A. DC. in DC., Prodr. 7: 369 (1839).
Lobelia parvifolia Raf., New Fl. N. Amer. 2: 18 (1837); non P. J. Bergius, Descr. Pl. Cap.: 345 (1767); nec R. Br., Prodr.: 564 (1810).
Lobelia algoensis A. DC. in DC., Prodr. 7: 369 (1839).
Lobelia bracteolata A. DC. in DC., Prodr. 7: 371 (1839). *Lobelia erinus* var. *bracteolata* (A. DC.) E. Wimm., Ann. Naturhist. Mus. Wien 56: 350 (1948).
Lobelia senegalensis A. DC. in DC., Prodr. 7: 372 (1839). *Lobelia kohautiana* Vatke, Linnaea 38: 719 (1874). *Dortmanna senegalensis* (A. DC.) Kuntze, Revis. Gen. Pl. 2: 973 (1891).
Lobelia erinus var. *grandiflora* Paxton, Paxton's Mag. Bot. 10: 75 (1843).
Lobelia lavendulacea Klotzsch in Peters, Naturw. Reise Mossambique 6: 302 (1861). *Dortmanna lavendulacea* (Klotzsch) Kuntze, Revis. Gen. Pl. 2: 972 (1891).
Lobelia natalensis var. *subulifolia* Sond. in Harv. & Sond., Fl. Cap. 3: 545 (1865).
Lobelia nuda Hemsl. in Oliv., Fl. Trop. Afr. 3: 469 (1877).
Lobelia erinus "section" *compacta* G. Nichols., Ill. Dict. Gard. 2: 290 (1885).

Lobelia erinus "section" *paxtoniana* G. Nichols., Ill. Dict. Gard. 2: 291 (1885).
Lobelia erinus "section" *pumila* G. Nichols., Ill. Dict. Gard. 2: 291 (1885).
Lobelia erinus "section" *ramosoides* G. Nichols., Ill. Dict. Gard. 2: 291 (1885).
Lobelia erinus "section" *speciosa* G. Nichols., Ill. Dict. Gard. 2: 291 (1885).
Lobelia pubescens var. *simplex* Kuntze, Jahrb. Königl. Bot. Gart. Berlin 4: 267 (1886).
Lobelia dortmannii Dammann, Wiener Ill. Gart.-Zeitung 19: 459 (1894).
Dortmanna debilis var. *natalensis* Kuntze, Revis. Gen. Pl. 3(2): 187 (1898). *Lobelia benguellensis* Hiern, Cat. Afr. Pl. 1: 626 (1898); non *Lobelia natalensis* A. DC. in DC., Prodr. 7: 369 (1839).
Lobelia chilawana Schinz, Mém. Herb. Boissier 10: 70 (1900).
Lobelia rosulata S. Moore, J. Bot. 41: 402 (1903). *Lobelia nuda* var. *rosulata* (S. Moore) E. Wimm., Kew Bull. 1952: 138 (1952).
Lobelia jugosa S. Moore, J. Linn. Soc., Bot. 40: 125 (1911).
Lobelia transvaalensis Schltr., Bot. Jahrb. Syst. 57: 622 (1922).
Lobelia trierarchii R. D. Good, J. Bot. 62: 139 (1924), as 'trierarchi'.
Lobelia turgida E. Wimm., Repert. Spec. Nov. Regni Veg. 38: 82 (1935). *Lobelia senegalensis* var. *turgida* (E. Wimm.) E. Wimm., Pflanzenr. IV.276b: 552 (1953).
Lobelia rosulata f. *hirtella* E. Wimm., Notizbl. Bot. Gart. Berlin-Dahlem 15: 634 (1941).
Lobelia turgida var. *peteri* E. Wimm., Notizbl. Bot. Gart. Berlin-Dahlem 15: 634 (1941).
Lobelia lydenburgensis E. Wimm., Ann. Naturhist. Mus. Wien 56: 342 (1948).
Lobelia nuda f. *hirtella* E. Wimm., Ann. Naturhist. Mus. Wien 56: 343 (1948).
Lobelia erinus f. *elata* E. Wimm., Ann. Naturhist. Mus. Wien 56: 350 (1948).
Lobelia erinus var. *subvillosa* E. Wimm., Ann. Naturhist. Mus. Wien 56: 350 (1948).
Lobelia filiformis f. *albiflora* E. Wimm., Ann. Naturhist. Mus. Wien 56: 355 (1948).
Lobelia filiformis f. *multipilis* E. Wimm., Ann. Naturhist. Mus. Wien 56: 355 (1948).
Lobelia filiformis f. *muzandazora* E. Wimm., Ann. Naturhist. Mus. Wien 56: 355 (1948).
Lobelia montaguensis E. Wimm., Ann. Naturhist. Mus. Wien 56: 357 (1948).
Lobelia altimontis E. Wimm., Kew Bull. 1952: 138 (1952).
Lobelia polyodon E. Wimm., Kew Bull. 1952: 140 (1952).
Lobelia wildii E. Wimm., Kew Bull. 1952: 141 (1952).
Lobelia wildii var. *arcana* E. Wimm., Kew Bull. 1952: 141 (1952).
Lobelia raridentata E. Wimm., Pflanzenr. IV.276b: 538 (1953).
Lobelia keilhackii E. Wimm., Pflanzenr. IV.276b: 543 (1953).
Lobelia nuzana E. Wimm., Pflanzenr. IV.276b: 780 (1953).

Lobelia erlangeriana Engl., Bot. Jahrb. Syst. 32: 118 (1902).
Ethiopia. 24 ETH. Ther.

Lobelia eryliae C. E. C. Fisch., Bull. Misc. Inform. Kew 1928: 141 (1928).
Sulawesi. 42 SUL. Nanophan.
Lobelia epilobioides E. Wimm., Repert. Spec. Nov. Regni Veg. 38: 79 (1935).
Lobelia epilobioides var. *sarasinorum* E. Wimm., Ann. Naturhist. Mus. Wien 56: 367 (1948).

Lobelia eurypoda E. Wimm., Ann. Naturhist. Mus. Wien 56: 362 (1948).
S. Africa. 27 NAT SWZ TVL. Cham. or hemicr.

Lobelia exaltata Pohl, Pl. Bras. Icon. Descr. 2: 101 (1831). *Rapuntium exaltatum* (Pohl) C. Presl, Prodr. Monogr. Lobel.: 24 (1836). *Haynaldia exaltata* (Pohl) Kanitz, Magyar Növényt. Lapok 1: 4 (1877). *Euhaynaldia exaltata* (Pohl) Borbás in Pall., Nagy Lex. 8: 779 (1894). *Dortmanna exaltata* (Pohl) Kuntze, Revis. Gen. Pl. 2: 972 (1891).
Brazil (Minas Gerais, Rio de Janeiro, São Paulo, Paraná). 84 BZL BZS. Herb. phan. or hemicr. $2n = 28$.
Haynaldia exaltata var. *ramosa* Kanitz in Mart., Fl. Bras. 6(4): 141 (1878).

Lobelia excelsa Bonpl., Descr. Pl. Malmaison: 112 (1816). *Rapuntium excelsum* (Bonpl.) C. Presl, Prodr. Monogr. Lobel.: 29 (1836). *Dortmanna excelsa* (Bonpl.) Kuntze, Revis. Gen. Pl. 2: 972 (1891).

NC. Chile (Elqui to Talca). 85 CLC CLN. Phan. or nanophan. $2n = 42$.
- *Lobelia gigantea* Sims, Bot. Mag. 32: tab. 1325 (1810); non Cav., Anales Hist. Nat. 2: 104 (1800). *Lobelia salicifolia* Sweet, Hort. Suburb. Lond.: 37 (1818). *Tupa salicifolia* (Sweet) G. Don, Gen. Hist. 3: 700 (1834).
- *Lobelia arguta* Lindl., Bot. Reg. 12: pl. 973 (1826). *Tupa arguta* (Lindl.) G. Don, Gen. Hist. 3: 700 (1834). *Dortmanna arguta* (Lindl.) Kuntze, Revis. Gen. Pl. 2: 972 (1891).
- *Lobelia neriifolia* Moris, Enum. Sem. Hort. Bot. Taurin. 1833: 20 (1833).
- *Tupa salicifolia* var. *purpurea* Vatke, Linnaea 38: 726 (1874).
- *Tupa glaucescens* Phil., Anales Univ. Chile 90: 187 (1895).

Lobelia exilis Hochst. ex A. Rich., Tent. Fl. Abyss. 2: 7 (1850).
Ethiopia. 24 ETH. Ther.
- *Lobelia tenerrima* Chiov., Ann. Bot. (Rome) 9: 78 (1911).
- *Lobelia exilis* var. *pusilla* E. Wimm., Ann. Naturhist. Mus. Wien 56: 347 (1948).
- *Lobelia exilis* var. *major* E. Wimm., Ann. Naturhist. Mus. Wien 56: 347 (1948).

Lobelia fangiana (E. Wimm.) S. Y. Hu, J. Arnold Arbor. 61: 90 (1980).
SC. China (Sichuan). 36 CHC. Cham. or hemicr.
- * *Pratia fangiana* E. Wimm., Repert. Spec. Nov. Regni Veg. 38: 3 (1935).
- *Lobelia omiensis* E. Wimm., Ann. Naturhist. Mus. Wien 56: 366 (1948).

Lobelia fastigiata Kunth in Humb., Bonpl. & Kunth, Nov. Gen. Sp. 3: 313 (quarto), 244 (folio) (1819). *Rapuntium fastigiatum* (Kunth) C. Presl, Prodr. Monogr. Lobel.: 15 (1836). *Dortmanna fastigiata* (Kunth) Kuntze, Revis. Gen. Pl. 2: 972 (1891).
Honduras to Ecuador, Paraguay & E. Brazil. 80 HON PAN 81 TRT 82 VEN 83 CLM ECU 84 BZC BZE 85 PAR. Ther.
- *Lobelia tenuifolia* Willd. ex Schult. in Roem. & Schult., Syst. Veg. 5: 56 (1819).
- *Lobelia trinitensis* Griseb., Fl. Brit. W.I.: 385 (1861). *Dortmanna trinitensis* (Kunth) Kuntze, Revis. Gen. Pl. 2: 380 (1891).
- *Lobelia gardneriana* Kanitz in Mart., Fl. Bras. 6(4): 138 (1878). *Dortmanna gardneriana* (Kanitz) Kuntze, Revis. Gen. Pl. 2: 972 (1891).
- *Lobelia gardneriana* var. *foliosa* Zahlbr., Bull. Herb. Boissier (ser. 2) 3: 922 (1903).

Lobelia fawcettii Urb., Symb. Antill. 1: 452 (1899). *Pratia fawcettii* (Urb.) McVaugh, N. Amer. Fl. 32A: 112 (1943).
Jamaica. 81 JAM. Nanophan. or cham.

Lobelia feayana A. Gray, Proc. Amer. Acad. Arts 12: 60 (1876). *Dortmanna feayana* (A. Gray) Kuntze, Revis. Gen. Pl. 2: 972 (1891).
SE. U.S.A. (Florida). 78 FLA. Hemicr. $2n = 14$.

Lobelia fenestralis Cav., Icon. 6: 8 (1800). *Rapuntium fenestrale* (Cav.) C. Presl, Prodr. Monogr. Lobel.: 13 (1836). *Dortmanna fenestralis* (Cav.) Kuntze, Revis. Gen. Pl. 2: 972 (1891).
SW. U.S.A. (Arizona, New Mexico, Texas) to S. Mexico. 76 ARI 77 ALL 79 MXC MXE MXS. Ther.
- *Lobelia crispa* Graham, Edinburgh New Philos. J. 1: 173 (1826).
- *Lobelia spicata* Ruiz & Pav. ex G. Don, Gen. Hist. 3: 705 (1834); non Lam., Encycl. 3: 587 (1792). *Rapuntium spicatum* C. Presl, Prodr. Monogr. Lobel.: 31 (1836).
- *Lobelia stricta* M. Martens & Galeotti, Bull. Acad. Roy. Sci. Bruxrelles 9(2): 43 (1842); non Sw., Prodr.: 117 (1788); nec R. Br., Prodr.: 564 (1810).
- *Lobelia pectinata* Engelm. in Wislizenus, Mem. Tour N. Mexico: 108 (1848). *Lobelia fenestralis* var. *pectinata* (Engelm.) E. Wimm., Ann. Naturhist. Mus. Wien 56: 358 (1948).
- *Lobelia spicata* Moç. in Sessé & Moç., Pl. Nov. Hisp.: 151 (1890); non Lam., Encycl. 3: 587 (1792); nec Ruiz & Pav. ex G. Don, Gen. Hist. 3: 705 (1834).

Lobelia fervens Thunb., Fl. Cap. 2: 46 (1818). *Rapuntium fervens* (Thunb.) C. Presl, Prodr. Monogr. Lobel.: 30 (1836). *Dortmanna fervens* (Thunb.) Kuntze, Revis. Gen. Pl. 2: 972 (1891). Angola to Ethiopia & Mozambique; Comoros, Madagascar & Mauritius; naturalized in Brazil. 23 BUR ZAI 24 ETH SOM 25 ALL 26 ANG MOZ ZAM ZIM 29 COM MAU MFG (84) bzl. Hemicr. or ther. $2n$ = 12, 14.

   subsp. **fervens**
      Ethiopia to Mozambique; Comoros, Madagascar, Mauritius; naturalized in Brazil. 23 BUR ZAI 24 ETH SOM 25 KEN TAN UGA 26 ANG MOZ ZAM ZIM 29 COM MAU MFG (84) bzl. Hemicr. or ther. $2n$ = 12, 14.
      Lobelia madagascariensis Schult. in Roem. & Schult., Syst. Veg. 5: 67 (1819). *Rapuntium platycaulum* C. Presl, Prodr. Monogr. Lobel.: 14 (1836). *Dortmanna madagascariensis* (Schult.) Kuntze, Revis. Gen. Pl. 2: 972 (1891).
      Lobelia pterocaulon Klotzsch in Peters, Naturw. Reise Mossambique 6: 299(1861).
      Lobelia subulata Klotzsch in Peters, Naturw. Reise Mossambique 6: 300 (1861); non Benth., Pl. Hartweg.: 137 (1844).
      Lobelia asperulata Klotzsch in Peters, Naturw. Reise Mossambique 6: 300 (1861). *Lobelia fervens* var. *asperulata* (Klotzsch) Sond. in Harv. & Sond., Fl. Cap. 3: 548 (1865). *Lobelia anceps* var. *asperulata* (Klotzsch) E. Wimm., Ann. Naturhist. Mus. Wien 56: 346 (1948).
      Lobelia humilis Klotzsch in Peters, Naturw. Reise Mossambique 6: 301 (1861).
      Lobelia petersiana Klotzsch in Peters, Naturw. Reise Mossambique 6: 302 (1861).
      Lobelia trialata var. *grandiflora* Chiov., Result. Scientif. Missione Stefanini-Paoli Somal. Ital. 1: 108 (1916).
      Lobelia mearnsii De Wild., Pl. Bequaert. 1: 293 (1922), as 'mearnsi'.
      Lobelia odontoptera Schltr., Bot. Jahrb. Syst. 57: 620 (1922).
      Lobelia anceps f. *pleniflora* E. Wimm., Kew Bull. 1952: 137 (1952).
      Lobelia odontoptera var. *depilis* E. Wimm., Pflanzenr. IV.276b: 468 (1953).

   subsp. **recurvata** (E. Wimm.) Thulin in Launert, Fl. Zambes. 7(1): 133 (1983).
      Angola to Zaïre, Kenya & Mozambique. 23 BUR ZAI 25 KEN TAN UGA 26 ANG MOZ ZAM ZIM. Hemicr. or ther.
      * *Lobelia anceps* var. *recurvata* E. Wimm., Notizbl. Bot. Gart. Berlin-Dahlem 12: 103 (1934).

Lobelia filicaulis (C. Presl) Schönland, Phan. Fl. Uitenhage: 97 (1919).
   Cape Provinces. 27 CPP. Ther.
      * *Mezleria filicaulis* C. Presl, Prodr. Monogr. Lobel.: 7 (1836).
      Rapuntium ecklonianum C. Presl in Eckl. & Zeyh., Enum. Pl. Afric. Austral.: 396 (1837). *Lobelia eckloniana* (C. Presl) D. Dietr., Syn. Pl. 1: 736 (1839). *Isolobus ecklonianus* (C. Presl) Sond. in Harv. & Sond., Fl. Cap. 3: 536 (1865). *Dortmanna eckloniana* (C. Presl) Kuntze, Revis. Gen. Pl. 2: 971 (1891).
      Isolobus sparsiflorus A. DC. in DC., Prodr. 7: 353 (1839).

Lobelia filiformis Lam., Encycl. 3: 588 (1792). *Rapuntium filiforme* (Lam.) C. Presl, Prodr. Monogr. Lobel.: 18 (1836). *Dortmanna filiformis* (Lam.) Kuntze, Revis. Gen. Pl. 2: 972 (1891). Madagascar, Réunion, Mauritius. 29 MAU MDG REU. Hemicr.
      Lobelia polymorpha Bory, Voy. Îles Afrique 2: 138 (1804). *Rapuntium boryanum* C. Presl, Prodr. Monogr. Lobel.: 24 (1836).
      Lobelia filiformis var. *subaquatilis* E. Wimm., Pflanzenr. IV.276b: 780 (1953).

Lobelia filipes E. Wimm., Pflanzenr. IV.276c: 863 (1968).
   Namibia. 27 NAM. Ther.

Lobelia fistulosa Vell., Fl. Flumin., Icon. 8: pl. 157 (1831). *Dortmanna fistulosa* (Vell.) Kuntze, Revis. Gen. Pl. 2: 972 (1891).
   Brazil (Minas Gerais, Rio de Janeiro, São Paulo, Rio Grande do Sul). 84 BZL BZS. Herb. phan. or cham. $2n$ = 28.

Lobelia uranocoma Cham., Linnaea 8: 321 (1833). *Rapuntium uranocomum* (Cham.) C. Presl, Prodr. Monogr. Lobel.: 24 (1836). *Haynaldia uranocoma* (Cham.) Kanitz, Magyar. Növényt. Lapok 1: 4 (1877). *Dortmanna uranocoma* (Cham.) Kuntze, Revis. Gen. Pl. 2: 973 (1891).

Lobelia flaccida (C. Presl) A. DC. in DC., Prodr. 7: 360 (1839).
Sudan to S. Africa. 24 SUD 25 KEN UGA 26 MOZ ZIM 27 CPP LES NAT OFS SWZ TVL. Hemicr. or ther.
* *Rapuntium flaccidum* C. Presl, Prodr. Monogr. Lobel.: 13 (1836).

subsp. **flaccida**
S. Africa. 27 CPP LES NAT OFS SWZ TVL. Hemicr. or ther.
Rapuntium scabripes C. Presl, Prodr. Monogr. Lobel.: 17 (1836). *Lobelia scabripes* (C. Presl) A. DC. in DC., Prodr. 7: 369 (1839). *Lobelia flaccida* var. *scabripes* (C. Presl) E. Wimm., Pflanzenr. IV.276b: 509 (1953).
Rapuntium bellidifolium var. *brevidens* C. Presl in Eckl. & Zeyh., Enum. Pl. Afr. Austral.: 396 (1837). *Lobelia bellidifolia* var. *brevidens* (C. Presl) A. DC. in DC., Prodr. 7: 785 (1839).
Lobelia bellidifolia var. *glabrata* C. Presl ex A. DC. in DC., Prodr. 7: 369 (1839).
Lobelia bellidifolia var. *hirsuta* C. Presl ex A. DC. in DC., Prodr. 7: 369 (1839). *Lobelia flaccida* var. *hirsuta* (C. Presl ex A. DC.) E. Wimm., Pflanzenr. IV.276b: 509 (1953).
Lobelia filiformis var. *krebsiana* E. Wimm., Pflanzenr. IV.276b: 542 (1953).
Lobelia bellidifolia f. *flexuosa* Zahlbr., Ann. K. K. Naturhist. Hofmus. 18: 407 (1903).
Lobelia bellidifolia f. *stricta* Zahlbr., Ann. K. K. Naturhist. Hofmus. 18: 407 (1903). *Lobelia flaccida* var. *stricta* (Zahlbr.) E. Wimm., Pflanzenr. IV.276b: 509 (1953).
Lobelia knysnensis Schltr., Bot. Jahrb. Syst. 57: 618 (1922).
Lobelia flaccida var. *caffra* E. Wimm., Pflanzenr. IV.276b: 509 (1953).
Lobelia flaccida f. *densifoliata* E. Wimm., Pflanzenr. IV.276b: 509 (1953).
Lobelia flaccida f. *laxifoliata* E. Wimm., Pflanzenr. IV.276b: 509 (1953).
Lobelia filiformis f. *rusticana* E. Wimm., Pflanzenr. IV.276b: 542 (1953).

subsp. **granvikii** (T. C. E. Fr.) Thulin, Nordic J. Bot. 3: 375 (1983).
Sudan, Kenya, Uganda. 24 SUD 25 KEN UGA. Hemicr. or ther.
* *Lobelia granvikii* T. C. E. Fr., Notizbl. Bot. Gart. Berlin-Dahlem 8: 408 (1923).
Lobelia anceps f. *ugandensis* E. Wimm., Kew Bull. 1952: 137 (1952).
Lobelia melleri var. *grossidens* E. Wimm., Kew Bull. 1952: 139 (1952).
Lobelia orbiculata E. Wimm., Kew Bull. 1952: 139 (1952).
Lobelia orbiculata f. *subcuneata* E. Wimm., Kew Bull. 1952: 140 (1952).

subsp. **mossiana** (R. D. Good) Thulin in Launert, Fl. Zambes. 7(1): 129 (1983).
Mozambique to S. Africa. 26 MOZ ZIM 27 NAT SWZ TVL.
* *Lobelia mossiana* R. D. Good, J. Bot. 62: 49 (1924).
Lobelia senegalensis var. *subaspera* E. Wimm., Pflanzenr. IV.276b: 552 (1953).

Lobelia flaccidifolia Small, Bull. Torrey Bot. Club 24: 338 (1897).
S. U.S.A. (Texas to Florida). 77 TEX 78 ALA FLA GEO LOU. Hemicr. $2n = 14$.
Lobelia ludoviciana A. Gray, Proc. Amer. Acad. Arts 12: 60 (1876); non A. W. Wood, Class-book Bot.: 476 (1861). *Dortmanna ludoviciana* Kuntze, Revis. Gen. Pl. 2: 972 (1891). *Lobelia halei* Small, Fl. S.E. U.S.: 1145 (1903).

Lobelia flexicaulis Rzed. & Calderón, Acta Bot. Mex. 55: 35 (2001).
S. Mexico (México State, Guerrero). 79 MXC MXS. Hemicr. or geophyte.

Lobelia flexuosa (C. Presl) A. DC. in DC., Prodr. 7: 378 (1839).
W. Mexico. 79 MXE MXS. Ther. $2n = 14$.

Rapuntium flexuosum C. Presl, Prodr. Monogr. Lobel.: 23 (1836); non C. Presl, Prodr. Monogr. Lobel.: 16 (1836). *Dortmanna mexicana* Kuntze, Revis. Gen. Pl. 2: 972 (1891); non *Dortmanna flexuosa* Kuntze, Revis. Gen. Pl. 2: 972 (1891). *Heterotoma flexuosa* (C. Presl) McVaugh, Bull. Torrey Bot. Club 67: 143 (1940). *Calcaratolobelia flexuosa* (C. Presl) Wilbur, Sida 17: 563 (1997).

subsp. **flexuosa**
W. Mexico (Nayarit, Jalisco, Oaxaca). 79 MXE MXS. Ther. *2n* = 14.
Lobelia arabidoides Hook. & Arn., Bot. Beechey Voy.: 301 (1838). *Heterotoma arabidoides* (Hook. & Arn.) Benth. & Hook. ex Hemsl., Biol. Cent.-Amer., Bot. 2: 269 (1881).
Heterotoma flexuosa var. *liebmanniana* E. Wimm., Pflanzenr. IV.276b: 718 (1953).

subsp. **intermedia** (Hemsl.) Lammers, Novon 16: 71 (2006).
W. Mexico (SW. Durango). 79 MXE MXS. Ther. *2n* = 14.
**Heterotoma intermedia* Hemsl., Biol. Cent.-Amer., Bot. 2: 269 (1881). *Heterotoma cordifolia* var. *intermedia* (Hemsl.) E. Wimm., Pflanzenr. IV.276b: 717 (1953). *Lobelia flexuosa* var. *intermedia* (Hemsl.) T. J. Ayers, Syst. Bot. 15: 321 (1990). *Calcaratolobelia flexuosa* var. *intermedia* (Hemsl.) Wilbur, Sida 17: 563 (1997).

Lobelia floridana Chapm., Bot. Gaz. (Crawfordsville) 3: 9 (1878). *Lobelia paludosa* var. *floridana* (Chapm.) A. Gray, Syn. Fl. N. Amer. 2(1): 394 (1878).
SE. U.S.A. (Texas to North Carolina). 77 TEX 78 ALA FLA GEO LOU MSI NCA. Hemicr. *2n* = 42.

Lobelia foliiformis T. J. Zhang & D. Y. Hong, Acta Phytotax. Sin. 30: 155 (1992).
SC. China (Yunnan). 36 CHC. Nanophan.

Lobelia galpinii Schltr., Bot. Jahrb. Syst. 57: 615 (1922).
S. Africa. 27 CPP LES. Hemicr.
**Lobelia aquatica* E. Phillips, Ann. S. African Mus. 16: 175 (1917); non Cham., Linnaea 8: 211 (1833).
Lobelia galpinii var. *detonsa* E. Wimm., Pflanzenr. IV.276c: 861 (1968).

Lobelia gattingeri A. Gray, Proc. Amer. Acad. Arts 17: 221 (1882). *Dortmanna gattingeri* (A. Gray) Kuntze, Revis. Gen. Pl. 2: 972 (1891). *Lobelia appendiculata* var. *gattingeri* (A. Gray) McVaugh, N. Amer. Fl. 32A: 69 (1943).
SE. U.S.A. (Kentucky to Alabama). 78 ALA KTY TEN. Ther. *2n* = 14.

Lobelia gaudichaudii A. DC. in DC., Prodr. 7: 384 (1839). *Dortmanna gaudichaudii* (A. DC.) Kuntze, Revis. Gen. Pl. 2: 972 (1891). *Galeatella gaudichaudii* (A. DC.) O. Deg. & I. Deg., Fl. Hawaiiensis, fam. 339 (1962).
Hawaiian Is. (O'ahu). 63 HAW. Nanophan.

subsp. **gaudichaudii**
Hawaiian Is. (S. Ko'olau Mts. of O'ahu). 63 HAW. Nanophan.
Lobelia gaudichaudii var. *coccinea* Rock, Bull. Torrey Bot. Club 44: 238 (1917).

subsp. **koolauensis** (Hosaka & Fosberg) Lammers, Syst. Bot. 13: 506 (1988).
Hawaiian Is. (N. Ko'olau Mts. of O'ahu). 63 HAW. Nanophan.
**Lobelia gaudichaudii* var. *koolauensis* Hosaka & Fosberg, Occas. Pap. Bernice Pauahi Bishop Mus. 14: 4 (1938). *Galeatella gaudichaudii* var. *koolauensis* (Hosaka & Fosberg) O. Deg. & I. Deg., Fl. Hawaiiensis, fam. 339 (1962).

Lobelia gelida F. Muell., Fragm. 4: 183 (1864). *Pratia gelida* (F. Muell.) Benth., Fl. Austral. 4: 132 (1868).
SE. Australia (New South Wales, Victoria). 50 NSW VIC. Hemicr. *2n* = 14.

Lobelia georgiana McVaugh, Bull. Torrey Bot. Club 67: 144 (1940).
 SE. U.S.A. (Louisiana to Virginia). 78 ALA FLA GEO LOU NCA SCA TEN VRG. Hemicr.
 $2n = 14$.
 * *Lobelia amoena* var. *glandulifera* A. Gray, Syn. Fl. N. Amer. 2(1): 4 (1878). *Lobelia
 glandulifera* (A. Gray) Small, Fl. S.E. U.S.: 1144 (1903); non (A. DC.) Kuntze, Revis.
 Gen. Pl. 2: 378 (1891).
 Lobelia amoena var. *obtusata* A. Gray, Syn. Fl. N. Amer. 2(1): 4 (1878). *Lobelia amoena* f.
 obtusata (A. Gray) Voss in Siebert & Voss, Vilm. Blumengärtn. (ed. 3) 1: 576 (1894).
 Lobelia georgiana f. *subhirta* E. Wimm., Pflanzenr. IV.276b: 429 (1953).

Lobelia ghiesbreghtii Decne., Rev. Hort. (ser. 3) 2: 341 (1848).
 SW. Mexico (Oaxaca). 79 MXS. Nanophan. or hemicr.
 Tupa crassicaulis Hook., Bot. Mag. 76: tab. 4505 (1850).
 Lobelia regalis Fernald, Proc. Amer. Acad. Arts. 36: 503 (1901).

Lobelia gibbosa Labill., Nov. Holl. Pl. 1: 50 (1805). *Rapuntium gibbosum* (Labill.) C. Presl,
 Prodr. Monogr. Lobel.: 13 (1836). *Dortmanna gibbosa* (Labill.) Kuntze, Revis. Gen. Pl. 2:
 972 (1891).
 Australia. 50 NSW QLD SOA TAS VIC WAU. Ther. $2n = 20$.
 Lobelia stricta R. Br., Prodr.: 564 (1810); non Sw., Prodr.: 117 (1788). *Lobelia browniana*
 Schult. in Roem. & Schult., Syst. Veg. 5: 71 (1819). *Rapuntium brownianum* (Schult.)
 C. Presl, Prodr. Monogr. Lobel.: 30 (1836). *Dortmanna browniana* (Schult.) Kuntze,
 Revis. Gen. Pl. 2: 972 (1891). *Lobelia gibbosa* var. *browniana* (Schult.) F. M. Bailey,
 Queensl. Fl.: 916 (1900).
 Lobelia microsperma F. Muell., Fragm. 10: 41 (1876). *Dortmanna microsperma* (F. Muell.)
 Kuntze, Revis. Gen. Pl. 2: 972 (1891). *Lobelia gibbosa* var. *microsperma* (F. Muell.) F.
 M. Bailey, Queensl. Fl.: 916 (1900).
 Lobelia toppii Luehm., Victoria Naturalist 17: 169 (1901).

Lobelia giberroa Hemsl. in Oliv., Fl. Trop. Afr. 3: 465 (1877).
 Sudan to Zaïre, Zambia & Tanzania. 23 BUR RWA ZAI 24 ETH SUD 25 ALL 26 MLW ZAM.
 Phan. $2n = 28$.
 * *Tupa schimperi* Hochst. ex A. Rich., Tent. Fl. Abyss. 2: 10 (1850); non *Lobelia schimperi*
 Hochst. ex A. Rich., Tent. Fl. Abyss. 2: 6 (1850). *Dortmanna giberroa* (Hemsl.) Kuntze,
 Revis. Gen. Pl. 2: 972 (1891); non *Dortmanna schimperi* (Hochst. ex A. Rich.) Kuntze,
 Revis. Gen. Pl. 2: 973 (1891).
 Lobelia volkensii Engl., Bot. Jahrb. Syst. 19 (Beibl. 47): 49 (1894). *Rapuntium volkensii*
 (Engl.) Kuntze, Deutsche Bot. Monatsschr. 21: 173 (1903). *Lobelia giberroa* var.
 volkensii (Engl.) Hauman, Mém. Inst. Roy. Colon. Belge, Sect. Sci. Nat. (8°) 2: 8 (1934).
 Lobelia volkensii var. *ulugurensis* Engl., Notizbl. Königl. Bot. Gart. Berlin 1: 106 (1895).
 Lobelia ulugurensis (Engl.) Engl. ex R. E. Fr. & T. C. E. Fr., Svensk Bot. Tidskr. 16: 398
 (1922). *Lobelia giberroa* var. *ulugurensis* (Engl.) Hauman, Mém. Inst. Roy. Colon.
 Belge, Sect. Sci. Nat. (8°) 2: 9 (1934).
 Lobelia squarrosa Baker f., Bull. Misc. Inform. Kew 1898: 157 (1898). *Lobelia giberroa*
 subsp. *squarrosa* (Baker f.) Mabb., Kew Bull. 29: 566 (1974).
 Lobelia usafuensis Engl., Bot. Jahrb. Syst. 30: 420 (1901). *Lobelia giberroa* var. *usafuensis*
 (Engl.) Hauman, Mém. Inst. Roy. Colon. Belge, Sect. Sci. Nat. (8°) 2: 7, 11 (1934).
 Lobelia giberroa var. *longibracteata* Hauman, Mém. Inst. Roy. Colon. Belge, Sect. Sci. Nat.
 (8°) 2: 10 (1934).
 Lobelia intermedia Hauman, Mém. Inst. Roy. Colon. Belge, Sect. Sci. Nat. (8°) 2: 31
 (1934). *Lobelia giberroa* var. *intermedia* (Hauman) W. Robyns, Fl. Sperm. Parc. Nat.
 Albert 2: 413 (1947).
 Lobelia giberroa var. *iringensis* E. Wimm., Notizbl. Bot. Gart. Berlin-Dahlem 12: 106
 (1934).
 Lobelia giberroa var. *mionandra* E. Wimm., Ann. Naturhist. Mus. Wien 56: 368
 (1948).

Lobelia gilgii Engl., Notizbl. Königl. Bot. Gart. Berlin 1: 108 (1895). *Rapuntium gilgii* (Engl.)
Kuntze, Deutsche Bot. Monatsschr. 21: 173 (1903).
Tanzania. 25 TAN. Hemicr.
 Lobelia unamata E. Wimm., Notizbl. Bot. Gart. Berlin-Dahlem 12: 107 (1934).

Lobelia gilletii De Wild., Ann. Mus. Congo, Bot. (ser. 5) 1: 85 (1904). *Pratia gilletii* (De
Wild.) E. Wimm., Pflanzenr. IV.276b: 115 (1943).
Gabon & Zaïre. 23 GAB ZAI. Hemicr.

Lobelia gladiaria McVaugh, J. Wash. Acad. Sci. 39: 157 (1949).
E. Mexico (Hidalgo). 79 MXE. Hemicr.

Lobelia glandulosa Walter, Fl. Carol.: 218 (1788). *Rapuntium glandulosum* (Walter) C. Presl,
Prodr. Monogr. Lobel.: 21 (1836). *Dortmanna glandulosa* (Walter) Kuntze, Revis. Gen. Pl. 2:
972 (1891).
SE. U.S.A. (Mississippi to Maryland). 78 ALA FLA GEO MRY MSI NCS SCA VRG. Hemicr.
 $2n = 28$.
 Lobelia crassiuscula Michx., Fl. Bor.-Amer. 2: 152 (1803).
 Lobelia puberula var. *glabella* Elliott, Sketch Bot. S. Carolina 1: 267 (1817).
 Lobelia glandulosa var. *laevicalyx* Fernald, Rhodora 49: 186 (1947).

Lobelia glaucescens E. Wimm., Pflanzenr. IV.276b: 514 (1953).
E. Mexico (Veracruz). 79 MXG. Ther.

Lobelia glazioviana Zahlbr., Vidensk. Meddel. Dansk Naturhist. Foren. Kjøbenhavn 1895:
69 (1895).
Brazil (Rio de Janeiro). 84 BZL. Herb. phan. or hemicr.
 Lobelia zahlbruckneri E. Wimm., Repert Spec. Nov. Regni Veg. 19: 387 (1924).

Lobelia gloria-montis Rock, Monogr. Stud. Haw. Lobelioid.: 117 (1919). *Lobelia gaudichaudii*
var. *gloria-montis* (Rock) H. St. John & Hosaka, Occas. Pap. Bernice Pauahi Bishop Mus. 14:
121 (1938). *Galeatella gloria-montis* (Rock) O. Deg. & I. Deg., Fl. Hawaiiensis, fam. 339 (1962).
Hawaiian Is. (Moloka'i, Maui). 63 HAW. Nanophan.
 Lobelia gaudichaudii var. *longibracteata* Rock, Indig. Trees Haw. Isl.: 78 (1913). *Lobelia
 gloria-montis* var. *longibracteata* (Rock) Rock, Monogr. Stud. Haw. Lobelioid.: 119
 (1919). *Galeatella gloria-montis* var. *longibracteata* (Rock) O. Deg. & I. Deg., Fl.
 Hawaiiensis, fam. 339 (1962). *Lobelia longibracteata* (Rock) E. Wimm., Pflanzenr.
 IV.276c: 889 (1968). *Galeatella longibracteata* (Rock) O. Deg. & I. Deg., Prodr.
 Galeatella Neowimmeria: 6 (1974).
 Lobelia gloria-montis var. *molokaiensis* O. Deg., Fl. Hawaiiensis, fam. 339 (1938).
 Galeatella gloria-montis var. *molokaiensis* (O. Deg.) O. Deg. & I. Deg., Fl. Hawaiiensis,
 fam. 339 (1962).
 Lobelia gaudichaudii f. *bryanii* H. St. John & Hosaka, Occas. Pap. Bernice Pauahi Bishop
 Mus. 14: 123 (1938). *Galeatella gloria-montis* var. *bryanii* (H. St. John & Hosaka) O.
 Deg. & I. Deg., Fl. Hawaiiensis, fam. 339 (1962).
 Lobelia gaudichaudii f. *kukuiensis* H. St. John & Hosaka, Occas. Pap. Bernice Pauahi
 Bishop Mus. 14: 123 (1938). *Galeatella gloria-montis* var. *kukuiensis* (H. St. John &
 Hosaka) O. Deg. & I. Deg., Fl. Hawaiiensis, fam. 339 (1962).
 Lobelia gaudichaudii f. *sanguinea* H. St. John & Hosaka, Occas. Pap. Bernice Pauahi
 Bishop Mus. 14: 123 (1938). *Galeatella gloria-montis* f. *sanguinea* (H. St. John &
 Hosaka) O. Deg. & I. Deg., Fl. Hawaiiensis, fam. 339 (1962).
 Lobelia gaudichaudii var. *albiflora* H. St. John & A. C. Medeiros, Phytologia 63: 366 (1987).

Lobelia goetzei Diels, Bot. Jahrb. Syst. 18: 501 (1900).
Tanzania to S. Africa. 25 TAN 26 MLW MOZ ZIM 27 TVL. Hemicr. or ther.
 Lobelia graciliflora E. Wimm., Notizbl. Bot. Gart. Berlin-Dahlem 12: 105 (1934).

Lobelia chamaedryfolia var. *confinis* E. Wimm., Kew Bull. 1952: 142 (1952).
Lobelia holstii var. *subhirsuta* E. Wimm., Kew Bull. 1952: 142 (1952).

Lobelia goldmanii (Fernald) T. J. Ayers, Syst. Bot. 15: 322 (1990), as 'goldmannii'.
NW. Mexico (Sonora, Sinaloa, Durango). 79 MXE MXN. Ther. $2n = 14$.
* *Heterotoma goldmanii* Fernald, Proc. Amer. Acad. Arts 36: 504 (1901). *Calcaratolobelia goldmanii* (Fernald) Wilbur, Sida 17: 563 (1997).

Lobelia gouldii W. Fitzg., Victoria Naturalist 18: 104 (1901).
SE. Australia (Victoria). 50 VIC. Ther.

Lobelia gracillima Welw. ex Hiern, Cat. Afr. Pl. 1: 625 (1898).
Angola. 26 ANG. Ther.
Lobelia pedicellata Diels, Bot. Jahrb. Syst. 26: 116 (1898).

Lobelia grandifolia Britton, Bull. Torrey Bot. Club 37: 359 (1910). *Pratia grandifolia* (Britton) McVaugh, N. Amer. Fl. 32A: 112 (1943).
Jamaica. 81 JAM. Cham. or hemicr.

Lobelia graniticola E. Wimm., Ann. Naturhist. Mus. Wien 56: 357 (1948).
Tanzania. 25 TAN. Hemicr.

Lobelia grayana E. Wimm., Ann. Naturhist. Mus. Wien 56: 369 (1948).
Hawaiian Is. (E. Maui). 63 HAW. Nanophan. $2n = 28$.
* *Lobelia neriifolia* A. Gray, Proc. Amer. Acad. Arts 5: 150 (1861); non Moris, Enum. Sem. Hort. Bot. Taurin.: 1833: 20 (1833). *Dortmanna neriifolia* Kuntze, Revis. Gen. Pl. 2: 973 (1891). *Neowimmeria grayana* (E. Wimm.) O. Deg. & I. Deg., Phytologia 12: 73 (1965).

Lobelia gregoriana Baker f., J. Bot. 32: 66 (1894). *Lobelia keniensis* R. E. Fr. & T. C. E. Fr., Svensk Bot. Tidskr. 16: 413 (1922). *Lobelia deckenii* subsp. *keniensis* Mabb., Kew Bull. 29: 575 (1974).
Kenya & Uganda. 25 KEN UGA. Nanophan. or herb. phan. $2n = 28$.

subsp. **elgonensis** (R. E. Fr. & T. C. E. Fr.) E. B. Knox, Contr. Univ. Michigan Herb. 19: 250 (1993).
Kenya & Uganda. 25 KEN UGA. Nanophan. or herb. phan.
* *Lobelia elgonensis* R. E. Fr. & T. C. E. Fr., Svensk Bot. Tidskr. 16: 411 (1922). *Lobelia deckenii* subsp. *elgonensis* (R. E. Fr. & T. C. E. Fr.) Mabb., Kew Bull. 29: 576 (1974).

subsp. **gregoriana**
Kenya. 25 KEN. Nanophan. or herb. phan. $2n = 28$.

subsp. **sattimae** (R. E. Fr. & T. C. E. Fr.) E. B. Knox, Contr. Univ. Michigan Herb. 19: 250 (1993).
Kenya. 25 KEN. Nanophan. or herb. phan.
* *Lobelia sattimae* R. E. Fr. & T. C. E. Fr., Svensk Bot. Tidskr. 16: 414 (1922). *Lobelia deckenii* subsp. *sattimae* (R. E. Fr. & T. C. E. Fr.) Mabb., Kew Bull. 29: 576 (1974).

Lobelia griffithii Hook. f. & Thomson, J. Proc. Linn. Soc., Bot. 2: 28 (1858). *Dortmanna griffithii* (Hook. f. & Thomson) Kuntze, Revis. Gen. Pl. 2: 380 (1891).
Myanmar to Vietnam & Malaya. 41 CBD LAO MYA THA VIE 42 MLY. Ther.
Lobelia dopatrioides Kurz, J. Asiat. Soc. Bengal, Pt. 2, Nat. Hist. 39: 77 (1870). *Lobelia griffithii* var. *dopatrioides* (Kurz) Kurz, J. Asiat. Soc. Bengal, Pt. 2, Nat. Hist. 46: 211 (1877).
Lobelia hosseusii E. Wimm., Repert. Spec. Nov. Regni Veg. 26: 2 (1929).
Lobelia hosseusii var. *villosa* Kerr in Craib, Fl. Siam. 2: 304 (1936).

Lobelia gruina Cav., Icon. 6: 8 (1800). *Rapuntium gruinum* (Cav.) C. Presl, Prodr. Monogr.
Lobel.: 25 (1836). *Dortmanna gruina* (Cav.) Kuntze, Revis. Gen. Pl. 2: 972 (1891).
Mexico (Nuevo León, San Luis Potosí, Durango, México State, Guerrero, Oaxaca). 79
MXC MXE MXS. Hemicr.

subsp. **gruina**
NE. Mexico (Nuevo León, San Luis Potosí, Durango, México State). 79 MXC MXE.
Hemicr.
Lobelia rapunculoides Kunth in Humb., Bonpl. & Kunth, Nov. Gen. Sp. 3: 312 (quarto),
243 (folio) (1819). *Rapuntium rapunculoides* (Kunth) C. Presl, Prodr. Monogr. Lobel.:
23 (1836). *Dortmanna rapunculoides* (Kunth) Kuntze, Revis. Gen. Pl. 2: 973 (1891), as
'rapunculodes'. *Lobelia gruina* var. *rapunculoides* (Kunth) E. Wimm., Ann. Naturhist.
Mus. Wien 56: 341 (1948).
Lobelia pauciflora Kunth in Humb., Bonpl. & Kunth, Nov. Gen. Sp. 3: 314 (quarto), 245
(folio) (1819). *Rapuntium pauciflorum* (Kunth) C. Presl, Prodr. Monogr. Lobel.: 15
(1836). *Dortmanna pauciflora* (Kunth) Kuntze, Revis. Gen. Pl. 2: 973 (1891).
Lobelia commutata Schult. in Roem. & Schult., Syst. Veg. 5: 73 (1819).
Lobelia orizabae M. Martens & Galeotti, Bull. Acad. Roy. Sci. Bruxelles 9(2): 44 (1842).
Dortmanna orizabae (M. Martens & Galeotti) Kuntze, Revis. Gen. Pl. 2: 973 (1891).
Lobelia gruina var. *orizabae* (M. Martens & Galeotti) E. Wimm., Ann. Naturhist. Mus.
Wien 56: 342 (1948).
Lobelia gruina var. *conferta* Fernald, Proc. Amer. Acad. Arts 36: 503 (1901). *Lobelia gruina*
f. *conferta* (Fernald) McVaugh, Amer. Midl. Natualist 24: 687 (1940).
Lobelia gruina var. *rouaixii* Conz. in M. Gamio, Población Valle Teotihuacán 1: 41 (1922).
Lobelia gruina f. *flava* E. Wimm., Ann. Naturhist. Mus. Wien 56: 341 (1948).

subsp. **peduncularis** (McVaugh) Lammers, Novon 16: 71 (2006).
SW. Mexico (Guerrero, Oaxaca). 79 MXS. Hemicr.
* *Lobelia gruina* var. *peduncularis* McVaugh, Amer. Midl. Natualist 24: 687 (1940).

Lobelia guatemalensis (B. L. Rob. ex Donn. Sm.) Wilbur, Sida 14: 564 (1991).
Guatemala & Honduras. 80 GUA HON. Cham. or hemicr.
* *Centropogon guatemalensis* B. L. Rob. ex Donn. Sm., Bot. Gaz. (Crawfordsville) 20: 4
(1895). *Pratia guatemalensis* (B. L. Rob. ex Donn. Sm.) E. Wimm., Repert. Spec. Nov.
Regni Veg. 29: 50 (1931).

Lobelia guerrerensis Eakes & Lammers, Novon 9: 381 (1999).
SW. Mexico (Guerrero). 79 MXS. Nanophan. or cham.

Lobelia gypsophila T. J. Ayers, Sida 13: 144 (1988).
NE. Mexico (Nuevo León). 79 MXE. Hemicr.
* *Heterotoma pringlei* B. L. Rob., Proc. Amer. Acad. Arts 44: 615 (1909); non *Lobelia pringlei*
S. Watson, Proc. Amer. Acad. Arts 25: 157 (1890). *Calcaratolobelia pringlei* (B. L. Rob.)
Wilbur, Sida 17: 564 (1997).

Lobelia hainanensis E. Wimm., Ann. Naturhist. Mus. Wien 56: 348 (1948).
Hainan. 36 CHH. Ther.

Lobelia harrisii Urb., Symb. Antill. 5: 520 (1908). *Pratia harrisii* (Urb.) McVaugh, N. Amer.
Fl. 32A: 110 (1943).
Jamaica. 81 JAM. Hemicr.

Lobelia hartlaubii Buchenau, Jahresber. Naturwiss. Vereines Bremen 7: 201 (1881), as
'hartlaubi'.
Nigeria to Uganda; Madagascar. 22 NGA 23 BUR CMN RWA ZAI 25 UGA 29 MDG. Hemicr.
Lobelia schaeferi Schltr., Bot. Jahrb. Syst. 57: 624 (1922).

Lobelia hartwegii A. DC. ex Benth., Pl. Hartweg.: 16 (1839), as 'hartwegi'. *Dortmanna hartwegii* (A. DC. ex Benth.) Kuntze, Revis. Gen. Pl. 2: 972 (1891).
SW. Mexico (México State to Oaxaca). 79 MXC MXS. Hemicr. or geophyte.
  Lobelia velutina M. Martens & Galeotti, Bull. Acad. Roy. Sci. Bruxelles 9(2): 41 (1842).
  Lobelia hartwegii var. *angusta* E. Wimm., Ann. Naturhist. Mus. Wien 56: 359 (1948).

Lobelia hassleri Zahlbr., Bull. Herb. Boissier (ser. 2) 7: 445 (1907).
S. Brazil (Paraná, Santa Catarina, Rio Grande do Sul) to NE. Argentina (Misiones, Corrientes). 84 BZS 85 AGE PAR. Cham. or hemicr. $2n = 28$.

Lobelia hederacea Cham., Linnaea 8: 212 (1833). *Pratia hederacea* (Cham.) G. Don, Gen. Hist. 3: 699 (1834).
Bolivia to SE. Brazil (Minas Gerais to Rio Grande do Sul), Uruguay & Argentina (Formosa to Misiones & Rio Negro). 83 BOL 84 BZL BZS 85 AGE AGS PAR URU. Hemicr.
  Lobelia odorata Graham, Edinburgh New Philos. J. 16: 178 (1834). *Lobelia hederacea* var. *elliptica* Hook. & Arn., J. Bot. (Hooker) 1: 277 (1834). *Pratia hederacea* var. *elliptica* (Hook. & Arn.) A. DC. in DC., Prodr. 7: 340 (1839). *Lobelia serpyllacea* var. *odorata* (Graham) Heynh., Nom. Bot. Hort. 1: 473 (1840). *Pratia hederacea* var. *odorata* (Graham) Steud., Nomencl. Bot. (ed. 2) 2: 392 (1841). *Pratia elliptica* (Hook. & Arn.) Hook. f., Fl. Antarct.: 43 (1844).
  Pratia serpyllacea C. Presl, Prodr. Monogr. Lobel.: 46 (1836). *Lobelia serpyllacea* (C. Presl) Heynh., Nom. Bot. Hort. 1: 473 (1840).

Lobelia henodon E. Wimm., Ann. Naturhist. Mus. Wien 56: 343 (1948).
Angola. 26 ANG. Hemicr.

Lobelia henricksonii M. C. Johnst., Nordic J. Bot. 2: 1 (1982).
NE. Mexico (Coahuila). 79 MXE. Hemicr.

Lobelia hereroensis Schinz, Vierteljahrsschr. Naturf. Ges. Zürich 61: 442 (1916).
Namibia. 27 NAM. Hemicr.?

Lobelia heteroclita McVaugh, Ann. Missouri Bot. Gard. 52: 404 (1965).
Colombia. 83 CLM. Nanophan. or cham.

Lobelia heterophylla Labill., Nov. Holl. Pl. 1: 52 (1805). *Rapuntium heterophyllum* (Labill.) C. Presl, Prodr. Monogr. Lobel.: 13 (1836). *Dortmanna heterophylla* (Labill.) Kuntze, Revis. Gen. Pl. 2: 972 (1891).
Australia. 50 NSW NTA QLD SOA WAU. Ther. $2n = 22$.
  Lobelia heterophylla f. *compacta* Voss in Siebert & Voss, Vilm. Blumengärtn. (ed. 3) 1: 575 (1894).

Lobelia heyneana Schult. in Roem. & Schult., Syst. Veg. 5: 50 (1819), as 'heyniana'.
Zaïre to Zambia & Ethiopia; Oman; India to SC. China (Yunnan), Philippines (Luzon) & New Guinea. 23 ZAI 24 ETH 25 TAN 26 ZAM 35 OMA 36 CHC 40 EHM IND NEP SRL 41 LAO MYA THA VIE 42 JAW LSI PHI SUM 43 NWG. Ther. $2n = 24$.
  Lobelia decurrens Roth, Nov. Pl. Sp.: 145 (1821); non Cav., Icon. 6: 13 (1800).
  Lobelia micrantha Hook., Exot. Fl. 1: 44 (1823); non Kunth in Humb., Bonpl. & Kunth, Nov. Gen. Sp. 3: 316 (quarto), 247 (folio) (1819).
  Lobelia trialata Buch.-Ham. ex D. Don, Prodr. Fl. Nepal.: 157 (1825). *Rapuntium trialatum* (Buch.-Ham. ex D. Don) C. Presl, Prodr. Monogr. Lobel.: 13 (1836). *Dortmanna trialata* (Buch.-Ham. ex D. Don) Kuntze, Revis. Gen. Pl. 2: 973 (1891).
  Rapuntium reinwardtianum C. Presl, Prodr. Monogr. Lobel.: 14 (1836). *Lobelia reinwardtiana* (C. Presl) A. DC. in DC., Prodr. 7: 367 (1839). *Dortmanna reinwardtiana* (C. Presl) Kuntze, Revis. Gen. Pl. 2: 973 (1891).

Rapuntium arenarioides C. Presl, Prodr. Monogr. Lobel.: 17 (1836). *Lobelia arenarioides*
   (C. Presl) A. DC. in DC., Prodr. 7: 367 (1839).
Lobelia subincisa Wall. ex A. DC. in DC., Prodr. 7: 367 (1839).
Lobelia dichotoma Miq., Fl. Ned. Ind.: 576 (1857).
Lobelia subracemosa Miq., Fl. Ned. Ind.: 576 (1857).
Lobelia subracemosa var. *rigidior* Miq., Fl. Ned. Ind.: 576 (1857).
Lobelia umbrosa Hochst. ex Hemsl. in Oliv., Fl. Trop. Afr. 3: 468 (1877). *Lobelia trialata*
   var. *umbrosa* (Hemsl.) Chiov., Result. Scientif. Missione Stefanini-Paoli Somal. Ital. 1:
   109 (1916).
Lobelia zeylanica var. *walkeri* C. B. Clarke in Hook. f., Fl. Brit. India 3: 425 (1881).
Lobelia trialata var. *lamiifolia* C. B. Clarke in Hook. f., Fl. Brit. India 3: 425 (1881).
   Lobelia heyneana var. *lamiifolia* (C. B. Clarke) E. Wimm., Ann. Naturhist. Mus. Wien
   56: 345 (1948).
Lobelia bialata Merr., Philipp. J. Sci. 7: 105 (1912).
Lobelia trialata var. *asiatica* Chiov., Result. Scientif. Missione Stefanini-Paoli Somal. Ital.
   1: 109 (1916).
Lobelia aligera Haines, J. & Proc. Asiat. Soc. Bengal 15: 316 (1920). *Lobelia zeylanica* var.
   aligera (Haines) Haines, Bot. Bihar Orissa 4: 501 (1922). *Lobelia dichotoma* var. *aligera*
   (Haines) E. Wimm., Ann. Naturhist. Mus. Wien 56: 345 (1948).
Lobelia zeylanica var. *parviflora* Danguy in Lecomte & Humbert, Fl. Indo-Chine 3: 679
   (1930). *Lobelia heyneana* var. *parviflora* (Danguy) E. Wimm., Ann. Naturhist. Mus.
   Wien 56: 345 (1948). *Lobelia heyneana* f. *parviflora* (Danguy) E. Wimm., Pflanzenr.
   IV.276c: 475 (1953).
Lobelia dichotoma var. *pilosella* E. Wimm., Repert. Spec. Nov. Regni Veg. 38: 78 (1935).
Lobelia heyneana var. *viridissima* E. Wimm., Pflanzenr. IV.276c: 858 (1968).

Lobelia hilaireana (Kanitz) E. Wimm., Revista Sudamer. Bot. 2: 106 (1935).
   SE. Brazil (Minas Gerais). 84 BZL. Cham. or hemicr.
      * *Haynaldia hilaireana* Kanitz in Mart., Fl. Bras. 6(4): 143 (1878). *Dortmanna hilaireana*
      (Kanitz) Kuntze, Revis. Gen. Pl. 2: 972 (1891).

Lobelia hillebrandii Rock, Monogr. Stud. Haw. Lobelioid.: 133 (1919). *Neowimmeria*
   hillebrandii (Rock) O. Deg. & I. Deg., Phytologia 12: 73 (1965).
   Hawaiian Is. (Maui). 63 HAW. Nanophan.

Lobelia hintoniorum B. L. Turner, Phytologia 79: 293 (1996).
   S. Mexico (Oaxaca). 79 MXS. Hemicr. or geophyte.

Lobelia hirtipes E. Wimm., Pflanzenr. IV.276b: 781 (1953).
   Madagascar. 29 MDG. Ther.
      Lobelia hirtipes var. *subpaludosa* E. Wimm., Pflanzenr. IV.276b: 781 (1953).

Lobelia holotricha E. Wimm., Repert. Spec. Nov. Regni Veg. 38: 87 (1935).
   Peru. 83 PER. Hemicr.

Lobelia holstii Engl., Bot. Jahrb. Syst. 19 (Beibl. 47): 51 (1894).
   Zaïre to Ethiopia & Tanzania. 23 BUR RWA ZAI 24 ETH 25 KEN TAN. Hemicr. $2n = 12$.
      Lobelia holstii f. *minor* Engl., Pflanzenw. Ost-Afrikas C: 402 (1895).
      Lobelia johnstonii C. H. Wright, Bull. Misc. Inform. Kew 1906: 164 (1906), as 'johnstoni'.
      Lobelia scioensis Chiov., Ann. Bot. (Rome) 10: 388 (1912).
      Lobelia holstii f. *alba* E. Wimm., Notizbl. Bot. Gart. Berlin-Dahlem 12: 106 (1934).

Lobelia homophylla E. Wimm., Repert. Spec. Nov. Regni Veg. 22: 194 (1926).
   SE. U.S.A. (Florida). 78 FLA. Ther.

Lobelia horombensis E. Wimm., Pflanzenr. IV.276b: 779 (1953).
   Madagascar. 29 MDG. Ther.

Lobelia hotteana Judd & Skean, Bull. Florida State Mus., Biol. Sci. 32: 139 (1987).
Haiti. 81 HAI. Nanophan.

Lobelia humistrata F. Muell. ex F. M. Bailey, Syn. Queensland Fl.: 282 (1883).
NE. Australia (Queensland). 50 QLD. Hemicr.
Lobelia humistrata var. *kingiana* E. Wimm., Ann. Naturhist. Mus. Wien 56: 346 (1948).

Lobelia humpatensis E. Wimm., Ann. Naturhist. Mus. Wien 56: 363 (1948).
Angola. 26 ANG. Nanophan. or cham.

Lobelia × hybrida Voss in Siebert & Voss, Vilm. Blumengärtn. (ed. 3) 1: 576 (1894).
L. amoena × L. syphilitica
Cult. Hemicr.

Lobelia hypnodes E. Wimm. ex McVaugh, Amer. Midl. Naturalist 24: 698 (1940).
E. Mexico (Veracruz). 79 MXG. Ther.

Lobelia hypoleuca Hillebr., Fl. Hawaiian Isl.: 238 (1888). *Neowimmeria hypoleuca* (Hillebr.)
O. Deg. & I. Deg., Phytologia 12: 73 (1965).
Hawaiian Is. (Kaua'i, O'ahu, Moloka'i, Lāna'i, Maui, Hawai'i). 63 HAW. Nanophan.
2n = 28.
Lobelia hypoleuca f. *macrophyta* Rock, Monogr. Stud. Haw. Lobel.: 125 (1919).
Neowimmeria hypoleuca var. *macrophyta* (Rock) O. Deg. & I. Deg., Prodr.
Neowimmeria Galeatella: 9 (1974).
Lobelia hypoleuca var. *rockii* H. St. John & Hosaka, Occas. Pap. Bernice Pauahi Bishop
Mus. 14: 125 (1938). *Neowimmeria rockii* (H. St. John & Hosaka) O. Deg. & I. Deg.,
Prodr. Neowimmeria Galeatella: 11 (1974).
Lobelia hypoleuca var. *heterocarpa* E. Wimm., Ann. Naturhist. Mus. Wien 56: 369 (1948).
Neowimmeria heterocarpa (E. Wimm.) O. Deg. & I. Deg., Prodr. Galeatella
Neowimmeria: 8 (1974).

Lobelia hypsibata E. Wimm., Ann. Naturhist. Mus. Wien 56: 349 (1948).
Cape Provinces. 27 CPP. Ther.

Lobelia illota McVaugh, Amer. Midl. Naturalist 24: 689 (1940).
S. Mexico (Chiapas). 79 MXT. Hemicr.

Lobelia imberbis (Griseb.) Urb., Symb. Antill. 1: 455 (1899).
Cuba. 81 CUB. Cham. or hemicr.
* *Tupa imberbis* Griseb., Mem. Amer. Acad. Arts (n.s.) 8: 516 (1862). *Dortmanna imberbis*
(Griseb.) Kuntze, Revis. Gen. Pl. 2: 972 (1891).
Lobelia piedrana Urb., Ark. Bot. 23A(5): 105 (1930).

Lobelia imperialis E. Wimm., Revista Sudamer. Bot. 2: 94 (1935).
SE. Brazil (Rio de Janeiro). 84 BZL. Cham. or hemicr. 2n = 28.

Lobelia inconspicua A. Rich., Tent. Fl. Abyss. 2: 8 (1850). *Dortmanna inconspicua* (A. Rich.)
Kuntze, Revis. Gen. Pl. 2: 972 (1891). *Lobelia heyneana* var. *inconspicua* (A. Rich.) E.
Wimm., Pflanzenr. IV.276b: 475 (1953).
Sierre Leone to Kenya & Malawi. 22 NGA SIE 23 BUR CMN ZAI 24 ETH 25 KEN TAN UGA
26 MLW. Ther.
Lobelia maranguensis Engl., Pflanzenw. Ost-Afrikas C: 401 (1895).
Lobelia ilysanthoides Schltr., Bot. Jahrb. Syst. 57: 617 (1922).

Lobelia inflata L., Sp. Pl.: 931 (1753). *Rapuntium inflatum* (L.) Mill., Gard. Dict. (ed. 8) (1768). *Dortmanna inflata* (L.) Kuntze, Revis. Gen. Pl. 2: 380 (1891).
  E. North America (Nova Scotia to Nebraska, Louisiana & Georgia); cult.; naturalized in Japan. (38) jap 72 NBR NSC ONT PEI QUE 74 ILL IOW KAN MIN MSO NEB OKL WIS 75 ALL 78 ALA ARK DEL KTY LOU GEO MRY MSI NCA SCA TEN VRG. Biennial or ther. $2n = 14$.
  Lobelia michauxii Nutt., Gen. N. Amer. Pl. 2: 76 (1818). *Rapuntium michauxii* (Nutt.) C. Presl, Prodr. Monogr. Lobel.: 25 (1836).
  Lobelia inflata var. *simplex* Raf. ex Millsp., Prelim. Cat. Fl. W. Virginia: 398 (1892).
  Lobelia inflata f. *albiflora* Moldenke, Castanea 9: 65 (1944).

Lobelia innominata Rendle, J. Bot. 73: 274 (1935). *Pratia innominata* (Rendle) McVaugh, N. Amer. Fl. 32A: 111 (1943). *Pratia acuminata* var. *innominata* (Rendle) E. Wimm., Pflanzenr. IV.276b: 767 (1953).
  Jamaica. 81 JAM. Nanophan. or cham.

Lobelia intercedens (E. Wimm.) Thulin in Launert, Fl. Zambes. 7(1): 136 (1983).
  Malawi, Zimbabwe, Zambia. 26 MLW ZAM ZIM. Ther.
   * *Lobelia heyneana* var. *intercedens* E. Wimm., Pflanzenr. IV.276b: 475 (1953).

Lobelia irasuensis Planch. & Oerst., Vidensk. Meddel. Dansk Naturhist. Foren. Kjøbenhavn 1857: 153 (1857). *Laurentia irasuensis* (Planch. & Oerst.) E. Wimm. in Standl., Fl. of Costa Rica: 1415 (1938), as 'irazuensis'. *Diastatea irasuensis* (Planch. & Oerst.) E. Wimm., Ann. Naturhist. Mus. Wien 56: 332 (1948), as 'irazuensis'.
  C. & W. Mexico; Costa Rica & Panama. 79 MXC MXE MXN 80 COS PAN. Hemicr. or geophyte.

  subsp. **fucata** (McVaugh) Lammers, Novon 16: 71 (2006).
    W. Mexico (Sinaloa, Durango). 79 MXE MXN. Hemicr. or geophyte
   Lobelia irasuensis var. *fucata* McVaugh, Amer. Midl. Naturalist 24: 697 (1940).

  subsp. **irasuensis**
    Costa Rica & Panama. 80 COS PAN. Hemicr. or geophyte.
    Lobelia warscewiczii Vatke, Linnaea 38: 718 (1874). *Dortmanna warscewiczii* (Vatke) Kuntze, Revis. Gen. Pl. 2: 973. 1891, as 'warczewiczii'.

  subsp. **picta** (B. L. Rob. & Seaton) Lammers, Novon 16: 71 (2006).
    C. Mexico (México State). 79 MXC. Hemicr. or geophyte.
   * *Lobelia picta* B. L. Rob. & Seaton, Proc. Amer. Acad. Arts 28: 112 (1893). *Lobelia irasuensis* var. *picta* (B. L. Rob. & Seaton) McVaugh, Amer. Midl. Naturalist 24: 697 (1940).

Lobelia irrigua R. Br., Prodr.: 563 (1810). *Rapuntium irriguum* (R. Br.) C. Presl, Prodr. Monogr. Lobel.: 30 (1836). *Pratia irrigua* (R. Br.) Benth., Fl. Austral. 4: 132 (1868).
  Tasmania. 50 TAS. Hemicr.

Lobelia iteophylla C. Y. Wu, Rep. Yunnan Trop. Subtrop. Fl. Res. Inst. 1: 93 (1965).
  SC. China (Yunnan). 36 CHC. Cham. or hemicr.

Lobelia jaliscensis McVaugh, Amer. Midl. Naturalist 24: 697 (1940).
  W. Mexico (Jalisco). 79 MXS. Hemicr.

Lobelia jasionoides (A. DC.) E. Wimm., Pflanzenr. IV.276b: 119 (1943).
  Cape Provinces. 27 CPP. Hemicr.
  Lobelia corymbosa Hook. ex Graham, Edinburgh New Philos. J. 1: 385 (1826); non P. J. Bergius, Descr. Pl. Cap.: 344 (1767). *Pratia corymbosa* G. Don, Gen. Hist. 3: 699 (1834). *Monopsis corymbosa* (G. Don) C. Presl, Prodr. Monogr. Lobel.: 11 (1836). *Rapuntium corymbosum* (G. Don) C. Presl in E. Mey., Comm. Pl. Afr. Austr.: 287 (1838). *Isolobus corymbosus* (G. Don) A. DC. in DC., Prodr. 7: 352 (1839). *Dortmanna corymbosa* (G. Don) Kuntze, Revis. Gen. Pl. 2: 972 (1891).

Monopsis corymbosa var. *pedicellaris* C. Presl in Eckl. & Zeyh., Enum. Pl. Afric. Austral.: 394 (1837).

Rapuntium corymbosum var. *longifolium* C. Presl in E. Mey., Comm. Pl. Afr. Austr.: 287 (1838).

* *Isolobus jasionoides* A. DC. in DC., Prodr. 7: 353 (1839). *Isolobus corymbosus* var. *jasionoides* (A. DC.) Sond. in Harv. & Sond., Fl. Cap. 3: 535 (1865).

Isolobus corymbosus var. *sparsiflorus* Sond. in Harv. & Sond., Fl. Cap. 3: 535 (1865). *Lobelia jasionoides* var. *sparsiflora* (Sond.) E. Wimm., Ann. Naturhist. Mus. Wien 56: 364 (1948).

Lobelia spathulata R. D. Good, J. Bot. 62: 50 (1924). *Isolobus ecklonianus* var. *spathulatus* (R. D. Good) Adamson, J. S. African Bot. 5: 55 (1939). *Lobelia goodii* E. Wimm., Ann. Naturhist. Mus. Wien 56: 347 (1948).

Lobelia kalmii L., Sp. Pl.: 930 (1753). *Rapuntium kalmii* (L.) C. Presl, Prodr. Monogr. Lobel.: 23 (1836). *Dortmanna kalmii* (L.) Kuntze, Revis. Gen. Pl. 2: 380 (1891).

Canada & N. U.S.A. (Washington to West Virginia & Maine). 70 NWT 71 ABT BRC MAN SAS 72 NBR NFL NSC ONT QUE 73 IDA MNT WAS 74 ILL IOW MIN NDA SDA WIS 75 CNT INI MAI MAS MIC NWH NWJ NWY OHI PEN VER WVA. Hemicr. $2n = 14$.

Lobelia falcata Raf., New Fl. N. Amer. 2: 18 (1837).

Lobelia kalmii var. *strictiflora* Rydb., Mem. New York Bot. Gard. 1: 378 (1900). *Lobelia strictiflora* (Rydb.) Lunell, Bull. Leeds Herb. 2: 8 (1908). *Petromarula strictiflora* (Rydb.) Nieuwl. & Lunell, Amer. Midl. Naturalist 5: 13 (1917).

Lobelia kalmii var. *capillaris* Farw., Amer. Midl. Naturalist 10: 217 (1927).

Lobelia kalmii var. *edithae* E. Wimm., Repert. Spec. Nov. Regni Veg. 38: 85 (1935).

Lobelia kalmii f. *leucantha* Rouleau, Naturaliste Canad. 71: 270 (1944).

Lobelia kalobaensis E. Wimm. ex Thulin, Nordic J. Bot. 3: 374 (1983).

Zaïre. 23 ZAI. Hemicr. or geophyte.

Lobelia kauaiensis (A. Gray) A. Heller, Minnesota Bot. Stud. 1: 911 (1897), as 'kauaensis'.

Hawaiian Is. (Kaua'i). 63 HAW. Nanophan.

* *Lobelia gaudichaudii* var. *kauaiensis* A. Gray, Proc. Amer. Acad. Arts 5: 150 (1861), as 'kauaensis'. *Galeatella kauaiensis* (A. Gray) O. Deg. & I. Deg., Fl. Hawaiiensis, fam. 339 (1962), as 'kauaensis'.

Lobelia gaudichaudii f. *hirsuta* H. St. John & Hosaka, Occas. Pap. Bernice Pauahi Bishop Mus. 14: 121 (1938). *Galeatella kauaiensis* var. *hirsuta* (H. St. John & Hosaka) O. Deg. & I. Deg., Fl. Hawaiiensis, fam. 339 (1962).

Lobelia kirkii R. E. Fr., Wiss. Ergebn. Schwed. Kongo-Rhodesia-Exped. 1: 318 (1914).

Angola to Zimbabwe. 26 ANG ZAM ZIM. Hemicr.

Lobelia kirkii var. *microphylla* Schltr., Bot. Jahrb. Syst. 57: 618 (1922).

Lobelia knoblochii T. J. Ayers, Brittonia 39: 420 (1987). *Calcaratolobelia knoblochii* (T. J. Ayers) Wilbur, Sida 17: 562 (1997).

N. Mexico (Chihuahua). 79 MXE. Geophyte. $2n = 14$.

Lobelia kraussii Graham, Edinburgh New Philos. J. 8: 379 (1830). *Rapuntium kraussii* (Graham) C. Presl, Prodr. Monogr. Lobel.: 27 (1836). *Siphocampylus kraussii* (Graham) G. Don in Sweet, Hort. Brit. (ed. 3): 424 (1839).

Windward Is. (Dominica, Martinique). 81 WIN. Hemicr.

Lobelia kundelungensis Thulin, Nordic J. Bot. 3: 378 (1983).

Zaïre. 23 ZAI. Hemicr.

Lobelia langeana Dusén, Ark. Bot. 9(15): 18 (1910).

S. Brazil (Paraná, Santa Catarina). 84 BZS. Cham. or hemicr. $2n = 28$.

Lobelia lasiocalycina E. Wimm., Repert. Spec. Nov. Regni Veg. 38: 81 (1935).
Zaïre to Malawi. 23 BUR ZAI 26 MLW ZAM. Ther.
Lobelia kassneri E. Wimm., Pflanzenr. IV.276b: 500 (1953).
Lobelia ingrata E. Wimm., Pflanzenr. IV.276c: 859 (1968).

Lobelia laurentioides Schltr., Bot. Jahrb. Syst. 27: 196 (1899).
Cape Provinces. 27 CPP. Ther.

Lobelia laxa MacOwan, J. Linn. Soc., Bot. 25: 392 (1890).
S. Africa. 27 CPP LES NAT TVL. Hemicr.
Lobelia krookii Zahlbr., Ann. K. K. Naturhist. Hofmus. 18: 407 (1903), as 'krooki'.
Lobelia tysonii E. Phillips, Ann. S. African Mus. 16: 176 (1917).

Lobelia laxiflora Kunth in Humb., Bonpl. & Kunth, Nov. Gen. Sp. 3: 311 (quarto), 242
(folio) (1819). *Rapuntium laxiflorum* (Kunth) C. Presl, Prodr. Monogr. Lobel.: 26 (1836).
Tupa laxiflora (Kunth) Planch. & Oerst., Vidensk. Meddel. Dansk Naturhist. Foren.
Kjøbenhavn 1857: 154 (1857). *Lobelia persicifolia* var. *laxiflora* (Kunth) Vatke, Linnaea 38:
722 (1874). *Dortmanna laxiflora* (Kunth) Kuntze, Revis. Gen. Pl. 2: 972 (1891).
SW. U.S.A. to Colombia; cult.; naturalized in Madeira. (21) mdr 76 ARI 79 MXC MXE MXN
MXS MXT 80 COS ELS GUA HON NIC PAN 83 CLM. Nanophan. or cham. $2n = 14$.

subsp. **angustifolia** (A. DC.) Eakes & Lammers, Novon 9: 384 (1999).
SW. U.S.A. (S. Arizona) & E. Mexico (Chihuahua to Oaxaca); cult. 76 ARI 79 MXC MXE
MXN MXS. Nanophan. or cham. $2n = 14$.
* *Lobelia laxiflora* var. *angustifolia* A. DC. in DC., Prodr. 7: 383 (1839). *Lobelia persicifolia*
var. *angustifolia* (A. DC.) Vatke, Linnaea 38: 723 (1874). *Lobelia laxiflora* f. *angustifolia*
(A. DC.) Voss in Siebert & Voss, Vilm. Blumengärtn. (ed. 3) 1: 576 (1894). *Lobelia*
angustifolia (A. DC.) Urbina, Cat. Pl. Mexican.: 201 (1897); non Cham., Linnaea 8:
219 (1833); nec Benth. in Endl., Fenzl, Benth. & Schott, Enum. Pl.: 74 (1837).
Lobelia dracunculoides Willd. ex Schult. in Roem. & Schult., Syst. Veg. 5: 56 (1819).
Rapuntium kunthianum C. Presl, Prodr. Monogr. Lobel.: 27 (1836). *Lobelia persicifolia* var.
amygdalina Vatke, Linnaea 38: 723 (1874).
Lobelia cavanillesii var. *lutea* F. Haage & K. Schmidt, Gartenflora 52: 577 (1903). *Lobelia*
laxiflora f. *lutea* (F. Haage & K. Schmidt) E. Wimm., Pflanzenr. IV.276b: 682 (1953);
non Standl. & Steyerm., Publ. Field Mus. Nat. Hist., Bot. Ser. 23: 98 (1944).
Lobelia nelsonii var. *fragilis* B. L. Rob. & Fernald, Proc. Amer. Acad. Arts 43: 27 (1907).
Lobelia laxiflora f. *fragilis* (B. L. Rob. & Fernald) E. Wimm., Pflanzenr. IV.276b:
682 (1953).
Lobelia laxiflora var. *brevipes* E. Wimm., Pflanzenr. IV.276b: 683 (1953).

subsp. **laxiflora**
N. Mexico to Colombia; cult.; naturalized in Madeira. (21) mdr 79 MXC MXE MXN
MXS MXT 80 COS ELS GUA HON NIC PAN 83 CLM. Nanophan. or cham. $2n = 14$.
Lobelia persicifolia Cav., Icon. 6: 12 (1800); non Lam., Encycl. 3: 584 (1792). *Lobelia*
cavanillesiana Schult. in Roem. & Schult., Syst. Veg. 5: 43 (1819). *Lobelia cavanillesii*
Mart., Ausw. Merkw. Pfl.: 12 (1830). *Rapuntium cavanillesianum* (Schult.) C. Presl,
Prodr. Monogr. Lobel.: 27 (1836). *Tupa persicifolia* G. Don in Sweet, Hort. Brit. (ed.
3): 424 (1839); non (Lam.) A. DC. in DC., Prodr. 7: 395 (1839). *Lobelia laxiflora* var.
cavanillesii Steud., Nomencl. Bot. (ed. 2) 2: 61 (Mar 1841).
Lobelia rigidula Kunth in Humb., Bonpl. & Kunth, Nov. Gen. Sp. 3: 311 (quarto),
243 (folio) (1819). *Rapuntium rigidulum* (Kunth) C. Presl, Prodr. Monogr. Lobel.:
26 (1836).
Lobelia fissa Willd. ex Schult. in Roem. & Schult., Syst. Veg. 5: 57 (1819).
Lobelia amygdalina Willd. ex Schult. in Roem. & Schult., Syst. Veg. 5: 57 (1819).
Rapuntium amygdalinum (Willd. ex Schult.) C. Presl, Prodr. Monogr. Lobel.: 27
(1836). *Dortmanna amygdalina* (Willd. ex Schult.) Kuntze, Revis. Gen. Pl. 2:
972 (1891).

Rapuntium haenkeanum C. Presl, Prodr. Monogr. Lobel.: 26 (1836). *Lobelia haenkeana* (C. Presl) A. DC. in DC., Prodr. 7: 382 (1839). *Dortmanna haenkeana* (C. Presl) Kuntze, Revis. Gen. Pl. 2: 972 (1891).

Lobelia prunifolia Humb. ex C. Presl, Prodr. Monogr. Lobel.: 37 (1836). *Siphocampylus prunifolius* (Humb. ex C. Presl) A. DC. in DC., Prodr. 7: 401 (1839).

Lobelia canescens C. Presl, Prodr. Monogr. Lobel.: 38 (1836). *Siphocampylus canescens* (C. Presl) A. DC. in DC., Prodr. 7: 402 (1839).

Siphocampylus bicolor D. Don in Sweet, Brit. Fl. Gard. 7: pl. 389 (1837). *Lobelia laxiflora* var. *bicolor* (D. Don) Endl., Cat. Horti Vindob. 1: 436 (1842). *Tupa bicolor* (D. Don) Planch., Hort. Donat.: 78 (1858).

Lobelia ovalifolia Hook. & Arn., Bot. Beechey Voy.: 300 (1838).

Lobelia angulatodentata Hook. & Arn., Bot. Beechey Voy.: 301 (1838), as 'angulato-dentata'.

Lobelia lanceolata Hook. & Arn., Bot. Beechey Voy.: 301 (1838); non (Gaudich.) Hook. & Arn., Bot. Beechey Voy.: 88 (1832). *Lobelia laxiflora* f. *lanceolata* E. Wimm., Pflanzenr. IV.276b: 683 (1953).

Lobelia concolor M. Martens & Galeotti, Bull. Acad. Roy. Sci. Bruxelles 9(2): 46 (1842); non R. Br., Prodr.: 563 (1810). *Dortmanna concolor* Kuntze, Revis. Gen. Pl. 2: 972 (1891). *Lobelia laxiflora* f. *concolor* (Kuntze) E. Wimm., Pflanzenr. IV.276b: 684 (1953).

Lobelia andina Benth., Pl. Hartweg.: 213 (1845).

Siphocampylus mollis Regel, Flora 33: 353 (1850); non Planch., Fl. Serres Jard. Eur. 6: 36 (1850). *Siphocampylus warscewiczii* Regel, Schweiz. Z. Gartenbau 1850: 131 (1851). *Lobelia persicifolia* var. *warscewiczii* (Regel) Vatke, Linnaea 38: 723 (1874).

Tupa costaricana Planch. & Oerst., Vidensk. Meddel. Dansk Naturhist. Foren. Kjøbenhavn 1857: 154 (1857). *Lobelia costaricana* (Planch. & Oerst.) E. Wimm., Ann. Naturhist. Mus. Wien 46: 239 (1932).

Tupa costaricana var. *stricta* Planch. & Oerst., Vidensk. Meddel. Dansk Naturhist. Foren. Kjøbenhavn 1857: 155 (1857). *Lobelia laxiflora* var. *stricta* (Planch. & Oerst.) McVaugh, N. Amer. Fl. 32A: 96 (1943).

Tupa costaricana var. *patula* Planch. & Oerst., Vidensk. Meddel. Dansk Naturhist. Foren. Kjøbenhavn 1857: 155 (1857). *Lobelia laxiflora* var. *patula* (Planch. & Oerst.) E. Wimm., Pflanzenr. IV.276b: 683 (1953).

Lobelia persicifolia var. *mollis* Vatke, Linnaea 38: 722 (1874). *Lobelia laxiflora* var. *mollis* (Vatke) Zahlbr., Repert. Spec. Nov. Regni Veg. 14: 185 (1916).

Lobelia patzquarensis Sessé & Moç., Pl. Nov. Hisp.: 152 (1890).

Lobelia nelsonii Fernald, Proc. Amer. Acad. Arts 36: 503 (1901). *Lobelia laxiflora* var. *nelsonii* (Fernald) McVaugh, Ann. Missouri Bot. Gard. 27: 349 (1940), as 'nelsoni'.

Lobelia laxiflora var. *brevifolia* Zahlbr., Repert. Spec. Nov. Regni Veg. 14: 185 (1916). *Lobelia laxiflora* f. *brevifolia* (Zahlbr.) E. Wimm., Pflanzenr. IV.276b: 684 (1953).

Lobelia laxiflora var. *foliosa* Zahlbr., Repert. Spec. Nov. Regni Nov. 14: 185 (1916).

Lobelia delessertiana E. Wimm., Repert. Spec. Nov. Regni Veg. 19: 386 (1924).

Lobelia loretensis M. E. Jones, Contr. W. Bot. 18: 68 (1933).

Lobelia costaricana var. *magna* E. Wimm., Repert. Spec. Nov. Regni Veg. 38: 85 (1935). *Lobelia laxiflora* f. *magna* (E. Wimm.) E. Wimm., Pflanzenr. IV.276b: 684 (1953).

Lobelia rensonii E. Wimm., Repert. Spec. Nov. Regni Veg. 38: 85 (1935).

Lobelia laxiflora f. *lutea* Standl. & Steyerm., Publ. Field Mus. Nat. Hist., Bot. Ser. 23: 98 (1944).

Lobelia haenkeana var. *panamensis* E. Wimm., Ann. Naturhist. Mus. Wien 56: 368 (1948).

Lobelia laxiflora var. *petiolata* E. Wimm., Ann. Naturhist. Mus. Wien 56: 369 (1948).

Lobelia laxiflora f. *flava* E. Wimm., Pflanzenr. IV.276b: 685 (1953).

Lobelia leichhardtii E. Wimm., Pflanzenr. IV.276b: 584 (1953).

NE. Australia (Queensland). 50 QLD. Hemicr.

Lobelia lepida E. Wimm., Ann. Naturhist. Mus. Wien 56: 342 (1948).
  Angola. 26 ANG. Ther.
    * *Lobelia pusilla* Welw. ex Hiern, Cat. Afr. Pl. 1: 626 (1898); non G. Don, Gen. Hist. 3: 711
      (1834); nec (C. Presl) A. DC. in DC., Prodr. 7: 379 (1839).

Lobelia leschenaultiana (C. Presl) Skottsb., Acta Horti Gothob. 4: 4 (1928).
  S. India (Tamil Nadu, Kerala) & Sri Lanka. 40 IND SRL. Nanophan. or cham.
    * *Lobelia excelsa* Lesch. ex Roxb., Fl. Ind. 2: 114 (1824); non Bonpl., Descr. Pl.
      Malmaison: 112 (1816). *Rapuntium leschenaultianum* C. Presl, Prodr. Monogr. Lobel.:
      24 (1836). *Dortmanna leschenaultiana* (C. Presl) Kuntze, Revis. Gen. Pl. 2: 972 (1891).
    Lobelia aromatica Moon ex Wight, Icon. Pl. Ind. Orient. 4: pl. 1172 (1848).
    Lobelia leschenaultiana f. *glabrior* Skottsb., Acta Horti Gothob. 4: 4 (1928).

Lobelia leucotos Albr., Austrobaileya 5: 706 (2000).
  NE. Australia (Queensland). 50 QLD. Hemicr.

Lobelia limosa (Adamson) E. Wimm., Pflanzenr. IV.276c: 878 (1968).
  Cape Provinces. 27 CPP. Hemicr.
    * *Lobelia depressa* var. *thunbergii* A. DC. in DC., Prodr. 7: 379 (1839). *Mezleria limosa*
      Adamson, J. S. African Bot. 7: 276 (1942).
    Lobelia capensis E. Wimm., Pflanzenr. IV.276b: 603 (1953).

Lobelia linarioides (C. Presl) A. DC. in DC., Prodr. 7: 371 (1839).
  Cape Provinces. 27 CPP. Cham. or hemicr.
    * *Rapuntium linarioides* C. Presl, Prodr. Monogr. Lobel.: 22 (1836). *Dortmanna linarioides*
      (C. Presl) Kuntze, Revis. Gen. Pl. 2: 972 (1891).

Lobelia lindblomii Mildbr., Notizbl. Bot. Gart. Berlin-Dahlem 8: 235 (1922).
  Kenya & Uganda. 25 KEN UGA. Hemicr. $2n = 26$.
    Lobelia lindblomii f. *nanella* Mildbr., Notizbl. Bot. Gart. Berlin-Dahlem 8: 235 (1922).

Lobelia linearis Thunb., Prodr. Pl. Cap.: 39 (1794). *Dortmanna linearis* (Thunb.) Kuntze,
  Revis. Gen. Pl. 2: 972 (1891).
  Cape Provinces. 27 CPP. Nanophan. or cham.
    Rapuntium spartioides C. Presl in Eckl. & Zeyh., Enum. Pl. Afric. Austral.: 395 (1837).
      Lobelia spartioides (C. Presl) D. Dietr., Syn. Pl. 1: 728 (1839). *Dortmanna spartioides*
      (C. Presl) Kuntze, Revis. Gen. Pl. 2: 972 (1891), as 'spartiodes'.
    Rapuntium tenuifolium C. Presl in Eckl. & Zeyh., Enum. Pl. Afric. Austral.: 395 (1837).
      Lobelia tenuifolia (C. Presl) D. Dietr., Syn. Pl. 1: 731 (1839); non Willd. ex Schult. in
      Roem. & Schult., Syst. Veg. 5: 56 (1819). *Lobelia ericetorum* A. DC. in DC., Prodr. 7:
      784 (1839).
    Rapuntium lasianthum C. Presl in E. Mey., Comm. Pl. Afr. Austr.: 288 (1838). *Lobelia
      lasiantha* (C. Presl) A. DC. in DC., Prodr. 7: 363 (1839). *Dortmanna lasiantha* (C.
      Presl) Kuntze, Revis. Gen. Pl. 2: 972 (1891).
    Lobelia linearis var. *pinnata* Schltr., Bot. Jahrb. Syst. 27: 196 (1899).
    Lobelia linearis var. *gloveri* E. Wimm., Pflanzenr. IV.276v: 877 (1968).

Lobelia lingulata E. Wimm., Pflanzenr. IV.276b: 537 (1953).
  Madagascar. 29 MDG. Hemicr.

Lobelia lisowskii Thulin, Bull. Jard. Bot. Belg. 53: 489 (1983).
  Zaïre. 23 ZAI. Ther.

Lobelia livingstoniana R. E. Fr., Wiss. Ergebn. Schwed. Rhodesia-Kongo-Exped. 1: 317 (1914).
  Angola & Zambia. 26 ANG ZAM. Hemicr. or geophyte.
    Lobelia borleana E. Wimm., Pflanzenr. IV.276c: 867 (1968).

Lobelia lobata E. Wimm., Pflanzenr. IV.276c: 864 (1968).
Zimbabwe & Transvaal. 26 ZIM 27 TVL. Ther.

Lobelia longicaulis Brandegee, Univ. Calif. Publ. Bot. 6: 73 (1914).
S. Mexico to Panama. 79 MXC MXS MXT 80 COS ELS GUA HON NIC PAN. Hemicr.
Lobelia neglecta Vatke, Linnaea 38: 720 (1874); non Schult. in Roem. & Schult., Syst.
Veg. 5: 633 (1819). *Dortmanna neglecta* Kuntze, Revis. Gen. Pl. 2: 973 (1891). *Lobelia urticifolia* E. Wimm., Repert. Spec. Nov. Regni Veg. 22: 195 (1926).
Lobelia plebeia E. Wimm., Repert. Spec. Nov. Regni Veg. 22: 195 (1926). *Lobelia longicaulis* var. *plebeia* (E. Wimm.) E. Wimm., Pflanzenr. IV.27b: 507 (1953).
Lobelia poasensis E. Wimm., Ann. Naturhist. Mus. Wien 46: 239 (1932).

Lobelia longipedicellata C. E. C. Fisch., Bull. Misc. Inform. Kew 1940: 298 (1941). *Pratia longipedicellata* (C. E. C. Fisch.) E. Wimm., Pflanzenr. IV.276c: 833 (1968).
E. Himalaya. 40 EHM. Cham. or hemicr.

Lobelia longisepala Engl., Bot. Jahrb. Syst. 32: 117 (1902).
Tanzania. 25 TAN. Phan.

Lobelia loochooensis Koidz., Bot. Mag. (Tokyo) 43: 406 (1929).
Nansei-shoto (Kume, Amami-Oshima). 38 NNS. Hemicr. $2n = 14$.

Lobelia lucayana Britton & Millsp., Bahama Fl.: 428 (1920).
Bahamas (Inaguas, San Salvador) & Turks-Caicos Is. 81 BAH TCI. Ther.

Lobelia lukwangulensis Engl., Notizbl. Königl. Bot. Gart. Berlin 1: 107 (1895). *Rapuntium lukwangulense* (Engl.) Kuntze, Deutsche Bot. Monatsschr. 21: 173 (1903).
Tanzania. 25 TAN. Phan. $2n = 28$.

Lobelia luruniensis E. Wimm., Pflanzenr. IV.276c: 862 (1968).
Bolivia. 83 BOL. Hemicr. or ther.

Lobelia luzoniensis (Pers.) Merr., Enum. Philipp. Fl. Pl. 3: 588 (1923).
Philippines. 42 PHI. Ther.
* *Lobelia filiformis* var. *luzoniensis* Pers., Syn. Pl. 2: 214 (1806).

Lobelia macdonaldii B. L. Turner, Phytologia 72: 34 (1992).
S. Mexico (Oaxaca). 79 MXS. Hemicr.

Lobelia macraeana E. Wimm., Ann. Naturhist. Mus. Wien 56: 361 (1948).
Sri Lanka. 40 SRL. Hemicr.
Dortmanna trigona var. *nummulariifolia* Kuntze, Revis. Gen. Pl. 2: 380 (1891), as 'nummulariaefolia'.

Lobelia macrocentron (Benth.) T. J. Ayers, Brittonia 39: 418 (1987).
NW. Mexico (?). 79 +. Geophyte.
* *Heterotoma macrocentron* Benth., Hooker's Icon. Pl. 12: pl. 1177 (1876). *Calcaratolobelia macrocentron* (Benth.) Wilbur, Sida 17: 562 (1997).

Lobelia macrodon (Hook. f.) Lammers, Novon 8: 34 (1998).
New Zealand. 51 NZS. Hemicr. $2n = 14$.
* *Pratia macrodon* Hook. f., Handb. New Zealand Fl.: 172 (1867).

Lobelia malowensis E. Wimm., Ann. Naturhist. Mus. Wien 56: 362 (1948).
S. Africa. 27 CPP NAT SWZ TVL. Cham. or hemicr.

Lobelia margarita E. Wimm., Ann. Naturhist. Mus. Wien 56: 355 (1948). *Calcaratolobelia margarita* (E. Wimm.) Wilbur, Sida 17: 564 (1997).
NE. Mexico (Nuevo León). 79 MXE. Hemicr.

Lobelia martagon (Griseb.) Hitchc., Annual Rep. Missouri Bot. Gard. 4: 103 (1893).
Jamaica. 81 JAM. Cham. or hemicr.
* *Tupa martagon* Griseb., Fl. Brit. W.I.: 386 (1861). *Dortmanna martagon* (Griseb.) Kuntze, Revis. Gen. Pl. 2: 972 (1891).

Lobelia mcvaughii T. J. Ayers, Brittonia 39: 421 (1987). *Calcaratolobelia mcvaughii* (T. J. Ayers) Wilbur, Sida 17: 562 (1997).
N. Mexico (Durango). 79 MXE. Geophyte.

Lobelia melliana E. Wimm., Akad. Wiss. Wien, Math.-Naturwiss. Kl., Anz. 61: 111 (1925).
SE. China (Zhejiang, Fujian, Jiangxi, Hunan, Guangdong). 36 CHS. Cham. or hemicr.

Lobelia membranacea R. Br., Prodr.: 563 (1810). *Rapuntium membranaceum* (R. Br.) C. Presl, Prodr. Monogr. Lobel.: 30 (1836). *Dortmanna membranacea* (R. Br.) Kuntze, Revis. Gen. Pl. 2: 972 (1891).
NE. Australia (Queensland). 50 QLD. Hemicr.

Lobelia mexicana E. Wimm., Ann. Naturhist. Mus. Wien 56: 358 (1948).
S. Mexico to Guatemala. 79 MXT 80 GUA. Hemicr.

Lobelia mezlerioides E. Wimm., Ann. Naturhist. Mus. Wien 56: 341 (1948).
NE. Australia (Queensland). 50 QLD. Ther.

Lobelia microcarpa C. B. Clarke in Hook. f., Fl. Brit. India 3: 424 (1881). *Dortmanna trigona* var. *microcarpa* (C. B. Clarke) Kuntze, Revis. Gen. Pl. 2: 380 (1891).
West Bengal to Laos, Sri Lanka, Andaman Is., Nicobar Is. & Jawa. 40 IND SRL 41 AND NCB MYA LAO 42 JAW. Ther.

Lobelia mildbraedii Engl. in Mildbr., Wiss. Erg. Deut. Zentr.-Afr. Exped., Bot. 2: 344 (1911).
Zaïre to Uganda & Malawi. 23 BUR RWA ZAI 25 TAN UGA 26 MLW. Nanophan. or herb. phan. $2n = 28$.
Lobelia utshungwensis R. E. Fr. & T. C. E. Fr., Svensk Bot. Tidskr. 16: 405 (1922).
Lobelia suavibracteata Hauman, Mém. Inst. Roy. Colon. Belge, Sect. Sci. Nat. (8°) 2: 33 (1934).
Lobelia mildbraedii var. *robynsii* E. Wimm., Ann. Naturhist. Mus. Wien 56: 368 (1948).
Lobelia mildbraedii f. *acutifolia* E. Wimm., Pflanzenr. IV.276c: 885 (1968).

Lobelia minutula Engl., Bot. Jahrb. Syst. 19 (Beibl. 47): 50 (1894).
Cameroon to Kenya & Tanzania. 23 BUR CMN RWA ZAI 25 KEN TAN UGA. Hemicr. $2n = 12$.
Lobelia kiwuensis Engl. in Mildbr., Wiss. Erg. Deut. Zentr.-Afr. Exped., Bot. 2: 345 (1911). *Lobelia minutula* var. *kiwuensis* (Engl.) E. Wimm., Ann. Naturhist. Mus. Wien 56: 347 (1948).
Lobelia rugegensis T. C. E. Fr., Notizbl. Bot. Gart. Berlin-Dahlem 8: 403 (1923). *Lobelia minutula* var. *rugegensis* (T. C. E. Fr.) E. Wimm., Ann. Naturhist. Mus. Wien 56: 347 (1948).
Lobelia nannae T. C. E. Fr., Notizbl. Bot. Gart. Berlin-Dahlem 8: 406 (1923).

Lobelia mishmica C. B. Clarke in Hook. f., Fl. Brit. India 3: 426 (1881). *Dortmanna mishmica* (C. B. Clarke) Kuntze, Revis. Gen. Pl. 2: 972 (1891).
E. Himalaya. 40 EHM. Ther.

Lobelia modesta Wedd., Chlor. Andina 2: 14 (1858). *Dortmanna modesta* (Wedd.) Kuntze,
Revis. Gen. Pl. 2: 972 (1891).
Venezuela & Colombia. 82 VEN 83 CLM. Hemicr.

Lobelia molleri Henriq., Bol. Soc. Brot. 10: 137 (1892).
São Tomé; Cameroon to Sudan, Tanzania, Malawi & Zimbabwe. 23 BUR CMN EQG GGI
ZAI 24 SUD 25 TAN UGA 26 MLW ZIM. Hemicr. or ther. *2n* = 28.
Lobelia thomensis Engl. & Diels, Bot. Jahrb. Syst. 26: 114 (1898).
Lobelia butaguensis De Wild., Rev. Zool. Afr. 8 (Suppl. Bot.): 25 (1920). *Lobelia molleri*
var. *butaguensis* (De Wild.) E. Wimm. ex W. Robyns, Fl. Sperm. Parc Nat. Albert 2:
409 (1947).
Lobelia dealbata E. Wimm., Notizbl. Bot. Gart. Berlin-Dahlem 12: 104 (1934).
Lobelia molleri f. *latifolia* E. Wimm., Kew Bull. 1952: 138 (1952).

Lobelia monostachya (Rock) Lammers, Syst. Bot. 13: 506 (1988).
Hawaiian Is. (O'ahu). 63 HAW. Nanophan.
* *Lobelia hillebrandii* var. *monostachya* Rock, Monogr. Stud. Haw. Lobelioid.: 135 (1919).
Neowimmeria monostachya (Rock) O. Deg. & I. Deg., Prodr. Galeatella Neowimmeria:
9 (1974).

Lobelia montana Reinw. ex Blume, Bijdr.: 728 (1826). *Pratia montana* (Reinw. ex Blume)
Hassk., Cat. Hort. Bot. Bogor.: 106 (1844). *Piddingtonia montana* (Reinw. ex Blume) Miq.,
Fl. Ned. Ind. 2: 573 (1857). *Speirema montanum* (Reinw. ex Blume) Hook. f. & Thomson, J.
Proc. Linn. Soc., Bot. 2: 27 (1858).
Nepal to SC. China (Yunnan), Vietnam & Jawa. 36 CHT CHC 40 ASS EHM IND NEP 41
MYA VIE 42 JAW MLY SUM. Cham. or hemicr.
Piddingtonia patens Miq., Fl. Ned. Ind. 2: 573 (1857).
Piddingtonia cyanocarpa Hassk., Bonplandia 7: 179 (1859). *Pratia montana* var.
cyanocarpa (Hassk.) E. Wimm., Pflanzenr. IV.276b: 117 (1943).
Pratia montana f. *variegata* Hochr., Candollea 5: 293 (1934).
Lobelia wardii C. E. C. Fisch., Bull. Misc. Inform. Kew 1940: 298 (1941). *Pratia wardii* (C.
E. C. Fisch.) E. Wimm., Pflanzenr. IV.276c: 833 (1968).
Lobelia deleiensis C. E. C. Fisch., Bull. Misc. Inform. Kew 1940: 297 (1941). *Pratia
montana* var. *deleiensis* (C. E. C. Fisch.) E. Wimm., Pflanzenr. IV.276c: 833 (1968).

Lobelia morogoroensis Knox & Pócs, Kew Bull. 47: 505 (1992).
Tanzania. 25 TAN. Phan. *2n* = 28.

Lobelia muscoides Cham., Linnaea 8: 215 (1833). *Rapuntium muscoides* (Cham.) C. Presl,
Prodr. Monogr. Lobel.: 22 (1836). *Dortmanna muscoides* (Cham.) Kuntze, Revis. Gen. Pl. 2:
972 (1891), as 'muscodes'.
Cape Provinces. 27 CPP. Ther.

Lobelia nana Kunth in Humb., Bonpl. & Kunth, Nov. Gen. Sp. 3: 317 (quarto), 247 (folio)
(1819). *Rapuntium nanum* (Kunth) C. Presl, Prodr. Monogr. Lobel.: 22 (1836). *Dortmanna
nana* (Kunth) Kuntze, Revis. Gen. Pl. 2: 973 (1891).
N. Mexico to Guatemala; Colombia to N. Argentina (Jujuy, Salta, Tucumán, Catamarca,
Córdoba). 79 MXE MXG MXS MXT 80 GUA 83 BOL CLM ECU PER 85 AGE AGW.
Hemicr.
Pratia boliviensis A. DC. in DC., Prodr. 7: 340 (1839).
Lobelia nana var. *flagelliformis* Wedd., Chlor. Andina 2: 13 (1858).
Lobelia cymbalaria Griseb., Abh. Königl. Ges. Wiss. Göttingen 19: 200 (1874).
Dortmanna cymbalaria (Griseb.) Kuntze, Revis. Gen. Pl. 2: 972 (1891).
Lobelia cymbalarioides Zahlbr., Bot. Jahrb. Syst. 37: 461 (1906); non Engl., Bot. Jahrb.
Syst. 19 (Beibl. 47): 50 (1894). *Lobelia nana* var. *cymbalarioides* E. Wimm. in J. F.
Macbr., Fl. Peru 6: 479 (1937).
Lobelia bellis E. Wimm., Repert. Spec. Nov. Regni Veg. 22: 193 (1926).

Lobelia neglecta Schult. in Roem. & Schult., Syst. Veg. 5: 633 (1819).
Cape Provinces. 27 CPP. Cham.
Rapuntium pedunculare C. Presl, Prodr. Monogr. Lobel.: 19 (1836). *Lobelia longipes* A. DC.
in DC., Prodr. 7: 361 (1839).
Rapuntium pedunculare var. *angustifolium* C. Presl in Eckl. & Zeyh., Enum. Pl. Afric.
Austral.: 398 (1837).
Rapuntium pedunculare var. *brevifolium* C. Presl in Eckl. & Zeyh., Enum. Pl. Afric.
Austral.: 398 (1837).
Rapuntium pedunculare var. *latifolium* C. Presl in Eckl. & Zeyh., Enum. Pl. Afric. Austral.:
398 (1837).

Lobelia neumannii T. C. E. Fr., Notizbl. Bot. Gart. Berlin-Dahlem 8: 409 (1923).
Cameroon to Ethiopia & Kenya. 23 CMN 24 ETH SUD 25 KEN. Ther. $2n = 12$.
Lobelia rubrimaris E. Wimm., Ann. Naturhist. Mus. Wien 56: 352 (1948).
Lobelia albomaculata E. Wimm., Pflanzenr. IV.276b: 499 (1953).
Lobelia kinangopia E. Wimm., Pflanzenr. IV.276b: 777 (1953).

Lobelia nicotianifolia Roth ex Schult. in Roem. & Schult., Syst. Veg. 5: 47 (1819), as
'nicotianaefolia'. *Rapuntium nicotianifolium* (Roth ex Schult.) C. Presl, Prodr. Monogr.
Lobel.: 24 (1836), as 'nicotianaefolium'. *Dortmanna nicotianifolia* (Roth ex Schult.) Kuntze,
Revis. Gen. Pl. 2: 973 (1891).
Sri Lanka; C. & S. India (Madhya Pradesh, Maharashtra, Karnataka, Kerala, Tamil Nadu) to
Thailand. 40 IND SRL 41 MYA THA. Nanophan. or herb. phan. $2n = 28$.
Lobelia trichandra Wight, Icon. Pl. Ind. Orient. 4: pl. 1171 (1848). *Lobelia nicotianifolia*
var. *trichandra* (Wight) C. B. Clarke in Hook. f., Fl. Brit. India 3: 427 (1881).
Lobelia nicotianifolia var. *macrostemon* Skottsb., Acta Horti Gothob. 4: 12 (1928).
Lobelia camptodon E. Wimm., Ann. Naturhist. Mus. Wien 56: 366 (1948).
Lobelia nicotianifolia var. *bibarbata* E. Wimm., Pflanzenr. IV.276b: 644 (1953).
Lobelia camptodon var. *brevipedicellata* E. Wimm., Pflanzenr. IV.276c: 881 (1968).
Lobelia courtallensis K. K. N. Nair, Proc. Indian Acad. Sci. 87B(5): 106 (1978).

Lobelia niihauensis H. St. John, Occas. Pap. Bernice Pauahi Bishop Mus. 9(14): 9 (1931).
Neowimmeria niihauensis (H. St. John) O. Deg. & I. Deg., Phytologia 12: 73 (1965).
Hawaiian Is. (Ni'ihau, Kaua'i, O'ahu). 63 HAW. Nanophan.
Lobelia tortuosa A. Heller, Minnesota Bot. Stud. 1: 912 (1897); non (Benth.) Kuntze,
Revis. Gen. Pl. 2: 378 (1891). *Neowimmeria tortuosa* O. Deg. & I. Deg., Phytologia 12:
73 (1965).
Lobelia tortuosa f. *glabrata* Skottsb., Acta Horti Gothob. 2: 263 (1926). *Neowimmeria*
tortuosa var. *glabrata* (Skottsb.) O. Deg. & I. Deg., Prodr. Galeatella Neowimmeria:
11 (1974).
Lobelia niihauensis var. *forbesii* H. St. John, Occas. Pap. Bernice Pauahi Bishop Mus. 15:
23 (1939). *Neowimmeria niihauensis* var. *forbesii* (H. St. John) O. Deg. & I. Deg., Prodr.
Galeatella Neowimmeria: 10 (1974).
Lobelia niihauensis var. *meridiana* H. St. John, Occas. Pap. Bernice Pauahi Bishop Mus.
15: 23 (1939). *Neowimmeria meridiana* (H. St. John) O. Deg. & I. Deg., Prodr.
Galeatella Neowimmeria: 9 (1974).
Lobelia tortuosa var. *intermedia* H. St. John, Occas. Pap. Bernice Pauahi Bishop Mus. 15:
24 (1939). *Neowimmeria intermedia* (H. St. John) O. Deg. & I. Deg., Prodr. Galeatella
Neowimmeria: 9 (1974).
Lobelia tortuosa var. *haupuensis* H. St. John, Phytologia 62: 435 (1987).

Lobelia nubicola McVaugh, N. Amer. Fl. 32A: 94 (1943).
Guatemala & Honduras. 80 GUA HON. Nanophan. or hemicr.

Lobelia nubigena J. Anthony, Notes Roy. Bot. Gard. Edinburgh 19: 175 (1936).
Bhutan. 40 EHM. Nanophan. or herb. phan.

Lobelia nugax E. Wimm., Ann. Naturhist. Mus. Wien 56: 364 (1948).
  Cape Provinces. 27 CPP. Ther.

Lobelia nummularia Lam., Encycl. 3: 589 (1792). *Rapuntium nummularia* (Lam.) C. Presl,
  Prodr. Monogr. Lobel.: 30 (1836), as 'nummularium'. *Piddingtonia nummularia* (Lam.) A.
  DC. in DC., Prodr. 7: 341 (1839). *Pratia nummularia* (Lam.) A. Braun & Asch., Index Sem.
  Hort. Berol., Appendix: 6 (1861).
  S. & E. India (Tamil Nadu, West Bengal) to SE. China (Hubei, Hunan, Guangxi), Taiwan,
    Philippines & New Guinea. 36 CHC CHS 38 TAI 40 ASS BAN EHM IND NEP SRL 41
    LAO MYA THA VIE 42 JAW LSI MLY PHI SUL SUM 43 NWG. Hemicr. $2n = 12$.
    Lobelia begoniifolia Wall., Asiat. Res. 13: 377 (1820), as 'begonifolia'. *Pratia begoniifolia*
      (Wall.) Lindl., Edwards' Bot. Reg. 16: pl. 1373 (1830), as 'begonifolia'.
    Lobelia javanica Thunb., Fl. Jav.: 9 (1825).
    Lobelia obliqua Ham. ex D. Don., Prodr. Fl. Nepal.: 158 (1825).
    Pratia zeylanica Hassk., Cat. Hort. Bot. Bogor.: 106 (1844).
    Lobelia horsfieldiana Miq., Fl. Ned. Ind.: 577 (1857).
    Pratia papuana S. Moore, Trans. Linn. Soc. London, Bot. (ser. 2) 9: 88 (1916). *Lobelia
      angulata* var. *papuana* (S. Moore) Gilli, Ann. Naturhist. Mus. Wien 83: 417 (1980).
    Pratia wollastonii S. Moore, Trans. Linn. Soc. London, Bot. (ser. 2) 9: 89 (1916).
    Lobelia arfakensis Gibbs, Fl. Arfak Mts.: 183 (1917).
    Pratia podenzanae S. Moore, J. Bot. 55: 306 (1917).
    Lobelia paradoxa E. Wimm., Repert. Spec. Nov. Regni Veg. 26: 2 (1929).
    Lobelia linnaeoides var. *brevipilis* E. Wimm., Pflanzenr. IV.276b: 483 (1953).

Lobelia nummularioides Cham., Linnaea 8: 209 (1833). *Rapuntium nummularioides*
  (Cham.) C. Presl, Prodr. Monogr. Lobel.: 21 (1836). *Dortmanna nummularioides* (Cham.)
  Kuntze, Revis. Gen. Pl. 2: 973 (1891), as 'nummulariodes'.
  SE. Brazil (Minas Gerais, São Paulo, Paraná) to NE. Argentina (Misiones, Corrientes). 84
    BZL BZS 85 AGE PAR. Ther.
    Lobelia prostrata Zahlbr., Bull. Herb. Boissier (ser. 2) 7: 447 (1907). *Lobelia nummularioides*
      var. *prostrata* (Zahlbr.) E. Wimm., Revista Sudamer. Bot. 2: 104 (1935).

Lobelia nuttallii Schult. in Roem. & Schult., Syst. Veg. 5: 39 (1819), as 'nuttalli'.
  SE. U.S.A. (Oklahoma to New York & Florida). 74 OKL 75 NWJ NWY PEN 78 ALA DEL
    FLA GEO KTY LOU MRY NCA SCA TEN VRG. Hemicr. $2n = 14, 28$.
    * *Lobelia gracilis* Nutt., Gen. N. Amer. Pl. 2: 77 (1818); non Salisb., Prodr. Stirp. Chap.
      Allerton: 128 (1796); nec Andrews, Bot. Repos. 5: 340 (1803). *Lobelia kalmii* var.
      gracilis W. P. C. Barton, Fl. N. Amer. 1: 122 (1821). *Rapuntium nuttallianum* C. Presl,
      Prodr. Monogr. Lobel.: 23 (1836). *Dortmanna nuttallii* (Schult.) Kuntze, Revis. Gen.
      Pl. 2: 973 (1891).
    Lobelia kalmii var. *caroliniana* Michx., Fl. Bor.-Amer. 2: 153 (1803).

Lobelia oahuensis Rock, Bull. Torrey Bot. Club 45: 137 (1918). *Neowimmeria oahuensis*
  (Rock) O. Deg. & I. Deg., Prodr. Galeatella Neowimmeria: 10 (1974).
  Hawaiian Is. (O'ahu). 63 HAW. Nanophan.

Lobelia obconica E. Wimm., Ann. Naturhist. Mus. Wien 56: 348 (1948).
  Mexico (?). 79 +. Hemicr.?

Lobelia occidentalis McVaugh & Huft, Contr. Univ. Michigan Herb. 11: 66 (1975).
  W. Mexico (Jalisco, Guerrero). 79 MXS. Hemicr.

Lobelia oligophylla (Wedd.) Lammers, Novon 9: 74 (1999).
  Ecuador to Tierra del Fuego. 83 BOL ECU PER 85 AGW AGS CLC CLN CLS. Hemicr.
    Lysipomia reniformis Kunth in Humb., Bonpl. & Kunth, Nov. Gen. Sp. 3: 320 (quarto),
      250 (folio) (1819); non *Lobelia reniformis* Cham., Linnaea 8: 210 (1833). *Hypsela
      reniformis* (Kunth) C. Presl, Prodr. Monogr. Lobel.: 45 (1836).

Pratia longiflora Hook. f., Fl. Antarct.: 325 (1846); non *Lobelia longiflora* L., Sp. Pl.: 930 (1753). *Hypsela longiflora* (Hook. f.) F. Phil., Cat. Pl. Vasc. Chil.: 188 (1881).
* *Pratia oligophylla* Wedd., Chlor. Andina 2: 10 (1858). *Hypsela oligophylla* (Wedd.) Benth. & Hook. f. ex Zahlbr., Bull. Torrey Bot. Club 24: 387 (1897).
  Pratia subsesslis Wedd., Chlor. Andina 2: 10 (1858). *Hypsela subsessilis* (Wedd.) Benth. & Hook. f. ex Zahlbr., Bull. Torrey Bot. Club 24: 387 (1897).
  Pratia atacamensis Phil., Fl. Atacam.: 34 (1860). *Hypsela atacamensis* (Phil.) F. Phil., Cat. Pl. Vasc. Chil.: 188 (1881).
  Pratia pencana Phil., Anales Univ. Chile 18: 53 (1861).

Lobelia oreas E. Wimm., Pflanzenr. IV.276b: 538 (1953).
  KwaZulu-Natal. 27 NAT. Hemicr.

Lobelia organensis Gardner, London J. Bot. 4: 128 (1845). *Haynaldia organensis* (Gardner) Kanitz in Mart., Fl. Bras. 6(4): 143 (1878). *Euhaynaldia organensis* (Gardner) Borbás in Pall., Nagy Lex. 8: 779 (1894). *Dortmanna organensis* (Gardner) Kuntze, Revis. Gen Pl. 2: 973 (1891).
  Brazil (Minas Gerais, Rio de Janeiro). 84 BZL. Cham. or hemicr.
    Haynaldia organensis var. *insignis* Kanitz in Mart., Fl. Bras. 6(4): 143 (1878).
    Lobelia imperialis var. *kanitzii* E. Wimm., Pflanzenr. IV.276b: 641 (1953).

Lobelia orientalis Rzed. & Calderón, Acta Bot. Méx. 40: 62 (1997).
  C. Mexico (Querétaro). 79 MXE. Hemicr.

Lobelia ovina E. Wimm., Notizbl. Bot. Gart. Berlin-Dahlem 12: 106 (1934).
  Tanzania to Zambia. 25 TAN 26 MLW ZAM. Hemicr.

Lobelia oxyphylla Urb., Symb. Antill. 7: 418 (1912).
  Cuba. 81 CUB. Nanophan. $2n = 28$.

Lobelia paludigena Thulin, Nordic J. Bot. 3: 376 (1983).
  Zaïre. 23 ZAI. Ther.

Lobelia paludosa Nutt., Gen. N. Amer. Pl. 2: 75 (1818). *Dortmanna paludosa* (Nutt.) G. Don, Gen. Hist. 3: 715 (1834). *Rapuntium paludosum* (Nutt.) C. Presl, Prodr. Monogr. Lobel.: 18 (1836).
  SE. U.S.A. (Alabama, Georgia, Florida). 78 ALA FLA GEO. Hemicr. $2n = 42$.
    Lobelia nudicaulis Raf., Atlantic J.: 147 (1832); non Lam., Encycl. 3: 590 (1792).

Lobelia paranaensis R. Braga, Tribuna Farm. 32: 5 (1964).
  Brazil (Paraná). 84 BZS. Ther.

Lobelia parva Badré & Cadet, Adansonia (n.s.) 11: 672 (1972).
  Réunion. 29 REU. Hemicr.

Lobelia parvidentata L. O. Williams, Ceiba 4: 41 (1953).
  Honduras. 80 HON. Nanophan. or hemicr.

Lobelia parvisepala E. Wimm., Pflanzenr. IV.276b: 498 (1953).
  Cape Provinces. 27 CPP. Ther.

Lobelia patula L. f., Suppl. Pl.: 395 (1782). *Rapuntium patulum* (L. f.) C. Presl, Prodr. Monogr. Lobel.: 30 (1836). *Dortmanna patula* (L. f.) Kuntze, Revis. Gen. Pl. 2: 973 (1891).
  Cape Provinces. 27 CPP. Hemicr.
    Rapuntium genistoides C. Presl in E. Mey., Comm. Pl. Afr. Austr.: 285 (1838). *Lobelia genistoides* (C. Presl) A. DC. in DC., Prodr. 7: 358 (1839).

Lobelia pedunculata R. Br., Prodr.: 563 (1810). *Rapuntium pedunculatum* (R. Br.) C. Presl, Prodr.
Monogr. Lobel.: 30 (1836). *Pratia pedunculata* (R. Br.) Benth., Fl. Austral. 4: 133 (1868).
E. Australia. 50 NSW QLD SOA TAS VIC. Hemicr. or geophyte. $2n = 14, 28$.

Lobelia pentheri E. Wimm., Ann. Naturhist. Mus. Wien 56: 353 (1948).
Cape Provinces. 27 CPP. Hemicr.

Lobelia perpusilla Hook. f., Fl. Nov.-Zel. 1: 158 (1852). *Pratia perpusilla* (Hook. f.) Hook. f.,
Handb. New Zealand Fl. 172 (1867).
New Zealand. 51 NZN NZS. Hemicr. $2n = 42$.

Lobelia perrieri E. Wimm., Pflanzenr. IV.276b: 779 (1953).
Madagascar. 29 MDG. Ther.

Lobelia persicifolia Lam., Encycl. 3: 584 (1792). *Rapuntium persicifolium* (Lam.) C. Presl,
Prodr. Monogr. Lobel.: 27 (1836). *Siphocampylus persicifolius* (Lam.) G. Don in Sweet,
Hort. Brit. (ed. 3): 424 (1839). *Tupa persicifolia* (Lam.) A. DC. in DC., Prodr. 7: 395 (1839),
as 'persicaefolia'. *Dortmanna persicifolia* (Lam.) Kuntze, Revis. Gen. Pl. 2: 973 (1891).
Leeward Is. (Guadeloupe). 81 LEE. Cham. or hemicr.

Lobelia petiolata Hauman, Mém. Inst. Roy. Colon. Belge, Sect. Sci. Nat. (8°) 2: 36 (1934).
Zaïre & Rwanda. 23 RWA ZAI. Phan. $2n = 28$.

Lobelia philippinensis Skottsb., Acta Horti Gothob. 4: 13 (1928).
Philippines. 42 PHI. Nanophan. or herb. phan.
 Lobelia nicotianifolia var. *mollis* Elmer, Leafl. Philipp. Bot. 9: 3180 (1934).
 Lobelia epilobioides var. *luzonica* E. Wimm., Pflanzenr. IV.276c: 882 (1968).

Lobelia physaloides A. Cunn., Ann. Nat. Hist. 2: 51 (1839). *Colensoa physaloides* (A. Cunn.)
Hook. f., Fl. Nov.-Zel. 1: 157 (1852). *Pratia physaloides* (A. Cunn.) Hemsl., Hooker's Icon.
Pl. 16: pl. 1532 (1886).
New Zealand. 51 NZN. Nanophan. or cham. $2n = 26$.

Lobelia pinifolia L., Sp. Pl.: 929 (1753). *Rapuntium pinifolium* (L.) C. Presl, Prodr. Monogr.
Lobel.: 19 (1836). *Dortmanna pinifolia* (L.) Kuntze, Revis. Gen. Pl. 2: 973 (1891).
S. Africa. 27 CPP NAT. Cham.
 Lobelia capillipes Schltr., Bot. Jahrb. Syst. 24: 448 (1897).
 Dortmanna pinifolia var. *comosa* Kuntze, Revis. Gen. Pl. 3(2): 188 (1898).
 Dortmanna pinifolia var. *laxa* Kuntze, Revis. Gen. Pl. 3(2): 188 (1898).
 Lobelia pinifolia var. *laricina* E. Wimm., Ann. Naturhist. Mus. Wien 56: 363 (1948).
 Lobelia pinifolia f. *rubriflora* Wilms ex E. Wimm., Ann. Naturhist. Mus. Wien 56:
  363 (1948).
 Lobelia pinifolia f. *parviflora* Wilms ex E. Wimm., Pflanzenr. IV.276b: 588 (1953).

Lobelia platycalyx (F. Muell.) F. Muell., Fragm. 4: 183 (1864).
SE. Australia. 50 SOA TAS VIC. Hemicr. $2n = 42$.
 * *Laurentia platycalyx* F. Muell., Trans. & Proc. Philos. Inst. Victoria 1: 39 (1855). *Pratia*
  platycalyx (F. Muell.) Benth., Fl. Austral. 4: 133 (1868).

Lobelia pleotricha Diels, Notes Roy. Bot. Gard. Edinburgh 5: 170 (1912). *Lobelia davidii* var.
pleotricha (Diels) E. Wimm., Pflanzenr. IV.276b: 658 (1953).
SC. China (Yunnan). 36 CHC. Nanophan. or cham.
 Lobelia handelii E. Wimm., Akad. Wiss. Wien, Math.-Naturwiss. Kl., Anz. 61: 109
  (1925). *Lobelia davidii* var. *handelii* (E. Wimm.) E. Wimm., Pflanzenr. IV.276b: 658
  (1953). *Lobelia pleotricha* var. *handelii* (E. Wimm.) C. Y. Wu, Rep. Yunnan Trop.
  Subtrop. Fl. Res. Inst. 1: 95 (1965).
 Lobelia pleotricha var. *cacuminiflora* Y. S. Lian, Fl. Reipubl. Popul. Sin. 73(2): 188 (1983).

Lobelia poetica E. Wimm., Repert. Spec. Nov. Regni Veg. 22: 195 (1926).
W. Mexico (Nayarit). 79 MXS. Hemicr. or geophyte.

Lobelia polyphylla Hook. & Arn., Bot. Beechey Voy.: 33 (1830). *Tupa polyphylla* (Hook. &
Arn.) G. Don, Gen. Hist. 3: 700 (1834). *Rapuntium polyphyllum* (Hook. & Arn.) C. Presl,
Prodr. Monogr. Lobel.: 28 (1836). *Dortmanna polyphylla* (Hook. & Arn.) Kuntze, Revis.
Gen. Pl. 2: 972 (1891).
NC. Chile (Copiapó to San Antonio). 85 CLC CLN. Nanophan. 2n = 42.
   Lobelia purpurea Lindl., Edwards' Bot. Reg. 16: pl. 1325 (1830); non Breiter, Hort.
      Breiter.: 249 (1817); nec Vellozo, Fl. Flumin.: 353 (1825). *Tupa purpurea* G. Don, Gen.
      Hist. 3: 700 (1834). *Rapuntium purpureum* (G. Don) C. Presl, Prodr. Monogr. Lobel.:
      29 (1836). *Dortmanna purpurea* (G. Don) Kuntze, Revis. Gen. Pl. 2: 972 (1891).
   Rapuntium besserianum C. Presl, Prodr. Monogr. Lobel.: 28 (1836). *Tupa besseriana* (C.
      Presl) A. DC. in DC., Prodr. 7: 393 (1839). *Tupa polyphylla* var. *besseriana* (C. Presl)
      Vatke, Linnaea 38: 727 (1874). *Dortmanna besseriana* (C. Presl) Kuntze, Revis. Gen.
      Pl. 2: 972 (1891). *Lobelia polyphylla* var. *besseriana* (C. Presl) Reiche, Anales Univ.
      Chile 117: 459 (1905).
   Rapuntium subdentatum C. Presl, Prodr. Monogr. Lobel.: 28 (1836). *Tupa subdentata* (C.
      Presl) A. DC. in DC., Prodr. 7: 393 (1839). *Dortmanna subdentata* (C. Presl) Kuntze,
      Revis. Gen. Pl. 2: 973 (1891). *Lobelia polyphylla* f. *subdentata* (C. Presl) E. Wimm.,
      Pflanzenr. IV.276b: 615 (1953).
   Rapuntium bracteosum C. Presl, Prodr. Monogr. Lobel.: 29 (1836). *Tupa bracteosa* (C.
      Presl) A. DC. in DC., Prodr. 7: 393 (1839). *Tupa polyphylla* var. *bracteosa* (C. Presl)
      Vatke, Linnaea 38: 727 (1874). *Dortmanna bracteosa* (C. Presl) Kuntze, Revis. Gen. Pl.
      2: 972 (1891). *Lobelia polyphylla* var. *bracteosa* (C. Presl) Reiche, Anales Univ. Chile
      117: 459 (1905). *Lobelia polyphylla* f. *bracteosa* (C. Presl) E. Wimm., Ann. Naturhist.
      Mus. Wien 56: 365 (1948).
   Rapuntium hyssopifolium C. Presl, Prodr. Monogr. Lobel.: 29 (1836). *Tupa hyssopifolia* (C.
      Presl) A. DC. in DC., Prodr. 7: 392 (1839). *Lobelia hyssopifolia* (C. Presl) Gay, Fl.
      Chile: 326 (1849). *Dortmanna hyssopifolia* (C. Presl) Kuntze, Revis. Gen. Pl. 2: 972
      (1891). *Lobelia polyphylla* f. *hyssopifolia* (C. Presl) E. Wimm., Ann. Naturhist. Mus.
      Wien 56: 365 (1948).
   Tupa polyphylla var. *angustifolia* Hook. & Arn. ex A. DC. in DC., Prodr. 7: 393 (1839).
      Lobelia polyphylla var. *angustifolia* (Hook. & Arn. ex A. DC.) Heynh., Nom. Bot. Hort.
      1: 473 (1840).
   Tupa polyphylla var. *latifolia* A. DC. in DC., Prodr. 7: 393 (1839). *Lobelia polyphylla* var.
      latifolia (A. DC.) Heynh., Nom. Bot. Hort. 1: 473 (1840).
   Tupa atropurpurea Vis., Ill. Piant. Nuov. 2: 23 (1844).
   Tupa ovata Phil., Anales Univ. Chile 43: 507 (1873); non G. Don, Gen. Hist. 3: 700
      (1834). *Lobelia ovata* Reiche, Anales Univ. Chile 117: 460 (1905).
   Tupa polyphylla var. *coquimbana* Vatke, Linnaea 38: 727 (1874). *Lobelia polyphylla* var.
      coquimbana (Vatke) Reiche, Anales Univ. Chile 117: 459 (1905).
   Tupa poeppigiana Phil., Anales Univ. Chile 90: 188 (1895).
   Tupa axilliflora Phil., Anales Univ. Chile 90: 188 (1895). *Lobelia axilliflora* (Phil.) Reiche,
      Anales Univ. Chile 117: 460 (1895).
   Tupa serrata Phil., Anales Univ. Chile 90: 189 (1895).
   Tupa linearifolia Phil., Anales Univ. Chile 90: 190 (1895). *Lobelia polyphylla* f. *linearifolia*
      (Phil.) E. Wimm., Ann. Naturhist. Mus. Wien 56: 365 (1948).
   Tupa gayana Phil., Anales Univ. Chile 90: 190 (1895).

Lobelia porphyrea Rzed. & Calderón, Acta Bot. Mex. 55: 33 (2001).
E. Mexico (Hidalgo). 79 MXE. Hemicr.

Lobelia portoricensis (Vatke) Urb., Symb. Antill. 1: 453 (1899).
Puerto Rico. 81 PUE. Cham. or hemicr. 2n = 14.
  * *Tupa portoricensis* Vatke, Linnaea 38: 727 (1874).

Lobelia pratiana Gaudich. ex Lammers, Novon 8: 34 (1998).
S. Argentina (Chubut, Santa Cruz, Tierra del Fuego), S. Chile (Llanquihue to Magallanes),
Falkland Isl. 85 AGS CLS 90 FAL. Hemicr.
* *Pratia repens* Gaudich., Ann. Sci. Nat. 5: 103 (1825); non *Lobelia repens* Thunb., Prodr.
Pl. Cap.: 40 (1794).
Pratia repens var. *urvilleana* A. DC. in DC., Prodr. 7: 341 (1839).

Lobelia pratioides Benth., Fl. Austral. 4: 131 (1868). *Dortmanna pratioides* (Benth.) Kuntze,
Revis. Gen. Pl. 2: 973 (1891).
SE. Australia. 50 NSW SOA TAS VIC. Hemicr. $2n = 14, 42$.

Lobelia preslii A. DC. in DC., Prodr. 7: 358 (1839).
S. Africa. 27 CPP LES NAT OFS. Hemicr.
* *Rapuntium cordatum* C. Presl in E. Mey., Comm. Pl. Afr. Austr.: 285 (1838); non *Lobelia
cordata* Willd. ex Schult. in Roem. & Schult., Syst. Veg. 5: 58 (1819). *Dortmanna
cordata* Kuntze, Revis. Gen. Pl. 2: 971 (1891).
Lobelia preslii f. *glabra* E. Wimm., Pflanzenr. IV.276b: 584 (1953).

Lobelia pringlei S. Watson, Proc. Amer. Acad. Arts 25: 157 (1890).
NE. Mexico (Nuevo León). 79 MXE. Hemicr.

Lobelia psilostoma E. Wimm., Pflanzenr. IV.276b: 564 (1953).
W. Australia. 50 WAU. Ther.

Lobelia pteropoda (C. Presl) A. DC. in DC., Prodr. 7: 364 (1839).
S. Africa. 26 MOZ 27 CPP NAT. Hemicr.
* *Rapuntium pteropodum* C. Presl, Prodr. Monogr. Lobel.: 21 (1836). *Lobelia patula* var.
pteropoda (C. Presl) Sond. in Harv. & Sond., Fl. Cap. 3: 544 (1865).

Lobelia puberula Michx., Fl. Bor.-Amer. 2: 152 (1803). *Rapuntium puberulum* (Michx.) C.
Presl, Prodr. Monogr. Lobel.: 23 (1836). *Dortmanna puberula* (Michx.) Kuntze, Revis. Gen.
Pl. 2: 973 (1891).
C. & SE. U.S.A. (Texas to Illinois, Ohio, New Jersey & Florida). 74 ILL MSO OKL 75 INI
NWJ OHI PEN WVA 77 TEX 78 ALL. Hemicr. $2n = 14, 28$.
Lobelia puberula var. *glabella* Hook., Bot. Mag. 61: tab. 3292 (1834); non Elliott, Sketch
Bot. S. Carolina 1: 267 (1817). *Lobelia puberula* var. *subglabra* Hook., Companion Bot.
Mag. 1: 100 (1835). *Rapuntium puberulum* var. *glabellum* C. Presl, Prodr. Monogr.
Lobel.: 23 (1836). *Lobelia puberula* f. *glabella* (C. Presl) Voss in Siebert & Voss, Vilm.
Blumengärtn. (ed. 3) 1: 576 (1894). *Lobelia puberula* var. *laeviuscula* C. Mohr, Contr.
U.S. Natl. Herb. 6: 750 (1901).
Lobelia glandulosa var. *obtusifolia* A. DC. in DC., Prodr. 7: 378 (1839). *Lobelia puberula*
var. *obtusifolia* (A. DC.) Fernald, Rhodora 49: 184 (1947).
Lobelia puberula var. *mineolana* E. Wimm., Repert. Spec. Nov. Regni Veg. 26: 4 (1929).
Lobelia puberula f. *candida* Fernald, Rhodora 41: 570 (1939).
Lobelia puberula var. *simulans* Fernald, Rhodora 49: 184 (1947). *Lobelia puberula* f.
simulans (Fernald) Bowden, Bull. Torrey Bot. Club 86: 98 (1959).
Lobelia puberula f. *argutidentata* E. Wimm., Pflanzenr. IV.276b: 427 (1953).

Lobelia pubescens Aiton, Hort. Kew. 3: 498 (1789). *Rapuntium pubescens* (Aiton) C. Presl,
Prodr. Monogr. Lobel.: 17 (1836). *Dortmanna pubescens* (Aiton) Kuntze, Revis. Gen. Pl. 2:
973 (1891).
Cape Provinces. 27 CPP. Hemicr. or ther.
Lobelia alyssifolia Salisb., Icon. Stirp. Rar.: 19 (1791) .
Rapuntium incisum C. Presl in Eckl. & Zeyh., Enum. Pl. Afric. Austral.: 397 (1837). *Lobelia
incisa* (C. Presl) D. Dietr., Syn. Pl. 1: 733 (1839); non Thunb., Prodr. Pl. Cap.: 39 (1794).
Lobelia pubescens var. *incisa* (C. Presl) Sond. in Harv. & Sond., Fl. Cap. 3: 547 (1865).

Lobelia pubescens var. *thunbergiana* Sond. in Harv. & Sond., Fl. Cap. 3: 546 (1865).
Lobelia pubescens var. *jacquiniana* Sond. in Harv. & Sond., Fl. Cap. 3: 547 (1865).
Lobelia pubescens var. *rotundifolia* E. Wimm., Ann. Naturhist. Mus. Wien 56: 351 (1948).
Lobelia pubescens var. *holopsida* E. Wimm., Kew Bull. 1952: 140 (1952).

Lobelia pulchella Vatke, Linnaea 38: 720 (1874). *Dortmanna pulchella* (Vatke) Kuntze, Revis. Gen. Pl. 2: 973 (1891).
　　SW. Mexico (Guerrero, Oaxaca). 79 MXS. Hemicr.
　　　Lobelia bryophila E. Wimm., Repert. Spec. Nov. Regni Veg. 19: 385 (1924)
　　　Lobelia setulosa E. Wimm., Repert. Spec. Nov. Regni Veg. 19: 386 (1924).
　　　Lobelia bryophila var. *fimbriosa* E. Wimm., Ann. Naturhist. Mus. Wien 56: 342 (1948).

Lobelia purpurascens R. Br., Prodr.: 563 (1810). *Rapuntium purpurascens* (R. Br.) C. Presl, Prodr. Monogr. Lobel.: 17 (1836). *Dortmanna purpurascens* (R. Br.) Kuntze, Revis. Gen. Pl. 2: 973 (1891). *Pratia purpurascens* (R. Br.) E. Wimm., Pflanzenr. IV.276b: 764 (1953).
　　E. Australia. 50 NSW QLD VIC. Hemicr. or geophyte. $2n = 56$.
　　　Lobelia ilicifolia Sims, Bot. Mag. 44: tab. 1896 (1817). *Lobelia purpurascens* var. *ilicifolia* (Sims) A. DC. in DC., Prodr. 7: 366 (1839).

Lobelia purpusii Brandegee, Univ. Calif. Publ. Bot. 4: 190 (1911).
　　E. Mexico (San Luis Potosí, Veracruz). 79 MXE MXG. Hemicr. $2n = 14$.

Lobelia pyramidalis Wall., Asiat. Res. 13: 376 (1820). *Rapuntium pyramidale* (Wall.) C. Presl, Prodr. Monogr. Lobel.: 23 (1836). *Dortmanna pyramidalis* (Wall.) Kuntze, Revis. Gen. Pl. 2: 380 (1891).
　　Nepal to S. China (Guangxi, Guizhou, Yunnan) & Myanmar. 36 CHC CHS 40 EHM IND NEP 41 MYA. Cham. or hemicr. $2n = 28$.
　　　Rapuntium wallichianum C. Presl, Prodr. Monogr. Lobel.: 24 (1836). *Lobelia pyramidalis* var. *wallichianum* (C. Presl) Steud., Nomencl. Bot. (ed. 2) 2: 62 (1841). *Lobelia wallichiana* (C. Presl) Hook. f. & Thomson, J. Proc. Linn. Soc. Bot. 2: 29 (1858).
　　　Lobelia seguinii var. *arakana* E. Wimm., Ann. Naturhist. Mus. Wien 56: 367 (1948).

Lobelia quadrangularis R. Br., Prodr.: 563 (1810). *Rapuntium quadrangulare* (R. Br.) C. Presl, Prodr. Monogr. Lobel.: 16 (1836). *Dortmanna quadrangularis* (R. Br.) Kuntze, Revis. Gen. Pl. 2: 973 (1891).
　　W. Australia. 50 WAU. Hemicr.
　　　Lobelia benthamiana S. Moore, J. Bot. 55: 305 (1917).

Lobelia quadrisepala (R. D. Good) E. Wimm., Pflanzenr. IV.276c: 878 (1968).
　　Cape Provinces. 27 CPP. Hemicr.
　　　Mezleria quadrisepala R. D. Good, J. Bot. 63: 172 (1925).

Lobelia quarreana E. Wimm., Repert. Spec. Nov. Regni Veg. 38: 80 (1935).
　　Zaïre & Zambia. 23 ZAI 26 ZAM. Ther. $2n = 12$.

Lobelia rarifolia E. Wimm., Ann. Naturhist. Mus. Wien 56: 360 (1948).
　　W. Australia. 50 WAU. Ther.
　　　Lobelia parvifolia R. Br., Prodr.: 564 (1810); non P. J. Bergius, Descr. Pl. Cap.: 345 (1767). *Rapuntium parvifolium* C. Presl, Prodr. Monogr. Lobel.: 31 (1836).

Lobelia reflexisepala Lammers, Novon 8: 34 (1998).
　　Tibet. 36 CHT. Cham. or hemicr.
　　　Pratia reflexa Y. S. Lian, Acta Phytotax. Sin. 17: 123 (1979); non *Lobelia reflexa* Stokes, Bot. Mat. Med. 1: 342 (1812).

Lobelia reinekeana E. Wimm., Pflanzenr. IV.276c: 858 (1968).
　　Transvaal. 27 TVL. Ther.

Lobelia remyi Rock, Monogr. Stud. Haw. Lobelioid.: 137 (1919). *Neowimmeria remyi* (Rock) O. Deg. & I. Deg., Phytologia 12: 73 (1965).
   Hawaiian Is. (O'ahu). 63 HAW. Nanophan.

Lobelia reniformis Cham., Linnaea 8: 210 (1833). *Rapuntium reniforme* (Cham.) C. Presl, Prodr. Monogr. Lobel.: 15 (1836). *Pratia reniformis* (Cham.) Kanitz in Mart., Fl. Bras. 6(4): 136 (1878).
   Brazil (?). 84+. Hemicr.

Lobelia reverchonii B. L. Turner, Field & Lab. 18: 46 (1950), as 'reverchoni'.
   SC. U.S.A. (Texas, Louisiana). 77 TEX 78 LOU. Hemicr. 2*n* = 14.
   * *Lobelia puberula* var. *pauciflora* Bush, Annual Rep. Missouri Bot. Gard. 17: 122 (1906);
      non *Lobelia pauciflora* Kunth in Humb., Bonpl. & Kunth, Nov. Gen. Sp. 3: 314
      (quarto), 245 (folio) (1819). *Lobelia puberula* subsp. *pauciflora* (Bush) Bowden, Bull.
      Torrey Bot. Club 86: 97 (1959).
   Lobelia vaughiana E. Wimm., Pflanzenr. IV.276b: 425 (1953).

Lobelia rhombifolia de Vriese in Lehm., Pl. Preiss.: 397 (1845). *Dortmanna rhombifolia* (de Vriese) Kuntze, Revis. Gen. Pl. 2: 973 (1891).
   S. Australia. 50 SOA TAS VIC WAU. Ther. 2*n* = 20.
   Lobelia rhombifolia var. *bidentula* E. Wimm., Pflanzenr. IV.276c: 869 (1968).
   Lobelia kochiana E. Wimm., Pflanzenr. IV.276c: 870 (1968).

Lobelia rhynchopetalum Hemsl. in Oliv., Fl. Trop. Afr. 3: 465 (1877).
   Ethiopia. 24 ETH. Phan. 2*n* = 28.
   * *Rhynchopetalum montanum* Fresen., Flora 21: 604 (1838); non *Lobelia montana* Reinw. ex
      Blume, Bijdr.: 728 (1826). *Tupa rhynchopetalum* Hochst. ex A. Rich., Tent. Fl. Abyss.
      2: 9 (1850). *Tupa montana* (Fresen.) Vatke, Linnaea 38: 726 (1874); non C. Wright ex
      Griseb., Cat. Pl. Cub.: 159 (1866); nec Phil., Anales Univ. Chile 43: 506 (1873).
      Dortmanna montana (Fresen.) Kuntze, Revis. Gen. Pl. 2: 972 (1891).
   Lobelia rhynchopetalum subsp. *callosomarginata* E. Wimm., Pflanzenr. IV.276c: 884 (1968).

Lobelia rhytidosperma Benth., Fl. Austral. 4: 126 (1868). *Dortmanna rhytidosperma* (Benth.) Kuntze, Revis. Gen. Pl. 2: 973 (1891).
   W. Australia. 50 WAU. Ther. 2*n* = 22.

Lobelia ritabeaniana E. B. Knox, Contr. Univ. Michigan Herb. 19: 251 (1993).
   Tanzania. 25 TAN. Phan.

Lobelia rivalis E. Wimm., Repert. Spec. Nov. Regni Veg. 38: 80 (1935).
   Zaïre. 23 ZAI. Ther.

Lobelia robusta Graham, Edinburgh New Philos. J. 11: 378 (1831). *Tylomium robustum* (Graham) C. Presl, Prodr. Monogr. Lobel.: 32 (1836). *Tupa robusta* (Graham) A. DC. in DC., Prodr. 7: 394 (1839). *Dortmanna robusta* (Graham) Kuntze, Revis. Gen. Pl. 2: 973 (1891).
   Cuba, Hispaniola, Puerto Rico. 81 CUB DOM HAI PUE. Nanophan.
   Tupa assurgens var. *portoricensis* A. DC. in DC., Prodr. 7: 394 (1839), as 'porto-ricensis'.
      Lobelia assurgens var. *portoricensis* (A. DC.) Urb., Symb. Antill. 1: 454 (1899). *Lobelia*
      robusta var. *portoricensis* (A. DC.) McVaugh, Bull. Torrey Bot. Club 67: 144 (1940), as
      'porto-ricensis'.
   Tupa conglobata var. *elatior* A. DC. in DC., Prodr. 7: 395 (1839), as 'oelatior'.
   Dortmanna acuminata var. *pubescens* Kuntze, Revis. Gen. Pl. 2: 379 (1891).

Lobelia × rogersii Bowden, Canad. J. Bot. 39: 1688 (1961). L. brevifolia × L. puberula
   SE. U.S.A. (Louisiana to Florida). 78 ALA FLA LOU MSI. Hemicr.

Lobelia rosea Wall. in Roxb., Fl. Ind. 2: 115 (1824). *Rapuntium roseum* (Wall.) C. Presl, Prodr. Monogr. Lobel.: 24 (1836). *Dortmanna rosea* (Wall.) Kuntze, Revis. Gen. Pl. 2: 973 (1891).
Nepal to Vietnam. 40 ASS BAN EHM IND NEP 41 LAO MYA THA VIE. Ther.

Lobelia rotundifolia Juss. ex A. DC. in DC., Prodr. 7: 383 (1839). *Dortmanna rotundifolia* (Juss. ex A. DC.) Kuntze, Revis. Gen. Pl. 2: 973 (1891).
Hispaniola & Puerto Rico. 81 DOM HAI PUE. Nanophan.
Tupa domingensis Vatke, Linnaea 38: 728 (1874).
Lobelia rotundifolia var. *angustifolia* Ekman, Repert. Spec. Nov. Regni Veg. 32: 94 (1933).
Lobelia praetervisa Borhidi, Acta Bot. Hung. 38: 192 (1994).

Lobelia roughii Hook. f., Handb. New Zealand Fl.: 171 (1867). *Dortmanna roughii* (Hook. f.) Kuntze, Revis. Gen. Pl. 2: 973 (1891).
New Zealand. 51 NZS. Hemicr. $2n = 14$.
Lobelia roughii var. *alces* E. Wimm., Pflanzenr. IV.276b: 607 (1953).

Lobelia rubescens De Wild., Rev. Zool. Afr. 8 (Suppl. Bot.): 27 (1920).
Sierra Leone to Tanzania. 22 NGA SIE 23 BUR CMN EQG RWA ZAI 25 TAN UGA. Hemicr. or ther.
Lobelia kamerunensis Engl. ex Hutch. & Dalziel, Fl. W. Trop. Afr. 2: 192 (1931).

Lobelia salicina Lam., Encyc. 3: 583 (1792).
Cuba & Hispaniola. 81 CUB DOM HAI. Cham. or hemicr.
Tupa ensifolia A. DC. in DC., Prodr. 7: 396 (1839). *Dortmanna ensifolia* (A. DC.) Kuntze, Revis. Gen. Pl. 2: 972 (1891). *Lobelia ensifolia* (A. DC.) Hitchc., Annual Rep. Missouri Bot. Gard. 4: 103 (1893).
Lobelia salicina var. *brachyantha* Urb., Symb. Antill. 9: 431 (1928). *Lobelia salicina* subsp. *brachyantha* (Urb.) Borhidi & Muñiz, Bot. Közlem. 58: 177 (1971).
Lobelia turckheimii Rendle, J. Bot. 73: 279 (1935).

Lobelia sancta Thulin, Kew Bull. 34: 815 (1980).
Tanzania. 25 TAN. Phan.

Lobelia santa-luciae Rendle, J. Bot. 75: 74 (1937).
Windward Is. (St. Lucia). 81 WIN. Cham. or hemicr.

Lobelia santos-limae Brade, Rodriguésia 10(20): 46 (1947).
SE. Brazil (Rio de Janeiro). 84 BZL. Cham. or hemicr.

Lobelia sapinii De Wild., Bull. Jard. Bot. État 3: 261 (1911), as 'sapini'.
Senegal to Tanzania & Mozambique. 22 GHA IVO NGA SEN SIE TOG 23 CAF CMN ZAI 25 TAN 26 MLW MOZ. Ther.
Lobelia ledermannii Schltr., Bot. Jahrb. Syst. 57: 619 (1922).
Lobelia lelyana E. Wimm., Pflanzenr. IV.276b: 465 (1953).

Lobelia sartorii Vatke, Linnaea 38: 721 (1874). *Dortmanna sartorii* (Vatke) Kuntze, Revis. Gen. Pl. 2: 973 (1891).
NE. Mexico to Guatemala. 79 MXC MXE MXS MXT 80 GUA. Hemicr.
Lobelia novella B. L. Rob., Proc. Amer. Acad. Arts 26: 167 (1891).
Lobelia piscinula E. Wimm., Repert. Spec. Nov. Regni Veg. 38: 84 (1935). *Lobelia sartorii* var. *piscinula* (E. Wimm.) E. Wimm., Pflanzenr. IV.276b: 517 (1953).
Lobelia sartorii var. *opulenta* E. Wimm., Pflanzenr. IV.276b: 517 (1953).

Lobelia scaevolifolia Roxb. in Beatson, Tracts St. Helena: 312 (1816). *Dortmanna scaevolifolia* (Roxb.) Kuntze, Revis. Gen. Pl. 2: 973 (1891). *Trimeris scaevolifolia* (Roxb.) Mabb., Kew Bull. 29: 579 (1974).
St. Helena. 28 STH. Nanophan.
    Trimeris oblongifolia C. Presl, Prodr. Monogr. Lobel.: 46 (1836).

Lobelia scebelii Chiov., Nuovo Giorn. Bot. Ital. (n.s.) 36: 365 (1929).
Ethiopia. 24 ETH. Hemicr.

Lobelia schimperi Hochst. ex A. Rich., Tent. Fl. Abyss. 2: 6 (1850). *Dortmanna schimperi* (Hochst. ex A. Rich.) Kuntze, Revis. Gen. Pl. 2: 973 (1891).
Ethiopia. 24 ETH. Hemicr.

Lobelia scrobiculata E. Wimm., Pflanzenr. IV.276b: 514 (1953).
Mexico (?). 79 +. Hemicr.

Lobelia seguinii H. Lév. & Vaniot, Repert. Spec. Nov. Regni Veg. 12: 186 (1913), as 'seguini'.
S. China (Hubei, Sichuan, Guizhou, Guangxi, Yunnan) & Taiwan. 36 CHC CHS 38 TAI. Cham.
    Lobelia seguinii f. *longisepala* E. Wimm., Akad. Wiss. Wien, Math.-Naturwiss. Kl., Anz. 61: 111 (1925).
    Lobelia seguinii f. *brevisepala* E. Wimm., Akad. Wiss. Wien, Math.-Naturwiss. Kl., Anz. 61: 111 (1925).

Lobelia serpens Lam., Encycl. 3: 588 (1792). *Rapuntium serpens* (Lam.) C. Presl, Prodr. Monogr. Lobel.: 15 (1836).
Madagascar, Réunion, Mauritius. 29 MAU MDG REU. Hemicr. or ther.
    Rapuntium cheiranthifolium C. Presl, Prodr. Monogr. Lobel.: 18 (1836). *Lobelia serpens* var. *cheiranthifolia* (C. Presl) A. DC. in DC., Prodr. 7: 368 (1839).
    Lobelia serpens var. *pedicellata* A. DC. in DC., Prodr. 7: 368 (1839).
    Lobelia serpens var. *bicolor* Steud., Nomencl. Bot. (ed. 2) 2: 62 (1841).
    Lobelia serpens var. *puberula* E. Wimm., Pflanzenr. IV.276b: 493 (1953).
    Lobelia serpens var. *sieberi* E. Wimm., Pflanzenr. IV.276b: 493 (1953).

Lobelia sessilifolia Lamb., Trans. Linn. Soc. London 10: 160 (1811). *Lobelia camtschatica* Pall. ex Spreng., Syst. Veg. 1: 712 (1824). *Rapuntium kamtschaticum* C. Presl, Prodr. Monogr. Lobel.: 24 (1836). *Dortmanna sessilifolia* (Lamb.) Kuntze, Revis. Gen. Pl. 2: 973 (1891).
E. Siberia to S. & E. China, Korea, Japan & Russian Far East. 30 BRY CTA YAK 31 ALL 36 CHC CHM CHN CHS 38 JAP KOR. Hemicr. or geophyte. $2n = 28$.
    Lobelia saligna Fisch., Mém. Soc. Imp. Naturalistes Moscou 3: 65 (1812).
    Lobelia salicifolia Fisch. ex Trautv., Increm. Fl. Phaenog. Ross.: 501 (1883); non Sweet, Hort. Suburb. Lond.: 37 (1818).
    Lobelia sessilifolia var. *latifolia* Nakai, Bot. Mag. (Tokyo) 34: 51 (1920).
    Lobelia sessilifolia f. *leuantha* H. Hara, Bot. Mag. (Tokyo) 51: 898 (1937).

Lobelia setacea Thunb. in Hoffm., Phytogr. Bl.: 21 (1803). *Rapuntium setaceum* (Thunb.) C. Presl, Prodr. Monogr. Lobel.: 19 (1836). *Dortmanna setacea* (Thunb.) Kuntze, Revis. Gen. Pl. 2: 973 (1891).
Cape Provinces. 27 CPP. Hemicr.
    Lobelia setacea var. *parviflora* Sond. in Harv. & Sond., Fl. Cap. 3: 541 (1865).
    Lobelia glaucoleuca Schltr., Bot. Jahrb. Syst. 24: 449 (1897).

Lobelia shaferi Urb., Symb. Antill. 7: 419 (1912).
Cuba. 81 CUB. Cham. or hemicr.
    Lobelia obtusata Urb., Ark. Bot. 23A(5): 105 (1930). *Lobelia shaferi* var. *obtusata* (Urb.) E. Wimm., Ann. Naturhist. Mus. Wien 56: 365 (1948).

Lobelia nipensis Urb., Ark. Bot. 23A(5): 106 (1930). *Lobelia shaferi* var. *nipensis* (Urb.) E. Wimm., Ann. Naturhist. Mus. Wien 56: 365 (1948).

Lobelia simplicicaulis R. Br., Prodr.: 564 (1810). *Rapuntium simplicicaule* (R. Br.) C. Presl, Prodr. Monogr. Lobel.: 30 (1836). *Lobelia gibbosa* var. *simplicicaulis* (R. Br.) F. M. Bailey, Queensl. Fl.: 916 (1900).
   E. Australia. 50 NSW QLD SOA TAS WAU. Hemicr. or ther.

Lobelia sinaloae Sprague, Bull. Misc. Inform. Kew 1929: 7 (1929).
   W. Mexico (Sinaloa). 79 MXN. Hemicr.

Lobelia siphilitica L., Sp. Pl.: 931 (1753). *Rapuntium siphiliticum* (L.) Mill., Gard. Dict. (ed. 8) (1768). *Lobelia reflexa* Stokes, Bot. Mater. Med. 1: 342 (1812). *Lobelia antisyphilitica* Hayne, Brandt & Ratzeb. in Hayne, Getreue Darstell. Gew. 13: pl. 9 (1837). *Dortmanna siphilitica* (L.) Kuntze, Revis. Gen. Pl. 2: 380 (1891), as 'syphilitica'.
   E. & C. North America (Manitoba to Colorado, Texas, Georgia & Maine); cult. 71 MAN 72 ONT 73 COL WYO 74 ALL 75 ALL 77 TEX 78 ALA ARK DEL GEO KTY LOU MRY MSI NCA SCA TEN VRG. Hemicr. $2n = 14$.
   Lobelia coelestis Nutt. ex Loudon, Hort. Brit. (ed. 2): 592 (1832) .
   Lobelia siphilitica var. *maculata* G. Don, Gen. Hist. 3: 706 (1834).
   Lobelia siphilitica var. *minor* Hook., Companion Bot. Mag. 1: 100 (1835).
   Lobelia siphilitica var. *alba* G. Don & Baxter in Loudon, Hort. Brit. (ed. 3) 646 (1839).
   Lobelia siphilitica var. *ludoviciana* A. DC. in DC., Prodr. 7: 377 (1839). *Lobelia siphilitica* f.
      ludoviciana (A. DC.) Voss in Siebert & Voss, Vilm. Blumengärtn. (ed. 3) 1: 576 (1894).
   Lobelia belgica de Jonghe, Rev. Hort. (ser. 2) 4: 39 (1845).
   Lobelia densiflora Rennie, Mag. Bot. Gard.1: pl. 67 (1833). *Lobelia siphilitica* var.
      densiflora (Rennie) E. Wimm., Ann. Naturhist. Mus. Wien 56: 340 (1948).
   Lobelia siphilitica var. *candida* A. W. Wood, Class-book Bot. 476 (1869), as 'candidus'.
   Lobelia siphilitica var. *rosea* N. Coleman, Kent Sci. Inst. Misc. Publ. 2: 24 (1874).
   Lobelia siphilitica f. *albiflora* Britton, Bull. Torrey Bot. Club 17: 125 (1890).
   Lobelia siphilitica f. *alba* Voss in Siebert & Voss, Vilm. Blumengärtn. (ed. 3) 1: 576 (1894).
   Lobelia siphilitica f. *purpurea* Voss in Siebert & Voss, Vilm. Blumengärtn. (ed. 3) 1:
      576 (1894).
   Lobelia bollii E. Wimm., Repert. Spec. Nov. Regni Veg. 26: 3 (1929). *Lobelia siphilitica*
      var. *bollii* (E. Wimm.) E. Wimm., Pflanzenr. IV.276b: 423 (1953).
   Lobelia siphilitica f. *laevicalyx* Fernald, Rhodora 45: 476 (1943).
   Lobelia siphilitica f. *grandis* E. Wimm., Pflanzenr. IV.276b: 422 (1953).
   Lobelia siphilitica f. *purpurea* E. J. Palmer & Steyerm., Brittonia 10: 118 (1958); non Voss
      in Siebert & Voss, Vilm. Blumengärtn. (ed. 3) 1: 576 (1894).

Lobelia solaris E. Wimm., Pflanzenr. IV.276b: 778 (1953).
   Madagascar. 29 MDG. Ther.

Lobelia sonderiana (Kuntze) Lammers, Novon 9: 386 (1999).
   Kenya to S. Africa. 25 KEN TAN 26 ANG MLW MOZ ZAM ZIM 27 BOT CPP LES NAM OFS TVL. Hemicr. or ther.
 * *Mezleria depressa* A. DC. in DC., Prodr. 7: 350 (1839); non (L. f.) C. Presl, Prodr. Monogr.
      Lobel.: 7 (1836); nec *Lobelia depressa* L. f., Suppl. Pl.: 395 (1782). *Mezleria dregeana*
      Sond. in Harv. & Sond., Fl. Cap. 3: 533 (1865); non *Lobelia dregeana* (C. Presl) A. DC.
      in DC., Prodr. 7: 731 (1839). *Dortmanna sonderiana* Kuntze, Revis. Gen. Pl. 2: 972
      (1891); non *Dortmanna dregeana* (C. Presl) Kuntze, Revis. Gen. Pl. 2: 972 (1891).
      Lobelia sonderi Zahlbr., Ann. K. K. Naturhist. Hofmus. 18: 404 (1903). *Lobelia depressa*
      var. *dregeana* (Sond.) E. Wimm., Notizbl. Bot. Gart. Berlin-Dahlem 15: 633 (1941).
   Lobelia angolensis Engl. & Diels, Bot. Jahrb. Syst. 26: 114 (1898).
   Lobelia lythroides Diels, Bot. Jahrb. Syst. 26: 113 (1899).

Lobelia dekindtiana Engl., Bot. Jahrb. Syst. 32: 147 (1902).

Lobelia minutidentata Engl. & Gilg in Warb., Kunene-Sambesi-Exped.: 397 (1903).

Lobelia seineri Schltr., Bot. Jahrb. Syst. 57: 616 (1922). *Lobelia depressa* var. *seineri* (Schltr.) E. Wimm., Notizbl. Bot. Gart. Berlin-Dahlem 15: 633 (1941).

Lobelia tsotsorogensis Bremek. & Oberm., Ann. Transvaal Mus. 16: 437 (1935).

Lobelia depressa var. *linearifolia* E. Wimm., Pflanzenr. IV.276c: 879 (1968).

Lobelia spathopetala Diels, Bot. Jahrb. Syst. 26: 115 (1898).
Madagascar. 29 MDG. Ther.

> *Lobelia spathopetala* var. *decaryana* E. Wimm., Mém. Inst. Sci. Madagascar, Sér. B, Biol. Vég. 5: 82 (1954).

Lobelia × speciosa Sweet, Brit. Fl. Gard. 5: pl. 174 (1833), pro sp. L. cardinalis × L. siphilitica. *Rapuntium speciosum* (Sweet) C. Presl, Prodr. Monogr. Lobel.: 19 (1836). *Lobelia siphilitica* var. *speciosa* (Sweet) Steud., Nomencl. Bot. (ed. 2) 2: 63 (1841).
C. North America (Ontario, Illinois, Missouri); cult. 72 ONT 74 ILL MSO. Hemicr.

> *Lobelia siphilitica* var. *hybrida* Hook., Bot. Mag. 64: tab. 3604 (1837). *Lobelia siphilitica* f. *hybrida* (Hook.) Voss in Siebert & Voss, Vilm. Blumengärtn. (ed. 3) 1: 576 (1894).
>
> *Lobelia cardinalis* var. *milleri* D. Don in Sweet, Brit. Fl. Gard. 7: pl. 372 (1837). *Lobelia cardinalis* f. *milleri* (D. Don) Voss in Siebert & Voss, Vilm. Blumengärtn. (ed. 3) 1: 575 (1894).
>
> *Lobelia milleri* Paxton, Paxton's Mag. Bot. 6: 166 (1839), as 'millerii'. *Lobelia siphilitica* var. *milleri* (Paxton) Steud., Nomencl. Bot. (ed. 2) 2: 63 (1841).
>
> *Lobelia formosa* de Jonghe, Rev. Hort. (ser. 2) 4: 39 (1845).
>
> *Lobelia makoyi* de Jonghe, Rev. Hort. (ser. 2) 4: 39 (1845).
>
> *Lobelia nova* de Jonghe, Rev. Hort. (ser. 2) 4: 39 (1845).
>
> *Lobelia gerardii* Sauvageau, Rev. Hort. 65: 519 (1893), as 'gerardi'.
>
> *Lobelia rivoirei* Bain, Gard. Chron. (ser. 3) 25: 232 (1898).
>
> *Lobelia gerardii* var. *lugdunensis* L. H. Bailey, Cycl. Amer. Hort.: 936 (1900).
>
> *Lobelia × speciosa* nothovar. *occidentalis* Bowden, Canad. J. Bot. 60: 2066 (1982), pro nothomorph.
>
> *Lobelia × speciosa* nothovar. *schneckii* Bowden, Canad. J. Bot. 60: 2065 (1982), pro nothomorph.

Lobelia spicata Lam., Encycl. 3: 587 (1792). *Dortmanna spicata* (Lam.) Kuntze, Revis. Gen. Pl. 2: 380 (1891).
E. & C. North America (Alberta to Texas, Georgia & Nova Scotia). 71 ABT MAN SAS 72 NBR NSC ONT PEI QUE 73 MNT 74 ALL 75 ALL 77 TEX 78 ALA ARK DEL GEO KTY LOU MRY MSI NCA SCA TEN VRG. Hemicr. $2n = 14$.

> *Lobelia claytoniana* Michx., Fl. Bor.-Amer. 2: 153 (1803). *Rapuntium claytonianum* (Michx.) C. Presl, Prodr. Monogr. Lobel.: 23 (1836). *Lobelia pallida* var. *claytoniana* (Michx.) Darby, Bot. South. States: 412 (1855).
>
> *Lobelia goodenioides* Willd., Hort. Berol.: pl. 30 (1804). *Lobelia pallida* Muhl., Cat. Pl. Amer. Sept.: 22 (1813).
>
> *Lobelia nivea* Raf., Ann. Nat.: 15 (1820).
>
> *Lobelia paniculata* Raf., New Fl. N. Amer. 2: 18 (1837); non L., Sp. Pl.: 930 (1753).
>
> *Lobelia leptostachys* A. DC. in DC., Prodr. 7: 376 (1839). *Dortmanna leptostachys* (A. DC.) Kuntze, Revis. Gen. Pl. 2: 972 (1891). *Lobelia spicata* var. *leptostachys* (A. DC.) Mack. & Bush, Fl. Jackson County Missouri: 183 (1902).
>
> *Lobelia spicata* var. *parviflora* A. Gray, Syn. Fl. N. Amer. 2(1): 6 (1878).
>
> *Lobelia spicata* var. *hirtella* A. Gray, Syn. Fl. N. Amer. 2(1): 6 (1878). *Lobelia hirtella* (A. Gray) Greene, Pittonia 3: 349 (1898). *Lobelia leptostachys* var. *hirtella* (A. Gray) Farw., Rep. Michigan Acad. Sci. 15: 188 (1913). *Petromarula hirtella* (A. Gray) Nieuwl. & Lunell, Amer. Midl. Naturalist 5: 13 (1917).
>
> *Lobelia bracteata* Small, Fl. S.E. U.S.: 1146 (1903).
>
> *Lobelia spicata* f. *albiflora* Ralph Hoffm., Proc. Boston Soc. Nat. Hist. 36: 334 (1922).

Lobelia spicata var. *campanulata* McVaugh, Rhodora 38: 316 (1936). *Lobelia spicata* f.
   campanulata (McVaugh) Bowden, Bull. Torrey Bot. Club 86: 96 (1959).
   Lobelia spicata var. *scaposa* McVaugh, Rhodora 38: 318 (1936).

Lobelia standleyi McVaugh, N. Amer. Fl. 32A: 47 (1943).
   Guatemala. 80 GUA. Hemicr.

Lobelia stellfeldii R. Braga, Tribuna Farm. 32: 6 (1964).
   Brazil (Paraná). 84 BZS. Cham. or hemicr.

Lobelia stenocarpa E. Wimm., Repert. Spec. Nov. Regni Veg. 38: 81 (1935).
   Madagascar. 29 MDG. Hemicr.

Lobelia stenodonta (Fernald) McVaugh, Bull. Torrey Bot. Club 67: 144 (1940).
   S. Mexico (Chiapas). 79 MXT. Hemicr.
      Heterotoma stenodonta Fernald, Proc. Amer. Acad. Arts 36: 504 (1901).

Lobelia stenophylla Benth., Fl. Austral. 4: 130 (1868). *Dortmanna stenophylla* (Benth.)
   Kuntze, Revis. Gen. Pl. 2: 973 (1891).
   N. Australia (Northern Territory, Queensland). 50 NTA QLD. Hemicr.
      Lobelia douglasiana F. M. Bailey, Queensland Agric. J. 1: 228 (1897).

Lobelia stenosiphon (Adamson) E. Wimm., Pflanzenr. IV.276c: 862 (1968).
   Cape Provinces. 27 CPP. Hemicr.
      Mezleria stenosiphon Adamson, J. S. African Bot. 12: 36 (1946).
      Lobelia isotomoides E. Wimm., Ann. Naturhist. Mus. Wien 56: 353 (1948).

Lobelia stolonifera Donn. Sm., Bot. Gaz. (Crawfordsville) 27: 338 (1899).
   Guatemala. 80 GUA. Hemicr. $2n = 14$.

Lobelia stricklandiae Gilliland, J. Bot. 73: 248 (1935), as 'stricklandae'.
   Tanzania to S. Africa. 25 TAN 26 MLW MOZ ZIM 27 TVL. Phan. $2n = 28$.
      Lobelia stricklandiae f. *uncinata* E. Wimm., Pflanzenr. IV.276c: 886 (1968).

Lobelia stricta Sw., Prodr.: 117 (1788). *Rapuntium strictum* (Sw.) C. Presl, Prodr. Monogr.
   Lobel.: 27 (1836). *Tupa stricta* (Sw.) A. DC. in DC., Prodr. 7: 395 (1839). *Dortmanna stricta*
   (Sw.) Kuntze, Revis. Gen. Pl. 2: 973 (1891).
   Lesser Antilles (St. Kitts-Nevis, Guadeloupe, Dominica, Martinique). 81 LEE WIN. Hemicr.
      Lobelia areolata A. Rich. ex Juss., Ann. Mus. Natl. Hist. Nat. 18: 3 (1811).
      Tupa flavescens A. DC. in DC., Prodr. 7: 395 (1839). *Dortmanna flavescens* (A. DC.)
         Kuntze, Revis. Gen. Pl. 2: 972 (1891). *Lobelia flavescens* (A. DC.) E. H. L. Krause,
         Beih. Bot. Centralbl. 32(2): 337 (1914).

Lobelia stuhlmannii Schweinf. ex Stuhlmann, Mit Emin Pasha in Afrika: 295 (1893).
   Zaïre to Uganda. 23 RWA ZAI 25 UGA. Phan. $2n = 28$.
      Lobelia lanuriensis De Wild., Rev. Zool. Afr. 8 (Suppl. Bot.): 29 (1920). *Lobelia
         stuhlmannii* var. *lanuriensis* (De Wild.) Hauman, Bull. Acad. Roy. Belgique, C. Sci. (ser.
         5) 19: 608 (1933).
      Lobelia karisimbensis R. E. Fr. & T. C. E. Fr., Svensk Bot. Tidskr. 16: 400 (1922). *Lobelia
         lanuriensis* var. *karisimbensis* (R. E. Fr. & T. C. E. Fr.) Hauman, Mém. Inst. Roy. Colon.
         Belge, Sect. Sci. Nat. (8°) 2: 26 (1934).
      Lobelia lanuriensis var. *ericeti* Hauman, Mém. Inst. Roy. Colon. Belge, Sect. Sci. Nat. (8°)
         2: 25 (1934).
      Lobelia lanuriensis var. *kahuzica* Humbert, Bull. Soc. Bot. France 81: 841 (1935).

Lobelia sublibera S. Watson, Proc. Amer. Acad. Arts 25: 157 (1890).
   NE. Mexico (Nuevo León, Tamaulipas). 79 MXE. Hemicr.

Lobelia subnuda Benth., Pl. Hartweg.: 44 (1840). *Dortmanna subnuda* (Benth.) Kuntze,
    Revis. Gen. Pl. 2: 973 (1891).
    E. Mexico (Hidalgo). 79 MXE. Ther.
        Lobelia discolor Klotzsch in Link, Klotzsch & Otto, Icon. Pl. Rar.: 3 (1840).

Lobelia subpubera Wedd., Chlor. Andina 2: 14 (1858). *Dortmanna subpubera* (Wedd.)
    Kuntze, Revis. Gen. Pl. 2: 973 (1891).
    Ecuador. 83 ECU. Hemicr.

Lobelia sumatrana Merr., Notul. Nat. Acad. Nat. Sci. Philadelphia 47: 9 (1940).
    Sumatera. 42 SUM. Cham. or hemicr.

Lobelia surrepens Hook. f., Fl. Tasman. 1: 237 (1856). *Dortmanna surrepens* (Hook. f.)
    Kuntze, Revis. Gen. Pl. 2: 973 (1891), as 'subrepens'. *Pratia surrepens* (Hook. f.) E. Wimm.,
    Pflanzenr. IV.276b: 108 (1943).
    SE. Australia. 50 NSW TAS VIC. Hemicr. $2n = 28, 42$.

Lobelia sutherlandii E. Wimm., Kew Bull. 1952: 141 (1952).
    Cape Provinces. 27 CPP. Hemicr. or geophyte.

Lobelia tarsophora Seaton ex Greenm., Proc. Amer. Acad. Arts 33: 489 (1898).
    E. Mexico (Hidalgo, Veracruz). 79 MXE MXG. Ther.

Lobelia tatea (E. Wimm.) E. Wimm., Pflanzenr. IV.276b: 119 (1943).
    S. Mexico to Nicaragua. 79 MXS 80 BLZ GUA HON NIC. Nanophan. or hemicr.
      * *Pratia tatea* E. Wimm., Repert. Spec. Nov. Regni Veg. 29: 51 (1931).

Lobelia telekii Schweinf. in L. Höhn., Rudolf- Stephanie-See: 861 (1892).
    Kenya & Uganda. 25 KEN UGA. Nanophan. or herb. phan. $2n = 28$.
        Lobelia fenniae T. C. E. Fr., Bot. Not. 1923: 296 (1923).

Lobelia telephioides (C. Presl) A. DC. in DC., Prodr. 7: 367 (1839).
    Réunion & Mauritius. 29 MAU REU. Hemicr.
      * *Rapuntium telephioides* C. Presl, Prodr. Monogr. Lobel.: 18 (1836).
        Lobelia telephioides f. *pilosella* E. Wimm., Pflanzenr. IV.276b: 494 (1953).

Lobelia tenera Kunth in Humb., Bonpl. & Kunth, Nov. Gen. Sp. 3: 314 (quarto), 245 (folio)
    (1819). *Rapuntium tenerum* (Kunth) C. Presl, Prodr. Monogr. Lobel.: 15 (1836). *Dortmanna
    tenera* (Kunth) Kuntze, Revis. Gen. Pl. 2: 973 (1891).
    Venezuela to Peru. 82 VEN 83 CLM ECU PER. Hemicr. or geophyte.
        Lobelia rupestris Kunth in Humb., Bonpl. & Kunth, Nov. Gen. Sp. 3: 313 (quarto), 244
            (folio) (1819). *Rapuntium rupestre* (Kunth) C. Presl, Prodr. Monogr. Lobel.: 15 (1836).
            Dortmanna rupestris (Kunth) Kuntze, Revis. Gen. Pl. 2: 973 (1891).
        Lobelia polygalifolia Willd. ex Schult. in Roem. & Schult., Syst. Veg. 5: 56 (1819), as
            'polygalaefolia'.
        Lobelia veronicifolia Willd. ex Schult. in Roem. & Schult., Syst. Veg. 5: 56 (1819), as
            'veronicaefolia'. *Rapuntium veronicifolium* (Willd. ex Schult.) C. Presl, Prodr. Monogr.
            Lobel.: 23 (1836), as 'veronicaefolium'.
        Lobelia weberbaueri Zahlbr., Bot. Jahrb. Syst. 37: 462 (1906). *Lobelia subpubera* var.
            weberbaueri (A. Zahlbr.) E. Wimm., Repert. Spec. Nov. Regni Veg. 29: 51 (1931).
        Lobelia tenera var. *belladonna* E. Wimm. in J. F. Macbr., Fl. Peru 6: 481 (1937).
        Lobelia tenera var. *tenuicaulis* E. Wimm., Ann. Naturhist. Mus. Wien 56: 356 (1948).

Lobelia tenuior R. Br., Prodr.: 564 (1810). *Rapuntium tenuius* (R. Br.) C. Presl, Prodr. Monogr.
    Lobel.: 30 (1836). *Dortmanna tenuior* (R. Br.) Kuntze, Revis. Gen. Pl. 2: 973 (1891).
    W. Australia; cult. 50 WAU. Ther. $2n = 18$.

Lobelia ramosa Benth. in Maund & Hensl., Botanist: pl. 93 (1838).
Lobelia heterophylla var. *major* Paxton, Paxton's Mag. Bot. 9: 101 (1842).
Lobelia longipedunculata de Vriese in Lehm., Pl. Preiss.: 394 (1845), as 'longepedunculata'.
Lobelia longipedunculata var. *gracilior* de Vriese in Lehm., Pl. Preiss.: 395 (1845).
Lobelia adscendens de Vriese in Lehm., Pl. Preiss.: 395 (1845).
Lobelia ciliata de Vriese in Lehm., Pl. Preiss.: 397 (1845).
Lobelia ramosa f. *alba* Voss in Siebert & Voss, Vilm. Blumengärtn. (ed. 3) 1: 575 (1894).
Lobelia ramosa f. *nana* Voss in Siebert & Voss, Vilm. Blumengärtn. (ed. 3) 1: 575 (1894).
Lobelia ramosa f. *rosea* Voss in Siebert & Voss, Vilm. Blumengärtn. (ed. 3) 1: 575 (1894).
Lobelia tenuior var. *glabra* E. Wimm., Pflanzenr. IV.276b: 566 (1953).

Lobelia terminalis C. B. Clarke in Hook. f., Fl. Brit. India 3: 424 (1881). *Dortmanna trigona* var. *terminalis* (C. B. Clarke) Kuntze, Revis. Gen. Pl. 2: 380 (1891).
C. & E. India (Madhya Pradesh, Bihar, West Bengal) to SC. China (Yunnan) & Vietnam. 36 CHC 40 ASS EHM IND 41 LAO THA VIE. Ther. $2n = 14$.
Lobelia terminalis var. *minuta* C. B. Clarke in Hook. f., Fl. Brit. India 3: 424 (1881).
Lobelia thorelii E. Wimm., Repert. Spec. Nov. Regni Veg. 26: 3 (1929).

Lobelia thapsoidea Schott ex Pohl, Pl. Bras. Icon. Descr. 2: 102 (1831). *Rapuntium thapsoideum* (Schott ex Pohl) C. Presl, Prodr. Monogr. Lobel.: 24 (1836). *Haynaldia thapsoidea* (Schott ex Pohl) Kanitz, Magyar. Növényt. Lapok 1: 4 (1877). *Dortmanna thapsoidea* (Schott ex Pohl) Kuntze, Revis. Gen. Pl. 2: 973 (1891), as 'thapsodea'. *Euhaynaldia thapsoidea* (Schott ex Pohl) Borbás in Pall., Nagy Lex. 8: 779 (1894).
Brazil (Goiás, Minas Gerais, Rio de Janeiro, São Paulo). 84 BZC BZL. Cham. or hemicr. $2n = 14$.
Geniostoma brasiliense Spreng., Syst. Veg. 1: 588 (1824); non *Lobelia brasiliensis* A. O. S. Vieira & G. J. Sheph., Novon 8: 457 (1998).

Lobelia thermalis Thunb., Prodr. Pl. Cap.: 40 (1794). *Rapuntium thermale* (Thunb.) C. Presl, Prodr. Monogr. Lobel.: 12 (1836). *Parastranthus thermalis* (Thunb.) A. DC. in DC., Prodr. 7: 354 (1839). *Dortmanna thermalis* (Thunb.) Kuntze, Revis. Gen. Pl. 2: 973 (1891).
S. Africa. 26 ANG ZIM 27 BOT CPP LES NAM OFS TVL. Hemicr. or ther.
Lobelia leptocarpa Griess., Linnaea 5: 419 (1830). *Rapuntium glabrifolium* C. Presl, Prodr. Monogr. Lobel.: 12 (1836).
Lobelia mundtiana Cham., Linnaea 8: 215 (1833). *Rapuntium mundtianum* (Cham.) C. Presl in Eckl. & Zeyh., Enum. Pl. Afric. Austral.: 395 (1837). *Rapuntium glabrifolium* var. *mundtianum* (Cham.) C. Presl in E. Mey., Comm. Pl. Afr. Austral.: 385 (1838).
Lobelia thermalis var. *pungens* E. Wimm., Pflanzenr. IV.276b: 568 (1953).

Lobelia thuliniana E. B. Knox, Contr. Univ. Michigan Herb. 19: 253 (1993).
Tanzania. 25 TAN. Nanophan.

Lobelia tibetica W. L. Zheng, Acta Phytotax. Sin. 36: 549 (1998).
Tibet. 36 CHT. Hemicr.

Lobelia tomentosa L. f., Suppl. Pl.: 394 (1782). *Rapuntium tomentosum* (L. f.) C. Presl, Prodr. Monogr. Lobel.: 19 (1836). *Dortmanna tomentosa* (L. f.) Kuntze, Revis. Gen. Pl. 2: 973 (1891). *Dortmanna coronopifolia* var. *tomentosa* (L. f.) Kuntze, Revis. Gen. Pl. 3(2): 187 (1898).
S. Africa. 27 CPP NAT. Cham.
Rapuntium paucidentatum C. Presl, Prodr. Monogr. Lobel.: 20 (1836). *Lobelia paucidentata* (C. Presl) A. DC. in DC., Prodr. 7: 362 (1839). *Lobelia tomentosa* var. *paucidentata* (C. Presl) Sond. in Harv. & Sond., Fl. Cap. 3: 543 (1865). *Dortmanna coronopifolia* var. *paucidentata* (C. Presl) Kuntze, Revis. Gen. Pl. 3(2): 187 (1898).
Rapuntium tomentosum var. *tenuifolium* C. Presl in Eckl. & Zeyh., Enum. Pl. Afric. Austral.: 399 (1837).

Lobelia tomentosa var. *multiflora* A. DC. in DC., Prodr. 7: 364 (1839). *Dortmanna coronopifolia* var. *multiflora* (A. DC.) Kuntze, Revis. Gen. Pl. 3(2): 187 (1898). *Dortmanna coronopifolia* f. *azurea* Kuntze, Revis. Gen. Pl. 3(2): 187 (1898); non Kuntze, Revis. Gen. Pl. 3(2): 187 (1898); nec Kuntze, Revis. Gen. Pl. 3(2): 187 (1898).

Lobelia trigonocaulis F. Muell., Fragm. 1: 18 (1858). *Dortmanna trigonocaulis* (F. Muell.) Kuntze, Revis. Gen. Pl. 2: 973 (1891).
E. Australia (Queensland, New South Wales). 50 NSW QLD. Hemicr. $2n = 20$.

Lobelia tripartita Thulin, Nordic J. Bot. 11: 525 (1991).
Ethiopia. 24 ETH. Hemicr. or ther.

Lobelia trullifolia Hemsl. in Oliv., Fl. Trop. Afr. 3: 466 (1877). *Dortmanna trullifolia* (Hemsl.) Kuntze, Revis. Gen. Pl. 2: 973 (1891).
Ethiopia to Zaïre & S. Africa. 23 ZAI 24 ETH 25 KEN TAN 26 MLW MOZ ZAM 27 SWZ TVL. Hemicr. or ther.

subsp. **delicatula** (Compton) Thulin, Nordic J. Bot. 3: 374 (1983).
S. Africa. 27 SWZ TVL. Hemicr. or ther.
Lobelia delicatula Compton, J. S. African Bot. 41: 50 (1975).

subsp. **minor** Thulin in Launert, Fl. Zambes. 7(1): 127 (1983).
Tanzania to Zambia. 25 TAN 26 MLW ZAM. Ther.

subsp. **pinnatifida** Thulin in Launert, Fl. Zambes. 7(1): 127 (1983).
Zambia. 26 ZAM. Ther.

subsp. **rhodesica** (R. E. Fr.) Thulin in Launert, Fl. Zambes. 7(1): 128 (1983).
Zambia. 26 ZAM. Hemicr. or ther.
Lobelia rhodesica R. E. Fr., Wiss. Ergebn. Schwed. Rhodesia-Kongo-Exped. 1: 316 (1914).

subsp. **trullifolia**
Ethiopia to Mozambique & Zambia. 24 ETH 25 KEN TAN 26 MLW MOZ ZAM. Hemicr. or ther.
Lobelia melleri Hemsl. in Oliv., Fl. Trop. Afr. 3: 468 (1877). *Dortmanna melleri* (Hemsl.) Kuntze, Revis. Gen. Pl. 2: 972 (1891).
Lobelia usambarensis Engl., Bot. Jahrb. Syst. 19 (Beibl. 47): 50 (1894).
Lobelia nyassae Engl., Pflanzenw. Ost-Afrikas C: 401 (1895).
Lobelia buchananii Baker f., Bull. Misc. Inform. Kew 1898: 156 (1898), as 'buchanani'.
Lobelia intertexta Baker f., Bull. Misc. Inform. Kew 1898: 157 (1898).
Lobelia nyikensis Baker f., Bull. Misc. Inform. Kew 1898: 157 (1898).
Lobelia wentzeliana Engl., Bot. Jahrb. Syst. 30: 420 (1901).
Lobelia dolichopus Schltr., Bot. Jahrb. Syst. 57: 617 (1922).
Lobelia saliensis E. Wimm., Notizbl. Bot. Gart. Berlin-Dahlem 12: 107 (1934). *Lobelia trullifolia* var. *saliensis* (E. Wimm.) E. Wimm., Pflanzenr. IV.276b: 490 (1953).
Lobelia usambarensis var. *hispidella* E. Wimm., Notizbl. Bot. Gart. Berlin-Dahlem 12: 107 (1934).
Lobelia brassiana E. Wimm., Kew Bull. 1952: 139 (1952).
Lobelia melleri f. *pilosula* E. Wimm., Pflanzenr. IV.276b: 529 (1953).
Lobelia melleri var. *pulchra* E. Wimm., Pflanzenr. IV.276b: 529 (1953).
Lobelia intertexta f. *arida* E. Wimm., Pflanzenr. IV.276b: 530 (1953).
Lobelia intertexta var. *diversifolia* E. Wimm., Pflanzenr. IV.276b: 530 (1953). *Lobelia trullifolia* f. *glabricalycina* E. Wimm., Pflanzenr. IV.276c: 861 (1968).
Lobelia melleri var. *trichella* E. Wimm., Pflanzenr. IV.276c: 865 (1968).
Lobelia zombaensis E. Wimm., Pflanzenr. IV.276c: 866 (1968).
Lobelia usambarensis var. *calantha* E. Wimm., Pflanzenr. IV.276c: 868 (1968).

Lobelia tupa L., Sp. Pl.: 929 (1753), as 'trapa'. *Tupa feuillei* G. Don, Gen. Hist. 3: 700 (1834).
Rapuntium tupa (L.) C. Presl, Prodr. Monogr. Lobel.: 28 (1836). *Dortmanna tupa* (L.)
Kuntze, Revis. Gen. Pl. 2: 972 (1891). *Lobelia feuillei* (G. Don) Voss in Siebert & Voss,
Vilm. Blumengärtn. (ed. 3) 1: 577 (1894).
SC. Chile (Cachapoal to Chiloé); cult. 85 CLC CLS. Cham. or hemicr.
    Lobelia mucronata Cav., Icon. 6:11 (1800). *Tupa cavanillesiana* G. Don, Gen. Hist. 3: 700
       (1834). *Rapuntium mucronatum* (Cav.) C. Presl, Prodr. Monogr. Lobel.: 29 (1836). *Tupa*
       mucronata (Cav.) A. DC. in DC., Prodr. 7: 392 (1839). *Tupa feuillei* var. *mucronata*
       (Cav.) Vatke, Linnaea 38: 726 (1874). *Dortmanna mucronata* (Cav.) Kuntze, Revis.
       Gen. Pl. 2: 972 (1891). *Lobelia tupa* var. *mucronata* (Cav.) Reiche, Anales Univ. Chile
       117: 458 (1905).
    Lobelia serrata Meyen, Reise 1: 300 (1834).
    Tupa berteroi A. DC. in DC., Prodr. 7: 392 (1839), as 'berterii'. *Tupa feuillei* var. *berteroi*
       (A. DC.) Vatke, Linnaea 38: 726 (1874), as 'berterii'. *Dortmanna berteroi* (A. DC.)
       Kuntze, Revis. Gen. Pl. 2: 972 (1891), as 'berterii'. *Lobelia tupa* var. *berteroi* (A. DC.)
       Reiche, Anales Univ. Chile 117: 458 (1905), as 'berterii'. *Lobelia mucronata* var.
       berteroi (A. DC.) E. Wimm., Pflanzenr. IV.276b: 614 (1953), as 'berterii'.
    Tupa mucronata var. *hookeri* A. DC. in DC., Prodr. 7: 392 (1839). *Lobelia mucronata* f.
       hookeri (A. DC.) E. Wimm., Pflanzenr. IV.276b: 614 (1953).
    Tupa montana Phil., Anales Univ. Chile 43: 506 (1873); non C. Wright ex Griseb., Cat.
       Pl. Cub.: 159 (1866). *Dortmanna philippiana* Kuntze, Revis. Gen. Pl. 2: 972 (1891);
       non *Dortmanna montana* (Fresen.) Kuntze, Revis. Gen. Pl. 2: 972 (1891). *Lobelia tupa*
       var. *montana* Reiche, Anales Univ. Chile 117: 459 (1905).
    Tupa feuillei var. *macrophylla* Vatke, Linnaea 38: 726 (1874).
    Dortmanna bicalcarata Kuntze, Revis. Gen. Pl. 3(2): 186 (1898). *Lobelia bicalcarata*
       (Kuntze) Zahlbr. ex K. Schum., Just's Bot. Jahresber. 26(1): 373 (1900). *Lobelia*
       tupa var. *bicalcarata* (Kuntze) E. Wimm., Ann. Naturhist. Mus. Wien 56:
       365 (1948).
    Lobelia mucronata f. *ovalifolia* E. Wimm., Pflanzenr. IV.276b: 614 (1953).
    Lobelia tupa var. *pavonii* E. Wimm., Pflanzenr. IV.276c: 881 (1968).

Lobelia udzungwensis Thulin, Kew Bull. 59: 190 (2004).
Tanzania. 25 TAN. Phan.

Lobelia uliginosa E. Wimm., Pflanzenr. IV.276c: 867 (1968).
Tanzania & Zambia. 25 TAN 26 ZAM. Ther.

Lobelia umbellifera McVaugh, Bull. Torrey Bot. Club 67: 144 (1940).
S. Mexico to Guatemala. 79 MXT 80 GUA. Hemicr. or geophyte.
  * *Lobelia fasciculata* Donn. Sm., Bot. Gaz. (Crawfordsville) 27: 338 (1899); non (Benth.)
    Kuntze, Revis. Gen. Pl. 2: 378 (1891).

Lobelia urens L., Sp. Pl.: 931 (1753). *Rapuntium urens* (L.) Mill., Gard. Dict. (ed. 8) (1768).
Lobelia verbenifolia Salisb., Prodr. Stirp. Chap. Allerton: 129 (1796), as 'verbenaefolia'.
Mecoschistum urens (L.) Dulac, Fl. Hautes-Pyrénées: 459 (1867). *Dortmanna urens* (L.)
Kuntze, Revis. Gen. Pl. 2: 973 (1891). Hemicr.
W. Europe, Morocco, Macaronesia. 10 GRB 11 BGM 12 FRA POR SPA 20 MOR 21 AZO MDR.
    Lobelia farsetia Vand., Fasc. Pl.: 19 (1771).
    Lobelia serrulata Schott ex Brot., Fl. Lusit. 1: 304 (1804). *Rapuntium serrulatum* (Schott
       ex Brot.) C. Presl, Prodr. Monogr. Lobel.: 14 (1836). *Lobelia urens* var. *serrulata*
       (Schott ex Brot.) Steud., Nomencl. Bot. (ed. 2) 2: 63 (1841).
    Lobelia urens var. *brevibracteata* Pérez Lara, Anales Soc. Esp. Hist. Nat. 18: 36 (1889).
    Lobelia urens var. *longibracteata* Pérez Lara, Anales Soc. Esp. Hist. Nat. 18: 36 (1889), as
       'longebracteata'.
    Lobelia urens var. *integra* Chabert, Bull. Herb. Boissier 3: 148 (1895).

Lobelia vagans Balf. f., J. Linn Soc., Bot. 16: 16 (1877). *Dortmanna vagans* (Balf. f.) Kuntze, Revis. Gen. Pl. 2: 973 (1891).
   Rodrigues. 29 ROD. Ther.

Lobelia valida L. Bolus, Bull. Misc. Inform. Kew 1934: 260 (1934).
   Cape Provinces. 27 CPP. Cham.

Lobelia vanreenensis (Kuntze) K. Schum., Just's Bot. Jahresber. 26(1): 373 (1900).
   S. Africa. 27 CPP LES NAT OFS TVL. Hemicr.
      * *Dortmanna vanreenensis* Kuntze, Revis. Gen. Pl. 3(2): 188 (1898).
      Lobelia vanreenensis var. *villosula* E. Wimm., Pflanzenr. IV.276b: 585 (1953).

Lobelia victoriensis P. Royen, Bot. J. Linn. Soc. 77: 118 (1978).
   New Guinea. 43 NWG. Hemicr. or ther.

Lobelia villaregalis T. J. Ayers, Brittonia 39: 419 (1987). *Calcaratolobelia villaregalis* (T. J. Ayers) Wilbur, Sida 17: 562 (1997).
   W. Mexico (Jalisco). 79 MXS. Geophyte.

Lobelia villosa (Rock) H. St. John & Hosaka, Occas. Pap. Bernice Pauahi Bishop Mus. 14: 124 (1938).
   Hawaiian Is. (Kaua'i). 63 HAW. Nanophan.
      * *Lobelia kauaiensis* var. *villosa* Rock, Bull. Torrey Bot. Club 44: 237 (1917). *Galeatella villosa* (Rock) O. Deg. & I. Deg., Fl. Hawaiiensis, fam. 339 (1962).

Lobelia viridiflora McVaugh, N. Amer. Fl. 32A: 91 (1943).
   Jamaica. 81 JAM. Cham. or hemicr.

Lobelia vivaldii Lammers & Proctor, Brittonia 46: 274 (1994).
   Puerto Rico (Isla de Mona). 81 PUE. Nanophan.

Lobelia volcanica T. J. Ayers, Syst. Bot. 15: 317 (1990).
   S. Mexico (Guerrero, México State, Morelos, Veracruz). 79 MXC MXG MXS. Ther.
      * *Heterotoma tenella* Turcz., Bull. Soc. Imp. Naturalistes Moscou 25(3): 175 (1852); non *Lobelia tenella* L., Mant. Pl.: 120 (1767); nec Burm. f., Prodr. Fl. Cap.: 29 (1768).
      Heterotoma cordifolia var. *tenella* (Turcz.) E. Wimm., Pflanzenr. IV.276b: 717 (1953).
      Calcaratolobelia tenella (Turcz.) Wilbur, Sida 17: 563 (1997).

Lobelia welwitschii Engl. & Diels, Bot. Jahrb. Syst. 26: 116 (1898).
   Cameroon to Ethiopia, Mozambique & Angola. 23 CMN ZAI 24 ETH 25 ALL 26 ANG MLW MOZ ZAM ZIM. Hemicr.
      Lobelia welwitschii var. *albiflora* Hiern, Cat. Afr. Pl. 1: 627 (1898).
      Lobelia stolzii Schltr., Bot. Jahrb. Syst. 57: 621 (1922).

Lobelia wilmsiana Diels, Bot. Jahrb. Syst. 26: 115 (1898).
   Transvaal. 27 TVL. Hemicr.

Lobelia winifrediae Diels, Bot. Jahrb. Syst. 35: 549 (1904), as 'winfridae'.
   W. Australia. 50 WAU. Ther.

Lobelia wollastonii Baker f., J. Linn. Soc., Bot. 38: 265 (1908), as 'wollastoni'.
   Zaïre to Uganda. 23 RWA ZAI 25 UGA. Phan.
      Lobelia wollastonii var. *scaettana* Hauman, Mém. Inst. Roy. Colon. Belge, Sect. Sci. Nat. (8°) 2: 28 (1934).

Lobelia xalapensis Kunth in Humb., Bonpl. & Kunth, Nov. Gen. Sp. 3: 315 (quarto), 246
(folio) (1819). *Rapuntium xalapense* (Kunth) C. Presl, Prodr. Monogr. Lobel.: 25 (1836).
Lobelia cliffortiana var. *xalapensis* (Kunth) A. Gray, Syn. Fl. N. Amer. 2(1): 7 (1878).
Dortmanna xalapensis (Kunth) Kuntze, Revis. Gen. Pl. 2: 973 (1891). *Dortmanna cliffortiana*
var. *xalapensis* (Kunth) Kuntze, Revis. Gen. Pl. 3(2): 187 (1898).
S. Mexico to Bolivia & N. Argentina (Jujuy, Salta, Tucumán, Santiago del Estero, Chaco,
Formosa, Missiones, Corrientes); Galápagos (Isabela, Santa Cruz, Santa María);
naturalized in Lesser Antilles (Martinique). 79 MXG MXS MXT 80 BLZ GUA HON NIC
COS PAN (81) win 82 VEN 83 BOL CLM ECU GAL PER 84 BZE BZL BZS 85 AGE AGW
PAR. Ther.
Lobelia monticola Kunth in Humb., Bonpl. & Kunth, Nov. Gen. Sp. 3: 316 (quarto), 246
(folio) (1819). *Rapuntium monticolum* (Kunth) C. Presl, Prodr. Monogr. Lobel.: 25
(1836). *Dortmanna monticola* (Kunth) Kuntze, Revis. Gen. Pl. 2: 972 (1891).
Lobelia palmaris Willd. ex Schult. in Roem. & Schult., Syst. Veg. 5: 56 (1819).
Lobelia mollis Graham, Edinburgh New Philos. J. 8: 185 (1830). *Rapuntium molle*
(Graham) C. Presl, Prodr. Monogr. Lobel.: 30 (1836). *Dortmanna mollis* (Graham)
Kuntze, Revis. Gen. Pl. 2: 972 (1891).
Lobelia ocimoides Kunze, Linnaea 24: 178 (1851). *Dortmanna ocimoides* (Kunze) Kuntze,
Revis. Gen. Pl. 2: 973 (1891), as 'ocimodes'.

Lobelia xongorolana E. Wimm., Repert. Spec. Nov. Regni Veg. 38: 79 (1935).
Angola. 26 ANG. Phan.

Lobelia yucatana E. Wimm., Repert. Spec. Nov. Regni Veg. 38: 83 (1935).
SE. Mexico (Yucatán) & Guatemala. 79 MXT 80 GUA. Ther.

Lobelia yuccoides Hillebr., Fl. Hawaiian Isl.: 237 (1888). *Dortmanna yuccoides* (Hillebr.)
Kuntze, Revis. Gen. Pl. 2: 973 (1891), as 'yuccodes'. *Neowimmeria yuccoides* (Hillebr.) O.
Deg. & I. Deg., Phytologia 12: 73 (1965).
Hawaiian Is. (Kaua'i, O'ahu). 63 HAW. Nanophan.

Lobelia zelayensis Wilbur, Sida 14: 563 (1991).
Nicaragua. 81 NIC. Nanophan. or hemicr.

Lobelia zeylanica L., Sp. Pl.: 932 (1753). *Rapuntium zeylanicum* (L.) C. Presl, Prodr. Monogr.
Lobel.: 13 (1836). *Dortmanna zeylanica* (L.) Kuntze, Revis. Gen. Pl. 2: 380 (1891).
C. & E. India (Madhya Pradesh, Bihar, West Bengal) to Nepal, S. China (Fujian,
Guangdong, Guangxi, Yunnan), Taiwan, Philippines & New Guinea; naturalized in Fiji.
36 CHC CHS 38 TAI 40 ASS BAN EHM IND NEP SRL 41 LAO MYA THA VIE 42 BOR
JAW MLY PHI SUL SUM 43 NWG (60) fij. Hemicr.
Lobelia hirta L., Sp. Pl.: 932 (1753) *Rapuntium hirtum* (L.) Mill., Gard. Dict. (ed. 8)
(1768). *Lobelia zeylanica* var. *hirta* (L.) Martyn in Mill., Gard. Dict. (ed. 9) (1797).
Lobelia succulenta Blume, Bijdr.: 728 (1826). *Rapuntium succulentum* (Blume) C. Presl,
Prodr. Monogr. Lobel.: 13 (1836). *Dortmanna succulenta* (Blume) Kuntze, Revis. Gen.
Pl. 2: 973 (1891).
Lobelia affinis Wall. ex G. Don, Gen. Hist. 3: 709 (1834); non Mirb., Hist. Nat. Pl. 14:
237 (1805). *Rapuntium affine* C. Presl, Prodr. Monogr. Lobel.: 13 (1836). *Dortmanna*
trigona var. *affinis* (C. Presl) Kuntze, Revis. Gen. Pl. 2: 380 (1891).
Lobelia subcuneata Miq., Fl. Ned. Ind.: 574 (1857). *Dortmanna subcuneata* (Miq.) Kuntze,
Revis. Gen. Pl. 2: 973 (1891).
Lobelia lobbiana Hook. f. & Thomson, J. Proc. Linn. Soc., Bot. 2: 28 (1858). *Lobelia*
affinis var. *lobbiana* (Hook. f. & Thomson) C. B. Clarke in Hook. f., Fl. Brit. India 3:
424 (1881). *Lobelia succulenta* var. *lobbiana* (Hook. f. & Thomson) E. Wimm., Ann.
Naturhist. Mus. Wien 56: 361 (1948). *Lobelia zeylanica* var. *lobbiana* (Hook. f. &
Thomson) Y. S. Lian, Fl. Reipubl. Popul. Sin. 73(2): 149 (1983).
Lobelia barbata Warb., Bot. Jahrb. Syst. 13: 444 (1891); non Cav., Icon. 6: 12 (1800).

Pratia torricellensis Lauterb. in K. Schum. & Lauterb., Fl. Schutzgeb. Südsee, Nachträge: 402 (1905).

Pratia ovata Elmer, Leafl. Philipp. Bot. 2: 593 (1909).

Lobelia zeylanica f. *viridis* Hochr., Candollea 5: 292 (1934).

Lobelia zeylanica f. *rubescens* Hochr., Candollea 5: 292 (1934).

Lobelia succulenta f. *glabra* E. Wimm., Ann. Naturhist. Mus. Wien 56: 361 (1948).

Lobelia zwartkopensis E. Wimm., Pflanzenr. IV.276b: 603 (1953).
  Cape Provinces. 27 CPP. Ther.

Synonyms:

Lobelia sect. *Colensoa* (Hook. f.) Murata === **Lobelia** L.

Lobelia sect. *Euhaynaldia* (Borbás) Zahlbr. === **Lobelia** L.

Lobelia sect. *Galeatella* E. Wimm. === **Lobelia** L.

Lobelia sect. *Isolobus* (A. DC.) C. B. Clarke === **Lobelia** L.

Lobelia sect. *Mezleria* (C. Presl) Schönl. === **Monopsis** Salisb.

Lobelia sect. *Palmerella* (A. Gray) McVaugh === **Palmerella** A. Gray

Lobelia sect. *Revolutella* E. Wimm. === **Lobelia** L.

Lobelia sect. *Rhynchopetalum* (Fresen.) Benth. === **Lobelia** L.

Lobelia sect. *Trimeris* (C. Presl) A. DC. === **Lobelia** L.

Lobelia sect. *Tupa* (G. Don) Benth. === **Lobelia** L.

Lobelia sect. *Tylomium* (C. Presl) Benth. === **Lobelia** L.

Lobelia subg. *Isolobus* (A. DC.) Y. S. Lian === **Lobelia** L.

Lobelia subg. *Mezleria* (C. Presl) E. Wimm. === **Monopsis** Salisb.

Lobelia subg. *Tupa* E. Wimm. === **Lobelia** L.

Lobelia [unranked] *Centropogon* (C. Presl) Heynh. === **Centropogon** C. Presl

Lobelia [unranked] *Enchysia* (C. Presl) Heynh. === **Lobelia** L.

Lobelia [unranked] *Isolobus* (A. DC.) Heynh. === **Lobelia** L.

Lobelia [unranked] *Isotoma* R. Br. === **Isotoma** (R. Br.) Lindl.

Lobelia [unranked] *Monopsis* (Salisb.) Heynh. === **Monopsis** Salisb.

Lobelia [unranked] *Parastranthus* (G. Don) Heynh. === **Monopsis** Salisb.

Lobelia [unranked] *Piddingtonia* (A. DC.) Heynh. === **Lobelia** L.

Lobelia [unranked] *Pratia* (Gaudich.) Heynh. === **Lobelia** L.

Lobelia [unranked] *Tupa* (G. Don) Heynh. === **Lobelia** L.

Lobelia acuminata (Gaudich.) Hook. & Arn. === **Cyanea acuminata** (Gaudich.) Hillebr.

Lobelia acutangula (C. Presl) A. DC. === **Lobelia erinus** L.

Lobelia adscendens de Vriese === **Lobelia tenuior** R. Br.

Lobelia affinis Mirb. === **Siphocampylus affinis** (Mirb.) McVaugh

Lobelia affinis Wall. ex G. Don === **Lobelia zeylanica** L.

Lobelia affinis var. *lobbiana* (Hook. f. & Thomson) C. B. Clarke === **Lobelia zeylanica** L.

Lobelia afromontana T. C. E. Fr. === **Lobelia cymbalarioides** Engl.

Lobelia alata Labill. === **Lobelia anceps** L. f.

Lobelia alata var. *cuneiformis* (Labill.) R. Br. === **Lobelia anceps** L. f.

Lobelia alata var. *longisepala* E. Wimm. === **Lobelia anceps** L. f.

Lobelia alata var. *minor* (Sond.) E. Wimm. === **Lobelia anceps** L. f.

Lobelia alata var. *ottoniana* (C. Presl) E. Wimm. === **Lobelia anceps** L. f.

Lobelia alata var. *rudatisii* (Schltr.) E. Wimm. === **Lobelia anceps** L. f.

Lobelia alata var. *stolonifera* Endl. === **Lobelia anceps** L. f.

Lobelia albomaculata E. Wimm. === **Lobelia neumannii** T. C. E. Fr.

Lobelia alexia E. Wimm. === **Lobelia acuminata** Sw.

Lobelia algoensis A. DC. === **Lobelia erinus** L.

Lobelia aligera Haines === **Lobelia heyneana** Schult.

Lobelia alpina Vell. === ?

Lobelia alsinoides f. *elongata* (Danguy) E. Wimm. === **Lobelia alsinoides** Lam.

Lobelia alsinoides subsp. *hancei* (H. Hara) Lammers === **Lobelia chevalieri** Danguy

Lobelia alsinoides var. *cantonensis* (E. Wimm. ex Danguy) E. Wimm. === **Lobelia alsinoides** Lam.

Lobelia alsinoides var. *hirta* (E. Wimm. ex Danguy) E. Wimm. === **Lobelia alsinoides** Lam.

Lobelia altimontis E. Wimm. === **Lobelia erinus** L.

Lobelia alyssifolia Salisb. === **Lobelia pubescens** Aiton

Lobelia amabilis M. E. Jones === **Lobelia aurita** (Brandegee) T. J. Ayers

Lobelia ambigua Cham. === **Cyanea lanceolata** (Gaudich.) Lammers, Givnish & Sytsma

Lobelia amoena var. *glandulifera* A. Gray === **Lobelia georgiana** McVaugh

Lobelia amoena f. *obtusata* (A. Gray) Voss === **Lobelia georgiana** McVaugh

Lobelia amoena var. *obtusata* A. Gray === **Lobelia georgiana** McVaugh

Lobelia amplexicaulis de Vriese === **Grammatotheca bergiana** (Cham.) C. Presl

Lobelia amygdalina Willd. ex Schult. === **Lobelia laxiflora** Kunth subsp. **laxiflora**

Lobelia anatina var. *riskindii* M. C. Johnst. === **Lobelia anatina** E. Wimm.

Lobelia anceps f. *pleniflora* E. Wimm. === **Lobelia fervens** Thunb. subsp. **fervens**

Lobelia anceps f. *ugandensis* E. Wimm. === **Lobelia flaccida** subsp. **granvikii** (T. C. E. Fr.) Thulin

Lobelia anceps var. *alata* (Labill.) Heynh. === **Lobelia anceps** L. f.

Lobelia anceps var. *asperulata* (Klotzsch) E. Wimm. === **Lobelia fervens** Thunb. subsp. **fervens**

Lobelia anceps var. *cuneiformis* (Labill.) A. DC. === **Lobelia anceps** L. f.

Lobelia anceps var. *minor* Sond. === **Lobelia anceps** L. f.

Lobelia anceps var. *recurvata* E. Wimm. === **Lobelia fervens** subsp. **recurvata** (E. Wimm.) Thulin

Lobelia anceps var. *vagans* R. D. Good === **Lobelia anceps** L. f.

Lobelia andina Benth. === **Lobelia laxiflora** Kunth subsp. **laxiflora**

Lobelia andropogon Cav. === **Centropogon cornutus** (L.) Druce

Lobelia androsacea Willd. ex Schult. === **Lysipomia acaulis** Kunth

Lobelia angolensis Engl. & Diels === **Lobelia sonderiana** (Kuntze) Lammers

Lobelia angulata var. *papuana* (S. Moore) Gilli === **Lobelia nummularia** Lam.

Lobelia angulatodentata Hook. & Arn. === **Lobelia laxiflora** Kunth subsp. **laxiflora**

Lobelia angustifolia (A. DC.) Urbina === **Lobelia laxiflora** subsp. **angustifolia** (A. DC.) Eakes & Lammers

Lobelia angustifolia Benth. === **Lobelia anceps** L. f.

Lobelia angustifolia Cham. === **Cyanea angustifolia** (Cham.) Hillebr.

Lobelia antisyphilitica Hayne, Brandt & Ratzeb. === **Lobelia siphilitica** L.

Lobelia aphylla Nutt. === **Apteria aphylla** (Nutt.) Barnhart (Burmanniaceae)

Lobelia appendiculata var. *gattingeri* (A. Gray) McVaugh === **Lobelia gattingeri** A. Gray

Lobelia aquatica E. Phillips === **Lobelia galpinii** Schltr.

Lobelia aquatica var. *gracilis* E. Wimm. === **Lobelia aquatica** Cham.

Lobelia arabidea (C. Presl) Steud. === **Wimmerella arabidea** (C. Presl) Serra, M. B. Crespo & Lammers

Lobelia arabidoides Hook. & Arn. === **Lobelia flexuosa** (C. Presl) A. DC. subsp. **flexuosa**

Lobelia arborea G. Forst. === **Sclerotheca arborea** (G. Forst.) A. DC.

Lobelia arenarioides (C. Presl) A. DC. === **Lobelia heyneana** Schult.

Lobelia areolata A. Rich. ex Juss. === **Lobelia stricta** Sw.

Lobelia arfakensis Gibbs === **Lobelia nummularia** Lam.

Lobelia arguta Lindl. === **Lobelia excelsa** Bonpl.

Lobelia aromatica Moon ex Wight === **Lobelia leschenaultiana** (C. Presl) Skottsb.

Lobelia asclepiadea Willd. ex Schult. === **Centropogon asclepiadeus** (Willd. ex Schult.) E. Wimm.

Lobelia aspera Spreng. === **Monopsis simplex** (L.) E. Wimm.

Lobelia asperulata Klotzsch === **Lobelia fervens** Thunb. subsp. **fervens**

Lobelia assurgens var. *jamaicensis* Urb. === **Lobelia assurgens** L.

Lobelia assurgens var. *portoricensis* (A. DC.) Urb. === **Lobelia robusta** Graham

Lobelia assurgens var. *santa-clarae* McVaugh === **Lobelia assurgens** L.

Lobelia ayavacensis Willd. ex Schult. === **Centropogon ayavacensis** (Willd. ex Schult.) Lammers

Lobelia axilliflora (Phil.) Reiche === **Lobelia polyphylla** Hook. & Arn.

Lobelia backhouseana Lem. === **Lobelia caerulea** Sims

Lobelia baoulensis A. Chev. === **Lobelia djurensis** Engl. & Diels

Lobelia barbata Cav. === **Centropogon ferrugineus** (L. f.) Gleason

Lobelia barbata Warb. === **Lobelia zeylanica** L.

Lobelia baumannii f. *alba* E. Wimm. === **Lobelia baumannii** Engl.

Lobelia baumannii var. *cuneata* E. Wimm. === **Lobelia baumannii** Engl.

Lobelia baumannii var. *kummeriae* (Engl. ex T. C. E. Fr.) E. Wimm. === **Lobelia baumannii** Engl.

Lobelia begoniifolia Wall. === **Lobelia nummularia** Lam.

Lobelia belgica de Jonghe === **Lobelia siphilitica** L.

Lobelia bellidifolia L. f. === **Lobelia erinus** L.

Lobelia bellidifolia f. *flexuosa* Zahlbr. === **Lobelia flaccida** (C. Presl) A. DC. subsp. **flaccida**

Lobelia bellidifolia f. *stricta* Zahlbr. === **Lobelia flaccida** (C. Presl) A. DC. subsp. **flaccida**

Lobelia bellidifolia var. *brevidens* (C. Presl) A. DC. === **Lobelia flaccida** (C. Presl) A. DC. subsp. **flaccida**

Lobelia bellidifolia var. *glabrata* C. Presl ex A. DC. === **Lobelia flaccida** (C. Presl) A. DC. subsp. **flaccida**

Lobelia bellidifolia var. *hirsuta* C. Presl ex A. DC. === **Lobelia flaccida** (C. Presl) A. DC. subsp. **flaccida**

Lobelia bellis E. Wimm. === **Lobelia nana** Kunth

Lobelia benguellensis Hiern === **Lobelia erinus** L.

Lobelia benthamiana S. Moore === **Lobelia quadrangularis** R. Br.

Lobelia bergiana Cham. === **Grammatotheca bergiana** (Cham.) C. Presl

Lobelia berlandieri var. *brachypoda* (A. Gray) McVaugh === **Lobelia berlandieri** A. DC.

Lobelia berlandieri var. *seleriana* (E. Wimm.) E. Wimm. === **Lobelia divaricata** Hook. & Arn.

Lobelia berteriana Spreng. === **Centropogon berterianus** (Spreng.) A. DC.

Lobelia beslerioides Kunth === **Centropogon beslerioides** (Kunth) A. DC.

Lobelia betulifolia Cham. === **Siphocampylus betulifolius** (Cham.) G. Don

Lobelia bialata Merr. === **Lobelia heyneana** Schult.

Lobelia bicalcarata (Kuntze) Zahlbr. ex K. Schum. === **Lobelia tupa** L.

Lobelia bicolor Sims === **Lobelia erinus** L.

Lobelia bifida Thunb. === **Wimmerella bifida** (Thunb.) Serra, M. B. Crespo & Lammers

Lobelia biserrata Cav. === **Siphocampylus biserratus** (Cav.) A. DC.

Lobelia biserrata var. *spicata* Hook. === **Siphocampylus biserratus** (Cav.) A. DC.

Lobelia bisserrata Sessé & Moç. === **Lobelia cliffortiana** L.

Lobelia bivonae Tineo === **Solenopsis bivonae** (Tineo) M. B. Crespo, Serra & Juan

Lobelia blanda (D. Don) Heynh. === **Lobelia bridgesii** Hook. & Arn.

Lobelia bollii E. Wimm. === **Lobelia siphilitica** L.

Lobelia boninensis f. *rubra* Nakai === **Lobelia boninensis** Koidz.

Lobelia boninensis f. *viridis* Nakai === **Lobelia boninensis** Koidz.

Lobelia bonplandiana Schult. === **Centropogon cornutus** (L.) Druce

Lobelia borleana E. Wimm. === **Lobelia livingstoniana** R. E. Fr.

Lobelia brachypoda (A. Gray) Small === **Lobelia berlandieri** A. DC.

Lobelia bracteata Small === **Lobelia spicata** Lam.

Lobelia bracteolata A. DC. === **Lobelia erinus** L.

Lobelia bracteolata Vatke === **Lobelia aquatica** Cham.

Lobelia brassiana E. Wimm. === **Lobelia trullifolia** Hemsl. subsp. **trullifolia**

Lobelia breynii Lam. === **Monopsis lutea** (L.) Urb.

Lobelia breynii var. *bragae* Engl. === **Monopsis decipiens** (Sond.) Thulin

Lobelia broussonetia Bory === **Wahlenbergia lobelioides** (L. f.) Schrad. ex Link subsp. **lobelioides**

Lobelia browniana Schult. === **Lobelia gibbosa** Labill.

Lobelia bryoides Willd. ex Schult. === **Arenaria dicranoides** Kunth (Caryophyllaceae)

Lobelia bryophila E. Wimm. === **Lobelia pulchella** Vatke

Lobelia bryophila var. *fimbriosa* E. Wimm. === **Lobelia pulchella** Vatke

Lobelia buchananii Baker f. === **Lobelia trullifolia** Hemsl. subsp. **trullifolia**

Lobelia bulbosa L. === **Cyphia bulbosa** (L.) P. J. Bergius

Lobelia burmaniana D. Dietr. === ?

Lobelia butaguensis De Wild. === **Lobelia molleri** Henriq.

Lobelia caerulea f. *procera* (C. Presl) E. Wimm. === **Lobelia caerulea** Sims

Lobelia caerulea var. *backhouseana* (Lem.) E. Wimm. === **Lobelia caerulea** Sims

Lobelia caerulea var. *glabrescens* (C. Presl) A. DC. === **Lobelia caerulea** Sims

Lobelia caerulea var. *latidens* (C. Presl) A. DC. === **Lobelia caerulea** Sims

Lobelia caerulea var. *macularis* (C. Presl) E. Wimm. === **Lobelia caerulea** Sims

Lobelia caespitosa Blume === **Lobelia chinensis** Lour.

Lobelia calcarata Bertol. === **Heterotoma lobelioides** Zucc.

Lobelia calcarea E. Wimm. === **Lobelia berlandieri** A. DC.

Lobelia calycina Cham. === **Cyanea calycina** (Cham.) Lammers

Lobelia campanulata Cav. === **Siphocampylus affinis** (Mirb.) McVaugh

Lobelia campanulata Lam. === **Monopsis debilis** (L. f.) C. Presl

Lobelia campanuloides Thunb. === **Lobelia chinensis** Lour.

Lobelia campanuloides Thunb. === **Wahlenbergia marginata** (Thunb. ex Murray) A. DC.

Lobelia camporum f. *angustifolia* E. Wimm. === **Lobelia camporum** Pohl

Lobelia camporum f. *minor* E. Wimm. === **Lobelia camporum** Pohl

Lobelia camporum var. *lundiana* A. DC. === **Lobelia camporum** Pohl

Lobelia camptodon E. Wimm. === **Lobelia nicotianifolia** Roth ex Schult.

Lobelia camptodon var. *brevipedicellata* E. Wimm. === **Lobelia nicotianifolia** Roth ex Schult.

Lobelia camtschatica Pall. ex Spreng. === **Lobelia sessilifolia** Lamb.

Lobelia caoutchouc Kunth === **Centropogon caoutchouc** (Kunth) E. Wimm.

Lobelia capensis E. Wimm. === **Lobelia limosa** (Adamson) E. Wimm

Lobelia capillifolia var. *velutina* A. DC. === **Lobelia capillifolia** (C. Presl) A. DC.

Lobelia capillipes Schltr. === **Lobelia pinifolia** L.

Lobelia capitata Burm. f. === **Lobelia comosa** L.

Lobelia cardamines Thunb. === **Cyphia incisa** (Thunb.) Willd.

Lobelia cardinalis f. *alba* (J. McNab) H. St. John === **Lobelia cardinalis** L.

Lobelia cardinalis f. *cordigera* (Cav.) Bowden === **Lobelia cardinalis** L.

Lobelia cardinalis f. *hispidula* (E. Wimm.) Bowden === **Lobelia cardinalis** L.

Lobelia cardinalis f. *milleri* (D. Don) Voss === **Lobelia** × **speciosa** Sweet

Lobelia cardinalis f. *rosea* H. St. John === **Lobelia cardinalis** L.

Lobelia cardinalis f. *splendens* (Humb. & Bonpl. ex Willd.) Bowden === **Lobelia cardinalis** L.

Lobelia cardinalis subsp. *graminea* (Lam.) McVaugh === **Lobelia cardinalis** L.

Lobelia cardinalis var. *alba* J. McNab === **Lobelia cardinalis** L.

Lobelia cardinalis var. *angustifolia* Vatke === **Lobelia cardinalis** L.

Lobelia cardinalis var. *candida* A. W. Wood === **Lobelia cardinalis** L.

Lobelia cardinalis var. *glandulosa* N. Coleman === **Lobelia cardinalis** L.

Lobelia cardinalis var. *graminea* (Lam.) McVaugh === **Lobelia cardinalis** L.

Lobelia cardinalis var. *hispidula* E. Wimm. === **Lobelia cardinalis** L.

Lobelia cardinalis var. *integerrima* A. W. Wood === **Lobelia cardinalis** L.

Lobelia cardinalis var. *meridionalis* Bowden === **Lobelia cardinalis** L.

Lobelia cardinalis var. *milleri* D. Don === **Lobelia** × **speciosa** Sweet

Lobelia cardinalis var. *multiflora* (Paxton) McVaugh === **Lobelia cardinalis** L.

Lobelia cardinalis var. *phyllostachya* (Engelm.) McVaugh === **Lobelia cardinalis** L.

Lobelia cardinalis var. *propinqua* (Paxton) Bowden === **Lobelia cardinalis** L.

Lobelia cardinalis var. *pseudosplendens* McVaugh === **Lobelia cardinalis** L.

Lobelia cardinalis var. *texensis* (Raf.) Rothr. === **Lobelia cardinalis** L.

Lobelia carnosula Hook. & Arn. === **Porterella carnosula** (Hook. & Arn.) Torr.

Lobelia cautschuk Willd. ex Schult. === **Centropogon caoutchouc** (Kunth) E. Wimm.

Lobelia cavaleriei H. Lév. === ?

Lobelia cavanillesiana Schult. === **Lobelia laxiflora** Kunth subsp. **laxiflora**

Lobelia cavanillesii Mart. === **Lobelia laxiflora** Kunth subsp. **laxiflora**

Lobelia cavanillesii var. *lutea* F. Haage & K. Schmidt === **Lobelia laxiflora** subsp. **angustifolia** (A. DC.) Eakes & Lammers

Lobelia ceratophylla (C. Presl) D. Dietr. === **Lobelia chamaepitys** Lam.

Lobelia ceratophylla var. *glabrata* (C. Presl) A. DC. === **Lobelia chamaepitys** Lam.

Lobelia chamaedryfolia var. *confinis* E. Wimm. === **Lobelia goetzei** Diels

Lobelia chamaepitys var. *ceratophylla* (C. Presl) E. Wimm. === **Lobelia chamaepitys** Lam.

Lobelia cheiranthus L. === **Manulea cheiranthus** (L.) L. (Scrophulariaceae)

Lobelia chenopodiifolia Wall. ex G. Don === **Lobelia cliffortiana** L.

Lobelia chilawana Schinz === **Lobelia erinus** L.

Lobelia chinensis f. *elongata* Danguy === **Lobelia alsinoides** Lam.

Lobelia chinensis f. *lactiflora* (Hisauti) H. Hara === **Lobelia chinensis** Lour.

Lobelia chinensis f. *plena* (Makino) H. Hara === **Lobelia chinensis** Lour.

Lobelia chinensis var. *albiflora* (E. Wimm.) E. Wimm. === **Lobelia chinensis** Lour.

Lobelia chinensis var. *cantonensis* E. Wimm. ex Danguy === **Lobelia alsinoides** Lam.

Lobelia chinensis var. *hirta* E. Wimm. ex Danguy === **Lobelia alsinoides** Lam.

Lobelia ciliata de Vriese === **Lobelia tenuior** R. Br.

Lobelia cinerea Thunb. === **Wahlenbergia albicaulis** (Sond.) Lammers

Lobelia cirsiifolia var. *opilionis* E. Wimm. === **Lobelia cirsiifolia** Lam.

Lobelia cladlomesa Raf. === **?**

Lobelia claytoniana Michx. === **Lobelia spicata** Lam.

Lobelia clermontiana Gaudich. ex D. Dietr. === **Clermontia persicifolia** Gaudich.

Lobelia cliffortiana var. *brachypoda* A. Gray === **Lobelia berlandieri** A. DC.

Lobelia cliffortiana var. *xalapensis* (Kunth) A. Gray === **Lobelia xalapensis** Kunth

Lobelia coccinea de Jonghe === **Lobelia cardinalis** L.

Lobelia coccinea Stokes === **Lobelia cardinalis** L.

Lobelia coddii Compton === **Monopsis malvacea** E. Wimm.

Lobelia coelestis Nutt. ex Loudon === **Lobelia siphilitica** L.

Lobelia colorata Sweet === **Lobelia amoena** Michx.

Lobelia colorata var. *baculus* E. Wimm. === **Lobelia colorata** Wall. subsp. **colorata**

Lobelia colorata var. *dsolinhoensis* E. Wimm. === **Lobelia colorata** Wall. subsp. **colorata**

Lobelia columneae L. f. === **Siphocampylus columneae** (L. f.) G. Don

Lobelia commutata Schult. === **Lobelia gruina** Cav.

Lobelia commutata Schult. === **Lobelia neglecta** Schult.

Lobelia comosa Cav. === **Siphocampylus comosus** G. Don

Lobelia comosa var. *foliosa* (E. Wimm.) E. Wimm. === **Lobelia comosa** L.

Lobelia comosa var. *microdon* (C. Presl) E. Wimm. === **Lobelia comosa** L.

Lobelia comosa var. *secundata* (Sond.) E. Wimm. === **Lobelia comosa** L.

Lobelia concolor M. Martens & Galeotti === **Lobelia laxiflora** Kunth subsp. **laxiflora**

Lobelia conraui R. E. Fr. & T. C. E. Fr. === **Lobelia columaris** Hook. f.

Lobelia convolvulacea Cham. === **Siphocampylus convolvulaceus** (Cham.) G. Don

Lobelia cordata Willd. ex Schult. === **Siphocampylus cordatus** (Willd. ex Schult.) E. Wimm.

Lobelia cordigera Cav. === **Lobelia cardinalis** L.

Lobelia cordigera var. *fatalis* E. Wimm. === **Lobelia cardinalis** L.

Lobelia cordigera var. *marryattae* (Paxton) E. Wimm. === **Lobelia cardinalis** L.

Lobelia cordigera var. *multiflora* (Paxton) E. Wimm. === **Lobelia cardinalis** L.

Lobelia cordigera var. *pyramidalis* (Paxton) E. Wimm. === **Lobelia cardinalis** L.

Lobelia cornuta L. === **Centropogon cornutus** (L.) Druce

Lobelia coronopifolia f. *albiflora* E. Wimm. === **Lobelia coronopifolia** L.

Lobelia coronopifolia var. *caerulea* (Sims) Sond. === **Lobelia caerulea** Sims

Lobelia coronopifolia var. *glabrescens* (C. Presl) Sond. === **Lobelia caerulea** Sims

Lobelia coronopifolia var. *macularis* (C. Presl) Sond. === **Lobelia caerulea** Sims

Lobelia coronopifolia var. *simsii* (Sweet) E. Wimm. === **Lobelia coronopifolia** L.

Lobelia coronopifolia var. *uniflora* A. DC. === **Lobelia coronopifolia** L.

Lobelia corymbosa Hook. ex Graham === **Lobelia jasionoides** (A. DC.) E. Wimm.

Lobelia corymbosa P. J. Bergius === **Lobelia comosa** L.

Lobelia costaricana (Planch. & Oerst.) E. Wimm. === **Lobelia laxiflora** Kunth subsp. **laxiflora**

Lobelia costaricana var. *magna* E. Wimm. === **Lobelia laxiflora** Kunth subsp. **laxiflora**

Lobelia costata E. Wimm. === **Lobelia dunbariae** subsp. **paniculata** (Rock) Lammers

Lobelia cotensis M. E. Jones === **Lobelia aurita** (Brandegee) T. J. Ayers

Lobelia courtallensis K. K. N. Nair === **Lobelia nicotianifolia** Roth ex Schult.
Lobelia crassiuscula Michx. === **Lobelia glandulosa** Walter
Lobelia crenata Thunb. === **Cyphia crenata** (Thunb.) C. Presl
Lobelia crispa (Gaudich.) Endl. === **Cyanea crispa** (Gaudich.) Lammers, Givnish & Sytsma
Lobelia crispa Graham === **Lobelia fenestralis** Cav.
Lobelia cuneifolia var. *ananda* E. Wimm. === **Lobelia cuneifolia** Link & Otto
Lobelia cuneifolia var. *hirsutum* (C. Presl) A. DC. === **Lobelia cuneifolia** Link & Otto
Lobelia cuneiformis Labill. === **Lobelia anceps** L. f.
Lobelia cymbalaria Griseb. === **Lobelia nana** Kunth
Lobelia cymbalarioides Zahlbr. === **Lobelia nana** Kunth
Lobelia cyphia J. F. Gmel. === **Cyphia bulbosa** (L.) P. J. Bergius
Lobelia davidii var. *dolichothyrsa* (Diels) E. Wimm. === **Lobelia davidii** Franch.
Lobelia davidii var. *glaberrima* E. Wimm. === **Lobelia davidii** Franch.
Lobelia davidii var. *handelii* (E. Wimm.) E. Wimm. === **Lobelia pleotricha** Diels
Lobelia davidii var. *kwangsiensis* (E. Wimm.) Y. S. Lian === **Lobelia davidii** Franch.
Lobelia davidii var. *pleotricha* (Diels) E. Wimm. === **Lobelia pleotricha** Diels
Lobelia davidii var. *sichuanensis* Y. S. Lian === **Lobelia davidii** Franch.
Lobelia dealbata E. Wimm. === **Lobelia molleri** Henriq.
Lobelia debilis L. f. === **Monopsis debilis** (L. f.) C. Presl
Lobelia decipiens Sond. === **Monopsis decipiens** (Sond.) Thulin
Lobelia deckenii subsp. *bequaertii* (De Wild.) Mabb. === **Lobelia bequaertii** De Wild.
Lobelia deckenii subsp. *burttii* (E. A. Bruce) Mabb. === **Lobelia burttii** E. A. Bruce
Lobelia deckenii subsp. *elgonensis* (R. E. Fr. & T. C. E. Fr.) Mabb. === **Lobelia gregoriana**
    subsp. **elgonensis** (R. E. Fr. & T. C. E. Fr.) E. B. Knox
Lobelia deckenii subsp. *keniensis* Mabb. === **Lobelia gregoriana** Baker f. subsp. **gregoriana**
Lobelia deckenii subsp. *sattimae* (R. E. Fr. & T. C. E. Fr.) Mabb. === **Lobelia gregoriana** subsp.
    sattimae (R. E. Fr. & T. C. E. Fr.) E. B. Knox
Lobelia deckenii var. *cacuminum* R. E. Fr. & T. C. E. Fr. === **Lobelia deckenii** (Asch.) Hemsl.
    subsp. **deckenii**
Lobelia deckenii var. *tayloriana* (Baker f.) E. Wimm. === **Lobelia deckenii** (Asch.) Hemsl.
    subsp. **deckenii**
Lobelia decumbens A. Rich. ex Schult. === **Siphocampylus decumbens** (A. Rich. ex Schult.)
    Juss. ex A. DC.
Lobelia decumbens Sims === **Lobelia anceps** L. f.
Lobelia decurrens Roth === **Lobelia heyneana** Schult.
Lobelia decurrens var. *foliosa* (Kunth) Heynh. === **Lobelia decurrens** Cav. subsp. **decurrens**
Lobelia decurrens var. *jaensis* E. Wimm. === **Lobelia decurrens** Cav. subsp. **decurrens**
Lobelia dekindtiana Engl. === **Lobelia sonderiana** (Kuntze) Lammers
Lobelia deleiensis C. E. C. Fisch. === **Lobelia montana** Reinw. ex Blume
Lobelia delessertiana E. Wimm. === **Lobelia laxiflora** Kunth subsp. **laxiflora**
Lobelia delicatula Compton === **Lobelia trullifolia** subsp. **delicatula** (Compton) Thulin
Lobelia delisseana Gaudich. ex D. Dietr. === **Cyanea acuminata** (Gaudich.) Hillebr.
Lobelia densiflora Rennie === **Lobelia siphilitica** L.
Lobelia depressa L. f. === **Monopsis debilis** (L. f.) C. Presl
Lobelia depressa var. *dregeana* (Sond.) E. Wimm. === **Lobelia sonderiana** (Kuntze) Lammers
Lobelia depressa var. *linearifolia* E. Wimm. === **Lobelia sonderiana** (Kuntze) Lammers
Lobelia depressa var. *seineri* (Schltr.) E. Wimm. === **Lobelia sonderiana** (Kuntze) Lammers
Lobelia depressa var. *thunbergii* A. DC. === **Lobelia limosa** (Adamson) E. Wimm.
Lobelia dichotoma Miq. === **Lobelia heyneana** Schult.
Lobelia dichotoma var. *aligera* (Haines) E. Wimm. === **Lobelia heyneana** Schult.
Lobelia dichotoma var. *pilosella* E. Wimm. === **Lobelia heyneana** Schult.
Lobelia digitalifolia var. *guadeloupensis* (Urb.) McVaugh === **Lobelia digitalifolia** (Griseb.) Urb.
Lobelia digitata Thunb. === **Cyphia bulbosa** (L.) P. J. Bergius
Lobelia discolor Klotzsch === **Lobelia subnuda** Benth.
Lobelia disperma E. Wimm. === **Unigenes humifusa** (A. DC.) E. Wimm.
Lobelia dissecta subsp. *humidulorum* Friis & Vollesen === **Lobelia dissecta** M. B. Moss

Lobelia diversifolia Willd. ex Scuhult. === ?
Lobelia dobrowskioides Diels === **Monopsis decipiens** (Sond.) Thulin
Lobelia dobrowskyana D. Dietr. === **Monopsis unidentata** (W. T. Aiton) E. Wimm. subsp. **unidentata**
Lobelia dodiana var. *radicans* (Schönland) E. Wimm. === **Lobelia dodiana** E. Wimm.
Lobelia dolichopus Schltr. === **Lobelia trullifolia** Hemsl. subsp. **trullifolia**
Lobelia dolichothyrsa Diels === **Lobelia davidii** Franch.
Lobelia domingensis A. DC. === **Lobelia aquatica** Cham.
Lobelia dopatrioides Kurz === **Lobelia griffithii** Hook. f. & Thomson
Lobelia dortmanna f. *terrestris* Sylvèn === **Lobelia dortmanna** L.
Lobelia dortmanna var. *decolorata* Lindb. f. === **Lobelia dortmanna** L.
Lobelia dortmannii von Damman === **Lobelia erinus** L.
Lobelia douglasiana F. M. Bailey === **Lobelia stenophylla** Benth.
Lobelia douglasii A. W. Wood === **Downingia elegans** (Dougl. ex Lindl.) Torr.
Lobelia draba Willd. ex Schult. === **Diastatea micrantha** (Kunth) McVaugh
Lobelia dracunculoides Willd. ex Schult. === **Lobelia laxiflora** subsp. **angustifolia** (A. DC.) Eakes & Lammers
Lobelia dubia de Vriese === **Wahlenbergia multicaulis** Benth.
Lobelia dunnii Greene === **Palmerella debilis** A. Gray
Lobelia dunnii var. *serrata* (A. Gray) McVaugh === **Palmerella debilis** subsp. **serrata** (A. Gray) Lammers
Lobelia eckloniana (C. Presl) D. Dietr. === **Lobelia filicaulis** (C. Presl) Schönland
Lobelia ehrenbergii var. *gracilens* (A. Gray) McVaugh === **Lobelia ehrenbergii** subsp. **gracilens** (A. Gray) Lammers
Lobelia ekmanii Urb. === **Siphocampylus caudatus** McVaugh
Lobelia elegans de Vriese === **Isotoma pusilla** Benth.
Lobelia elgonensis R. E. Fr. & T. C. E. Fr. === **Lobelia gregoriana** subsp. **elgonensis** (R. E. Fr. & T. C. E. Fr.) E. B. Knox
Lobelia elliptica Willd. ex Schult. === **Siphocampylus ellipticus** (Willd. ex Schult.) Vatke
Lobelia ensifolia (A. DC.) Hitchc. === **Lobelia salicina** Lam.
Lobelia ensiformis Vell. === ?
Lobelia epilobioides E. Wimm. === **Lobelia eryliae** C. E. C. Fisch.
Lobelia epilobioides var. *luzonica* E. Wimm. === **Lobelia philippinensis** Skottsb.
Lobelia epilobioides var. *sarasinorum* E. Wimm. === **Lobelia eryliae** C. E. C. Fisch.
Lobelia erecta de Vriese === **Lobelia anceps** L. f.
Lobelia erecta Hook. f. & Thomson === **Lobelia erectiuscula** H. Hara
Lobelia ericetorum A. DC. === **Lobelia linearis** Thunb.
Lobelia ericoides (C. Presl) D. Dietr. === **Monopsis lutea** (L.) Urb.
Lobelia erinifolia Salisb. === **Lobelia erinus** L.
Lobelia erinoides L. === **Lobelia erinus** L.
Lobelia erinus f. *bicolor* (Sims) Zahlbr. === **Lobelia erinus** L.
Lobelia erinus f. *elata* E. Wimm. === **Lobelia erinus** L.
Lobelia erinus "section" *compacta* G. Nichols. === **Lobelia erinus** L.
Lobelia erinus "section" *paxtoniana* G. Nichols. === **Lobelia erinus** L.
Lobelia erinus "section" *pumila* G. Nichols. === **Lobelia erinus** L.
Lobelia erinus "section" *ramosoides* G. Nichols. === **Lobelia erinus** L.
Lobelia erinus "section" *speciosa* G. Nichols. === **Lobelia erinus** L.
Lobelia erinus var. *bellidifolia* (L. f.) Sond. === **Lobelia erinus** L.
Lobelia erinus var. *bracteolata* (A. DC.) E. Wimm. === **Lobelia erinus** L.
Lobelia erinus var. *grandiflora* Paxton === **Lobelia erinus** L.
Lobelia erinus var. *microdon* (C. Presl) Sond. === **Lobelia comosa** L.
Lobelia erinus var. *schrankii* (Sweet) E. Wimm. === **Lobelia erinus** L.
Lobelia erinus var. *subvillosa* E. Wimm. === **Lobelia erinus** L.
Lobelia esquirolii H. Lév. === **Mazus pumilus** (Burm. f.) Steenis (Scrophulariaceae)
Lobelia esterhuyseniae E. Wimm. === **Lobelia comptonii** E. Wimm.
Lobelia eurypoda var. *fissurarum* E. Wimm. === **Lobelia eurypoda** E. Wimm.
Lobelia excelsa Lesch. ex Roxb. === **Lobelia leschenaultiana** (C. Presl) Skottsb.

Lobelia exilis var. *major* E. Wimm. === **Lobelia exilis** Hochst. ex A. Rich.
Lobelia exilis var. *pusilla* E. Wimm. === **Lobelia exilis** Hochst. ex A. Rich.
Lobelia falcata Raf. === **Lobelia kalmii** L.
Lobelia farsetia Vand. === **Lobelia urens** L.
Lobelia fasciculata Donn. Sm. === **Lobelia umbellifera** McVaugh
Lobelia fenestralis var. *pectinata* (Engelm.) E. Wimm. === **Lobelia fenestralis** Cav.
Lobelia fenniae T. C. E. Fr. === **Lobelia telekii** Schweinf.
Lobelia ferruginea L. f. === **Centropogon ferrugineus** (L. f.) Gleason
Lobelia fervens var. *asperulata* (Klotzsch) Sond. === **Lobelia fervens** Thunb. subsp. **fervens**
Lobelia feuillei (G. Don) Voss === **Lobelia tupa** L.
Lobelia filiformis f. *albiflora* E. Wimm. === **Lobelia erinus** L.
Lobelia filiformis f. *multipilis* E. Wimm. === **Lobelia erinus** L.
Lobelia filiformis f. *muzandazora* E. Wimm. === **Lobelia erinus** L.
Lobelia filiformis f. *rusticana* E. Wimm. === **Lobelia flaccida** (C. Presl) A. DC. subsp. **flaccida**
Lobelia filiformis var. *krebsiana* E. Wimm. === **Lobelia flaccida** (C. Presl) A. DC. subsp. **flaccida**
Lobelia filiformis var. *luzoniensis* Pers. === **Lobelia luzoniensis** (Pers.) Merr.
Lobelia filiformis var. *natalensis* (A. DC.) E. Wimm. === **Lobelia erinus** L.
Lobelia filiformis var. *subaquatilis* E. Wimm. === **Lobelia filiformis** Lam.
Lobelia fissa Willd. ex Schult. === **Lobelia laxiflora** Kunth subsp. **laxiflora**
Lobelia fistulosa Raf. === ?
Lobelia flaccida f. *densifoliata* E. Wimm. === **Lobelia flaccida** (C. Presl) A. DC. subsp. **flaccida**
Lobelia flaccida f. *laxifoliata* E. Wimm. === **Lobelia flaccida** (C. Presl) A. DC. subsp. **flaccida**
Lobelia flaccida var. *caffra* E. Wimm. === **Lobelia flaccida** (C. Presl) A. DC. subsp. **flaccida**
Lobelia flaccida var. *hirsuta* (C. Presl ex A. DC.) E. Wimm. === **Lobelia flaccida** (C. Presl)
    A. DC. subsp. **flaccida**
Lobelia flaccida var. *scabripes* (C. Presl) E. Wimm. === **Lobelia flaccida** (C. Presl) A. DC.
    subsp. **flaccida**
Lobelia flaccida var. *stricta* (Zahlbr.) E. Wimm. === **Lobelia flaccida** (C. Presl) A. DC.
    subsp. **flaccida**
Lobelia flava (C. Presl) D. Dietr. === **Monopsis flava** (C. Presl) E. Wimm.
Lobelia flavescens (A. DC.) E. H. L. Krause === **Lobelia stricta** Sw.
Lobelia flexuosa var. *intermedia* (Hemsl.) T. J. Ayers === **Lobelia flexuosa** subsp. **intermedia**
    (Hemsl.) Lammers
Lobelia fluminensis Vell. === **Siphocampylus fluminensis** (Vell.) E. Wimm.
Lobelia fluviatilis R. Br. === **Isotoma fluviatilis** (R. Br.) F. Muell. ex Benth.
Lobelia foetida Kunth === **Centropogon foetidus** (Kunth) Zahlbr.
Lobelia foliosa Kunth === **Lobelia decurrens** Cav. subsp. **decurrens**
Lobelia fonticola Engl. & Gilg. === **Monopsis zeyheri** (Sond.) Thulin
Lobelia formosa de Jonghe === **Lobelia × speciosa** Sweet
Lobelia fossarum E. Wimm. === **Lobelia colorata** subsp. **taliensis** (Diels) T. J. Zhang &
    D. Y. Hong
Lobelia fulgens Humb. & Bonpl. ex Willd. === **Lobelia cardinalis** L.
Lobelia fulgens f. *atrosanguinea* (Hook.) Voss === **Lobelia cardinalis** L.
Lobelia fulgens f. *maculata* Voss === **Lobelia cardinalis** L.
Lobelia fulgens f. *rosea* Voss === **Lobelia cardinalis** L.
Lobelia fulgens var. *glabriuscula* Klotzsch ex Walp. === **Lobelia cardinalis** L.
Lobelia fulgens var. *marryattae* (Paxton) Paxton === **Lobelia cardinalis** L.
Lobelia fulgens var. *multiflora* Paxton === **Lobelia cardinalis** L.
Lobelia fulgens var. *propinqua* Paxton === **Lobelia cardinalis** L.
Lobelia fulgens var. *pyramidalis* Paxton === **Lobelia cardinalis** L.
Lobelia galeopsoides Engl. & Diels === **Dielsantha galeopsoides** (Engl. & Diels) E. Wimm.
Lobelia galpinii var. *detonsa* E. Wimm. === **Lobelia galpinii** Schltr.
Lobelia gardneriana Kanitz === **Lobelia fastigiata** Kunth
Lobelia gardneriana var. *foliosa* Zahlbr. === **Lobelia fastigiata** Kunth
Lobelia gasparrinii Tineo === **Solenopsis laurentia** (L.) C. Presl
Lobelia gaudichaudii f. *bryanii* H. St. John & Hosaka === **Lobelia gloria-montis** Rock

Lobelia gaudichaudii f. *hirsuta* H. St. John & Hosaka === **Lobelia kauaiensis** (A. Gray) A. Heller

Lobelia gaudichaudii f. *kukuiensis* H. St. John & Hosaka === **Lobelia gloria-montis** Rock

Lobelia gaudichaudii f. *sanguinea* H. St. John & Hosaka === **Lobelia gloria-montis** Rock

Lobelia gaudichaudii var. *albiflora* H. St. John & A. C. Medeiros === **Lobelia gloria-montis** Rock

Lobelia gaudichaudii var. *coccinea* Rock === **Lobelia gaudichaudii** A. DC. subsp. **gaudichaudii**

Lobelia gaudichaudii var. *gloria-montis* (Rock) H. St. John & Hosaka === **Lobelia gloria-montis** Rock

Lobelia gaudichaudii var. *kauaiensis* A. Gray === **Lobelia kauaiensis** (A. Gray) A. Heller

Lobelia gaudichaudii var. *koolauensis* Hosaka & Fosberg === **Lobelia gaudichaudii** subsp. **koolauensis** (Hosaka & Fosberg) Lammers

Lobelia gaudichaudii var. *longibracteata* Rock === **Lobelia gloria-montis** Rock

Lobelia genistoides (C. Presl) A. DC. === **Lobelia patula** L. f.

Lobelia georgiana var. *azurea* E. Wimm. === **Lobelia elongata** Small

Lobelia gerardii Sauvageau === **Lobelia × speciosa** Sweet

Lobelia gerardii var. *lugdunensis* L. H. Bailey === **Lobelia × speciosa** Sweet

Lobelia ghiesbreghtii Lem. === ?

Lobelia gibbosa var. *browniana* (Schult.) F. M. Bailey === **Lobelia gibbosa** Labill.

Lobelia gibbosa var. *microsperma* (F. Muell.) F. M. Bailey === **Lobelia gibbosa** Labill.

Lobelia gibbosa var. *simplicicaulis* (R. Br.) F. M. Bailey === **Lobelia simplicicaulis** R. Br.

Lobelia giberroa subsp. *squarrosa* (Baker f.) Mabb. === **Lobelia giberroa** Hemsl.

Lobelia giberroa var. *intermedia* (Hauman) W. Robyns === **Lobelia giberroa** Hemsl.

Lobelia giberroa var. *iringensis* E. Wimm. === **Lobelia giberroa** Hemsl.

Lobelia giberroa var. *longibracteata* Hauman === **Lobelia giberroa** Hemsl.

Lobelia giberroa var. *mionandra* E. Wimm. === **Lobelia giberroa** Hemsl.

Lobelia giberroa var. *ulugurensis* (Engl.) Hauman === **Lobelia giberroa** Hemsl.

Lobelia giberroa var. *usafuensis* (Engl.) Hauman === **Lobelia giberroa** Hemsl.

Lobelia giberroa var. *volkensii* (Engl.) Hauman === **Lobelia giberroa** Hemsl.

Lobelia giftbergensis E. Phillips === **Wimmerella giftbergensis** (E. Phillips) Serra, M. B. Crespo & Lammers

Lobelia gigantea Cav. === **Siphocampylus giganteus** (Cav.) G. Don

Lobelia gigantea Sims === **Lobelia excelsa** Bonpl.

Lobelia glabrata Kunth. === **Burmeistera glabrata** (Kunth) Benth. & Hook. f. ex B. D. Jacks.

Lobelia glandulifera (A. Gray) Small === **Lobelia georgiana** McVaugh

Lobelia glandulosa var. *glabra* A. DC. === **Lobelia elongata** Small

Lobelia glandulosa var. *laevicalyx* Fernald === **Lobelia glandulosa** Walter

Lobelia glandulosa var. *obtusifolia* A. DC. === **Lobelia puberula** Michx.

Lobelia glaucoleuca Schltr. === **Lobelia setacea** Thunb.

Lobelia gloria-montis var. *longibracteata* (Rock) Rock === **Lobelia gloria-montis** Rock

Lobelia gloria-montis var. *molokaiensis* O. Deg. === **Lobelia gloria-montis** Rock

Lobelia goodenioides Willd. === **Lobelia spicata** Lam.

Lobelia goodii E. Wimm. === **Lobelia jasionoides** (A. DC.) E. Wimm.

Lobelia gracilens A. Gray === **Lobelia ehrenbergii** subsp. **gracilens** (A. Gray) Lammers

Lobelia graciliflora E. Wimm. === **Lobelia goetzei** Diels

Lobelia gracilis Andrews === **Lobelia andrewsii** Lammers

Lobelia gracilis Nutt. === **Lobelia nuttallii** Schult.

Lobelia gracilis Salisb. === **Solenopsis laurentia** (L.) C. Presl

Lobelia gracilis f. *rosea* (J. W. Loudon) E. Wimm. === **Lobelia andrewsii** Lammers

Lobelia gracilis var. *major* Benth. === **Lobelia andrewsii** Lammers

Lobelia gracilis var. *rosea* J. W. Loudon === **Lobelia andrewsii** Lammers

Lobelia graminea Lam. === **Lobelia cardinalis** L.

Lobelia graminea f. *kerneri* (L. Nagy) E. Wimm. === **Lobelia cardinalis** L.

Lobelia graminea var. *phyllostachya* (Engelm.) E. Wimm. === **Lobelia cardinalis** L.

Lobelia graminea var. *pseudosplendens* (McVaugh) E. Wimm. === **Lobelia cardinalis** L.

Lobelia graminea var. *intermedia* E. Wimm. === **Lobelia cardinalis** L.

Lobelia grandidentata Schlecht. === **Centropogon grandidentatus** (Schlecht.) Zahlbr.

Lobelia grandiflora (Gaudich.) Endl. === **Clermontia grandiflora** Gaudich.

Lobelia grandis L. f. === **Centropogon grandis** (L. f.) C. Presl

Lobelia granvikii T. C. E. Fr. === **Lobelia flaccida** subsp. **granvikii** (T. C. E. Fr.) Thulin

Lobelia griffithii var. *dopatrioides* (Kurz) Kurz === **Lobelia griffithii** Hook. f. & Thomson

Lobelia grimesiana (Gaudich.) Hook. & Arn. === **Cyanea grimesiana** Gaudich.

Lobelia gruina f. *conferta* (Fernald) McVaugh === **Lobelia gruina** Cav.

Lobelia gruina f. *flava* E. Wimm. === **Lobelia gruina** Cav.

Lobelia gruina var. *conferta* Fernald === **Lobelia gruina** Cav.

Lobelia gruina var. *peduncularis* McVaugh === **Lobelia gruina** Cav.

Lobelia gruina var. *orizabae* (M. Martens & Galeotti) E. Wimm. === **Lobelia gruina** Cav.

Lobelia gruina var. *rapunculoides* (Kunth) E. Wimm. === **Lobelia gruina** Cav.

Lobelia gruina var. *rouaixii* Conz. === **Lobelia gruina** Cav.

Lobelia guadeloupensis Urb. === **Lobelia digitalifolia** (Griseb.) Urb.

Lobelia haenkeana (C. Presl) A. DC. === **Lobelia laxiflora** Kunth subsp. **laxiflora**

Lobelia haenkeana var. *panamensis* E. Wimm. === **Lobelia laxiflora** Kunth subsp. **laxiflora**

Lobelia halei Small === **Lobelia flaccidifolia** Small

Lobelia hancei H. Hara === **Lobelia chevalieri** Danguy

Lobelia handelii E. Wimm. === **Lobelia pleotricha** Diels

Lobelia hartwegii var. *angusta* E. Wimm. === **Lobelia hartwegii** A. DC. ex Benth.

Lobelia hederacea A. DC. === **Lobelia cuneifolia** Link & Otto

Lobelia hederacea var. *elliptica* Hook. & Arn. === **Lobelia hederacea** Cham.

Lobelia heterodonta Sprague === **Lobelia cirsiifolia** Lam.

Lobelia heteromalla Schrad. === **Lobelia comosa** L.

Lobelia heterophylla f. *compacta* Voss === **Lobelia heterophylla** Labill.

Lobelia heterophylla var. *major* Paxton === **Lobelia tenuior** R. Br.

Lobelia heyneana f. *parviflora* (Danguy) E. Wimm. === **Lobelia heyneana** Schult.

Lobelia heyneana var. *inconspicua* (A. Rich.) E. Wimm. === **Lobelia inconspicua** A. Rich.

Lobelia heyneana var. *intercedens* E. Wimm. === **Lobelia intercedens** (E. Wimm.) Thulin

Lobelia heyneana var. *lamiifolia* (C. B. Clarke) E. Wimm. === **Lobelia heyneana** Schult.

Lobelia heyneana var. *parviflora* (Danguy) E. Wimm. === **Lobelia heyneana** Schult.

Lobelia heyneana var. *viridissima* E. Wimm. === **Lobelia heyneana** Schult.

Lobelia hillebrandii var. *monostachya* Rock === **Lobelia monostachya** (Rock) Lammers

Lobelia hillebrandii var. *paniculata* Rock === **Lobelia dunbariae** subsp. **paniculata** (Rock) Lammers

Lobelia hirsuta L. === **Gnidia hirsuta** (L.) Thulin (Thymeleaceae)

Lobelia hirta Cav. === **Centropogon hirtus** (Cav.) C. Presl

Lobelia hirtella (A. Gray) Greene === **Lobelia spicata** Lam.

Lobelia hirtipes var. *subpaludosa* E. Wimm. === **Lobelia hirtipes** E. Wimm.

Lobelia holstii f. *alba* E. Wimm. === **Lobelia holstii** Engl.

Lobelia holstii f. *minor* Engl. === **Lobelia holstii** Engl.

Lobelia holstii var. *subhirsuta* E. Wimm. === **Lobelia goetzei** Diels

Lobelia horsfieldiana Miq. === **Lobelia nummularia** Lam.

Lobelia hortensis A. DC. === **Lobelia amoena** Michx.

Lobelia hosseusii E. Wimm. === **Lobelia griffithii** Hook. f. & Thomson

Lobelia hosseusii var. *villosa* Kerr === **Lobelia griffithii** Hook. f. & Thomson

Lobelia humboldtiana Schult. === **Triodanis perfoliata** subsp. **biflora** (Ruiz & Pav.) Lammers

Lobelia humilis Klotzsch === **Lobelia fervens** Thunb. subsp. **fervens**

Lobelia humistrata var. *kingiana* E. Wimm. === **Lobelia humistrata** F. Muell. ex F. M. Bailey

Lobelia hybrida C. Y. Wu === **Lobelia colorata** subsp. **taliensis** (Diels) T. J. Zhang & D. Y. Hong

Lobelia hypocrateriformis R. Br. === **Isotoma hypocrateriformis** (R. Br.) Druce

Lobelia hypoleuca f. *macrophyta* Rock === **Lobelia hypoleuca** Hillebr.

Lobelia hypoleuca var. *heterocarpa* E. Wimm. === **Lobelia hypoleuca** Hillebr.

Lobelia hypoleuca var. *rockii* H. St. John & Hosaka === **Lobelia hypoleuca** Hillebr.

Lobelia hyssopifolia (C. Presl) Gay === **Lobelia polyphylla** Hook. & Arn.

Lobelia ignea Paxton === **Lobelia cardinalis** L.

Lobelia ignea Vell. === **Siphocampylus corymbiferus** Pohl

Lobelia ilicifolia Sims === **Lobelia purpurascens** R. Br.

Lobelia ilysanthoides Schltr. === **Lobelia inconspicua** A. Rich.

Lobelia imbricata Cham. === **Siphocampylus imbricatus** (Cham.) G. Don

Lobelia imperialis var. *kanitzii* E. Wimm. === **Lobelia organensis** Gardner

Lobelia incana Ruiz & Pav. ex B. D. Jacks. === **Siphocampylus obovatus** (G. Don) E. Wimm.

Lobelia incisa (C. Presl) D. Dietr. === **Lobelia pubescens** Aiton

Lobelia incisa Thunb. === **Cyphia incisa** (Thunb.) Willd.

Lobelia incurva Raf. === ?

Lobelia infesta (Griseb.) Urb. === **Lobelia cirsiifolia** Lam.

Lobelia inflata f. *albiflora* Moldenke === **Lobelia inflata** L.

Lobelia inflata var. *simplex* Raf. ex Millsp. === **Lobelia inflata** L.

Lobelia ingrata E. Wimm. === **Lobelia lasiocalycina** E. Wimm.

Lobelia intermedia Hauman === **Lobelia giberroa** Hemsl.

Lobelia intertexta Baker f. === **Lobelia trullifolia** Hemsl. subsp. **trullifolia**

Lobelia intertexta f. *arida* E. Wimm. === **Lobelia trullifolia** Hemsl. subsp. **trullifolia**

Lobelia intertexta var. *diversifolia* E. Wimm. === **Lobelia trullifolia** Hemsl. subsp. **trullifolia**

Lobelia inundata R. Br. === **Isotoma fluviatilis** (R. Br.) F. Muell. ex Benth. subsp. **fluviatilis**

Lobelia irasuensis var. *fucata* McVaugh === **Lobelia irasuensis** subsp. **fucata** (McVaugh) Lammers

Lobelia irasuensis var. *picta* (B. L. Rob. & Seaton) McVaugh === **Lobelia irasuensis** subsp. **picta** (B. L. Rob. & Seaton) Lammers

Lobelia isotomoides E. Wimm. === **Lobelia stenosiphon** (Adamson) E. Wimm.

Lobelia jamaicensis (Urb.) Urb. === **Lobelia assurgens** L.

Lobelia japonica F. Dietr. === **Lobelia chinensis** Lour.

Lobelia jasionoides var. *sparsiflora* (Sond.) E. Wimm. === **Lobelia jasionoides** (A. DC.) E. Wimm.

Lobelia javanica Thunb. === **Lobelia nummularia** Lam.

Lobelia johnstonii C. H. Wright === **Lobelia holstii** Engl.

Lobelia jugosa S. Moore === **Lobelia erinus** L.

Lobelia kakeana (Meyen) Endl. === **Clermontia kakeana** Meyen

Lobelia kalmii f. *leucantha* Rouleau === **Lobelia kalmii** L.

Lobelia kalmii var. *capillaris* Farw. === **Lobelia kalmii** L.

Lobelia kalmii var. *caroliniana* Michx. === **Lobelia nuttallii** Schult.

Lobelia kalmii var. *edithae* E. Wimm. === **Lobelia kalmii** L.

Lobelia kalmii var. *gracilis* W. P. C. Barton === **Lobelia nuttallii** Schult.

Lobelia kalmii var. *strictiflora* Rydb. === **Lobelia kalmii** L.

Lobelia kamerunensis Engl. ex Hutch. & Dalziel === **Lobelia rubescens** De Wild.

Lobelia karisimbensis R. E. Fr. & T. C. E. Fr. === **Lobelia stuhlmannii** Schweinf. ex Stuhlmann

Lobelia kassneri E. Wimm. === **Lobelia lasiocalycina** E. Wimm.

Lobelia kauaiensis var. *villosa* Rock === **Lobelia villosa** (Rock) H. St. John & Hosaka

Lobelia keilhackii E. Wimm. === **Lobelia erinus** L.

Lobelia keniensis R. E. Fr. & T. C. E. Fr. === **Lobelia gregoriana** Baker f. subsp. **gregoriana**

Lobelia kerii (A. DC.) Heynh. === **Lobelia chinensis** Lour.

Lobelia kerneri L. Nagy === **Lobelia cardinalis** L.

Lobelia kilimandsharica Engl. === **Wahlenbergia pusilla** Hochst. ex A. Rich.

Lobelia kinangopia E. Wimm. === **Lobelia neumannii** T. C. E. Fr.

Lobelia kirkii var. *microphylla* Schltr. === **Lobelia kirkii** R. E. Fr.

Lobelia kiwuensis Engl. === **Lobelia minutula** Engl.

Lobelia knysnensis Schltr. === **Lobelia flaccida** (C. Presl) A. DC. subsp. **flaccida**

Lobelia kochiana E. Wimm. === **Lobelia rhombifolia** de Vriese

Lobelia kohautiana Vatke === **Lobelia erinus** L.

Lobelia krebsiana C. Presl ex A. DC. === **Lobelia flaccida** (C. Presl) A. DC. subsp. **flaccida**

Lobelia krookii Zahlbr. === **Lobelia laxa** MacOwan

Lobelia kummeriae Engl. ex T. C. E. Fr. === **Lobelia baumannii** Engl.

Lobelia kwangsiensis E. Wimm. === **Lobelia davidii** Franch.

Lobelia laciniata Lam. === **Siphocampylus sonchifolius** (Sw.) McVaugh

Lobelia lactescens Martyn === **Sclerotheca arborea** (G. Forst.) A. DC.

Lobelia lacustris Salisb. === **Lobelia dortmanna** L.

Lobelia laevigata L. f. === **Centropogon cornutus** (L.) Druce

Lobelia lanceolata (Gaudich.) Hook. & Arn. === **Cyanea lanceolata** (Gaudich.) Lammers, Givnish & Sytsma

Lobelia lanceolata Hook. & Arn. === **Lobelia laxiflora** Kunth subsp. **laxiflora**

Lobelia lanuriensis De Wild. === **Lobelia stuhlmannii** Schweinf. ex Stuhlmann

Lobelia lanuriensis var. *ericeti* Hauman === **Lobelia stuhlmannii** Schweinf. ex Stuhlmann

Lobelia lanuriensis var. *kahuzica* Humbert === **Lobelia stuhlmannii** Schweinf. ex Stuhlmann

Lobelia lanuriensis var. *karisimbensis* (R. E. Fr. & T. C. E. Fr.) Hauman === **Lobelia stuhlmannii** Schweinf. ex Stuhlmann

Lobelia laricina A. Spreng. === **Prismatocarpus diffusus** (L. f.) A. DC.

Lobelia lasiantha (C. Presl) A. DC. === **Lobelia linearis** Thunb.

Lobelia laurentia L. === **Solenopsis laurentia** (L.) C. Presl

Lobelia laurentia var. *nana* Hoffmanns. & Link === **Solenopsis laurentia** (L.) C. Presl

Lobelia lavendulacea Klotzsch === **Lobelia erinus** L.

Lobelia laxiflora f. *angustifolia* (A. DC.) Voss === **Lobelia laxiflora** subsp. **angustifolia** (A. DC.) Eakes & Lammers

Lobelia laxiflora f. *brevifolia* (Zahlbr.) E. Wimm. === **Lobelia laxiflora** Kunth subsp. **laxiflora**

Lobelia laxiflora f. *concolor* (Kuntze) E. Wimm. === **Lobelia laxiflora** Kunth subsp. **laxiflora**

Lobelia laxiflora f. *flava* E. Wimm. === **Lobelia laxiflora** Kunth subsp. **laxiflora**

Lobelia laxiflora f. *fragilis* (B. L. Rob. & Fernald) E. Wimm. === **Lobelia laxiflora** subsp. **angustifolia** (A. DC.) Eakes & Lammers

Lobelia laxiflora f. *lanceolata* E. Wimm. === **Lobelia laxiflora** Kunth subsp. **laxiflora**

Lobelia laxiflora f. *lutea* (F. Haage & K. Schmidt) E. Wimm. === **Lobelia laxiflora** subsp. **angustifolia** (A. DC.) Eakes & Lammers

Lobelia laxiflora f. *lutea* Standl. & Steyerm. === **Lobelia laxiflora** Kunth subsp. **laxiflora**

Lobelia laxiflora f. *magna* (E. Wimm.) E. Wimm. === **Lobelia laxiflora** Kunth subsp. **laxiflora**

Lobelia laxiflora var. *angustifolia* A. DC. === **Lobelia laxiflora** subsp. **angustifolia** (A. DC.) Eakes & Lammers

Lobelia laxiflora var. *bicolor* (D. Don) Endl. === **Lobelia laxiflora** Kunth subsp. **laxiflora**

Lobelia laxiflora var. *brevifolia* Zahlbr. === **Lobelia laxiflora** Kunth subsp. **laxiflora**

Lobelia laxiflora var. *brevipes* E. Wimm. === **Lobelia laxiflora** subsp. **angustifolia** (A. DC.) Eakes & Lammers

Lobelia laxiflora var. *cavanillesii* Steud. === **Lobelia laxiflora** Kunth subsp. **laxiflora**

Lobelia laxiflora var. *foliosa* Zahlbr. === **Lobelia laxiflora** Kunth subsp. **laxiflora**

Lobelia laxiflora var. *insignis* Donn. Sm. === **Lobelia aguana** E. Wimm.

Lobelia laxiflora var. *mollis* (Vatke) Zahlbr. === **Lobelia laxiflora** Kunth subsp. **laxiflora**

Lobelia laxiflora var. *nelsonii* (Fernald) McVaugh === **Lobelia laxiflora** Kunth subsp. **laxiflora**

Lobelia laxiflora var. *patula* (Planch. & Oerst.) E. Wimm. === **Lobelia laxiflora** Kunth subsp. **laxiflora**

Lobelia laxiflora var. *petiolata* E. Wimm. === **Lobelia laxiflora** Kunth subsp. **laxiflora**

Lobelia laxiflora var. *stricta* (Planch. & Oerst.) McVaugh === **Lobelia laxiflora** Kunth subsp. **laxiflora**

Lobelia ledermannii Schltr. === **Lobelia sapinii** De Wild.

Lobelia lehmannii de Vriese === **Isotoma hypocrateriformis** (R. Br.) Druce

Lobelia lelyana E. Wimm. === **Lobelia sapinii** De Wild.

Lobelia leptocarpa Griess. === **Lobelia thermalis** Thunb.

Lobelia leptostachys A. DC. === **Lobelia spicata** Lam.

Lobelia leptostachys var. *hirtella* (A. Gray) Farw. === **Lobelia spicata** Lam.

Lobelia leschenaultiana f. *glabrior* Skottsb. === **Lobelia leschenaultiana** (C. Presl) Skottsb.

Lobelia leucanthera Kerr === **Lobelia colorata** Wall. subsp. **colorata**

Lobelia limburgensis de Jonghe === ?

Lobelia limoselloides Humb. & Bonpl. ex Schult. === **Lysipomia montioides** Kunth

Lobelia lindblomii f. *nanella* Mildbr. === **Lobelia lindblomii** Mildbr.

Lobelia linearis var. *gloveri* E. Wimm. === **Lobelia linearis** Thunb.

Lobelia linearis var. *pinnata* Schltr. === **Lobelia linearis** Thunb.

Lobelia linifolia Willd. ex Schult. === **Lobelia collina** Kunth

Lobelia linnaeoides (Hook. f.) Petrie === **Lobelia angulata** G. Forst.

Lobelia linnaeoides var. *brevipilis* E. Wimm. === **Lobelia nummularia** Lam.

Lobelia littoralis R. Cunn. ex A. Cunn. === **Lobelia angulata** G. Forst.

Lobelia lobbiana Hook. f. & Thomson === **Lobelia zeylanica** L.

Lobelia lobelioides (Zucc.) Koopman & T. J. Ayers === **Heterotoma lobelioides** Zucc.

Lobelia longibracteata (Rock) E. Wimm. === **Lobelia gloria-montis** Rock

Lobelia longicaulis var. *plebeia* (E. Wimm.) E. Wimm. === **Lobelia longicaulis** Brandegee

Lobelia longiflora L. === **Hippobroma longiflora** (L.) G. Don

Lobelia longifolia (C. Presl) A. DC. === **Lobelia cardinalis** L.

Lobelia longipedunculata de Vriese === **Lobelia tenuior** R. Br.

Lobelia longipedunculata var. *gracilior* de Vriese === **Lobelia tenuior** R. Br.

Lobelia longipes A. DC. === **Lobelia neglecta** Schult.

Lobelia longipilosa E. Wimm. === **Lobelia cobaltica** S. Moore

Lobelia longiracemosa R. D. Good === **Lobelia anceps** L. f.

Lobelia loretensis M. E. Jones === **Lobelia laxiflora** Kunth subsp. **laxiflora**

Lobelia loxensis Willd. ex Schult. === **Siphocampylus loxensis** (Willd. ex Schult.) Vatke

Lobelia lucaeanum (C. Presl) A. DC. === **Lobelia bridgesii** Hook. & Arn.

Lobelia ludoviciana A. Gray === **Lobelia flaccidifolia** Small

Lobelia ludoviciana A. W. Wood === **Lobelia brevifolia** Nutt. ex A. DC.

Lobelia lutea L. === **Monopsis lutea** (L.) Urb.

Lobelia lycioides Cham. === **Siphocampylus lycioides** (Cham.) G. Don.

Lobelia lydenburgensis E. Wimm. === **Lobelia erinus** L.

Lobelia lythroides Diels === **Lobelia sonderiana** (Kuntze) Lammers

Lobelia macrocarpa de Vriese === **Grammatotheca bergiana** (Cham.) C. Presl

Lobelia macrocarpa var. *genistoides* de Vriese === **Grammatotheca bergiana** (Cham.) C. Presl

Lobelia macropoda Thunb. === **Siphocampylus macropodus** (Thunb.) G. Don

Lobelia macrostachys Hook. & Arn. === **Trematolobelia macrostachys** (Hook. & Arn.)
Zahlbr. ex Rock

Lobelia macularis (C. Presl) A. DC. === **Lobelia caerulea** Sims

Lobelia macularis var. *procera* (C. Presl) A. DC. === **Lobelia caerulea** Sims

Lobelia madagascariensis Schult. === **Lobelia fervens** Thunb. subsp. **fervens**

Lobelia makoyi de Jonghe === **Lobelia** × **speciosa** Sweet

Lobelia maranguensis Engl. === **Lobelia inconspicua** A. Rich.

Lobelia marryattae Paxton === **Lobelia cardinalis** L.

Lobelia mearnsii De Wild. === **Lobelia fervens** Thunb. subsp. **fervens**

Lobelia megapotamica Spreng. === **Wahlenbergia linarioides** (Lam.) A. DC.

Lobelia melleri Hemsl. === **Lobelia trullifolia** Hemsl. subsp. **trullifolia**

Lobelia melleri f. *pilosula* E. Wimm. === **Lobelia trullifolia** Hemsl. subsp. **trullifolia**

Lobelia melleri var. *grossidens* E. Wimm. === **Lobelia flaccida** subsp. **granvikii** (T. C. E. Fr.)
Thulin

Lobelia melleri var. *pulchra* E. Wimm. === **Lobelia trullifolia** Hemsl. subsp. **trullifolia**

Lobelia melleri var. *trichella* E. Wimm. === **Lobelia trullifolia** Hemsl. subsp. **trullifolia**

Lobelia michauxii Nutt. === **Lobelia inflata** L.

Lobelia michelii (A. DC.) Colmeiro === **Solenopsis laurentia** (L.) C. Presl

Lobelia michelii var. *integrifolia* (Lange) Colmeiro === **Solenopsis laurentia** (L.) C. Presl

Lobelia micrantha (E. Mey.) Heynh. === **Wimmerella pygmaea** (Thunb.) Serra, M. B. Crespo
& Lammers

Lobelia micrantha Hook. === **Lobelia heyneana** Schult.

Lobelia micrantha Kunth === **Diastatea micrantha** (Kunth) McVaugh

Lobelia micrantha var. *subtilis* (Kunth) Vatke === **Diastatea micrantha** (Kunth) McVaugh

Lobelia microdon (C. Presl) A. DC. === **Lobelia comosa** L.

Lobelia microphylla Raf. === ?

Lobelia microsperma F. Muell. === **Lobelia gibbosa** Labill.

Lobelia mildbraedii f. *acutifolia* E. Wimm. === **Lobelia mildbraedii** Engl.

Lobelia mildbraedii var. *robynsii* E. Wimm. === **Lobelia mildbraedii** Engl.

Lobelia milleri Paxton === **Lobelia × speciosa** Sweet
Lobelia minima Sims === ?
Lobelia minuta L. === **Solenopsis minuta** (L.) C. Presl
Lobelia minuta var. *minima* (Sims) Heynh. === ?
Lobelia minutidentata Engl. & Gilg === **Lobelia sonderiana** (Kuntze) Lammers
Lobelia minutiflora Kunze === **Diastatea micrantha** (Kunth) McVaugh
Lobelia minutiflora Pau === **Wahlenbergia lobeliioides** subsp. **nutabunda** (Guss.) Murb.
Lobelia minutula var. *kiwuensis* (Engl.) E. Wimm. === **Lobelia minutula** Engl.
Lobelia minutula var. *rugegensis* (T. C. E. Fr.) E. Wimm. === **Lobelia minutula** Engl.
Lobelia molleri f. *latifolia* E. Wimm. === **Lobelia molleri** Henriq.
Lobelia molleri var. *butaguensis* (De Wild.) E. Wimm. ex W. Robyns === **Lobelia molleri** Henriq.
Lobelia mollis Graham === **Lobelia xalapensis** Kunth
Lobelia monadelpha Larran. === ?
Lobelia monanthos de Vriese === **Isotoma scapigera** (R. Br.) G. Don
Lobelia montaguensis E. Wimm. === **Lobelia erinus** L.
Lobelia monticola Kunth === **Lobelia xalapensis** Kunth
Lobelia mossiana R. D. Good === **Lobelia flaccida** subsp. **mossiana** (R. D. Good) Thulin
Lobelia mucronata Cav. === **Lobelia tupa** L.
Lobelia mucronata Engelm. === **Lobelia cardinalis** L.
Lobelia mucronata f. *hookeri* (A. DC.) E. Wimm. === **Lobelia tupa** L.
Lobelia mucronata f. *ovalifolia* E. Wimm. === **Lobelia tupa** L.
Lobelia mucronata var. *berterii* (A. DC.) E. Wimm. === **Lobelia tupa** L.
Lobelia mukuluensis De Wild. === **Monopsis stellarioides** (C. Presl) Urb. subsp. **stellarioides**
Lobelia multicaulis (C. Presl) Steud. === ?
Lobelia multiflora Knowles & Westc. === ?
Lobelia mundtiana Cham. === **Lobelia thermalis** Thunb.
Lobelia mutisiana Kunth === **Burmeistera mutisiana** (Kunth) E. Wimm.
Lobelia nana var. *cymbalarioides* E. Wimm. === **Lobelia nana** Kunth
Lobelia nana var. *flagelliformis* Wedd. === **Lobelia nana** Kunth
Lobelia nannae T. C. E. Fr. === **Lobelia minutula** Engl.
Lobelia natalensis A. DC. === **Lobelia erinus** L.
Lobelia natalensis var. *subulifolia* Sond. === **Lobelia erinus** L.
Lobelia neglecta Vatke === **Lobelia longicaulis** Brandegee
Lobelia nelsonii Fernald === **Lobelia laxiflora** Kunth subsp. **laxiflora**
Lobelia nelsonii var. *fragilis* B. L. Rob. & Fernald === **Lobelia laxiflora** subsp. **angustifolia** (A. DC.) Eakes & Lammers
Lobelia neriifolia A. Gray === **Lobelia grayana** E. Wimm.
Lobelia neriifolia Moris === **Lobelia excelsa** Bonpl.
Lobelia nicotianifolia var. *bibarbata* E. Wimm. === **Lobelia nicotianifolia** Roth ex Schult.
Lobelia nicotianifolia var. *macrostemon* Skottsb. === **Lobelia nicotianifolia** Roth ex Schult.
Lobelia nicotianifolia var. *mollis* Elmer === **Lobelia philippinensis** Skottsb.
Lobelia nicotianifolia var. *trichandra* (Wight) C. B. Clarke === **Lobelia nicotianifolia** Roth ex Schult.
Lobelia niihauensis var. *forbesii* H. St. John === **Lobelia niihauensis** H. St. John
Lobelia niihauensis var. *meridiana* H. St. John === **Lobelia niihauensis** H. St. John
Lobelia nipensis Urb. === **Lobelia shaferi** Urb.
Lobelia nivea Raf. === **Lobelia spicata** Lam.
Lobelia nivea Willd. ex Schult. === **Siphocampylus niveus** (Willd. ex Schult.) Vatke
Lobelia nova de Jonghe === **Lobelia × speciosa** Sweet
Lobelia novella B. L. Rob. === **Lobelia sartorii** Vatke
Lobelia nuda Hemsl. === **Lobelia erinus** L.
Lobelia nuda f. *hirtella* E. Wimm. === **Lobelia erinus** L.
Lobelia nuda var. *rosulata* (S. Moore) E. Wimm. === **Lobelia erinus** L.
Lobelia nudicaulis Lam. === **Cyphia phyteuma** (L.) Willd.
Lobelia nudicaulis Raf. === **Lobelia paludosa** Nutt.
Lobelia nummularioides var. *prostrata* (Zahlbr.) E. Wimm. === **Lobelia nummularioides** Cham.

Lobelia nuzana E. Wimm. === **Lobelia erinus** L.

Lobelia nyassae Engl. === **Lobelia trullifolia** Hemsl. subsp. **trullifolia**

Lobelia nyikensis Baker f. === **Lobelia trullifolia** Hemsl. subsp. **trullifolia**

Lobelia obliqua Ham. ex D. Don. === **Lobelia nummularia** Lam.

Lobelia oblongifolia (Gaudich.) Endl. === **Clermontia oblongifolia** Gaudich.

Lobelia obscura L. === **Centropogon cornutus** (L.) Druce

Lobelia obtusata Urb. === **Lobelia shaferi** Urb.

Lobelia obtusifolia Willd. ex Schult. === **Siphocampylus scandens** (Kunth) G. Don

Lobelia ocimoides Kunze === **Lobelia xalapensis** Kunth

Lobelia odontoptera Schltr. === **Lobelia fervens** Thunb. subsp. **fervens**

Lobelia odontoptera var. *depilis* E. Wimm. === **Lobelia fervens** Thunb. subsp. **fervens**

Lobelia odorata Graham === **Lobelia hederacea** Cham.

Lobelia oligantha C. Y. Wu === **Lobelia davidii** Franch.

Lobelia omiensis E. Wimm. === **Lobelia fangiana** (E. Wimm.) S. Y. Hu

Lobelia omphalodoides Schltr. === **Lobelia dregeana** (C. Presl) A. DC.

Lobelia ophiocephala de Vriese === **Isotoma scapigera** (R. Br.) G. Don

Lobelia orbiculata E. Wimm. === **Lobelia flaccida** subsp. **granvikii** (T. C. E. Fr.) Thulin

Lobelia orbiculata f. *subcuneata* E. Wimm. === **Lobelia flaccida** subsp. **granvikii** (T. C. E. Fr.) Thulin

Lobelia orizabae M. Martens & Galeotti === **Lobelia gruina** Cav.

Lobelia ottoniana (C. Presl) A. DC. === **Lobelia anceps** L. f.

Lobelia ovalifolia Hook. & Arn. === **Lobelia laxiflora** Kunth subsp. **laxiflora**

Lobelia ovata Reiche === **Lobelia polyphylla** Hook. & Arn.

Lobelia oxyphylla subsp. *cacuminis* (Britton & P. Wilson) Borhidi === **Lobelia cacuminis** Britton & P. Wilson

Lobelia oxyphylla var. *cacuminis* (Britton & P. Wilson) E. Wimm. === **Lobelia cacuminis** Britton & P. Wilson

Lobelia pallida Muhl. === **Lobelia spicata** Lam.

Lobelia pallida var. *claytoniana* (Michx.) Darby === **Lobelia spicata** Lam.

Lobelia palmaris Willd. ex Schult. === **Lobelia xalapensis** Kunth

Lobelia palmeri Greene === **Diastatea tenera** (A. Gray) McVaugh

Lobelia paludosa var. *floridana* (Chapm.) A. Gray === **Lobelia floridana** Chapm.

Lobelia palustris Kerr === **Lobelia colorata** Wall. subsp. **colorata**

Lobelia paniculata L. === **Wahlenbergia exilis** A. DC.

Lobelia paniculata Raf. === **Lobelia spicata** Lam.

Lobelia paradoxa E. Wimm. === **Lobelia nummularia** Lam.

Lobelia parviflora M. Martens & Galeotti === **Diastatea micrantha** (Kunth) McVaugh

Lobelia parvifolia P. J. Bergius === **Wahlenbergia parvifolia** (P. J. Bergius) Lammers

Lobelia parvifolia Raf. === **Lobelia erinus** L.

Lobelia parvifolia R. Br. === **Lobelia rarifolia** E. Wimm.

Lobelia patula var. *pteropoda* (C. Presl) Sond. === **Lobelia pteropoda** (C. Presl) A. DC.

Lobelia patzquarensis Sessé & Moç. === **Lobelia laxiflora** Kunth subsp. **laxiflora**

Lobelia paucidentata (C. Presl) A. DC. === **Lobelia tomentosa** L. f.

Lobelia pauciflora Kunth === **Lobelia gruina** Cav.

Lobelia paulista E. Wimm. === **Lobelia camporum** Pohl

Lobelia pectinata Engelm. === **Lobelia fenestralis** Cav.

Lobelia pedicellata Diels === **Lobelia gracillima** Welw. ex Hiern

Lobelia pedunculata Sims === **Lobelia coronopifolia** L.

Lobelia persicifolia Cav. === **Lobelia laxiflora** Kunth subsp. **laxiflora**

Lobelia persicifolia (Gaudich.) Endl. === **Clermontia persicifolia** Gaudich.

Lobelia persicifolia var. *amygdalina* Vatke === **Lobelia laxiflora** subsp. **angustifolia** (A. DC.) Eakes & Lammers

Lobelia persicifolia var. *angustifolia* (A. DC.) Vatke === **Lobelia laxiflora** subsp. **angustifolia** (A. DC.) Eakes & Lammers

Lobelia persicifolia var. *laxiflora* (Kunth) Vatke === **Lobelia laxiflora** Kunth subsp. **laxiflora**

Lobelia persicifolia var. *mollis* Vatke === **Lobelia laxiflora** Kunth subsp. **laxiflora**

Lobelia persicifolia var. *warscewiczii* (Regel) Vatke === **Lobelia laxiflora** Kunth subsp. **laxiflora**
Lobelia petersiana Klotzsch === **Lobelia fervens** Thunb. subsp. **fervens**
Lobelia phoenicea de Jonghe === ?
Lobelia phyllostachya Engelm. === **Lobelia cardinalis** L.
Lobelia phyteuma L. === **Cyphia phyteuma** (L.) Willd.
Lobelia phyteumoides Willd. ex Schult. === **Lobelia collina** Kunth
Lobelia picta B. L. Rob. & Seaton === **Lobelia irasuensis** subsp **picta** (B. L. Rob. & Seaton)
   Lammers
Lobelia piedrana Urb. === **Lobelia imberbis** (Griseb.) Urb.
Lobelia pilosa Schltr. === **Lobelia cuneifolia** Link & Otto
Lobelia pinifolia f. *parviflora* Wilms ex E. Wimm. === **Lobelia pinifolia** L.
Lobelia pinifolia f. *rubriflora* Wilms ex E. Wimm. === **Lobelia pinifolia** L.
Lobelia pinifolia var. *laricina* E. Wimm. === **Lobelia pinifolia** L.
Lobelia pinnata Lam. === ?
Lobelia pinnatifida Cham. === **Cyanea pinnatifida** (Cham.) E. Wimm.
Lobelia piscinula E. Wimm. === **Lobelia sartorii** Vatke
Lobelia plebeia E. Wimm. === **Lobelia longicaulis** Brandegee
Lobelia pleotricha var. *cacuminiflora* Y. S. Lian === **Lobelia pleotricha** Diels
Lobelia pleotricha var. *handelii* (E. Wimm.) C. Y. Wu === **Lobelia pleotricha** Diels
Lobelia poasensis E. Wimm. === **Lobelia longicaulis** Brandegee
Lobelia polygalifolia Willd. ex Schult. === **Lobelia tenera** Kunth
Lobelia polymorpha Bory === **Lobelia filiformis** Lam.
Lobelia polyodon E. Wimm. === **Lobelia erinus** L.
Lobelia polyphylla f. *bracteosa* (C. Presl) E. Wimm. === **Lobelia polyphylla** Hook. & Arn.
Lobelia polyphylla f. *hyssopifolia* (C. Presl) E. Wimm. === **Lobelia polyphylla** Hook. & Arn.
Lobelia polyphylla f. *linearifolia* (Phil.) E. Wimm. === **Lobelia polyphylla** Hook. & Arn.
Lobelia polyphylla f. *subdentata* (C. Presl) E. Wimm. === **Lobelia polyphylla** Hook. & Arn.
Lobelia polyphylla var. *angustifolia* (Hook. & Arn. ex A. DC.) Heynh. === **Lobelia polyphylla**
   Hook. & Arn.
Lobelia polyphylla var. *besseriana* (C. Presl) Reiche === **Lobelia polyphylla** Hook. & Arn.
Lobelia polyphylla var. *bracteosa* (C. Presl) Reiche === **Lobelia polyphylla** Hook. & Arn.
Lobelia polyphylla var. *coquimbana* (Vatke) Reiche === **Lobelia polyphylla** Hook. & Arn.
Lobelia polyphylla var. *latifolia* (A. DC.) Heynh. === **Lobelia polyphylla** Hook. & Arn.
Lobelia porphyrantha Decne. ex Groenland === **Lobelia cardinalis** L.
Lobelia preslii f. *glabra* E. Wimm. === **Lobelia preslii** A. DC.
Lobelia princeps Otto & A. Dietr. === **Lobelia cardinalis** L.
Lobelia praetervisa Borhidi === **Lobelia rotundifolia** Juss. ex A. DC.
Lobelia procumbens J. Forbes === **Lobelia erinus** L.
Lobelia prostrata Zahlbr. === **Lobelia nummarioides** Cham.
Lobelia pterocaulon Klotzsch === **Lobelia fervens** Thunb. subsp. **fervens**
Lobelia puberula f. *argutidentata* E. Wimm. === **Lobelia puberula** Michx.
Lobelia puberula f. *candida* Fernald === **Lobelia puberula** Michx.
Lobelia puberula f. *glabella* (C. Presl) Voss === **Lobelia puberula** Michx.
Lobelia puberula f. *simulans* (Fernald) Bowden === **Lobelia puberula** Michx.
Lobelia puberula subsp. *pauciflora* (Bush) Bowden === **Lobelia reverchonii** B. L. Turner
Lobelia puberula var. *glabella* Elliott === **Lobelia glandulosa** Walter
Lobelia puberula var. *glabella* Hook. === **Lobelia puberula** Michx.
Lobelia puberula var. *laeviuscula* C. Mohr === **Lobelia puberula** Michx.
Lobelia puberula var. *mineolana* E. Wimm. === **Lobelia puberula** Michx.
Lobelia puberula var. *obtusifolia* (A. DC.) Fernald === **Lobelia puberula** Michx.
Lobelia puberula var. *pauciflora* Bush === **Lobelia reverchonii** B. L. Turner
Lobelia puberula var. *simulans* Fernald === **Lobelia puberula** Michx.
Lobelia puberula var. *subglabra* Hook. === **Lobelia puberula** Michx.
Lobelia pubescens var. *holopsida* E. Wimm. === **Lobelia pubescens** Aiton
Lobelia pubescens var. *incisa* (C. Presl) Sond. === **Lobelia pubescens** Aiton
Lobelia pubescens var. *jacquiniana* Sond. === **Lobelia pubescens** Aiton

Lobelia pubescens var. *rotundifolia* E. Wimm. === **Lobelia pubescens** Aiton
Lobelia pubescens var. *simplex* Kuntze === **Lobelia erinus** L.
Lobelia pubescens var. *thunbergiana* Sond. === **Lobelia pubescens** Aiton
Lobelia pulverulenta Pers. === **Siphocampylus affinis** (Mirb.) McVaugh
Lobelia pumila Burm. f. === **Mazus pumilus** (Burm. f.) Steenis (Scrophulariaceae)
Lobelia pumila Salisb. === **Solenopsis minuta** (L.) C. Presl subsp. **minuta**
Lobelia punicea Otto & A. Dietr. === **Lobelia cardinalis** L.
Lobelia punicea var. *kerneri* (L. Nagy) E. Wimm. === **Lobelia cardinalis** L.
Lobelia purpurascens var. *ilicifolia* (Sims) A. DC. === **Lobelia purpurascens** R. Br.
Lobelia purpurea Lindl. === **Lobelia polyphylla** Hook. & Arn.
Lobelia purpurea Vell. === **Centropogon cornutus** (L.) Druce
Lobelia pusilla (C. Presl) A. DC. === **Lobelia divaricata** Hook. & Arn.
Lobelia pusilla G. Don === **Wimmerella pygmaea** (Thunb.) Serra, M. B. Crespo & Lammers
Lobelia pusilla Welw. ex Hiern === **Lobelia lepida** E. Wimm.
Lobelia pygmaea Thunb. === **Wimmerella pygmaea** (Thunb.) Serra, M. B. Crespo & Lammers
Lobelia pyramidalis var. *wallichianum* (C. Presl) Steud. === **Lobelia pyramidalis** Wall.
Lobelia racemosa Sims === **Lobelia cirsiifolia** Lam.
Lobelia racemosa subsp. *ryanii* (Rendle) Sastre === **Lobelia cirsiifolia** Lam.
Lobelia racemosa Vest ex Schult. === ?
Lobelia radicans Thunb. === **Lobelia chinensis** Lour.
Lobelia radicans f. *lactiflora* Hisauti === **Lobelia chinensis** Lour.
Lobelia radicans f. *plena* Makino === **Lobelia chinensis** Lour.
Lobelia radicans var. *albiflora* E. Wimm. === **Lobelia chinensis** Lour.
Lobelia ramosa Benth. === **Lobelia tenuior** R. Br.
Lobelia ramosa Burb. === **Lobelia cardinalis** L.
Lobelia ramosa f. *alba* Voss === **Lobelia tenuior** R. Br.
Lobelia ramosa f. *nana* Voss === **Lobelia tenuior** R. Br.
Lobelia ramosa f. *rosea* Voss === **Lobelia tenuior** R. Br.
Lobelia ramosissima M. Martens & Galeotti === **Diastatea virgata** Scheidw.
Lobelia rapunculoides Kunth === **Lobelia gruina** Cav.
Lobelia raridentata E. Wimm. === **Lobelia erinus** L.
Lobelia recurvifolia Spreng. === **Lobelia coronopifolia** L.
Lobelia reflexa Stokes === **Lobelia siphilitica** L.
Lobelia regalis Fernald === **Lobelia ghiesbreghtii** Decne.
Lobelia reinwardtiana (C. Presl) A. DC. === **Lobelia heyneana** Schult.
Lobelia reniformis Cham. === **Lobelia oligophylla** (Wedd.) Lammers
Lobelia rensonii E. Wimm. === **Lobelia laxiflora** Kunth subsp. **laxiflora**
Lobelia repanda Martyn === **Lobelia angulata** G. Forst.
Lobelia repens Thunb. === **Lobelia anceps** L. f.
Lobelia reticulata Willd. ex Schult. === **Siphocampylus reticulatus** (Willd. ex Schult.)
    Klotzsch & H. Karst. ex Vatke
Lobelia retrorsa Willd. ex Schult. === **Siphocampylus retrorsus** (Willd. ex Schult.) Vatke
Lobelia rhizophyta Spreng. === **Lobelia anceps** L. f.
Lobelia rhombifolia var. *bidentula* E. Wimm. === **Lobelia rhombifolia**
Lobelia rhynchopetalum subsp. *callosomarginata* E. Wimm. === **Lobelia rhynchopetalum** Hemsl.
Lobelia rhynchopetalum var. *acrochilus* E. Wimm. === **Lobelia acrochilus** (E. Wimm.) E. B. Knox
Lobelia rhodesica R. E. Fr. === **Lobelia trullifolia** subsp. **rhodesica** (R. E. Fr.) Thulin
Lobelia rigidula Kunth === **Lobelia laxiflora** Kunth subsp. **laxiflora**
Lobelia rivoirei Bain === **Lobelia** × **speciosa** Sweet
Lobelia robusta var. *portoricensis* (A. DC.) McVaugh === **Lobelia robusta** Graham
Lobelia rollandia Gaudich. ex D. Dietr. === **Cyanea lanceolata** (Gaudich.) Lammers, Givnish
    & Sytsma
Lobelia rosulata S. Moore === **Lobelia erinus** L.
Lobelia rosulata f. *hirtella* E. Wimm. === **Lobelia erinus** L.
Lobelia rothrockii Greene === **Palmerella debilis** subsp. **serrata** (A. Gray) Lammers
Lobelia rotundifolia var. *angustifolia* Ekman === **Lobelia rotundifolia** Juss. ex A. DC.

Lobelia roughii var. *alces* E. Wimm. === **Lobelia roughii** Hook. f.
Lobelia roxburgiana (A. DC.) Heynh. === **Lobelia chinensis** Lour.
Lobelia rubrimaris E. Wimm. === **Lobelia neumannii** T. C. E. Fr.
Lobelia rudatisii Schltr. === **Lobelia anceps** L. f.
Lobelia ruderalis Willd. ex Schult. === **Diastatea micrantha** (Kunth) McVaugh
Lobelia rugegensis T. C. E. Fr. === **Lobelia minutula** Engl.
Lobelia rugulosa Graham === **Lobelia angulata** G. Forst.
Lobelia rupestris Kunth === **Lobelia tenera** Kunth
Lobelia rupincola Bertero ex Colla === **Lobelia anceps** L. f.
Lobelia ryanii Rendle === **Lobelia cirsiifolia** Lam.
Lobelia salicifolia Fisch. ex Trautv. === **Lobelia sessilifolia** Lamb.
Lobelia salicifolia Sweet === **Lobelia excelsa** Bonpl.
Lobelia salicina subsp. *brachyantha* (Urb.) Borhidi & Muñiz === **Lobelia salicina** Lam.
Lobelia salicina var. *brachyantha* Urb. === **Lobelia salicina** Lam.
Lobelia saliensis E. Wimm. === **Lobelia trullifolia** Hemsl. subsp. **trullifolia**
Lobelia saligna Fisch. === **Lobelia sessilifolia** Lamb.
Lobelia salviifolia A. Rich. === **Siphocampylus manettiiflorus** Hook.
Lobelia salzmanniana C. Presl === **Solenopsis laurentia** (L.) C. Presl
Lobelia sartorii var. *opulenta* E. Wimm. === **Lobelia sartorii** Vatke
Lobelia sartorii var. *piscinula* (E. Wimm.) E. Wimm. === **Lobelia sartorii** Vatke
Lobelia sattimae R. E. Fr. & T. C. E. Fr. === **Lobelia gregoriana** subsp. **sattimae** (R. E. Fr. & T. C. E. Fr.) E. B. Knox
Lobelia saxicola de Vriese === **Lobelia anceps** L. f.
Lobelia scabra Spreng. === **Monopsis simplex** (L.) E. Wimm.
Lobelia scabra Thunb. === **Monopsis scabra** (Thunb.) Urb.
Lobelia scabripes (C. Presl) A. DC. === **Lobelia flaccida** (C. Presl) A. DC. subsp. **flaccida**
Lobelia scandens Kunth === **Siphocampylus scandens** (Kunth) G. Don
Lobelia scapigera R. Br. === **Isotoma scapigera** (R. Br.) G. Don
Lobelia scapigera var. *biuncialis* R. Br. === **Isotoma scapigera** (R. Br.) G. Don
Lobelia scapigera var. *pusilla* R. Br. === **Isotoma scapigera** (R. Br.) G. Don
Lobelia schaeferi Schltr. === **Lobelia hartlaubii** Buchenau
Lobelia schmitzii E. Wimm. === **Lobelia divaricata** Hook. & Arn.
Lobelia schrankii Sweet === **Lobelia erinus** L.
Lobelia scioensis Chiov. === **Lobelia holstii** Engl.
Lobelia sebae A. DC. in DC. === **Monopsis debilis** (L. f.) C. Presl
Lobelia secunda L. f. === **Wimmerella secunda** (L. f.) Serra, M. B. Crespo & Lammers
Lobelia seguinii f. *brevisepala* E. Wimm. === **Lobelia seguinii** H. Lév. & Vaniot
Lobelia seguinii f. *longisepala* E. Wimm. === **Lobelia seguinii** H. Lév. & Vaniot
Lobelia seguinii var. *arakana* E. Wimm. === **Lobelia pyramidalis** Wall.
Lobelia seguinii var. *doniana* (Skottsb.) E. Wimm. === **Lobelia doniana** Skottsb.
Lobelia seineri Schltr. === **Lobelia sonderiana** (Kuntze) Lammers
Lobelia seleriana E. Wimm. === **Lobelia divaricata** Hook. & Arn.
Lobelia senecioides A. Cunn. ex Sims === **Isotoma axillaris** Lindl.
Lobelia senecionis Spreng. === **Isotoma axillaris** Lindl.
Lobelia senegalensis A. DC. === **Lobelia erinus** L.
Lobelia senegalensis var. *subaspera* E. Wimm. === **Lobelia flaccida** subsp. **mossiana** (R. D. Good) Thulin
Lobelia senegalensis var. *turgida* (E. Wimm.) E. Wimm. === **Lobelia erinus** L.
Lobelia serpens var. *bicolor* Steud. === **Lobelia serpens** Lam.
Lobelia serpens var. *cheiranthifolia* (C. Presl) A. DC. === **Lobelia serpens** Lam.
Lobelia serpens var. *pedicellata* A. DC. === **Lobelia serpens** Lam.
Lobelia serpens var. *puberula* E. Wimm. === **Lobelia serpens** Lam.
Lobelia serpens var. *sieberi* E. Wimm. === **Lobelia serpens** Lam.
Lobelia serpyllacea (C. Presl) Heynh. === **Lobelia hederacea** Cham.
Lobelia serpyllacea var. *odorata* (Graham) Heynh. === **Lobelia hederacea** Cham.
Lobelia serrata Meyen === **Lobelia tupa** L.

Lobelia serrulata Schott ex Brot. === **Lobelia urens** L.
Lobelia sessilifolia f. *leuantha* H. Hara === **Lobelia sessilifolia** Lamb.
Lobelia sessilifolia var. *latifolia* Nakai === **Lobelia sessilifolia** Lamb.
Lobelia setacea Sm. === **Solenopsis bivonae** (Tineo) M. B. Crespo, Serra & Juan
Lobelia setacea var. *dissectifolia* E. Wimm. === **Lobelia setacea** Thunb.
Lobelia setacea var. *parviflora* Sond. === **Lobelia setacea** Thunb.
Lobelia setulosa E. Wimm. === **Lobelia pulchella** Vatke
Lobelia shaferi var. *nipensis* (Urb.) E. Wimm. === **Lobelia shaferi** Urb.
Lobelia shaferi var. *obtusata* (Urb.) E. Wimm. === **Lobelia shaferi** Urb.
Lobelia simplex L. === **Monopsis simplex** (L.) E. Wimm.
Lobelia simsii Sweet === **Lobelia coronopifolia** L.
Lobelia siphilitica f. *alba* Voss === **Lobelia siphilitica** L.
Lobelia siphilitica f. *albiflora* Britton === **Lobelia siphilitica** L.
Lobelia siphilitica f. *grandis* E. Wimm. === **Lobelia siphilitica** L.
Lobelia siphilitica f. *hybrida* (Hook.) Voss === **Lobelia × speciosa** Sweet
Lobelia siphilitica f. *laevicalyx* Fernald === **Lobelia siphilitica** L.
Lobelia siphilitica f. *ludoviciana* (A. DC.) Voss === **Lobelia siphilitica** L.
Lobelia siphilitica f. *purpurea* E. J. Palmer & Steyerm. === **Lobelia siphilitica** L.
Lobelia siphilitica f. *purpurea* Voss === **Lobelia siphilitica** L.
Lobelia siphilitica var. *alba* G. Don & Baxter === **Lobelia siphilitica** L.
Lobelia siphilitica var. *bollii* (E. Wimm.) E. Wimm. === **Lobelia siphilitica** L.
Lobelia siphilitica var. *candida* A. Wood === **Lobelia siphilitica** L.
Lobelia siphilitica var. *densiflora* (Rennie) E. Wimm. === **Lobelia siphilitica** L.
Lobelia siphilitica var. *hybrida* Hook. === **Lobelia × speciosa** Sweet
Lobelia siphilitica var. *ludoviciana* A. DC. === **Lobelia siphilitica** L.
Lobelia siphilitica var. *maculata* G. Don === **Lobelia siphilitica** L.
Lobelia siphilitica var. *milleri* (Paxton) Steud.=== **Lobelia × speciosa** Sweet
Lobelia siphilitica var. *minor* Hook. === **Lobelia siphilitica** L.
Lobelia siphilitica var. *rosea* N. Coleman === **Lobelia siphilitica** L.
Lobelia siphilitica var. *speciosa* (Sweet) Steud. === **Lobelia × speciosa** Sweet
Lobelia sonchifolia Sw. === **Siphocampylus sonchifolius** (Sw.) McVaugh
Lobelia sonderi Zahlbr. === **Lobelia sonderiana** (Kuntze) Lammers
Lobelia spartioides (C. Presl) D. Dietr. === **Lobelia linearis** Thunb.
Lobelia spathopetala var. *decaryana* E. Wimm. === **Lobelia spathopetala** Diels
Lobelia spathulata R. D. Good === **Lobelia jasionoides** (A. DC.) E. Wimm.
Lobelia × speciosa nothovar. *occidentalis* Bowden === **Lobelia × speciosa** Sweet
Lobelia × speciosa nothovar. *schneckii* Bowden === **Lobelia × speciosa** Sweet
Lobelia spectabilis Kunth === **Centropogon cornutus** (L.) Druce
Lobelia speculum Andr. === **Monopsis debilis** (L. f.) C. Presl
Lobelia spicata Moç. === **Lobelia fenestralis** Cav.
Lobelia spicata Ruiz & Pav. ex G. Don === **Lobelia fenestralis** Cav.
Lobelia spicata f. *albiflora* Ralph Hoffm. === **Lobelia spicata** Lam.
Lobelia spicata f. *campanulata* (McVaugh) Bowden === **Lobelia spicata** Lam.
Lobelia spicata var. *campanulata* McVaugh === **Lobelia spicata** Lam.
Lobelia spicata var. *hirtella* A. Gray === **Lobelia spicata** Lam.
Lobelia spicata var. *leptostachys* (A. DC.) Mack. & Bush === **Lobelia spicata** Lam.
Lobelia spicata var. *parviflora* A. Gray === **Lobelia spicata** Lam.
Lobelia spicata var. *scaposa* McVaugh === **Lobelia spicata** Lam.
Lobelia splendens Humb. & Bonpl. ex Willd. === **Lobelia cardinalis** L.
Lobelia splendens f. *ignea* (Hook.) Voss === **Lobelia cardinalis** L.
Lobelia splendens var. *atrosanguinea* Hook. === **Lobelia cardinalis** L.
Lobelia splendens var. *fulgens* (Humb. & Bonpl. ex Willd.) S. Watson === **Lobelia cardinalis** L.
Lobelia splendens var. *ignea* Hook. === **Lobelia cardinalis** L.
Lobelia squarrosa Baker f. === **Lobelia giberroa** Hemsl.
Lobelia stellarioides (C. Presl) Benth. & Hook. f. ex Hemsl. === **Monopsis stellarioides** (C. Presl) Urb.

Lobelia stenotheca F. Muell. === **Grammatotheca bergiana** (Cham.) C. Presl
Lobelia stipularis Roth ex Schult. === **Lobelia alsinoides** Lam.
Lobelia stolzii Schltr. === **Lobelia welwitschii** Engl. & Diels
Lobelia stricklandiae f. *uncinata* E. Wimm. === **Lobelia stricklandiae** Gilliland
Lobelia stricta R. Br. === **Lobelia gibbosa** Labill.
Lobelia stricta de Vriese === **Lobelia anceps** L. f.
Lobelia stricta M. Martens & Galeotti === **Lobelia fenestralis** Cav.
Lobelia strictiflora (Rydb.) Lunell === **Lobelia kalmii** L.
Lobelia stuhlmannii var. *lanuriensis* (De Wild.) Hauman === **Lobelia stuhlmannii** Schweinf. ex Stuhlmann
Lobelia suavibracteata Hauman === **Lobelia mildbraedii** Engl.
Lobelia subcuneata Miq. === **Lobelia zeylanica** L.
Lobelia subincisa Wall. ex A. DC. === **Lobelia heyneana** Schult.
Lobelia submersa R. Cunn. ex A. Cunn. === **Glossostigma elatinoides** Benth. (Scrophulariaceae)
Lobelia subpubera var. *weberbaueri* (A. Zahlbr.) E. Wimm. === **Lobelia tenera** Kunth
Lobelia subracemosa Miq. === **Lobelia heyneana** Schult.
Lobelia subracemosa var. *rigidior* Miq. === **Lobelia heyneana** Schult.
Lobelia subtilis Kunth === **Diastatea micrantha** (Kunth) McVaugh
Lobelia subulata Benth. === **Lysipomia laricina** E. Wimm.
Lobelia subulata Klotzsch === **Lobelia fervens** Thunb. subsp. **fervens**
Lobelia succulenta Blume === **Lobelia zeylanica** L.
Lobelia succulenta f. *glabra* E. Wimm. === **Lobelia zeylanica** L.
Lobelia succulenta var. *lobbiana* (Hook. f. & Thomson) E. Wimm. === **Lobelia zeylanica** L.
Lobelia superba Cham. === **Cyanea superba** (Cham.) A. Gray
Lobelia surinamensis L. === **Centropogon cornutus** (L.) Druce
Lobelia sylvatica Fourc. === **Lobelia cuneifolia** Link & Otto
Lobelia sylvicola Lejoly & Lisowski === **Dielsantha galeopsoides** (Engl. & Diels) E. Wimm.
Lobelia taliensis Diels === **Lobelia colorata** subsp. **taliensis** (Diels) T. J. Zhang & D. Y. Hong
Lobelia tayloriana Baker f. === **Lobelia deckenii** (Asch.) Hemsl. subsp. **deckenii**
Lobelia telephioides f. *pilosella* E. Wimm. === **Lobelia telephioides** (C. Presl) A. DC.
Lobelia tenella Biv. === **Solenopsis bivonae** (Tineo) M. B. Crespo, Serra & Juan
Lobelia tenella Burm. f. === ?
Lobelia tenella L. === **Wahlenbergia parvifolia** (P. J. Bergius) Lammers
Lobelia tenera var. *belladonna* E. Wimm. === **Lobelia tenera** Kunth
Lobelia tenera var. *tenuicaulis* E. Wimm. === **Lobelia tenera** Kunth
Lobelia tenerrima Chiov. === **Lobelia exilis** Hochst. ex A. Rich.
Lobelia tenuifolia (C. Presl) D. Dietr. === **Lobelia linearis** Thunb.
Lobelia tenuifolia Willd. ex Schult. === **Lobelia fastigiata** Kunth
Lobelia tenuior var. *glabra* E. Wimm. === **Lobelia tenuior** R. Br.
Lobelia terminalis var. *minuta* C. B. Clarke === **Lobelia terminalis** C. B. Clarke
Lobelia tetragona Blume === **Lobelia cliffortiana** L.
Lobelia texensis Raf. === **Lobelia cardinalis** L.
Lobelia thermalis var. *pungens* E. Wimm. === **Lobelia thermalis** Thunb.
Lobelia thorelii E. Wimm. === **Lobelia terminalis** C. B. Clarke
Lobelia thomensis Engl. & Diels === **Lobelia molleri** Henriq.
Lobelia thunbergii Sweet === **Lobelia coronopifolia** L.
Lobelia tomentosa var. *ceratophylla* (C. Presl) Sond. === **Lobelia chamaepitys** Lam.
Lobelia tomentosa var. *multiflora* A. DC. === **Lobelia tomentosa** L. f.
Lobelia tomentosa var. *paucidentata* (C. Presl) Sond. === **Lobelia tomentosa** L. f.
Lobelia toppii Luehm. === **Lobelia gibbosa** Labill.
Lobelia tortuosa A. Heller === **Lobelia niihauensis** H. St. John
Lobelia tortuosa f. *glabrata* Skottsb. === **Lobelia niihauensis** H. St. John
Lobelia tortuosa var. *haupuensis* H. St. John === **Lobelia niihauensis** H. St. John
Lobelia tortuosa var. *intermedia* H. St. John === **Lobelia niihauensis** H. St. John
Lobelia transvaalensis Schltr. === **Lobelia erinus** L.

Lobelia trialata Buch.-Ham. ex D. Don === **Lobelia heyneana** Schult.

Lobelia trialata var. *asiatica* Chiov. === **Lobelia heyneana** Schult.

Lobelia trialata var. *grandiflora* Chiov. === **Lobelia fervens** Thunb. subsp. **fervens**

Lobelia trialata var. *lamiifolia* C. B. Clarke === **Lobelia heyneana** Schult.

Lobelia trialata var. *umbrosa* (Hemsl.) Chiov. === **Lobelia heyneana** Schult.

Lobelia triangulata Roxb. === **Lobelia alsinoides** Lam.

Lobelia trichandra Wight === **Lobelia nicotianifolia** Roth ex Schult.

Lobelia trierarchii R. D. Good === **Lobelia erinus** L.

Lobelia trigona Roxb. === **Lobelia alsinoides** Lam.

Lobelia trinitensis Griseb. === **Lobelia fastigiata** Kunth

Lobelia triquetra L. === **Lobelia comosa** L.

Lobelia triquetra var. *alba* G. Don === **Lobelia comosa** L.

Lobelia triquetra var. *comosa* (L.) Schult. === **Lobelia comosa** L.

Lobelia triquetra var. *microdon* (C. Presl) E. Wimm. === **Lobelia comosa** L.

Lobelia triquetra var. *secundata* Sond. === **Lobelia comosa** L.

Lobelia triquetra var. *foliosa* E. Wimm. === **Lobelia comosa** L.

Lobelia trivialis E. Wimm. === **Lobelia divaricata** Hook. & Arn.

Lobelia trullifolia f. *glabricalycina* E. Wimm. === **Lobelia trullifolia** Hemsl. subsp. **trullifolia**

Lobelia trullifolia var. *saliensis* (E. Wimm.) E. Wimm. === **Lobelia trullifolia** Hemsl. subsp. **trullifolia**

Lobelia tsotsorogensis Bremek. & Oberm. === **Lobelia sonderiana** (Kuntze) Lammers

Lobelia tuberosa Burm. f. === **Cyphia bulbosa** (L.) P. J. Bergius

Lobelia tupa var. *berterii* (A. DC.) Reiche === **Lobelia tupa** L.

Lobelia tupa var. *bicalcarata* (Kuntze) E. Wimm. === **Lobelia tupa** L.

Lobelia tupa var. *montana* Reiche === **Lobelia tupa** L.

Lobelia tupa var. *mucronata* (Cav.) Reiche === **Lobelia tupa** L.

Lobelia tupa var. *pavonii* E. Wimm. === **Lobelia tupa** L.

Lobelia turckheimii Rendle === **Lobelia salicina** Lam.

Lobelia turgida E. Wimm. === **Lobelia erinus** L.

Lobelia turgida var. *peteri* E. Wimm. === **Lobelia erinus** L.

Lobelia tyrianthina J. Forbes === ?

Lobelia tysonii E. Phillips === **Lobelia laxa** MacOwan

Lobelia ulugurensis (Engl.) Engl. ex R. E. Fr. & T. C. E. Fr. === **Lobelia giberroa** Hemsl.

Lobelia umbellata Kunth === **Siphocampylus umbellatus** (Kunth) G. Don

Lobelia umbellata Vest ex Schult. === ?

Lobelia umbrosa Hemsl. === **Lobelia heyneana** Schult.

Lobelia unamata E. Wimm. === **Lobelia gilgii** Engl.

Lobelia uncinata de Vriese === **Lobelia anceps** L. f.

Lobelia unidentata W. T. Aiton === **Monopsis unidentata** (W. T. Aiton) E. Wimm.

Lobelia unidentata Spreng. === **Monopsis unidentata** (W. T. Aiton) E. Wimm.

Lobelia uranocoma Cham. === **Lobelia fistulosa** Vell.

Lobelia urens var. *brevibracteata* Pérez Lara === **Lobelia urens** L.

Lobelia urens var. *integra* Chabert === **Lobelia urens** L.

Lobelia urens var. *longibracteata* Pérez Lara === **Lobelia urens** L.

Lobelia urens var. *serrulata* (Schott ex Brot.) Steud. === **Lobelia urens** L.

Lobelia urticifolia E. Wimm. === **Lobelia longicaulis** Brandegee

Lobelia usafuensis Engl. === **Lobelia giberroa** Hemsl.

Lobelia usambarensis Engl. === **Lobelia trullifolia** Hemsl. subsp. **trullifolia**

Lobelia usambarensis var. *calantha* E. Wimm. === **Lobelia trullifolia** Hemsl. subsp. **trullifolia**

Lobelia usambarensis var. *hispidella* E. Wimm. === **Lobelia trullifolia** Hemsl. subsp. **trullifolia**

Lobelia utshungwensis R. E. Fr. & T. C. E. Fr. === **Lobelia mildbraedii** Engl.

Lobelia valdiviana (Phil.) Vatke === **Legenere valdiviana** (Phil.) E. Wimm.

Lobelia vanreenensis var. *villosula* E. Wimm. === **Lobelia vanreenensis** (Kuntze) K. Schum.

Lobelia variifolia Sims === **Monopsis variifolia** (Sims) Urb.
Lobelia vaughiana E. Wimm. === **Lobelia reverchonii** B. L. Turner
Lobelia velutina M. Martens & Galeotti === **Lobelia hartwegii** A. DC. ex Benth.
Lobelia verbenifolia Salisb. === **Lobelia urens** L.
Lobelia veronicifolia Willd. ex Schult. === **Lobelia tenera** Kunth
Lobelia verticillata Cham. === **Siphocampylus verticillatus** (Cham.) G. Don
Lobelia victorialis E. Wimm. === **Lobelia chireensis** A. Rich.
Lobelia violaceo-aurantiaca De Wild. === **Monopsis stellarioides** (C. Presl) Urb. subsp. **stellarioides**
Lobelia virescens Benth. === **Burmeistera virescens** (Benth.) Benth. & Hook. ex Hemsl.
Lobelia volkensii Engl. === **Lobelia giberroa** Hemsl.
Lobelia volkensii var. *ulugurensis* Engl. === **Lobelia giberroa** Hemsl.
Lobelia volubilis Burm. f. === **Cyphia volubilis** (Burm. f.) Willd.
Lobelia volubilis Kunth === **Siphocampylus cordatus** (Willd. ex Schult.) E. Wimm.
Lobelia wallichiana (C. Presl) Hook. f. & Thomson === **Lobelia pyramidalis** Wall.
Lobelia wardii C. E. C. Fisch. === **Lobelia montana** Reinw. ex Blume
Lobelia warscewiczii Vatke === **Lobelia irasuensis** Planch. & Oerst. subsp. **irasuensis**
Lobelia weberbaueri Zahlbr. === **Lobelia tenera** Kunth
Lobelia welwitschii var. *albiflora* Hiern === **Lobelia welwitschii** Engl. & Diels
Lobelia wentzeliana Engl. === **Lobelia trullifolia** Hemsl. subsp. **trullifolia**
Lobelia westiniana Thunb. === **Siphocampylus westinianus** (Thunb.) Pohl
Lobelia wildii E. Wimm. === **Lobelia erinus** L.
Lobelia wildii var. *arcana* E. Wimm. === **Lobelia erinus** L.
Lobelia willdenowiana Schult. === **Lysipomia aretioides** Kunth
Lobelia wollastonii var. *scaettana* Hauman === **Lobelia wollastonii** Baker f.
Lobelia zahlbruckneri E. Wimm. === **Lobelia glazioviana** Zahlbr.
Lobelia zeyheri Sond. === **Monopsis zeyheri** (Sond.) Thulin
Lobelia zeylanica f. *rubescens* Hochr. === **Lobelia zeylanica** L.
Lobelia zeylanica f. *viridis* Hochr. === **Lobelia zeylanica** L.
Lobelia zeylanica var. *aligera* (Haines) Haines === **Lobelia heyneana** Schult.
Lobelia zeylanica var. *lobbiana* (Hook. f. & Thomson) Y. S. Lian === **Lobelia zeylanica** L.
Lobelia zeylanica var. *parviflora* Danguy === **Lobelia heyneana** Schult.
Lobelia zeylanica var. *walkeri* C. B. Clarke === **Lobelia heyneana** Schult.
Lobelia zombaensis E. Wimm. === **Lobelia trullifolia** Hemsl. subsp. **trullifolia**

Lobelia

The name *Lobelia* originated with Plumier (1703), who coined the polynomial *Lobelia frutescens, portulacae folia* for the species of Goodeniaceae now known as *Scaevola plumieri* (L.) Vahl. Linnaeus (1753, 1754, under the first **Lobelia**) took up the genus (calling this species *L. plumieri*), but added to it many other species treated by earlier authors as *Dortmanna* or *Rapuntium* (q.v.). Numerous authors, including Miller (1754) and Adanson (1763), objected to this expanded circumscription as unnatural, and restricted *Lobelia* to the original goodeniaceous species, assigning the campanulaceous interlopers to *Dortmanna, Laurentia, Rapuntium,* etc. Eventually, Linnaeus (1771) concurred with the criticism that his *Lobelia* was unnatural. However, instead of removing the numerous campanulaceous species and leaving *Lobelia* goodeniaceous, he simply removed *L. plumieri* to a new genus, *Scaevola,* leaving the campanulaceous species in place. Objection to this move continued into the early 20th Century (e.g., Safford 1905), reaching a crescendo with Kuntze (1891) and his absolutist views of priority. Type [by monotypy]: *Lobelia frutescens* Mill.

Plumier, C. (1703). Nova plantarum Americanarum genera. Paris: J. Boudot. La. — Pre-Linnaean origin of name *Lobelia* for species of Goodeniaceae now called *Scaevola plumieri* (L.) Vahl.

Miller, P. (1754). The gardeners dictionary ... abridged from the last folio edition (ed. 4). London: J. & J. Rivington. En. — Restriction of *Lobelia* to a species of Goodeniaceae, thus establishing later homonym of *Lobelia* L.

Adanson, M. (1763). Les chevrefeuilles. *Caprifolia*. In Familles des plantes 2: 153-159, 571. Paris: Vincent. Fr. — Defines *Lobelia* in same sense as Miller (1754).

Miller, P. (1768). The gardeners dictionary (ed. 8). London: J. & F. Rivington. En. — Validation of requisite binomial.

Linnaeus, C. (1771). Mantissa plantarum altera. Stockholm: Laurentius salvius. La. — Establishment of genus *Scaevola* for the goodeniaceous *Lobelia*.

Kuntze, O. (1891). Goodeniaceae. In Revisio generum plantarum vascularium omnium atque cellularium multarum secundum leges nomenclaturae internationales 2: 377-378. Leipzig: Arthur Felix. Ge. — Advocates use of *Lobelia* (attributed to Adanson) for *Scaevola*.

Safford, W. E. (1905). The useful plants of the island of Guam. Contr. U.S. Natl. Herb. 9: 1-416, illus. En. — Supports Kuntze's (1891) view, arguing for use of this name over *Scaevola*.

Synonyms:

Lobelia Mill. === **Scaevola** L., nom. cons. (Goodeniaceae)

Lobelia aemula (R. Br.) Kuntze === **Scaevola aemula** R. Br. (Goodeniaceae)

Lobelia amblyanthera (F. Muell.) Kuntze === **Scaevola amblyanthera** F. Muell. (Goodeniaceae)

Lobelia anchusifolia (Benth.) Kuntze === **Scaevola anchusifolia** Benth. (Goodeniaceae)

Lobelia angulata (R. Br.) Kuntze === **Scaevola angulata** R. Br. (Goodeniaceae)

Lobelia apterantha (F. Muell.) Kuntze === **Scaevola ramosissima** (Sm.) K. Krause (Goodeniaceae)

Lobelia atriplicina (F. Muell.) Kuntze === **Scaevola tomentosa** Gaudich. (Goodeniaceae)

Lobelia attenuata (R. Br.) Kuntze === **Scaevola nitida** R. Br. (Goodeniaceae)

Lobelia auriculata (Benth.) Kuntze === **Scaevola auriculata** Benth. (Goodeniaceae)

Lobelia brookeana (F. Muell.) Kuntze === **Scaevola brookeana** F. Muell. (Goodeniaceae)

Lobelia calendulacea (J. Kenn.) Kuntze === **Scaevola calendulacea** (J. Kenn.) Druce (Goodeniaceae)

Lobelia canescens (Benth.) Kuntze === **Scaevola canescens** Benth. (Goodeniaceae)

Lobelia chamissoniana (Gaudich.) Kuntze === **Scaevola chamissoniana** Gaudich. (Goodeniaceae)

Lobelia collaris (F. Muell.) Kuntze === **Scaevola collaris** F. Muell. (Goodeniaceae)

Lobelia coriacea (Nutt.) Kuntze === **Scaevola coriacea** Nutt. (Goodeniaceae)

Lobelia crassifolia (Labill.) Kuntze === **Scaevola crassifolia** Labill. (Goodeniaceae)

Lobelia cuneiformis (Labill.) Kuntze === **Scaevola cuneiformis** Labill. (Goodeniaceae)

Lobelia cunninghamii (DC.) Kuntze === **Scaevola cunninghamii** DC. (Goodeniaceae)

Lobelia cylindrocarpa (Hillebr.) Kuntze === **Scaevola chamissoniana** Gaudich. (Goodeniaceae)

Lobelia depauperata (R. Br.) Kuntze === **Scaevola depauperata** R. Br. (Goodeniaceae)

Lobelia enantophylla (F. Muell.) Kuntze === **Scaevola enantophylla** F. Muell. (Goodeniaceae)

Lobelia fasciculata (Benth.) Kuntze === **Goodenia fasciculata** (Benth.) Carolin (Goodeniaceae)

Lobelia frutescens Mill. === **Scaevola plumieri** (L.) Vahl (Goodeniaceae)

Lobelia gaudichaudii (Hook. & Arn.) Kuntze === **Scaevola gaudichaudii** Hook. & Arn. (Goodeniaceae)

Lobelia glabra (Hook. & Arn.) Kuntze === **Scaevola glabra** Hook. & Arn. (Goodeniaceae)

Lobelia glandulifera (DC.) Kuntze === **Scaevola glandulifera** DC. (Goodeniaceae)

Lobelia globulifera (Labill.) Kuntze === **Scaevola globulifera** Labill. (Goodeniaceae)

Lobelia gracilis (Hook. f.) Kuntze === **Scaevola gracilis** Hook. f. (Goodeniaceae)

Lobelia hispida (Cav.) Kuntze === **Scaevola ramosissima** (Sm.) K. Krause (Goodeniaceae)

Lobelia holosericea (de Vriese) Kuntze === **Scaevola anchusifolia** Benth. (Goodeniaceae)

Lobelia hookeri (F. Muell.) Kuntze === **Scaevola hookeri** F. Muell (Goodeniaceae)

Lobelia humifusa (de Vriese) Kuntze === **Scaevola humifusa** de Vriese (Goodeniaceae)

Lobelia humilis (R. Br.) Kuntze === **Scaevola humilis** R. Br. (Goodeniaceae)

Lobelia koenigii (Vahl) W. Wight === **Scaevola taccada** (Gaertn.) Roxb. (Goodeniaceae)

Lobelia lanceolata (Benth.) Kuntze === **Scaevola lanceolata** Benth. (Goodeniaceae)

Lobelia linearis (R. Br.) Kuntze === **Scaevola linearis** R. Br. (Goodeniaceae)
Lobelia longifolia (de Vriese) Kuntze === **Scaevola lanceolata** Benth. (Goodeniaceae)
Lobelia longiscapa de Vriese === **Goodenia pulchella** Benth. (Goodeniaceae)
Lobelia macrophylla (de Vriese) Kuntze === **Scaevola macrophylla** (de Vriese) Benth. (Goodeniaceae)
Lobelia macrostachya (de Vriese) Kuntze === **Scaevola macrostachya** (de Vriese) Benth. (Goodeniaceae)
Lobelia micrantha (C. Presl) Kuntze === **Scaevola micrantha** C. Presl (Goodeniaceae)
Lobelia microcarpa (Cav.) Kuntze === **Scaevola albida** (Sm.) Druce (Goodeniaceae)
Lobelia microphylla (de Vriese) Kuntze === **Scaevola microphylla** (de Vriese) Benth. (Goodeniaceae)
Lobelia mollis (Hook. & Arn.) Kuntze === **Scaevola mollis** Hook. & Arn. (Goodeniaceae)
Lobelia myrtifolia (de Vriese) Kuntze === **Scaevola myrtifolia** (de Vriese) K. Krause (Goodeniaceae)
Lobelia nitida (R. Br.) Kuntze === **Scaevola nitida** R. Br. (Goodeniaceae)
Lobelia oldfieldii (F. Muell.) Kuntze === **Scaevola oldfieldii** F. Muell. (Goodeniaceae)
Lobelia oppositifolia (Roxb.) Kuntze === **Scaevola oppositifolia** Roxb. (Goodeniaceae)
Lobelia ovalifolia (R. Br.) Kuntze === **Scaevola ovalifolia** R. Br. (Goodeniaceae)
Lobelia oxyclona (F. Muell.) Kuntze === **Scaevola oxyclona** F. Muell. (Goodeniaceae)
Lobelia paludosa (R. Br.) Kuntze === **Scaevola paludosa** R. Br. (Goodeniaceae)
Lobelia parvifolia (F. Muell. ex Benth.) Kuntze === **Scaevola parvifolia** F. Muell. ex Benth. (Goodeniaceae)
Lobelia phlebopetala (F. Muell.) Kuntze === **Scaevola phlebopetala** F. Muell. (Goodeniaceae)
Lobelia piliplena (Miq.) Kuntze === **Scaevola taccada** (Gaertn.) Roxb. (Goodeniaceae)
Lobelia pilosa (Benth.) Kuntze === **Scaevola pilosa** Benth. (Goodeniaceae)
Lobelia platyphylla (Lindl.) Kuntze === **Scaevola platyphylla** Lindl. (Goodeniaceae)
Lobelia plumieri L. === **Scaevola plumieri** (L.) Vahl (Goodeniaceae)
Lobelia porocarya (F. Muell.) Kuntze === **Scaevola porocarya** F. Muell. (Goodeniaceae)
Lobelia procera (Hillebr.) Kuntze === **Scaevola procera** Hillebr. (Goodeniaceae)
Lobelia restiacea (Benth.) Kuntze === **Scaevola restiacea** Benth. (Goodeniaceae)
Lobelia revoluta (R. Br.) Kuntze === **Scaevola revoluta** R. Br. (Goodeniaceae)
Lobelia sericea (Vahl) Kuntze === **Scaevola taccada** (Gaertn.) Roxb. (Goodeniaceae)
Lobelia sericea var. *koenigii* (Vahl) Kuntze === **Scaevola taccada** (Gaertn.) Roxb. (Goodeniaceae)
Lobelia sericophylla (F. Muell. ex Benth.) Kuntze === **Scaevola sericophylla** F. Muell. ex Benth. (Goodeniaceae)
Lobelia spinescens (R. Br.) Kuntze === **Scaevola spinescens** R. Br. (Goodeniaceae)
Lobelia stenophylla (F. Muell.) Kuntze === **Goodenia stenophylla** F. Muell. (Goodeniaceae)
Lobelia striata (R. Br.) Kuntze === **Scaevola striata** R. Br. (Goodeniaceae)
Lobelia taccada Gaertn. === **Scaevola taccada** (Gaertn.) Roxb. (Goodeniaceae)
Lobelia thesioides (Benth.) Kuntze === **Scaevola thesioides** Benth. (Goodeniaceae)
Lobelia tomentosa (Gaudich.) Kuntze === **Scaevola tomentosa** Gaudich. (Goodeniaceae)
Lobelia tortuosa (Benth.) Kuntze === **Scaevola tortuosa** Benth. (Goodeniaceae)
Lobelia velutina (C. Presl) Kuntze === **Scaevola taccada** (Gaertn.) Roxb. (Goodeniaceae)

Lobelia

Presl (1836) removed all of the original Linnaean species from *Lobelia* L., leaving behind only species described later (the majority of which are referable to *Siphocampylus*). By this action, he created a later homonym. The typification effected here makes this name homotypic with *Siphocampylus*. Type [designated here]: *Lobelia westiniana* Thunb.

Presl, C. (1836). *Centropogon*. In Prodromus monographiae Lobeliacearum: 33-40. Prague: Theophilus Haase. La. — Establishment of genus.

Synonyms:
Lobelia C. Presl === **Siphocampylus** Pohl
Lobelia attenuata C. Presl === **Siphocampylus attenuatus** (C. Presl) A. DC.
Lobelia cana (Pohl) C. Presl === **Siphocampylus macropodus** (Thunb.) G. Don
Lobelia canescens C. Presl === **Lobelia laxiflora** Kunth subsp. **laxiflora**
Lobelia cardiophylla (Pohl) C. Presl === **Siphocampylus corymbiferus** Pohl
Lobelia corymbifera (Pohl) C. Presl === **Siphocampylus corymbiferus** Pohl
Lobelia crenata C. Presl === **Siphocampylus macropodus** (Thunb.) G. Don
Lobelia dombeyana C. Presl === **Centropogon dombeyanus** (C. Presl) E. Wimm.
Lobelia fusca (G. Don) C. Presl === **Centropogon fuscus** (G. Don) E. Wimm.
Lobelia humboldtiana C. Presl === **Siphocampylus humboldtianus** A. DC.
Lobelia jamesoniana C. Presl === **Siphocampylus affinis** (Mirb.) McVaugh
Lobelia kunthiana C. Presl === **Siphocampylus umbellatus** (Kunth) G. Don
Lobelia laciniata (G. Don) C. Presl === ?
Lobelia macrophylla (G. Don) C. Presl === **Centropogon macrophyllus** (G. Don) E. Wimm.
Lobelia macrostemon C. Presl === **Siphocampylus macrostemon** (C. Presl) A. DC.
Lobelia nitida (Pohl) C. Presl === **Siphocampylus nitidus** Pohl
Lobelia pedicellaris C. Presl === **Siphocampylus longipedunculatus** Pohl
Lobelia pendulifolia C. Presl === **Siphocampylus ovatus** (G. Don) E. Wimm.
Lobelia prunifolia Humb. ex C. Presl === **Lobelia laxiflora** Kunth subsp. **laxiflora**
Lobelia psilophylla (Pohl) C. Presl === **Siphocampylus psilophyllus** Pohl
Lobelia rosmarinifolia (G. Don) Dombey ex C. Presl === **Siphocampylus rosmarinifolius** G. Don
Lobelia rugosa C. Presl === **Siphocampylus dependens** G. Don
Lobelia scabra C. Presl === **Siphocampylus longipedunculatus** Pohl
Lobelia triphylla C. Presl === **Siphocampylus duploserratus** Pohl
Lobelia verbascifolia C. Presl === **Centropogon verbascifolius** (C. Presl) Gleason
Lobelia villosula (Pohl) C. Presl === **Siphocampylus macropodus** (Thunb.) G. Don

Loreia

Rafinesque (1837) segregated this species from *Campanula* as part of his attempt to create a more natural circumscription for that genus. In his view, *Campanula* consisted solely of species with exappendiculate calyx, campanulate corolla, and smooth three-locular poricidal capsules. This species (assigned to *Campanula* sect. *Rapunculus* Dumort. by most authors) differed in its rotate corolla and sulcate hypanthium. Type [designated by the author]: *Loreia baldensis* (Balb.) Raf.

> Rafinesque, C. S. (1837). Flora Telluriana, pars secunda. Philadelphia: published privately. En. — Establishment of genus as distinct from *Campanula*.

Synonyms:
Loreia Raf. === **Campanula** L.
Loreia baldensis (Balb.) Raf. === **Campanula ramosissima** Sm.

Lysipomia

Lobelioideae, 30 species, Andes of South America, from Venezuela to Bolivia; the greatest concentration occurs in Ecuador and Peru. The genus is divided into subg. *Lysipomia* and subg. *Rhizocephalum* (Wedd.) E. Wimm. on the basis of floral morphology (Wimmer 1953, McVaugh 1955); phylogenetic analysis of DNA data supports this classification (Ayers 1999). Despite their marked difference in habit and stature, *Lysipomia* and the largely sympatric *Centropogon-Burmeistera-Siphocampylus* clade appear to be sister groups, based on recent analyses of molecular data (E. Knox, pers. comm.; A. Antonelli, pers. comm.). The treatment here follows the monograph of McVaugh (1955), as amended by McVaugh (1965), Jeppesen (1981), and Ayers (1997). $2n = 20$. Type [designated by McVaugh, Brittonia 8: 72 (1955)]: *Lysipomia acaulis* Kunth.

Humboldt, A., A. Bonpland & C. S. Kunth (1819). *Lysipomia*. In Nova genera et species plantarum 3: 318-321 (quarto), 248-250 (folio), illus. Paris: Libraria Graeco-Latini-Germanica. La. — Establishment of genus as distinct from *Lobelia*.

- Wimmer, F. E. (1953). *Lysipomia* Humb., *Dominella* E. Wimm. In A. Engler & L. Diels, Das Pflanzenreich IV. 276b: 745-754, illus. Berlin: Akademie-Verlag. Ge., La. — Monograph with key, descriptions, infrageneric classification, full nomenclature, and specimen citations.
- McVaugh, R. (1955). A revision of *Lysipomia* (Campanulaceae, Lobelioideae). Brittonia 8: 69-105, illus. En. — Monograph with key, descriptions, infrageneric classification, full nomenclature, and specimen citations.

McVaugh, R. (1965). South American Lobelioideae new to science. Ann. Missouri Bot. Gard. 52: 399-409, illus. En. — Supplement to McVaugh (1955).

- Wimmer, F. E. (1968). *Lysipomia* HBK. In A. Engler & L. Diels, Das Pflanzenreich IV. 276c: 893-900, illus. Berlin: Akademie-Verlag. Ge., La. — Supplement to Wimmer (1953), with new key to species.

van der Hammen, T. & A. M. Cleef (1978). Pollen morphology of *Lysipomia* H.B.K. and *Rhizocephalum* Wedd. (Campanulaceae), and the revision of the pollen determination "*Valeriana*" *stenophylla* Killip. Rev. Paleobot. Palynol. 25: 367-376, illus. En. — Palynological data support merger of *Lysipomia* and *Rhizocephalum*.

- Jeppesen, S. (1981). *Lysipomia* H. B. K. In G. Harling & B. Sparre (eds.), Flora of Ecuador 14: 136-151, illus. Stockholm: Swedish Natural Science Research Council. En. — Flora with keys, descriptions, full nomenclature, and specimen citations; type of *Dominella* transferred to genus.

Ayers, T. J. (1997). Three new species of *Lysipomia* (Lobeliaceae) endemic to the páramos of southern Ecuador. Brittonia 49: 433-440, illus. En. — Revised key to all Ecuadorean species.

- Ayers, T. J. (1999). Biogeography of *Lysipomia* (Campanulaceae), a high elevation endemic: an illustration of species richness at the Huancabamba Depression, Peru. Arnaldoa 6(2): 13-28. En. — Overview of genus, its biogeography and evolution, based on phylogenetic analysis.

Lysipomia Kunth in Humb., Bonpl. & Kunth, Nov. Gen. Sp. 3: 318 (quarto), 248 (folio) (1819).
Andean South America. 82 83.
 Rhizocephalum Wedd., Chlor. Andina 2: 11 (1858).
 Dominella E. Wimm., Pflanzenr. IV.276b: 754 (1953).

Lysipomia acaulis Kunth in Humb., Bonpl. & Kunth, Nov. Gen. Sp. 3: 321 (quarto), 250 (folio) (1819).
Ecuador & Peru. 83 ECU PER. Cham.
 Lobelia androsacea Willd. ex Schult. in Roem. & Schult., Syst. Veg. 5: 41 (1819).

Lysipomia aretioides Kunth in Humb., Bonpl. & Kunth, Nov. Gen. Sp. 3: 321 (quarto), 250 (folio) (1819).
Ecuador & Peru. 83 ECU PER. Cham.
 Lobelia willdenowiana Schult. in Roem. & Schult., Syst. Veg. 5: 634 (1819).

Lysipomia bilineata McVaugh, Brittonia 8: 88 (1955).
Ecuador. 83 ECU. Cham.

Lysipomia bourgoinii Ernst, Revista Ci. Mens. Univ. Centr. Venezuela 1887: 134 (1887), as 'bourgoini'.
Venezuela. 82 VEN. Cham.
Lysipomia lycopodioides Goebel, Pflanzenbiol. Schilderungen 2: 50 (1891).

Lysipomia brachysiphonia (Zahlbr.) E. Wimm. in J. F. Macbr., Fl. Peru 6: 484 (1937).
Peru. 83 PER. Cham.

Rhizocephalum brachysiphonium Zahlbr., Bot. Jahrb. Syst. 37: 461 (1906).
   Rhizocephalum brachysiphonium var. *brevifolium* Zahlbr., Bot. Jahrb. Syst. 37: 461
      (1906).*Lysipomia brachysiphonia* var. *brevifolia* (A. Zahlbr.) E. Wimm. in J. F. Macbr.,
      Fl. Peru 6: 484 (1937).

Lysipomia caespitosa T. J. Ayers, Brittonia 49: 434 (1997).
   Ecuador. 83 ECU. Cham.

Lysipomia crassimarginata (E. Wimm.) Jeppesen in Harling & Sparre, Fl. Ecuador 14: 140
   (1981), as 'crassomarginata'.
   Ecuador or Peru. 83 +. Hemicr.
    **Dominella crassimarginata* E. Wimm., Pflanzenr. IV.276b: 754 (1953), as 'crassomarginata'.

Lysipomia cuspidata McVaugh, Brittonia 8: 90 (1955).
   Ecuador. 83 ECU. Cham.

Lysipomia cylindrocarpa T. J. Ayers, Brittonia 49: 435 (1997).
   Ecuador. 83 ECU. Cham.

Lysipomia glandulifera (Wedd.) Schlecht. ex E. Wimm. in J. F. Macbr., Fl. Peru 6: 484 (1937).
   Peru & Bolivia. 83 BOL PER. Cham.
    **Pratia glandulifera* Wedd., Chlor. Andina 2: 11 (1858).

   subsp. **glandulifera**
      Peru & Bolivia. 83 BOL PER. Cham.

   subsp. **globulifera** McVaugh, Brittonia 8: 93 (1955).
      Peru. 83 PER. Cham.

Lysipomia globularis E. Wimm. in J. F. Macbr., Fl. Peru 6: 485 (1937).
   Peru. 83 PER. Cham.

Lysipomia gracilis (E. Wimm.) E. Wimm. in J. F. Macbr., Fl. Peru 6: 485 (1937).
   Peru. 83 PER. Cham.
    **Rhizocephalum gracile* E. Wimm., Repert. Spec. Nov. Regni Veg. 26: 4 (1929).

Lysipomia hirta E. Wimm., Repert. Spec. Nov. Regni Veg. 38: 87 (1935).
   Peru. 83 PER. Cham.

Lysipomia hutchisonii McVaugh, Ann. Missouri Bot. Gard. 52: 408 (1965).
   Peru. 83 PER. Cham.

Lysipomia laciniata A. DC. in DC., Prodr. 7: 349 (1839). *Rhizocephalum candollei* Wedd.,
   Chlor. Andina 2: 12 (1858), as 'candollii'.
   Venezuela to Bolivia. 82 VEN 83 BOL CLM PER. Hemicr. $2n = 20$.

   subsp. **fissicalyx** McVaugh, Brittonia 8: 102 (1955).
      Colombia. 83 CLM. Hemicr.

   subsp. **laciniata**
      Peru & Bolivia. 83 BOL PER. Hemicr. $2n = 20$.
      Rhizocephalum candollei var. *vulgare* Wedd., Chlor. Andina 2: 12 (1858). *Lysipomia*
         laciniata var. *vulgaris* (Wedd.) E. Wimm. in J. F. Macbr., Fl. Peru 6: 486 (1937).

   subsp. **linearifolia** (E. Wimm.) McVaugh, Brittonia 8: 96 (1955).
      Peru & Bolivia. 83 BOL PER. Hemicr.
       **Lysipomia linearifolia* E. Wimm. in J. F. Macbr., Fl. Peru 6: 487 (1937).
         Rhizocephalum candollei var. *ciliatum* Wedd., Chlor. Andina 2: 12 (1858). *Lysipomia*
            laciniata var. *ciliata* (Wedd.) McVaugh, Brittonia 8: 96 (1955).

subsp. **meridensis** McVaugh, Brittonia 8: 97 (1955).
Venezuela. 82 VEN. Hemicr.

subsp. **microsperma** McVaugh, Brittonia 8: 101 (1955).
Colombia. 83 CLM. Hemicr.
Rhizocephalum candollei var. *pubescens* Wedd., Chlor. Andina 2: 12 (1858). *Lysipomia laciniata* var. *pubescens* (Wedd.) E. Wimm., Pflanzenr. IV.276b: 753 (1953).

Lysipomia laricina E. Wimm. in J. F. Macbr., Fl. Peru 6: 486 (1937).
Ecuador. 83 ECU. Cham.
* *Lobelia subulata* Benth., Pl. Hartweg.: 137 (1844); non *Lysipomia subulata* G. Don, Gen. Hist. 3: 717 (1834).

Lysipomia lehmannii Hieron. ex Zahlbr., Repert. Spec. Nov. Regni Veg. 14: 185 (1915), as 'lehmanni'.
Ecuador. 83 ECU. Cham.

Lysipomia montioides Kunth in Humb., Bonpl. & Kunth, Nov. Gen. Sp. 3: 320 (quarto), 249 (folio) (1819).
Colombia to Peru. 83 CLM ECU PER. Cham.
Lobelia limoselloides Humb. & Bonpl. ex Schult. in Roem. & Schult., Syst. Veg. 5: 41 (1819).

Lysipomia multiflora McVaugh, Brittonia 8: 83 (1955).
Peru. 83 PER. Cham.

Lysipomia muscoides Hook. f., London J. Bot. 6: 286 (1847).
Colombia to Peru. 83 CLM ECU PER. Cham.

subsp. **delicatula** McVaugh, Ann. Missouri Bot. Gard. 52: 406 (1965).
Peru. 83 PER. Cham.

subsp. **muscoides**
Colombia & Ecuador. 83 CLM ECU. Cham.

subsp. **simulans** McVaugh, Brittonia 8: 87 (1955).
Colombia. 83 CLM. Cham.

Lysipomia oellgaardii Jeppesen in Harling & Sparre, Fl. Ecuador 14: 144 (1981).
Ecuador. 93 ECU. Hemicr.

Lysipomia pumila (Wedd.) E. Wimm. in J. F. Macbr., Fl. Peru 6: 488 (1937).
Peru & Bolivia. 83 BOL PER. Hemicr.
* *Rhizocephalum pumilum* Wedd., Chlor. Andina 2: 13 (1858).

Lysipomia rhizomata McVaugh, Brittonia 8: 84 (1955).
Ecuador. 83 ECU. Cham.

Lysipomia sparrei Jeppesen in Harling & Sparre, Fl. Ecuador 14: 145 (1981).
Ecuador. 83 ECU. Cham.

Lysipomia speciosa T. J. Ayers, Brittonia 49: 437 (1997).
Ecuador. 83 ECU. Cham.

Lysipomia sphagnophila Griseb. ex Wedd., Chlor. Andina 2: 15 (1858), as 'sphagnophilum'.
Colombia to Peru. 83 CLM ECU PER. Cham.

subsp. **acuta** (E. Wimm.) McVaugh, Brittonia 8: 79 (1955).
Peru. 83 PER. Cham.
Lysipomia acuta E. Wimm. in J. F. Macbr., Fl. Peru 6: 483 (1937).

subsp. **angelensis** Jeppesen in Harling & Sparre, Fl. Ecuador 14: 147 (1981).
Ecuador. 83 ECU. Cham.

subsp. **minor** McVaugh, Brittonia 8: 78 (1955).
Colombia. 83 CLM. Cham.
Lysipomia obliqua E. Wimm., Ann. Naturhist. Mus. Wien 56: 374 (1948), as 'oblinqua'.

subsp. **sphagnophila**
Peru. 83 PER. Cham.

subsp. **variabilis** McVaugh, Brittonia 8: 80 (1955). *Lysipomia variabilis* (McVaugh) E.
Wimm., Pflanzenr. IV.276b: 898 (1968).
Ecuador. 83 ECU. Cham.

Lysipomia subpeltata McVaugh, Ann. Missouri Bot. Gard. 52: 407 (1965).
Peru. 83 PER. Hemicr.

Lysipomia tubulosa McVaugh, Brittonia 8: 88 (1955).
Ecuador. 83 ECU. Cham.

Lysipomia vitreola McVaugh, Brittonia 8: 90 (1955).
Ecuador. 83 ECU. Cham.

Lysipomia wurdackii McVaugh, Ann. Missouri Bot. Gard. 52: 409 (1965).
Peru. 83 PER. Cham.

Synonyms:
Lysipomia sect. *Hypsela* (C. Presl) A. DC. === **Lobelia** L.
Lysipomia acuta E. Wimm. === **Lysipomia sphagnophila** subsp. **acuta** (E. Wimm.) McVaugh
Lysipomia brachysiphonia var. *brevifolia* (A. Zahlbr.) E. Wimm. === **Lysipomia
brachysiphonia** (Zahlbr.) E. Wimm.
Lysipomia laciniata var. *ciliata* (Wedd.) McVaugh === **Lysipomia laciniata** subsp.
linearifolia (E. Wimm.) McVaugh
Lysipomia laciniata var. *pubescens* (Wedd.) E. Wimm. === **Lysipomia laciniata** subsp.
microsperma McVaugh
Lysipomia laciniata var. *vulgaris* (Wedd.) E. Wimm. === **Lysipomia laciniata** A. DC.
subsp. **laciniata**
Lysipomia linearifolia E. Wimm. === **Lysipomia laciniata** subsp. **linearifolia** (E. Wimm.)
McVaugh
Lysipomia lycopodioides Goebel === **Lysipomia bourgoinii** Ernst
Lysipomia obliqua E. Wimm. === **Lysipomia sphagnophila** subsp. **minor** McVaugh
Lysipomia reniformis Kunth === **Lobelia oligophylla** (Wedd.) Lammers
Lysipomia subulata G. Don === ?
Lysipomia variabilis (McVaugh) E. Wimm. === **Lysipomia sphagnophila** subsp. **variabilis**
McVaugh

Macrochilus

This genus was distinguished from *Cyanea* and *Delissea* largely on the mistaken belief that
the calyx lobes were imbricate, a condition not found in the family. Type [by monotypy]:
Macrochilus superbus (Cham.) C. Presl.

Presl, C. (1836). *Macrochilus*. In Prodromus monographiae Lobeliacearum: 47. Prague:
Theophilus Haase. La. — Establishment of genus as distinct from *Lobelia*.

Gray, A. (1861). Notes on Lobeliaceae, Goodeniaceae, &c. of the collections of the U.S. South Pacific Exploring Expedition. Proc. Amer. Acad. Arts 5: 146-152. En. — Genus subsumed into *Cyanea*.

Synonyms:
Macrochilus C. Presl === **Cyanea** Gaudich.
Macrochilus superbus (Cham.) C. Presl === **Cyanea superba** (Cham.) A. Gray

Marianthemum

This genus was created to accomodate species of *Campanula* with reflexed appendages between the calyx lobes. Its type is referable to sect. *Sibiricae* (Fomin) Kharadze, but the other species that have been assigned here are quite diverse, belonging to sect. *Cordifoliae* (Fomin) Kharadze, sect. *Dasystigma* (Fed.) Victorov, sect. *Involucrata* (Fomin) Kharadze, sect. *Quinqueloculares* (Boiss.) Phitos, and sect. *Scapiflorae* (Boiss.) Kharadze. Type [by monotypy]: *Marianthemum pannonicum* Schrank.

von Schrank, F. (1822). Bemerkungen über einige seltnere Pflanzen des k. botanischen Gartens zu München. Denkschr. Königl.-Baier. Bot. Ges. Regensburg 2: 21-72. Ge. — Establishment of genus as distinct from *Campanula*.

Synonyms:
Marianthemum Schrank === **Campanula** L.
Marianthemum aggregatum (Willd.) Schrank === **Campanula glomerata** L. subsp. **glomerata**
Marianthemum alpinum (Jacq.) Schur === **Campanula alpina** Jacq.
Marianthemum betonicifolium Schrank === **Campanula sarmatica** Ker Gawl.
Marianthemum divergens (Waldst. & Kit.) Schur === **Campanula sibirica** subsp. **divergens** (Waldst. & Kit. ex Willd.) Nyman
Marianthemum longifolium (Pers.) Schrank === **Campanula speciosa** Pourr. subsp. **speciosa**
Marianthemum medium (L.) Schur === **Campanula medium** L.
Marianthemum pannonicum Schrank === **Campanula sibirica** subsp. **divergens** (Waldst. & Kit. ex Willd.) Nyman
Marianthemum sibiricum (L.) Schur === **Campanula sibirica** L.
Marianthemum spathulatum Schur === **Campanula sibirica** subsp. **divergens** (Waldst. & Kit. ex Willd.) Nyman

Mecoschistum

In the protologue, *Lobelia* was cited as a synonym and the description did not differentiate the new genus from the elder. For this reason, it seems that Dulac (1867) was coining a new name for all of *Lobelia*, rather than merely segregating *L. urens* from it. Perhaps he believed (like Miller 1754, under **Lobelia**) that the name *Lobelia* properly applied to a genus in the Goodeniaceae. Based on: *Lobelia* L.

Dulac, J. (1867). Flore du département des Hautes-Pyrénées. Paris: F. Savy. Fr. — Establishment of name.

Synonyms:
Mecoschistum Dulac === **Lobelia** L.
Mecoschistum urens (L.) Dulac === **Lobelia urens** L.

Medium

This pre-Linnaean generic name was resurrected to accomodate species of *Campanula* with an appendiculate calyx. As with *Marianthemum* (q.v.), the included species are a heterogeneous lot. The type selected here belongs to sect. *Quinqueloculares* (Boiss.) Phitos, while the remaining species are referable to sect. *Cordifoliae* (Fomin) Kharadze, sect. *Scapiflorae* (Boiss.) Kharadze, and sect. *Sibiricae* (Fomin) Kharadze. Type [designated here]: *Medium grandiflorum* Spach.

> Spach, É. (1838). Les Campanulacées. – Campanulaceae. Histoire naturelle des végétaux, Phanérogames 9: 534-569. Paris: Roret. Fr. — Establishment of genus as distinct from *Campanula*.

Synonyms:
Medium Spach === **Campanula** L.
Medium alliariifolium (Willd.) Spach === **Campanula alliariifolia** Willd.
Medium barbatum (L.) Spach === **Campanula barbata** L.
Medium grandiflorum Spach === **Campanula medium** L.
Medium gummiferum (Willd. ex Schult.) Spach === **Campanula sarmatica** Ker Gawl.
   subsp. **sarmatica**
Medium sibiricum (L.) Spach === **Campanula sibirica** L.

Medium

This genus was one of two created by Opiz (in Berchtold 1839) to accomodate species of *Campanula* with an appendiculate calyx. Those with a three-loculed capsule were assigned to *Nenningia*, those with five were assigned here. As such, the genus is roughly equivalent to *Campanula* sect. *Quinqueloculares* (Boiss) Phitos. Type [designated here]: *Campanula medium* L.

> Berchtold, F. (1839). Oekonomisch-technische flora Böhmens, band 2, abteil. 2. Prague: Thomas Thabor. Ge. — Establishment of genus as distinct from *Campanula*.

Synonyms:
Medium Opiz === **Campanula** L.

Merciera

Campanuloideae, 6 species, Cape Provinces of South Africa. No species has been included yet in a molecular phylogeny, but one would assume on the basis of morphology and geography that it would be part of the "wahlenbergioid" clade that includes *Craterocapsa* and *Roella* (Eddie et al. 2003, under **General**). The treatment here follows Cupido (2003). Type [designated by Pfeiff., Nomencl. Bot. 2: 284 (1874)]: *Trachelium tenuifolium* L. f.

> de Candolle, A. (1830). *Merciera*. In Monographie des Campanulées: 369-371. Paris: Veuve Desray. La. — Establishment of genus as distinct from *Trachelium*.
>
> Adamson, R. S. (1950). *Merciera* A. DC. In R. S. Adamson & T. M. Salter, Flora of the Cape Peninsula: 745. Cape Town: Juta. En. — Flora with description.
>
> • Adamson, R. S. (1955). The genus *Merciera* A. DC. J. S. African Bot. 20: 157-163. — En. Monograph with key, descriptions, full nomenclature, and specimen citations.
>
> Dyer, R. A. (1975). *Merciera* A. DC. In The genera of southern African flowering plants 1: 639. Pretoria: Department of Agricultural Technical Services. En. — Generic description.
>
> Goldblatt, P. & J. Manning (2000). Cape plants. A conspectus of the Cape flora of South Africa. *Merciera*. Strelitzia 9: 391-392. En. — Checklist for Cape of Good Hope, with brief species desciptions.
>
> Welman, W. G. (2000). *Merciera* A. DC. In O. A. Leistner (ed.), Seed plants of southern Africa: families and genera. Strelitzia 10: 201. En. — Generic description.
>
> • Cupido, C. N. (2003). Systematic studies in the genus *Merciera* (Campanulaceae): a reassessment of species boundaries, illus. Adansonia (sér. 3) 25: 33-44. En. — Synopsis of classification of genus, based on morphometric analyses.

Merciera A. DC., Monogr. Campan.: 369 (1830).
   S. Africa. 27.

Merciera azurea Schltr., Bot. Jahrb. Syst. 24: 447 (1897). *Merciera tenuifolia* var. *azurea*
   (Schltr.) Adamson, J. S. African Bot. 20: 160 (1955).
   Cape Provinces. 27 CPP. Nanophan.

Merciera brevifolia A. DC., Monogr. Campan.: 371 (1830).
   Cape Provinces. 27 CPP. Nanophan.

Merciera eckloniana H. Buek in Eckl. & Zeyh., Enum. Pl. Afric. Austral.: 387 (1837).
   Merciera tenuifolia var. *eckloniana* (H. Buek) Sond. in Harv. & Sond., Fl. Cap. 3: 596 (1865).
   Cape Provinces. 27 CPP. Nanophan.

Merciera leptoloba A. DC., Monogr. Campan.: 371 (1830). *Merciera brevifolia* var. *leptoloba*
   (A. DC.) Sond. in Harv. & Sond., Fl. Cap. 3: 596 (1865).
   Cape Provinces. 27 CPP. Nanophan.

Merciera tenuifolia (L. f.) A. DC., Monogr. Campan.: 370 (1830).
   Cape Provinces. 27 CPP. Nanophan.
      * *Trachelium tenuifolium* L. f., Suppl. Pl.: 143 (1782). *Roella tenuifolia* (L. f.) Thunb., Fl.
         Cap. (ed. 2): 174 (1823).
      Merciera tenuifolia var. *candolleana* Sond. in Harv. & Sond., Fl. Cap. 3: 596 (1865).
      Merciera tenuifolia var. *thunbergiana* Sond. in Harv. & Sond., Fl. Cap. 3: 596 (1865).

Merciera tetraloba Cupido, Bothalia 32: 74 (2002).
   Cape Provinces. 27 CPP. Nanophan.

Synonyms:
Merciera brevifolia var. *leptoloba* (A. DC.) Sond. === **Merciera leptoloba** A. DC.
Merciera heteromorpha H. Buek === **Carpacoce heteromorpha** (H. Buek) L. Bolus (Rubiaceae)
Merciera tenuifolia var. *azurea* (Schltr.) Adamson === **Merciera azurea** Schltr.
Merciera tenuifolia var. *candolleana* Sond. === **Merciera tenuifolia** (L. f.) A. DC.
Merciera tenuifolia var. *eckloniana* (H. Buek) Sond. === **Merciera brevifolia** A. DC.
Merciera tenuifolia var. *thunbergiana* Sond. === **Merciera tenuifolia** (L. f.) A. DC.
Merciera vaginata Adamson === Rubiaceae

Mezleria

Although this name typically has been used for a subgenus of *Lobelia* (e.g., Wimmer 1953,
Murata 1995, both under **General**), its type is actually referable to the genus *Monopsis*
(Thulin 1983, under **Africa**; Lammers 1999b, under **Special**). Type [designated by Pfeiff.,
Nomencl. Bot. 2: 298 (1874)]: *Lobelia depressa* L. f.

   Presl, C. (1836). *Mezleria*. In Prodromus monographiae Lobeliacearum: 7-8. Prague:
      Theophilus Haase. La. — Establishment of genus as distinct from *Lobelia*.
   Adamson, R. S. (1950). *Mezleria* Presl. In R. S. Adamson & T. M. Salter, Flora of the Cape
      Peninsula: 753-754. Cape Town: Juta. En. — One of few 20th Century works to
      recognize genus.

Synonyms:
Mezleria C. Presl === **Monopsis** Salisb.
Mezleria depressa (L. f.) C. Presl === **Monopsis debilis** (L. f.) C. Presl
Mezleria dregeana Sond. === **Lobelia sonderiana** (Kuntze) Lammers
Mezleria filicaulis C. Presl === **Lobelia filicaulis** (C. Presl) Schönl.

Mezleria humifusa A. DC. === **Unigenes humifusa** (A. DC.) E. Wimm.
Mezleria limosa Adamson === **Lobelia limosa** (Adamson) E. Wimm.
Mezleria quadrisepala R. Good === **Lobelia quadrisepala** (R. Good) E. Wimm.
Mezleria stenosiphon Adamson === **Lobelia stenosiphon** (Adamson) E. Wimm.
Mezleria valdiviana Phil. === **Legenere valdiviana** (Phil.) E. Wimm.

Michauxia

Campanuloideae, 7 species, endemic to southwestern Asia, from Turkey and Lebanon to Iran. The name *Mindium* has been used for this genus by some authors (e.g., Rechinger & Schiman-Czeika 1965), but *Michauxia* has been formally conserved; at the same time, the name *Mindium* was typified such that it became a homotypic synonym of *Canarina* and then formally rejected in favor of that name (Meikle 1972, McVaugh 1974, Stafleu & Voss 1975). The single species included in a molecular phylogeny (Eddie et al. 2003, under **General**) formed a clade with several species of *Campanula* sect. *Campanula* within the "Campanulaceae s. str. clade". There is no monograph of *Michauxia*, but the floristic treatments for Turkey (Damboldt 1978) and Iran (Rechinger and Schiman-Czeika 1965) taken together account for all seven species. $2n$ = 28, 30, 34. Type [conserved]: *Michauxia campanuloides* L'Hér.

> L'Héritier de Brutelle, C. L. (1788). *Michauxia.* (Octandria monogynia). 3 pp. + 2 pl., illus. Paris: published privately. La. — Establishment of genus.
> Hooker, J. D. (1900). *Michauxia tchihatcheffii.* Bot. Mag. 126: tab. 7742 + 2 pp., illus. En. — Portrait with description.
> Fedorov, A. A. (1957). *Michauxia* L'Hér. In V. L. Komarov, Flora URSS 24: 383-387, illus. Leningrad: Academia Scientiae URSS. Ru. — Flora for former Soviet Union, with description and full nomenclature.
> • Rechinger, K. H. & H. Schiman-Czeika (1965). *Mindium.* In K. H. Rechinger (ed.), Flora Iranica 13: 47- 49 + tab. 12, illus. Graz: Akademische Druck- und Verlagsanstalt. Ge., La. — Flora with keys, descriptions, full nomenclature, and specimen citations.
> Meikle, R. D. (1972). Proposal to conserve the generic name 8651 *Michauxia* L'Hérit. (1788) (Campanulaceae) against *Mindium* Adanson (1763). Taxon 21: 541-542. En. — Attempt to maintain and clarify use of name.
> McVaugh, R. (1974). Report of the Committee for Spermatophyta. Conservation of generic names, XVII. Taxon 23: 819-824. En. — Recommendation to reject *Mindium* so that it threatens neither *Canarina* or *Michauxia*.
> Stafleu, F. A. & E. G. Voss (1975). Synopsis of proposals on botanical nomenclature Leningrad 1975. Taxon 24: 201-251. En. — General committee recommends conservation.
> • Damboldt, J. (1978). *Michauxia* L'Hérit., nom. cons. In P. H. Davis (ed.), Flora of Turkey and the East Aegean Islands 6: 81-83. Edinburgh: University Press. En. — Flora with keys, descriptions, full nomenclature, and specimen citations.
> Kolakovskii, A. A. (1982). [Biological "mechanisms" of *Michauxia*.] Soobšč. Akad. Nauk Gruzinsk. SSR 106: 369-372, illus. Ru. — Interpretation of floral and fruiting structure and function; *M. laevigata* obligately autogamous.
> Shear, W. A. (1994). Marvelous *Michauxia.* Flower & Garden 38(3): 60, illus. En. — Brief popular account from horticultural perspective.
> Al-zein, M. S. & L. J. Musselman (2004). *Michauxia* (Campanulaceae): a western Asian genus honoring a North American pioneer botanist. Castanea Occas. Pap. 2: 200-205, illus. En. — Overview of genus and its nomenclatural history, with notes on pollination biology in *M. campanuloides.*

Michauxia L'Hér., Michauxia (1788), nom. cons.
W. Asia. 33 34.

Michauxia campanuloides L'Hér., Michauxia (1788). *Mindium rhazis* Rauwolff ex Juss.,
Gen. Pl.: 164 (1789). *Mindium spicatum* J. F. Gmel., Syst. Nat.: 618 (1791). *Campanula
lyrifolia* Salisb., Prodr. Stirp. Chap. Allerton: 127 (1796), as 'lyraefolia'. *Michauxia strigosa*
Pers., Syn. Pl. 1: 418 (1805). *Mindium campanuloides* (L'Hér.) Rech. f. & Schiman-Czeika in
Rech. f., Fl. Iran. 13: 47 (1965).
Turkey to Israel. 34 LBS PAL TUR. Biennial. $2n$ = 30, 34.
   Mindium isauricum Contandr., Quézel & Pamukç., Ann. Univ. Provence Sci. 46: 55 (1972).

Michauxia koeieana Rech. f., Dansk Bot. Ark. 15(4): 41 (1954). *Mindium koeieanum* (Rech.
f.) Rech. f. & Schiman-Czeika in Rech. f., Fl. Iran. 13: 48 (1965).
Iran. 34 IRN. Biennial.

Michauxia laevigata Vent., Descr. Pl. Nouv.: pl. 81 (1802). *Mindium laevigatum* (Vent.)
Rech. f. & Schiman-Czeika in Rech. f., Fl. Iran. 13: 47 (1965). $2n$ = 30, 34.
Turkey & Caucasus (Armenia, Nakhichevan, Azerbaijan) to Iran. 33 TCS 34 IRN IRQ
TUR. Biennial.
   Michauxia laevigata var. *setosa* Wettst., Denkshr. Kaiserl. Akad. Wiss., Math.-Naturwiss.
   Kl. 50: 69 (1885).

Michauxia nuda A. DC. in DC., Prodr. 7: 457 (1839). *Michauxia laevigata* var. *nuda* (A. DC.)
Parsa, Fl. Iran 3: 857 (1949). *Mindium nudum* (A. DC.) Rech. f. & Schiman-Czeika in Rech.
f., Fl. Iran. 13: 48 (1965).
Syria & Iraq. 34 IRQ LBS. Biennial.

Michauxia stenophylla Boiss. & Hausskn. in Boiss., Fl. Orient. 3: 891 (1875). *Mindium
stenophyllum* (Boiss. & Hausskn.) Rech. f. & Schiman-Czeika in Rech. f., Fl. Iran. 13:
48 (1965).
Iran. 34 IRN. Biennial.

Michauxia tchihatcheffii Fisch. & C. A. Mey., Ann. Sci. Nat., Bot. (ser. 4) 1: 32 (1854).
Turkey; cult. 34 TUR. Biennial. $2n$ = 28, 30.

Michauxia thyrsoidea Boiss. & Heldr. in Boiss., Diagn. Pl. Orient. (ser. 1) 11: 61 (1849).
Turkey. 34 TUR. Biennial. $2n$ = 30.

Synonyms:
Michauxia laevigata var. *nuda* (A. DC.) Parsa === **Michauxia nuda** A. DC.
Michauxia laevigata var. *setosa* Wettst. === **Michauxia laevigata** Vent.
Mindium spicatum J. F. Gmel. === **Michauxia campanuloides** L'Hér.
Michauxia strigosa Pers. === **Michauxia campanuloides** L'Hér.

Microcodon

Campanuloideae, 4 species, Cape Provinces of South Africa. The genus is presumably part of
the "wahlenbergioid" clade (Eddie et al. 2003, under **General**), though none of its species
has been included in a phylogenetic analysis. The most comprehensive treatment available
is the floristic account by Adamson (1950), which covers all species save one. Type
[designated by Pfeiff., Nomencl. Bot. 2: 304 (1874)]: *Microcodon glomeratus* A. DC.

   de Candolle, A. (1830). *Microcodon*. In Monographie des Campanulées: 127-129. Paris: Veuve
      Desray. La. — Establishment of genus as distinct from *Campanula* and *Wahlenbergia*.
• Adamson, R. S. (1950). *Microcodon* A. DC. In R. S. Adamson & T. M. Salter, Flora of the
      Cape Peninsula: 745-746. Cape Town: Juta. En. — Flora with key and descriptions.
   Dyer, R. A. (1975). *Microcodon* A. DC. In The genera of southern African flowering
      plants 1: 642-643. Pretoria: Department of Agricultural Technical Services. En. —
      Generic description.

Goldblatt, P. & J. Manning (2000). Cape plants. A conspectus of the Cape flora of South Africa. *Microcodon*. Strelitzia 9: 392. En. — Checklist for Cape of Good Hope, with brief species desciptions.

Welman, W. G. (2000). *Microcodon* A. DC. In O. A. Leistner (ed.), Seed plants of southern Africa: families and genera. Strelitzia 10: 201. En. — Generic description.

Microcodon A. DC., Monogr. Campan.: 127 (1830). *Campanula* sect. *Microcodon* (A. DC.) D. Dietr., Syn. Pl. 1: 757 (1839).
S. Africa. 27.

Microcodon glomeratus A. DC., Monogr. Campan.: 127 (1830), as 'glomeratum'. *Campanula microcodon* D. Dietr., Syn. Pl. 1: 757 (1839); non *Campanula glomerata* L., Sp. Pl.: 166 (1753).
Cape Provinces. 27 CPP. Ther.

> *Microcodon brevibracteatus* H. Buek in Eckl. & Zeyh., Enum. Pl. Afric. Austral.: 377 (1837), as 'brevibracteatum'. *Campanula brevibracteata* (H. Buek) D. Dietr., Syn. Pl. 1: 757 (1839). *Microcodon glomeratus* var. *brevibracteatus* (H. Buek) Sond. in Harv. & Sond., Fl. Cap. 3: 565 (1865), as 'brevibracteatum'.
> *Microcodon pygmaeus* H. Buek in Eckl. & Zeyh., Enum. Pl. Afr. Austral.: 377 (1837), as 'pygmaeum'. *Campanula pygmaea* (H. Buek) D. Dietr., Syn. Pl. 1: 757 (1839); non DC. in Lam. & DC., Fl. Franç. (ed. 3) 3: 705 (1805). *Microcodon glomeratus* var. *pygmaeus* (H. Buek) Sond. in Harv. & Sond., Fl. Cap. 3: 565 (1865), as 'pygmaeum'.
> *Microcodon candolleanus* H. Buek in Eckl. & Zeyh., Enum. Pl. Afric. Austral.: 377 (1837), as 'candollianum'.
> *Microcodon glomeratus* var. *singuliflorus* Sond. in Harv. & Sond., Fl. Cap. 3: 565 (1865), as 'singuliflorum'.

Microcodon hispidulus (L. f.) Sond. in Harv. & Sond., Fl. Cap. 3: 565 (1865), as 'hispidulum'.
Cape Provinces. 27 CPP. Ther.

> * *Campanula hispidula* L. f., Suppl. Pl.: 142 (1782). *Wahlenbergia hispidula* (L. f.) Schrad., Index Seminum Horti Acad. Gott. 1821: 4 (1821).
> *Microcodon depressus* A. DC. in DC., Prodr. 7: 422 (1839), as 'depressum'.

Microcodon linearis (L. f.) H. Buek in Eckl. & Zeyh., Enum. Pl. Afric. Austral.: 378 (1837), as 'lineare'.
Cape Provinces. 27 CPP. Ther.

> * *Campanula linearis* L. f., Suppl. Pl.: 140 (1782). *Wahlenbergia linearis* (L. f.) A. DC., Monogr. Campan.: 137 (1830).
> *Wahlenbergia diffusa* A. DC., Monogr. Campan.: 137 (1830). *Campanula diffusa* (A. DC.) D. Dietr., Syn. Pl. 1: 752 (1839); non (L. f.) D. Dietr., Syn. Pl. 1: 756 (1839). *Microcodon linearis* var. *diffusus* (A. DC.) Sond. in Harv. & Sond., Fl. Cap. 3: 564 (1865), as 'diffusa'.

Microcodon sparsiflorus A. DC., Monogr. Campan.: 128 (1830), as 'sparsiflorum'.
Campanula sparsiflora (A. DC.) D. Dietr., Syn. Pl. 1: 757 (1839).
Cape Provinces. 27 CPP. Ther.

Synonyms:
Microcodon brevibracteatus H. Buek === **Microcodon glomeratus** A. DC.
Microcodon candolleanus H. Buek === **Microcodon glomeratus** A. DC.
Microcodon depressus A. DC. === **Microcodon hispidulus** (L. f.) Sond.
Microcodon glomeratus var. *brevibracteatus* (H. Buek) Sond. === **Microcodon glomeratus** A. DC.
Microcodon glomeratus var. *pygmaeus* (H. Buek) Sond. === **Microcodon glomeratus** A. DC.
Microcodon glomeratus var. *singuliflorus* Sond. === **Microcodon glomeratus** A. DC.
Microcodon linearis var. *diffusus* (A. DC.) Sond. === **Microcodon linearis** (L. f.) H. Buek
Microcodon longibracteatus H. Buek === **Treichelia longibracteata** (H. Buek) Vatke
Microcodon pygmaeus H. Buek === **Microcodon glomeratus** A. DC.

Mindium

As originally circumscribed by Adanson (1763), this genus encompassed elements referable to both *Canarina* and *Michauxia* (q.v.). Although some modern authors have typified the name to allow its use for the latter, the earliest typification dictates the opposite. *Mindium* does have priority over *Canarina*, but was formally rejected in favor of that name (Meikle 1972, McVaugh 1974, Stafleu & Voss 1975). Type [designated by Raf., Fl. Tellur. 2: 79 (1837)]: *Mindium canariense* (L.) Raf.

> Adanson, M. (1763). Les campanules. *Campanulae*. In Familles des Plantes 2: 132-134. Paris: Vincent. Fr. — Establishment of genus as distinct from *Campanula*.
>
> Rafinesque, C. S. (1837). Flora Telluriana, pars secunda. Philadelphia: published privately. En. — First to recognize the mixed nature of genus; explicitly restricted name to plants treated today as *Canarina*.
>
> Meikle, R. D. (1972). Proposal to conserve the generic name 8656 *Canarina* L. (1771) (Campanulaceae) against *Mindium* Adanson (1763). Taxon 21: 542-543. En. — Formal motion to reject name.
>
> McVaugh, R. (1974). Report of the Committee for Spermatophyta. Conservation of generic names, XVII. Taxon 23: 819-824. En. — Recommendation to reject *Mindium* so that it threatens neither *Canarina* or *Michauxia*.
>
> Stafleu, F. A. & E. G. Voss (1975). Synopsis of proposals on botanical nomenclature Leningrad 1975. Taxon 24: 201-251. En. — General Committee recommends rejection.

Synonyms:

Mindium Adans. === **Canarina** L.

Mindium campanuloides (L'Hér.) Rech. f. & Schiman-Czeika === **Michauxia campanuloides** L'Hér.

Mindium canariense (L.) Raf. === **Canarina canariensis** (L.) Vatke

Mindium isauricum Contandr., Quézel & Pamukç. === **Michauxia campanuloides** L'Hér.

Mindium koeieanum (Rech. f.) Rech. f. & Schiman-Czeika === **Michauxia koeieana** Rech. f.

Mindium laevigatum (Vent.) Rech. f. & Schiman-Czeika === **Michauxia laevigata** Vent.

Mindium nudum (A. DC.) Rech. f. & Schiman-Czeika === **Michauxia nuda** A. DC.

Mindium rhazis Rauwolff ex Juss. === **Michauxia campanuloides** L'Hér.

Mindium stenophyllum (Boiss. & Hausskn.) Rech. f. & Schiman-Czeika === **Michauxia stenophylla** Boiss. & Hausskn.

Monopsis

Lobelioideae, 15 species, tropical and southern Africa. The genus is divided (Wimmer 1953) into sect. *Dobrowskya* (C. Presl) Urb., sect. *Monopsis* (including *Mezleria*, q.v.), and sect. *Parastranthus* (G.Don) E. Wimm. (although a combination based on *Rapuntium* sect. *Xanthomeria* C. Presl would have priority for the last). In a phylogenetic analysis of DNA data (E. Knox, pers. comm.), the genus was shown to be monophyletic and sister to a group of African species belonging to *Lobelia* sect. *Delostemon* (E. Wimm.) J. Murata. Furthermore, its division into three sections was largely supported: the species of sect. *Dobrowskya* were sister to a clade comprising sect. *Monopsis* and sect. *Parastranthus*, each of which was supported as monophyletic. The most recent monograph was that of Wimmer (1953, 1968); the species of tropical Africa were covered subsequently by Thulin (1979), while some of those in southern Africa were treated by Phillipson (1986b, 1989). 2n = 28. Type [designated by Pfeiff., Nomencl. Bot. 2: 349 (1874)]: *Lobelia speculum* Andrews.

> Sims, J. (1810). *Lobelia lutea*. Yellow lobelia. Bot. Mag. 32: tab. 1319 + 2 pp., illus. En. — Portrait and description of *M. lutea*.
>
> Sims, J. (1815). *Lobelia variifolia*. Various-leaved lobelia. Bot. Mag. 41: tab. 1692 + 1 pg., illus. En. — Portrait and description of *M. variifolia*.

Salisbury, R. A. (1817). On the cultivation of the *Monopsis conspicua,* a beautiful annual, with a figure of it, and a short account of another species. Trans. Hort. Soc. London 2: 37-41, illus. En. — Establishment of genus as distinct from *Lobelia,* with horticultural commentary.

- Urban, I. (1881). Die Bestäubungseinrichtungen bei den Lobeliaceen nebst einer Monographie der afrikanischen Lobeliaceen-Gattung *Monopsis.* Jahrb. Königl. Bot. Gart. Berlin 1: 260-277, illus. Ge., La. — Monograph with key, descriptions, infrageneric classification, and full nomenclature; *Dobrowskya* (including *Parastranthus*) included at sectional rank.

Marloth, R. (1932). Campanulaceae. In The flora of South Africa 3: 208-215 + pl. 54, illus. Capetown: Darter. En. — Generalized floristic account of family; *Dobrowskya, Mezleria, Monopsis,* and *Parastranthus* all subsumed into *Lobelia.*

Adamson, R. S. (1950). *Monopsis* Salisb., *Parastranthus* G. Don. In R. S. Adamson & T. M. Salter, Flora of the Cape Peninsula: 754-755, 758-759. Cape Town: Juta. En. — Flora with keys and descriptions; genus includes *Dobrowskya* but not *Parastranthus.*

- Wimmer, F. E. (1953). *Monopsis* Salisb. In A. Engler & L. Diels, Das Pflanzenreich IV. 276b: 698-713, 783-784, illus. Berlin: Akademie-Verlag. Ge., La. — Monograph with key, descriptions, infrageneric classification, full nomenclature, and specimen citations.

- Wimmer, F. E. (1968). *Monopsis* Salisb. In A. Engler & L. Diels, Das Pflanzenreich IV. 276c: 889-891 + Taf. 28, illus. Berlin: Akademie-Verlag. Ge., La. — Supplement to Wimmer (1953).

Dyer, R. A. (1975). *Monopsis* Salisb. In The genera of southern African flowering plants 1: 645. Pretoria: Department of Agricultural Technical Services. En. — Generic description.

- Thulin, M. (1979). *Monopsis* (Lobeliaceae) in tropical Africa. Bot. Not. 132: 131-137, illus., maps. En. — Monograph of species outside S. Africa, with key, descriptions, and synonymy.

- Thulin, M. (1983). *Monopsis* Salisb. In E. Launert (ed.), Flora Zambesiaca 7(1): 145-148, illus. Kew: Royal Botanic Garden. En. — Flora of Mozambique, Malawi, Zambia, Zimbabwe and Botswana, with keys, descriptions, full nomenclature and specimen citations.

Barker, W. R. (1984). Notes on the South Australian flora: *Monopsis simplex* (L.) E. Wimmer (Campanulaceae), a new generic record for South Australia. J. Adelaide Bot. Gard. 7: 135-136. En. — Naturalized in Australia.

Phillipson, P. B. (1986a). Lectotypification of *Monopsis* Salisb. (Lobeliaceae). Taxon 35: 361-363. En. — Typification of two original species.

- Phillipson, P. B. (1986b). Taxonomy of *Monopsis* (Lobeliaceae): *M. simplex, M. debilis* and a new species. Bot. J. Linn. Soc. 93: 329-341, illus., maps. En. — Treatment of three closely related species, with descriptions, full nomenclature, and specimen citations.

- Phillipson, P. B. (1989). Taxonomy of *Monopsis* (Lobeliaceae): *M. scabra* and *M. unidentata.* Bot. J. Linn. Soc. 99: 255-272, illus., maps. En. — Treatment of two closely related species, with descriptions, full nomenclature, and specimen citations.

Gess, S. K. & F. W. Gess (1994). Love among the flowers. The small annual, *Monopsis debilis,* and its pollinator, the bee, *Haplomelitta oliviei.* Veld & Flora 80: 18-19, illus. En. — Popular account of pollination biology.

Goldblatt, P. & J. Manning (2000). Cape plants. A conspectus of the Cape flora of South Africa. *Monopsis.* Strelitzia 9: 392. En. — Checklist for Cape of Good Hope, with brief species descriptions.

Welman, W. G. (2000). *Monopsis* Salisb. In O. A. Leistner (ed.), Seed plants of southern Africa: families and genera. Strelitzia 10: 340. En. — Generic description.

Monopsis Salisb., Trans. Hort. Soc. London 2: 37 (1817). *Lobelia* [unranked] *Monopsis* (Salisb.) Heynh., Nom. Bot. Hort. 1: 473 (1840). *Rapuntium* sect. *Monopsis* (Salisb.) Kuntze in T. Post & Kuntze, Lex. Gen. Phan.: 478 (1903).

Trop. & S. Africa; naturalized in Australia. 23 24 25 26 27 29 (50).

Parastranthus G. Don, Gen. Hist. 3: 716 (1834). *Lobelia* [unranked] *Parastranthus* (G. Don) Heynh., Nom. Bot. Hort. 1: 473 (1840). *Rapuntium* sect. *Parastranthus* (G. Don.) Kuntze in T. Post & Kuntze, Lex. Gen. Phan.: 478 (1903). *Monopsis* sect. *Parastranthus* (G. Don) E. Wimm., Pflanzenr. IV.276b: 709 (1953).
Mezleria C. Presl, Prodr. Monogr. Lobel.: 7 (1836). *Lobelia* sect. *Mezleria* (C. Presl) Schönl. in Engl. & Prantl, Nat. Pflanzenfam. IV.5: 67 (1889). *Rapuntium* sect. *Mezleria* (C. Presl) Kuntze in T. Post & Kuntze, Lex. Gen. Phan.: 477 (1903). *Lobelia* subg. *Mezleria* (C. Presl) E. Wimm., Ann. Naturhist. Mus. Wien 56: 364 (1948).
Dobrowskya C. Presl, Prodr. Monogr. Lobel.: 8 (1836). *Monopsis* sect. *Dobrowskya* (C. Presl) Urb., Jahrb. Königl. Bot. Gart. Berlin 1: 271 (1881).

Monopsis acrodon E. Wimm., Ann. Naturhist. Mus. Wien 56: 370 (1948).
Cape Provinces. 27 CPP. Hemicr.
Monopsis acrodon subsp. *herois* E. Wimm., Pflanzenr. IV.276c: 890 (1968).

Monopsis alba Phillipson, Bot. J. Linn. Soc. 93: 334 (1986).
Cape Provinces. 27 CPP. Hemicr.

Monopsis belliflora E. Wimm., Pflanzenr. IV.276b: 705 (1953).
KwaZulu-Natal. 27 NAT. Hemicr.

Monopsis debilis (L. f.) C. Presl, Prodr. Monogr. Lobel.: 11 (1836).
Cape Provinces; cult. 27 CPP. Ther.
* *Lobelia debilis* L. f., Suppl. Pl.: 395 (1782). *Dortmanna debilis* (L. f.) Kuntze, Revis. Gen. Pl. 3(2): 187 (1898).
Lobelia depressa L. f., Suppl. Pl.: 395 (1782). *Mezleria depressa* (L. f.) C. Presl, Prodr. Monogr. Lobel.: 7 (1836). *Dortmanna depressa* (L. f.) Kuntze, Revis. Gen. Pl. 2: 972 (1891). *Monopsis debilis* var. *depressa* (L. f.) Phillipson, Bot. J. Linn. Soc. 93: 338 (1986).
Lobelia campanulata Lam., Encycl. 3: 588 (1792). *Monopsis campanulata* (Lam.) Sond. in Harv. & Sond., Fl. Cap. 3: 534 (1865).
Lobelia speculum Andrews, Bot. Repos. 10: pl. 644 (1811). *Monopsis speculum* (Andrews) A. DC. in DC., Prodr. 7: 351 (1839).
Monopsis inconspicua Salisb., Trans. London Hort. Soc. 2: 40 (1817).
Monopsis conspicua var. *gracilis* C. Presl in E. Mey., Comm. Pl. Afr. Austr.: 284 (1838). *Monopsis gracilis* (C. Presl) A. DC. in DC., Prodr. 7: 352 (1839). *Monopsis debilis* var. *gracilis* (C. Presl) Phillipson, Bot. J. Linn. Soc. 93: 340 (1986).
Dobrowskya serratifolia A. DC. in DC., Prodr. 7: 356 (1839).
Lobelia sebae A. DC. in DC., Prodr. 7: 386 (1839).
Monopsis litigiosa Fisch. & C. A. Mey., Index Seminum Hort. Petrop. 7: 53 (1841) .
Monopsis monantha E. Wimm., Ann. Naturhist. Mus. Wien 56: 370 (1948).

Monopsis decipiens (Sond.) Thulin, Bot. Not. 132: 134 (1979).
S. Africa. 26 MOZ ZIM 27 BOT CPP LES NAT OFS SWZ TVL. Hemicr. or ther.
* *Lobelia decipiens* Sond. in Harv. & Sond., Fl. Cap. 3: 540 (1865). *Dortmanna decipiens* (Sond.) Kuntze, Revis. Gen. Pl. 2: 972 (1891).
Lobelia breynii var. *bragae* Engl., Pflanzenw. Ost-Afrikas C: 402 (1895).
Lobelia dobrowskioides Diels, Bot. Jahrb. Syst. 26: 117 (1898).

Monopsis flava (C. Presl) E. Wimm., Ann. Naturhist. Mus. Wien 56: 371 (1948).
Cape Provinces. 27 CPP. Hemicr.
* *Rapuntium flavum* C. Presl in Eckl. & Zeyh., Enum. Pl. Afric. Austral.: 394 (1837). *Parastranthus flavus* (C. Presl) A. DC. in DC., Prodr. 7: 785 (1839). *Lobelia flava* (C. Presl) D. Dietr., Syn. Pl. 1: 734 (1839).
Monopsis flava var. *vitellina* E. Wimm., Pflanzenr. IV.276b: 710 (1953).

Monopsis kowynensis E. Wimm., Pflanzenr. IV.276b: 784 (1953).
Northern Provinces. 27 TVL. Hemicr.

Monopsis lutea (L.) Urb., Jahrb. Königl. Bot. Gart. Berlin 1: 276 (1881).
Cape Provinces. 27 CPP. Hemicr.
Lobelia lutea L., Sp. Pl.: 932 (1753). *Rapuntium luteum* (L.) C. Presl, Prodr. Monogr.
Lobel.: 11 (1836). *Parastranthus simplex* G. Don, Gen. Hist. 3: 716 (1834).
Parastranthus luteus (L.) A. DC. in DC., Prodr. 7: 354 (1839). *Dortmanna lutea* (L.)
Kuntze, Revis. Gen. Pl. 3(2): 187 (1898).
Lobelia breynii Lam., Encycl. 3: 588 (1792). *Rapuntium breynii* (Lam.) C. Presl, Prodr.
Monogr. Lobel.: 19 (1836). *Dortmanna breynii* (Lam.) Kuntze, Revis. Gen. Pl. 2: 972
(1891).
Rapuntium capitatum C. Presl, Prodr. Monogr. Lobel.: 11 (1836). *Parastranthus capitatus*
(C. Presl) A. DC. in DC., Prodr. 7: 354 (1839). *Parastranthus luteus* var. *capitatus* (C.
Presl) Sond. in Harv. & Sond., Fl. Cap. 3: 536 (1865).
Rapuntium ericoides C. Presl, Prodr. Monogr. Lobel.: 12 (1836). *Lobelia ericoides* (C. Presl)
D. Dietr., Syn. Pl. 1: 733 (1839). *Parastranthus ericoides* (C. Presl) A. DC. in DC.,
Prodr. 7: 354 (1839). *Parastranthus luteus* var. *ericoides* (C. Presl) Sond. in Harv. &
Sond., Fl. Cap. 3: 536 (1865). *Monopsis lutea* var. *ericoides* (C. Presl) Urb., Jahrb.
Königl. Bot. Gart. Berlin 1: 276 (1881).
Parastranthus luteus var. *euphrasioides* Sond. in Harv. & Sond., Fl. Cap. 3: 536 (1865).
Monopsis lutea var. *euphrasioides* (Sond.) Urb., Jahrb. Königl. Bot. Gart. Berlin 1:
276 (1881). *Dortmanna lutea* var. *euphrasioides* (Sond.) Kuntze, Revis. Gen. Pl. 3(2):
188 (1898).
Monopsis lutea var. *subcoerulea* Zahlbr., Ann. K. K. Naturhist. Hofmus. 18: 408 (1903).
Monopsis arenaria E. Wimm., Ann. Naturhist. Mus. Wien 56: 371 (1948).

Monopsis malvacea E. Wimm., Kew Bull. 1952: 142 (1952).
Swaziland. 27 SWZ. Hemicr.
Lobelia coddii Compton, J. S. African Bot. 41: 49 (1975).

Monopsis scabra (Thunb.) Urb., Jahrb. Königl. Bot. Gart. Berlin 1: 274 (1881).
Cape Provinces. 27 CPP. Cham. or hemicr.
Lobelia scabra Thunb. in Hoffm., Phytogr. Bl.: 21 (1803). *Dobrowskya thunbergiana* C. Presl,
Prodr. Monogr. Lobel.: 9 (1836). *Dobrowskya scabra* (Thunb.) A. DC. in DC., Prodr. 7:
355 (1839). *Dortmanna scabra* (Thunb.) Kuntze, Revis. Gen. Pl. 3(2): 188 (1898).
Dobrowskya dregeana C. Presl, Prodr. Monogr. Lobel.: 9 (1836). *Dobrowskya scabra* var.
dregeana (C. Presl) Sond. in Harv. & Sond., Fl. Cap. 3: 549 (1865).
Dobrowskya polyphylla C. Presl, Prodr. Monogr. Lobel.: 9 (1836).
Dortmanna scabra var. *bicolor* Kuntze, Revis. Gen. Pl. 3(2): 188 (1898). *Monopsis scabra*
var. *bicolor* (Kuntze) E. Wimm., Pflanzenr. IV.276b: 709 (1953).
Monopsis scabra var. *subhispida* E. Wimm., Pflanzenr. IV.276b: 708 (1953).

Monopsis simplex (L.) E. Wimm., Ann. Naturhist. Mus. Wien 56: 370 (1948).
Cape Provinces; naturalized in S. Australia. 27 CPP (50) nsw soa vic wau. Hemicr. or ther.
$2n = 28$.
Lobelia simplex L., Mant. Pl.: 291 (1771). *Rapuntium simplex* (L.) C. Presl, Prodr. Monogr.
Lobel.: 17 (1836).
Lobelia scabra Spreng., Neue Entd. 1: 272 (1820); non Thunb. in Hoffm., Phytogr. Bl.:
21 (1803). *Lobelia aspera* Spreng., Neue Entd. 3: 222 (1822). *Dobrowskya eckloniana* C.
Presl in Eckl. & Zeyh., Enum. Pl. Afric. Austral.: 393 (1837). *Dobrowskya aspera*
(Spreng.) A. DC. in DC., Prodr. 7: 356 (1839). *Monopsis aspera* (Spreng.) Urb., Jahrb.
Königl. Bot. Gart. Berlin 1: 274 (1881).
Wahlenbergia chondrophylla H. Buek in Eckl. & Zeyh., Enum. Pl. Afric. Austral.: 381
(1837). *Campanula chondrophylla* (H. Buek) D. Dietr., Syn. Pl. 1: 753 (1839).

Monopsis stellarioides (C. Presl) Urb., Jahrb. Königl. Bot. Gart. Berlin 1: 275 (1881).
Trop. & S. Africa; Comoros. 23 BUR CMN EQG GGI RWA ZAI 24 ETH 25 KEN TAN UGA
26 MLW ZAM 27 CPP NAT SWZ TVL 29 COM. Hemicr. or ther. $2n = 28$.

Dobrowskya stellarioides C. Presl, Prodr. Monogr. Lobel.: 10 (1836). *Parastranthus stellarioides* (C. Presl) Vatke, Linnaea 38: 717 (1874). *Lobelia stellarioides* (C. Presl) Benth. & Hook. f. ex Hemsl. in Oliv., Fl. Trop. Afr. 3: 470 (1877). *Dortmanna stellarioides* (C. Presl) Kuntze, Revis. Gen. Pl. 3(2): 188 (1898), as 'stellariodes'.

subsp. **schimperiana** (Urb.) Thulin, Bot. Notiser 132: 132 (1979).
Trop. & S. Africa; Comoros. 23 BUR CMN EQG GGI-BI RWA ZAI 24 ETH 25 KEN TAN UGA 26 MLW ZAM 29 COM. Hemicr. or ther. *2n* = 28.
Monopsis schimperiana Urb., Jahrb. Königl. Bot. Gart. Berlin 1: 275 (1881). *Monopsis stellarioides* var. *schimperiana* (Urb.) E. Wimm., Pflanzenr. IV.276b: 704 (1953).
Lobelia mukuluensis De Wild., Rev. Zool. Afr. 8 (Suppl. Bot.): 26 (1920), as 'mokuluensis'.
Lobelia violaceo-aurantiaca De Wild., Rev. Zool. Afr. 8 (Suppl. Bot.): 28 (1920). *Monopsis stellarioides* f. *violaceo-aurantiaca* (De Wild.) E. Wimm., Pflanzenr. IV.276b: 704 (1953).
Monopsis brevicalyx T. C. E. Fr., Notizbl. Bot. Gart. Berlin-Dahlem 8: 411 (1923).
Monopsis schimperiana var. *brevifolia* Chiov., Racc. Bot. 73 (1935).

subsp. **stellarioides**
S. Africa. 27 CPP NAT SWZ TVL. Hemicr. or ther.

Monopsis unidentata (W. T. Aiton) E. Wimm., Pflanzenr. IV.276b: 705 (1953).
S. Africa. 27 CPP NAT. Cham. or hemicr.
Lobelia unidentata W. T. Aiton, Hortus Kew. 1: 356 (1810). *Parastranthus unidentatus* (W. T. Aiton) G. Don, Gen. Hist. 3: 716 (1834), as 'unidentata'. *Dobrowskya massoniana* C. Presl, Prodr. Monogr. Lobel.: 9 (1836). *Dobrowskya unidentata* (W. T. Aiton) A. DC. in DC., Prodr. 7: 355 (1839).
Lobelia unidentata Spreng., Novi Provent.: 26 (1818); non W. T. Aiton, Hortus Kew. 1: 356 (1810).

subsp. **intermedia** Phillipson, Bot. J. Linn. Soc. 99: 268 (1989).
Cape Provinces. 27 CPP. Cham. or hemicr.

subsp. **laevicaulis** (C. Presl) Phillipson, Bot. J. Linn. Soc. 99: 270 (1989).
KwaZulu-Natal & Cape Provinces. 27 NAT CPP. Cham. or hemicr.
Dobrowskya laevicaulis C. Presl, Prodr. Monogr. Lobel.: 10 (1836). *Monopsis laevicaulis* (C. Presl) E. Wimm., Ann. Naturhist. Mus. Wien 56: 371 (1948).
Dobrowskya scabra var. *glabrata* Sond. in Harv. & Sond., Fl. Cap. 3: 549 (1865). *Monopsis scabra* var. *glabrata* (Sond.) Schönl., Phan. Fl. Uitenhage: 98 (1919).
Monopsis scabra var. *robusta* Diels, Bot. Jahrb. Syst. 26: 118 (1899).

subsp. **unidentata**
Cape Provinces. 27 CPP. Cham. or hemicr.
Dobrowskya stricta C. Presl ex Eckl. & Zeyh., Enum. Pl. Afric. Austral.: 393 (1837). *Lobelia dobrowskyana* D. Dietr., Syn. Pl. 1: 732 (1839); non *Lobelia stricta* Sw., Prodr. Veg. Ind. Occid.: 117 (1787); nec R. Br., Prodr.: 564 (1810). *Monopsis stricta* (C. Presl ex Eckl. & Zeyh.) E. Wimm., Ann. Naturhist. Mus. Wien 56: 371 (1948).
Dortmanna scabra var. *violacea* Kuntze, Revis. Gen. Pl. 3(2): 188 (1898).
Dortmanna scabra var. *albiflora* Kuntze, Revis. Gen. Pl. 3(2): 188 (1898). *Monopsis scabra* var. *albiflora* (Kuntze) E. Wimm., Pflanzenr. IV.276b: 709 (1953).

Monopsis variifolia (Sims) Urb., Jahrb. Königl. Bot. Gart. Berlin 1: 277 (1881).
Cape Provinces. 27 CPP. Cham. or hemicr.
Lobelia variifolia Sims, Bot. Mag. 41: tab. 1692 (1814). *Parastranthus variifolius* (Sims) G. Don, Gen. Hist. 3: 716 (1834), as 'variifolia'. *Rapuntium variifolium* (Sims) C. Presl, Prodr. Monogr. Lobel.: 12 (1836).

Monopsis zeyheri (Sond.) Thulin, Bot. Not. 132: 135 (1979).
Kenya to S. Africa. 25 KEN TAN 26 MLW ZAM ZIM 27 NAM TVL. Ther.

Lobelia zeyheri Sond. in Harv. & Sond., Fl. Cap. 3: 539 (1865).

Lobelia fonticola Engl. & Gilg. in Warb., Kunene-Sambesi-Exped.: 398 (1903).

Cephalostigma nanellum R. E. Fr., Wiss. Ergebn. Schwed. Rhodesia-Kongo-Exped. 1: 315(1914), as 'nanella'.

Synonyms:

Monopsis sect. *Dobrowskya* (C. Presl) Urb. === **Monopsis** Salisb.

Monopsis sect. *Parastranthus* (G. Don) E. Wimm. === **Monopsis** Salisb.

Monopsis [unranked] *Holostigma* (G. Don) Endl. === **Lobelia** L.

Monopsis [unranked] *Isolobus* (A. DC.) Endl. === **Lobelia** L.

Monopsis acrodon subsp. *herois* E. Wimm. === **Monopsis acrodon** E. Wimm.

Monopsis arenaria E. Wimm. === **Monopsis lutea** (L.) Urb.

Monopsis aspera (Spreng.) Urb. === **Monopsis simplex** (L.) E. Wimm.

Monopsis brevicalyx T. C. E. Fr. === **Monopsis stellarioides** subsp. **schimperiana** (Urb.) Thulin

Monopsis campanulata (Lam.) Sond. === **Monopsis debilis** (L. f.) C. Presl

Monopsis conspicua Salisb. === **Lobelia erinus** L.

Monopsis conspicua var. *gracilis* C. Presl === **Monopsis debilis** (L. f.) C. Presl

Monopsis corymbosa (G. Don) C. Presl === **Lobelia jasionoides** (A. DC.) E. Wimm.

Monopsis corymbosa var. *pedicellaris* C. Presl === **Lobelia jasionoides** (A. DC.) E. Wimm.

Monopsis debilis var. *conspicua* Sond. === **Lobelia erinus** L.

Monopsis debilis var. *depressa* (L. f.) Phillipson === **Monopsis debilis** (L. f.) C. Presl

Monopsis debilis var. *gracilis* (C. Presl) Phillipson === **Monopsis debilis** (L. f.) C. Presl

Monopsis dioica (R. Br.) C. Presl === **Lobelia dioica** R. Br.

Monopsis flava var. *vitellina* E. Wimm. === **Monopsis flava** (C. Presl) E. Wimm.

Monopsis gracilis (C. Presl) A. DC. === **Monopsis debilis** (L. f.) C. Presl

Monopsis inconspicua Salisb. === **Monopsis debilis** (L. f.) C. Presl

Monopsis laevicaulis (C. Presl) E. Wimm. === **Monopsis unidentata** subsp. **laevicaulis** (C. Presl) Phillipson

Monopsis litigiosa Fisch. & C. Mey. === **Monopsis debilis** (L. f.) C. Presl

Monopsis lutea var. *ericoides* (C. Presl) Urb. === **Monopsis lutea** (L.) Urb.

Monopsis lutea var. *euphrasioides* (C. Presl ex Sond.) Urb. === **Monopsis lutea** (L.) Urb.

Monopsis lutea var. *subcoerulea* Zahlbr. === **Monopsis lutea** (L.) Urb.

Monopsis monantha E. Wimm. === **Monopsis debilis** (L. f.) C. Presl

Monopsis scabra var. *albiflora* (Kuntze) E. Wimm. === **Monopsis unidentata** (W. T. Aiton) E. Wimm. subsp. **unidentata**

Monopsis scabra var. *bicolor* (Kuntze) E. Wimm. === **Monopsis scabra** (Thunb.) Urb.

Monopsis scabra var. *glabrata* (Sond.) Schönl. === **Monopsis unidentata** subsp. **laevicaulis** (C. Presl) Phillipson

Monopsis scabra var. *robusta* Diels === **Monopsis unidentata** subsp. **laevicaulis** (C. Presl) Phillipson

Monopsis scabra var. *subhispida* E. Wimm. === **Monopsis scabra** (Thunb.) Urb.

Monopsis schimperiana Urb. === **Monopsis stellarioides** subsp. **schimperiana** (Urb.) Thulin

Monopsis schimperiana var. *brevifolia* Chiov. === **Monopsis stellarioides** subsp. **schimperiana** (Urb.) Thulin

Monopsis simplex var. *conspicua* (Sond.) E. Wimm. === **Lobelia erinus** L.

Monopsis speculum (Andr.) A. DC. === **Monopsis debilis** (L. f.) C. Presl

Monopsis stellarioides f. *violaceo-aurantiaca* (De Wild.) E. Wimm. === **Monopsis stellarioides** subsp. **schimperiana** (Urb.) Thulin

Monopsis stellarioides var. *schimperiana* (Urb.) E. Wimm. === **Monopsis stellarioides** subsp. **schimperiana** (Urb.) Thulin

Monopsis stricta (C. Presl ex Eckl. & Zeyh.) E. Wimm. === **Monopsis unidentata** (W. T. Aiton) E. Wimm. subsp. **unidentata**

Monopsis tenella (L.) Urb. === **Wahlenbergia parvifolia** (P. J. Bergius) Lammers

Muehlbergella

Campanuloideae, 1 species, eastern Caucasus. Although it has been treated as part of the southeastern European genus *Edraianthus* by some authors (e.g., Fedorov 1957), it seems more closely allied on the basis of morphology to two other Caucasian endemics, *Sachokiella* and *Theodorovia* (q.v.). However, all these genera have yet to be represented in a phylogeny based on molecular data. Type [by monotypy]: *Muehlbergella oweriniana* (Rupr.) Feer.

Feer, H. (1890). Beiträge zur Systematik und Morphologie der Campanulaceen. II.A. Drei neue monotype Gattungen. Bot. Jahrb. Syst. 12: 610-617 + Taf. VII-VIII, illus. Ge. — Establishment of genus as distinct from *Edraianthus*.

Fedorov, A. A. (1957). *Edrajanthus* [sic] A. DC. In V. L. Komarov, Flora URSS 24: 442-444, illus. Leningrad: Academia Scientiae URSS. Ru. — Flora of former Soviet Union, with description and full nomenclature.

Muehlbergella Feer, Bot. Jahrb. Syst. 12: 615 (1890).
Caucasus. 33.

Muehlbergella oweriniana (Rupr.) Feer, Bot. Jahrb. Syst. 12: 616 (1890).
Caucasus (Dagestan). 33 NCS. Cham.
    * *Edraianthus owerinianus* Rupr., Bull. Acad. Imp. Sci. Saint-Pétersbourg 11: 203 (1867).

Musschia

Campanuloideae, 2 species, Madeira. In a molecular phylogenetic analysis (Eddie et al. 2003, under **General**), the type species formed a clade with the types of *Echinocodonia* and *Gadellia* plus *Campanula peregrina;* this was one of three "transitional" clades forming a pentachotomy with the "Campanulaceae s. str. clade" and the "Rapunculus clade". The most recent taxonomic treatment is the floristic account by Tebbs (1994). $2n = 32$. Type [by monotypy]: *Musschia aurea* (L. f.) Dumort.

Ker-Gawler, J. B. (1815). *Campanula aurea*. α. Broad-leaved golden bell-flower. Edwards' Bot. Reg. 1: pl. 57 + 2 pp., illus. En. — Portrait of *M. aurea*, with description.

Dumortier, B. (1822). *Musschia*. In Commentationes botanicae: 28-29. Tournay: Ch. Casterman-Dien. La. — Establishment of genus as distinct from *Campanula*.

Johnson, J. Y. (1857). Notes on some rare and little-known plants of Madeira. Hooker's J. Bot. Kew Gard. Misc. 9: 161-165. En. — Descriptive account.

Hooker, J. D. (1866). *Musschia wollastoni* [sic]. Mr. Wollaston's musschia. Bot. Mag. 92: tab. 5606 + 2 pp., illus. En. — Portrait with description.

Hooker, J. D. (1881). *Musschia aurea*. Bot. Mag. 92: tab. 6556 + 2 pp., illus. En. — Portrait with description.

Elvers, I. (1978). The Madeiran lizard-flower connection observed in a natural habitat. Bot. Not. 131: 159-160, illus. En. — Pollination of *M. aurea* by reptile *Lacerta muralis dugesii*.

Perez de Paz, J. (1978). Estudio palinologico de las Campanulaceas endemicas de la region Macaronesica. Bot. Macarones. 6: 27-42, illus. Por. — Descriptive account of pollen morphology in both species.

Vogel, S., C. Westerkamp, B. Thiel & K. Gessner (1984). Ornithophilie auf den Canarischen Inseln. Pl. Syst. Evol. 146: 225-248. Ge. — Ornithophilous pollination.

Oleson, J. M. (1985). The Macaronesian bird-flower element and its relation to bird and bee opportunists. Bot. J. Linn. Soc. 91: 395-414. En. — Ornithophilous pollination.

• Turland, N. J. (1994). *Musschia* Dumort. In J. R. Press and M. J. Short (eds.), Flora of Madeira: 331 + pl. 42, illus. London: HMSO. En. — Flora with key and descriptions.

Musschia Dumort., Comment. Bot.: 28 (1822). *Chrysangia* Link, Handbuch 1: 632 (1829).
Benaurea Raf., Fl. Tellur. 2: 77 (1837). *Campanula* sect. *Musschia* (Dumort.) D. Dietr., Syn.
Pl. 1: 758 (1839).
Macaronesia. 21.

Musschia aurea (L. f.) Dumort., Comment. Bot.: 28 (1822).
Madeira, Deserta Grande, Doca. 21 MDR. Cham. 2*n* = 32.
* *Campanula aurea* L. f., Suppl. Pl.: 141 (1782). *Chrysangia aurea* (L. f.) Link, Handbuch 1:
632 (1829). *Benaurea sempervirens* Raf., Fl. Tellur. 2: 77 (1837).
Campanula aurea var. *angustifolia* Ker Gawl., Bot. Reg. 1: pl. 57 (1815). *Musschia aurea*
var. *angustifolia* (Ker Gawl.) Dumort., Comment. Bot.: 29 (1822).

Musschia wollastonii Lowe, Hooker's J. Bot. Kew Gard. Misc. 8: 298 (1856), as 'wollastoni'.
Madeira. 21 MDR. Nanophan.

Synonyms:
Musschia aurea var. *angustifolia* (Ker Gawl.) A. DC. === **Musschia aurea** (L. f.) Dumort.

Myopsia

Presl (1836) apparently was unaware of the name *Heterotoma* when he described this genus.
Type [by monotypy]: *Myopsia mexicana* C. Presl.

Presl, C. (1836). *Myopsia*. In Prodromus monographiae Lobeliacearum: 8. Prague:
Theophilus Haase. La. — Establishment of genus.
Reichenbach, H. G. L. (1841). Der Deutsche Botaniker. Erster Band. Das Herbarienbuch.
Ge. — Realized name was synonymous with *Heterotoma* but used *Myopsia* to avoid
homonymy with the earlier insect genus *Heterotoma* Lepeletier & Serville (Heteroptera:
Miridae); because of the independence of botanical and zoological nomenclature
(ICBN Principle I), this not a consideration.
Turczaninow, N. (1852). Decas septima. Generum plantarum non descriptorum adjectis
descriptionis nonnullarum specierum. Bull. Soc. Imp. Naturalistes Moscou 25(3): 138-
181. La. — Suggested that his *H. tenella* (*Lobelia volcanica,* q.v.) might be identical with
type of *Myopsia,* but not certain.

Synonyms:
Myopsia C. Presl === **Heterotoma** Zucc.
Myopsia mexicana C. Presl === **Heterotoma lobelioides** Zucc.

Mzymtella

This genus was based upon minor characters of the fruit which have not seemed significant
to subsequent authors. Its type is a species of *Campanula* sect. *Cordifoliae* (Fomin) Kharadze.
Type [designated by the author]: *Mzymtella sclerophylla* Kolak.

Kolakovskii, A. A. (1981). [Another two new monotypical genera of Campanulaceae for
the flora of the USSR.] Soobšč. Akad. Nauk Gruzinsk. SSR 103: 149-152, illus. Ru. —
Establishment of genus as distinct from *Campanula.*

Synonyms:
Mzymtella Kolak. === **Campanula** L.
Mzymtella sclerophylla Kolak. === **Campanula sclerophylla** (Kolak.) Ogan.

Namacodon

Campanuloideae, 1 species, southern Africa. Based on morphology, the genus is related to *Prismatocarpus*. Type [designated by the author]: *Namacodon schinzianus* (Markgr.) Thulin.

> Thulin, M. (1974). *Gunillaea* and *Namacodon*. Two new genera of Campanulaceae in Africa. Bot. Not. 127: 165-182, illus., maps. En. — Establishment of genus as distinct from *Prismatocarpus*.
> Dyer, R. A. (1975). *Namacodon* Thulin. In The genera of southern African flowering plants 1: 734. Pretoria: Department of Agricultural Technical Services. En. — Generic description.
> Welman, W. G. (2000). *Namacodon* Thulin. In O. A. Leistner (ed.), Seed plants of southern Africa: families and genera. Strelitzia 10: 201. En. — Generic description.

Namacodon Thulin, Bot. Not. 127: 173 (1974).
> S. Africa. 27.

Namacodon schinzianus (Markgr.) Thulin, Bot. Not. 127: 173 (1974), as 'schinzianum'.
> Namibia & Cape Provinces. 27 CPP NAM. Nanophan. or cham.
> * *Prismatocarpus junceus* Schinz, Mém. Herb. Boissier 20: 35 (1900); non H. Buek in Eckl.& Zeyh., Enum. Pl. Afric. Austral.: 383 (1837). *Prismatocarpus schinzianus* Markgr., Notizbl. Bot. Gart. Berlin-Dahlem 15: 465 (1941).

Nemacladus

Nemacladoideae, 13 species, western North America, from Oregon and Idaho to Sonora and Baja California; all species occur in California. The genus is most closely allied to *Parishella* and *Pseudonemacladus;* the three are treated as subfamily Nemacladoideae (Lammers 1998, 2006, under **General**). The relationships of this subfamily are as yet uncertain. A phylogenetic analysis of DNA data (Cosner et al. 1994, under **General**) showed *Nemacladus* to be the sister-group of the Campanuloideae, while a morphological phylogeny (Gustafsson & Bremer 1995, under **General**) showed it to be the sister-group of a clade comprising Cyphioideae, Cyphocarpoideae, and Lobelioideae; a subsequent analysis combining molecular and morphological data (Lundberg & Bremer 2003, under **General**) placed it as the sister group of Lobelioideae. The treatment here follows that by Morin and Milburn (1993). $2n$ = 18. Type [by monotypy]: *Nemacladus ramosissimus* Nutt.

> Nuttall, T. (1842). Description and notices of new or rare plants in the natural orders Lobeliaceae, Campanulaceae, Vaccinieae, Ericaceae, collected in a journey over the continent of North America, and during a visit to the Sandwich Islands, and Upper California. Trans. Amer. Philos. Soc. (n.s.) 8: 251-272. En. — Establishment of genus and family Nemacladaceae.
> • Munz, P. A. (1924). A revision of the genus *Nemacladus* (Campanulaceae). Amer. J. Bot. 11: 233-248 + pl. IX-X, illus. En. — Monograph with key, descriptions, full nomenclature, and specimen citations.
> McVaugh, R. (1939). Some taxonomic realignments in the genus *Nemacladus*. Amer. Midl. Naturalist 22: 521-550, illus., maps. En. — Revision of Munz (1924), with key, descriptions, full nomenclature, and some specimen citations.
> • McVaugh, R. (1943). *Nemacladus* Nutt. In North American Flora 32A: 4-12. New York: New York Botanical Garden. En. — Flora with key, descriptions, and full nomenclature.
> Robbins, G. T. (1958). Notes on the genus *Nemacladus*. Aliso 4: 139-147, illus. En. — Taxonomic notes on select species, updating McVaugh (1939).
> Munz, P. A. & D. D. Keck (1970). *Nemacladus* Nutt. In A California flora: 1066-1068. Berkeley: University of California Press. En. — Flora with key, descriptions, and synonymy.
> Dunbar, A. & H. G. Wallentinus (1976). On pollen of Campanulaceae III. A numerical taxonomic investigation. Bot. Not. 129: 69-72. En. — Cluster analysis of palynological data suggests Cyphioideae s. l. not a natural group.

Cronquist, A., A. H. Holmgren, N. H. Holmgren, J. L. Reveal & P. K. Holmgren (1984). *Nemacladus* Nutt. In Intermountain flora 4: 524-527, illus. New York: New York Botanical Garden. En. — Flora with key, descriptions, and full nomenclature.
* Morin, N. R. & J. Milburn (1993). *Nemacladus*. In J. C. Hickman (ed.), The Jepson manual, higher plants of California: 465-468, illus. Berkeley: University of California Press. En. — Flora with key and descriptions; covers all species in genus.

Nemacladus Nutt., Trans. Amer. Philos. Soc. (n.s.) 8: 254 (1842).
W. North America. 73 76 77 79.

Nemacladus capillaris Greene, Bull. Calif. Acad. Sci. 1: 196 (1885). *Nemacladus rigidus* var. *capillaris* (Greene) Munz, Amer. J. Bot. 11: 244 (1924).
California & Oregon. 73 ORE 76 CAL. Ther. $2n = 18$.

Nemacladus glanduliferus Jeps., Man. Fl. Pl. Calif.: 975 (1925).
California to Utah, New Mexico, Sonora & Baja California. 76 ARI CAL NEV UTA 77 NWM 79 MXN. Ther. $2n = 18$.
Nemacladus rigidus var. *australis* Munz, Amer. J. Bot. 11: 242 (1924). *Nemacladus glanduliferus* var. *australis* (Munz) McVaugh, Amer. Midl. Naturalist 22: 540 (1939).
Nemacladus glanduliferus var. *orientalis* McVaugh, Amer. Midl. Naturalist 22: 540 (1939).

Nemacladus gracilis Eastw., Bull. Torrey Bot. Club 30: 500 (1903). *Nemacladus ramosissimus* var. *gracilis* (Eastw.) Munz, Amer. J. Bot. 11: 240 (1924).
California to Nevada, Arizona & Baja California. 76 ARI CAL NEV 79 MXN. Ther.

Nemacladus interior (Munz) G. T. Robbins, Aliso 4: 146 (1958).
California & Oregon. 73 ORE 76 CAL. Ther.
** Nemacladus rigidus* var. *interior* Munz, Amer. J. Bot. 11: 243 (1924). *Nemacladus rubescens* var. *interior* (Munz) McVaugh, Amer. Midl. Naturalist 22: 537 (1939).

Nemacladus longiflorus A. Gray, Proc. Amer. Acad. Arts 12: 60 (1877).
California, Arizona, Baja California. 76 ARI CAL 79 MXN. Ther.
Nemacladus longiflorus var. *breviflorus* McVaugh, Amer. Midl. Naturalist 22: 526 (1939).

Nemacladus montanus Greene, Bull. Calif. Acad. Sci. 1: 197 (1885). *Nemacladus ramosissimus* var. *montanus* (Greene) A. Gray, Syn. Fl. N. Amer. (ed. 2) 2(1): 393 (1886).
Nemacladus rigidus var. *montanus* (Greene) Munz, Amer. J. Bot. 11: 243 (1924).
N. California. 76 CAL. Ther. $2n = 18$.

Nemacladus pinnatifidus Greene, Bull. Calif. Acad. Sci. 1: 197 (1885). *Nemacladus ramosissimus* var. *pinnatifidus* (Greene) A. Gray, Syn. Fl. N. Amer. (ed. 2) 2(1): 393 (1886).
California & Baja California. 76 CAL 79 MXN. Ther.

Nemacladus ramosissimus Nutt., Trans. Amer. Philos. Soc. (n.s.) 8: 254 (1842).
California & Baja California. 76 CAL 79 MXN. Ther.
Nemacladus tenuissimus Greene, Bull. Calif. Acad. Sci. 1: 198 (1885).

Nemacladus rigidus Curran, Bull. Calif. Acad. Sci. 1: 154 (1885).
California to Oregon, Idaho & Nevada. 73 IDA ORE 76 CAL NEV. Ther.

Nemacladus rubescens Greene, Bull. Calif. Acad. Sci. 1: 197 (1885). *Nemacladus rigidus* var. *rubescens* (Greene) Munz, Amer. J. Bot. 11: 245 (1924).
California to Utah, Arizona & Baja California. 76 ARI CAL NEV UTA 79 MXN. Ther.
Nemacladus adenophorus Parish, Bull. S. Calif. Acad. Sci. 2: 28 (1903).
Nemacladus rubescens var. *tenuis* McVaugh, Amer. Midl. Naturalist 22: 536 (1939).

Nemacladus secundiflorus G. T. Robbins, Aliso 4: 142 (1958).
C. California. 76 CAL. Ther.

Nemacladus sigmoideus G. T. Robbins, Aliso 4: 144 (1958).
California to Nevada, Arizona & Baja California. 76 ARI CAL NEV 79 MXN. Ther.

Nemacladus twisselmannii J. T. Howell, Leafl. W. Bot. 10: 45 (1963).
S. California (N. Kern Co.). 76 CAL. Ther.

Synonyms:
Nemacladus sect. *Baclea* Kuntze === **Pseudonemacladus** McVaugh
Nemacladus adenophorus Parish === **Nemacladus rubescens** Greene
Nemacladus glanduliferus var. *australis* (Munz) McVaugh === **Nemacladus glanduliferus** Jeps.
Nemacladus glanduliferus var. *orientalis* McVaugh === **Nemacladus glanduliferus** Jeps.
Nemacladus longiflorus var. *breviflorus* McVaugh === **Nemacladus longiflorus** A. Gray
Nemacladus oppositifolia B. L. Rob. === **Pseudonemacladus oppositifolius** (B. L. Rob.)
McVaugh
Nemacladus ramosissimus var. *gracilis* (Eastw.) Munz === **Nemacladus gracilis** Eastw.
Nemacladus ramosissimus var. *montanus* (Greene) A. Gray === **Nemacladus montanus** Greene
Nemacladus ramosissimus var. *pinnatifidus* (Greene) A. Gray === **Nemacladus pinnatifidus**
Greene
Nemacladus rigidus var. *australis* Munz === **Nemacladus glanduliferus** Jeps.
Nemacladus rigidus var. *capillaris* (Greene) Munz === **Nemacladus capillaris** Greene
Nemacladus rigidus var. *interior* Munz === **Nemacladus interior** (Munz) G. T. Robbins
Nemacladus rigidus var. *montanus* (Greene) Munz === **Nemacladus montanus** Greene
Nemacladus rigidus var. *rubescens* (Greene) Munz === **Nemacladus rubescens** Greene
Nemacladus rubescens var. *interior* (Munz) McVaugh === **Nemacladus interior** (Munz)
G. T. Robbins
Nemacladus rubescens var. *tenuis* McVaugh === **Nemacladus rubescens** Greene
Nemacladus tenuissimus Greene === **Nemacladus ramosissimus** Nutt.

Nenningia

Like *Medium* (q.v.), this genus was created by Opiz (in Berchtold 1839) to accomodate
species of *Campanula* with an appendiculate calyx; those with a five-loculed capsule were
assigned to *Medium*, those with three were placed here. Its type is a species of *Campanla* sect.
Sibiricae (Fomin) Kharadze. Type [designated by Opiz, Seznam: 68 (1852)]: *Nenningia
paniculata* Opiz.

Berchtold, F. (1839). Oekonomisch-technische flora Böhmens, band 2, abteil. 2. Prague:
Thomas Thabor. Ge. — Establishment of the genus as distinct from *Campanula*.

Synonyms:
Nenningia Opiz === **Campanula** L.
Nenningia paniculata Opiz === **Campanula sibirica** L.

Neocodon

This genus was segregated from *Campanula* to accomodate species that lack appendages
between the calyx lobes and that differ in details of capsule structure. All species referable
here that were included in a molecular phylogeny (Eddie et al. 2003, under **General**) fell
into the "Rapunculus clade". The type of this name is also the type (under ICBN Art. 22.6)
of *Campanula* sect. *Rapunculus* Dumort. Type [designated by the authors]: *Neocodon
rapunculus* (L.) Kolak. & Serdyuk.

Kolakovskii, A. A. & L. Serdyukova (1984). De genere *Rapunculus* Fourr. (Campanulaceae). Zametki Sist. Geogr. Rast. 40: 26-29, illus. Ru. — Establishment of genus as distinct from *Campanula* and *Hemisphaera*.

Synonyms:

Neocodon Kolak. & Serdyuk. === **Campanula** L.

Neocodon abietinus (Griseb. & Schenk) Kolak. & Serdyuk. === **Campanula patula** subsp. **abietina** (Griseb. & Schenk) Simonk.
(Griseb. & Schenk) Kolak. & Serdyuk.

Neocodon alberti (Trautv.) Kolak. & Serdyuk. === **Campanula stevenii** subsp. **alberti** (Trautv.) Victorov

Neocodon altaicus (Ledeb.) Kolak. & Serdyuk. === **Campanula stevenii** subsp. **altaica** (Ledeb.) Fed.

Neocodon beauverdianus (Fomin) Kolak. & Serdyuk. === **Campanula beauverdiana** Fomin

Neocodon carpaticus (Jacq.) Kolak. & Serdyuk. === **Campanula carpatica** Jacq.

Neocodon hemscinicus (K. Koch) Kolak. & Serdyuk. === **Campanula olympica** Boiss.

Neocodon lambertianus (A. DC.) Kolak. & Serdyuk. === **Campanula rapunculus** subsp. **lambertiana** (A. DC.) Rech. f.

Neocodon patulus (L.) Kolak. & Serdyuk. === **Campanula patula** L.

Neocodon persicifolius (L.) Kolak. & Serdyuk. === **Campanula persicifolia** L.

Neocodon ponticus (Albov) Kolak. & Serdyuk. === **Campanula pontica** Albov

Neocodon rapunculus (L.) Kolak. & Serdyuk. === **Campanula rapunculus** L.

Neocodon sparsus (Friv.) Kolak. & Serdyuk. === **Campanula sparsa** Friv.

Neocodon spatulatus (Sm.) Kolak. & Serdyuk. === **Campanula spatulata** Sm.

Neocodon stevenii (M. Bieb.) Kolak. & Serdyuk. === **Campanula stevenii** M. Bieb.

Neocodon turczaninovii (Fed.) Kolak. & Serdyuk. === **Campanula stevenii** subsp. **turczaninovii** (Fed.) Victorov

Neocodon wolgensis (P. A. Smirn.) Kolak. & Serdyuk. === **Campanula stevenii** subsp. **wolgensis** (P. A. Smirn.) Fed.

Neowimmeria

As with *Galeatella* (q.v.), Degener and Degener's (1963, 1965) basis for according this taxon generic rank largely was left unstated. Two species referable here were included in a preliminary phylogenetic analysis of DNA data (E. Knox, pers. comm.), where they formed the sister-group of *Trematolobelia* and *Lobelia* sect. *Galeatella* E. Wimm. within the Hawaiian clade. Based on: *Lobelia* sect. *Revolutella* E. Wimm. Type [designated by O. Deg. & I. Deg., Phytologia 12: 73 (1965)]: *Lobelia grayana* E. Wimm.

Degener, O. & I. Degener (1963). *Neowimmeria* Deg. & Deg. In Flora Hawaiiensis, contents of sixth century and important notes: K16. Honolulu: published privately. En. — First attempt to elevate *Lobelia* sect. *Revolutella* to generic rank, but without citation of type.

Melville, R. (1965). Book reviews. Flora of Hawaii. Kew Bull. 19: 206. En. — Chastises Degeners for way in which *Neowimmeria* proposed.

Degener, O. & I. Degener (1965). The Hawaiian genus *Neowimmeria* (Lobeliaceae). Phytologia 12: 73. En. — Validation of generic name.

Stone, B. C. (1967). A review of the endemic genera of Hawaiian plants. *Neowimmeria* Deg. & Deg. Bot. Rev. (Lancaster) 33: 247-248. En. — Discusses merits of recognizing genus as distinct from *Lobelia*.

Degener, O. & I. Degener (1974). Prodromus of *Galeatella* and *Neowimmeria*. Honolulu: published privately. En. — Checklist of species, with full nomenclature.

Synonyms:

Neowimmeria O. Deg. & I. Deg. === **Lobelia** L.

Neowimmeria costata (E. Wimm.) O. Deg. & I. Deg. === **Lobelia dunbariae** subsp. **paniculata** (Rock) Lammers

Neowimmeria dunbariae (Rock) O. Deg. & I. Deg. === **Lobelia dunbariae** Rock subsp. **dunbariae**
Neowimmeria grayana (E. Wimm.) O. Deg. & I. Deg. === **Lobelia grayana** E. Wimm.
Neowimmeria heterocarpa (E. Wimm.) O. Deg. & I. Deg. === **Lobelia hypoleuca** Hillebr.
Neowimmeria hillebrandii (Rock) O. Deg. & I. Deg. === **Lobelia hillebrandii** Rock
Neowimmeria hypoleuca (Hillebr.) O. Deg. & I. Deg. === **Lobelia hypoleuca** Hillebr.
Neowimmeria hypoleuca var. *macrophyta* (Rock) O. Deg. & I. Deg. === **Lobelia hypoleuca**
  Hillebr.
Neowimmeria intermedia (H. St. John) O. Deg. & I. Deg. === **Lobelia niihauensis** H. St. John
Neowimmeria meridiana (H. St. John) O. Deg. & I. Deg. === **Lobelia niihauensis** H. St. John
Neowimmeria monostachya (Rock) O. Deg. & I. Deg. === **Lobelia monostachya** (Rock)
  Lammers
Neowimmeria niihauensis (H. St. John) O. Deg. & I. Deg. === **Lobelia niihauensis** H. St. John
Neowimmeria niihauensis var. *forbesii* (H. St. John) O. Deg. & I. Deg. === **Lobelia niihauensis**
  H. St. John
Neowimmeria oahuensis (Rock) O. Deg. & I. Deg. === **Lobelia oahuensis** Rock
Neowimmeria paniculata (Rock) O. Deg. & I. Deg. === **Lobelia dunbariae** subsp. **paniculata**
  (Rock) Lammers
Neowimmeria remyi (Rock) O. Deg. & I. Deg. === **Lobelia remyi** Rock
Neowimmeria rockii (H. St. John & Hosaka) O. Deg. & I. Deg. === **Lobelia hypoleuca** Hillebr.
Neowimmeria tortuosa O. Deg. & I. Deg. === **Lobelia niihauensis** H. St. John
Neowimmeria tortuosa var. *glabrata* (Skottsb.) O. Deg. & I. Deg. === **Lobelia niihauensis**
  H. St. John
Neowimmeria yuccoides (Hillebr.) O. Deg. & I. Deg. === **Lobelia yuccoides** Hillebr.

Nesocodon

Campanuloideae, 1 species, Mauritius in the western Indian Ocean. On the basis of morphology and biogeography, the genus is allied to *Heterochaenia* of Réunion (Thulin 1980, Wyse Jackson 1990, Weberling & Barthlott 1998). $2n = 34$. Type [designated by the author]: *Nesocodon mauritianus* (I. Richardson) Thulin.

  Richardson, I. B. K. (1979). A distinctive new species of *Wahlenbergia* (Campanulaceae) from Mauritius. Kew Bull. 33: 547-550, illus. En. — Discovery and description of type species.
• Thulin, M. (1980). *Nesocodon,* a new genus in Campanulaceae. Kew Bull. 34: 813-814, illus. En. — Establishment of genus as distinct from *Wahlenbergia* and *Heterochaenia.*
• Wyse Jackson, P. S. (1990). *Nesocodon mauritianus.* Kew Mag. 7: 113-117 + pl. 152, illus. En. — Portrait with detailed description, including chromosome count and comments on cultivation.
  Lorence, D. H. (1991). A Mascarene Island interlude. Bull. Natl. Trop. Bot. Gard. 21(4): 1-7, illus. En. — Popular account of the species, emphasizing conservation.
• Weberling, F. & W. Barthlott (1998). Lateral inflorescences — a distinguishing character of *Nesocondon* [sic] Thulin (Campanulaceae)? Beitr. Biol. Pflanzen 71: 41-44, illus. En. — Reinterpretation of inflorescence as terminal; recommends (but does not effect) transfer of type to *Heterochaenia.*

Nesocodon Thulin, Kew Bull. 34: 813 (1980).
  Mascarenes. 29.

Nesocodon mauritianus (I. Richardson) Thulin, Kew Bull. 34: 813 (1980).
  Mauritius. 29 MAU. Nanophan. $2n = 34$.
    ** Wahlenbergia mauritiana* I. Richardson, Kew Bull. 33: 547 (1979).

Numaeacampa

The identity of this scandent herb is a mystery. Gagnepain (1948) assigned it to
Campanulaceae, and stated that it was closely related to *Campanumoea* (here included in
Codonopsis) on the basis of its opposite leaves, solitary flowers, and ovary "au fond du
calice". However, the apically porose bisporangiate anthers are unlike anything else in the
family, and the undivided style and 10-lobed calyx are likewise quite unusual. Type [by
monotypy]: *Numaeacampa kerrii* Gagnep.

> Gagnepain, F. (1948). Genres nouveaux, espèces nouvelles d'Indochine. Bull. Soc. Bot.
> France 95: 26-34. Fr. — Establishment of genus as distinct from *Campanumoea*.

Synonyms:
Numaeacampa Gagnep. === ?
Numaeacampa kerrii Gagnep. === ?

Ostrowskia

Campanuloideae, 1 species, Tadzhikistan and Kirgizstan. It has not been included in a
phylogenetic analysis, but is thought be be related to platycodonoid genera such as
Canarina and *Codonopsis* on the basis of morphology, despite the lateral dehiscence of its
capsules. 2*n* = 34. Type [by monotypy]: *Ostrowskia magnifica* Regel.

> Regel, E. (1884). Descriptiones plantarum novarum et minus cognitarum, fasc. IX. Trudy
> Imp. S.-Peterburgsk. Bot. Sada 8: 641-702, illus. La. — Establishment of genus.
> Anonymous (1888). *Ostrowskya* [sic] *magnifica*. Vick's Monthly Mag. 11: 305-306, illus.
> En. — Popular description.
> Fedorov, A. A. (1957). *Ostrowskia* Rgl. In V. L. Komarov, Flora URSS 24: 382-383, illus.
> Leningrad: Academia Scientiae URSS. Ru. — Flora of former Soviet Union, with
> description and full nomenclature.
> Tarasevich, V. F. & K. K. Shrestha (1992). [Palynological data on the position of the genus
> *Ostrowskia* within the Campanulaceae family.] Bot. Žurn. 77(9): 27-30. Ru. — Pollen
> morphology.
> Kamelina, O. P. & N. A. Zhinkina (1998). [On the embryology of *Ostrowskia magnifica*
> (Campanulaceae). The ovule and seed.] Bot. Žurn. 83(3): 9-20, illus. Ru. —
> Embryological data suggest isolated and primitive position within family.

Ostrowskia Regel, Trudy Imp. S. Peterburgsk. Bot. Sada 8: 686 (1884).
Middle Asia. 32.

Ostrowskia magnifica Regel, Trudy Imp. S. Peterburgsk. Bot. Sada 8: 686 (1884).
Tadzhikistan & Kirgizstan. 32 KGZ TZK. Geophyte. 2*n* = 34.

Ovilla

This pre-Linnaean name has been substituted for *Jasione* by those who did not recognize
1753 as the beginning of botanical nomenclature. Based on: *Jasione* L.

> Adanson, M. (1763). Les campanules. *Campanulae.* In Familles des plantes 2: 132-134.
> Paris: Vincent. Fr. — Validation of name as replacement for *Jasione*.
> Ruprecht, F. J. (1856). Flora Ingrica sive historia plantarum gubernii petropolitani, vol.
> 1. St. Petersburg: Eggers. La. — Accepts substitution of *Ovilla* for *Jasione* on grounds
> of priority.
> Bubani, P. (1899). Flora pyrenaea per ordines naturales gradatim digesta, vol. 2. Milan:
> Ulrich Hoepli. La. — Accepts substitution of *Ovilla* for *Jasione* on grounds of priority.

Synonyms:
Ovilla Adans. === **Jasione** L.
Ovilla amethystina (Lag. & Rodr.) Bubani === **Jasione crispa** subsp. **tristis** (Bory) G. Lopéz
Ovilla globulariiflora Rupr. === **Jasione montana** L.
Ovilla humilis (Pers.) Bubani === **Jasione crispa** (Pourr.) Samp.
Ovilla perennis (L. f.) Bubani === **Jasione laevis** Lam. subsp. **laevis**

Palmerella

Lobelioideae, 1 species, western North America, in southern California and northern Baja California. Although some authors (e.g., McVaugh 1940a, Wimmer 1953, Serra & Crespo 1997) have emphasized a relationship to *Laurentia* s.l. (q.v.), preliminary molecular analysis (E. Knox, pers. comm.) places this genus as the sister-group of a clade comprising several other western North American endemics: *Downingia, Howellia, Legenere,* and *Porterella.* $2n$ = 14. Type [by monotypy]: *Palmerella debilis* A. Gray.

> Gray, A. (1876). Miscellaneous botanical contributions. Proc. Amer. Acad. Arts 11: 71-104. En. — Establishment of genus.
>
> Greene, E. L. (1889). Analogies and affinities. I. Pittonia 1: 293-300. En. — Subsumed into *Lobelia;* contends that George W. Dunn was actual discoverer of type species.
>
> Palmer, E. (1890). *Palmerella.* West Amer. Sci. 7(5): 8-9. En. — Rebuttal to Greene (1889), supporting Palmer as discoverer.
>
> • McVaugh, R. (1940a). A revision of "Laurentia" and allied genera in North America. Bull. Torrey Bot. Club 67: 778-798. En. — Subsumed into *Isotoma* (as *Lobelia* sect. *Isotoma*); includes key and descriptions of infraspecific taxa.
>
> McVaugh, R. (1940b). A key to the North American species of *Lobelia* (sect. *Hemipogon*). Amer. Midl. Naturalist 24: 681-702, illus. En. — Assigned sectional rank under *Lobelia.*
>
> McVaugh, R. (1943). [*Lobelia*] Section *Palmerella* (A. Gray) McVaugh. In North American Flora 32A: 65-66. New York: New York Botanical Garden. En. — Flora with description and full nomenclature.
>
> • Wimmer, F. E. (1953). [*Laurentia* (Mich.) Adans.] Sectio II. *Palmerella* (A. Gray) E. Wimm. In A. Engler & L. Diels, Das Pflanzenreich IV. 276b: 397-398, illus. Berlin: Akademie-Verlag. Ge., La. — Monograph with description, full nomenclature, and specimen citations.
>
> Ayers, T. (1993). *Lobelia.* In J. C. Hickman (ed.), The Jepson manual, higher plants of California: 465, illus. Berkeley: University of California Press. En. — Flora with description.
>
> Serra, L. & M. B. Crespo (1997). An outline revision of the subtribe Siphocampylinae (Lobeliaceae). Lagascalia 19: 881-888, illus., maps. En. — Key to distinguish genus from alleged allies.

Palmerella A. Gray, Proc. Amer. Acad. Arts 11: 80 (1876). *Lobelia* sect. *Palmerella* (A. Gray) McVaugh, Amer. Midl. Naturalist 24: 682 (1940). *Laurentia* sect. *Palmerella* (A. Gray) E. Wimm., Ann. Naturhist. Mus. Wien 56: 335 (1948).
W. North America. 76 79.

Palmerella debilis A. Gray, Proc. Amer. Acad. Arts 11: 80 (1876). *Lobelia dunnii* Greene, Pittonia 1: 297 (1889); non *Lobelia debilis* L. f., Suppl. Pl.: 395 (1782). *Laurentia debilis* (A. Gray) McVaugh, Bull. Torrey Bot. Club 67: 144 (1940).
SW. U.S.A. & NW. Mexico. 76 CAL 79 MXN. Hemicr. $2n$ = 14.

subsp. **debilis**
NW. Mexico (N. Baja California). 79 MXN. Hemicr.

subsp. **serrata** (A. Gray) Lammers, Novon 16: 72 (2006).
SW. U.S.A. (S. California) & NW. Mexico (N. Baja California). 76 CAL 79 MXN. Hemicr. $2n$ = 14.

* *Palmerella debilis* var. *serrata* A. Gray in S. Watson, Bot. California 1: 620 (1876). *Lobelia rothrockii* Greene, Pittonia 1: 297 (1889); non *Lobelia serrata* Meyen, Reise 1: 300 (1834). *Laurentia debilis* var. *serrata* (A. Gray) McVaugh, Bull. Torrey Bot. Club 67: 144 (1940). *Lobelia dunnii* var. *serrata* (A. Gray) McVaugh, Bull. Torrey Bot. Club 67: 795 (1940).

Synonyms:
Palmerella debilis var. *serrata* A. Gray === **Palmerella debilis** subsp. **serrata** (A. Gray) Lammers
Palmerella tenera A. Gray === **Diastatea tenera** (A. Gray) McVaugh

Parastranthus

This genus was segregated from *Lobelia* on the basis of its non-resupinate flowers. It is treated by most authors (e.g., Wimmer 1953, under **Monopsis**) as *Monopsis* sect. *Parastranthus* (G. Don) E. Wimm., although a combination based on *Rapuntium* sect. *Xanthomeria* C. Presl has priority at that rank. The monophyly of the group and its relationship to the rest of *Monopsis* (q.v.) are strongly supported by DNA sequence data (E. Knox, pers. comm.). Type [designated by Pfeiff., Nomencl. Bot. 2: 588 (1874)]: *Lobelia lutea* L.

> Don, G. (1834). *Parastranthus*. In A general history of the dichlamydeous plants: 697-698, 716-717. London: J. G. & F. Rivington. En. — Establishment of genus as distinct from *Lobelia*.
> Urban, I. (1881). Die Bestäubungseinrichtungen bei den Lobeliaceen nebst einer Monographie der afrikanischen Lobeliaceen-Gattung *Monopsis*. Jahrb. Königl. Bot. Gart. Berlin 1: 260-277, illus. Ge., La. — Subsumed into *Monopsis* sect. *Dobrowskya*.
> Adamson, R. S. (1950). *Parastranthus* G. Don. In R. S. Adamson & T. M. Salter, Flora of the Cape Peninsula: 758-759. Cape Town: Juta. En. — Flora with key and descriptions; one of few 20th Century floras to recognize genus.

Synonyms:
Parastranthus G. Don === **Monopsis** Salisb.
Parastranthus capitatus (C. Presl) A. DC. === **Monopsis lutea** (L.) Urb.
Parastranthus ericoides (C. Presl) A. DC. === **Monopsis lutea** (L.) Urb.
Parastranthus flavus (C. Presl) A. DC. === **Monopsis flava** (C. Presl) E. Wimm.
Parastranthus luteus (L.) A. DC. === **Monopsis lutea** (L.) Urb.
Parastranthus luteus var. *capitatus* (C. Presl) Sond. === **Monopsis lutea** (L.) Urb.
Parastranthus luteus var. *ericoides* (C. Presl) Sond. === **Monopsis lutea** (L.) Urb.
Parastranthus luteus var. *euphrasioides* Sond. === **Monopsis lutea** (L.) Urb.
Parastranthus simplex G. Don === **Monopsis lutea** (L.) Urb.
Parastranthus stellarioides (C. Presl) Vatke === **Monopsis stellarioides** (C. Presl) Urb.
Parastranthus thermalis (Thunb.) A. DC. === **Lobelia thermalis** Thunb.
Parastranthus unidentatus (W. T. Aiton) G. Don === **Monopsis unidentata** (W. T. Aiton) E. Wimm.
Parastranthus variifolius (Sims) G. Don === **Monopsis variifolia** (Sims) Urb.

Parishella

Nemacladoideae, 1 species, southern California. Allied to *Nemacladus* and *Pseudonemacladus*, the other genera of the subfamily. Type [by monotypy]: *Parishella californica* A. Gray.

> Gray, A. (1882). *Parishella californica*. Bot. Gaz. (Crawfordsville) 7: 94-95. En. — Establishment of genus.
> • McVaugh, R. (1943). *Parishella* A. Gray. In North American Flora 32A: 13. New York: New York Botanical Garden. En. — Flora with description and full nomenclature.

- Wimmer, F. E. (1968). *Parishella* A. Gray. In A. Engler & L. Diels, Das Pflanzenreich IV. 276c: 923. Berlin: Akademie-Verlag. Ge., La. — Monograph with description, full nomenclature, and specimen citations.

  Munz, P. A. & D. D. Keck (1970). *Parishella* Gray. In A California flora: 1068-1069. Berkeley: University of California Press. En. — Flora with description.

  Dunbar, A. (1984). Pollen morphology in Campanulaceae IV. Nord. J. Bot. 4: 1-19, illus. En. — Grains unlike most Campanuloideae and all Lobelioideae and Cyphocarpoideae surveyed: 6-colpate with spinules and a perforated tectum

  Morin, N. (1993). *Parishella*. In J. C. Hickman (ed.), The Jepson manual, higher plants of California: 468, illus. Berkeley: University of California Press. En. — Flora with description.

Parishella A. Gray, Bot. Gaz. (Crawfordsville) 7: 94 (1882).
   W. North America. 76.

Parishella californica A. Gray, Bot. Gaz. (Crawfordsville) 7: 94 (1882).
   C. & S. California. 76 CAL. Ther.

Pentagonia

Kuntze (1891) revived this pre-Linnaean name for *Specularia* (q.v.) as part of his absolutist views of priority. Under the current ICBN, it not only lacks priority, it is also a later homonym of *Pentagonia* Heister ex Fabr., nom. rej. (Solanaceae) and *Pentagonia* Benth., nom. cons. (Rubiaceae). Based on: *Specularia* Heister ex A. DC.

  Kuntze, O. (1891). *Pentagonia*. In Revisio generum plantarum vascularium omnium atque cellularium multarum secundum leges nomenclaturae internationales 2: 381. Leipzig: Arthur Felix. Ge. — Validation of name.

Synonyms:
Pentagonia Moehring ex Kuntze === **Legousia** Durande
Pentagonia biflora (Ruiz & Pav.) Kuntze === **Triodanis perfoliata** subsp. **biflora** (Ruiz & Pav.) Lammers
Pentagonia coa (A. DC.) Kuntze === **Legousia pentagonia** (L.) Thell.
Pentagonia coloradoensis (Buckley) Kuntze === **Triodanis coloradoensis** (Buckley) McVaugh
Pentagonia falcata (Ten.) Kuntze === **Legousia falcata** (Ten.) Fritsch ex Janch.
Pentagonia hybrida (L.) Kuntze === **Legousia hybrida** (L.) Delarbre
Pentagonia leptocarpa (Nutt.) Kuntze === **Triodanis leptocarpa** (Nutt.) Nieuwl.
Pentagonia perfoliata Kuntze === **Triodanis perfoliata** (L.) Nieuwl.

Pentropis

Rafinesque (1837) accorded this species of *Campanula* sect. *Quinqueloculares* (Boiss.) Phitos generic rank on the basis of its appendiculate calyx and five-locular five-ribbed globose hypanthium. In his view, *Campanula* consisted solely of species with exappendiculate calyx, campanulate corolla, and smooth three-locular hypanthium. Type [designated by the author]: *Pentropis saxatilis* (L.) Raf.

  Rafinesque, C. S. (1837). Flora Telluriana, pars secunda. Philadelphia: published privately. En. — Estsablishment of genus as distinct from *Campanula*.

Synonyms:
Pentropis Raf. === **Campanula** L.
Pentropis saxatilis (L.) Raf. === **Campanula saxatilis** L.

Peracarpa

Campanuloideae, 1 speces, eastern Asia, from Sakhalin to Himalaya and as far south as New Guinea. The genus has not yet been included in a molecular phylogeny, but is thought to be related to *Homocodon* (likewise unsampled yet) on the basis of similarities in habit and fruit morphology. The treatment here follows Barnesky and Lammers (1997). 2*n* = 28, 30. Type [by monotypy]: *Peracarpa carnosa* (Wall.) Hook. f. & Thomson.

Hooker, J. D. & T. Thomson (1858). Praecursores ad floram Indicam: being sketches of the natural families of Indian plants, with remarks on their distribution, structure, and affinities. Series I. Stylidieae, Goodenovieae, et Campanulaceae (including Lobeliaceae). En. — Establishment of genus as distinct from *Campanula*.

Hara, H. (1947). Annotationes miscellaneae ad plantas Asiae-orientalis (II). J. Jap. Bot. 21: 14-21. La. — Division of species into five geographic varieties.

Fedorov, A. A. (1957). *Peracarpa* Hook. f. & Thoms. In V. L. Komarov, Flora URSS 24: 380-381, illus. Leningrad: Academia Scientiae URSS. Ru. — Flora of former Soviet Union, with description and full nomenclature.

• Hara, H. (1966). Taxonomic comparison between corresponding taxa of Spermatophyta in eastern Himalaya and Japan. In H. Hara (ed.), The Flora of eastern Himalaya: 627-657, illus., map. Tokyo: University of Tokyo Press. En. — Brief discussion of geographically correlated morphological variation, reducing varieties to three.

Hong, D. Y. (1983). *Peracarpa* Hook. f. et Thoms. In Flora Reipublicae Popularis Sinicae 73(2): 142-143. Beijing: Science Press. Ch. — Flora for China, with description and full nomenclature.

• Barnesky, A. L. & T. G. Lammers (1997). Revision of the endemic Asian genus *Peracarpa* (Campanulaceae: Campanuloideae) via numerical phenetics. Bot. Bull. Acad. Sin. 38: 49-56. En. — Morphometric analysis failed to support recognition of infraspecific taxa.

Peracarpa Hook. f. & Thomson, J. Proc. Linn Soc., Bot. 2: 26 (1858).
Temp. & trop. Asia. 31 36 38 40 41 42 43.

Peracarpa carnosa (Wall.) Hook. f. & Thomson, J. Proc. Linn Soc., Bot. 2: 26 (1858).
Sakhalin to NE. India (Uttar Pradesh, West Bengal), Thailand, Philippines & New Guinea. 31 SAK 36 CHC CHS CHT 38 JAP KOR TAI 40 ASS EHM IND NEP 41 MYA THA 42 PHI 43 NWG. Hemicr. 2*n* = 28, 30.
  * *Campanula carnosa* Wall. in Roxb., Fl. Ind. 2: 102 (1824).
    Wahlenbergia ovata D. Don, Prodr. Fl. Nepal.: 156 (1825). *Campanula ovata* (D. Don) Spreng., Syst. Veg. 4(2): 78 (1827).
    Campanula circaeoides F. Schmidt, Reis. Amur-Land., Bot.: 154 (1868). *Peracarpa circaeoides* (F. Schmidt) Feer, Bot. Jahrb. Syst. 12: 621 (1890). *Peracarpa carnosa* var. *circaeoides* (F. Schmidt) Makino, Ill. Fl. Nippon: 82 (1940).
    Peracarpa luzonica Rolfe, Bull. Misc. Inform. Kew 1906: 201 (1906).
    Peracarpa carnosa var. *formosana* H. Hara, J. Jap. Bot. 21: 19 (1947).
    Peracarpa carnosa var. *kiusiana* H. Hara, J. Jap. Bot. 21: 20 (1947).
    Peracarpa carnosa f. *macrantha* Nakai ex H. Hara, J. Jap. Bot. 21: 20 (1947).
    Peracarpa carnosa var. *pumila* H. Hara, J. Jap. Bot. 21: 21 (1947).

Synonyms:
Peracarpa carnosa f. *macrantha* Nakai ex H. Hara === **Peracarpa carnosa** (Wall.) Hook. f. & Thomson
Peracarpa carnosa f. *pumila* H. Hara === **Peracarpa carnosa** (Wall.) Hook. f. & Thomson
Peracarpa carnosa var. *circaeoides* (Fr. Schmidt) Makino === **Peracarpa carnosa** (Wall.) Hook. f. & Thomson
Peracarpa carnosa var. *formosana* H. Hara === **Peracarpa carnosa** (Wall.) Hook. f. & Thomson
Peracarpa carnosa var. *kiusiana* H. Hara === **Peracarpa carnosa** (Wall.) Hook. f. & Thomson
Peracarpa circaeoides (Fr. Schmidt) Feer === **Peracarpa carnosa** (Wall.) Hook. f. & Thomson
Peracarpa luzonica Rolfe === **Peracarpa carnosa** (Wall.) Hook. f. & Thomson

Pernetya

This name was proposed as a substitute for *Canarina* L. to avoid confusion with the earlier *Canarium* L. (Burseraceae); however, these names are not considered homonyms under today's ICBN (Art. 53 Ex. 10). In 1966, *Pernetya* was formally rejected in favor of *Pernettya* Gaudich. (Ericaceae), which is its homonym. Based on: *Canarina* L.

> Scopoli, J. A. (1777). Introductio ad historiam naturalem. Prague: Wolfgang Gerle. La. — Establishment of genus.

Synonyms:
Pernetya Scop. === **Canarina** L.

Petalostima

Rafinesque (1837) seemed unaware that this species had been assigned to *Wahlenbergia*, commenting only on its differences from *Campanula*. Type [by monotypy]: *Petalostima capensis* (L.) Raf.

> Rafinesque, C. S. (1837). Flora Telluriana, pars secunda. Philadelphia: published privately. En. — Establishment of genus as distinct from *Campanula*.

Synonyms:
Petalostima Raf. === **Wahlenbergia** Schrad. ex Roth
Petalostima capensis (L.) Raf. === **Wahlenbergia capensis** (L.) A. DC.

Petkovia

This genus was segregated from *Campanula* on the basis of its indehiscent fruit, reminiscent of that seen in *Peracarpa*. Type [by monotypy]: *Petkovia orphanidea* (Boiss.) Stef.

> Stefanoff, B. (1936). [Über die systematische Stellung einiger Arten der Familie Campanulaceae.] God. Sofiisk. Univ. Agron.-Lesoved. Fak. 14: 93-104. Bu. — Establishment of genus as distinct from *Campanula*.

Synonyms:
Petkovia Stef. === **Campanula** L.
Petkovia orphanidea (Boiss.) Stef. === **Campanula orphanidea** Boiss.

Petromarula

Campanuloideae, 1 species, Kriti in the eastern Mediterranean. In a phylogeny based on DNA sequences (Eddie et al. 2003, under **General**), its type was sister to a clade comprising a species of *Asyneuma* and the type species of *Phyteuma* and *Physoplexis,* within the "Rapunculus clade". $2n = 30$. Type [designated by A. DC., Monogr. Campan. 209 (1830)]: *Petromarula pinnata* (L.) A. DC.

> Hedwig, R. (1806). Genera plantarum secundum characteres differentiales. Leipzig: I. H. Reclam. La. — Establishment of genus as distinct from *Phyteuma.*
> • Tutin, T. G. (1976). *Petromarula* Vent. ex Hedwig fil. In T. G. Tutin, V. H. Heywood, N., A. Burges, D. M. Moore, D. H. Valentine, S. M. Walters & D. A. Webb (eds.), Flora Europaea 4: 95. Cambridge: Cambridge University Press. En. — Flora of Europe with description and full nomenclature.
> Gudkova, I. Y. & G. P. Borshchenko (1991). [The serological study of the Campanulaceae. The phylogenetic relations in the tribe Phyteumateae.] Bot. Žurn. 76(6): 809-817. Ru. — Protein serology supports close relationship to *Asyneuma, Cryptocodon, Phyteuma,* and *Sergia,* and a somewhat more distant relationship to *Cylindrocarpa.*

Igersheim A. (1993). Floral development and secondary pollen presentation in *Petromarula* Vent. ex Hedwig f. (Campanulaceae). Bot. Jahrb. Syst. 115: 301-313, illus. En. — Floral anatomy and morphology in relation to pollination.

Petromarula Vent. ex R. Hedw., Gen. Pl. 139 (1806).
E. Mediterranean. 13.
Phyteuma [unranked] *Petromarula* Pers., Syn. Pl. 1: 192 (1805). *Phyteuma* sect.
Petromarula (Pers.) Kuntze in T. Post & Kuntze, Lex. Gen. Phan.: 437 (1904).

Petromarula pinnata (L.) A. DC., Monogr. Campan.: 209 (1830).
Kriti. 13 KRI. Hemicr. $2n = 30$.
* *Phyteuma pinnatum* L., Sp. Pl.: 171 (1753), as 'pinnata'.
Petromarula pinnata var. *pubescens* A. DC., Monogr. Campan.: 209 (1830).
Petromarula oxyloba Gand., Fl. Cret.: 69 (1916). *Petromarula pinnata* f. *oxyloba* (Gand.)
Hayek, Repert. Spec. Nov. Regni Veg. Beih. 30(2): 553 (1930).

Synonyms:
Petromarula oxyloba Gand. === **Petromarula pinnata** (L.) A. DC.
Petromarula pinnata f. *oxyloba* (Gand.) Hayek === **Petromarula pinnata** (L.) A. DC.
Petromarula pinnata var. *pubescens* A. DC. === **Petromarula pinnata** (L.) A. DC.

Petromarula

Embracing an absolutist view of priority, Nieuwland and Lunell (in Lunell 1917) resurrected a pre-Linnaean name for *Lobelia*. Under the current ICBN, not only does it lack priority over *Lobelia*, it is also a later homonym of *Petromarula* Vent. ex R. Hedw. (Campanulaceae). Fortunately, only two of the many possible combinations were ever effected. Type [designated here]: *Petromarula hirtella* (A. Gray) Nieuwl. & Lunell.

Lunell, J. (1917). Enumerantur plantae Dakotae Septentrionalis vasculares. – X. Amer. Midl. Naturalist 5: 1-13. En. — Establishment of name as a substitute for *Lobelia*.

Synonyms:
Petromarula Belli ex Nieuwl. & Lunell === **Lobelia** L.
Petromarula hirtella (A. Gray) Nieuwl. & Lunell === **Lobelia spicata** Lam.
Petromarula strictiflora (Rydb.) Nieuwl. & Lunell === **Lobelia kalmii** L.

Phyllocharis

Diels (1917) overlooked the fact that this name had been used earlier for a genus of folicolous lichens (*Phyllocharis* Fée). When a bid at conservation failed (Bakhuizen van den Brink et al. 1961, Rickett 1963), it was renamed *Ruthiella* (q.v.). Type [designated by Steenis, Blumea 13: 127 (1965)]: *Phyllocharis schlechteri* Diels.

Diels, F. L. (1917). Neue Campanulaceen von Papuasien. Bot. Jahrb. Syst. 55: 121-125. Ge. — Establishment of genus.
• Wimmer, F. E. (1953). *Phyllocharis* Diels. In A. Engler & L. Diels, Das Pflanzenreich IV. 276b: 724-725. Berlin: Akademie-Verlag. Ge., La. — Monograph with key, descriptions, full nomenclature, and specimen citations.
• Tuyn, P. (1960). *Phyllocharis*. In C. G. G. J. van Steenis (ed.), Flora Malesiana (ser. I) 6(1): 137-139, illus. Djakarta: Noordhoff-Kolff. En. — Flora with key, descriptions, and full nomenclature.
Bakhuizen van den Brink, R. C., R. A. Maas Geesteranus & C. G. G. J. van Steenis (1961). Proposal to conserve the generic name *Phyllocharis* Diels (Lobeliaceae). Taxon 10: 246. En. — Formal proposal to protect name.

Rickett, H. W. (1963). Report of the Committee for Spermatophyta. Conservation of generic names V. Taxon 12: 235-238. En. — Declined to conserve name.

Synonyms:
Phyllocharis Diels === **Ruthiella** Steenis
Phyllocharis lamiifolia E. Wimm. === **Ruthiella subcordata** (Merr. & L. M. Perry) Steenis
Phyllocharis oblongifolia Diels === **Ruthiella oblongifolia** (Diels) Steenis
Phyllocharis saxicola P. Royen === **Ruthiella saxicola** (P. Royen) Steenis
Phyllocharis schlechteri Diels === **Ruthiella schlechteri** (Diels) Steenis
Phyllocharis subcordata Merr. & L. M. Perry === **Ruthiella subcordata** (Merr. & L. M. Perry) Steenis

Physoplexis

Campanuloideae, 1 species, southeastern Europe. This species sometimes has been included in *Phyteuma* (e.g., Hooker 1880), a position supported by a very preliminary phylogenetic analysis (Eddie et al. 2003, under **General**). There, it formed an unresolved trichotomy with the type and a second species of *Phyteuma*. 2*n* = 34, 68. Based on: *Phyteuma* [unranked] *Physoplexis* Endl. Type [by monotypy]: *Phyteuma comosum* L.

Schur, F. (1853). Sertum florae Transsilvaniae. Hermannstadt: Georg von Closius. La. — Elevation of basionym to generic rank.
Hooker, J. D. (1880). *Phyteuma comosum*. Bot. Mag. 106: tab. 6478 + 2 pp., illus. En. — Portrait of *Physoplexis comosa,* with description.
Schulz, R. (1904). Monographie der Gattung *Phyteuma*. Geisenheim: Johann Schneck. Ge. — Elevation to generic rank under the name *Synotoma.*
• Damboldt, J. (1976). *Physoplexis* (Endl.) Schur. In T. G. Tutin, V. H. Heywood, N., A. Burges, D. M. Moore, D. H. Valentine, S. M. Walters & D. A. Webb (eds.), Flora Europaea 4: 98. Cambridge: Cambridge University Press. En. — Flora of Europe with description and full nomenclature.

Physoplexis (Endl.) Schur, Sert. Fl. Transsilv.: 47 (1853).
   S. Alps. 11 13.
     * *Phyteuma* [unranked] *Physoplexis* Endl., Gen. Pl.: 517 (1838).
     Phyteuma sect. *Synotoma* G. Don, Gen. Hist. 3: 746 (1834). *Synotoma* (G. Don) Rich. Schulz, Monogr. Phyteuma: 18 (1904).

Physoplexis comosa (L.) Schur, Sert. Fl. Transsilv.: 47 (1853).
   N. Italy, Austria, Slovenia. 11 AUT 13 ITA YUG. Biennial. 2*n* = 34, 68.
     * *Phyteuma comosum* L., Sp. Pl.: 171 (1753), as 'comosa'. *Rapunculus comosus* (L.) Mill., Gard. Dict. (ed. 8) (1768). *Synotoma comosum* (L.) Dalla Torre & Sarnth., Fl. Tirol 6(3): 458 (1911).
     Phyteuma comosum var. *pubescens* Facchini ex Murr, Allg. Bot. Z. Syst. 4: 7 (1898). *Synotoma comosum* var. *pubescens* (Facchini ex Murr) Dalla Torre & Sarnth., Fl. Tirol 6(3): 458 (1911).

Phyteuma

Campanuloideae, 22 species, Europe, with one extending into North Africa; the greatest concentration of species is found in the Alps. In a preliminary molecular phylogeny (Eddie et al. 2003, under **General**), the type of the genus and a congener formed a clade with the types of *Petromarula* and *Physoplexis,* plus the one sampled species of *Asyneuma;* this clade was then part of the "Rapunculus clade". The last monograph was that by Schulz (1904); the best modern account is the floristic treatment by Damboldt (1976). 2*n* = 16, 18, 20, 22, 24, 26, 28, 36. Type [designated by Hitchc. & M. L. Green in Anon., Nomencl. Prop. Brit. Bot. 131 (1929)]: *Phyteuma spicatum* L.

Linnaeus, C. (1753). *Phyteuma*. In Species plantarum: 170-171. Stockholm: Laurentius Salvius. La. — Establishment of genus in Pentandria Monogynia.

Linnaeus, C. (1754). *Phyteuma*. In Genera plantarum: 78. Stockholm: Laurentius Salvius. La. — Description of genus to accompany preceding.

Sims, J. (1812). *Phyteuma cordata*. Horned rampion. Bot. Mag. 36: tab. 1466 + 1 pg., illus. En. — Portrait of *P. orbiculare*, with description.

Sims, J. (1819). *Phyteuma betonicifolium*. Betony-leaved rampion. Bot. Mag. 46: tab. 2066 + 2 pp., illus. En. — Portrait of *P. scorzonerifolia*, with description.

Sims, J. (1821). *Phyteuma scorzonerifolia*. Scorzonera-leaved rampion. Bot. Mag. 48: tab. 2271 + 2 pp., illus. En. — Portrait with description.

Sims, J. (1822). *Phyteuma spicatum*. Spiked rampion. Bot. Mag. 49: tab. 2347 + 2 pp., illus. En. — Portrait with description.

Murr, J. (1896). Über Hybride der Gattung *Phyteuma*. Deutsche Bot. Monatsschr. 14: 116-120. Ge. — Description of several interspecific hybrids.

• Schulz, R. (1904). Monographie der Gattung *Phyteuma*. Geisenheim: J. Schneck. Ge. — Monograph with keys, descriptions, infrageneric classification, full nomenclature, and specimen citations.

Chouard, C. (1952). Espèces en formation chez les *Phyteuma* du groupe *hemisphaericum*, particulièrement dans les Pyrénées. Bull. Soc. Bot. France 99: 25-29, illus. Fr. — Brief regional account of problematic species complex.

Fedorov, A. A. (1957). *Phyteuma* L. In V. L. Komarov, Flora URSS 24: 388-396, illus. Leningrad: Academia Scientiae URSS. Ru. — Flora of former Soviet Union, with key, descriptions, infrageneric classification, and full nomenclature.

Kovanda, M. (1971). Observations on *Phyteuma tenerum* R. Schulz in England. Watsonia 8: 385-389. En. — Taxonomic notes, including chromosome counts.

• Damboldt, J. (1976). *Phyteuma* L. In T. G. Tutin, V. H. Heywood, N., A. Burges, D. M. Moore, D. H. Valentine, S. M. Walters & D. A. Webb (eds.), Flora Europaea 4: 95-98. Cambridge: Cambridge University Press. En. — Flora of Europe with key, descriptions, and full nomenclature.

Palmer, R. C. (1977). *Phyteuma scheuchzeri* All. naturalized in Oxford. Watsonia 11: 254-255. En. — Species of S. Europe naturalized in Britain.

Martini, F. (1978). Distribuzione di *Phyteuma betonicifolium* Vill. e *P. zahlbruckneri* Vest nelle Alpi sudorientali. Giorn. Bot. Ital. 112: 53-62, map. It. — Biogeography of pair of closely related species.

• Kovanda, M. (1981). Studies in *Phyteuma*. Preslia 53: 211-238. En. — Biosystematic and cytological studies of species in Czechoslovakia.

Brunerye, L. (1989). Note sur les *Phyteuma* du groupe *spicatum* s.l. de la flore de France. Bull. Soc. Bot. Centre-Ouest (n.s.) 20: 13-21, illus. Fr. — Descriptive floristic account of problematic species complex, with keys.

Weeda, E. J. (1989). *Phyteuma nigrum* F. W. Schmidt en *P. spicatum* L. in Nederland. Gorteria 15: 6-27, maps. Du. — Morphometric and vegetational analysis on local scale supports treatment as conspecific subspecies.

Ayers, T. J. (1991). The cultivated species of *Phyteuma* and *Asyneuma* (Campanulaceae). Baileya 23: 126-138, illus. En. — Descriptive account and key for species used in horticulture.

Gudkova, I. Y. & G. P. Borshchenko (1991). [The serological study of the Campanulaceae. The phylogenetic relations in the tribe Phyteumateae.] Bot. Žurn. 76(6): 809-817. Ru. — Protein serology supports close relationship to *Asyneuma, Cryptocodon, Petromarula,* and *Sergia,* and a somewhat more distant relationship to *Cylindrocarpa.*

Cafferty, S. & F. Sales (1999). Proposal to reject the name *Phyteuma pauciflorum* (Campanulaceae). Taxon 48: 601-602. En. — Formal proposal to reject a Linnaean name used in contradictory ways by different authors.

Wheeler, B. R. & M. J. Hutchings (1999). The history and distribution of *Phyteuma spicatum* L. (Campanulaceae) in Britain. Watsonia 22: 387-395, maps. En. — Biogeographic notes.

Sales, F. & I. C. Hedge (2000). *Phyteuma* L. (Campanulaceae): some taxonomic notes. Anales Jard. Bot. Madrid 57: 474-477. En. — Miscellaneous taxonomic comments preparatory to Sales & Hedge (2001).

Brummitt, R. K. (2001). Report of the Committee for Spermatophyta: 51. Taxon 50: 559-568. En. — Recommended rejecting the name *Phyteuma pauciflorum*.

Sales, F. & I. C. Hedge (2001). *Phyteuma* L. In S. Castroviejo (ed.), Flora Iberica 14: 143-150, illus. Madrid: Real Jardín Botánico. Sp. — Flora of Spain with key, descriptions, and full nomenclature.

Wheeler, B.R. & M. J. Hutchings (2002). Biological flora of the British Isles No. 223 *Phyteuma spicatum* L. J. Ecol. 90: 581-591, illus., maps. En. — Ecological life history of widespread polymorphic species.

Phyteuma L., Sp. Pl.: 170 (1753). *Rapunculus* Mill., Gard. Dict. (abr. ed. 4) (1754).
Europe & N. Africa. 11 12 13 14 20.

Phyteuma × **adulterinum** Wallr., Erst. Beitr. Fl. Hercyn.: 188 (1840). P. nigrum × P. spicatum
Germany. 11 GER. Hemicr.

Phyteuma betonicifolium Vill., Hist. Pl. Dauphiné 2: 518 (1787), as 'betonicaefolia'.
Phyteuma spicatum var. *betonicifolium* (Vill.) Lapeyr., Hist. Pl. Pyrénées: 112 (1813), as 'betonicaefolia'. *Phyteuma michelii* subsp. *betonicifolium* (Vill.) Nyman, Consp. Fl. Eur.: 484 (1879), as 'betonicaefolium'.
S. Germany to S. France & Slovenia. 11 AUT GER SWI 12 FRA 13 ITA YUG. Hemicr.
   $2n = 24$.

subsp. **betonicifolium**
   S. Germany to S. France & Slovenia. 11 AUT GER SWI 12 FRA 13 ITA YUG. Hemicr.
      $2n = 24$.
   Phyteuma betonicifolium var. *pubescens* A. DC., Monogr. Campan.: 194 (1830). *Phyteuma betonicifolium* f. *pubescens* (A. DC.) Rich. Schulz, Monogr. Phyteuma: 95 (1904).
   Phyteuma betonicifolium var. *sessilifolium* A. DC., Monogr. Campan.: 194 (1830). *Phyteuma michelii* var. *sessilifolium* (A. DC.) Rouy, Fl. France 10: 86 (1908).
   Phyteuma veronicifolium Schrad. ex A. DC., Monogr. Campan.: 196 (1830), as 'veronicaefolium'. *Phyteuma michelii* var. *veronicifolium* (Schrad. ex A. DC.) Nyman, Consp. Fl. Eur.: 484 (1879), as 'veronicaefolium'.
   Phyteuma betonicifolium f. *glabrum* Rich. Schulz, Monogr. Phyteuma: 95 (1904).
   Phyteuma betonicifolium f. *alpestre* Rich. Schulz, Monogr. Phyteuma: 95 (1904).
   Phyteuma betonicifolium var. *lanceolatum* Rich. Schulz, Monogr. Phyteuma: 95 (1904).
   Phyteuma betonicifolium f. *vulgare* Rich. Schulz, Monogr. Phyteuma: 95 (1904).
   Phyteuma betonicifolium f. *rhaeticum* Rich. Schulz, Monogr. Phyteuma: 95 (1904).

subsp. **scaposum** (Rich. Schulz) Pign., Giorn. Bot. Ital. 111: 54 (1977).
   SE. France to Austria. 11 AUT SWI 12 FRA 13 ITA. Hemicr. $2n = 24$.
   Phyteuma elegans Hegetschw., Fl. Schweiz: 228 (1838-39).
  Phyteuma scaposum Rich. Schulz, Monogr. Phyteuma: 96 (1904). *Phyteuma michelii* subsp. *scaposum* (Rich. Schulz) P. Fourn., Quatre Fl. France: 918 (1939).
   Phyteuma scaposum f. *glabrum* Rich. Schulz, Monogr. Phyteuma: 98 (1904).
   Phyteuma scaposum f. *cordifolium* Rich. Schulz, Monogr. Phyteuma: 98 (1904).

Phyteuma charmelii Vill., Hist. Pl. Dauphiné 2: 516 (1787). *Phyteuma reniforme* Dulac, Fl. Hautes-Pyrénées: 456 (1867). *Phyteuma scheuchzeri* subsp. *charmelii* (Vill.) Nyman, Consp. Fl. Eur.: 484 (1879). *Phyteuma longibracteatum* St. Lag. in Cariot, Étude Fl. (ed. 8) 2: 543 (1889).
Morocco; Spain to N. Italy; S. Appennines. 12 FRA SPA 13 ITA 20 MOR. Hemicr. $2n = 18$.
   Rapunculus scheuchzeri Bubani, Fl. Pyren. 2: 24 (1899).
   Phyteuma villarsii Rich. Schulz, Monogr. Phyteuma: 143 (1904).
   Phyteuma charmelii f. *microcephala* Sennen, Diagn. Nouv.: 22 (1936).

Phyteuma confusum A. Kern., Z. Ferdinandeums Tirol (ser. 3) 15: 247 (1870), as 'confusa'.
  Phyteuma orbiculare subvar. *confusum* (A. Kern.) Nyman, Consp. Fl. Eur.: 484 (1879).
  Phyteuma latifolium Dalla Torre in Hartinger, Atlas Alpenfl.: 160 (1884). *Phyteuma
  hemisphaericum* subsp. *confusum* (A. Kern.) Nyman, Consp. Fl. Eur., Suppl. 2: 212 (1889).
  Austria to Albania & Bulgaria. 11 AUT 13 ALB BUL ROM YUG. Hemicr. *2n* = 28.

Phyteuma cordatum Balb., Mém. Acad. Sci. Turin 16: 208 (1809). *Phyteuma balbisii* A. DC.,
  Monogr. Campan.: 200 (1830).
  S. France & NE. Italy. 12 FRA 13 ITA. Hemicr. *2n* = 24.
    Phyteuma balbisii var. *petraeum* A. DC., Monogr. Campan.: 200 (1830).
    Phyteuma michelii subsp. *alpini* Ces ex Arcang., Comp. Fl. Ital.: 449 (1882).
    Phyteuma cordatum f. *obtusatum* Rich. Schulz, Monogr. Phyteuma: 109 (1904).

Phyteuma gallicum Rich. Schulz, Monogr. Phyteuma: 88 (1904).
  SC. France. 12 FRA. Hemicr.

Phyteuma globulariifolium Sternb. & Hoppe, Denkschr. Königl.-Baier. Bot. Ges.
  Regensburg 1(2): 100 (1818), as 'globulariaefolium'. *Phyteuma pauciflorum* var.
  macrophyllum Schur, Enum. Pl. Transsilv.: 428 (1866). *Phyteuma pauciflorum* subsp.
  globulariifolium (Sternb. & Hoppe) Nyman, Consp. Fl. Eur.: 485 (1879), as
  'globulariaefolium'.
  Pyrénées & Alps. 11 AUT SWI 12 FRA SPA 13 ITA. Hemicr. *2n* = 24, 28.
    Phyteuma pauciflorum L., Sp. Pl.: 170 (1753), as 'pauciflora', nom. rejic. *Rapunculus
      pauciflorus* (L.) Mill., Gard. Dict. (ed. 8) (1768), nom. rejic.
    Phyteuma capituliforme Rochel, Pl. Banat. Rar.: 6 (1828).
    Phyteuma nanum Schur, Verh. Mitth. Siebenbürg. Vereins Naturwiss. Hermannstadt 3:
      88 (1852). *Phyteuma pauciflorum* var. *nanum* (Schur) Schur, Enum. Pl. Transsilv.:
      428 (1866).
    Phyteuma hemisphaericum var. *transsilvanicum* Schur, Enum. Pl. Transsilv.: 429 (1866).
    Phyteuma pauciflorum f. *nanum* Rich. Schulz, Monogr. Phyteuma: 160 (1904).
    Phyteuma pauciflorum f. *asteranthum* Rich. Schulz, Monogr. Phyteuma: 160 (1904).
    Phyteuma pauciflorum f. *albiflorum* Rich. Schulz, Monogr. Phyteuma: 160 (1904).
    Phyteuma globulariifolium var. *tirolense* Rich. Schulz, Monogr. Phyteuma: 163 (1904).
    Phyteuma globulariifolium f. *integrum* Rich. Schulz, Monogr. Phyteuma: 163 (1904).
    Phyteuma globulariifolium var. *acutifolium* Rich. Schulz, Monogr. Phyteuma: 163 (1904).
    Phyteuma globulariifolium f. *vulgare* Rich. Schulz, Monogr. Phyteuma: 163 (1904).
    Phyteuma globulariifolium f. *simplex* Rich. Schulz, Monogr. Phyteuma: 163 (1904).
    Phyteuma globulariifolium var. *nanum* Rich. Schulz, Monogr. Phyteuma: 163 (1904).
    Phyteuma pedemontanum Rich. Schulz, Monogr. Phyteuma: 163 (1904). *Phyteuma
      pauciflorum* subsp. *pedemontanum* (Rich. Schulz) P. Fourn., Quatre Fl. France: 917
      (1939).
    Phyteuma pedemontanum f. *intermedium* Rich. Schulz, Monogr. Phyteuma: 168 (1904).
    Phyteuma pedemontanum f. *humillimum* Rich. Schulz, Monogr. Phyteuma: 168 (1904).

Phyteuma hedraianthifolium Rich. Schulz, Monogr. Phyteuma: 150 (1904).
  Switzerland & N. Italy. 11 SWI 13 ITA. Hemicr.
    Phyteuma hedraianthifolium f. *graminifolium* Rich. Schulz, Monogr. Phyteuma: 152
      (1904).

Phyteuma hemisphaericum L., Sp. Pl.: 170 (1753), as 'hemisphaerica'. *Rapunculus
  hemisphaericus* (L.) Mill., Gard. Dict. (ed. 8) (1768), as 'hemisphericus'.
  Spain to E. Austria. 11 AUT GER SWI 12 FRA SPA. Hemicr. *2n* = 16, 24, 28.
    Phyteuma graminifolium Sieber, Flora 5: 648 (1822). *Phyteuma hemisphaericum* var.
      graminifolium (Sieber) Schur, Enum. Pl. Transsilv.: 429 (1866).
    Phyteuma hemisphaericum var. *setaceum* Hegetschw., Reis. Glarus Graubünden: 147
      (1825).

Phyteuma intermedium Hegetschw., Reis. Glarus Graubünden: 147 (1825).
Phyteuma hemisphaericum var. *latifolium* Schur, Enum. Pl. Transsilv.: 429 (1866).
Phyteuma hemisphaericum var. *albiflorum* Schur, Enum. Pl. Transsilv.: 429 (1866).
Phyteuma hemisphaericum f. *vulgare* Rich. Schulz, Monogr. Phyteuma: 149 (1904).
Phyteuma hemisphaericum f. *albiflorum* Rich. Schulz, Monogr. Phyteuma: 149 (1904).
Phyteuma hemisphaericum var. *carinthiacum* Rich. Schulz, Monogr. Phyteuma: 149 (1904).
Phyteuma hemisphaericum var. *platyphyllum* Rich. Schulz, Monogr. Phyteuma: 149 (1904).
Phyteuma hemisphaericum var. *subacaule* Rouy, Fl. France 10: 89 (1908).
Phyteuma gaussenii Chouard, Bull. Soc. Bot. France 99: 26 (1952).
Phyteuma hemisphaericum var. *hedraianthoides* Chouard, Bull. Soc. Bot. France 99: 26 (1952).
Phyteuma serratoides Chouard, Bull. Soc. Bot. France 99: 28 (1952).

Phyteuma humile Schleich. ex Gaudin in Murith, Guide Bot. Voy. Valais: 84 (1810), as 'humilis'.
SE. France to N. Italy. 11 SWI 12 FRA 13 ITA. Hemicr.
Phyteuma carestiae Biroli, Giorn. Fis. (ser. 2) 1: 37 (1818) .
Phyteuma humile var. *humillimum* A. DC., Monogr. Campan.: 186 (1830).
Phyteuma linearifolium Biroli ex Colla, Herb. Pedem. 4: 34 (1835).

Phyteuma × huteri Murr, Progr. Oberrealschule Innsbruck 1890-91: 56 (1891). P. betonicifolium × P. ovatum
Austria. 11 AUT. Hemicr.
Phyteuma × murrianum Borbás ex Murr, Progr. Oberrealschule Innsbruck 1890-91: 56 (1891).
Phyteuma × khekii Murr, Deutsche Bot. Monatsschr. 14: 119 (1896).
Phyteuma × hellwegeri Murr, Deutsche Bot. Monatsschr. 14: 121 (1896).

Phyteuma michelii All., Fl. Pedem. 1: 115 (1785). *Phyteuma scorzonerifolium* var. *angustissimum* St.-Lag. in Cariot, Étude Fl. (ed. 8) 2: 542 (1889).
SE. France & N. Italy. 12 FRA 13 ITA. Hemicr.

Phyteuma nigrum F. W. Schmidt, Fl. Boëm. 2: 87 (1794). *Phyteuma spicatum* var. *caerulescens* Godr., Fl. Lorraine 2: 91 (1843). *Phyteuma spicatum* var. *nigrum* (F. W. Schmidt) Kuntze, Taschen-Fl. Leipzig: 126 (1867). *Phyteuma halleri* subsp. *nigrum* (F. W. Schmidt) Arcang., Comp. Fl. Ital.: 450 (1882). *Phyteuma spicatum* subsp. *nigrum* (F. W. Schmidt) Weeda, Gorteria 15: 24 (1989).
Belgium to E. Austria. 11 AUT BGM CZE GER 12 FRA. Hemicr. 2*n* = 22, 26.
Phyteuma ovatum F. W. Schmidt, Fl. Boëm. 2: 87 (1794); non Honck., Verz. Gew. Teutsch.: 653 (1782), as 'ovata'.
Phyteuma atropurpureum Hoppe, Bot. Taschenb. 1802: 27 (1802). *Phyteuma spicatum* var. *atropurpureum* (Hoppe) Steud., Nomencl. Bot.: 618 (1821), as 'atropurpurea'.
Phyteuma nigrum f. *longibracteatum* Rich. Schulz, Monogr. Phyteuma: 86 (1904).
Phyteuma nigrum var. *integrifolium* Rich. Schulz, Monogr. Phyteuma: 86 (1904).
Phyteuma nigrum f. *vulgare* Rich. Schulz, Monogr. Phyteuma: 86 (1904).
Phyteuma nigrum f. *latibracteatum* Rich. Schulz, Monogr. Phyteuma: 86 (1904).
Phyteuma nigrum f. *interruptum* Rich. Schulz, Monogr. Phyteuma: 86 (1904).
Phyteuma nigrum var. *acuminatum* Rich. Schulz, Monogr. Phyteuma: 86 (1904).
Phyteuma nigrum var. *coeruleum* Rich. Schulz, Monogr. Phyteuma: 86 (1904).

Phyteuma × obornyanum Hayek, Fl. Steiermark 2(1): 468 (1912), pro sp. P. confusum × P. globulariifolium
Austria. 11 AUT. Hemicr.

Phyteuma orbiculare L., Sp. Pl.: 170 (1753), as 'orbicularis'. *Rapunculus orbicularis* (L.) Mill., Gard. Dict. (ed. 8) (1768), as 'orbicularus'.

S. Great Britain to Latvia, Ukraine, Greece & Spain; cult. 10 GRB 11 AUT BGM CZE GER
HUN POL SWI 12 FRA SPA 13 ALB GRC ITA ROM YUG 14 BLR BLT UKR. Hemicr. $2n =$
22, 24.

Phyteuma cordifolium Vill., Hist. Pl. Dauphiné 2: 517 (1787), as 'cordifolia'. *Phyteuma
orbiculare* var. *cordifolium* (Vill.) Griseb., Spic. Fl. Rumel. 2: 291 (1846), as 'cordatum'.
Phyteuma orbiculare var. *cordatum* Gren. & Godr., Fl. Fr. 2: 402 (1853).

Phyteuma ellipticifolium Vill., Hist. Pl. Dauphiné 2: 517 (1787), as 'ellipticifolia'.
Phyteuma orbiculare var. *ellipticum* Pers., Syn. Pl. 1: 194 (1805). *Phyteuma orbiculare*
subsp. *ellipticum* Gaudin, Fl. Helv. 2: 174 (1828). *Phyteuma orbiculare* var. *giganteum*
A. DC., Monogr. Campan.: 188 (1830). *Phyteuma orbiculare* var. *ellipticifolium* (Vill.)
Nyman, Consp. Fl. Eur.: 484 (1879), as 'ellipticifolia'. *Phyteuma orbiculare* subsp.
ellipticifolium (Vill.) Arcang., Comp. Fl. Ital.: 449 (1882). *Phyteuma delphinense* var.
ellipticifolium (Vill.) Dalla Torre & Sarnth., Fl. Tirol. 6(3): 469 (1911).

Phyteuma lanceolatum Vill., Hist. Pl. Dauphiné 2: 517 (1787), as 'lanceolata'. *Phyteuma
orbiculare* var. *lanceolatum* (Vill.) Pers., Syn. Pl. 1: 194 (1805), as 'lanceolata'.
Phyteuma orbiculare subsp. *lanceolatum* (Vill.) Arcang., Comp. Fl. Ital.: 449 (1882).

Phyteuema inaequatum Kit. ex Schult., Oesterr. Fl. (ed. 2) 1: 398 (1814). *Phyteuma
orbiculare* var. *inaequatum* (Kit. ex Schult.) Nyman, Consp. Fl. Eur.: 484 (1879).

Phyteuma brevifolium Schleich., Cat. Pl. Helvet. (ed. 4): 25 (1821), as 'brevifolia'.
Phyteuma pilosum Hegetschw., Reis. Glarus Graubünden: 149 (1835). *Phyteuma
orbiculare* subsp. *decipiens* Gaudin, Fl. Helv. 2: 176 (1828). *Phyteuma orbiculare* var.
decipiens (Gaudin) A. DC., Monogr. Campan.: 188 (1830). *Phyteuma orbiculare* var.
brevifolium (Schleich.) Steud., Nomencl. Bot. (ed. 2) 2: 331 (1841). *Phyteuma
orbiculare* f. *pilosum* Rich. Schulz, Monogr. Phyteuma: 114 (1904).

Phyteuma orbiculare var. *comosum* Steud., Nomencl. Bot.: 618 (1821), as 'comosa'.

Phyteuma fistulosum Rchb., Flora 5: 534 (1822). *Phyteuma orbiculare* var. *fistulosum*
(Rchb.) Steud., Nomencl. Bot. (ed. 2) 2: 331 (1841). *Phyteuma orbiculare* subsp.
fistulosum (Rchb.) Nyman, Consp. Fl. Eur.: 484 (1879).

Phyteuma ellipticifolium var. *pauciflorum* Hegetschw., Reis. Glarus Graubünden: 148
(1825).

Phyteuma orbiculare subsp. *cordatum* Gaudin, Fl. Helv. 2: 174 (1828).

Phyteuma orbiculare subsp. *cinerascens* Gaudin, Fl. Helv. 2: 175 (1828).

Phyteuma orbiculare subsp. *lancifolium* Gaudin, Fl. Helv. 2: 176 (1828).

Phyteuma orbiculare var. *columnae* A. DC., Monogr. Campan.: 188 (1830).

Phyteuma angustatum Wender., Schriften Ges. Beförd. Gesammten Naturwiss. Marburg
2: 247 (1831), as 'angustata'.

Phyteuma bovelinii Hegetschw., Fl. Schweiz: 224 (1838-39), as 'bovelini'.

Phyteuma hispidum Hegetschw., Fl. Schweiz: 224 (1838-39). *Phyteuma orbiculare* f.
hispidum (Hegetschw.) Rich. Schulz, Monogr. Phyteuma: 115 (1904).

Phyteuma longifolium Hegetschw., Fl. Schweiz: 225 (1838-39).

Phyteuma michelii Hegetschw., Fl. Schweiz: 228 (1838-39); non All., Fl. Pedem. 1: 115
(1785).

Phyteuma orbiculare var. *alpinum* Schur, Enum. Pl. Transsilv.: 429 (1866).

Phyteuma pseudorbiculare Pant., Verh. Vereins Natur-Heilk. Presburg (n.s.) 2: 53 (1874),
as 'pseudoorbiculare'. *Phyteuma orbiculare* var. *pseudorbiculare* (Pant.) Nyman, Consp.
Fl. Eur.: 484 (1879).

Phyteuma austriacum Beck, Verh. K. K. Zool.-Bot. Ges. Wien 32: 179 (1883). *Phyteuma
orbiculare* subsp. *austriacum* (Beck) Nyman, Consp. Fl. Eur., Suppl. 2: 211 (1889).

Phyteuma orbiculare var. *angustifolium* St.-Lag. in Cariot, Étude Fl. (ed. 8) 2: 543 (1889).

Phyteuma austriacum var. *vestitum* Murr, Oesterr. Bot. Z. 47: 250 (1897). *Phyteuma
delphinense* var. *vestitum* (Murr) Dalla Torre & Sarnth., Fl. Tirol 6(3): 469 (1911).

Rapunculus sylvestris Tragi ex Bubani, Fl. Pyren. 2: 24 (1899).

Phyteuma orbiculare subsp. *pratense* Rich. Schulz, Monogr. Phyteuma: 113 (1904).
Phyteuma orbiculare var. *pratense* (Rich. Schulz) Hayek & Hegi in Hegi, Ill. Fl. Mitt.-
Eur. 6: 379 (1916).

Phyteuma orbiculare var. *patens* Rich. Schulz, Monogr. Phyteuma: 113 (1904).

Phyteuma orbiculare f. *glabrescens* Rich. Schulz, Monogr. Phyteuma: 113 (1904).

Phyteuma orbiculare f. *hirsutum* Rich. Schulz, Monogr. Phyteuma: 113 (1904).

Phyteuma orbiculare subsp. *montanum* Rich. Schulz, Monogr. Phyteuma: 113 (1904).
  Phyteuma montanum (Rich. Schulz) Dalla Torre & Sarnth., Fl. Tirol 6(3): 468 (1911);
  non C. K. Spreng., Entd. Geheimn. Nat.: 115 (1793). *Phyteuma orbiculare* var.
  montanum (Rich. Schulz) Hayek & Hegi in Hegi, Ill. Fl. Mitt.-Eur. 6: 379 (1916).

Phyteuma orbiculare var. *suffultum* Rich. Schulz, Monogr. Phyteuma: 114 (1904).
  Phyteuma montanum var. *suffultum* (Rich. Schulz) Dalla Torre & Sarnth., Fl. Tirol 6(3):
  468 (1911).

Phyteuma orbiculare f. *glabriusculum* Rich. Schulz, Monogr. Phyteuma: 114 (1904).

Phyteuma orbiculare var. *exinvolucratum* Rich. Schulz, Monogr. Phyteuma: 114 (1904).
  Phyteuma montanum var. *exinvolucratum* (Rich. Schulz) Dalla Torre & Sarnth., Fl. Tirol
  6(3): 468 (1911).

Phyteuma orbiculare f. *glabrum* Rich. Schulz, Monogr. Phyteuma: 114 (1904).

Phyteuma orbiculare f. *pilosiusculum* Rich. Schulz, Monogr. Phyteuma: 114 (1904).

Phyteuma orbiculare var. *vulgare* Rich. Schulz, Monogr. Phyteuma: 114 (1904).

Phyteuma orbiculare f. *minus* Rich. Schulz, Monogr. Phyteuma: 114 (1904).

Phyteuma orbiculare f. *majus* Rich. Schulz, Monogr. Phyteuma: 114 (1904).

Phyteuma orbiculare subsp. *delphinense* Rich. Schulz, Monogr. Phyteuma: 114 (1904).
  Phyteuma delphinense (Rich. Schulz) Dalla Torre & Sarnth., Fl. Tirol 6(3): 469 (1911).

Phyteuma orbiculare f. *glabratum* Rich. Schulz, Monogr. Phyteuma: 114 (1904).

Phyteuma orbiculare f. *hispidulum* Rich. Schulz, Monogr. Phyteuma: 114 (1904).

Phyteuma orbiculare f. *nanum* Rich. Schulz, Monogr. Phyteuma: 114 (1904).

Phyteuma orbiculare f. *alpestre* Rich. Schulz, Monogr. Phyteuma: 115 (1904). *Phyteuma
  delphinense* f. *alpestre* (Rich. Schulz) Dalla Torre & Sarnth., Fl. Tirol 6(3): 469 (1911),
  as 'alpestris'.

Phyteuma orbiculare f. *hispidum* Hegetschw. ex Rich. Schulz, Monogr. Phyteuma: 115
  (1904). *Phyteuma delphinense* f. *hispidum* (Hegetschw. ex Rich. Schulz) Dalla Torre &
  Sarnth., Fl. Tirol 6(3): 469 (1911), as 'hispida'.

Phyteuma orbiculare f. *stellulatum* Rich. Schulz, Monogr. Phyteuma: 115 (1904)

Phyteuma orbiculare subsp. *depauperatum* Rich. Schulz, Monogr. Phyteuma: 115 (1904).

Phyteuma orbiculare var. *liguricum* Rich. Schulz, Monogr. Phyteuma: 115 (1904).

Phyteuma orbiculare f. *nudum* Rich. Schulz, Monogr. Phyteuma: 115 (1904).

Phyteuma orbiculare f. *pubescens* Rich. Schulz, Monogr. Phyteuma: 115 (1904).

Phyteuma orbiculare f. *humile* Rich. Schulz, Monogr. Phyteuma: 115 (1904).

Phyteuma orbiculare subsp. *flexuosum* Rich. Schulz, Monogr. Phyteuma: 115 (1904).

Phyteuma orbiculare var. *carpaticum* Rich. Schulz, Monogr. Phyteuma: 115 (1904).

Phyteuma orbiculare var. *hungaricum* Rich. Schulz, Monogr. Phyteuma: 115 (1904).

Phyteuma tenerum Rich. Schulz, Monogr. Phyteuma: 122 (1904). *Phyteuma orbiculare*
  subsp. *tenerum* (Rich. Schulz) P. Fourn., Quatre Fl. France: 918 (1939).

Phyteuma tenerum subsp. *anglicum* Rich. Schulz, Monogr. Phyteuma: 124 (1904).
  Phyteuma orbiculare subsp. *anglicum* (Rich. Schulz) P. Fourn., Quatre Fl. France:
  918 (1939).

Phyteuma tenerum var. *tenerrimum* Rich. Schulz, Monogr. Phyteuma: 124 (1904).

Phyteuma tenerum subvar. *brevifolium* Rich. Schulz, Monogr. Phyteuma: 125 (1904).

Phyteuma tenerum f. *glabrum* Rich. Schulz, Monogr. Phyteuma: 125 (1904).

Phyteuma tenerum f. *hirsutum* Rich. Schulz, Monogr. Phyteuma: 125 (1904).

Phyteuma tenerum subvar. *longifolium* Rich. Schulz, Monogr. Phyteuma: 125 (1904).

Phyteuma tenerum f. *glabrescens* Rich. Schulz, Monogr. Phyteuma: 125 (1904).

Phyteuma tenerum f. *pilosum* Rich. Schulz, Monogr. Phyteuma: 125 (1904).

Phyteuma tenerum var. *ellipticum* Rich. Schulz, Monogr. Phyteuma: 125 (1904).

Phyteuma tenerum var. *anomalum* Rich. Schulz, Monogr. Phyteuma: 125 (1904).

Phyteuma tenerum subsp. *ibericum* Rich. Schulz, Monogr. Phyteuma: 125 (1904).
  Phyteuma orbiculare subsp. *ibericum* (Rich. Schulz) P. Fourn., Quatre Fl. France:
  918 (1939).

Phyteuma tenerum var. *microphyllum* Rich. Schulz, Monogr. Phyteuma: 125 (1904).

Phyteuma tenerum var. *macrophyllum* Rich. Schulz, Monogr. Phyteuma: 125 (1904).
Phyteuma hispanicum Rich. Schulz, Monogr. Phyteuma: 127 (1904).
Phyteuma pseudorbiculare f. *angustifolium* Rich. Schulz, Monogr. Phyteuma: 132 (1904).
Phyteuma scorzonerifolium var. *eynense* Sennen, Bol. Soc. Ibér. Ci. Nat. 27: 173 (1929).
  Phyteuma eynense (Sennen) Sennen, Diagn. Nouv.: 22 (1936).
Phyteuma sallei Sennen & Elias, Bol. Soc. Ibér. Ci. Nat. 28: 172 (1930).
Phyteuma orbiculare [unranked] *crucis* Sennen, Diagn. Nouv.: 32 (1936).
Phyteuma orbiculare var. *ciliata* Gajiç, Glasn. Prir. Mus. Beogradu, Ser. B, Biol. 28: 62 (1973).

Phyteuma ovatum Honck., Verz. Gew. Teutsch.: 653 (1782), as 'ovata'.
Pyrénées to Slovenia. 11 AUT GER SWI 12 FRA 13 ITA YUG. Hemicr. 2*n* = 22, 26.

subsp. **ovatum**
Pyrénées to Slovenia. 11 AUT GER SWI 12 FRA 13 ITA YUG. Hemicr. 2*n* = 22, 26.
Phyteuma halleri All., Fl. Pedem. 1: 116 (1785), as 'hallerii'. *Phyteuma urticifolium* Clairv.,
Man. Herbor. Suisse: 63 (1811), as 'urticaefolium'. *Phyteuma spicatum* var. *alpestre*
Godr., Fl. Lorraine 2: 91 (1843). *Phyteuma spicatum* race *alpestre* (Godr.) Rouy, Fl.
France 10: 85 (1908). *Phyteuma spicatum* subsp. *alpestre* (Godr.) Kerguélen, Index
Synonym. Fl. France: xv (1993).
Phyteuma ovale Hoppe, Bot. Taschenb. 1794: 83 (1794). *Phyteuma spicatum* var.
rapunculus Pers., Syn. Pl. 1: 194 (1805).
Phyteuma halleri var. *caerulescens* Bonnet, Bull. Soc. Bot. France 27: xiii (1880). *Phyteuma*
halleri subvar. *caerulescens* (Bonnet) Rouy, Fl. France 10: 84 (1908).
Phyteuma halleri var. *pseudonigrum* Murr, Deutsche Bot. Monatsschr. 17: 151 (1899).
Phyteuma halleri f. *longibracteatum* Rich. Schulz, Monogr. Phyteuma: 74 (1904).
Phyteuma halleri f. *brevibracteatum* Rich. Schulz, Monogr. Phyteuma: 74 (1904).
Phyteuma halleri f. *pilosum* Rich. Schulz, Monogr. Phyteuma: 74 (1904).
Phyteuma halleri f. *umbrosum* Rich. Schulz, Monogr. Phyteuma: 74 (1904).
Phyteuma halleri var. *cordifolium* Rich. Schulz, Monogr. Phyteuma: 74 (1904).
Phyteuma halleri f. *macrophyllum* Rich. Schulz, Monogr. Phyteuma: 74 (1904).
Phyteuma halleri f. *microphyllum* Rich. Schulz, Monogr. Phyteuma: 74 (1904).
Phyteuma halleri f. *pubescens* Rich. Schulz, Monogr. Phyteuma: 74 (1904).
Phyteuma halleri f. *silvaticum* Rich. Schulz, Monogr. Phyteuma: 74 (1904).
Phyteuma halleri var. *coeruleum* Rich. Schulz, Monogr. Phyteuma: 75 (1904).
Phyteuma halleri var. *glabriflora* Rouy, Fl. France 10: 84 (1908).

subsp. **pseudospicatum** Pign., Giorn. Bot. Ital. 111: 54 (1977).
N. Italy. 13 ITA. Hemicr.

Phyteuma × pyrenaeum Sennen, Bull. Soc. Bot. France 74: 386 (1927), pro sp. P. ovatum ×
P. spicatum
Pyrénées. 12 FRA. Hemicr.

Phyteuma rupicola Braun-Blanq., Commun. Stat. Int. Géobot. Médit. Montpellier 87: 231
(1945). *Phyteuma globulariifolium* subsp. *rupicola* (Braun-Blanquet) O. Bolòs & Vigo,
Collect. Bot. (Barcelona) 14: 102 (1983).
Pyrénées. 12 FRA. Hemicr.

Phyteuma scheuchzeri All., Auct. Syn. Stirp. Horti Taur.: 11 (1773). *Phyteuma ovatum* Lam.,
Tabl. Encycl. 2: 66 (1796), as 'ovata;' non Honck., Verz. Gew. Teutsch.: 653 (1782), as
'ovata;' nec F. W. Schmidt, Fl. Boëm. 2: 87 (1794). *Phyteuma corniculatum* Clairv., Man.
Herbor. Suisse: 63 (1811).
S. Alps & N. Appennines; naturalized in Great Britain. (10) grb 11 SWI 12 FRA 13 ITA
YUG. Hemicr. 2*n* = 26, 36.

subsp. **columnae** (Thomas) Bech., Vierteljahrsschr. Naturf. Ges. Zürich 68: 471 (1923).
S. Alps & N. Appennines. 11 SWI 13 ITA YUG. Hemicr.

Phyteuma columnae Thomas, Cat. Pl. Suiss.: 22 (1818). *Phyteuma corniculatum* subsp. *columnae* (Thomas) Gaudin, Fl. Helv. 2: 178 (1828).

Phyteuma charmelioides Biroli, Mem. Reale Accad. Sci. Torino 24: 577 (1820). *Phyteuma corniculatum* subsp. *charmelioides* (Biroli) Rich. Schulz, Monogr. Phyteuma: 137 (1904). *Phyteuma scheuchzeri* subsp. *charmelioides* (Biroli) Hayek in Hegi, Ill. Fl. Mitt.-Eur. 6: 381 (1916).

Phyteuma scheuchzeri var. *serratum* W. D. J. Koch, Syn. Fl. Germ. Helv.: 534 (1837). *Phyteuma corniculatum* var. *serratum* (W. D. J. Koch) Rich. Schulz, Monogr. Phyteuma: 139 (1904). *Phyteuma charmelii* var. *serratum* (W. D. J. Koch) Rouy, Fl. France 10: 88 (1908).

Phyteuma corniculatum var. *petraeum* Rich. Schulz, Monogr. Phyteuma: 137 (1904).

subsp. **scheuchzeri**

S. Alps; naturalized in Great Britain. (10) grb 11 SWI 12 FRA 13 ITA. Hemicr. 2*n* = 26, 36.

Phyteuma scheuchzeri var. *michelii* Schult. in Roem. & Schult., Syst. Veg. 5: 77 (1819), as 'micheli'.

Phyteuma humile Roth, Enum. Pl. Phaen. Germ. 1: 696 (1827); non Schleich. ex Gaudin in Murith, Guide Bot. Voy. Valais: 84 (1810), as 'humilis'.

Phyteuma corniculatum var. *angustifolium* Gaudin, Fl. Helv. 2: 177 (1828).

Phyteuma scheuchzeri var. *leucanthum* Schur, Enum. Pl. Transsilv.: 429 (1866).

Phyteuma corniculatum var. *vulgare* Rich. Schulz, Monogr. Phyteuma: 137 (1904).

Phyteuma scorzonerifolium Vill., Hist. Pl. Dauphiné 2: 519 (1787), as 'scorzonerifolia'. *Phyteuma scheuchzeri* var. *scorzonerifolium* (Vill.) Pers., Syn. Pl. 1: 193 (1805). *Phyteuma michelii* subsp. *scorzonerifolium* (Vill.) Arcang., Comp. Fl. Ital.: 449 (1882), as 'scorzoneraefolium'. *Phyteuma michelii* var. *scorzonerifolium* (Vill.) Nyman, Consp. Fl. Eur.: 484 (1879).

S. Alps & N. Appennines. 11 SWI 12 FRA 13 ITA. Hemicr.

Phyteuma scorzonerifolium subsp. *laxiflorum* Beyer, Jahreskat. Wiener Bot. Tauschanst. ?? (1898). *Phyteuma scorzonerifolium* f. *laxiflorum* (Beyer) Rich. Schulz, Monogr. Phyteuma: 101 (1904).

Phyteuma scorzonerifolium f. *albiflorum* Rich. Schulz, Monogr. Phyteuma: 101 (1904).

Phyteuma scorzonerifolium var. *tenue* Rich. Schulz, Monogr. Phyteuma: 101 (1904).

Phyteuma michelii f. *umbrofilum* Fiori, Nuov. Giorn. Bot. Ital. (n.s.) 21: 74 (1914).

Phyteuma serratum Viv., Fl. Cors. Prodr., Appendix: 1 (1825).

Corse. 12 COR. Hemicr. 2*n* = 26, 28.

Phyteuma serratum var. *nanum* Rich. Schulz, Monogr. Phyteuma: 146 (1904).

Phyteuma sieberi Spreng., Pl. Min. Cogn. Pug. 1: 15 (1813), as 'sieberii'.

N. Italy to Slovenia. 11 AUT 13 ITA YUG. Hemicr. 2*n* = 20.

Phyteuma sieberi f. *glabrum* Rich. Schulz, Monogr. Phyteuma: 135 (1904).

Phyteuma sieberi f. *pilosum* Rich. Schulz, Monogr. Phyteuma: 135 (1904).

Phyteuma sieberi var. *alpinum* Rich. Schulz, Monogr. Phyteuma: 135 (1904).

Phyteuma sieberi f. *glabrescens* Rich. Schulz, Monogr. Phyteuma: 135 (1904).

Phyteuma sieberi f. *pubescens* Rich. Schulz, Monogr. Phyteuma: 135 (1904).

Phyteuma sieberi f. *pygmaeum* Bolzon, Nuov. Giorn. Bot. Ital. (n.s.) 21: 154 (1914).

Phyteuma spicatum L., Sp. Pl.: 171 (1753), as 'spicata'. *Rapunculus spicatus* (L.) Mill., Gard. Dict. (ed. 8) (1768).

Great Britain to S. Norway, Estonia, Ukraine, Montenegro & N. Spain; naturalized in Finland & Sweden. 10 GRB DEN fin NOR swe 11 AUT BGM CZE GER HUN NET POL SWI 12 FRA SPA 13 ALB ITA ROM YUG 14 BLR BLT UKR. Hemicr. 2*n* = 22, 36.

Phyteuma rapunculus Pers., Syst. Veg. (ed. 15): 220 (1797).

Phyteuma spicatum var. *caeruleum* Hegetschw., Reis. Glarus Graubünden: 149 (1825), as 'coeruleum'.

Phyteuma spicatum var. *bracteatum* A. DC., Monogr. Campan.: 198 (1830).

Phyteuma elongatum Hegetschw., Fl. Schweiz: 227 (1838-39).

Phyteuma spicatum var. *halleri* Steud., Nomencl. Bot. (ed. 2) 2: 331 (1841).

Phyteuma angustifolium Ledeb., Fl. Ross. 2: 874 (1846).

Phyteuma spicatum var. *caeruleum* Gren. & Godr., Fl. Fr. 2: 403 (1853); non Hegetschw., Reis. Glarus Graubünden: 149 (1825), as 'coeruleum'. *Phyteuma spicatum* subvar. *caeruleum* Rouy, Fl. France 10: 85 (1908).

Phyteuma spicatum var. *caeruleum* Gremli, Excursionsfl. Schweiz (ed. 4): 291 (1881), as 'coeruleum;' non Hegetschw., Reis. Glarus Graubünden: 149 (1825), as 'coeruleum;' nec Gren. & Godr., Fl. Fr. 2: 403 (1853). *Phyteuma caeruleum* Dalla Torre & Sarnth., Fl. Tirol. 6(3): 461 (1911), as 'coeruleum'.

Rapunculus ovatus Bubani, Fl. Pyren. 2: 26 (1899).

Phyteuma spicatum subvar. *longibracteatum* Rich. Schulz, Monogr. Phyteuma: 67 (1904).

Phyteuma spicatum f. *crenatum* Rich. Schulz, Monogr. Phyteuma: 67 (1904).

Phyteuma spicatum f. *grossidentatum* Rich. Schulz, Monogr. Phyteuma: 67 (1904).

Phyteuma spicatum subvar. *brevibracteatum* Rich. Schulz, Monogr. Phyteuma: 67 (1904).

Phyteuma spicatum f. *bicrenatum* Rich. Schulz, Monogr. Phyteuma: 67 (1904).

Phyteuma spicatum f. *macrodon* Rich. Schulz, Monogr. Phyteuma: 67 (1904).

Phyteuma spicatum subvar. *macrophyllum* Rich. Schulz, Monogr. Phyteuma: 67 (1904).

Phyteuma spicatum f. *microdon* Rich. Schulz, Monogr. Phyteuma: 67 (1904).

Phyteuma spicatum f. *incisum* Rich. Schulz, Monogr. Phyteuma: 67 (1904).

Phyteuma spicatum subvar. *microphyllum* Rich. Schulz, Monogr. Phyteuma: 67 (1904).

Phyteuma spicatum f. *crenatoserratum* Rich. Schulz, Monogr. Phyteuma: 67 (1904), as 'crenato-serratum'.

Phyteuma spicatum f. *fissum* Rich. Schulz, Monogr. Phyteuma: 67 (1904).

Phyteuma spicatum subsp. *jurassicum* Rich. Schulz, Monogr. Phyteuma: 67 (1904). *Phyteuma spicatum* var. *jurassicum* (Rich. Schulz) Hayek & Hegi in Hegi, Ill. Fl. Mitt.-Eur. 6: 373 (1916).

Phyteuma spicatum var. *bracteatum* Rich. Schulz, Monogr. Phyteuma: 67 (1904); non A. DC., Monogr. Campan.: 198 (1830).

Phyteuma spicatum var. *vulgare* Rich. Schulz, Monogr. Phyteuma: 67 (1904).

Phyteuma spicatum subsp. *occidentale* Rich. Schulz, Monogr. Phyteuma: 67 (1904).

Phyteuma spicatum var. *glabrum* Rich. Schulz, Monogr. Phyteuma: 68 (1904).

Phyteuma spicatum var. *pilosum* Rich. Schulz, Monogr. Phyteuma: 68 (1904).

Phyteuma spicatum subsp. *caeruleum* Rich. Schulz, Monogr. Phyteuma: 68 (1904), as 'coeruleum'.

Phyteuma spicatum var. *alpinum* Rich. Schulz, Monogr. Phyteuma: 68 (1904).

Phyteuma spicatum f. *involucratum* Rich. Schulz, Monogr. Phyteuma: 68 (1904).

Phyteuma spicatum f. *ebracteatum* Rich. Schulz, Monogr. Phyteuma: 68 (1904).

Phyteuma spicatum f. *cordatum* Rich. Schulz, Monogr. Phyteuma: 68 (1904).

Phyteuma spicatum f. *divaricatum* Rich. Schulz, Monogr. Phyteuma: 68 (1904).

Phyteuma pyrenaicum Rich. Schulz, Monogr. Phyteuma: 79 (1904). *Phyteuma spicatum* var. *pyrenaicum* (Rich. Schulz) O. Bolòs & Vigo, Collect. Bot. (Barcelona) 14: 102 (1983).

Phyteuma pyrenaicum subsp. *cordifolium* Rich. Schulz, Monogr. Phyteuma: 82 (1904).

Phyteuma pyrenaicum var. *involucratum* Rich. Schulz, Monogr. Phyteuma: 82 (1904).

Phyteuma pyrenaicum f. *nudum* Rich. Schulz, Monogr. Phyteuma: 82 (1904).

Phyteuma pyrenaicum f. *pilosum* Rich. Schulz, Monogr. Phyteuma: 82 (1904).

Phyteuma pyrenaicum var. *brevibracteatum* Rich. Schulz, Monogr. Phyteuma: 82 (1904).

Phyteuma pyrenaicum f. *glabrum* Rich. Schulz, Monogr. Phyteuma: 82 (1904).

Phyteuma pyrenaicum f. *pubescens* Rich. Schulz, Monogr. Phyteuma: 82 (1904).

Phyteuma pyrenaicum subsp. *betonicoides* Rich. Schulz, Monogr. Phyteuma: 82 (1904).

Phyteuma pyrenaicum var. *bracteatum* Rich. Schulz, Monogr. Phyteuma: 82 (1904).

Phyteuma pyrenaicum f. *glabrescens* Rich. Schulz, Monogr. Phyteuma: 82 (1904).

Phyteuma pyrenaicum f. *pilosiusculum* Rich. Schulz, Monogr. Phyteuma: 82 (1904).

Phyteuma pyrenaicum var. *ebracteatum* Rich. Schulz, Monogr. Phyteuma: 82 (1904).

Phyteuma pyrenaicum f. *glabriusculum* Rich. Schulz, Monogr. Phyteuma: 82 (1904).

Phyteuma pyrenaicum f. *hirsutum* Rich. Schulz, Monogr. Phyteuma: 82 (1904).
Phyteuma spicatum subsp. *ambigens* Rouy, Fl. France 10: 85 (1908).
Phyteuma bracteatum Losa, Bol. Soc. Ibér. Ci. Nat. 29: 98 (1931).
Phyteuma abelis Sennen, Diagn. Nouv.: 22 (1936).
Phyteuma spicatum var. *roseum* Degen & P. Rossi in Degen, Fl. Veleb. 3: 107 (1938).
Phyteuma spicatum var. *pseudohalleri* Font Quer ex O. Bolòs & Vigo, Collect. Bot. (Barcelona) 14: 102 (1983).

Phyteuma tetramerum Schur, Sert. Fl. Transsilv.: 47 (1853), as 'tetramerium'. *Phyteuma spicatum* var. *tetramerum* (Schur) Nyman, Consp. Fl. Eur.: 484 (1879), as 'tetramerium'.
E. & C. Carpathians. 13 ROM 14 UKR. Hemicr.

Phyteuma vagneri A. Kern., Sched. Fl. Exs. Austro-Hung. 3: 107 (1884). *Phyteuma atropurpureum* Schur, Verh. Mitth. Siebenbürg. Vereins Naturwiss. Hermannstadt 3: 88 (1852); non Hoppe, Bot. Taschenb. 1802: 27 (1802). *Phyteuma nigrum* var. *atropurpureum* Schur, Enum. Pl. Transsilv.: 430 (1866).
E. & S. Carpathians. 13 ROM 14 UKR. Hemicr. $2n = 24$.
Phyteuma vagneri f. *brevibracteatum* Rich. Schulz, Monogr. Phyteuma: 78 (1904).
Phyteuma vagneri f. *grossidentatum* Rich. Schulz, Monogr. Phyteuma: 78 (1904).
Phyteuma vagneri f. *latibracteatum* Rich. Schulz, Monogr. Phyteuma: 78 (1904).
Phyteuma vagneri f. *alpinum* Rich. Schulz, Monogr. Phyteuma: 79 (1904).

Phyteuma zahlbruckneri Vest, Steiermärk. Z. 3: 159 (1821). *Phyteuma betonicifolium* subsp. *zahlbruckneri* (Vest) Hayek, Repert. Spec. Nov. Regni Veg. Beih. 30(2): 558 (1930).
N. Italy to Slovenia. 11 AUT 13 ITA YUG. Hemicr. $2n = 24$.
Phyteuma persicifolium Hoppe ex A. DC., Monogr. Campan.: 196 (1830), as 'persicaefolium'. *Phyteuma betonicifolium* var. *persicifolium* (Hoppe ex A. DC.) Steud., Nomencl. Bot. (ed. 2) 2: 330 (1841). *Phyteuma michelii* var. *persicifolium* (Hoppe ex A. DC.) Nyman, Consp. Fl. Eur.: 484 (1879). *Phyteuma scorzonerifolium* var. *persicifolium* (Hoppe ex A. DC.) St.-Lag. in Cariot, Étude Fl. (ed. 8) 2: 542 (1889).
Phyteuma persicifolium f. *carniolicum* Rich. Schulz, Monogr. Phyteuma: 106 (1904).
Phyteuma persicifolium var. *lineare* Rich. Schulz, Monogr. Phyteuma: 106 (1904).
Phyteuma persicifolium var. *latifolium* Rich. Schulz, Monogr. Phyteuma: 106 (1904).
Phyteuma persicifolium var. *angustissimum* Rich. Schulz, Monogr. Phyteuma: 106 (1904).
Phyteuma persicifolium var. *laxifolium* Rich. Schulz, Monogr. Phyteuma: 106 (1904).
Phyteuma persicifolium var. *alpestre* Rich. Schulz, Monogr. Phyteuma: 106 (1904).
Phyteuma persicifolium var. *albiflorum* Rich. Schulz, Monogr. Phyteuma: 106 (1904).

Synonyms:
Phyteuma sect. *Cylindrocarpa* (Regel) Kuntze === **Cylindrocarpa** Regel.
Phyteuma sect. *Petromarula* (Pers.) Kuntze === **Petromarula** Vent. ex R. Hedw.
Phyteuma sect. *Podanthum* G. Don === **Asyneuma** Griseb. & Schenk.
Phyteuma sect. *Synotoma* G. Don === **Physoplexis** (Endl.) Schur
Phyteuma [unranked] *Petromarula* Pers. === **Petromarula** Vent. ex R. Hedw.
Phyteuma abelis Sennen === **Phyteuma spicatum** L.
Phyteuma aizoon (Hausskn. ex Bornm.) Hausskn. ex Bornm. === **Asyneuma virgatum** (Labill.) Bornm. subsp. **virgatum**
Phyteuma americanum Hill === **Campanula americana** L.
Phyteuma amplexicaule Willd. === **Asyneuma amplexicaule** (Willd.) Hand.-Mazz.
Phyteuma amplexicaule var. *majus* K. Koch === **Asyneuma amplexicaule** (Willd.) Hand.-Mazz. subsp. **amplexicaule**
Phyteuma amplexicaule var. *minus* K. Koch === **Asyneuma amplexicaule** (Willd.) Hand.-Mazz. subsp. **amplexicaule**
Phyteuma angustatum Wender. === **Phyteuma orbiculare** L.
Phyteuma angustifolium Ledeb. === **Phyteuma spicatum** L.
Phyteuma anthericoides (Janka) Nyman === **Asyneuma anthericoides** (Janka) Bornm.

Phyteuma argutum Regel === **Asyneuma argutum** (Regel) Bornm.

Phyteuma asperum Boiss. === **Asyneuma persicum** (A. DC.) Bornm.

Phyteuma atropurpureum Hoppe === **Phyteuma nigrum** F. W. Schmidt

Phyteuma atropurpureum Schur === **Phyteuma vagneri** A. Kern.

Phyteuma attenuatum Franch. === **Asyneuma argutum** (Regel) Bornm. subsp. **argutum**

Phyteuma aucheri A. DC. === **Asyneuma amplexicaule** subsp. **aucheri** (A. DC.) Bornm.

Phyteuma austriacum Beck === **Phyteuma orbiculare** L.

Phyteuma austriacum var. *vestitum* Murr === **Phyteuma orbiculare** L.

Phyteuma balbisii A. DC. === **Phyteuma cordatum** Balb.

Phyteuma balbisii var. *petraeum* A. DC. === **Phyteuma cordatum** Balb.

Phyteuma barrelieri Vill. === ?

Phyteuma begoniifolium Roxb. ex Jack === **Pentaphragma begoniifolium** (Roxb. ex Jack) G. Don (Pentaphragmataceae)

Phyteuma betonicifolium f. *alpestre* Rich. Schulz === **Phyteuma betonicifolium** Vill. subsp. **betonicifolium**

Phyteuma betonicifolium f. *glabrum* Rich. Schulz === **Phyteuma betonicifolium** Vill. subsp. **betonicifolium**

Phyteuma betonicifolium f. *pubescens* (A. DC.) Rich. Schulz === **Phyteuma betonicifolium** Vill. subsp. **betonicifolium**

Phyteuma betonicifolium f. *rhaeticum* Rich. Schulz=== **Phyteuma betonicifolium** Vill. subsp. **betonicifolium**

Phyteuma betonicifolium f. *vulgare* Rich. Schulz === **Phyteuma betonicifolium** Vill. subsp. **betonicifolium**

Phyteuma betonicifolium subsp. *zahlbruckneri* (Vest) Hayek === **Phyteuma zahlbruckneri** Vest

Phyteuma betonicifolium var. *lanceolatum* Rich. Schulz === **Phyteuma betonicifolium** Vill. subsp. **betonicifolium**

Phyteuma betonicifolium var. *persicifolium* (Hoppe ex A. DC.) Steud. === **Phyteuma zahlbruckneri** Vest

Phyteuma betonicifolium var. *pubescens* A. DC. === **Phyteuma betonicifolium** Vill. subsp. **betonicifolium**

Phyteuma betonicifolium var. *sessilifolium* A. DC. === **Phyteuma betonicifolium** Vill. subsp. **betonicifolium**

Phyteuma bipinnata Lour. === **Sambucus ebuloides** Desv. ex DC. (Caprifoliaceae)

Phyteuma bovelinii Hegetschw. === **Phyteuma orbiculare** L.

Phyteuma bracteatum Losa === **Phyteuma spicatum** L.

Phyteuma brevifolium Schleich. === **Phyteuma orbiculare** L.

Phyteuma caeruleum Dalla Torre & Sarnth. === **Phyteuma spicatum** L.

Phyteuma campanuloides M. Bieb. ex Sims === **Asyneuma campanuloides** (M. Bieb. ex Sims) Bornm.

Phyteuma campanuloides var. *amplexifolium* K. Koch === **Asyneuma campanuloides** (M. Bieb. ex Sims) Bornm.

Phyteuma campanuloides var. *sibthorpianum* (Schult.) A. DC. === **Asyneuma rigidum** subsp. **sibthorpianum** (Schult.) Damboldt

Phyteuma canescens Waldst. & Kit. === **Asyneuma canescens** (Waldst. & Kit.) Griseb. & Schenk

Phyteuma canescens f. *laevis* (Rohlena) Rohlena === **Asyneuma canescens** (Waldst. & Kit.) Griseb. & Schenk subsp. **canescens**

Phyteuma canescens subsp. *foliosum* Nyman === **Asyneuma canescens** (Waldst. & Kit.) Griseb. & Schenk subsp. **canescens**

Phyteuma canescens var. *humile* K. Koch === **Asyneuma rigidum** (Willd.) Grossh. subsp. **rigidum**

Phyteuma canescens var. *salicifolium* Steud. === **Asyneuma canescens** (Waldst. & Kit.) Griseb. & Schenk subsp. **canescens**

Phyteuma canescens var. *salignum* (Waldst. & Kit. ex Besser) Nyman === **Asyneuma canescens** (Waldst. & Kit.) Griseb. & Schenk subsp. **canescens**

Phyteuma capensis Burm. f. === ?

Phyteuma capituliforme Rochel === **Phyteuma globulariifolium** Sternb. & Hoppe

Phyteuma cappadocicum Boiss. === **Asyneuma virgatum** (Labill.) Bornm. subsp. **virgatum**

Phyteuma carestiae Biroli === **Phyteuma humile** Schleich. ex Gaudin

Phyteuma charmelii f. *microcephala* Sennen === **Phyteuma charmelii** Vill.

Phyteuma charmelii var. *serratum* (W. D. J. Koch) Rouy === **Phyteuma scheuchzeri** subsp. **columnae** (Thomas) Bech.

Phyteuma charmelioides Biroli === **Phyteuma scheuchzeri** subsp. **columnae** (Thomas) Bech.

Phyteuma cichoriiforme Boiss. === **Asyneuma virgatum** subsp. **cichoriiforme** (Boiss.) Damboldt

Phyteuma cochinchinensis Lour. === **Sambucus phyteumoides** DC. (Caprifoliaceae)

Phyteuma collinum Guss. === **Asyneuma limonifolium** (L.) Janch. subsp. **limoniifolium**

Phyteuma columnae Thomas === **Phyteuma scheuchzeri** subsp. **columnae** (Thomas) Bech.

Phyteuma comosum L. === **Physoplexis comosa** (L.) Schur

Phyteuma comosum var. *pubescens* Facchini ex Murr === **Physoplexis comosa** (L.) Schur

Phyteuma cordatum f. *obtusatum* Rich. Schulz === **Phyteuma cordatum** Balb.

Phyteuma cordifolium Vill. === **Phyteuma orbiculare** L.

Phyteuma corniculatum Clairv. === **Phyteuma scheuchzeri** All.

Phyteuma corniculatum subsp. *charmelioides* (Biroli) Rich. Schulz === **Phyteuma scheuchzeri** subsp. **columnae** (Thomas) Bech.

Phyteuma corniculatum subsp. *columnae* (Thomas) Gaudin === **Phyteuma scheuchzeri** subsp. **columnae** (Thomas) Bech.

Phyteuma corniculatum var. *angustifolium* Gaudin === **Phyteuma scheuchzeri** All. subsp. **scheuchzeri**

Phyteuma corniculatum var. *petraeum* Rich. Schulz === **Phyteuma scheuchzeri** subsp. **columnae** (Thomas) Bech.

Phyteuma corniculatum var. *serratum* (W. D. J. Koch) Rich. Schulz === **Phyteuma scheuchzeri** subsp. **columnae** (Thomas) Bech.

Phyteuma corniculatum var. *vulgare* Rich. Schulz === **Phyteuma scheuchzeri** All. subsp. **scheuchzeri**

Phyteuma crispum Pourr. === **Jasione crispa** (Pourr.) Samp.

Phyteuma degenii Gáyer === ?

Phyteuma delphinense (Rich. Schulz) Dalla Torre & Sarnth. === **Phyteuma orbiculare** L.

Phyteuma delphinense f. *alpestre* (Rich. Schulz) Dalla Torre & Sarnth. === **Phyteuma orbiculare** L.

Phyteuma delphinense f. *hispidum* (Hegetschw. ex Rich. Schulz) Dalla Torre & Sarnth. === **Phyteuma orbiculare** L.

Phyteuma delphinense var. *ellipticifolium* (Vill.) Dalla Torre & Sarnth. === **Phyteuma orbiculare** L.

Phyteuma delphinense var. *vestitum* (Murr) Dalla Torre & Sarnth. === **Phyteuma orbiculare** L.

Phyteuma eldivenum (Czeczott) Czeczott === **Asyneuma rigidum** (Willd.) Grossh. subsp. **rigidum**

Phyteuma elegans Hegetschw. === **Phyteuma betonicifolium** subsp. **scaposum** (Rich. Schulz) Pign.

Phyteuma ellipticum Sm. === **Asyneuma rigidum** subsp. **sibthorpianum** (Schult.) Damboldt

Phyteuma ellipticifolium Vill. === **Phyteuma orbiculare** L.

Phyteuma ellipticifolium var. *pauciflorum* Hegetschw. === **Phyteuma orbiculare** L.

Phyteuma elongatum Hegetschw. === **Phyteuma spicatum** L.

Phyteuma eynense (Sennen) Sennen === **Phyteuma orbiculare** L.

Phyteuma fistulosum Rchb. === **Phyteuma orbiculare** L.

Phyteuma foliosum Bellardi ex Colla === ?

Phyteuma foliosum Kit. === **Asyneuma canescens** (Waldst. & Kit.) Griseb. & Schenk subsp. **canescens**

Phyteuma gaussenii Chouard === **Phyteuma hemisphaericum** L.

Phyteuma giganteum (Boiss.) Nyman === **Asyneuma giganteum** (Boiss.) Bornm.

Phyteuma globulariifolium f. *integrum* Rich. Schulz === **Phyteuma globulariifolium** Sternb. & Hoppe

Phyteuma globulariifolium f. *simplex* Rich. Schulz === **Phyteuma globulariifolium** Sternb. & Hoppe

Phyteuma globulariifolium f. *vulgare* Rich. Schulz === **Phyteuma globulariifolium** Sternb. & Hoppe

Phyteuma globulariifolium subsp. *rupicola* (Braun-Blanquet) O. Bolòs & Vigo === **Phyteuma rupicola** Braun-Blanq.

Phyteuma globulariifolium var. *acutifolium* Rich. Schulz === **Phyteuma globulariifolium** Sternb. & Hoppe

Phyteuma globulariifolium var. *nanum* Rich. Schulz === **Phyteuma globulariifolium** Sternb. & Hoppe

Phyteuma globulariifolium var. *tirolense* Rich. Schulz === **Phyteuma globulariifolium** Sternb. & Hoppe

Phyteuma gracile Boiss. & Heldr. === **Asyneuma lobelioides** (Willd.) Hand.-Mazz.

Phyteuma graminifolium Sieber === **Phyteuma hemisphaericum** L.

Phyteuma grossheimii Karjagin ex Grossh. === **Asyneuma rigidum** (Willd.) Grossh. subsp. **rididum**

Phyteuma halleri All. === **Phyteuma ovatum** Honck. subsp. **ovatum**

Phyteuma halleri f. *brevibracteatum* Rich. Schulz=== **Phyteuma ovatum** Honck. subsp. **ovatum**

Phyteuma halleri f. *longibracteatum* Rich. Schulz=== **Phyteuma ovatum** Honck. subsp. **ovatum**

Phyteuma halleri f. *macrophyllum* Rich. Schulz === **Phyteuma ovatum** Honck. subsp. **ovatum**

Phyteuma halleri f. *microphyllum* Rich. Schulz === **Phyteuma ovatum** Honck. subsp. **ovatum**

Phyteuma halleri f. *pilosum* Rich. Schulz === **Phyteuma ovatum** Honck. subsp. **ovatum**

Phyteuma halleri f. *pubescens* Rich. Schulz === **Phyteuma ovatum** Honck. subsp. **ovatum**

Phyteuma halleri f. *silvaticum* Rich. Schulz === **Phyteuma ovatum** Honck. subsp. **ovatum**

Phyteuma halleri f. *umbrosum* Rich. Schulz=== **Phyteuma ovatum** Honck. subsp. **ovatum**

Phyteuma halleri subsp. *nigrum* (F. W. Schmidt) Arcang. === **Phyteuma nigrum** F. W. Schmidt

Phyteuma halleri subvar. *caerulescens* (Bonnet) Rouy === **Phyteuma ovatum** Honck. subsp. **ovatum**

Phyteuma halleri var. *caerulescens* Bonnet === **Phyteuma ovatum** Honck. subsp. **ovatum**

Phyteuma halleri var. *coeruleum* Rich. Schulz === **Phyteuma ovatum** Honck. subsp. **ovatum**

Phyteuma halleri var. *cordifolium* Rich. Schulz === **Phyteuma ovatum** Honck. subsp. **ovatum**

Phyteuma halleri var. *glabriflora* Rouy === **Phyteuma ovatum** Honck. subsp. **ovatum**

Phyteuma halleri var. *pseudonigrum* Murr === **Phyteuma ovatum** Honck. subsp. **ovatum**

Phyteuma hedraianthifolium f. *graminifolium* Rich. Schulz === **Phyteuma hedraianthifolium** Rich. Schulz

Phyteuma × *hellwegeri* Murr === **Phyteuma** × **huteri** Murr

Phyteuma hemisphaericum f. *albiflorum* Rich. Schulz === **Phyteuma hemisphaericum** L.

Phyteuma hemisphaericum f. *vulgare* Rich. Schulz === **Phyteuma hemisphaericum** L.

Phyteuma hemisphaericum subsp. *confusum* (A. Kern.) Nyman === **Phyteuma confusum** A. Kern.

Phyteuma hemisphaericum var. *albiflorum* Schur === **Phyteuma hemisphaericum** L.

Phyteuma hemisphaericum var. *carinthiacum* Rich. Schulz === **Phyteuma hemisphaericum** L.

Phyteuma hemisphaericum var. *graminifolium* (Sieber) Schur === **Phyteuma hemisphaericum** L.

Phyteuma hemisphaericum var. *hedraianthoides* Chouard === **Phyteuma hemisphaericum** L.

Phyteuma hemisphaericum var. *latifolium* Schur === **Phyteuma hemisphaericum** L.

Phyteuma hemisphaericum var. *platyphyllum* Rich. Schulz === **Phyteuma hemisphaericum** L.

Phyteuma hemisphaericum var. *setaceum* Hegetschw. === **Phyteuma hemisphaericum** L.

Phyteuma hemisphaericum var. *subacaule* Rouy === **Phyteuma hemisphaericum** L.

Phyteuma hemisphaericum var. *transsilvanicum* Schur === **Phyteuma globulariifolium** Sternb. & Hoppe

Phyteuma hispanicum Rich. Schulz === **Phyteuma orbiculare** L.

Phyteuma hispidum Hegetschw. === **Phyteuma orbiculare** L.

Phyteuma humile Roth === **Phyteuma scheuchzeri** All. subsp. **scheuchzeri**

Phyteuma humile var. *humillimum* A. DC. === **Phyteuma humile** Schleich. ex Gaudin

Phyteuma inaequatum Kit. ex Schult. === **Phyteuma orbiculare** L.

Phyteuma intermedium Hegetschw. === **Phyteuma hemisphaericum** L.

Phyteuma jacquinii Sieber === **Campanula jacquinii** (Sieber) A. DC.

Phyteuma japonicum Miq. === **Asyneuma japonicum** (Miq.) Briq.

Phyteuma × *khekii* Murr === **Phyteuma** × **huteri** Murr

Phyteuma kotschyi Boiss. === **Asyneuma amplexicaule** (Willd.) Hand.-Mazz. subsp. **amplexicaule**

Phyteuma kotschyi var. *obtusicrenum* Boiss. === **Asyneuma amplexicaule** (Willd.) Hand.-Mazz. subsp. **amplexicaule**

Phyteuma lanceolatum Desf.=== **Asyneuma lobelioides** (Willd.) Hand.-Mazz.

Phyteuma lanceolatum Vill. === **Phyteuma orbiculare** L.

Phyteuma lanceolatum Willd. === **Asyneuma rigidum** (Willd.) Grossh. subsp. **rigidum**

Phyteuma lanceolatum var. *canescens* (Waldst. & Kit.) Pers. === **Asyneuma canescens** (Waldst. & Kit.) Griseb. & Schenk

Phyteuma latifolium Dalla Torre === **Phyteuma confusum** A. Kern.

Phyteuma leianthum Trautv. === **Asyneuma virgatum** (Labill.) Bornm. subsp. **virgatum**

Phyteuma limoniifolium (L.) Sm. === **Asyneuma limoniifolium** (L.) Janch.

Phyteuma limoniifolium var. *strictum* (Sims) K. Koch === **Asyneuma limonifolium** (L.) Janch. subsp. **limoniifolium**

Phyteuma linearifolium Biroli ex Colla === **Phyteuma humile** Schleich. ex Gaudin

Phyteuma linifolium Boiss. & Heldr. === **Asyneuma linifolium** (Boiss. & Heldr.) Bornm.

Phyteuma lobelioides Willd. === **Asyneuma lobelioides** (Willd.) Hand.-Mazz.

Phyteuma longibracteatum St. Lag. === **Phyteuma charmelii** Vill.

Phyteuma longifolium Hegetschw. === **Phyteuma orbiculare** L.

Phyteuma lycium Boiss. === **Asyneuma lycium** (Boiss.) Bornm.

Phyteuma michelii Hegetschw. === **Phyteuma orbiculare** L.

Phyteuma michelii f. *umbrofilum* Fiori === **Phyteuma scorzonerifolium** Vill.

Phyteuma michelii subsp. *betonicifolium* (Vill.) Nyman === **Phyteuma betonicifolium** Vill.

Phyteuma michelii subsp. *scaposum* (Rich. Schulz) P. Fourn. === **Phyteuma betonicifolium** subsp. **scaposum** (Rich. Schulz) Pign.

Phyteuma michelii subsp. *alpini* Ces ex Arcang. === **Phyteuma cordatum** Balb.

Phyteuma michelii subsp. *scorzonerifolium* (Vill.) Arcang. === **Phyteuma scorzonerifolium** Vill.

Phyteuma michelii var. *involucrata* Schur === ?

Phyteuma michelii var. *persicifolium* (Hoppe ex A. DC.) Nyman === **Phyteuma zahlbruckneri** Vest

Phyteuma michelii var. *scorzonerifolium* (Vill.) Nyman === **Phyteuma scorzonerifolium** Vill.

Phyteuma michelii var. *sessilifolium* (A. DC.) Rouy === **Phyteuma betonicifolium** Vill. subsp. **betonicifolium**

Phyteuma michelii var. *veronicifolium* (Schrad. ex A. DC.) Nyman === **Phyteuma betonicifolium** Vill. subsp. **betonicifolium**

Phyteuma minutum Schult. === **Jasione foliosa** Cav. subsp. **foliosa**

Phyteuma monocephalum (Trautv.) Pavlov === **Cryptocodon monocephalus** (Trautv.) Fed.

Phyteuma montanum C. K. Spreng. === ?

Phyteuma montanum (Rich. Schulz) Dalla Torre & Sarnth. === **Phyteuma orbiculare** L.

Phyteuma montanum var. *exinvolucratum* (Rich. Schulz) Dalla Torre & Sarnth. === **Phyteuma orbiculare** L.

Phyteuma montanum var. *suffultum* (Rich. Schulz) Dalla Torre & Sarnth. === **Phyteuma orbiculare** L.

Phyteuma multicaule Franch. === **Sergia regelii** (Trautv.) Fed.

Phyteuma × *murrianum* Borbás ex Murr === **Phyteuma** × **huteri** Murr

Phyteuma nanum Schur === **Phyteuma globulariifolium** Sternb. & Hoppe

Phyteuma nigrum f. *interruptum* Rich. Schulz === **Phyteuma nigrum** F. W. Schmidt

Phyteuma nigrum f. *latibracteatum* Rich. Schulz === **Phyteuma nigrum** F. W. Schmidt

Phyteuma nigrum f. *longibracteatum* Rich. Schulz === **Phyteuma nigrum** F. W. Schmidt

Phyteuma nigrum f. *vulgare* Rich. Schulz === **Phyteuma nigrum** F. W. Schmidt

Phyteuma nigrum var. *acuminatum* Rich. Schulz === **Phyteuma nigrum** F. W. Schmidt

Phyteuma nigrum var. *atropurpureum* Schur === **Phyteuma vagneri** A. Kern.

Phyteuma nigrum var. *coeruleum* Rich. Schulz === **Phyteuma nigrum** F. W. Schmidt

Phyteuma nigrum var. *integrifolium* Rich. Schulz === **Phyteuma nigrum** F. W. Schmidt
Phyteuma obtusifolium Freyn === ?
Phyteuma obtusifolium (Hausskn. ex Bornm.) Hausskn. ex Bornm. === **Asyneuma virgatum** (Labill.) Bornm. subsp. **virgatum**
Phyteuma occultans Popov & Vved. === **Cryptocodon monocephalus** (Trautv.) Fed.
Phyteuma orbiculare f. *alpestre* Rich. Schulz === **Phyteuma orbiculare** L.
Phyteuma orbiculare f. *glabratum* Rich. Schulz === **Phyteuma orbiculare** L.
Phyteuma orbiculare f. *glabrescens* Rich. Schulz === **Phyteuma orbiculare** L.
Phyteuma orbiculare f. *glabriusculum* Rich. Schulz === **Phyteuma orbiculare** L.
Phyteuma orbiculare f. *glabrum* Rich. Schulz === **Phyteuma orbiculare** L.
Phyteuma orbiculare f. *hirsutum* Rich. Schulz === **Phyteuma orbiculare** L.
Phyteuma orbiculare f. *hispidulum* Rich. Schulz === **Phyteuma orbiculare** L.
Phyteuma orbiculare f. *hispidum* (Hegetschw.) Rich. Schulz === **Phyteuma orbiculare** L.
Phyteuma orbiculare f. *humile* Rich. Schulz === **Phyteuma orbiculare** L.
Phyteuma orbiculare f. *majus* Rich. Schulz === **Phyteuma orbiculare** L.
Phyteuma orbiculare f. *minus* Rich. Schulz === **Phyteuma orbiculare** L.
Phyteuma orbiculare f. *nanum* Rich. Schulz === **Phyteuma orbiculare** L.
Phyteuma orbiculare f. *nudum* Rich. Schulz === **Phyteuma orbiculare** L.
Phyteuma orbiculare f. *pilosiusculum* Rich. Schulz === **Phyteuma orbiculare** L.
Phyteuma orbiculare f. *pilosum* Rich. Schulz === **Phyteuma orbiculare** L.
Phyteuma orbiculare f. *pubescens* Rich. Schulz === **Phyteuma orbiculare** L.
Phyteuma orbiculare f. *stellulatum* Rich. Schulz === **Phyteuma orbiculare** L.
Phyteuma orbiculare subsp. *anglicum* (Rich. Schulz) P. Fourn. === **Phyteuma orbiculare** L.
Phyteuma orbiculare subsp. *austriacum* (Beck) Nyman === **Phyteuma orbiculare** L.
Phyteuma orbiculare subsp. *cinerascens* Gaudin === **Phyteuma orbiculare** L.
Phyteuma orbiculare subsp. *cordatum* Gaudin === **Phyteuma orbiculare** L.
Phyteuma orbiculare subsp. *decipiens* Gaudin === **Phyteuma orbiculare** L.
Phyteuma orbiculare subsp. *delphinense* Rich. Schulz === **Phyteuma orbiculare** L.
Phyteuma orbiculare subsp. *depauperatum* Rich. Schulz === **Phyteuma orbiculare** L.
Phyteuma orbiculare subsp. *ellipticifolium* (Vill.) Arcang. === **Phyteuma orbiculare** L.
Phyteuma orbiculare subsp. *ellipticum* Gaudin === **Phyteuma orbiculare** L.
Phyteuma orbiculare subsp. *fistulosum* (Rchb.) Nyman === **Phyteuma orbiculare** L.
Phyteuma orbiculare subsp. *flexuosum* Rich. Schulz === **Phyteuma orbiculare** L.
Phyteuma orbiculare subsp. *ibericum* (Rich. Schulz) P. Fourn. === **Phyteuma orbiculare** L.
Phyteuma orbiculare subsp. *lanceolatum* (Vill.) Arcang. === **Phyteuma orbiculare** L.
Phyteuma orbiculare subsp. *lancifolium* Gaudin === **Phyteuma orbiculare** L.
Phyteuma orbiculare subsp. *montanum* Rich. Schulz === **Phyteuma orbiculare** L.
Phyteuma orbiculare subsp. *pratense* Rich. Schulz === **Phyteuma orbiculare** L.
Phyteuma orbiculare subsp. *tenerum* (Rich. Schulz) P. Fourn. === **Phyteuma orbiculare** L.
Phyteuma orbiculare subvar. *confusum* (A. Kern.) Nyman === **Phyteuma confusum** A. Kern.
Phyteuma orbiculare [unranked] *crucis* Sennen === **Phyteuma orbiculare** L.
Phyteuma orbiculare var. *alpinum* Schur === **Phyteuma orbiculare** L.
Phyteuma orbiculare var. *angustifolium* St.-Lag. === **Phyteuma orbiculare** L.
Phyteuma orbiculare var. *brevifolium* (Schleich.) Steud. === **Phyteuma orbiculare** L.
Phyteuma orbiculare var. *carpaticum* Rich. Schulz === **Phyteuma orbiculare** L.
Phyteuma orbiculare var. *ciliata* Gajiç === **Phyteuma orbiculare** L.
Phyteuma orbiculare var. *comosum* Steud. === **Phyteuma orbiculare** L.
Phyteuma orbiculare var. *cordatum* Gren. & Godr. === **Phyteuma orbiculare** L.
Phyteuma orbiculare var. *cordifolium* (Vill.) Griseb. === **Phyteuma orbiculare** L.
Phyteuma orbiculare var. *decipiens* (Gaudin) A. DC. === **Phyteuma orbiculare** L.
Phyteuma orbiculare var. *ellipticifolium* (Vill.) Nyman === **Phyteuma orbiculare** L.
Phyteuma orbiculare var. *ellipticum* Pers. === **Phyteuma orbiculare** L.
Phyteuma orbiculare var. *exinvolucratum* Rich. Schulz === **Phyteuma orbiculare** L.
Phyteuma orbiculare var. *fistulosum* (Rchb.) Steud. === **Phyteuma orbiculare** L.
Phyteuma orbiculare var. *giganteum* A. DC. === **Phyteuma orbiculare** L.
Phyteuma orbiculare var. *hungaricum* Rich. Schulz === **Phyteuma orbiculare** L.

Phyteuma orbiculare var. *inaequatum* (Kit. ex Schult.) Nyman === **Phyteuma orbiculare** L.

Phyteuma orbiculare var. *lanceolatum* (Vill.) Pers. === **Phyteuma orbiculare** L.

Phyteuma orbiculare var. *liguricum* Rich. Schulz === **Phyteuma orbiculare** L.

Phyteuma orbiculare var. *montanum* (Rich. Schulz) Hayek & Hegi === **Phyteuma orbiculare** L.

Phyteuma orbiculare var. *patens* Rich. Schulz === **Phyteuma orbiculare** L.

Phyteuma orbiculare var. *pratense* (Rich. Schulz) Hayek & Hegi === **Phyteuma orbiculare** L.

Phyteuma orbiculare var. *pseudorbiculare* (Pant.) Nyman === **Phyteuma orbiculare** L.

Phyteuma orbiculare var. *suffultum* Rich. Schulz === **Phyteuma orbiculare** L.

Phyteuma otites (Boiss.) Trautv. === **Asyneuma limonifolium** (L.) Janch. subsp. **limoniifolium**

Phyteuma ovale Hoppe === **Phyteuma ovatum** Honck. subsp. **ovatum**

Phyteuma ovatum Lam. === **Phyteuma scheuchzeri** All.

Phyteuma ovatum F. W. Schmidt === **Phyteuma nigrum** F. W. Schmidt

Phyteuma pancicii Rohlena === ?

Phyteuma pauciflorum L. === **Phyteuma globulariifolium** Sternb. & Hoppe

Phyteuma pauciflorum f. *asteranthum* Rich. Schulz === **Phyteuma globulariifolium** Sternb. & Hoppe

Phyteuma pauciflorum f. *albiflorum* Rich. Schulz === **Phyteuma globulariifolium** Sternb. & Hoppe

Phyteuma pauciflorum f. *nanum* Rich. Schulz === **Phyteuma globulariifolium** Sternb. & Hoppe

Phyteuma pauciflorum subsp. *globulariifolium* (Sternb. & Hoppe) Nyman === **Phyteuma globulariifolium** Sternb. & Hoppe

Phyteuma pauciflorum subsp. *pedemontanum* (Rich. Schulz) P. Fourn. === **Phyteuma globulariifolium** Sternb. & Hoppe

Phyteuma pauciflorum var. *macrophyllum* Schur === **Phyteuma globulariifolium** Sternb. & Hoppe

Phyteuma pauciflorum var. *nanum* (Schur) Schur === **Phyteuma globulariifolium** Sternb. & Hoppe

Phyteuma pedemontanum Rich. Schulz === **Phyteuma globulariifolium** Sternb. & Hoppe

Phyteuma pedemontanum f. *humillimum* Rich. Schulz === **Phyteuma globulariifolium** Sternb. & Hoppe

Phyteuma pedemontanum f. *intermedium* Rich. Schulz === **Phyteuma globulariifolium** Sternb. & Hoppe

Phyteuma persicifolium Hoppe ex A. DC. === **Phyteuma zahlbruckneri** Vest

Phyteuma persicifolium f. *carniolicum* Rich. Schulz === **Phyteuma zahlbruckneri** Vest

Phyteuma persicifolium var. *alpestre* Rich. Schulz === **Phyteuma zahlbruckneri** Vest

Phyteuma persicifolium var. *albiflorum* Rich. Schulz === **Phyteuma zahlbruckneri** Vest

Phyteuma persicifolium var. *angustissimum* Rich. Schulz === **Phyteuma zahlbruckneri** Vest

Phyteuma persicifolium var. *latifolium* Rich. Schulz === **Phyteuma zahlbruckneri** Vest

Phyteuma persicifolium var. *laxifolium* Rich. Schulz === **Phyteuma zahlbruckneri** Vest

Phyteuma persicifolium var. *lineare* Rich. Schulz === **Phyteuma zahlbruckneri** Vest

Phyteuma pestalozzae Boiss. === **Asyneuma limonifolium** subsp. **pestalozzae** (Boiss.) Damboldt

Phyteuma pilosum Hegetschw. === **Phyteuma orbiculare** L.

Phyteuma pinnatum L. === **Petromarula pinnata** (L.) A. DC.

Phyteuma pseudorbiculare Pant. === **Phyteuma orbiculare** L.

Phyteuma pseudorbiculare f. *angustifolium* Rich. Schulz === **Phyteuma orbiculare** L.

Phyteuma pulchellum Fisch. & C. A. Mey. === **Asyneuma pulchellum** (Fisch. & C. A. Mey.) Bornm.

Phyteuma pyrenaicum Rich. Schulz === **Phyteuma spicatum** L.

Phyteuma pyrenaicum Sennen === ?

Phyteuma pyrenaicum f. *glabrescens* Rich. Schulz === **Phyteuma spicatum** L.

Phyteuma pyrenaicum f. *glabriusculum* Rich. Schulz === **Phyteuma spicatum** L.

Phyteuma pyrenaicum f. *glabrum* Rich. Schulz === **Phyteuma spicatum** L.

Phyteuma pyrenaicum f. *hirsutum* Rich. Schulz === **Phyteuma spicatum** L.

Phyteuma pyrenaicum f. *nudum* Rich. Schulz === **Phyteuma spicatum** L.

Phyteuma pyrenaicum f. *pilosiusculum* Rich. Schulz === **Phyteuma spicatum** L.

Phyteuma pyrenaicum f. *pilosum* Rich. Schulz === **Phyteuma spicatum** L.

Phyteuma pyrenaicum f. *pubescens* Rich. Schulz === **Phyteuma spicatum** L.

Phyteuma pyrenaicum subsp. *betonicoides* Rich. Schulz === **Phyteuma spicatum** L.

Phyteuma pyrenaicum subsp. *cordifolium* Rich. Schulz === **Phyteuma spicatum** L.

Phyteuma pyrenaicum var. *brevibracteatum* Rich. Schulz === **Phyteuma spicatum** L.

Phyteuma pyrenaicum var. *bracteatum* Rich. Schulz === **Phyteuma spicatum** L.

Phyteuma pyrenaicum var. *ebracteatum* Rich. Schulz === **Phyteuma spicatum** L.

Phyteuma pyrenaicum var. *involucratum* Rich. Schulz === **Phyteuma spicatum** L.

Phyteuma rapunculus Pers. === **Phyteuma spicatum** L.

Phyteuma regelii Trautv. === **Sergia regelii** (Trautv.) Fed.

Phyteuma reniforme Dulac === **Phyteuma charmelii** Vill.

Phyteuma repandum Sm. === **Asyneuma limonifolium** (L.) Janch. subsp. **limoniifolium**

Phyteuma rigidifolium Dufour ex Schult. === **Jasione foliosa** Cav. subsp. **foliosa**

Phyteuma rigidum Willd. === **Asyneuma rigidum** (Willd.) Grossh.

Phyteuma rumelianum (Hampe) Griseb. === **Campanula rumeliana** (Hampe) Vatke

Phyteuma salicifolium Waldst. & Kit. ex Schult. === **Asyneuma canescens** (Waldst. & Kit.) Griseb. & Schenk subsp. **canescens**

Phyteuma salignum Waldst. & Kit. ex Besser === **Asyneuma canescens** (Waldst. & Kit.) Griseb. & Schenk subsp. **canescens**

Phyteuma sallei Sennen & Elias === **Phyteuma orbiculare** L.

Phyteuma scaposum Rich. Schulz === **Phyteuma betonicifolium** subsp. **scaposum** (Rich. Schulz) Pign.

Phyteuma scaposum f. *cordifolium* Rich. Schulz === **Phyteuma betonicifolium** subsp. **scaposum** (Rich. Schulz) Pign.

Phyteuma scaposum f. *glabrum* Rich. Schulz === **Phyteuma betonicifolium** subsp. **scaposum** (Rich. Schulz) Pign.

Phyteuma scheuchzeri subsp. *charmelii* (Vill.) Nyman === **Phyteuma charmelii** Vill.

Phyteuma scheuchzeri subsp. *charmelioides* (Biroli) Hayek === **Phyteuma scheuchzeri** subsp. **columnae** (Thomas) Bech.

Phyteuma scheuchzeri var. *leucanthum* Schur === **Phyteuma scheuchzeri** All. subsp. **scheuchzeri**

Phyteuma scheuchzeri var. *michelii* Schult. === **Phyteuma scheuchzeri** All. subsp. **scheuchzeri**

Phyteuma scheuchzeri var. *scorzonerifolium* (Vill.) Pers. === **Phyteuma scorzonerifolium** Vill.

Phyteuma scheuchzeri var. *serratum* W. D. J. Koch === **Phyteuma scheuchzeri** subsp. **columnae** (Thomas) Bech.

Phyteuma schlatteri Chenev. === ?

Phyteuma scorzonerifolium f. *laxiflorum* (Beyer) Rich. Schulz === **Phyteuma scorzonerifolium** Vill.

Phyteuma scorzonerifolium subsp. *laxiflorum* Beyer === **Phyteuma scorzonerifolium** Vill.

Phyteuma scorzonerifolium var. *angustissimum* St.-Lag. === **Phyteuma michelii** All.

Phyteuma scorzonerifolium var. *eynense* Sennen === **Phyteuma orbiculare** L.

Phyteuma scorzonerifolium var. *persicifolium* (Hoppe ex A. DC.) St.-Lag. === **Phyteuma zahlbruckneri** Vest

Phyteuma serratoides Chouard === **Phyteuma hemisphaericum** L.

Phyteuma serratum var. *nanum* Rich. Schulz === **Phyteuma serratum** Viv.

Phyteuma sewerzowii Regel === **Cylindrocarpa sewerzowii** (Regel) Regel

Phyteuma sibiricum Vest ex Schult. === ?

Phyteuma sibthorpianum Schult. === **Asyneuma rigidum** subsp. **sibthorpianum** (Schult.) Damboldt

Phyteuma sibthorpianum var. *campanuloides* (M. Bieb. ex Sims) Steud. === **Asyneuma campanuloides** (M. Bieb. ex Sims) Bornm.

Phyteuma sieberi f. *glabrescens* Rich. Schulz === **Phyteuma sieberi** Spreng.

Phyteuma sieberi f. *glabrum* Rich. Schulz === **Phyteuma sieberi** Spreng.

Phyteuma sieberi f. *pilosum* Rich. Schulz === **Phyteuma sieberi** Spreng.

Phyteuma sieberi f. *pubescens* Rich. Schulz === **Phyteuma sieberi** Spreng.

Phyteuma sieberi f. *pygmaeum* Bolzon === **Phyteuma sieberi** Spreng.

Phyteuma sieberi var. *alpinum* Rich. Schulz === **Phyteuma sieberi** Spreng.

Phyteuma sinai A. DC. === **Asyneuma rigidum** subsp. **sinai** (A. DC.) Damboldt

Phyteuma sintenisii (Hausskn. ex Bornm.) Hausskn. ex Bornm. === **Asyneuma virgatum** (Labill.) Bornm.

Phyteuma spicatum f. *bicrenatum* Rich. Schulz === **Phyteuma spicatum** L.

Phyteuma spicatum f. *cordatum* Rich. Schulz === **Phyteuma spicatum** L.

Phyteuma spicatum f. *crenatoserratum* Rich. Schulz === **Phyteuma spicatum** L.

Phyteuma spicatum f. *crenatum* Rich. Schulz === **Phyteuma spicatum** L.

Phyteuma spicatum f. *divaricatum* Rich. Schulz === **Phyteuma spicatum** L.

Phyteuma spicatum f. *ebracteatum* Rich. Schulz === **Phyteuma spicatum** L.

Phyteuma spicatum f. *fissum* Rich. Schulz === **Phyteuma spicatum** L.

Phyteuma spicatum f. *grossidentatum* Rich. Schulz === **Phyteuma spicatum** L.

Phyteuma spicatum f. *incisum* Rich. Schulz === **Phyteuma spicatum** L.

Phyteuma spicatum f. *involucratum* Rich. Schulz === **Phyteuma spicatum** L.

Phyteuma spicatum f. *macrodon* Rich. Schulz === **Phyteuma spicatum** L.

Phyteuma spicatum f. *microdon* Rich. Schulz === **Phyteuma spicatum** L.

Phyteuma spicatum race *alpestre* (Godr.) Rouy === **Phyteuma ovatum** Honck. subsp. **ovatum**

Phyteuma spicatum subsp. *alpestre* (Godr.) Kerguélen === **Phyteuma ovatum** Honck. subsp. **ovatum**

Phyteuma spicatum subsp. *ambigens* Rouy === **Phyteuma spicatum** L.

Phyteuma spicatum subsp. *caeruleum* Rich. Schulz === **Phyteuma spicatum** L.

Phyteuma spicatum subsp. *jurassicum* Rich. Schulz === **Phyteuma spicatum** L.

Phyteuma spicatum subsp. *nigrum* (F. W. Schmidt) Weeda === **Phyteuma nigrum** F. W. Schmidt

Phyteuma spicatum subsp. *occidentale* Rich. Schulz === **Phyteuma spicatum** L.

Phyteuma spicatum subvar. *brevibracteatum* Rich. Schulz === **Phyteuma spicatum** L.

Phyteuma spicatum subvar. *caeruleum* Rouy === **Phyteuma spicatum** L.

Phyteuma spicatum subvar. *longibracteatum* Rich. Schulz === **Phyteuma spicatum** L.

Phyteuma spicatum subvar. *macrophyllum* Rich. Schulz === **Phyteuma spicatum** L.

Phyteuma spicatum subvar. *microphyllum* Rich. Schulz === **Phyteuma spicatum** L.

Phyteuma spicatum var. *alpestre* Godr. === **Phyteuma ovatum** Honck. subsp. **ovatum**

Phyteuma spicatum var. *alpinum* Rich. Schulz === **Phyteuma spicatum** L.

Phyteuma spicatum var. *atropurpureum* (Hoppe) Steud. === **P. nigrum** F. W. Schmidt

Phyteuma spicatum var. *betonicifolium* (Vill.) Lapeyr. === **Phyteuma betonicifolium** Vill.

Phyteuma spicatum var. *bracteatum* A. DC. === **Phyteuma spicatum** L.

Phyteuma spicatum var. *bracteatum* Rich. Schulz === **Phyteuma spicatum** L.

Phyteuma spicatum var. *caerulescens* Godr. === **Phyteuma nigrum** F. W. Schmidt

Phyteuma spicatum var. *caeruleum* Gremli === **Phyteuma spicatum** L.

Phyteuma spicatum var. *caeruleum* Gren. & Godr. === **Phyteuma spicatum** L.

Phyteuma spicatum var. *caeruleum* Hegetschw. === **Phyteuma spicatum** L.

Phyteuma spicatum var. *glabrum* Rich. Schulz === **Phyteuma spicatum** L.

Phyteuma spicatum var. *halleri* Steud. === **Phyteuma spicatum** L.

Phyteuma spicatum var. *jurassicum* (Rich. Schulz) Hayek & Hegi === **Phyteuma spicatum** L.

Phyteuma spicatum var. *nigrum* (F. W. Schmidt) Kuntze === **Phyteuma nigrum** F. W. Schmidt

Phyteuma spicatum var. *pilosum* Rich. Schulz === **Phyteuma spicatum** L.

Phyteuma spicatum var. *pseudohalleri* Font Quer ex O. Bolòs & Vigo === **Phyteuma spicatum** L.

Phyteuma spicatum var. *pyrenaicum* (Rich. Schulz) O. Bolòs & Vigo === **Phyteuma spicatum** L.

Phyteuma spicatum var. *rapunculus* Pers. === **Phyteuma ovatum** Honck. subsp. **ovatum**

Phyteuma spicatum var. *roseum* Degen & P. Rossi === **Phyteuma spicatum** L.

Phyteuma spicatum var. *tetramerum* (Schur) Nyman === **Phyteuma tetramerum** Schur

Phyteuma spicatum var. *vulgare* Rich. Schulz === **Phyteuma spicatum** L.

Phyteuma strictum Sims === **Asyneuma limonifolium** (L.) Janch. subsp. **limoniifolium**

Phyteuma stylidioides Boiss. === **Asyneuma limonifolium** (L.) Janch. subsp. **limoniifolium**

Phyteuma stylosum Schrank === **Asyneuma limonifolium** (L.) Janch. subsp. **limoniifolium**

Phyteuma supina (Sieber ex Spreng.) G. Don === **Jasione supina** Sieber ex Spreng.

Phyteuma tenerum Rich. Schulz === **Phyteuma orbiculare** L.

Phyteuma tenerum f. *glabrescens* Rich. Schulz === **Phyteuma orbiculare** L.

Phyteuma tenerum f. *glabrum* Rich. Schulz === **Phyteuma orbiculare** L.
Phyteuma tenerum f. *hirsutum* Rich. Schulz === **Phyteuma orbiculare** L.
Phyteuma tenerum f. *pilosum* Rich. Schulz === **Phyteuma orbiculare** L.
Phyteuma tenerum subsp. *anglicum* Rich. Schulz === **Phyteuma orbiculare** L.
Phyteuma tenerum subsp. *ibericum* Rich. Schulz === **Phyteuma orbiculare** L.
Phyteuma tenerum subvar. *brevifolium* Rich. Schulz === **Phyteuma orbiculare** L.
Phyteuma tenerum subvar. *longifolium* Rich. Schulz === **Phyteuma orbiculare** L.
Phyteuma tenerum var. *anomalum* Rich. Schulz === **Phyteuma orbiculare** L.
Phyteuma tenerum var. *ellipticum* Rich. Schulz === **Phyteuma orbiculare** L.
Phyteuma tenerum var. *macrophyllum* Rich. Schulz === **Phyteuma orbiculare** L.
Phyteuma tenerum var. *microphyllum* Rich. Schulz === **Phyteuma orbiculare** L.
Phyteuma tenerum var. *tenerrimum* Rich. Schulz === **Phyteuma orbiculare** L.
Phyteuma tenuifolium A. DC. === **Asyneuma limonifolium** (L.) Janch. subsp. **limoniifolium**
Phyteuma thomsonii (Hook. f.) C. B. Clarke === **Asyneuma thomsonii** (Hook. f.) Bornm.
Phyteuma trautvetteri (B. Fedtsch.) B. Fedtsch === **Asyneuma argutum** (Regel) Bornm.
    subsp. **argutum**
Phyteuma trichocalycinum (Ten.) Tanfani === **Asyneuma trichocalycinum** (Ten.) K. Malý
Phyteuma tricolor Molina === **Salpiglossis sinuata** Ruiz & Pav. (Solanaceae)
Phyteuma urticifolium Clairv. === **Phyteuma ovatum** Honck. subsp. **ovatum**
Phyteuma vagneri f. *alpinum* Rich. Schulz=== **Phyteuma vagneri** A. Kern.
Phyteuma vagneri f. *brevibracteatum* Rich. Schulz === **Phyteuma vagneri** A. Kern.
Phyteuma vagneri f. *grossidentatum* Rich. Schulz=== **Phyteuma vagneri** A. Kern.
Phyteuma vagneri f. *latibracteatum* Rich. Schulz=== **Phyteuma vagneri** A. Kern.
Phyteuma veronicifolium Schrad. ex A. DC. === **Phyteuma betonicifolium** Vill. subsp.
    betonicifolium
Phyteuma villarsii Rich. Schulz === **Phyteuma charmelii** Vill.
Phyteuma virgatum (Labill.) Willd. === **Asyneuma virgatum** (Labill.) Bornm.

Piddingtonia

Candolle (1839) created this name for an unranked taxon within *Pratia* upon its elevation to generic rank, to avoid having two genera named for the same person (both *Pratia* and *Bernonia* commemorate a French naval officer named Prat-Bernon). In a preliminary phylogenetic analysis (E. Knox, pers. comm.), its type was part of a largely Australasian clade that included some species of *Isotoma* (but not the type); and of *Lobelia* sect. *Dioica* (E. Wimm.), sect. *Isolobus* (A. DC.) C. B. Clarke, sect. *Paramezleria* E. Wimm., and sect. *Pratia* (Gaudich.) J. Murata. Based on: *Pratia* [unranked] *Bernonia* Endl. Type [by monotypy]: *Lobelia begoniifolia* Wall.

> de Candolle, A. (1839). *Piddingtonia*. In A. P. de Candolle, Prodromus systematis naturalis regni vegetabilis 7: 341. Paris: Treuttel & Würtz. La. — Establishment of genus as distinct from *Lobelia* and *Pratia*.
>
> Hooker, J. D. (1852). *Pratia*, Gaud. In The botany of the antarctic voyage of H.M. discovery ships *Erebus* and *Terror* in the years 1839-1843. II. Flora Novae-Zelandiae 1: 156-157. London: Reeve Bros. En. — Suggests *Piddingtonia* not distinct from *Pratia*.
>
> Bentham, G. & F. Mueller (1868). *Pratia*, Gaudich. In Flora Australiensis: a description of the plants of the Australian Territory 4: 131-134. London: L.. Reeve. En. — Suggests *Piddingtonia* not distinct from *Pratia*.

Synonyms:
Piddingtonia A. DC. === **Lobelia** L.
Piddingtonia cyanocarpa Hassk. === **Lobelia montana** Reinw. ex Blume
Piddingtonia montana (Reinw. ex Blume) Miq. === **Lobelia montana** Reinw. ex Blume
Piddingtonia nummularia (Lam.) A. DC. === **Lobelia nummularia** Lam.
Piddingtonia palliardii Lehm. === **Lobelia angulata** G. Forst.
Piddingtonia patens Miq. === **Lobelia montana** Reinw. ex Blume

Pilorea

Rafinesque (1837) segregated this species from *Campanula* on the basis of its capitate inflorescence and two-locular ovary. Though it has priority over *Edraianthus* (Lammers 1996), *Pilorea* has been formally rejected to protect that name (Brummitt 1999). Type [designated by the author]: *Pilorea graminifolia* (L.) Raf.

> Rafinesque, C. S. (1837). Flora Telluriana, pars secunda. Philadelphia: published privately. En. — Establishment of genus as distinct from *Campanula*.
> Lammers, T. G. (1996). Proposal to conserve the name *Edraianthus* against *Pilorea* (Campanulaceae). Taxon 45: 563-564. En. — Formal proposal to reject name.
> Brummitt, R. K. (1999). Report of the Committee fro Spermatophyta: 48. Taxon 48: 359-371. En. — Committee recommended rejection.

Synonyms:
Pilorea Raf. === **Edraianthus** A. DC.
Pilorea graminifolia (L.) Raf. === **Edraianthus graminifolius** (L.) A. DC.

Platycodon

Campanuloideae, 1 species, eastern Asia and widely cultivated. In a preliminary phylogenetic analysis based on nucleic acid data (Eddie et al. 2003, under **General**), *Platycodon* was embedded in a clade that otherwise comprised the nine sampled species of *Codonopsis* subg. *Codonopsis,* including the type species of *Campanumoea* and *Leptocodon*. $2n = 18, 36$. Type [designated by Pfeiff., Nomencl. Bot. 2: 746 (1874)]: *Campanula grandiflora* Jacq.

> de Candolle, A. (1830). *Platycodon.* In Monographie des Campanulées: 125-127. Paris: Veuve Desray. La. — Establishment of genus as distinct from *Campanula* and *Wahlenbergia*.
> Decaisne, J. (1848). *Platycodon autumnale* Dne. Rev. Hort. (sér. 3) 2: 361-362 + fig. 19, illus. Fr. — Portrait of *P. grandiflorus,* with description.
> Lindley, J. & J. Paxton (1851). The Chinese platycode. In Paxton's flower garden 2: 121-122 + pl. 61, illus. En. — Portrait of *P. grandiflorus,* with description.
> Boynton, K. R. (1920). *Platycodon grandiflorum.* Addisonia 5: 13-14 + pl. 167, illus. En. — Portrait with description.
> Bailey, L. H. (1953). *Platycodon.* In The garden of bellflowers in North America: 132-133, illus. New York: MacMillan. En. — Account of species in cultivation.
> Fedorov, A. A. (1957). *Platycodon* A. DC. In V. L. Komarov, Flora URSS 24: 439-441, illus. Leningrad: Academia Scientiae URSS. Ru. — Flora of former Soviet Union, with description and full nomenclature.
> • Hong, D. Y. (1983). *Platycodon* A. DC. In Flora Reipublicae Popularis Sinicae 73(2): 76-77. Beijing: Science Press. Ch. — Flora for China, with description and full nomenclature.
> Liou, M. Y. & C. X. Fu (1985). [Study on the biology of *Platycodon grandiflorum* (Jacq.) A. DC.] Bull. Bot. Res., Harbin 5: 71-80, illus. Ch. — Ecological life history.
> Foster, S. & C. X. Yue (1992). Balloonflower. In Herbal Emissaries. Bringing Chinese Herbs to the West: 155-158. Rochester: Healing Arts Press. En. — Pharmaceutical use in Asia.
> Li, P. (1995). [Brief report of the investigation on the comparative morphology of *Platycodon* from Zhejiang, Liaoning and Sichuan.] J. Sichuan Univ. Nat. Sci. Ed. 32 (special issue): 9-16, illus. Ch. — Analysis of morphological variation among populations in S. & E. China.
> • Eddie, W. M. M. (2000). *Platycodon* de Candolle. In J. Cullen, J. C. M. Alexander, C. D. Brickell, J. R. Edmondson, P. S. Green, V. H. Heywood, P. M. Jørgensen, S. L. Jury, S. G. Knees, H. S. Maxwell, D. M. Willer, N. K. B. Robson, S. M. Walters & P. F. Yeo (eds.), The European garden flora 6: 495-496, illus. Cambridge: Cambridge University Press. En. — Account of species in cultivation.

Kim, Y. S., J. S. Kim, S. U. Choi, J. S. Kim, H. S. Lee, S. H. Roh, Y. C. Jeong, Y. K. Kim & S. Y. Ryu (2005). Isolation of a new saponin and cytotoxic effect of saponins from the root of *Platycodon grandiflorum* [sic] on human tumor cell lines. Planta Med. 71: 566-568. En. — Saponins significantly inhibit tumor proliferation *in vitro*.

Platycodon A. DC., Monogr. Campan.: 125 (1830).
Temp. E. Asia. 30 31 36 38.

Platycodon grandiflorus (Jacq.) A. DC., Monogr. Campan.: 125 (1830), as 'grandiflorum'.
SE. Siberia to Japan & S. China; cult. 30 BRY CTA 31 AMU KHA PRM 36 CHC CHI CHM CHN CHS 38 JAP KOR. Hemicr. or geophyte. $2n = 18, 36$.
* *Campanula grandiflora* Jacq., Hort. Bot. Vindob. 3: 4 (1776). *Campanula gentianoides* Lam., Encycl. 1: 581 (1785).
Campanula glauca Thunb. ex Murray, Syst. Veg. (ed. 14): 211 (1784). *Platycodon glaucus* (Thunb. ex Murray) Nakai, Bot. Mag. (Tokyo) 38: 301 (1924), as 'glaucum'. *Platycodon grandiflorus* var. *glaucus* (Thunb.) Sieb. & Zucc., Abh. Math.-Phys. Cl. Königl. Bayer. Akad. Wiss. 4: 179 (1846).
Platycodon autumnalis Decne., Rev. Hort. (ser. 3) 2: 361 (1848), as 'autumnale'. *Platycodon grandiflorus* var. *autumnalis* (Decne.) Voss in Siebert & Voss, Vilm. Blumengärtn. (ed. 3) 1: 572 (1894).
Platycodon chinensis Lindl. & Paxton, Paxt. Fl. Gard. 2: 121 (1851), as 'chinense'.
Platycodon sinensis Lem., Jard. Fleur.: pl. 250 (1853), as 'sinense'.
Platycodon grandiflorus var. *mariesii* Lynch, Garden (London) 27: 216 (1885).
Platycodon mariesii Wittm., Gartenfl. 41: 655 (1892).
Platycodon mariesii var. *albus* Wittm., Gartenfl. 41: 655 (1892), as 'album'. *Platycodon grandiflorus* f. *albonanus* H. Hara, Enum. Sperm. Jap. 2: 102 (1952), as 'albonanum'.
Platycodon grandiflorus var. *albus* Stubenrauch in L. H. Bailey, Cycl. Amer. Hort.: 1370 (1901), as 'album'.
Platycodon grandiflorus var. *japonicus* Stubenrauch in L. H. Bailey, Cycl. Amer. Hort.: 1370 (1901), as 'japonicum'.
Platycodon grandiflorus var. *semiplenus* Stubenrauch in L. H. Bailey, Cycl. Amer. Hort.: 1370 (1901), as 'semi-plenum'.
Platycodon grandiflorus var. *striatus* Stubenrauch in L. H. Bailey, Cycl. Amer. Hort.: 1370 (1901), as 'striatum'. *Platycodon grandiflorus* f. *striatus* (Stubenrauch) H. Hara, Enum. Sperm. Jap. 2: 101 (1952), as 'striatum'.
Platycodon grandiflorus var. *duplex* Makino, Bot. Mag. (Tokyo) 22: 157 (1908).
Platycodon grandiflorus var. *pentapetalus* Makino, Bot. Mag. (Tokyo) 23: 21 (1909). *Platycodon glaucus* var. *pentapetalus* (Makino) Makino, J. Jap. Bot. 11: 44 (1926).
Platycodon glaucus f. *albus* Makino, J. Jap. Bot. 11: 43 (1926). *Platycodon grandiflorus* f. *leucanthus* H. Hara, Enum. Sperm. Jap. 2: 102 (1952), as 'leucanthum'.
Platycodon glaucus f. *bicolor* Makino, J. Jap. Bot. 11: 43 (1926). *Platycodon grandiflorus* f. *bicolor* (Makino) H. Hara, Enum. Sperm. Jap. 2: 102 (1952).
Platycodon glaucus f. *violaceus* Makino, J. Jap. Bot. 11: 44 (1926).
Platycodon glaucus f. *albiflorus* Honda, Bot. Mag. (Tokyo) 51: 858 (1937), as 'albiflorum'. *Platycodon grandiflorus* f. *albiflorus* (Honda) H. Hara, Enum. Sperm. Jap. 2: 101 (1952), as 'albiflorum'.
Platycodon glaucus f. *subasepalus* Honda, Bot. Mag. (Tokyo) 52: 517 (1938), as 'subasepalum'. *Platycodon glaucus* var. *subasepalus* (Honda) Nakai, J. Jap. Bot. 15: 686 (1939), as 'subasepalum'. *Platycodon grandiflorus* f. *subasepalus* (Honda) H. Hara, Enum. Sperm. Jap. 2: 101 (1952), as 'subasepalum'.
Platycodon glaucus var. *monanthus* Nakai, J. Jap. Bot. 15: 186 (1939), as 'monanthum'. *Platycodon grandiflorus* f. *monanthus* (Nakai) Kim, Fl. Coreana 6: 97 (1976), as 'monanthum'.
Platycodon glaucus var. *planicorollatus* Makino, Zissai-Engei 26: 461 (1950), as 'planicorollatum'. *Platycodon grandiflorus* var. *planicorollatus* (Makino) H. Hara, Enum. Sperm. Jap. 2: 102 (1952), as 'planicorollatum'.

Platycodon glaucus var. *rugosus* Makino, Zissai-Engei 26: 461 (1950), as 'rugosum'.
Platycodon grandiflorus var. *rugosus* (Makino) H. Hara, Enum. Sperm. Jap. 2: 102 (1952), as 'rugosum'.

Synonyms:

Platycodon autumnalis Decne. === **Platycodon grandiflorus** (Jacq.) A. DC.
Platycodon chinensis Lindl. & Paxton === **Platycodon grandiflorus** (Jacq.) A. DC.
Platycodon glaucus (Thunb. ex Murray) Nakai === **Platycodon grandiflorus** (Jacq.) A. DC.
Platycodon glaucus f. *albiflorus* Honda === **Platycodon grandiflorus** (Jacq.) A. DC.
Platycodon glaucus f. *albus* Makino === **Platycodon grandiflorus** (Jacq.) A. DC.
Platycodon glaucus f. *bicolor* Makino === **Platycodon grandiflorus** (Jacq.) A. DC.
Platycodon glaucus f. *subasepalus* Honda === **Platycodon grandiflorus** (Jacq.) A. DC.
Platycodon glaucus f. *violaceus* Makino === **Platycodon grandiflorus** (Jacq.) A. DC.
Platycodon glaucus var. *monanthum* Nakai === **Platycodon grandiflorus** (Jacq.) A. DC.
Platycodon glaucus var. *pentapetalus* (Makino) Makino === **Platycodon grandiflorus** (Jacq.) A. DC.
Platycodon glaucus var. *planicorollatus* Makino === **Platycodon grandiflorus** (Jacq.) A. DC.
Platycodon glaucus var. *rugosus* Makino === **Platycodon grandiflorus** (Jacq.) A. DC.
Platycodon glaucus var. *subasepalus* (Honda) Nakai === **Platycodon grandiflorus** (Jacq.) A. DC.
Platycodon grandiflorus f. *albiflorus* (Honda) H. Hara === **Platycodon grandiflorus** (Jacq.) A. DC.
Platycodon grandiflorus f. *albonanus* H. Hara === **Platycodon grandiflorus** (Jacq.) A. DC.
Platycodon grandiflorus f. *bicolor* (Makino) H. Hara === **Platycodon grandiflorus** (Jacq.) A. DC.
Platycodon grandiflorus f. *leucanthus* H. Hara === **Platycodon grandiflorus** (Jacq.) A. DC.
Platycodon grandiflorus f. *monanthus* (Nakai) Kim === **Platycodon grandiflorus** (Jacq.) A. DC.
Platycodon grandiflorus f. *striatus* (Stubenrauch) H. Hara === **Platycodon grandiflorus** (Jacq.) A. DC.
Platycodon grandiflorus var. *albus* Stubenrauch === **Platycodon grandiflorus** (Jacq.) A. DC.
Platycodon grandiflorus var. *autumnalis* (Decne.) Voss === **Platycodon grandiflorus** (Jacq.) A. DC.
Platycodon grandiflorus var. *duplex* Makino === **Platycodon grandiflorus** (Jacq.) A. DC.
Platycodon grandiflorus var. *glaucus* (Thunb.) Sieb. & Zucc. === **Platycodon grandiflorus** (Jacq.) A. DC.
Platycodon grandiflorus var. *japonicus* Stubenrauch === **Platycodon grandiflorus** (Jacq.) A. DC.
Platycodon grandiflorus var. *mariesii* Lynch === **Platycodon grandiflorus** (Jacq.) A. DC.
Platycodon grandiflorus var. *pentapetalus* Makino === **Platycodon grandiflorus** (Jacq.) A. DC.
Platycodon grandiflorus var. *planicorollatus* (Makino) H. Hara === **Platycodon grandiflorus** (Jacq.) A. DC.
Platycodon grandiflorus var. *rugosus* (Makino) H. Hara === **Platycodon grandiflorus** (Jacq.) A. DC.
Platycodon grandiflorus var. *semiplenus* Stubenrauch === **Platycodon grandiflorus** (Jacq.) A. DC.
Platycodon grandiflorus var. *striatus* Stubenrauch === **Platycodon grandiflorus** (Jacq.) A. DC.
Platycodon homallanthinus (Ledeb.) A. DC. === **Campanula expansa** Rudolph
Platycodon homallanthinus var. *angustifolius* A. DC. === **Campanula expansa** Rudolph
Platycodon mariesii Wittm. === **Platycodon grandiflorus** (Jacq.) A. DC.
Platycodon mariesii var. *albus* Wittm. === **Platycodon grandiflorus** (Jacq.) A. DC.
Platycodon sinensis Lem. === **Platycodon grandiflorus** (Jacq.) A. DC.

Podanthum

This name has been used for *Asyneuma,* though under the current ICBN, the latter has priority. Based on: *Phyteuma* sect. *Podanthum* G. Don. Type [designated here]: *Phyteuma canescens* Waldst. & Kit.

Boissier, P. E. (1875). *Podanthum*. In Flora orientalis sive enumeratio plantarum in Oriente a Graecia et Aegypto ad Indiae fines hucusque observatum 3: 945-957. Basel: H. Georg. La. — Elevation of *Phyteuma* sect. *Podanthum* to generic rank.

Synonyms:

Podanthum (G. Don) Boiss. === **Asyneuma** Griseb. & Schenk.

Podanthum aizoon Hausskn. ex Bornm. === **Asyneuma virgatum** (Labill.) Bornm. subsp. **virgatum**

Podanthum amplexicaule (Willd.) Boiss. === **Asyneuma amplexicaule** (Willd.) Hand.-Mazz.

Podanthum amplexicaule var. *angustifolium* Boiss. === **Asyneuma amplexicaule** (Willd.) Hand.-Mazz. subsp. **amplexicaule**

Podanthum amplexicaule var. *aucheri* (A. DC.) Bornm. === **Asyneuma amplexicaule** subsp. **aucheri** (A. DC.) Bornm.

Podanthum amplexicaule var. *kotschyi* (Boiss.) Boiss. === **Asyneuma amplexicaule** (Willd.) Hand.-Mazz. subsp. **amplexicaule**

Podanthum anthericoides Janka === **Asyneuma anthericoides** (Janka) Bornm.

Podanthum anthericoides f. *glabrata* Stoj. & Acht. === **Asyneuma anthericoides** (Janka) Bornm.

Podanthum anthericoides f. *villosa* Stoj. & Acht. === **Asyneuma anthericoides** (Janka) Bornm.

Podanthum anthericoides var. *grandiflorum* (Velen.) Stoj. & Stef. === **Asyneuma anthericoides** (Janka) Bornm.

Podanthum argutum (Regel) O. Fedtsch. & B. Fedtsch. === **Asyneuma argutum** (Regel) Bornm.

Podanthum argutum var. *baldshuanicum* O. Fedtsch. === **Asyneuma argutum** subsp. **baldshuanicum** (O. Fedtsch.) Damboldt

Podanthum argutum var. *elegans* O. Fedtsch. === **Asyneuma argutum** (Regel) Bornm. subsp. **argutum**

Podanthum argutum var. *foliosum* O. Fedtsch. === **Asyneuma argutum** (Regel) Bornm. subsp. **argutum**

Podanthum asperum (Boiss.) Boiss. === **Asyneuma persicum** (A. DC.) Bornm.

Podanthum attenuatum (Franch.) O. Fedtsch. & B. Fedtsch. === **Asyneuma argutum** (Regel) Bornm. subsp. **argutum**

Podanthum aurasiacum Batt. & Trabut === **Asyneuma rigidum** subsp. **aurasiacum** (Batt. & Trabut) Damboldt

Podanthum brachylobum Boiss. === **Asyneuma virgatum** (Labill.) Bornm. subsp. **virgatum**

Podanthum brachylobum var. *sintensii* (Hausskn. ex Bornm.) Bornm. === **Asyneuma virgatum** (Labill.) Bornm. subsp. **virgatum**

Podanthum campanuloides (M. Bieb. ex Sims) Boiss. === **Asyneuma campanuloides** (M. Bieb. ex Sims) Bornm.

Podanthum canescens (Waldst. & Kit.) Boiss. === **Asyneuma canescens** (Waldst. & Kit.) Griseb. & Schenk

Podanthum canescens f. *laevis* Rohlena === **Asyneuma canescens** (Waldst. & Kit.) Griseb. & Schenk subsp. **canescens**

Podanthum canescens subsp. *rhodopeum* Formánek === **Asyneuma canescens** (Waldst. & Kit.) Griseb. & Schenk subsp. **canescens**

Podanthum canescens var. *nudiflorum* Davidov === **Asyneuma canescens** (Waldst. & Kit.) Griseb. & Schenk subsp. **canescens**

Podanthum canescens var. *salicifolium* (Steud.) Fomin === **Asyneuma canescens** (Waldst. & Kit.) Griseb. & Schenk subsp. **canescens**

Podanthum cappadocicum (Boiss.) Boiss. === **Asyneuma virgatum** (Labill.) Bornm. subsp. **virgatum**

Podanthum cichoriiforme (Boiss.) Boiss. === **Asyneuma virgatum** subsp. **cichoriiforme** (Boiss.) Damboldt

Podanthum controversum (Boiss.) Boiss. === **Asyneuma rigidum** (Willd.) Grossh. subsp. **rigidum**

Podanthum floribundum Stapf === **Asyneuma limonifolium** subsp. **pestalozzae** (Boiss.) Damboldt

Podanthum giganteum Boiss. === **Asyneuma giganteum** (Boiss.) Bornm.

Podanthum grandiflorum Velen. === **Asyneuma anthericoides** (Janka) Bornm.

Podanthum kellerianum Stef. === **Asyneuma limoniifolium** (L.) Janch. subsp. **limoniifolium**

Podanthum lanceolatum Boiss. === **Asyneuma rigidum** (Willd.) Grossh. subsp. **rigidum**

Podanthum lanceolatum var. *alpinum* Boiss. === **Asyneuma rigidum** subsp. **sinai** (A. DC.) Damboldt

Podanthum lanceolatum var. *controversum* (Boiss.) Bornm. === **Asyneuma rigidum** (Willd.) Grossh. subsp. **rigidum**

Podanthum lanceolatum var. *flagellatum* Hausskn. & Bornm. ex Bornm. === **Asyneuma rigidum** (Willd.) Grossh. subsp. **rigidum**

Podanthum lanceolatum var. *rigidum* (Willd.) Boiss. === **Asyneuma rigidum** (Willd.) Grossh.

Podanthum leianthum (Trautv.) Boiss. === **Asyneuma virgatum** (Labill.) Bornm. subsp. **virgatum**

Podanthum limonifolium (L.) Boiss. === **Asyneuma limonifolium** (L.) Janch. subsp. **limonifolium**

Podanthum limonifolium f. *heterophyllum* Rohlena === **Asyneuma limonifolium** (L.) Janch. subsp. **limonifolium**

Podanthum limonifolium var. *alpinum* Boiss. === **Asyneuma limonifolium** (L.) Janch. subsp. **limonifolium**

Podanthum limonifolium var. *canescens* Boiss. === **Asyneuma limonifolium** (L.) Janch. subsp. **limonifolium**

Podanthum limonifolium var. *ramosum* Hausskn. === **Asyneuma limonifolium** (L.) Janch. subsp. **limonifolium**

Podanthum limonifolium var. *repandum* (Sm.) Halácsy === **Asyneuma limonifolium** (L.) Janch. subsp. **limonifolium**

Podanthum linifolium (Boiss. & Heldr.) Boiss. === **Asyneuma linifolium** (Boiss. & Heldr.) Bornm.

Podanthum lobelioides (Willd.) Boiss. === **Asyneuma lobelioides** (Willd.) Hand.-Mazz.

Podanthum lobelioides f. *stenophylla* Bornm. === **Asyneuma limonifolium** subsp. **pestalozzae** (Boiss.) Damboldt

Podanthum lobelioides var. *urceolatum* Fomin === **Asyneuma lobelioides** (Willd.) Hand.-Mazz.

Podanthum lycium (Boiss.) Boiss. === **Asyneuma lycium** (Boiss.) Bornm.

Podanthum macrodon Boiss. & Hausskn. === **Asyneuma macrodon** (Boiss. & Hausskn.) Bornm.

Podanthum obtusifolium Hausskn. ex Bornm. === **Asyneuma virgatum** (Labill.) Bornm. subsp. **virgatum**

Podanthum otites Boiss. === **Asyneuma limonifolium** (L.) Janch. subsp. **limonifolium**

Podanthum persicum (A. DC.) Boiss. === **Asyneuma persicum** (A. DC.) Bornm.

Podanthum persicum f. *subsimplex* Bornm. === **Asyneuma persicum** (A. DC.) Bornm.

Podanthum persicum var. *asperum* (Boiss.) Bornm. === **Asyneuma persicum** (A. DC.) Bornm.

Podanthum persicum var. *multicaule* (Boiss.) Bornm. === **Asyneuma persicum** (A. DC.) Bornm.

Podanthum persicum var. *pumilum* Boiss. === **Asyneuma persicum** (A. DC.) Bornm.

Podanthum psaridis Heldr. ex Halácsy === **Asyneuma limonifolium** (L.) Janch. subsp. **limonifolium**

Podanthum psilostachyum (Boiss. & Kotschy) Boiss. === **Campanula psilostachya** Boiss. & Kotschy

Podanthum pulchellum (Fisch. & C. A. Mey.) Boiss. === **Asyneuma pulchellum** (Fisch. & C. A. Mey.) Bornm.

Podanthum regelii (Trautv.) O. Fedtsch. & B. Fedtsch. === **Sergia regelii** (Trautv.) Fed.

Podanthum rhodopaeum Davidov === **Asyneuma limonifolium** (L.) Janch. subsp. **limonifolium**

Podanthum salicifolium Rupr. === **Asyneuma canescens** (Waldst. & Kit.) Griseb. & Schenk subsp. **canescens**

Podanthum scoparium Boiss. & Hausskn. === **Campanula scoparia** (Boiss. & Hausskn.) Damboldt

Podanthum serbicum Formánek === **Asyneuma anthericoides** (Janka) Bornm.

Podanthum sibthorpianum (Schult.) Boiss. === **Asyneuma rigidum** subsp. **sibthorpianum**
(Schult.) Damboldt

Podanthum sibthorpianum var. *tmoleum* Boiss. === **Asyneuma rigidum** subsp.
sibthorpianum (Schult.) Damboldt

Podanthum sintensii Hausskn. ex Bornm. === **Asyneuma virgatum** (Labill.) Bornm.
subsp. **virgatum**

Podanthum strigillosum (Boiss.) Boiss. === **Campanula strigillosa** Boiss.

Podanthum supinum Wettst. === **Asyneuma rigidum** (Willd.) Grossh.

Podanthum tenuifolium (A. DC.) Boiss. === **Asyneuma limonifolium** (L.) Janch. subsp.
limonifolium

Podanthum trautvetteri B. Fedtsch. === **Asyneuma argutum** (Regel) Bornm.

Podanthum trichocalycinum (Ten.) Boiss. === **Asyneuma trichocalycinum** (Ten.) K. Malý

Podanthum trichocalycinum var. *densiflora* Hausskn. === **Asyneuma trichocalycinum** (Ten.)
K. Malý

Podanthum trichostegium Boiss. === **Asyneuma trichostegium** (Boiss.) Bornm.

Podanthum urceolatum (Fomin) Schischk. === **Asyneuma lobelioides** (Willd.) Hand.-Mazz.

Podanthum virgatum (Labill.) Boiss. === **Asyneuma virgatum** (Labill.) Bornm.

Podanthum woronowii Fomin === **Asyneuma lobelioides** (Willd.) Hand.-Mazz.

Popoviocodonia

This genus was established for a pair of species from northeastern Asia that were said to be
intermediate between *Adenophora* and *Campanula,* but which differ from *Campanula* in no
discernible fashion; if anything, the deeply cut corolla is more reminiscent of *Asyneuma*
than *Adenophora.* A relationship to *Campanula prenanthoides* (q.v.) of western North
America should be investigated. Type [designated by the author]: *Popoviocodonia uyemurae*
(Kudô) Fed.

> Fedorov, A. A. (1957). *Popoviocodonia* Fed., gen. nov. In V. L. Komarov, Flora URSS 24: 372-
> 374, 470-471, illus. Leningrad: Academia Scientiae URSS. Ru. — Establishment of genus
> as distinct from *Adenophora* and *Campanula.*

Synonyms:

Popoviocodonia Fed. === **Campanula** L.

Popoviocodonia stenocarpa (Trautv. & C. A. Mey.) Fed. === **Campanula stenocarpa** Trautv. &
C. A. Mey.

Popoviocodonia uyemurae (Kudô) Fed. === **Campanula uyemurae** (Kudô) Miyabe & Tatew.

Porterella

Lobelioideae, 1 species, western United States. Although Wimmer (1953) included it in
Laurentia sect. *Laurentia* (as "sect. Solenopsis") with species of *Solenopsis* and *Wimmerella,*
this is not supported by preliminary analyses of DNA data (E. Knox, pers. comm.). In Knox's
phylogeny, *Porterella* is part of a clade that also includes *Downingia, Howellia, Legenere,* and
Palmerella, while *Solenopsis* and *Wimmerella* fall into two other clades. $2n = 22, 24$. Type [by
monotypy]: *Porterella carnosula* (Hook. & Arn.) Torr.

> Porter, T. C. (1872). Catalogue of plants collected during the expedition to the headwaters
> of the Yellowstone River in 1871. In F. V. Hayden, Preliminary report of the United
> States Geological Survey of Montana and portions of adjacent territories, being a fifth
> annual report of progress: 477-498. Washington: Government Printing Office. En. —
> establishment of genus as distinct from *Lobelia.*

> • McVaugh, R. (1940). A revision of "Laurentia" and allied genera in North America. Bull.
> Torrey Bot. Club 67: 778-798. En. — Monograph with description and full
> nomenclature.

McVaugh, R. (1943). *Porterella* Torr. In North American Flora 32A: 25-26. New York: New York Botanical Garden. En. — Flora with description and full nomenclature.
- Wimmer, F. E. (1953). [*Laurentia* (Mich.) Adans.] Sectio I. *Solenopsis* Endl. In A. Engler & L. Diels, Das Pflanzenreich IV. 276b: 387-397. Berlin: Akademie-Verlag. Ge., La. — Monograph with description, full nomenclature, and specimen citations; placed in same taxon as species here referred to *Solenopsis* and *Wimmerella*.

Correll, D. S. & H. B. Correll (1972). *Porterella* Torr. In Aquatic and wetland plants of southwestern United States: 1583-1585, illus. Washington: Environmental Protection Agency. En. — Flora with description.

Cronquist, A., A. H. Holmgren, N. H. Holmgren, J. L. Reveal & P. K. Holmgren (1984). *Porterella* Torr. In Intermountain flora 4: 520-521, illus. New York: New York Botanical Garden. En. — Flora with description and full nomenclature.

Morin, N. (1993). *Porterella*. In J. C. Hickman (ed.), The Jepson manual, higher plants of California: 468, illus. Berkeley: University of California Press. En. — Flora with description.

Serra, L. & M. B. Crespo (1997). An outline revision of the subtribe Siphocampylinae (Lobeliaceae). Lagascalia 19: 881-888, illus., maps. En. — Key to distinguish genus from alleged allies.

Porterella Torr. in Hayden, Prelim. Rep. Geol. Surv. Montana: 488 (1872).
W. North America. 73 76.

Porterella carnosula (Hook. & Arn.) Torr. in Hayden, Prelim. Rep. Geol. Surv. Montana: 488 (1872), as 'carnulosa'.
Oregon to Wyoming, Arizona & California. 73 IDA ORE WYO 76 ARI CAL NEV UTA. Ther. $2n = 22, 24$.
* *Lobelia carnosula* Hook. & Arn., Bot. Beechey Voy.: 362 (1839). *Laurentia carnosula* (Hook. & Arn.) Benth. ex A. Gray in S. Watson, Bot. California 1: 444 (1876).
Porterella eximia A. Nelson, Bull. Torrey Bot. Club 27: 270 (1900). *Laurentia eximia* (A. Nelson) A. Nelson in J. M. Coult. & A. Nelson, New Man. Bot. Centr. Rocky Mts.: 475 (1909).

Synonyms:
Porterella eximia A. Nelson === **Porterella carnosula** (Hook. & Arn.) Torr.

Pratia

Although *Pratia* has been recognized by numerous workers over the years (e.g., Wimmer 1943), others (e.g., Moeliono 1960, Adams 1972) have questioned the wisdom of distinguishing it from *Lobelia* on the basis of a single character, its fleshy fruit. A molecular phylogeny (E. Knox, pers. comm.) supports the latter view, as the sampled species fall into two discrete clades. The majority are scattered among species referable to *Isotoma* and to *Lobelia* sect. *Dioica* (E. Wimm.) J. Murata, sect. *Isolobus* (A. DC.) C. B. Clarke, and sect. *Paramezleria* E. Wimm. Two species, however, belong to a clade that otherwise includes the type of *Isotoma* plus various Australasian species of *Lobelia* sect. *Delostemon* (E. Wimm.) J. Murata. The type of *Pratia* has yet to be sampled. Type [by monotypy]: *Pratia repens* Gaudich.

Gaudichaud, C. (1825). Rapport sur la flore des Îles Malouines. Ann. Sci. Nat. 5: 89-110. Fr. — Establishment of genus.

Hooker, J. D. (1844). *Pratia,* Gaud. In The botany of the antarctic voyage of H.M. discovery ships *Erebus* and *Terror* in the years 1839-1843. I. Flora antarctica: 41-44, 325 + tab. XXIX, illus. London: Reeve Bros. En. — Synopsis of genus with descriptions and full nomenclature.

- Bentham, G. & F. Mueller (1868). *Pratia,* Gaudich. In Flora Australiensis: a description of the plants of the Australian Territory 4: 131-134. London: L. Reeve. En. — Flora with key, descriptions, full nomenclature, and specimen citations; comments that genus not truly distinct from *Lobelia.*

    Wimmer, F. E. (1943). *Pratia* Gaudich. In A. Engler & L. Diels, Das Pflanzenreich IV. 276b: 104-119, illus. Leipzig: Wilhelm Engelmann. Ge., La. — Monograph with key, descriptions, infrageneric classification, full nomenclature, and specimen citations.

    Wimmer, F. E. (1953). *Pratia.* In A. Engler & L. Diels, Das Pflanzenreich IV. 276b: 763-767. Berlin: Akademie-Verlag. Ge., La. — Supplement to Wimmer (1943).

    Moeliono, B. (1960). *Lobelia.* In C. G. G. J. van Steenis (ed.), Flora Malesiana (ser. I) 6(1): 121-136, illus., maps. Djakarta: Noordhoff-Kolff. En. — Presents evidence supporting merger of genus with *Lobelia.*

    Wimmer, F. E. (1968). *Pratia* Gaudich. In A. Engler & L. Diels, Das Pflanzenreich IV. 276c: 832-834 + Taf. 12, illus. Berlin: Akademie-Verlag. Ge., La. — Supplement to Wimmer (1943, 1953).

    Adams, C. D. (1972). Campanulaceae. In Flowering plants of Jamaica: 734-737. Mona: University of the West Indies. En. — Subsumes endemic species into *Lobelia,* stating that fruit distinction is unclear and that all seem to form natural group.

- Murata, J. (1995). A revision of infrageneric classification of *Lobelia* (Campanulaceae-Lobelioideae) with special reference to seed coat morphology. J. Fac.Sci. Univ. Tokyo, Sect. 3, Bot. 15: 349-371, illus. En. — *Pratia* accorded sectional rank within *Lobelia.*

    Chiapella, J. (1996). Nota sobre la identidad de *Pratia repens* (Campanulaceae: Lobelioideae). Bol. Soc. Argent. Bot. 32: 123-135, illus. Sp. — Morphometric analysis supports conspecificity of type species of *Pratia* and *Hypsela.*

    Lammers, T. G. (1998). New names and new combinations in Campanulaceae. Novon 8: 31-35. En. — Transfer of last remaining species to *Lobelia.*

Synonyms:

Pratia Gaudich. === **Lobelia** L.

Pratia [unranked] *Bernonia* Endl. === **Lobelia** L.

Pratia sect. *Colensoa* (Hook. f.) Baill. === **Lobelia** L.

Pratia sect. *Hypsela* (C. Presl) Baill. === **Lobelia** L.

Pratia sect. *Speirema* (Hook. f.) Baill. === **Lobelia** L.

Pratia acuminata (Sw.) McVaugh === **Lobelia acuminata** Sw.

Pratia acuminata var. *innominata* (Rendle) E. Wimm. === **Lobelia innominata** Rendle

Pratia angulata (G. Forst.) Hook. f === **Lobelia angulata** G. Forst.

Pratia angulata var. *arenaria* (Hook. f.) Hook. f. === **Lobelia angulata** G. Forst.

Pratia angulata var. *minor* Carse === **Lobelia angulata** G. Forst.

Pratia angulata var. *obovata* E. Wimm. === **Lobelia angulata** G. Forst.

Pratia archboldiana Merr. & L. M. Perry === **Lobelia archboldiana** (Merr. & L. M. Perry) Moeliono

Pratia arenaria Hook. f. === **Lobelia angulata** G. Forst.

Pratia atacamensis Phil. === **Lobelia oligophylla** (Wedd.) Lammers

Pratia begoniifolia (Wall.) Lindl. === **Lobelia nummularia** Lam.

Pratia boliviensis A. DC. === **Lobelia nana** Kunth

Pratia borneensis Hemsl. === **Lobelia borneensis** (Hemsl.) Moeliono

Pratia borneensis var. *grandiflora* Stapf === **Lobelia borneensis** (Hemsl.) Moeliono

Pratia brachyantha (Merr. & L. M. Perry) E. Wimm. === **Lobelia brachyantha** Merr. & L. M. Perry

Pratia brevisepala Y. S. Lian === **Lobelia brevisepala** (Y. S. Lian) Lammers

Pratia calochlamys (Donn. Sm.) E. Wimm. === **Lobelia calochlamys** (Donn. Sm.) Wilbur

Pratia concolor (R. Br.) Druce === **Lobelia concolor** R. Br.

Pratia concolor var. *longipes* (Hook. f.) E. Wimm. === **Lobelia concolor** R. Br.

Pratia corymbosa G. Don === **Lobelia jasionoides** (A. DC.) E. Wimm.

Pratia cunninghamii (A. DC.) Hook. f. === **Lobelia concolor** R. Br.

Pratia cunninghamii var. *longipes* Hook. f. === **Lobelia concolor** R. Br.

Pratia conferta (Merr. & L. M. Perry) E. Wimm. === **Lobelia conferta** Merr. & L. M. Perry
Pratia darlingensis E. Wimm. === **Lobelia darlingensis** (E. Wimm.) Albr.
Pratia elliptica (Hook. & Arn.) Hook. f. === **Lobelia hederacea** Cham.
Pratia erecta Gaudich. === **Lobelia concolor** R. Br.
Pratia fangiana E. Wimm. === **Lobelia fangiana** (E. Wimm.) S. Y. Hu
Pratia fawcettii (Urb.) McVaugh === **Lobelia fawcettii** Urb.
Pratia gelida (F. Muell.) Benth. === **Lobelia gelida** F. Muell.
Pratia gilletii (De Wild.) E. Wimm. === **Lobelia gilletii** De Wild.
Pratia glandulifera Wedd. === **Lysipomia glandulifera** (Wedd.) Schlecht. ex E. Wimm.
Pratia grandifolia (Britt.) McVaugh === **Lobelia grandifolia** Britt.
Pratia guatemalensis (B. L. Rob.) E. Wimm. === **Lobelia guatemalensis** (B. L. Rob.) Wilbur
Pratia harrisii (Urb.) McVaugh === **Lobelia harrisii** Urb.
Pratia hederacea (Cham.) G. Don === **Lobelia hederacea** Cham.
Pratia hederacea var. *elliptica* A. DC. === **Lobelia hederacea** Cham.
Pratia hederacea var. *odorata* (Graham) Steud. === **Lobelia hederacea** Cham.
Pratia innominata (Rendle) McVaugh === **Lobelia innominata** Rendle
Pratia irrigua (R. Br.) Benth. === **Lobelia irrigua** R. Br.
Pratia linnaeoides Hook. f. === **Lobelia angulata** G. Forst.
Pratia longiflora Hook. f. === **Lobelia oligophylla** (Wedd.) Lammers
Pratia longipedicellata (C. E. C. Fisch.) E. Wimm. === **Lobelia longipedicellata** C. E. C. Fisch.
Pratia macrodon Hook. f. === **Lobelia macrodon** (Hook. f.) Lammers
Pratia montana (Reinw. ex Blume) Hassk. === **Lobelia montana** Reinw. ex Blume
Pratia montana f. *variegata* Hochr. === **Lobelia montana** Reinw. ex Blume
Pratia montana var. *cyanocarpa* (Hassk.) E. Wimm. === **Lobelia montana** Reinw. ex Blume
Pratia montana var. *deleiensis* (C. E. C. Fischer) E. Wimm. === **Lobelia montana** Reinw. ex Blume
Pratia nummularia (Lam.) A. Braun & Asch. === **Lobelia nummularia** Lam.
Pratia oligophylla Wedd. === **Lobelia oligophylla** (Wedd.) Lammers
Pratia ovata Elmer === **Lobelia zeylanica** L.
Pratia papuana S. Moore === **Lobelia nummularia** Lam.
Pratia pedunculata (R. Br.) Benth. === **Lobelia pedunculata** R. Br.
Pratia pencana Phil. === **Lobelia oligophylla** (Wedd.) Lammers
Pratia perpusilla (Hook. f.) Hook. f. === **Lobelia perpusilla** Hook. f.
Pratia physaloides (A. Cunn.) Hemsl. === **Lobelia physaloides** A. Cunn.
Pratia platycalyx (F. Muell.) Benth. === **Lobelia platycalyx** (F. Muell.) F. Muell.
Pratia podenzanae S. Moore === **Lobelia nummularia** Lam.
Pratia prostrata E. Wimm. === **Lobelia australiensis** Lammers
Pratia puberula Benth. === **Lobelia benthamii** F. Muell.
Pratia puberula f. *arguta* E. Wimm. === **Lobelia benthamii** F. Muell.
Pratia purpurascens (R. Br.) E. Wimm. === **Lobelia purpurascens** R. Br.
Pratia radicans G. Don === **Lobelia chinensis** Lour.
Pratia reflexa Y. S. Lian === **Lobelia reflexisepala** Lammers
Pratia reniformis (Cham.) Kanitz === **Lobelia reniformis** Cham.
Pratia repens Gaudich. === **Lobelia pratiana** Gaudich. ex Lammers
Pratia repens var. *urvilleana* A. DC. === **Lobelia pratiana** Gaudich. ex Lammers
Pratia rugulosa (Graham) G. Don === **Lobelia angulata** G. Forst.
Pratia serpyllacea C. Presl === **Lobelia hederacea** Cham.
Pratia subsesslis Wedd. === **Lobelia oligophylla** (Wedd.) Lammers
Pratia surrepens (Hook. f.) E. Wimm. === **Lobelia surrepens** Hook. f.
Pratia tatea E. Wimm. === **Lobelia tatea** (E. Wimm.) E. Wimm.
Pratia thunbergii G. Don === **Lobelia chinensis** Lour.
Pratia torricellensis Lauterb. === **Lobelia zeylanica** L.
Pratia wardii (C. E. C. Fisch.) E. Wimm. === **Lobelia montana** Reinw. ex Blume
Pratia wollastonii S. Moore === **Lobelia nummularia** Lam.
Pratia zeylanica Hassk. === **Lobelia nummularia** Lam.

Prismatocarpus

Campanuloideae, 29 species, Cape Provinces of South Africa. The European genus *Legousia* (q.v.) was included in synonymy at the time *Prismatocarpus* was created. Because that name had priority and should have been used, the name *Prismatocarpus* was illegitimate (ICBN Art. 52.1). Even though Candolle (1830) removed the European species, it remained so until formally conserved a century and a half later (Thulin 1985, Brummitt 1988). The genus is divided (Adamson 1952) into subg. *Prismatocarpus* and subg. *Afrotrachelium* Adamson; the former comprises ser. *Prismatocarpus,* ser. *Stricti* Adamson, and ser. *Nitidi* Adamson. Although none of its species have been included in recent molecular studies, one suspects on the basis of morphology and biogeography that the genus would be part of the "wahlenbergioid" clade (Eddie et al. 2003, under **General**) with *Craterocapsa* and *Roella.* The most recent monograph is that by Adamson (1952). *2n* = 34. Type [conserved]: *Prismatocarpus paniculatus* L'Hér.

> L'Héritier de Brutelle, C. L. (1789). Sertum anglicum. Paris: Didot. La. — Establishment of genus as distinct from *Campanula,* but encompassing *Legousia.*
> Hooker, W. J. (1827). *Campanula prismatocarpus.* Angular-fruited Cape Campanula. Bot. Mag. 54: tab. 2733 + 2 pp., illus. En. — Portrait of *P. nitidus,* with description.
> • Adamson, R. S. (1950). *Prismatocarpus* L'Hérit. In R. S. Adamson & T. M. Salter, Flora of the Cape Peninsula: 741-742. Cape Town: Juta. En. — Flora with key and descriptions.
> • Adamson, R. S. (1952). A revision of the genera *Prismatocarpus* and *Roella.* J. S. African Bot. 17: 93-166. En. — Monograph with keys, descriptions, infrageneric classification, full nomenclature, and specimen citations.
> Dyer, R. A. (1975). *Prismatocarpus* L'Hérit. In The genera of southern African flowering plants 1: 640. Pretoria: Department of Agricultural Technical Services. En. — Generic description.
> Thulin, M. (1985). Proposal to conserve 8663 *Prismatocarpus* (Campanulaceae) with a conserved type. Taxon 34: 538-539. En. — Propose to typify name so no longer based on *Legousia.*
> Brummitt, R. K. (1988). Report of the Committee for Spermatophyta: 35. Taxon 37: 444-450. En. — Conservation recommended.
> Goldblatt, P. & J. Manning (2000). Cape plants. A conspectus of the Cape flora of South Africa. *Prismatocarpus.* Strelitzia 9: 393-394, 706. En. — Checklist for Cape of Good Hope, with brief species desciptions.
> Welman, W. G. (2000). *Prismatocarpus* L'Hér. In O. A. Leistner (ed.), Seed plants of southern Africa: families and genera. Strelitzia 10: 201. En. — Generic description.

Prismatocarpus L'Hér., Sert. Angl.: 1 (1789), nom. cons. *Campanula* sect. *Prismatocarpus* (L'Hér.) Schult. in Roem. & Schult., Syst. Veg. 5: 152 (1819).
S. Africa. 27.
> *Codiphus* Raf., Fl. Tellur. 2: 77 (1837).
> *Concilium* Raf., Fl. Tellur. 2: 78 (1837).

Prismatocarpus alpinus (Bond) Adamson, J. S. African Bot. 17: 106 (1952).
Cape Provinces. 27 CPP. Nanophan. or cham.
> ** Roella alpina* Bond, J. S. African Bot. 6: 61 (1940).

Prismatocarpus altiflorus L'Hér., Sert. Angl.: 2 (1789). *Campanula altiflora* (L'Hér.) Pers., Syn. Pl. 1: 192 (1805).
Cape Provinces. 27 CPP. Nanophan. or cham.

Prismatocarpus brevilobus A. DC. in DC., Prodr. 7: 443 (1839).
Cape Provinces. 27 CPP. Nanophan. or cham.

Prismatocarpus campanuloides (L. f.) Sond. in Harv. & Sond., Fl. Cap. 3: 589 (1865).
Cape Provinces. 27 CPP. Nanophan. or cham. *2n* = 34.

* *Polemonium campanuloides* L. f., Suppl. Pl.: 139 (1782). *Retzia campanuloides* (L. f.)
    Spreng., Syst. Veg. 1: 589 (1824).
Prismatocarpus ecklonii A. DC., Monogr. Campan.: 168 (1830), as 'eklonii'.
Prismatocarpus linariifolius A. DC., Monogr. Campan.: 169 (1830), as 'linariaefolius'.
    Campanula linariifolia (A. DC.) D. Dietr., Syn. Pl. 1: 756 (1839), as 'lineareifolia'.
Prismatocarpus strictus A. DC., Monogr. Campan.: 169 (1830).
Prismatocarpus campanuloides var. *dentatus* Adamson, J. S. African Bot. 17: 111 (1952).

Prismatocarpus candolleanus Cham., Linnaea 8: 197 (1833). *Campanula candolleana* (Cham.)
    D. Dietr., Syn. Pl. 1: 756 (1839).
    Cape Provinces. 27 CPP. Nanophan. or cham.
    Prismatocarpus virgatus Fourc., Trans. Roy. Soc. South Africa 21: 83 (1932).

Prismatocarpus cliffortioides Adamson, J. S. African Bot. 17: 114 (1952).
    Cape Provinces. 27 CPP. Nanophan. or cham.

Prismatocarpus cordifolius Adamson, J. S. African Bot. 17: 121 (1952).
    Cape Provinces. 27 CPP. Hemicr.

Prismatocarpus crispus L'Hér., Sert. Angl.: 2 (1789). *Campanula plicata* Pers., Syn. Pl. 1: 193
    (1805); non *Campanula crispa* Lam., Encycl. 1: 581 (1785). *Campanula longirostris* Banks ex
    D. Dietr., Syn. Pl. 1: 756 (1839).
    Cape Provinces. 27 CPP. Ther.

Prismatocarpus debilis Bolus ex Adamson, J. S. African Bot. 12: 37 (1946).
    Cape Provinces. 27 CPP. Hemicr.
    Prismatocarpus nitidus var. *ovatus* Adamson, J. S. African Bot. 17: 120 (1952).
    Prismatocarpus debilis var. *elongatus* Adamson, J. S. African Bot. 17: 122 (1952).

Prismatocarpus decurrens Adamson, J. S. African Bot. 17: 105 (1952).
    Cape Provinces. 27 CPP. Nanophan. or cham.

Prismatocarpus diffusus (L. f.) A. DC., Monogr. Campan.: 164 (1830).
    Cape Provinces. 27 CPP. Nanophan. or cham.
    * *Trachelium diffusum* L. f., Suppl. Pl.: 143 (1782). *Campanula diffusa* (L. f.) D. Dietr., Syn.
        Pl. 1: 756 (1839); non (A. DC.) D. Dietr., Syn. Pl. 1: 752 (1839).
    Lobelia laricina A. Spreng., Tent. Suppl. Syst. Veg.: 9 (1828). *Prismatocarpus laricinus* (A.
        Spreng.) C. Presl, Prodr. Monogr. Lobel.: 52 (1836).

Prismatocarpus fastigiatus C. Presl ex A. DC. in DC., Prodr. 7: 442 (1839).
    Cape Provinces. 27 CPP. Nanophan. or cham.

Prismatocarpus fruticosus (L.) L'Hér., Sert. Angl.: 2 (1789).
    Cape Provinces. 27 CPP. Nanophan. or cham.
    * *Campanula fruticosa* L., Sp. Pl.: 168 (1753). *Concilium pedunculare* Raf., Fl. Tellur. 2: 78
        (1837), as 'peduncularis'. *Lightfootia fruticosa* (L.) Druce, Bot. Exch. Club Soc. Brit.
        Isles 3: 420 (1914).
    Campanula subulata Thunb., Prodr. Pl. Cap.: 38 (1794). *Lightfootia ciliata* Spreng., Syst.
        Veg. 1: 809 (1824); non *Lightfootia subulata* L'Hér., Sert. Angl.: 4 (1789).
        Prismatocarpus subulatus (Thunb.) A. DC., Monogr. Campan.: 166 (1830).
    Prismatocarpus bergianus Cham., Linnaea 8: 199 (1833). *Campanula bergiana* (Cham.) D.
        Dietr., Syn. Pl. 1: 757 (1839). *Prismatocarpus subulatus* var. *bergianus* (Cham.) Sond. in
        Harv. & Sond., Fl. Cap. 3: 587 (1865).
    Ireon ciliatum Raf., Sylva Tellur.: 138 (1838).
    Prismatocarpus subulatus var. *penicillatus* A. DC. in DC., Prodr. 7: 444 (1839), as
        'penicillata'.

Prismatocarpus hildebrandtii Vatke, Linnaea 38: 701 (1874).
Cape Provinces. 27 CPP. Ther.

Prismatocarpus hispidus Adamson, J. S. African Bot. 17: 112 (1952).
Cape Provinces. 27 CPP. Nanophan. or cham.

Prismatocarpus implicatus Adamson, J. S. African Bot. 17: 128 (1952).
Cape Provinces. 27 CPP. Hemicr.

Prismatocarpus lasiophyllus Adamson, J. S. African Bot. 17: 120 (1952).
Cape Provinces. 27 CPP. Hemicr.

Prismatocarpus lycioides Adamson, J. S. African Bot. 17: 123 (1952).
Cape Provinces. 27 CPP. Nanophan. or cham.

Prismatocarpus lycopodioides A. DC. in DC., Prodr. 7: 443 (1839).
Cape Provinces. 27 CPP. Nanophan. or cham.
Prismatocarpus subulatus var. *pauciflorus* Sond. in Harv. & Sond., Fl. Cap. 3: 587 (1865).
Prismatocarpus lycopodioides var. *hispidus* Adamson, J. S. African Bot. 17: 103 (1952).

Prismatocarpus nitidus L'Hér., Sert. Angl.: 2 (1789). *Campanula prismatocarpus* Aiton, Hort.
Kew. 1: 224 (1789); non *Campanula nitida* Aiton, Hort. Kew. 1: 221 (1789). *Codiphus
nitidus* (L'Hér.) Raf., Fl. Tellur. 2: 77 (1837).
Cape Provinces. 27 CPP. Hemicr.
Prismatocarpus hookeri Sweet, Hort. Brit. (ed. 2): 328 (1830).

Prismatocarpus pauciflorus Adamson, J. S. African Bot. 17: 126 (1952).
Cape Provinces. 27 CPP. Nanophan. or cham.

Prismatocarpus pedunculatus (P. J. Bergius) A. DC. in DC., Prodr. 7: 443 (1839).
Cape Provinces. 27 CPP. Nanophan. or cham.
Campanula fruticosa Hill, Veg. Syst. 8: 9 (1765); non L., Sp. Pl.: 168 (1753).
* *Roella pedunculata* P. J. Bergius, Descr. Pl. Cap.: 42 (1767). *Prismatocarpus roelloides* var.
pedunculatus (P. J. Bergius) Zahlbr., Ann. K. K. Naturhist. Hofmus. 18: 402 (1903).
Polemonium roelloides L. f., Suppl. Pl.: 139 (1782). *Retzia roelloides* (L. f.) Spreng., Syst.
Veg. 1: 589 (1824). *Prismatocarpus roelloides* (L. f.) Sond. in Harv. & Sond., Fl. Cap. 3:
586 (1865).
Prismatocarpus paniculatus L'Hér., Sert. Angl.: 2 (1789). *Campanula pedunculata* J. F.
Gmel., Syst. Nat. 2: 352 (1791); non *Campanula paniculata* L. f., Suppl. Pl.: 139 (1782).
Prismatocarpus interruptus L'Hér., Sert. Angl.: 2 (1789). *Campanula interrupta* (L'Hér.)
Pers., Syn. Pl. 1: 192 (1805). *Prismatocarpus roelloides* var. *interruptus* (L'Hér.) Sond. in
Harv. & Sond., Fl. Cap. 3: 587 (1865).
Campanula ericoides Lam., Tabl. Encycl. 2: 63 (1796). *Roella ericoides* (Lam.) Spreng.,
Syst. Veg. 1: 723 (1824).
Prismatocarpus grandiflorus E. Mey. ex A. DC. in DC., Prodr. 7: 443 (1839).
Prismatocarpus roelloides var. *grandiflorus* (E. Mey. ex A. DC.) Sond. in Harv. & Sond.,
Fl. Cap. 3: 587 (1865).

Prismatocarpus pilosus Adamson, J. S. African Bot. 17: 127 (1952).
Cape Provinces. 27 CPP. Nanophan. or cham.

Prismatocarpus rogersii Fourc., Trans. Roy. Soc. South Africa 21: 83 (1932).
Cape Provinces. 27 CPP. Nanophan. or cham.

Prismatocarpus schlechteri Adamson, J. S. African Bot. 17: 111 (1952).
Cape Provinces. 27 CPP. Nanophan. or cham.

Prismatocarpus sessilis Eckl. ex A. DC., Monogr. Campan.: 171 (1830). *Campanula sessilis*
(Eckl. ex A. DC.) D. Dietr., Syn. Pl. 1: 756 (1839).
Cape Provinces. 27 CPP. Hemicr.
Prismatocarpus acerosus Schinz, Bull. Herb. Boissier 2: 217 (1894).
Wahlenbergia filicaulis R. D. Good, J. Bot. 62: 48 (1924).
Prismatocarpus sessilis var. *macrocarpus* Adamson, J. S. African Bot. 17: 117 (1952).

Prismatocarpus spinosus Adamson, J. S. African Bot. 17: 115 (1952).
Cape Provinces. 27 CPP. Nanophan. or cham.

Prismatocarpus tenellus Oliv., Hooker's Icon. Pl. 15: pl. 1460 (1884).
Cape Provinces. 27 CPP. Hemicr.

Prismatocarpus tenerrimus H. Buek in Eckl. & Zeyh., Enum. Pl. Afric. Austral.: 384 (1837).
Campanula tenerrima (H. Buek) D. Dietr., Syn. Pl. 1: 757 (1839).
Cape Provinces. 27 CPP. Hemicr.
Prismatocarpus burchelii Vatke, Linnaea 38: 701 (1874).

Synonyms:

Prismatocarpus acerosus Schinz === **Prismatocarpus sessilis** Eckl. ex A. DC.
Prismatocarpus bergianus Cham. === **Prismatocarpus fruticosus** (L.) L'Hér.
Prismatocarpus burchelii Vatke === **Prismatocarpus tenerrimus** H. Buek
Prismatocarpus campanuloides var. *dentatus* Adamson === **Prismatocarpus campanuloides**
(L. f.) Sond.
Prismatocarpus confertus Moench === **Legousia hybrida** (L.) Delarbre
Prismatocarpus cordatus (Vis.) Rchb. === **Legousia speculum-veneris** (L.) Durande ex Vill.
Prismatocarpus debilis var. *elongatus* Adamson === **Prismatocarpus debilis** Bolus ex Adamson
Prismatocarpus ecklonii A. DC. === **Prismatocarpus campanuloides** (L. f.) Sond.
Prismatocarpus falcatus Ten. === **Legousia falcata** (Ten.) Fritsch ex Janch.
Prismatocarpus grandiflorus E. Mey. ex A. DC. === **Prismatocarpus pedunculatus** (P. J.
Bergius) A. DC.
Prismatocarpus hirtus Ten. === **Legousia speculum-veneris** (L.) Durande ex Vill.
Prismatocarpus hookeri Sweet === **Prismatocarpus nitidus** L'Hér.
Prismatocarpus hybridus (L.) L'Hér. === **Legousia hybrida** (L.) Delarbre
Prismatocarpus interruptus L'Hér. === **Prismatocarpus pedunculatus** (P. J. Bergius) A. DC.
Prismatocarpus junceus H. Buek === **Wahlenbergia juncea** (H. Buek) Lammers
Prismatocarpus junceus Schinz === **Namacodon schinzianus** (Markgr.) Thulin
Prismatocarpus laricinus (A. Spreng.) C. Presl === **Prismatocarpus diffusus** (L. f.) A. DC.
Prismatocarpus linariifolius A. DC. === **Prismatocarpus campanuloides** (L. f.) Sond.
Prismatocarpus lycopodioides var. *hispidus* Adamson === **Prismatocarpus lycopodioides** A. DC.
Prismatocarpus nitidus var. *ovatus* Adamson === **Prismatocarpus debilis** Adamson
Prismatocarpus paniculatus L'Hér. === **Prismatocarpus pedunculatus** (P. J. Bergius) A. DC.
Prismatocarpus pentagonia (L.) L'Hér. === **Legousia pentagonia** (L.) Thell.
Prismatocarpus perfoliatus (L.) Sweet === **Triodanis perfoliata** (L.) Nieuwl.
Prismatocarpus rhodesicus Adamson === **Gunillaea rhodesica** (Adamson) Thulin
Prismatocarpus roelloides (L. f.) Sond. === **Prismatocarpus pedunculatus** (P. J. Bergius) A. DC.
Prismatocarpus roelloides var. *grandiflorus* (E. Mey. ex A. DC.) Sond. === **Prismatocarpus
pedunculatus** (P. J. Bergius) A. DC.
Prismatocarpus roelloides var. *interruptus* (L'Hér.) Sond. === **Prismatocarpus pedunculatus** (P.
J. Bergius) A. DC.
Prismatocarpus roelloides var. *pedunculatus* (P. J. Bergius) Zahlbr. === **Prismatocarpus
pedunculatus** (P. J. Bergius) A. DC.
Prismatocarpus scaber Lowe === **Legousia scabra** (Lowe) Gamisans
Prismatocarpus schinzianus Markgr. === **Namacodon schinzianus** (Markgr.) Thulin
Prismatocarpus sessilis var. *macrocarpus* Adamson === **Prismatocarpus sessilis** Eckl. ex A. DC.
Prismatocarpus speculum-veneris (L.) L'Hér. === **Legousia speculum-veneris** (L.) Durande ex Vill.

Prismatocarpus speculum-veneris var. *hirtus* K. Koch === **Legousia speculum-veneris** (L.) Durande ex Vill.

Prismatocarpus strictus A. DC. === **Prismatocarpus campanuloides** (L. f.) Sond.

Prismatocarpus subulatus (Thunb.) A. DC. === **Prismatocarpus fruticosus** (L.) L'Hér.

Prismatocarpus subulatus var. *bergianus* (Cham.) Sond. === **Prismatocarpus fruticosus** (L.) L'Hér.

Prismatocarpus subulatus var. *pauciflorus* Sond. === **Prismatocarpus lycopodioides** A. DC.

Prismatocarpus subulatus var. *penicillatus* A. DC. === **Prismatocarpus fruticosus** (L.) L'Hér.

Prismatocarpus virgatus Fourc. === **Prismatocarpus candolleanus** Cham.

Protoedraianthus

This genus was established by Lakušić (1988) to accomodate species of *Edraianthus* (q. v.) that differ in several features, including lateral capsule dehiscence via pores or valves as in *Campanula* (vs. irregularly at apex) and white or yellowish (vs. blue or violet) corollas. Unfortunately, he failed to comply with various provisions of the ICBN, and all but one of the species assigned here do not have a valid name in the genus. Type [by monotypy]: *Protoedraianthus majae* Lakušić.

Lakušić, R. (1988). *Protoedraianthus* Lakušić status nov. (syn.: *Edraianthus* DC. subgen. *Protoedraianthus* Lakušić subgen. nov.) In Zbnornik Referata Naučnog skupa "Minerali, Stijene, Izumrli i živi svijet Bosne i Hercegovine:" 263-272, illus. Sarajevo: Zemaljski Muzej Bosne i Hercegovine. Bos. — Establishment of genus; several names not validly published under ICBN Art. 33.3 or Art. 37.1.

Synonyms:
Protoedraianthus Lakušić === **Edraianthus** A. DC.
Protoedraianthus majae Lakušić === **Edraianthus glisicii** Černjavski & Soška

Pseudocampanula

This genus was segregated from *Campanula* largely on the basis of the "hardness" of its capsule. It is treated most often as *Campanula* subg. *Pseudocampanula* (Kolak.) Ogan. or sect. *Pseudocampanula* (Kolak.) Victorov. Type [designated by the author]: *Pseudocampanula dzaaku* (Albov) Kolak.

Kolakovskii, A. A. (1980). [*Pseudocampanula* — a new monotypical genus from Kolkheti limestones.] Soobšč. Akad. Nauk Gruzinsk. SSR 97: 413-415. Ru. — Establishment of genus.

Synonyms:
Pseudocampanula Kolak. === **Campanula** L.
Pseudocampanula dzaaku (Albov) Kolak. === **Campanula dzaaku** Albov

Pseudonemacladus

Nemacladoideae, 1 species, eastern Mexico. Allied to *Nemacladus* and *Parishella*, the other genera of the subfamily (Haberle & Ayers 1997). Although the name *Baclea* Greene (q.v.) has priority, it is a later homonym of *Baclea* Fourn. ex Baill. (Asclepiadaceae). Based on: *Baclea* Greene.

McVaugh, R. (1943). *Pseudonemacladus* McVaugh, nom. nov. In North American Flora 32A: 3-4. New York: New York Botanical Garden. En. — Validation of replacement name; flora with description and full nomenclature.

- Wimmer, F. E. (1968). *Pseudonemacladus* McVaugh. In A. Engler & L. Diels, Das Pflanzenreich IV. 276c: 935 + Taf. 31, illus. Berlin: Akademie-Verlag. Ge., La. — Monograph with description, full nomenclature, and specimen citations.

  Haberle, R. C. & T. J. Ayers (1997). Systematics of *Pseudonemacladus* (Nemacladaceae) [abstract]. Amer. J. Bot. 84 (6, Suppl.): 200. En. — Cytological, morphological, and molecular data support relationship to *Nemacladus* and *Parishella*.

Pseudonemacladus McVaugh, N. Amer. Fl. 32A: 3 (1943).
  Mexico. 79.
  * *Baclea* Greene, Erythrea 1: 238 (1893); non Fourn. ex Baill., Dict. Bot. 1: 338 (1877).
      Nemacladus sect. *Baclea* Kuntze in T. Post & Kuntze, Lex. Gen. Phan.: 385 (1904).

Pseudonemacladus oppositifolius (B. L. Rob.) McVaugh, N. Amer. Fl. 32A: 3 (1943).
  E. Mexico (San Luis Potosí, Hidalgo). 79 MXE. Hemicr.
  * *Nemacladus oppositifolia* B. L. Rob., Proc. Amer. Acad. Arts 26: 168 (1891). *Baclea oppositifolia* (B. L. Rob.) Greene, Erythrea 1: 238 (1893).

Quinquelocularia

This genus was established to accomodate a species of *Campanula* with a five-locular ovary and unappendaged calyx. Its type is a species of *Campanula* sect. *Quinqueloculares* (Boiss.) Phitos. Type [by monotypy]: *Quinquelocularia crispa* (Lam.) K. Koch.

  Koch, K. (1850). Beiträge zu einer Flora des Orients, VI. Linnaea 23: 577-713. Ge. — Segregation of genus from *Campanula*.

Synonyms:
Quinquelocularia K. Koch === **Campanula** L.
Quinquelocularia crispa (Lam.) K. Koch === **Campanula crispa** Lam.

Rapunculus

This is a pre-Linnaean name for *Phyteuma*, used by authors with absolutist views of priority. Type [designated here]: *Rapunculus spicatus* (L.) Mill.

  Miller, P. (1754). The gardeners dictionary ... abridged from the last folio edition (ed. 4). London: J. & J. Rivington. En. — First use of pre-Linnaean generic name after 1753.

  Miller, P. (1768). The gardeners dictionary (ed. 8). London: J. & F. Rivington. En. — Validation of requisite binomials.

  Bubani, P. (1899). Flora pyrenaea per ordines naturales gradatim digesta, vol. 2. Milan: Ulrich Hoepli. La. — Accepts substitution of *Rapunculus* for *Phyteuma* on grounds of priority.

Synonyms:
Rapunculus Mill. === **Phyteuma** L.
Rapunculus caesalpini Bubani === ?
Rapunculus comosus (L.) Mill. === **Physoplexis comosa** (L.) Schur
Rapunculus hemisphericus (L.) Mill. === **Phyteuma hemisphaericum** L.
Rapunculus orbicularis (L.) Mill. === **Phyteuma orbiculare** L.
Rapunculus ovatus Bubani === **Phyteuma spicatum** L.
Rapunculus pauciflorus (L.) Mill. === **Phyteuma globulariifolium** Sternb. & Hoppe
Rapunculus scheuchzeri Bubani === **Phyteuma charmelii** Vill.
Rapunculus spicatus (L.) Mill. === **Phyteuma spicatum** L.
Rapunculus sylvestris Tragi ex Bubani === **Phyteuma orbiculare** L.

Rapuntia

This genus, like *Marianthemum* and *Medium* (q.v.) was erected to accomodate species of *Campanula* with reflexed calyx appendages. Its type is a member of *Campanula* sect. *Quinqueloculares* (Boiss.) Phitos. Type [by monotypy]: *Rapuntia medium* (L.) Chevall.

> Chevallier, F. F. (1836). Flore générale des environs de Paris, selon la méthode naturelle, ed. 2, vol. 2. Paris: Ferra. Fr. — Segregation of genus from *Campanula*.

Synonyms:
Rapuntia Chevall. === **Campanula** L.
Rapuntia medium (L.) Chevall. === **Campanula medium** L.

Rapuntium

As with *Dortmanna* (q.v.), this pre-Linnaean name for *Lobelia* was used by later authors who considered the proper type of *Lobelia* to be *L. plumieri* L., a species of Goodeniaceae now called *Scaevola plumieri* (L.) Vahl. However, one major proponent of substituting *Rapuntium* for *Lobelia* (Presl 1836) inexplicably circumscribed *Lobelia* so as to exclude *every* species placed there by Linnaeus, including *L. plumieri,* assigning to it species that are today assigned to *Siphocampylus*. Type [designated here]: *Rapuntium cardinale* (L.) Mill.

> Miller, P. (1754). The gardeners dictionary ... abridged from the last folio edition (ed. 4). London: J. & J. Rivington. En. — First use of pre-Linnaean generic name after 1753.
> Miller, P. (1768). The gardeners dictionary (ed. 8). London: J. & F. Rivington. En. — Validation of requisite binomials.
> Moench, C. (1794). *Rapuntium*. In Methodus plantas horti botanici et agri Marburgensis, a staminum situ describendi: 655-656. Marburg: Nova Libraria Academiae. La. — Affirmation of use of this name for *Lobelia*.
> Moench, C. (1802). *Rapuntium*. In Supplementum ad methodum plantas a staminum situ describendi: 277-278. Marburg: Nova Libraria Academiae. La. — Supplement to Moench (1794).
> Presl, C. (1836). *Rapuntium*. Tourn. Mill. Gaertn. Moench. In Prodromus monographiae Lobeliacearum: 11-31. Prague: Theophilus Haase. La. — Affirmation of use of this name for *Lobelia*.

Synonyms:
Rapuntium Mill. === **Lobelia** L.
Rapuntium sect. *Grammatotheca* (C. Presl) Kuntze === **Grammatotheca** C. Presl
Rapuntium sect. *Haynaldia* Kuntze === **Lobelia** L.
Rapuntium sect. *Isolobus* Kuntze === **Lobelia** L.
Rapuntium sect. *Mezleria* (C. Presl) Kuntze === **Monopsis** Salisb.
Rapuntium sect. *Monopsis* (Salisb.) Kuntze === **Monopsis** Salisb.
Rapuntium sect. *Parastranthus* (Salisb.) Kuntze === **Monopsis** Salisb.
Rapuntium sect. *Trematocarpus* Kuntze === **Trematolobelia** Zahlbr. ex Rock
Rapuntium sect. *Trimeris* (C. Presl) Kuntze === **Lobelia** L.
Rapuntium sect. *Tupa* (G. Don) Kuntze === **Lobelia** L.
Rapuntium sect. *Tylomium* (C. Presl) Kuntze === **Lobelia** L.
Rapuntium acuminatum (Sw.) C. Presl === **Lobelia acuminata** Sw.
Rapuntium acutangulum C. Presl === **Lobelia erinus** L.
Rapuntium affine C. Presl === **Lobelia zeylanica** L.
Rapuntium alatum (L. f.) C. Presl === **Lobelia anceps** L. f.
Rapuntium alsinoides (Lam.) C. Presl === **Lobelia alsinoides** Lam.
Rapuntium amoenum (Michx.) C. Presl === **Lobelia amoena** Michx.
Rapuntium amygdalinum (Willd. ex Schult.) C. Presl === **Lobelia laxiflora** Kunth
  subsp. **laxiflora**

Rapuntium anceps (L. f.) C. Presl === **Lobelia anceps** L. f.

Rapuntium angulatum (G. Forst.) C. Presl === **Lobelia angulata** G. Forst.

Rapuntium aquaticum (Cham.) C. Presl === **Lobelia aquatica** Cham.

Rapuntium arabideum C. Presl === **Wimmerella arabidea** (C. Presl) Serra, M. B. Crespo & Lammers

Rapuntium arenarioides C. Presl === **Lobelia heyneana** Schult.

Rapuntium bellidifolium (L. f.) C. Presl === **Lobelia erinus** L.

Rapuntium bellidifolium var. *brevidens* C. Presl === **Lobelia flaccida** (C. Presl) A. DC. subsp. **flaccida**

Rapuntium besserianum C. Presl === **Lobelia polyphylla** Hook. & Arn.

Rapuntium bicolor (Sims) C. Presl === **Lobelia erinus** L.

Rapuntium bifidum (Thunb.) C. Presl === **Wimmerella bifida** (Thunb.) Serra, M. B. Crespo & Lammers

Rapuntium blandum (D. Don) C. Presl === **Lobelia bridgesii** Hook. & Arn.

Rapuntium boryanum C. Presl === **Lobelia filiformis** Lam.

Rapuntium bracteosum C. Presl === **Lobelia polyphylla** Hook. & Arn.

Rapuntium breynii (Lam.) C. Presl === **Monopsis lutea** (L.) Urb.

Rapuntium bridgesii (Hook. & Arn.) C. Presl === **Lobelia bridgesii** Hook. & Arn.

Rapuntium brownianum (Schult.) C. Presl === **Lobelia gibbosa** Labill.

Rapuntium burmanianum C. Presl === ?

Rapuntium caeruleum (Sims) C. Presl === **Lobelia caerulea** Sims

Rapuntium caeruleum var. *glabrescens* C. Presl === **Lobelia caerulea** Sims

Rapuntium caeruleum var. *latidens* C. Presl === **Lobelia caerulea** Sims

Rapuntium caespitosum (Blume) C. Presl === **Lobelia chinensis** Lour.

Rapuntium campanuloides (Thunb.) C. Presl === **Lobelia chinensis** Lour.

Rapuntium camporum (Pohl) C. Presl === **Lobelia camporum** Pohl

Rapuntium capillifolium C. Presl === **Lobelia capillifolia** (C. Presl) A. DC.

Rapuntium capitatum C. Presl === **Monopsis lutea** (L.) Urb.

Rapuntium cardinale (L.) Mill. === **Lobelia cardinalis** L.

Rapuntium cavanillesianum (Schult.) C. Presl === **Lobelia laxiflora** Kunth subsp. **laxiflora**

Rapuntium ceratophyllum C. Presl === **Lobelia chamaepitys** Lam.

Rapuntium ceratophyllum var. *glabratum* C. Presl === **Lobelia chamaepitys** Lam.

Rapuntium chamaedryifolium C. Presl === **Lobelia chamaedryifolia** (C. Presl) A. DC.

Rapuntium chamaepitys (Lam.) C. Presl === **Lobelia chamaepitys** Lam.

Rapuntium cheiranthifolium C. Presl === **Lobelia serpens** Lam.

Rapuntium chenopodiifolium (Wall. ex G. Don) C. Presl === **Lobelia cliffortiana** L.

Rapuntium chinense (Lour.) C. Presl === **Lobelia chinensis** Lour.

Rapuntium cinereum (Thunb.) C. Presl === **Wahlenbergia albicaulis** (Sond.) Lammers

Rapuntium circaeoides C. Presl === **Lobelia circaeoides** (C. Presl) A. DC.

Rapuntium cirsiifolium (Lam.) C. Presl === **Lobelia cirsiifolia** Lam.

Rapuntium clifortianum (L.) Mill. === **Lobelia cliffortiana** L.

Rapuntium claytonianum (Michx.) C. Presl === **Lobelia spicata** Lam.

Rapuntium coccineum Moench === **Lobelia cardinalis** L.

Rapuntium collinum (Kunth) C. Presl === **Lobelia collina** Kunth

Rapuntium coloratum C. Presl === **Lobelia amoena** Michx.

Rapuntium coloratum (Wall.) C. Presl === **Lobelia colorata** Wall.

Rapuntium cordatum C. Presl === **Lobelia preslii** A. DC.

Rapuntium cordifolium Moench === **Lobelia cliffortiana** L.

Rapuntium cordigerum (Cav.) C. Presl === **Lobelia cardinalis** L.

Rapuntium cornopifolium (L.) C. Presl === **Lobelia coronopifolia** L.

Rapuntium coronopifolium var. *latifolium* C. Presl === **Lobelia coronopifolia** L.

Rapuntium corymbosum (G. Don) C. Presl === **Lobelia jasionoides** (A. DC.) E. Wimm.

Rapuntium corymbosum var. *longifolium* C. Presl === **Lobelia jasionoides** (A. DC.) E. Wimm.

Rapuntium cuneiforme (Labill.) C. Presl === **Lobelia anceps** L. f.

Rapuntium decurrens (Cav.) C. Presl === **Lobelia decurrens** Cav.

Rapuntium dentatum (Cav.) C. Presl === **Lobelia dentata** Cav.

Rapuntium dortmanna (L.) C. Presl === **Lobelia dortmanna** L.

Rapuntium dregeanum C. Presl === **Lobelia dregeana** (C. Presl) A. DC.

Rapuntium ecklonianum C. Presl === **Lobelia filicaulis** (C. Presl) Schönl.

Rapuntium ericoides C. Presl === **Monopsis lutea** (L.) Urb.

Rapuntium erinoides (L.) Mill. === **Lobelia erinus** L.

Rapuntium erinus (L.) Mill. === **Lobelia erinus** L.

Rapuntium exaltatum (Pohl) C. Presl === **Lobelia exaltata** Pohl

Rapuntium excelsum (Bonpl.) C. Presl === **Lobelia excelsa** Bonpl.

Rapuntium fastigiatum (Kunth) C. Presl === **Lobelia fastigiata** Kunth

Rapuntium fenestrale (Cav.) C. Presl === **Lobelia fenestralis** Cav.

Rapuntium fervens (Thunb.) C. Presl === **Lobelia fervens** Thunb.

Rapuntium filiforme (Lam.) C. Presl === **Lobelia filiformis** Lam.

Rapuntium flaccidum C. Presl === **Lobelia flaccida** (C. Presl) A. DC.

Rapuntium flavum C. Presl ex Eckl. & Zeyh. === **Monopsis flava** (C. Presl) E. Wimm.

Rapuntium flexuosum C. Presl === **Lobelia erinus** L.

Rapuntium flexuosum C. Presl === **Lobelia flexuosa** (C. Presl) A. DC.

Rapuntium fluviatile (R. Br.) C. Presl === **Isotoma fluviatilis** (R. Br.) F. Muell. ex Benth.

Rapuntium foliosum (Kunth) C. Presl === **Lobelia decurrens** Cav. subsp. **decurrens**

Rapuntium fulgens (Humb. & Bonpl. ex Willd.) C. Presl === **Lobelia cardinalis** L.

Rapuntium genistoides C. Presl === **Lobelia patula** L. f.

Rapuntium gibbosum (Labill.) C. Presl === **Lobelia gibbosa** Labill.

Rapuntium gilgii (Engl.) Kuntze === **Lobelia gilgii** Engl.

Rapuntium glabrifolium C. Presl === **Lobelia thermalis** Thunb.

Rapuntium glabrifolium var. *mundtianum* (Cham.) C. Presl === **Lobelia thermalis** Thunb.

Rapuntium glandulosum (Walt.) C. Presl === **Lobelia glandulosa** Walt.

Rapuntium gracile C. Presl === **Lobelia andrewsii** Lammers

Rapuntium gramineum (Lam.) C. Presl === **Lobelia cardinalis** L.

Rapuntium gruinum (Cav.) C. Presl === **Lobelia gruina** Cav.

Rapuntium haenkeanum C. Presl === **Lobelia laxiflora** Kunth subsp. **laxiflora**

Rapuntium heteromallum (Schrad.) C. Presl === **Lobelia comosa** L.

Rapuntium heterophyllum (Labill.) C. Presl === **Lobelia heterophylla** Labill.

Rapuntium hirsutum (L.) Moench === **Gnidia hirsuta** (L.) Thulin (Thymeleaceae)

Rapuntium hyssopifolium C. Presl === **Lobelia polyphylla** Hook. & Arn.

Rapuntium incisum C. Presl === **Lobelia pubescens** Aiton

Rapuntium inflatum (L.) Mill. === **Lobelia inflata** L.

Rapuntium inundatum (R. Br.) C. Presl === **Isotoma fluviatilis** (R. Br.) F. Muell. ex Benth. subsp. **fluviatilis**

Rapuntium irriguum (R. Br.) C. Presl === **Lobelia irrigua** R. Br.

Rapuntium kalmii (L.) C. Presl === **Lobelia kalmii** L.

Rapuntium kamtschaticum C. Presl === **Lobelia sessilifolia** Lamb.

Rapuntium kraussii (Graham) C. Presl === **Lobelia kraussii** Graham

Rapuntium kunthianum C. Presl === **Lobelia laxiflora** subsp. **angustifolia** (A. DC.) Eakes & Lammers

Rapuntium laciniatum (Lam.) C. Presl === **Siphocampylus sonchifolius** (Sw.) McVaugh

Rapuntium lasianthum C. Presl === **Lobelia linearis** Thunb.

Rapuntium laxiflorum (Kunth) C. Presl === **Lobelia laxiflora** Kunth

Rapuntium leschenaultianum C. Presl === **Lobelia leschenaultiana** (C. Presl) Skottsb.

Rapuntium linarioides C. Presl === **Lobelia linarioides** (C. Presl) A. DC.

Rapuntium longiflorum (L.) Mill. === **Hippobroma longiflora** (L.) G. Don

Rapuntium longifolium C. Presl === **Lobelia cardinalis** L.

Rapuntium lucaeanum C. Presl === **Lobelia bridgesii** Hook. & Arn.

Rapuntium lukwangulense (Engl.) Kuntze === **Lobelia lukwangulensis** Engl.

Rapuntium luteum (L.) C. Presl === **Monopsis lutea** (L.) Urb.

Rapuntium maculare C. Presl === **Lobelia caerulea** Sims

Rapuntium maculare var. procerum C. Presl === **Lobelia caerulea** Sims

Rapuntium membranaceum (R. Br.) C. Presl === **Lobelia membranacea** R. Br.
Rapuntium michauxii (Nutt.) C. Presl === **Lobelia inflata** L.
Rapuntium micranthum (Kunth) C. Presl === **Diastatea micrantha** (Kunth) McVaugh
Rapuntium microdon C. Presl === **Lobelia comosa** L.
Rapuntium molle (Graham) C. Presl === **Lobelia xalapensis** Kunth
Rapuntium monticolum (Kunth) C. Presl === **Lobelia xalapensis** Kunth
Rapuntium mucronatum (Cav.) C. Presl === **Lobelia tupa** L.
Rapuntium multicaule C. Presl === ?
Rapuntium muscoides (Cham.) C. Presl === **Lobelia muscoides** Cham.
Rapuntium mundtianum (Cham.) C. Presl === **Lobelia thermalis** Thunb.
Rapuntium nanum (Kunth) C. Presl === **Lobelia nana** Kunth
Rapuntium nicotianifolium (Roth ex Schult.) C. Presl === **Lobelia nicotianifolia** Roth
    ex Schult.
Rapuntium nummularioides (Cham.) C. Presl === **Lobelia nummularioides** Cham.
Rapuntium nummularia (Lam.) C. Presl === **Lobelia nummularia** Lam.
Rapuntium nuttallii (Schult.) C. Presl === **Lobelia nuttallii** Schult.
Rapuntium ottonianum C. Presl === **Lobelia anceps** L. f.
Rapuntium obovatum (G. Don) C. Presl === **Siphocampylus obovatus** (G. Don) E. Wimm.
Rapuntium ovatum C. Presl === **Lobelia cuneifolia** Link & Otto
Rapuntium ovatum (G. Don) C. Presl === **Siphocampylus ovatus** (G. Don) E. Wimm.
Rapuntium ovatum var. *hirsutum* C. Presl === **Lobelia cuneifolia** Link & Otto
Rapuntium paludosum (Nutt.) C. Presl === **Lobelia paludosa** Nutt.
Rapuntium parvifolium C. Presl === **Lobelia rarifolia** E. Wimm.
Rapuntium patulum (L. f.) C. Presl === **Lobelia patula** L. f.
Rapuntium paucidentatum C. Presl === **Lobelia tomentosa** L. f.
Rapuntium pauciflorum (Kunth) C. Presl === **Lobelia gruina** Cav.
Rapuntium pedunculare C. Presl === **Lobelia neglecta** Schult.
Rapuntium pedunculare var. *angustifolium* C. Presl === **Lobelia neglecta** Schult.
Rapuntium pedunculare var. *brevifolium* C. Presl === **Lobelia neglecta** Schult.
Rapuntium pedunculare var. *latifolium* C. Presl === **Lobelia neglecta** Schult.
Rapuntium pedunculatum (R. Br.) C. Presl === **Lobelia pedunculata** R. Br.
Rapuntium persicifolium (Lam.) C. Presl === **Lobelia persicifolia** Lam.
Rapuntium pinifolium (L.) C. Presl === **Lobelia pinifolia** L.
Rapuntium platycaulum C. Presl === **Lobelia fervens** Thunb. subsp. **fervens**
Rapuntium polyphyllum (Hook. & Arn.) C. Presl === **Lobelia polyphylla** Hook. & Arn.
Rapuntium pteropodum C. Presl === **Lobelia pteropoda** (C. Presl) A. DC.
Rapuntium puberulum (Michx.) C. Presl === **Lobelia puberula** Michx.
Rapuntium puberulum var. *glabellum* C. Presl === **Lobelia puberula** Michx.
Rapuntium pubescens (Aiton) C. Presl === **Lobelia pubescens** Aiton
Rapuntium purpurascens (R. Br.) C. Presl === **Lobelia purpurascens** R. Br.
Rapuntium purpureum (G. Don) C. Presl === **Lobelia polyphylla** Hook. & Arn.
Rapuntium pusillum C. Presl === **Lobelia divaricata** Hook. & Arn.
Rapuntium pygmaeum (Thunb.) C. Presl === **Wimmerella pygmaea** (Thunb.) Serra, M. B.
    Crespo & Lammers
Rapuntium pyramidale (Wall.) C. Presl === **Lobelia pyramidalis** Wall.
Rapuntium quadrangulare (R. Br.) C. Presl === **Lobelia quadrangularis** R. Br.
Rapuntium racemosum C. Presl === **Lobelia cirsiifolia** Lam.
Rapuntium radicans (Thunb.) C. Presl === **Lobelia chinensis** Lour.
Rapuntium rapunculoides (Kunth) C. Presl === **Lobelia gruina** Cav.
Rapuntium reinwardtianum C. Presl === **Lobelia heyneana** Schult.
Rapuntium reniforme (Cham.) C. Presl === **Lobelia reniformis** Cham.
Rapuntium richardianum C. Presl === **Siphocampylus decumbens** (A. Rich. ex Schult.) Juss.
    ex A. DC.
Rapuntium rigidulum (Kunth) C. Presl === **Lobelia laxiflora** Kunth subsp. **laxiflora**
Rapuntium roseum (Wall.) C. Presl === **Lobelia rosea** Wall.
Rapuntium rugulosum (Graham) C. Presl === **Lobelia angulata** G. Forst.

Rapuntium rupestre (Kunth) C. Presl === **Lobelia tenera** Kunth

Rapuntium scabripes C. Presl === **Lobelia flaccida** (C. Presl) A. DC. subsp. **flaccida**

Rapuntium secundum (Ruiz & Pav. ex G. Don) C. Presl === **Siphocampylus pavonis** E. Wimm.

Rapuntium serpens (Lam.) C. Presl === **Lobelia serpens** Lam.

Rapuntium serrulatum (Schott ex Brot.) C. Presl === **Lobelia urens** L.

Rapuntium setaceum (Thunb.) C. Presl === **Lobelia setacea** Thunb.

Rapuntium simplex (L.) C. Presl === **Monopsis simplex** (L.) E. Wimm.

Rapuntium simplicicaule (R. Br.) C. Presl === **Lobelia simplicicaulis** R. Br.

Rapuntium simsii (Sweet) C. Presl === **Lobelia coronopifolia** L.

Rapuntium siphiliticum (L.) Mill. === **Lobelia siphilitica** L.

Rapuntium sonchifolium (Sw.) C. Presl === **Siphocampylus sonchifolius** (Sw.) McVaugh

Rapuntium spartioides C. Presl === **Lobelia linearis** Thunb.

Rapuntium speciosum (Sweet) C. Presl === **Lobelia** × **speciosa** Sweet

Rapuntium spicatum C. Presl === **Lobelia fenestralis** Cav.

Rapuntium splendens (Humb. & Bonpl. ex Willd.) C. Presl === **Lobelia cardinalis** L.

Rapuntium strictum (Sw.) C. Presl === **Lobelia stricta** Sw.

Rapuntium subdentatum C. Presl === **Lobelia polyphylla** Hook. & Arn.

Rapuntium subtile (Kunth) C. Presl === **Diastatea micrantha** (Kunth) McVaugh

Rapuntium succulentum (Blume) C. Presl === **Lobelia zeylanica** L.

Rapuntium telephioides C. Presl === **Lobelia telephioides** (C. Presl) A. DC.

Rapuntium tenerum (Kunth) C. Presl === **Lobelia tenera** Kunth

Rapuntium tenuifolium C. Presl === **Lobelia linearis** Thunb.

Rapuntium tenuius (R. Br.) C. Presl === **Lobelia tenuior** R. Br.

Rapuntium thapsoideum (Schott ex Pohl) C. Presl === **Lobelia thapsoidea** Schott ex Pohl

Rapuntium thermale (Thunb.) C. Presl === **Lobelia thermalis** Thunb.

Rapuntium thunbergii (Sweet) C. Presl === **Lobelia coronopifolia** L.

Rapuntium tomentosum (L. f.) C. Presl === **Lobelia tomentosa** L. f.

Rapuntium tomentosum var. *tenuifolium* C. Presl === **Lobelia tomentosa** L. f.

Rapuntium trialatum (Buch.-Ham. ex D. Don) C. Presl === **Lobelia heyneana** Schult.

Rapuntium triquetrum (L.) C. Presl === **Lobelia comosa** L.

Rapuntium tupa (L.) C. Presl === **Lobelia tupa** L.

Rapuntium umbellatum C. Presl === ?

Rapuntium uranocomum (Cham.) C. Presl === **Lobelia fistulosa** Vell.

Rapuntium urens (L.) Mill. === **Lobelia urens** L.

Rapuntium variifolium (Sims) C. Presl === **Monopsis variifolia** (Sims) Urb.

Rapuntium veronicifolium (Willd. ex Schult.) C. Presl === **Lobelia tenera** Kunth

Rapuntium volkensii (Engl.) Kuntze === **Lobelia giberroa** Hemsl.

Rapuntium wallichianum C. Presl === **Lobelia pyramidalis** Wall.

Rapuntium xalapense (Kunth) C. Presl === **Lobelia xalapensis** Kunth

Rapuntium zeylanicum (L.) C. Presl === **Lobelia zeylanica** L.

Rhigiophyllum

Campanuloideae, 1 species, Cape Provinces of South Africa. Although the genus has not been included in any DNA phylogeny, on the basis of morphology it appears to be related to *Siphocodon* (q.v.); one suspects that both would be part of the "wahlenbergioid" clade (Eddie et al. 2003, under **General**) with *Craterocapsa* and *Roella*. Type [by monotypy]: *Rhigiophyllum squarrosum* Hochst.

Hochstetter, C. F. (1842). Nova genera plantarum Africae tum australis tum tropicae boreralis. Flora 25: 225-240. La. — Establishment of genus.

Dyer, R. A. (1975). *Rhigiophyllum* Hochst. In The genera of southern African flowering plants 1: 640. Pretoria: Department of Agricultural Technical Services. En. — Description.

Goldblatt, P. & J. Manning (2000). Cape plants. A conspectus of the Cape flora of South Africa. *Rhigiophyllum*. Strelitzia 9: 395. En. — Brief desciption.

Welman, W. G. (2000). *Rhigiophyllum* Hochst. In O. A. Leistner (ed.), Seed plants of southern Africa: families and genera. Strelitzia 10: 201. En. — Description.

Rhigiophyllum Hochst., Flora 25: 232 (1842).
S. Africa. 27.

Rhigiophyllum squarrosum Hochst., Flora 25: 232 (1842).
Cape Provinces. 27 CPP. Nanophan. or cham.

Rhizocephalum

This is treated by most authors (Wimmer 1953, McVaugh 1955) as *Lysipomia* subg. *Rhizocephalum* (Wedd.) E. Wimm. Type [designated by McVaugh, Brittonia 8: 70 (1955)]: *Lysipomia laciniata* A. DC.

Weddell, H. A. (1858). *Rhizocephalum*. In Chloris andina. Essai d'une flore de la région alpine des cordillères de l'Amérique du Sud 2: 11-13. Paris: Bertrand. Fr. — Segregation of genus from *Lysipomia*.
Baillon, H. (1885). *Lysipoma* [sic] H.B.K. In Histoire des plantes 8: 365. Paris: L. Hachette. Fr. — Genus merged with *Lysipomia*.
Wimmer, F. E. (1953). *Lysipomia* Humb. In A. Engler & L. Diels, Das Pflanzenreich IV. 276b: 745-753, illus. Berlin: Akademie-Verlag. Ge., La. — Recognition as subgenus within *Lysipomia*.
• McVaugh, R. (1955). A revision of *Lysipomia* (Campanulaceae, Lobelioideae). Brittonia 8: 69-105, illus. En. — Recognition as subgenus within *Lysipomia*.
van der Hammen, T. & A. M. Cleef (1978). Pollen morphology of *Lysipomia* H.B.K. and *Rhizocephalum* Wedd. (Campanulaceae), and the revision of the pollen determination "*Valeriana*" *stenophylla* Killip. Rev. Paleobot. Palynol. 25: 367-376, illus. En. — Palynological data support merger of *Lysipomia* and *Rhizocephalum*..

Synonyms:
Rhizocephalum Wedd. === **Lysipomia** Kunth
Rhizocephalum brachysiphonium Zahlbr. === **Lysipomia brachysiphonia** (Zahlbr.) E. Wimm.
Rhizocephalum brachysiphonium var. *brevifolium* Zahlbr. === **Lysipomia brachysiphonia** (Zahlbr.) E. Wimm.
Rhizocephalum candollei Wedd. === **Lysipomia laciniata** A. DC. subsp. **laciniata**
Rhizocephalum candollei var. *ciliata* Wedd. === **Lysipomia laciniata** subsp. **linearifolia** (E. Wimm.) McVaugh
Rhizocephalum candollei var. *pubescens* Wedd. === **Lysipomia laciniata** subsp. **microsperma** McVaugh
Rhizocephalum candollei var. *vulgare* Wedd. === **Lysipomia laciniata** A. DC. subsp. **laciniata**
Rhizocephalum gracilis E. Wimm. === **Lysipomia gracilis** (E. Wimm.) E. Wimm.
Rhizocephalum pumilum Wedd. === **Lysipomia pumila** (Wedd.) E. Wimm.

Rhynchopetalum

This is treated by modern authors (e.g., Wimmer 1953, Murata 1995, under **Lobelia**) as *Lobelia* sect. *Rhynchopetalum* (Fresen.) Benth. Type [by monotypy]: *Rhynchopetalum montanum* Fresen.

Fresenius, G. (1838). Diagnoses generum speciarumque novarum in Abyssinia a cl. Rueppell detectarum. Flora 21(2): 601-616. La. — Establishment of genus.
Bentham, G. (1876). *Lobelia*, Linn. In G. Bentham & J. D. Hooker, Genera plantarum 2: 551-553. London: L. Reeve. La. — Transferred to sectional rank within *Lobelia*.

Synonyms:
Rhynchopetalum Fresen. === **Lobelia** L.
Rhynchopetalum montanum Fresen. === **Lobelia rhynchopetalum** Hemsl.

Roella

Campanuloideae, 20 species, Cape Provinces of South Africa, with one species extending into KwaZulu-Natal. The genus is divided (Adamson 1952) into ser. *Muscosae* Adamson, ser. *Prostratae* Adamson, ser. *Roella*, ser. *Spicatae* Adamson, and ser. *Squarrosae* Adamson. In a molecular phylogeny (Eddie et al. 2003, under **General**), the type formed a clade (the "wahlenbergioids") with the sole included species of *Craterocapsa;* this clade was sister to the "campanuloid clade". The most recent monograph is that of Adamson (1952). Type [designated by Hitchc. & M. L. Green in Anon., Nomencl. Prop. Brit. Bot. 131 (1929)]: *Roella ciliata* L.

> Linnaeus, C. (1753). *Roella.* In Species plantarum: 170. Stockholm: Laurentius Salvius. La. — Establishment of genus in Pentandria Monogynia.
> Linnaeus, C. (1754). *Roella.* In Genera plantarum: 77. Stockholm: Laurentius Salvius. La. — Description of genus to accompany preceding.
> Adamson, R. S. (1946). *Roella reticulata.* J. S. African Bot. 12: 107-109. En. — Name actually refers to species of Asteraceae.
> Adamson, R. S. (1950). *Roella* L. In R. S. Adamson & T. M. Salter, Flora of the Cape Peninsula: 742-745. Cape Town: Juta. En. — Flora with key and descriptions.
> Adamson, R. S. (1952). A revision of the genera *Prismatocarpus* and *Roella.* J. S. African Bot. 17: 93-166. — Monograph, with keys, descriptions, infrageneric classification, full nomenclature, and specimen citations.
> Dyer, R. A. (1975). *Roella* L. In The genera of southern African flowering plants 1: 639. Pretoria: Department of Agricultural Technical Services. En. — Generic description.
> Belyaev, A. A. (1986). [The features of anatomy and surface ultrastructure of the seed coat in some representatives of critical genera of the Campanulaceae family.] Bot. Žurn. 71: 1371-1376 + 2 pl., illus. Ru. — Description of testa in *R. ciliata.*
> Goldblatt, P. & J. Manning (2000). Cape plants. A conspectus of the Cape flora of South Africa. *Roella.* Strelitzia 9: 395-396, 706-707. En. — Checklist for Cape of Good Hope, with brief species desciptions; circumscriptions of several species revised.
> Welman, W. G. (2000). *Roella* L. In O. A. Leistner (ed.), Seed plants of southern Africa: families and genera. Strelitzia 10: 201. En. — Generic description.

Roella L., Sp. Pl.: 170 (1753).
S. Africa. 27.

Roella amplexicaulis Wolley-Dod, J. Bot. 39: 400 (1901), as 'amplexicaule'.
Cape Provinces. 27 CPP. Nanophan. or cham.

Roella arenaria Schltr., Bot. Jahrb. Syst. 27: 193 (1899).
Cape Provinces. 27 CPP. Nanophan. or cham.

Roella bryoides H. Buek in Eckl. & Zeyh., Enum. Pl. Afric. Austral.: 386 (1837).
Cape Provinces. 27 CPP. Nanophan. or cham.
Roella ciliata var. *elongata* H. Buek in Eckl. & Zeyh., Enum. Pl. Afric. Austral.: 384 (1837).
Roella minor Eckl. ex H. Buek in Eckl. & Zeyh., Enum. Pl. Afric. Austral.: 385 (1837).
Roella ciliata var. *minor* (Eckl. ex H. Buek) Sond. in Harv. & Sond., Fl. Cap. 3: 591 (1865).
Roella gracilis H. Buek in Eckl. & Zeyh., Enum. Pl. Afric. Austral.: 385 (1837).

Roella ciliata L., Sp. Pl.: 170 (1753). *Campanula ciliaris* Salisb., Prodr. Stirp. Chap. Allerton: 127 (1796).
Cape Provinces. 27 CPP. Nanophan. or cham.

Roella compacta Schltr., Bot. Jahrb. Syst. 27: 193 (1899).
Cape Provinces. 27 CPP. Nanophan. or cham.
Roella cuspidata Adamson, J. S. African Bot. 17: 151 (1952).
Roella cuspidata var. *hispida* Adamson, J. S. African Bot. 17: 151 (1952).

Roella decurrens L'Hér., Sert. Angl.: 4 (1789).
Cape Provinces. 27 CPP. Hemicr. or ther.
Roella glabra Poir. in Lam., Encycl. 6: 232 (1804).

Roella dregeana A. DC. in DC., Prodr. 7: 446 (1839). *Roella ciliata* var. *dregeana* (A. DC.)
Sond. in Harv. & Sond., Fl. Cap. 3: 591 (1865).
Cape Provinces. 27 CPP. Nanophan. or cham.
Roella leptosepala Sond. in Harv. & Sond., Fl. Cap. 3: 593 (1865).
Roella nitida Schltr., Bot. Jahrb. Syst. 24: 446 (1897). *Roella dregeana* var. *nitida* (Schltr.)
Adamson, J. S. African Bot. 17: 139 (1952).
Roella psammophila Schltr., Bot. Jahrb. Syst. 24: 446 (1897).

Roella dunantii A. DC., Monogr. Campan.: 175 (1830). *Roella reticulata* var. *dunantii* (A.
DC.) Sond. in Harv. & Sond., Fl. Cap. 3: 593 (1865).
Cape Provinces. 27 CPP. Nanophan. or cham.

Roella glomerata A. DC. in DC., Prodr. 7: 447 (1839).
Cape Provinces & KwaZulu-Natal. 27 NAT CPP. Nanophan. or cham.

Roella goodiana Adamson, J. S. African Bot. 5: 55 (1939).
Cape Provinces. 27 CPP. Nanophan. or cham.
* *Roella ericoides* R. D. Good, J. Bot. 62: 48 (1924); non Spreng., Syst. Veg. 1: 723 (1824);
nec H. Buek in Eckl. & Zeyh., Enum. Pl. Afric. Austral.: 385 (1837).

Roella incurva Banks ex A. DC., Monogr. Campan.: 172 (1830). *Roella ciliata* var. *incurva*
(Banks ex A. DC.) Sond. in Harv. & Sond., Fl. Cap. 3: 591 (1865).
Cape Provinces. 27 CPP. Nanophan. or cham.
Roella rhodantha Adamson, J. S. African Bot. 17: 135 (1952).

Roella latiloba A. DC. in DC., Prodr. 7: 447 (1839).
Cape Provinces. 27 CPP. Nanophan. or cham.

Roella maculata Adamson, J. S. African Bot. 17: 136 (1952).
Cape Provinces. 27 CPP. Nanophan. or cham.

Roella muscosa L. f., Suppl. Pl.: 143 (1782).
Cape Provinces. 27 CPP. Hemicr.

Roella prostrata E. Mey. ex A. DC. in DC., Prodr. 7: 447 (1839). *Roella reticulata* var. *prostrata*
(E. Mey. ex A. DC.) Sond. in Harv. & Sond., Fl. Cap. 3: 592 (1865).
Cape Provinces. 27 CPP. Nanophan. or cham.
Roella reticulata var. *ternifolia* Sond. in Harv. & Sond., Fl. Cap. 3: 592 (1865).
Roella incurva var. *rigida* Adamson, J. S. African Bot. 17: 135 (1952).

Roella recurvata A. DC. in DC., Prodr. 7: 447 (1839).
Cape Provinces. 27 CPP. Nanophan. or cham.

Roella secunda H. Buek in Eckl. & Zeyh., Enum. Pl. Afric. Austral.: 386 (1837).
Cape Provinces. 27 CPP. Nanophan. or cham.
Roella glauca H. Buek in Eckl. & Zeyh., Enum. Pl. Afric. Austral.: 386 (1837).

Roella spicata L. f., Suppl. Pl.: 143 (1782).
Cape Provinces. 27 CPP. Nanophan. or cham.
Campanula bracteata Thunb. in Hoffm., Phytogr. Bl.: 20 (1803). *Roella bracteata*
(Thunb.) A. DC., Monogr. Campan.: 179 (1830).
Roella eckloniana H. Buek in Eckl. & Zeyh., Enum. Pl. Afric. Austral.: 385 (1837).
Roella ericoides H. Buek in Eckl. & Zeyh., Enum. Pl. Afric. Austral.: 385 (1837); non
(Lam.) Spreng., Syst. Veg. 1: 723 (1824).
Roella campestris A. DC. in DC., Prodr. 7: 447 (1839).
Roella eckloniana var. *pubescens* Sond. in Harv. & Sond., Fl. Cap. 3: 593 (1865).
Roella lightfootioides Schltr., Bot. Jahrb. Syst. 24: 445 (1897).
Roella spicata var. *burchellii* Adamson, J. S. African Bot. 17: 150 (1952).

Roella squarrosa P. J. Bergius, Descr. Pl. Cap. 42 (1767). *Roella filiformis* Lam., Tab. Encycl. 2:
66 (1796).
Cape Provinces. 27 CPP. Nanophan. or cham.

Roella triflora (R. D. Good) Adamson, J. S. African Bot. 17: 137 (1952).
Cape Provinces. 27 CPP. Nanophan. or cham.
* *Roella ciliata* var. *triflora* R. D. Good, J. Bot. 62: 48 (1924).

Synonyms:
Roella alpina Bond === **Prismatocarpus alpinus** (Bond) Adamson
Roella angustifolia Roxb. === **Wahlenbergia angustifolia** (Roxb.) A. DC.
Roella bracteata (Thunb.) A. DC. === **Roella spicata** L. f.
Roella campestris A. DC. === **Roella spicata** L. f.
Roella ciliata var. *dregeana* (A. DC.) Sond. === **Roella dregeana** A. DC.
Roella ciliata var. *elongata* H. Buek === **Roella bryoides** H. Buek
Roella ciliata var. *incurva* (Banks ex A. DC.) Sond. === **Roella incurva** Banks ex A. DC.
Roella ciliata var. *minor* (Eckl. ex H. Buek) Sond. === **Roella bryoides** H. Buek
Roella ciliata var. *triflora* R. D. Good === **Roella triflora** (R. D. Good) Adamson
Roella cinerea (L. f.) A. DC. === **Wahlenbergia cinerea** (L. f.) Lammers
Roella cuspidata Adamson === **Roella compacta** Schltr.
Roella cuspidata var. *hispida* Adamson === **Roella compacta** Schltr.
Roella decurrens Andrews === **Wahlenbergia capensis** (L.) A. DC.
Roella dregeana var. *nitida* (Schltr.) Adamson === **Roella dregeana** A. DC.
Roella eckloniana H. Buek === **Roella spicata** L. f.
Roella eckloniana var. *pubescens* Sond. === **Roella spicata** L. f.
Roella elegans Paxton === Gesneriaceae
Roella ericoides H. Buek === **Roella spicata** L. f.
Roella ericoides R. D. Good === **Roella goodiana** Adamson
Roella ericoides (Lam.) Spreng. === **Prismatocarpus pedunculatus** (P. J. Bergius) A. DC.
Roella filiformis Lam. === **Roella squarrosa** P. J. Bergius
Roella glabra Poir. === **Roella decurrens** L'Hér.
Roella glauca H. Buek === **Roella secunda** H. Buek
Roella gracilis H. Buek === **Roella bryoides** H. Buek
Roella incurva var. *rigida* Adamson === **Roella prostrata** E. Mey. ex A. DC.
Roella insizwae Zahlbr. === **Craterocapsa insizwae** (Zahlbr.) Hilliard & B. L. Burtt
Roella leptosepala Sond. === **Roella dregeana** A. DC.
Roella lightfootioides Schltr. === **Roella spicata** L. f.
Roella linifolia Roxb. === **Wahlenbergia linifolia** (Roxb.) A. DC.
Roella minor Eckl. ex H. Buek === **Roella bryoides** H. Buek
Roella nitida Schltr. === **Roella dregeana** A. DC.
Roella paniculata Roxb. === **Wahlenbergia roxburghii** A. DC.
Roella pedunculata P. J. Bergius === **Prismatocarpus pedunculatus** (P. J. Bergius) A. DC.
Roella psammophila Schltr. === **Roella dregeana** A. DC.
Roella reticulata L. === **Cullumia reticulata** (L.) Greuter, M. V. Agab. & Wagenitz (Asteraceae)

Roella reticulata var. *dunantii* (A. DC.) Sond. === **Roella dunantii** A. DC.
Roella reticulata var. *prostrata* (E. Mey. ex A. DC.) Sond. === **Roella prostrata** E. Mey. ex
    A. DC.
Roella reticulata var. *ternifolia* Sond. === **Roella prostrata** E. Mey. ex A. DC.
Roella rhodantha Adamson === **Roella incurva** Banks ex A. DC.
Roella spicata var. *burchellii* Adamson === **Roella spicata** L. f.
Roella tenuifolia (L. f.) Thunb. === **Merciera tenuifolia** (L. f.) A. DC.
Roella thunbergii (Schult.) A. DC. === **Wahlenbergia thunbergii** (Schult.) B. Nord.

Rollandia

Although this genus was recognized for many years (e.g., Hillebrand 1888, Rock 1919,
Lammers 1990), molecular phylogenetic analyses revealed that it was embedded among
species of *Cyanea* (Lammers et al. 1993). Type [by monotypy]: *Rollandia lanceolata* Gaudich.

Gaudichaud, C. (1826). *Rollandia*. In Voyage autour du monde, entrepris par order du roi,
    ... exécuté sur les corvettes de S. M. *l'Uranie* et *la Physicienne*, pendant les années 1817,
    1818, 1819 et 1820. Botanique: 458-459 + pl. 74, illus. Paris: Pillet-ainé. La., Fr. —
    Establishment of genus.
Gray, A. (1861). Notes on Lobeliaceae, Goodeniaceae, &c. of the collections of the U.S.
    South Pacific Exploring Expedition. Proc. Amer. Acad. Arts 5: 146-152. En. — Adnation
    of staminal column believed erroneous; species divided between *Cyanea* and *Delissea*.
Baillon, H. (1885). *Delissea* Gaudich. In Histoire des plantes 8: 364. Paris: L. Hachette. Fr.
    — Treated as section of *Delissea*.
Hillebrand, W. (1888). *Rollandia,* Gaud. In Flora of the Hawaiian Islands: 244-248.
    London: Williams & Norgate. En. — Flora with keys, descriptions, full nomenclature,
    and specimen citations.
Rock, J. F. (1919). A monographic study of the Hawaiian species of the tribe Lobelioideae
    family Campanulaceae, *Rollandia* Gaudichaud. Mem. Bernice Pauahi Bishop Mus. 7(2):
    363-383, illus. En. — Monograph with key, descriptions, full nomenclature, and
    specimen citations.
Wimmer, F. E. (1943). *Rollandia* Gaudich. In A. Engler & L. Diels, Das Pflanzenreich IV.
    276b: 96-104, illus. Leipzig: Wilhelm Engelmann. Ge., La. — Monograph with key,
    descriptions, full nomenclature, and specimen citations.
Lammers, T. G. (1990). *Rollandia* Gaud. In W. L. Wagner, D. R. Herbst & S. H. Sohmer,
    Manual of the flowering plants of Hawai'i: 480-485, illus. Honolulu: University of
    Hawaii Press. En. — Flora with key and descriptions.
Lammers, T. G., T. J. Givnish & K. J. Sytsma (1993). Merger of the endemic Hawaiian
    genera *Cyanea* and *Rollandia* (Campanulaceae: Lobelioideae). Novon 3: 437-441. En. —
    Rollandia subsumed into *Cyanea* on basis of molecular phylogeny.

Synonyms:
Rollandia Gaudich. === **Cyanea** Gaudich.
Rollandia alba H. St. John & W. N. Takeuchi === **Cyanea pinnatifida** (Cham.) E. Wimm.
Rollandia ambigua (Cham.) G. Don === **Cyanea lanceolata** (Gaudich.) Lammers, Givnish
    & Sytsma
Rollandia angustifolia (Hillebr.) Rock === **Cyanea koolauensis** Lammers, Givnish & Sytsma
Rollandia angustifolia var. *ochreata* E. Wimm. === **Cyanea koolauensis** Lammers, Givnish
    & Sytsma
Rollandia bidentata H. St. John === **Cyanea calycina** (Cham.) Lammers
Rollandia calycina (Cham.) G. Don === **Cyanea calycina** (Cham.) Lammers
Rollandia calycina var. *kaalae* (Wawra) E. Wimm. === **Cyanea calycina** (Cham.) Lammers
Rollandia crispa Gaudich. === **Cyanea crispa** (Gaudich.) Lammers, Givnish & Sytsma
Rollandia crispa var. *muricata* Rock === **Cyanea crispa** (Gaudich.) Lammers, Givnish
    & Sytsma

Rollandia degeneriana E. Wimm. === **Cyanea longiflora** (Wawra) Lammers, Givnish & Sytsma
Rollandia delessertiana Gaudich. === **Cyanea lanceolata** (Gaudich.) Lammers, Givnish & Sytsma
Rollandia fauriei H. Lév. === **Cyanea spathulata** (Hillebr.) A. Heller
Rollandia grandifolia (A. DC.) Hillebr. === **Cyanea lanceolata** (Gaudich.) Lammers, Givnish & Sytsma
Rollandia humboldtiana Gaudich. === **Cyanea humboldtiana** (Gaudich.) Lammers, Givnish & Sytsma
Rollandia humboldtiana f. *albida* H. St. John === **Cyanea humboldtiana** (Gaudich.) Lammers, Givnish & Sytsma
Rollandia humboldtiana var. *tomentella* Wawra === **Cyanea calycina** (Cham.) Lammers
Rollandia kaalae Wawra === **Cyanea calycina** (Cham.) Lammers
Rollandia lanceolata Gaudich. === **Cyanea lanceolata** (Gaudich.) Lammers, Givnish & Sytsma
Rollandia lanceolata f. *rockii* E. Wimm. === **Cyanea lanceolata** (Gaudich.) Lammers, Givnish & Sytsma
Rollandia lanceolata [unranked] *tomentosa* Hillebr. === **Cyanea calycina** (Cham.) Lammers
Rollandia lanceolata subsp. *calycina* (Cham.) Lammers === **Cyanea calycina** (Cham.) Lammers
Rollandia lanceolata var. *brevipes* E. Wimm. === **Cyanea longiflora** (Wawra) Lammers, Givnish & Sytsma
Rollandia lanceolata var. *glaberrima* E. Wimm. === **Cyanea lanceolata** (Gaudich.) Lammers, Givnish & Sytsma
Rollandia lanceolata var. *grandifolia* A. DC. === **Cyanea lanceolata** (Gaudich.) Lammers, Givnish & Sytsma
Rollandia lanceolata var. *kipapaensis* Hosaka ex Fosberg === **Cyanea lanceolata** (Gaudich.) Lammers, Givnish & Sytsma
Rollandia lanceolata var. *tomentella* (Wawra) E. Wimm. === **Cyanea calycina** (Cham.) Lammers
Rollandia lanceolata var. *tomentosa* (Hillebr.) Drake === **Cyanea calycina** (Cham.) Lammers
Rollandia lanceolata var. *viridiflora* Rock === **Cyanea lanceolata** (Gaudich.) Lammers, Givnish & Sytsma
Rollandia longiflora Wawra === **Cyanea longiflora** (Wawra) Lammers, Givnish & Sytsma
Rollandia longiflora var. *angustifolia* Hillebr. === **Cyanea koolauensis** Lammers, Givnish & Sytsma
Rollandia obatae H. St. John === **Cyanea calycina** (Cham.) Lammers
Rollandia parviflora C. N. Forbes === **Cyanea parvifolia** (C. N. Forbes) Lammers, Givnish & Sytsma
Rollandia pedunculosa Wawra === **Cyanea humboldtiana** (Gaudich.) Lammers, Givnish & Sytsma
Rollandia pinnatifida (Cham.) G. Don === **Cyanea pinnatifida** (Cham.) E. Wimm.
Rollandia purpurellifolia Rock === **Cyanea purpurellifolia** (Rock) Lammers, Givnish & Sytsma
Rollandia racemosa (H. Mann) Hillebr. === **Cyanea humboldtiana** (Gaudich.) Lammers, Givnish & Sytsma
Rollandia scabra Wawra === **Cyanea calycina** (Cham.) Lammers
Rollandia sessilifolia O. Deg. === **Cyanea sessilifolia** (O. Deg.) Lammers
Rollandia st.-johnii Hosaka === **Cyanea st.-johnii** (Hosaka) Lammers, Givnish & Sytsma
Rollandia st.-johnii var. *obtusisepala* E. Wimm. === **Cyanea st.-johnii** (Hosaka) Lammers, Givnish & Sytsma
Rollandia truncata Rock === **Cyanea truncata** (Rock) Rock
Rollandia waianaeensis H. St. John === **Cyanea calycina** (Cham.) Lammers

Rotantha

This genus was created for a pair of Floridean species of *Campanula* with rotate deeply cleft corollas. However, the two do not appear to be closely related. While Shetler (1963) suggested that annual *C. robinsiae* was related to *C. reverchonii* of Texas, seed morphology (Shetler & Morin 1986) argues that it occupies a highly isolated position within the genus.

Similarly, Shetler (1963) suggested that perennial *C. floridana* was a derivative of northern *C. aparinoides*, but its seed morphology (Shetler & Morin 1986) is far more similar to that of Appalachian *C. divaricata*. In any event, the name *Rotantha* cannot be used for these species, as it is a later homonym of *Rotantha* Baker (Lythraceae). Type [designated here]: *Rotantha robinsiae* (Small) Small.

> Small, J. K. (1933). *Rotantha* Small. In Manual of the southeastern flora: 1289-1290, illus. New York: published privately. En. — Segregation of genus from *Campanula*.
> Shetler, S. G. (1963). A checklist and key to the species of *Campanula* native or commonly naturalized in North America. Rhodora 65: 319-337. En. — Discusses disparate relationships of the two species included here.
> Shetler, S. G. & N. R. Morin (1986). Seed morphology in North American Campanulaceae. Ann. Missouri Bot. Gard. 73: 653-688, illus. En. — Detailed study of seed coat structure and its taxonomic implications.

Synonyms:
Rotantha Small === **Campanula** L.
Rotantha floridana (S. Watson ex A. Gray) Small === **Campanula floridana** S. Watson ex
  A. Gray
Rotantha robinsiae (Small) Small=== **Campanula robinsiae** Small

Roucela

This genus was established for annual species of *Campanula* with an unappendaged calyx. Some authors accept it (e.g., Kharadze 1976), while others (e.g., Oganesian 1995, under **Campanula**) treat it as *Campanula* subg. *Roucela* (Dumort.) Damboldt. In a DNA phylogeny (Eddie et al. 2003, under **General**), its type plus *C. mollis* and the types of *Azorina* and *Feeria* formed a clade within the "Campanulaceae s. str. clade". Type [designated by Pfeiff., Nomencl. Bot. 2: 994 (1874)]: *Campanula erinus* L.

> Dumortier, B. (1822). *Roucela*. In Commentationes botanicae: 14-15. Tournay: Ch. Casterman-Dien. La. — Establishment of genus as distinct from *Campanula*.
> Kharadze, A. (1976). Genus *Campanula* L. s.l. in Caucaso (conspectus). Zametki Sist. Geogr. Rast. 32: 45-56. Ru. — Modern revival of genus.

Synonyms:
Roucela Dumort. === **Campanula** L.
Roucela drabifolia (Sm.) Dumort. === **Campanula drabifolia** Sm.
Roucela erinus (L.) Dumort. === **Campanula erinus** L.
Roucela hederacea (L.) Dumort. === **Wahlenbergia hederacea** (L.) Rchb.
Roucela propinqua (Fisch. & C. A. Mey.) Kharadze === **Campanula propinqua** Fisch. &
  C. A. Mey.

Ruthiella

Lobelioideae, 4 species, New Guinea. The name *Phyllocharis* Diels (q.v.) was used for this genus for many years, but is a later homonym of the lichen genus *Phyllocharis* Fée. Based on: *Phyllocharis* Diels.

- Wimmer, F. E. (1953). *Phyllocharis* Diels. In A. Engler & L. Diels, Das Pflanzenreich IV. 276b: 724-725. Berlin: Akademie-Verlag. Ge., La. — Monograph with key, descriptions, full nomenclature, and specimen citations.
- Tuyn, P. (1960). *Phyllocharis*. In C. G. G. J. van Steenis (ed.), Flora Malesiana (ser. I) 6(1): 137-139, illus. Djakarta: Noordhoff-Kolff. En. — Flora with key, descriptions, and full nomenclature.

Steenis, C. G. G. (1965). *Ruthiella,* a new generic name for *Phyllocharis* Diels, 1917, non Fée, 1824 (Campanulaceae). Blumea 13: 127-128. En. — Provision of replacement name.

Steenis, C. G. G. J. (1972). Campanulaceae. In C. G. G. J. van Steenis (ed.), Flora Malesiana (ser. I) 6(6): 928. Djakarta: Noordhoff-Kolff. En. — Modification of Tuyn (1960) in light of Steenis (1965).

Ruthiella Steenis, Blumea 13: 127 (1965).
Papuasia. 43.
* *Phyllocharis* Diels, Bot. Jahrb. Syst. 55: 122 (1917); non Fée, Essai Crypt. Écorc. lix, xciv, xcix (1824).

Ruthiella oblongifolia (Diels) Steenis, Blumea 13: 127 (1965).
New Guinea. 43 NWG. Ther.
* *Phyllocharis oblongifolia* Diels, Bot. Jahrb. Syst. 55: 124 (1917).

Ruthiella saxicola (P. Royen) Steenis, Blumea 13: 127 (1965).
New Guinea. 43 NWG. Ther.
* *Phyllocharis saxicola* P. Royen in Steenis, Fl. Males. (ser. 1) 6: 139 (1960).

Ruthiella schlechteri (Diels) Steenis, Blumea 13: 127 (1965).
New Guinea. 43 NWG. Ther.
* *Phyllocharis schlechteri* Diels, Bot. Jahrb. Syst. 55: 124 (1917).

Ruthiella subcordata (Merr. & L. M. Perry) Steenis, Blumea 13: 127 (1965).
New Guinea. 43 NWG. Ther.
* *Phyllocharis subcordata* Merr. & L. M. Perry, J. Arnold Arbor. 22: 387 (1941).
Phyllocharis lamiifolia E. Wimm., Ann. Naturhist. Mus. Wien 56: 372 (1948).

Sachokiella

Campanuloideae, 1 species, western Asia. Segregated from *Campanula* on the basis of its unique capsule, which splits from the base into five elongate segments, and its large erose-winged spongy verrucose seeds. Possible relationships with *Muehlbergella* and *Theodorovia* (q.v.) should be explored. Type [designated by the author]: *Sachokiella macrochlamys* (Boiss. & A. Huet) Kolak.

Kolakovskii, A. A. (1985). [*Sachokiella:* a new monotypical genus of Campanulaceae.] Soobšč. Akad. Nauk Gruzinsk. SSR 118: 593-596, illus. Ru. — Establishment of genus.

Sachokiella Kolak., Soobšč. Akad. Nauk Gruzinsk. SSR 118: 595 (1985).
W. Asia. 33 34.

Sachokiella macrochlamys (Boiss. & A. Huet) Kolak., Soobšč. Akad. Nauk Gruzinsk. SSR 118: 595 (1985), as 'macrochlamis'.
Turkey & Caucasus (?). 33+ 34 TUR. Biennial.
* *Campanula macrochlamys* Boiss. & A. Huet in Boiss., Diagn. Pl. Orient. (ser. 2) 3: 111 (1856).
Campanula glomerata var. *artwinensis* Kuntze, Trudy Imp. S.-Peterburgsk. Bot. Sada 10: 206 (1887).

Schultesia

This name was created to accomodate the sole European representative of *Wahlenbergia*, but is a later homonym of *Schultesia* Spreng., nom. rej. (Poaceae); *Schultesia* Schrad. (Amaranthaceae); and *Schultesia* Mart., nom. cons. (Gentianaceae). Its avowed substitute

Aikinia (q.v.) is likewise illegitimate. Preliminary molecular data (Eddie et al. 2003, under **General**) suggest that its type might indeed be but distantly related to the other species of *Wahlenbergia*. If segregation does prove desirable, a new name will be needed. Type [by monotypy]: *Schultesia hederacea* (L.) Roth.

> Roth, A. W. (1827). Enumeratio plantarum phaenogamarum in Germania sponte nascentium, pars prima. Leipzig: Gleditsch. La. — Establishment of genus as distinct from *Campanula*.

Synonyms:
Schultesia Roth === **Wahlenbergia** Schrad. ex Roth
Schultesia hederacea (L.) Roth === **Wahlenbergia hederacea** (L.) Rchb.

Sclerotheca

Lobelioideae, 6 species, Polynesia; five are endemic to Tahiti in the Society Islands, the sixth to Rarotonga in the Cook Islands. A preliminary phylogeny based on DNA sequence data (E. Knox, pers. comm.) places two of its species within the Hawaiian clade, as sister to the clade comprising *Clermontia* and *Cyanea*. On the basis of morphology and geography, one suspects that *Apetahia* (q.v.) is even more closely related. The genus was last monographed by Wimmer (1953). Type [by monotypy]: *Sclerotheca arborea* (G. Forst.) A. DC.

> de Candolle, A. (1839). *Sclerotheca*. In A. P. de Candolle, Prodromus systematis naturalis regni vegetabilis 7: 356-367. Paris: Treuttel & Würtz. La. — Establishment of genus as distinct from *Lobelia* and *Delissea*.
>
> Brown, F. B. H. (1935). Flora of southeastern Polynesia III. Dicotyledons. *Sclerotheca* De Candolle. Bernice P. Bishop Mus. Bull. 130: 325-328, illus. En. — Emendation of generic description to encompass some species here assigned to *Apetahia*.
>
> • Wimmer, F. E. (1953). *Sclerotheca* A. De Candolle. In A. Engler & L. Diels, Das Pflanzenreich 276b: 754-756, illus. Berlin: Akademie-Verlag. Ge., La. — Monograph with key, descriptions, full nomenclature, and specimen citations; rejected Brown's (1935) expanded circumscription.
>
> Thorne, R. F. (1972). Major disjunctions in the geographic ranges of seed plants. Quart. Rev. Biol. 47: 365-411, maps. En. — *Sclerotheca* one of just 14 genera confined to Pacific Basin, which show major distributional disjunctions within that region.
>
> • Florence, J. (1997). New species of *Plakothira* (Loasaceae), *Melicope* (Rutaceae), and *Apetahia* (Campanulaceae) from the Marquesas Islands. Allertonia 7: 238-253, illus. En. — Clarification of distinctions seperating *Apetahia* and *Sclerotheca*.

Sclerotheca A. DC. in DC., Prodr. 7: 356 (1839).
  SC. Pacific. 61.

Sclerotheca arborea (G. Forst.) A. DC. in DC., Prodr. 7: 357 (1839).
  Society Is. (Tahiti). 61 SCI. Phan. or nanophan.
    * *Lobelia arborea* G. Forst., Fl. Ins. Austr.: 58 (1786). *Lobelia lactescens* Martyn in Mill., Gard. Dict. (ed. 9) (1797). *Delissea arborea* (G. Forst.) C. Presl, Prodr. Monogr. Lobel.: 47 (1836).

Sclerotheca forsteri Drake, Ill. Fl. Ins. Pacif.: 26 (1886).
  Society Is. (Tahiti). 61 SCI. Nanophan.

Sclerotheca jayorum J. Raynal, Adansonia (n.s.) 16: 380 (1976).
  Society Is. (Tahiti). 61 SCI. Phan. or nanophan.

Sclerotheca magdalenae J. Florence, Bull. Mus. Hist. Nat., B, Adansonia 18: 97 (1996).
  Society Is. (Tahiti). 61 SCI. Phan or nanophan.

Sclerotheca oreades E. Wimm., Ann. Naturhist. Mus. Wien 56: 373 (1948).
Society Is. (Tahiti). 61 SCI. Phan. or nanophan.

Sclerotheca viridiflora Cheeseman, Trans. Linn. Soc. London, Bot. (ser. 2) 6: 285 (1903).
Cook Islands (Rarotonga). 61 COO. Nanophan.

Synonyms:
Sclerotheca longistigmata F. Br. === **Apetahia longistigmata** (F. Br.) E. Wimm.
Sclerotheca margaretae F. Br. === **Apetahia margaretae** (F. Br.) E. Wimm.

Sergia

Campanuloideae, 2 species, Tadzhikstan and Kirghizstan. Although neither species has been included in a molecular phylogeny, on the basis of morphology and serology (Gudkova & Borshchenko 1991), one suspects that *Sergia* is part of the clade that comprises *Asyneuma, Petromarula, Physoplexis,* and *Phyteuma* within the "Rapunculus clade" (Eddie et al. 2003, under **General**). Type [designated by the author]: *Sergia sewerzowii* (Regel) Fed.

> Fedorov, A. A. (1957). *Sergia* Fed., gen. nov. In V. L. Komarov, Flora URSS 24: 420-423, 474, illus. Leningrad: Academia Scientiae URSS. Ru. — Establishment of genus as distinct from *Asyneuma, Phyteuma,* and *Campanula.*
> Belyaev, A. A. (1986). [The features of anatomy and surface ultrastructure of the seed coat in some representatives of critical genera of the Campanulaceae family.] Bot. Žurn. 71: 1371-1376 + 2 pl., illus. Ru. — Description of testa in *S. regelii.*
> Gudkova, I. Y. & G. P. Borshchenko (1991). [The serological study of the Campanulaceae. The phylogenetic relations in the tribe Phyteumateae.] Bot. Žurn. 76(6): 809-817. Ru. — Protein serology supports close relationship to *Asyneuma, Cryptocodon, Petromarula,* and *Phyteuma,* and somewhat more distant relationship to *Cylindrocarpa.*

Sergia Fed. in Kom., Fl. URSS 24: 474 (1957).
Middle Asia. 32.

Sergia regelii (Trautv.) Fed. in Kom., Fl. URSS 24: 421 (1957).
Tadzhikistan. 32 TZK. Hemicr.
  * *Phyteuma regelii* Trautv., Trudy Imp. S.-Peterburgsk. Bot. Sada 6: 53 (1879). *Podanthum regelii* (Trautv.) O. Fedtsch. & B. Fedtsch., Bot. Žurn. (St. Petersburg) 1: 53 (1906). *Asyneuma regelii* (Trautv.) Bornm., Beih. Bot. Centralbl. 38(2): 348 (1921). *Phyteuma multicaule* Franch., Mission Capus Pl. Turkestan: 115 (1883).

Sergia sewerzowii (Regel) Fed. in Kom., Fl. URSS 24: 420 (1957).
Kirgizstan. 32 KGZ. Hemicr.
  * *Campanula sewerzowii* Regel, Bull. Soc. Imp. Naturalistes Moscou 40(3): 188 (1867), as 'sewerzowi'.

Sicyocodon

This species was segregated from *Campanula* on the basis of its annual duration, dichasial inflorescence, appendiculate calyx, and long-exsert style. Most recent authors (e.g., Sáez & Aldasoro 2003) treat it as *Campanula* subg. *Sicyocodon* (Feer) Damboldt. Type [by monotypy]: *Sicyocodon macrostylus* (Boiss. & Heldr.) Feer.

> Feer, H. (1890). Beiträge zur Systematik und Morphologie der Campanulaceen. II.A. Drei neue monotype Gattungen. Bot. Jahrb. Syst. 12: 610-617 + Taf. VII-VIII, illus. Ge. — Establishment of genus as distinct from *Campanula.*

Sáez, L. & J. J. Aldasoro (2003). A taxonomic revision of *Campanula* L. subgenus *Sicyocodon* (Feer) Damboldt and subgenus *Megalocalyx* Damboldt (Campanulaceae). Bot. J. Linn. Soc. 141: 215-241, illus., maps. En. — Monograph as subgenus of *Campanula*.

Synonyms:
Sicyocodon Feer === **Campanula** L.
Sicyocodon macrostylus (Boiss. & Heldr.) Feer === **Campanula macrostyla** Boiss. & Heldr.

Siphocampylus

Lobelioideae, 231 species, Neotropics, from Costa Rica to Argentina and in the Greater Antilles; the greatest diversity is in the northern Andes. The genus has been divided (Wimmer 1953, 1968) into sect. *Siphocampylus* (three subsections) and sect. *Brachysiphon* E. Wimm. (four subsections). Preliminary analyses of molecular data (E. Knox, pers. comm.; A. Antonelli, pers. comm.) support a close relationship to *Burmeistera* and *Centropogon,* but cast doubt on the strict monophyly of the genus. The most recent monograph is that of Wimmer (1953, 1968). $2n$ = 28. Type [designated by McVaugh, N. Amer. Fl. 32A: 100 (1943)]: *Siphocampylus westinianus* (Thunb.) Pohl.

Pohl, J. E. (1831). *Siphocampylus*. In Plantarum Brasiliae icones et descriptiones hactenus ineditae 2: 104-115 + tab. 168-177, illus. Vienna: published privately. La. — Establishment of genus as distinct from *Lobelia*.

Presl, C. (1836). *Lobelia*. In Prodromus monographiae Lobeliacearum: 33-40. Prague: Theophilus Haase. La. — All original Linnaean species removed from *Lobelia* and assigned elsewhere (primarily to *Rapuntium*), and that name inexplicably used for *Siphocampylus*.

Hooker, W. J. (1842). *Siphocampylus betulaefolius*. Birch-leaved siphocampylus. Bot. Mag. 69: tab. 3973 + 2 pp., illus. En. — Portrait with description.

Hooker, W. J. (1843). *Siphocampylos* [sic] *longepedunculatus*. Long flower-stalked siphocampylos. Bot. Mag. 69: tab. 4015 + 2 pp., illus. En. — Portrait with description.

Hooker, W. J. (1844). *Siphocampylus lantanifolius*. Lantana-leaved siphocampylus. Bot. Mag. 70: tab. 5105 + 2 pp., illus. En. — Portrait of *S. reticulatus,* with description.

Otto, F. & A. Dietrich (1844). Ueber die *Siphocampylus*-arten, welche in unsern Gärten kultivirt werden. Allg. Gartenzeitung 12: 369-372. Ge., La. — Account of species in cultivation at that time, with descriptions and full nomenclature.

Hooker, W. J. (1845). *Siphocampylus giganteus,* Cav. Icon. Pl. 8: pl. 716 + 1 pg., illus. En. — Portrait with description.

Morren, C. (1846). *Siphocampylus nitidus*. De Jonghe. (siphocampyle brillant). Ann. Soc. Roy. Agric. Gand 2: 319-320 + pl. 78, illus. Fr. — Portrait with description.

Hooker, W. J. (1847). *Siphocampylos* [sic] *microstoma*. Small-mouthed siphocampylos. Bot. Mag. 73: tab. 4286 + 2 pp., illus. En. — Portrait with description.

Hooker, W. J. (1848). *Siphocampylos* [sic] *manettiaeflorus*. Manettia-flowered siphocampylos. Bot. Mag. 74: tab. 4403 + 2 pp., illus. En. — Portrait with description.

Paxton, J. (1849). *Siphocampylos* [sic] *manettiaeflorus*. (Manettia-flowered siphocampylos). Paxton's Mag. Bot. 15: 267 + plate, illus. En. — Portrait with description.

Planchon, J. E. & L. van Houtte (1850). *Siphocampylus orbignyanus*. Syphocampylus de D'Orbigny. Fl. Serres Jard. Eur. 6: 15 + pl. 544, illus. — Portrait with description.

Lindley, J. & J. Paxton (1851). The small-mouthed siphocampyl. (*Siphocampylus microstoma*). Paxton's Fl. Gard. 2: 33-34 + pl. 44, illus. En. — Portrait of *S. lindleyi,* with description.

Lemaire, C. (1852). *Siphocampylus lindleyi*. Syphocampyle de Lindley. Jard. Fleur. 2: pl. 142 + 2 pp., illus. Fr. — Portrait with description, clarifying Lindley & Paxton's (1851) misidentification of their plant.

Hooker, W. J. (1853). *Syphocampylus* [sic] *orbignianus*. D'Orbigny's syphocampylus. Bot. Mag. 74: tab. 4403 + 2 pp., illus. En. — Portrait with description.

Dombrain, H. H. (1866). *Siphocampylus fulgens*. Fl. Mag. (London) 5: pl. 320 + 2 pp., illus. En. — Portrait with description.

Hooker, J. D. (1867). *Siphocampylus humboldtianus*. Humboldt's siphocampylus. Bot. Mag. 93: tab. 5631 + 2 pp., illus. En. — Portrait with description.

Gleason, H. A. (1921). A rearrangement of the Bolivian species of *Centropogon* and *Siphocampylus*. Bull. Torrey Bot. Club 48: 189-201, illus. En. — Floristic treatment, with key, brief descriptions, and full nomenclature.

McVaugh, R. (1943). *Siphocampylus* Pohl. In North American Flora 32A: 100-110. New York: New York Botanical Garden. En. — Flora with key, descriptions, infrageneric classification, and full nomenclature.

• McVaugh, R. (1949). Studies in South American Lobelioideae (Campanulaceae) with special reference to Colombian species. II. A new subsection of *Siphocampylus* sect. *Eusiphocampylus*. Brittonia 6: 453-458. En. — Establishment of subsect. *Nivei,* including key and descriptions for its species.

Garello, F. (1950). Contribucion al estudio del *Siphocampylus foliosus* Gris. Arch. Farm. Bioquím. Tucumán 5: 5-46 + lam. I-VIII, illus. Sp. — Detailed anatomical and phytochemical study; isolation and purification of new alkaloiid, siphocampilin.

• Wimmer, F. E. (1953). *Siphocampylus* Pohl. In A. Engler & L. Diels, Das Pflanzenreich IV. 276b: 264-380, 775 + 7 pl., illus. Berlin: Akademie-Verlag. Ge., La. — Monograph with key, descriptions, infrageneric classification, full nomenclature, and specimen citations.

• Wimmer, F. E. (1968). *Siphocampylus* Pohl. In A. Engler & L. Diels, Das Pflanzenreich IV. 276c: 845-852 + Taf. 16-19, illus. Berlin: Akademie-Verlag. Ge., La. — Supplement to Wimmer (1953).

Vogel, S. (1969). Chiropterophilie in der neotropischen Flora, neue Mitteilungen II. Flora (abt. B) 158: 185-222, illus. Ge. — Bat pollination in *S. tunicatus*.

Belem, C. I. F. (1976). Descrição palinológica de espécies de Campanulaceae dos gêneros *Centropogon* e *Siphocampylus*. Revista Brasil. Biol. 36: 861-870, illus. Por. — Palynological survey of Brazilian spp.

Borhidi, A. & Z. Kereszty (1979). New names and new species in the flora of Cuba resp. Antilles. Acta Bot. Acad. Sci. Hung. 25: 1-37. En. — Discussion of *S. glaber*.

• Jeppesen, S. (1981). *Siphocampylus* Pohl. In G. Harling & B. Sparre (eds.), Flora of Ecuador 14: 151-170, illus. Stockholm: Swedish Natural Science Research Council. En. — Flora with keys, descriptions, full nomenclature, and specimen citations.

Stein, B. A. (1987). *Siphocampylus oscitans* (Campanulaceae: Lobelioideae), a new name for *Burmeistera weberbaueri* from Peru. Ann. Missouri Bot. Gard. 74: 491-493. En. — Discussion of species with open anther orifice like that of *Burmeistera*.

Galetto, L., L. M. Bernardello & H. R. Juliani (1993). Estructura del nectario, composition quimica del nectar y mecanismo de polinizacion en tres especies de *Siphocampylus* (Campanulaceae). Kurtziana 22: 81-96, illus. Sp. — Pollination biology in three Argentinian species; nectar sucrose dominant, hummingbirds observed.

Sazima, M., I. Sazima & S. Buzato (1994). Nectar by day and night: *Siphocampylus sulfureus* (Lobeliaceae) pollinated by hummingbirds and bats. Pl. Syst. Evol. 191: 237-246, illus. En. — Documentation of both chiropterophily and ornithology in single Brazilian species.

Trentin, A. P., A. R. S. Santos, O. G. Miguel, M. G. Pizzolatti, R. A. Yunes & J. B. Calixto (1997). Mechanisms involved in the antinociceptive effect in mice of the hydroalcoholic extract of *Siphocampylus verticillatus*. J. Pharm. Pharmacol. 49: 567-572. En. — Detailed biochemical study of analgesic potential of plant used in folk medicine to treat asthma.

Santos, A. R. S., O. G. Miguel, R. A. Yunes & J. B. Calixto (1997). Antinociceptive properties of the new alkaloid, cis-8,10-di-*n*-propyllobelidiol hydrochloride dihydrate isolated from *Siphocampylus verticillatus:* evidence for the mechanism of action. J. Pharmacol. Exp. Therap. 289: 417-426. En. — Detailed biochemical study of potential new analgesic.

Serra, L. & M. B. Crespo (1997). An outline revision of the subtribe Siphocampylinae (Lobeliaceae). Lagascalia 19: 881-888, illus., maps. En. — Key to distinguish genus from alleged allies.

- Lammers, T. G. (1998). Review of the neotropical endemics *Burmeistera, Centropogon,* and *Siphocampylus* (Campanulaceae: Lobeliodeae), with description of 18 new species and a new section. Brittonia 50: 233-262. En. — Includes summary of current knowledge of genus.
  Saburi, W. Jr., L. R. Reato & S. A. Pires de Godoy (2005). Floral venation patterns in *Siphocampylus* (Campanulaceae). Amer. J. Bot. 92: 797-801, illus. En. — Anatomical study in eight Brazilian species supports appendicular origin of hypanthium.

Siphocampylus Pohl, Pl. Bras. Icon. Descr. 2: 104 (1831). *Lobelia* C. Presl, Prodr. Monogr. Lobel.: 33 (1836); non L., Sp. Pl.: 929 (1753); nec Mill., Gard. Dict. (abr. ed. 4) (1754). Trop. America. 80 81 82 83 84 85.
  Canonanthus G. Don, Gen. Hist. 3: 718 (1834).
  Byrsanthes C. Presl, Prodr. Monogr. Lobel.: 41 (1836), nom. rejic.; non *Byrsanthus* Guill. in Deless., Icon. Select. Pl. 3: 30 (1838), nom. cons. *Siphocampylus* sect. *Byrsanthes* (C. Presl) Schönl. in Engl. & Prantl, Naturl. Pflanzenfam. IV.5: 66 (1889), nom. rejic.
  Cremochilus Turcz., Bull. Soc. Imp. Naturalistes Moscou 25(3): 174 (1852). *Siphocampylus* sect. *Cremochilus* (Turcz.) Schönl. in Engl. & Prantl, Naturl. Pflanzenfam. IV.5: 66 (1889). *Siphocampylus* grex *Cremochilus* (Turcz.) E. Wimm., Ann. Naturhist. Mus. Wien 56: 329 (1948).
  Campylosiphon St.-Lag., Ann. Soc. Bot. Lyon 7: 135 (1880).
  Juchia Neck. ex Kuntze in T. Post & Kuntze, Lex. Gen. Phan.: 303 (1903); non Roem., Fam. Nat. Syn. Monogr. 2: 11 (1846).

Siphocampylus actinothrix E. Wimm., Repert. Spec. Nov. Regni Veg. 19: 253 (1924). Peru. 83 PER. Nanophan.
  Siphocampylus actinothrix f. *minor* E. Wimm., Pflanzenr. IV.276b: 367 (1953).

Siphocampylus adhaerens Lammers, Novon 12: 224 (2002). Venezuela. 82 VEN. Nanophan. or cham.

Siphocampylus affinis (Mirb.) McVaugh, Brittonia 6: 455 (1949). Ecuador. 83 ECU. Nanophan. or cham. $2n = 28$.
  * *Lobelia campanulata* Cav., Anales Hist. Nat. 2: 107 (1800); non Lam., Encycl. 3: 588 (1791). *Lobelia affinis* Mirb., Hist. Nat. Pl. 14: 237 (1805). *Lobelia pulverulenta* Pers., Syn. Pl. 2: 212 (1806). *Canonanthus campanulatus* G. Don, Gen. Hist. 3: 719 (1834). *Siphocampylus campanulatus* A. DC. in DC., Prodr. 7: 401 (1839).
  Lobelia jamesoniana C. Presl, Prodr. Monogr. Lobel.: 36 (1836). *Siphocampylus jamesonianus* (C. Presl) A. DC. in DC., Prodr. 7: 402 (1839).
  Byrsanthes halliana C. Presl, Prodr. Monogr. Lobel.: 42 (1836). *Siphocampylus hallianus* (C. Presl) Vatke, Linnaea 38: 732 (1874).

Siphocampylus albiguttur McVaugh, N. Amer. Fl. 32A: 109 (1943). Panama. 80 PAN. Cham. or hemicr.

Siphocampylus albus E. Wimm., Repert. Spec. Nov. Regni Veg. 26: 17 (1929). Peru. 83 PER. Phan. or nanophan.

Siphocampylus amalfiensis E. Wimm., Repert. Spec. Nov. Regni Veg. 22: 206 (1926). Colombia. 83 CLM. Cl. nanophan.

Siphocampylus ambivalens Lammers, Novon 12: 226 (2002). Bolivia. 83 BOL. (Cl.) nanophan.

Siphocampylus amoenus Planch., Fl. Serres Jard. Eur. 6: 273 (1851). Brazil (?). 84+. Nanophan. or cham.

Siphocampylus andinus Britton, Bull. Torrey Bot. Club 19: 373 (1892).
Bolivia. 83 BOL. Cl. nanophan.
    Siphocampylus andinus var. *elegantissimus* E. Wimm., Repert. Spec. Nov. Regni Veg. 38: 22 (1935).
    Siphocampylus andinus var. *solemnis* E. Wimm., Repert. Spec. Nov. Regni Veg. 38: 23 (1935).

Siphocampylus angustiflorus Schltdl. ex Zahlbr., Bull. Torrey Bot. Club 24: 379 (1897).
Peru & Bolivia. 83 BOL PER. Cl. nanophan.

Siphocampylus antioquianus E. Wimm., Repert. Spec. Nov. Regni Veg. 29: 80 (1931).
Colombia. 83 CLM. Cl. nanophan.

Siphocampylus apricus E. Wimm., Repert. Spec. Nov. Regni Veg. 22: 206 (1926).
Peru. 83 PER. Nanophan. or cham.

Siphocampylus arachnes E. Wimm., Notizbl. Bot. Gart. Berlin-Dahlem 10: 743 (1929).
Peru. 83 PER. Nanophan. or cham.

Siphocampylus argentinus (Griseb.) Hieron. ex E. Wimm., Revista Sudamer. Bot. 2: 99 (1935).
Bolivia & NW. Argentina (Jujuy to Catamarca). 83 BOL 85 AGW. Cham.
    * *Centropogon argentinus* Griseb., Abh. Königl. Ges. Wiss. Göttingen 24: 219 (1879).
    Siphocampylus cuspidatus E. Wimm., Repert. Spec. Nov. Regni Veg. 38: 20 (1935).
    Siphocampylus argentinus var. *cuspidatus* (E. Wimm.) E. Wimm., Ann. Naturhist. Mus. Wien 56: 326 (1948).
    Siphocampylus argentinus var. *hirticorollinus* E. Wimm., Pflanzenr. IV.276c: 847 (1968).

Siphocampylus argutus Zahlbr., Bull. Torrey Bot. Club 24: 383 (1897).
Bolivia. 83 BOL. Cham.

Siphocampylus asplundii Jeppesen in Harling & Sparre, Fl. Ecuador 14: 154 (1981).
Ecuador. 83 ECU. Cl. nanophan.

Siphocampylus attenuatus (C. Presl) A. DC. in DC., Prodr. 7: 398 (1839).
Peru. 83 PER. Nanophan.
    * *Lobelia attenuata* C. Presl, Prodr. Monogr. Lobel.: 34 (1836).

Siphocampylus aureus Rusby, Mem. Torrey Bot. Club 6: 72 (1896).
Bolivia. 83 BOL. Cl. nanophan.
    Siphocampylus aureus var. *latior* Zahlbr., Bull. Torrey Bot. Club 24: 378 (1897).

Siphocampylus aurocinctus E. Wimm., Repert. Spec. Nov. Regni Veg. 19: 243 (1924).
Peru. 83 PER. Nanophan.

Siphocampylus avicularis E. Wimm., Repert. Spec. Nov. Regni Veg. 38: 20 (1935).
Colombia. 83 CLM. Cl. nanophan.

Siphocampylus ayersiae Lammers, Brittonia 50: 236 (1998).
Bolivia. 83 BOL. Cl. nanophan.

Siphocampylus baracoensis Vict., Contr. Inst. Bot. Univ. Montreál 49: 6 (1944).
Cuba. 81 CUB. Nanophan. or cham.

Siphocampylus benthamianus Walp., Repert. Bot. Syst. 6: 379 (1846).
Colombia. 83 CLM. Nanophan.
    * *Siphocampylus cordifolius* Benth., Pl. Hartweg.: 213 (1845); non Otto & A. Dietr., Allg. Gartenzeitung 12: 371 (1844).

Siphocampylus betulifolius (Cham.) G. Don, Gen. Hist. 3: 703 (1834), as 'betulaefolius'.
SE. Brazil (Rio de Janeiro, São Paulo, Paraná, Santa Catarina). 84 BZL BZS. Nanophan.
* *Lobelia betulifolia* Cham., Linnaea 8: 204 (1833), as 'betulaefolia'.
Siphocampylus cordifolius Otto & A. Dietr., Allg. Gartenzeitung 12: 371 (1844).
Siphocampylus betulifolius var. *cordifolius* (Otto & A. Dietr.) E. Wimm., Repert. Spec. Nov. Regni Veg. 22: 207 (1926).
Siphocampylus betulifolius var. *uleanus* E. Wimm., Repert. Spec. Nov. Regni Veg. 22: 207 (1926).

Siphocampylus bichromatus E. Wimm., Repert. Spec. Nov. Regni Veg. 19: 253 (1924).
Peru. 83 PER. Nanophan.

Siphocampylus bilabiatus Zahlbr., Bull. Torrey Bot. Club 24: 382 (1897).
Bolivia. 83 BOL. Nanophan. or cham.
Siphocampylus bilabiatus var. *glabratus* Lauterb. ex Buchtien, Contrib. Fl. Bolivia 1: 187 (1910).
Siphocampylus subcordatus var. *dives* E. Wimm. in J. F. Macbr., Fl. Peru 6: 470 (1937). *Siphocampylus bilabiatus* var. *dives* (E. Wimm.) E. Wimm., Pflanzenr. IV.276b: 343 (1953).

Siphocampylus biserratus (Cav.) A. DC. in DC., Prodr. 7: 397 (1839).
Peru. 83 PER. Cham.
* *Lobelia biserrata* Cav., Icon. 6: 10 (1800). *Siphocampylus cavanillesianus* G. Don, Gen. Hist. 3: 702 (1834).
Lobelia biserrata var. *spicata* Hook., Bot. Misc. 2: 221 (1831). *Siphocampylus biserratus* var. *spicatus* (Hook.) E. Wimm., Pflanzenr. IV.276b: 299 (1953).
Siphocampylus biserratus var. *latifolius* A. DC. in DC., Prodr. 7: 397 (1839), as 'latifolia'.
Siphocampylus biserratus var. *petiolaris* E. Wimm., Notizbl. Bot. Gart. Berlin-Dahlem 10: 737 (1929).

Siphocampylus bogotensis E. Wimm., Repert. Spec. Nov. Regni Veg. 26: 12 (1929).
Colombia. 83 CLM. Cham.

Siphocampylus boliviensis Zahlbr., Ann. K. K. Naturhist. Hofmus. 6: 443 (1891).
Peru & Bolivia. 83 BOL PER. Nanophan. or cham.

Siphocampylus bonplandianus Zahlbr., Repert. Spec. Nov. Regni Veg. 14: 183 (1915).
Colombia. 83 CLM. Nanophan.

Siphocampylus brevicalyx E. Wimm., Ann. Naturhist. Mus. Wien 56: 326 (1948).
Ecuador. 83 ECU. Cl. nanophan.

Siphocampylus brevidens Regel, Index Sem. Hort. Petrop. 1868: 84 (1868).
Brazil (?). 84+. Cl. nanophan.

Siphocampylus brevipedicellatus E. Wimm., Repert. Spec. Nov. Regni Veg. 38: 22 (1935).
Colombia. 83 CLM. Cl. nanophan.

Siphocampylus buesii E. Wimm., Repert. Spec. Nov. Regni Veg. 38: 17 (1935).
Peru. 83 PER. Cl. nanophan.

Siphocampylus bullatus McVaugh, Brittonia 6: 457 (1949).
Colombia. 83 CLM. Nanophan. or cham.

Siphocampylus calodontus E. Wimm., Repert. Spec. Nov. Regni Veg. 29: 87 (1931).
Peru. 83 PER. Cl. nanophan.

Siphocampylus candollei E. Wimm., Repert. Spec. Nov. Regni Veg. 22: 208 (1926).
   Peru. 83 PER. Cham.
   Siphocampylus candollei var. *breviflorus* E. Wimm., Notizbl. Bot. Gart. Berlin-Dahlem 10: 746 (1929).
   Siphocampylus candollei var. *illustris* E. Wimm., Repert. Spec. Nov. Regni Veg. 38: 75 (1935).

Siphocampylus capribarba E. Wimm., Repert. Spec. Nov. Regni Veg. 19: 254 (1924).
   S. Brazil (?). 84+. Cham.

Siphocampylus caudatus McVaugh, N. Amer. Fl. 32A: 107 (1943).
   Haiti. 81 HAI. Nanophan.
   * *Lobelia ekmanii* Urb., Ark. Bot. 20A(5): 63 (1926); non *Siphocampylus ekmanii* Urb., Symb. Antill. 9: 429 (1925).

Siphocampylus cernuus Griseb., Cat. Pl. Cub.: 159 (1866).
   Cuba. 81 CUB. Nanophan. or cham.
   Siphocampylus cernuus var. *nipensis* Urb., Symb. Antill. 9: 430 (1925). *Siphocampylus cernuus* subsp. *nipensis* (Urb.) Borhidi, Bot. Közlem. 58: 177 (1971).

Siphocampylus chloroleucus E. Wimm., Repert. Spec. Nov. Regni Veg. 38: 17 (1935).
   Ecuador & Peru. 83 ECU PER. Cl. nanophan.

Siphocampylus citrinus E. Wimm., Repert. Spec. Nov. Regni Veg. 38: 75 (1935).
   Peru. 83 PER. Nanophan.

Siphocampylus clotho E. Wimm., Repert. Spec. Nov. Regni Veg. 19: 254 (1924).
   Peru. 83 PER. Cl. nanophan.
   Siphocampylus clotho var. *calvescens* E. Wimm., Repert. Spec. Nov. Regni Veg. 19: 255 (1924).

Siphocampylus coltinya E. Wimm., Repert. Spec. Nov. Regni Veg. 29: 84 (1931).
   Peru. 83 PER. Nanophan.

Siphocampylus columneae (L. f.) G. Don, Gen. Hist. 3: 701 (1834), as 'columnae'.
   Columbia & Ecuador. 83 CLM ECU. Nanophan.
   * *Lobelia columneae* L. f., Suppl. Pl.: 393 (1782).

Siphocampylus comosus G. Don, Gen. Hist. 3: 702 (1834).
   Peru & Bolivia. 83 BOL PER. Nanophan.
   * *Lobelia comosa* Cav., Icon 6: 9 (1800); non L., Sp. Pl.: 933 (1753).
   Siphocampylus virgatus A. DC. in DC., Prodr. 7: 398 (1839).
   Siphocampylus comosus var. *atrichus* E. Wimm. in J. F. Macbr., Fl. Peru 6: 448 (1937).

Siphocampylus convolvulaceus (Cham.) G. Don, Gen. Hist. 3: 703 (1834).
   SE. Brazil (Rio de Janeiro, São Paulo, Paraná). 84 BZL BZS. Cl. nanophan.
   * *Lobelia convolvulacea* Cham., Linnaea 8: 205 (1833).
   Siphocampylus weirianus E. Wimm., Repert. Spec. Nov. Regni Veg. 29: 79 (1931). *Siphocampylus convolvulaceus* var. *weirianus* (E. Wimm.) E. Wimm., Pflanzenr. IV.276b: 331 (1953).

Siphocampylus cordatus (Willd. ex Schult.) E. Wimm., Pflanzenr. IV.276b: 346 (1953).
   Venezuela & Colombia. 82 VEN 83 CLM. Cl. nanophan.
   * *Lobelia cordata* Willd. ex Schult. in Roem. & Schult., Syst. Veg. 5: 58 (1819).
   Lobelia volubilis Kunth in Humb., Bonpl. & Kunth, Nov. Gen. Sp. 3: 308 (quarto), 240 (folio) (1819); non Burm. f., Prodr. Fl. Cap.: 29 (1768). *Siphocampylus volubilis* G. Don, Gen. Hist. 3: 703 (1834).

Siphocampylus volubilis var. *demissus* E. Wimm., Repert. Spec. Nov. Regni Veg. 22: 215
(1926).

Siphocampylus volubilis var. *obtusus* E. Wimm., Repert. Spec. Nov. Regni Veg. 22: 215
(1926). *Siphocampylus cordatus* var. *obtusus* (E. Wimm.) E. Wimm., Pflanzenr.
IV.276b: 347 (1953).

Siphocampylus coronatus Gleason, Bull. Torrey Bot. Club 52: 72 (1925).
Colombia. 83 CLM. Cl. nanophan.

Siphocampylus correoides Zahlbr., Bull. Torrey Bot. Club 24: 382 (1897).
Peru & Bolivia. 83 BOL PER. Cl. nanophan.

Siphocampylus corymbiferus Pohl, Pl. Bras. Icon. Descr. 2: 112 (1831). *Lobelia corymbifera*
(Pohl) C. Presl, Prodr. Monogr. Lobel.: 37 (1836). Nanophan. or cham.
Peru to SE. Brazil (Mato Grosso, Minas Gerais, Rio de Janeiro, São Paulo). 83 BOL PER 84
BZC BZL.

Siphocampylus cardiophyllus Pohl, Pl. Bras. Icon. Descr. 2: 110 (1831). *Lobelia cardiophylla*
(Pohl) C. Presl, Prodr. Monogr. Lobel.: 37 (1836).

Lobelia ignea Vell., Fl. Flumin., Icon. 8: pl. 158 (1831). *Siphocampylus igneus* (Vell.) E.
Wimm., Repert. Spec. Nov. Regni Veg. 22: 211 (1926); non Urb., Symb. Antill. 1:
452 (1899).

Siphocampylus gracilis Britton, Bull. Torrey Bot. Club 19: 374 (1892). *Siphocampylus
corymbiferus* var. *gracilis* (Britton) Zahlbr., Bull. Torrey Bot. Club 24: 384 (1897).
Siphocampylus igneus var. *gracilis* (Britton) E. Wimm., Repert. Spec. Nov. Regni Veg.
22: 211 (1926).

Siphocampylus gracilis var. *glabris* Britton, Bull. Torrey Bot. Club 19: 374 (1892).

Siphocampylus corynellus Gleason, Torreya 25: 93 (1925).
Peru. 83 PER. Nanophan. or cham.

Siphocampylus raimondii E. Wimm., Notizbl. Bot. Gart. Berlin-Dahlem 10: 741 (1929).

Siphocampylus corynoides E. Wimm., Repert. Spec. Nov. Regni Veg. 19: 255 (1924).
Peru. 83 PER. Nanophan.

Siphocampylus corynoides f. *fortunatus* E. Wimm., Repert Spec. Nov. Regni Veg. 38: 20
(1935).

Siphocampylus crenatus (E. Wimm.) E. Wimm., Pflanzenr. IV.276b: 11 (1943).
Bolivia. 83 BOL. Cl. nanophan.

**Siphocampylus oblongifolius* var. *crenatus* E. Wimm., Repert. Spec. Nov. Regni Veg. 38:
19 (1935).

Siphocampylus cutervensis Zahlbr., Ann. K. K. Naturhist. Hofmus. 6: 442 (1891).
Peru. 83 PER. Nanophan. or cham.

Siphocampylus megalandrus E. Wimm., Repert. Spec. Nov. Regni Veg. 19: 259 (1924).

Siphocampylus darienensis Wilbur, Ann. Missouri Bot. Gard. 63: 648 (1977).
Panama. 80 PAN. Nanophan. or cham.

Siphocampylus declinatus Rusby, Descr. S. Amer. Pl.: 145 (1920). *Centropogon declinatus*
(Rusby) E. Wimm., Pflanzenr. IV.276b: 218 (1943).
Colombia. 83 CLM. Nanophan. or cham.

Siphocampylus decumbens (A. Rich. ex Schult.) Juss. ex A. DC. in DC., Prodr. 7: 397 (1839).
Haiti. 81 HAI. Cham. or hemicr.

**Lobelia decumbens* A. Rich. ex Schult. in Roem. & Schult., Syst. Veg. 5: 67 (1819).

Rapuntium richardianum C. Presl, Prodr. Monogr. Lobel.: 27 (1836).

Siphocampylus densidentatus E. Wimm., Repert. Spec. Nov. Regni Veg. 29: 79 (1931).
S. Brazil (Santa Catarina). 84 BZS. Nanophan. or cham.

Siphocampylus densiflorus Planch., Fl. Serres Jard. Eur. 6: 18 (1850).
Colombia. 83 CLM. Nanophan. or cham.

Siphocampylus dentatus Gleason, Bull. Torrey Bot. Club 52: 69 (1925).
Colombia. 83 CLM. Cl. cham.
Siphocampylus clarus E. Wimm., Repert. Spec. Nov. Regni Veg. 29: 81 (1931). *Siphocampylus dentatus* var. *clarus* (E. Wimm.) E. Wimm., Pflanzenr. IV.276b: 346 (1953).

Siphocampylus denticulosus Planch., Fl. Serres Jard. Eur. 6: 19 (1850).
Colombia. 83 CLM. Cl. nanophan.
Siphocampylus eupeplus E. Wimm., Repert. Spec. Nov. Regni Veg. 19: 256 (1924).
Siphocampylus phyllobotrys E. Wimm., Repert. Spec. Nov. Regni Veg. 19: 256 (1924).
Siphocampylus phyllobotrys f. *leucanthus* E. Wimm., Pflanzenr. IV.276b: 348 (1953).

Siphocampylus dependens G. Don, Gen. Hist. 3: 704 (1834).
Peru. 83 PER. Cl. nanophan.
Lobelia rugosa C. Presl, Prodr. Monogr. Lobel.: 34 (1836). *Siphocampylus rugosus* (C. Presl) A. DC. in DC., Prodr. 7: 399 (1839).

Siphocampylus domingensis A. DC. in DC., Prodr. 7: 397 (1839).
Haiti. 81 HAI. Cham. or hemicr.

Siphocampylus dossennus E. Wimm., Notizbl. Bot. Gart. Berlin-Dahlem 10: 742 (1929).
Peru. 83 PER. Nanophan.

Siphocampylus dubius Zahlbr., Bull. Torrey Bot. Club 24: 385 (1897). *Centropogon dubius* (Zahlbr.) E. Wimm., Pflanzenr. IV.276b: 196 (1943).
Bolivia. 83 BOL. Nanophan. or cham.

Siphocampylus duploserratus Pohl, Pl. Bras. Icon. Descr. 2: 114 (1831).
SE. Brazil (Minas Gerais, Rio de Janeiro, São Paulo). 84 BZL. Cham. or hemicr.
Lobelia triphylla C. Presl, Symb. Bot. 1: 63 (1831).
Siphocampylus duploserratus var. *infundibularis* E. Wimm., Ann. Naturhist. Mus. Wien 56: 323 (1948).

Siphocampylus ecuadorensis E. Wimm., Repert. Spec. Nov. Regni Veg. 22: 209 (1926).
Ecuador. 83 ECU. Cham.
Siphocampylus exuberans McVaugh, Ann. Missouri Bot. Gard. 52: 403 (1965).

Siphocampylus eichleri Kanitz in Mart., Fl. Bras. 6(4): 148 (1878).
Brazil (Goiás, São Paulo). 84 BZC BZL. Cham. or hemicr.

Siphocampylus elegans Planch., Fl. Serres Jard. Eur. 6: 19 (1850).
Colombia. 83 CLM. Cl. nanophan.
Siphocampylus elegans var. *obtusatus* E. Wimm., Ann. Naturhist. Mus. Wien 56: 326 (1948).

Siphocampylus elfriedii E. Wimm., Repert Spec. Nov. Regni Veg. 38: 75 (1935), as 'elfriedi'.
Peru. 83 PER. Nanophan.

Siphocampylus ellipticus (Willd. ex Schult.) Vatke, Linnaea 38: 724 (1874).
Colombia. 83 CLM. Nanophan. or cham.
* *Lobelia elliptica* Willd. ex Schult. in Roem. & Schult., Syst. Veg. 5: 57 (1819).

Siphocampylus funckeanus Planch., Fl. Serres Jard. Eur. 6: 19 (1850). *Siphocampylus ellipticus* var. *funckeanus* (Planch.) E. Wimm., Pflanzenr. IV.276b: 341 (1953). *Siphocampylus funckeanus* var. *rugosus* Planch., Fl. Serres Jard. Eur. 6: 19 (1850). *Siphocampylus peritornus* E. Wimm., Repert. Spec. Nov. Regni Veg. 29: 82 (1931). *Siphocampylus ellipticus* var. *peritornus* (E. Wimm.) E. Wimm., Pflanzenr. IV.276b: 341 (1953).

Siphocampylus elsworthii E. Wimm., Repert. Spec. Nov. Regni Veg. 38: 17 (1935).
Colombia. 83 CLM. Nanophan.

Siphocampylus fallax Lammers, Brittonia 50: 238 (1998).
Peru. 83 PER. Nanophan.

Siphocampylus fiebrigii E. Wimm., Repert. Spec. Nov. Regni Veg. 22: 209 (1926).
Bolivia & NW. Argentina (Jujuy to Salta). 83 BOL 85 AGW. Nanophan.
Siphocampylus fiebrigii var. *intermedius* E. Wimm., Pflanzenr. IV.276b: 310 (1953).

Siphocampylus fimbriatus Regel, Gartenflora 17: 355 (1868).
Bolivia, Argentina (?), S. Brazil (Santa Catarina). 83 BOL 84 BZS 85+. Cl. nanophan.

Siphocampylus fissus Gleason, Torreya 25: 95 (1925).
Peru. 83 PER. Cl. nanophan.

Siphocampylus flagelliformis Zahlbr., Bull. Torrey Bot. Club 24: 380 (1897).
Bolivia. 83 BOL. Cl. nanophan.
Siphocampylus altiscandens Gleason, Bull. Torrey Bot. Club 48: 198 (1921).
Siphocampylus flagelliformis var. *glaber* E. Wimm., Ann. Naturhist. Mus. Wien 56: 324 (1948).

Siphocampylus floribundus Zahlbr., Bot. Jahrb. Syst. 37: 460 (1906).
Peru. 83 PER. Cl. nanophan.

Siphocampylus fluminensis (Vell.) E. Wimm., Repert. Spec. Nov. Regni Veg. 22: 210 (1926).
SE. Brazil (São Paulo). 84 BZL. Nanophan. or cham.
* *Lobelia fluminensis* Vell., Fl. Flumin., Icon. 8: pl. 159 (1831).
Siphocampylus fluminensis var. *oppositifolius* E. Wimm., Ann. Naturhist. Mus. Wien 56: 323 (1948).

Siphocampylus foliosus Griseb., Abh. Königl. Ges. Wiss. Göttingen 19: 201 (1874).
Colombia to N. Argentina (Jujuy to Córdoba). 83 BOL CLM PER 85 AGE AGW.
Nanophan. 2*n* = 28.
Siphocampylus foliosus var. *minor* Zahlbr. ex Kuntze, Revis. Gen. Pl. 3(2): 189 (1898).
Siphocampylus tupiformis var. *diversifolius* Hicken, Darwiniana 1: 142 (1924).
Siphocampylus foliosus var. *diversifolius* (Hicken) Pontiroli in Cabrera, Fl. Prov. Jujuy 9: 528 (1993).
Siphocampylus foliosus var. *colombinus* E. Wimm., Repert. Spec. Nov. Regni Veg. 29: 85 (1931).
Siphocampylus foliosus var. *glabratus* E. Wimm., Revista Sudamer. Bot. 2: 93 (1935).
Siphocampylus foliosus var. *lessenii* E. Wimm., Revista Sudamer. Bot. 2: 93 (1935).
Siphocampylus foliosus var. *subcanus* Zahlbr. ex E. Wimm., Revista Sudamer. Bot. 2: 100 (1935).

Siphocampylus fruticosus E. Wimm., Pflanzenr. IV.276c: 846 (1968).
Ecuador. 83 ECU. Nanophan.

Siphocampylus fulgens Dombrain, Fl. Mag. (London) 5: pl. 320 (1866).
   S. Brazil (Paraná). 84 BZS. Nanophan. or cham.

Siphocampylus funiculosus E. Wimm., Repert. Spec. Nov. Regni Veg. 29: 81 (1931).
   Colombia. 83 CLM. Cl. nanophan.

Siphocampylus furax E. Wimm., Repert. Spec. Nov. Regni Veg. 29: 86 (1931).
   Ecuador. 83 ECU. Nanophan.

Siphocampylus giganteus (Cav.) G. Don, Gen. Hist. 3: 702 (1834).
   Colombia to Peru. 83 CLM ECU PER. Phan. or nanophan. $2n = 28$.
      * *Lobelia gigantea* Cav., Anales Hist. Nat. 2: 104 (1800).
      Siphocampylus giganteus var. *latifolius* Vatke, Linnaea 38: 734 (1874).
      Siphocampylus giganteus var. *angustifolius* Vatke, Linnaea 38: 734 (1874).
      Siphocampylus giganteus f. *brachyodon* E. Wimm., Ann. Naturhist. Mus. 56: 331 (1948).

Siphocampylus glaber McVaugh, N. Amer. Fl. 32A: 103 (1943). *Siphocampylus subglaber* var.
   glaber (McVaugh) E. Wimm., Pflanzenr. IV.276b: 271 (1953).
   Cuba. 81 CUB. Nanophan.

Siphocampylus glareosus Zahlbr., Repert. Spec. Nov. Regni Veg. 14: 182 (1915).
   Colombia. 83 CLM. Nanophan. or cham.

Siphocampylus goebelii E. Wimm., Repert. Spec. Nov. Regni Veg. 38: 76 (1935).
   Peru. 83 PER. Nanophan. or cham.

Siphocampylus grandiflorus E. Wimm., Repert. Spec. Nov. Regni Veg. 19: 257 (1924).
   Peru. 83 PER. Cl. nanophan.

Siphocampylus heliades E. Wimm., Notizbl. Bot. Gart. Berlin-Dahlem 10: 744 (1929).
   Peru. 83 PER. Nanophan. or cham.

Siphocampylus helmutii E. Wimm., Notizbl. Bot. Gart. Berlin-Dahlem 10: 745 (1929),
   as 'helmuti'.
   Peru. 83 PER. Cl. nanophan.
      Siphocampylus helmutii var. *pilosifolius* E. Wimm., Pflanzenr. IV.276b: 775 (1953).

Siphocampylus hispidus Benth., Pl. Hartweg.: 214 (1845).
   Colombia. 83 CLM. Nanophan.

Siphocampylus humboldtianus A. DC. in DC., Prodr. 7: 398 (1839).
   Ecuador & Peru. 83 ECU PER. Nanophan.
      * *Lobelia humboldtiana* C. Presl, Prod. Monogr. Lobel.: 35 (1836); non Schult. in Roem. &
         Schult., Syst. Veg. 5: 68 (1819).
      Siphocampylus pubescens Benth., Pl. Hartweg.: 139 (1845).
      Centropogon griseus Gleason, Bull. Torrey Bot. Club 52: 62 (1925). *Siphocampylus griseus*
         (Gleason) E. Wimm., Pflanzenr. IV.276b: 260 (1943).
      Siphocampylus humboldtianus var. *ovatus* E. Wimm. in J. F. Macbr., Fl. Peru 6: 455 (1937).

Siphocampylus humilis E. Wimm., Pflanzenr. IV.276b: 332 (1953).
   SE. Brazil (Rio de Janeiro). 84 BZL. Hemicr.

Siphocampylus hypoleucus E. Wimm., Pflanzenr. IV.276c: 849 (1968).
   Peru. 83 PER. Nanophan.

Siphocampylus hypopogon E. Wimm., Revista Sudamer. Bot. 2: 91 (1935).
   NE. Argentina (Misiones). 85 AGE. Nanophan. or cham.

Siphocampylus hypsophilus E. Wimm., Repert. Spec. Nov. Regni Veg. 26: 14 (1929).
Colombia. 83 CLM. Cl. nanophan.

Siphocampylus igneus Urb., Symb. Antill. 1: 452 (1899). *Siphocampylus urbanii* E. Wimm.,
Repert. Spec. Nov. Regni Veg. 22: 214 (1926).
Dominican Rep. 81 DOM. Cham. or hemicr.

Siphocampylus imbricatus (Cham.) G. Don, Gen. Hist. 3: 703 (1834).
E. Brazil (Bahia, Minas Gerais). 84 BZE BZL. Nanophan.
  * *Lobelia imbricata* Cham., Linnaea 8: 206 (1833).
    Siphocampylus thomesianus Moric., Pl. Nouv. Amér.: 142 (1847).
    Siphocampylus imbricatus var. *casarettoi* E. Wimm., Repert. Spec. Nov. Regni Veg. 26:
      16 (1929).
    Siphocampylus imbricatus var. *glabratus* E. Wimm., Ann. Naturhist. Mus. Wien 56: 327
      (1948).

Siphocampylus isochilus E. Wimm., Brittonia 8: 111 (1955).
Colombia. 83 CLM. Nanophan.

Siphocampylus jelskii Zahlbr., Ann. K. K. Naturhist. Hofmus. 6: 441 (1891).
Ecuador & Peru. 83 ECU PER. Herb. phan. or nanophan.
    Siphocampylus superbus Zahlbr., Bot. Jahrb. Syst. 37: 455 (1906).
    Siphocampylus jelskii f. *eugenius* E. Wimm., Notizbl. Bot. Gart. Berlin-Dahlem 10:
      740 (1929).

Siphocampylus keissleri E. Wimm., Repert. Spec. Nov. Regni Veg. 22: 211 (1926),
as 'keisslerii'.
Colombia to Peru. 83 CLM ECU PER. (Cl.) nanophan.
    Siphocampylus laetus E. Wimm., Repert. Spec. Nov. Regni Veg. 22: 212 (1926).

Siphocampylus krauseanus E. Wimm., Repert. Spec. Nov. Regni Veg. 26: 16 (1929).
Peru. 83 PER. Nanophan.

Siphocampylus kuntzeanus Zahlbr., Bull. Torrey Bot. Club 24: 378 (1897).
Bolivia. 83 BOL. Nanophan. or cham.

Siphocampylus laevigatus Planch., Fl. Serres Jard. Eur. 6: 34 (1850).
Colombia. 83 CLM. Cl. nanophan.

Siphocampylus lasiandrus Planch., Fl. Serres Jard. Eur. 6: 36 (1850).
Colombia. 83 CLM. Nanophan.

Siphocampylus lauroanus Handro & M. Kuhlm., Arq. Bot. Estado São Paulo (n.s.) 3:
263 (1962).
SE. Brazil (São Paulo). 84 BZL. Nanophan.

Siphocampylus lecomtei E. Wimm., Repert. Spec. Nov. Regni Veg. 26: 15 (1929).
Colombia. 83 CLM. Cl. nanophan.

Siphocampylus leptophyllus Urb., Ark. Bot. 23A(5): 104 (1930).
Haiti. 81 HAI. Cham. or hemicr.

Siphocampylus lindleyi Lem., Jard. Fleur. 2: pl. 142 (1852).
Colombia. 83 CLM. Nanophan. or cham.
    Siphocampylus pubiflorus E. Wimm., Repert. Spec. Nov. Regni Veg. 19: 261 (1924).

Siphocampylus lindleyi var. *dissitiflorus* E. Wimm., Repert. Spec. Nov. Regni Veg. 29: 83 (1931).

Siphocampylus lobbii Zahlbr., Bot. Jahrb. Syst. 37: 457 (1906).
Peru. 83 PER. Nanophan.
 Siphocampylus lobbii var. *megodontus* E. Wimm., Repert. Spec. Nov. Regni Veg. 29: 88 (1931).

Siphocampylus longibracteolatus E. Wimm., Repert. Spec. Nov. Regni Veg. 19: 258 (1924), as 'longebracteolatus'.
Colombia. 83 CLM. Cl. nanophan.

Siphocampylus longior Lammers, Novon 12: 227 (2002).
Peru. 83 PER. Cl. nanophan.

Siphocampylus longipedunculatus Pohl, Pl. Bras. Icon. Descr. 2: 109 (1831), as 'longepedunculatus'. *Lobelia pedicellaris* C. Presl, Prodr. Monogr. Lobel.: 34 (1836).
SE. Brazil (Minas Gerais, Rio de Janeiro, São Paulo). 84 BZL. Nanophan.
 Lobelia scabra C. Presl, Prodr. Monogr. Lobel.: 33 (1836); non Thunb. in Hoffm., Phytogr. Bl.: 21 (1803); nec Spreng., Neue Entd. 1: 272 (1820). *Siphocampylus scaber* A. DC. in DC., Prodr. 7: 399 (1839).
 Siphocampylus longipedunculatus var. *meiopodus* E. Wimm., Pflanzenr. IV.276b: 332 (1953).
 Siphocampylus longipedunculatus var. *trichophyllus* E. Wimm., Pflanzenr. IV.276b: 332 (1953).

Siphocampylus lorentzii E. Wimm., Repert. Spec. Nov. Regni Veg. 29: 85 (1931).
Bolivia & N. Argentina (Córdoba). 83 BOL 85 AGE. Nanophan. or cham.

Siphocampylus loxensis (Willd. ex Schult.) Vatke ex E. Wimm. in J. F. Macbr., Fl. Peru 6: 457 (1937).
Ecuador. 83 ECU. Nanophan.
 * *Lobelia loxensis* Willd. ex Schult. in Roem. & Schult., Syst. Veg. 5: 72 (1819). *Dortmanna loxensis* (Willd. ex Schult.) Kuntze, Revis. Gen. Pl. 2: 972 (1891).

Siphocampylus lucidus E. Wimm., Repert. Spec. Nov. Regni Veg. 19: 391 (1924).
Ecuador. 83 ECU. Nanophan. or cham. $2n = 28$.

Siphocampylus lucifer E. Wimm., Ann. Naturhist. Mus. Wien 56: 324 (1948).
Colombia. 83 CLM. Cl. nanophan.
 * *Siphocampylus acuminatus* E. Wimm., Repert. Spec. Nov. Regni Veg. 22: 205 (1926); non (Sw.) G. Don in Sweet, Hort. Brit. (ed. 3): 424 (1839).

Siphocampylus lycioides (Cham.) G. Don., Gen. Hist. 3: 703 (1834).
E. Brazil (Goiás, Paraná). 84 BZC BZS. Cham. or hemicr.
 * *Lobelia lycioides* Cham., Linnaea 8: 207 (1833). *Campylosiphon lycioides* (Cham.) St.-Lag., Ann. Soc. Bot. Lyon 7: 135 (1880), as 'lycioideus'.

Siphocampylus macropodoides Zahlbr., Bot. Jahrb. Syst. 37: 458 (1906).
Peru. 83 PER. Nanophan.

Siphocampylus macropodus (Thunb.) G. Don, Gen. Hist. 3: 702 (1834).
Brazil (Mato Grosso, Minas Gerais, Rio de Janeiro, São Paulo, Paraná). 84 BZC BZL BZS. (Cl.) nanophan. $2n = 28$.
 * *Lobelia macropoda* Thunb., Pl. Bras.: 6 (1817).
 Siphocampylus canus Pohl, Pl. Bras. Icon. Descr. 2: 106 (1831). *Lobelia cana* (Pohl) C. Presl, Prodr. Monogr. Lobel.: 37 (1836).
 Siphocampylus crenatifolius Pohl, Pl. Bras. Icon. Descr. 2: 107 (1831). *Lobelia crenata* C. Presl, Prodr. Monogr. Lobel.: 37 (1836); non Thunb., Prodr. Pl. Cap.: 39 (1794).

Siphocampylus villosulus Pohl, Pl. Bras. Icon. Descr. 2: 108 (1831). *Lobelia villosula* (Pohl)
C. Presl, Prodr. Monogr. Lobel.: 38 (1836).
Siphocampylus cinerascens E. Wimm., Pflanzenr. IV.276c: 847 (1968).

Siphocampylus macrostemon (C. Presl) A. DC. in DC., Prodr. 7: 403 (1839).
Peru. 83 PER. Cham. or hemicr.
* *Lobelia macrostemon* C. Presl, Prodr. Monogr. Lobel.: 36 (1836). *Centropogon
macrostemon* (C. Presl) Zahlbr., Repert. Spec. Nov. Regni Veg. 14: 181 (1915).

Siphocampylus manettiiflorus Hook., Bot. Mag. 74: tab. 4403 (1848), as 'manettiaeflorus'.
Cuba. 81 CUB. Nanophan. or cham.
Siphocampylus nitidus de Jonghe ex C. Morren, Ann. Soc. Roy. Agric. Bot. Gand 2: 319
(1846); non Pohl, Pl. Bras. Icon. Descr. 2: 111 (1831).
Lobelia salviifolia A. Rich. in Sagra, Hist. Fis. Cub. 11: 69 (1850), as 'salviaefolia'.
Siphocampylus impressus Urb., Symb. Antill. 7: 417 (1912).
Siphocampylus undulatus Urb., Symb. Antill. 9: 428 (1925).
Siphocampylus libanensis Urb., Symb. Antill. 9: 429 (1925). *Siphocampylus manettiiflorus*
var. *libanensis* (Urb.) E. Wimm., Ann. Naturhist. Mus. Wien 56: 322 (1948).
Siphocampylus ekmanii Urb., Symb. Antill. 9: 429 (1925).

Siphocampylus matthiaei A. DC. in DC., Prodr. 7: 405 (1839). *Burmeistera matthiaei* (A.
DC.) Benth. & Hook. f. ex B. D. Jacks., Index Kew. 1: 361 (1893).
Peru. 83 PER. Nanophan.
Siphocampylus matthewsii E. Wimm., Repert. Spec. Nov. Regni Veg. 19: 258 (1924), as
'mathewsii'.

Siphocampylus maxonis E. Wimm., Repert. Spec. Nov. Regni Veg. 19: 259 (1924).
Panama. 80 PAN. Cham. or hemicr.

Siphocampylus megalanthus Zahlbr., Repert. Spec. Nov. Regni Veg. 14: 184 (1915).
Costa Rica & Colombia. 80 COS. 83 CLM. Nanophan.
Siphocampylus megalanthus var. *orthosepalus* E. Wimm., Repert. Spec. Nov. Regni Veg.
26: 15 (1929).
Siphocampylus megalanthus var. *costaricanus* E. Wimm., Pflanzenr. IV.276c: 849 (1968).

Siphocampylus megastoma E. Wimm., Notizbl. Bot. Gart. Berlin-Dahlem 10: 739 (1929).
Peru. 83 PER. Nanophan.

Siphocampylus membranaceus Britton, Bull. Torrey Bot. Club 19: 372 (1892).
Bolivia. 83 BOL. Nanophan. or cham.

Siphocampylus microstoma Hook., Bot. Mag. 73: tab. 4286 (1847).
Colombia. 83 CLM. Nanophan.
Siphocampylus microstoma var. *sparsus* E. Wimm., Pflanzenr. IV.276b: 284 (1953).

Siphocampylus mirabilis E. Wimm., Repert. Spec. Nov. Regni Veg. 29: 80 (1931).
Colombia. 83 CLM. Cl. nanophan.

Siphocampylus moritzianus E. Wimm., Repert. Spec. Nov. Regni Veg. 19: 260 (1924).
Venezuela. 82 VEN. Cl. nanophan.

Siphocampylus nematosepalus (Donn. Sm.) E. Wimm., Repert. Spec. Nov. Regni Veg. 38:
22 (1935).
Costa Rica. 80 COS. Cham. or hemicr.
* *Centropogon nematosepalus* Donn. Sm., Bot. Gaz. (Crawfordsville) 44: 114 (1907).

Siphocampylus nemoralis Griseb., Abh. Königl. Ges. Wiss. Göttingen 19: 201 (1874).
Bolivia & NW. Argentina (Jujuy to Catamarca). 83 BOL 85 AGW. Nanophan.
Siphocampylus nemoralis var. *grisebachii* E. Wimm., Revista Sudamer. Bot. 2: 92 (1935).
Siphocampylus nemoralis var. *tarijanus* E. Wimm., Revista Sudamer. Bot. 2: 92 (1935).
Siphocampylus nemoralis f. *hieronymi* E. Wimm., Revista Sudamer. Bot. 2: 100 (1935).

Siphocampylus neurotrichus E. Wimm., Repert. Spec. Nov. Regni Veg. 38: 24 (1935).
Bolivia. 83 BOL. Cham. or hemicr.

Siphocampylus nitidus Pohl, Pl. Bras. Icon. Descr. 2: 111 (1831). *Lobelia nitida* (Pohl) C.
Presl, Prodr. Monogr. Lobel.: 37 (1836).
E. Brazil (Minas Gerais). 84 BZL. Cl. nanophan.
Siphocampylus nitidus var. *nitidissimus* E. Wimm., Repert. Spec. Nov. Regni Veg. 22:
212 (1926).
Siphocampylus nitidus var. *pleotrichus* E. Wimm., Repert. Spec. Nov. Regni Veg. 22:
212 (1926).
Siphocampylus nitidus f. *acuminatus* E. Wimm., Pflanzenr. IV.276b: 314 (1953).

Siphocampylus niveus (Willd. ex Schult.) Vatke, Linnaea 38: 732 (1874).
Colombia. 83 CLM. Nanophan.
* *Lobelia nivea* Willd. ex Schult. in Roem. & Schult., Syst. Veg. 5: 58 (1819). *Byrsanthes
humboldtiana* C. Presl, Prodr. Monogr. Monogr. Lobel.: 42 (1836). *Byrsanthes nivea*
(Willd. ex Schult.) A. DC. in DC., Prodr. 7: 408 (1839).
Siphocampylus lanatus Benth., Pl. Hartweg. 214 (1845).
Siphocampylus pennellii Gleason, Bull. Torrey Bot. Club 52: 65 (1925).

Siphocampylus nobilis E. Wimm., Repert. Spec. Nov. Regni Veg. 26: 12 (1929).
Peru. 83 PER. Cl. nanophan.

Siphocampylus nummularius E. Wimm., Repert. Spec. Nov. Regni Veg. 38: 18 (1935).
Bolivia. 83 BOL. Cl. nanophan.

Siphocampylus oblongifolius Rusby, Mem. Torrey Bot. Club 6: 73 (1896).
Bolivia. 83 BOL. Cl. nanophan.

Siphocampylus obovatus (G. Don) E. Wimm., Repert. Spec. Nov. Regni Veg. 38: 19 (1935).
Peru. 83 PER. Nanophan. or cham.
* *Tupa obovata* G. Don, Gen. Hist. 3: 700 (1834). *Rapuntium obovatum* (G. Don) C. Presl,
Prodr. Monogr. Lobel.: 25 (1836). *Dortmanna obovata* (G. Don) Kuntze, Revis. Gen.
Pl. 2: 973 (1891). *Lobelia incana* Ruiz & Pav. ex B. D. Jacks., Index Kew. 2: 104 (1894).

Siphocampylus obtusus E. Wimm., Repert. Spec. Nov. Regni Veg. 19: 261 (1924).
Colombia. 83 CLM. Nanophan.

Siphocampylus odontosepalus Vatke, Linnaea 38: 729 (1874).
Venezuela. 82 VEN. Nanophan. or cham.

Siphocampylus onagrius E. Wimm., Repert. Spec. Nov. Regni Veg. 22: 212 (1926).
Peru. 83 PER. Cl. nanophan.
Siphocampylus dependens var. *undulatus* E. Wimm., Repert. Spec. Nov. Regni Veg. 22: 209
(1926). *Siphocampylus onagrius* var. *undulatus* (E. Wimm.) E. Wimm., Pflanzenr.
IV.276b: 342 (1953).
Siphocampylus onagrius f. *longisepalus* E. Wimm., Pflanzenr. IV.276b: 343 (1953).

Siphocampylus orbignianus A. DC. in DC., Prodr. 7: 405 (1839).
Bolivia & NW. Argentina (Jujuy to Salta). 83 BOL 85 AGW. Nanophan.

Siphocampylus orbignianus var. *discurrens* E. Wimm., Ann. Naturhist. Mus. Wien 56:
323 (1948).

Siphocampylus oscitans B. A. Stein, Ann. Missouri Bot. Gard. 74: 492 (1987).
Peru. 83 PER. Nanophan.
* *Burmeistera weberbaueri* Zahlbr., Bot. Jahrb. Syst. 37: 451 (1906); non *Siphocampylus
weberbaueri* Zahlbr., Bot. Jahrb. Syst. 37: 456 (1906).

Siphocampylus ovatus (G. Don) E. Wimm., Repert. Spec. Nov. Regni Veg. 38: 18 (1935).
Peru. 83 PER. Cl. nanophan.
* *Tupa ovata* G. Don, Gen. Hist. 3: 700 (1834). *Rapuntium ovatum* (G. Don) C. Presl,
Prodr. Monogr. Lobel.: 29 (1836); non C. Presl, Prodr. Monogr. Lobel.: 16 (1836).
Dortmanna ovata (G. Don) Kuntze, Revis. Gen. Pl. 2: 973 (1891).
Lobelia pendulifolia C. Presl, Prodr. Monogr. Lobel.: 35 (1836). *Siphocampylus
pendulifolius* (C. Presl) A. DC. in DC., Prodr. 7: 398 (1839).

Siphocampylus palilloanus E. Wimm., Repert. Spec. Nov. Regni Veg. 19: 261 (1924).
Peru. 83 PER. Nanophan. or cham.

Siphocampylus pallidus E. Wimm. in J. F. Macbr., Fl. Peru 6: 462 (1937).
Peru. 83 PER. Nanophan. or cham.

Siphocampylus paramicola McVaugh, Brittonia 6: 458 (1949).
Colombia. 83 CLM. Nanophan. or cham.

Siphocampylus parvifolius E. Wimm. in J. F. Macbr., Fl. Peru 6: 463 (1937).
Peru. 83 PER. Nanophan. or cham.

Siphocampylus parvilobus E. Wimm., Repert. Spec. Nov. Regni Veg. 38: 21 (1935).
Peru. 83 PER. Cl. nanophan.
* *Siphocampylus brevidens* E. Wimm., Repert. Spec. Nov. Regni Veg. 29: 89 (1931); non
Regel, Index Sem. Hort. Petrop. 1868: 84 (1868).

Siphocampylus patens Griseb., Cat. Pl. Cub.: 159 (1866).
Cuba. 81 CUB. Cham. or hemicr. $2n = 28$.

Siphocampylus pavonis E. Wimm., Repert. Spec. Nov. Regni Veg. 29: 89 (1931).
Peru. 83 PER. Cl. nanophan.
* *Tupa secunda* G. Don, Gen. Hist. 3: 700 (1834); non *Siphocampylus secundus* E. Wimm.,
Repert. Spec. Nov. Regni Veg. 19: 262 (1924). *Rapuntium secundum* (G. Don) C. Presl,
Prodr. Monogr. Lobel.: 29 (1836). *Dortmanna doniana* Kuntze, Revis. Gen. Pl. 2: 971
(1891). *Siphocampylus donianus* E. Wimm., Pflanzenr. IV.276b: 364 (1953).

Siphocampylus penduliflorus Decne. ex Planch., Fl. Serres Jard. Eur. 8: 35 (1852).
Venezuela & Peru. 82 VEN 83 PER. Cl. nanophan.
Siphocampylus penduliflorus var. *asperatulus* E. Wimm., Repert. Spec. Nov. Regni Veg. 29:
88 (1931).

Siphocampylus peruvianus A. DC. in DC., Prodr. 7: 401 (1839).
Peru. 83 PER. Cl. nanophan.

Siphocampylus phaeton E. Wimm., Repert. Spec. Nov. Regni Veg. 29: 83 (1931).
Peru. 83 PER. Nanophan. or cham.

Siphocampylus pilosus Gleason, Bull. Torrey Bot. Club 52: 70 (1925).
Colombia. 83 CLM. Cham. or hemicr.

Siphocampylus platysiphon Lammers, Brittonia 45: 28 (1993).
Peru. 83 PER. Nanophan.

Siphocampylus plegmatocaulis Lammers, Novon 12: 228 (2002).
Peru. 83 PER. Cl. nanophan.

Siphocampylus polyanthus E. Wimm., Pflanzenr. IV.276c: 848 (1968).
Colombia. 83 CLM. Nanophan. or cham.

Siphocampylus polycladus E. Wimm., Notizbl. Bot. Gart. Berlin-Dahlem 10: 745 (1929).
Ecuador & Peru. 83 ECU PER. Nanophan.
Siphocampylus polycladus var. *rubicundus* E. Wimm., Notizbl. Bot. Gart. Berlin-Dahlem 10: 746 (1929).

Siphocampylus polyphyllus Planch., Fl. Serres Jard. Eur. 6: 36 (1850). *Siphocampylus planchonis* var. *polyphyllus* (Planch.) E. Wimm., Ann. Naturhist. Mus. Wien 56: 327 (1948).
Venezuela & Colombia. 82 VEN 83 CLM. Nanophan.
Siphocampylus mollis Planch., Fl. Serres Jard. Eur. 6: 36 (1850); non Regel, Flora 33: 353 (1850). *Siphocampylus planchonis* E. Wimm., Repert. Spec. Nov. Regni Veg. 22: 213 (1926).
Siphocampylus planchonis var. *daweanus* E. Wimm., Repert. Spec. Nov. Regni Veg. 29: 90 (1931).
Siphocampylus planchonis f. *flaviflorus* E. Wimm., Ann. Naturhist. Mus. Wien 56: 327 (1948).
Siphocampylus planchonis var. *truxillensis* E. Wimm., Ann. Naturhist. Mus. Wien 56: 327 (1948).
Siphocampylus planchonis f. *glabrifolia* E. Wimm., Pflanzenr. IV.276c: 848 (1968).

Siphocampylus popayanensis Benth., Pl. Hartweg.: 213 (1845). *Centropogon popayanensis* (Benth.) Benth. & Hook. f. ex B. D. Jacks., Index Kew. 2: 1274 (1895).
Colombia & Ecuador. 83 CLM ECU. Cl. nanophan.
Siphocampylus trichodontus E. Wimm., Repert. Spec. Nov. Regni Veg. 19: 262 (1924).

Siphocampylus pozuzensis E. Wimm., Repert. Spec. Nov. Regni Veg. 29: 87 (1931).
Peru. 83 PER. Cl. nanophan.

Siphocampylus praevaricator Lammers, Novon 12: 230 (2002).
Venezuela. 82 VEN. Nanophan. or cham.

Siphocampylus psilophyllus Pohl, Pl. Bras. Icon. Descr. 2: 113 (1831). *Lobelia psilophylla* (Pohl) C. Presl, Prodr. Monogr. Lobel.: 36 (1836).
SE. Brazil (Minas Gerais, Rio de Janeiro). 84 BZL. Nanophan.
Siphocampylus psilophyllus var. *longisepalus* E. Wimm., Revista Sudamer. Bot. 2: 93 (1935).

Siphocampylus puberulus E. Wimm., Repert. Spec. Nov. Regni Veg. 26: 13 (1929).
Bolivia. 83 BOL. Nanophan. or cham.

Siphocampylus purdieanus Planch., Fl. Serres Jard. Eur. 6: 18 (1850), as 'purdiaeanus'.
Colombia. 83 CLM. Nanophan. or cham.

Siphocampylus pyriformis Zahlbr., Repert. Spec. Nov. Regni Veg. 14: 183 (1915).
Colombia. 83 CLM. Cl. nanophan.
Siphocampylus obovoideus Gleason, Bull. Torrey Bot. Club 52: 71 (1925).
Siphocampylus pyriformis var. *reflexisepalus* E. Wimm., Brittonia 8: 111 (1955).

Siphocampylus queluzensis E. Wimm., Revista Sudamer. Bot. 2: 91 (1935).
E. Brazil (Minas Gerais). 84 BZL. Nanophan.

Siphocampylus quetamensis E. Wimm., Repert. Spec. Nov. Regni Veg. 29: 90 (1931).
Colombia. 83 CLM. Cl. nanophan.

Siphocampylus quindioros E. Wimm., Pflanzenr. IV.276c: 848 (1968).
Colombia. 83 CLM. Cham. or hemicr.

Siphocampylus radiatus Rusby, Mem. Torrey Bot. Club 6: 73 (1896).
Bolivia. 83 BOL. Nanophan. or cham.
Centropogon brittonianus var. *brevidentatus* Zahlbr. & Rech., Meded. Rijks.-Herb. 19: 51
(1913). *Siphocampylus radiatus* var. *brevidentatus* (Zahlbr. & Rech.) E. Wimm., Ann.
Naturhist. Mus. Wien 56: 331 (1948).
Siphocampylus radiatus var. *minor* Zahlbr., Bull. Torrey Bot. Club 24: 376 (1897).

Siphocampylus ravidus E. Wimm., Repert. Spec. Nov. Regni Veg. 26: 13 (1929).
Colombia. 83 CLM. Cl. phan. or cl. nanophan.

Siphocampylus rectiflorus Rusby, Descr. S. Amer. Pl.: 145 (1920).
Colombia. 83 CLM. Phan. or nanophan.
Siphocampylus lilacinus E. Wimm., Repert. Spec. Nov. Regni Veg. 19: 257 (1924).

Siphocampylus reflexus Rusby, Bull. New York Bot. Gard. 4: 403 (1907).
Bolivia. 83 BOL. Cl. nanophan.
Siphocampylus elegans var. *cordatus* Zahlbr., Bull. Torrey Bot. Club 24: 381 (1897).

Siphocampylus reticulatus (Willd. ex Schult.) Klotzsch & H. Karst. ex Vatke, Linnaea 38:
728 (1874).
Venezuela. 82 VEN. Nanophan. or cham.
* *Lobelia reticulata* Willd. ex Schult. in Roem. & Schult., Syst. Veg. 5: 58 (1819).
Dortmanna reticulata (Willd. ex Schult.) Kuntze, Revis. Gen. Pl. 2: 973 (1891).
Siphocampylus lantanifolius A. DC. in DC., Prodr. 7: 399 (1839).
Siphocampylus revolutus Graham, Edinburgh New Philos. J. 30: 429 (1841).
Siphocampylus lantanifolius var. *glabriusculus* Hook., Bot. Mag. 70: tab. 4105 (1844).
Siphocampylus reticulatus var. *glabriusculus* (Hook.) E. Wimm., Pflanzenr. IV.276b: 286
(1953).
Siphocampylus hamatus Wendl., Allg. Gartenzeitung 18: 138 (1850). *Siphocampylus
reticulatus* var. *hamatus* (Wendl.) E. Wimm., Ann. Naturhist. Mus. Wien 56: 324 (1948).
Siphocampylus tepuiensis Gleason, Brittonia 3: 194 (1939).

Siphocampylus retrorsus (Willd. ex Schult.) Vatke, Linnaea 38: 724 (1874).
Colombia. 83 CLM. Cl. nanophan.
* *Lobelia retrorsa* Willd. ex Schult. in Roem. & Schult., Syst. Veg. 5: 57 (1819). *Dortmanna
retrorsa* (Willd. ex Schult.) Kuntze, Revis. Gen. Pl. 2: 973 (1891).
Siphocampylus asper Benth., Pl. Hartweg.: 213 (1845). *Siphocampylus retrorsus* var. *asper*
(Benth.) E. Wimm., Repert. Spec. Nov. Regni Veg. 26: 18 (1929).
Siphocampylus eximius Planch., Fl. Serres Jard. Eur. 6: 16 (1850). *Siphocampylus retrorsus*
var. *eximius* (Planch.) E. Wimm., Repert. Spec. Nov. Regni Veg. 26: 19 (1929).
Siphocampylus reflexifolius Zahlbr., Repert. Spec. Nov. Regni Veg. 14: 181 (1915).
Siphocampylus retrorsus var. *semiasper* E. Wimm., Repert. Spec. Nov. Regni Veg. 26:
19 (1929).

Siphocampylus rictus Lammers, Brittonia 50: 239 (1998).
Peru. 83 PER. Nanophan.

Siphocampylus rosmarinifolius G. Don, Gen. Hist. 3: 704 (1834). *Lobelia rosmarinifolia*
(G. Don) Dombey ex C. Presl, Prodr. Monogr. Lobel.: 36 (1836).
Peru. 83 PER. Nanophan.

Siphocampylus rostratus E. Wimm., Pflanzenr. IV.276c: 846 (1968).
Ecuador. 83 ECU. Nanophan.

Siphocampylus ruber Alain, Contr. Ocas. Mus. Hist. Nat. Colegio "De la Salle" 18: 2 (1960).
Cuba. 81 CUB. Nanophan.

Siphocampylus rupestris E. Wimm., Repert. Spec. Nov. Regni Veg. 19: 391 (1924).
Ecuador. 83 ECU. Nanophan.

Siphocampylus rusbyanus Britton, Bull. Torrey Bot. Club 19: 372 (1892).
Peru & Bolivia. 83 BOL PER. Nanophan.
Siphocampylus rusbyanus var. *subtervestitus* Zahlbr., Bot. Jahrb. Syst. 37: 459 (1906), as 'subtervestita'.
Centropogon weddellii E. Wimm., Repert. Spec. Nov. Regni Veg. 26: 11 (1929).
Burmeistera tricolorata E. Wimm., Repert. Spec. Nov. Regni Veg. 30: 22 (1932).

Siphocampylus salviifolius E. Wimm., Notizbl. Bot. Gart. Berlin-Dahlem 10: 743 (1929).
Peru. 83 PER. Nanophan.

Siphocampylus sanchezii Lammers, Brittonia 50: 237 (1998).
Peru. 83 PER. Cl. nanophan.

Siphocampylus sandemanii E. Wimm., Pflanzenr. IV.276b: 313 (1953).
Colombia. 83 CLM. Cham.

Siphocampylus sanguineus Zahlbr., Bot. Jahrb. Syst. 37: 456 (1906).
Ecuador & Peru. 83 ECU PER. Cl. nanophan. $2n = 28$.
Centropogon hitchcockii Gleason, Bull. Torrey Bot. Club 52: 63 (1925). *Siphocampylus hitchcockii* (Gleason) E. Wimm., Repert. Spec. Nov. Regni Veg. 29: 78 (1931). *Siphocampylus sanguineus* var. *hitchcockii* (Gleason) E. Wimm., Ann. Naturhist Mus. Wien 56: 327 (1948).

Siphocampylus scandens (Kunth) G. Don, Gen. Hist. 3: 703 (1834).
Colombia & Ecuador. 83 CLM ECU. Cl. nanophan.
* *Lobelia scandens* Kunth in Humb., Bonpl. & Kunth, Nov. Gen. Sp. 3: 309 (quarto), 241 (folio) (1819).
Lobelia obtusifolia Willd. ex Schult. in Roem. & Schult., Syst. Veg. 5: 57 (1819).
Siphocampylus divaricatus Benth., Pl. Hartweg.: 138 (1845).
Siphocampylus subcarnosus Benth., Pl. Hartweg.: 138 (1845).
Siphocampylus penduliflorus var. parvicalicinus E. Wimm., Repert. Spec. Nov. Regni Veg. 29: 88 (1931).
Siphocampylus rotundifolius E. Wimm., Pflanzenr. IV.276c: 846 (1968).

Siphocampylus sceptrum Decne. in Linden, Établ. Linden, Prix-courant 5: 7 (1850).
Venezuela & Colombia. 83 VEN 83 CLM. Nanophan. or cham.
Cremochilus meridensis Turcz., Bull. Soc. Imp. Naturalistes Moscou 25(3): 174 (1852). *Siphocampylus meridensis* (Turcz.) Zahlbr., Ann. K. K. Naturhist. Hofmus. 6: 440 (1891).
Siphocampylus sessilifolius Vatke, Linnaea 38: 731 (1874). *Siphocampylus sceptrum* var. *latifolius* E. Wimm., Ann. Naturhist. Mus. Wien 56: 330 (1948).
Siphocampylus sceptrum var. *denudatus* E. Wimm., Ann. Naturhist. Mus. Wien 56: 330 (1948).
Siphocampylus sceptrum var. *tachirensis* E. Wimm., Pflanzenr. IV.276c: 851 (1968).

Siphocampylus schizandrus E. Wimm., Pflanzenr. IV.276b: 3 (1943).
Colombia. 83 CLM. Nanophan. or cham.

Siphocampylus schlimianus Planch., Fl. Serres Jard. Eur. 6: 34 (1850), as 'schlimmianus'.
  Venezuela & Colombia. 82 VEN 83 CLM. Cl. nanophan.

Siphocampylus schultzeanus E. Wimm., Repert. Spec. Nov. Regni Veg. 26: 14 (1929).
  Colombia. 83 CLM. Cl. nanophan.

Siphocampylus secundus E. Wimm., Repert. Spec. Nov. Regni Veg. 19: 262 (1924).
  Peru. 83 PER. Cl. nanophan.

Siphocampylus sissi E. Wimm., Pflanzenr. IV.276c: 849 (1968).
  Colombia. 83 CLM. Cl. nanophan.

Siphocampylus smilax Lammers, Brittonia 50: 240 (1998).
  Bolivia. 83 BOL. Nanophan. or cham.

Siphocampylus sonchifolius (Sw.) McVaugh, N. Amer. Fl. 32A: 105 (1943).
  Hispaniola. 81 DOM HAI. Cham. or hemicr.
    Lobelia laciniata Lam., Encycl. 3: 584 (1792). *Rapuntium laciniatum* (Lam.) C. Presl,
      Prodr. Monogr. Lobel.: 27 (1836). *Siphocampylus lamarckii* A. DC. in DC., Prodr. 7:
      397 (1839); non *Siphocampylus laciniatus* G. Don, Gen. Hist. 3: 704 (1834).
      Siphocampylus laciniatus (Lam.) Urb., Symb. Antill. 1: 451 (1899); non G. Don, Gen.
      Hist. 3: 704 (1834). *Siphocampylus sonchifolius* var. *laciniatus* (Lam.) McVaugh, N.
      Amer. Fl. 32A: 105. 1943.
    * *Lobelia sonchifolia* Sw., Fl. Ind. Occid.: 1947 (1806). *Rapuntium sonchifolium* (Sw.) C.
      Presl, Prodr. Monogr. Lobel.: 27 (1836). *Tupa sonchifolia* (Sw.) Griseb., Fl. Brit. W.I.:
      388 (1861). *Dortmanna sonchifolia* (Sw.) Kuntze, Revis. Gen. Pl. 2: 973 (1891).
    Siphocampylus tuerckheimii Urb., Symb. Antill. 7: 416 (1912). *Siphocampylus sonchifolius*
      var. *tuerckheimii* (Urb.) McVaugh, N. Amer. Fl. 32A: 106 (1943). *Siphocampylus lamarckii*
      var. *tuerckheimii* (Urb.) E. Wimm., Ann. Naturhist. Mus. Wien 56: 322 (1948).
    Siphocampylus pinnatisectus Gleason, Bull. Torrey Bot. Club 50: 56 (1923).
    Siphocampylus linearifolius Leonard, J. Wash. Acad. Sci. 14: 417 (1924).
    Siphocampylus lamarckii var. *sinuatus* E. Wimm., Ann. Naturhist. Mus. Wien 56:
      321 (1948).

Siphocampylus soraticus E. Wimm., Repert. Spec. Nov. Regni Veg. 38: 23 (1935).
  Peru & Bolivia. 83 BOL PER. Cl. nanophan.
    Siphocampylus elegans var. *boliviensis* Zahlbr., Bull. Torrey Bot. Club 24: 381 (1897).
    Siphocampylus soraticus var. *angustatus* E. Wimm., Repert. Spec. Nov. Regni Veg. 38: 23
      (1935).

Siphocampylus sparsipilus E. Wimm., Repert. Spec. Nov. Regni Veg. 19: 388 (1924).
  Bolivia. 83 BOL. Cl. nanophan.

Siphocampylus splendens (E. Wimm.) Jeppesen ex B. A. Stein, Ann. Missouri Bot. Gard.
  74: 493 (1987).
  Colombia. 83 CLM. Nanophan.
  * *Burmeistera splendens* E. Wimm., Pflanzenr. IV.276c: 836 (1968).

Siphocampylus spruceanus Zahlbr., Ann. K. K. Naturhist. Hofmus. 6: 443 (1891).
  Peru. 83 PER. Nanophan.

Siphocampylus stenolobus E. Wimm., Repert. Spec. Nov. Regni Veg. 38: 24 (1935).
  Peru. 83 PER. Cl. nanophan.

Siphocampylus subcordatus Rusby, Bull. New York Bot. Gard. 8: 121 (1912).
  Bolivia. 83 BOL. Nanophan.

Siphocampylus subglaber Urb., Symb. Antill. 7: 418 (1912).
Cuba. 81 CUB. Nanophan.

Siphocampylus sulfureus E. Wimm., Repert. Spec. Nov. Regni Veg. 22: 213 (1926).
SE. Brazil (Minas Gerais, Rio de Janeiro, São Paulo, Paraná) & NE. Argentina (Misiones).
84 BZL BZS 85 AGE. Cham.
Siphocampylus verticillatus var. *glaber* Zahlbr., Vidensk. Meddel. Dansk Naturhist. Foren.
Kjøbenhavn 1895: 68 (1895). *Siphocampylus sulfureus* var. *glaber* (Zahlbr.) E. Wimm.,
Repert. Spec. Nov. Regni Veg. 22: 214 (1926).

Siphocampylus tenuisepalus E. Wimm., Pflanzenr. IV.276c: 845 (1968).
Bolivia. 83 BOL. Nanophan. or cham.

Siphocampylus tillettii Steyerm., Brittonia 30: 50 (1978).
Venezuela. 82 VEN. Nanophan. or cham.

Siphocampylus tolimanus E. Wimm., Trab. Mus. Nac. Ci. Nat., Ser. Bot. 26: 28 (1933).
Colombia. 83 CLM. Nanophan.

Siphocampylus tortuosus Zahlbr., Bot. Jahrb. Syst. 37: 459 (1906).
Peru. 83 PER. Cl. nanophan.

Siphocampylus trianae E. Wimm., Repert. Spec. Nov. Regni Veg. 26: 17 (1929).
Colombia. 83 CLM. Nanophan. or cham.

Siphocampylus tuberculatus E. Wimm., Ann. Naturhist. Mus. Wien 56: 324 (1948).
Colombia. 83 CLM. Nanophan.
Siphocampylus tuberculatus var. *carmesinus* E. Wimm., Ann. Naturhist. Mus. Wien 56:
325 (1948).

Siphocampylus tunarensis Zahlbr., Bull. Torrey Bot. Club 24: 376 (1897).
Bolivia. 83 BOL. Phan. or nanophan.

Siphocampylus tunicatus Zahlbr. ex Kuntze, Revis. Gen. Pl. 3(2): 189 (1898).
Bolivia. 83 BOL. Phan. or nanophan.

Siphocampylus tupiformis Zahlbr., Ann. K. K. Naturhist. Hofmus. 6: 440 (1891), as
'tupaeformis'.
Peru & Bolivia. 83 BOL PER. Nanophan. or cham.
Siphocampylus flavoruber Gleason, Torreya 25: 94 (1925).
Siphocampylus tupiformis var. *reduncus* E. Wimm. ex Gleason, Torreya 25: 95 (1925).
Siphocampylus tupiformis var. *dulcis* E. Wimm., Repert. Spec. Nov. Regni Veg. 29: 84 (1931).
Siphocampylus tupiformis var. *stenophyllus* E. Wimm., Repert. Spec. Nov. Regni Veg. 29:
84 (1931).
Siphocampylus tupiformis f. *glabricorollinus* E. Wimm., Pflanzenr. IV.276b: 301 (1953).

Siphocampylus umbellatus (Kunth) G. Don, Gen. Hist. 3: 702 (1834).
Bolivia & SE. Brazil (Minas Gerais, Rio de Janeiro). 83 BOL 84 BZL. Phan. or nanophan.
* *Lobelia umbellata* Kunth in Humb., Bonpl. & Kunth, Nov. Gen. Sp. 3: 304 (quarto), 237
(folio) (1819).
Lobelia kunthiana C. Presl, Prodr. Monogr. Lobel.: 39 (1836). *Siphocampylus umbellatus*
var. *chamissonianus* A. DC. in DC., Prodr. 7: 407 (1839). *Siphocampylus umbellatus* var.
kunthianus (C. Presl) Steud., Nomencl. Bot. (ed. 2) 2: 592 (1841). *Centropogon*
chamissonianus (A. DC.) Kanitz in Mart., Fl. Bras. 6(4): 133 (1878).
Siphocampylus umbellatus var. *wettsteinii* E. Wimm., Ann. Naturhist. Mus. Wien 56:
331 (1948).

Siphocampylus uncipes McVaugh, J. Wash. Acad. Sci. 39: 158 (1949).
Ecuador. 83 ECU. Nanophan.

Siphocampylus vatkeanus Zahlbr., Bull. Torrey Bot. Club 24: 377 (1897).
Peru & Bolivia. 83 BOL PER. Nanophan.

Siphocampylus venosus Gleason, Bull. Torrey Bot. Club 52: 71 (1925).
Colombia. 83 CLM. Cl. nanophan.
Siphocampylus venosus var. *neivanus* E. Wimm., Repert. Spec. Nov. Regni Veg. 38: 21 (1935).

Siphocampylus venustus E. Wimm., Repert. Spec. Nov. Regni Veg. 19: 263 (1924).
Peru. 83 PER. Nanophan. or cham.

Siphocampylus versicolor E. Wimm., Repert. Spec. Nov. Regni Veg. 19: 263 (1924).
Peru. 83 PER. Nanophan.

Siphocampylus verticillatus (Cham.) G. Don, Gen. Hist. 3: 703 (1834).
Brazil (Minas Gerais, São Paulo, Paraná, Rio Grande do Sul) to NE. Argentina (Misiones).
84 BZL BZS 85 AGE PAR URU. Hel. *2n* = 28.
* *Lobelia verticillata* Cham., Linnaea 8: 202 (1833).
Siphocampylus verticillatus var. *grandiflorens* E. Wimm., Repert. Spec. Nov. Regni Veg. 22: 214 (1926).

Siphocampylus veteranus E. Wimm., Repert. Spec. Nov. Regni Veg. 19: 264 (1924).
Peru. 83 PER. Nanophan.
Siphocampylus veteranus f. *laevigatus* E. Wimm., Ann. Naturhist. Mus. Wien 56: 329 (1948).

Siphocampylus violaceus E. Wimm., Repert. Spec. Nov. Regni Veg. 19: 264 (1924).
Venezuela. 82 VEN. Cl. nanophan.

Siphocampylus viscidus E. Wimm., Repert. Spec. Nov. Regni Veg. 22: 215 (1926).
SE. Brazil (Rio de Janeiro). 84 BZL. Cham. or hemicr.

Siphocampylus warmingii Kanitz in Mart., Fl. Bras. 6(4): 148 (1878).
SE. Brazil (Minas Gerais, São Paulo). 84 BZL. Cham. or hemicr.

Siphocampylus weberbaueri Zahlbr., Bot. Jahrb. Syst. 37: 456 (1906).
Peru. 83 PER. Nanophan.

Siphocampylus werdermannii E. Wimm., Repert. Spec. Nov. Regni Veg. 38: 76 (1935).
Bolivia. 83 BOL. Nanophan.

Siphocampylus westinianus (Thunb.) Pohl, Pl. Bras. Icon. Descr. 2: 115 (1831).
SE. Brazil (Minas Gerais, Rio de Janeiro, São Paulo). 84 BZL. Nanophan. or cham.
* *Lobelia westiniana* Thunb., Pl. Bras.: 5 (1817).
Siphocampylus westinianus var. *brevipedicellatus* A. DC. in DC., Prodr. 7: 405 (1839), as 'brevipedicellata'.
Siphocampylus westinianus var. *chamissonianus* E. Wimm., Repert. Spec. Nov. Regni Veg. 22: 216 (1926).
Siphocampylus westinianus var. *sellowianus* E. Wimm., Repert. Spec. Nov. Regni Veg. 22: 216 (1926).
Siphocampylus westinianus var. *sparsifoliatus* E. Wimm., Repert. Spec. Nov. Regni Veg. 22: 216 (1926).
Siphocampylus westinianus f. *longiflorus* E. Wimm., Repert. Spec. Nov. Regni Veg. 22: 216 (1926).
Siphocampylus westinianus var. *kanitzii* E. Wimm., Revista Sudamer. Bot. 2: 93 (1935).

Siphocampylus williamsii Rusby, Bull. New York Bot. Gard. 8: 122 (1912).
Peru & Bolivia. B3 BOL PER. Nanophan. or cham.

Siphocampylus yerbalensis E. Wimm., Repert. Spec. Nov. Regni Veg. 22: 216 (1926).
NE. Argentina (Misiones). 85 AGE. Nanophan.

Siphocampylus yumuriensis Vict., Contr. Inst. Bot. Univ. Montréal 49: 4 (1944).
Cuba. 81 CUB. Cham. or hemicr.

Synonyms:
Siphocampylus sect. *Byrsanthes* (C. Presl) Schönl. === **Siphocampylus** Pohl
Siphocampylus grex *Cremochilus* (Turcz.) E. Wimm. === **Siphocampylus** Pohl
Siphocampylus sect. *Cremochilus* (Turcz.) Schönl. === **Siphocampylus** Pohl
Siphocampylus actinothrix f. *minor* E. Wimm. === **Siphocampylus actinothrix** E. Wimm.
Siphocampylus acuminatus E. Wimm. === **Siphocampylus lucifer** E. Wimm.
Siphocampylus acuminatus (Sw.) G. Don === **Lobelia acuminata** Sw.
Siphocampylus aggregatus Rusby === **Centropogon granulosus** C. Presl
Siphocampylus altiscandens Gleason === **Siphocampylus flagelliformis** Zahlbr.
Siphocampylus andinus var. *elegantissimus* E. Wimm. === **Siphocampylus andinus** Britton
Siphocampylus andinus var. *solemnis* E. Wimm. === **Siphocampylus andinus** Britton
Siphocampylus andropogon (Cav.) G. Don === **Centropogon cornutus** (L.) Druce
Siphocampylus argentinus var. *cuspidatus* (E. Wimm.) E. Wimm. === **Siphocampylus
    argentinus** (Griseb.) Hieron. ex E. Wimm.
Siphocampylus argentinus var. *hirticorollinus* E. Wimm. === **Siphocampylus argentinus**
    (Griseb.) Hieron. ex E. Wimm.
Siphocampylus asper Benth. === **Siphocampylus retrorsus** (Willd. ex Schult.) Vatke
Siphocampylus aureus var. *latior* Zahlbr. === **Siphocampylus aureus** Rusby
Siphocampylus barbatus (Cav.) G. Don === **Centropogon ferrugineus** (L. f.) Gleason
Siphocampylus berterianus (Spreng.) G. Don === **Centropogon berterianus** (Spreng.) A. DC.
Siphocampylus beslerioides (Kunth) G. Don === **Centropogon beslerioides** (Kunth) A. DC.
Siphocampylus betulifolius var. *cordifolius* (Otto & A. Dietr.) E. Wimm. === **Siphocampylus
    betulifolius** (Cham.) G. Don
Siphocampylus betulifolius var. *uleanus* E. Wimm. === **Siphocampylus betulifolius** (Cham.)
    G. Don
Siphocampylus bicolor D. Don === **Lobelia laxiflora** Kunth subsp. **laxiflora**
Siphocampylus bilabiatus var. *dives* (E. Wimm.) E. Wimm. === **Siphocampylus bilabiatus**
    Zahlbr.
Siphocampylus bilabiatus var. *glabratus* Lauterb. ex Buchtien === **Siphocampylus bilabiatus**
    Zahlbr.
Siphocampylus biserratus var. *latifolius* A. DC. === **Siphocampylus biserratus** (Cav.) A. DC.
Siphocampylus biserratus var. *petiolaris* E. Wimm. === **Siphocampylus biserratus** (Cav.) A. DC.
Siphocampylus biserratus var. *spicatus* (Hook.) E. Wimm. === **Siphocampylus biserratus**
    (Cav.) A. DC.
Siphocampylus bracteatus Decne. ex Linden === ?
Siphocampylus brevidens E. Wimm. === **Siphocampylus parvilobus** E. Wimm.
Siphocampylus campanulatus A. DC. === **Siphocampylus affinis** (Mirb.) McVaugh
Siphocampylus candollei var. *breviflorus* E. Wimm. === **Siphocampylus candollei** E. Wimm.
Siphocampylus candollei var. *illustris* E. Wimm. === **Siphocampylus candollei** E. Wimm.
Siphocampylus canescens (C. Presl) A. DC. === **Lobelia laxiflora** Kunth subsp. **laxiflora**
Siphocampylus canus Pohl === **Siphocampylus macropodus** (Thunb.) G. Don
Siphocampylus caoutchouc (Kunth) G. Don === **Centropogon caoutchouc** (Kunth) E. Wimm.
Siphocampylus cardiophyllus Pohl === **Siphocampylus corymbiferus** Pohl
Siphocampylus cavanillesianus G. Don === **Siphocampylus biserratus** (Cav.) A. DC.
Siphocampylus cernuus subsp. *nipensis* (Urb.) Borhidi === **Siphocampylus cernuus** Griseb.
Siphocampylus cernuus var. *nipensis* Urb. === **Siphocampylus cernuus** Griseb.
Siphocampylus cinerascens E. Wimm. === **Siphocampylus macropodus** (Thunb.) G. Don

Siphocampylus cirsiifolius (Lam.) G. Don === **Lobelia cirsiifolia** Lam.
Siphocampylus clarus E. Wimm. === **Siphocampylus dentatus** Gleason
Siphocampylus clotho var. *calvescens* E. Wimm. === **Siphocampylus clotho** E. Wimm.
Siphocampylus coccineus Hook. === **Centropogon coccineus** (Hook.) Regel ex B. D. Jacks.
Siphocampylus coccineus var. *leucostomus* Van Houtte === **Centropogon coccineus** (Hook.)
    Regel ex B. D. Jacks.
Siphocampylus coleoides Vatke === **Burmeistera coleoides** (Vatke) E. Wimm.
Siphocampylus comosus var. *atrichus* E. Wimm. === **Siphocampylus comosus** G. Don
Siphocampylus convolvulaceus var. *weirianus* (E. Wimm.) E. Wimm. === **Siphocampylus
    convolvulaceus** (Cham.) G. Don
Siphocampylus cordatus var. *obtusus* (E. Wimm.) E. Wimm. === **Siphocampylus cordatus**
    (Willd. ex Schult.) E. Wimm.
Siphocampylus cordifolius Benth. === **Siphocampylus benthamianus** Walp.
Siphocampylus cordifolius Otto & A. Dietr. === **Siphocampylus betulifolius** (Cham.) G. Don
Siphocampylus corymbiferus var. *gracilis* (Britton) Zahlbr. === **Siphocampylus corymbiferus**
    Pohl
Siphocampylus corynoides f. *fortunatus* E. Wimm. === **Siphocampylus corynoides** E. Wimm.
Siphocampylus costaricae Vatke === **Centropogon costaricae** (Vatke) McVaugh
Siphocampylus crenatifolius Pohl === **Siphocampylus macropodus** (Thunb.) G. Don
Siphocampylus cubensis A. Rich. === **Lobelia assurgens** L.
Siphocampylus cuspidatus E. Wimm. === **Siphocampylus argentinus** (Griseb.) Hieron. ex
    E. Wimm.
Siphocampylus cylindricus Gleason === **Centropogon ayavacensis** subsp. **cylindricus**
    (Gleason) Lammers
Siphocampylus dentatus var. *clarus* (E. Wimm.) E. Wimm. === **Siphocampylus dentatus**
    Gleason
Siphocampylus dependens var. *undulatus* E. Wimm. === **Siphocampylus onagrius** E. Wimm.
Siphocampylus discolor Donn. Sm. === **Centropogon smithii** E. Wimm.
Siphocampylus divaricatus Benth. === **Siphocampylus scandens** (Kunth) G. Don
Siphocampylus dombeyanus (C. Presl) A. DC. === **Centropogon dombeyanus** (C. Presl)
    E. Wimm.
Siphocampylus donianus E. Wimm. === **Siphocampylus pavonis** E. Wimm.
Siphocampylus duploserratus var. *infundibularis* E. Wimm. === **Siphocampylus duploserratus**
    Pohl
Siphocampylus ekmanii Urb. === **Siphocampylus manettiiflorus** Hook.
Siphocampylus elegans var. *boliviensis* Zahlbr. === **Siphocampylus soraticus** E. Wimm.
Siphocampylus elegans var. *cordatus* Zahlbr. === **Siphocampylus reflexus** Rusby
Siphocampylus elegans var. *obtusatus* E. Wimm. === **Siphocampylus elegans** Planch.
Siphocampylus ellipticus var. *funckeanus* (Planch.) E. Wimm. === **Siphocampylus ellipticus**
    (Willd. ex Schult.) Vatke
Siphocampylus ellipticus var. *peritornus* (E. Wimm.) E. Wimm. === **Siphocampylus ellipticus**
    (Willd. ex Schult.) Vatke
Siphocampylus erianthus Benth. === **Centropogon erianthus** (Benth.) Benth. & Hook. f.
    ex Drake
Siphocampylus eupeplus E. Wimm. === **Siphocampylus denticulosus** Planch.
Siphocampylus eximius Planch. === **Siphocampylus retrorsus** (Willd. ex Schult.) Vatke
Siphocampylus exuberans McVaugh === **Siphocampylus ecuadorensis** E. Wimm.
Siphocampylus ferrugineus (L. f.) G. Don === **Centropogon ferrugineus** (L. f.) Gleason
Siphocampylus fiebrigii var. *intermedius* E. Wimm. === **Siphocampylus fiebrigii** E. Wimm.
Siphocampylus flagelliformis var. *glaber* E. Wimm. === **Siphocampylus flagelliformis** Zahlbr.
Siphocampylus flavoruber Gleason === **Siphocampylus tupiformis** Zahlbr.
Siphocampylus floccosus Planch. ex Linden === ?
Siphocampylus fluminensis var. *oppositifolius* E. Wimm. === **Siphocampylus fluminensis**
    (Vell.) E. Wimm.
Siphocampylus foetidus (Kunth) G. Don === **Centropogon foetidus** (Kunth) Zahlbr.
Siphocampylus foliosus var. *colombinus* E. Wimm. === **Siphocampylus foliosus** Griseb.

Siphocampylus foliosus var. *diversifolius* (Hicken) Pontiroli === **Siphocampylus foliosus** Griseb.
Siphocampylus foliosus var. *glabratus* E. Wimm. === **Siphocampylus foliosus** Griseb.
Siphocampylus foliosus var. *lessenii* E. Wimm. === **Siphocampylus foliosus** Griseb.
Siphocampylus foliosus var. *minor* Zahlbr. ex Kuntze === **Siphocampylus foliosus** Griseb.
Siphocampylus foliosus var. *subcanus* Zahlbr. ex E. Wimm. === **Siphocampylus foliosus** Griseb.
Siphocampylus funckeanus Planch. === **Siphocampylus ellipticus** (Willd. ex Schult.) Vatke
Siphocampylus funckeanus var. *rugosus* Planch. === **Siphocampylus ellipticus** (Willd. ex Schult.) Vatke
Siphocampylus fuscus G. Don === **Centropogon fuscus** (G. Don) E. Wimm.
Siphocampylus giganteus f. *brachyodon* E. Wimm. === **Siphocampylus giganteus** (Cav.) G. Don
Siphocampylus giganteus var. *angustifolius* Vatke === **Siphocampylus giganteus** (Cav.) G. Don
Siphocampylus giganteus var. *latifolius* Britton === **Centropogon brittonianus** Zahlbr.
Siphocampylus giganteus var. *latifolius* Vatke === **Siphocampylus giganteus** (Cav.) G. Don
Siphocampylus glabratus (Kunth) G. Don === **Burmeistera glabrata** (Kunth) Benth. & Hook. f. ex B. D. Jacks.
Siphocampylus glandulosus Hook. === **Centropogon glandulosus** (Hook.) Decne.
Siphocampylus glandulosus var. *ovatus* E. Wimm. === **Centropogon glandulosus** (Hook.) Decne.
Siphocampylus gloriosus Britton === **Centropogon gloriosus** (Britton) Zahlbr.
Siphocampylus gracilis Britton === **Siphocampylus corymbiferus** Pohl
Siphocampylus gracilis var. *glabris* Britton === **Siphocampylus corymbiferus** Pohl
Siphocampylus grandidentatus (Schltdl.) D. Dietr. === **Centropogon grandidentatus** (Schltdl.) Zahlbr.
Siphocampylus grandis (L. f.) G. Don === **Centropogon grandis** (L. f.) C. Presl
Siphocampylus griseus (Gleason) E. Wimm. === **Siphocampylus humboldtianus** A. DC.
Siphocampylus guatemalensis Vatke === **Centropogon cordifolius** Benth.
Siphocampylus gutierrezii Planch. & Oerst. === **Centropogon gutierrezii** (Planch. & Oerst.) E. Wimm.
Siphocampylus hartwegii Benth. === **Centropogon hartwegii** (Benth.) Benth. & Hook. f. ex B. D. Jacks.
Siphocampylus hazenii Gleason === **Centropogon hazenii** (Gleason) E. Wimm.
Siphocampylus hirtus (Cav.) G. Don === **Centropogon hirtus** (Cav.) C. Presl
Siphocampylus humboldtianus var. ovatus E. Wimm. === **Siphocampylus humboldtianus** A. DC.
Siphocampylus hallianus (C. Presl) Vatke === **Siphocampylus affinis** (Mirb.) McVaugh
Siphocampylus hamatus Wendl. === **Siphocampylus reticulatus** (Willd. ex Schult.) Klotzsch & H. Karst. ex Vatke
Siphocampylus helmutii var. *pilosifolius* E. Wimm. === **Siphocampylus helmutii** E. Wimm.
Siphocampylus hitchcockii (Gleason) E. Wimm. === **Siphocampylus sanguineus** Zahlbr.
Siphocampylus igneus (Vell.) E. Wimm. === **Siphocampylus corymbiferus** Pohl
Siphocampylus igneus var. *gracilis* (Britton) E. Wimm. === **Siphocampylus corymbiferus** Pohl
Siphocampylus imbricatus var. *casarettoi* E. Wimm. === **Siphocampylus imbricatus** (Cham.) G. Don
Siphocampylus imbricatus var. *glabratus* E. Wimm. === **Siphocampylus imbricatus** (Cham.) G. Don
Siphocampylus impressus Urb. === **Siphocampylus manettiiflorus** Hook.
Siphocampylus incanus Britton === **Centropogon incanus** (Britton) Zahlbr.
Siphocampylus jamesonianus (C. Presl) A. DC. === **Siphocampylus affinis** (Mirb.) McVaugh
Siphocampylus jelskii f. *eugenius* E. Wimm. === **Siphocampylus jelskii** Zahlbr.
Siphocampylus kraussii (Graham) G. Don === **Lobelia kraussii** Graham
Siphocampylus laciniatus (Lam.) Urb. === **Siphocampylus sonchifolius** (Sw.) McVaugh
Siphocampylus laetus E. Wimm. === **Siphocampylus keissleri** E. Wimm.
Siphocampylus lamarckii A. DC. === **Siphocampylus sonchifolius** (Sw.) McVaugh
Siphocampylus lamarckii var. *sinuatus* E. Wimm. === **Siphocampylus sonchifolius** (Sw.) McVaugh
Siphocampylus lamarckii var. *tuerckheimii* (Urb.) E. Wimm. === **Siphocampylus sonchifolius** (Sw.) McVaugh

Siphocampylus lanatus Benth. === **Siphocampylus niveus** (Willd. ex Schult.) Vatke

Siphocampylus lantanifolius A. DC. === **Siphocampylus reticulatus** (Willd. ex Schult.) Klotzsch & H. Karst. ex Vatke

Siphocampylus lantanifolius var. *glabriusculus* Hook. === **Siphocampylus reticulatus** (Willd. ex Schult.) Klotzsch & H. Karst. ex Vatke

Siphocampylus libanensis Urb. === **Siphocampylus manettiiflorus** Hook.

Siphocampylus lilacinus E. Wimm. === **Siphocampylus rectiflorus** Rusby

Siphocampylus lindleyi var. *dissitiflorus* E. Wimm. === **Siphocampylus lindleyi** Lem.

Siphocampylus linearifolius Leonard === **Siphocampylus sonchifolius** (Sw.) McVaugh

Siphocampylus lobbii var. *megodontus* E. Wimm. === **Siphocampylus lobbii** Zahlbr.

Siphocampylus longipedunculatus var. *meiopodus* E. Wimm. === **Siphocampylus longipedunculatus** Pohl

Siphocampylus longipedunculatus var. *trichophyllus* E. Wimm. === **Siphocampylus longipedunculatus** Pohl

Siphocampylus longipes (Regel) Vatke === **Centropogon solanifolius** Benth.

Siphocampylus macranthus Pohl === **Centropogon cornutus** (L.) Druce

Siphocampylus macrophyllus G. Don === **Centropogon macrophyllus** (G. Don) E. Wimm.

Siphocampylus manettiiflorus var. *libanensis* (Urb.) E. Wimm. === **Siphocampylus manettiiflorus** Hook.

Siphocampylus matthewsii E. Wimm. === **Siphocampylus matthiaei** A. DC.

Siphocampylus megalandrus E. Wimm. === **Siphocampylus cutervensis** Zahlbr.

Siphocampylus megalanthus var. *orthosepalus* E. Wimm. === **Siphocampylus megalanthus** Zahlbr.

Siphocampylus megalanthus var. *costaricanus* E. Wimm. === **Siphocampylus megalanthus** Zahlbr.

Siphocampylus meridensis (Turcz.) Zahlbr. === **Siphocampylus sceptrum** Decne.

Siphocampylus microstoma var. *sparsus* E. Wimm. === **Siphocampylus microstoma** Hook.

Siphocampylus mollis Planch. === **Siphocampylus polyphyllus** Planch.

Siphocampylus mollis Regel === **Lobelia laxiflora** Kunth subsp. **laxiflora**

Siphocampylus mutisiana (Kunth) G. Don === **Burmeistera mutisiana** (Kunth) E. Wimm.

Siphocampylus nemoralis f. *hieronymi* E. Wimm. === **Siphocampylus nemoralis** Griseb.

Siphocampylus nemoralis var. *grisebachii* E. Wimm. === **Siphocampylus nemoralis** Griseb.

Siphocampylus nemoralis var. *tarijanus* E. Wimm. === **Siphocampylus nemoralis** Griseb.

Siphocampylus nitidus de Jonghe ex C. Morren === **Siphocampylus manettiiflorus** Hook.

Siphocampylus nitidus f. *acuminatus* E. Wimm. === **Siphocampylus nitidus** Pohl

Siphocampylus nitidus var. *nitidissimus* E. Wimm. === **Siphocampylus nitidus** Pohl

Siphocampylus nitidus var. *pleotrichus* E. Wimm. === **Siphocampylus nitidus** Pohl

Siphocampylus oaxacanus E. Wimm. === **Centropogon grandidentatus** (Schltdl.) Zahlbr.

Siphocampylus oblongifolius var. *crenatus* E. Wimm. === **Siphocampylus crenatus** (E. Wimm.) E. Wimm.

Siphocampylus obovoideus Gleason === **Siphocampylus pyriformis** Zahlbr.

Siphocampylus onagrius var. *undulatus* (E. Wimm.) E. Wimm. === **Siphocampylus onagrius** E. Wimm.

Siphocampylus onagrius f. *longisepalus* E. Wimm. === **Siphocampylus onagrius** E. Wimm.

Siphocampylus orbignianus var. *discurrens* E. Wimm. === **Siphocampylus orbignianus** A. DC.

Siphocampylus penduliflorus var. *asperatulus* E. Wimm. === **Siphocampylus penduliflorus** Decne. ex Planch.

Siphocampylus penduliflorus var. *parvicalicinus* E. Wimm. === **Siphocampylus scandens** (Kunth) G. Don

Siphocampylus pendulifolius (C. Presl) A. DC. === **Siphocampylus ovatus** (G. Don) E. Wimm.

Siphocampylus pennellii Gleason === **Siphocampylus niveus** (Willd. ex Schult.) Vatke

Siphocampylus peritornus E. Wimm. === **Siphocampylus ellipticus** (Willd. ex Schult.) Vatke

Siphocampylus persicifolius (Lam.) G. Don === **Lobelia persicifolia** Lam.

Siphocampylus phyllobotrys E. Wimm. === **Siphocampylus denticulosus** Planch.

Siphocampylus phyllobotrys f. *leucanthus* E. Wimm. === **Siphocampylus denticulosus** Planch.

Siphocampylus pinnatisectus Gleason === **Siphocampylus sonchifolius** (Sw.) McVaugh

Siphocampylus planchonis E. Wimm. === **Siphocampylus polyphyllus** Planch.
Siphocampylus planchonis f. *flaviflorus* E. Wimm. === **Siphocampylus polyphyllus** Planch.
Siphocampylus planchonis f. *glabrifolia* E. Wimm. === **Siphocampylus polyphyllus** Planch.
Siphocampylus planchonis var. *daweanus* E. Wimm. === **Siphocampylus polyphyllus** Planch.
Siphocampylus planchonis var. *polyphyllus* (Planch.) E. Wimm. === **Siphocampylus polyphyllus** Planch.
Siphocampylus planchonis var. *truxillensis* E. Wimm. === **Siphocampylus polyphyllus** Planch.
Siphocampylus polycladus var. *rubicundus* E. Wimm. === **Siphocampylus polycladus** E. Wimm.
Siphocampylus prunifolius (Humb. ex C. Presl) A. DC. === **Lobelia laxiflora** Kunth subsp. **laxiflora**
Siphocampylus psilophyllus var. *longisepalus* E. Wimm. === **Siphocampylus psilophyllus** Pohl
Siphocampylus pubescens Benth. === **Siphocampylus humboldtianus** A. DC.
Siphocampylus pubiflorus E. Wimm. === **Siphocampylus lindleyi** Lem.
Siphocampylus pulchellus Planch. & Linden === **?**
Siphocampylus pyriformis var. *reflexisepalus* E. Wimm. === **Siphocampylus pyriformis** Zahlbr.
Siphocampylus radiatus var. *brevidentatus* (Zahlbr. & Rech.) E. Wimm. === **Siphocampylus radiatus** Rusby
Siphocampylus radiatus var. *minor* Zahlbr. === **Siphocampylus radiatus** Rusby
Siphocampylus radicans Kuntze === **Centropogon coccineus** (Hook.) Regel ex B. D. Jacks.
Siphocampylus raimondii E. Wimm. === **Siphocampylus corynellus** Gleason
Siphocampylus reflexifolius Zahlbr. === **Siphocampylus retrorsus** (Willd. ex Schult.) Vatke
Siphocampylus regelii Vatke === **Centropogon ferrugineus** (L. f.) Gleason
Siphocampylus regelii var. *umbrosus* Vatke === **Centropogon ferrugineus** (L. f.) Gleason
Siphocampylus reticulatus var. *glabriusculus* (Hook.) E. Wimm. === **Siphocampylus reticulatus** (Willd. ex Schult.) Klotzsch & H. Karst. ex Vatke
Siphocampylus reticulatus var. *hamatus* (Wendl.) E. Wimm. === **Siphocampylus reticulatus** (Willd. ex Schult.) Klotzsch & H. Karst. ex Vatke
Siphocampylus retrorsus var. *asper* (Benth.) E. Wimm. === **Siphocampylus retrorsus** (Willd. ex Schult.) Vatke
Siphocampylus retrorsus var. *eximius* (Planch.) E. Wimm. === **Siphocampylus retrorsus** (Willd. ex Schult.) Vatke
Siphocampylus retrorsus var. *semiasper* E. Wimm. === **Siphocampylus retrorsus** (Willd. ex Schult.) Vatke
Siphocampylus revolutus Graham === **Siphocampylus reticulatus** (Willd. ex Schult.) Klotzsch & H. Karst. ex Vatke
Siphocampylus roseus Donn. Sm. === **Centropogon coccineus** (Hook.) Regel ex B. D. Jacks.
Siphocampylus rotundifolius E. Wimm. === **Siphocampylus scandens** (Kunth) G. Don
Siphocampylus rugosus (C. Presl) A. DC. === **Siphocampylus dependens** G. Don
Siphocampylus rugosus Poit. & Vilm. === **Centropogon coccineus** (Hook.) Regel ex B. D. Jacks.
Siphocampylus rusbyanus var. *subtervestitus* Zahlbr. === **Siphocampylus rusbyanus** Britton
Siphocampylus sanguineus var. *hitchcockii* (Gleason) E. Wimm. === **Siphocampylus sanguineus** Zahlbr.
Siphocampylus scaber A. DC. === **Siphocampylus longipedunculatus** Pohl
Siphocampylus sceptrum var. *denudatus* E. Wimm. === **Siphocampylus sceptrum** Decne.
Siphocampylus sceptrum var. *latifolius* E. Wimm. === **Siphocampylus sceptrum** Decne.
Siphocampylus sceptrum var. *tachirensis* E. Wimm. === **Siphocampylus sceptrum** Decne.
Siphocampylus sessilifolius Vatke === **Siphocampylus sceptrum** Decne.
Siphocampylus sonchifolius var. *laciniatus* (Lam.) McVaugh === **Siphocampylus sonchifolius** (Sw.) McVaugh
Siphocampylus sonchifolius var. *tuerckheimii* (Urb.) McVaugh === **Siphocampylus sonchifolius** (Sw.) McVaugh
Siphocampylus soraticus var. *angustatus* E. Wimm. === **Siphocampylus soraticus** E. Wimm.
Siphocampylus spectabilis (Kunth) G. Don === **Centropogon cornutus** (L.) Druce
Siphocampylus stellatus Gleason === **Centropogon ayavacensis** (Willd. ex Schult.) Lammers subsp. **ayavacensis**
Siphocampylus subcarnosus Benth. === **Siphocampylus scandens** (Kunth) G. Don

Siphocampylus subcordatus var. *dives* E. Wimm. === **Siphocampylus bilabiatus** Zahlbr.
Siphocampylus subglaber var. *glaber* (McVaugh) E. Wimm. === **Siphocampylus glaber** McVaugh
Siphocampylus sulfureus var. *glaber* (Zahlbr.) E. Wimm. === **Siphocampylus sulfureus** E. Wimm.
Siphocampylus superbus Zahlbr. === **Siphocampylus jelskii** Zahlbr.
Siphocampylus surinamensis (L.) G. Don === **Centropogon cornutus** (L.) Druce
Siphocampylus tepuiensis Gleason === **Siphocampylus reticulatus** (Willd. ex Schult.)
    Klotzsch & H. Karst. ex Vatke
Siphocampylus thomesianus Moric. === **Siphocampylus imbricatus** (Cham.) G. Don
Siphocampylus thysanopetalus Vatke=== **Centropogon gutierrezii** (Planch. & Oerst.) E. Wimm.
Siphocampylus trichodontus E. Wimm. === **Siphocampylus popayanensis** Benth.
Siphocampylus tuberculatus var. *carmesinus* E. Wimm. === **Siphocampylus tuberculatus**
    E. Wimm.
Siphocampylus tuerckheimii Urb. === **Siphocampylus sonchifolius** (Sw.) McVaugh
Siphocampylus tupiformis f. *glabricorollinus* E. Wimm. === **Siphocampylus tupiformis** Zahlbr.
Siphocampylus tupiformis var. *diversifolius* Hicken === **Siphocampylus foliosus** Griseb.
Siphocampylus tupiformis var. *dulcis* E. Wimm. === **Siphocampylus tupiformis** Zahlbr.
Siphocampylus tupiformis var. *reduncus* E. Wimm. ex Gleason === **Siphocampylus tupiformis**
    Zahlbr.
Siphocampylus tupiformis var. *stenophyllus* E. Wimm. === **Siphocampylus tupiformis** Zahlbr.
Siphocampylus umbellatus var. *ayavacensis* (Willd. ex Schult.) Steud. === **Centropogon
    ayavacensis** (Willd. ex Schult.) Lammers
Siphocampylus umbellatus var. *chamissonianus* A. DC. === **Siphocampylus umbellatus**
    (Kunth) G. Don
Siphocampylus umbellatus var. *kunthianus* (C. Presl) Steud. === **Siphocampylus umbellatus**
    (Kunth) G. Don
Siphocampylus umbellatus var. *wettsteinii* E. Wimm. === **Siphocampylus umbellatus** (Kunth)
    G. Don
Siphocampylus umbellatus var. *willdenowianus* A. DC. === **Centropogon ayavacensis** (Willd.
    ex Schult.) Lammers
Siphocampylus unduavensis Britton === **Centropogon unduavensis** (Britton) Zahlbr.
Siphocampylus undulatus Urb. === **Siphocampylus manettiiflorus** Hook.
Siphocampylus urbanii E. Wimm. === **Siphocampylus igneus** Urb.
Siphocampylus venosus var. *neivanus* E. Wimm. === **Siphocampylus venosus** Gleason
Siphocampylus verbascifolius (C. Presl) A. DC. === **Centropogon verbascifolius** (C. Presl)
    Gleason
Siphocampylus verticillatus var. *glaber* Zahlbr. === **Siphocampylus sulfureus** E. Wimm.
Siphocampylus verticillatus var. *grandiflorens* E. Wimm. === **Siphocampylus verticillatus**
    (Cham.) G. Don
Siphocampylus veteranus f. *laevigatus* E. Wimm. === **Siphocampylus veteranus** E. Wimm.
Siphocampylus villosulus Pohl === **Siphocampylus macropodus** (Thunb.) G. Don
Siphocampylus virgatus A. DC. === **Siphocampylus comosus** G. Don
Siphocampylus volubilis G. Don === **Siphocampylus cordatus** (Willd. ex Schult.) E. Wimm.
Siphocampylus volubilis var. *demissus* E. Wimm. === **Siphocampylus cordatus** (Willd. ex
    Schult.) E. Wimm.
Siphocampylus volubilis var. *obtusus* E. Wimm. === **Siphocampylus cordatus** (Willd. ex
    Schult.) E. Wimm.
Siphocampylus warscewiczii Regel === **Lobelia laxiflora** Kunth subsp. **laxiflora**
Siphocampylus weirianus E. Wimm. === **Siphocampylus convolvulaceus** (Cham.) G. Don
Siphocampylus westinianus f. *longiflorus* E. Wimm. === **Siphocampylus westinianus**
    (Thunb.) Pohl
Siphocampylus westinianus var. *brevipedicellatus* A. DC. === **Siphocampylus westinianus**
    (Thunb.) Pohl
Siphocampylus westinianus var. *chamissonianus* E. Wimm. === **Siphocampylus westinianus**
    (Thunb.) 'Pohl
Siphocampylus westinianus var. *kanitzii* E. Wimm. === **Siphocampylus westinianus**
    (Thunb.) Pohl

Siphocampylus westinianus var. *sellowianus* E. Wimm. === **Siphocampylus westinianus**
   (Thunb.) Pohl
Siphocampylus westinianus var. *sparsifoliatus* E. Wimm. === **Siphocampylus westinianus**
   (Thunb.) Pohl

Siphocodon

Campanuloideae, 2 species, Cape Provinces of South Africa. Although neither species has been included in a molecular analysis (Eddie et al. 2003, under **General**), on the basis of morphology and biogeography, the genus appears to be related to *Rhigiophyllum* (q.v.); one suspects that both would be part of the "wahlenbergioid clade" with *Craterocapsa* and *Roella*. Type [by monotypy]: *Siphocodon spartioides* Turcz.

> Turczaninow, N. (1852). Decas septima. Generum plantarum non descriptorum adjectis descriptionis nonnullarum specierum. Bull. Soc. Imp. Naturalistes Moscou 25(3): 138-181. La. — Establishment of genus.
> Dyer, R. A. (1975). *Siphocodon* Trucz. [sic]. In The genera of southern African flowering plants 1: 641. Pretoria: Department of Agricultural Technical Services. En. — Generic description.
> Goldblatt, P. & J. Manning (2000). Cape plants. A conspectus of the Cape flora of South Africa. *Siphocodon*. Strelitzia 9: 396. En. — Checklist for Cape of Good Hope, with brief species desciptions.
> Welman, W. G. (2000). *Siphocodon* Turcz. In O. A. Leistner (ed.), Seed plants of southern Africa: families and genera. Strelitzia 10: 201. En. — Generic description.

Siphocodon Turcz., Bull. Soc. Imp. Naturalistes Moscou 25(3): 175 (1852).
   S. Africa. 27.

Siphocodon debilis Schltr., Bot. Jahrb. Syst. 24: 447 (1897).
   Cape Provinces. 27 CPP. Nanophan. or cham.

Siphocodon spartioides Turcz., Bull. Soc. Imp. Naturalistes Moscou 25(3): 176 (1852).
   Cape Provinces. 27 CPP. Nanophan. or cham.

Solenopsis

Lobelioideae, 6 species, Mediterranean. These species were known by the generic name *Laurentia* (q.v.) until Meikle (1979) noted that this name was illegitimate under ICBN Art. 52.1, and resurrected *Solenopsis* as a replacement. The genus is not related to *Wimmerella* (q.v.), with which it was included under the name *Laurentia*, but rather is sister to a clade containing all sampled North American species of *Lobelia* (E. Knox, pers. comm.; A. Antonelli, pers. comm.). The treatment here follows the recent monograph by Crespo et al. (1998), as modified by Aldasoro et al. (2001). $2n$ = 22, 28. Type [designated by Pfeiff., Nomencl. Bot. 2: 1192 (1874)]: *Lobelia minuta* L.

> Presl, C. (1836). *Solenopsis*. In Prodromus monographiae Lobeliacearum: 32-33. Prague: Theophilus Haase. La. — Establishment of genus as distinct from *Lobelia*.
> Turrill, W. B. (1949). *Laurentia* in the Balkan Peninsula. Kew Bull. 1949: 22. En. — Brief floristic and taxonomic note.
> • Wimmer, F. E. (1953). [*Laurentia* (Mich.) Adans.] Sectio I. *Solenopsis* Endl. In A. Engler & L. Diels, Das Pflanzenreich 276b: 387-397, illus. Berlin: Akademie-Verlag. Ge., La. — Monograph with key, descriptions, full nomenclatrure, and specimen citations; circumscribed to include *Porterella* and *Wimmerella*.
> • Wimmer, F. E. (1968). *Laurentia* (Mich.) Adams. [sic] sect. I. *Laurentia*. In A. Engler & L. Diels, Das Pflanzenreich IV. 276c: 853-855. Berlin: Akademie-Verlag. Ge., La. — Supplement to Wimmer (1953).

Meikle, R. D. (1979). Some notes on *Laurentia* Adanson (Campanulaceae). Kew Bull. 34: 373-375. En. — Argued that *Laurentia* was proposed as an avowed substitute for *Lobelia* and thus is illegitimate; advocated use of next available name, *Solenopsis*.

Lammers, T. G. (1997). Proposal to conserve the name *Laurentia* (Campanulaceae) with a conserved type. Taxon 46: 795-799. En. — Argued that *Laurentia* not based on *Lobelia* but on *Lobelia laurentia,* and thus *is* correct name for genus.

Serra, L. & M. B. Crespo (1997). An outline revision of the subtribe Siphocampylinae (Lobeliaceae). Lagascalia 19: 881-888, illus., maps. En. — Key to distinguish genus from alleged allies.

• Crespo, M. B., L. Serra & A. Juan (1998). *Solenopsis* (Lobeliaceae): a genus endemic in the Mediterranean region. Pl. Syst. Evol. 210: 211-229, illus., maps. En. — Monograph with key, descriptions, full nomenclature, and specimen citations.

Brummitt, R. K. (2000). Report of the Committee for Spermatophyta: 49. Taxon 49: 261-278. En. — Committee agreed with Meikle (1979) that *Laurentia* illegitimate; proposal to conserve not recommended.

Aldasoro, J. J., S. Castroviejo, F. Sales & I. C. Hedge (2001). ¿Qué es *Lobelia minima* Sims? Anales Jard. Bot. Madrid 59: 173. Sp. — Disagree with Crespo et al. (1998) that *S. minima* applies to the Balearic species; replace it with *S. balearica* and consider *S. minima* doubtful.

Solenopsis C. Presl, Prodr. Monogr. Lobel.: 32 (1836). *Laurentia* [unranked] *Solenopsis* (C. Presl) Endl., Gen. Pl.: 512 (1838). *Laurentia* subg. *Solenopsis* (C. Presl) Peterm., Pflanzenreich: 444 (1845).
Mediterranean. 12 13 20 21 34.

Solenopsis balearica (E. Wimm.) Aldasoro, Castrov., Sales & Hedge, Anales Jard. Bot. Madrid 59: 173 (2001).
Baleares. 12 BAL. Hemicr.
  * *Laurentia minuta* f. *balearica* E. Wimm., Ann. Naturhist. Mus. Wien 56: 334 (1948). *Solenopsis minuta* subsp. *balearica* (E. Wimm.) Meikle, Kew Bull. 34: 374 (1979).

Solenopsis bicolor (Batt.) Greuter & Burdet, Willdenowia 11: 41 (1981).
Algeria & Tunisia. 20 ALG TUN. Ther.
  * *Laurentia michelii* var. *bicolor* Batt., Contr. Fl. Atl.: 59 (1919). *Laurentia bicolor* (Batt.) Maire & T. Stephenson, Bull. Soc. Hist. Nat. Afrique N. 21: 49 (1930).

Solenopsis bivonae (Tineo) M. B. Crespo, Serra & Juan, Pl. Syst. Evol. 210: 219 (1998).
Sardegna, Sicilia, Cyprus. 12 SAR 13 SIC 34 CYP. Hemicr. *2n* = 28.
  * *Lobelia tenella* Biv., Sicul. Pl. 1: 53 (1806); non L., Mant. Pl.: 120 (1767); nec Burm. f., Prodr. Fl. Cap.: 29 (1768). *Lobelia bivonae* Tineo, Cat. Pl. Hort. Panorm.: 279 (1827). *Solenopsis bivonaeana* C. Presl, Prodr. Monogr. Lobel.: 32 (1836). *Laurentia tenella* A. DC. in DC., Prodr. 7: 410 (1839). *Laurentia bivonae* (Tineo) Pign., Giorn. Bot. Ital. 111: 54 (1977). *Laurentia gasparinii* subsp. *tenella* O. Bolòs & Vigo, Collect. Bot. (Barcelona) 14: 102 (1983). *Solenopsis laurentia* subsp. *tenella* (O. Bolòs & Vigo) O. Bolòs, Vigo, Masalles & Ninot, Fl. Man. Països Catalans: 1215 (1990).
  Lobelia setacea Sm. in Sibth. & Sm., Fl. Graec. Prodr. 1: 145 (1806); non Thunb. in Hoffm., Phytogr. Bl.: 21 (1803).
  Laurentia minuta f. *nobilis* E. Wimm., Ann. Naturhist. Mus. Wien 56: 333 (1948). *Solenopsis minuta* subsp. *nobilis* (E. Wimm.) Meikle, Kew Bull. 34: 374 (1979).

Solenopsis corsica (Meikle) M. B. Crespo, Serra & Juan, Pl. Syst. Evol. 210: 221 (1998).
Corse & Sardegna. 12 COR SAR. Hemicr.
  * *Solenopsis minuta* subsp. *corsica* Meikle, Kew Bull. 34: 374 (1979).

Solenopsis laurentia (L.) C. Presl, Prodr. Monogr. Lobel.: 32 (1836).
Canary Is.; Portugal to N. Africa, Lebanon & Turkey. 12 COR POR SAR SPA FRA 13 ALB GRC ITA KRI SIC 20 ALG MOR TUN 21 CNY 34 LBS TUR. Ther. *2n* = 22.

Lobelia laurentia L., Sp. Pl.: 931 (1753). *Lobelia gracilis* Salisb., Prodr. Stirp. Chap. Allerton: 129 (1796). *Laurentia michelii* A. DC. in DC., Prodr. 7: 409 (1839). *Lobelia michelii* (A. DC.) Colmeiro, Enum. Pl. Penins. Hispano-Lusit. 3: 493 (1887).
Lobelia laurentia var. *nana* Hoffmanns. & Link, Fl. Portug. 2: 21 (1813-20).
Lobelia gasparrinii Tineo, Cat. Pl. Hort. Panorm.: 279 (1827), as 'gasparrini'. *Laurentia michelii* var. *gasparrinii* (Tineo) Nyman, Consp. Fl. Eur.: 487 (1879), as 'gasparrini'. *Laurentia gasparrinii* (Tineo) Strobl, Flora 66: 547 (1883), as 'gasparrini'.
Lobelia salzmanniana C. Presl, Symb. Bot. 1: 31 (1831). *Solenopsis salzmanniana* (C. Presl) C. Presl, Prodr. Monogr. Lobel.: 32 (1836).
Solenopsis canariensis (C. Presl) C. Presl, Prodr. Monogr. Lobel.: 32 (1836). *Laurentia canariensis* (C. Presl) A. DC. in DC., Prodr. 7: 410 (1839).
Laurentia michelii var. *integrifolia* Lange, Pug. Pl. Hispan.: 155 (1861). *Lobelia michelii* var. *integrifolia* (Lange) Colmeiro, Enum. Pl. Penins. Hispano-Lusit. 3: 493 (1887).
Laurentia michelii var. *subacaulis* Pomel ex Batt., Fl. Algérie, Dicotyl.: 569 (1889). *Laurentia gasparrinii* var. *subacaulis* (Pomel ex Batt.) E. Wimm., Ann. Naturhist. Mus. Wien 56: 334 (1948).
Laurentia michelii f. *albiflora* Merino, Fl. Galicia 2: 303 (1906). *Laurentia gasparrinii* f. *albiflora* (Merino) E. Wimm., Ann. Naturhist. Mus. Wien 56: 334 (1948).
Laurentia michelii var. *perpusilla* Maire, Cavanillesia 4: 15 (1931).

Solenopsis minuta (L.) C. Presl, Prodr. Monogr. Lobel.: 32 (1836).
Sardegna, Sicilia, Kriti. 12 SAR 13 KRI SIC. Hemicr. or ther.
Lobelia minuta L., Mant. Pl.: 292 (1771). *Lobelia pumila* Salisb., Prodr. Stirp. Chap. Allerton: 129 (1796); non Burm. f., Fl. Indica: 186 (1786). *Laurentia minuta* (L.) A. DC. in DC., Prodr. 7: 410 (1839).

subsp. **annua** Greuter, Matthäs & Risse, Willdenowia 14: 30 (1984).
Kriti. 13 KRI. Ther.

subsp. **minuta**
Sardegna, Sicilia, Kriti. 12 SAR 13 KRI SIC. Hemicr.

Synonyms:
Solenopsis bivonaeana C. Presl === **Solenopsis bivonae** (Tineo) M. B. Crespo, Serra & Juan
Solenopsis canariensis (C. Presl) C. Presl === **Solenopsis laurentia** (L.) C. Presl
Solenopsis dregeana C. Presl === **Wimmerella bifida** (Thunb.) Serra, M. B. Crespo & Lammers
Solenopsis laurentia subsp. *tenella* (O. Bolòs & Vigo) O. Bolòs, Vigo, Masalles & Ninot === **Solenopsis bivonae** (Tineo) M. B. Crespo, Serra & Juan
Solenopsis longiflora (L.) M. R. Almeida === **Hippobroma longiflora** (L.) G. Don
Solenopsis minima (Sims) M. B. Crespo, Serra & Juan === ?
Solenopsis minuta subsp. *balearica* (E. Wimm.) Meikle === **Solenopsis balearica** (E. Wimm.) Aldasoro, Castrov., Sales & Hedge
Solenopsis minuta subsp. *corsica* Meikle === **Solenopsis corsica** (Meikle) M. B. Crespo, Serra & Juan
Solenopsis minuta subsp. *nobilis* (E. Wimm.) Meikle === **Solenopsis bivonae** (Tineo) M. B. Crespo, Serra & Juan
Solenopsis minuta var. *minima* (Sims) C. Presl === ?
Solenopsis salzmanniana (C. Presl) C. Presl === **Solenopsis laurentia** (L.) C. Presl

Specularia

This pre-Linnaean name was resurrected by Candolle (1830) for species then placed in *Prismatocarpus* (q.v.) that were not from southern Africa. The name was applied for many years to the species here called *Legousia* (q.v.), but it does not have priority. Based on: *Legousia* Durande.

de Candolle, A. (1830). *Specularia*. In Monographie des Campanulées: 344-352. Paris: Veuve Desray. La. — Establishment of genus as distinct from *Campanula* and *Prismatocarpus*.

Synonyms:

Specularia Heister ex A. DC. === **Legousia** Durande

Specularia sect. *Campylocera* (Nutt.) A. Gray === **Triodanis** Raf.

Specularia sect. *Hesperocodon* Webb & Berth. === **Triodanis** Raf.

Specularia sect. *Triodanis* (Raf.) Torr. === **Triodanis** Raf.

Specularia [unranked] *Dysmicodon* Endl. === **Triodanis** Raf.

Specularia americana (L.) Morgan ex J. James === **Campanula americana** L.

Specularia arvensis Montandon === **Legousia speculum-veneris** (L.) Durande ex Vill.

Specularia biflora (Ruiz & Pav.) Fisch. & C. A. Mey. === **Triodanis perfoliata** subsp. **biflora** (Ruiz & Pav.) Lammers

Specularia californica (Nutt.) Vatke === **Triodanis perfoliata** subsp. **biflora** (Ruiz & Pav.) Lammers

Specularia calycina Lojac. === **Legousia hybrida** (L.) Delarbre

Specularia castellana Lange === **Legousia scabra** (Lowe) Gamisans

Specularia castellana var. *grandiflora* Willk. === **Legousia scabra** (Lowe) Gamisans

Specularia coa A. DC. === **Legousia pentagonia** (L.) Thell.

Specularia coloradoensis (Buckley) Small === **Triodanis coloradoensis** (Buckley) McVaugh

Specularia conferta E. H. L. Krause === **Legousia hybrida** (L.) Delarbre

Specularia cordata (Vis.) Heynh. === **Legousia speculum-veneris** (L.) Durande ex Vill.

Specularia falcata (Ten.) A. DC. === **Legousia falcata** (Ten.) Fritsch ex Janch.

Specularia falcata f. *scabra* (Lowe) H. Lindb. === **Legousia scabra** (Lowe) Gamisans

Specularia falcata subsp. *castellana* (Lange) Nyman === **Legousia scabra** (Lowe) Gamisans

Specularia falcata var. *pusilla* Boiss. === **Legousia falcata** (Ten.) Fritsch ex Janch.

Specularia falcata var. *scabra* (Lowe) A. DC. === **Legousia scabra** (Lowe) Gamisans

Specularia ghilanensis (Pall. ex Schult.) A. DC. === **Campanula ghilanensis** Pall. ex Schult.

Specularia hirta (Ten.) Gand. === **Legousia speculum-veneris** (L.) Durande ex Vill.

Specularia holzingeri (McVaugh) Fernald === **Triodanis holzingeri** McVaugh

Specularia hybrida (L.) A. DC. === **Legousia hybrida** (L.) Delarbre

Specularia julianii Batt. === **Legousia julianii** (Batt.) Briq.

Specularia lamprosperma (McVaugh) Fernald === **Triodanis lamprosperma** McVaugh

Specularia leptocarpa (Nutt.) A. Gray === **Triodanis leptocarpa** (Nutt.) Nieuwl.

Specularia lindheimeri Vatke === **Triodanis coloradoensis** (Buckley) McVaugh

Specularia linsecomia Buckley === **Triodanis leptocarpa** (Nutt.) Nieuwl.

Specularia ovata (Nutt.) Vatke === **Triodanis perfoliata** subsp. **biflora** (Ruiz & Pav.) Lammers

Specularia parviflora (Gray) Bubani === **Legousia hybrida** (L.) Delarbre

Specularia parviflora St.-Lag. === **Legousia hybrida** (L.) Delarbre

Specularia pentagonia (L.) A. DC. === **Legousia pentagonia** (L.) Thell.

Specularia pentagonia var. *coa* (A. DC.) Nyman === **Legousia pentagonia** (L.) Thell.

Specularia pentagonia var. *pubescens* A. DC. === **Legousia pentagonia** (L.) Thell.

Specularia perfoliata (L.) A. DC. === **Triodanis perfoliata** (L.) Nieuwl.

Specularia perfoliata f. *alba* (J. W. Voigt) Steyerm. === **Triodanis perfoliata** (L.) Nieuwl. subsp. **perfoliata**

Specularia perfoliata f. *ramosa* Arechav. === **Triodanis perfoliata** subsp. **biflora** (Ruiz & Pav.) Lammers

Specularia perfoliata f. *rigida* Arechav. === **Triodanis perfoliata** subsp. **biflora** (Ruiz & Pav.) Lammers

Specularia polypiflora David. === **Legousia speculum-veneris** (L.) Durande ex Vill.

Specularia rariflora (Nutt.) McVaugh === **Heterocodon rariflorum** Nutt.

Specularia speculum-veneris (L.) A. DC. === **Legousia speculum-veneris** (L.) Durande ex Vill.

Specularia speculum-veneris f. *plena* Voss === **Legousia speculum-veneris** (L.) Durande ex Vill.

Specularia speculum-veneris f. *procumbens* Voss === **Legousia speculum-veneris** (L.) Durande ex Vill.

Specularia speculum-veneris subsp. *hirta* (Ten.) Arcang. === **Legousia speculum-veneris** (L.) Durande ex Vill.

Specularia speculum-veneris subvar. *pubescens* (A. DC.) Rouy === **Legousia speculum-veneris** (L.) Durande ex Vill.

Specularia speculum-veneris var. *calycina* A. DC. === **Legousia speculum-veneris** (L.) Durande ex Vill.

Specularia speculum-veneris var. *cordata* (Vis.) Nyman === **Legousia speculum-veneris** (L.) Durande ex Vill.

Specularia speculum-veneris var. *hirta* (Ten.) Nyman === **Legousia speculum-veneris** (L.) Durande ex Vill.

Specularia speculum-veneris var. *libanotica* A. DC. === **Legousia speculum-veneris** (L.) Durande ex Vill.

Specularia speculum-veneris var. *procumbens* Anonymous === **Legousia speculum-veneris** (L.) Durande ex Vill.

Specularia speculum-veneris var. *pubescens* A. DC. === **Legousia speculum-veneris** (L.) Durande ex Vill.

Specularia speculum-veneris var. *racemosa* Boiss. === **Legousia speculum-veneris** (L.) Durande ex Vill.

Specularia speculum-veneris var. *stricta* Griseb. === **Legousia speculum-veneris** (L.) Durande ex Vill.

Specularia texana (McVaugh) Fernald === **Triodanis texana** McVaugh

Specularia vulgaris Kitt. === **Legousia speculum-veneris** (L.) Durande ex Vill.

Speirema

This species was segregated from *Lobelia* primarily on the basis of its large robust stature and membranous or somewhat fleshy berry, rupturing irregularly. It is currently assigned to *Lobelia* sect. *Colensoa* (Hook. f.) J. Murata. Type [by monotypy]: *Speirema montanum* (Reinw. ex Blume) Hook. f. & Thomson.

> Hooker, J. D. & T. Thomson (1858). Praecursores ad floram Indicam: being sketches of the natural families of Indian plants, with remarks on their distribution, structure, and affinities. Series I. Stylidieae, Goodenovieae, et Campanulaceae (including Lobeliaceae). J. Proc. Linn. Soc., Bot. 2: 1-29. En. — Establishment of genus as distinct from *Lobelia*.
>
> Baillon, H. (1885). *Pratia* Gaudich. In Histoire des plantes 8: 366. Paris: L. Hachette. Fr. — Included in *Pratia* as one of four sections.
>
> Hemsley, W. B. (1886). *Pratia borneensis,* Hemsl. Hooker's Icon. Pl. 16: pl. 1532 + 1 pg., illus. En. — Included in *Pratia* as one of three sections.

Synonyms:
Speirema Hook. f. & Thomson === **Lobelia** L.
Speirema montanum (Reinw. ex Blume) Hook. f. & Thomson === **Lobelia montana** Reinw. ex Blume

Streleskia

Originally described as a genus of Lobelioideae, but the stated characters are perfectly at home in the campanuloid genus *Wahlenbergia*. Type [by monotypy]: *Streleskia montana* Hook. f.

> Hooker, J. D. (1847). Flora Tasmaniae spicilegium. London J. Bot. 6: 106-125, 265-286, 461-479. La. — Establishment of genus as allied to *Isotoma*.

Hooker, J. D. (1852). *Wahlenbergia,* Schrad. In The botany of the antarctic voyage of H.M. discovery ships *Erebus* and *Terror* in the years 1839-1843. II. Flora Novae-Zelandiae 1: 159-160. London: Reeve Bros. En. — Notes that type of *Streleskia montana* is a specimen of *Wahlenbergia saxicola.*

Synonyms:
Streleskia Hook. f. === **Wahlenbergia** Schrad. ex Roth
Streleskia montana Hook. f. === **Wahlenbergia saxicola** (R. Br.) A. DC.

Symphyandra

This name was used for many years (e.g., Fedorov 1957, Tutin 1976) for species that resemble *Campanula* except for their connate stamens. Today, most are treated as part of *Campanula* sect. *Symphyandriformes* (Fomin) Kharadze (e.g., Kolakovskii & Serdyukova 1980, Oganesian 2001, Viktorov 2002). This is supported by a preliminary phylogenetic analysis (Eddie et al. 2003, under **General**), in which the three sampled species referable here fell into two different clades within the "Campanulaceae s. str. clade", among numerous species of *Campanula.* Type [designated by Fed. in Kom., Fl. URSS 24: 332 (1957)]: *Symphyandra cretica* A. DC.

de Candolle, A. (1830). *Symphyandra.* In Monographie des Campanulées: 365-368. Paris: Veuve Desray. La. — Establishment of genus as distinct from *Campanula.*
Fedorov, A. A. (1957). *Symphyandra* A. DC. In V. L. Komarov, Flora URSS 24: 331-341, illus. Leningrad: Academia Scientiae URSS. Ru. — Flora for former Soviet Union, with key, descriptions, infrageneric classifications, and full nomenclature.
Tutin, T. G. (1976). *Symphyandra* A. DC. In T. G. Tutin, V. H. Heywood, N., A. Burges, D. M. Moore, D. H. Valentine, S. M. Walters & D. A. Webb (eds.), Flora Europaea 4: 93. Cambridge: Cambridge University Press. En. — Flora with key, descriptions, and full nomenclature.
Crook, H. C. (1977). The genus *Symphyandra.* Bull. Alpine Gard. Soc. Gr. Brit. 45: 246-254, illus. En. — Popular account with brief descriptions, published posthumously.
Kolakovskii, A. A. & L. B. Serdyukova (1980). [Ad systematicam specierum Caucasicarum generis *Campanula* L. notitiae novae.] Zametki Sist. Geogr. Rast. 36: 44-57, illus. Ru. — Transfer of Caucasian species to *Campanula.*
Greuter, W. (1981). Med-Checklist notulae, 3. Willdenowia 11: 23-43. En. — Transfer of type species to *Campanula.*
Oganesian, M. E. (2001). [Review of the sections *Symphyandriformes* and *Cordifoliae* of the subgenus *Campanula* (*Campanula,* Campanulaceae).] Bot. Žurn. 86(11): 134-150. Ru. — Synopsis of group of species in W. Asia, with key.
Viktorov, V. (2002). [Conspectus taxonomicus generis *Campanula* L. (Campanulaceae) Rossiae et civitatum collimitanearum.] Novosti Sist. Vyssh. Rast. 34: 197-234. Ru. — Species in former Soviet Union treated as *Campanula.*

Synonyms:
Symphyandra A. DC. === **Campanula** L.
Symphyandra antiqua Kolak. === **Campanula pendula** M. Bieb.
Symphyandra armena (Steven) A. DC. === **Campanula armena** Steven
Symphyandra armena subsp. *daralaghezica* (Grossh.) Fed. === **Campanula armena** Steven
Symphyandra armena var. *serratosepala* Fomin === **Campanula armena** Steven
Symphyandra asiatica Nakai === **Hanabusaya asiatica** (Nakai) Nakai
Symphyandra cretica A. DC. === **Campanula cretica** (A. DC.) D. Dietr.
Symphyandra cretica subsp. *samothracica* (Degen) Hayek === **Campanula samothracica** (Degen) Greuter & Burdet
Symphyandra cretica subsp. *sporadum* (Halácsy) Hayek === **Campanula samothracica** subsp. **sporadum** (Halácsy) Greuter & Burdet

Symphyandra cretica var. *samothracica* Degen === **Campanula samothracica** (Degen) Greuter & Burdet

Symphyandra daralaghezica Grossh. === **Campanula armena** Steven

Symphyandra finitima (Fomin) Fomin === **Campanula betulifolia** K. Koch

Symphyandra hofmannii Pant. === **Campanula hofmannii** (Pant.) Greuter & Burdet

Symphyandra lazica Boiss. & Balansa === **Campanula lazica** (Boiss. & Balansa) Kharadze

Symphyandra lezgina Alex. === **Campanula lezgina** (Alex.) Kolak. & Serdyuk.

Symphyandra odontosepala (Boiss.) Esfand. === **Campanula odontosepala** Boiss.

Symphyandra ossetica (M. Bieb.) A. DC. === **Campanula ossetica** M. Bieb.

Symphyandra pangaea Heldr. & Charrel === **Campanula orphanidea** Boiss.

Symphyandra pendula (M. Bieb.) A. DC. === **Campanula pendula** M. Bieb.

Symphyandra pendula var. *transcaucasica* Sommier & Levier === **Campanula pendula** M. Bieb.

Symphyandra repens Karpiss. === **Campanula odontosepala** Boiss.

Symphyandra samothracica (Degen) Halácsy === **Campanula samothracica** (Degen) Greuter & Burdet

Symphyandra sporadum Halácsy === **Campanula samothracica** subsp. **sporadum** (Halácsy) Greuter & Burdet

Symphyandra transcaucasica (Sommier & Levier) Grossh. === **Campanula pendula** M. Bieb.

Symphyandra wanneri (Rochel) Heuff. === **Campanula wanneri** Rochel

Symphyandra zangezura Lipsky === **Campanula zangezura** (Lipsky) Kolak. & Serdyuk.

Synotoma

This name applies to the same taxon as *Physoplexis*, but does not have priority. Based on: *Phyteuma* sect. *Synotoma* G. Don. Type [by monotypy]: *Phyteuma comosum* L.

Schulz, R. (1904). Monographie der Gattung *Phyteuma*. Geisenheim: J. Schneck. Ge. — Elevation of *Phyteuma* sect. *Synotoma* to generic rank.

Synonyms:

Synotoma (G. Don) Rich. Schulz === **Physoplexis** (Endl.) Schur

Synotoma comosum (L.) Dalla Torre & Sarnth. === **Physoplexis comosa** (L.) Schur

Synotoma comosum var. *pubescens* (Facchini ex Murr) Dalla Torre & Sarnth. === **Physoplexis comosa** (L.) Schur

Theilera

Campanuloideae, 2 species, Cape Provinces of South Africa. Although neither species has been included in a molecular phylogeny (Eddie et al. 2003, under **General**), on the basis of morphology and biogeography, one suspects that the genus would be part of the "wahlenbergioid clade". Type [designated by the author]: *Theilera guthriei* (L. Bolus) E. Phillips.

Phillips, E. P. (1926). *Theilera*, Phillips. In The genera of South African flowering plants: 606. Cape Town: Cape Times. En. — Establishment of genus as distinct from *Wahlenbergia*.

Dyer, R. A. (1975). *Theilera* Phillips. In The genera of southern African flowering plants 1: 643. Pretoria: Department of Agricultural Technical Services. En. — Generic description.

Goldblatt, P. & J. Manning (2000). Cape plants. A conspectus of the Cape flora of South Africa. *Wahlenbergia*. Strelitzia 9: 396-401, 707-708. En. — Argues for return of *Theilera* to *Wahlenbergia*.

Welman, W. G. (2000). *Theilera* E. Phillips. In O. A. Leistner (ed.), Seed plants of southern Africa: families and genera. Strelitzia 10: 202. En. — Generic description.

Theilera E. Phillips, Gen. S. Afr. Fl. Pl.: 606 (1926).
S. Africa. 27.

Theilera capensis D. Y. Hong, Taxon 51: 732 (2003).
Cape Provinces. 27 CPP. Nanophan. or cham.

Theilera guthriei (L. Bolus) E. Phillips, Gen. S. Afr. Fl. Pl. 606 (1926).
Cape Provinces. 27 CPP. Nanophan. or cham.
* *Wahlenbergia guthriei* L. Bolus, Ann. Bolus Herb. 1: 193 (1915).

Theodorovia

Campanuloideae, 1 species, Caucasus. Judging from morphology and biogeography, the genus may be related to *Muehlbergella* and *Sachokiella* (q.v.). Based on: *Fedorovia* Kolak., non Yakovlev.

Kolakovskii, A. A. (1980). [Fedorovia – a new monotypical genus from limestone of South Transcaucasia.] Soobšč. Akad. Nauk Gruzinsk. SSR 97: 685-688, illus. Ru. — Establishment of genus as distinct from *Campanula*, under an illegitimate name.
Kolakovskii, A. A. (1985). [Fruit types in the Campanulaceae.] Bot. Žurn. 70(1): 3-11, illus. Ru. — Validation of substitute name.
Oganesian, M. (1995). Proposal to conserve *Campanula karakuschensis* Grossh. against *Campanula minsteriana* Grossh. (Campanulaceae). Taxon 44: 241. En. — Formal proposal to protect name of type species from competitor with priority.
Brummitt, R. K. (1997). Report of the Committee for Spermatophyta: 45. Taxon 46: 323-328. En. — Proposal for conservation not recommended.
Nicolson, D. H. (1999). Report of the General Committee: 8. Taxon 48: 373-378. En. — Proposal for conservation "on hold".

Theodorovia Kolak., Bot. Žurn. (Moscow & Leningrad) 70: 7 (1985).
W. Asia. 33 34.
* *Fedorovia* Kolak., Soobšč. Akad. Nauk Gruzinsk. SSR 97: 687 (1980); non Yakovlev, Bot. Žurn. (Moscow & Leningrad) 56: 656 (1971). *Campanula* subg. *Theodorovia* (Kolak.) Ogan., Candollea 50: 299 (1995). *Campanula* sect. *Theodorovia* (Kolak.) Victorov, Novosti Sist. Vyssh. Rast. 34: 227 (2002).

Theodorovia karakuschensis (Grossh.) Kolak., Bot. Žurn. (Moscow & Leningrad) 70: 7 (1985).
Caucasus (Armenia, Nakhichevan), Turkey, Iran. 33 TCS 34 IRN TUR. Hemicr. or geophyte.
Campanula minsteriana Grossh., Trudy Bot. Inst. Akad. Nauk SSSR Azerbajdžansk. Fil. 2: 257 (1936), nom. rejic. prop.
* *Campanula karakuschensis* Grossh., Izv. Azerbajdžansk. Fil. 1-2: 118 (1939), nom. conserv. prop. *Fedorovia karakuschensis* (Grossh.) Kolak., Soobšč. Akad. Nauk Gruzinsk. SSR 97: 687 (1980).
Campanula hakkiarica P. H. Davis, Notes Roy. Bot. Gard. Edinburgh 24: 32 (1962).

Trachanthelium

This name has the same basis as *Podanthum* (q.v.), and like it, lacks priority over *Asyneuma*. Based on: *Phyteuma* sect. *Podanthum* G. Don. Type [designated under **Podanthum**]: *Phyteuma canescens* Waldst. & Kit.

Schur, F. (1866). *Trachanthelium* (Kittel) Schur. In Enumeratio plantarum Transsilvaniae: 431. Vienna: Braumüller. Ge. — Elevation of *Phyteuma* sect. *Podanthum* to generic rank under a new name.

Synonyms:

Trachanthelium Kit. ex Schur. === **Asyneuma** Griseb. & Schenk.
Trachanthelium canescens (Waldst. & Kit.) Kit. ex Schur. === **Asyneuma canescens** (Waldst. & Kit.) Griseb. & Schenk
Trachanthelium canescens var. *transsilvanicum* Schur === **Asyneuma canescens** (Waldst. & Kit.) Griseb. & Schenk subsp. **canescens**

Trachelioides

Opiz (in Berchtold 1839) established this genus for those species of *Campanula* with exappendiculate calyx, long-campanulate corolla, three-locular five-angled ovary, and basally dehiscent capsules. When Opiz next mentioned the name (Opiz 1852), he spelled it "Tracheliopsis;" this is interpreted here as an orthographic error. Type [designated here]: *Campanula trachelium* L.

> Berchtold, F. (1839). Oekonomisch-technische flora Böhmens, band 2, abteil. 2. Prague: Thomas Thabor. Ge. — Segregation of genus from *Campanula*.
> Opiz, F. M. (1852). Seznam rostlin květeny České. Prague: Řiunáce. Cz. — Name maintained but misspelled.

Synonyms:
Trachelioides Opiz === **Campanula** L.

Tracheliopsis

Like *Diosphaera* (q.v.), this genus has been recognized by some workers, but considered too ill-defined for distinction by others. It seems best treated as *Campanula* sect. *Tracheliopsis* (Buser) Kuntze (Damboldt 1976). One suspects that if any of its species were to be included in a molecular phylogeny, they would fall near *Campanula rumeliana* in the "Campanulaceae s. str. clade" (cf. Eddie et al. 2003, under **General**). Type [designated here]: *Tracheliopsis postii* (Boiss.) Buser.

> Buser, R. (1894). Contributions a la connaissance des Campanulacées. I. *Trachelium* L. revisum. Bull. Herb. Boissier 2: 501-532 + pl. XV-XIX, illus. Fr. — Segregation of genus from *Trachelium*.
> Engler, A. (1897). Campanulaceae. In A. Engler & K. Prantl, Die natürlichen Pflanzenfamilien, Nachtr.: 319-320. Leipzig: Wilhelm Engelmann. Ge. — Places genus in synonymy of *Campanula*.
> Davis, P. H. (1950). *Tracheliopsis antilibanotica* P. H. Davis. Hooker's Icon. Pl. 35: tab. 3498 + 5pp., illus. En. — Doubts merit of recognizing genus.
> Davis, P. H. (1972). *Campanula myrtifolia*. Bot. Mag. 179: tab. 637 + 3 pp., illus. En. — Update of Davis (1950), advocating inclusion in *Campanula*.
> Damboldt, J. (1976). Materials for a flora of Turkey XXXII: Campanulaceae. Notes Roy. Bot. Gard. Edinburgh 35: 39-52, illus. En. — Genus reduced to sectional rank within *Campanula*.

Synonyms:
Tracheliopsis Buser === **Campanula** L.
Tracheliopsis albicans Buser === **Campanula petraea** L.
Tracheliopsis antilibanotica P. H. Davis === **Campanula antilibanotica** (P. H. Davis) Greuter & Burdet
Tracheliopsis fruticulosa O. Schwarz & P. H. Davis === **Campanula fruticulosa** (O. Schwarz & P. H. Davis) Damboldt
Tracheliopsis myrtifolia (Boiss. & Heldr.) O. Schwarz & P. H. Davis === **Campanula myrtifolia** Boiss. & Heldr.
Tracheliopsis petraea (L.) Buser === **Campanula petraea** L.
Tracheliopsis postii (Boiss.) Buser === **Campanula postii** (Boiss.) Engl.
Tracheliopsis postii subsp. *shepardii* (Post) P. H. Davis === **Campanula postii** (Boiss.) Engl.
Tracheliopsis pubicalyx P. H. Davis === **Campanula pubicalyx** (P. H. Davis) Damboldt
Tracheliopsis taurica Contandr., Quézel & Pamukç. === **Campanula myrtifolia** Boiss. & Heldr.
Tracheliopsis tubulosa (Boiss.) Buser === **Campanula buseri** Damboldt
Tracheliopsis tubulosa var. *berytensis* Buser === **Campanula buseri** Damboldt
Tracheliopsis tubulosa var. *taurica* Buser === **Campanula buseri** Damboldt
Tracheliopsis tubulosa var. *libanotica* Buser === **Campanula buseri** Damboldt

Trachelium

Campanuloideae, 2 species, Macaronesia and western Mediterranean. In a preliminary DNA phylogeny (Eddie et al. 2003, under **General**), its type was sister to the other members of the "Campanulaceae s. str. clade", including all sampled species of *Azorina, Edraianthus, Feeria,* and *Michauxia,* plus many species of *Campanula.* 2n = 32, 34. Type [by monotypy]: *Trachelium caeruleum* L.

Linnaeus, C. (1753). *Trachelium.* In Species plantarum: 171. Stockholm: Laurentius Salvius. La. — Establishment of genus in Pentandria Monogynia.

Linnaeus, C. (1754). *Trachelium.* In Genera plantarum: 78. Stockholm: Laurentius Salvius. La. — Description of genus to accompany preceding.

- Buser, R. (1894). Contributions a la connaissance des Campanulacées. I. Genus *Trachelium* L. revisum. Bull. Herb. Boissier 2: 501-532 + pl. XV-XIX, illus. Fr. — Removal to segregate genera of most species previouslly assigned to genus; monograph with descriptions, complete nomenclature, and specimen citations.

Sandwith, N. Y. (1940). *Trachelium lanceolatum.* Campanulaceae. Bot. Mag. 163: tab. 9603 + 4 pp., illus. En. — Portrait with description; advocates specific rank for this taxon and hybrid status of *T.* × *halteratum.*

Turrill, W. B. (1963). *Trachelium caeruleum.* Campanulaceae. Bot. Mag. 174: tab. 427 + 4pp., illus. En. — Portrait with description and notes on cultivation; supports specific rank for *T. lanceolatum.*

- Tutin, T. G. (1976). *Trachelium.* In T. G. Tutin, V. H. Heywood, N., A. Burges, D. M. Moore, D. H. Valentine, S. M. Walters & D. A. Webb (eds.), Flora Europaea 4: 94-95. Cambridge: Cambridge University Press. En. — Flora with key, descriptions, and full nomenclature; circumscribed to include *Diosphaera.*

Cosner, M. E., R. K. Jansen, J. D. Palmer & S. R. Downie (1997). The highly rearranged chloroplast genome of *Trachelium caeruleum* (Campanulaceae): multiple inversions, inverted repeat expansion and contrqaction, transposition, insertion/deletions, and several repeat families. Curr. Genet. 31: 419-429. En. — Chloroplast DNA has suffered numerous structural mutations.

Eddie, W. M. M. (2000). *Trachelium* Linnaeus. In J. Cullen, J. C. M. Alexander, C. D. Brickell, J. R. Edmondson, P. S. Green, V. H. Heywood, P. M. Jørgensen, S. L. Jury, S. G. Knees, H. S. Maxwell, D. M. Willer, N. K. B. Robson, S. M. Walters & P. F. Yeo (eds.), The European garden flora 6: 481-482, illus. Cambridge: Cambridge University Press. En. — Account of species in cultivation, with description.

Trachelium L., Sp. Pl.: 171 (1753). *Campanula* sect. *Trachelium* (L.) Kuntze in T. Post & Kuntze, Lex. Gen. Phan.: 95 (1904).
Macaronesia & W. Mediterranean; naturalized in British Isles & Australasia. (10) 12 13 20 21 (51).

Trachelium caeruleum L., Sp. Pl. 171 (1753).
Azores; Madeira; Portugal to Italy & N. Africa; cult.; naturalized in Great Britain, Channel Is. (Guernsey, Jersey) & New Zealand. (10) grb 12 fra POR SPA 13 ITA SIC 20 ALG MOR 21 AZO MDR (51) nzn. Cham. or hemicr. 2n = 32, 34.
Trachelium azureum Gouan, Hortus Monsp.: 100 (1762).
Trachelium caeruleum f. *isabellinum* Voss in Siebert & Voss, Vilm. Blumengärtn. (ed. 3) 1: 563 (1894).
Trachelium caeruleum f. *brevidentatum* Sennen & Mauricio in Sennen, Diagn. Nouv.: 236 (1936), as 'brevedentatum'.
Trachelium caeruleum f. *longiflorum* Sennen & Mauricio in Sennen, Diagn. Nouv.: 236 (1936).

Trachelium × **halteratum** (Bianca ex Ces., Pass. & Gibelli) Sandwith, Bot. Mag. 163: tab. 9603 (1940). T. caeruleum × T. lanceolatum
Sicilia. 13 SIC. Cham. or hemicr.

* *Trachelium caeruleum* var. *halteratum* Bianca ex Ces., Pass. & Gibelli, Comp. Fl. Ital.: 429 (1877). *Trachelium caeruleum* subsp. *halteratum* (Bianca ex Ces., Pass. & Gibelli) Nyman, Consp. Fl. Eur.: 485 (1879), as 'alteratum'.

Trachelium lanceolatum Guss., Fl. Sicul. Syn. 1: 252 (1843). *Trachelium caeruleum* var. *lanceolatum* (Guss.) Ces., Pass. & Gibelli, Comp. Fl. Ital.: 429 (1877). *Trachelium caeruleum* subsp. *lanceolatum* (Guss.) Arcang., Comp. Fl. Ital.: 457 (1882).
Sicilia. 13 SIC. Cham. or hemicr. $2n = 34$.

Synonyms:

Trachelium angustifolium Schousb. === **Feeria angustifolia** (Schousb.) Buser

Trachelium asperuloides Boiss. & Orph. === **Campanula asperuloides** (Boiss. & Orph.) Engl.

Trachelium azureum Gouan === **Trachelium caeruleum** L.

Trachelium caeruleum f. *brevidentatum* Sennen & Mauricio === **Trachelium caeruleum** L.

Trachelium caeruleum f. *isabellinum* Voss === **Trachelium caeruleum** L.

Trachelium caeruleum f. *longiflorum* Sennen & Mauricio === **Trachelium caeruleum** L.

Trachelium caeruleum subsp. *halteratum* (Bianca ex Ces., Pass. & Gibelli) Nyman === **Trachelium × halteratum** (Bianca ex Ces., Pass. & Gibelli) Sandwith.

Trachelium caeruleum var. *halteratum* Bianca ex Ces., Pass. & Gibelli === **Trachelium × halteratum** (Bianca ex Ces., Pass. & Gibelli) Sandwith.

Trachelium caeruleum subsp. *lanceolatum* (Guss.) Arcang. === **Trachelium lanceolatum** Guss.

Trachelium caeruleum var. *lanceolatum* (Guss.) Ces., Pass. & Gibelli === **Trachelium lanceolatum** Guss.

Trachelium diffusum L. f. === **Prismatocarpus diffusus** (L. f.) A. DC.

Trachelium jacquinii (Sieber) Boiss. === **Campanula jacquinii** (Sieber) A. DC.

Trachelium jacquinii subsp. *dalgiciorum* Özhatay & Dane === **Campanula rumeliana** (Hampe) Vatke subsp. **rumeliana**

Trachelium jacquinii subsp. *rumelianum* (Hampe) Tutin === **Campanula rumeliana** (Hampe) Vatke

Trachelium myrtifolium (Boiss. & Heldr.) Boiss. === **Campanula myrtifolia** Boiss. & Heldr.

Trachelium postii Boiss. === **Campanula postii** (Boiss.) Engl.

Trachelium rumelianum Hampe === **Campanula rumeliana** (Hampe) Vatke

Trachelium rumelianum var. *cinerascens* Vandas === **Campanula rumeliana** (Hampe) Vatke subsp. **rumeliana**

Trachelium taygeteum Quézel & Contandr. === **Campanula asperuloides** subsp. **taygetea** (Quézel & Contandr.) Greuter & Burdet

Trachelium tenuifolium L. f. === **Merciera tenuifolia** (L. f.) A. DC.

Trachelium tubulosum Boiss. === **Campanula buseri** Damboldt

Treichelia

Campanuloideae, 1 species, Cape Provinces of South Africa. The genus presumably is allied to other "wahlenbergioid" genera in that region, though it has not yet been included in a molecular phylogeny (Eddie et al. 2003, under **General**). Based on: *Leptocodon* Sond., non (Hook. f.) Lem.

Sonder, O. W. (1865). *Leptocodon*, Sond. In W. H. Harvey & O. W. Sonder, Flora Capensis 3: 584-585. Dublin: Hodges Smith. En. — Establishment of genus under illegitimate name.

Vatke, W. (1874). Notulae in Campanulaceas herbarii regii berolinensis. Linnaea 38: 699-735. La. — Validation of replacement name.

Adamson, R. S. (1950). *Treichelia* Vatke. In R. S. Adamson & T. M. Salter, Flora of the Cape Peninsula: 745. Cape Town: Juta. En. — Flora with description.

Dyer, R. A. (1975). *Treichelia* Vatke. In The genera of southern African flowering plants 1: 640. Pretoria: Department of Agricultural Technical Services. En. — Generic description.

Goldblatt, P. & J. Manning (2000). Cape plants. A conspectus of the Cape flora of South
Africa. *Treichelia*. Strelitzia 9: 396. En. — Brief species desciption.
Welman, W. G. (2000). *Treichelia* Vatke. In O. A. Leistner (ed.), Seed plants of southern
Africa: families and genera. Strelitzia 10: 202. En. — Generic description.

Treichelia Vatke, Linnaea 38: 700 (1874).
S. Africa. 27.
* *Leptocodon* Sond. in Harv. & Sond., Fl. Cap. 3: 584 (1865); non (Hook. f.) Lem., Ill. Hort. 3
(Misc.): 49 (1856).

Treichelia longibracteata (H. Buek) Vatke, Linnaea 38: 700 (1874), as 'longebracteata'.
Cape Provinces. 27 CPP. Ther.
* *Microcodon longibracteatus* H. Buek in Eckl. & Zeyh., Enum. Pl. Afric. Austral.: 378
(1837), as 'longebracteatum'. *Campanula longibracteata* (H. Buek) D. Dietr., Syn. Pl.
1: 757 (1839), as 'longebracteata'. *Leptocodon longibracteatus* (H. Buek) Sond. in Harv.
& Sond., Fl. Cap. 3: 585 (1865), as 'longebracteatum'.
Wahlenbergia depressa Wolley-Dod, J. Bot. 39: 400 (1901); non J. M. Wood & M. S.
Evans, J. Bot. 35: 489 (1897).

Trematocarpus

Though it is earlier, this name cannot be used for *Trematolobelia* because it is a later
homonym of *Trematocarpus* Kütz., the name of a genus of red algae. Type [by monotypy]:
Trematocarpus macrostachys (Hook. & Arn) Zahlbr.

Zahlbruckner, A. (1891). Ueber einige Lobeliaceen des Wiener Herbariums. *Trematocarpus*
A. Zahlbr. nov. gen. Ann. K. K. Naturhist. Hofmus. 6: 430-432, illus. Ge. —
Establishment of genus as distinct from *Lobelia* on basis of its poricidal fruit.
Hemsley, W. B. (1892). *Trematocarpus*. Ann. Bot. 6: 154. En. — Hemsley opines that
poricidal fruit characterizing new genus due to insect damage.
Anonymous (1893). Botanischer Discussions- und Literaturabend am 20. Jänner 1893.
Verh. K. K. Zool.-Bot. Ges. Wien 43 (Sitzungsber.): 6-7. Ge. — Report of oral
presentation by Zahlbruckner on his controversy with Hemsley over the fruit of
Trematocarpus.
Zahlbruckner, A. (1893). The genus *Trematocarpus*. Ann. Bot. 7: 289-290. En. —
Zahlbruckner defends correctness of his description, followed by brief response from
Hemsley.
Stapf, O. (1893). The genus *Trematocarpus*. Ann. Bot. 7: 396-398. En. — Resolution of
Zahlbruckner-Hemsley debate through examination of anatomy of fruit.

Synonyms:
Trematocarpus Zahlbr. === **Trematolobelia** Zahlbr. ex Rock
Trematocarpus macrostachys (Hook. & Arn) Zahlbr. === **Trematolobelia macrostachys**
(Hook. & Arn.) Zahlbr. ex Rock

Trematolobelia

Lobelioideae, 4 species, Hawaiian Islands. Based on morphology (Carlquist 1962) and
molecular biology (E. Knox, pers. comm.), the genus is most closely allied to *Lobelia* sect.
Galeatella E. Wimm., a taxon likewise endemic to the Hawaiian Islands. The treatment
adopted here is that of Lammers (1990). $2n = 28$. Based on: *Trematocarpus* Zahlbr., non Kütz.

Rock, J. F. (1913). New species of Hawaiian plants. Coll. Hawaii Publ. Bull. 2: 39-47 + pl.
IX-XII, illus. En. — Validation of replacement name.

- Rock, J. F. (1919). A monographic study of the Hawaiian species of the tribe Lobelioideae family Campanulaceae, *Trematolobelia* Zahlbruckner. Mem. Bernice Pauahi Bishop Mus. 7(2): 139-148, illus. En. — Monograph with key, descriptions, full nomenclature, and specimen citations.

    Selling, O. H. (1947). Studies in Hawaiian pollen statistics II. The pollens of the Hawaiian phanerogams. Lobeliaceae. Spec. Publ. Bernice Pauahi Bishop Mus. 38: 318-320 + pl. 49, illus. En. — Descriptive palynology of one species.

- Wimmer, F. E. (1953). *Trematolobelia* A. Zahlb. In A. Engler & L. Diels, Das Pflanzenreich IV. 276b: 756-758. Berlin: Akademie-Verlag. Ge., La. — Monograph with key, descriptions, full nomenclature, and specimen citations.

- Carlquist, S. (1962). *Trematolobelia:* seed dispersal; anatomy of fruit and seeds. Pacific Sci. 16: 126-134, illus. En. — Detailed anatomical study, in light of ecology; confirms unique poricidal nature of fruit.

    Stone, B. C. (1967). A review of the endemic genera of Hawaiian plants. *Trematolobelia* Zahlbruckner ex Rock. Bot. Rev. (Lancaster) 33: 246. En. — Brief descriptive treatment with full nomenclature.

- Wimmer, F. E. (1968). *Trematolobelia* A. Zahlbr. In A. Engler & L. Diels, Das Pflanzenreich IV. 276c: 901 + Taf. 29, illus. Berlin: Akademie-Verlag. Ge., La. — Supplement to Wimmer (1953), with revised key.

- St. John, H. (1983). Monograph of *Trematolobelia* (Lobeliaceae). Hawaiian plant studies 107. Pacific Sci. 36: 483-506, illus., maps. En. — Monograph with key, descriptions, full nomenclature, and specimen citations.

- Lammers, T. G. (1990). *Trematolobelia* A. Zahlbr. In W. L. Wagner, D. R. Herbst & S. H. Sohmer, Manual of the flowering plants of Hawai'i: 485-488, illus. Honolulu: University of Hawaii Press. En. — Flora with key and descriptions.

    Duvall, F. II (1991). Cultivation of *Clermontia, Cyanea, Lobelia,* and *Trematolobelia* from seed. Hawaii's Forests & Wildlife 6(3): 12-13, illus. En. — Practical methods for cultivation of endangered rainforest species.

Trematolobelia Zahlbr. ex Rock, Coll. Hawaii Publ. Bull. 2: 45 (1913).
Hawaiian Is. 63 HAW.
* *Trematocarpus* Zahlbr., Ann. K. K. Naturhist. Hofmus. 6: 430 (1891); non Kütz., Phycol. General.: 410 (1843). *Rapuntium* sect. *Trematocarpus* Kuntze in T. Post & Kuntze, Lex. Gen. Phan.: 478 (1903).

Trematolobelia grandifolia (Rock) O. Deg., Fl. Hawaiiensis, fam. 339 (1934).
Hawaiian Is. (Hawai'i). 63 HAW. Phan. or nanophan.
* *Trematolobelia macrostachys* var. *grandifolia* Rock, Coll. Hawaii Publ. Bull. 2: 46 (1913). *Trematolobelia wimmeri* O. Deg. & I. Deg., Phytologia 17: 370 (1968). *Trematolobelia kohalaensis* H. St. John, Pacific Sci. 36: 494 (1983). *Trematolobelia rockii* var. *hawaiiensis* H. St. John, Pacific Sci. 36: 501 (1983).

Trematolobelia kauaiensis (Rock) Skottsb., Acta Horti Gothob. 10: 189 (1936).
Hawaiian Is. (Kaua'i). 63 HAW. Nanophan. $2n = 28$.
* *Trematolobelia macrostachys* var. *kauaiensis* Rock, Coll. Hawaii Publ. Bull. 2: 46 (1913).

Trematolobelia macrostachys (Hook. & Arn.) Zahlbr. ex Rock, Coll. Hawaii Publ. Bull. 2: 45 (1913).
Hawaiian Is. (O'ahu, Moloka'i, Lāna'i, Maui, Hawai'i). 63 HAW. Phan. or nanophan. $2n = 28$.
* *Lobelia macrostachys* Hook. & Arn., Bot. Beechey Voy.: 88 (1832). *Delissea macrostachys* (Hook. & Arn.) C. Presl, Prodr. Monogr. Lobel.: 47 (1836). *Dortmanna macrostachys* (Hook. & Arn.) Kuntze, Revis. Gen. Pl. 2: 972. (1891). *Trematocarpus macrostachys* (Hook. & Arn) Zahlbr., Ann. K. K. Naturhist. Hofmus. 6: 430 (1891). *Trematolobelia sandwicensis* O. Deg., Fl. Hawaiiensis, fam. 339 (1934).

Trematolobelia sandwicensis var. *kaalae* O. Deg., Fl. Hawaiiensis, fam. 339 (1936).
  Trematolobelia macrostachys var. *kaalae* (O. Deg.) E. Wimm., Pflanzenr. IV.276b: 758 (1953).
Trematolobelia macrostachys var. *haleakalaensis* H. St. John, Pacific Sci. 25: 67 (1971).
  Trematolobelia haleakalaensis (H. St. John) O. Deg. & I. Deg., Pl. Haw. Natl. Parks (rev. ed.): 461 (1983).
Trematolobelia auriculata H. St. John, Paciific Sci. 36: 486 (1983).
Trematolobelia forbesii H. St. John, Pacific Sci. 36: 488 (1983).
Trematolobelia lustriatilis H. St. John, Pacific Sci. 36: 495 (1983), 'lustrialis'.
Trematolobelia macrostachys var. *ligulata* H. St. John, Pacific Sci. 36: 499 (1983).
Trematolobelia rockii H. St. John, Pacific Sci. 36: 499 (1983).

Trematolobelia singularis H. St. John, Pacific Sci. 36: 501 (1983).
  Hawaiian Is. (O'ahu). 63 HAW. Nanophan. $2n = 28$.

Synonyms:
Trematolobelia auriculata H. St. John === **Trematolobelia macrostachys** (Hook. & Arn.) Zahlbr. ex Rock
Trematolobelia forbesii H. St. John === **Trematolobelia macrostachys** (Hook. & Arn.) Zahlbr. ex Rock
Trematolobelia haleakalaensis (H. St. John) O. Deg. & I. Deg. === **Trematolobelia macrostachys** (Hook. & Arn.) Zahlbr. ex Rock
Trematolobelia kauaiensis (Rock) Skottsb. === **Trematolobelia kauaiensis** (Rock) Skottsb.
Trematolobelia kohalaensis H. St. John === **Trematolobelia grandifolia** (Rock) O. Deg.
Trematolobelia lustriatilis H. St. John === **Trematolobelia macrostachys** (Hook. & Arn.) Zahlbr. ex Rock
Trematolobelia macrostachys var. *grandifolia* Rock === **Trematolobelia grandifolia** (Rock) O. Deg.
Trematolobelia macrostachys var. *haleakalaensis* H. St. John === **Trematolobelia macrostachys** (Hook. & Arn.) Zahlbr. ex Rock
Trematolobelia macrostachys var. *kaalae* (O. Deg.) E. Wimm. === **Trematolobelia macrostachys** (Hook. & Arn.) Zahlbr. ex Rock
Trematolobelia macrostachys var. *kauaiensis* Rock === **Trematolobelia kauaiensis** (Rock) Skottsb.
Trematolobelia macrostachys var. *ligulata* H. St. John === **Trematolobelia macrostachys** (Hook. & Arn.) Zahlbr. ex Rock
Trematolobelia rockii H. St. John === **Trematolobelia macrostachys** (Hook. & Arn.) Zahlbr. ex Rock
Trematolobelia rockii var. *hawaiiensis* H. St. John === **Trematolobelia grandifolia** (Rock) O. Deg.
Trematolobelia sandwicensis O. Deg. === **Trematolobelia macrostachys** (Hook. & Arn.) Zahlbr. ex Rock
Trematolobelia sandwicensis var. *kaalae* O. Deg. === **Trematolobelia macrostachys** (Hook. & Arn.) Zahlbr. ex Rock
Trematolobelia wimmeri O. Deg. & I. Deg. === **Trematolobelia grandifolia** (Rock) O. Deg.

Trimeris

Mabberley (1974) advocated recognition of the endemic lobelioid on St. Helena as a distinct genus. However, it differs in no substantive way from *Lobelia* and is probably best assigned to *Lobelia* sect. *Colensoa* (Hook. f.) J. Murata. Type [by monotypy]: *Trimeris oblongifolia* C. Presl.

Roxburgh, W. (1816). An alphabetical list of plants, seen by Dr. Roxburgh growing on the island of St. Helena, in 1813-14. In A. Beatson, Tracts relative to the island of St. Helena: 295-326. London: W. Bulmer. En. — Description of sole species.
Presl, C. (1836). *Trimeris*. In Prodromus monographiae Lobeliacearum: 46. Prague: Theophilus Haase. La. — Establishment of genus.

Bentham, G. (1876). *Lobelia,* Linn. In G. Bentham & J. D. Hooker, Genera plantarum 2: 551-553. London: L. Reeve. La. — Transferred to sectional rank within *Lobelia.*

Turrill, W. B. (1949). On the flora of St. Helena. Kew Bull. 1948: 358-362. En. — Discussion of conservation status of endemic flora; *Lobelia scaevolifolia* still present.

Mabberley, D. J. (1974). The pachycaul lobelias of Africa and St. Helena. Kew Bull. 29: 535-584, illus. En. — Resurrection of *Trimeris;* suggests relationship to *Apetahia* and *Sclerotheca* of Polynesia.

Goodenough, S. (1985). St. Helena and its endemic plants – a conservation status. Kew Mag. 2: 369-379, illus. En. — *Lobelia scaevolifolia* still extant.

Cronk, Q. C. B. (1987). The history of endemic flora of St Helena: a relictual series. New Phytol. 105: 509-520. En. — Accepts genus; related to genera of South America and Pacific, not Africa.

Synonyms:

Trimeris C. Presl === **Lobelia** L.

Trimeris oblongifolia C. Presl === **Lobelia scaevolifolia** Roxb.

Trimeris scaevolifolia (Roxb.) Mabb. === **Lobelia scaevolifolia** Roxb.

Triodanis

Campanuloideae, 6 species, North America, with one subspecies disjunct in temperate South America. It has been included by some authors in the otherwise Old World genus *Legousia,* generally under the illegitimate name *Specularia* (q.v.). However, a preliminary phylogenetic analysis of DNA data (Eddie et al. 2003, under **General**) indicates that *Legousia* circumscribed to include *Triodanis* would also have to include *Campanula americana* and *C. divaricata.* The most recent monograph is that of McVaugh (1945), who included *Legousia falcata* (q.v.) of the Old World in the genus. $2n = 28, 56, 60.$ Type [designated by McVaugh, Wrightia 1: 23 (1945)]: *Triodanis rupestris* Raf.

Rafinesque, C. S. (1838). New flora and botany of North America, part 4. Philadelphia: published privately. En. — Establishment of genus as allied to *Legousia.*

Gray, A. (1876). Miscellaneous botanical contributions, *Specularia* Heister. Proc. Amer. Acad. Arts 11: 81-82. En. — Synopsis of "Specularia Americanae", dividing them between two sections, and providing descriptions, full nomenclature, and specimen citations; description of cleistogamous flowers.

Trent, J. A. (1940). Floral variation in *Specularia perfoliata* (L.) A. DC. Amer. Midl. Naturalist 23: 448-454, illus. En. — Variation in floral morphology and its effects on breeding system.

• McVaugh, R. (1945). The genus *Triodanis* Rafinesque, and its relationships to *Specularia* and *Campanula.* Wrightia 1: 13-52, maps. En. — Monograph with key, descriptions, full nomenclature, and specimen citations; discussion of generic boundaries in family and general criteria for drawing these boundaries.

Fernald, M. L. (1946). *Triodanis* versus *Specularia.* Rhodora 48: 209-214 + pl. 1049-1050, illus. En. — Criticizes recognition of *Triodanis* as distinct from *Legousia.*

McVaugh, R. (1948). Generic status of *Triodanis* and *Specularia.* Rhodora 50: 38-49. En. — Rebuttal to Fernald (1946), with additional comments on generic issues.

Bradley, T. R. (1975). Hybridization between *Triodanis perfoliata* and *Triodanis biflora* (Campanulaceae). Brittonia 27: 110-114, illus. En. — Experimental and natural hybridization suggests these two taxa not specifically distinct.

Shetler, S. G. & N. R. Morin (1986). Seed morphology in North American Campanulaceae. Ann. Missouri Bot. Gard. 73: 653-688, illus. En. — Seed coat morphology suggests that Texas endemic *Campanula reverchonii* (q.v.) may belong to *Triodanis;* seeds of *T. texana* unique within family.

Shen, B. A. (1988). A new species of *Asyneuma* Griseb. et Schenk from China. Acta Phytotax. Sin. 26: 463-464, illus. Ch. — Description of naturalized plants of *T. perfoliata* subsp. *biflora* as endemic species of *Asyneuma.*

Gara, B. & G. Muenchow (1990). Chasmogamy/cleistogamy in *Triodanis perfoliata* (Campanulaceae): some CH/CL comparisons in fitness parameters. Amer. J. Bot. 77: 1-6. En. — Experimental studies on breeding system.

Chen, L. J., Z. Y. Li & D. Y. Hong (1992). *Triodanis* Raf. — a new recorded genus of Campanulaceae in China. Acta Phytotax. Sin. 30: 473-475. Ch. — Both subspecies of *T. perfoliata* naturalized in China.

Peng, C. I. & T. G. Lammers (1998). *Triodanis* Raf. (Campanulaceae: Campanuloideae), a new generic record for the flora of Taiwan. Bot. Bull. Acad. Sin. 39: 213-216, illus. En. — *T. perfoliata* subsp. *biflora* naturalized in N. Taiwan.

Triodanis Raf., New Fl. N. Amer. 4: 67 (1838). *Specularia* sect. *Triodanis* (Raf.) Torr., Fl. New York 1: 428 (1843), as 'Triodallus'.

Temp. North America & South America; naturalized in Asia, Australia & West Indies. (36) (38) (50) 71 72 73 74 75 76 77 78 79 80 (81) 83 84 85.

Specularia [unranked] *Dysmicodon* Endl., Gen. Pl.: 518 (1838). *Dysmicodon* (Endl.) Nutt., Trans. Amer. Philos. Soc. (n.s.) 8: 255 (1842). *Specularia* sect. *Hesperocodon* Webb & Berth., Hist. Nat. Iles Canaries 2(3): 6 (1844).

Campylocera Nutt., Trans. Amer. Philos. Soc. (n.s.) 8: 257 (1842). *Specularia* sect. *Campylocera* (Nutt.) A. Gray, Proc. Amer. Acad. Arts 11: 82 (1876). *Legousia* sect. *Campylocera* (Nutt.) Hayek & Hegi in Hegi, Ill. Fl. Mitt.-Eur. 6: 366 (1916).

Triodanis coloradoensis (Buckley) McVaugh, Wrightia 1: 25 (1945).

SC. Texas (Edwards Plateau). 77 TEX. Ther.

* *Campanula coloradoensis* Buckley, Proc. Acad. Nat. Sci. Philadelphia 13: 460 (1861), as 'coloradoense'. *Pentagonia coloradoensis* (Buckley) Kuntze, Revis. Gen. Pl. 2: 381 (1891). *Legousia coloradoensis* (Buckley) A. Heller, Bot. Explor. S. Texas: 99 (1895), as 'coloradoense'. *Specularia coloradoensis* (Buckley) Small, Fl. S.E. U.S.: 1142 (1903).

Specularia lindheimeri Vatke, Linnaea 38: 713 (1874).

Triodanis holzingeri McVaugh, Wrightia 1: 45 (1945). *Specularia holzingeri* (McVaugh) Fernald, Rhodora 48: 214 (1946).

Wyoming to Tennessee, Texas & Arizona. 73 COL WYO 74 KAN MSO NEB OKL 76 ARI 77 TEX 78 ARK TEN. Ther.

Triodanis lamprosperma McVaugh, Wrightia 1: 42 (1945). *Specularia lamprosperma* (McVaugh) Fernald, Rhodora 48: 214 (1946).

Kansas to Louisiana & Texas. 74 KAN MSO OKL 77 TEX 78 ARK LOU. Ther.

Triodanis leptocarpa (Nutt.) Nieuwl., Amer. Midl. Naturalist 3: 192 (1914).

Montana to Minnesota, Indiana & Texas. 73 COL MNT WYO 74 ILL IOW KAN MIN MSO NDA NEB OKL SDA 75 INI 77 TEX 78 ARK. Ther.

* *Campylocera leptocarpa* Nutt., Trans. Amer. Philos. Soc. (n.s.) 8: 257 (1842). *Specularia leptocarpa* (Nutt.) A. Gray, Proc. Amer. Acad. Arts 11: 82 (1876). *Pentagonia leptocarpa* (Nutt.) Kuntze, Revis. Gen. Pl. 2: 381 (1891). *Legousia leptocarpa* (Nutt.) Britt., Mem. Torrey Bot. Club 5: 309 (1894).

Campylocera leptocarpa var. *glabella* Nutt., Trans. Amer. Philos. Soc. (n.s.) 8: 257 (1842). *Specularia linsecomia* Buckley, Proc. Acad. Nat. Sci. Philadelphia 13: 460 (1861).

Triodanis perfoliata (L.) Nieuwl., Amer. Midl. Naturalist 3: 192 (1914).

S. Canada to Guatemala; Ecuador to SE. Brazil, S. Argentina & C. Chile; naturalized in E. Asia, SE. Australia & West Indies. (36) chs (38) tai (50) nsw 71 BRC 72 ONT QUE 73 COL IDA MNT ORE WAS WYO 74 ILL IOW KAN MIN MSO NDA NEB OKL SDA WIS 75 CNT INI MAI MAS MIC NWH NWJ NWY OHI PEN RHO VER WVA 76 ARI CAL NEV UTA 77 NWM TEX 78 ALA ARK DEL FLA GEO KTY LOU MRY MSI NCA SCA TEN VRG 79 MXC MXE MXG MXI MXN 80 GUA (81) dom jam 83 BOL ECU PER 84 BZL BZS 85 AGE AGS AGW PAR URU. Ther. 2*n* = 28, 56.

* *Campanula perfoliata* L., Sp. Pl.: 169 (1753). *Campanula amplexicaulis* Michx., Fl. Bor.-
Amer. 1: 108 (1803). *Prismatocarpus perfoliatus* (L.) Sweet, Hort. Brit.: 251 (1826).
Specularia perfoliata (L.) A. DC., Monogr. Campan.: 351 (1830). *Dysmicodon perfoliatus*
(L.) Nutt., Trans. Amer. Philos. Soc. (n.s.) 8: 256 (1842), as 'perfoliatum'. *Pentagonia
perfoliata* Kuntze, Revis. Gen. Pl. 2: 381 (1891). *Legousia perfoliata* (L.) Britt., Mem.
Torrey Bot. Club 5: 309 (1894).

subsp. **perfoliata**
S. Canada (British Columbia to Quebec) to Guatemala; naturalized in New South
Wales, SE. China (Fujian), Dominican Rep. & Jamaica. (36) chs (50) nsw 71 BRC 72
ONT QUE 73 COL IDA MNT ORE WAS WYO 74 ILL IOW KAN MIN MSO NDA NEB
OKL SDA WIS 75 CNT INI MAI MAS MIC NWH NWJ NWY OHI PEN RHO VER WVA
76 ARI CAL NEV UTA 77 NWM TEX 78 ALA ARK DEL FLA GEO KTY LOU MRY MSI
NCA SCA TEN VRG 79 MXC MXE MXG 80 GUA (81) dom jam. Ther. *2n* = 56.
Campanula verticillata Hill, Veg. Syst. 8: 10 (1765).
Campanula flagellaris Kunth in Humb., Bonpl. & Kunth, Nov. Gen. Sp. 3: 301 (quarto),
234 (folio) (1819).
Triodanis rupestris Raf., New Fl. N. Amer. 4: 67 (1838).
Triodanis perfoliata f. *alba* J. W. Voigt in Mohlenbr. & J. W. Voigt, Fl. S. Illinois: 325
(1959). *Specularia perfoliata* f. *alba* (J. W. Voigt) Steyerm., Rhodora 62: 131 (1960).

subsp. **biflora** (Ruiz & Pav.) Lammers, Novon 16: 72 (2006).
U.S.A. (Oregon to New York) to N. Mexico (Baja California, Nueva León, San Luis
Potosí); Guadeloupe I.; Ecuador to SE. Brazil (Rio de Janeiro, Rio Grande do Sul),
Argentina (Jujuy to Chubut) & C. Chile; naturalized in SE. China (Anhui, Fujian,
Zhejiang) and Taiwan. (36) chs (38) tai 73 ORE 74 ILL KAN MSO NEB OKL 75 PEN
NWY 76 ARI CAL 77 NWM TEX 78 ALA ARK FLA GEO KTY LOU MSI NCA SCA TEN
VRG 79 MXE MXI MXN 83 BOL ECU PER 84 BZL BZS 85 AGE AGS AGW PAR URU.
Ther. *2n* = 28, 56.
* *Campanula biflora* Ruiz & Pav., Fl. Peruv. 2: 55 (1799). *Specularia biflora* (Ruiz & Pav.)
Fisch. & C. A. Mey., Index Sem. Hort. Petrop. 2: 22 (1836). *Pentagonia biflora* (Ruiz &
Pav.) Kuntze, Revis. Gen. Pl. 2: 381 (1891). *Legousia biflora* (Ruiz & Pav.) Britt., Mem.
Torrey Bot. Club 5: 309 (1894). *Triodanis biflora* (Ruiz & Pav.) Greene, Man. Bot. San
Francisco: 230 (1894). *Triodanis perfoliata* var. *biflora* (Ruiz & Pav.) T. R. Bradley,
Brittonia 27: 114 (1975).
Campanula angulata Raf., Fl. Ludov.: 55 (1817).
Lobelia humboldtiana Schult. in Roem. & Schult., Syst. Veg. 5: 68 (1819).
Campanula montevidensis Spreng., Syst. Veg. 1: 738 (1824).
Dysmicodon ovatus Nutt., Trans. Amer. Philos. Soc. (n.s.) 8: 256 (1842), as 'ovatum'.
Specularia ovata (Nutt.) Vatke, Linnaea 38: 713 (1874).
Dysmicodon californicus Nutt., Trans. Amer. Philos. Soc. (n.s.) 8: 256 (1842), as
'californicum'. *Specularia californica* (Nutt.) Vatke, Linnaea 38: 714 (1874).
Campanula ludoviciana Riddell, New Orleans Med. Surg. J. 9: 609 (1853).
Heterocodon minimus Kellogg, Proc. Calif. Acad. Sci. 7: 111 (1877), as 'minimum'.
Specularia perfoliata f. *ramosa* Arechav., Anales Mus. Nac. Montevideo 7: 14 (1909).
Specularia perfoliata f. *rigida* Arechav., Anales Mus. Nac. Montevideo 7: 14 (1909).
Asyneuma anhuiense B. A. Shen, Acta Phytotax. Sin. 26: 463 (1988).

Triodanis texana McVaugh, Wrightia 1: 43 (1945). *Specularia texana* (McVaugh) Fernald,
Rhodora 48: 214 (1946).
E. Texas. 77 TEX. Ther.

Synonyms:
Triodanis biflora (Ruiz & Pav.) Greene === **Triodanis perfoliata** subsp. **biflora** (Ruiz & Pav.)
Lammers
Triodanis falcata (Ten.) McVaugh === **Legousia falcata** (Ten.) Fritsch ex Janch.
Triodanis perfoliata f. *alba* J. W. Voigt === **Triodanis perfoliata** (L.) Nieuwl. subsp. **perfoliata**

Triodanis perfoliata var. *biflora* (Ruiz & Pav.) T. R. Bradley === **Triodanis perfoliata** subsp. **biflora** (Ruiz & Pav.) Lammers
Triodanis rupestris Raf. === **Triodanis perfoliata** (L.) Nieuwl. subsp. **perfoliata**
Triodanis scabra Raf. === ?

Trochocodon

This was described as a new genus allied to *Trachelium,* but is considered a synonym of *Asyneuma limoniifolium* (L.) Janch. subsp. *limoniifolium* (Greuter et al. 1984). Type [by monotypy]: *Trochocodon spicatus* Candargy.

Candargy, T. (1897). *Trochocodon* (aff. *Trachelio*) *spicatus*. Bull. Soc. Bot. France 44: 148-149. La. — Establishment of the genus.

Damboldt, J. (1978). *Campanula* L. In P. H. Davis (ed.), Flora of Turkey and the East Aegean Islands 6: 2-64, illus., maps. Edinburgh: University Press. En. — Identity of type unknown.

Greuter, W., H. M. Burdet & G. Long (1984). Campanulaceae. In Med-checklist 1: 120-154, map. Genève: Conservatoire et Jardin Botanique. En. — Name treated as synonym of *Asyneuma limoniifolium* (L.) Janch. subsp. *limoniifolium*

Trochocodon Candargy === **Asyneuma** Griseb. & Schenk.
Trochocodon spicatus Candargy === **Asyneuma limoniifolium** (L.) Janch. subsp. **limoniifolium**

Tupa

This genus was segregated from *Lobelia* on the basis of its robust herbaceoous habit and unilabiate corolla with lobes coherent at apex. It is treated by most current authors as *Lobelia* subg. *Tupa* (G. Don) E. Wimm. and *Lobelia* sect. *Tupa* (G. Don) Benth. Type [designated by Pfeiff., Nomencl. Bot. 2: 1509 (1874)]: *Lobelia tupa* L.

Don, G. (1834). *Tupa.* In A general history of the dichlamydeous plants: 697, 700. London: J. G. & F. Rivington. En. — Segregation of genus from *Lobelia.*

Bentham, G. (1876). *Lobelia,* Linn. In G. Bentham & J. D. Hooker, Genera plantarum 2: 551-553. London: L. Reeve. La. — Transferred to sectional rank within *Lobelia.*

Lammers, T. G. (2000). Revision of *Lobelia* sect. *Tupa* (Campanulaceae: Lobelioideae). Sida 19: 87-110. En. — Monograph with key, descriptions, full nomenclature, and specimen citations.

Synonyms:
Tupa G. Don === **Lobelia** L.
Tupa sect. *Rhynchopetalum* (Fresen.) A. Rich. === **Lobelia** L.
Tupa sect. *Tylomium* (C. Presl) A. DC. === **Lobelia** L.
Tupa acuminata (Sw.) A. DC. === **Lobelia acuminata** Sw.
Tupa arguta (Lindl.) G. Don === **Lobelia excelsa** Bonpl.
Tupa assurgens (L.) A. DC. === **Lobelia assurgens** L.
Tupa assurgens var. *portoricensis* A. DC. === **Lobelia robusta** Graham
Tupa atropurpurea Vis. === **Lobelia polyphylla** Hook. & Arn.
Tupa axilliflora Phil. === **Lobelia polyphylla** Hook. & Arn.
Tupa berteroi A. DC. === **Lobelia tupa** L.
Tupa besseriana (C. Presl) A. DC. === **Lobelia polyphylla** Hook. & Arn.
Tupa bicolor (D. Don) Planch. === **Lobelia laxiflora** Kunth subsp. **laxiflora**
Tupa blanda D. Don === **Lobelia bridgesii** Hook. & Arn.
Tupa bracteosa (C. Presl) A. DC. === **Lobelia polyphylla** Hook. & Arn.
Tupa bridgesii (Hook. & Arn.) A. DC. === **Lobelia bridgesii** Hook. & Arn.
Tupa caudata Griseb. === **Lobelia caudata** (Griseb.) Urb.

Tupa cavanillesiana G. Don === **Lobelia tupa** L.
Tupa cirsiifolia (Lam.) A. DC. === **Lobelia cirsiifolia** Lam.
Tupa columnaris (Hook. f.) Vatke === **Lobelia columaris** Hook. f.
Tupa conglobata (Lam.) A. DC. === **Lobelia conglobata** Lam.
Tupa conglobata var. *elatior* A. DC. === **Lobelia robusta** Graham
Tupa costaricana Planch. & Oerst. === **Lobelia laxiflora** Kunth subsp. **laxiflora**
Tupa costaricana var. *patula* Planch. & Oerst. === **Lobelia laxiflora** Kunth subsp. **laxiflora**
Tupa costaricana var. *stricta* Planch. & Oerst. === **Lobelia laxiflora** Kunth subsp. **laxiflora**
Tupa crassicaulis Hook. === **Lobelia ghiesbreghtii** Decne.
Tupa deckenii Asch. === **Lobelia deckenii** (Asch.) Hemsl.
Tupa decurrens (Cav.) G. Don === **Lobelia decurrens** Cav.
Tupa digitalifolia Griseb. === **Lobelia digitalifolia** (Griseb.) Urb.
Tupa domingensis Vatke === **Lobelia rotundifolia** Juss. ex A. DC.
Tupa ensifolia A. DC. === **Lobelia salicina** Lam.
Tupa feuillei G. Don === **Lobelia tupa** L.
Tupa feuillei var. *berteroi* (A. DC.) Vatke === **Lobelia tupa** L.
Tupa feuillei var. *macrophylla* Vatke === **Lobelia tupa** L.
Tupa feuillei var. *mucronata* (Cav.) Vatke === **Lobelia tupa** L.
Tupa flavescens A. DC. === **Lobelia stricta** Sw.
Tupa gayana Phil. === **Lobelia polyphylla** Hook. & Arn.
Tupa glaucescens Phil. === **Lobelia excelsa** Bonpl.
Tupa hyssopifolia (C. Presl) A. DC. === **Lobelia polyphylla** Hook. & Arn.
Tupa ignescens Payer === ?
Tupa imberbis Griseb. === **Lobelia imberbis** (Griseb.) Urb.
Tupa infesta Griseb. === **Lobelia cirsiifolia** Lam.
Tupa kerstenii Vatke === **Lobelia deckenii** (Asch.) Hemsl.
Tupa kingii Phil. === **Lobelia excelsa** Bonpl. × **Lobelia polyphylla** Hook. & Arn.
Tupa laxiflora (Kunth) Planch. & Oerst. === **Lobelia laxiflora** Kunth
Tupa linearifolia Phil. === **Lobelia polyphylla** Hook. & Arn.
Tupa martagon Griseb. === **Lobelia martagon** (Griseb.) Hitchc.
Tupa montana C. Wright ex Griseb. === **Lobelia cubana** Urb.
Tupa montana (Fresen.) Vatke === **Lobelia rhynchopetalum** Hemsl.
Tupa montana Phil. === **Lobelia tupa** L.
Tupa mucronata (Cav.) A. DC. === **Lobelia tupa** L.
Tupa mucronata var. *hookeri* A. DC. === **Lobelia tupa** L.
Tupa obovata G. Don === **Siphocampylus obovatus** (G. Don) E. Wimm.
Tupa ovata G. Don === **Siphocampylus ovatus** (G. Don) E. Wimm.
Tupa ovata Phil. === **Lobelia polyphylla** Hook. & Arn.
Tupa persicifolia G. Don === **Lobelia laxiflora** Kunth subsp. **laxiflora**
Tupa persicifolia (Lam.) A. DC. === **Lobelia persicifolia** Lam.
Tupa poeppigiana Phil. === **Lobelia polyphylla** Hook. & Arn.
Tupa polyphylla (Hook. & Arn.) G. Don === **Lobelia polyphylla** Hook. & Arn.
Tupa polyphylla var. *angustifolia* Hook. & Arn. ex A. DC. === **Lobelia polyphylla** Hook. & Arn.
Tupa polyphylla var. *besseriana* (C. Presl) Vatke === **Lobelia polyphylla** Hook. & Arn.
Tupa polyphylla var. bracteosa (C. Presl) Vatke === **Lobelia polyphylla** Hook. & Arn.
Tupa polyphylla var. *coquimbana* Vatke === **Lobelia polyphylla** Hook. & Arn.
Tupa polyphylla var. *latifolia* A. DC. === **Lobelia polyphylla** Hook. & Arn.
Tupa portoricensis Vatke === **Lobelia portoricensis** (Vatke) Urb.
Tupa purpurea G. Don === **Lobelia polyphylla** Hook. & Arn.
Tupa racemosa (C. Presl) A. DC. === **Lobelia cirsiifolia** Lam.
Tupa rhynchopetalum Hochst. ex A. Rich. === **Lobelia rhynchopetalum** Hemsl.
Tupa robusta (Graham) A. DC. === **Lobelia robusta** Graham
Tupa salicifolia (Sweet) G. Don === **Lobelia excelsa** Bonpl.
Tupa salicifolia var. *purpurea* Vatke === **Lobelia excelsa** Bonpl.
Tupa schimperi Hochst. ex A. Rich. === **Lobelia giberroa** Hemsl.
Tupa secunda Ruiz & Pav. ex G. Don === **Siphocampylus pavonis** E. Wimm.

Tupa serrata Phil. === **Lobelia polyphylla** Hook. & Arn.
Tupa sonchifolia (Sw.) Griseb. === **Siphocampylus sonchifolius** (Sw.) McVaugh
Tupa stricta (Sw.) A. DC. === **Lobelia stricta** Sw.
Tupa subdentata (C. Presl) A. DC. === **Lobelia polyphylla** Hook. & Arn.

Tylomium

This genus was segregated from *Lobelia* on the basis of its deciduous corolla and stamens, leaving a prominent raised rim atop the hypanthium. It is treated by most authors (e.g., McVaugh 1943, Wilbur 1991) as *Lobelia* sect. *Tylomium* (C. Presl) Benth. Type [designated by Pfeiff., Nomencl. Bot. 2: 1515 (1874)]: *Lobelia assurgens* L.

> Presl, C. (1836). *Tylomium.* In Prodromus monographiae Lobeliacearum: 31-32. Prague: Theophilus Haase. La. — Segregation of genus from *Lobelia.*
> Bentham, G. (1876). *Lobelia,* Linn. In G. Bentham & J. D. Hooker, Genera plantarum 2: 551-553. London: L. Reeve. La. — Transferred to sectional rank within *Lobelia.*
> McVaugh, R. (1943). *Lobelia* L., *Pratia* Gaud. In North American Flora 32A: 35-99, 110-114. New York: New York Botanical Garden. En. — Monograph of section, with keys, descriptions, infrageneric classification, and full nomenclature.
> Wilbur, R. L. (1991). Synopsis of the Mexican and Central American representatives of *Lobelia* section *Tylomium* (Campanulaceae: Lobelioideae). Sida 14: 555-567. En. — Floristic treatment, including key, of species on North American mainland.

Synonyms:
Tylomium C. Presl === **Lobelia** L.
Tylomium assurgens (L.) C. Presl === **Lobelia assurgens** L.
Tylomium conglobatum (Lam.) C. Presl === **Lobelia conglobata** Lam.
Tylomium robustum (Graham) C. Presl === **Lobelia robusta** Graham

Unigenes

Lobelioideae, 1 species, Cape Provinces of South Africa. In a preliminary phylogenetic analysis (E. Knox, pers. comm.), it was part of a clade comprising several southern African species of *Lobelia* sect. *Heyneana* J. Murata and sect. *Isolobus* (A. DC.) C. B. Clarke. Type [by monotypy]: *Unigenes humifusa* (A. DC.) E. Wimm.

> Wimmer, F. E. (1948). Vorarbeiten zur Monographie der Campanulaceae-Lobelioideae: II. Trib. Lobelieae. Ann. Naturhist. Mus. Wien 56: 317-374. Ge. — Establishment of genus.
> Adamson, R. S. (1950). *Mezleria* Presl. In R. S. Adamson & T. M. Salter, Flora of the Cape Peninsula: 753-754. Cape Town: Juta. En. — Type of *Unigenes* assigned to *Mezleria* (here treated as a synonym of *Monopsis*).
> • Wimmer, F. E. (1953). *Unigenes* E. Wimm. In A. Engler & L. Diels, Das Pflanzenreich IV. 276b: 726-728, illus. Berlin: Akademie-Verlag. Ge., La. — Monograph with description, full nomenclature, and specimen citations.
> Goldblatt, P. & J. Manning (2000). Cape plants. A conspectus of the Cape flora of South Africa. *Lobelia.* Strelitzia 9: 390-391. En. — Brief species desciption; argues for inclusion of *Unigenes* in *Lobelia,* but does not effect requisite new combination.

Unigenes E. Wimm., Ann. Naturhist. Mus. Wien 56: 373 (1948).
S. Africa. 27.

Unigenes humifusa (A. DC.) E. Wimm., Ann. Naturhist. Mus. Wien 56: 374 (1948).
Cape Provinces. 27 CPP. Ther.
> * *Mezleria humifusa* A. DC. in DC., Prodr. 7: 351 (1839). *Dortmanna humifusa* (A. DC.) Kuntze, Revis. Gen. Pl. 2: 972 (1891).
> *Lobelia disperma* E. Wimm., Pflanzenr. IV.276b: 606 (1953).

Urumovia

This genus was segregated from *Jasione* on the basis of its ebracteate flowers, cordate anthers, dilated filament bases, and reduced number of ovules. Type [by monotypy]: *Urumovia foliosa* (Cav.) Stef.

> Stefanoff, B. (1936). [Über die systematische Stellung einiger Arten der Familie Campanulaceae.] God. Sofiisk. Univ. Agron.-Lesoved. Fak. 14: 93-104. Bu. — Establishment of genus as distinct from *Jasione.*

Synonyms:
Urumovia Stef. === **Jasione** L.
Urumovia foliosa (Cav.) Stef. === **Jasione foliosa** Cav.

Valvinterlobus

In a situation parallel to *Mecoschistum* (q.v.), it seems that Dulac coined a new name for all of *Wahlenbergia* rather than just segregating one species. If *W. hederacea* indeed proves to be only distantly related to its congeners, as suggested a a recent phylogeny (Eddie et al. 2003, under **General**), a new generic name will be required, as none is available. Based on: *Wahlenbergia* Schrad. ex Roth.

> Dulac, J. (1867). Flore du Département des Hautes-Pyrénées. Paris: F. Savy. Fr. — Validation of name.

Synonyms:
Valvinterlobus Dulac === **Wahlenbergia** Schrad. ex Roth
Valvinterlobus filiformis Dulac === **Wahlenbergia hederacea** (L.) Rchb.

Wahlenbergia

Campanuloideae, 260 species, throughout the Southern Hemisphere and extending north to western Europe, southern and eastern Asia, and northern South America; fully 80% of the species are African. Though a number of infrageneric names have been validated, no current comprehensive classification of the genus is available. Just one species, European *W. hederacea,* has been included in a molecular phylogeny (Eddie et al. 2003, under **General**), where it formed one of the three "transitional taxa" forming a pentachotomy with the "Campanulaceae s. str". and "Rapunculus" clades. One suspects that if more typical members of the genus were to be sequenced, they would belong to the "wahlenbergioid" clade comprising *Roella* and *Craterocapsa.* The last comprehensive monograph was that of Candolle (1839, under **General**). $2n$ = 14, 16, 18, 22, 36, 42, 54, 72, 90. Type [by monotypy]: *Wahlenbergia elongata* (Willd.) Schrad. ex Roth.

> Roth, A. W. (1821). Novae plantarum species praesertim Indiae orientalis. Halberstadt: Vogler. La. — Establishment of genus.
>
> Decaisne, J. (1849). *Wahlenbergia vincaeflora,* Dne. Rev. Hort. (sér. 3) 3: 41-42 + fig. 5, illus. Fr. — Portrait of *W. gracilis,* with description.
>
> Hooker, W. J. (1851). *Wahlenbergia albomarginata,* Hook. Icon. Pl. 9: tab 818 + 1 pg., illus. En. — Portrait with description.
>
> Hooker, J. D. (1875). *Wahlenbergia tuberosa.* Bot. Mag. 101: tab. 6155 + 2 pp., illus. En. — Portrait with description.
>
> Hooker, J. D. (1882). *Wahlenbergia saxicola.* Bot. Mag. 108: tab. 6613 + 2 pp., illus. En. — Portrait with description.
>
> Hooker, J. D. (1891). *Wahlenbergia undulata.* Bot. Mag. 117: tab. 7174 + 2 pp., illus. En. — Portrait with description.

Wright, C. H. (1900). *Lightfootia leptophylla,* C. H. Wright. Icon. Pl. 27: pl. 2659 + 1 pg., illus. En. — Portrait of *W. collomoides,* with description.

Brown, N. E. (1913). The wahlenbergias of Australia and New Zealand. Gard. Chron. (ser. 3) 54: 316-317, 336-337, 354-355. En. — Attempt to clarify nomenclature of species in cultivation, with descriptions and full nomenclature.

Skottsberg, C. (1921). *Wahlenbergia.* In The natural history of Juan Fernandez and Easter Island 2: 175-179, illus. Uppsala: Almqvist & Wiksells. En., La. — Floristic treatment, with key and descriptions.

McVaugh, R. (1945). Notes on North American Campanulaceae. 2. The genus *Wahlenbergia* in the North American flora. Bartonia 23: 36-40. En. — First report of genus as naturalized in North America.

Lothian, N. (1947a). Remarks on Victorian bluebells (*Wahlenbergia*). Vict. Naturalist 63: 229-235, illus. En. — Floristic account with keys and descriptions.

Lothian, N. (1947b). Critical notes on the genus *Wahlenbergia* Schrader; with descriptions of new species in the Australian region. Proc. Linn. Soc. New South Wales 71: 201-235, illus. En. — Diverse taxonomic and nomenclatural comments on genus in general and in Australia.

Lothian, N. (1948). *Wahlenbergia limenophylax.* An unintentional orthgraphic error. Proc. Linn. Soc. New South Wales 72: 366. En. — Correction of Lothian (1947b).

Turrill, W. B. (1949). On the flora of St. Helena. Kew Bull. 1948: 358-362. En. — Endemic *W. angustifolia* and *W. linifolia* abundant but *W. burchellii* not found despite repeated searches.

Adamson, R. S. (1950). *Wahlenbergia* Schrad., *Lightfootia* L'Herit. In R. S. Adamson & T. M. Salter, Flora of the Cape Peninsula: 746-750. Cape Town: Juta. En. — Flora with keys and descriptions.

Stearn, W. T. (1951). *Wahlenbergia trichogyna.* Gard. Chron. (ser. 3) 130: 169. En. — Attempt to clarify identity issues.

Adamson, R. S. (1953). Notes on nomenclature in *Lightfootia.* J. S. African Bot. 19: 157-159. En. — Clarification of application of questionable names.

• Adamson, R. S. (1955). The South African species of *Lightfootia.* J. S. African Bot. 21: 155-218. En. — Regional monograph, with keys, descriptions, infrageneric classification, full nomenclature, and specimen citations.

Lothian, N. (1956). Alpine bluebells. Some further comments and descriptions of two new species of *Wahlenbergia.* Vict. Naturalist 72: 165-169, illus. En. — Supplement to Lothian (1947a, 1947b).

Melville, R. (1959). *Wahlenbergia consimilis.* Bot. Mag. 172: tab. 343 + 4 pp., illus. En. — Portrait of *W. stricta,* with description and notes on cultivation.

• Tuyn, P. (1960). *Wahlenbergia.* In C. G. G. J. van Steenis (ed.), Flora Malesiana (ser. I) 6(1): 110-118, illus., map. Djakarta: Noordhoff-Kolff. En. — Flora with key, descriptions, and full nomenclature; advocates inclusion of *Cephalostigma* and *Lightfootia.*

Lambinon, J. & P. Duvigneaud (1961). Étude systématique et phytogéographique d'un groupe de *Lightfootia* à inflorescence contractée. Bull. Soc. Roy. Bot. Belgique 93: 41-53, illus., map. Fr. — Monograph of group in tropical Africa, with key, full nomenclature, and specimen citations.

Godfrey, R. K. (1963). *Wahlenbergia linarioides* (Campanulaceae) in Florida: a second adventive species for the United States. Sida 1: 185-186. En. — First report for North America.

Carolin, R. C. (1964). Taxonomic and nomenclatural notes on the genus *Wahlenbergia* in Australia. Proc. Linn. Soc. New South Wales 89: 235-240. En. — Miscellaneous nomenclatural changes.

Dyer, R. A. (1975). *Wahlenbergia* Schrad ex Roth, *Cephalostigma* A. DC., *Lightfootia* L'Hérit. In The genera of southern African flowering plants 1: 641-642. Pretoria: Department of Agricultural Technical Services. En. — Generic descriptions.

• Thulin, M. (1975). The genus *Wahlenbergia* s. lat. (Campanulaceae) in tropical Africa and Madagascar. Symb. Bot. Upsal. 21: 1-223, illus., maps. En. — Regional monograph with keys, descriptions, informal infrageneric classafication, full nomenclature, and specimen citations; advocates inclusion of *Cephalostigma* and *Lightfootia.*

- Thulin, M. (1976). *Wahlenbergia*. In Flora of tropical east Africa, Campanulaceae: 4-32, illus. Rotterdam: Balkema. En. — Flora with keys, descriptions, full nomenclature, and specimen citations.
- Thulin, M. (1977). *Wahlenbergia* Schrad. ex Roth, nom. conserv. In Flora d'Afrique centrale (Zaire-Rwanda-Burundi), Spermatophytes, Campanulaceae: 4-41, illus. Meise: Jardin Botanique National de Belgique. Fr. — Flora with keys, descriptions, full nomenclature, and specimen citations.
- Thulin, M. (1978). *Wahlenbergia* Schrader ex Roth. In Flore de Madagascar et des Comores. 187. Campanulacées: 4-21, illus. Fr. Paris: Muséum National d'Histoire Naturelle. — Flora with keys, descriptions, full nomenclature, and specimen citations.

Bamps, P. (1982). Deux *Wahlenbergia* (Campanulaceae) nouveaux pour le Burundi. Bull. Jard. Bot. Belg. 52: 484-485. Fr. — First report of *W. hookeri* and *W. huttonii* from Burundi.

Murthy, G. V. S. (1982). Pollen morphology of Indian *Wahlenbergia* (Campanulaceae). Int. Quart. J. Pl. Sci. Res. 7: 199-201, illus. En. — Regional palynological survey.

Hong, D. Y. (1983). *Wahlenbergia* Schrad. ex Roth, *Cephalostigma* A. DC. In Flora Reipublicae Popularis Sinicae 73(2): 28-31, illus. Ch. — Flora of China, with descriptions and full nomenclature.

- Thulin, M. (1983). *Wahlenbergia* Schrader ex Roth. In E. Launert (ed.), Flora Zambesiaca 7(1): 88-111, illus. Kew: Royal Botanic Garden. En. — Flora of Mozambique, Malawi, Zambia, Zimbabwe and Botswana, with keys, descriptions, full nomenclature and specimen citations.

Cronk, Q. C. B. (1987). The history of endemic flora of St Helena: a relictual series. New Phytol. 105: 509-520. En. — Brief note on the four endemic species; two extinct.

Crawford, D. J., T. F. Stuessy, T. G. Lammers, M. Silva O. & P. Pacheco (1990). Allozyme variation and evolutionary relationships among three species of *Wahlenbergia* (Campanulaceae) in the Juan Fernandez Islands. Bot. Gaz. (Crawfordsville) 151: 119-124. En. — Enzyme electrophoresis supports hypothesis that *W. berteroi* and *W. masafuerae* were derived from *W. fernandeziana*.

- Smith, P. J. (1992). A revision of the genus *Wahlenbergia* (Campanulaceae) in Australia. Telopea 5: 91-175, illus., maps. En. — Floristic treatment with keys, descriptions, full nomenclature, and specimen citations.

Ricci, M. & L. Eaton (1994). The rescue of *Wahlenbergia larrainii* in Robinson Crusoe Island, Chile. Biol. Conserv. 68: 89-93, map. En. — *Ex situ* conservation in Juan Fernández Islands.

Lammers, T. G. (1995). Transfer of the southern African species of *Lightfootia*, nom. illeg., to *Wahlenbergia* (Campanulaceae, Campanuloideae). Taxon 44: 333-339. En. — Transfer of remaining species of *Lightfootia* to *Wahlenbergia*.

Petterson, J. A., E. G. Williams & M. I. Dawson (1995). Contributions to a chromosome atlas of the New Zealand flora – 34. *Wahlenbergia* (Campanulaceae). New Zealand J. Bot. 33: 489-496, illus. En. — Rhizomatous species tetraploid, radicate ones hexaploid or more commonly octoploid.

- Haridasan, V. K. & P. K. Mukherjee (1996). *Wahlenbergia*. In P. K. Hajra & M. Sanjappa (eds.), Fascicles of flora of India 22: 104-114. Calcutta: Botanical Survey of India. En. — Flora with keys, descriptions, and full nomenclature.
- Lammers, T. G. (1996). Phylogeny, biogeography, and systematics of the *Wahlenbergia fernandeziana* complex (Campanulaceae: Campanuloideae). Syst. Bot. 21: 397-415, illus., maps. En. — Monograph of species endemic to Juan Fernández Is., with key, descriptions, full nomenclature, and specimen citations.
- Petterson, J. A. (1997). Revision of the genus *Wahlenbergia* (Campanulaceae) in New Zealand.New Zealand J. Bot. 35: 9-54, illus., maps. En. — Floristic treatment with keys, descriptions, full nomenclature, and specimen citations.

Petterson, J. A. (1997). Identity of the original *Wahlenbergia gracilis* (Campanulaceae) and allied species. An historical review of early collections. New Zealand J. Bot. 35: 55-78, illus. En. — Clarification of application of names in Australasia.

Goldblatt, P. & J. Manning (2000). Cape plants. A conspectus of the Cape flora of South Africa. *Wahlenbergia*. Strelitzia 9: 396-401. En. — Checklist for Cape of Good Hope, with brief species desciptions; advocates inclusion of *Theilera*.

Welman, W. G. (2000). *Wahlenbergia* Schrad. ex Roth. In O. A. Leistner (ed.), Seed plants of southern Africa: families and genera. Strelitzia 10: 202. En. — Generic description.

Anderson, G. J., G. Bernardello, T. F. Stuessy & D. J. Crawford (2001). Breeding system and pollination of selected plants endemic to Juan Fernández Islands. Amer. J. Bot. 88: 220-233. En. — *W. berteroi* and *W. fernandeziana* facultatively autogamous despite protandry.

Rix, M. (2004). *Wahlenbergia hederacea*. Curtis's Bot. Mag. 21: 61-64 + pl. 488, illus. En. — Portrait with description and notes on cultivation.

Wahlenbergia Schrad. ex Roth, Nov. Pl. Sp.: 399 (1821), nom. cons. *Campanula* sect. *Wahlenbergia* (Schrad. ex Roth) D. Dietr., Syn. Pl. 1: 751 (1839). *Valvinterlobus* Dulac, Fl. Hautes-Pyrénées: 459 (1867).

Trop. & S. Temp. Zones; naturalized in SE. U.S.A. & Hawaiian Is. 10 11 12 13 20 21 22 23 24 25 26 27 28 29 35 36 38 40 41 42 43 50 51 60 (63) (77) (78) 83 84 85.

Ireon Scop., Intr. Hist. Nat.: 229 (1777); non Burm. f., Prodr. Fl. Cap.: 6 (1768).

Lightfootia L'Hér., Sert. Angl.: 4 (1789); non Sw., Prodr.: 5, 83 (1788).

Campanula sect. *Campanopsis* R. Br., Prodr.: 561 (1810). *Campanula* sect. *Codonia* Colla, Herb. Pedem. 4: 20 (1835). *Campanopsis* (R. Br.) Kuntze, Revis. Gen. Pl. 2: 378 (1891).

Cervicina Delile, Descr. Égypte, Hist. Nat. 2: 150 (1813-1814), nom. rejic. *Wahlenbergia* sect. *Cervicina* (Delile) A. DC., Monogr. Campan.: 156 (1830), nom. rejic. *Campanopsis* sect. *Cervicina* (Delile) Kuntze in T. Post & Kuntze, Lex. Gen. Phan.: 95 (1904), nom. rejic.

Schultesia Roth, Enum. Pl. Phaen. Germ. 1: 690 (1827); non Spreng., Pl. Pugill. 2: 176 (1815), nom. rejic.; nec Schrad., Gött. Gel. Anz. 1821: 708 (1821); nec Mart., Nova Gen. Sp. Pl. 2: 103 (1827), nom. cons. *Aikinia* Salisb. ex Fourr., Ann. Soc. Linn. Lyon (n.s.) 17: 112 (1869); non Wall., Pl. Asiat. Rar. 3: pl. 273 (1832); nec R. Br. in Wall., Pl. Asiat. Rar. 3: 65 (1832).

Cephalostigma A. DC., Monogr. Campan.: 117 (1830).

Hecale Raf., Fl. Tellur. 2: 79 (1837).

Petalostima Raf., Fl. Tellur. 2: 79 (1837).

Streleskia Hook. f., London J. Bot. 6: 266 (1847).

Wahlenbergia abyssinica (Hochst. ex A. Rich.) Thulin, Symb. Bot. Upsal. 21: 160 (1975).

Somalia to Angola & KwaZulu-Natal; Madagascar. 23 BUR CMN ZAI 24 ETH SOM 25 KEN TAN 26 ANG MLW MOZ ZAM ZIM 27 NAT 29 MDG. Hemicr. or ther. $2n = 16$.

** Lightfootia abyssinica* Hochst. ex A. Rich., Tent. Fl. Abyss. 2: 1 (1850).

subsp. **abyssinica**

Somalia to Angola & KwaZulu-Natal; Madagascar. 23 BUR ZAI 24 ETH SOM 25 KEN TAN 26 ANG MLW MOZ ZAM ZIM 27 NAT 29 MDG. Hemicr. $2n = 16$.

Lightfootia madagascariensis A. DC., Monogr. Campan.: 116 (1830); non *Wahlenbergia madagascariensis* A. DC., Monogr. Campan.: 139 (1830).

Lightfootia abyssinica var. *tenuis* Oliv., J. Linn. Soc., Bot. 21: 401 (1885).

Lightfootia sodenii Engl., Bot. Jahrb. Syst. 19 (Beibl. 47): 52 (1894), as 'sodeni'.

Lightfootia madagascariensis var. *glabra* Engl., Pflanzenw. Ost-Afrikas C: 400 (1895).

Lightfootia rupestris Engl., Bot. Jahrb. Syst. 30: 419 (1901).

Lightfootia divaricata Engl., Bot. Jahrb. Syst. 32: 117 (1902); non H. Buek in Eckl.& Zeyh., Enum. Pl. Afric. Austral.: 375 (1837).

Lightfootia grandifolia Engl., Bot. Jahrb. Syst. 40: 48 (1907).

Lightfootia subulata Engl., Bot. Jahrb. Syst. 40: 48 (1907); non L'Hér., Sert. Angl.: 4 (1789).

Lightfootia ellenbeckii Engl., Bot. Jahrb. Syst. 40: 49 (1907).

Lightfootia elata Chiov., Racc. Bot.: 74 (1935).

Lightfootia arenicola Meikle, Kew Bull. 1948: 466 (1949).

subsp. **parvipetala** Thulin, Symb. Bot. Upsal. 21: 167 (1975).
    Kenya & Tanzania. 25 KEN TAN. Ther. 2*n* = 16.

Wahlenbergia acaulis E. Mey. ex A. DC. in DC., Prodr. 7: 430 (1839). *Campanopsis acaulis*
(E. Mey. ex A. DC.) Kuntze, Revis. Gen. Pl. 2: 379 (1891).
    Cape Provinces. 27 CPP. Ther.

Wahlenbergia acicularis Brehmer, Bot. Jahrb. Syst. 53: 90 (1915).
    KwaZulu-Natal. 27 NAT. Hemicr.?

Wahlenbergia acuminata Brehmer, Bot. Jahrb. Syst. 53: 87 (1915).
    S. Africa. 27 CPP LES OFS. Hemicr.?

Wahlenbergia adamsonii Lammers, Taxon 44: 334 (1995).
    Cape Provinces. 27 CPP. Nanophan. or cham.
      * *Lightfootia multicaulis* Adamson, J. S. African Bot. 21: 165 (1955); non *Wahlenbergia
        multicaulis* Benth. in Endl., Enum. Pl.: 75 (1837).

Wahlenbergia adpressa (L. f.) Sond. in Harv. & Sond., Fl. Cap. 3: 583 (1865).
    Cape Provinces. 27 CPP. Nanophan. or cham.
      * *Campanula adpressa* L. f., Suppl. Pl.: 140 (1782). *Lightfootia adpressa* (L. f.) A. DC.,
        Monogr. Campan.: 110 (1830). *Campanopsis adpressa* (L. f.) Kuntze, Revis. Gen. Pl. 2:
        379 (1891).

Wahlenbergia akaroa Petterson, New Zealand J. Bot. 35: 24 (1997).
    New Zealand (South I.). 51 NZS. Hemicr. 2*n* = 72.

Wahlenbergia albens (Spreng. ex A. DC.) Lammers, Taxon 44: 335 (1995).
    S. Africa. 27 CPP LES OFS. Nanophan. or cham. 2*n* = 16.
      * *Lightfootia albens* Spreng. ex A. DC., Monogr. Campan.: 110 (1830).
        Lightfootia laricina H. Buek in Eckl. & Zeyh., Enum. Pl. Afric. Austral.: 373 (1837).

Wahlenbergia albicaulis (Sond.) Lammers, Taxon 44: 335 (1995).
    Cape Provinces. 27 CPP. Ther.
      * *Lobelia cinerea* Thunb., Prodr. Fl. Cap.: 40 (1794); non *Wahlenbergia cinerea* (L. f.)
        Lammers, Taxon 44: 335 (1995). *Rapuntium cinereum* (Thunb.) C. Presl, Prodr. Monogr.
        Lobel.: 30 (1836). *Lightfootia albicaulis* Sond. in Harv. & Sond., Fl. Cap. 3: 556 (1865);
        non *Lightfootia cinerea* (L. f.) Sond. in Harv. & Sond., Fl. Cap. 3: 557 (1865).

Wahlenbergia albomarginata Hook., Icon. Pl. 9: pl. 818 (1851).
    New Zealand (South I.). 51 NZS. Geophyte. 2*n* = 36.

  subsp. **albomoarginata**
      New Zealand (South I.). 51 NZS. Geophyte. 2*n* = 36.
          Wahlenbergia brockiei J. A. Hay in Allan, Fl. New Zealand 1: 792, 976 (1961).

  subsp. **decora** Petterson, New Zealand J. Bot. 35: 39 (1997).
      New Zealand (South I.). 51 NZS. Geophyte. 2*n* = 36.

  subsp. **flexilis** (Petrie) Petterson, New Zealand J. Bot. 35: 36 (1997).
      New Zealand (South I.). 51 NZS. Geophyte. 2*n* = 36.
      * *Wahlenbergia flexilis* Petrie, Trans. & Proc. New Zealand Inst. 49: 51 (1916).
        Wahlenbergia simpsonii J. A. Hay in Allan, Fl. New Zealand 1: 793, 976 (1961).

  subsp. **laxa** (G. Simpson) Petterson, New Zealand J. Bot. 35: 38 (1997).
      New Zealand (South I.). 51 NZS. Geophyte. 2*n* = 36.
      * *Wahlenbergia laxa* G. Simpson, Trans. & Proc. Roy. Soc. New Zealand 79: 430 (1952).
        Wahlenbergia pygmaea var. *laxa* (G. Simpson) H. H. Allan, Fl. New Zealand 1:
        791 (1961).

subsp. **olivina** Petterson, New Zealand J. Bot. 35: 36 (1997).
New Zealand (South I.). 51 NZS. Geophyte. *2n* = 36.

Wahlenbergia androsacea A. DC., Monogr. Campan.: 150 (1830). *Campanula androsacea*
(A. DC.) D. Dietr., Syn. Pl. 1: 753 (1839). *Campanopsis androsacea* (A. DC.) Kuntze, Revis.
Gen. Pl. 2: 379 (1891). Ther. *2n* = 18.
S. Africa. 26 MOZ ZIM 27 BOT CPP LES NAM NAT OFS TVL.
Wahlenbergia nudicaulis A. DC., Monogr. Campan.: 149 (1830). *Campanula nudicaulis*
(A. DC.) D. Dietr., Syn. Pl. 1: 753 (1839). *Campanopsis nudicaulis* (A. DC.) Kuntze,
Revis. Gen. Pl. 2: 379 (1891).
Wahlenbergia arenaria A. DC. in DC., Prodr. 7: 436 (1839). *Campanopsis arenaria* (A.
DC.) Kuntze, Revis. Gen. Pl. 2: 379 (1891). *Campanopsis androsacea* var. *arenaria* (A.
DC.) Kuntze, Revis. Gen. Pl. 3(2): 185 (1898).
Wahlenbergia sphaerocarpa A. DC. in DC., Prodr. 7: 436 (1839). *Wahlenbergia arenaria*
var. *sphaerocarpa* (A. DC.) Sond. in Harv. & Sond., Fl. Cap. 3: 581 (1865).
Wahlenbergia verreauxii A. DC. in DC., Prodr. 7: 436 (1839). *Wahlenbergia arenaria* var.
verreauxii (A. DC.) Sond. in Harv. & Sond., Fl. Cap. 3: 581 (1865).
Wahlenbergia semiglabra A. DC. in DC., Prodr. 7: 436 (1839).
Wahlenbergia inhambanensis Klotzsch in Peters, Naturw. Reise Mossambique 6: 303
(1861). *Campanopsis inhambanensis* (Klotzsch) Kuntze, Revis. Gen. Pl. 2: 379 (1891).
Cephalostigma fockeanum Schinz, Verh. Bot. Vereins Prov. Brandenburg 30: 257 (1888).
Wahlenbergia inhambanensis var. *erecta* Brehmer, Bot. Jahrb. Syst. 53: 108 (1915).
Wahlenbergia perennis Brehmer, Bot. Jahrb. Syst. 53: 137 (1915).
Wahlenbergia androsacea var. *multicaulis* Brehmer, Bot. Jahrb. Syst. 53: 137 (1915).
Wahlenbergia rosulata Brehmer, Bot. Jahrb. Syst. 53: 138 (1915).
Wahlenbergia glandulosa Brehmer, Bot. Jahrb. Syst. 53: 140 (1915).

Wahlenbergia angustifolia (Roxb.) A. DC., Monogr. Campan.: 162 (1830).
St. Helena. 28 STH. Nanophan. or cham.
* *Roella angustifolia* Roxb. in Beatson, Tracts St. Helena: 320 (1816). *Campanopsis
angustifolia* (Roxb.) Kuntze, Revis. Gen. Pl. 2: 379 (1891).
Wahlenbergia clivosa A. DC., Monogr. Campan.: 161 (1830). *Campanula clivosa* (A. DC.)
D. Dietr., Syn. Pl. 1: 755 (1839).

Wahlenbergia annua (A. DC.) Thulin, Symb. Bot. Upsal. 21: 206 (1975).
Angola. 26 ANG. Ther. *2n* = 16.
* *Lightfootia annua* A. DC., Ann. Sci. Nat., Bot. (ser. 5) 6: 329 (1866).

Wahlenbergia annularis A. DC. in DC., Prodr. 7: 437 (1839). *Campanopsis annularis* (A.
DC.) Kuntze, Revis. Gen. Pl. 2: 379 (1891).
S. Africa. 27 NAM CPP. Ther.
Wahlenbergia annularis var. *bolusiana* Brehmer, Bot. Jahrb. Syst. 53: 140 (1915).

Wahlenbergia annuliformis Brehmer, Bot. Jahrb. Syst. 53: 118 (1915).
Cape Provinces. 27 CPP. Hemicr.?

Wahlenbergia appressifolia Hilliard & B. L. Burtt, Notes Roy. Bot. Gard. Edinburgh 43:
350 (1986).
S. Africa. 27 LES NAT. Geophyte.

Wahlenbergia arcta Thulin, Symb. Bot. Upsal. 21: 128 (1975).
Angola. 26 ANG. Hemicr. or ther.
* *Lightfootia gracilis* A. DC., Ann. Sci. Nat., Bot. (ser. 5) 6: 330 (1866); non (G. Forst.)
Miq., Fl. Ned. Ind. 2: 567 (1857); nec *Wahlenbergia gracilis* (G. Forst.) A. DC.,
Monogr. Campan.: 142 (1830).

Wahlenbergia aridicola P. J. Sm. in Jessop & Toelken, Fl. S. Australia 3: 1377 (1986).
  SE. Australia. 50 NSW SOA. Hemicr. $2n = 18, 36$.

Wahlenbergia asparagoides (Adamson) Lammers, Taxon 44: 335 (1995).
  Cape Provinces. 27 CPP. Nanophan. or cham.
   * *Lightfootia asparagoides* Adamson, J. S. African Bot. 21: 206 (1955).

Wahlenbergia asperifolia Brehmer, Bot. Jahrb. Syst. 53: 90 (1915).
  Cape Provinces. 27 CPP. Hemicr.?

Wahlenbergia axillaris (Sond.) Lammers, Taxon 44: 335 (1995).
  Cape Provinces. 27 CPP. Nanophan. or cham.
   * *Lightfootia axillaris* Sond. in Harv. & Sond., Fl. Cap. 3: 558 (1865).

Wahlenbergia banksiana A. DC., Monogr. Campan.: 154 (1830). *Campanula banksiana* (A.
  DC.) D. Dietr., Syn. Pl. 1: 754 (1839). *Campanopsis banksiana* (A. DC.) Kuntze, Revis. Gen.
  Pl. 2: 379 (1891).
  S. Africa. 26 ANG MOZ ZAM ZIM 27 BOT NAM NAT SWZ TVL. Cham. or hemicr.
    Wahlenbergia leucantha Engl. & Gilg in Warb., Kunene-Sambesi-Exped.: 396 (1903).
    Wahlenbergia mashonica N. E. Br., Bull. Misc. Inform. Kew 1906: 165 (1906).
    Wahlenbergia okavangensis N. E. Br., Bull. Misc. Inform. Kew 1909: 118 (1909).
    Wahlenbergia rhodesiana S. Moore, J. Linn. Soc., Bot. 40: 125 (1911).
    Wahlenbergia saginoides S. Moore, J. Bot. 49: 153 (1911).
    Wahlenbergia multiflora Conrath, Bull. Misc. Inform. Kew 1914: 134 (1914).
    Wahlenbergia mashonica var. *junodis* Brehmer, Bot. Jahrb. Syst. 53: 78 (1915).
    Wahlenbergia foliosa Brehmer, Bot. Jahrb. Syst. 53: 84 (1915); non A. DC., Monogr.
      Campan.: 160 (1830).
    Wahlenbergia gracillima S. Moore, J. Bot. 56: 9 (1918).

Wahlenbergia bernardii Leredde, Bull. Soc. Hist. Nat. Afrique N. 44: 246 (1953), as 'bernardi'.
  Algeria & Libya. 20 ALG LBY. Ther.

Wahlenbergia berteroi Hook. & Arn., J. Bot. (Hooker) 1: 279 (1834). *Campanopsis berteroi*
  (Hook. & Arn.) Kuntze, Revis. Gen. Pl. 2: 379 (1891). Cham. or hemicr.
  Juan Fernández Is. (Masatierra, Santa Clara). 85 JNF. $2n = 22$.
    Campanula gracilis var. *revoluta* Colla, Mem. Reale Accad. Sci. Torino 38: 119 (1835).

Wahlenbergia bolusiana Schltr. & Brehmer, Bot. Jahrb. Syst. 53: 88 (1915).
  Cape Provinces. 27 CPP. Hemicr.?

Wahlenbergia bowkerae Sond. in Harv. & Sond., Fl. Cap. 3: 577 (1865), as 'bowkeriae'.
  Campanopsis bowkerae (Sond.) Kuntze, Revis. Gen. Pl. 2: 379 (1891).
  Cape Provinces. 27 CPP. Ther.

Wahlenbergia brachiata (Adamson) Lammers, Taxon 44: 335 (1995).
  Northern Provinces. 27 TVL. Nanophan. or cham.
   * *Lightfootia brachiata* Adamson, J. S. African Bot. 21: 204 (1955).

Wahlenbergia brachycarpa Schltr., J. Bot. 35: 429 (1897).
  Cape Provinces. 27 CPP. Ther.
    Wahlenbergia brachycarpa var. *pilosa* Brehmer, Bot. Jahrb. Syst. 53: 86 (1915).

Wahlenbergia brachyphylla (Adamson) Lammers, Taxon 44: 335 (1995).
  Cape Provinces. 27 CPP. Nanophan. or cham.
   * *Lightfootia brachyphylla* Adamson, J. S. African Bot. 21: 164 (1955).

Wahlenbergia brasiliensis Cham., Linnaea 8: 318 (1833). *Campanopsis brasiliensis* (Cham.)
Kuntze, Revis. Gen. Pl. 2: 379 (1891).
Brazil (Minas Gerais, São Paulo). 84 BZL. Hemicr.

Wahlenbergia brehmeri Lammers, Novon 8: 35 (1998).
Cape Provinces. 27 CPP. Ther.
* *Wahlenbergia rotundifolia* Brehmer, Bot. Jahrb. Syst. 53: 124 (1915); non (Benth.) A. DC.
in DC., Prodr. 7: 425 (1839).

Wahlenbergia brevisquamifolia Brehmer, Bot. Jahrb. Syst. 53: 120 (1915).
Northern Provinces. 27 TVL. Hemicr.?

Wahlenbergia burchelii A. DC. in DC., Prodr. 7: 438 (1839). *Campanopsis burchelii* (A. DC.)
Kuntze, Revis. Gen. Pl. 2: 379 (1891).
St. Helena. 28 STH. Nanophan.

Wahlenbergia buseriana Schltr. & Brehmer, Bot. Jahrb. Syst. 53: 84 (1915).
Cape Provinces. 27 CPP. Hemicr.?

Wahlenbergia calcarea (Adamson) Lammers, Taxon 44: 335 (1995).
Cape Provinces. 27 CPP. Nanophan. or cham.
* *Lightfootia calcarea* Adamson, J. S. African Bot. 21: 188 (1955).

Wahlenbergia calycina Schltdl. ex Griseb., Abh. Königl. Ges. Wiss. Göttingen 24: 219 (1879).
Colombia to NE. Argentina (Salta). 83 BOL CLM PER 85 AGW. Hemicr.

Wahlenbergia campanuloides (Delile) Vatke, Linnaea 38: 700 (1874).
Egypt; Senegal; S. Africa. 20 EGY 22 SEN 26 ZAM 27 CPP NAM. Ther.
* *Cervicina campanuloides* Delile, Descr. Egypte, Hist. Nat. 2: 150 (1813-1814).
Wahlenbergia cervicina A. DC., Monogr. Campan.: 156 (1830). *Campanula cervicina* D.
Dietr., Syn. Pl. 1: 754 (1839). *Campanopsis campanuloides* (Delile) Kuntze, Revis. Gen.
Pl. 2: 379 (1891), as 'campanulodes'.
Campanula elongata Willd., Enum. Pl., Suppl.: 10 (1814). *Wahlenbergia elongata* (Willd.)
Schrad. ex Roth, Nov. Pl. Sp.: 399 (1821).
Wahlenbergia humifusa Markgr., Notizbl. Bot. Gart. Berlin-Dahlem 15: 760 (1942).

Wahlenbergia candolleana (Hiern) Thulin, Symb. Bot. Upsal. 21: 205 (1975).
Angola. 26 ANG. Hemicr. or ther.
* *Lightfootia paniculata* A. DC., Ann. Sci. Nat., Bot. (ser. 5) 6: 331 (1866); non Sond. in
Harv. & Sond., Fl. Cap. 3: 560 (1865). *Cephalostigma candolleanum* Hiern, Cat. Afr.
Pl. 1: 629 (1898); non *Cephalostigma paniculatum* Wall. ex A. DC., Monogr. Campan.:
117 (1830); nec *Wahlenbergia paniculata* (L. f.) A. DC., Monogr. Campan.: 153
(1830). *Wahlenbergia humpatensis* Brehmer, Bot. Jahrb. Syst. 53: 128 (1915).

Wahlenbergia candollei Tuyn in Steenis, Fl. Males. (ser. 1) 6: 114 (1960).
Myanmar. 41 MYA. Ther.
* *Cephalostigma paniculatum* A. DC., Monogr. Campan.: 117 (1830); non *Wahlenbergia
paniculata* (L. f.) A. DC., Monogr. Campan.: 153 (1830).

Wahlenbergia capensis (L.) A. DC., Monogr. Campan.: 136 (1830).
Cape Provinces; naturalized in W. Australia. 27 CPP (50) wau. Ther.
* *Campanula capensis* L., Sp. Pl.: 169 (1753). *Petalostima capensis* (L.) Raf., Fl. Tellur. 2: 79
(1837). *Campanopsis capensis* (L.) Kuntze, Revis. Gen. Pl. 2: 379 (1891).
Roella decurrens Andrews, Bot. Repos. 4: pl. 238 (1802); non L'Hér., Sert. Angl.: 4 (1789).
Wahlenbergia capensis var. *leiocalycina* Zahlbr., Ann. K. K. Naturhist. Hofmus. 18:
402 (1903).

Wahlenbergia capillacea (L. f.) A. DC., Monogr. Campan.: 156 (1830).
  Kenya to S. Africa. 23 BUR 25 KEN TAN 26 MLW MOZ ZIM 27 CPP LES NAT SWZ.
    Hemicr. 2n = 16.
    * *Campanula capillacea* L. f., Suppl. Pl.: 139 (1782). *Campanula setacea* D. Dietr., Syn. Pl.
      1: 754 (1839). *Campanopsis capillacea* (L. f.) Kuntze, Revis. Gen. Pl. 2: 379 (1891).

  subsp. **capillacea**
    S. Africa. 27 CPP LES NAT SWZ. Hemicr.
    Wahlenbergia pinnata var. *simplicifolia* Compton, J. S. African Bot. 33: 299 (1967).

  subsp. **tenuior** (Engl.) Thulin, Symb. Bot. Upsal. 21: 113 (1975).
    Kenya to Mozambique & Zimbabwe. 23 BUR 25 KEN TAN 26 MLW MOZ ZIM. Hemicr.
      2n = 16.
    Wahlenbergia kilimandscharica Engl., Hochgebirgsfl. Afrika: 412 (1892).
    Wahlenbergia oliveri Schweinf. ex Engl., Pflanzenw. Ost-Afrikas C: 400 (1895).
    * *Wahlenbergia capillacea* var. *tenuior* Engl., Bot. Jahrb. Syst. 30: 418 (1901).
    Wahlenbergia kilimandscharica var. *intermedia* Brehmer, Bot. Jahrb. Syst. 53: 78 (1915).
    Wahlenbergia aberdarica T. C. E. Fr., Notizbl. Bot. Gart. Berlin-Dahlem 8: 395 (1923).

Wahlenbergia capillaris (Lodd.) Sweet, Hort. Brit. (ed. 2): 593 (1830).
  New Guinea & Australia. 43 NWG 50 NSW NTA QLD SOA VIC WAU. Hemicr. 2n = 16,
    18, 36.
    Campanula gracilis var. *capillaris* R. Br., Prodr.: 561 (1810). *Wahlenbergia gracilis* var.
      capillaris (R. Br.) A. DC., Monogr. Campan.: 142 (1830).
    * *Campanula capillaris* Lodd., Bot. Cab. 15: pl. 1406 (1829).
    Wahlenbergia multicaulis var. *dispar* N. E. Br., Gard. Chron. (ser. 3) 54: 337 (1913).
    Wahlenbergia communis Carolin, Proc. Linn. Soc. New South Wales 89: 237 (1964).

Wahlenbergia capillata Brehmer, Bot. Jahrb. Syst. 53: 77 (1915).
  S. Africa. 27 NAT OFS. Hemicr.?

Wahlenbergia capillifolia E. Mey. ex Brehmer, Bot. Jahrb. Syst. 53: 58 (1915).
  Cape Provinces. 27 CPP. Hemicr.?
    Wahlenbergia capillifolia var. *conferta* Brehmer, Bot. Jahrb. Syst. 53: 58 (1915).

Wahlenbergia capitata (Baker) Thulin, Symb. Bot. Upsal. 21: 139 (1975).
  Zaïre to Uganda & Mozambique. 23 BUR RWA ZAI 24 TAN UGA 26 MLW MOZ ZAM ZIM.
    Hemicr. or ther. 2n = 16.
    * *Lightfootia capitata* Baker, Bull. Misc. Inform. Kew 1898: 158 (1898). *Lightfootia
      glomerata* var. *capitata* (Baker) Lambinon, Bull. Soc. Roy. Bot. Belgique 93: 47 (1961).
    Lightfootia glomerata var. *subspicata* Engl., Pflanzenw. Ost-Afrikas C: 400 (1895).
    Lightfootia bequaertii De Wild. & Ledoux. in De Wild., Contr. Fl. Katanga, Suppl. 3: 145
      (1930), as 'bequaerti'.
    Lightfootia elegans Gilli, Ann. Maturhist. Mus. Wien 77: 56 (1973).

Wahlenbergia cartilaginea Hook. f., Handb. New Zealand Fl.: 170 (1867). *Campanopsis
  cartilaginea* (Hook. f.) Kuntze, Revis. Gen. Pl. 2: 379 (1891).
  New Zealand (South I.). 51 NZS. Geophyte. 2n = 36.

Wahlenbergia caryophylloides P. J. Sm., Nuytsia 7: 63 (1989).
  N. Australia. 50 NTA QLD WAU. Ther.

Wahlenbergia celata P. I. Forst., Austrobaileya 5: 661 (2000).
  NE. Australia (Queensland). 50 QLD. Hemicr.

Wahlenbergia cephalodina Thulin, Symb. Bot. Upsal. 21: 171 (1975).
  Zambia. 26 ZAM. Ther.

Wahlenbergia ceracea Lothian, Victoria Naturalist 72: 166 (1956).
SE. Australia. 50 NSW TAS VIC. Geophyte. $2n = 18$.

Wahlenbergia cerastioides Thulin, Symb. Bot. Upsal. 21: 108 (1975).
Madagascar. 29 MDG. Hemicr.

Wahlenbergia cernua (Thunb.) A. DC., Monogr. Campan.: 148 (1830).
Cape Provinces. 27 CPP. Ther.
* *Campanula cernua* Thunb., Prodr. Fl. Cap.: 39 (1794). *Campanopsis cernua* (Thunb.)
    Kuntze, Revis. Gen. Pl. 2: 379 (1891).
  Wahlenbergia cernua var. *minor* H. Buek in Eckl. & Zeyh., Enum. Pl. Afric. Austral.:
    379 (1837).
  Wahlenbergia cernua var. *minor* Sond. in Harv. & Sond., Fl. Cap. 3: 579 (1865); non H.
    Buek in Eckl. & Zeyh., Enum. Pl. Afric. Austral.: 379 (1837).
  Wahlenbergia ciliolata A. DC. in DC., Prodr. 7: 436 (1839).
  Wahlenbergia cernua var. *cuspidata* Brehmer, Bot. Jahrb. Syst. 53: 103 (1915).
  Wahlenbergia maculata Brehmer, Bot. Jahrb. Syst. 53: 103 (1915).
  Wahlenbergia maculata var. *nuda* Brehmer, Bot. Jahrb. Syst. 53: 103 (1915).
  Wahlenbergia clavatula Brehmer, Bot. Jahrb. Syst. 53: 104 (1915).

Wahlenbergia cinerea (L. f.) Lammers, Taxon 44: 335 (1995).
Cape Provinces. 27 CPP. Nanophan. or cham.
* *Campanula cinerea* L. f., Suppl. Pl.: 139 (1782). *Roella cinerea* (L. f.) A. DC., Monogr.
    Campan.: 175 (1830). *Lightfootia cinerea* (L. f.) Sond. in Harv. & Sond., Fl. Cap. 3:
    557 (1865).
  Lightfootia grisea H. Buek in Eckl. & Zeyh., Enum. Pl. Afric. Austral.: 374 (1837).
  Lightfootia albanensis Sond. in Harv. & Sond., Fl. Cap. 3: 557 (1865).

Wahlenbergia clavata Brehmer, Bot. Jahrb. Syst. 53: 93 (1915).
Cape Provinces. 27 CPP. Hemicr.?

Wahlenbergia collomoides (A. DC.) Thulin, Symb. Bot. Upsal. 21: 168 (1975).
Central African Rep. to Angola & Mozambique. 23 CAF CON ZAI 25 TAN 26 ANG MLW
    MOZ ZAM. Hemicr. or ther.
* *Lightfootia collomioides* A. DC., Ann. Sci. Nat., Bot. (ser. 5) 6: 328 (1866).
  Lightfootia leptophylla C. H. Wright, Hooker's Icon. Pl. 27: pl. 2659 (1900).
  Lightfootia collomioides subsp. *katangensis* Lambinon, Bull. Soc. Roy. Bot. Belgique 93:
    49 (1961).

Wahlenbergia compacta Brehmer, Bot. Jahrb. Syst. 53: 79 (1915).
Cape Provinces. 27 CPP. Hemicr.?

Wahlenbergia confusa Merr. & L. M. Perry, J. Arnold Arbor. 22: 383 (1941).
New Guinea. 43 NWG. Hemicr.

Wahlenbergia congesta (Cheeseman) N. E. Br., Gard. Chron. (ser. 3) 54: 336 (1913).
New Zealand (South I.). 51 NZS. Geophyte. $2n = 36$.
* *Wahlenbergia saxicola* var. *congesta* Cheeseman, Man. New Zealand Fl.: 403 (1906).

subsp. **congesta**
  New Zealand (South I.). 51 NZS. Geophyte. $2n = 36$.
  Wahlenbergia morganii Petrie, Trans. & Proc. New Zealand Inst. 46: 34 (1914), as
    'morgani'.

subsp. **haastii** Petterson, New Zealand J. Bot. 35: 32 (1997).
  New Zealand (South I.). 51 NZS. Geophyte. $2n = 36$.

Wahlenbergia congestifolia Brehmer, Bot. Jahrb. Syst. 53: 125 (1915).
S. Africa. 27 CPP LES NAT. Hemicr.?
Wahlenbergia congestifolia var. *glabra* Brehmer, Bot. Jahrb. Syst. 53: 126 (1915).
Wahlenbergia congestifolia var. *laxa* Brehmer, Bot. Jahrb. Syst. 53: 126 (1915).

Wahlenbergia constricta Brehmer, Bot. Jahrb. Syst. 53: 97 (1915).
Cape Provinces. 27 CPP. Hemicr.

Wahlenbergia cooperi Brehmer, Bot. Jahrb. Syst. 53: 115 (1915).
S. Africa. 27 LES CPP. Hemicr.?

Wahlenbergia cordata (Adamson) Lammers, Taxon 44: 335 (1995).
Cape Provinces. 27 CPP. Cham.
* *Lightfootia cordata* Adamson, J. S. African Bot. 21: 169 (1955).

Wahlenbergia costata A. DC. in DC., Prodr. 7: 427 (1839). *Campanopsis costata* (A. DC.)
Kuntze, Revis. Gen. Pl. 2: 379 (1891).
Cape Provinces. 27 CPP. Ther.

Wahlenbergia cuspidata Brehmer, Bot. Jahrb. Syst. 53: 126 (1915).
S. Africa. 27 NAT CPP. Hemicr.?
Wahlenbergia dentifera Brehmer, Bot. Jahrb. Syst. 53: 117 (1915).
Wahlenbergia furcata Brehmer, Bot. Jahrb. Syst. 53: 124 (1915).

Wahlenbergia debilis H. Buek in Eckl. & Zeyh., Enum. Pl. Afric. Austral.: 378 (1837).
Campanula debilis (H. Buek) D. Dietr., Syn. Pl. 1: 752 (1839). *Campanopsis debilis* (H. Buek)
Kuntze, Revis. Gen. Pl. 2: 379 (1891).
Cape Provinces. 27 CPP. Ther.
Wahlenbergia ramulosa E. Mey. ex A. DC. in DC., Prodr. 7: 432 (1839). *Campanopsis
ramulosa* (E. Mey. ex A. DC.) Kuntze, Revis. Gen. Pl. 2: 379 (1891).
Wahlenbergia lobata Brehmer, Bot. Jahrb. Syst. 53: 83 (1915).

Wahlenbergia decipiens A. DC. in DC., Prodr. 7: 427 (1839). *Campanopsis decipiens* (A. DC.)
Kuntze, Revis. Gen. Pl. 2: 379 (1891).
Cape Provinces. 27 CPP. Nanophan. or cham.
Wahlenbergia longisepala Brehmer, Bot. Jahrb. Syst. 53: 77 (1915).

Wahlenbergia densicaulis Brehmer, Bot. Jahrb. Syst. 53: 129 (1915).
Namibia. 27 NAM. Ther.
Wahlenbergia sabulosa Brehmer, Bot. Jahrb. Syst. 53: 82 (1915).
Wahlenbergia densicaulis var. *angusta* Brehmer, Bot. Jahrb. Syst. 53: 129 (1915).

Wahlenbergia densifolia Lothian, Victoria Naturalist 72: 167 (1956).
SE. Australia. 50 NSW VIC. Geophyte. $2n = 18$.

Wahlenbergia dentata Brehmer, Bot. Jahrb. Syst. 53: 79 (1915).
Cape Provinces. 27 CPP. Hemicr.?

Wahlenbergia denticulata (Burch.) A. DC., Monogr. Campan.: 152 (1830).
Zaïre to Kenya & S. Africa. 23 ZAI 25 KEN TAN 26 MLW ZAM ZIM 27 BOT CPP LES NAM
OFS TVL. Hemicr. or ther.
* *Campanula denticulata* Burch., Trav. S. Africa 1: 371 (1822). *Lightfootia denticulata*
(Burch.) Sond. in Harv. & Sond., Fl. Cap. 3: 559 (1865).
Wahlenbergia denticulata var. *scabra* A. DC., Monogr. Campan.: 152 (1830).
Lightfootia tenuifolia A. DC., Ann. Sci. Nat., Bot. (ser. 5) 6: 327 (1866).
Wahlenbergia spinulosa Engl., Bot. Jahrb. Syst. 10: 271 (1889); non A. DC., Monogr.
Campan.: 155 (1830).

Lightfootia goetzeana Engl., Bot. Jahrb. Syst. 30: 419 (1901).
Lightfootia laricifolia Engl. & Gilg in Warb., Kunene-Sambesi-Exped.: 397 (1903).
Lightfootia dinteri Engl. ex Dinter, Repert. Spec. Nov. Regni Veg. 18: 438 (1922).
Lightfootia intricata Dinter & Markgr., Notizbl. Bot. Gart. Berlin-Dahlem 15: 467 (1941).
Lightfootia denticulata var. *podanthoides* Markgr., Notizbl. Bot. Gart. Berlin-Dahlem 15: 466 (1941).
Lightfootia denticulata var. *transvaalensis* Adamson, J. S. African Bot. 21: 172 (1955).
   Wahlenbergia denticulata var. *transvaalensis* (Adamson) Welman, Bothalia 26: 157 (1996).

Wahlenbergia denudata A. DC., Monogr. Campan.: 147 (1830). *Campanula denudata* (A. DC.) D. Dietr., Syn. Pl. 1: 753 (1839). *Campanopsis denudata* (A. DC.) Kuntze, Revis. Gen. Pl. 2: 379 (1891). *Cervicina denudata* (A. DC.) S. Moore, J. Bot. 41: 402 (1903).
Cape Provinces. 27 CPP. Ther.
   Wahlenbergia denudata var. *mutata* Brehmer, Bot. Jahrb. Syst. 53: 125 (1915).

Wahlenbergia depressa J. M. Wood & M. S. Evans, J. Bot. 35: 489 (1897). *Cervicina depressa* (J. M. Wood & M. S. Evans) S. Moore, J. Bot. 41: 402 (1903). Ther.
S. Africa. 27 LES OFS TVL.

Wahlenbergia desmantha Lammers, Taxon 44: 335 (1995).
Cape Provinces. 27 CPP. Nanophan. or cham.
   * *Campanula fasciculata* L. f., Suppl. Pl.: 139 (1782); non *Wahlenbergia fasciculata* Brehmer, Bot. Jahrb. Syst. 53: 92 (1915). *Lightfootia fasciculata* (L. f.) Spreng., Syst. Veg. 1: 809 (1824). *Lightfootia tenella* var. *fasciculata* (L. f.) Sond. in Harv. & Sond., Fl. Cap. 3: 562 (1865).

Wahlenbergia dichotoma A. DC. in DC., Prodr. 7: 437 (1839). *Campanopsis dichotoma* (A. DC.) Kuntze, Revis. Gen. Pl. 2: 379 (1891).
Cape Provinces. 27 CPP. Ther.
   Wahlenbergia parviflora A. DC. in DC., Prodr. 7: 437 (1839).
   Wahlenbergia dichotoma var. *simplex* Sond. in Harv. & Sond., Fl. Cap. 3: 581 (1865).

Wahlenbergia dieterlenii (E. Phillips) Lammers, Taxon 44: 336 (1995).
S. Africa. 27 LES OFS TVL. Nanophan. or cham.
   Lightfootia caledonica Sond. in Harv. & Sond., Fl. Cap. 3: 559 (1865); non *Wahlenbergia caledonica* Sond. in Harv. & Sond., Fl. Cap. 3: 579 (1865).
   * *Lightfootia dieterlenii* E. Phillips, Ann. S. African Mus. 16: 178 (1917).

Wahlenbergia dilatata Brehmer, Bot. Jahrb. Syst. 53: 122 (1915).
Cape Provinces. 27 CPP. Hemicr.?

Wahlenbergia distincta Brehmer, Bot. Jahrb. Syst. 53: 109 (1915).
Cape Provinces. 27 CPP. Hemicr.?

Wahlenbergia divergens A. DC. in DC., Prodr. 7: 427 (1839). *Campanopsis divergens* (A. DC.) Kuntze, Revis. Gen. Pl. 2: 379 (1891).
Cape Provinces. 27 CPP. Ther.

Wahlenbergia doleritica Hilliard & B. L. Burtt, Notes Roy. Bot. Gard. Edinburgh 43: 196 (1986).
KwaZulu-Natal. 27 NAT. Hemicr.

Wahlenbergia dunantii A. DC., Monogr. Campan.: 152 (1830). *Campanula dunantii* (A. DC.) D. Dietr., Syn. Pl. 1: 754 (1839).
Cape Provinces. 27 CPP. Ther.
   Wahlenbergia dunantii var. *glabrata* Brehmer, Bot. Jahrb. Syst. 53: 73 (1915).

Wahlenbergia ecklonii H. Buek in Eckl. & Zeyh., Enum. Pl. Afric. Austral.: 380 (1837).
Campanula ecklonii (H. Buek) D. Dietr., Syn. Pl. 1: 753 (1839). *Campanopsis ecklonii* (H.
Buek) Kuntze, Revis. Gen. Pl. 2: 379 (1891).
Cape Provinces. 27 CPP. Hemicr.
   Wahlenbergia swellendamensis H. Buek in Eckl. & Zeyh., Enum. Pl. Afric. Austral.: 381
     (1837). *Campanula swellendamensis* (H. Buek) D. Dietr., Syn. Pl. 1: 753 (1839).
   Wahlenbergia turbinata A. DC. in DC., Prodr. 7: 427 (1839).
   Wahlenbergia saxifraga A. DC. in DC., Prodr. 7: 427 (1839)
   Wahlenbergia pentamera A. DC. in DC., Prodr. 7: 428 (1839). *Wahlenbergia ecklonii* var.
     pilosa Sond. in Harv. & Sond., Fl. Cap. 3: 570 (1865).
   Wahlenbergia dregeana A. DC. in DC., Prodr. 7: 428 (1839). *Wahlenbergia ecklonii* var.
     gracilis Sond. in Harv. & Sond., Fl. Cap. 3: 570 (1865).
   Wahlenbergia ecklonii var. *brevisepala* Brehmer, Bot. Jahrb. Syst. 53: 74 (1915).
   Wahlenbergia macra Schltr. & Brehmer, Bot. Jahrb. Syst. 53: 91 (1915).

Wahlenbergia effusa (Adamson) Lammers, Taxon 44: 336 (1995).
Cape Provinces. 27 CPP. Nanophan. or cham.
   * *Lightfootia effusa* Adamson, J. S. African Bot. 21: 206 (1955).

Wahlenbergia epacridea Sond. in Harv. & Sond., Fl. Cap. 3: 584 (1865). *Campanopsis
epacridea* (Sond.) Kuntze, Revis. Gen. Pl. 2: 379 (1891). Hemicr.
S. Africa. 27 NAT SWZ TVL.
   Wahlenbergia epacridea var. *glabrata* Sond. in Harv. & Sond., Fl. Cap. 3: 584 (1865)

Wahlenbergia erecta (Roth ex Schult.) Tuyn in Steenis, Fl. Males. (ser. 1) 6: 113 (1960).
Nigeria to Angola, Zimbabwe & Ethiopia; India; Sumatera. 22 NGA 24 ETH SUD 25 KEN
TAN 26 ANG MLW ZAM ZIM 40 ASS IND 42 SUM. Ther.
   * *Dentella erecta* Roth ex Schult. in Roem. & Schult., Syst. Veg. 5: 25 (1819).
     Cephalostigma erectum (Roth ex Schult.) Vatke, Linnaea 38: 699 (1874).
   Dentella perotifolia Willd. ex Schult. in Roem. & Schult., Syst. Veg. 5: 25 (1819).
     Wahlenbergia perotifolia (Willd. ex Schult.) Wight & Arn., Prodr. Fl. Ind. Orient.: 405
     (1834). *Cephalostigma perotifolium* (Willd. ex Schult.) Hutch. & Dalziel, Fl. W. Trop.
     Afr. 2: 190 (1931). *Lightfootia perotifolia* (Willd. ex Schult) E. Wimm. ex Agnew,
     Upland Kenya Wildflowers: 511 (1974).
   Cephalostigma schimperi Hochst. ex A. Rich., Tent. Fl. Abyss. 2: 2 (1850). *Wahlenbergia
     schimperi* (A. Rich.) Schweinf. & Asch., Beitr. Fl. Aethiop. 1: 282 (1867).
   Lightfootia arenaria A. DC., Ann. Sci. Nat., Bot. (ser. 5) 6: 329 (1866).

Wahlenbergia ericoidella (P. A. Duvign. & Denaeyer) Thulin, Symb. Bot. Upsal. 21:
185 (1975).
Zaïre. 23 ZAI. Hemicr.
   * *Lightfootia ericoidella* P. A. Duvign. & Denaeyer, Bull. Soc. Roy. Bot. Belgique 96:
     132 (1963).

Wahlenbergia erophiloides Markgr., Notizbl. Bot. Gart. Berlin-Dahlem 15: 468 (1941).
Namibia. 27 NAM. Ther.
   Wahlenbergia pavida Launert, Mitt. Bot. Staatssamml. München 2: 306 (1957).

Wahlenbergia exilis A. DC., Monogr. Campan.: 151 (1830). *Campanula exilis* (A. DC.) D.
Dietr., Syn. Pl. 1: 754 (1839). *Campanopsis exilis* (A. DC.) Kuntze, Revis. Gen. Pl. 2:
379 (1891).
Cape Provinces. 27 CPP. Ther.
   Lobelia paniculata L., Sp. Pl.: 930 (1753); non *Wahlenbergia paniculata* (L. f.) A. DC.,
     Monogr. Campan.: 153 (1830).
   Wahlenbergia exilis var. *major* Sond. in Harv. & Sond., Fl. Cap. 3: 577 (1865).

Wahlenbergia fasciculata Brehmer, Bot. Jahrb. Syst. 53: 92 (1915).
S. Africa. 27 CPP LES NAT OFS. Cham.
Wahlenbergia fasciculata var. *pilosa* Brehmer, Bot. Jahrb. Syst. 53: 92 (1915).

Wahlenbergia fernandeziana A. DC., Monogr. Campan.: 160 (1830). *Campanula fernandeziana* (A. DC.) D. Dietr., Syn. Pl. 1: 754 (1839). *Campanopsis fernandeziana* (A. DC.) Kuntze, Revis. Gen. Pl. 2: 379 (1891).
Juan Fernández Is. (Masatierra). 85 JNF. Nanophan. $2n = 22$.
Campanula larrainii Bertero in Colla, Mem. Reale Accad. Sci. Torino 38: 118 (1835), as 'larraini'. *Wahlenbergia larrainii* (Bertero) Skottsb., Nat. Hist. Juan Fernandez 2: 176 (1921), as 'larraini'.
Wahlenbergia fernandeziana f. *elata* Skottsb., Nat. Hist. Juan Fernandez 2: 177 (1921).

Wahlenbergia filipes Brehmer, Bot. Jahrb. Syst. 53: 89 (1915).
Cape Provinces. 27 CPP. Ther.
Wahlenbergia filipes var. *dentata* Brehmer, Bot. Jahrb. Syst. 53: 89 (1915).

Wahlenbergia fistulosa Brehmer, Bot. Jahrb. Syst. 53: 119 (1915).
KwaZulu-Natal. 27 NAT. Hemicr.?

Wahlenbergia flexuosa (Hook. f. & Thomson) Thulin, Symb. Bot. Upsal. 21: 158 (1975).
Nigeria to Ethiopia & Malawi; Oman; S. India (Karnataka, Maharashtra). 22 NGA 23 CAF CMN ZAI 24 ETH 25 TAN UGA 26 MLW 35 OMA 40 IND. Ther.
* *Cephalostigma flexuosum* Hook. f. & Thomson, J. Proc. Linn. Soc., Bot. 2: 9 (1858).
Cephalostigma erectum var. *luteum* Chiov., Ann. Bot. (Rome) 9: 79 (1911).

Wahlenbergia floribunda Schltr. & Brehmer, Bot. Jahrb. Syst. 53: 136 (1915).
Cape Provinces. 27 CPP. Hemicr.?

Wahlenbergia fluminalis (J. M. Black) E. Wimm. ex H. Eichler, Taxon 12: 297 (1963), as 'fluminale'.
E. Australia. 50 NSW QLD SOA VIC. Hemicr. $2n = 36$.
* *Cephalostigma fluminale* J. M. Black, Trans. & Proc. Roy. Soc. South Australia 58: 184 (1934).

Wahlenbergia fruticosa Brehmer, Bot. Jahrb. Syst. 53: 98 (1915).
Cape Provinces. 27 CPP. Nanophan. or cham.

Wahlenbergia galpiniae Schltr., J. Bot. 35: 343 (1897).
Cape Provinces. 27 CPP. Ther.?
Wahlenbergia galpiniae var. *excedens* Brehmer, Bot. Jahrb. Syst. 53: 121 (1915).

Wahlenbergia glabra P. J. Sm., Telopea 5: 113 (1992).
E. Australia (McPherson Range). 50 NSW QLD. Hemicr. $2n = 18$.

Wahlenbergia glandulifera Brehmer, Bot. Jahrb. Syst. 53: 117 (1915).
KwaZulu-Natal. 27 NAT. Hemicr.?

Wahlenbergia globularis E. Wimm. in J. F. Macbr., Fl. Peru 6: 386 (1937).
Peru. 83 PER. Hemicr.

Wahlenbergia gloriosa Lothian, Proc. Linn. Soc. New South Wales 71: 224 (1947).
SE. Australia. 50 NSW VIC. Geophyte. $2n = 18$.
Wahlenbergia marginata var. *grandiflora* Tuyn in Steenis, Fl. Males. (ser. 1) 6: 118 (1960).

Wahlenbergia gracilenta Lothian, Proc. Linn. Soc. New South Wales 71: 217 (1947).
S. Australia. 50 NSW SOA TAS VIC WAU. Ther. $2n = 18, 36$.

Wahlenbergia gracilis (G. Forst.) A. DC., Monogr. Campan.: 142 (1830).
Norfolk I., Lord Howe I., New Caledonia, Tonga ('Eua). 50 NFK 60 NWC TON. Hemicr.
$2n = 72$.
* *Campanula gracilis* G. Forst., Fl. Ins. Austr.: 15 (1786). *Campanula vinciflora* Vent., Jard.
Malmaison: 12 (1803), as 'vincaeflora'. *Wahlenbergia vinciflora* Decne., Rev. Hort. (ser.
3) 3: 41 (1849), as 'vincaeflora'. *Lightfootia gracilis* (G. Forst.) Miq., Fl. Ned. Ind. 2:
567 (1857). *Cervicina gracilis* (G. Forst.) Britten in Banks & Sol., Ill. Austral. Pl. Cook's
Voy.: 56 (1901).
Wahlenbergia marginata var. *neocaledonica* Lothian, Proc. Linn. Soc. New South Wales
71: 214 (1947).

Wahlenbergia grahamiae Hemsl., Rep. Challenger, Bot. 1(3): 46 (1884), as 'grahamae'.
Juan Fernández Is. (Masatierra). 85 JNF. Nanophan.

Wahlenbergia grandiflora Brehmer, Bot. Jahrb. Syst. 53: 115 (1915).
S. Africa. 27 CPP NAT TVL. Cham.?
Wahlenbergia grandiflora var. *fissa* Brehmer, Bot. Jahrb. Syst. 53: 116 (1915).
Wahlenbergia grandiflora var. *lanceolata* Brehmer, Bot. Jahrb. Syst. 53: 116 (1915).
Wahlenbergia grandiflora var. *lata* Brehmer, Bot. Jahrb. Syst. 53: 116 (1915).
Wahlenbergia grandiflora var. *undulata* Brehmer, Bot. Jahrb. Syst. 53: 117 (1915).

Wahlenbergia graniticola Carolin, Proc. Linn. Soc. New South Wales 89: 239 (1964).
E. Australia. 50 NSW QLD VIC. Hemicr. $2n = 18, 36$.

Wahlenbergia hederacea (L.) Rchb., Icon. Bot. Pl. Crit. 5: 47 (1827).
British Isles; Portugal to Germany. 10 GRB IRE 11 BGM GER NET 12 FRA POR SPA.
Hemicr. $2n = 36$.
* *Campanula hederacea* L., Sp. Pl.: 169 (1753). *Campanula hederifolia* Salisb., Prodr. Stirp.
Chap. Allerton: 127 (1796), as 'hederaefolia'. *Schultesia hederacea* (L.) Roth, Enum. Pl.
Phaen. Germ. 1: 690 (1827). *Roucela hederacea* (L.) Dumort., Fl. Belg.: 58 (1827).
Campanula pentagonophylla Vuk., Linnaea 26: 325 (1854). *Valvinterlobus filiformis*
Dulac, Fl. Hautes-Pyrénées: 459 (1867). *Aikinia hederacea* (L.) Fourr., Ann. Soc. Linn.
Lyon (n.s.) 17: 112 (1869). *Campanopsis hederacea* (L.) Kuntze, Revis. Gen. Pl. 2: 379
(1891). *Cervicina hederacea* (L.) Druce, Fl. Berkshire: 324 (1898). *Wahlenbergia
hederifolia* Bubani, Fl. Pyren. 2: 18 (1899).

Wahlenbergia hirsuta (Edgew.) Tuyn in Steenis, Fl. Males. (ser. 1) 6: 113 (1960).
Senegal to Namibia, Zimbabwe & Ethiopia; Madagascar; Comoros; Yemen; Himalaya &
NE. India (Bihar, Uttar Pradesh). 22 GHA NGA SEN 23 BUR CAF CMN RWA ZAI 24
SUD ETH 25 KEN TAN UGA 26 ANG MLW ZAM ZIM 27 NAM 29 COM MDG 35 YEM
40 EHM IND NEP WHM. Ther. $2n = 42$.
* *Cephalostigma hirsutum* Edgew., Trans. Linn. Soc. London 20: 81 (1846), as 'hirsuta'.
Lightfootia hirsuta (Edgew.) E. Wimm. ex Hepper, Kew Bull. 15: 61 (1961).
Cephalostigma erectum var. *coeruleum* Chiov., Ann. Bot. (Rome) 9: 79 (1911).

Wahlenbergia hookeri (C. B. Clarke) Tuyn in Steenis, Fl. Males. (ser. 1) 6: 114 (1960).
Cameroon to Ethiopia & Tanzania; India to SC. China (Yunnan) & Thailand; Jawa. 23
BUR CMN ZAI 24 ETH 25 KEN TAN 36 CHC-YN 40 ASS IND 41 THA 42 JAW. Ther.
* *Cephalostigma hookeri* C. B. Clarke in Hook. f., Fl. Brit. India 3: 429 (1881).

Wahlenbergia humbertii Thulin, Symb. Bot. Upsal. 21: 182 (1975).
Madagascar. 29 MDG. Hemicr.

Wahlenbergia huttonii (Sond.) Thulin, Symb. Bot. Upsal. 21: 129 (1975).
Burundi & Tanzania to S. Africa. 23 BUR 25 TAN 26 MLW 27 CPP NAT SWZ TVL. Cham.
or hemicr.

Lightfootia huttonii Sond. in Harv. & Sond., Fl. Cap. 3: 556 (1865), as 'huttoni'.
  Lightfootia corymbosa Kuntze, Revis. Gen. Pl. 3(2): 188 (1898).
  Lightfootia lycopodioides Mildbr., Notizbl. Bot. Gart. Berlin-Dahlem 11: 686 (1932); non
    A. DC., Monogr. Campan.: 114 (1830).

Wahlenbergia ingrata A. DC. in DC., Prodr. 7: 432 (1839). *Campanopsis ingrata* (A. DC.)
  Kuntze, Revis. Gen. Pl. 2: 379 (1891).
  Cape Provinces. 27 CPP. Hemicr.

Wahlenbergia insulae-howei Lothian, Proc. Linn. Soc. New South Wales 71: 234 (1947).
  Lord Howe I. 50 NFK-LH. Hemicr. or geophyte.
    Wahlenbergia limenophylax Lothian, Proc. Linn. Soc. New South Wales 71: 233 (1947),
      as 'limnophalyx'.

Wahlenbergia intermedia Zahlbr., Verh. K. K. Zool.-Bot. Ges. Wien 49: 518 (1899).
  SE. Brazil (Minas Gerais). 84 BZL. Hemicr.

Wahlenbergia islensis P. J. Sm., Telopea 5: 114 (1992).
  NE. Australia (Queensland). 50 QLD. Hemicr. $2n = 18$.

Wahlenbergia juncea (H. Buek) Lammers, Taxon 44: 336 (1995).
  Cape Provinces. 27 CPP. Cham. or hemicr.
    * *Prismatocarpus junceus* H. Buek in Eckl. & Zeyh., Enum. Pl. Afric. Austral.: 383 (1837).
      Campanula juncea (H. Buek) D. Dietr., Syn. Pl. 1: 756 (1839); non Hill, Syst. Veg. 8:
      14 (1765). *Lightfootia juncea* (H. Buek) Sond. in Harv. & Sond., Fl. Cap. 3: 563 (1865).
    Wahlenbergia spicata E. Mey. ex A. DC. in DC., Prodr. 7: 441 (1839).

Wahlenbergia kowiensis R. A. Dyer, Bull. Misc. Inform. Kew 1934: 265 (1934).
  Cape Provinces. 27 CPP. Ther.

Wahlenbergia krebsii Cham., Linnaea 8: 195 (1833). *Campanula krebsii* (Cham.) D. Dietr.,
  Syn. Pl. 1: 753 (1839). *Campanopsis krebsii* (Cham.) Kuntze, Revis. Gen. Pl. 3(2): 185 (1898).
  Bioko; Cameroon to Ethiopia & Tanzania; S. Africa. 23 BUR CMN EQG GGI RWA ZAI 24
  ETH 25 KEN TAN UGA 27 CPP LES NAT OFS SWZ TVL. Hemicr. $2n = 14$.

subsp. **arguta** (Hook. f.) Thulin, Symb. Bot. Upsal. 21: 99 (1975).
  Bioko; Cameroon to Ethiopia & Tanzania. 23 BUR CMN EQG GGI RWA ZAI 24 ETH 25
  KEN TAN UGA. Hemicr. $2n = 14$.
    * *Wahlenbergia arguta* Hook. f., J. Proc. Linn. Soc., Bot. 6: 15 (1862). *Campanopsis arguta*
      (Hook. f.) Kuntze, Revis. Gen. Pl. 2: 379 (1891).
    Lightfootia arabidifolia Engl., Bot. Jahrb. Syst. 19 (Beibl. 47): 53 (1894). *Wahlenbergia*
      arabidifolia (Engl.) Brehmer, Bot. Jahrb. Syst. 53: 99 (1915).
    Campanopsis krebsii var. *transvalica* Kuntze, Revis. Gen. Pl. 3(2): 185 (1898).
    Wahlenbergia coerulea H. J. P. Winkl., Bot. Jahrb. Syst. 41: 285 (1908).
    Wahlenbergia arguta var. *parvilocula* Brehmer, Bot. Jahrb. Syst. 53: 100 (1915).
    Wahlenbergia arguta var. *longifusiformis* Brehmer, Bot. Jahrb. Syst. 53: 100
    Wahlenbergia sarmentosa T. C. E. Fr., Notizbl. Bot. Gart. Berlin-Dahlem 8: 396 (1923).

subsp. **krebsii**
  S. Africa. 27 CPP LES NAT OFS SWZ TVL. Hemicr.
    Wahlenbergia zeyheri H. Buek in Eckl. & Zeyh., Enum. Pl. Afric. Austral.: 379 (1837).
      Campanula zeyheri (H. Buek) D. Dietr., Syn. Pl. 1: 753 (1839). *Campanopsis zeyheri* (H.
      Buek) Kuntze, Revis. Gen. Pl. 2: 379 (1891).
    Wahlenbergia variabilis E. Mey. ex A. DC. in DC., Prodr. 7: 436 (1839).
    Wahlenbergia variabilis var. *krebsiana* A. DC. in DC., Prodr. 7: 436 (1839). *Wahlenbergia*
      zeyheri var. *krebsiana* (A. DC.) Sond. in Harv. & Sond., Fl. Cap. 3: 580 (1865).
      Wahlenbergia zeyheri var. *linearis* Sond. in Harv. & Sond., Fl. Cap. 3: 580 (1865).

Wahlenbergia zeyheri var. *natalensis* Sond. in Harv. & Sond., Fl. Cap. 3: 580 (1865).
Wahlenbergia zeyheri var. *pyriformis* Brehmer, Bot. Jahrb. Syst. 53: 102 (1915).
Wahlenbergia zeyheri var. *lanceolata* Brehmer, Bot. Jahrb. Syst. 53: 102 (1915).

Wahlenbergia lasiocarpa Schltr. & Brehmer, Bot. Jahrb. Syst. 53: 113 (1915).
Cape Provinces. 27 CPP. Hemicr.

Wahlenbergia laxiflora (Sond.) Lammers, Taxon 44: 336 (1995).
Cape Provinces. 27 CPP. Cham. or hemicr.
 * *Lightfootia laxiflora* Sond. in Harv. & Sond., Fl. Cap. 3: 564 (1865).

Wahlenbergia levynsiae Lammers, Taxon 44: 336 (1995).
Cape Provinces. 27 CPP. Nanophan. or cham.
 * *Lightfootia squarrosa* Adamson, J. S. African Bot. 21: 189 (1955); non *Wahlenbergia squarrosa* Brehmer, Bot. Jahrb. Syst. 53: 131 (1915).

Wahlenbergia linarioides (Lam.) A. DC., Monogr. Campan.: 158 (1830).
Ecuador to SE. Brazil (Minas Gerais), Argentina (Rio Negro), and S. Chile (Llanquihue); naturalized in SE. U.S.A. (Florida). (78) fla 83 BOL ECU PER 84 BZL BZS 85 AGE AGS AGW CLC CLS PAR URU. Cham. or hemicr.
 * *Campanula linarioides* Lam., Encycl. 1: 580 (1785). *Campanopsis linarioides* (Lam.) Kuntze, Revis. Gen. Pl. 2: 378 (1891), as 'linariodes'.
 Campanula arida Kunth in Humb., Bonpl. & Kunth, Nov. Gen. Sp. 3: 301 (quarto), 235 (folio) (1819). *Wahlenbergia linarioides* var. *arida* (Kunth) A. DC., Monogr. Campan.: 158 (1830). *Wahlenbergia arida* (Kunth) Griseb., Abh. Königl. Ges. Wiss. Göttingen 19: 200 (1874).
 Campanula filiformis Ruiz & Pav., Fl. Peruv. 2: 55 (1799). *Campanula chilensis* Mol., Sag. Stor. Nat. Chili (ed. 2): 281 (1810). *Wahlenbergia linarioides* var. *filiformis* (Ruiz & Pav.) A. DC., Monogr. Campan.: 158 (1830).
 Breweria linifolia Spreng., Syst. Veg. 1: 614 (1824).
 Lobelia megapotamica Spreng., Syst. Veg. 4(2): 75 (1827).
 Wahlenbergia linarioides var. *latifolia* A. DC. in DC., Prodr. 7: 441 (1839).
 Wahlenbergia linarioides var. *macrantha* Phil., Linnaea 33: 171 (1864).
 Wahlenbergia linarioides var. *micrantha* Phil., Linnaea 33: 172 (1864).
 Campanopsis linarioides var. *longifolia* Kuntze, Revis. Gen. Pl. 3(2): 185 (1891).

Wahlenbergia linifolia (Roxb.) A. DC., Monogr. Campan.: 162 (1830).
St. Helena. 28 STH. Nanophan.
 * *Roella linifolia* Roxb. in Beatson, Tracts St. Helena: 321 (1816). *Campanopsis linifolia* (Roxb.) Kuntze, Revis. Gen. Pl. 2: 379 (1891).
 Wahlenbergia foliosa A. DC., Monogr. Campan.: 160 (1830). *Campanula st.-helenae* D. Dietr., Syn. Pl. 1: 755 (1839); non *Campanula foliosa* Ten., Fl. Napol. 1: 71 (1811).

Wahlenbergia littoralis (Labill.) Sweet, Hort. Brit. (ed. 2): 593 (1830).
SE. Australia. 50 SOA TAS VIC. Geophyte. $2n = 18$.
 * *Campanula littoralis* Labill., Nov. Holl. Pl. 1: 49 (1805*). Campanula gracilis* var. *littoralis* (Labill.) R. Br., Prodr.: 561 (1810). *Wahlenbergia gracilis* var. *littoralis* (Labill.) A. DC., Monogr. Campan.: 142 (1830). *Wahlenbergia gracilis* f. *littoralis* (Labill.) Wawra, Itin. Princ. S. Coburgi 1: 133 (1883). *Wahlenbergia vinciflora* var. *littoralis* (Labill.) N. E. Br., Gard. Chron. (ser. 3) 54: 336 (1913). *Wahlenbergia marginata* var. *littoralis* (Labill.) Hochr., Candollea 5: 291 (1934). *Wahlenbergia billardieri* Lothian, Proc. Linn. Soc. New South Wales 71: 226 (1947).
 Wahlenbergia gymnoclada Lothian, Proc. Linn. Soc. New South Wales 71: 227 (1947).

Wahlenbergia littoricola P. J. Sm. in Jessop & Toelken, Fl. S. Australia 3: 1380 (1986), as 'litticola'.
S. Australia. 50 NSW NSW SOA TAS VIC WAU. Hemicr. $2n = 54, 72$.

Wahlenbergia lobelioides (L. f.) Schrad. ex Link, Handbuch 1: 632 (1829).
Macaronesia; S. Europe to trop. W. Africa; Socotra; Saudi Arabia. 12 SAR SPA 13 ITA SIC 20
ALG EGY MOR TUN 21 CNY CVI MDR SEL 22 GNB GUI MLI NGA SEN 23 CMN 24
CHA ETH SOC SUD 25 KEN 35 SAU. Ther. $2n = 18$.
* *Campanula lobelioides* L. f., Suppl. Pl.: 140 (1782). *Campanula parviflora* Salisb., Prodr.
Stirp. Chap. Allerton: 126 (1796); non Lam., Encycl. 1: 588 (1785). *Wahlenbergia
pendula* Schrad., Blumenbachia: 38 (1827). *Hecale lobelioides* (L. f.) Raf., Fl. Tellur. 2:
79 (1837). *Campanopsis lobelioides* (L. f.) Kuntze, Revis. Gen. Pl. 2: 379 (1891).
Cervicina pendula Druce, Bot. Soc. Exch. Club Brit. Isles 6: 32 (1921). *Cervicina
lobelioides* (L. f.) Druce, Bot. Soc. Exch. Club Brit. Isles 8: 543, 547 (1928).

subsp. **lobelioides**
Madeira, Selvagens, Canary Is., Cape Verde. 21 CNY CVI MDR SEL. Ther. $2n = 18$.
Lobelia broussonetia Bory, Ess. Isl. Fortunées: 330 (1803).
Wahlenbergia lobelioides var. *macilenta* Webb & Berth., Hist. Nat. Iles Canaries 2(3):
4 (1844).

subsp. **nutabunda** (Guss.) Murb., Acta Univ. Lund. 33(12): 115 (1897).
Italy to Spain, Morocco & Ethiopia; Socotra; Saudi Arabia. 12 SAR SPA 13 ITA SIC 20
ALG EGY MOR TUN 24 ETH SOC SUD 35 SAU. Ther.
* *Campanula nutabunda* Guss. in Ten., Fl. Neapol. Prodr. (App. 5): 8 (1826). *Wahlenbergia
nutabunda* (Guss.) A. DC., Monogr. Campan.: 151 (1830). *Wahlenbergia lobelioides*
var. *gussonei* Webb & Berth., Hist. Nat. Iles Canaries 2(3): 4 (1844).
Laurentia etbaica Schweinf., Verh. K. K. Zool.-Bot. Ges. Wien 18: 683 (1868).
Wahlenbergia etbaica (Schweinf.) Vatke, Linnaea 38: 700 (1874). *Campanopsis etbaica*
(Schweinf.) Kuntze, Revis. Gen. Pl. 2: 379 (1891).
Lobelia minutiflora Pau, Bol. Soc. Aragonesa Ci. Nat. 1: 30 (1902); non Kunze, Linnaea
16: 318 (1842), as 'minutiflorum'.
Wahlenbergia nutabunda var. *erythraeae* Chiov., Ann. Bot. (Rome) 10: 390 (1912).
Wahlenbergia riparia var. *etbaica* Brehmer, Bot. Jahrb. Syst. 53: 110 (1915).

subsp. **riparia** (A. DC.) Thulin, Symb. Bot. Upsal. 21: 94 (1975).
Senegal to Chad & Cameroon. 22 GNB GUI MLI NGA SEN 23 CMN 24 CHA. Ther.
* *Wahlenbergia riparia* A. DC., Monogr. Campan.: 146 (1830). *Campanula riparia* (A. DC.)
Leprieur & Perottet ex D. Dietr., Syn. Pl. 1: 753 (1839). *Campanopsis riparia* (A. DC.)
Kuntze, Revis. Gen. Pl. 2: 379 (1891).
Wahlenbergia humilis A. DC., Monogr. Campan.: 147 (1830). *Campanula humilis* (A.
DC.) D. Dietr., Syn. Pl. 1: 753 (1839). *Campanopsis humilis* (A. DC.) Kuntze, Revis.
Gen. Pl. 2: 379 (1891).
Wahlenbergia riparia var. *clavata* Brehmer, Bot. Jahrb. Syst. 53: 110 (1915).
Wahlenbergia riparia var. *segregata* Brehmer, Bot. Jahrb. Syst. 53: 110 (1915).
Wahlenbergia riparia var. *virgulta* Brehmer, Bot. Jahrb. Syst. 53: 110 (1915).

Wahlenbergia lobulata Brehmer, Bot. Jahrb. Syst. 53: 87 (1915).
Cape Provinces. 27 CPP. Cham.?
Wahlenbergia galpinii E. Phillips, Ann. S. African Mus. 16: 182 (1917).
Wahlenbergia monotropa Killick, Bothalia 8: 164 (1964).

Wahlenbergia longifolia (A. DC.) Lammers, Taxon 44: 336 (1995).
Cape Provinces. 27 CPP. Nanophan. or cham.
* *Lightfootia longifolia* A. DC., Monogr. Campan.: 108 (1830).
Lightfootia longifolia var. *lanuginosa* Cham., Linnaea 8: 193 (1833).
Lightfootia longifolia var. *oppositifolia* Sond. in Harv. & Sond., Fl. Cap. 3: 558 (1865).
Lightfootia erecta R. D. Good, J. Bot. 62: 48 (1924).
Lightfootia longifolia var. *corymbosa* Adamson, J. S. African Bot. 21: 177 (1955).
Wahlenbergia longifolia var. *corymbosa* (Adamson) Welman, Bothalia 26:
157 (1996).

Wahlenbergia longisquamifolia Brehmer, Bot. Jahrb. Syst. 53: 119 (1915).
Cape Provinces. 27 CPP. Hemicr.?

Wahlenbergia luteola P. J. Sm. in Jessop & Toelken, Fl. S. Australia 3: 1380 (1986).
SE. Australia. 50 NSW SOA VIC. Hemicr. $2n = 18, 36$.

Wahlenbergia lycopodioides Schltr. & Brehmer, Bot. Jahrb. Syst. 53: 91 (1915).
S. Africa. 27 LES NAT OFS TVL. Ther.?

Wahlenbergia macrostachys (A. DC.) Lammers, Taxon 44: 336 (1995).
Cape Provinces. 27 CPP. Nanophan. or cham.
   Lightfootia spicata H. Buek in Eckl. & Zeyh., Enum. Pl. Afric. Austral.: 374 (1837); non
      Wahlenbergia spicata E. Mey. ex A. DC. in DC., Prodr. 7: 441 (1839).
   * *Lightfootia macrostachys* A. DC. in DC., Prodr. 7: 787 (1839).

Wahlenbergia madagascariensis A. DC., Monogr. Campan.: 139 (1830). *Campanula
madagascariensis* (A. DC.) D. Dietr., Syn. Pl. 1: 752 (1839). *Campanopsis madagascariensis*
(A. DC.) Kuntze, Revis. Gen. Pl. 2: 379 (1891).
Malawi to S. Africa; Madagascar. 26 MLW MOZ ZIM 27 CPP NAT SWZ TVL 29 MDG.
   Hemicr.
   Wahlenbergia hilsenbergii A. DC. in DC., Prodr. 7: 429 (1839). *Campanopsis hilsenbergii*
      (A. DC.) Kuntze, Revis. Gen. Pl. 2: 379 (1891).
   Wahlenbergia oppositifolia A. DC. in DC., Prodr. 7: 429 (1839). *Campanopsis oppositifolia*
      (A. DC.) Kuntze, Revis. Gen. Pl. 2: 379 (1891). *Campanopsis procumbens* var.
      oppositifolia (A. DC.) Kuntze, Revis. Gen. Pl. 3(2): 185 (1898).
   Wahlenbergia oppositifolia var. *crispa* A. DC. in DC., Prodr. 7: 429 (1839).
   Wahlenbergia oppositifolia var. *crenata* Brehmer, Bot. Jahrb. Syst. 53: 134 (1915).

Wahlenbergia magaliesbergensis Lammers, Taxon 44: 336 (1995).
Northern Provinces. 27 TVL. Hemicr.
   * *Lightfootia paniculata* Sond. in Harv. & Sond., Fl. Cap. 3: 560 (1865); non *Wahlenbergia
      paniculata* (L. f.) A. DC., Monogr. Campan.: 153 (1830).

Wahlenbergia malaissei Thulin, Nordic J. Bot. 7: 265 (1987).
Zaïre. 23 ZAI. Ther.

Wahlenbergia marginata (Thunb. ex Murray) A. DC., Monogr. Campan.: 143 (1830).
India to Japan, Nansei-shoto, Taiwan, Philippines & New Guinea; naturalized in SE.
   U.S.A. & Hawaiian Is. 36 CHS 38 JAP NNS KOR TAI 40 ASS EHM IND NEP SRL WHM
   41 LAO MYA VIE 42 JAW LSI MOL PHI SUL 43 NWG (63) haw (77) tex (78) ala ark fla
   geo lou msi nca sca. Hemicr. $2n = 18, 36, 54, 72, 90$.
   * *Campanula marginata* Thunb. ex Murray, Syst. Veg. (ed. 14): 211 (1784). *Campanopsis
      marginata* (Thunb. ex Murray) Kuntze, Revis. Gen. Pl. 2: 378 (1891).
   Lobelia campanuloides Thunb., Trans. Linn. Soc. London 2: 331 (1794).
   Campanula dehiscens Roxb., Asiat. Res. 12: 571 (1819). *Wahlenbergia dehiscens* (Roxb.) A.
      DC., Monogr. Campan.: 145 (1830).
   Campanula agrestis Wall. in Roxb., Fl. Ind. 2: 97 (1824). *Wahlenbergia agrestis* (Wall.) A.
      DC., Monogr. Campan.: 145 (1830).
   Campanula lavandulifolia Reinw. ex Blume, Bijdr.: 726 (1826). *Wahlenbergia
      lavandulifolia* (Reinw. ex Blume) A. DC., Monogr. Campan.: 144 (1830), as
      'lavandulaefolia'. *Lightfootia gracilis* var. *lavandulifolia* (Reinw. ex Blume) Miq., Fl.
      Ned. Ind. 2: 567 (1857), as 'lavandulaefolia'.
   Wahlenbergia indica A. DC., Monogr. Campan.: 146 (1830). *Campanula indica* (A. DC.)
      D. Dietr., Syn. Pl. 1: 753 (1839).
   Wahlenbergia gracilis var. *misera* Hemsl., J. Linn. Soc., Bot. 26: 4 (1889).
   Campanopsis marginata var. *rigida* Kuntze, Revis. Gen. Pl. 2: 378 (1891).

Campanopsis marginata var. *polymorpha* Kuntze, Revis. Gen. Pl. 2: 378 (1891).
Wahlenbergia bivalvis Merr., Philipp. J. Sci. 1(Suppl.): 242 (1906).

Wahlenbergia marunguensis Thulin, Nordic J. Bot. 7: 264 (1987).
Zaïre. 23 ZAI. Hemicr.

Wahlenbergia masafuerae (Phil.) Skottsb., Kongl. Svenska Vitenskapsakad. Handl. (n.s.) 51(9): 6 (1914).
Juan Fernández Is. (Masafuera). 85 JNF. Cham. or hemicr. $2n = 22$.
* *Euphorbia masafuerae* Phil., Bot. Zeitung (Berlin) 14: 647 (1856).
Wahlenbergia masafuerae f. *rosea* Skottsb., Nat. Hist. Juan Fernandez 2: 178 (1921).

Wahlenbergia massonii A. DC., Monogr. Campan.: 153 (1830). *Campanula massonii* (A. DC.) D. Dietr., Syn. Pl. 1: 754 (1839). *Wahlenbergia paniculata* var. *massonii* (A. DC.) Sond. in Harv. & Sond., Fl. Cap. 3: 575 (1865).
Cape Provinces. 27 CPP. Ther.

Wahlenbergia matthewsii Cockayne, Trans. & Proc. New Zealand Inst. 47: 113 (1915).
New Zealand (South I.). 51 NZS. Cham. or hemicr. $2n = 36$.

Wahlenbergia meyeri A. DC. in DC., Prodr. 7: 439 (1839). *Campanopsis meyeri* (A. DC.) Kuntze, Revis. Gen. Pl. 2: 379 (1891).
Cape Provinces. 27 CPP. Ther.
Wahlenbergia meyeri var. *lanceolata* Brehmer, Bot. Jahrb. Syst. 53: 127 (1915).
Wahlenbergia meyeri var. *subacaulis* Brehmer, Bot. Jahrb. Syst. 53: 127 (1915).

Wahlenbergia microphylla (Adamson) Lammers, Taxon 44: 336 (1995).
Cape Provinces. 27 CPP. Nanophan. or cham.
* *Lightfootia microphylla* Adamson, J. S. African Bot. 21: 201 (1955).

Wahlenbergia minuta Brehmer, Bot. Jahrb. Syst. 53: 79 (1915).
Cape Provinces. 27 CPP. Hemicr.?

Wahlenbergia mollis Brehmer, Bot. Jahrb. Syst. 53: 74 (1915).
Cape Provinces. 27 CPP. Hemicr.?

Wahlenbergia multicaulis Benth. in Endl., Enum. Pl.: 75 (1837).
S. Australia. 50 NSW SOA TAS VIC WAU. Hemicr. $2n = 36, 54, 72$.
Wahlenbergia simplicicaulis de Vriese in Lehm., Pl. Preiss. 2: 241 (1848)
Lobelia dubia de Vriese in Lehm., Pl. Preiss. 2: 242 (1848).
Wahlenbergia tagdellii Lothian, Proc. Linn. Soc. New South Wales 71: 228 (1947).

Wahlenbergia namaquana Sond. in Harv. & Sond., Fl. Cap. 3: 582 (1865). *Campanopsis namaquana* (Sond.) Kuntze, Revis. Gen. Pl. 2: 379 (1891).
Cape Provinces. 27 CPP. Ther.

Wahlenbergia nana Brehmer, Bot. Jahrb. Syst. 53: 135 (1915).
S. Africa. 27 NAM OFS. Ther.

Wahlenbergia napiformis (A. DC.) Thulin, Symb. Bot. Upsal. 21: 133 (1975).
Central African Rep. to Ethiopia, Mozambique & Namibia. 23 BUR CAF CON RWA ZAI 24 ETH SUD 25 KEN TAN UGA 26 ANG MLW MOZ ZAM ZIM 27 BOT NAM. Hemicr. or biennial. $2n = 16$.
Lightfootia marginata A. DC., Ann. Sci. Nat., Bot. (ser. 5) 6: 326 (1866); non *Wahlenbergia marginata* (Thunb. ex Murray) A. DC., Monogr. Campan.: 143 (1830).
* *Lightfootia napiformis* A. DC., Ann. Sci. Nat., Bot. (ser. 5) 6: 328 (1866).

Lightfootia glomerata Engl., Bot. Jahrb. Syst. 19 (Beibl. 47): 52 (1894).
Lightfootia abyssinica var. *glaberrima* Engl., Pflanzenw. Ost-Afrikas C: 400 (1895).
Lightfootia abyssinica var. *cinerea* Engl. & Gilg in Warb., Kunene-Sambesi-Exped.: 397 (1903).
Lightfootia campestris Engl. in Mildbr., Wiss. Erg. Deut. Zentr.-Afr.-Exped., Bot.: 343 (1911).
Lightfootia kagerensis S. Moore, J. Linn. Soc., Bot. 37: 176 (1905).
Lightfootia graminicola M. B. Scott, Bull. Misc. Inform. Kew 1915: 45 (1915).
Lightfootia marginata var. *lucens* Lambinon, Bull. Soc. Roy. Bot. Belgique 93: 46 (1961).

Wahlenbergia neorigida Lammers, Taxon 44: 337 (1995).
Cape Provinces. 27 CPP. Nanophan. or cham.
* *Lightfootia rigida* Adamson, J. S. African Bot. 21: 199 (1955); non *Wahlenbergia rigida* Bernh., Flora 27: 820 (1844).

Wahlenbergia neostricta Lammers, Taxon 44: 337 (1995).
Cape Provinces. 27 CPP. Nanophan. or cham.
* *Lightfootia stricta* Adamson, J. S. African Bot. 21: 179 (1955); non *Wahlenbergia stricta* (Sm. ex R. Br.) Sweet, Hort. Brit. (ed. 2): 593 (1830).

Wahlenbergia nodosa (H. Buek) Lammers, Taxon 44: 337 (1995).
S. Africa. 27 CPP OFS. Nanophan. or cham.
* *Lightfootia nodosa* H. Buek in Eckl. & Zeyh., Enum. Pl. Afric. Austral.: 376 (1837).
Lightfootia tenella var. *longivalvis* A. DC. in DC., Prodr. 7: 418 (1839).
Lightfootia tenella var. *rigida* Sond. in Harv. & Sond., Fl. Cap. 3: 562 (1865).
Lightfootia tenella var. *microphylla* Sond. in Harv. & Sond., Fl. Cap. 3: 562 (1865).

Wahlenbergia obovata Brehmer, Bot. Jahrb. Syst. 53: 111 (1915).
Cape Provinces. 27 CPP. Ther.
* *Wahlenbergia arenaria* var. *lasiocarpa* Sond. in Harv. & Sond., Fl. Cap. 3: 581 (1865); non *Wahlenbergia lasiocarpa* Schltr. & Brehmer, Bot. Jahrb. Syst. 53: 113 (1915).
Wahlenbergia obovata var. *cernua* Brehmer, Bot. Jahrb. Syst. 53: 112 (1915).
Wahlenbergia obovata var. *fissa* Brehmer, Bot. Jahrb. Syst. 53: 112 (1915).
Wahlenbergia obovata var. *lata* Brehmer, Bot. Jahrb. Syst. 53: 112 (1915).

Wahlenbergia oligantha Lammers, Taxon 44: 337 (1995).
Cape Provinces. 27 CPP. Nanophan. or cham.
* *Lightfootia pauciflora* Adamson, J. S. African Bot. 21: 201 (1955); non *Wahlenbergia pauciflora* A. DC. in DC., Prodr. 7: 437 (1839).

Wahlenbergia oligotricha Schltr. & Brehmer, Bot. Jahrb. Syst. 53: 140 (1915).
Cape Provinces. 27 CPP. Ther.?
Wahlenbergia oligotricha var. *hispidula* Brehmer, Bot. Jahrb. Syst. 53: 141 (1915).

Wahlenbergia oocarpa Sond. in Harv. & Sond., Fl. Cap. 3: 576 (1865). *Campanopsis oocarpa* (Sond.) Kuntze, Revis. Gen. Pl. 2: 379 (1891). Ther.
Cape Provinces. 27 CPP.

Wahlenbergia orae Lammers, Novon 11: 67 (2001).
Cape Provinces. 27 CPP. Ther.?
* *Wahlenbergia littoralis* Schltr. & Brehmer, Bot. Jahrb. Syst. 53: 127 (1915); non (Labill.) Sweet, Hort. Brit. (ed. 2): 593 (1830).

Wahlenbergia oxyphylla A. DC. in DC., Prodr. 7: 427 (1839). *Campanopsis oxyphylla* (A. DC.) Kuntze, Revis. Gen. Pl. 2: 379 (1891).
S. Africa. 27 CPP NAM. Ther.

Wahlenbergia pallidiflora Hilliard & B. L. Burtt, Notes Roy. Bot. Gard. Edinburgh 43: 351 (1986).
KwaZulu-Natal. 27 NAT. Hemicr.

Wahlenbergia paludicola Thulin, Symb. Bot. Upsal. 21: 202 (1975).
Cameroon to Tanzania, Zimbabwe & Angola; Madagascar. 23 CMN ZAI 25 TAN UGA 26 ANG MLW ZAM ZIM 29 MDG. Ther.
   Lightfootia gracillima R. E. Fr., Wissensch. Ergebn. Schwed. Rhodesia-Kongo-Exped. 1: 316 (1916); non *Wahlenbergia gracillima* S. Moore, J. Bot. 56: 9 (1918).

Wahlenbergia paniculata (L. f.) A. DC., Monogr. Campan.: 153 (1830).
S. Africa. 27 CPP LES NAM OFS TVL. Ther.
   Campanula paniculata L. f., Suppl. Pl.: 139 (1782). *Campanopsis paniculata* (L. f.) Kuntze, Revis. Gen. Pl. 2: 379 (1891).
   Wahlenbergia rudis E. Mey. ex A. DC. in DC., Prodr. 7: 431 (1839). *Wahlenbergia paniculata* var. *rudis* (E. Mey. ex A. DC.) Sond. in Harv. & Sond., Fl. Cap. 3: 575 (1865).
   Wahlenbergia divaricata E. Mey. ex A. DC., Prodr. 7: 432 (1839).

Wahlenbergia papuana P. Royen, Bot. J. Linn. Soc. 77: 121 (1978).
New Guinea. 43 NWG. Hemicr.

Wahlenbergia parvifolia (P. J. Bergius) Lammers, Taxon 44: 337 (1995).
Cape Provinces. 27 CPP. Nanophan. or cham.
   Lobelia parvifolia P. J. Bergius, Descr. Fl. Cap.: 345 (1767). *Ireon parvifolia* (P. J. Bergius) Scop., Intr. Hist. Nat.: 229 (1777), as 'parviflora'. *Lightfootia parvifolia* (P. J. Bergius) Adamson, J. S. African Bot. 7: 273 (1942).
   Lobelia tenella L., Mant. Pl.: 120 (1767); non *Wahlenbergia tenella* (L. f.) Lammers, Taxon 44: 338 (1995). *Dobrowskya tenella* (L.) Sond. in Harv. & Sond., Fl. Cap. 3: 549 (1865). *Monopsis tenella* (L.) Urb., Jahrb. Königl. Bot. Gart. Berlin 1: 273 (1881).
   Lightfootia oxycoccoides L'Hér., Sert. Angl.: 4 (1789).
   Campanula ottoniana Schult. in Roem. & Schult., Syst. Veg. 5: 113 (1819).

Wahlenbergia patula A. DC. in DC., Prodr. 7: 436 (1839). *Campanopsis patula* (A. DC.) Kuntze, Revis. Gen. Pl. 2: 379 (1891).
S. Africa. 27 CPP NAM. Ther.

Wahlenbergia paucidentata Schinz, Bull. Herb. Boissier 3: 422 (1895).
S. Africa. 27 CPP LES NAT. Hemicr.
   Wahlenbergia paucidentata var. *tysonii* Schinz, Bull. Herb. Boissier 3: 423 (1895).

Wahlenbergia pauciflora A. DC. in DC., Prodr. 7: 437 (1839). *Campanopsis pauciflora* (A. DC.) Kuntze, Revis. Gen. Pl. 2: 379 (1891).
Cape Provinces. 27 CPP. Ther.

Wahlenbergia peduncularis (Wall. ex A. DC.) Hook. f. & Thomson, J. Proc. Linn Soc., Bot. 2: 22 (1858).
Himalaya & N. India (Uttar Pradesh). 40 IND NEP WHM. Hemicr.
   Campanula peduncularis Wall. ex A. DC. in DC., Prodr. 7: 483 (1839). *Campanopsis peduncularis* (Wall. ex A. DC.) Kuntze, Revis. Gen. Pl. 2: 379 (1891).

Wahlenbergia perrieri Thulin, Symb. Bot. Upsal. 21: 130 (1975).
Madagascar. 29 MDG. Cham. or hemicr.

Wahlenbergia perrottetii (A. DC.) Thulin, Symb. Bot. Upsal. 21: 199 (1975).
French Guiana; E. Brazil (Bahia); Senegal to Chad, Malawi & Angola; Madagascar; Comoros. 22 BEN GHA GNB GUI IVO LBR MLI NGA SEN SIE TOG 23 BUR CAF CMN CON EQG GAB RWA ZAI 24 CHA 25 TAN UGA 26 ANG MLW 29 COM MDG 82 FRG 84 BZE. Ther. $2n = 16$.

Cephalostigma perrottetii A. DC., Monogr. Campan.: 118 (1830).
  Cephalostigma prieurei A. DC., Monogr. Campan.: 118 (1830).
  Cephalostigma bahiense A. DC. in DC., Prodr. 7: 421 (1839).
  Cephalostigma bahiense var. *major* A. DC. in DC., Prodr. 7: 421 (1839).

Wahlenbergia persimilis Thulin, Symb. Bot. Upsal. 21: 178 (1975).
  Zaïre. 23 ZAI. Ther.

Wahlenbergia peruviana A. Gray, Proc. Amer. Acad. Arts 5: 152 (1861).
  Peru to NW. Argentina (Salta). 83 BOL PER 85 AGW. Hemicr.

Wahlenbergia petraea Thulin, Nordic J. Bot. 7: 262 (1987).
  Burundi. 23 BUR. Hemicr.

Wahlenbergia pilosa H. Buek in Eckl. & Zeyh., Enum. Pl. Afric. Austral.: 381 (1837).
  Campanula buekii D. Dietr., Syn. Pl. 1: 753 (1839), as 'buchii;' non *Campanula pilosa* Pall.
  ex Schult. in Roem. & Schult., Syst. Veg. 5: 148 (1819). *Campanopsis pilosa* (H. Buek)
  Kuntze, Revis. Gen. Pl. 2: 379 (1891).
  Cape Provinces. 27 CPP. Ther.

Wahlenbergia pinifolia N. E. Br., Bull. Misc. Inform. Kew 1895: 148 (1895).
  KwaZulu-Natal. 27 NAT. Hemicr.

Wahlenbergia pinnata Compton, J. S. African Bot. 33: 299 (1967).
  S. Africa. 27 NAT SWZ. Hemicr.

Wahlenbergia planiflora P. J. Sm., Telopea 5: 142 (1992).
  SE. Australia. 50 NSW QLD VIC. Hemicr. $2n = 18, 36, 54$.

  subsp. **longipila** Carolin ex P. J. Sm., Telopea 5: 144 (1992).
    SE. Australia. 50 NSW QLD. Hemicr. $2n = 18, 36$.

  subsp. **planiflora**
    SE. Australia. 50 NSW VIC. Hemicr. $2n = 36, 54$.

Wahlenbergia polyantha Lammers, Taxon 44: 337 (1995).
  Cape Provinces. 27 CPP. Nanophan. or cham.
    Lightfootia multiflora Adamson, J. S. African Bot. 21: 180 (1955); non *Wahlenbergia
    multiflora* Conrath, Bull. Misc. Inform. Kew 1914: 134 (1914).

Wahlenbergia polycephala (Mildbr.) Thulin, Symb. Bot. Upsal. 21: 173 (1975).
  Tanzania. 25 TAN. Hemicr. $2n = 16$.
    Lightfootia polycephala Mildbr., Notizbl. Bot. Gart. Berlin-Dahlem 11: 686 (1932).

Wahlenbergia polyclada A. DC. in DC., Prodr. 7: 789 (1839).
  Cape Provinces. 27 CPP. Ther.

Wahlenbergia polyphylla Thulin, Nordic J. Bot. 7: 264 (1987).
  Zaïre. 23 ZAI. Hemicr.

Wahlenbergia polytrichifolia Schltr., J. Bot. 34: 392 (1896).
  S. Africa. 27 CPP LES NAT. Hemicr.

  subsp. **dracomontana** Hilliard & B. L. Burtt, Notes Roy. Bot. Gard. Edinburgh 43:
    197 (1986).
    Lesotho & KwaZulu-Natal. 27 LES NAT. Hemicr.

  subsp. **polytrichifolia**
    Cape Provinces. 27 CPP. Hemicr.

Wahlenbergia preissii de Vriese in Lehm., Pl. Preiss. 2: 241 (1848).
SW. Australia. 50 SOA WAU. Ther.

Wahlenbergia procumbens (L. f.) A. DC., Monogr. Campan.: 140 (1830).
S. Africa. 27 CPP NAM. Hemicr.
  * *Campanula procumbens* L. f., Suppl. Pl.: 141 (1782). *Campanopsis procumbens* (L. f.)
    Kuntze, Revis. Gen. Pl. 2: 379 (1891).
  Wahlenbergia repens Schrad., Blumenbachia: 37 (1827).
  Wahlenbergia diversifolia A. DC., Monogr. Campan.: 139 (1830). *Campanula diversifolia*
    (A. DC.) D. Dietr., Syn. Pl. 1: 752 (1839); non Dumort., Fl. Belg.: 58 (1827).
    Wahlenbergia procumbens var. *diversifolia* (A. DC.) Sond. in Harv. & Sond., Fl. Cap. 3:
    573 (1865).
  Wahlenbergia procumbens var. *foliosa* A. DC., Monogr. Campan.: 140 (1830).
  Wahlenbergia procumbens var. *intermedia* Brehmer, Bot. Jahrb. Syst. 53: 133 (1915).
  Wahlenbergia saxifragoides Brehmer, Bot. Jahrb. Syst. 53: 133 (1915).

Wahlenbergia prostrata E. Mey. ex A. DC. in DC., Prodr. 7: 431 (1839). *Campanopsis
prostrata* (E. Mey. ex A. DC.) Kuntze, Revis. Gen. Pl. 2: 379 (1891). Ther.
S. Africa. 27 CPP NAM.

Wahlenbergia psammophila Schltr., Bot. Jahrb. Syst. 27: 192 (1899).
Cape Provinces. 27 CPP. Ther.
  Wahlenbergia psammophila var. *longisepala* Brehmer, Bot. Jahrb. Syst. 53: 82 (1915).

Wahlenbergia pseudoandrosacea Brehmer, Bot. Jahrb. Syst. 53: 135 (1915).
Cape Provinces. 27 CPP. Hemicr.?

Wahlenbergia pseudoinhambanensis Brehmer, Bot. Jahrb. Syst. 53: 110 (1915).
Cape Provinces. 27 CPP. Hemicr.?

Wahlenbergia pseudonudicaulis Brehmer, Bot. Jahrb. Syst. 53: 139 (1915).
Cape Provinces. 27 CPP. Hemicr.?
  Wahlenbergia pseudonudicaulis var. *diversa* Brehmer, Bot. Jahrb. Syst. 53: 139 (1915).

Wahlenbergia pulchella Thulin, Symb. Bot. Upsal. 21: 147 (1975).
Zaïre to Tanzania & Zambia. 23 BUR RWA ZAI 25 TAN 26 ZAM. Ther.

subsp. **laurentii** Thulin, Symb. Bot. Upsal. 21: 150 (1975).
Zaïre. 23 ZAI. Ther.

subsp. **mbalensis** Thulin, Symb. Bot. Upsal. 21: 150 (1975).
Tanzania & Zambia. 25 TAN 26 ZAM. Ther.

subsp. **michelii** Thulin, Symb. Bot. Upsal. 21: 154 (1975).
Zaïre, Rwanda, Burundi. 23 BUR RWA ZAI. Ther.

subsp. **paradoxa** Thulin, Symb. Bot. Upsal. 21: 153 (1975).
Tanzania & Zambia. 25 TAN 26 ZAM. Ther.

subsp. **pedicellata** Thulin, Symb. Bot. Upsal. 21: 153 (1975).
Rwanda, Burundi, Tanzania. 23 BUR RWA 25 TAN. Ther.

subsp. **pulchella**.
Burundi. 23 BUR. Ther.

Wahlenbergia pulvillus-gigantis Hilliard & B. L. Burtt, Notes Roy. Bot. Gard. Edinburgh
31: 2 (1971).
S. Africa. 27 LES NAT. Cham. or hemicr.

Wahlenbergia pusilla Hochst. ex A. Rich., Tent. Fl. Abyss. 2: 2 (1850). *Campanopsis pusilla* (Hochst. ex A. Rich.) Kuntze, Revis. Gen. Pl. 2: 379 (1891).
Ethiopia to Tanzania. 24 ETH 25 KEN TAN. Hemicr. $2n = 14$.
   Lobelia kilimandsharica Engl., Bot. Jahrb. Syst. 19 (Beibl. 47): 52 (1894).

Wahlenbergia pygmaea Colenso, Trans. & Proc. New Zealand Inst. 31: 273 (1899). *Wahlenbergia albomarginata* var. *pygmaea* (Colenso) N. E. Br., Gard. Chron. (ser. 3) 54: 336 (1913).
New Zealand (North I.). 51 NZN. Geophyte. $2n = 36$.

   subsp. **drucei** Petterson, New Zealand J. Bot. 35: 44 (1997).
   New Zealand (North I.). 51 NZN. Geophyte. $2n = 36$.

   subsp. **pygmaea**
   New Zealand (North I.). 51 NZN. Geophyte. $2n = 36$.

   subsp. **tararua** Petterson, New Zealand J. Bot. 35: 43 (1997).
   New Zealand (North I.). 51 NZN. Geophyte. $2n = 36$.

Wahlenbergia pyrophila Lammers, Taxon 44: 337 (1995).
Cape Provinces. 27 CPP. Nanophan. or cham.
   * *Lightfootia tenuis* Adamson, J. S. African Bot. 7: 202 (1941); non *Wahlenbergia tenuis* A. DC. in DC., Prodr. 7: 440 (1839).

Wahlenbergia quadrifida (R. Br.) A. DC., Monogr. Campan.: 144 (1830).
E. Australia. 50 NSW NTA QLD SOA TAS VIC. Hemicr. $2n = 54$.
   * *Campanula quadrifida* R. Br., Prodr.: 561 (1810).
   Wahlenbergia sieberi A. DC., Monogr. Campan.: 144 (1830). *Campanula sieberi* (A. DC.) D. Dietr., Syn. Pl. 1: 752 (1839).
   Wahlenbergia eurycarpa Domin, Biblioth. Bot. 89: 638 (1929).
   Wahlenbergia dioica Domin, Biblioth. Bot. 89: 639 (1929).

Wahlenbergia queenslandica Carolin ex P. J. Sm. in Jessop & Toelken, Fl. S. Australia 3: 1381 (1986).
N. Australia. 50 NSW NTA QLD SOA WAU. Hemicr. $2n = 18, 36$.

Wahlenbergia ramifera Brehmer, Bot. Jahrb. Syst. 53: 73 (1915).
Cape Provinces. 27 CPP. Hemicr.?

Wahlenbergia ramosa G. Simpson, Trans. & Proc. Roy. Soc. New Zealand 75: 196 (1945).
New Zealand. 51 NZN NZS. Hemicr. $2n = 72$.

Wahlenbergia ramosissima (Hemsl.) Thulin, Symb. Bot. Upsal. 21: 187 (1975).
Nigeria to Tanzania, Zimbabwe & Namibia. 22 NGA 23 CMN ZAI 25 TAN 26 ANG MLW ZAM ZIM 27 BOT NAM. Ther. $2n = 16$.
   * *Cephalostigma ramosissimum* Hemsl. in Oliv., Fl. Trop. Afr. 3: 472 (1877). *Lightfootia ramosissima* (Hemsl.) E. Wimm. ex Hepper, Kew Bull. 15: 61 (1961).

   subsp. **centiflora** Thulin, Symb. Bot. Upsal. 21: 194 (1975).
   Angola to Zaïre, Tanzania & Malawi. 23 ZAI 25 TAN 26 ANG MLW ZAM. Ther.

   subsp. **lateralis** (Brehmer) Thulin, Symb. Bot. Upsal. 21: 193 (1975).
   Angola, Namibia, Botswana. 26 ANG 27 BOT NAM. Ther.
      * *Lightfootia exilis* A. DC., Ann. Sci. Nat., Bot. (ser. 5) 6: 330 (1866); non *Wahlenbergia exilis* A. DC., Monogr. Campan.: 151 (1830). *Wahlenbergia lateralis* Brehmer, Bot. Jahrb. Syst. 53: 128 (1915).
      Lightfootia debilis A. DC., Ann. Sci. Nat., Bot. (ser. 5) 6: 331 (1866).
      Cephalostigma pyramidale Schinz, Vierteljahrsschr. Naturf. Ges. Zürich 60: 421(1915).

subsp. **oldenlandioides** Thulin, Symb. Bot. Upsal. 21: 196 (1975).
Tanzania & Zambia. 25 TAN 26 ZAM. Ther. $2n = 16$.

subsp. **ramosissima**
Nigeria & Cameroon. 22 NGA 23 CMN. Ther.

subsp. **richardsiae** Thulin, Symb. Bot. Upsal. 21: 197 (1975).
Zambia. 26 ZAM. Ther.

subsp. **subcapitata** Thulin, Symb. Bot. Upsal. 21: 192 (1975).
Zaïre & Tanzania. 23 ZAI 25 TAN. Ther.

subsp. **zambiensis** Thulin, Symb. Bot. Upsal. 21: 191 (1975).
Zaïre & Zambia. 23 ZAI 26 ZAM. Ther.

Wahlenbergia rara Schltr. & Brehmer, Bot. Jahrb. Syst. 53: 82 (1915).
Cape Provinces. 27 CPP. Hemicr.?

Wahlenbergia rhytidosperma Thulin, Bot. Not. 131: 67 (1978).
Northern Provinces. 27 TVL. Hemicr.

Wahlenbergia riversdalensis Lammers, Taxon 44: 337 (1995).
Cape Provinces. 27 CPP. Hemicr.
* *Lightfootia planifolia* Adamson, J. S. African Bot. 21: 202 (1955); non *Wahlenbergia planifolia* Gand., Bull. Soc. Bot. France 65: 54 (1918).

Wahlenbergia rivularis Diels, Bot. Jahrb. Syst. 26: 111 (1898).
S. Africa. 27 CPP NAT OFS TVL. Hemicr.
Wahlenbergia tysonii Zahlbr., Ann. K. K. Naturhist. Hofmus. 18: 403 (1903), as 'tysoni'.
Wahlenbergia rivularis var. *oblonga* Brehmer, Bot. Jahrb. Syst. 53: 118 (1915).

Wahlenbergia robusta (A. DC.) Sond. in Harv. & Sond., Fl. Cap. 3: 584 (1865).
Cape Provinces. 27 CPP. Nanophan. or cham.
* *Lightfootia robusta* A. DC. in DC., Prodr. 7: 420 (1839). *Campanopsis robusta* (A. DC.) Kuntze, Revis. Gen. Pl. 2: 379 (1891).
Wahlenbergia rigida Bernh., Flora 27: 820 (1844).

Wahlenbergia roelliflora Schltr. & Brehmer, Bot. Jahrb. Syst. 53: 96 (1915).
Cape Provinces. 27 CPP. Hemicr.

Wahlenbergia roxburghii A. DC., Monogr. Campan.: 162 (1830).
St. Helena. 28 STH. Nanophan.
* *Roella paniculata* Roxb. in Beatson, Tracts St. Helena: 320 (1816); non *Wahlenbergia paniculata* (L. f.) A. DC., Monogr. Campan.: 153 (1830). *Campanopsis roxburghii* (A. DC.) Kuntze, Revis. Gen. Pl. 2: 379 (1891); non *Campanopsis paniculata* (L. f.) Kuntze, Revis. Gen. Pl. 2: 379 (1891).

Wahlenbergia rubens (H. Buek) Lammers, Taxon 44: 337 (1995).
Cape Provinces. 27 CPP. Nanophan. or cham.
* *Lightfootia rubens* H. Buek in Eckl. & Zeyh., Enum. Pl. Afric. Austral.: 373 (1837).
Lightfootia angustifolia A. DC. in DC., Prodr. 7: 418 (1839).
Lightfootia rubens var. *brachyphylla* Adamson, J. S. African Bot. 21: 188 (1955).
Wahlenbergia rubens var. *brachyphylla* (Adamson) Welman, Bothalia 26: 157 (1996).

Wahlenbergia rubioides (A. DC.) Lammers, Taxon 44: 337 (1995).
Cape Provinces. 27 CPP. Nanophan. or cham.
* *Lightfootia rubioides* A. DC., Monogr. Campan.: 116 (1830).
Lightfootia rubioides var. *stokoei* Adamson, J. S. African Bot. 21: 164 (1955). *Wahlenbergia rubioides* var. *stokoei* (Adamson) Welman, Bothalia 26: 157 (1996).

Wahlenbergia rupestris G. Simpson, Trans. & Proc. Roy. Soc. New Zealand 79: 431 (1952).
New Zealand. 51 NZN NZS. Hemicr. $2n = 72$.

Wahlenbergia saxicola (R. Br.) A. DC., Monogr. Campan.: 144 (1830).
Tasmania. 50 TAS. Geophyte. $2n = 72$.
  * *Campanula saxicola* R. Br., Prodr.: 561 (1810). *Campanopsis saxicola* (R. Br.) Kuntze,
     Revis. Gen. Pl. 2: 379 (1891).
     Streleskia montana Hook. f., London J. Bot. 6: 267 (1847).

Wahlenbergia schistacea Brehmer, Bot. Jahrb. Syst. 53: 100 (1915).
Cape Provinces. 27 CPP. Hemicr.?

Wahlenbergia schlechteri Brehmer, Bot. Jahrb. Syst. 53: 113 (1915).
Cape Provinces. 27 CPP. Hemicr.?

Wahlenbergia schwackeana Zahlbr., Verh. K. K. Zool.-Bot. Ges. Wien 49: 517 (1899).
SE. Brazil (Minas Gerais). 84 BZL. Ther.

Wahlenbergia scopella Brehmer, Bot. Jahrb. Syst. 53: 85 (1915).
Cape Provinces. 27 CPP. Hemicr.?
  Wahlenbergia scopella var. *rotundata* Brehmer, Bot. Jahrb. Syst. 53: 86 (1915).

Wahlenbergia scopulicola Carolin ex P. J. Sm., Telopea 5: 111 (1992).
E. Australia (McPherson Range). 50 NSW QLD. Hemicr. $2n = 18$.

Wahlenbergia scottii Thulin, Symb. Bot. Upsal. 21: 145 (1975).
Kenya & Tanzania. 25 KEN TAN. Hemicr.
  * *Lightfootia cartilaginea* M. B. Scott, Bull. Misc. Inform. Kew 1915: 45 (1915); non
     Wahlenbergia cartilaginea Hook. f., Handb. New Zealand Fl.: 170 (1867).

Wahlenbergia serpentina Brehmer, Bot. Jahrb. Syst. 53: 134 (1915).
Northern Provinces. 27 TVL. Ther.?

Wahlenbergia sessiliflora Brehmer, Bot. Jahrb. Syst. 53: 72 (1915).
Cape Provinces. 27 CPP. Hemicr.?
  Wahlenbergia sessiliflora var. *dentata* Brehmer, Bot. Jahrb. Syst. 53: 72 (1915).

Wahlenbergia silenoides Hochst. ex A. Rich., Tent. Fl. Abyss. 2: 3 (1850). *Campanopsis
silenoides* (Hochst. ex A. Rich.) Kuntze, Revis. Gen. Pl. 2: 379 (1891), as 'silenodes'.
Bioko & São Tomé; Nigeria to Ethiopia & Tanzania. 22 NGA 23 BUR CMN EQG GGI ZAI
  24 ETH SUD 25 KEN TAN UGA. Hemicr. $2n = 18$.
  Wahlenbergia polyclada Hook. f., J. Proc. Linn. Soc., Bot. 6: 15 (1862); non A. DC. in
     DC., Prodr. 7: 789 (1839). *Wahlenbergia mannii* Vatke, Linnaea 38: 700 (1874).
     Campanopsis mannii (Vatke) Kuntze, Revis. Gen. Pl. 2: 379 (1891).
  Wahlenbergia silenoides var. *elongata* Brehmer, Bot. Jahrb. Syst. 53: 128 (1915).
  Wahlenbergia mannii var. *intermedia* Brehmer, Bot. Jahrb. Syst. 53: 129 (1915).
  Wahlenbergia mannii var. *virgulta* Brehmer, Bot. Jahrb. Syst. 53: 129 (1915).

Wahlenbergia sonderi Lammers, Taxon 44: 338 (1995).
Cape Provinces. 27 CPP. Ther.
  * *Lightfootia namaquana* Sond. in Harv. & Sond., Fl. Cap. 3: 556 (1865); non *Wahlenbergia
     namaquana* Sond. in Harv. & Sond., Fl. Cap. 3: 582 (1865).

Wahlenbergia songeana Thulin, Symb. Bot. Upsal. 21: 86 (1975).
Tanzania. 25 TAN. Hemicr.

Wahlenbergia sphaerica Brehmer, Bot. Jahrb. Syst. 53: 76 (1915).
Cape Provinces. 27 CPP. Ther.
Wahlenbergia sphaerica var. *longifolia* Brehmer, Bot. Jahrb. Syst. 53: 76 (1915).

Wahlenbergia squamifolia Brehmer, Bot. Jahrb. Syst. 53: 94 (1915).
S. Africa. 27 NAT OFS SWZ TVL. Cham.
Wahlenbergia squamifolia var. *tenuis* Brehmer, Bot. Jahrb. Syst. 53: 96 (1915).

Wahlenbergia squarrosa Brehmer, Bot. Jahrb. Syst. 53: 131 (1915).
Namibia. 27 NAM. Cham.?

Wahlenbergia stellarioides Cham., Linnaea 8: 196 (1833). *Campanula stellarioides* (Cham.)
D. Dietr., Syn. Pl. 1: 753 (1839). *Campanopsis stellarioides* (Cham.) Kuntze, Revis. Gen. Pl.
2: 379 (1891), as 'stellariodes'.
Cape Provinces. 27 CPP. Hemicr.
Wahlenbergia integrifolia A. DC. in DC., Prodr. 7: 428 (1839). *Campanopsis stellarioides*
var. *integrifolia* (A. DC.) Kuntze, Revis. Gen. Pl. 3(2): 185 (1898).
Wahlenbergia stellarioides var. *major* Sond. in Harv. & Sond., Fl. Cap. 3: 572 (1865).
Wahlenbergia stellarioides var. *angusta* Sond. in Harv. & Sond., Fl. Cap. 3: 572 (1865).
Campanopsis stellarioides var. *angusta* (Sond.) Kuntze, Revis. Gen. Pl. 3(2): 185 (1898).

Wahlenbergia stricta (R. Br.) Sweet, Hort. Brit. (ed. 2): 593 (1830).
SE. Australia; naturalized farther west. 50 NSW NSW QLD TAS VIC wau. Hemicr. $2n = 18, 36$.
* *Campanula stricta* Sm., Exot. Bot. pl. 45 (1805); non L., Sp. Pl. (ed. 2): 238 (1762).
Campanula gracilis var. *stricta* R. Br., Prodr.: 561 (1810). *Wahlenbergia gracilis* var.
stricta (R. Br.) A. DC., Monogr. Campan.: 142 (1830). *Wahlenbergia gracilis* f. *stricta*
(R. Br.) Voss in Siebert & Voss, Vilm. Blumengärtn. (ed. 3) 1: 572 (1894).
Wahlenbergia bicolor Lothian, Proc. Linn. Soc. New South Wales 71: 230 (1947).

subsp. **alterna** P. J. Sm., Telopea 5: 128 (1992).
SE. Australia. 50 NSW QLD. Hemicr. $2n = 18, 36$.

subsp. **stricta**
SE. Australia; naturalized farther west. 50 NSW QLD TAS VIC wau. Hemicr. $2n = 18, 36$.
Wahlenbergia consimilis Lothian, Proc. Linn. Soc. New South Wales 71: 223 (1947).
Wahlenbergia trichogyna Stearn, Gard. Chron. (ser. 3) 130: 169 (1951). *Wahlenbergia
marginata* subvar. *trichogyna* (Stearn) Tuyn in Steenis, Fl. Males. (ser. 1) 6: 118 (1960).
Wahlenbergia vinciflora f. *eriocalyx* Domin, Biblioth. Bot. 22: 1192 (1929).

Wahlenbergia subaphylla (Baker) Thulin, Symb. Bot. Upsal. 21: 124 (1975).
Zaïre to Tanzania & Mozambique; Madagascar. 23 ZAI 25 TAN 26 MLW MOZ ZAM ZIM 29
MDG. Hemicr.
* *Lightfootia subaphylla* Baker, J. Linn. Soc., Bot. 20: 193 (1883).

subsp. **subaphylla**
Madagascar. 29 MDG. Hemicr.

subsp. **scoparia** (Wild) Thulin, Symb. Bot. Upsal. 21: 127 (1975).
Zambia to Mozambique. 26 MOZ ZAM ZIM. Hemicr.
* *Lightfootia scoparia* Wild, Kirkia 4: 161 (1964).

subsp. **thesioides** Thulin, Symb. Bot. Upsal. 21: 126 (1975).
Zaïre, Tanzania, Malawi. 23 ZAI 25 TAN 26 MLW. Hemicr.

Wahlenbergia subfusiformis Brehmer, Bot. Jahrb. Syst. 53: 80 (1915).
Northern Provinces. 27 TVL. Hemicr.?
Wahlenbergia subfusiformis var. *involuta* Brehmer, Bot. Jahrb. Syst. 53: 81 (1915).

Wahlenbergia subpilosa Brehmer, Bot. Jahrb. Syst. 53: 88 (1915).
Cape Provinces. 27 CPP. Hemicr.?
* *Wahlenbergia paniculata* var. *glabrata* Sond. in Harv. & Sond., Fl. Cap. 3: 575 (1865).

Wahlenbergia subrosulata Brehmer, Bot. Jahrb. Syst. 53: 101 (1915).
S. Africa. 27 CPP NAM. Hemicr.?
Wahlenbergia subrosulata var. *grandifolia* Brehmer, Bot. Jahrb. Syst. 53: 101 (1915).

Wahlenbergia subtilis Brehmer, Bot. Jahrb. Syst. 53: 109 (1915).
Cape Provinces. 27 CPP. Hemicr.?

Wahlenbergia subulata (L'Hér.) Lammers, Taxon 44: 338 (1995).
Cape Provinces. 27 CPP. Nanophan. or cham.
Campanula sessiliflora L. f., Suppl. Pl.: 139 (1782); non *Wahlenbergia sessiliflora*
Brehmer, Bot. Jahrb. Syst. 53: 72 (1915). *Lightfootia sessiliflora* (L. f.) Spreng., Syst.
Veg. 1: 809 (1824).
* *Lightfootia subulata* L'Hér., Sert. Angl.: 4 (1789).
Lightfootia subulata var. *tenuifolia* Adamson, J. S. African Bot. 21: 161 (1955).
Wahlenbergia subulata var. *tenuifolia* (Adamson) Welman, Bothalia 26: 157 (1996).
Lightfootia subulata var. *congesta* Adamson, J. S. African Bot. 21: 162 (1955).
Wahlenbergia subulata var. *congesta* (Adamson) Welman, Bothalia 26: 157 (1996).

Wahlenbergia subumbellata Markgr., Notizbl. Bot. Gart. Berlin-Dahlem 15: 761 (1942).
Namibia. 27 NAM. Ther.

Wahlenbergia tenella (L. f.) Lammers, Taxon 44: 338 (1995).
Cape Provinces. 27 CPP. Nanophan. or cham.
* *Campanula tenella* L. f., Suppl. Pl.: 141 (1782). *Lightfootia tenella* (L. f.) A. DC., Monogr.
Campan.: 111 (1830); non Lodd., Bot. Cab. 11: pl. 1038 (1825).
Lightfootia diffusa H. Buek in Eckl. & Zeyh., Enum. Pl. Afric. Austral.: 376 (1837).
Lightfootia tenella var. *diffusa* (H. Buek) Zahlbr., Ann. K. K. Naturhist. Hofmus. 18:
403 (1903).
Lightfootia tenella var. *brevivalvis* A. DC. in DC., Prodr. 7: 419 (1839).
Lightfootia tenella f. *flaviflora* Kuntze, Revis. Gen. Pl. 3(2): 188 (1898).
Lightfootia diffusa var. *palustris* Adamson, J. S. African Bot. 21: 193 (1955). *Wahlenbergia*
tenella var. *palustris* (Adamson) Welman, Bothalia 26: 157 (1996).
Lightfootia diffusa var. *stokoei* Adamson, J. S. African Bot. 21: 194 (1955). *Wahlenbergia*
tenella var. *stokoei* (Adamson) Welman, Bothalia 26: 157 (1996).

Wahlenbergia tenerrima (H. Buek) Lammers, Taxon 44: 338 (1995).
Cape Provinces. 27 CPP. Nanophan. or cham.
Lightfootia tenella Lodd., Bot. Cab. 11: pl. 1038 (1825); non *Wahlenbergia tenella* (L. f.)
Lammers, Taxon 44: 338 (1995). *Lightfootia loddigesii* A. DC., Monogr. Campan.:
114 (1830).
* *Lightfootia tenerrima* H. Buek in Eckl. & Zeyh., Enum. Pl. Afric. Austral.: 375 (1837).
Lightfootia tenella var. *tenerrima* (H. Buek) Sond. in Harv. & Sond., Fl. Cap. 3:
562 (1865).
Lightfootia tenella f. *coerulescens* Kuntze, Revis. Gen. Pl. 3(2): 188 (1898).
Lightfootia tenella var. *montana* Adamson, J. S. African Bot. 21: 199 (1955). *Wahlenbergia*
tenella var. *montana* (Adamson) Welman, Bothalia 26: 157 (1996).

Wahlenbergia tenuiloba Thulin, Bot. Not. 128: 355 (1976).
Zaïre. 23 ZAI. Ther.
* *Wahlenbergia congesta* Thulin, Symb. Bot. Upsal. 21: 209 (1975); non (Cheeseman) N. E.
Br., Gard. Chron. (ser. 3) 54: 336 (1913).

Wahlenbergia tenuis A. DC. in DC., Prodr. 7: 440 (1839). *Campanopsis tenuis* (A. DC.) Kuntze, Revis. Gen. Pl. 2: 379 (1891).
Cape Provinces. 27 CPP. Ther.

Wahlenbergia tetramera Thulin, Nordic J. Bot. 7: 261 (1987).
KwaZulu-Natal. 27 NAT. Hemicr. or ther.

Wahlenbergia thulinii Lammers, Novon 11: 68 (2001).
Cape Provinces. 27 CPP. Nanophan. or cham.
   * *Lightfootia oppositifolia* A. DC., Monogr. Campan.: 115 (1830); non *Wahlenbergia oppositifolia* A. DC. in DC., Prodr. 7: 429 (1839).
   Lightfootia capillaris H. Buek in Eckl. & Zeyh., Enum. Pl. Afric. Austral.: 372 (1837). *Wahlenbergia capillaris* (H. Buek) Lammers, Taxon 44: 335 (1995); non (Lodd.) Sweet, Hort. Brit. (ed. 2): 593 (1830).

Wahlenbergia thunbergiana (H. Buek) Lammers, Taxon 44: 338 (1995).
S. Africa. 27 CPP NAM. Nanophan. or cham.
   * *Lightfootia thunbergiana* H. Buek in Eckl. & Zeyh., Enum. Pl. Afric. Austral.: 376 (1837).
   Lightfootia anomala A. DC. in DC., Prodr. 7: 418 (1839).

Wahlenbergia thunbergii (Schult.) B. Nord., Taxon 52: 821 (2003).
Cape Provinces. 27 CPP. Nanophan. or cham.
   * *Campanula ciliata* Thunb., Mém. Acad. Imp. Sci. St. Pétersbourg Hist. Acad. 4: 358 (1813); non Steven, Mém. Soc. Imp. Nat. Moscou 3: 256 (1812). *Campanula thunbergii* Schult. in Roem. & Schult., Syst. Veg. 5: 135 (1819), as 'thunbergi'. *Lightfootia ciliata* Sond. in Harv. & Sond., Fl. Cap. 3: 561 (1865); non Spreng., Syst. Veg. 1: 809 (1824). *Roella thunbergii* (Schult.) A. DC., Monogr. Campan.: 174 (1830).
   Lightfootia uitenhagensis H. Buek in Eckl. & Zeyh., Enum. Pl. Afric. Austral.: 373 (1837). *Wahlenbergia uitenhagensis* (H. Buek) Lammers, Taxon 44: 338 (1995).
   Lightfootia laricina H. Buek in Eckl. & Zeyh., Enum. Pl. Afric. Austral.: 373 (1837).
   Lightfootia thymifolia H. Buek in Eckl. & Zeyh., Enum. Pl. Afric. Austral.: 373 (1837).
   Lightfootia intermedia H. Buek in Eckl. & Zeyh., Enum. Pl. Afric. Austral.: 374 (1837); non *Wahlenbergia intermedia* Zahlbr., Verh. K. K. Zool.-Bot. Ges. Wien 49: 518 (1899).
   Lightfootia divaricata H. Buek in Eckl. & Zeyh., Enum. Pl. Afric. Austral.: 375 (1837); non *Wahlenbergia divaricata* E. Mey. ex A. DC. in DC., Prodr. 7: 432 (1839).
   Lightfootia mucronulata H. Buek in Eckl. & Zeyh., Enum. Pl. Afric. Austral.: 375 (1837).
   Lightfootia pubescens A. DC. in DC., Prodr. 7: 419 (1839). *Lightfootia ciliata* var. *pubescens* (A. DC.) Sond. in Harv. & Sond., Fl. Cap. 3: 561 (1865), as 'pubescent'.
   Lightfootia buekii Sond. in Harv. & Sond., Fl. Cap. 3: 560 (1865).
   Lightfootia ciliata var. *major* Sond. in Harv. & Sond., Fl. Cap. 3: 561 (1865).
   Lightfootia ciliata var. *debilis* Sond. in Harv. & Sond., Fl. Cap. 3: 561 (1865). *Lightfootia divaricata* var. *debilis* (Sond.) Adamson, J. S. African Bot. 21: 182 (1955). *Wahlenbergia uitenhagensis* var. *debilis* (Sond.) Welman, Bothalia 27: 140 (1997). *Wahlenbergia thunbergii* var. *debilis* (Sond.) B. Nord., Taxon 52: 821 (2003).
   Lightfootia divaricata var. *filifolia* Adamson, J. S. African Bot. 21: 183 (1955). *Wahlenbergia uitenhagensis* var. *filifolia* (Adamson) Welman, Bothalia 26: 157 (1996). *Wahlenbergia thunbergii* var. *filifolia* (Adamson) B. Nord., Taxon 52: 821 (2003).

Wahlenbergia tibestica Quézel, Bull. Soc. Hist. Nat. Afrique N. 48: 98 (1957).
Chad. 24 CHA. Ther.

Wahlenbergia tomentosula Brehmer, Bot. Jahrb. Syst. 53: 86 (1915).
Cape Provinces. 27 CPP. Hemicr.?

Wahlenbergia tortilis Brehmer, Bot. Jahrb. Syst. 53: 81 (1915).
Northern Provinces. 27 TVL. Hemicr.?

Wahlenbergia transvaalensis Brehmer, Bot. Jahrb. Syst. 53: 104 (1915).
   S. Africa. 27 CPP TVL. Hemicr.?

Wahlenbergia tsaratananae Thulin, Symb. Bot. Upsal. 21: 109 (1975).
   Madagascar. 29 MDG. Hemicr.

Wahlenbergia tuberosa Hook. f., Bot. Mag. 101: tab. 6155 (1875). *Campanopsis tuberosa*
   (Hook. f.) Kuntze, Revis. Gen. Pl. 2: 379 (1891).
   Juan Fernández Is. (Masafuera). 85 JNF. Cham. or hemicr. $2n = 22$.

Wahlenbergia tumida Brehmer, Bot. Jahrb. Syst. 53: 80 (1915).
   Cape Provinces. 27 CPP. Ther.?
      Wahlenbergia tumida var. *gracilis* Brehmer, Bot. Jahrb. Syst. 53: 80 (1915).

Wahlenbergia tumidifructa P. J. Sm. in Jessop & Toelken, Fl. S. Australia 3: 1383 (1986).
   Australia. 50 NSW NTA QLD SOA VIC WAU. Hemicr. or ther. $2n = 54$.

Wahlenbergia umbellata (Adamson) Lammers, Taxon 44: 339 (1995).
   Cape Provinces. 27 CPP. Nanophan. or cham.
      * *Lightfootia umbellata* Adamson, J. S. African Bot. 21: 194 (1955).

Wahlenbergia undulata (L. f.) A. DC., Monogr. Campan.: 148 (1830).
   Tanzania to S. Africa; Madagascar; cult. 25 TAN 26 MLW MOZ ZAM ZIM 27 BOT CPP LES
      NAM NAT OFS SWZ TVL 29 MDG. Hemicr. or ther.
      * *Campanula undulata* L. f., Suppl. Pl.: 142 (1782). *Campanopsis undulata* (L. f.) Kuntze,
         Revis. Gen. Pl. 2: 379 (1891). *Cervicina undulata* (L. f.) Skeels, U.S.D.A. Bur. Pl.
         Industr. Bull. 208: 17 (1911).
      Wahlenbergia chamissoniana G. Don, Gen. Hist. 3: 740 (1834).
      Wahlenbergia chamissoniana var. *macrantha* Cham. ex G. Don, Gen. Hist. 3: 740 (1834).
         Wahlenbergia undulata var. *macrantha* (Cham. ex G. Don) Sond. in Harv. & Sond., Fl.
         Cap. 3: 579 (1865).
      Wahlenbergia chamissoniana var. *micrantha* Cham. ex G. Don, Gen. Hist. 3: 740 (1834).
      Wahlenbergia bojeri A. DC. in DC., Prodr. 7: 435 (1839). *Campanopsis bojeri* (A. DC.)
         Kuntze, Revis. Gen. Pl. 2: 379 (1891).
      Wahlenbergia bilocularis A. DC. in DC., Prodr. 7: 439 (1839). *Wahlenbergia denudata* var.
         brevisepala Brehmer, Bot. Jahrb. Syst. 53: 125 (1915).
      Wahlenbergia striata A. DC. in DC., Prodr. 7: 439 (1839). *Wahlenbergia undulata* var.
         glabrata Sond. in Harv. & Sond., Fl. Cap. 3: 579 (1865). *Wahlenbergia undulata* var.
         striata (A. DC.) Zahlbr., Ann. K. K. Naturhist. Mus. 18: 402 (1903), as 'stricta'.
      Wahlenbergia caledonica Sond. in Harv. & Sond., Fl. Cap. 3: 579 (1865). *Campanopsis
         caledonica* (Sond.) Kuntze, Revis. Gen. Pl. 2: 379 (1891).
      Wahlenbergia oatesii Rolfe in Oates, Matabele Land Victoria Falls (ed. 2) appendix 5:
         402 (1889).
      Campanopsis undulata f. *albiflora* Kuntze, Revis. Gen. Pl. 3(2): 185 (1898).
      Campanopsis undulata f. *azurea* Kuntze, Revis. Gen. Pl. 3(2): 185 (1898).
      Campanopsis undulata f. *flavida* Kuntze, Revis. Gen. Pl. 3(2): 185 (1898).
      Campanopsis undulata f. *lilacina* Kuntze, Revis. Gen. Pl. 3(2): 185 (1898).
      Wahlenbergia cyanea Engl. & Gilg in Warb., Kunene-Sambesi-Exped.: 395 (1903).
         Wahlenbergia caledonica var. *cyanea* (Engl. & Gilg) Brehmer, Bot. Jahrb. Syst. 53:
         105 (1915).
      Wahlenbergia engleri Brehmer, Bot. Jahrb. Syst. 53: 105 (1915).
      Wahlenbergia dinteri Brehmer, Bot. Jahrb. Syst. 53: 106 (1915).
      Wahlenbergia dinteri var. *elongata* Brehmer, Bot. Jahrb. Syst. 53: 107 (1915).
      Wahlenbergia dinteri var. *paucilaciniata* Brehmer, Bot. Jahrb. Syst. 53: 107 (1915).
      Wahlenbergia dinteri var. *rotundicapsula* Brehmer, Bot. Jahrb. Syst. 53: 107 (1915).
      Wahlenbergia dinteri var. *virgulta* Brehmer, Bot. Jahrb. Syst. 53: 107 (1915).

Wahlenbergia scoparia Brehmer, Bot. Jahrb. Syst. 53: 108 (1915).
Wahlenbergia scoparia var. *obovata* Brehmer, Bot. Jahrb. Syst. 53: 108 (1915).
Wahlenbergia undulata var. *latisepala* Brehmer, Bot. Jahrb. Syst. 53: 122 (1915).
Wahlenbergia undulata var. *rotundifolia* Brehmer, Bot. Jahrb. Syst. 53: 122 (1915).
Wahlenbergia polychotoma Brehmer, Bot. Jahrb. Syst. 53: 123 (1915).

Wahlenbergia unidentata (L. f.) Lammers, Taxon 44: 339 (1995).
Cape Provinces. 27 CPP. Nanophan. or cham.
** Campanula unidentata* L. f., Suppl. Pl.: 139 (1782). *Lightfootia unidentata* (L. f.) A. DC.,
Monogr. Campan.: 109 (1830).
Lightfootia lycopodioides A. DC., Monogr. Campan.: 114 (1830). *Lightfootia unindentata*
var. *lycopodioides* (A. DC.) Sond. in Harv. & Sond., Fl. Cap. 3: 559 (1865).
Lightfootia unidentata var. *pubescens* Sond. in Harv. & Sond., Fl. Cap. 3: 559 (1865),
as 'pubescent'.

Wahlenbergia upembenesis Thulin, Symb. Bot. Upsal. 21: 184 (1975).
Zaïre & Zambia. 23 ZAI 26 ZAM. Hemicr.

Wahlenbergia urcosensis E. Wimm. in J. F. Macbr., Fl. Peru 6: 388 (1937).
Peru. 83 PER. Hemicr.

Wahlenbergia verbascoides Thulin, Symb. Bot. Upsal. 21: 144 (1975).
Angola. 26 ANG. Nanophan. or cham.

Wahlenbergia vernicosa Petterson, New Zealand J. Bot. 35: 26 (1997).
New Zealand (North I.), Kermadec Is., Chatham Is., Tonga. 51 CTM KER NZN 60 TON.
Hemicr. $2n = 54$.
Wahlenbergia gracilis var. *polymorpha* A. DC., Monogr. Campan.: 142 (1830).
Wahlenbergia marginata var. *polymorpha* (A. DC.) Hochr., Candollea 5: 290 (1934).

Wahlenbergia victoriensis P. J. Sm., Telopea 5: 158 (1992).
SE. Australia. 50 NSW VIC. Ther.

Wahlenbergia violacea Petterson, New Zealand J. Bot. 35: 17 (1997).
New Zealand. 51 NZN NZS. Hemicr. $2n = 72$.

Wahlenbergia virgata Engl., Pflanzenw. Ost-Afrikas C: 400 (1895).
Sudan & Ethiopia to S. Africa. 23 BUR 24 ETH SUD 25 KEN TAN UGA 26 MLW MOZ ZAM
ZIM 27 LES NAT OFS SWZ TVL. Hemicr. $2n = 18$.
Wahlenbergia subnuda Conrath, Bull. Misc. Inform. Kew 1908: 225 (1908).
Wahlenbergia sparticula Chiov., Ann. Bot. (Rome) 10: 389 (1912).
Wahlenbergia recurvata Brehmer, Bot. Jahrb. Syst. 51: 232 (1914).
Wahlenbergia virgata var. *longisepala* Brehmer, Bot. Jahrb. Syst. 53: 122 (1915).
Wahlenbergia virgata var. *tenuis* Brehmer, Bot. Jahrb. Syst. 53: 122 (1915).
Wahlenbergia virgata var. *valida* Brehmer, Bot. Jahrb. Syst. 53: 122 (1915).

Wahlenbergia virgulta Brehmer, Bot. Jahrb. Syst. 53: 93 (1915).
Cape Provinces. 27 CPP. Hemicr.?

Wahlenbergia welwitschii (A. DC.) Thulin, Symb. Bot. Upsal. 21: 208 (1975).
Angola. 26 ANG. Ther.
** Lightfootia welwitschii* A. DC., Ann. Sci. Nat., Bot. (ser. 5) 6: 332 (1866).

Wahlenbergia wittei Thulin, Symb. Bot. Upsal. 21: 180 (1975).
Zaïre. 23 ZAI. Ther.

Wahlenbergia wyleyana Sond. in Harv. & Sond., Fl. Cap. 3: 583 (1865). *Campanopsis wyleyana* (Sond.) Kuntze, Revis. Gen. Pl. 2: 379 (1891).
Cape Provinces. 27 CPP. Ther.

Synonyms:

Wahlenbergia sect. *Cervicina* (Delile) A. DC. === **Wahlenbergia** Schrad. ex Roth
Wahlenbergia sect. *Edraiantha* A. DC. === **Edraianthus** A. DC.
Wahlenbergia sect. *Heterochaenia* (A. DC.) Cordem. === **Heterochaenia** A. DC.
Wahlenbergia aberdarica T. C. E. Fr. === **Wahlenbergia capillacea** subsp. **tenuior** (Engl.)
   Thulin
Wahlenbergia agrestis (Wall.) A. DC. === **Wahlenbergia marginata** (Thunb. ex Murray)
   A. DC.
Wahlenbergia albomarginata var. *pygmaea* (Colenso) N. E. Br. === **Wahlenbergia pygmaea**
   Colenso
Wahlenbergia androsacea var. *multicaulis* Brehmer === **Wahlenbergia androsacea** A. DC.
Wahlenbergia annularis var. *bolusiana* Brehmer === **Wahlenbergia annularis** A. DC.
Wahlenbergia arabidifolia (Engl.) Brehmer === **Wahlenbergia krebsii** subsp. **arguta** (Hook.
   f.) Thulin
Wahlenbergia arenaria A. DC. === **Wahlenbergia androsacea** A. DC.
Wahlenbergia arenaria var. *lasiocarpa* Sond. === **Wahlenbergia obovata** Brehmer
Wahlenbergia arenaria var. *sphaerocarpa* (A. DC.) Sond. === **Wahlenbergia androsacea** A. DC.
Wahlenbergia arenaria var. *verreauxii* (A. DC.) Sond. === **Wahlenbergia androsacea** A. DC.
Wahlenbergia arguta Hook. f. === **Wahlenbergia krebsii** subsp. **arguta** (Hook. f.) Thulin
Wahlenbergia arguta var. *longifusiformis* Brehmer === **Wahlenbergia krebsii** subsp. **arguta**
   (Hook. f.) Thulin
Wahlenbergia arguta var. *parvilocula* Brehmer === **Wahlenbergia krebsii** subsp. **arguta**
   (Hook. f.) Thulin
Wahlenbergia arida (Kunth) Griseb. === **Wahlenbergia linarioides** (Lam.) A. DC.
Wahlenbergia arvatica (Lag.) Sweet === **Campanula arvatica** Lag.
Wahlenbergia baikalensis Freyn === ?
Wahlenbergia basutica E. Phillips === **Craterocapsa montana** (A. DC.) Hilliard & B. L. Burtt
Wahlenbergia bicolor Lothian === **Wahlenbergia stricta** (R. Br.) Sweet
Wahlenbergia billardieri Lothian === **Wahlenbergia littoralis** (Labill.) Sweet
Wahlenbergia bilocularis A. DC.   === **Wahlenbergia undulata** (L. f.) A. DC.
Wahlenbergia bivalvis Merr. === **Wahlenbergia marginata** (Thunb. ex Murray) A. DC.
Wahlenbergia bojeri A. DC. === **Wahlenbergia undulata** (L. f.) A. DC.
Wahlenbergia brachycarpa var. *pilosa* Brehmer === **Wahlenbergia brachycarpa** Schltr.
Wahlenbergia brevipes Hemsl. === **Homocodon brevipes** (Hemsl.) D. Y. Hong
Wahlenbergia brockiei J. A. Hay === **Wahlenbergia albomarginata** Hook. subsp.
   albomarginata
Wahlenbergia caledonica Sond. === **Wahlenbergia undulata** (L. f.) A. DC.
Wahlenbergia caledonica var. *cyanea* (Engl. & Gilg) Brehmer === **Wahlenbergia undulata**
   (L. f.) A. DC.
Wahlenbergia californica Kellogg === **Campanula californica** (Kellogg) A. Heller
Wahlenbergia capensis var. *leiocalycina* Zahlbr. === **Wahlenbergia capensis** (L.) A. DC.
Wahlenbergia capillacea var. *tenuior* Engl. === **Wahlenbergia capillacea** subsp. **tenuior**
   (Engl.) Thulin
Wahlenbergia capillaris (H. Buek) Lammers === **Wahlenbergia thulinii** Lammers
Wahlenbergia capillifolia var. *conferta* Brehmer === **Wahlenbergia capillifolia** E. Mey.
   ex Brehmer
Wahlenbergia caudata (Vis.) Vatke === **Edraianthus dalmaticus** (A. DC.) A. DC.
Wahlenbergia cernua var. *cuspidata* Brehmer === **Wahlenbergia cernua** (Thunb.) A. DC.
Wahlenbergia cernua var. *minor* H. Buek === **Wahlenbergia cernua** (Thunb.) A. DC.
Wahlenbergia cernua var. *minor* Sond. === **Wahlenbergia cernua** (Thunb.) A. DC.
Wahlenbergia cervicina A. DC. === **Wahlenbergia campanuloides** (Delile) Vatke
Wahlenbergia chamissoniana G. Don === **Wahlenbergia undulata** (L. f.) A. DC.

Wahlenbergia chamissoniana var. *macrantha* Cham. ex G. Don === **Wahlenbergia undulata** (L. f.) A. DC.

Wahlenbergia chamissoniana var. *micrantha* Cham. ex G. Don === **Wahlenbergia undulata** (L. f.) A. DC.

Wahlenbergia chondrophylla H. Buek === **Monopsis simplex** (L.) E. Wimm.

Wahlenbergia ciliolata A. DC. === **Wahlenbergia cernua** (Thunb.) A. DC.

Wahlenbergia clavatula Brehmer === **Wahlenbergia cernua** (Thunb.) A. DC.

Wahlenbergia clematidea Schrenk === **Codonopsis clematidea** (Schrenk) C. B. Clarke

Wahlenbergia clivosa A. DC. === **Wahlenbergia angustifolia** (Roxb.) A. DC.

Wahlenbergia coerulea H. J. P. Winkl. === **Wahlenbergia krebsii** subsp. **arguta** (Hook. f.) Thulin

Wahlenbergia colensoi N. E. Br. === ?

Wahlenbergia communis Carolin === **Wahlenbergia capillaris** (Lodd.) Sweet

Wahlenbergia congesta Thulin === **Wahlenbergia tenuiloba** Thulin

Wahlenbergia congestifolia var. *glabra* Brehmer === **Wahlenbergia congestifolia** Brehmer

Wahlenbergia congestifolia var. *laxa* Brehmer === **Wahlenbergia congestifolia** Brehmer

Wahlenbergia consimilis Lothian === **Wahlenbergia stricta** (R. Br.) Sweet subsp. **stricta**

Wahlenbergia crispa Nees === ?

Wahlenbergia croaticus (A. Kern.) Vatke === **Edraianthus graminifolius** (L.) A. DC. subsp. **graminifolius**

Wahlenbergia cyanea Engl. & Gilg === **Wahlenbergia undulata** (L. f.) A. DC.

Wahlenbergia cylindrica Pax & K. Hoffm. === **Campanula aristata** Wall.

Wahlenbergia dalmatica A. DC. === **Edraianthus dalmaticus** (A. DC.) A. DC.

Wahlenbergia dehiscens (Roxb.) A. DC. === **Wahlenbergia marginata** (Thunb. ex Murray) A. DC.

Wahlenbergia densicaulis var. *angusta* Brehmer === **Wahlenbergia densicaulis** Brehmer

Wahlenbergia denticulata var. *scabra* A. DC. === **Wahlenbergia denticulata** (Burch.) A. DC.

Wahlenbergia denticulata var. *transvaalensis* (Adamson) Welman === **Wahlenbergia denticulata** (Burch.) A. DC.

Wahlenbergia dentifera Brehmer === **Wahlenbergia cuspidata** Brehmer

Wahlenbergia denudata var. *brevisepala* Brehmer === **Wahlenbergia undulata** (L. f.) A. DC.

Wahlenbergia denudata var. *mutata* Brehmer === **Wahlenbergia denudata** A. DC.

Wahlenbergia depressa Wolley-Dod === **Treichelia longibracteata** (H. Buek) Vatke

Wahlenbergia dicentrifolia C. B. Clarke === **Codonopsis dicentrifolia** (C. B. Clarke) W. W. Sm.

Wahlenbergia dichotoma var. *simplex* Sond. === **Wahlenbergia dichotoma** A. DC.

Wahlenbergia diffusa A. DC. === **Microcodon linearis** (L. f.) H. Buek

Wahlenbergia dinteri Brehmer === **Wahlenbergia undulata** (L. f.) A. DC.

Wahlenbergia dinteri var. *elongata* Brehmer === **Wahlenbergia undulata** (L. f.) A. DC.

Wahlenbergia dinteri var. *paucilaciniata* Brehmer === **Wahlenbergia undulata** (L. f.) A. DC.

Wahlenbergia dinteri var. *rotundicapsula* Brehmer === **Wahlenbergia undulata** (L. f.) A. DC.

Wahlenbergia dinteri var. *virgulta* Brehmer === **Wahlenbergia undulata** (L. f.) A. DC.

Wahlenbergia dioica Domin === **Wahlenbergia quadrifida** (R. Br.) A. DC.

Wahlenbergia divaricata E. Mey. ex A. DC. === **Wahlenbergia paniculata** (L. f.) A. DC.

Wahlenbergia diversifolia A. DC. === **Wahlenbergia procumbens** (L. f.) A. DC.

Wahlenbergia dregeana A. DC. === **Wahlenbergia ecklonii** H. Buek

Wahlenbergia dunantii var. *glabrata* Brehmer === **Wahlenbergia dunantii** A. DC.

Wahlenbergia ecklonii var. *brevisepala* Brehmer === **Wahlenbergia ecklonii** H. Buek

Wahlenbergia ecklonii var. *gracilis* Sond. === **Wahlenbergia ecklonii** H. Buek

Wahlenbergia ecklonii var. *pilosa* Sond. === **Wahlenbergia ecklonii** H. Buek

Wahlenbergia elatines (L.) Link === **Campanula elatines** L.

Wahlenbergia elongata (Willd.) Schrad. ex Roth === **Wahlenbergia campanuloides** (Delile) Vatke

Wahlenbergia emirnensis A. DC. === **Gunillaea emirnensis** (A. DC.) Thulin

Wahlenbergia engleri Brehmer === **Wahlenbergia undulata** (L. f.) A. DC.

Wahlenbergia ensifolia (Lam.) A. DC. === **Heterochaenia ensifolia** (Lam.) A. DC.

Wahlenbergia epacridea var. *glabrata* Sond. === **Wahlenbergia epacridea** Sond.

Wahlenbergia erinus (L.) Link === **Campanula erinus** L.

Wahlenbergia etbaica (Schweinf.) Vatke === **Wahlenbergia lobelioides** subsp. **nutabunda** (Guss.) Murb.

Wahlenbergia eurycarpa Domin === **Wahlenbergia quadrifida** (R. Br.) A. DC.

Wahlenbergia exilis var. *major* Sond. === **Wahlenbergia exilis** A. DC.

Wahlenbergia expansa (Rudolph) Steffen === **Campanula expansa** Rudolph

Wahlenbergia fasciculata var. *pilosa* Brehmer === **Wahlenbergia fasciculata** Brehmer

Wahlenbergia fernandeziana f. *elata* Skottsb. === **Wahlenbergia fernandeziana** A. DC.

Wahlenbergia filicaulis R. D. Good === **Prismatocarpus sessilis** Eckl. ex A. DC.

Wahlenbergia filipes M. E. Jones === ?

Wahlenbergia filipes var. *dentata* Brehmer === **Wahlenbergia filipes** Brehmer

Wahlenbergia flaccida A. DC. === ?

Wahlenbergia flaccida C. Presl === **Campanula garganica** Ten. subsp. **garganica**

Wahlenbergia flexilis Petrie === **Wahlenbergia albomarginata** subsp. **flexilis** (Petrie) Petterson

Wahlenbergia foliosa A. DC. === **Wahlenbergia linifolia** (Roxb.) A. DC.

Wahlenbergia foliosa Brehmer === **Wahlenbergia banksiana** A. DC.

Wahlenbergia furcata Brehmer === **Wahlenbergia cuspidata** Brehmer

Wahlenbergia galpiniae var. *excedens* Brehmer === **Wahlenbergia galpiniae** Schltr.

Wahlenbergia galpinii E. Phillips === **Wahlenbergia lobulata** Brehmer

Wahlenbergia garganica G. Don === **Campanula garganica** Ten. subsp. **garganica**

Wahlenbergia glandulosa Brehmer === **Wahlenbergia androsacea** A. DC.

Wahlenbergia gracilis f. *littoralis* (Labill.) Wawra === **Wahlenbergia littoralis** (Labill.) Sweet

Wahlenbergia gracilis var. *capillaris* (R. Br.) A. DC. === **Wahlenbergia capillaris** (Lodd.) Sweet

Wahlenbergia gracilis var. *integerrima* Brehmer === ?

Wahlenbergia gracilis var. *littoralis* (Labill.) A. DC. === **Wahlenbergia littoralis** (Labill.) Sweet

Wahlenbergia gracilis var. *misera* Hemsl. === **Wahlenbergia marginata** (Thunb. ex Murray) A. DC.

Wahlenbergia gracilis var. *polymorpha* A. DC. === **Wahlenbergia vernicosa** Petterson

Wahlenbergia gracilis var. *stricta* (R. Br.) A. DC. === **Wahlenbergia stricta** (R. Br.) Sweet

Wahlenbergia gracillima S. Moore === **Wahlenbergia banksiana** A. DC.

Wahlenbergia graminifolia (L.) A. DC. === **Edraianthus graminifolius** (L.) A. DC.

Wahlenbergia grandiflora var. *fissa* Brehmer === **Wahlenbergia grandiflora** Brehmer

Wahlenbergia grandiflora var. *lanceolata* Brehmer === **Wahlenbergia grandiflora** Brehmer

Wahlenbergia grandiflora var. *lata* Brehmer === **Wahlenbergia grandiflora** Brehmer

Wahlenbergia grandiflora var. *undulata* Brehmer === **Wahlenbergia grandiflora** Brehmer

Wahlenbergia guthriei L. Bolus === **Theilera guthriei** (L. Bolus) E. Phillips

Wahlenbergia gymnoclada Lothian === **Wahlenbergia littoralis** (Labill.) Sweet.

Wahlenbergia hederacea var. *arvatica* (Lag.) Steud. === **Campanula arvatica** Lag.

Wahlenbergia hederifolia Bubani === **Wahlenbergia hederacea** (L.) Rchb.

Wahlenbergia hercegovina (K. Malý) K. Malý === **Edraianthus hercegovinus** K. Malý

Wahlenbergia hilsenbergii Bojer ex A. DC. === **Wahlenbergia madagascariensis** A. DC.

Wahlenbergia hispidula (L. f.) Schrad. === **Microcodon hispidulum** (L. f.) Sond.

Wahlenbergia homallanthina (Ledeb.) A. DC. === **Campanula expansa** Rudolph

Wahlenbergia homallanthina var. *angustifolia* A. DC. === **Campanula expansa** Rudolph

Wahlenbergia huillana A. DC. === **Gunillaea emirnensis** (A. DC.) Thulin

Wahlenbergia huillana var. *pusilla* A. DC. === **Gunillaea emirnensis** (A. DC.) Thulin

Wahlenbergia humifusa Markgr. === **Wahlenbergia campanuloides** (Delile) Vatke

Wahlenbergia humilis A. DC. === **Wahlenbergia lobelioides** subsp. **riparia** (Leprieur & Perottet ex A. DC.) Thulin

Wahlenbergia humpatensis Brehmer === **Wahlenbergia candolleana** (Hiern) Thulin

Wahlenbergia inconspicua A. DC. === ?

Wahlenbergia indica A. DC. === **Wahlenbergia marginata** (Thunb. ex Murray) A. DC.

Wahlenbergia inhambanensis Klotzsch === **Wahlenbergia androsacea** A. DC.

Wahlenbergia inhambanensis var. *erecta* Brehmer === **Wahlenbergia androsacea** A. DC.

Wahlenbergia integrifolia A. DC. === **Wahlenbergia stellarioides** Cham.

Wahlenbergia kilimandscharica Engl. === **Wahlenbergia capillacea** subsp. **tenuior** (Engl.) Thulin

Wahlenbergia kilimandscharica var. *intermedia* Brehmer === **Wahlenbergia capillacea** subsp. **tenuior** (Engl.) Thulin

Wahlenbergia kitaibelii A. DC. === **Edraianthus graminifolius** (L.) A. DC. subsp. **graminifolius**

Wahlenbergia larrainii (Bertero) Skottsb. === **Wahlenbergia fernandeziana** A. DC.

Wahlenbergia lateralis Brehmer === **Wahlenbergia ramosissima** subsp. **lateralis** (Brehmer) Thulin

Wahlenbergia lavandulifolia (Reinw. ex Blume) A. DC. === **Wahlenbergia marginata** (Thunb. ex Murray) A. DC.

Wahlenbergia laxa G. Simpson === **Wahlenbergia albomarginata** subsp. **laxa** (G. Simpson) Petterson

Wahlenbergia leucantha Engl. & Gilg === **Wahlenbergia banksiana** A. DC.

Wahlenbergia limenophylax Lothian === **Wahlenbergia insulae-howei** Lothian

Wahlenbergia linarioides var. *arida* (Kunth) A. DC. === **Wahlenbergia linarioides** (Lam.) A. DC.

Wahlenbergia linarioides var. *filiformis* (Ruiz & Pav.) A. DC. === **Wahlenbergia linarioides** (Lam.) A. DC.

Wahlenbergia linarioides var. *latifolia* A. DC. === **Wahlenbergia linarioides** (Lam.) A. DC.

Wahlenbergia linarioides var. *macrantha* Phil. === **Wahlenbergia linarioides** (Lam.) A. DC.

Wahlenbergia linarioides var. *micrantha* Phil. === **Wahlenbergia linarioides** (Lam.) A. DC.

Wahlenbergia linearis (L. f.) A. DC. === **Microcodon lineare** (L. f.) H. Buek

Wahlenbergia littoralis Schltr. & Brehmer === **Wahlenbergia orae** Lammers

Wahlenbergia lobata Brehmer === **Wahlenbergia debilis** H. Buek

Wahlenbergia lobelioides var. *gussonei* Webb & Berth. === **Wahlenbergia lobelioides** subsp. **nutabunda** (Guss.) Murb.

Wahlenbergia lobelioides var. *macilenta* Webb & Berth. === **Wahlenbergia lobelioides** (L. f.) Schrad. ex Link subsp. **lobelioides**

Wahlenbergia longifolia var. *corymbosa* (Adamson) Welman === **Wahlenbergia longifolia** (A. DC.) Lammers

Wahlenbergia longisepala Brehmer === **Wahlenbergia decipiens** A. DC.

Wahlenbergia lurida (Lindl.) G. Manetti === **Codonopsis rotundifolia** Benth.

Wahlenbergia macra Schltr. & Brehmer === **Wahlenbergia ecklonii** H. Buek

Wahlenbergia maculata Brehmer === **Wahlenbergia cernua** (Thunb.) A. DC.

Wahlenbergia madagascariensis A. DC. === **Wahlenbergia abyssinica** (Hochst. ex A. Rich.) Thulin

Wahlenbergia mairei H. Lév. === **Cyananthus flavus** C. Marquand

Wahlenbergia mannii Vatke === **Wahlenbergia silenoides** Hochst. ex A. Rich.

Wahlenbergia mannii var. *intermedia* Brehmer === **Wahlenbergia silenoides** Hochst. ex A. Rich.

Wahlenbergia mannii var. *virgulta* Brehmer === **Wahlenbergia silenoides** Hochst. ex A. Rich.

Wahlenbergia marginata subvar. *trichogyna* (Stearn) Tuyn === **Wahlenbergia stricta** (R. Br.) Sweet subsp. **stricta**

Wahlenbergia marginata var. *grandiflora* Tuyn === **Wahlenbergia gloriosa** Lothian

Wahlenbergia marginata var. *littoralis* (Labill.) Hochr. === **Wahlenbergia littoralis** (Labill.) Sweet

Wahlenbergia marginata var. *neocaledonica* Lothian === **Wahlenbergia gracilis** (G. Forst.) A. DC.

Wahlenbergia marginata var. *polymorpha* (A. DC.) Hochr. === **Wahlenbergia vernicosa** Petterson

Wahlenbergia masafuerae f. *rosea* Skottsb. === **Wahlenbergia masafuerae** (Phil.) Skottsb.

Wahlenbergia mashonica N. E. Br. === **Wahlenbergia banksiana** A. DC.

Wahlenbergia mashonica var. *junodis* Brehmer === **Wahlenbergia banksiana** A. DC.

Wahlenbergia mauritiana I. Richardson === **Nesocodon mauritianus** (I. Richardson) Thulin

Wahlenbergia meyeri var. *lanceolata* Brehmer === **Wahlenbergia meyeri** A. DC.

Wahlenbergia meyeri var. *subacaulis* Brehmer === **Wahlenbergia meyeri** A. DC.

Wahlenbergia micrantha E. Mey. === **Wimmerella pygmaea** (Thunb.) Serra, M. B. Crespo & Lammers

Wahlenbergia monantha H. J. P. Winkl. ex H. Limpr. === **Homocodon brevipes** (Hemsl.) D. Y. Hong

Wahlenbergia monotropa Killick === **Wahlenbergia lobulata** Brehmer

Wahlenbergia montana A. DC. === **Craterocapsa montana** (A. DC.) Hilliard & B. L. Burtt

Wahlenbergia montana var. *angustisepala* Brehmer === **Craterocapsa tarsodes** Hilliard & B. L. Burtt

Wahlenbergia montana var. *glabrata* Sond. === **Craterocapsa tarsodes** Hilliard & B. L. Burtt

Wahlenbergia morganii Petrie === **Wahlenbergia congesta** (Cheeseman) N. E. Br. subsp. **congesta**

Wahlenbergia multicaulis var. *dispar* N. E. Br. === **Wahlenbergia capillaris** (Lodd.) Sweet

Wahlenbergia multiflora Conrath === **Wahlenbergia banksiana** A. DC.

Wahlenbergia nudicaulis A. DC. === **Wahlenbergia androsacea** A. DC.

Wahlenbergia nutabunda (Guss.) A. DC. === **Wahlenbergia lobelioides** subsp. **nutabunda** (Guss.) Murb.

Wahlenbergia nutabunda var. *erythraeae* Chiov. === **Wahlenbergia lobelioides** subsp. **nutabunda** (Guss.) Murb.

Wahlenbergia oatesii Rolfe === **Wahlenbergia undulata** (L. f.) A. DC.

Wahlenbergia obovata var. *cernua* Brehmer === **Wahlenbergia obovata** Brehmer

Wahlenbergia obovata var. *fissa* Brehmer === **Wahlenbergia obovata** Brehmer

Wahlenbergia obovata var. *lata* Brehmer === **Wahlenbergia obovata** Brehmer

Wahlenbergia okavangensis N. E. Br. === **Wahlenbergia banksiana** A. DC.

Wahlenbergia oligotricha var. *hispidula* Brehmer === **Wahlenbergia oligotricha** Schltr. & Brehmer

Wahlenbergia oliveri Schweinf. === **Wahlenbergia capillacea** subsp. **tenuior** (Engl.) Thulin

Wahlenbergia oppositifolia A. DC. === **Wahlenbergia madagascariensis** A. DC.

Wahlenbergia oppositifolia var. *crenata* Brehmer === **Wahlenbergia madagascariensis** A. DC.

Wahlenbergia oppositifolia var. *crispa* A. DC. === **Wahlenbergia madagascariensis** A. DC.

Wahlenbergia ovalis Brehmer === **Craterocapsa insizwae** (Zahlbr.) Hilliard & B. L. Burtt

Wahlenbergia ovata D. Don === **Percarpa carnosa** (Wall.) Hook. f. & Thomson

Wahlenbergia paniculata var. *glabrata* Sond. === **Wahlenbergia subpilosa** Brehmer

Wahlenbergia paniculata var. *massonii* (A. DC.) Sond. === **Wahlenbergia massonii** A. DC.

Wahlenbergia paniculata var. *rudis* (E. Mey. ex A. DC.) Sond. === **Wahlenbergia paniculata** (L. f.) A. DC.

Wahlenbergia parviflora A. DC. === **Wahlenbergia dichotoma** A. DC.

Wahlenbergia paucidentata var. *tysonii* Schinz === **Wahlenbergia paucidentata** Schinz

Wahlenbergia pavida Launert === **Wahlenbergia erophiloides** Markgr.

Wahlenbergia pendula Schrad. === **Wahlenbergia lobelioides** (L.) Schrad. ex Link subsp. **lobelioides**

Wahlenbergia pentamera A. DC. === **Wahlenbergia ecklonii** H. Buek

Wahlenbergia perennis Brehmer === **Wahlenbergia androsacea** A. DC.

Wahlenbergia perotifolia (Willd. ex Schult.) Wight & Arn. === **Wahlenbergia erecta** (Roth ex Schult.) Tuyn

Wahlenbergia pinnata var. *simplicifolia* Compton === **Wahlenbergia capillacea** subsp. **tenuior** (Engl.) Thulin

Wahlenbergia planifolia Gand. === ?

Wahlenbergia polychotoma Brehmer === **Wahlenbergia undulata** (L. f.) A. DC.

Wahlenbergia polyclada Hook. f. === **Wahlenbergia silenoides** Hochst. ex A. Rich.

Wahlenbergia procumbens var. *diversifolia* (A. DC.) Sond. === **Wahlenbergia procumbens** (L. f.) A. DC.

Wahlenbergia procumbens var. *foliosa* A. DC. === **Wahlenbergia procumbens** (L. f.) A. DC.

Wahlenbergia procumbens var. *intermedia* Brehmer === **Wahlenbergia procumbens** (L. f.) A. DC.

Wahlenbergia psammophila var. *longisepala* Brehmer === **Wahlenbergia psammophila** Schltr.

Wahlenbergia pseudonudicaulis var. *diversa* Brehmer === **Wahlenbergia pseudonudicaulis** Brehmer

Wahlenbergia pumilio (Portenschlag ex Schult.) A. DC. === **Edraianthus pumilio** (Portenschlag ex Schult.) A. DC.

Wahlenbergia purpurea (Wall.) A. DC. === **Codonopsis purpurea** Wall.

Wahlenbergia pygmaea var. *laxa* (G. Simpson) H. H. Allan === **Wahlenbergia albomarginata** subsp. **laxa** (G. Simpson) Petterson

Wahlenbergia ramulosa E. Mey. ex A. DC. === **Wahlenbergia debilis** H. Buek

Wahlenbergia recurvata Brehmer === **Wahlenbergia virgata** Engl.

Wahlenbergia repens Schrad. === **Wahlenbergia procumbens** (L. f.) A. DC.

Wahlenbergia rhodesiana S. Moore === **Wahlenbergia banksiana** A. DC.

Wahlenbergia rigida Bernh. === **Wahlenbergia robusta** (A. DC.) Sond.

Wahlenbergia riparia Leprieur & Perottet ex A. DC. === **Wahlenbergia lobelioides** subsp. **riparia** (Leprieur & Perottet ex A. DC.) Thulin

Wahlenbergia riparia var. *clavata* Brehmer === **Wahlenbergia lobelioides** subsp. **riparia** (Leprieur & Perottet ex A. DC.) Thulin

Wahlenbergia riparia var. *etbaica* Brehmer === **Wahlenbergia lobelioides** subsp. **nutabunda** (Guss.) Murb.

Wahlenbergia riparia var. *segregata* Brehmer === **Wahlenbergia lobelioides** subsp. **riparia** (Leprieur & Perottet ex A. DC.) Thulin

Wahlenbergia riparia var. *virgulta* Brehmer === **Wahlenbergia lobelioides** subsp. **riparia** (Leprieur & Perottet ex A. DC.) Thulin

Wahlenbergia rivularis var. *oblonga* Brehmer === **Wahlenbergia rivularis** Diels

Wahlenbergia rosulata Brehmer === **Wahlenbergia androsacea** A. DC.

Wahlenbergia rotundifolia Brehmer === **Wahlenbergia brehmeri** Lammers

Wahlenbergia rotundifolia (Benth.) A. DC. === **Codonopsis rotundifolia** Benth.

Wahlenbergia roylei A. DC. === **Codonopsis ovata** Benth.

Wahlenbergia rubens var. *brachyphylla* (Adamson) Welman === **Wahlenbergia rubens** (H. Buek) Lammers

Wahlenbergia rubioides var. *stokoei* (Adamson) Welman === **Wahlenbergia rubioides** (Banks ex A. DC.) Lammers

Wahlenbergia rudis E. Mey. ex A. DC. === **Wahlenbergia paniculata** (L. f.) A. DC.

Wahlenbergia rutenbergiana Vatke === ?

Wahlenbergia sabulosa Brehmer === **Wahlenbergia densicaulis** Brehmer

Wahlenbergia saginoides S. Moore === **Wahlenbergia banksiana** A. DC.

Wahlenbergia sarmentosa T. C. E. Fr. === **Wahlenbergia krebsii** subsp. **arguta** (Hook. f.) Thulin

Wahlenbergia saxicola var. *congesta* Cheeseman === **Wahlenbergia congesta** (Cheeseman) N. E. Br.

Wahlenbergia saxifragoides Brehmer === **Wahlenbergia procumbens** (L. f.) A. DC.

Wahlenbergia schimperi (A. Rich.) Schweinf. & Asch. === **Wahlenbergia erecta** (Roth ex Schult.) Tuyn

Wahlenbergia scoparia Brehmer === **Wahlenbergia undulata** (L. f.) A. DC.

Wahlenbergia scoparia var. *obovata* Brehmer === **Wahlenbergia undulata** (L. f.) A. DC.

Wahlenbergia scopella var. *rotundata* Brehmer === **Wahlenbergia scopella** Brehmer

Wahlenbergia semiglabra A. DC. === **Wahlenbergia androsacea** A. DC.

Wahlenbergia serpyllifolia (Vis.) Vatke === **Edraianthus serpyllifolius** (Vis.) A. DC.

Wahlenbergia sessiliflora var. *dentata* Brehmer === **Wahlenbergia sessiliflora** Brehmer

Wahlenbergia sieberi A. DC. === **Wahlenbergia quadrifida** (R. Br.) A. DC.

Wahlenbergia silenoides var. *elongata* Brehmer === **Wahlenbergia silenoides** Hochst. ex A. Rich.

Wahlenbergia simplicicaulis de Vriese === **Wahlenbergia multicaulis** Benth.

Wahlenbergia simpsonii J. A. Hay === **Wahlenbergia albomarginata** subsp. **flexilis** (Petrie) Petterson

Wahlenbergia solitaria Brehmer === ?

Wahlenbergia sparticula Chiov. === **Wahlenbergia virgata** Engl.

Wahlenbergia sphaerica var. *longifolia* Brehmer === **Wahlenbergia sphaerica** Brehmer

Wahlenbergia sphaerocarpa A. DC. === **Wahlenbergia androsacea** A. DC.

Wahlenbergia spicata E. Mey. ex A. DC. === **Wahlenbergia juncea** (H. Buek) Lammers

Wahlenbergia spinulosa A. DC. === ?

Wahlenbergia spinulosa Engl. === **Wahlenbergia denticulata** (Burch.) A. DC.

Wahlenbergia squamifolia var. *tenuis* Brehmer === **Wahlenbergia squamifolia** Brehmer

Wahlenbergia stellarioides var. *angusta* Sond. === **Wahlenbergia stellarioides** Cham.

Wahlenbergia stellarioides var. *major* Sond. === **Wahlenbergia stellarioides** Cham.

Wahlenbergia striata A. DC. === **Wahlenbergia undulata** (L. f.) A. DC.

Wahlenbergia subnuda Conrath === **Wahlenbergia virgata** Engl.

Wahlenbergia subfusiformis var. *involuta* Brehmer === **Wahlenbergia subfusiformis** Brehmer

Wahlenbergia subrosulata var. *grandifolia* Brehmer === **Wahlenbergia subrosulata** Brehmer

Wahlenbergia subulata var. *congesta* (Adamson) Welman === **Wahlenbergia subulata** (L'Hér.) Lammers

Wahlenbergia subulata var. *tenuifolia* (Adamson) Welman === **Wahlenbergia subulata** (L'Hér.) Lammers

Wahlenbergia swellendamensis H. Buek === **Wahlenbergia ecklonii** H. Buek

Wahlenbergia tagdellii Lothian === **Wahlenbergia multicaulis** Benth.

Wahlenbergia tenella var. *montana* (Adamson) Welman === **Wahlenbergia tenerrima** (H. Buek) Lammers

Wahlenbergia tenella var. *palustris* (Adamson) Welman === **Wahlenbergia tenella** (L. f.) Lammers

Wahlenbergia tenella var. *stokoei* (Adamson) Welman === **Wahlenbergia tenella** (L. f.) Lammers

Wahlenbergia tenuifolia (Waldst. & Kit.) A. DC. === **Edraianthus tenuifolius** (Waldst. & Kit.) A. DC.

Wahlenbergia thalictrifolia (Wall.) A. DC. === **Codonopsis thalictrifolia** Wall.

Wahlenbergia thunbergii var. *debilis* (Sond.) B. Nord. === **Wahlenbergia thunbergii** (Schult.) B. Nord.

Wahlenbergia thunbergii var. *filifolia* (Adamson) B. Nord. === **Wahlenbergia thunbergii** (Schult.) B. Nord.

Wahlenbergia trichogyna Stearn === **Wahlenbergia stricta** (R. Br.) Sweet subsp. **stricta**

Wahlenbergia tumida var. *gracilis* Brehmer === **Wahlenbergia tumida** Brehmer

Wahlenbergia turbinata A. DC. === **Wahlenbergia ecklonii** H. Buek

Wahlenbergia tysonii Zahlbr. === **Wahlenbergia rivularis** Diels

Wahlenbergia uitenhagensis (H. Buek) Lammers === **Wahlenbergia thunbergii** (Schult.) B. Nord.

Wahlenbergia uitenhagensis var. *debilis* (Sond.) Welman === **Wahlenbergia thunbergii** (Schult.) B. Nord.

Wahlenbergia uitenhagensis var. *filifolia* (Adamson) Welman === **Wahlenbergia thunbergii** (Schult.) B. Nord.

Wahlenbergia undulata var. *glabrata* Sond. === **Wahlenbergia undulata** (L. f.) A. DC.

Wahlenbergia undulata var. *latisepala* Brehmer === **Wahlenbergia undulata** (L. f.) A. DC.

Wahlenbergia undulata var. *macrantha* (Cham. ex G. Don) Sond. === **Wahlenbergia undulata** (L. f.) A. DC.

Wahlenbergia undulata var. *rotundifolia* Brehmer === **Wahlenbergia undulata** (L. f.) A. DC.

Wahlenbergia undulata var. *striata* (A. DC.) Zahlbr. === **Wahlenbergia undulata** (L. f.) A. DC.

Wahlenbergia variabilis E. Mey. ex A. DC. === **Wahlenbergia krebsii** Cham. subsp. **krebsii**

Wahlenbergia variabilis var. *krebsiana* A. DC. === **Wahlenbergia krebsii** Cham. subsp. **krebsii**

Wahlenbergia verreauxii A. DC. === **Wahlenbergia androsacea** A. DC.

Wahlenbergia vinciflora Decne. === **Wahlenbergia gracilis** (G. Forst.) A. DC.

Wahlenbergia vinciflora f. *eriocalyx* Domin === **Wahlenbergia stricta** (R. Br.) Sweet subsp. **stricta**

Wahlenbergia vinciflora var. *littoralis* (Labill.) N. E. Br. === **Wahlenbergia littoralis** (Labill.) Sweet

Wahlenbergia vinciflora var. *rosulata* J. M. Black === ?

Wahlenbergia virgata var. *longisepala* Brehmer === **Wahlenbergia virgata** Engl.

Wahlenbergia virgata var. *tenuis* Brehmer === **Wahlenbergia virgata** Engl.
Wahlenbergia virgata var. *valida* Brehmer === **Wahlenbergia virgata** Engl.
Wahlenbergia viridis (Wall.) A. DC. === **Codonopsis viridis** Wall.
Wahlenbergia zeyheri H. Buek === **Wahlenbergia krebsii** Cham. subsp. **krebsii**
Wahlenbergia zeyheri var. *krebsiana* (A. DC.) Sond. === **Wahlenbergia krebsii** Cham.
   subsp. **krebsii**
Wahlenbergia zeyheri var. *lanceolata* Brehmer === **Wahlenbergia krebsii** Cham. subsp. **krebsii**
Wahlenbergia zeyheri var. *linearis* Sond. === **Wahlenbergia krebsii** Cham. subsp. **krebsii**
Wahlenbergia zeyheri var. *natalensis* Sond. === **Wahlenbergia krebsii** Cham. subsp. **krebsii**
Wahlenbergia zeyheri var. *pyriformis* Brehmer === **Wahlenbergia krebsii** Cham. subsp. **krebsii**

Weitenwebera

This genus was established for those species of *Campanula* with exappendiculate calyx, short-campanulate corolla, three-locular 10-angled ovary, and basally dehiscent capsules. Type [designated here]: *Weitenwebera cervicaria* (L.) Opiz.

   Berchtold, F. (1839). Oekonomisch-technische flora Böhmens, band 2, abteil. 2. Prague: Thomas Thabor. Ge. — Segregation of genus from *Campanula*.

Synonyms:
Weitenwebera Opiz === **Campanula** L.
Weitenwebera cervicaria (L.) Opiz === **Campanula cervicaria** L.
Weitenwebera glomerata (L.) Opiz === **Campanula glomerata** L.
Weitenwebera glomerata var. *mollis* (Tausch) Opiz === ?
Weitenwebera glomerata var. *nana* Opiz === ?
Weitenwebera glomerata var. *salviifolia* (Wallr.) Opiz === ?

Wimmerella

Lobelioideae, 10 species, Cape Provinces of South Africa, with one species extending into KwaZulu-Natal. These species were assigned to *Solenopsis* (under the illegitimate name *Laurentia;* q.v.) for many years, until segregated as *Enchysia* (Serra & Crespo 1997; Crespo et al. 1998). However, that name properly is a synonym of *Lobelia,* and it was necessary to create the genus *Wimmerella* to house these species. In a preliminary molecular phylogeny (E. Knox, pers. comm.), all sampled species were members of a single clade that also included *Lobelia anceps.* Type [designated by the authors]: *Wimmerella secunda* (L. f.) Serra, M. B. Crespo & Lammers.

   Adamson, R. S. (1950). *Laurentia* Neck. In R. S. Adamson & T. M. Salter, Flora of the Cape Peninsula: 752-753. Cape Town: Juta. En. — Flora with key and descriptions.
 • Wimmer, F. E. (1953). [*Laurentia* (Mich.) Adans.] Sectio I. *Solenopsis* Endl. In A. Engler & L. Diels, Das Pflanzenreich 276b: 387-397, illus. Berlin: Akademie-Verrlag. Ge., La. — Monograph with key, descriptions, full nomenclatrure, and specimen citations.
 • Wimmer, F. E. (1968). *Laurentia* (Mich.) Adams. [sic] sect. I. *Laurentia.* In A. Engler & L. Diels, Das Pflanzenreich IV. 276c: 853-855. Berlin: Akademie-Verrlag. Ge., La. — Supplement to Wimmer (1953).
   Dyer, R. A. (1975). *Laurentia* Michx. [sic] ex Adans. In The genera of southern African flowering plants 1: 646. Pretoria: Department of Agricultural Technical Services. En. — Generic description.
   Serra, L. & M. B. Crespo (1997). An outline revision of the subtribe Siphocampylinae (Lobeliaceae). Lagascalia 19: 881-888, illus., map. En. — Name *Enchysia* used for the South African species of *Laurentia* s.l.
   Crespo, M. B., L. Serra & A. Juan (1998). *Solenopsis* (Lobeliaceae): a genus endemic in the Mediterranean Region. Pl. Syst. Evol. 210: 211-229, illus., maps. En. — Name *Enchysia* used for the South African species of *Laurentia* s.l.

- Serra, L., M. B. Crespo & T. G. Lammers (1999). *Wimmerella,* a new South African genus of Lobelioideae (Campanulaceae). Novon 9: 414-418. En. — Establishment of genus; includes key to distinguish it from allied genera and key to its species.

  Goldblatt, P. & J. Manning (2000). Cape plants. A conspectus of the Cape flora of South Africa. *Wimmerella.* Strelitzia 9: 401-402. En. — Checklist for Cape of Good Hope, with brief species desciptions.

  Welman, W. G. (2000). *Wimmerella* L. Serra, M. B. Crespo & Lammers. In O. A. Leistner (ed.), Seed plants of southern Africa: families and genera. Strelitzia 10: 340. En. — Generic description.

Wimmerella Serra, M. B. Crespo & Lammers, Novon 9: 415 (1999).
S. Africa. 27.

Wimmerella arabidea (C. Presl) Serra, M. B. Crespo & Lammers, Novon 9: 416 (1999).
Cape Provinces & KwaZulu-Natal. 27 CPP NAT. Hemicr.
* *Rapuntium arabideum* C. Presl, Prodr. Monogr. Lobel.: 18 (1836). *Laurentia arabidea* (C. Presl) A. DC. in DC., Prodr. 7: 410 (1839). *Lobelia arabidea* (C. Presl) Steud., Nomencl. Bot. (ed. 2) 2: 59 (1841).

Wimmerella bifida (Thunb.) Serra, M. B. Crespo & Lammers, Novon 9: 416 (1999).
Cape Provinces. 27 CPP. Hemicr.
* *Lobelia bifida* Thunb., Prodr. Fl. Cap.: 40 (1794). *Rapuntium bifidum* (Thunb.) C. Presl, Prodr. Monogr. Lobel.: 30 (1836). *Laurentia bifida* (Thunb.) Sond. in Harv. & Sond., Fl. Cap. 3: 552 (1865).
  Solenopsis dregeana C. Presl in E. Mey., Comm. Pl. Afr. Austr.: 290 (1838). *Laurentia dregeana* (C. Presl) A. DC. in DC., Prodr. 7: 411 (1839).

Wimmerella frontidentata (E. Wimm.) Serra, M. B. Crespo & Lammers, Novon 9: 417, 418 (1999).
Cape Provinces. 27 CPP. Hemicr.
* *Laurentia frontidentata* E. Wimm., Pflanzer. IV.276c: 855 (1968).

Wimmerella giftbergensis (E. Phillips) Serra, M. B. Crespo & Lammers, Novon 9: 417 (1999).
Cape Provinces. 27 CPP. Hemicr.
* *Lobelia giftbergensis* E. Phillips, Ann. S. African Mus. 9: 121 (1913). *Laurentia giftbergensis* (E. Phillips) E. Wimm., Repert. Spec. Nov. Regni Veg. 38: 77 (1935).

Wimmerella hederacea (Sond.) Serra, M. B. Crespo & Lammers, Novon 9: 417 (1999).
Cape Provinces. 27 CPP. Hemicr.
* *Laurentia hederacea* Sond. in Harv. & Sond., Fl. Cap. 3: 553 (1865).

Wimmerella hedyotidea (Schltr.) Serra, M. B. Crespo & Lammers, Novon 9: 417 (1999).
Cape Provinces. 27 CPP. Hemicr.
* *Laurentia hedyotidea* Schltr., Bot. Jahrb. Syst. 27: 197 (1899).
  Laurentia hedyotidea var. *major* E. Wimm., Pflanzenr. IV.276c: 854 (1968).

Wimmerella longitubus (E. Wimm.) Serra, M. B. Crespo & Lammers, Novon 9: 417 (1999).
Cape Provinces. 27 CPP. Hemicr.
  Laurentia longiflora Schltr., Bot. Jahrb. Syst. 57: 625 (1922); non (L.) Peterm., Pflanzenreich: 444 (1845).
* *Laurentia longitubus* E. Wimm., Repert. Spec. Nov. Regni Veg. 22: 193 (1926).

Wimmerella mariae (E. Wimm.) Serra, M. B. Crespo & Lammers, Novon 9: 417, 418 (1999).
Cape Provinces. 27 CPP. Hemicr.
* *Laurentia mariae* E. Wimm., Pflanzenr. IV.276c: 854 (1968).

Wimmerella pygmaea (Thunb.) Serra, M. B. Crespo & Lammers, Novon 9: 417 (1999).
Cape Provinces. 27 CPP. Hemicr.
  * *Lobelia pygmaea* Thunb., Prodr. Fl. Cap.: 40 (1794). *Rapuntium pygmaeum* (Thunb.) C.
    Presl, Prodr. Monogr. Lobel.: 22 (1836). *Laurentia pygmaea* (Thunb.) Sond. in Harv. &
    Sond., Fl. Cap. 3: 553 (1865).
  Lobelia pusilla G. Don, Gen. Hist. 3: 711 (1834).
  Wahlenbergia micrantha E. Mey. in Drège, Cat. Pl. Afr. Austral.: 9 (1837). *Laurentia
    micrantha* (E. Mey.) A. DC. in DC., Prodr. 7: 411 (1839). *Lobelia micrantha* (E. Mey.)
    Heynh., Nom. Hort. Bot. 1: 471 (1840); non Kunth in Humb., Bonpl. & Kunth, Nov.
    Gen. Sp. 3: 316 (quarto), 247 (folio) (1819); nec Hook., Exot. Fl. 1: 44 (1823).
  Laurentia pygmaea var. *glabra* Sond. in Harv. & Sond., Fl. Cap. 3: 553 (1865).
  Laurentia pygmaea var. *obtusiloba* Sond. in Harv. & Sond., Fl. Cap. 3: 553 (1865).

Wimmerella secunda (L. f.) Serra, M. B. Crespo & Lammers, Novon 9: 417 (1999).
Cape Provinces. 27 CPP. Hemicr.
  * *Lobelia secunda* L. f., Suppl. Pl.: 395 (1782). *Enchysia secunda* (L. f.) Sond. in Harv. &
    Sond., Fl. Cap. 3: 551 (1865). *Laurentia secunda* (L. f.) Kuntze, Revis. Gen. Pl. 3(2):
    188 (1898).
  Enchysia erecta A. DC. in DC., Prodr. 7: 409 (1839). *Enchysia secunda* var. *glabrata* Sond.
    in Harv. & Sond., Fl. Cap. 3: 551 (1865). *Laurentia erecta* (A. DC.) Schönl. in Engl. &
    Prantl, Natürl. Pflanzenfam. IV.5: 45 (1894). *Laurentia erinoides* f. *erecta* (A. DC.) E.
    Wimm., Ann. Naturhist. Mus. Wien 56: 334 (1948). *Laurentia secunda* f. *erecta* (A.
    DC.) E. Wimm., Pflanzenr. IV.276b: 395 (1953).
  Enchysia dentata A. DC. in DC., Prodr. 7: 409 (1839). *Enchysia secunda* var. *dentata* (A.
    DC.) Sond. in Harv. & Sond., Fl. Cap. 3: 551 (1865). *Laurentia erinoides* f. *dentata* (A.
    DC.) E. Wimm., Ann. Naturhist. Mus. Wien 56: 334 (1948). *Laurentia secunda* f.
    dentata (A. DC.) E. Wimm., Pflanzenr. IV.276b: 396 (1953).
  Enchysia repens var. *reflexa* Kunze, Ind. Sem. Hort. Lips. ? (1846). *Enchysia secunda* var.
    reflexa (Kunze) Sond. in Harv. & Sond., Fl. Cap. 3: 551 (1865).

Wittea

Like *Bolelia* and *Gynampsis* (q.v.), this name was proposed as an avowed substitute for the
illegitimate name *Clintonia*. Based on: *Clintonia* Dougl. ex Lindl., non Raf.

  Torrey, J. (1857). Report on the botany of the Expedition. In Explorations and surveys for a
    railroad route from the Mississippi River to the Pacific Ocean. Route near the thirty-fifth
    parallel, explored by Lieutenant A. W. Whipple, topographical engineers, in 1853 and
    1854. Washington: Government Printing Office. En. — Torrey explicitly declined to
    take up *Wittea,* on grounds that it is improper to name two genera after single person.

Synonyms:
Wittea Kunth === **Downingia** Torr.

Zeugandra

Campanuloideae, 2 species, Iran. On the basis of morphology, it is believed to be closely
related to *Campanula*. Type [by monotypy]: *Zeugandra iranica* P. H. Davis.

  Davis, P. H. (1950). *Zeugandra* P. H. Davis, gen. nov. Hooker's Icon. Pl. 35: pl. 3497 + 2 pp.
    text, illus. En. — Establishment of genus as allied to *Campanula*.
  Rechinger, K. H. & H. Schiman-Czeika (1965). *Zeugandra*. In K. H. Rechinger (ed.), Flora
    Iranica 13: 2-3. Graz: Akademische Druck- und Verlagsanstalt. Ge., La. — Floristic
    treatment with description, full nomenclature, and specimen citations.

Zeugandra P. H. Davis, Hooker's Icon. Pl. 35: pl. 3497 (1950).
   Western Asia. 34.

Zeugandra iranica P. H. Davis, Hooker's Icon. Pl. 35: pl. 3497 (1950).
   Iran. 34 IRN. Hemicr.

Zeugandra iranshahrii Esfand., Notes Roy. Bot. Gard. Edinburgh 38: 447 (1980).
   Iran. 34 IRN. Hemicr.

Summary of names of uncertain application

Some validly published names unfortunately cannot be equated with an accepted taxon. These all are summarized here, together with the full bibliographic citation for each. The names also are included in the lists of synonyms following each genus. In both places, a question mark replaces the name of an accepted taxon after the equal sign (===). Some names appear here largely because they have been forgotten or ignored since their initial publication. In other cases, subsequent workers have attempted to ascertain the identity of the plant represented, but failed. In these cases, the interrogatory is followed by a bracketed citation of the pertinent reference.

Adenophora polymorpha f. **integrisepala** Herder, Trudy Imp. S.-Peterburgsk. Bot. Sada 1: 308 (1873).
=== ?

Adenophora polymorpha f. **raddeana** Herder, Trudy Imp. S.-Peterburgsk. Bot. Sada 1: 307 (1873).
=== ?

Adenophora polymorpha var. **alternifolia** Franch. & Sav., Enum. Pl. Jap. 2: 422 (1879).
=== ?

Adenophora polymorpha var. **calicina** Franch. & Sav. , Enum. Pl. Jap. 2: 422 (1879).
=== ?

Adenophora polymorpha var. **chanetii** H. Lév., Repert. Spec. Nov. Regni Veg. 12: 22 (1913).
=== ? [D. F. Chamb., Notes Roy. Bot. Gard. Edinburgh 35: 248 (1977).]

Adenophora polymorpha var. **urticifolia** Franch. & Sav., Enum. Pl. Jap. 2: 422 (1879), as 'urticaefolia'.
=== ?

Adenophora prenanthoides Prain ex Diels, Notes Roy. Bot. Gard. Edinburgh 5: 176 (1912).
=== ?

Adenophora reticulata G. Don ex W. H. Baxter & Wooster, Suppl. Hort. Brit.: 471 (1850).
=== ?

Adenophora tatewakiana Hurus., Bot. Mag. (Tokyo) 62: 47 (1949).
=== ?

Adenophora verbenifolia Bunge ex Heynh., Alph. Aufz. Gew.: 9 (1846), as 'verbenaefolia'.
=== ?

Campanopsis caffra (A. DC.) Kuntze, Revis. Gen. Pl. 2: 379 (1891). Basionym: *Wahlenbergia caffra* A. DC.
=== ?

Campanopsis flaccida (A. DC.) Kuntze, Revis. Gen. Pl. 2: 379 (1891). Basionym: *Wahlenbergia flaccida* A. DC.
=== ?

Campanopsis inconspicua (A. DC.) Kuntze, Revis. Gen. Pl. 2: 379 (1891). Basionym: *Wahlenbergia inconspicua* A. DC.
=== ?

Campanopsis procumbens var. **bolusiana** Kuntze, Revis. Gen. Pl. 2: 379 (1891).
=== ?

Campanopsis spinulosa (A. DC.) Kuntze, Revis. Gen. Pl. 2: 379 (1891). Basionym: *Wahlenbergia spinulosa* A. DC.
=== ?

Campanula adscendens Vest ex Schult. in Roem. & Schult., Syst. Veg. 5: 89 (1819).
=== ? [Fed. in Kom., Fl. URSS 24: 323-324 (1957).]

Campanula aggregata Hegetschw., Fl. Schweiz: 235 (1838-1839); non Willd., Enum. Pl., Suppl.: 10 (1814).
=== ?

Campanula ajugifolia Sest. ex Schult. in Roem. & Schult., Syst. Veg. 5: 132 (1819), as 'ajugaefolia'.
=== ? [A. DC., Monogr. Campan.: 341 (1830).]

Campanula amasiae Post, J. Linn. Soc., Bot. 24: 435 (1888).
=== ? [Damboldt in P. H. Davis, Fl. Turkey 6: 63 (1978).]

Campanula amoris Sennen, Diagn. Nouv.: 283 (1936).
=== ?

Campanula arctolinifolia Hruby, Magyar Bot. Lapok 29: 231 (1930).
=== ? [Podlech, Feddes Repert. 71: 161 (1965).]

Campanula arenaria Formánek, Verh. Naturf. Vereines Brünn 32: 290 (1894).
=== ?

Campanula aspera Moench, Suppl. Meth.: 188 (1802); non (Boiss.) Boiss., Diagn. P. Orient. (ser. 1) 11: 77 (1849).
=== ? [A. DC. in DC., Prodr. 7: 485 (1839).]

Campanula asperrima Zuccagni, Cent. Observ. Bot. 1: 16 (1806).
=== ? [A. DC., Monogr. Campan.: 341 (1830).]

Campanula barrelieri Marnock, Floric. Mag. & Misc. Gard. 1: 161 (1836); non C. Presl, Symb. Bot. 1: 30 (1831).
=== ?

Campanula bulgarica Davidov, Trav. Soc. Bulg. Sci. Nat. 8: 91 (1915); non Witasek, Magyar Bot. Lapok. 5: 244 (1906).
=== ?

Campanula camtschatica Pall. ex Schult. in Roem. & Schult., Syst. Veg. 5: 152 (1819).
=== ? [Fed. in Kom., Fl. URSS 24: 324-326 (1957).]

Campanula cephalaria Vuk., Linnaea 26: 333 (1854).
=== ?

Campanula cephalaria var. **cardiophylla** Vuk., Linnaea 26: 334 (1854).
=== ?

Campanula cephalaria var. **ovaliphylla** Vuk., Linnaea 26: 335 (1854).
=== ?

Campanula cephalaria var. **polyanthemos** Vuk., Linnaea 26: 334 (1854).
=== ?

Campanula chrysogonii Sennen, Diagn. Nouv.: 283 (1936), as 'chrysogoni'.
=== ?

Campanula cochleariifolia subsp. **septentrionalis** Hruby, Magyar Bot. Lapok 29: 238 (1930).
=== ? [Podlech, Feddes Repert. 71: 161 (1965).]

Campanula cordata Peterm., Fl. Lips. Excurs.: 187 (1838); non Vis., Stirp. Dalmat. Spec.: 5 (1826).
=== ?

Campanula cuatrecasasii Sennen & Losa in Sennen, Diagn. Nouv.: 275 (1936), as 'cuatrecasi;' non Pau, Trab. Mus. Ci. Nat. Barcelona 12: 442 (1929).
=== ?

Campanula davurica Siev., Briefe Siberien: ? (1796).
=== ?

Campanula decloetiana Ortmann in Weitenw., Beiträge 3: 220 (1838).
=== ? [Podlech, Feddes Repert. 71: 161 (1965).]

Campanula densifoliata Sennen & Losa in Sennen, Diagn. Nouv.: 275 (1936).
=== ?

Campanula draba Burm. f., Prodr. Fl. Cap.: 5 (1768).
=== ?

Campanula elata Colla, Herb. Pedem. 4: 26 (1835).
=== ?

Campanula erinoides L., Mant. Pl.: 44 (1767).
=== ?

Campanula exigua Formánek, Verh. Naturf. Vereines Brünn 32: 153 (1894); non Rattan, Bot. Gaz (Crawfordsville) 11: 339 (1886).
=== ?

Campanula exul Schott in Schott, Nyman & Kotschy, Analect. Bot.: 9 (1854).
=== ?

Campanula ficarioides var. **major** Timb.-Lagr., Mém. Acad. Sci. Toulouse (ser. 7) 5: 274 (1873). Based on: *Campanula rohdii* Loisel.
=== ? [Podlech, Feddes Repert. 71: 161 (1965).]

Campanula flaccida (A. DC.) D. Dietr., Syn. Pl. 1: 752 (1839). Basionym: *Wahlenbergia flaccida* A. DC.
=== ?

Campanula flagellaris Halácsy, Denkschr. Kaiserl. Akad. Wiss., Math.-Naturwiss. Cl. 61: 246 (1846).
=== ?

Campanula glomerata Hegetschw., Fl. Schweiz: 235 (1838-39).
=== ?

Campanula glomerata subsp. **fatrae** (Borbás) Sóo, Acta Bot. Acad. Sci. Hung. 17: 124 (1972). Basionym: *Campanula glomerata* var. *fatrae* Borbás.
=== ?

Campanula glomerata subsp. **salviifolia** (Wallr.) O. Schwarz, Mitt. Thüring. Bot. Ges. 1: 118 (1949), as 'salviaefolia'. Basionym: *Campanula glomerata* var. *salviifolia* Wallr.
=== ?

Campanula glomerata var. **capituliformis** K. Koch, Linnaea 23: 639 (1850).
=== ?

Campanula glomerata var. **cordifolia** Rohlena, Allg. Bot. Z. Syst. 5: 94 (1899).
=== ?

Campanula glomerata var. **coriacea** Fomin in N. M. Kusn., N. Bush & Fomin, Mater. Fl. Kavkaza 4(6): 104 (1906).
=== ?

Campanula glomerata var. **fatrae** Borbás, V. K. Reálisk. Ért.: 22 (1898).
=== ?

Campanula glomerata var. **latibracteata** Sommier & Levier, Enum. Pl. Cauc.: 326 (1900).
=== ?

Campanula glomerata var. **mollis** Tausch, ?
=== ?

Campanula glomerata var. **nana** Peterm., Fl. Lips. Excurs.: ? (1838); non Noulet, Fl. Bass. Sous-Pyrén.: 408 (1837).
=== ?

Campanula glomerata var. **salviifolia** Wallr., Sched. Crit.: 90 (1822), as 'salviaefolia'.
=== ?

Campanula gracilis Avé-Lall., Pl. Ital. Bor.: 10 (1829); non G. Forst., Fl. Ins. Austr.: 15 (1786).
=== ? [Podlech, Feddes Repert. 71: 161 (1965).]

Campanula grammosepala Vuk., Linnaea 26: 323 (1854).
=== ?

Campanula grammosepala var. **cardiophylla** Vuk., Linnaea 26: 324 (1854).
=== ?

Campanula grammosepala var. **spathiphylla** Vuk., Linnaea 26: 324 (1854).
=== ?

Campanula grisebachiana Gand., Contrib. Fl. Terr. Slav. Merid. 1: 21 (1883).
=== ?

Campanula hirsuta Pant., Verh. Vereins Natur- Heilk. Presburg (n.s.) 2: 54 (1871-72).
=== ?

Campanula hirta Hegetschw., Fl. Schweiz: 235 (1838-39).
=== ?

Campanula hostii Baumg., Enum. Stirp. Transsilv. 3: 342 (1817).
=== ? [Podlech, Feddes Repert. 71: 161 (1965).]

Campanula hostii var. **uniflora** A. DC., Monogr. Campan.: 277 (1830).
=== ? [Podlech, Feddes Repert. 71: 161 (1965).]

Campanula inconcessa Schott in Schott, Nyman & Kotschy, Analect. Bot.: 10 (1854).
=== ? [Podlech, Feddes Repert. 71: 161 (1965).]

Campanula intermedia Gand., Contrib. Fl. Terr. Slav. Merid. 1: 20 (1883); non Schult. in Roem. & Schult., Syst. Veg. 5: 110 (1819).
=== ?

Campanula italica Arv.-Touv., Bull. Soc. Dauphin. Échange Pl. 14: 584 (1887), as 'italicum'.
=== ?

Campanula juncea Hill, Veg. Syst. 8: 14 (1765).
=== ?

Campanula lagranensis Sennen & Losa in Sennen, Diagn. Nouv.: 274 (1936).
=== ?

Campanula lanceolata Hegetschw., Fl. Schweiz: 230 (1838-39); non Lapeyr., Hist. Pl. Pyrénées: 105 (1815).
=== ?

Campanula ligularis Lam., Encycl. 1: 585 (1785).
=== ? [A. DC., Monogr. Campan.: 341 (1830).]

Campanula linifolia Hegetschw., Fl. Schweiz: 230 (1838-39); non L., Fl. Monsp.: 12 (1756); nec Scop., Annus Hist.-Nat. 2: 47 (1769); nec Lam., Encycl. 1: 579 (1785).
=== ?

Campanula macrorrhiza Vuk., Linnaea 26: 329 (1854).
=== ?

Campanula megaphylla Gand., Contrib. Fl. Terr. Slav. Merid. 1: 20 (1883).
=== ?

Campanula minuta Savi, Due Cent. Piante: 54 (1804).
=== ? [Podlech, Feddes Repert. 71: 161 (1965).]

Campanula moldavica Gand., Contrib. Fl. Terr. Slav. Merid. 1: 21 (1883).
=== ?

Campanula molineri Colla, Herb. Pedem. 4: 17 (1835).
=== ?

Campanula montana Delarbre, Fl. Auvergne (ed. 2): 40 (1800).
=== ? [A. DC. in DC., Prodr. 7: 485 (1839).]

Campanula multicaulis Witasek, Meddeland. Soc. Fauna Fl. Fenn. 29: 209 (1904); non Boiss., Diagn. Pl. Orient. (ser. 1) 7: 19 (1846).
=== ? [Podlech, Feddes Repert. 71: 161 (1965).]

Campanula murrayana Karsch, Phan.-Fl. Westfalen: 770 (1853).
=== ?

Campanula muscosa Mazziari, Ionios Antologi 2: 204 (1835).
=== ?

Campanula oxyphylla Vuk., Linnaea 26: 331 (1854).
  === ?

Campanula paniculata Turra ex Sacc., Atti Reale Ist. Veneto Sci. (ser. 3) 9: 486 (1863-64);
  non L. f., Suppl. Pl.: 139 (1782); nec Pohl, Tent. Fl. Bohem. 1: 207 (1809).
  === ?

Campanula plasonii Formánek, Verh. Narurf. Vereines Brünn 35: 155 (1897).
  === ?

Campanula polyantha Schult. in Roem. & Schult., Syst. Veg. 5: 130 (1819).
  === ? [Fed. in Kom., Fl. URSS 24: 330 (1957).]

Campanula pourretii Jeanb. & Timb.-Lagr., Bull. Soc. Sci. Phys. Nat. Toulouse 6: 153
  (1883-84).
  === ? [Podlech, Feddes Repert. 71: 161 (1965).]

Campanula praecox Miégev., Bull. Soc. Bot. France 40: 304 (1894).
  === ?

Campanula preissii de Vriese in Lehm., Pl. Preiss. 2: 241 (1848).
  === ?

Campanula pruinosa Mazziari, Ionios Antologi 2: 204 (1835).
  === ?

Campanula pulliformis Rouy, Monde Pl. 4: 49 (1894).
  === ? [Podlech, Feddes Repert. 71: 161 (1965).]

Campanula pusilla Hegetschw., Fl. Schweiz: 231 (1838-39); non Haenke in Jacq., Collect. 2:
  79 (1789).
  === ?

Campanula racemosa Vuk., Linnaea 26: 332 (1854).
  === ?

Campanula reniformis Schur, Enum. Pl. Transsilv.: 441 (1866).
  === ?

Campanula rigescens Pall. ex Schult. in Roem. & Schult., Syst. Veg. 5: 102 (1819).
  === ? [Fed. in Kom., Fl. URSS 24: 330 (1957).]

Campanula rohdii Loisel., Fl. Gall. (ed. 2) 1: 140 (1828).
  === ? [Podlech, Feddes Repert. 71: 161 (1965).]

Campanula rosea Noronha, Verh. Batav. Genootsch. Kunsten 5 (Art. IV): 12 (1790).
  === ?

Campanula rotundifolia subsp. **linifolia** Lapeyr., Hist. Pl. Pyrénées: 104 (1813).
  === ?

Campanula rotundifolia var. **decloetiana** (Ortmann) Nyman, Consp. Fl. Eur.: 479 (1879).
  Basionym: *Campanula decloetiana* Ortmann.
  === ? [Podlech, Feddes Repert. 71: 161 (1965).]

Campanula rotundifolia var. **hostii** (Baumg.) Nyman, Consp. Fl. Eur.: 479 (1879).
  Basionym: *Campanula hostii* Baumg.
  === ? [Podlech, Feddes Repert. 71: 161 (1965).]

Campanula rotundifolia var. **montana** Lecoq & Lamotte, Cat. Pl. Plateau Central:
  260 (1848).
  === ? [Podlech, Feddes Repert. 71: 161 (1965).]

Campanula rumenica Gand., Contrib. Fl. Terr. Slav. Merid. 1: 20 (1883).
  === ?

Campanula scabrida Hochst. in Lorent, Wanderungen Morgenlande: 331 (1845).
  === ?

Campanula scandens Pall. ex Schult. in Roem. & Schult., Syst. Veg. 5: 120 (1819).
  === ? [Fed. in Kom., Fl. URSS 24: 330 (1957).]

Campanula scheuchzeri race **pourretii** (Jeanb. & Timb.-Lagr.) Rouy, Fl. France 10: 78 (1908).
Basionym: *Campanula pourretii* Jeanb. & Timb.-Lagr.
=== ? [Podlech, Feddes Repert. 71: 161 (1965).]

Campanula scheuchzeri var. **inconcessa** (Schott) Nyman, Consp. Fl. Eur.: 479 (1879).
Basionym: *Campanula inconcessa* Schott.

=== ? [Podlech, Feddes Repert. 71: 161 (1965).]

Campanula scheuchzeri var. **rohdii** (Loisel.) Rouy, Fl. France 10: 77 (1908), as 'rhodii'.
Basionym: *Campanula rohdii* Loisel.
=== ? [Podlech, Feddes Repert. 71: 161 (1965).]

Campanula serratifolia Vuk., Linnaea 26: 335 (1854).
=== ?

Campanula seticalyx Gand., Contrib. Fl. Terr. Slav. Merid. 1: 19 (1883).
=== ?

Campanula setosa J. C. Wendl., Bot. Beob.: 42 (1798).
=== ?

Campanula sibirica f. **nana** Fomin in N. M. Kusn., N. Bush & Fomin, Mater. Fl. Kavkaza
4(6): 27 (1904).
=== ?

Campanula sibirica var. **saxicola** K. Koch, Linnaea 23: 636 (1850).
=== ?

Campanula spinulosa (A. DC.) D. Dietr., Syn. Pl. 1: 754 (1839). Basionym: *Wahlenbergia
spinulosa* A. DC.
=== ?

Campanula stenopoda Gand., Contrib. Fl. Terr. Slav. Merid. 1: 19 (1883).
=== ?

Campanula stolonifera Mign., Bull. Soc. Bot. France 12: 342 (1865).
=== ? [Podlech, Feddes Repert. 71: 161 (1965).]

Campanula subuniflora Lam., Tabl. Encycl. 2: 53 (1796).
=== ? [Podlech, Feddes Repert. 71: 161 (1965).]

Campanula thessala Gand., Contrib. Fl. Terr. Slav. Merid. 1: 21 (1883).
=== ?

Campanula toletana Sennen, Diagn. Nouv.: 283 (1936).
=== ?

Campanula triangularis Parsa, Kew Bull. 1948: 209 (1948).
=== ? [Rech. f. & Schiman-Czeika in Rech. f., Fl. Iran. 13: 38 (1965).]

Campanula urticifolia Hegetschw., Fl. Schweiz: 236 (1838-39); non Turra, Fl. Ital. Prodr.: 64
(1780); nec All., Fl. Pedem. 1: 110 (1785); nec Salisb., Prodr. Stirp. Chap. Allerton: 127 (1796).
=== ?

Campanula verbenifolia Sieber, Ann. Fl. Pomone 1838: 375 (1838), as 'verbenaefolia'.
=== ?

Campanula vulgaris Gueldenst., Reis. Russland 2: 10 (1791).
=== ?

Canarina zanguebar Lour., Fl. Cochinch.: 195 (1790).
=== ? [Hedberg, Svensk. Bot. Tidskr. 55: 57 (1961).]

Centropogon laciniatus (G. Don) Zahlbr., Bull. Herb. Boissier 3: 623 (1895). Basionym:
Siphocampylus laciniatus G. Don.
=== ? [E. Wimm. in Engl. Pflanzenr. IV.276b: 330 (1953).]

Centropogon ovalifolius var. **glabristamineus** E. Wimm., Repert. Spec. Nov. Regni Veg. 19:
64 (1931).
=== ?

Centropogon pilosulus var. **quindiuensis** E. Wimm., Repert. Spec. Nov. Regni Veg. 29:
64 (1931).
=== ?

Centropogon scandens Planch. & Oerst., Vidensk. Meddel. Dansk Naturhist. Foren.
Kjøbenhavn 1857: 157 (1857).
=== ? [E. Wimm. in Engl., Pflanzenr. IV.276b: 260 (1943).]

Clermontia carinifera H. Lév., Repert. Spec. Nov. Regni Veg. 12: 505 (1913).
=== ? [Lammers, Syst. Bot. Monogr. 32: 83 (1991).]

Clermontia fulva H. Lév., Repert. Spec. Nov. Regni Veg. 12: 506 (1913).
=== ? [Lammers, Syst. Bot. Monogr. 32: 83 (1991).]

Codonopsis draco Pampan., Nuovo Giorn. Bot. Ital. (n.s.) 17: 733 (1910).
=== ? [Anthony, Notes Roy. Bot. Gard. Edinburgh 15: 187 (1926).]

Codonopsis japonica Miq., Ann. Mus. Bot. Lugduno-Batavum 2: 192 (1866), as 'iaponica'.
=== ? [Chipp, J. Linn. Soc. Bot. 38: 390 (1908).]

Cyananthus incanus var. **bicolor** Cowan, New Fl. & Silva 10: 185 (1938).
=== ?

Cyananthus incanus var. **nudicalyx** Cowan, New Fl. & Silva 10: 185 (1938).
=== ?

Cyananthus inflatus var. **rufus** Franch., J. Bot. (Morot) 1: 281 (1887).
=== ?

Cyananthus inflatus var. **sylvestris** C. Marquand, Bull. Misc. Inform. Kew 1924: 249 (1924).
=== ?

Cyananthus microrhombeus var. **leiocalyx** C. Y. Wu, Rep. Yunnan Trop. Subtrop. Fl. Res.
Inst. 1: 86 (1965).
=== ?

Cyananthus pedunculatus var. **crenatus** C. Marquand, New Fl. & Silva 8: 208 (1936).
=== ?

Cyanea bonita Rock, Monogr. Stud. Haw. Lobelioid.: 51 (1919). Based on: *Delissea
kunthiana* Gaudich.
=== ? [Lammers, Syst. Bot. Monogr. 73: 55 (2005).]

Cyanea holophylla var. **obovata** Rock, Monogr. Stud. Haw. Lobelioid.: 263 (1919).
=== ? [Lammers, Syst. Bot. Monogr. 73: 54 (2005).]

Cyanea scabra f. **sinuata** (Rock) E. Wimm., Pflanzenr. IV.276b: 66 (1943). Basionym:
Cyanea scabra var. *sinuata* Rock
=== ?

Cyanea scabra var. **sinuata** Rock, Monogr. Stud. Haw. Lobelioid.: 259 (1919).
=== ?

Cyphia pinnata (Lam.) Schult. in Roem. & Schult., Syst. Veg. 5: 477 (1819). Basionym:
Lobelia pinnata Lam.
=== ?

Delissea acuta H. St. John, Phytologia 64: 165 (1988).
=== ? [Lammers, Syst. Bot. Monogr. 73: 44 (2005).]

Delissea holophylla var. **obovata** (Rock) H. St. John, Phytologia 63: 84 (1987). Basionym:
Cyanea holophylla var. *obovata* Rock
=== ? [Lammers, Syst. Bot. Monogr. 73: 54 (2005).]

Delissea kunthiana Gaudich., Voy. Bonite, Bot.: tab. 77 (1844).
=== ? [Lammers, Syst. Bot. Monogr. 73: 55 (2005).]

Delissea muriculata H. St. John, Phytologia 64: 166 (1988).
=== ? [Lammers, Syst. Bot. Monogr. 73: 59 (2005).]

Diastatea ghiesbreghtii (Kuntze) E. Wimm., Ann. Naturhist. Mus. Wien 56: 333 (1948).
Basionym: *Dortmanna ghiesbreghtii* Kuntze
=== ? [McVaugh, N. Amer. Fl. 32A: 99 (1943).]

Diastatea lemairei E. Wimm., Pflanzenr. IV.276b: 368 (1953). Based on: *Lobelia ghiesbreghtii*
Lem. non Decne.
=== ? [E. Wimm. in Engl., Pflanzenr. IV.276c: 853 (1968).]

Dortmanna alpina (Vell.) Kuntze, Revis. Gen. Pl. 2: 972 (1891). Basionym: *Lobelia alpina* Vell.
=== ? [E. Wimm. in Engl., Pflanzenr. IV.276b: 694 (1953).]

Dortmanna diversifolia (Willd. ex Schult.) Kuntze, Revis. Gen. Pl. 2: 972 (1891). Basionym:
Lobelia diversifolia Willd. ex Schult.
=== ? [E. Wimm. in Engl., Pflanzenr. IV.276b: 695 (1953).]

Dortmanna ensiformis (Vell.) Kuntze, Revis. Gen. Pl. 2: 972 (1891). Basionym: *Lobelia
ensiformis* Vell.
=== ? [E. Wimm. in Engl., Pflanzenr. IV.276b: 694 (1953).]

Dortmanna ghiesbreghtii Kuntze, Revis. Gen. Pl. 2: 972 (1891). Based on: *Lobelia
ghiesbreghtii* Lem. non Decne.
=== ? [E. Wimm. in Engl., Pflanzenr. IV.276c: 853 (1968).]

Dortmanna trigona var. **intermedia** Kuntze, Revis. Gen. Pl. 2: 380 (1891).
=== ?

Heterotoma riparia E. Wimm., Ann. Naturhist. Mus. Wien 56: 372 (1948).
=== ? [Ayers, Syst. Bot. 15: 325-326 (1990).]

Jasione amethystina f. **glaberrima** H. Lindb., Acta Soc. Sci. Fenn., ser. B, Opera Biol. 1(2):
150 (1932).
=== ?

Jasione amethystina f. **glabra** Schmeja, Beih. Bot. Centralbl. 48(2): 23 (1931); non
Schmeja, Beih. Bot. Centralbl. 48(2): 23 (1931).
=== ?

Jasione amethystina f. **glabra** Schmeja, Beih. Bot. Centralbl. 48(2): 23 (1931); non
Schmeja, Beih. Bot. Centralbl. 48(2): 23 (1931).
=== ?

Jasione amethystina f. **hirsuta** Schmeja, Beih. Bot. Centralbl. 48(2): 23 (1931); non
Schmeja, Beih. Bot. Centralbl. 48(2): 23 (1931).
=== ?

Jasione amethystina f. **hirsuta** Schmeja, Beih. Bot. Centralbl. 48(2): 23 (1931); non
Schmeja, Beih. Bot. Centralbl. 48(2): 23 (1931).
=== ?

Jasione amethystina var. **intermedia** Willk. in Willk. & Lange, Prodr. Fl. Hispan. 2: 283 (1868).
=== ? [Sales & Hedge, Anales Jard. Bot. Madrid 59: 172 (2001).]

Jasione caespitosa Roth, Enum. Pl. Phaen. Germ. 1: 774 (1827).
=== ?

Jasione genisticola Gand., Bull. Soc. Bot. France 45: 594 (1898).
=== ? [Sales & Hedge, Anales Jard. Bot. Madrid 59: 172 (2001).]

Jasione montana f. **glabra** Schmeja, Beih. Bot. Centralbl. 48(2): 31 (1931); non Schmeja,
Beih. Bot. Centralbl. 48(2): 32 (1931); nec Schmeja, Beih. Bot. Centralbl. 48(2): 33 (1931).
=== ?

Jasione montana f. **glabra** Schmeja, Beih. Bot. Centralbl. 48(2): 32 (1931); non Schmeja,
Beih. Bot. Centralbl. 48(2): 31 (1931); nec Schmeja, Beih. Bot. Centralbl. 48(2): 33 (1931).
=== ?

Jasione montana f. **glabra** Schmeja, Beih. Bot. Centralbl. 48(2): 33 (1931); non Schmeja,
Beih. Bot. Centralbl. 48(2): 31 (1931); nec Schmeja, Beih. Bot. Centralbl. 48(2): 32 (1931).
=== ?

Jasione montana f. **hirsuta** Schmeja, Beih. Bot. Centralbl. 48(2): 31 (1931); non Schmeja, Beih. Bot. Centralbl. 48(2): 32 (1931); nec Schmeja, Beih. Bot. Centralbl. 48(2): 33 (1931).
=== ?

Jasione montana f. **hirsuta** Schmeja, Beih. Bot. Centralbl. 48(2): 32 (1931); non Schmeja, Beih. Bot. Centralbl. 48(2): 31 (1931); nec Schmeja, Beih. Bot. Centralbl. 48(2): 33 (1931).
=== ?

Jasione montana f. **hirsuta** Schmeja, Beih. Bot. Centralbl. 48(2): 33 (1931); non Schmeja, Beih. Bot. Centralbl. 48(2): 31 (1931); nec Schmeja, Beih. Bot. Centralbl. 48(2): 32 (1931).
=== ?

Jasione montana f. **rossularis** Schmeja, Beih. Bot. Centralbl. 48(2): 33 (1931).
=== ?

Jasione montana var. **glaberrima** Merino, Brotéria, Sér. Bot. 12: 114 (1914).
=== ? [Sales & Hedge, Anales Jard. Bot. Madrid 59: 172 (2001).]

Jasione montana var. **umbellata** Lapeyr., Hist. Pl. Pyrénées: 102 (1813).
=== ? [Sales & Hedge, Anales Jard. Bot. Madrid 59: 172 (2001).]

Jasione perennis f. **carpetana** Schmeja, Beih. Bot. Centralbl. 48(2): 26 (1931).
=== ?

Jasione perennis f. **megacarpaea** Coustur. & Gand., Bull. Soc. Bot. France 60: 551 (1913).
=== ? [Sales & Hedge, Anales Jard. Bot. Madrid 59: 172 (2001).]

Jasione perennis f. **rosularis** Schmeja, Beih. Bot. Centralbl. 48(2): 26 (1931).
=== ?

Jasione perennis var. **alpestris** Willk., Flora 35: 198 (1852).
=== ? [Sales & Hedge, Anales Jard. Bot. Madrid 59: 172 (2001).]

Jasione perennis var. **intermedia** Coss., Notes Pl. Crit.:121 (1851).
=== ? [Sales & Hedge, Anales Jard. Bot. Madrid 59: 172 (2001).]

Laurentia commutata Tod., Index Sem. Panorm.: ?? (1872).
=== ?

Laurentia minuta var. **minima** (Sims) A. DC. in DC., Prodr. 7: 410 (1839). Basionym: *Lobelia minima* Sims.
=== ? [Aldasoro, Castrov., Sales & Hedge, Anales Jard. Bot. Madrid 59: 173 (2001).]

Legousia balearica Sennen, Bol. Soc. Ibér. Ci. Nat. 28: 171 (1930).
=== ?

Legousia speculum-veneris var. **maroccana** Pau & Font Quer in Font Quer, Iter Marocc.: 639 (1927).
=== ?

Lobelia alpina Vell., Fl. Flumin., Icon. 8: pl. 154 (1831).
=== ? [E. Wimm. in Engl., Pflanzenr. IV.276b: 694 (1953).]

Lobelia burmaniana D. Dietr., Syn. Pl. 1: 728 (1839), as 'burmanniana'.
=== ? [E. Wimm. in Engl., Pflanzenr. IV.276b: 695 (1953).]

Lobelia cavaleriei H. Lév., Repert. Spec. Nov. Regni Veg. 9: 455 (1911).
=== ? [E. Wimm. in Engl., Pflanzenr. IV.276b: 694 (1953).]

Lobelia cladlomesa Raf., New Fl. N. Amer. 2: 17 (1837).
=== ? [McVaugh, N. Amer. Fl. 32A: 99 (1943).]

Lobelia diversifolia Willd. ex Schult. in Roem. & Schult, Syst. Veg. 5: 68 (1819).
=== ? [E. Wimm. in Engl., Pflanzenr. IV.276b: 695 (1953).]

Lobelia ensiformis Vell., Fl. Flumin., Icon. 8: pl. 155 (1831).
=== ? [E. Wimm. in Engl., Pflanzenr. IV.276b: 694 (1953).]

Lobelia fistulosa Raf., New. Fl. N. Amer. 2: 18 (1837); non Vell., Fl. Flumin., Icon. 8: pl. 157 (1831).
=== ? [McVaugh, N. Amer. Fl. 32A: 99 (1943).]

Lobelia ghiesbreghtii Lem., Ill. Hort. 1: pl. 34 (1854); non Decne., Rev. Hort. (ser. 3) 2: 341 (1848).
=== ? [McVaugh, N. Amer. Fl. 32A: 99 (1943).]

Lobelia incurva Raf., New. Fl. N. Amer. 2: 18 (1837).
=== ? [McVaugh, N. Amer. Fl. 32A: 99 (1943).]

Lobelia laciniata (G. Don) C. Presl, Prodr. Monogr. Lobel.: 40 (1836); non Lam., Encycl. 3: 584 (1792).   Basionym: *Siphocampylus laciniatus* G. Don.
=== ? [E. Wimm. in Engl. Pflanzenr. IV.276b: 330 (1953).]

Lobelia limburgensis de Jonghe, Rev. Hort. (ser. 2) 4: 39 (1845).
=== ?

Lobelia microphylla Raf., Atlantic J.: 147 (1832).
=== ? [McVaugh, N. Amer. Fl. 32A: 68 (1943).]

Lobelia minima Sims, Bot. Mag. 46: pl. 2077 (1819).
=== ? [Aldasoro, Castrov., Sales & Hedge, Anales Jard. Bot. Madrid 59: 173 (2001).]

Lobelia minuta var. **minima** (Sims) Heynh., Nom. Bot. Hort. 1: 471 (1840). Basionym: *Lobelia minima* Sims.
=== ? [Aldasoro, Castrov., Sales & Hedge, Anales Jard. Bot. Madrid 59: 173 (2001).]

Lobelia monadelpha Larran., Escritos 1: 16 (1922).
=== ? [E. Wimm. in Engl. Pflanzenr. IV.276b: 694 (1953).]

Lobelia multicaulis (C. Presl) Steud., Nomencl. Bot. (ed. 2) 2: 61 (1841), as 'multicaule'. Basionym: *Rapuntium multicaule* C. Presl.
=== ? [E. Wimm. in Engl. Pflanzenr. IV.276b: 694 (1953).]

Lobelia multiflora Knowles & Westc., Fl. Cab. 3: 126 (1839).
=== ? [E. Wimm. in Engl. Pflanzenr. IV.276b: 694 (1953).]

Lobelia phoenicea de Jonghe, Rev. Hort. (ser. 2) 4: 39 (1845).
=== ?

Lobelia pinnata Lam., Encycl. 3: 591 (1792).
=== ? [E. Wimm. in Engl. Pflanzenr. IV.276b: 695 (1953).]

Lobelia racemosa Vest ex Schult. in Roem. & Schult., Syst. Veg. 5: 58 (1819).
=== ? [E. Wimm. in Engl. Pflanzenr. IV.276b: 694 (1953).]

Lobelia tenella Burm. f., Prodr. Fl. Cap.: 29 (1768); non L., Mant. Pl.: 120 (1767).
=== ?

Lobelia tyrianthina J. Forbes, Hort. Woburn.: 33 (1833).
=== ?

Lobelia umbellata Vest ex Schult. in Roem. & Schult., Syst. Veg. 5: 58 (1819); non Kunth in Humb., Bonpl. & Kunth, Nov. Gen. Sp. 3: 304 (quarto), 237 (folio) (1819).
=== ? [E. Wimm. in Engl. Pflanzenr. IV.276b: 694 (1953).]

Lysipomia subulata G. Don, Gen. Hist. 3: 717 (1834).
=== ? [McVaugh, Brittonia 8: 104 (1955).]

Numaeacampa Gagnep., Bull. Soc. Bot. France 95: 33 (1948).
=== ?

Numaeacampa kerrii Gagnep., Bull. Soc. Bot. France 95: 33 (1948).
=== ?

Phyteuma barrelieri Vill., Cat. Pl. Jard.: 183 (1807).
=== ?

Phyteuma capensis Burm. f., Prodr. Fl. Cap.: 5 (1768).
=== ? [Rich. Schulz, Monogr. Phyteuma: 180 (1904).]

Phyteuma degenii Gáyer, Magyar Bot. Lapok. 25: 82 (1927), as 'degeni'.
=== ?

Phyteuma foliosum Bellardi ex Colla, Herb. Pedem. 4: 33 (1835).
=== ?

Phyteuma michelii var. **involucrata** Schur, Verh. Mitth. Siebenbürg. Vereins Naturwiss. Hermanstadt 10: 139 (1859).
=== ?

Phyteuma montanum C. K. Spreng., Entd. Geheimn. Nat.: 115 (1793).
=== ?

Phyteuma obtusifolium Freyn, Oesterr. Bot. Z. 41: 56 (1891).
=== ?

Phyteuma pancicii Rohlena, Sitzungsber. Königl. Böhm. Ges. Wiss. Prag, Math.-Naturwiss. Cl. 1: 83 (1913).
=== ?

Phyteuma pyrenaicum Sennen, Bol. Soc. Ibér. Ci. Nat. 28: 172 (1930); non Rich. Schulz, Monogr. Phyteuma: 79 (1904).
=== ?

Phyteuma schlatteri Chenev., Bull. Trav. Soc. Bot. Genève 9: 126 (1902).
=== ?

Phyteuma sibiricum Vest ex Schult. in Roem. & Schult., Syst. Veg. 5: 77 (1819).
=== ? [Fed. in Kom., Fl. URSS 24: 395 (1957).]

Rapunculus caesalpini Bubani, Fl. Pyren. 2: 25 (1899).
=== ? [Rich. Schulz, Monogr. Phyteuma: 180 (1904).]

Rapuntium burmanianum C. Presl, Prodr. Monogr. Lobel.: 31 (1836), as 'burmannianum'.
=== ?

Rapuntium multicaule C. Presl, Prodr. Monogr. Lobel.: 31 (1836).
=== ? [E. Wimm. in Engl. Pflanzenr. IV.276b: 694 (1953).]

Rapuntium umbellatum C. Presl, Prodr. Monogr. Lobel.: 31 (1836). Basionym: *Lobelia umbellata* Vest ex Schult. non Kunth.
=== ? [E. Wimm. in Engl. Pflanzenr. IV.276b: 694 (1953).]

Siphocampylus bracteatus Decne. ex Linden, Établ. Linden, Prix-courant 5: 7 (1850).
=== ? [E. Wimm. in Engl. Pflanzenr. IV.276b: 379 (1953).]

Siphocampylus floccosus Planch. ex Linden, Établ. Linden, Prix-courant 5: 7 (1850).
=== ? [E. Wimm. in Engl. Pflanzenr. IV.276b: 379 (1953).]

Siphocampylus laciniatus G. Don, Gen. Hist. 3: 704 (1834).
=== ? [E. Wimm. in Engl. Pflanzenr. IV.276b: 330 (1953).]

Siphocampylus pulchellus Planch. & Linden in Linden, Prix-courant 10: 7 (1855).
=== ? [E. Wimm. in Engl. Pflanzenr. IV.276b: 379 (1953).]

Solenopsis minima (Sims) M. B. Crespo, Serra & Juan, Pl. Syst. Evol. 210: 224 (1998). Basionym: *Lobelia minima* Sims.
=== ? [Aldasoro, Castrov., Sales & Hedge, Anales Jard. Bot. Madrid 59: 173 (2001).]

Solenopsis minuta var. **minima** (Sims) C. Presl, Prodr. Monogr. Lobel.: 32 (1836). Basionym: *Lobelia minima* Sims.
=== ? [Aldasoro, Castrov., Sales & Hedge, Anales Jard. Bot. Madrid 59: 173 (2001).]

Triodanis scabra Raf., New. Fl. N. Amer. 4: 67 (1838).
=== ? [McVaugh, Wrightia 1: 50 (1945).]

Tupa ignescens Payer, Traité Organogén. Fl.: 644 (1857).
=== ?

Wahlenbergia baikalensis Freyn, Oesterr. Bot. Z. 40: 46 (1890).
=== ?

Wahlenbergia caffra A. DC. in DC., Prodr. 7: 432 (1839).
=== ?

Wahlenbergia colensoi N. E. Br., Gard. Chron. (ser. 3) 54: 317 (1913).
=== ? [Petterson, New Zealand J. Bot. 35: 77 (1997).]

Wahlenbergia crispa Nees, Del. Sem. Hort. Vratislav. 1850: 3 (1850).
=== ?

Wahlenbergia filipes M. E. Jones, Contr. W. Bot. 18: 66 (1933); non Brehmer, Bot. Jahrb.
Syst. 53: 89 (1915).
=== ? [C. V. Morton, Contr. U.S. Natl. Herb. 29: 114 (1945).]

Wahlenbergia flaccida A. DC., Monogr. Campan.: 138 (1830).
=== ?

Wahlenbergia gracilis var. **integerrima** Brehmer, Bot. Jahrb. Syst. 53: 112 (1915).
=== ?

Wahlenbergia inconspicua A. DC. in DC., Prodr. 7: 430 (1839).
=== ?

Wahlenbergia planifolia Gand., Bull. Soc. Bot. France 65: 54 (1918).
=== ?

Wahlenbergia rutenbergiana Vatke, Abh. Naturwiss. Vereine Bremen 9: 123 (1885).
=== ?

Wahlenbergia solitaria Brehmer, Bot. Jahrb. Syst. 53: 132 (1915).
=== ?

Wahlenbergia spinulosa A. DC., Monogr. Campan.: 155 (1830).
=== ?

Wahlenbergia vinciflora var. **rosulata** J. M. Black, Trans. & Proc. Roy. Soc. South Australia
58: 183 (1934).
=== ? [P. J. Sm., Telopea 5: 163 (1992).]

Weitenwebera glomerata var. **mollis** (Tausch) Opiz, Seznam: 105 (1852). Basionym:
Campanula glomerata var. *mollis* Tausch.
=== ?

Weitenwebera glomerata var. **nana** Opiz, Seznam: 105 (1852). Based on: *Campanula
glomerata* var. *nana* Peterm. non Noulet
=== ?

Weitenwebera glomerata var. **salviifolia** (Wallr.) Opiz, Seznam: 105 (1852), as 'salviaefolia'.
Basionym: *Campanula glomerata* var. *salviifolia* Wallr.
=== ?

Summary of excluded names

Some names assigned to genera of Campanulaceae in fact represent species that belong to other families. These all are summarized here, together with the full bibliographic citation for each; they also are included in the lists of synonyms following each genus. In both places, the family to which the name properly belongs is given parenthetically. Here only, this is followed by a bracketed reference that justifies the indicated identity.

Campanula porosa L. f., Suppl. Pl.: 142 (1782).
=== Samolus valerandi L. (Primulaceae) [R. Br., Prodr. 561 (1810).]

Campanula pygmaea DC. in Lam. & DC., Fl. Franç. (ed. 3) 3: 705 (1805).
=== Borago laxiflora DC. (Boraginaceae) [DC., Prodr. 10: 34 (1846).]

Campanula repens Lour., Fl. Cochinch., 1: 139 (1790).
=== Dentella repens (L.) J. R. Forst. & G. Forst. (Rubiaceae) [Merr., Trans. Amer. Philos. Soc. (n.s.) 24(2): 155 (1935).]

Campanula stellata Thunb. in Hoffm., Phytogr. Bl.: 20 (1803).
=== Viola decumbens L. (Violaceae) [Sond. in Harv. & Sond., Fl. Cap. 3: 597 (1865).]

Cyrtandroidea F. Br., Fl. S.E. Polynesia: 323 (1935).
=== Cyrtandra J. R. Forst. & G. Forst. (Gesneriaceae) [B. L. Burtt, Notes Roy. Bot. Gard. Edinburgh 28: 217-218 (1967).]

Cyrtandroidea jonesii F. Br., Fl. S.E. Polynesia: 324 (1935).
=== Cyrtandra jonesii (F. Br.) G. W. Gillett (Gesneriaceae) [G. W. Gillett, Univ. Calif. Publ. Bot. 66: 55 (1973).]

Dortmanna hirsuta (L.) Kuntze, Revis. Gen. Pl. 2: 972 (1891). Basionym: *Lobelia hirsuta* L.
=== Gnidia hirsuta (L.) Thulin (Thymeleaceae) [Thulin, Taxon 35: 724 (1986).]

Jasione ausfeldii Regel, Gartenflora 10: 356 (1861).
=== Brunonia australis Sm. ex R. Br. (Goodeniaceae) [Schmeja, Beih. Bot. Centralbl. 48(2): 39 (1931).]

Jasione capensis P. J. Bergius, Nova Acta Regiae Soc. Sci. Upsal. 3: 187 (1780).
=== Alpidea capensis (P. J. Bergius) R. A. Dyer (Apiaceae) [R. A. Dyer, Veget. Albany Bathurst: 128 (1937).]

Lobelia Mill., Gard. Dict. (abr. ed. 4) (1754); non L., Sp. Pl.: 929 (1753).
=== Scaevola L., nom. cons. (Goodeniaceae)

Lobelia aemula (R. Br.) Kuntze, Revis. Gen. Pl. 2: 378 (1891). Basionym: *Scaevola aemula* R. Br. (Goodeniaceae)
=== Scaevola aemula R. Br. (Goodeniaceae) [Carolin, Rajput & D. A. Morrison, Fl. Australia 35: 127 (1992).]

Lobelia amblyanthera (F. Muell.) Kuntze, Revis. Gen. Pl. 2: 378 (1891). Basionym: *Scaevola amblyanthera* F. Muell. (Goodeniaceae)
=== Scaevola amblyanthera F. Muell. (Goodeniaceae) [Carolin, Rajput & D. A. Morrison, Fl. Australia 35: 128 (1992).]

Lobelia anchusifolia (Benth.) Kuntze, Revis. Gen. Pl. 2: 378 (1891). Basionym: *Scaevola anchusifolia* Benth. (Goodeniaceae)
=== Scaevola anchusifolia Benth. (Goodeniaceae) [Carolin, Rajput & D. A. Morrison, Fl. Australia 35: 113 (1992).]

Lobelia angulata (R. Br.) Kuntze, Revis. Gen. Pl. 2: 378 (1891); non G. Forst., Fl. Ins. Austr.: 58 (1786). Basionym: *Scaevola angulata* R. Br. (Goodeniaceae)
=== Scaevola angulata R. Br. (Goodeniaceae) [Carolin, Rajput & D. A. Morrison, Fl. Australia 35: 144 (1992).]

Lobelia aphylla Nutt., Amer. J. Sci. 5: 297 (1822).
=== Apteria aphylla (Nutt.) Barnhart (Burmanniaceae) [E. Wimm. in Engl., Pflanzenr. IV.276b: 694 (1953).]

Lobelia apterantha (F. Muell.) Kuntze, Revis. Gen. Pl. 2: 378 (1891). Basionym: *Scaevola apterantha* F. Muell. (Goodeniaceae)
=== Scaevola ramosissima (Sm.) K. Krause (Goodeniaceae) [Carolin, Rajput & D. A. Morrison, Fl. Australia 35: 134 (1992).]

Lobelia atriplicina (F. Muell.) Kuntze, Revis. Gen. Pl. 2: 378 (1891). Basionym: *Scaevola atriplicina* F. Muell. (Goodeniaceae)
=== Scaevola tomentosa Gaudich. (Goodeniaceae) [Carolin, Rajput & D. A. Morrison, Fl. Australia 35: 100 (1992).]

Lobelia attenuata (R. Br.) Kuntze, Revis. Gen. Pl. 2: 378 (1891); non C. Presl, Prodr. Lobel.: 34 (1836). Basionym: *Scaevola attenuata* R. Br. (Goodeniaceae)
=== Scaevola nitida R. Br. (Goodeniaceae) [Carolin, Rajput & D. A. Morrison, Fl. Australia 35: 104 (1992).]

Lobelia auriculata (Benth.) Kuntze, Revis. Gen. Pl. 2: 378 (1891). Basionym: *Scaevola auriculata* Benth. (Goodeniaceae)
=== Scaevola auriculata Benth. (Goodeniaceae) [Carolin, Rajput & D. A. Morrison, Fl. Australia 35: 132 (1992).]

Lobelia brookeana (F. Muell.) Kuntze, Revis. Gen. Pl. 2: 378 (1891), as 'brooksiana'. Basionym: *Scaevola brookeana* F. Muell. (Goodeniaceae)
=== Scaevola brookeana F. Muell. (Goodeniaceae) [Carolin, Rajput & D. A. Morrison, Fl. Australia 35: 116 (1992).]

Lobelia bryoides Willd. ex Schult. in Roem. & Schult., Syst. Veg. 5: 41 (1819).
=== Arenaria dicranoides Kunth (Caryophyllaceae) [E. Wimm. in Engl., Pflanzenr. IV.276b: 695 (1953).]

Lobelia calendulacea (J. Kenn.) Kuntze, Revis. Gen. Pl. 2: 378 (1891). Basionym: *Goodenia calendulacea* J. Kenn. (Goodeniaceae)
=== Scaevola calendulacea (J. Kenn.) Druce (Goodeniaceae) [Carolin, Rajput & D. A. Morrison, Fl. Australia 35: 115 (1992).]

Lobelia canescens (Benth.) Kuntze, Revis. Gen. Pl. 2: 378 (1891); non C. Presl, Prodr. Monogr. Lobel.: 38 (1836). Basionym: *Scaevola canescens* Benth. (Goodeniaceae)
=== Scaevola canescens Benth. (Goodeniaceae) [Carolin, Rajput & D. A. Morrison, Fl. Australia 35: 117 (1992).]

Lobelia chamissoniana (Gaudich.) Kuntze, Revis. Gen. Pl. 2: 378 (1891). Basionym: *Scaevola chamissoniana* Gaudich. (Goodeniaceae)
=== Scaevola chamissoniana Gaudich. (Goodeniaceae) [R. Patt. in W. L. Wagner, D. R. Herbst & Sohmer, Man. Fl. Pl. Hawai'i: 783 (1990).]

Lobelia cheiranthus L., Sp. Pl.: 933 (1753).
=== Manulea cheiranthus (L.) L. (Scrophulariaceae) [E. Wimm. in Engl., Pflanzenr. IV.276b: 695 (1953).

Lobelia collaris (F. Muell.) Kuntze, Revis. Gen. Pl. 2: 378 (1891). Basionym: *Scaevola collaris* F. Muell. (Goodeniaceae)
=== Scaevola collaris F. Muell. (Goodeniaceae) [Carolin, Rajput & D. A. Morrison, Fl. Australia 35: 103 (1992).]

Lobelia coriacea (Nutt.) Kuntze, Revis. Gen. Pl. 2: 378 (1891). Basionym: *Scaevola coriacea* Nutt. (Goodeniaceae)
=== Scaevola coriacea Nutt. (Goodeniaceae) [R. Patt. in W. L. Wagner, D. R. Herbst & Sohmer, Man. Fl. Pl. Hawai'i: 784 (1990).]

Lobelia crassifolia (Labill.) Kuntze, Revis. Gen. Pl. 2: 378 (1891). Basionym: *Scaevola crassifolia* Labill. (Goodeniaceae)
=== Scaevola crassifolia Labill. (Goodeniaceae) [Carolin, Rajput & D. A. Morrison, Fl. Australia 35: 103 (1992).]

Lobelia cuneiformis (Labill.) Kuntze, Revis. Gen. Pl. 2: 378 (1891); non Labill., Nov. Holl. Pl.: 51 (1805). Basionym: *Scaevola cuneiformis* Labill. (Goodeniaceae)
=== Scaevola cuneiformis Labill. (Goodeniaceae) [Carolin, Rajput & D. A. Morrison, Fl. Australia 35: 126 (1992).]

Lobelia cunninghamii (DC.) Kuntze, Revis. Gen. Pl. 2: 378 (1891). Basionym: *Scaevola cunninghamii* DC. (Goodeniaceae)
=== Scaevola cunninghamii DC. (Goodeniaceae) [Carolin, Rajput & D. A. Morrison, Fl. Australia 35: 107 (1992).]

Lobelia cylindrocarpa (Hillebr.) Kuntze, Revis. Gen. Pl. 2: 378 (1891). Basionym: *Scaevola cylindrocarpa* Hillebr. (Goodeniaceae)
=== Scaevola chamissoniana Gaudich. (Goodeniaceae) [R. Patt. in W. L. Wagner, D. R. Herbst & Sohmer, Man. Fl. Pl. Hawai'i: 783 (1990).]

Lobelia depauperata (R. Br.) Kuntze, Revis. Gen. Pl. 2: 378 (1891). Basionym: *Scaevola depauperata* R. Br. (Goodeniaceae)
=== Scaevola depauperata R. Br. (Goodeniaceae) [Carolin, Rajput & D. A. Morrison, Fl. Australia 35: 142 (1992).]

Lobelia enantophylla (F. Muell.) Kuntze, Revis. Gen. Pl. 2: 378 (1891). Basionym: *Scaevola enantophylla* F. Muell. (Goodeniaceae)
=== Scaevola enantophylla F. Muell. (Goodeniaceae) [Carolin, Rajput & D. A. Morrison, Fl. Australia 35: 101 (1992).]

Lobelia esquirolii H. Lév., Fl. Kouy-Tchéou: 58 (1915).
=== Mazus pumilus (Burm. f.) Steenis (Scrophulariaceae) [E. Wimm. in Engl., Pflanzenr. IV.276b: 695 (1953).]

Lobelia fasciculata (Benth.) Kuntze, Revis. Gen. Pl. 2: 378 (1891). Basionym: *Scaevola fasciculata* Benth. (Goodeniaceae)
=== Goodenia fasciculata (Benth.) Carolin (Goodeniaceae) [Carolin, Rajput & D. A. Morrison, Fl. Australia 35: 169 (1992).]

Lobelia frutescens Mill., Gard. Dict. (ed. 8) (1768). Based on: *Lobelia plumieri* L. (Goodeniaceae)
=== Scaevola plumieri (L.) Vahl (Goodeniaceae) [C. Jeffrey, Kew Bull. 34: 540 (1979).]

Lobelia gaudichaudii (Hook. & Arn.) Kuntze, Revis. Gen. Pl. 2: 378 (1891); non A. DC. in DC., Prodr. 7: 384 (1839). Basionym: *Scaevola gaudichaudii* Hook. & Arn. (Goodeniaceae)
=== Scaevola gaudichaudii Hook. & Arn. (Goodeniaceae) [R. Patt. in W. L. Wagner, D. R. Herbst & Sohmer, Man. Fl. Pl. Hawai'i: 784 (1990).]

Lobelia glabra (Hook. & Arn.) Kuntze, Revis. Gen. Pl. 2: 378 (1891). Basionym: *Scaevola glabra* Hook. & Arn. (Goodeniaceae)
=== Scaevola glabra Hook. & Arn. (Goodeniaceae) [R. Patt. in W. L. Wagner, D. R. Herbst & Sohmer, Man. Fl. Pl. Hawai'i: 786 (1990).]

Lobelia glandulifera (DC.) Kuntze, Revis. Gen. Pl. 2: 378 (1891). Basionym: *Scaevola glandulifera* DC. (Goodeniaceae)
=== Scaevola glandulifera DC. (Goodeniaceae) [Carolin, Rajput & D. A. Morrison, Fl. Australia 35: 111 (1992).]

Lobelia globulifera (Labill.) Kuntze, Revis. Gen. Pl. 2: 378 (1891). Basionym: *Scaevola globulifera* Labill. (Goodeniaceae)
=== Scaevola globulifera Labill. (Goodeniaceae) [Carolin, Rajput & D. A. Morrison, Fl. Australia 35: 106 (1992).]

Lobelia gracilis (Hook. f.) Kuntze, Revis. Gen. Pl. 2: 378 (1891); non Salisb., Prodr. Stirp. Chap. Allerton: 129 (1796); nec Andrews, Bot. Repos. 5: pl. 340 (1803); nec Nutt., Gen. N. Amer. Pl. 2: 77 (1818). Basionym: *Scaevola gracilis* Hook. f. (Goodeniaceae)
=== Scaevola gracilis Hook. f. (Goodeniaceae) [Sykes, New Zealand J. Bot. 36: 672 (1998).]

Lobelia hirsuta L., Sp. Pl.: 932 (1753).
=== Gnidia hirsuta (L.) Thulin (Thymeleaceae) [Thulin, Taxon 35: 724 (1986).]

Lobelia hispida (Cav.) Kuntze, Revis. Gen. Pl. 2: 378 (1891). Basionym: *Scaevola hispida* Cav. (Goodeniaceae)
=== Scaevola ramosissima (Sm.) K. Krause (Goodeniaceae) [Carolin, Rajput & D. A. Morrison, Fl. Australia 35: 134 (1992).]

Lobelia holosericea (de Vriese) Kuntze, Revis. Gen. Pl. 2: 378 (1891). Basionym: *Scaevola holosericea* de Vriese (Goodeniaceae)
=== Scaevola anchusifolia Benth. (Goodeniaceae) [Carolin, Rajput & D. A. Morrison, Fl. Australia 35: 113 (1992).]

Lobelia hookeri (F. Muell.) Kuntze, Revis. Gen. Pl. 2: 378 (1891). Basionym: *Scaevola hookeri* F. Muell. (Goodeniaceae)
=== Scaevola hookeri F. Muell (Goodeniaceae) [Carolin, Rajput & D. A. Morrison, Fl. Australia 35: 135 (1992).]

Lobelia humifusa (de Vriese) Kuntze, Revis. Gen. Pl. 2: 378 (1891). Basionym: *Scaevola humifusa* de Vriese (Goodeniaceae)
=== Scaevola humifusa de Vriese (Goodeniaceae) [Carolin, Rajput & D. A. Morrison, Fl. Australia 35: 120 (1992).]

Lobelia humilis (R. Br.) Kuntze, Revis. Gen. Pl. 2: 378 (1891); non Klotzsch in Peters, Naturw. Reise Mossambique 6: 301 (1861). Basionym: *Scaevola humilis* R. Br. (Goodeniaceae)
=== Scaevola humilis R. Br. (Goodeniaceae) [Carolin, Rajput & D. A. Morrison, Fl. Australia 35: 130 (1992).]

Lobelia koenigii (Vahl) W. Wight, Contr. U.S. Natl. Herb. 9: 310 (1905). Basionym: *Scaevola koenigii* Vahl (Goodeniaceae)
=== Scaevola taccada (Gaertn.) Roxb. (Goodeniaceae) [P. S. Green, Taxon 40: 121 (1991).]

Lobelia lanceolata (Benth.) Kuntze, Revis. Gen. Pl. 2: 378 (1891); non (Gaudich.) Hook. & Arn., Bot. Beechey Voy.: 88 (1832); nec Hook. & Arn., Bot. Beechey Voy.: 301 (1838). Basionym: *Scaevola lanceolata* Benth. (Goodeniaceae)
=== Scaevola lanceolata Benth. (Goodeniaceae) [Carolin, Rajput & D. A. Morrison, Fl. Australia 35: 114 (1992).]

Lobelia linearis (R. Br.) Kuntze, Revis. Gen. Pl. 2: 378 (1891); non Thunb., Prodr. Fl. Cap.: 39 (1794). Basionym: *Scaevola linearis* R. Br. (Goodeniaceae)
=== Scaevola linearis R. Br. (Goodeniaceae) [Carolin, Rajput & D. A. Morrison, Fl. Australia 35: 120 (1992).]

Lobelia longifolia (de Vriese) Kuntze, Revis. Gen. Pl. 2: 378 (1891); non (C. Presl) A. DC. in DC., Prodr. 7: 382 (1839). Basionym: *Scaevola longifolia* de Vriese (Goodeniaceae)
=== Scaevola lanceolata Benth. (Goodeniaceae) [Carolin, Rajput & D. A. Morrison, Fl. Australia 35: 114 (1992).]

Lobelia longiscapa de Vriese in Lehm., Pl. Preiss.: 398 (1845).
=== Goodenia pulchella Benth. (Goodeniaceae) [Carolin, Rajput & D. A. Morrison, Fl. Australia 35: 249 (1992).]

Lobelia macrophylla (de Vriese) Kuntze, Revis. Gen. Pl. 2: 378 (1891); non (G. Don) C. Presl, Prodr. Monogr. Lobel.: 39 (1836). Basionym: *Molkenboeria macrophylla* de Vriese (Goodeniaceae)
=== Scaevola macrophylla (de Vriese) Benth. (Goodeniaceae) [Carolin, Rajput & D. A. Morrison, Fl. Australia 35: 133 (1992).]

Lobelia macrostachya (de Vriese) Kuntze, Revis. Gen. Pl. 2: 378 (1891). Basionym: *Scaevola macrostachya* (de Vriese) Benth. (Goodeniaceae)
=== Scaevola macrostachya (de Vriese) Benth. (Goodeniaceae) [Carolin, Rajput & D. A. Morrison, Fl. Australia 35: 111 (1992).]

Lobelia micrantha (C. Presl) Kuntze, Revis. Gen. Pl. 2: 378 (1891); non Kunth in Humb., Bonpl. & Kunth, Nov. Gen. Sp. 3: 316 (quarto), 247 (folio) (1819); nec Hook., Exot. Fl. 1: 44 (1823); nec (E. Mey.) Heynh., Nom. Bot. Hort. 1: 471 (1840). Basionym: *Scaevola micrantha* C. Presl (Goodeniaceae)
=== Scaevola micrantha C. Presl (Goodeniaceae) [Leenh. in Steenis, Fl Males. (ser. I) 5: 342 (1957).]

Lobelia microcarpa (Cav.) Kuntze, Revis. Gen. Pl. 2: 378 (1891); non C. B. Clarke in Hook. f., Fl. Brit. India 3: 424 (1881). Basionym: *Scaevola microcarpa* Cav. (Goodeniaceae)
=== Scaevola albida (Sm.) Druce (Goodeniaceae) [Carolin, Rajput & D. A. Morrison, Fl. Australia 35: 122 (1992).]

Lobelia microphylla (de Vriese) Kuntze, Revis. Gen. Pl. 2: 378 (1891). Basionym: *Scaevola microphylla* (de Vriese) Benth. (Goodeniaceae)
=== Scaevola microphylla (de Vriese) Benth. (Goodeniaceae) [Carolin, Rajput & D. A. Morrison, Fl. Australia 35: 132 (1992).]

Lobelia mollis (Hook. & Arn.) Kuntze, Revis. Gen. Pl. 2: 378 (1891); non Graham, Edinburgh New Philos. J. 8: 185 (1830). Basionym: *Scaevola mollis* Hook. & Arn. (Goodeniaceae)
=== Scaevola mollis Hook. & Arn. (Goodeniaceae) [R. Patt. in W. L. Wagner, D. R. Herbst & Sohmer, Man. Fl. Pl. Hawai'i: 786 (1990).]

Lobelia myrtifolia (de Vriese) Kuntze, Revis. Gen. Pl. 2: 378 (1891). Basionym: *Scaevola myrtifolia* (de Vriese) K. Krause (Goodeniaceae)
=== Scaevola myrtifolia (de Vriese) K. Krause (Goodeniaceae) [Carolin, Rajput & D. A. Morrison, Fl. Australia 35: 100 (1992).]

Lobelia nitida (R. Br.) Kuntze, Revis. Gen. Pl. 2: 378 (1891). Basionym: *Scaevola nitida* R. Br. (Goodeniaceae)
=== Scaevola nitida R. Br. (Goodeniaceae) [Carolin, Rajput & D. A. Morrison, Fl. Australia 35: 104 (1992).]

Lobelia oldfieldii (F. Muell.) Kuntze, Revis. Gen. Pl. 2: 378 (1891). Basionym: *Scaevola oldfieldii* F. Muell.
=== Scaevola oldfieldii F. Muell. (Goodeniaceae) [Carolin, Rajput & D. A. Morrison, Fl. Australia 35: 119 (1992).]

Lobelia oppositifolia (Roxb.) Kuntze, Revis. Gen. Pl. 2: 378 (1891). Basionym: *Scaevola oppositifolia* Roxb. (Goodeniaceae)
=== Scaevola oppositifolia Roxb. (Goodeniaceae) [Leenh. in Steenis, Fl Males. (ser. I) 5: 342 (1957).]

Lobelia ovalifolia (R. Br.) Kuntze, Revis. Gen. Pl. 2: 378 (1891); non Hook. & Arn., Bot. Beechey Voy.: 300 (1838). Basionym: *Scaevola ovalifolia* R. Br. (Goodeniaceae)
=== Scaevola ovalifolia R. Br. (Goodeniaceae) [Carolin, Rajput & D. A. Morrison, Fl. Australia 35: 123 (1992).]

Lobelia oxyclona (F. Muell.) Kuntze, Revis. Gen. Pl. 2: 378 (1891). Basionym: *Scaevola oxyclona* F. Muell. (Goodeniaceae)
=== Scaevola oxyclona F. Muell. (Goodeniaceae) [Carolin, Rajput & D. A. Morrison, Fl. Australia 35: 144 (1992).]

Lobelia paludosa (R. Br.) Kuntze, Revis. Gen. Pl. 2: 378 (1891); non Nutt., Gen. N. Amer. Pl. 2: 76 (1818). Basionym: *Scaevola paludosa* R. Br. (Goodeniaceae)
=== Scaevola paludosa R. Br. (Goodeniaceae) [Carolin, Rajput & D. A. Morrison, Fl. Australia 35: 121 (1992).]

Lobelia parvifolia (F. Muell. ex Benth.) Kuntze, Revis. Gen. Pl. 2: 378 (1891); non P. J. Bergius, Descr. Pl. Cap.: 345 (1767); nec R. Br., Prodr.: 564 (1810); nec Raf., New Fl. N. Amer. 2: 18 (1837). Basionym: *Scaevola parvifolia* F. Muell. ex Benth. (Goodeniaceae)
=== Scaevola parvifolia F. Muell. ex Benth. (Goodeniaceae) [Carolin, Rajput & D. A. Morrison, Fl. Australia 35: 141 (1992).]

Lobelia phlebopetala (F. Muell.) Kuntze, Revis. Gen. Pl. 2: 378 (1891). Basionym: *Scaevola phlebopetala* F. Muell. (Goodeniaceae)
=== Scaevola phlebopetala F. Muell. (Goodeniaceae) [Carolin, Rajput & D. A. Morrison, Fl. Australia 35: 138 (1992).]

Lobelia piliplena (Miq.) Kuntze, Revis. Gen. Pl. 2: 378 (1891). Basionym: *Scaevola piliplena* Miq.
=== Scaevola taccada (Gaertn.) Roxb. (Goodeniaceae) [Leenh. in Steenis, Fl Males. (ser. I) 5: 339 (1957).]

Lobelia pilosa (Benth.) Kuntze, Revis. Gen. Pl. 2: 378 (1891). Basionym: *Scaevola pilosa* Benth. (Goodeniaceae)
=== Scaevola pilosa Benth. (Goodeniaceae) [Carolin, Rajput & D. A. Morrison, Fl. Australia 35: 135 (1992).]

Lobelia platyphylla (Lindl.) Kuntze, Revis. Gen. Pl. 2: 378 (1891). Basionym: *Scaevola platyphylla* Lindl. (Goodeniaceae)
=== Scaevola platyphylla Lindl. (Goodeniaceae) [Carolin, Rajput & D. A. Morrison, Fl. Australia 35: 133 (1992).]

Lobelia plumieri L., Sp. Pl.: 929 (1753).
=== Scaevola plumieri (L.) Vahl (Goodeniaceae)) [C. Jeffrey, Kew Bull. 34: 538-539 (1979).]

Lobelia porocarya (F. Muell.) Kuntze, Revis. Gen. Pl. 2: 378 (1891). Basionym: *Scaevola porocarya* F. Muell. (Goodeniaceae)
=== Scaevola porocarya F. Muell. (Goodeniaceae) [Carolin, Rajput & D. A. Morrison, Fl. Australia 35: 107 (1992).]

Lobelia procera (Hillebr.) Kuntze, Revis. Gen. Pl. 2: 378 (1891). Basionym: *Scaevola procera* Hillebr. (Goodeniaceae)
=== Scaevola procera Hillebr. (Goodeniaceae) [R. Patt. in W. L. Wagner, D. R. Herbst & Sohmer, Man. Fl. Pl. Hawai'i: 788 (1990).]

Lobelia pumila Burm. f., Fl. Indica: 186 (1768).
=== Mazus pumilus (Burm f.) Steenis (Scrophulariaceae) [Steenis, Nova Guinea (n.s.) 9: 31 (1958).]

Lobelia restiacea (Benth.) Kuntze, Revis. Gen. Pl. 2: 378 (1891). Basionym: *Scaevola restiacea* Benth. (Goodeniaceae)
=== Scaevola restiacea Benth. (Goodeniaceae) [Carolin, Rajput & D. A. Morrison, Fl. Australia 35: 143 (1992).]

Lobelia revoluta (R. Br.) Kuntze, Revis. Gen. Pl. 2: 378 (1891). Basionym: *Scaevola revoluta* R. Br. (Goodeniaceae)
=== Scaevola revoluta R. Br. (Goodeniaceae) [Carolin, Rajput & D. A. Morrison, Fl. Australia 35: 109 (1992).]

Lobelia sericea (Vahl) Kuntze, Revis. Gen. Pl. 2: 378 (1891). Basionym: *Scaevola sericea* Vahl (Goodeniaceae)
=== Scaevola taccada (Gaertn.) Roxb. (Goodeniaceae) [P. S. Green, Taxon 40: 121 (1991).]

Lobelia sericea var. **koenigii** (Vahl) Kuntze, Revis. Gen. Pl. 2: 378 (1891). Basionym: *Scaevola koenigii* Vahl (Goodeniaceae)
=== Scaevola taccada (Gaertn.) Roxb. (Goodeniaceae) [P. S. Green, Taxon 40: 121 (1991).]

Lobelia sericophylla (F. Muell. ex Benth.) Kuntze, Revis. Gen. Pl. 2: 378 (1891). Basionym: *Scaevola sericophylla* F. Muell. ex Benth. (Goodeniaceae)
=== Scaevola sericophylla F. Muell. ex Benth. (Goodeniaceae) [Carolin, Rajput & D. A. Morrison, Fl. Australia 35: 117 (1992).]

Lobelia spinescens (R. Br.) Kuntze, Revis. Gen. Pl. 2: 378 (1891). Basionym: *Scaevola spinescens* R. Br. (Goodeniaceae)
=== Scaevola spinescens R. Br. (Goodeniaceae) [Carolin, Rajput & D. A. Morrison, Fl. Australia 35: 97 (1992).]

Lobelia stenophylla (F. Muell.) Kuntze, Revis. Gen. Pl. 2: 378 (1891); non Benth., Fl.
Austral. 4: 130 (1868). Basionym: *Goodenia stenophylla* F. Muell. (Goodeniaceae)
=== Goodenia stenophylla F. Muell. (Goodeniaceae) [Carolin, Rajput & D. A. Morrison, Fl.
Australia 35: 170 (1992).]

Lobelia striata (R. Br.) Kuntze, Revis. Gen. Pl. 2: 378 (1891). Basionym: *Scaevola striata* R. Br.
(Goodeniaceae)
=== Scaevola striata R. Br. (Goodeniaceae) [Carolin, Rajput & D. A. Morrison, Fl. Australia
35: 136 (1992).]

Lobelia submersa R. Cunn. ex A. Cunn., Ann. Nat. Hist. 2: 50 (1839).
=== Glossostigma elatinoides Benth. (Scrophulariaceae) [E. Wimm. in Engl., Pflanzenr.
IV.276b: 695 (1953).]

Lobelia taccada Gaertn., Fruct. Sem. Pl. 1: 119 (1788).
=== Scaevola taccada (Gaertn.) Roxb. (Goodeniaceae) [P. S. Green, Taxon 40: 121 (1991).]

Lobelia thesioides (Benth.) Kuntze, Revis. Gen. Pl. 2: 378 (1891). Basionym: *Scaevola
thesioides* Benth. (Goodeniaceae)
=== Scaevola thesioides Benth. (Goodeniaceae) [Carolin, Rajput & D. A. Morrison, Fl.
Australia 35: 105 (1992).]

Lobelia tomentosa (Gaudich.) Kuntze, Revis. Gen. Pl. 2: 378 (1891); non L. f., Suppl. Pl.:
394 (1782). Basionym: *Scaevola tomentosa* Gaudich. (Goodeniaceae)
=== Scaevola tomentosa Gaudich. (Goodeniaceae) [Carolin, Rajput & D. A. Morrison, Fl.
Australia 35: 100 (1992).]

Lobelia tortuosa (Benth.) Kuntze, Revis. Gen. Pl. 2: 378 (1891). Basionym: *Scaevola tortuosa*
Benth.
=== Scaevola tortuosa Benth. (Goodeniaceae) [Carolin, Rajput & D. A. Morrison, Fl.
Australia 35: 145 (1992).]

Lobelia velutina (C. Presl) Kuntze, Revis. Gen. Pl. 2: 378 (1891); non M. Martens &
Galeotti, Bull. Acad. Roy. Sci. Bruxelles 9(2): 41 (1842). Basionym: *Scaevola velutina* C.
Presl (Goodeniaceae)
=== Scaevola taccada (Gaertn.) Roxb. (Goodeniaceae) [Leenh. in Steenis, Fl Males. (ser. I)
5: 339 (1957).]

Merciera heteromorpha H. Buek in Eckl. & Zeyh., Enum. Pl. Afric. Austral.: 387 (1837).
=== Carpacoce heteromorpha (H. Buek) L. Bolus (Rubiaceae) [Adamson, J. S. African Bot.
20: 163 (1955).]

Merciera vaginata Adamson, J. S. African Bot. 20: 162 (1955).
=== ? (Rubiaceae) [Cupido, Adansonia (sér. 3) 25: 34 (2003).]

Phyteuma begoniifolium Roxb. ex Jack, Malayan Misc. 1(1): 5 (1820), as 'begonifolium'.
=== Pentaphragma begoniifolium (Roxb. ex Jack) G. Don (Pentaphragmataceae) [G. Don,
Gen. Hist. 3: 731 (1834).]

Phyteuma bipinnata Lour., Fl. Cochinch. 1: 138 (1790).
=== Sambucus ebuloides Desv. ex DC. (Caprifoliaceae) [DC., Prodr. 4: 323 (1830).]

Phyteuma cochinchinensis Lour., Fl. Cochinch. 1: 139 (1790).
=== Sambucus phyteumoides DC. (Caprifoliaceae) [DC., Prodr. 4: 323 (1830).]

Phyteuma tricolor Molina, Sag. Stor. Nat. Chili (ed. 2): 281 (1810).
=== Salpiglossis sinuata Ruiz & Pav. (Solanaceae) [Phil., Anales Univ. Chile 22: 710 (1863).]

Rapuntium hirsutum (L.) Moench, Methodus: 656 (1794). Basionym: *Lobelia hirsuta* L.
=== Gnidia hirsuta (L.) Thulin (Thymeleaceae) [Thulin, Taxon 35: 724 (1986).]

Roella elegans Paxton, Paxton's Mag. Bot. 6: 27 (1839).
=== ? (Gesneriaceae) [Adamson, J. S. African Bot. 17: 156 (1952).]

Roella reticulata L., Sp. Pl.: 170 (1753).
=== Cullumia reticulata (L.) Greuter, M. V. Agab. & Wagenitz (Asteraceae) [Greuter, M. V.
Agab. & Wagenitz, Taxon 54: 155 (2005).]

Summary of names based on fossils

Only a single name based on fossil material has been published in Campanulaceae.

Campanula palaeopyramidalis Lańc.-Środ., Acta Paleobot. 18: 38 (1977).
=== seeds; Miocene of Poland.

Summary of named hybrids without a legitimate name

Hybridization is known or suspected to occur in some genera of Campanulaceae. Some interspecific hybrids have been named, while others (not included in this checklist) are merely known by formulae. A few of the named hybrids unfortunately lack a legitimate name; for convenience, they are listed here.

Campanula decumbens var. **pseudospecularioides** G. Lopéz, Bol. Soc. Brot. (ser. 2) 53: 301 (1980). C. decumbens × C. lusitanica subsp. specularioides [L. Sáez & Aldasoro in Paiva et al., Fl. Iberica 14: 131 (2001).]
S. Spain. 12 SPA. Hemicr. or ther.?

Campanula hausmannii Rchb. f., Flora 60: 31 (1877), as 'hausmanni'. C. barbata L. × Phyteuma hemisphaericum L. [Wehrh., Gartenstauden 2: 985 (1931).]
Austria. 11 AUT. Hemicr.

Campanula × **vrtacensis** Ravnik, Phyton 12: 171 (1967). C. cochleariifiolia × Favratia zoysii
Slovenia. 13 YUG. Hemicr.

Centropogon × **lucyanus** Houllet, Rev. Hort. 40: 291 (1868). C. cornutus × Siphocampylus betulifolius
Cult. Nanophan. or cham. [Esson, Addisonia 23(2): 3 (1955).]

Cyanea atra var. **lobata** Rock, Indig. Trees Haw. Isl.: 511 (1913). *Delissea atra* var. *lobata* (Rock) H. St. John, Phytologia 63: 81 (1987). C. horrida × C. macrostegia [Lammers, Syst. Bot. Monogr. 73: 46 (2005).]
Hawaiian Is. (E. Maui). 63 HAW. Phan.

Cyanea macrostegia var. **viscosa** Rock, Monogr. Stud. Haw. Lobelioid.: 183 (1919). *Delissea macrostegia* var. *viscosa* (Rock) H. St. John, Phytologia 63: 85 (1987). C. aculeatiflora × C. macrostegia [Lammers, Syst. Bot. Monogr. 73: 57-58 (2005).]
Hawaiian Is. (E. Maui). 63 HAW. Phan.

Tupa kingii Phil., Anales Univ. Chile 90: 189 (1895), as 'kingi'. Lobelia excelsa × L. polyphylla
C. Chile (Valaparaíso). 85 CLC. Nanophan. [Lammers, Sida 19: 105-106 (2000).]

Outline of a revised classification of the Campanulaceae

The provisional classification proffered here is intended to integrate this checklist of species with the recent generic summary of the family (Lammers 2007, under **General**). The latter embodied a classification of the family that was reflected in the sequence of the genera but not otherwise articulated, due to constraints of format. That underlying classification is presented here in its entirety. While recent molecular phylogenetic studies have already pointed to weaknesses in the scheme, such research has not as yet included enough taxa to permit a comprehensive revision. Thus, this arrangement may stand as a summary of our understanding of the family at this point in time.

Family **Campanulaceae** Juss., Gen. Pl.: 163 (1789), nom. cons. Limbaceae Dulac, Fl. Hautes-Pyrénées: 459 (1867). Type: *Campanula* L.

The monophyly of the family as circumscribed here has been supported by numerous studies (e.g., Cosner et al. 1994, Gustafsson & Bremer 1995, Bremer & Gustafsson 1997, Eddie et al. 2003, Lundberg & Bremer 2003, under **General**).

Subfamily I. **Campanuloideae** Burnett, Outl. Bot.: 942, 1094, 1110 (1835), as 'Campanulidae'. Type: *Campanula* L.

The same studies that support the monophyly of the family (above) support the monophyly of this subfamily, which comprises 1045 species, 279 subspecies, and 21 named interspecific hybrids. Its division into tribes is reasonably consistent with the phylogeny of Eddie et al. (2003, under **General**). Tribe 1 corresponds to their "platycodonoids" and tribe 2 to their "wahlenbergioids", while tribes 3 through 9 represent the "campanuloids".

Tribe 1. **Cyanantheae** Meisn., Pl. Vasc. Gen.: Tab. Diagn. 273, Comm. 180 (1840). Cyananthaceae J. Agardh, Theoria Syst. Pl.: 384 (1858), as 'Cyanantheae'. Type: *Cyananthus* Wall. ex Benth.
Canarineae Webb. & Berth., Hist. Nat. Iles Canaries 2(3): 1 (1844). Canarinoideae Kolak., Bot. Žurn. (Moscow & Leningrad) 12: 1574 (1987). Type: *Canarina* L.
Platycodoninae Schönl. in Engl. & Prantl, Nat. Pflanzenfam. IV.5: 48 (1889). Platycodoneae Yeo, Taxon 42: 109 (1993). Type: *Platycodon* A. DC.
Campanumoeae Kolak., Bot. Žurn. (Moscow & Leningrad) 79: 111 (1994). Type: *Campanumoea* Blume.
Ostrowskieae Fed. in Kom., Fl. URSS 24: 471 (1957). Ostrowskioideae Takht., Diversity Classific. Fl. Pl.: 408 (1997). Type: *Ostrowskia* Regel.

This tribe corresponds to the "platycodonoid" clade of Eddie et al. (2003, under **General**); and to the *Cyananthus*, *Platycodon* and *Ostrowskia* Groups of Hong (1995, under **General**). Included genera: *Cyananthus*, *Platycodon*, *Codonopsis*, *Cyclocodon*, *Echinocodon*, *Ostrowskia*, and *Canarina*, totalling 90 species. Distribution: Macaronesia; E. Africa; Asia (Iran to Russian Far East & Papuasia). 21 23 24 25 26 30 31 32 34 36 37 38 40 41 42 43. Chromosome numbers: $2n = 14, 16, 18, 34, 36$.

Tribe 2. **Wahlenbergieae** Endl., Gen. Pl.: 514 (1838). Wahlenbergiinae Sond. in Harv. & Sond., Fl. Cap. 3: 531 (1865), as 'Wahlenbergieae'. Wahlenbergioideae Kolak., Bot. Žurn. (Moscow & Leningrad) 12: 1574 (1987). Type: *Wahlenbergia* Schrad. ex Roth
Lightfootiinae Endl., Gen. Pl.: 514 (1838), as 'Lighfootieae'. Type: *Lightfootia* L'Hér. non Sw.
Prismatocarpinae Endl., Gen. Pl.: 516 (1838), as 'Prismatocarpeae'. Prismatocarpoideae Kolak., Bot. Žurn. (Moscow & Leningrad) 12: 1574 (1987). Type: *Prismatocarpus* L'Hér.
Merciereae Meisn., Pl. Vasc. Gen.: Tab. Diagn. 238, 242, Comm. 150 (1839). Mercierinae Sond. in Harv. & Sond., Pl. Cap. 3: 531 (1865), as 'Mercierieae'. Type: *Merciera* A. DC.
Siphocodoninae Sond. in Harv. & Sond., Fl. Cap. 3: 531 (1865), as 'Siphocodeae'. Type: *Siphocodon* Turcz.

This tribe corresponds to the "wahlenbergioid" clade of Eddie et al. (2003, under **General**) and the *Wahlenbergia* Group of Hong (1995, under **General**). Included genera: *Wahlenbergia, Nesocodon, Heterochaenia, Berenice, Theilera, Namacodon, Craterocapsa, Roella, Gunillaea, Prismatocarpus, Treichelia, Microcodon, Merciera, Siphocodon,* and *Rhigiophyllum,* totalling 338 species. Distribution: S. Hemisphere north to W. Europe, Arabian Pen., E. Asia & N. South America, with the greatest concentration of diversity in S. Africa. 10 11 12 13 20 21 22 23 24 25 26 27 28 29 35 36 38 40 41 42 43 50 51 60 83 85. Chromosome numbers: $2n$ = 14, 16, 18, 22, 34, 36, 42, 54, 72, 90.

Tribe 3. **Edraiantheae** Fed. in Kom., Fl. URSS 24: 475 (1957). Type: *Edraianthus* A. DC. Michauxieae Fed. in Kom., Fl. URSS 24: 472 (1957). Type: *Michauxia* L'Hér.

These genera were members of the "Campanulaceae s. str. clade" in the phylogeny of Eddie et al. (2003, under **General**), and the *Campanula* group of Hong (1995, under **General**). Included genera: *Edraianthus, Feeria, Michauxia,* and *Trachelium,* totalling 23 species. Distribution: Mediterranean, from Macaronesia to W. Asia. 12 13 20 21 33 34. Chromosome numbers: $2n$ = 28, 30, 32, 34.

Tribe 4. **Jasioneae** Dumort, Fl. Belg.: 59 (1827). Jasionaceae Dumort., Anal. Fam. Pl.: 28, 30 (1829), as 'Jasionideae'. Jasioninae Endl., Gen. Pl.: 514 (1838), as 'Jasioneae'. Type: *Jasione* L.

This tribe is identical in composition to one of the three "transitional taxa" recognized by Eddie et al. (2003, under **General**), and to the *Jasione* Group of Hong (1995, under **General**). Sole genus: *Jasione,* with 15 species. Distribution: Europe to N. Africa & W. Asia. 10 11 12 13 14 20 34. Chromosome numbers: $2n$ = 12, 14, 18, 24, 36, 48, 60.

Tribe 5. **Musschieae** Kolak., Bot. Žurn. (Moscow & Leningrad) 12: 1575 (1987). Type: *Musschia* Dumort.

Sole genus: *Musschia,* with 2 species. Macaronesia. 21. This genus belonged to another of the "transitional taxa" recognized by Eddie et al. (2003, under **General**), and to the *Campanula* Group of Hong (1995, under **General**). Chromosome number: $2n$ = 32.

Tribe 6. **Campanuleae** Dumort., Fl. Belg.: 58 (1827). Campanulinae Schönl. in Engl. & Prantl, Nat. Pflanzenfam. IV.5: 48 (1889). Type: *Campanula* L.
Azorineae Kolak., Bot. Žurn. (Moscow & Leningrad) 12: 1575 (1987). Type: *Azorina* Feer
Echinocodoneae Kolak., Bot. Žurn. (Moscow & Leningrad) 12: 1575 (1987).
    Echinocodonieae Kolak., Bot. Žurn. (Moscow & Leningrad) 79: 114 (1994). Type: *Echinocodonia* Kolak.
Annaeae Kolak., Bot. Žurn. (Moscow & Leningrad) 12: 1575 (1987), as 'Annaea'. Type: *Annaea* Kolak.
Gadellieae Kolak., Bot. Žurn. (Moscow & Leningrad) 12: 1576 (1987). Type: *Gadellia* Schulkina
Neocodoneae Kolak., Bot. Žurn. (Moscow & Leningrad) 12: 1577 (1987). Type: *Neocodon* Kolak. & Serdyuk.
Mzymtelleae Kolak., Bot. Žurn. (Moscow & Leningrad) 12: 1578 (1987), as 'Mzymteleae'. Type: *Mzymtella* Kolak.
Pseudocampanuleae Kolak., Bot. Žurn. (Moscow & Leningrad) 79: 115 (1994). Type: *Pseudocampanula* Kolak.

This is the one patently unnatural tribe recognized in this subfamily, due largely to the polyphyly of *Campanula* as currently circumscribed. Elements of this tribe are found in the "Campanulaceae s. str. clade", the "Rapunculus clade", and in the "transitional" taxon that includes *Musschia* (Eddie et al. 2003, under **General**). All were among the

members of the *Campanula* Group of Hong (1995, under **General**). Included genera:
Azorina, Campanula, Favratia, Zeugandra, Adenophora, and *Hanabusaya,* totalling 492
species. Distribution: N. Hemisphere south to trop. Africa, S. Asia & N. Mexico, with the
greatest concentration of diversity from the Mediterranean to the Caucasus. 10 11 12 13
14 20 21 23 24 25 30 31 32 33 34 35 36 37 38 40 41 70 71 72 73 74 75 76 77 78 79.
Chromosome numbers: $2n$ = 14, 16, 18, 20, 22, 24, 26, 28, 30, 32, 34, 36, 40, 46, 48, 50,
52, 54, 56, 58, 60, 68, 70, 72, 80, 84, 90, 102.

Tribe 7. **Theodorovieae** Kolak., Bot. Žurn. (Moscow & Leningrad) 12: 1575 (1987). Type:
Theodorovia Kolak.
Muehlbergelleae Kolak., Bot. Žurn. (Moscow & Leningrad) 12: 1575 (1987). Type:
  Muehlbergella Feer
Sachokielleae Kolak., Bot. Žurn. (Moscow & Leningrad) 12: 1578 (1987), as 'Sachokieleae'.
  Type: *Sachokiella* Kolak.

No representatives of these genera have yet been included in a phylogenetic analysis of
the family, nor are any chromosome numbers known. All were included in the
Campanula Group of Hong (1995, under **General**). Included genera: *Theodorovia,*
Muehlbergella, and *Sachokiella,* each unispecific. Distribution: Caucasus. 33 34.

Tribe 8. **Peracarpeae** Fed. in Kom., Fl. URSS 24: 471 (1957). Type: *Peracarpa* Hook. f.
& Thomson

These genera did not form a clade in the phylogeny by Eddie et al. (2003, under
General), but all (so far as sampled) were members of the "Rapunculus clade". All were
members of the *Campanula* Group of Hong (1995, under **General**), while Takhtajan
(1996, under **General**) removed the four non-Asian genera to the following tribe.
Included genera: *Peracarpa, Homocodon, Legousia, Triodanis, Heterocodon,* and *Githopsis,*
totalling 21 species. Distribution: Northern Hemisphere south to N. Africa, S. Asia & S.
South America, with the greatest concentration of diversity in North America & E. Asia.
10 11 12 13 14 20 21 31 32 33 34 35 36 38 40 41 42 43 71 72 73 74 75 76 77 78 79 80 83
84 85. Chromosome numbers: $2n$ = 14, 16, 18, 20, 26, 28, 30, 36, 38, 40, 56, 60.

Tribe 9. **Phyteumeae** Dumort, Fl. Belg: 59 (1827). Phyteuminae Caruel, Epit. Fl. Eur. 2: 248
(1894), as 'Phyteumeae'. Type: *Phyteuma* L.
    Sergieae Kolak., Bot. Žurn. (Moscow & Leningrad) 12: 1577 (1987). Type: *Sergia* Fed.

These genera (so far as sampled) formed a clade within the "Rapunculus clade" (Eddie
et al. 2003, under **General**), and were assigned to the *Campanula* Group by Hong
(1995, under **General**). Included genera: *Asyneuma, Cryptocodon, Petromarula,*
Cylindrocarpa, Sergia, Phyteuma, and *Physoplexis,* totalling 61 species. Distribution:
Europe, N. Africa, Asia (Turkey to Russian Far East, Japan & Indo-China), with the
greatest concentration of diversity around the Mediterranean. 11 12 13 14 20 31 32 33
34 36 38 40 41. Chromosome numbers: $2n$ = 16, 18, 20, 22, 24, 26, 28, 30, 32, 34, 36,
48, 56, 68.

Subfamily II. **Nemacladoideae** Lammers, Novon 8: 37 (1998). Nemacladaceae Nutt., Trans.
Amer. Philos. Soc. (n.s.) 8: 254 (1842). Type: *Nemacladus* Nutt.

Segregation of this subfamily from Cyphioideae s.l. has been supported by several studies
(e.g., Cosner et al. 1994, Bremer & Gustafsson 1997, Lundberg & Bremer 2003, under
General). Included genera: *Pseudonemacladus, Nemacladus,* and *Parishella,* totalling 15
species. Distribution: W. North America, with the greatest concentration of diversity in
California. 73 76 77 79. Chromosome number: $2n$ = 18.

Subfamily III. **Lobelioideae** Burnett, Outl. Bot.: 942, 1094, 1110 (1835), as 'Lobelidae'.
Lobeliaceae Juss. ex Bonpl., Descr. Pl. Malmaison: 19, ad t. 7 (1813), nom. cons.
Ciliovallaceae Dulac, Fl. Hautes-Pyrénées: 459 (1867). Type: *Lobelia* L.
Dortmannaceae Rupr., Fl. Ingr. 1: 649 (1856), as 'Dortmanniaceae'. Type: *Dortmanna* Hill

The same studies that support the monophyly of the family (see above) support the monophyly of this subfamily, which comprises 1192 species, 110 subspecies, and 6 named interspecific hybrids.

Tribe 1. **Lobelieae** Rchb., Fl. Germ. Excurs. 1(3): 296 (1831). Lobeliinae Sond. in Harv. & Sond., Fl. Cap. 3: 531 (1865), as 'Lobelieae'. Type: *Lobelia* L.
Grammatotheceae A. Gray, J. Linn. Soc., Bot. 14: 29 (1873). Type: *Grammatotheca* C. Presl.
Phyllocharinae E. Wimm., Ann. Naturhist. Mus. Wien 56: 319 (1948). Type: *Phyllocharis* Diels non Fée.
Unigeninae E. Wimm., Ann. Naturhist. Mus. Wien 56: 373 (1948). Type: *Unigenes* E. Wimm.

This is the one patently unnatural tribe recognized in this subfamily, due largely to the polyphyly of *Lobelia* as currently circumscribed. Elements of this tribe are found throughout the phylogeny (E. Knox, pers. comm.; A. Antonelli, pers. comm.). Included genera: *Dialypetalum, Lobelia, Solenopsis, Wimmerella, Grammatotheca, Dielsantha, Monopsis, Unigenes, Isotoma, Ruthiella, Diastatea, Hippobroma,* and *Heterotoma,* totalling 469 species. Distribution: Almost cosmopolitan, on all six continents and several archipelagoes. 10 11 12 13 14 20 21 22 23 24 25 26 27 28 29 30 31 34 35 36 38 40 41 42 43 50 51 61 63 70 71 72 73 74 75 76 77 78 79 80 81 82 83 84 85 90. Chromosome numbers: $2n$ = 12, 14, 18, 22, 24, 26, 28, 38, 42, 56, 70, 140.

Tribe 2. **Downingieae** Lammers, nom. nov. Based on: Clintonieae C. Presl, Prodr. Monogr. Lobel.: 43 (1836), nom. illeg. sub ICBN Art. 19.5. Type: *Clintonia* Douglas ex Lindl. non Raf.
Howelliinae E. Wimm., Ann. Naturhist. Mus. Wien 56: 319 (1948). Type: *Howellia* A. Gray

This tribe is monophyletic (E. Knox, pers. comm.; A. Antonelli, pers. comm.). Included genera: *Palmerella, Porterella, Legenere, Howellia,* and *Downingia,* totalling 17 species. Distribution: W. North America & S. South America. 71 73 76 79 85. Chromosome numbers: $2n$ = 12, 14, 16, 18, 20, 22, 24.

Tribe 3. **Lysipomieae** A. DC. in DC., Prodr. 7: 348 (1839). Lysipomiinae E. Wimm. in J. F. Macbr., Fl. Peru 6: 481 (1937). Type: *Lysipomia* Kunth
Burmeisterinae E. Wimm. in Engl., Pflanzenr. IV.276b: 104 (1943). Type: *Burmeistera* Triana.
Siphocampylinae E. Wimm., Ann. Naturhist. Mus. Wien 56: 318 (1948). Type: *Siphocampylus* Pohl.

This tribe is monophyletic (E. Knox, pers. comm.; A. Antonelli, pers. comm.). Included genera: *Lysipomia, Siphocampylus, Centropogon,* and *Burmeistera,* totalling 575 species. Distribution: Neotropics (Mexico to Argentina; West Indies). 79 80 81 82 83 84 85. Chromosome numbers: $2n$ = 20, 28.

Tribe 4. **Delisseeae** C. Presl, Prodr. Monogr. Lobel.: 46 (1836). Type: *Delissea* Gaudich.
Clermontieae Horan., Char. Ess. Fam.: 100 (1847). Type: *Clermontia* Gaudich.
Cyaneinae E. Wimm. in Engl., Pflanzenr. IV.276b: 41 (1943). Type: *Cyanea* Gaudich.
Apetahiinae E. Wimm., Ann. Naturhist. Mus. Wien 56: 320 (1948). Type: *Apetahia* Baill.
Brighamiinae E. Wimm., Ann. Naturhist. Mus. Wien 56: 320 (1948). Type: *Brighamia* A. Gray.
Sclerothecinae E. Wimm., Ann. Naturhist. Mus. Wien 56: 320 (1948). Type: *Sclerotheca* A. DC.

This tribe with the addition of a few species of woody *Lobelia* would represent a
monophyletic group (E. Knox, pers. comm.; A. Antonelli, pers. comm.). Included genera:
Sclerotheca, Apetahia, Trematolobelia, Brighamia, Delissea, Cyanea, and *Clermontia,* totalling
131 species. Distribution: Polynesia. 61 63. Chromosome number: $2n = 28$.

Subfamily IV. **Cyphocarpoideae** Miers, London J. Bot. 7: 61 (1848), as 'Cyphocarpaceae'.
Cyphocarpaceae Reveal & Hoogland, Phytologia 79: 72 (1996). Type: *Cyphocarpus* Miers.

Segregation of this subfamily from Cyphioideae s.l. has been supported by several studies
(e.g., Cosner et al. 1994, Bremer & Gustafsson 1997, under **General**). Sole genus:
Cyphocarpus, with 3 species. Distribution: N. Chile. 85.

Subfamily V. **Cyphioideae** Walp., Ann. Bot. Syst. 2: 1037 (1852), as 'Cyphiaceae'.
Cyphiaceae A. DC. in DC., Prodr. 7: 497 (1839). Cyphieae Sond. in Harv. & Sond., Fl.
Cap. 3: 531 (1865). Type: *Cyphia* P. J. Bergius.

The narrow circumscription employed here has been supported by several studies (e.g.,
Cosner et al. 1994, Bremer & Gustafsson 1997, Lundberg & Bremer 2003, under
General). Sole genus: *Cyphia,* with 64 species and 2 subspecies. Distribution: Trop. & S.
Africa. 23 24 25 26 27. Chromosome number: $2n = 18$.

Printed in the United Kingdom by
Lightning Source UK Ltd., Milton Keynes
137446UK00001B/137-140/A

9 781842 461860